The Changing World Religion Map

Stanley D. Brunn
Editor

The Changing World Religion Map

Sacred Places, Identities, Practices and Politics

Volume IV

Donna A. Gilbreath
Assistant Editor

Editor
Stanley D. Brunn
Department of Geography
University of Kentucky
Lexington, KY, USA

Assistant Editor
Donna A. Gilbreath
UK Markey Cancer Center
Research Communications Office
Lexington, KY, USA

ISBN 978-94-017-9375-9 ISBN 978-94-017-9376-6 (eBook)
DOI 10.1007/978-94-017-9376-6

Library of Congress Control Number: 2014960060

Springer Dordrecht Heidelberg New York London

Printed on acid-free paper

Springer Science+Business Media B.V. Dordrecht is part of Springer Science+Business Media (www.springer.com)

Preface: A Continuing Journey

Religion has always been a part of my life. I am a Presbyterian PK (preacher's kid). From my father I inherited not only an interest in the histories and geographies of religions, not just Christianity, but also a strong sense of social justice, a thread that has been part of my personal and professional (teaching, research, service) life. My mother was raised as a Quaker and from her I also learned much about social justice, peace and reconciliation and being a part of an effective voice calling for ends to war, social discrimination of various types, and other injustices that seem to be a continual part of daily life on the planet. My father had churches mostly in the rural Upper Middle West. These were open country and small town congregations in Illinois, Wisconsin, Minnesota, South Dakota, Nebraska, and Missouri. The members of these congregations were Germans, Czech, Scandinavians (Norwegians, and Swedes), and English. Perhaps or probably because of these experiences, I had friendships with many young people who comprised the mosaic of the rural Middle West. Our family moved frequently when I was living at home, primarily because my father's views on social issues were often not popular with the rural farming communities. (He lost his church in northwest Missouri in 1953 because he supported the Supreme Court's decision on desegregation of schools. By the time I graduated from high school in a small town in southeastern Illinois, I had attended schools in a half-dozen states; these include one-room school house experiences as well as those in small towns.

During my childhood days my interests in religion were, of course, important in the views I had about many subjects about those of different faiths and many places on the planet. I was born in a Catholic hospital, which I always attribute to the beginning of my ecumenical experiences. The schools I attended mixes of Catholics and Protestants; I had few experiences with Native Americans, Jews and African, and Asian Americans before entering college. But that background changed, as I will explain below. My father was always interested in missionaries and foreign missions and once I considered training for a missionary work. What fascinated me most about missionaries were that they were living in distant lands, places that I just longed to know about; an atlas was always my favorite childhood book, next to a dictionary. I was always glad when missionaries visited our churches and stayed in our homes. The fascination extended to my corresponding with missionaries in Africa, Asia, and Latin America. I was curious what kind of work they did. I also found them a source for stamps, a hobby that I have pursued since primary school. Also I collected the call letters of radio stations, some which were missionary stations, especially in Latin America. (Some of these radio stations are still broadcasting.)

When I enrolled as an undergraduate student at Eastern Illinois University, a small regional university in east central Illinois, I immediately requested roommates from different countries. I very much wanted to make friends with students from outside the United States and learn about their culture. During my 3 years at EIU, I had roommates from Jordan, Samoa, Costa Rica, Ethiopia, and South Korea; these were very formative years in helping me understand cross-cultural, and especially, religious diversity. On reflection, I think that most of the Sunday services I attended were mostly Presbyterian and Methodist, not Catholic, Lutheran, or Baptist. When I entered the University of Wisconsin, Madison, for the M.A. degree, I was again exposed to some different views about religion. The Madison church that fascinated me the most was the Unitarian church, a building designed by Frank Lloyd Wright. I remember how different the services and sermons were from Protestant churches, but intellectually I felt at home. My father was not exactly pleased I found the Unitarian church a good worship experience. The UW-Madison experience also introduced me to the study of geography and religion. This was brought home especially in conversations with my longtime and good friend, Dan Gade, but also a cultural geography course I audited with Fred Simoons, whose new book on religion and food prejudices just appeared and I found fascinating. Also I had conversations with John Alexander, who eventually left the department to continue in his own ministry with the Inter Varsity Christian Fellowship. A seminar on Cultural Plant Geography co-taught with Fred Simoons, Jonathan Sauer, and Clarence Olmstead provided some opportunities to explore cultural and historical dimensions of religion, which were the major fields where geographers could study religion. The geographers I knew who were writing about religion were Pierre Deffontaines, Eric Issac, and Xavier de Phanol. That narrow focus, has, of course, changed in the past several decades, as I will discuss below.

The move to Ohio State University for my doctoral work did not have the strong religious threads that had emerged before. I attended a variety of Protestant churches, especially Presbyterian, Congregational, and Methodist. I took no formal courses in geography that dealt with religion, although I was very interested when Wilbur Zelinsky's lengthy article on church membership patterns appeared in the *Annals of the Association of American Geographers* in 1961. I felt then that this was, and would be, a landmark study in American human geography, as the many maps of denominational membership patterns plus extensive references would form the basis for future scholars interested in religion questions, apart from historical and cultural foci which were the norm at that time. My first article on religion was on religious town names; I wrote it when I was at Ohio State with another longtime friend, Jim Wheeler, who had little interest in religion. I can still remember using my knowledge of biblical place names and going through a Rand McNally atlas with Jim identifying these town names. This study appeared in *Names*, which cultural geographers acknowledge is one of the premier journals concerned with names and naming processes. Even though my dissertation on changes in the central place functions in small towns in northwest Ohio and southeast Ohio (Appalachia) did not look specifically at churches, I did tabulate the number and variety during extensive fieldwork in both areas.

My first teaching job was at the University of Florida in fall 1966. I decided once I graduated from OSU that I wanted to live in a different part of the United States where I could learn about different regional cultures and politics. I was discouraged by some former teachers about teaching in Florida, especially about the region's segregation history, recent civil rights struggles in the South and also the John Birch Society (which was also active in Columbus when I lived there). The 3 years (1966–1969) in Gainesville were also very rewarding years. These were also very formative years in developing my interests in the social geography, a new field that was just beginning to be studied in the mid-1960s. Included in the forefront of this emerging field of social geographers were Anne Buttimer, Paul Claval, Yi-Fu Tuan, Dick Morrill, Richard Peet, Bill Bunge, Wilbur Zelinsky, David Harvey, and David Smith, all who were challenging geographers to study the social geographies of race, employment, school and housing discrimination, but also poverty, environmental injustice, inequities in federal and state programs promoting human welfare, the privileges of whiteness and the minorities' participation in the voting/political process. Living in northern Florida in the late 1960s or "Wallace years" could not help but alert one to the role that religion was playing in rural and urban areas in the South. Gainesville had distinct racial landscapes. I was definitely a "northerner" and carpetbagger who was an outcast in many ways in southern culture. One vivid memory is attending a University of Florida football game (a good example of regional pride and nationalism) and being about the only person seated while the band played "Dixie." I joined a Congregational/United Church of Christ church which was attended by a small number of "northern faculty" who were supportive of initiatives to end discriminatory practices at local, university, and broader levels. At this time I also was learning about the role of the Southern Baptist Church, a bastion of segregation that was very slow to accommodate to the wishes of those seeking ends to all kinds of overt and subtle discrimination (gender, race, class) practices. The term Bible Belt was also a label that rang true; it represented, as it still does, those who adhere to a literal interpretation of the Bible, a theological position I have never felt comfortable. I soon realized that if one really wanted to make a difference in the lives of those living with discrimination, poverty, and ending racial disenfranchisement in voting, religion was a good arena to express one's feelings and work with others on coordinated efforts. Published research that emerged from my Florida experiences included studies on poverty in the United States (with Jim Wheeler), the geographies of federal outlays to states, an open housing referendum in Flint, Michigan (with Wayne Hoffman), and school levies in cities that illustrated social inequities (with Wayne Hoffman and Gerald Romsa). My Florida years also provided me the first opportunity to travel in the developing world; that was made possible with a summer grant where I visited nearly 15 different Caribbean capitals where I witnessed housing, social, and infrastructure gaps. This experience provided my first experiences with the developing world and led to a Cities of the World class I taught at Michigan State University and also co-edited several editions with the same title of a book with Jack Williams.

The 11 years at Michigan State University did not result in any major research initiatives related to religion, although it did broaden my horizons about faiths other than Christianity. I began to learn about Islam, especially from graduate students in the department from Saudi Arabia, Libya, Kuwait, and Iran. Many of these I advised on religion topics about their own cultures, especially those dealing with pilgrimages and sacred sites. Probably the main gain from living in Michigan was support for and interest in an emerging secular society. The religious "flavors" of Michigan's religious landscape ran the full gamut from those who were very traditional and conservative to those who were globally ecumenical, interfaith, and even agnostic. I continued to be active in Presbyterian and United Churches of Christ, both which were intellectually and spiritually challenging places for adult classes and singing in a choir.

When I moved to the University of Kentucky in 1980, I knew that living in the Bluegrass State would be different from Michigan in at least two respects. One is that Kentucky was considered a moderate to progressive state with many strong traditional and conservative churches, especially the Southern Baptist denomination. Zelinsky's map accurately portrayed this region as having a dominance of conservative and evangelical Protestantism. Second, I realized that for anyone interested in advancing social issues related to race and gender equality or environmental quality (especially strip mining in eastern Kentucky), there would likely be some conflicts. I also understood before coming to Kentucky that alcoholic beverage consumption was a big issue in some countries; that fact was evident in an innovative regional map Fraser Hart prepared in a small book about the South. And then there was the issue about science and religion in school curricula. With this foreknowledge, I was looking forward to living in a region where the cross-currents of religion and politics meshed, not only experiencing some of these social issues or schisms firsthand, but also having an opportunity to study them, as I did.

I realized when I moved to Lexington, it was in many ways and still is a slowly progressing socially conscious city. Southern Baptist churches, Christian churches, and Churches of Christ were dominant in the landscape and in their influences on social issues. One could not purchase alcoholic beverages on Sunday in restaurants until a couple referenda were passed in the mid-1980s that permitted sales. I think 90 of the state's 120 counties were officially dry, although everyone living in a dry county knew where to purchase liquor. One could not see the then-controversial "The Last Temptation of Christ" movie when it appeared unless one would drive three hours to Dayton. "Get Right with God" signs were prominent along rural highways. The University of Kentucky chimes in Memorial Hall on campus played religious hymns until this practice stopped sometime in the middle of the decade; I am not exactly sure why. Public schools had prayers before athletic events; some still do. Teachers in some public schools could lose their jobs if they taught evolution. Creationism was (and still is) alive and well. I was informed by university advisors that the five most "dangerous" subjects to new UK students were biology, anthropology, astronomy, geology, and physical geography. Students not used to other than literal biblical interpretations were confused and confounded by evolutional science. Betting on horses was legal, even though gambling was frowned on by some religious leaders. Cock fighting and snake handling still existed (and still do) in pockets in rural eastern Kentucky. In many ways living in Kentucky was like living "on the dark side of gray." Lexington in many ways was and still is an island or outlier. Desegregation was a slow moving process in a city with a strong southern white traditional heritage. Athletic programs were also rather slow to integrate, especially UK basketball. In short, how could one not study religion in such an atmosphere. Living in Kentucky is sort of the antipode to living in agnostic-thriving New England and Pacific Northwest. I would expect that within 100 miles of Lexington one would discover one of the most diverse religious denominational and faith belief landscapes in the United States. There are the old regular mainline denominations, new faiths that have come into the Bluegrass and also many one-of-a-kind churches, especially in rural eastern, southern, and southeastern Kentucky.

I have undertaken a number of studies related to religion in Kentucky and the South in the past three decades. Some of these have been single-authored projects, others with students and faculty at UK and elsewhere. Some were presentations at professional meetings; some resulted in publications. The topics that fascinated me were ones that I learned from my geography colleagues and those in other disciplines that were understudied. These include the history and current patterns of wet/dry counties in Kentucky, a topic that appears in local and statewide media with communities deciding whether to approve the sale of alcoholic beverages. This study I conducted with historian Tom Appleton. With regularity there were clergy of some fundamentalist denominations who decried the sale of such drinks; opposing these clergy and their supporters were often those interested in promoting tourism and attracting out-of-state traffic on interstates. Also I looked into legislation that focused on science/education interfaces in the public schools and on the types of religious books (or avoidance of such, such as dealing with Marx, Darwin, and interfaith relations) in county libraries. Craig Campbell and I published an article in *Political Geography* on Cristo Redentor (Christ of the Andes statue) as an example of differential locational harmony. At the regional level I investigated with Esther Long the mission statements of seminaries in the South, a study that led to some interesting variations not only in their statements, but course offerings and visual materials on websites. I published with Holly Barcus two articles in *Great Plains Research* about denominational changes in the Great Plains. Missionaries have also been relatively neglected in geography, so I embarked on a study with Elizabeth Leppman that looked at the contents of a leading Quaker journal in the early part of the past century. Religions magazines, as we acknowledged in our study, were (and probably still are) a very important medium for educating the public about places and cultures, especially those where most Americans would have limited first-hand knowledge. The music/religion interface has long fascinated me, not only as a regular choir member, but for the words used to convey messages about spirituality, human welfare and justice, religious traditions and promises of peace and hope.

After 11 September 2001, I collected information from a number of churches in eastern Kentucky about how that somber event was celebrated and also what hymns they sung on the tenth anniversary. As expected, some were very somber and dignified, others had words about hope, healing, and reaching across traditional religious boundaries that separate us. I also co-authored an article (mostly photos) in *Focus* on the Shankill-Falls divided between Catholic and Protestant areas of Belfast with three students in my geography of religion class at the National University of Ireland in Maynooth. The visualization theme was integral to a paper published in *Geographica Slovenica* on ecumenical spaces and the web pages of the World Council of Churches and papers I delivered how cartoonists depicted the controversial construction of a mosque at Ground Zero. How cartoonists depicted God-Nature themes (the 2011 Haitian earthquake and Icelandic volcanic eruption) were the focus of an article in *Mitteilungen der Ősterreichsten Geographischen Gesellschaft*. I published in *Geographical Review* an article how the renaissance of religion in Russia is depicted on stamp issues since 1991. A major change in my thinking about the subject of religion in the South was the study that I worked on with Jerry Webster and Clark Archer, a study that appeared in the *Southeastern Geographer* in late 2011. We looked at the definition and concept of the Bible Belt as first discussed by Charles Heatwole (who was in my classes when I taught at Michigan State University) in 1978 in the *Journal of Geography*. We wanted to update his study and learn what has happened to the Bible Belt (or Belts) since this pioneering effort. What we learned using the Glenmary Research Center's county data on adherents for the past several decades was that the "buckle" has relocated. As our maps illustrated, the decline in those counties with denominations adhering to a literal interpretation of the Bible in western North Carolina and eastern Tennessee and a shift to the high concentration of Bible Belt counties in western Oklahoma and panhandle Texas. In this study using Glenmary data for 2000, we also looked at the demographic and political/voting characteristics of these counties. (In this volume we look at the same phenomenon using 2010 data and also discuss some of the visual features of the Bible Belt landscapes.)

What also was instrumental in my thinking about religion and geography interfaces were activities outside my own research agenda. As someone who has long standing interests in working with others at community levels on peace and justice issues, I worked with three other similarly committed adults in Lexington to organize the Central Kentucky Council for Peace and Justice. CKCPJ emerged in 1983 as an interfaith and interdenominational group committed to working on peace and justice issues within Lexington, in Central Kentucky especially, but also with national and global interests. The other three who were active in this initiative were Betsy Neale (from the Friends), Marylynne Flowers (active in a local Presbyterian church) and Ernie Yanarella (political scientist, Episcopalian, longtime friend, and also contributor to a very thoughtful essay on Weber in this volume). This organization is a key agent in peace/justice issues in the Bluegrass; it hosts meetings, fairs, conferences, and other events for people of all ages, plans annual marches on Martin Luther King Jr. holiday, and is an active voice on issues related to capital punishment, gun control, gay/lesbian issues, fair trade and employment, environmental responsibility and stewardship, and the rights of women, children, and minorities.

I also led adult classes at Maxwell Street Presbyterian Church where we discussed major theologians and religious writers, including William Spong, Marcus Borg, Joseph Campbell, Philip Jenkins, Diane Eck, Kathleen Norris, Diana Butler Bass, Francis Collins, Sam Harris, Paul Alan Laughlin, James Kugel, Dorothy Bass, and Garry Wills. We discuss issues about science, secularism, death and dying, interfaith dialogue, Christianity in the twenty-first century, images of God, missions and missionaries, and more. I also benefitted from attending church services in the many countries I have traveled, lived, and taught classes in the past three decades. These include services in elaborate, formal, and distinguished cathedrals in Europe, Russian Orthodox services in Central Asia, and services in a black township and white and interracial mainline churches in Cape Town. Often I would attend services where I understood nothing or little, but that did not diminish the opportunity to worship with youth and elders (many more) on Sunday mornings and listen to choirs sing in multiple languages. These personal experiences also became part of my religion pilgrimage.

While religion has been an important part of my personal life, it was less important as part of my teaching program. Teaching classes on the geography of religion are few and far between in the United States; I think the subject was accepted much more in the instructional and research arenas among geographers in Europe. I think that part of my reluctance to pursue a major book project on religion was that for a long time I considered the subject too narrowly focused, especially on cultural and historical geography. From my reading of the geography and religion literature, there were actually few studies done before 1970s. (See the bibliography at the end of Chap. 1). I took some renewed interest in the subject in the mid-1980s when a number of geographers began to examine religion/nature/environment issues. The pioneering works of Yi-Fu Tuan and Anne Buttimer were instrumental in steering the study of values, ethics, spirituality, and religion into some new and productive directions. These studies paved the way for a number of other studies by social geographers (a field that was not among the major fields until the 1970s and early 1980s). The steady stream of studies on geography and religion continued with the emergence of GORABS (Geography of Religion and Belief Systems) as a Specialty Group of the Association of American Geographers. The publication of more articles and special journal issues devoted to the geography of religion continued into the last decade of the twentieth century and first decade of this century. The synthetic works of Lily Kong that have regularly appeared in *Progress in Human Geography* further supported those who wanted to look at religion from human/environmental perspectives. These reviews not only introduced the study of religion within geography, but also to those in related scholarly disciplines.

As more and more research appeared in professional journals and more conferences included presentations on religion from different fields and subfields, it became increasing apparent that the time was propitious for a volume that looked at religion/geography interfaces from a number of different perspectives. From my own vantage point, the study of religion was one that could, should, might, and would benefit from those who have theoretical and conceptual training in many of the discipline's major subfields. The same applied to those who were regional specialists; there were topics meriting study from those who looked a political/religion issues in Southeast Asia or Central America as well as cultural/historical themes in southern Africa and continental Europe and symbolic/architectural features and built environments of religions landscapes in California, southeast Australia, and southwest Asia. Studying religions topics would not have to be limited to those in human geography, but could be seen as opportunities for those studying religion/natural disaster issues in East Asia and southeast United States as well as the spiritual roots of early and contemporary religious thinking in Central Asia, East Asia, Russia, and indigenous groups in South America. For those engaged in the study of gender, law, multicultural education, and media disciplines, there were also opportunities to contribute to the study of this emerging field. In short, there were literally "gold mines" of potential research topics in rural and urban areas everywhere on the continent.

About 7 years ago I decided to offer a class on the geography of religion in the Department of Geography at the University of Kentucky. The numbers were never larger (less than 15), but these were always enlightening and interesting, because of the views expressed by students. Their views about religion ran the gamut from very conservative to very liberal and also agnostic and atheist, which made, as one would expect, some very interesting exchanges. Students were strongly encouraged (not required, as I could not do this in a public university) to attend a half dozen different worship services during the semester. This did not mean attending First Baptist, Second Baptist, Third Baptist, etc., but different kinds of experiences. For some this course component was the first some had ever attended a Jewish synagogue, Catholic mass, Baptist service, an African American church, Unitarian church, or visited a mosque. Some students used this opportunity to attend Wiccan services, or visit a Buddhist and Hindu temple. Their write-ups about these experiences and the ensuing discussion were one of the high spots of the weekly class. In addition, the classes discussed chapters in various books and articles from the geography literature about the state of studying religion. And we always discussed current news items, using materials from the RNS (Religion News Service) website.

Another ingredient that stimulated my decision to edit a book on the geography of religion emerged from geography of religion conferences held in Europe in recent years. These were organized by my good friends Ceri Peach (Oxford), Reinhard Henkel (Heidelberg), and also Martin Baumann and Andreas Tunger-Zanetti (University of Lucerne). These miniconferences, held in Oxford, Lucerne, and Gottingen, usually attracted 20–40 junior and senior geographers and other religious scholars, and were a rich source of ideas for topics that might be studied. The opportunities for small group discussions, the field trips, and special events were conducive to learning about historical and contemporary changes in the religious landscapes of the European continent and beyond. A number of authors contributing to this volume presented papers at one or more of these conferences. Additional names came from those attending sessions at annual meetings of the Association of American Geographers.

Some of my initial thoughts and inspiration about a book came from the course I taught, conversations with friends who studied and did not study religion, and also the book I edited on megaengineering projects. This three-volume, 126-chapter book, *Engineering Earth: The Impacts of Megaengineering Projects*, was published in 2011 by Springer. There were only a few chapters in this book that had a religious content, one on megachurches, another on liberation theologians fighting megadevelopment projects in the Philippines and Guatemala. When I approached Evelien Bakker and Bernadette Deelen-Mans, my first geography editors at Springer, about a religion book, they were excited and supportive, as they have been since day one. They gave me the encouragement, certainly the latitude (and probably the longitude) to pursue the idea, knowing that I would identify significant cutting-edge topics about religion and culture and society in all major world regions. The prospectus I developed was for an innovative book that would include the contributions of scholars from the social sciences and humanities, those from different counties and those from different faiths. For their confidence and support, I am very grateful. The reviews they obtained of the prospectus were encouraging and acknowledging that there was a definite need for a major international, interdisciplinary, and interfaith volume. Springer also saw this book as an opportunity to emphasize its new directions in the social sciences and humanities. I also want to thank Stefan Einarson who came on board late in the project and shepherded the project to its completion with the usual Springer traits of professionalism, kindness, and commitment to the project's publication. And I wish to thank Chitra Sundarajan and her staff for helpful professionalism in preparing the final manuscript for publication.

The organization of the book, which is discussed in Chapter One, basically reflects the way I look at religion from a geographical perspective. I look at the subject as more than simply investigations into human geography's fields and subfields, including cultural and historical, but also economic, social, and political geography, but also human/environmental geography (dealing with human values, ethics, behavior, disasters, etc.). I also look at the study of religious topics and phenomena with respects to major concepts we use in geography; these include landscapes, networks, hierarchies, scales, regions, organization of space, the delivery of services, and virtual religion. I started contacting potential authors in September 2010. Since then I have sent or received over 15,000 emails related the volume.

I am deeply indebted to many friends for providing names of potential authors. I relied on my global network of geography colleagues in colleges and universities around the world, who not only recommended specific individuals, but also topics they deemed worthy of inclusion. Some were geographers, but many were not; some taught in universities, others in divinity schools and departments of religion around the world. Those I specifically want to acknowledge include: Barbara Ambrose, Martin Checa Artasu, Martin Baumann, John Benson, Gary Bouma, John Benson, Dwight Billings, Marion Bowman, John D. Brewer, David Brunn, David Butler, Ron Byars, Heidi Campbell, Caroline Creamer, Janel Curry, David Eicher, Elizabeth Ferris, Richard Gale, Don Gross, Wayne Gnatuk, Martin Haigh, Dan Hofrenning, Wil Holden, Hannah Holtschneider, Monica Ingalls, Nicole Karapanagiotis, Aharon Kellerman, Judith Kenny, Jean Kilde, Ted Levin, James

Munder, Alec Murphy, Tad Mutersbuagh, Garth Myers, Lionel Obidah, Sam Otterstrom, Francis Owusu, Maria Paradiso, Ron Pen, Ivan Petrella, Adam Possamai, Leonard Primiano, Craig Revels, Heinz Scheifinger, Anna Secor, Ira Sheskin, Doug Slaymaker, Patricia Solis, Anita Stasulne, Jill Stevenson, Robert Strauss, Tristan Sturm, Greg Stump, Karen Till, Andreas Tunger-Zenetti, Gary Vachicouras, Viera Vlčkova, Herman van der Wusten, Stanley Waterman, Mike Whine, Don Zeigler, Shangyi Zhou, and Matt Zook.

And I want to thank John Kostelnick who provided the GORABS Working Bibliography; most of the entries, except dissertations and theses, are included in Chap. 1 bibliography. Others who helped him prepare this valuable bibliography also need to be acknowledged: John Bauer, Ed Davis, Michael Ferber, Julian Holloway, Lily Kong, Elizabeth Leppman, Carolyn Prorock, Simon Potter, Thomas Rumney, Rana P.B. Singh, and Robert Stoddard. These are scholars who devoted their lifetimes to advancing research on geography and religion.

Finally I want to thank Donna Gilbreath for another splendid effort preparing all the chapters for Springer. She formatted the chapters and prepared all the tables and illustrations per the publisher's guidelines. Donna is an invaluable and skilled professional who deserves much credit for working with multiple authors and the publisher to ensure that all text materials were correct and in order. Also I am indebted to her husband, Richard Gilbreath, for helping prepare some of the maps and graphics for authors without cartographic services and making changes on others. As Director of the Gyula Pauer Center for Cartography and GIS, Dick's work is always first class. And, finally, thanks are much in order to Natalya Tyutenkova for her interest, support, patience, and endurance in the past several years working on this megaproject, thinking and believing it would never end.

The journey continues.

February 2014 Stanley D. Brunn

Contents of Volume IV

Part XI Culture: Museums, Drama, Fashion, Food, Music, Sports and Science Fiction

Part XII Organizations

Contributors

Jamaine Abidogun Department of History, Missouri State University, Springfield, MO, USA

Afe Adogame School of Divinity, University of Edinburgh, Edinburgh, Scotland, UK

Christopher A. Airriess Department of Geography, Ball State University, Muncie, IN, USA

Kaarina Aitamurto Aleksanteri Institute, University of Helsinki, Helsinki, Finland

Mikael Aktor Institute of History, Study of Religions, University of Southern Denmark, Odense, Denmark

Elizabeth Allison Department of Philosophy and Religion, California Institute of Integral Studies, San Francisco, CA, USA

Johan Andersson Department of Geography, King's College London, London, UK

Stephen W. Angell Earlham College, School of Religion, Richmond, IN, USA

J. Clark Archer Department of Geography, School of Natural Resources, University of Nebraska-Lincoln, Lincoln, NE, USA

Ian Astley Asian Studies, University of Edinburgh, Edinburgh, UK

Steven M. Avella Professor of History, Marquette University, Milwaukee, WI, USA

Yulier Avello COPEXTEL S.A., Ministry of Informatics and Communications, Havana, Cuba

Erica Baffelli School of Arts, Languages and Cultures, University of Manchester, Manchester, UK

Bakama BakamaNume Division of Social Work, Behavioral and Political Science, Prairie View A&M University, Prairie View, TX, USA

Josiah R. Baker Methodist University, Fayetteville, NC, USA

Economics and Geography, Methodist University, Fayetteville, USA

Holly R. Barcus Department of Geography, Macalester College, St. Paul, MN, USA

David Bassens Department of Geography, Free University Brussels, Brussels, Belgium

Ramon Bauer Wittgenstein Centre for Demography and Global Human Capital (IIASA, VID/ŐAW, WU), Vienna Institute of Demography/Austrian Academy of Sciences, Vienna, Austria

Whitney A. Bauman Department of Religious Studies, Florida International University, Miami, FL, USA

Gwilym Beckerlegge Department of Religious Studies, The Open University, Milton Keynes, UK

Michael Bégin Department of Global Studies, Pusan National University, Pusan, Republic of Korea

Demyan Belyaev Collegium de Lyon/Institute of Advanced Studies, Lyon, France

Alexandre Benod Research Division, Department of Japanese Studies, Université de Lyon, Lyon, France

John Benson School of Teaching and Learning, Minnesota State University, Moorhead, MN, USA

Sigurd Bergmann Department of Philosophy and Religious Studies, Norwegian University of Science and Technology, Trondheim, Norway

Rachel Berndtson Department of Geographical Sciences, University of Maryland, College Park, MD, USA

Martha Bettis Gee Compassion, Peace and Justice, Peace and Justice Ministries, Presbyterian Mission Agency, Presbyterian Church (USA), Louisville, KY, USA

Warren Bird Research Division, Leadership Network, Dallas, TX, USA

Andrew Boulton Department of Geography, University of Kentucky, Lexington, KY, USA

Humana, Inc., Louisville, KY, USA

Kathleen Braden Department of Political Science and Geography, Seattle Pacific University, Seattle, WA, USA

Namara Brede Department of Geography, Macalester College, St. Paul, MN, USA

John D. Brewer Institute for the Study of Conflict Transformation and Social Justice, Queen's University Belfast, Belfast, UK

Laurie Brinklow School of Geography and Environmental Studies, University of Tasmania, Hobart, Australia

Interim Co-ordinator, Master of Arts in Island Studies Program, University of Prince Edward Island, Charlottetown, PE Canada

Dave Brunn Language and Linguistics Department, New Tribes Missionary Training Center, Camdenton, MO, USA

Stanley D. Brunn Department of Geography, University of Kentucky, Lexington, KY, USA

David J. Butler Department of Geography, University of Ireland, Cork, Ireland

Anne Buttimer Department of Geography, University College Dublin, Dublin, Ireland

Éric Caron Malenfant Demography Division, Statistics Canada, Ottawa, Canada

Lori Carter-Edwards Gillings School of Global Public Health, Public Health Leadership Program, University of North Carolina, Chapel Hill, NC, USA

Clemens Cavallin Department of Literature, History of Ideas and Religion, University of Gothenburg, Göteborg, Sweden

Martin M. Checa-Artasu Department of Sociology, Universidad Autónoma Metropolitana, Unidad Iztapalapa, Mexico, DF, Mexico

Richard Cimino Department of Anthropology and Sociology, University of Richmond, Richmond, VA, USA

Paul Claval Department of Geography, University of Paris-Sorbonne, Paris, France

Paul Cloke Department of Geography, Exeter University, Exeter, UK

Kevin Coe Department of Communication, University of Utah, Salt Lake City, UT, USA

Noga Collins-Kreiner Department of Geography and Environmental Studies, Centre for Tourism, Pilgrimage and Recreation, University of Haifa, Haifa, Israel

Louise Connelly Institute for Academic Development, University of Edinburgh, Edinburgh, Scotland, UK

Thia Cooper Department of Religion, Gustavus Adolphus College, St. Peter, MN, USA

Catherine Cottrell Department of Geography and Earth Sciences, Aberystwyth University, Aberystwyth, UK

Thomas W. Crawford Department of Geography, East Carolina University, Greenville, NC, USA

Janel Curry Provost, Gordon College, Wenham, MA, USA

Seif Da'Na Sociology and Anthropology Department, University of Wisconsin-Parkside, Kenosha, WI, USA

Erik Davis Department of Religious Studies, Rice University, Houston, TX, USA

Jenny L. Davis Department of American Indian Studies, University of Illinois, Urbana-Champaign, Urbana, USA

Kiku Day Department of Ethnomusicology, Aarhus University, Aarhus, Denmark

Renée de la Torre Castellanos Centro de Investigaciones y Estudios Superiores en Antropologia Social-Occidente, Guadalajara, Jalisco, Mexico

Frédéric Dejean Institut de recherche sur l'intégration professionnelle des immigrants, Collège de Maisonneuve, Montréal (Québec), Canada

Veronica della Dora Department of Geography, Royal Holloway University of London, UK

Sergio DellaPergola The Avraham Harman Institute of Contemporary Jewry, The Hebrew University of Jerusalem, Mt. Scopus, Jerusalem, Israel

Antoinette E. DeNapoli Religious Studies Department, University of Wyoming, Laramie, WY, USA

Matthew A. Derrick Department of Geography, Humboldt State University, Arcata, CA, USA

C. Nathan DeWall Department of Psychology, University of Kentucky, Lexington, KY, USA

Jualynne Dodson Department of Sociology, American and African Studies Program, Michigan State University, East Lansing, MI, USA

David Domke Department of Communication, University of Washington, Seattle, WA, USA

Katherine Donohue M.A. Diplomacy and International Commerce, Patterson School of Diplomacy and International Commerce, University of Kentucky, Lexington, KY, USA

Lizanne Dowds Northern Ireland Life and Times Survey, University of Ulster, Belfast, UK

Kevin M. Dunn School of Social Sciences and Psychology, University of Western Sydney, Penrith, NSW, Australia

Claire Dwyer Department of Geography, University College London, London, UK

Patricia Ehrkamp Department of Geography, University of Kentucky, Lexington, KY, USA

Paul Emerson Teusner School of Media and Communication, RMIT University, Melbourne, VIC, Australia

Chad F. Emmett Department of Geography, Brigham Young University, Provo, UT, USA

Ghazi-Walid Falah Department of Public Administration and Urban Studies, University of Akron, Akron, OH, USA

Yasser Farrés Department of Philosophy, University of Zaragoza, Pedro Cerbuna, Zaragoza, Spain

Timothy Joseph Fargo Department of City Planning, City of Los Angeles, Los Angeles, CA, USA

Michael P. Ferber Department of Geography, The King's University College, Edmonton, AB, Canada

Tatiana V. Filosofova Department of World Languages, Literatures, and Cultures, University of North Texas, Denton, TX, USA

John T. Fitzgerald Department of Theology, University of Notre Dame, Notre Dame, IN, USA

Colin Flint Department of Political Science, Utah State University, Logan, UT, USA

Daniel W. Gade Department of Geography, University of Vermont, Burlington, VT, USA

Armando Garcia Chiang Department of Sociology, Universidad Autónoma Metropolitana Iztapalapa, Iztapalapa, Mexico

Jeff Garmany King's Brazil Institute, King's College London, London, UK

Martha Geores Department of Geographical Sciences, University of Maryland, College Park, MD, USA

Hannes Gerhardt Department of Geosciences, University of West Georgia, Carrolton, GA, USA

Christina Ghanbarpour History Department, Saddleback College, Mission Viejo, CA, USA

Danilo Giambra Department of Theology and Religion, University of Otago-Te Whare Wänanga o Otägo, Dunedin, New Zealand/Aotearoa

Banu Gökarıksel Department of Geography, University of North Carolina, Chapel Hill, NC, USA

Margaret M. Gold London Guildhall Faculty of Business and Law, London Metropolitan University, London, UK

Anton Gosar Faculty of Tourism Studies, University of Primorska, Portorož, Slovenia

Anne Goujon Wittgenstein Centre for Demography and Global Human Capital (IIASA, VID/ŐAW, WU), International Institute for Applied Systems Analysis (IIASA), Laxenburg, Austria

Vienna Institute of Demography/Austrian Academy of Sciences, Vienna, Austria

Alyson L. Greiner Department of Geography, Oklahoma State University, Stillwater, OK, USA

Daniel Jay Grimminger Faith Lutheran Church, Kent State University, Millersburg, OH, USA

School of Music, Kent State University, Kent, OH, USA

Zeynep B. Gürtin Department of Sociology, University of Cambridge, Cambridge, UK

Cristina Gutiérrez Zúñiga Centro Universitario de Ciencias Sociales y Humanidades, El Colegio de Jalisco, Zapopan, Jalisco, Mexico

Martin J. Haigh Department of Social Sciences, Oxford Brookes University, Oxford, UK

Anna Halafoff Centre for Citizenship and Globalisation, Deakin University, Burwood, VIC, Australia

Airen Hall Department of Theology, Georgetown University, Washington, DC, USA

Randolph Haluza-DeLay Department of Sociology, The Kings University, Edmonton, AB, Canada

Tomáš Havlíček Faculty of Science, Department of Social Geography and Regional Development, Charles University, Prague 2, Czechia

C. Michael Hawn Sacred Music Program, Perkins School of Theology, Southern Methodist University, Dallas, TX, USA

Bernadette C. Hayes Department of Sociology, University of Aberdeen, Aberdeen, Scotland, UK

Peter J. Hemming School of Social Sciences, Cardiff University, Cardiff, Wales, UK

William Holden Department of Geography, University of Calgary, Calgary, AB, Canada

Edward C. Holland Havighurst Center for Russian and Post-Soviet Studies, Miami University, Oxford, OH, USA

Beverly A. Howard School of Music, California Baptist University, Riverside, CA, USA

Martina Hupková Faculty of Science, Department of Social Geography and Regional Development, Charles University, Prague 2, Czechia

Tim Hutchings Post Doc, St. John's College, Durham University, Durham, UK

Ronald Inglehart Institute of Social Research, University of Michigan, Ann Arbor, MI, USA

World Values Survey Association, Madrid, Spain

Marcia C. Inhorn Anthropology and International Affairs, Yale University, New Haven, CT, USA

Adrian Ivakhiv Environmental Program, University of Vermont, Burlington, VT, USA

Maria Cristina Ivaldi Dipartimento di Scienze Politiche "Jean Monnet", Seconda Università degli Studi di Napoli, Caserta, Italy

Thomas Jablonsky Professor of History, Marquette University, Milwaukee, WI, USA

Maria Jaschok International Gender Studies Centre, Lady Margaret Hall, Oxford University, Norham Gardens, UK

Philip Jenkins Institute for the Study of Religion, Baylor University, Waco, TX, USA

Wesley Jetton Student, University of Kentucky, Lexington, KY, USA

Shui Jingjun Henan Academy of Social Sciences, Zhengzhou, Henan Province, China

Mark D. Johns Department of Communication, Luther College, Decorah, IA, USA

James H. Johnson Jr. Kenan-Flagler Business School and Urban Investment Strategies Center, University of North Carolina, Chapel Hill, NC, USA

Lucas F. Johnston Department of Religion and Environmental Studies, Wake Forest University, Winston-Salem, NC, USA

Peter Jordan Austrian Academy of Sciences, Institute of Urban and Regional Research, Wien, Austria

Yakubu Joseph Geographisches Institut, University of Tübingen, Tübingen, Germany

Deborah Justice Yale Institute of Sacred Music, Yale University, New Haven, CT, USA

Akel Ismail Kahera College of Architecture, Art and Humanities, Clemson University, Clemson, SC, USA

P.P. Karan Department of Geography, University of Kentucky, Lexington, KY, USA

Sya Buryn Kedzior Department of Geography and Environmental Planning, Towson State University, Towson, MD, USA

Kevin D. Kehrberg Department of Music, Warren Wilson College, Asheville, NC, USA

Laura J. Khoury Department of Sociology, Birzeit University, West Bank, Palestine

Hans Knippenberg Department of Geography, Planning and International Development Studies, University of Amsterdam, Velserbroek, The Netherlands

Katherine Knutson Department of Political Science, Gustavus Adolphus College, St. Peter, MN, USA

Miha Koderman Science and Research Centre of Koper, University of Primorska, Koper-Capodistria, Slovenia

Lily Kong Department of Geography, National University of Singapore, Singapore, Singapore

Igor Kotin Museum of Anthropology and Ethnography, Russian Academy of Sciences, St. Petersburg, Russia

Katharina Kunter Faculty of Theology, University of Bochum, Bochum, Germany

Lisa La George International Studies, The Master's College, Santa Clarita, CA, USA

Shirley Lal Wijesinghe Faculty of Humanities, University of Kelaniya, Kelaniya, Sri Lanka

Ibrahim Badamasi Lambu Department of Geography, Faculty of Earth and Environmental Sciences, Bayero University Kano, Kano, Nigeria

Michelle Gezentsvey Lamy Comparative Education Research Unit, Ministry of Education, Wellington, New Zealand

Justin Lawson Health, Nature and Sustainability Research Group, School of Health and Social Development, Deakin University, Burwood, VIC, Australia

Deborah Lee Department of Geography, National University of Singapore, Singapore, Singapore

Karsten Lehmann Senior Lecturer, Science des Religions, Bayreuth University, Fribourg, Switzerland

Reina Lewis London College of Fashion, University of the Arts, London, UK

Micah Liben Judaic Studies, Kellman Brown Academy, Voorhees, NJ, USA

Edmund B. Lingan Department of Theater, University of Toledo, Toledo, OH, USA

Rubén C. Lois-González Departamento de Xeografía, Universidade de Santiago de Compostela, Galiza, Spain

Naomi Ludeman Smith Learning and Women's Initiatives, St. Paul, MN, USA

Katrín Anna Lund Department of Geography and Tourism, Faculty of Life and Environmental Sciences, University of Iceland, Reykjavik, Iceland

Avril Maddrell Department of Geography and Environmental Sciences, University of West England, Bristol, UK

Juraj Majo Department of Human Geography and Demography, Faculty of Sciences, Comenius University in Bratislava, Bratislava, Slovak Republic

Virginie Mamadouh Department of Geography, Planning and International Development Studies, University of Amsterdam, Amsterdam, The Netherlands

Mariana Mastagar Department of Theology, Trinity College, University of Toronto, Toronto, Canada

Alberto Matarán Department of Urban and Spatial Planning, University of Granada, Granada, Spain

René Matlovič Department of Geography and Applied Geoinformatics, Faculty of Humanities and Natural Sciences, University of Prešov, Prešov, Slovakia

Kvetoslava Matlovičová Department of Geography and Applied Geoinformatics, Faculty of Humanities and Natural Sciences, University of Prešov, Prešov, Slovakia

Hannah Mayne Department of Anthropology, University of Florida, Gainesville, FL, USA

Shampa Mazumdar Department of Sociology, University of California, Irvine, CA, USA

Sanjoy Mazumdar Department of Planning, Policy and Design, University of California, Irvine, CA, USA

Andrew M. McCoy Center for Ministry Studies, Hope College, Holland, MI, USA

Daniel McGowin Department of Geology and Geography, Auburn University, Auburn, AL, USA

James F. McGrath Department of Philosophy and Religion, Butler University, Indianapolis, IN, USA

Nick Megoran Department of Geography, University of Newcastle-upon-Tyne, Newcastle, UK

Amy Messer Department of Sociology, University of Kentucky, Lexington, KY, USA

Sarah Ann Deardorff Miller Researcher, Refugee Studies Centre, Oxford, UK

Kelly Miller Centre for Integrative Ecology, School of Life and Environmental Sciences, Deakin University, Burwood, VIC, Australia

Nathan A. Mosurinjohn Center for Social Research, Calvin College, Grand Rapids, MI, USA

Sven Müller Institute for Transport Economics, University of Hamburg, Hamburg, Germany

Erik Munder Institut für Vergleichende Kulturforschung - Kultur- u. Sozialanthropologie und Religionswissenschaft, Universität Marburg, Marburg, Germany

David W. Music School of Music, Baylor University, Waco, TX, USA

Kathleen Nadeau Department of Anthropology, California State University, San Bernadino, CA, USA

Caroline Nagel Department of Geography, University of South Carolina, Columbia, SC, USA

Pippa Norris John F. Kennedy School of Government, Harvard University, Cambridge, MA, USA

Government and International Relations, University of Sydney, Sydney, Australia

Orville Nyblade Makumira University College, Usa River, Tanzania

Lionel Obadia Department of Anthropology, Université de Lyon, Lyon, France

Daniel H. Olsen Department of Geography, Brigham Young University, Provo, UT, USA

Samuel M. Otterstrom Department of Geography, Brigham Young University, Provo, UT, USA

Barbara Palmquist Department of Geography, University of Kentucky, Lexington, KY, USA

Grigorios D. Papathomas Faculty of Theology, University of Athens, Athens, Greece

Nikos Pappas Musicology, University of Alabama, Tuscaloosa, AL, USA

Mohammad Aslam Parvaiz Islamic Foundation for Science and Environment (IFSE), New Delhi, India

Valerià Paül Departamento de Xeografía, Universidade de Santiago de Compostela, Galiza, Spain

Miguel Pazos-Otón Departamento de Xeografía, Universidade de Santiago de Compostela, Galiza, Spain

David Pereyra Toronto School of Theology, University of Toronto, Toronto, Canada

Bruce Phillips Loucheim School of Judaic Studies at the University of Southern California, Hebrew Union College-Jewish Institute of Religion, Los Angeles, CA, USA

Awais Piracha School of Social Sciences and Psychology, University of Western Sydney, Penrith, NSW, Australia

Linda Pittman Department of Geography, Richard Bland College of the College of William and Mary, Petersburg, VA, USA

Richard S. Pond Department of Psychology, University of North Carolina, Wilmington, NC, USA

Carolyn V. Prorok Independent Scholar, Slippery Rock, PA, USA

Steven M. Radil Department of Geography, University of Idaho, Moscow, ID, USA

Esther Long Ratajeski Independent Scholar, Lexington, KY, USA

Daniel Reeves Faculty of Science, Department of Social Geography and Regional Development, Charles University, Prague 2, Czechia

Arthur Remillard Department of Religious Studies, St. Francis University, Loretto, PA, USA

Claire M. Renzetti Department of Sociology, University of Kentucky, Lexington, KY, USA

Friedlind Riedel Department of Musicology, Georg-August-University of Göttingen, Göttingen, Germany

Sandra Milena Rios Oyola Department of Sociology and the Compromise after Conflict Research Programme, University of Aberdeen, Aberdeen, Scotland, UK

C.K. Robertson Presiding Bishop, The Episcopal Church, New York, NY, USA

Arsenio Rodrigues School of Architecture, Prairie View A&M University, Prairie View, TX, USA

Andrea Rota Institute for the Study of Religion, University of Bern, Bern, Switzerland

Rainer Rothfuss Geographisches Institut, University of Tübingen, Tübingen, Germany

Jeanmarie Rouhier-Willoughby Department of Modern and Classical Languages, Literatures and Cultures, University of Kentucky, Lexington, KY, USA

Rex J. Rowley Department of Geography-Geology, Illinois State University, Normal, IL, USA

Bradley C. Rundquist Department of Geography, University of North Dakota, Grand Forks, ND, USA

Simon Runkel Department of Geography, University of Bonn, Bonn, Germany

Joanna Sadgrove Research Staff, United Society, London, UK

Michael Samers Department of Geography, University of Kentucky, Lexington, KY, USA

Åke Sander Department of Literature, History of Ideas and Religion, University of Gothenburg, Göteborg, Sweden

Xosé M. Santos Departamento de Xeografía, Universidade de Santiago de Compostela, Galiza, Spain

Alessandro Scafi Medieval and Renaissance Cultural History, The Warburg Institute, University of London, London, UK

Anthony Schmidt Department of Communication Studies, Edmonds Community College, Edmonds, WA, USA

Mallory Schneuwly Purdie Institut de sciences sociales des religions contemporaines, Observatoire des religions en Suisse, Université de Lausanne – Anthropole, Lausanne, Switzerland

Anna J. Secor Department of Geography, University of Kentucky, Lexington, KY, USA

Hafid Setiadi Department of Geography, University of Indonesia, Depok, West Java, Indonesia

Fred M. Shelley Department of Geography and Environmental Sustainability, University of Oklahoma, Norman, OK, USA

Ira M. Sheskin Department of Geography and Regional Studies, University of Miami, Coral Gables, FL, USA

Lia Dong Shimada Conflict Mediator, Methodist Church in Britain, London, UK

Caleb Kwang-Eun Shin ABD, Korea Baptist Theological Seminary, Daejeon, Republic of Korea

J. Matthew Shumway Department of Geography, Brigham Young University, Provo, UT, USA

Dmitrii Sidorov Department of Geography, California State University, Long Beach, Long Beach, CA, USA

Caleb Simmons Religious Studies Program, University of Arizona, Tucson, AZ, USA

Devinder Singh Department of Geography, University of Jammu, Jammu, Jammu and Kashmir, India

Rana P.B. Singh Department of Geography, Banaras Hindu University, Varanasi, UP, India

Nkosinathi Sithole Department of English, University of Zululand, KwaZulu-Natal, South Africa

Vegard Skirbekk Wittgenstein Centre for Demography and Global Human Capital (IIASA, VID/ŐAW, WU), International Institute for Applied Systems Analysis, Laxenburg, Austria

Alexander Thomas T. Smith Department of Sociology, University of Warwick, Coventry, UK

Christopher Smith Independent Scholar, Tecumseh, OK, USA

Ryan D. Smith Compassion, Peace and Justice Ministries, Presbyterian Ministry at the U.N., Presbyterian Mission Agency, Presbyterian Church (USA), New York, NY, USA

Sara Smith Department of Geography, University of North Carolina, Chapel Hill, NC, USA

Leslie E. Sponsel Department of Anthropology, University of Hawaii, Honolulu, HI, USA

Chloë Starr Asian Christianity and Theology, Yale Divinity School, New Haven, CT, USA

Jeffrey Steller Public History, Northern Kentucky University, Highland Heights, KY, USA

Christopher Stephens Southlands College, University of Roehampton, London, UK

Jill Stevenson Department of Theater Arts, Marymount Manhattan College, New York, NY, USA

Anna Rose Stewart Department of Religious Studies, University of Kent, Canterbury, UK

Nancy Palmer Stockwell Senior Contract Administrator, Enerfin Resources, Houston, TX, USA

Robert Strauss President and CEO, Worldview Resource Group, Colorado Springs, CO, USA

Tristan Sturm School of Geography, Archaeology and Palaeoecology, Queen's University Belfast, Belfast, UK

Edward Swenson Department of Anthropology, University of Toronto, Toronto, ON, Canada

Anna Swynford Duke Divinity School, Duke University, Durham, NC, USA

Jonathan Taylor Department of Geography, California State University, Fullerton, CA, USA

Francis Teeney Institute for the Study of Conflict Transformation and Social Justice, Queen's University Belfast, Belfast, UK

Mary C. Tehan Stirling College, University of Divinity, Melbourne, Australia

Andrew R.H. Thompson The School of Theology, The University of the South, Sewanee, TN, USA

Scott L. Thumma Professor, Department of Sociology, Hartford Seminary, Hartford, CT, USA

Meagan Todd Department of Geography, University of Colorado, Boulder, CO, USA

Soraya Tremayne Fertility and Reproduction Studies Group, Institute of Social and Cultural Anthropology, University of Oxford, Oxford, UK

Gill Valentine Faculty of Social Sciences, University of Sheffield, Sheffield, UK

Inge van der Welle Department of Geography, Planning and International Development Studies, University of Amsterdam, Amsterdam, The Netherlands

Herman van der Wusten Department of Geography, Planning and International Development Studies, University of Amsterdam, Amsterdam, The Netherlands

Robert M. Vanderbeck Department of Geography, University of Leeds, West Yorkshire, UK

Jason E. VanHorn Department of Geography, Calvin College, Grand Rapids, MI, USA

Viera Vlčková Department of Public Administration and Regional Development, Faculty of National Economy, University of Economics in Bratislava, Bratislava, Slovakia

Geoffrey Wall Department of Geography and Environmental Management, University of Waterloo, Waterloo, ON, Canada

Robert H. Wall Counsel, Spilman Thomas & Battle, Winston-Salem, NC, USA

Kevin Ward School of Theology and Religious Studies, University of Leeds, West Yorkshire, UK

Barney Warf Department of Geography, University of Kansas, Lawrence, KS, USA

Stanley Waterman Department of Geography and Environmental Studies, University of Haifa, Haifa, Israel

Robert H. Watrel Department of Geography, South Dakota State University, Brookings, SD, USA

Gerald R. Webster Department of Geography, University of Wyoming, Laramie, WY, USA

Paul G. Weller Research, Innovation and Academic Enterprise, University of Derby, Derby, UK

Oxford Centre for Christianity and Culture, University of Oxford, Oxford, UK

Cynthia Werner Department of Anthropology, Texas A&M University, College Station, TX, USA

Geoff Wescott Centre for Integrative Ecology, School of Life and Environmental Sciences, Deakin University, Burwood, VIC, Australia

Carroll West Center for Historic Preservation, Middle Tennessee State University, Murfreesboro, TN, USA

Gerald West School of Religion, Philosophy and Classics, University of KwaZulu-Natal, Scottsville, South Africa

Mark Whitaker Department of Anthropology, University of Kentucky, Lexington, KY, USA

Thomas A. Wikle Department of Geography, Oklahoma State University, Stillwater, OK, USA

Justin Wilford Department of Geography, University of California, Los Angeles, CA, USA

Joseph Witt Department of Philosophy and Religion, Mississippi State University, Mississippi State, MS, USA

John D. Witvliet Calvin Institute of Christian Worship, Calvin College and Calvin Theological Seminary, Grand Rapids, MI, USA

Teresa Wright Department of Political Science, California State University, Long Beach, CA, USA

Ernest J. Yanarella Department of Political Science, University of Kentucky, Lexington, KY, USA

Yukio Yotsumoto College of Asia Pacific Studies, Ritsumeikan Asia Pacific University, Beppu, Oita, Japan

Samuel Zalanga Department of Anthropology, Sociology and Reconciliation Studies, Bethel University, St. Paul, MN, USA

Donald J. Zeigler Department of Geography, Old Dominion University, Virginia Beach, VA, USA

Shangyi Zhou School of Geography, Beijing Normal University, Beijing, China

Teresa Zimmerman-Liu Departments of Asian/Asian-American Studies and Sociology, California State University, Long Beach, CA, USA

Part IX
Secularization

Chapter 111
Secularization and Transformation of Religion in Post-War Europe

Hans Knippenberg

111.1 The Exceptional Case

In 2002, the British sociologist Grace Davie published a study on the "parameters of faith in the modern world" under the telling title *Europe: The exceptional case*. In a world in which religion has taken a prominent part, Europe is the scene of ongoing secularization in the meaning of the decline of traditional religion both at the societal and the individual level. State and church are becoming more separated and increasingly more people do not adhere to the traditional churches and religious communities, do not visit church on a regular basis and do not believe in God. That does not mean that people are no longer interested in spiritual affairs. Traditional religion is transforming into new, often hybrid forms of religiosity and spirituality such as the New Age movement, which has been described as the searching for the divine self.

Both processes, secularization and transformation of religion, will be discussed in this chapter in their geographical variety. Before doing so, we shall spend a few words on the background of the concept secularization, followed by an overview of the main theories on the subject.

111.2 Secularization

The term secularization derives from the Latin *saecularis*, the adjective of the noun *saeculum*, which originally meant a long period of time, but in Christian times also became used in the meaning of "the world in which we live," a world that is

H. Knippenberg (✉)
Department of Geography, Planning and International Development Studies, University of Amsterdam, Schoener 49, 1991 XA Velserbroek, The Netherlands
e-mail: hh.knippenberg@uva.nl

S.D. Brunn (ed.), *The Changing World Religion Map*,
DOI 10.1007/978-94-017-9376-6_111

characterised by sin and the rejection of God (Swatos and Christiano 1999; Bremmer 2008). Monastic priests, for example, were distinguished from the parish priests "in the world," who were called secular. In sixteenth century France, *séculariser* meant the transfer of goods from the possession of the Church into that of the world. In seventeenth century Germany, the Latin *saecularisatio* referred to the closure of monasteries or the liquidation of goods of the Catholic Church. In the eighteenth century the German terms *Secularisirung* and *Secularisation* were used in the same meaning, but in the nineteenth century, Karl Marx used *Säkularisation* in a more or less modern meaning, although German thinkers virtually always use the term *Verweltlichung*. At the turn of the century, Max Weber, followed by Ernst Troeltsch, introduced secularization into sociology, but the idea did not became a central concept in his sociology of religion. It was not until the early 1960s that the term was accepted as a major sociological concept due to the works of a generation of theologically interested sociologists and sociologically interested theologians such as Peter Berger (1967), Thomas Luckmann (1967), Bryan Wilson (1966) and David Martin (1969).

The most simple definition of secularization in the modern meaning is the decline of religion be it at the societal or the individual level. That means that secularization is always related to the definition of religion, and that the 'truth' of secularization depends on historical evidence, to wit, that people or societies were more religious in the past than in the present (Swatos and Christiano 1999: 213–214). However, the situation becomes complicated, when the character of religion changes over time. And that is what many social scientists observe to-day: on the one hand, traditional, organized religion is declining and on the other hand, religion is transforming into new, often privatized forms of religiosity and spirituality. In one way or the other, we must, therefore, try to differentiate between these two processes.

111.3 Theories of Secularization

For a long time, modernization was supposed to be the driving force behind the secularization in the Western part of the World, based on ideas of Max Weber concerning a growing rationality (*die Entzauberung der Welt*=the disenchantment or "de-magi-fication" of the world; Swatos and Christiano 1999: 212), or from Emile Durkheim on functional differentiation, that is the fragmentation of social life as specialized institutions are created to handle functions previously carried out by one institution, in this case the church (see for the classical secularization paradigm Wilson 1969, 1982, 1998; Bruce 2002, 2011). Once, religion covered the whole of society as a kind of *sacred canopy* (Berger 1967), now religion has been reduced to only one of many domains of modern society and has been privatized to such a large extent that it became "invisible" (Luckmann 1967). Also the rise of new secular ideologies such as socialism and nationalism was considered to have contributed to a declining importance of religion, both on the individual and societal level.

How plausible these explanations were in the European context, they could not explain why a modern nation such as the United States remained so religious. According to recent polls only 12–16 % of the Americans do not belong to a religious denomination (Bruce 2011: 158). That is why American sociologists developed a different kind of theory based on the principles of a market (Stark and Brainbridge 1985, 1987; Stark and Iannaccone 1994; Finke et al. 1996; Finke and Stark 1998; Stark 1999; Stark and Finke 2000). In fact, they assumed a stable demand of religious "products." They explained differences in church going and religiosity by differences in the supply side of the religious market. A strong regulation of the religious market (for instance in case of a state church, that by its monopoly prohibits the freedom of choice of individuals) and a small diversity of the supply side (that is, no competition between the religious "firms") should conduce secularization. Because the United States are characterized by a wide variety of churches and sects and, moreover, (in contrast to most European countries) there is no state church of otherwise dominant church, an optimal situation of something for everybody has developed, that prevented secularization.

Although these market theories experienced much empirical support in the American context, there still was much criticism both on theoretical and empirical grounds (Bruce 1999, 2002, 2011; Norris and Inglehart 2004; Wunder 2005; Gooren 2006; Pollack 2008). A first point of criticism concerned the rational choice paradigm that was behind it. Like many economic theories also the religious market model is based on the assumption that individuals make rational choices in order to reach an optimum between its profits and losses. In practice, this assumption does not hold, because no one is able to consider all relevant factors and because the knowledge on which the choice has been made is selective and strongly influenced by tradition and culture.

A second point of criticism concerns the one-sided orientation on the supply side of the religious market. By assuming a constant demand of religion individuals are reduced to "passive recipients of collectively created goods" (Bankston 2003: 165). From social science research we know that religious choices are affected by several individual and group characteristics such as education level, occupation, racial or ethnic minority status, socialization by family and peers, individual life events, etc. None of these characteristics are distributed in a constant fashion across history or geography.

Apart from its theoretical content, the religious market theories have been criticized on empirical grounds. A main hypothesis that was derived from these theories is that the more diversity of the supply side of the religious market, the more religiosity and religious participation. However, this hypothesis is not supported in the European context (Chaves and Gorski 2001; Wunder 2005: 175–186). The main feature of the changing religious landscape of Europe is the combination of growing religious plurality caused by immigration and crumbling religious monopolies on the one side, and continuing secularization on the other (Henkel and Knippenberg 2005). Thus in time perspective, growing diversity correlates with growing secularization, just the opposite of what religious market theories expect.

Also the assumption of the market theorists that there is a constant need for religion misses a firm empirical basis. In their courageous effort to analyze statistically the worldwide differences in religiosity, Norris and Inglehart (2004) make plausible that their data do not support a supply theory on secularization. Consequently, they reject the notion of a stable demand of religion in all parts of the world. Instead, they find a diminishing demand of religion in the more developed parts of the world, and a constant high level or even growing level of religiosity in other parts. In order to explain these differences, they developed their theory of existential security. Growing up in a society with a large amount of insecurity would lead to a relatively high level of religiosity and vice-versa. The relatively high level of secularization in the modern (West) European welfare states can thus be explained by the relatively high level of existential security.

Casanova (2001: 426) has pointed at processes of territorialisation in the past, which included a territorial embedding of religion. This territorialisation had to do with the rise of a system of modern nation states. The territorial embedding of religion was most influential in states where state churches developed. The well-known principle of the Augsburg Peace Treaty of 1555, that was confirmed by the Westphalian Peace Treaty in 1648, *cuius regio, eius religio* was the basis. *Cuius regio, eius religio* means that the rulers decide on the religion of the people on their territory.

This territorialization has three dimensions: c*losure, control* and *identity* (Dijkink 2008: 181). *Closure* represents the delimitation of territory, including all arrangements that define the relations with the outside world; *control* represents the influence over persons, activities and resources on the territory; *identity* represents the ideology that united the people and legitimates the power and authority over the territory. The first two dimensions concern state formation, while the latter concerns nation building.

State formation and nation building are crucial for an understanding of the geography of religions and its decline in Europe. Each country has developed its own mode of state-church/religion relations (Robbers 1995), that also influenced the level and speed of the secularization of the different societies. Two major divides are important in this respect: the divide between Western and Eastern Christianity, that originated in the schism between Rome and Byzantium in 1054, but had also influenced the conditions for state formation after 1500, and the North-south divide between Protestant and Roman-Catholic Europe after the Reformation of 1517, that had strongly influenced the conditions for nation building (Madeley 2003; Knippenberg 2006a). The West-east divide corresponds with Huntington's (1996) divide between Western and Orthodox civilizations and is also seen by Davie (2000: 3) as "a far more fundamental division … than the relatively recent opposition between communist and non-communist Europe in the post-war period."

The Reformation can be seen as the first major step toward the definition of territorial nations. Lutherans and Calvinists broke with the supra-territoriality of the Roman Church and merged the ecclesiastical bureaucracies with the secular territorial ones, thus accentuating the cultural and religious significance of the territorial borders and contributing to rapid cultural integration and nation building at state level in the North (Rokkan 1979: 80; Flora et al. 1999: 144). In the South, the Roman Catholic Church continued its supra-territorial power.

In eighteenth and nineteenth century Western Europe, the Enlightenment and French Revolution challenged the existing bond between "Crown and Altar" and brought (more) separation between church and state. Political power had to be legitimized by the nation instead of God. The confessional state changed into a secular or religiously neutral state and religious freedom ended the implementation of the *cuius regio, eius religio*-principle, although it took some time before this freedom of religion was realized completely. This greater freedom also included the freedom not to believe. A strict separation, as prescribed by the French 1905 Act (Bertrand 2006), was quite rare, however. The cultural revolutions of the 1960s further challenged the traditional state-church and nation-religion relations followed by mass secularization.

In Eastern Europe, the Russian Revolution brought an abrupt end to the close bond between Tsar and Orthodox Church. For the first time, the *cuius regio, eius religio*-principle was implemented from an atheistic viewpoint. The Orthodox Church lost its possessions and most of its rights and a 'scientifically based' atheism was instructed from Kindergarten to university. After the Second World War, a comparable anti-religious policy with slight differences in every state was initiated in the other states of the Communist bloc. This resulted in the strong secularization of most (but certainly not all) East European societies, until the velvet revolutions of 1989–1991 opened the door to more religious freedom and even a religious revival in some cases.

Notwithstanding these general trends, the levels of religiosity and secularization are differing considerably from country to country, as we shall see in the next section.

111.4 The Decline of Traditional Religion

Anyone, who wants to analyze comparable data on (the decline of) religion at different time cuts in the countries of post-War Europe meets immense problems. In many cases, comparable data are simply missing and as far as data are available, these are not always reliable or consistent and vary a great amount depending of the source (for instance census or surveys) or the kind of questioning.

An important data source are the World Values Surveys (WVS) and European Values Surveys (EVS), which both started in 1981. A tremendous effort was made by Norris and Inglehart (2004), who also incorporated older surveys and sources in their worldwide comparison on religion and politics. At the time of writing, the EVS provided the most recent data (2008; see Halman et al. 2012 for an overview). Therefore, we are using EVS data in this chapter and, if possible, add some data derived from older sources. The 1981 EVS were held in 15 (West) European countries, the 2008 EVS in 48 European countries.

With these remarks in mind we present some tables and maps on secularization issues in the countries of Europe. These issues concern three classical dimensions of religious involvement: belonging, practicing and believing. Each subsection starts with longitudinal data to analyze the process of declining traditional religion and concludes with a recent map of the level of religious involvement. Germany is split up in a west and an east part (the former GDR), and the United Kingdom in Great Britain and Northern Ireland. Kosovo and Turkish Cyprus are separate units. There are no data for Andorra, Liechtenstein, Monaco and San Marino.

111.4.1 Religious Belonging

Religious belonging is measured by the proportion of the population that answered "no" when asked "do you belong to a religious denomination?" The 1981 EVS included only West European countries. We, therefore, provide separate tables for West European and East European (=post-Communist and/or Orthodox) countries. Comparing both tables allows us to analyze whether the countries that experienced a Communist regime followed a different secularization path. European countries without longitudinal data on religion are not recorded in the tables.

The general trend in the West European countries is increasing non-denominationalism from about 10 % to more than a quarter of the population on average (Table 111.1). Comparing the results of the 1981 EVS with the 2008 EVS, all 15 countries had lower levels of non-denominationalism in 1981 than in 2008. In the Netherlands and France the decline of religious belonging had gone so far that even the majority of the population does not belong to a religious denomination

Table 111.1 Percent of the West European population that does not belong to a religious denomination, 1960–2008

West European country	1960/61	1970/71	1981	1990	1999	2008
Austria				14.6	11.9	17.2
Belgium			16.3	31.7	36.5	42.5
Denmark			5.6	8.4	10.0	12.3
Finland			7.8	11.1	11.9	23.8
France			26.3	38.5	42.5	50.7
Germany West	2.8	3.9	8.9	10.7	13.4	15.7
Great Britain			9.4	42.3	16.6	44.7
Iceland			1.3	2.1	4.3	8.2
Ireland			1.3	3.9	8.6	12.9
Italy			6.4	14.7	17.8	18.3
Malta			0.0	2.5	1.5	2.5
Netherlands	18.3	23.6	36.5	49.3	55.2	51.5
Northern Ireland			3.4	9.5	14.0	21.7
Norway			4.1	9.8		19.9
Portugal				27.6	11.0	18.8
Spain			8.9	13.1	18.0	25.4
Sweden			6.9	18.2	24.2	33.5
Average 15			9.5	17.7		25.6
Average 17				18.1		26.2

Data sources: European Values Surveys 1981–2008, Henkel (2001: 268), Knippenberg (1992: 276)

any more. Belgium and Great Britain almost reached the same level. The Netherlands have a long tradition of declining religious belonging that started already in the 1880s and reached a level of non-denominationalism of 14 % in 1930 (Knippenberg 1998). Following the 1789 French Revolution, the 1795 Batavian Revolution was a watershed in the relationship between church and state. The Calvinist Church lost its privileged position, all citizens (including Catholics and Jews) got equal rights independent of their religion and church and state became more separated (Knippenberg 2006b). The cultural revolution of the 1960s crumbled the religious "pillars," that had supported organized religion between 1930 and 1960, and accelerated the decline of religious belonging from 18 % in 1960 to 52 % in 2008, or even more than 60 % according to other surveys (Bernts et al. 2007: 14).

Great Britain shows a remarkable time series, probably due to a different kind of questioning in 1990 and 2008 compared to 1981 and 1999. However, there is no doubt that the trend is declining religious belonging (Davie 1994). Even very religious Ireland and Northern Ireland are secularizing, be it on a much lower level. Catholicism became an important element in the national identity of the Irish in their struggle for independence against the (Protestant) British rulers, which went on in Northern Ireland until fairly recently.

Unfortunately, reliable representative data on religious affiliations in the countries of Eastern Europe during the communist period are missing (Norris and Inglehart 2004: 111–112). Therefore, the analysis of long-term trends starts in 1990. Table 111.2 presents the available EVS data. There are remarkable differences with the secularization trends in Table 111.1. In contrast to the West European nations,

Table 111.2 Percent of the East European population that does not belong to a religious denomination, 1990–2008

East European country	1990	1999	2008
Bulgaria	65.9	30.0	26.2
Belarus		47.8	28.4
Croatia		11.1	16.8
Czech Republic	60.3	66.3	71.0
Estonia	87.2	75.2	68.8
Germany East	61.6	66.5	77.0
Greece		4.0	3.6
Hungary	41.7	42.9	45.3
Latvia	63.5	40.7	34.8
Lithuania	36.7	18.7	14.0
Poland	3.7	4.3	4.5
Romania	5.9	2.4	2.1
Russian Federation		49.5	38.1
Slovakia	28.2	23.2	23.1
Slovenia	26.4	30.0	28.8
Ukraine		43.6	23.4
Average 11	43.7	36.4	36.0
Average 16		34.8	31.6

Data source: European Values Survey 1990–2008

Bulgaria, Romania, Belarus, the Baltic states, Russia, Slovakia and Ukraine, all display a rise in religious belonging. Still, there is no general religious revival (in the sense of belonging) in post-Communist Europe, because Czechia, Hungary, Slovenia, Croatia, Poland and East Germany followed the Western trend of declining religious belonging. It is probably not a coincidence that the last group includes the most Western oriented nations and none of the Orthodox countries.

The differences between the Eastern European nations in the level of religious belonging are significant. Religious belonging is extremely low in Estonia, Czechia and East Germany, where more than two-thirds of the population does not belong to a religious denomination. Extremely high levels are found in Poland, Greece and Romania, where less than 6 % of the population did not belong to a religious denomination during the 1990–2008 period. Two elements are important in the explanation of these differences: the type of religion, whether it be Orthodox, Catholic or Protestant, and the relationship between religion and the "national myth" (Martin 2011: 135–148). Orthodoxy (and also Islam) is less vulnerable for secularization or even atheist state policies than Protestantism, because Orthodoxy is less differentiated at the national and communal level and more oriented on ritual acts that can be done at home and do not require assent as in Catholicism or personal commitment as in Protestantism. In the classical secularization paradigm, Protestantism is connected with individualism, social and structural differentiation, and rationality, all major dimensions of secularization (Bruce 2011: 27). That means that in general (all other things being equal) Protestant nations will secularize more than Catholic nations, which will in their turn secularize more than Orthodox (and Muslim) nations.

Concerning the relationship between religion and national identity, resistance against secularization is dependent on the role of the church to alien rule. In Poland the (Catholic) Church played a major role in the resistance against and bringing down of the atheist communist regime, in particular after the Pontificate of (the Polish) John Paul II. Also Greece knows a strong bond between nation and (Orthodox) religion, which goes back to the resistance against Muslim rulers in the past, but showed up again recently in the Greek resentment against the (supposed) secular agenda of the European Union (Halikiopoulou 2010). Very religious Romania was "a Latin isle in a Slav sea" and has been threatened in the past from three sides, by (predominantly) Orthodox/atheist Russia, (predominantly) Muslim Turkey and (predominantly) Catholic Austria-Hungary. The Romanian Orthodox Church remained the only vehicle of a continuing Romanian identity, in contrast to Slavic Bulgaria, where the Orthodox Church identifies with Russia (Martin 2011: 135–148).

The high levels of secularization in Czechia and Estonia have also to do with resistance against a foreign ruler with in this cases the same religion. In Estonia religion is associated with (Lutheran) German dominance in the past. In Czechia (Bohemia in particular), religion was associated with the Hapsburg re-Catholicization after the defeat of the Protestant Czechs in 1621. Following the creation of independent Czechoslovakia in 1918, the Catholic church lost its privileged position and was separated from state power (Havlíček 2006). The forced migration into Germany of German-speaking Catholics after World War II, but also relatively high levels of industrialization and urbanization further strengthened the secularization of the Czech Lands. The region of current Slovakia, that remained Catholic in the past, has not developed such a difficult relationship with Catholicism and shows a much higher level of religious belonging in 2008.

East (then Central) Germany, which has now the lowest level of religious belonging, had already a distinctive form of secularization well before the First World War (Martin 2011: 151–153). As the most Protestant region of Germany (Henkel 2001: 264), it had experienced a relatively great degree of religious individualization and weakening of the communal tie still retained in Catholicism. Secularist and anti-clerical movements and "free-thinking" organizations emerged early in that part of Germany. The ties of denominational culture were further weakened by National Socialism which was relatively strong in some parts of the region. As a result East Germany was already fairly secularized, when the Communists took over state power (Froese and Pfaff 2005). These Communists not only carried out an anti-religious policy, which undermined religious institutions, but also offered an ideological exchange whereby East Germans became the innocent proletarian victims of Nazism on the condition that they accepted their assigned role in the progress of Communism and its atheist worldview. 'National identity, which provided a base for resistance in Poland, could not exercise that role in East Germany because it had been so thoroughly discredited between 1933 and 1945 by the Nazi regime' (Martin 2011: 152).

The current state of religious belonging of all the European countries is recorded in Fig. 111.1 that maps the proportion of the population that does not belong to a

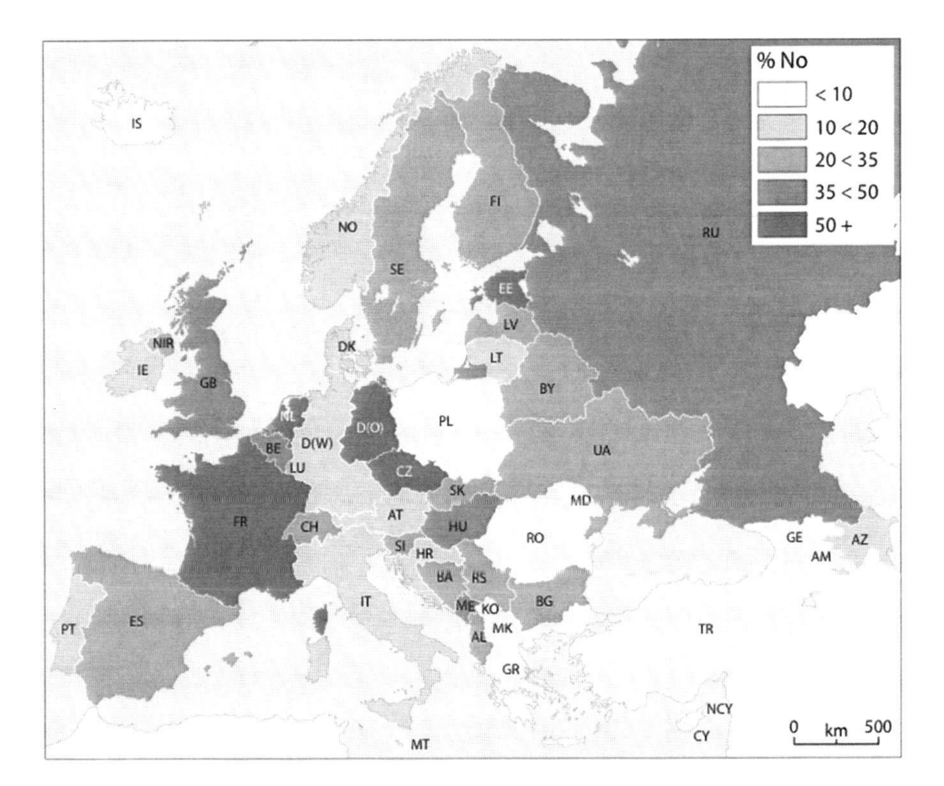

Fig. 111.1 Percent of the European population that does not belong to a religious denomination, 2008 (Map by Hans Knippenberg, data from European Values Survey 2008)

religious denomination based on the 2008 EVS. It appears that large parts of the European population do not belong to any religious denomination. The average level of non-denominationalism is almost a quarter of the population, but the differences between the European countries are tremendous, varying from <1 % in Cyprus to 77 % in East Germany. In general, the Protestant North is more secularized than the Catholic South, which is more secularized than the Orthodox East and the Muslim South-East. However, despite being strongly secularized in many other aspects, the Scandinavian countries have a relatively high level of belonging to the Lutheran denomination. Being Lutheran is more an expression of their national identity, than of their religious identity (see for instance Raivo 2005). Within the Catholic bloc, republican France and its northern neighbour Belgium have lower levels of religious belonging than the Southern partners, such as Spain, Portugal and Italy, which had a closer connection of state and (Catholic) church for a long time. In Spain, for example, the bond between state and (Catholic) church reached its height during the Franco regime (1939–1975), when the Roman-Catholic Church "obtained such a power that the religious sphere was hard to distinguish from the political sphere, a situation called National-Catholicism" and "Catholicism became state religion" (De Busser 2006: 286).

111.4.2 Religious Practice

Table 111.3 shows the West European trends in the second dimension of secularization: declining religious practice, measured by the percentage of the population that is attending religious services once a week or more according to their own report. Note, that actual church going is in general lower than self reported church going. The 1981–2008 data are derived again from the EVS; the 1970/71 data stem from the Mannheim Eurobarometer Trend File (Norris and Inglehart 2004: 72).

The general trend is a (strongly) declining religious practice, in particular when the 1970/71 data are drawn into the comparison. The average proportion of weekly church goers of the eight countries for which 1970/71 data are available diminished from 41 % in 1970/71 to 16 % in 2008. Even very religious Ireland underwent a strong decline from 91 to 40 %. The only exemptions are a few countries that had already very low levels of church going in 1981 such as Iceland and Norway. The Scandinavian countries that (as we saw) had a relatively high level of religious belonging appear to have a very low level of religious practice. In 2008, 92 % of the Icelanders said they belonged to the Lutheran denomination, but only 4 % said that they go to church on a weekly basis, illustrating that being Lutheran is more part of their national than of their religious identity. The average level of the 14 EVS countries in 1981 declined from 30 % in 1981 to 20 % in 2008.

In contrast to Western Europe, there is no clear general trend in Eastern Europe between 1990 en 1999, the first decennium after the collapse of communism (Table 111.4). In some countries (Latvia, Bulgaria, Romania and Slovakia), religious practice has increased, in others (Hungary, Poland, Slovenia, Czechia and East Germany), church going has declined. For the other countries relevant data are

Table 111.3 Percent of the West European population that visits church at least once a week, 1970–2008

West European country	1970/1971	1981	1990	1999	2008
Austria			25.0	22.8	16.6
Belgium	55	30.3	22.9	17.5	10.8
Denmark	5*	3.2	2.5	2.7	2.6
Finland			3.8	5.4	4.3
France	25	11.9	10.2	7.6	6.2
Germany West	34	21.8	18.6	15.6	10.0
Great Britain	16*	13.8	13.0	14.4	12.5
Iceland		2.5	2.4	3.2	3.9
Ireland	91*	82.4	80.9	59.0	40.0
Italy	57	36.0	40.6	40.5	33.5
Malta		92.3	88.0	82.2	76.9
Netherlands	45	26.9	20.7	13.9	15.2
Northern Ireland		52.6	50.0	48.6	40.6
Norway		5.6	5.1		5.7
Portugal			33.3	36.4	23.2
Spain		41.2	33.4	25.5	17.2
Sweden		13.8	13.0	14.4	12.5
Average 8	41.0	28.3	26.2	21.4	16.4
Average 14		31.0	28.7		20.5

Data sources: European Values Surveys 1981–2008, Norris and Inglehart (2004: 72) (1970/1971). *1973

Table 111.4 Percent of the East European population that visits church at least once a week, 1990–2008

East European country	1990	1999	2008
Bulgaria	5.5	8.9	5.6
Belarus		5.6	7.1
Croatia		31.6	24.6
Czech Republic	8.4	7.0	8.3
Estonia		3.7	3.7
Germany East	8.8	5.6	3.6
Greece		14.0	18.5
Hungary	14.0	10.5	9.0
Latvia	3.2	6.6	5.9
Lithuania		17.6	11.8
Poland	65.6	59.0	53.1
Romania	18.7	24.9	28.3
Russian Federation		3.2	5.3
Slovakia	31.7	40.6	34.6
Slovenia	22.9	17.2	17.1
Ukraine		9.1	12.4
Average 9	19.9	20.0	18.4
Average 16		16.6	15.6

Data source: European Values Surveys 1990–2008

missing. Between 1999 and 2008, the declining trend is dominating, but most Orthodox countries (Greece, Romania, Russia, Belarus and Ukraine) witnessed an increase in church going. As far as data are available, only East Germany, Hungary and Poland show a steady decline in church going during the whole 1990–2008 period, whereas Romania is the only country with a steady increase. In general East Europeans are visiting church less frequently than West Europeans.

Although declining, Poland still has one of the highest levels of church going of the whole of Europe. More than half of the Poles say they go to church once a week or more. The other East European countries are far below that level and on average are visiting church less than the countries of Western Europe. Only in Slovakia, Croatia and Romania does a quarter or more of the population attend religious services on a weekly (or more) basis.

The current state of religious practice in all the European countries is recorded in Fig. 111.2 based on the 2008 EVS. It appears that the great majority of the European population does not attend religious services on a weekly basis. The average level of weekly (or more) church going in the countries of Europe in 2008 is about 17 %. However, the differences between the European countries are again tremendous, varying from 77 % in the Isle of Malta to 3 % in Denmark. Denmark again illustrates the civil character of belonging to the Lutheran church.

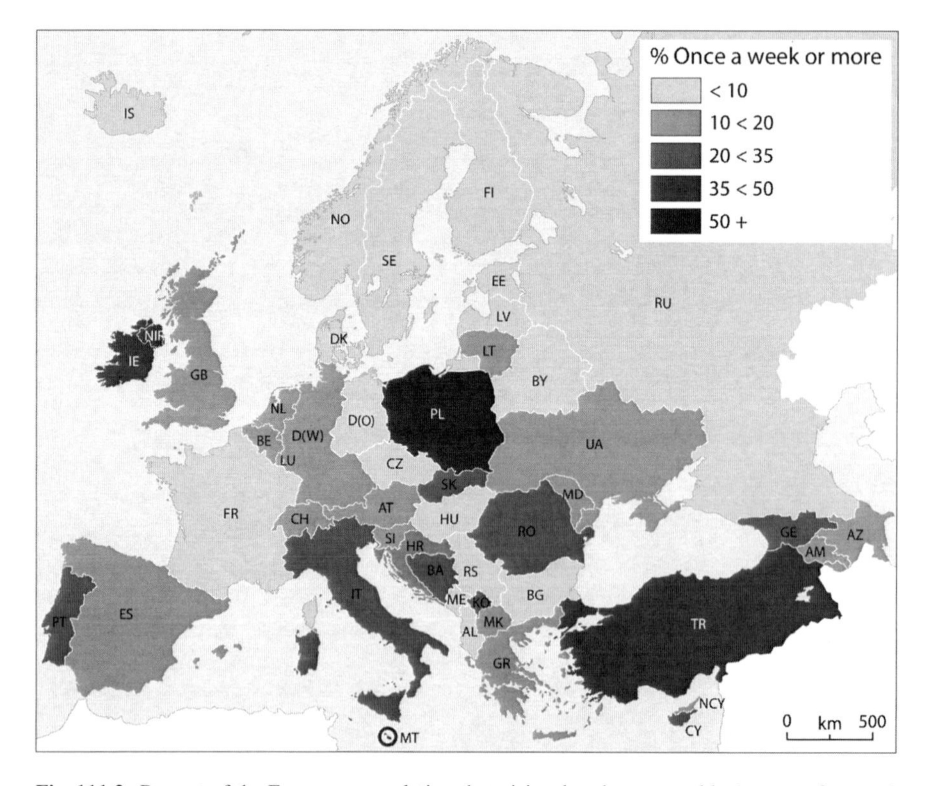

Fig. 111.2 Percent of the European population that visits church on a weekly (or more frequent) basis, 2008 (Map by Hans Knippenberg, data from European Values Survey 2008)

In most European countries less than 18 % attend religious services at least once a week. More than one-third is even below a level of 10 % regular church goers. In the Protestant North, including all Scandinavian countries, we found very low levels of church going, but also in Central European countries (East Germany, Czechia and Hungary), and East European Orthodox countries (Bulgaria, Belarus, Montenegro, Russia and Serbia). Relatively high levels of attending religious services are found in the Catholic countries such as Malta, Poland, Ireland, Slovakia, Italy, Catholic/Protestant mixed Northern Ireland (but not in republican France and Belgium), and in (predominantly) Islamic countries: Turkey, Kosovo and Bosnia-Herzegovina.

111.4.3 Religious Belief

The third dimension of religious involvement is belief. Table 111.5 provides some relevant information for a limited number of West European countries. Only countries are included with longitudinal data that started in 1981 or earlier. Recorded is the percentage of the respondents that answered "yes" when asked: "do you believe in God?" There certainly was a decline in the belief in God in the West European countries in the post-War period. However, that decline was not as significant as the decline in religious belonging and practising. The average level of belief in God of the six countries, for which 1947 data are available, has declined from 79 % in 1947 to 57 % in 2008. The eight countries with 1968 data have declined from 76 % in 1968 to 62 % on average in 2008.

Table 111.5 Percent of the West European population that believes in God, 1947–2008

West European country	1947	1968	1981	1990	1999	2008
Austria		85		85.7	86.8	79.6
Belgium			86.8	69.1	71.0	61.4
Denmark	80		68.2	64.3	68.9	62.9
Finland	83	83		75.9	82.5	69.0
France	66	73	68.0	61.9	61.5	52.4
Germany West		81	81.9	78.1	76.9	73.4
Great Britain		77	82.6	78.1	71.8	64.6
Iceland			81.3	85.1	84.4	72.1
Ireland			97.1	97.6	95.7	89.9
Italy			89.5	90.6	93.5	90.7
Malta			100.0	99.6	99.5	98.6
Netherlands	80	79	72.2	64.5	59.6	57.6
Norway	84	73	75.5	65.0		56.0
Spain			91.9	86.3	86.7	76.3
Sweden	80	60	60.4	45.2	53.4	45.6
Northern Ireland			96.6	96.7	93.2	91.4
Average 6	78.8			62.8		57.3
Average 8		76.4		69.3		62.3

Data sources: European Values Surveys 1981–2008, Norris and Inglehart (2004: 90) (1947, 1968)

The strongest decline was in Sweden, from 80 % in 1947 to 46 % in 2008. That means that the majority of the Swedish population does not believe in God any more. Hardly any decline occurred in a few traditional Catholic countries such as Italy, Malta and Ireland, where a very large majority of 90–99 % still believes in God. The same holds true for Northern Ireland. France had already a relatively low level of religious belief in 1947, but even that level was significantly lower in 2008, so that now almost half of the population does not believe in God any more. In general, the Protestant countries are more secularized in their beliefs than the Catholic ones.

Still, in 2008 on average almost 80 % of all the European nations said they believed in God. What they mean by belief in God, is not clear, however. That becomes more clear when we make use of another EVS question concerning belief. The EVS surveys also let the respondents choose between four statements and asked them which statement comes closest to their beliefs:

1. there is a personal God
2. there is some sort of spirit or life force
3. I don't really know what to think
4. I don't really think there is any sort of spirit, God or life force

A choice for option 1 will be considered as *traditional* belief, option 3 as *agnosticism* and option 4 as *atheism*. A choice for option 2 is the least clear and reflects maybe what Voas (2009) once labelled as "fuzzy fidelity." Since we are dealing with secularization in the sense of declining traditional religion, we first present data on the belief in a personal God.

Table 111.6 shows that between 1981 and 2008 Western Europe witnessed a clear decline in traditional belief. All nations, including very religious Malta, Ireland and Northern Ireland, have lower proportions of traditional believers in 2008 than in 1981. The biggest relative decline was in Belgium, where traditional belief has halved, followed by Norway, France and the Netherlands, the smallest in West Germany. The average level of traditional belief in the 12 countries with 1981 data has declined from 45 % in 1981 to 34 % in 2008. In 2008, the Protestant countries have in general lower levels of traditional belief than the Catholic countries (with the exception of France and Belgium).

Looking at the nations of Eastern Europe, the picture is very different (Table 111.7). Data for 1981 are missing, so we can only compare 2008 with 1990 or 1999. Probably the 1990 data were still influenced by the anti-religious culture of the communist period. During that period, responses to survey questions about religiosity may have been constrained by fear of governmental sanctions (Norris and Inglehart 2004: 111). Nevertheless, most East European nations display a rise in traditional belief between 1990 and 1999. Only in Czechia and Latvia was a decline. Between 1999 and 2008, there was no general trend. In some countries (Belarus and East Germany) traditional belief was declining sharply, whereas in others (Russia and Ukraine), the belief in a personal God was increasing.

Figure 111.3 gives a cartographic overview of the current level of traditional belief in all European countries according to the 2008 EVS. Most secularized are East Germany and Latvia with only 9 % traditional believers. Also Czechia, France, Serbia, Estonia and the Scandinavian countries have very low levels of traditional

Table 111.6 Percent of the West European population that believes in a personal God, 1981–2008

West European country	1981	1990	1999	2008
Austria		28.8	31.5	26.1
Belgium	45.4	31.5	28.6	21.7
Denmark	26.5	20.1	24.9	21.8
Finland		28.6	49.9	35.5
France	28.0	21.8	21.9	17.8
Germany West	27.4	25.1	38.5	25.9
Great Britain	31.6	32.9	31.0	25.8
Iceland	[a]	51.3	51.3	40.1
Ireland	75.8	67.2	65.0	65.0
Italy	[a]	67.0	70.7	61.6
Malta	78.9	72.4	77.9	67.2
Netherlands	37.1	28.3	23.6	24.0
Northern Ireland	71.7	66.4	61.4	58.7
Norway	39.2	29.8		22.9
Portugal		61.7	78.7	60.8
Spain	57.2	51.7	49.1	44.3
Sweden	20.0	15.5	16.1	14.7
Average 12	44.9	38.6		34.2
Average 17		41.2		37.3

Data source: European Values Survey, 2008
[a]1981 data of Iceland and Italy are not comparable

Table 111.7 Percent of the East European population that believes in a personal God, 1990–2008

East European country	1990	1999	2008
Bulgaria	10.2	35.8	33.8
Belarus		63.6	27.1
Croatia		41.4	42.4
Czech Republic	11.7	6.5	11.4
Estonia	6.7	15.9	18.5
Germany East	14.0	20.1	9.4
Greece		65.9	69.9
Hungary	39.5	44.7	41.2
Latvia	10.4	7.9	9.3
Lithuania	21.0	50.8	43.5
Poland	79.4	82.7	80.5
Romania	36.4	37.3	36.4
Russian Federation		32.1	45.8
Slovakia	36.0	35.3	41.9
Slovenia	22.3	24.1	22.9
Ukraine		41.6	56.6
Average 11	26.1	32.8	31.7
Average 16		37.9	36.9

Data source: European Values Survey 1990–2008

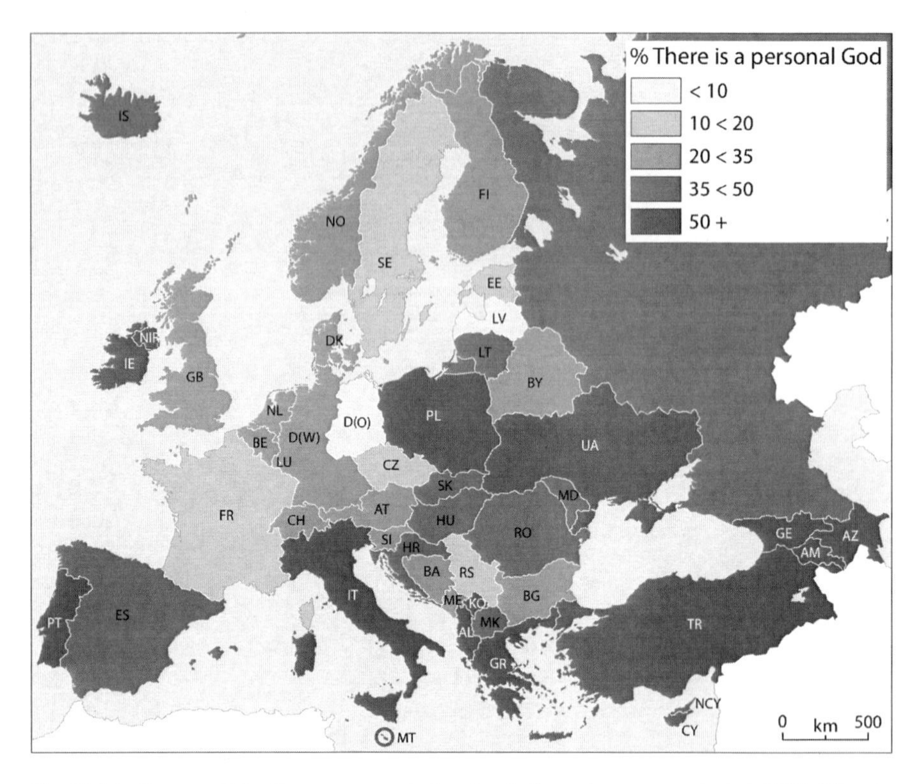

Fig. 111.3 Percent of the European population that believes in a personal God, 2008 (Map by Hans Knippenberg, data from European Values Survey 2008)

belief. In general Protestant countries have lower levels of traditional belief than Catholic and Orthodox countries. A high level of traditional belief can be found in Muslim countries such as Turkey (91 %), Kosovo (80 %) and Azerbaijan (76 %), but also in Christian countries as Orthodox Georgia (96 %) and Greece (70 %), and Catholic Poland (80 %). The average proportion of traditional believers in the European nations was 43 %.

A decline in traditional belief does not automatically implicate an increase in atheism or agnosticism as Tables 111.8 and 111.9 make clear. Unfortunately, pre-1981 (in Eastern Europe even pre-1990) data on these issues are missing. Data from other surveys in the Netherlands suggest a continuing increase in the proportion of combined agnostics and atheists since 1966 (Bernts et al. 2007: 40), but according to the EVS, West European nations such as Denmark, Iceland, Italy, West Germany and Malta, saw a decline in combined atheism and agnosticism between 1981 and 2008. All other West European countries, however, witnessed an increase in the same period; in Ireland the proportion of agnostics and atheists has even doubled.

Table 111.8 Combined percentage of agnostics and atheists in West European countries, 1981–2008

West European country	1981	1990	1999	2008
Austria		22.1	16.4	25.1
Belgium	26.7	46.4	34.4	40.7
Denmark	47.2	46.4	37.0	43.0
Finland		20.4	18.1	30.5
France	44.1	43.9	46.0	48.4
Germany West	31.9	29.7	26.0	27.7
Great Britain	28.5	25.8	29.0	34.7
Iceland	24.2	16.6	15.5	21.8
Ireland	8.0	8.6	10.8	16.7
Italy	18.4	11.0	9.8	14.5
Malta	7.9	8.4	3.4	6.6
Netherlands	31.4	29.8	27.5	32.5
Northern Ireland	9.6	13.4	12.9	15.5
Norway	27.7	33.9		36.9
Portugal		19.8	6.0	16.9
Spain	18.4	20.4	22.7	29.5
Sweden	38.8	38.5	30.0	40.5
Average 14	25.9	26.6		29.2
Average 17		25.6		28.3

Data source: European Values Survey 1981–2008

Table 111.9 Combined percentage of agnostics and atheists in East-European countries, 1990–2008

East European country	1990	1999	2008
Bulgaria	54.1	25.2	19.7
Belarus		27.6	23.6
Croatia		15.3	18.0
Czech Republic	50.5	43.3	58.7
Estonia	39.9	34.7	40.1
Germany East	65.5	63.4	72.9
Greece		11.4	7.0
Hungary	51.4	40.1	36.1
Latvia	34.1	25.2	23.8
Lithuania	34.1	25.2	23.8
Poland	15.2	7.2	8.9
Romania	20.9	17.1	17.1
Russian Federation		49.5	34.1
Slovakia	36.4	24.0	26.6
Slovenia	34.2	24.7	26.7
Ukraine		33.2	23.2
Average 11	39.7	30.0	32.2
Average 16		29.2	28.8

Data source: European Values Survey 1990–2008

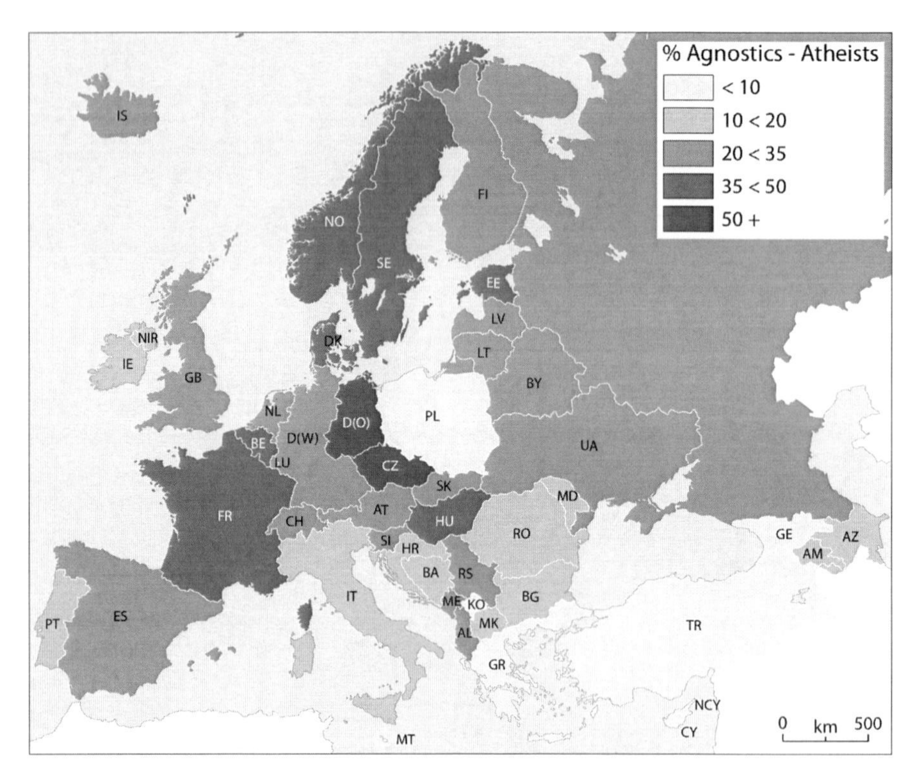

Fig. 111.4 Combined percentage of agnostics and atheists, European countries 2008 (Map by Hans Knippenberg, data from European Values Survey 2008)

In Eastern Europe, only in Czechia and East Germany agnosticism and atheism grew between 1990 and 2008, which resulted in unprecedentedly high levels in these countries. In former East Germany even the majority of the population (57 %) can now be considered as atheists; the combined proportion of agnostics and atheists concerns almost three-quarters of the population. Whereas Estonia remained at the same level, all other countries of Eastern Europe witnessed a rather large decline in the combined proportion of agnostics and atheists, which indicates a recovery of religious belief after the fall of the Berlin Wall.

Figure 111.4 records the current situation of combined agnosticism and atheism in Europe based on the 2008 EVS. On average about a quarter of the population in the European countries can be considered as agnostic or atheist, but the differences between the countries are large. The high levels in East Germany and Czechia contrast strongly with very low levels (below 10 %) in some Orthodox countries (only 1 % in Georgia, Cyprus and Greece), most Muslim countries (Turkey, Kosovo and Northern Cyprus) and a few Catholic countries (Malta and Poland). All Protestant countries have higher levels, varying from about one-fifth (Iceland, Switzerland) to almost three-quarters (East Germany) of the population.

111.5 Transforming Religion

Apart from the rise of agnosticism or atheism, the decline of traditional (Christian) religion as described in the previous section has given way to alternative forms of religion and spirituality.

111.5.1 Privatizing Religion

Already in 1967 Luckmann pointed at the "privatization" of religion in modern society. In his *The Invisible Religion*, he argued that religion did not disappear, but just changed its appearance. Dogmatic systems as well as ecclesiastical guidelines were no longer accepted and people tended to make up their own religions, which often take the form of a syncretistic patchwork. The "consumer orientation" and "sense of autonomy" of the individual in modern society changed his/her attitude to the sacred. He/she became a "buyer" who may choose from the assortment of "ultimate" meanings as he/she sees fit (Luckmann 1967: 98–99).

In a similar way of thinking, Grace Davie summarised her findings on religion in Britain and the European continent in the formula "believing without belonging," suggesting that secularization in the sense of declining religious belonging and practising does not mean that people are loosing their faith (Davie 1994, 2000). She coined the phrase "vicarious religion" (literally "substitutionary religion"), meaning "the notion of religion performed by an active minority but on behalf of a much larger number, who (implicitly at least) not only understand, but, quite clearly, approve of what the minority is doing" (Davie 2008: 169), in that way suggesting that secular societies are more religious than they seem (see Bruce 2011: 81–90 for a critical review). In her analysis of religion in modern Europe, she borrows from the work of the French sociologist Danièle Hervieu-Léger (1993) who approaches religion as collective memory. Concerning the secularization of modern European societies Hervieu-Léger argues that these societies are not less religious because they are increasingly rational, but because they are less capable of maintaining the collective memory which lies at the heart of their religious existence. In that way, they are "amnesic" societies.

111.5.2 A Spiritual Turn

Other sociologists of religion experience a more radical transformation of religion and speak about a "spiritual turn" or a "spiritual revolution." They refer to the rise of "post-Christian spirituality" or the New Age movement (Hanegraaff 1996; Heelas 1996; Heelas and Woodhead 2005; Houtman and Aupers 2007, 2008). Although often seen as an incoherent collection of ideas and practices, and referred to as "do-it-yourself religion" (Baerveldt 1996) or "pick-and-mix religion" (Hamilton 2000), they underscore that this post-Christian spirituality includes a commonly held belief, that "in the deepest layers of the self the "divine spark" – to borrow a term from ancient Gnosticism – is still smouldering, waiting to be stirred up, and succeed the socialized self. Getting in touch with this "true," "deeper," or "divine" self is ... understood as a long-term process [of personal growth]" (Houtman and Aupers 2007: 307). This immanent conception of the sacred differs considerably from the traditional Christian belief that the truth is "out there" rather than within and that the divine is transcendent rather than immanent (Heelas and Woodhead 2005: 22). Post-Christian spirituality rejects both scientific reasoning and religious faith as ways of finding the ultimate truth, because in the last resort there is no other authority than personal, inner experience (Hanegraaff 1996: 519). "We are all Gods" as a Dutch New Age teacher once put it (cited in Aupers 2005: 184).

Apart from these ideas of the "divine self," New Age spirituality often includes ideas that can be considered as millenarianism (the arrival of a New Age, characterised by harmony throughout the world, instead of conflict and struggle for domination) and holism (everything is permeated by a universal, interconnected energy, as opposed to the analytical methods of modern science) (Flere and Kirbiš 2009: 162).

Due to the heterogeneous and non-institutionalized nature of the New Age phenomenon, it is very difficult to measure its scope in terms of the number of "followers" or "participants." We could get an indication by counting the proportion of the respondents in the EVS that believe in some sort of spirit or life force, but does not belong to a religious denomination.

In the West European countries there has been a considerable increase in post-Christian spirituality between 1981 and 2008 (Table 111.10). The average proportion more than tripled. The Netherlands strikes the eye by its relatively high level of spirituality during this whole period, but Sweden had the biggest relative growth: from 1 % in 1981 to 13 % in 2008. The growing popularity of New Age spirituality has to do with a reaction to the rationalization of modern society that could be seen as most efficient from an institutional perspective, but no longer provided meaning for the individual, who experienced a feeling of alienation, of 'homelessness' (Berger et al. 1974). Analyzing New Age in the Netherlands the Dutch sociologist Stef Aupers (2005: 197–198) agrees that ongoing rationalisation and the feelings of alienation it generated triggered the emergence of self-spirituality, but also observed that, paradoxically, self-spirituality has blended in very well with the rational institutions it supposedly rejects: "In the days of the counterculture, the market of economy was seen as an evil force, transforming individuals into materialistic consumers. Nowadays, most New Age teachers accept the media, advertisements and marketing techniques as efficient tools to sell their New Age ideology and products. By doing so, they have become strong competitors to other providers of meaning on the "market of ultimate significance," most notably to the Christian churches and secular psychologists...The ethics of self-spirituality, authenticity and "being true to oneself" are important selling values in the new "experience economy"."

The East European countries show a mixed picture of increasing and declining post-Christian spirituality (Table 111.11). When analyzing only countries for which

Table 111.10 Percentage of the West European population that believes in a spirit or life force, but does not belong to a religious denomination, 1981–2008

West European country	1981	1990	1999	2008
Austria		6.1	5.6	6.4
Belgium	4.6	2.4	12.9	13.8
Denmark	1.3	3.2	2.9	3.2
Finland		5.4	4.0	7.2
France	5.8	11.8	11.4	15.4
Germany West	2.8	3.6	3.2	3.9
Great Britain	3.5	17.1	5.0	15.7
Iceland	0.4	0.3	1.5	2.8
Ireland	0.6	1.1	2.2	3.7
Italy	0.6	1.1	2.2	3.7
Malta	0.0	0.6	0.6	0.6
Netherlands	13.1	23.9	29.7	23.4
Northern Ireland	1.4	4.9	5.8	7.7
Norway	1.4	3.5		6.7
Portugal		7.1	3.4	5.2
Spain	2.3	3.9	5.1	6.8
Sweden	1.2	7.7	11.4	12.8
Average 13	2.8	6.1		8.6
Average 17		6.1		8.2

Data source: European Values Survey 1981–2008

Table 111.11 Percentage of the East European population that believes in a spirit or life force, but does not belong to a religious denomination, 1981–2008

East European country	1990	1999	2008
Bulgaria	17.6	10.1	12.2
Belarus		3.9	11.6
Croatia		6.1	6.7
Czech Republic	20.7	29.6	14.4
Estonia	46.0	36.2	26.9
Germany East	5.9	5.2	9.0
Greece		1.6	1.6
Hungary	4.2	6.3	10.5
Latvia	31.7	20.9	19.2
Lithuania	12.3	4.7	2.5
Poland	0.5	1.3	0.7
Romania	1.7	0.8	0.6
Russian Federation		7.7	7.2
Slovakia	5.5	6.4	3.6
Slovenia	11.6	14.5	13.6
Ukraine		11.8	6.1
Average 11	14.3	12.4	10.3
Average 16		10.4	9.2

Data source: European Values Survey 1981–2008

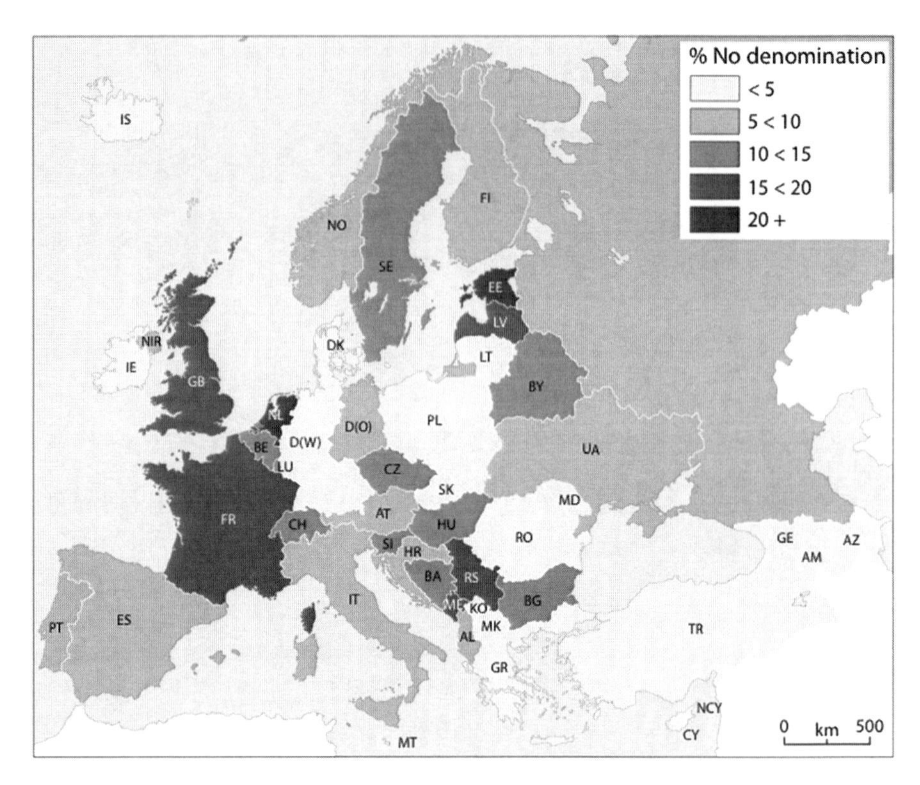

Fig. 111.5 Percentage of the European population that believes in a spirit or life force, but does not belong to a religious denomination, 2008 (Map by Hans Knippenberg, data from European Values Survey 2008)

1999 EVS data are available, Hungary, Finland, East Germany and Poland resemble the West European pattern of growing spirituality, whereas in the Baltic states, Bulgaria, Rumania, Czechia and Slovakia post-Christian spirituality is declining.

Figure 111.5 shows the current situation in its geographical variety. On average 8 % of the European nations believe in some sort of spirit or life force without belonging to a religious denomination, varying from <1 % in Cyprus, Turkey, Georgia, Malta, Romania, Poland and Macedonia to more than 20 % in the Netherlands and Estonia.

111.6 Conclusion

Europe is a fascinating continent as far as religion is concerned. At the eve of the French Revolution, once religiously united Europe presented the picture of a totally disunited continent divided into enemy religious denominations (Rémond 1999: 20–23). Apart from the Jewish and Muslim minorities, at least three entangled religious Europes can be distinguished: (a) Roman Catholic, (b) Protestant (among others divided in Lutheran, Calvinist and Anglican variants) and (c) Orthodox Europe. European states had a religion, just like individuals, as a consequence of the application of the *cuius regio, eius religio* principle in the past (Rémond 1999: 30–37).

The French Revolution ended this *ancien régime* and paved the way for a separation of church and state, religion and nation. A rich variety of church-state relations developed from strict separation such as in France to strong bonds such as in the Scandinavian states with their Lutheran state churches or Catholic states such as Spain and Portugal where the bond between "Crown and Altar" was continued for a long time. In multi-religious states, religion could become a major element in awakening national consciousness, such as the Catholic Poles witness in their struggle for independence against Lutheran Prussia and Orthodox Russia or the Catholic Irish in their struggle against Protestant British domination.

These historic processes of state formation and nation building proved to be very important not only for the development of religion, but also for its decline. Every European nation followed its own path in the secularization of Europe. That does not alter the fact that there was a general trend in post-War Europe of declining traditional religion in the sense of declining belonging, practice and even belief, although the decline of belief was less strong. That decline was only partially compensated by an increase in what has been called post-Christian spirituality or New Age religion. So, there was transformation of religion into a more privatized form, but there was also a general decline of religiosity.

That general decline was stronger in Protestant than in Catholic and Orthodox countries, since Protestantism encouraged individualism, fragmentation and rationality. The few Muslim countries hardly witnessed any decline of traditional religion.

Post-communist Europe followed a somewhat different path than the rest of Europe. In general the atheist policies during the communist period had a negative effect on the level of religious involvement, leading to very high levels of secularization. After the fall of the Berlin Wall and the following "velvet revolutions," the Orthodox nations in particular saw some religious recovery, but others, such as former East Germany, Czechia and Estonia continued their secularization path to the lowest levels of religious involvement in all its dimensions, even lower than many West European countries. The very strong secularization of the last group of East European countries contradicts in fact both modernization and market theories, because these countries are not more modern than West European countries and witnessed an increase in the supply of "religious goods" after the fall of the Berlin Wall. They do point to the significance of specific state formation and nation building processes in the past and the resulting state-church and nation-religion relationships.

Therein lies also the explanation of the striking difference between strongly secularized Europe and the relatively religious United States. The formation of the United States can be characterized as a kind of territorializing *from below*, which was strongly influenced by the religious, and in particular dissident Protestant culture of the first groups of settlers (Knippenberg 2010: 47–50). That culture was further elaborated in the civil American Creed, that emphasised individual freedom and responsibility, and distrusted government interference and in general hierarchical relationships, including collective welfare arrangements (Huntington 2004: 66–69). That culture also had its impact on the relationship between state and religion, which were formally separated, but not in order to free the state from religion (as in France and other French inspired European states), but in order to free religion from state interference and create maximal freedom of religion, which offered an ideal opportunity for religious entrepreneurs to start their own religious firm and extend it in mutual competition (Berger et al. 2008: 48–50).

In Europe, on the contrary, there was territorialisation *from above*. After the Roman Catholic Church had lost its monopoly during the Reformation, religion became embedded territorially according to the *cuius regio, eius religio* principle. Territorially organized state churches or at least dominant churches followed, which were functioning as a kind of public utilities, that gradually lost its functions to the rising welfare states (Berger et al. 2008: 35–36). Secular ideologies such as socialism and nationalism inspired the resistance against the vested order and also against the privileged churches that were connected with that order. Secularization was the consequence.

Appendix: Country Abbreviations That Have Been Used in the Figures

AL	Albania
AM	Armenia
AT	Austria
AZ	Azerbaijan
BA	Bosnia Herzegovina
BE	Belgium
BG	Bulgaria
BY	Belarus
CH	Switzerland
CY	Cyprus
CZ	Czechia
D(O)	Germany (East)
D(W)	Germany (West)
DK	Denmark
EE	Estonia
ES	Spain
FI	Finland
FR	France
GB	Great Britain
GE	Georgia
GR	Greece

HR Croatia
HU Hungary
IE Ireland
IS Iceland
IT Italy
KO Kosovo
LT Lithuania
LU Luxembourg
LV Latvia
MD Moldova
ME Montenegro
MK Macedonia
MT Malta
NCY Northern Cyprus
NIR Northern Ireland
NL Netherlands
NO Norway
PL Poland
PT Portugal
RO Romania
RS Serbia
RU Russia
SE Sweden
SL Slovenia
SK Slovakia
TR Turkey
UA Ukraine

References

Aupers, S. (2005). 'We are all gods.' New Age in the Netherlands 1960–2000. In E. Sengers (Ed.), *The Dutch and their gods. Secularization and transformation of religion in the Netherlands since 1950* (pp. 181–201). Hilversum: Verloren.

Baerveldt, C. (1996). New Age-religiositeit als individueel constructieproces. In M. Moerland (Ed.), *De kool en de geit in de nieuwe tijd: Wetenschappelijke reflecties op new age* (pp. 19–31). Utrecht: Jan van Arkel.

Bankston, C. L. (2003). Rationality, choice, and the religious economy: Individual and collective rationality in supply and demand. *Review of Religious Research, 45*, 155–171.

Berger, P. L. (1967). *The sacred canopy*. Garden City: Doubleday.

Berger, P. L., Berger, B., & Kellner, H. (1974). *The homeless mind: Modernization and consciousness*. Harmondsworth: Penguin.

Berger, P., Davie, G., & Fokas, E. (2008). *Religious America, secular Europe?* Farnham: Ashgate.

Bernts, T., Dekker, G., & De Hart, J. (2007). *God in Nederland 1996–2006*. Kampen: Ten Have.

Bertrand, J. R. (2006). State and church in France: Regulation and negotiation. *GeoJournal, 67*, 295–306.

Bremmer, J. (2008). Secularization: Notes toward a genealogy. In H. de Vries (Ed.), *Religion: Beyond a concept* (pp. 432–437). New York: Fordham University Press.

Bruce, S. (1999). *Choice and religion: A critique of rational choice theory*. Oxford: Oxford University Press.

Bruce, S. (2002). *God is dead. Secularization in the West*. Oxford: Blackwell.

Bruce, S. (2011). *Secularization: In defence of an unfashionable theory*. Oxford: Oxford University Press.

Casanova, J. (2001). Religion, the new millennium, and globalization. *Sociology of Religion, 62*, 415–441.

Chaves, M., & Gorski, P. S. (2001). Religious pluralism and religious participation. *Annual Review of Sociology, 27*, 261–281.

Davie, G. (1994). *Religion in Britain since 1945: Believing without belonging*. Oxford: Blackwell.

Davie, G. (2000). *Religion in modern Europe. A memory mutates*. Oxford: Oxford University Press.

Davie, G. (2008). From believing without belonging to vicarious religion. In D. Pollack & D. V. A. Olson (Eds.), *The role of religion in modern societies* (pp. 165–176). New York/London: Routledge.

De Busser, C. (2006). Church-state relations in Spain: Variations on a National-Catholic theme? *GeoJournal, 67*, 283–294.

Dijkink, G. J. (2008). *Territorial shock. Closure, control and identity in world affairs*. Unpublished manuscript, University of Amsterdam, Department of Geography, Planning and International Development Studies.

Finke, R., & Stark, R. (1998). Religious choice and competition. *American Sociological Review, 63*, 761–766.

Finke, R., Guest, A., & Stark, R. (1996). Mobilizing local religious markets: Religious pluralism in the empire state, 1855–1865. *American Sociological Review, 61*, 203–218.

Flere, S., & Kirbiš, A. (2009). New age, religiosity, and traditionalism: A cross-cultural comparison. *Journal for the Scientific Study of Religion, 48*, 161–184.

Flora, P., Kuhnle, S., & Urwin, D. (Eds.). (1999). *State formation, nation-building, and mass politics in Europe: The theory of Stein Rokkan*. Oxford: Oxford University Press.

Froese, P., & Pfaff, S. (2005). Explaining a religious anomaly: A historical analysis of secularization in Eastern Germany. *Journal for the Scientific Study of Religion, 44*, 397–422.

Gooren, H. (2006). The religious market model and conversion: Towards a new approach. *Exchange, 35*, 39–60.

Halikiopoulou, D. (2010). *Patterns of secularization: Church, state and nation in Greece and the Republic of Ireland*. Aldershot: Ashgate.

Halman, L., Sieben, I., & van Zundert, M. (2012). *Atlas of European values. Trends and traditions at the turn of the century*. Leiden: Brill.

Hamilton, M. (2000). An analysis of the festival for the mind-body-spirit, London. In S. Sutcliffe & M. Bowman (Eds.), *Beyond new age: Exploring alternative spirituality* (pp. 188–200). Edinburgh: Edinburgh University Press.

Hanegraaff, W. J. (1996). *New age religion and western culture: Esotericism in the mirror of secular thought*. Leiden: Brill.

Havlíček, T. (2006). Church-state relations in Czechia. *GeoJournal, 67*, 331–340.

Heelas, P. (1996). *The new age movement: The celebration of the self and the sacralisation of modernity*. Oxford: Blackwell.

Heelas, P., & Woodhead, L. (2005). *The spiritual revolution: Why religion is giving way to spirituality*. Oxford: Blackwell.

Henkel, R. (2001). *Atlas der Kirchen und der anderen Religionsgemeinschaften in Deutschland. Eine Religionsgeographie*. Stuttgart/Berlin/Köln: Kohlhammer.

Henkel, R., & Knippenberg, H. (2005). Secularisation and the rise of religious pluralism. In H. Knippenberg (Ed.), *The changing religious landscape of Europe* (pp. 1–13). Amsterdam: Het Spinhuis.

Hervieu-Léger, D. (1993). *La religion pour mémoire*. Paris: Le Cerf.

Houtman, D., & Aupers, S. (2007). The spiritual turn and the decline of tradition: The spread of post-Christian spirituality in 14 Western countries, 1981–2000. *Journal for the Scientific Study of Religion, 46*, 305–320.

Houtman, D., & Aupers, S. (2008). In H. de Vries (Ed.), *Religion: Beyond a concept* (pp. 798–812). New York: Fordham University Press.

Huntington, S. P. (1996). *The clash of civilizations and the remaking of world order*. London: Simon & Schuster.

Huntington, S. P. (2004). *Who are we? The challenges to American national identity*. London: Simon & Schuster.

Knippenberg, H. (1992). *De religieuze kaart van Nederland. Omvang en geografische spreiding van de godsdienstige gezindten vanaf de Reformatie tot heden*. Assen: Van Gorcum.

Knippenberg, H. (1998). Secularization in the Netherlands in its historical and geographical dimensions. *GeoJournal, 45*, 209–220.

Knippenberg, H. (2006a). The political geography of religion: Historical state-church relations in Europe and recent challenges. *GeoJournal, 67*, 253–265.

Knippenberg, H. (2006b). The changing relationship between state and church/religion in the Netherlands. *GeoJournal, 67*, 317–330.

Knippenberg, H. (2010). Secular Europe versus religious America: Geographical reflections on religion in Europe and the United States. *Folia Geographica 15*, Ročník XL, Presov, 42–54.

Luckmann, T. (1967). *The invisible religion: The problem of religion in modern society*. New York: Macmillan.

Madeley, J. T. S. (2003). A framework for the comparative analysis of church-state relations in Europe. *West European Politics, 26*, 23–50.

Martin, D. (1969). *The religious and the secular*. London: Routledge and Kegan Paul.

Martin, D. (2011). *The future of Christianity. Reflections on violence and democracy, religion and secularization*. Farnham: Ashgate.

Norris, P., & Inglehart, R. (2004). *Sacred and secular. Religion and politics worldwide*. Cambridge: Cambridge University Press.

Pollack, D. (2008). Introduction. Religious change in modern societies – Perspectives offered by the sociology of religion. In D. Pollack & D. V. A. Olson (Eds.), *The role of religion in modern societies* (pp. 1–21). New York/London: Routledge.

Raivo, P. J. (2005). Finland: Lutheran identity, orthodox heritage and a secular way of life. In H. Knippenberg (Ed.), *The changing religious landscape of Europe* (pp. 107–119). Amsterdam: Het Spinhuis.

Rémond, R. (1999). *Religion and society in modern Europe*. Oxford: Blackwell.

Robbers, G. (Ed.). (1995). *Staat und Kirche in der Europäischen Union*. Baden-Baden: Nomos.

Rokkan, S. (1979). Cities, states, nations: A dimensional for the study of contrast in development. In S. N. Eisenstad & S. Rokkan (Eds.), *Building states and nations* (pp. 73–97). Beverley Hills: Sage.

Stark, R. (1999). Secularization, R.I.P. *Sociology of Religion, 60*, 249–273.

Stark, R., & Brainbridge, W. S. (1985). *The future of religion. Secularization, revival and cult formation*. Berkeley/Los Angeles: University of California Press.

Stark, R., & Brainbridge, W. S. (1987). *A theory of religion*. Bern/New York: Peter Lang.

Stark, R., & Finke, R. (2000). *Acts of faith: Explaining the human side of religion*. Berkeley/Los Angeles: University of California Press.

Stark, R., & Iannaccone, L. R. (1994). A supply-side reinterpretation of the 'secularization' of Europe. *Journal for the Scientific Study of Religion, 33*, 230–252.

Swatos, W. H., & Christiano, K. J. (1999). Secularization theory: The course of a concept. *Sociology of Religion, 60*, 209–228.

Voas, D. (2009). The rise and fall of fuzzy fidelity in Europe. *European Sociological Review, 25*, 155–168.

Wilson, B. R. (1966). *Religion in secular society*. London: C.A. Watts.

Wilson, B. (1969). *Religion in secular society: A sociological comment*. Harmondsworth: Penguin.

Wilson, B. (1982). *Religion in sociological perspective*. Oxford: Oxford University Press.

Wilson, B. (1998). The secularization thesis: Criticisms and rebuttals. In B. Wilson, J. Billiet, & R. Laermans (Eds.), *Secularization and social integration: Papers in honour of Karel Dobbelaere* (pp. 45–66). Leuven: Leuven University Press.

Wunder, E. (2005). *Religion in der postkonfessionellen Gesellschaft. Ein Beitrag zur sozialwissenschaftlichen Theorieentwicklung in der Religionsgeographie*. München: Franz Steiner.

Chapter 112
Visualizing Secularization Through Changes in Religious Stamp Issues in Three Catholic European Countries

Stanley D. Brunn

112.1 Introduction

One distinctive feature of Europe's contemporary human landscapes is the display of secularism. This feature is evident in both visible and invisible dimensions. Visible secularism is most apparent in the empty, or nearly empty, church services on worship days (except major holidays) in churches that have closed and been recycled for other uses (offices, community gatherings, museums, etc.) and in the growing interests in popular culture that are substituted for religions activities (sports, entertainment and consumer shopping are examples). Invisible secularism emerges from the public opinion polls about religion and a religious society and experiences share the youth, middle aged and even among the elderly cohorts the about religion and religious institutions. These surveys reveal less interest in and less commitment to formal established institutions than previous generations, greater acceptance of tolerance and diversity towards others, weaker and questioning beliefs about Christian traditions among those in all ages and less support for state religious-funded institutions. While there are differences in these religious views and expressions of religion and religious life from western to eastern Europe and from northern to southern Europe, the secular landscape is a dominant feature across much of the continent.

The scholarly literature on European secularism is noteworthy; it discusses the causes, the results of opinion surveys about religious beliefs and sentiments and its impact on contemporary attitudes of youth, voters and political parties (Berger 1967; Luckmann 1967; Swaatos and Christiano 1999; Casanova 2001; Davie 2002; Knippenberg 2006; Berger et al. 2008; Bruce 2011). The contributors include historians, philosophers, sociologists, geographers and also religious scholars.

S.D. Brunn (✉)
Department of Geography, University of Kentucky, Lexington, KY 40506, USA
e-mail: brunn@uky.edu

S.D. Brunn (ed.), *The Changing World Religion Map*,
DOI 10.1007/978-94-017-9376-6_112

Major themes of the extant literature include post-World War II changes in the economy and wealth of Europe as well as changing church-state relations, institutional disestablishment and the Weberian views of a secular society and state.

While texts and analyses of texts provide useful and valuable insights into the impacts of secularism, there are additional methodologies and perspectives that might be describe this phenomenon. This chapter examines postage stamp issues as an additional measure or barometer to measure the emergence of a secular society.

112.2 Stamps as Texts

Postage stamps can be considered as one way to measure secularization is occurring within a state. The reasoning is as follows. Postage stamps are products of a state, that is, the state decides the subjects or themes and topics of stamp issues as well as the designs, colors and denominations. Also state philatelic divisions or bureaus decide the classes of mail for the stamps (internal or external use), the dates of issue and the total number of stamps to be printed. Stamps are also "visible products" of the state that become and are "political messengers" informing both inside and outside a country's territorial boundaries about what it wishes viewers/users to see and how it wishes to present this information. That is, stamps are major visible products of the state, but they are not the only popular visible state product; others are currency (Pointon 1988; Unwin and Hewitt 2001; Raento et al. 2004; Gilbert 2007; Hymans 2010), although currency is basically seen only with a state's territorial jurisdiction, flags, monuments, memorials, school maps and texts, official national maps (Zeigler 2002; Moisio 2008), and also a state's web pages (Brunn and Cottle 1997; Jacobson and Purcell 1997).

These products might be considered as ways in which states build nationalism (Boulding 1959) and imagined communities (Anderson 1991) and forms of banal nationalism (Billig 1995). A state's iconographies have been studied by a number of social, cultural, historical and political geographers (Dodds 2006). Stamps are a part of this extant literature; previous studies have looked at the themes and topics of first issues of newly independent states, the history of a state's stamps, state boosterism and nationalism (or philatelic nationalism) and how stamps can be used to measure political changes occurring within a country. See, for example, Leith (1971) in china, Grant (1995) on early Soviet stamps, Laurintzen (1988) on stamps of the Third Reich, McQueen (1988) on Australia, O'Sullivan (1988 on Ireland and South Africa) Reed (1993 on Iraq, Brunn (2002) on political stamps of the Arab world, Dobson (2002 on Japan), Brunn (2000) on the first issues of newly independent states, Deans on Taiwan (2005), Deans and Dobson (2005) on East Asia, Dobson on Japan (2005) and the United Kingdom (2005), Raento and Brunn (2005, 2008) and Raento (2006) on Finland's philatelic history, Johnson (2005) on the two Koreas, Covington and Brunn (2006) on music heritage, Kevane (2008) on Sudan and Burkina Faso and Hammett (2012) on South Africa's recent philatelic history. Many of the aforementioned studies not only look at a state's stamp policies (Stamp 1966; Altman 1991; Scott 1992, 1995; Reid 1984), but also at elements of a visual culture that accentuate the importance of semiotics, familiar symbols and designs that have various sensory appeals, especially visual (Barthes 1972; Mirzoeff 1998; Panofsky 1982; Hall 1997; Rose (1996, 2001, 2011).

The previous study that has most bearing on this discussion is by Brunn (2011) which examined the final years of Soviet stamps and the first years of Russian stamps which focuses on the concept as stamps being "messengers of change" in a country experiencing transition. His hypothesis was that the last stamps of the Soviet Union would be different in topics and designs than those appearing in the first years of the Russian state, that is, beginning in January 1992. One of the major findings of this detailed investigation on Soviet/Russian issues was that the transition marked the end of Soviet era stamp themes and the emergence of a new set of topics and themes that were primarily national (that is, Russian) in content rather than Soviet (former Soviet Union) themes.

112.3 The Overriding Hypothesis: The Emergence of Secularism in Europe

The overriding hypothesis in this study is that we can examine the visual content of stamp programs in three major countries in Europe that have a strong Catholic heritage to discern the emergence of secularism or the downplaying of religious stamp themes, topics and total stamp issues. The three countries I have selected are Ireland, Portugal and Spain, all European countries with a very strong Catholic heritage that extended into the 1980s and 1990s. In each state the Catholic church was considered a major social and political influence, not only in the observance of the church attending population on important of religious holidays, but in their school instruction and teacher training as well as the influence of legislation to prohibit abortions and divorces. Major political leaders were closely tied to the Catholic church leadership. However, in contemporary Ireland, Portugal and Spain, the support for the Catholic church has declined in recent decades as evidenced in lower attendance at masses, increased popularity and acceptance of divorce, support for abortions, loss of confidence and trust in Catholic clergy and declining support for church-state ties both inside and outside political parties.

112.4 Methodology

In order to discern if there were any differences in the number and topics of religious stamps in Ireland, Portugal and Spain, I examined the issues 10 years before and after a major change in political leadership or a major change in the religious life of a country. For Portugal I examined the decades before and after the death of Portuguese President A. Salazar in 1973; for Spain it was the decade before and after President Franco's death in 1975 and for Ireland it marked the 10 years before the first pedophile priest scandal in 1994 and 10 years later. The specific years for Portugal were 1964–1973 and 1974–1983, for Spain 1966–1975 and 1976–1985 and for Ireland 1984–1993 and 1994–2003. The deaths of Salazar and Franco were considered watershed events in recent Portuguese and Spanish history. Following their deaths there was more discussion about divorce and abortion in political parties as well as intellectual circles and calls for liberalizing the Catholic church's policies and views towards social issues. The pedophile priest scandal in Ireland was a similar watershed event where following the abuses priests inflicted on children, the Catholic church hierarchy was under intense criticism not only for sheltering priests, but not properly protecting the country's children.

112.5 Stamps as Visual Secularization

Stamps, as noted above, are texts, but they are a different kind of text than "word texts." They are "visual texts" in the sense that they convey through familiar images, symbols and icons important messages to the viewer. Each stamp contains one or more of these qualities that the producing agent, in this case the state, wishes the viewer/observer to take note. Both the casual viewer of a stamp issue and also a seasoned professional stamp purchaser or artist will look for familiar objects with which to associate or identify with the content. It may be the focus is on what is in the center of the illustration or stamp, but also may be background information, including colors, hues, and familiar objects or symbols that evoke a positive response. On a religious stamp, one may look for familiar objects that strike harmony with the viewer; they may be Christian crosses, a pilgrimage event, an image of a famous saint, a major landmark (building or memorial) feature or an abstract depiction of Jesus, Mary or the Holy Family in a painting.

I introduce the term "visual secularization" to relate to the reduction in religious stamp issues, images and designs and the replacement with other subjects of a non-religious or secular nature. These may be the emergence of topics related to an increasing secular society, such as sports, entertainment (film and television), literature and popular culture (popular stars in film and sports). When these become dominant or emerge in popularity in a country's stamp issues, they may reveal a social shift away from religious to increasingly secular or contemporary institutions. Their depiction on stamp issues at the same time religious stamp issues may be in decline, informs the casual and disciplined view of stamp issues that shifts underway are revealed in the stamps issued and used.

112.6 Data Set and Methodology

The stamp issues for Ireland, Portugal and Spain are listed in the *Scott's Standard Stamp Catalogue* (Snee et al. 2012). This source lists by date the stamp or stamps issued; illustrations are provided for most stamps. A brief summary of the subject is given and also the color and estimated cost/value. For example, one can find out the names of individuals on stamps issued for a group of talented poets and musicians as well as the names of edifices of famous religions sites (cathedrals, hospitals and pilgrimage destinations).

The procedure used here to classify the stamps into specific categories was used previously by Brunn (2000, 2002, 2011) and by Raento and Brunn (2005, 2008) and Raento (2006). Each stamp was carefully examined for one or more dominant themes and then placed into a category where one theme dominated. As anyone knows in observing stamp issues there are often multiple themes on an individual stamp; they may be nature, religion, and literature or a noted individual, monument and piece of art. Often in undertaking this "content analysis" of stamps, one shifts a stamp from one category to another, knowing all along that there is some fuzziness in any classification process as well as the placement of individual stamps into one category or another.

A religious stamp was defined as an individual stamp that contained one or more themes of a religious content. It may be a holiday (Christmas and Easter), a saint, a noted religious figure (author, artist, etc.), a religious building (cathedral or monastery), a religious site (apparition or birthplace of a noted saint), a map of pilgrimage routes, a religious event (pilgrimage, festival, holy day), a religious institution (the Vatican, for example) or some religious heritage event, such as the commemoration of a missionary organization, religious order or some religious/military battle.

112.7 Results

112.7.1 Number of Issues

The number of total issues and religious stamps varied among the three countries. For Ireland the number of total issues from 1984 to 1993 was 328; only 43 or 13 % had a religious theme. In the decade after the first pedophile priest scandal there were 631 stamps issued, but only 57 or 9 % had a religious theme. For Portugal in the decade before Salazar's death there were 267 stamps issued, but only 15 or 5 % had any religious theme. In the decade after Salazar's death, 1974–1983, there were many more stamps issued, 370 total, but only 22 or 9 % had any religious themes. The results for Spain were different than Ireland and Portugal in two significant ways, one there were more total stamps issued and more stamps with religious theme. In the decade before Franco's death, 1966–1975, there were 588 stamps issued of which 177 or 30 % had a religious theme. The decade after Franco's death, 1976–1985, there were also 588 stamps issued but only 80 religious stamps, which was 16 % the total.

112.7.2 Religious Stamps from Ireland

In the case of Ireland, as noted above, there were not many religious stamps issued in the 10 years before the pedophile scandal racked the Catholic church; only 43 stamps conveyed a religious theme. The most common theme was different paintings

Fig. 112.1 Ireland: Secular stamp issues, 1984–1993 (Source: Collection of Stanley D. Brunn)

depicting the Christmas season (Fig. 112.1). In the 10 years after 1983 there were many more religious stamps issued (57), but they comprised a smaller percentage of all stamps issued than the previous decade (9 %). Christmas stamps, which were annual issues, where high on the list, but so were stamps commemorating St. Patrick's Day and also the election of Pope John Paul II (Fig. 112.2). Examples of religious issues include:

Fig. 112.2 Ireland: Secular stamp issues, 1994–2003 (Source: Collection of Stanley D. Brunn)

Examples of Stamps: 1984–1993

- St. Vincent Hospital sesquicentennial (1984)
- Medical Missionaries of Mary (1987)
- Martyred Missionaries (1989)
- Irish Missionaries (1992)
- Christmas stamps were issued each year.

Examples of Stamps: 1994–2003

- Voyages of St. Brenadan (1994)
- Irish Society of Friends Soup Kitchen (1995)
- John Wesley, founder or Methodism (1997)
- Canonization of St. Pie (2002)
- Election of Pope John Paul II 25th Anniversary (2003)
- Christmas stamps were issued each year

112.7.3 Religious Stamps from Portugal

Of the three countries examined here, Portugal issued the fewest religious stamps and also the lowest percentages. The last 10 years of Salazar's regime were marked by stamps commemorating various Catholic saints, the victories over the Moors, Catholic shrines and a religious congress (Fig. 112.3). After 1973, Pope John Paul and Pope John XXIII and Pope John were on a number of stamps (Fig. 112.4). The total number of religious stamps issued were only 15 before 1973 and only 22 in the following decade. Both numbers were less than 6 % of all Portuguese issues. Some examples of religious issues are:

Examples of Stamps: 1964–1973

- Century of the Shrine of Our Lady of Mt. Sameiro, Braga (1964)
- 9th Century of the Capture of the City of Colmbra from the Moors (1965)
- 50th Anniversary of the Apparition of Virgin Mary (1967)
- 360th Anniversary of the death of Bento de Goes, Jesuit explorer of a route to China (1968)
- 400th Anniversary of the martyrdom of a group of missionaries en route to Brazil (1971)

Examples of Stamps: 1974–1983

- Pope XXI (the only Portuguese pope) (1977)
- St. Anthony of Lisbon (1981)
- 800th Birthday of St. Francis of Assisi (1982)
- Visit of Pope John Paul and cathedral (1982)
- Christmas stamps in 5 years

112.7.4 Religious Stamps from Spain

Spain issued the most religious stamps and also had the highest percentage of religious issues. In the decade before Franco's death in 1975, religious stamps were popular: 177 out of 588 (or 30 %) of all issues. In the decade following

Fig. 112.3 Portugal: Secular stamp issues, 1964–1973 (Source: Collection of Stanley D. Brunn)

Franco's death, there were only 80 (out of 588 stamps issued) or less than 14 %. Many religious themes appeared on Spanish stamps. Before Franco's death there where the annual Christmas stamps, but also stamps for pilgrimages, cathedrals, famous architecture and religious tourism (Fig. 112.5). Some of these same themes continued after Franco's death, including Christmas and cathedrals but also others

Fig. 112.4 Portugal: Secular stamp issues, 1974–1983 (Source: Collection of Stanley D. Brunn)

for heritage events, paintings and Pope John Paul II (Fig. 112.6). Examples of religious stamps include:

Examples of Stamps 1966–1975

- St. Mary Carthusian Monastery (1966)
- Map of Capuchia Missions along the Orinoco River (1968)

Fig. 112.5 Spain: Secular stamp issues, 1966–1975 (Source: Collection of Stanley D. Brunn)

- Bicentenary of San Diego, CA and Father Junipero Serra (1969)
- Explorations and Development of Chile (1969)
- Map of Main Pilgrimage Routs in Europe (1970)
- Virgin of Cabeza Sanctuary
- Scenes from the Apocalypse
- Christmas stamps every year

Fig. 112.6 Spain: Secular stamp issues, 1976–1985 (Source: Collection of Stanley D. Brunn)

Examples of Stamps: 1976–1985

- Holy Year of St. James of Compostala, Patron Saint (1976)
- Nicoya Church, Costa Rica: Spain's Links (1976)
- Century of the Founding of the Society of St. Theresa of Jesus (1977)
- Institute of Christian Brothers (1979)
- St. Benedict, Patron Saint of Europe (1981)
- Stained Glass windows (1983)
- Journey to the Holy Land – painting (1984)
- Christmas stamps every year

112.8 Discussion and Conclusions

We have observed from a careful examination of the religious content of the stamps of three previously strong European Catholic countries before and after the death of a major political leader with ties to the Catholic church (Salazar in Portugal and Franco in Spain) and a major religious watershed event in Ireland (first pedophile priest scandal in the mid 1990s) that stamps indeed can be read as social barometers. Basically, the frequencies and percentages of religious stamps were greater during the times of Salazar and Franco and when the Catholic church was the dominant religious institution in Ireland. The passing of Franco and Salazar and decline in support for the Catholic church in Ireland signaled a decline in attendance in Catholic churches, weaker church-state ties related to life issues (marriage, divorce and abortion) and the emergence of a secular society. In each of these countries religion remains an important institution, but one with much less influence and clout in the daily lives of young and middle aged cohorts especially. What is replacing religion in public and private life are areas of a popular culture (sports, the arts, lifelong learning, etc.) but also a wealthy consuming culture (food, entertainment, personal lifestyle purchases). Many of the youthful and middle aged generations will continue to subscribe to the respect, acceptance and toleration of others that were long associated with formal institutions religious. However, their daily and weekly livelihoods do not include the practices of formal attendance in religious services or engaging in specific formal religious practices. Rather, their lives could best summarized as secular.

Fewer religious stamps and fewer stamp themes can be used to illustrate the declining influence of religion within a state. All three European countries have become more secular within the past couple decades as a number of chapters in this volume point out. However, a few selected religious themes continue to be of major importance in all three countries, specifically, the celebration of Christmas and Easter. Declining in importance are stamps honoring popes, saints, festivals and major religious events in a nation's history.

This study suggests a number of follow-up research projects that might look at the topic of "stamp secularism." Let me suggest four. *First,* it would be useful to examine the religious stamp issues in east European countries since the end of the Cold War. Poland would be especially interesting because of its strong Catholic heritage and also the birthplace of Pope John XXIII. How does Poland's religious revival compare with that in Slovakia, Czech Republic, Ukraine and also former Yugoslavia, especially Slovenia, Croatia and Serbia, all with Catholic heritages? *Second,* how are the Protestant heritages of northern Europe reflected on recent stamps of Norway, Sweden, Finland, Denmark and Iceland? *Third,* since central Europe is the scene of an increased number of migrants from Muslim North Africa and southeast Europe as well as east Europe and Russia, how are these new religious mixes reflected in a state's stamp issues? Are there stamps commemorating Muslim holidays and showing mosques and Islamic architecture, for example, that did not appear before 1990? Which countries are taking the lead in displaying (on stamps) the continent's emerging multifaith diasporas? *Finally,* what are the dominant secular themes that have appeared on European stamps in the past decade, not only in former Catholic strongholds of Italy, Spain, Portugal and Ireland, but also Germany, the Netherlands, France, Belgium and northern Europe? If indeed, Europe has become more secular, as other authors in this volume discuss, then this trend should also be evident in stamp issues on contemporary sports, sports cars, music, theater, museums, literature, cartoon characters, food and clothing. Since stamps are "windows of the state," they merit further study to discern both the glaring and subtle changes that are taking place within a state's changing demographics, its social and political institutions and its relations with a wider world, whether a regional neighbor or an integral part of an increasingly globalized world.

References

Altman, D. (1991). *Paper ambassadors: The politics of stamps*. North Ryde: Angus and Robertson.

Anderson, B. (1991). *Imagined communities. Reflection on the origin and spread of nationalism*. London: Verso.

Barthes, R. (1972). *Mythologies* (trans: Lavers, A.). New York: Hill and Wang.

Berger, P. L. (1967). *The sacred canopy*. Garden City: Doubleday.

Berger, P. L., Davie, G., & Fokas, E. (2008). *Religions America, secular Europe*. Farnham: Ashgate.

Billig, M. (1995). *Banal nationalism*. London: Sage.

Boulding, K. E. (1959). National images and international systems. *Journal of Conflict Resolution, 2*, 120–131.

Bruce, S. (2011). *Secularization: In defence of an unfashionable theory*. Oxford: Oxford University Press.

Brunn, S. D. (2000). Stamps as iconography: Celebrating the independence of new European and Central Asian states. *GeoJournal, 52*, 315–323.

Brunn, S. D. (2002). Political stamps of the Arab world: 1950–1999. In C. Schofield, D. Newman, & A. Drysdale (Eds.), *The razor's edge. International boundaries and political geography. Essays in honour of professor Gerald Blake* (pp. 77–105). The Hague: Kluwer.

Brunn, S. D. (2011). Stamps as messengers of political transition. *Geographical Review, 101*, 19–36.

Brunn, S. D., & Cottle, C. (1997). Small states and cyberboosterism. *Geographical Review, 87*, 249–258.

Casanova, J. (2001). Religion, the new millennium and globalization. *Sociology of Religion, 62*, 415–441.

Covington, K., & Brunn, S. D. (2006). Celebrating a nation's heritage on music stamps. Constructing an international community. *GeoJournal, 65*, 125–135.

Davie, G. (2002). *Europe: The exceptional case. Parameters of faith in the modern world*. London: Darton, Longman and Todd.

Deans, P. (2005). Isolation, identity and Taiwanese stamps as vehicles for regime legitimation. *East Asia. An International Quarterly, 22*, 8–30.

Deans, P., & Dobson, N. (2005). East Asian postage stamps as socio-political artefacts. *East Asia. An International Quarterly, 22*, 3–7.

Dobson, H. (2002). Japanese postage stamps. Propaganda and decision making. *Japan Forum, 14*, 21–39.

Dobson, H. (2005). The stamp of approval. Decision-making processes and politics in Japan and the UK. *East Asia. An International Quarterly, 22*, 56–76.

Dodds, K. (2006). *Geopolitics in a changing world*. Harlow: Prentice-Hall.

Gilbert, E. (2007). Money, territoriality and the proposals for a North American Monetary Union. *Political Geography, 26*, 141–156.

Grant, J. (1995). The socialist construction of philately in the early Soviet era. *Comparative Studies in Society and History: An International Quarterly, 37*, 476–493.

Hall, S. (Ed.). (1997). *Representation: Cultural representations and signifying practices*. London: Sage.

Hammett, D. (2012). Envisaging the nation: The philatelic iconography of transforming South African national narratives. *Geopolitics, 17*, 526–552.

Hymans, J. (2010). East is east and west is west? Currency iconography as national-branding in the wider Europe. *Political Geography, 29*, 97–108.

Jacobson, M. H., & Purcell, D. (1997). Politics and media richness in the World Wide Web: Representations in the Former Yugoslavia. *Geographical Review, 87*, 249–258.

Johnson, G. (2005). The two Koreas' societies reflect in stamps. *East Asia. An International Quarterly, 22*, 77–95.

Kevane, M. (2008). Official representations of the nation: Comparing the postage stamps of Sudan and Burkina Faso. *African Studies Quarterly, 10*, Online: http://africa.ufl.edu/asq/v10/v10ila3.htm

Knippenberg, H. (2006). The political geography of religion: Historical state-church relations in Europe and recent challenges. *GeoJournal, 67*, 253–265.

Lauritzen, R. (1988). Propaganda art in the postage stamps of the Third Reich. *The Journal of Decorative and Propaganda Art, 10*(Autumn), 62–79.

Leith, J. A. (1971). Postage stamps and ideology in Communist China. *Queen's Quarterly, 78*, 176–186.

Luckmann, T. (1967). *The invisible religion: The problem of religion in modern society*. New York: Macmillan.

McQueen, H. (1988). The Australian stamp: Image, design and ideology. *Arena, 84*, 78–93.

Mirzoeff, N. (1998). What is visual culture? In N. Mirzoeff (Ed.), *The visual cultural reader* (pp. 3–13). London: Routledge.

Moisio, S. (2008). Finlandisation versus westernization: Political recognition and Finland's European Union membership debate. *National Identities, 10*, 77–93.

O'Sullivan, C. J. (1988). Impressions on Irish and South African national identity on government issued postage stamps. *Éire-Ireland, 23*, 104–115.

Panofsky, E. (1982). *Meaning in the visual arts*. Chicago: University of Chicago Press.

Pointon, M. R. (1988). Money and nationalism. In G. Cobbitt (Ed.), *Imaging nations* (pp. 229–254). Manchester/New York: Manchester University Press.

Raento, P. (2006). Communicating geopolitics through postage stamps: The case of Finland. *Geopolitics, 11*, 601–629.

Raento, P., & Brunn, S. D. (2005). Visualizing Finland: Postage stamps as political messengers. *Geografiska Annaler B, 87*, 145–163.

Raento, P., & Brunn, S. D. (2008). Picturing a nation: Finland on postage stamps, 1916–2000. *National Identities, 10*, 49–75.

Raento, P., Hãmmãnlãinen, A., Ikonen, H., & Mikkonen, N. (2004). Striking stories: A political geography of euro coinage. *Political Geography, 23*, 929–956.

Reed, D. (1984). The symbolism of postage stamps: A source for the historian. *Journal of Contemporary History, 19*, 223–249.

Reed, D. (1993). The postage stamp: A window of Saddam Hussein's Iraq. *The Middle East Journal, 47*, 77–89.

Rose, G. (1996). Teaching visualised geographies: Towards a methodology for the interpretation of visual methods. *Journal of Geography in Higher Education, 20*, 291–313.

Rose, G. (2001). *Visual methodologies. An introduction to the interpretation of visual materials*. London: Sage.

Rose, G. (2011). *Visual methodologies. An introduction to Researching visual materials*. London: Sage.

Scott, D. (1992). National icons: The semiotics of the French stamp. *French Cultural Studies, 3*, 215–234.

Scott, D. (1995). *European stamp designs. A semiotic approach to designing messages*. London: Wiley Academics.
Snee, C., & Kloetzel, J. E. et al. (2012). *Scott standard postage stamp catalogue*. Sidney: Scott Publishing Company.
Stamp, L. D. (1966). Philatelic cartography. A critical study of maps on stamps with special reference to the Commonwealth. *Geography, 51*, 179–197.
Swaatos, W. H., & Christiano, K. J. (1999). Secularization theory: The course of a concept. *Sociology of Religion, 60*, 209–228.
Unwin, T., & Hewitt, V. (2001). Banknotes and national identity in central and eastern Europe. *Political Geography, 20*, 1005–1028.
Zeigler, D. (2002). Post-communist East Europe and the cartography of independence. *Political Geography, 21*, 671–686.

Chapter 113
Demographic Forces Shaping the Religious Landscape of Vienna

Anne Goujon and Ramon Bauer

113.1 Introduction

It is usual for capital cities to undergo relatively rapid demographic change, whether in terms of absolute population, or age-specific population, compared to the rest of the country. Capital cities are usually the home of economic, social, educational, and cultural facilities, and therefore are poles of attraction for vital human capital (Harvey 1973, 1982). Another feature of major cities is that the socioeconomic compositions of their population can also evolve quite rapidly, as work opportunities make people migrate both on an international and national-regional scale, people of different social, ethnic, cultural and religious background. Since cities are the centers of change, they are also key actors in achieving social cohesion (Ranci 2011) as it is within cities that rapid increase of socio-economic and spatial disparities is most noticed (Cassiers and Kesteloot 2011).

Only a century ago, Vienna was one of the most populous cities of the world. Today, the Austrian capital is nowhere near the top in global urban population rankings; however it is still the second largest German-speaking city. After centuries of almost continuous growth, the population of Vienna peaked in the early 1900s,

A. Goujon (✉)
Wittgenstein Centre for Demography and Global Human Capital (IIASA, VID/ÖAW, WU), International Institute for Applied Systems Analysis (IIASA), Schlossplatz 1, 2361 Laxenburg, Austria

Vienna Institute of Demography/Austrian Academy of Sciences, Wohllebengasse 12-14, 1040 Vienna, Austria
e-mail: goujon@iiasa.ac.at

R. Bauer
Wittgenstein Centre for Demography and Global Human Capital (IIASA, VID/ŐAW, WU), Vienna Institute of Demography/Austrian Academy of Sciences, Wohllebengasse 12-14, 1040 Vienna, Austria
e-mail: ramon.bauer@oeaw.ac.at

S.D. Brunn (ed.), *The Changing World Religion Map*,
DOI 10.1007/978-94-017-9376-6_113

followed by nearly a century of decline and stagnation. Out of all European capital cities, only Vienna, together with Copenhagen, Berlin, and London were losing population between 1950 and 1975, and Budapest in the period from 1975 to 2010 (United Nations 2011). Contrasting with a stagnating absolute population of Vienna and surrounding areas – Vienna only grew by 7 % in 50 years between 1951 and 2001, from 1.45 to 1.55 million, yet its religious composition changed dramatically as more and more Christian inhabitants left the Church. Not until the late 1980s, Vienna's population numbers started to rise again and recently increasing growth rates have made Vienna one of the strongest growing European cities[1] during the first decade of the twenty-first century (Eurostat Database on Metropolitan regions)[2] which again had a large impact on the religious composition of the city.

Austria was in the past – as it still remains today – predominantly Roman Catholic, though several religious minorities, in particular Protestant and Jewish communities, have existed there for a number of centuries. Until recently, most changes in the religious landscape of Austria occurred through the enforcement of doctrines by the religious authorities, for example, the Counter-Reformation in the seventeenth century and/or by the political power in place, for example, the pogrom and holocaust of Jewish communities before and during the Second World War.

Since 1970 the homogeneity in terms of the domination in the population of one religious group, the Roman Catholic Church in the case of Austria, has been slowly fading away through two main forces: increased secularization and immigration of people belonging to other religions (Goujon et al. 2007). This transformation is quite unique in the history of religions but common to most "modern" societies. Moreover, migrant women, especially from Turkey, Serbia and Montenegro, and Macedonia have a higher number of children compared to native women, which is reinforcing the increase in religious pluralism due to migration. These phenomena lead to the diversification of the religious landscape in Austria and is exacerbated in the city of Vienna where both forces of secularization and migration are stronger than in any of the other Austrian federal provinces. Whereas from 1971 to 2001, the share of Roman Catholics decreased from 87 to 74 % in the whole country, it changed from 78 to 49 % in Vienna. During the same period, the share of those without religious affiliation rose from 4 to 12 % in Austria and from 10 to 26 % in Vienna. The share of the Muslim community, one of the fastest growing religions, rose from being close to 0 % in 1971 at the national as well as capital city level to 4 % in Austria and 8 % in Vienna in 2001. If half of the Muslims living in Vienna were born in Turkey, 30 % were actually born in Austria, and 18 % in former Yugoslavia. Vienna is home to 20 % of the Austrian population, but hosts a larger share of the population with a migratory background – 40 % – which makes 42 %

[1]The population of the city of Vienna increased from 1.55 to 1.66 million between 2002 and 2007 according to the Eurostat database (see note 3) – hence a 7.3 % gain which can be fully attributed to international immigration. The whole metropolitan Viennese area including suburbs and exurbs is inhabited by about 3.1 million people.

[2]Eurostat database (Metropolitan regions) available at http://epp.eurostat.ec.europa.eu/portal/page/portal/region_cities/metropolitan_regions/data_metro/database_sub3 [2/11/2012].

of its population: 74 % of those are migrants of the first generation. The difference in the religious composition of the population between Vienna and the city of Austria is shown in Fig. 113.1.

Side by side with the fading of church-related religiosity, individual beliefs such as the belief in God or self-assessed religiosity have weakened as well. During the last decade they have decreased most notably among young people and in rural areas (Zulehner and Polak 2009). Religious socialization in the family, an important precondition for future faith, is less common at present than 20 years ago. Baptism, a religious wedding and especially a religious funeral, on the other hand, remain widespread. When European countries are ranked by their level of religiosity, Austria is located at the lower end of the more religious half of countries, close to, for example, Switzerland or Slovenia (Voas 2009).

If secularization and international migration have been the main factors shaping Vienna's changing religious landscape, more restrictive and selective migration as well as the resurgence of religion (Kaufmann 2010) may mean that the extent of fertility differentials by religious denominations and of exogamy could play the major roles in determining the future religious composition of Vienna.

Overall, the relative sizes of secular and religious populations belong to the most important social characteristics of any society. In the wake of religious change, family behavior, including marriage and childbearing, is likely to be altered. European demographic trends, including that of low fertility and progressively later childbearing are

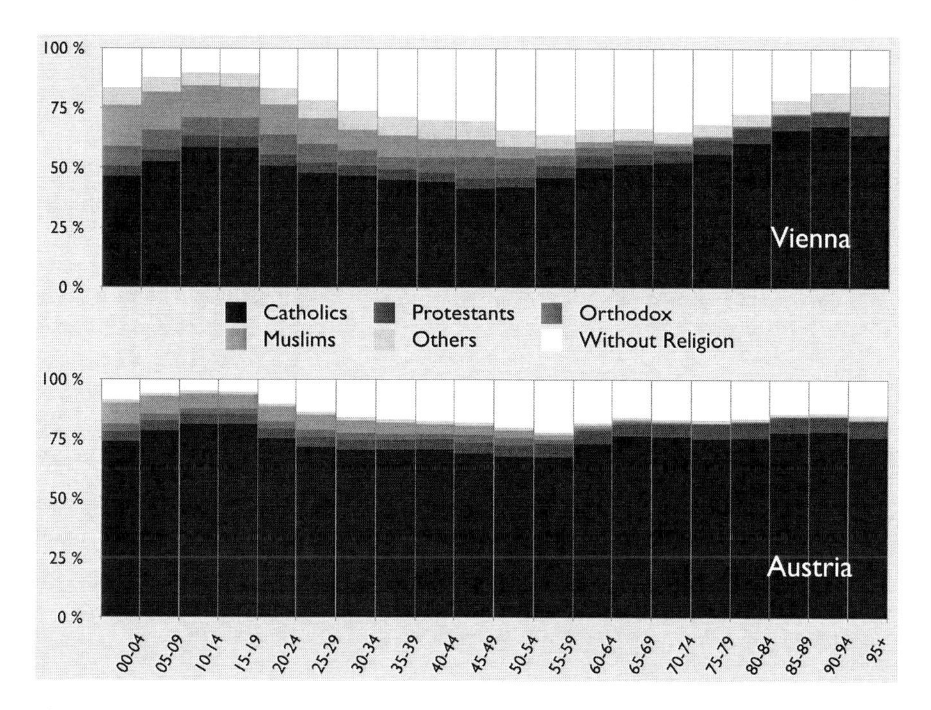

Fig. 113.1 Share of the population by religion and age in 2001 in Austria and Vienna (Source: A. Goujon and R. Bauer, data from 2001 population census, Statistics Austria)

also likely to be affected when there is a growth of distinct religious groups with high fertility and with low rates of conversion and secularization. The changing religious distribution of the population can also have wider social and political ramifications, affecting the level of social cohesion, voting preferences, and potentially leading to an increase in segregation of specific minorities at a level of urban districts (Borooah 2004; Lehrer 2005; Morgan et al. 2002). One of the primary challenges for the Austrian government and the city council of Vienna will be to safeguard religious freedoms and to ensure a fair voice for foreign-origin populations and religious minorities including Muslims, while combating extremism, supporting the integration of minorities, and adapting European societies to diverse religious communities. At the city level, the changes in the distribution of the religious communities between the 23 districts of Vienna could have problematic implications in terms of segregation and ghettoization.

Most datasets on religion are based on surveys, and very few contain detailed information for the whole population, which make the Austrian and Vienna Census-based datasets unique,[3] as well as other register-based data on marriages, births, divorces, deaths, in- and out-flows (to the main religious denominations) that have been collected for many years by Austrian statistical authorities. On the other hand, census data, which include religious denomination but no information on the degree of religiosity, may conceal differences in religious intensity between religious groups. This chapter does not, however, address the issue of religiosity, but purely on the different religious denominations and their size and demographic behavior.

In our research within the framework of the WIREL project[4] – WI for Wien/Vienna and REL for RELigion, that aims at addressing the role of religions in shaping the social and demographic structure of the population of Vienna, and their implications over a period ranging from 1951 to 2051. We investigate the changes that can be observed in the religious distribution of the population within Vienna and the different demographic forces that have been shaping this religious composition at the city level, namely, migration, differential fertility, and religious conversion. These findings will be used in the evaluation of the potential for the future evolution of these forces and the resulting religious landscape of Vienna until the middle of the century.

This chapter is divided into two main parts; the first section gives a brief history of the city of Vienna, the second shows the increasing religious diversity and in the third, we comment on the evolution of the different forces that were mainly shaping the city until now, namely migration and secularization, and the emerging demographic forces that will also be key to determine the future religious landscape, namely mixed unions and fertility.

---

[3] Unfortunately, 2001 was the last census by enumeration and Austria moved to new register-based census which does not include religion.

[4] The WIREL project on "Past, present and future religious prospects in Vienna, 1950–2050" received a grant from the WWTF (Vienna Science and Technology Fund) in its 2010 Diversity-Identity Call.

113.2 A Short History About the Population of Vienna

Throughout history, cities always were bound to attract migrants and this kept their population numbers from falling, since the number of deaths generally exceeded the number of births (Flinn 1981). This was also the case in Vienna, but recurring wars and plagues kept the population from growing substantially until the late seventeenth century. After Vienna became the capital of the expanding Habsburg Empire and thus experienced more than two centuries of unprecedented urban population growth, almost entirely driven by immigration: from estimated 125,000 in 1700 (Weigl 2000a, b) to more than half a million by 1850, and up to two million in 1910. Natural population growth was generally negative until the second half of the eighteenth century and never exceeded the demographic gains achieved by migration, even after mortality rates dropped significantly in the course of the demographic transition during the second half of the nineteenth century (Weigl 2000a). During the period of strongest growth, between 1850 and 1910, more than half of Vienna's population was born outside of the city. The many immigrants were not only attracted from the rural hinterland, but also – and in the wake of the Industrial Revolution (between 1830 and 1860 in Austria) predominately – from non-German-speaking parts of the multi-national Habsburg and Austro-Hungarian Empire: from Lombardy to Galicia and from Bohemia to Bosnia. By the turn of the twentieth century, Vienna was not only the fourth largest European city; it was also the second largest Czech city as well as the third largest Jewish city in Europe (Eppel 1996).

Vienna reached a peak population of 2.2 million in the aftermath of World War I, but found itself from now on as a capital of a small German-speaking country of 6.5 million, instead of being the administrative, economic and cultural center of a former multi-national Empire with more than 50 million people. Birth rates in Vienna tumbled down during war times and remained negative for the most part of the twentieth century. As accurately described by Weigl (2000a, b), between 1910 (just before World War I) and 1951 (when the first census after World War II was conducted), fertility rates strongly decreased far below the replacement level reaching absolute low-points during the Great Depression – the all-time low of 0.6 children per woman was achieved in 1934 (Lutz and Hanika 1989; Lutz et al. 2003). Some hundred thousand left Vienna in the years succeeding the war heading to the many successor states, but due to immigration from surrounding rural regions the population of Vienna stabilized during the 1920s and 1930s between 1.8 and 1.9 million. During World War II, Vienna lost almost its entire Jewish population of nearly 170,000 just before the war due to extensive expulsion in 1939 and the holocaust during the Nazi regime (Weigl 2000a). With more than 10 % of its population belonging to the Jewish denomination in 1923, Vienna was the third largest Jewish city in Europe (Lappin 1996). During World War II, the city's population also suffered great losses among adult men at war service as well as among the civil residents during the final stage of the war. As a consequence, Vienna's population decreased to merely 1.6 million after the end of the war and the city's demographic structure was shaped by a skewed age – due to fertility and migration forces – and sex distribution caused by war casualties among men. In 1951, 13 % of the population was below age 15 and 20 % above the age of 60, and there were 1.2 women for each men living in Vienna.

As a result of these historical events and the associated demographic dynamics during the first half of the twentieth century, post-war Vienna was inhabited to some extent by a residual population: the former multi-cultural Vienna had become a distinctly heterogeneous city with respect to the ethnic and cultural background, as well as religious denomination of the population.

After the drastic population losses during World War II, Vienna's population stabilized in numbers over the course of the 1950s and 1960s only because of considerable internal migration gains. International immigration remained on low levels, consisting mainly of refugee flows from neighboring socialist *Eastern bloc* countries: that is, Hungary in 1956, the former Czechoslovakia in 1968, and Poland in the early 1980s (Eppel 1996). However, the natural population balance still remained negative, even during the baby-boom years of the late 1950s and early 1960s. These demographic trends, in association with increases in life expectancy, turned Vienna into a city populated literally by old women. In 1971, more than a quarter of the population was above age 60, out of which two-thirds were women (Lutz and Hanika 1989).

The 1970s marked the reversal of several trends. First of all, internal migration became negative as more people moved to the suburbs of Vienna, and international migration increased substantially as a political effort to counteract the persistent shortage of labor with the recruitment of workers mainly from Yugoslavia and Turkey, the so-called *Gastarbeiter* (guestworkers). If the combination of both trends led the city to still lose about 5 % of its population between 1971 and 1981, it also triggered a process of (re-)diversification and rejuvenation in the long run. After 1971, the share of the elderly population gradually decreased and the share of the foreign-born population increased. The *Gastarbeiter* did not return home as originally intended, but rather brought their families in order to settle in the long-term, while more workers and their families followed their trails (Fassmann and Münz 1995).

In 1988, population numbers reached a low point of 1.48 million. After the fall of the Iron Curtain, Vienna became an attractive destination again for immigration from the neighboring Central European countries, and from Yugoslavia during the early 1990s, especially in the course of the Bosnian War. In 1995, Austria became member of the European Union, and Vienna started attracting as well EU citizens, especially from Germany. During the 1990s, international migration became the undisputed driver of Vienna's regained population growth. The 2001 census count for Vienna was 1.55 million and 1.71 million in 2011: a 10 % increase, the result of the persistent inflow of international immigrants and natural increases. Beyond that, the accelerated globalization established new migration regimes and further diversified the geographic origins of the new immigrants and, hence, also stimulated the ethnical, cultural and religious heterogeneity of today's population of Vienna. A research team at the University of Vienna is in the process of investigating the 800 or more places of cult in the city that have until now not been systematically enumerated.[5]

113.3 Increasing Religious Diversity

In 1951, the religious landscape of Vienna was characterized by the dominance of the Roman Catholic Church. About 82 % of the population were Catholics and, adding the 8 % Protestants as well as a few members of Eastern Orthodox Churches, more than 90 % of Vienna's population was Christian by denomination. The rest of the population split up between 2 % with other religions – including a small Jewish community that survived the *Shoa* and stayed in Vienna – and a considerable share of up to 8 % without any religious denomination. This relatively high share of people without religion in 1951, compared to the rest of Austria, and to presumably other European cities at the same time, find roots in the strong social-democratic tradition in Vienna (das "Rote Wien"), which goes back to the late nineteenth century and was predominately anti-clerical orientated – and also anti-Semitic to some extent.

Figure 113.2 shows that the religious distribution remained stable over the course of the 1950s and 1960s, although Catholics lost a few percentage points, while the share of secular people with no religious denomination increased slightly. Some other small shifts are visible – the share of Orthodox (included in "other" in the figure) and Muslims slightly increased and a Muslim community re-emerged since the arrival of the first *Gastarbeiter* from Yugoslavia and Turkey in the late 1950s.

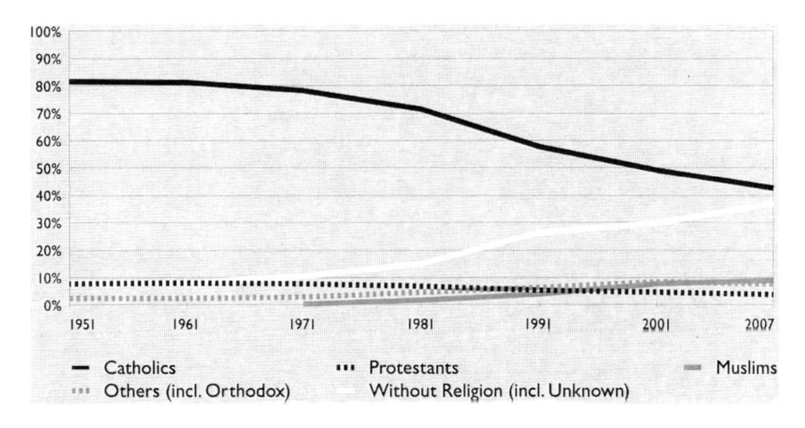

Fig. 113.2 Increasing religious diversity in Vienna, 1951–2007 (Source: A. Goujon and R. Bauer, data from Statistics Austria censuses for 1951–2001; estimates for 2007 from Goujon et al. 2007)

[5] A map of all religious places collected by the "Kartographie der Religionen in Wien" is available here: http://kartrel.univie.ac.at/?page_id=251 [16/11/2012].

The 1971 census clearly revealed that Vienna was still a largely Catholic city: almost four out of five professed to be Roman Catholic.

Between the 1971 and the 2001 census – when religious denominations were surveyed for the last time[3] – the previously rather homogeneous religious landscape of Vienna became considerably more diverse. By 2001, Catholics became a minority in Vienna (49.2 %), while the secularized population without any religious denomination strongly increased (from 10.5 % in 1971 to 29.5 %) and is about to become the strongest group (37 % in 2007 versus 42.6 % Catholics – see also Fig. 113.2). Like the Catholics, also the share of Protestants decreased by more than a third during the last three decades of the twentieth century (from 6.9 to 4.7 %). Besides those major shifts, the share of Muslims increased from less than a half per cent in 1971 (0.4 %) to nearly 8 % (7.8 %) and, hence, already outnumbered the city's Orthodox population (6 %). All in all, the 2001 census distinguished between 47 different religious denominations, including some established religions like Jews and Old Catholics (around 0.45 % each) and many previously unseen or unrepresented religions like Buddhism (0.3 %), Hinduism and Sikhism (both together 0.27 %) and also Mormonism and Baha'ism (both below 0.1 %) among others. At the turn of the millennium, Vienna had truly become a city of religious diversity.

Figure 113.3 shows the spatial dimension of Vienna's religious diversity in 2001 at the neighborhood level. The map is based on the share of the largest religious

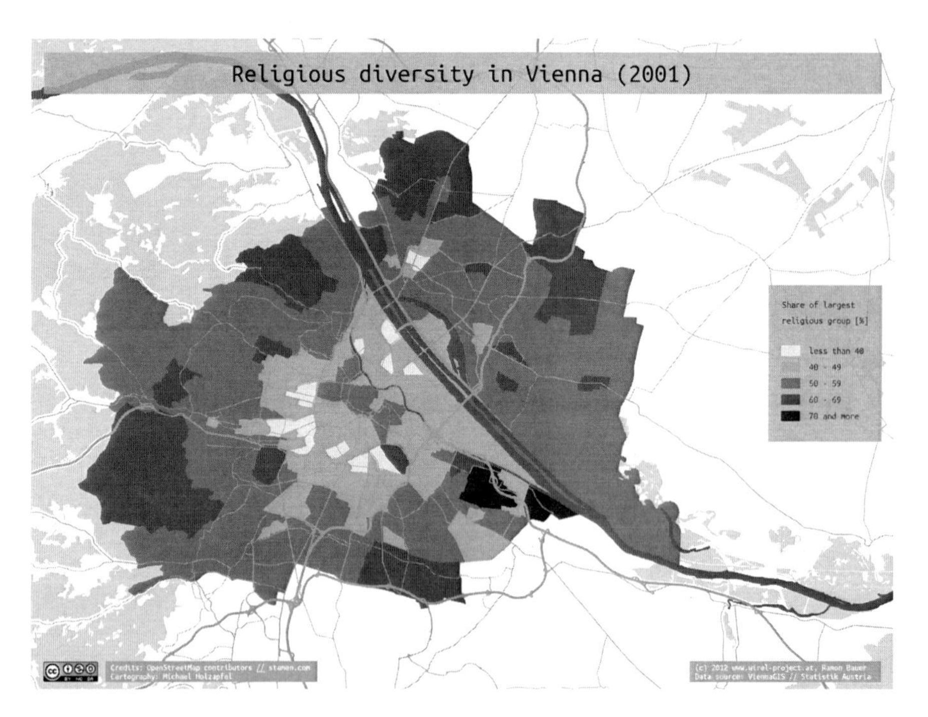

Fig. 113.3 Neighborhood diversity by religion in Vienna, 2001 (Map by A. Goujon and R. Bauer with data from 2001 population census 2001, Statistics Austria)

group (usually Catholic), that is, the smaller the share of the largest group, the more diverse is the neighborhood. Hence neighborhood diversity is assessing the mix of various groups rather than the concentration of one particular group, In Vienna, the most diverse neighborhoods are clustered along a heavy traffic ring street called "Gürtel" (German for "belt"), which surrounds the inner more urban districts and separates them from the outer ones. On one hand, these areas are characterized by high shares and hence concentration of ethnic minorities with a higher prevalence of migrant minority religions such as Islamic and Orthodox denominations; on the other hand, some of these areas are rather gentrified neighborhoods with a high share of secular population.

113.4 Demographic Drivers of Religious Change

113.4.1 Religious Mobility

One particularity of Austria is that members of the Catholic and Protestant Church have to pay a yearly fee directly to the Church – 1.1 % of the net pre-tax income of Catholic members and 1.5 % of the self-reported income of Protestants.[6] Hence religious mobility in terms of exit and entrance to the Christian churches are well known, and are shown in Fig. 113.4. Overall since the 1960s, the Catholic and Protestant Churches have been losing members, in a similar pattern but at a different scale, with less and less compensation coming from new entrants – data on entrance are only available until 1984). The period 1979–1983 was a time of major losses for the Catholic Church in Vienna and it peaked in 1983 with 17,000 members cancelling their Church membership – a record until now. Since then, the trend in terms of losses has been reversed, with clearly less exits by the year, but also more erratic patterns with peaks every now and then. Most of the peak losses (1995, 1999, 2004, and 2010) for the Catholic Church correspond to church-related sexual abuse scandals in Austrian Catholic Institutions (schools or Boy's choir) or elsewhere, notably in Belgium, Netherlands, Germany, and Ireland. Beyond the acts of child abuse themselves, the opinion in Austria was particularly scandalized by the attempt of the Church to cover up the stories, like in the case of the Archbishop of Vienna, Cardinal H.H. Groer, against whom charges of sexual misconduct were pronounced in 1995, but were ignored by the Church hierarchy until early 1998 when a papal investigation finally commenced. Finally a few months later, Cardinal Groer had to relinquish all ecclesiastical duties and privileges.

Although at first sight, the Protestant Church, which consists only of the Evangelical Church of the Augsburg and Helvetic Confessions (often abbreviated as Evangelical AB and HB) – other Protestant confessions are not counted under this group–, seems to be experiencing the same history of membership exits as the

[6] Other religions present in Austria do not pay taxes, but are encouraged to dedicate gifts to their religious community.

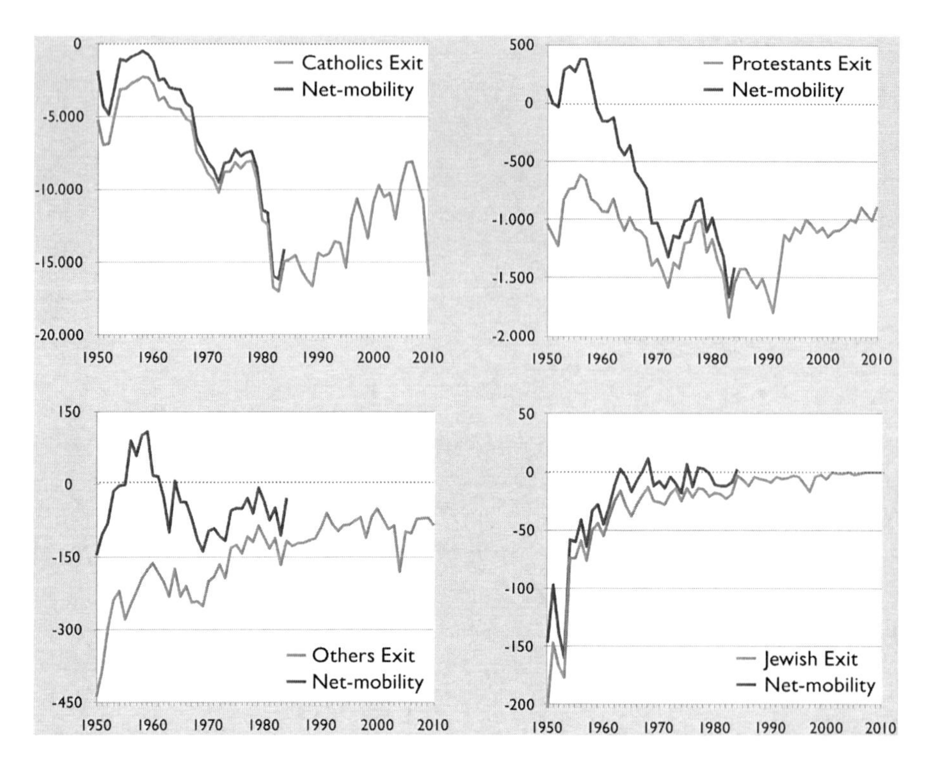

Fig. 113.4 Religious mobility in Vienna, 1950–2010 (Source: A. Goujon and R. Bauer, data from City of Vienna Statistical Yearbooks 1950–2011)

Catholic one, the Protestant pattern is different in three ways. First of all, the Protestants were able to compensate the loss of members through the entrance of new members much more than the Catholics. For instance in 1963, for one new member in the Catholic Church, there were three exiting members (1:3), whereas for the Protestants, the ratio was two to 3 (2:3). The difference between the two Churches in terms of gains and losses diminished over time. The second difference is that the Protestant Church is much less affected by the scandals than the Catholic one, and is, therefore, experiencing less membership fluctuations. Finally, and this is obvious looking at Fig. 113.5, whereas there is more mobility among the Catholic men, at least until the beginning of the twenty-first century, for the Protestants, and with a few exceptions, the majority of quitters are women. In former times, women have always been the ones in charge of religions in families, responsible for religious feelings, religious education, and religious practice, which would explain the higher retention for women among the Catholic Church. However, Protestant women in the 1950s–1970s, who were often marrying outside of their religion, mostly a Catholic husband, were more prone to change their religion in obedience to their spouse. Within the last few decades women started to look for new ways of religious experience, mostly in order to revitalize religion inside and outside of

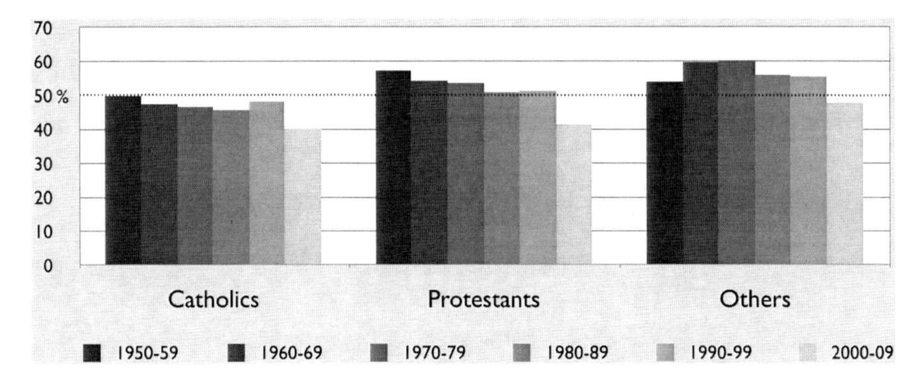

Fig. 113.5 Share of women in religious mobility in Vienna, 1950–2011 (Source: A. Goujon and R. Bauer, data from City of Vienna Statistical Yearbooks 1950–2011)

Churches, which would explain the stronger mobility in the most recent periods as well as the fact that women became more independent. For the other religions where data are available (old Catholic, Jewish, and other Christian denominations), there are also more women than men among the religiously mobile population which would confirm the minority hypothesis.

Trends in the age at which the members decide to leave the Church bring interesting information. Most of the transitions occur before the age of 60, with a peak around the age of 21–40. However the occurrence of the peak seems to differ greatly according to the decades of observations with no recognizable patterns, especially the 1960s are exhibiting some strange patterns potentially due to the absence of data for the years around 1968. Unfortunately, the age breakdown is not available after 1983.

113.4.2 International Migration Flows by Religion, Vienna (1992–2010)

Although very detailed data are available on religion in Vienna and the rest of Austria, very little is known about the religious denomination in migrant flows, meaning of the immigrants to, and emigrants from Austria. These data have to be imputed using the random migration hypothesis – meaning that migrants are assumed to have the same religion as the population in the country of origin. Although this is a strong assumption to make, and can be detrimental to the analysis for some particular religious groups in some particular years, this method has been shown to lead to quite reasonable estimates of the religious distribution of the migrants (see Goujon et al. 2012; Skirbekk et al. 2010). What is clearly visible from Fig. 113.6 is that immigration to Vienna was very high at the beginning of the 1990s, after the breakup of Yugoslavia. At that time, three-quarters of the international migrants to Vienna were Muslims from Turkey and Former Yugoslavia. In the

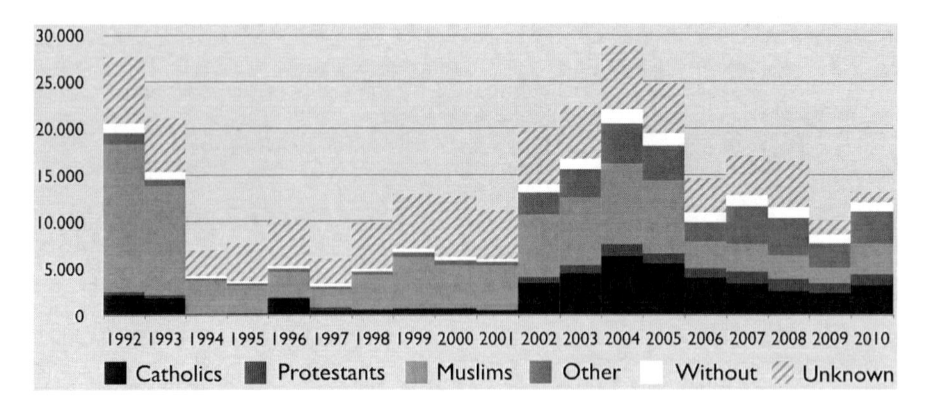

Fig. 113.6 International migration flows by religion, Vienna 1992–2010 (Note: migration flows of Austrian citizens are not included) (Source: A. Goujon and R. Bauer, data from City of Vienna Statistical Yearbooks 1993–2011)

early 2000s, following the entry of Austria in the European Union in 1995, international immigration to Vienna was equally high, but the share of Muslims has declined and recent migration flows are equally driven by the Catholic-, other-(mostly Orthodox) and Muslim-denominations. In 2011, the ten top countries in terms of net-migration to Vienna were Slovakia, Bulgaria, Turkey, Russia, Bosnia and Herzegovina, Afghanistan, Italy, Serbia (incl. Montenegro and Kosovo), Ukraine, and India. Migration is noticeably acting as a counterweight to secularization in the native population as the share of the population without religion seems to be lower as in the total population. However little is known about the secularization behavior of those immigrants to Austria.

If the religious status of migrants (flows) is not reported, more can be said about the foreign born population in Vienna (stock), based on censuses. In 2001, the foreign born population residing in Vienna was approximately shared between Roman Catholics (24 %), Muslims (23 %), Orthodox (19 %) and those without religion (20 %). As can be seen from Fig. 113.7 the largest group of foreign born population in Vienna, originated from Former Yugoslavia (particularly Serbia and Montenegro, and Bosnia and Herzegovina).

113.5 Total Fertility Rates by Religion, Vienna (1971–2010)

The fertility in Vienna has been overwhelmingly stable during the last 40 years as can be seen from Fig. 113.8, oscillating between 1.3 and 1.5, and actually most women in different religious denominations are having low and stable fertility behavior, between 1.2 and 1.5 – that is at least the case of the two main Christian denominations: the Catholics and the Protestants. Although the data on the Orthodox population only became available at the time of the 2001 census, it seems that this

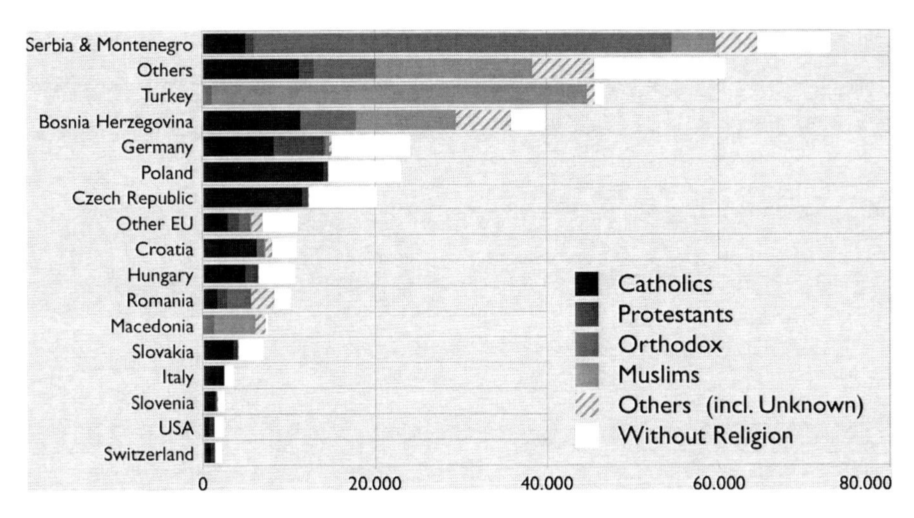

Fig. 113.7 Population by religion and country of birth (excluding Austrian citizens) in Vienna, 2001 (Source: 2001 population census 2001, Statistics Austria)

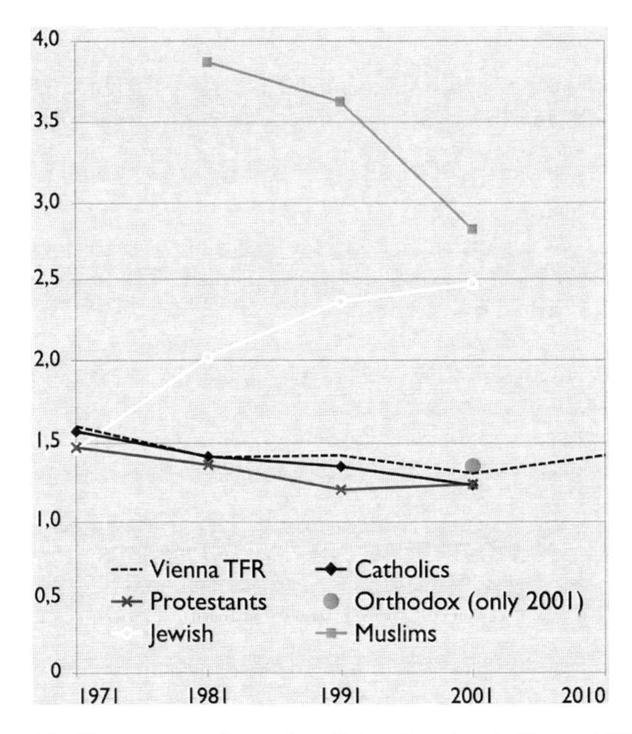

Fig. 113.8 Total fertility rates according to the religion of mothers in Vienna, 1971–2010 (Note: estimates at census years are based on the total population, the female population in childbearing ages, and the general fertility rate) (Source: A. Goujon and R. Bauer, data from census 1971–2011, Statistics Austria and City of Vienna Statistical Yearbooks 1972–2010)

Table 113.1 Births to Muslim women, by country of origin, 2008–2011

	Share of births		Average number of births per woman	
Country	Austria (%)	Vienna (%)	Austria	Vienna
Turkey	38	39	2.13	2.17
Austria	14	14	1.85	1.77
Bosnia and Herzegovina	12	7	1.86	1.83
Kosovo	6	4	2.08	2.05
Russian Federation	6	5	3.01	2.86
Serbia	4	3	2.11	1.97
Macedonia	4	5	1.98	1.93
Egypt	3	5	3.00	3.02
Afghanistan	2	3	2.58	2.59
Germany	1	1	1.79	1.74
Other	10	14	2.15	2.15
Total	100	100	2.13	2.15

Data source: Sobotka et al. (2012)

group is also following low patterns of fertility (with a TFR estimated at 1.35 in 2001). Two groups stand out: the Jewish, dominated by Orthodox Jews in Vienna, and the Muslims whose fertility started very high in the 1980s (3.9) as a result mostly of family reunification and changing countries of origin in the early 1990s and was around 2.8 in 2001. Further estimates of fertility by scholars (in Saunders 2012) point at further declines in Muslim fertility as Austria's Muslims had a recorded fertility rate of 3.09 children per mother in 1981, 2.77 in 1991, and 2.3 in 2001 (see also Goujon et al. 2007).

The decline in Muslim fertility, evident from birth records shown in Table 113.1, both in Vienna and in Austria is due to two phenomena. First of all, migration from Muslim countries is diminishing and hence the weight of second generation migrants converging to native (Austrian) low fertility patters in the overall Muslim population becomes more important. Secondly, in the main countries of origin of Muslim migrants, the fertility is experiencing strong declines – In Turkey, the total fertility is below the replacement level of 2.1, where as it was above 4 in the 1970s.

113.6 Marriages and Divorces Within Religions, Vienna 1950–2008

Catholics, especially men still marry in majority inside their religion whereas Protestants, in minority, tend to marry outside (Fig. 113.9). Both sexes in both religions tend to marry less and less inside their own religion. The trend is rather the opposite for Muslims and Orthodox. For Muslims, exogamy is rather typical of men whose wife will most likely convert to Islam. The transmission of religion from parents to children of those children born to mixed marriages will become increasingly important. Lutz and Uljas-Lutz (1998) estimated that only

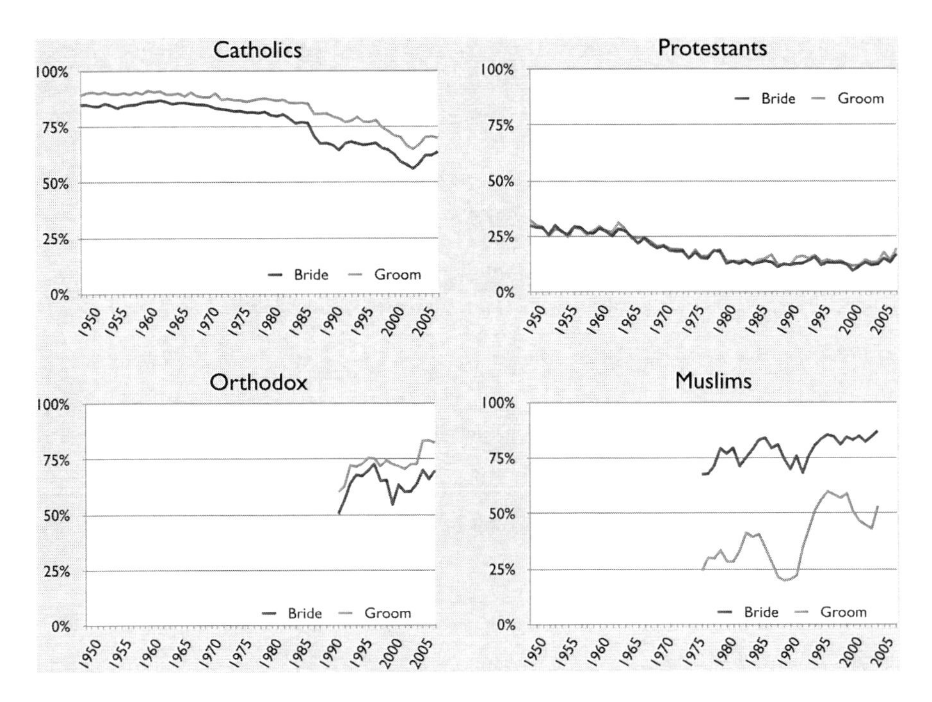

Fig. 113.9 Marriages within religions, Vienna 1950–2008 (Source: A. Goujon and R. Bauer, data from City of Vienna Statistical Yearbooks 1950–2011)

half of the children born to mixed couples with one spouse being Protestant became Protestant. Lutz (1985) shows that the religion of the mother is more important than the religion of the father for the transmission of religion from parents to children.

Not unrelated to the different patterns of mixed marriages across religions are those of divorces between sexes, which show that the majority of divorces among Catholics and Protestants were attributed to men, until 1985 when data were available, whereas there were more divorces among married women partners with other religions and/or no religion than among their male counterparts that could be due to the fact that these women had a higher status – mostly educational and financial – than the Catholic and Protestant women which allowed them more freedom to move out of a union.

113.7 Conclusions and Outlook

Vienna, a city that was distinctly homogenous in terms of population composition in the mid-twentieth century, became a city of ethnic, cultural and religious diversity. This can be studied demographically as all changes related to changes in the composition of the religious landscape of the city of Vienna are related to demographic events: migration and fertility. Other drivers of change such as secularization or religious mobility express also a form of migration from one religious status to another. We have shown in this chapter that secularization and international migration have been the main factors shaping Vienna's changing religious landscape, as this was the case in most countries in Europe, and North America. More restrictive and selective migration as well as the resurgence of religion (Kaufmann 2010) may mean that the extent of fertility differentials by religious denominations and of exogamy could play the major roles in determining the future religious composition of Vienna. The WIREL project will further evaluate the potential of future pathways of these forces and the resulting religious landscape of Vienna until the mid-twenty-first century.

Acknowledgments We are greatly indebted to Daniel Bichl for recovering and computing the data on religion from historical sources. Our sincere appreciation goes to Caroline Berghammer, Katrin Fliegenschnee, Richard Gisser, Regina Polak, Marcin Stonawski, and Krystof Zeman, for their help in solving problems and answering crucial questions along the way.

References

Borooah, V. (2004). The politics of demography: A study of inter-community fertility differences in India. *European Journal of Political Economy, 20*(3), 551–578.

Cassiers, T., & Kesteloot, C. (2011). Socio-spatial inequalities and social cohesion in European cities. *Urban Studies, 49*(9), 1909–1924.

Eppel, P. (1996). *Wir: zur Geschichte und Gegenwart der Zuwanderung nach Wien*. Vienna: Eigenverlag der Museen der Stadt Wien.

Fassmann, H., & Münz, R. (1995). *Einwanderungsland Österreich? Historische Migrationsmuster, aktuelle Trends und politische Maßnahmen*. Vienna: Jugend & Volk.

Flinn, M. W. (1981). *The European demographic system, 1500–1820*. Baltimore: The John Hopkins University Press.

Goujon, A., Skirbekk, V., Fliegenschnee, K., & Strzelecki, P. (2007). New times, old beliefs: Projecting the future size of religions in Austria. *Vienna yearbook of population research 2007, 5*, 237–270.

Goujon, A., Caron Malenfant, E., & Skirbekk, V. (2012). *Towards a Catholic North America? Projections of religion in Canada and the U.S. beyond the mid-21st century* (VID working paper 04/2008). Vienna: Vienna Institute of Demography, Austrian Academy of Sciences.

Harvey, D. (1973). *Social justice and the city*. London: Arnold.

Harvey, D. (1982). *The limits to capital*. Oxford: Blackwell Publishing.

Kaufmann, E. (2010). *Shall the religious inherit the earth? Demography and politics in the twenty-first century*. London: Profile Books Ltd.

Lappin, E. (1996). Juden in Wien. In P. Eppel (Ed.), *Wir: Zur Geschichte und Gegenwart der Zuwanderung nach Wien*. Vienna: Eigenverlag der Museen der Stadt Wien.

Lehrer, E. (2005). *Religious affiliation and participation as determinants of women's educational attainment and wages* (IZA discussion paper 1725). Bonn: Institute for the Study of Labor.

Lutz, W. (1985). Gemischt-konfessionelle Familien in Österreich. Analyse bevölkerungsstatistischer Daten [Families with mixed religious denominations in Austria, an analysis of population statistics]. In *Demographische Informationen* (pp. 77–80). Vienna Institute of Demography.

Lutz, W., & Hanika, A. (1989). Vienna: A city beyond aging. *Bulletin of the American Academy of Arts and Sciences, 12*(4), 14–21.

Lutz, W., & Uljas-Lutz, J. (1998). Konfessions-verbindende Familien als vordringliches Lernfeld einer Minderheitskirche. [Religious denomination connections families as important field for learning for a minority church]. In M. Bünker & T. Krobath (Eds.), *Festschrift für Johannes Dantine zum 60*. Geburtstag. Innsbruck-Wien: Tyrolia Verlag.

Lutz, W., Scherbov, S., & Hanika, A. (2003). Vienna: A city beyond aging – Revisited and revised. *Vienna yearbook of population research 2003, 1*, 181–195.

Morgan, P., Stash, S., Smith, H., & Mason, K. O. (2002). Muslim and non-Muslim differences in female fertility: Evidence from four Asian countries. *Population and Development Review, 28*(3), 515–537.

Ranci, C. (2011). Competitiveness and social cohesion in western European cities. *Urban Studies, 48*(13), 2789–2804.

Saunders, D. (2012). *The myth of the Muslim tide*. Toronto: Knopf Canada.

Skirbekk, V., Kaufmann, E., & Goujon, A. (2010). Secularism, fundamentalism, or Catholicism? The religious composition of the United States to 2043. *Journal for the Scientific Study of Religion, 49*(2), 293–310.

Sobotka, T., Zeman, K., & Winkler-Dworak, M. (2012). *Geburtenbarometer: Monitoring of fertility in Austria and Vienna*. Available at www.oeaw.ac.at/vid/Barometer. Accessed 19 Nov 2012.

United Nations, Department of Economic and Social Affairs, Population Division. (2011). *World population prospects: The 2010 revision*. New York: United Nations.

Voas, D. (2009). The rise and fall of fuzzy fidelity in Europe. *European Sociological Review, 25*, 155–168.

Weigl, A. (2000a). *Demographischer Wandel und Modernisierung in Wien*. Vienna: Wiener Stadt- und Landesarchiv.

Weigl, A. (2000b). *Die Wiener Bevölkerung in den letzten Jahrhunderten* (Statistische Mitteilungen der Stadt Wien 4/2000). Vienna: Magistrat der Stadt Wien.

Zulehner, P., & Polak, R. (2009). Von der "Wiederkehr der Religion" zur fragilen Pluralität. In C. Friesl, R. Polak, & U. Hamachers-Zuba (Eds.), *Die Österreicher innen: Wertewandel 1990–2008* (pp. 143–206). Vienna: Czernin.

Chapter 114
Secularization in Mexico City as a Constant, Current Paradigm

Armando García Chiang

114.1 Introduction

In Mexico the concept of secularization is one of the two dominant topics in the analysis of religion and, as in the Western world, a small but influential group of researchers considered the loss of the centrality of religion to be an inevitable constant. During the first two decades of this century the paradigm of secularization began to be questioned and the idea of a return of religion or a re-enchantment of the world began to emerge. It is possible to speculate that in large Mexican cities, especially in Mexico City, the process of secularization remains constant only in members of a middle class who can be considered carriers of an international subculture; these are people who have received a Western-style higher education, particularly in the humanities and social sciences.

In this context, the ideas of Peter Berger (2001), in which he stresses that an error in the theories of secularization is the belief that modernization inevitably leads to a loss of the importance of religion and that paradigm should be replaced by the analysis of the interaction between the forces of secularization and a counter-secularization, seem an appropriate manner to approach the study of religious issues in the larger Mexican cities.

A. García Chiang (✉)
Department of Sociology, Universidad Autónoma Metropolitana Iztapalapa,
Iztapalapa, Mexico
e-mail: agch@xanum.uam.mx

S.D. Brunn (ed.), *The Changing World Religion Map*,
DOI 10.1007/978-94-017-9376-6_114

114.2 The Study of Religions in Mexico

There are several precedents that must be considered in any analysis of religions in Mexico. First, it is important to emphasize the need that the Mexican state had to develop a feeling of national identity, based, significantly, in an identification with the indigenous past. This fact allows us an opportunity to understand that the anthropological and historical analyses played a major role in the development of the social science in Mexico other than the sociological, historical or geographical studies.

Secondly, it is possible to acknowledge that formerly the Berger (1971). dominant paradigm in the study of religions in Mexico was that of the secularization, a paradigm that, as noted above, only began to be questioned in the last two decades of the twentieth century, when the idea of a return of the religious or of a "re-enchantment" of the world began to emerge.

In this respect, Peter Berger (1967/2001) considers that, in opposition to the idea of a disenchantment of the world, religion, instead of disappearing seems to be recovering with great vitality and that the process of secularization continues being a constant only in Europe among the members of an international subculture of those who have received an education of Western type, especially in humanities and social sciences.

Berger's ideas can also apply to Mexico in the sense that the concept of secularization was one of two dominant axes in the development of the analysis of religious phenomena. The second major axis of this type of secularization studies has been the Catholic Church. It is evident that this development has both concealed the diversity and plurality of Mexican religious expressions in spite of the church's rich heritage. At present the monopoly of the sacred that belonged to the Roman Catholic Church has become lost and we are facing a multiplication of the religious offerings, where the Protestant churches, the sects, the non-Christian groups and the new religious movements are growing in spectacular numbers (García-Chiang 2007).

Based in the works of Roberto Blancarte (1992), Rodolfo Casillas (1996) and Cristián Parker Gumucio (1994) it can be affirmed that in Mexico, as in most of the Western world, anthropology, history and sociology have analyzed the study of religions from their own disciplinary perspectives. In the Mexican context it is important to emphasize the need that the State had to focus on a national identity that was based mainly on an identification with the indigenous past. This situation allows us to understand major developments in anthropological and historical analyses compared to contemporaneous developments in sociology. Likewise it is worth noting that the development of research on religion was very restricted until the 1970s.

Concerning anthropological studies, it should be noted that the Mexican state encouraged the development of anthropology to the extent that this discipline could provide materials that would help in the search for national identity. Moreover, the existence of a large number of indigenous groups (which compose about 10 % of the total population) and the peculiarities of their habits and customs are elements that help explain why a significant percentage of the work related to religious issues is anthropological and ethnographic, focused, above all, on analyzing the religious practices of indigenous peoples and on detailed descriptions of their practices and rituals (García-Chiang 2004).

Regarding historical works, it should be emphasized that in Latin America, including Mexico, the predominance of the Catholic Church was so intense that up to the nineteenth century, historical studies were devoted to an analysis of Christianity or more precisely of Roman Catholicism. This situation began to change when the Protestant churches, which had been politically excluded, arrived in Latin-American territory in 1821 in the case of Mexico, even though their influence did not begin to be felt until the start of the twentieth century.

For the sociological studies, one point to be emphasized is the difficulty of finding articles related to religious themes that were published in the 1930s Robertson (1980). But their real development occurred in the 1960s with the use of categories of sociological analysis by progressive sectors of the Catholic Church in their analysis of everyday life.

In this sense, it can be said that in Mexico there existed a religious or pastoralist sociology that never had a clear description about its focus as it did in the United States, where Protestant clergy took active roles in the birth of general sociology (Fukuyama 1963: 739–756), or in France, where researchers arose from the members of the Catholic Church and founded the Group of Sociology of the Religions. Nevertheless, the existence of this Mexican sociological current in religion contributed significantly to the development of Mexican sociology.

Finally, with respect to geography, an essential aspect to emphasize is the fact that the religious construction of space does not obey universal schemes. It is very appropriate to investigate the role of religion as a factor in transforming the environment and landscapes (Racine and Walther 2006). However, it should also be noted that while religion plays a prominent role in the contemporary political and cultural landscape, relatively few geographers have contributed to a better appreciation of this phenomenon (Proctor 2006: 165, Garcia-Chiang 2011).

114.3 Towards a Definition of the Concept of Secularization

In the sociology of religion, the concept of secularization has long been a paradigm that can be defined, despite its many interpretations, as a process related to the loss of the social influence of religion in modern societies.

In 1966, Wilson, approaching the question of the secularization, notes;

> In this private sphere, religion often continues, and even acquires new forms of expression, many of them much less related to others aspects of culture than were the religions of the past. (This lack of relation is also an evidence of the inconsequentiality of religion for the social system).... In this sense, religion remains an alternative culture, observed as unthreatening to the modern social system, in much the same way that entertainment is seen as unthreatening. It offers another world to explore as an escape from the rigors of technological order and the ennui that is the incidental by-product of an increasingly programmed world. (Wilson 1966: XIV)

A year later Peter L. Berger in his work *The Sacred Canopy: Elements of a Sociological Theory of Religion* offered a definition of the secularization that would become classic. For this author, secularization is the process by means of which sectors of the society and of the culture are removed from the authority of the religious institutions and symbols (1967: 174).

In spite of its ideological connotations in which the progress of modernity must necessarily go hand-in-hand with the loss of the importance of the religion, and the fact that the United States have never fit this scheme easily, the paradigm of secularization has been seen to be validated by empirical information that shows the decline of the affiliations and religious practices in Europe. However, this decline has not prevented critical discussion of this concept from the 1960s onward in which it has even been seen as "a sociological myth." This evaluation of the paradigm has increased from the 1990s to the present and it is even considered by some to be a "European exception" Berger (1971, 1999).

According to Mark Chaves (1991: 96) "secularization occurs by a weakening of religious authority, not by a decline in religion Chaves (1994, 2007)." In the same perspective Roy Wallis and Steve Bruce (1992, 1995, 1997, 1999, 2001, 2002, 2006), published a set of contributions of sociologists and historians who were discussing the classic theory of the secularization, which can be defined as a model that explains the decline of the social meaning of the religion because of three key factors linked to the modernization: (1) social differentiation, (2) socialization, and (3) rationalization that changes the way people conceive of the world.

The consequences of the functional differentiation of institutions and of activities tied to religion that provoke the secularization is not just one among the other social activities (such as the economy, education, health, politics (policy), etc.) that continues its own development. Following its proper logic it is not no more beholden or chained to the religious representations. This situation provokes a diversification of experiences in the life that carries a diversification of the central conceptions of life itself.

From Wilson's point of view, socialization is the passage of the community to a "becoming" society. The fact that life is increasingly organized on a societal scale, especially that of the national societies leaves behind the local scale and provokes the abandonment of the level at which religion was celebrating and experiencing difficulty Wilson (1985), Barker et al. (1993).

The consequences of these two combined processes ensure the decline of a moral and religious system, the privatization of the religion and its push towards the margins and interstices of the social order. The third secularization production process is that of rationalization, which changes the way that the peoples conceive the world. For Jean Paul Willaime (1977, 1996, 2004a, c, 2006a, b) this classic thesis was perfected and completed by Karel Dobbelaere (1981) and David Martin (1965). Dobbelaere developed a different level of secularization using three levels: micro, meso and macro. Martin paid greater attention to the national and religious contexts, differentiating the process of secularization according to dominant religions and the type of relation that exists among the State and the religious institutions.

In 1981, Dobbelaere distinguished three different dimensions of the secularization: the laicization[1] at the macro social level (the place of the religion in the society), the

[1]The French concept of laïcite denotes the absence of religious involvement in government affairs as well as absence of government involvement in religious affairs. The linked process called laicization has the same meaning in the Mexican context.

religious change at the meso social level (the evolution of the religious institutions themselves) and the religious implications at the individual scale (the evolution of the practices and individual attitudes). Reacting to this triple level distinction Chaves names the first dimension "social secularization", the second "organizational secularization" and the third "individual secularization." Dobbelaere utilizes this proposition and conceives laicization as a subcategory of societal and organizational secularization (2002: 14). Likewise, Dobbelaere employs a proposition of Jean Baubérot (1994a, b: 12–14) to differentiate secularization and laicization. In this proposition, laicization is tied to the tensions between different social forces that could be religious, cultural, political or even military and that can take the form of an open conflict. In his context, the stake is control of the State facilities, or at least, a strong influence on them in order to force the State to intervene as a social actor capable of offering or even imposing a solution to the matters related to the religion as a social institution.

The secularization thesis, on the contrary, would constitute a process of slow and progressive loss of social relevancy of religion to the predominant trends of the social dynamics are without a major clash between the political realm and the religious jurisdiction. Said differently, the economic, political, religious or scientific mutations that can lead, inside every field, not only to internal tensions but to serious dissonances, do not exist among the internal mutations of the religious. In this way religion can, in interaction with other fields, modify or restrict its social pretensions, or in certain cases is even able to provoke a loss of its own social influence.

114.3.1 Current Trends in the Analysis of Secularization

Secularization is far from representing an axiomatic manifestation between modernity and a worldwide secularization (Davie 1990, 1994; Bhargava 1998). In fact, it can be considered, following Grace Davie (1990, 1994, 2001), as a "European phenomenon with an European explanation" (2001: 120). The thesis of the European exception also occupies an important part in Berger's approach; he writes about the "secular Euro-culture" (2001: 25). But Davie (2004) has been particularly supportive of the thesis that Europe, where secularized society shows a low religious vitality and should be considered an exception, in contrast with the American situation where the religious question has always been present. Nevertheless, the difference between the United States and Europe must be questioned. Indeed, there are different aspects of secularity that must be included in the debate within the context of the globalization process, a dimension that has been strangely absent of the debates on the topic which Tschannen (1992: 371) calls "the most serious deficiency of the paradigm."

In her book *Europe: The Exceptional Case,* Davie (2002) positions Europe in a global context. Basing her arguments mainly on information relative to the diverse expressions of the Christianity, the author shows that in Latin America, Africa and Far East there does not exist a significant loss of the influence of the Christianity and it is possible to consider that in these regions, religion is experiencing great vitality.

About, Europe, Davie states that the Western Europeans are much more "unchurched" than "secular." Even though many Europeans have stopped being tied to a religious institution, she argues that this does not mean that they have cast aside their religious beliefs. Davie is also sensitive to the reconfigurations of the Jewish and Moslem religious minorities and to the appearance of the new and very diverse religious movements. She develops the concept of "vicarious religion" and wonders: "Could it be that Europeans are not so much less religious than populations in other parts of the world, but – quite simply – differently so?" (2002: 5). In her argument she emphasizes the diversity of the religious attitudes of the Europeans and the fact that it is useful to keep in mind the notion of believing without belonging. Using ideas of Eisenstadt (2000), Davie notes that the notion of "multiple modernities" implies not only the identification of western Europe as a particular version of modernity, but also differences inside Europe, especially the relationship of the State with the religion.

Again using the different elements that habitually have defined modernity, Yves Lambert shows, based on quantitative information of European surveys on the values, that far from having unilateral effects in the direction of decreasing religious commitments of individuals, modernity can also have positive religious effects. In this direction, individualization, instead of weakening religiousness, can lead to very personal emergencies of faith or even to an autonomous faith with regard to religions (2004: 331). Economic development can be translated into a religious value related to personal development. As to the idea of that progress in science necessarily implies the decline of the importance of religion, Isambert (1976) notes that it is not true that scientific progress necessarily makes it more difficult to believe in God. Isambert's assertion would falsify one of the theses of the secularization, that the "evolution of the humanity" would lead inevitably to a rational world where the gods did not exist.

On the other hand, regarding pluralism, the integration of pluralism is evident in the way of relating to religious truth. Some beliefs tend to grow, especially among younger people. As for those who declare themselves "without religion," nowadays it is necessary to distinguish between those "religious believers without religion" and "religious non-believers," in that declaring oneself as "without religion" does not necessarily mean an absence of beliefs or lack of interest in the spiritual matter (Davie 1990; Diotallevi 2000; Hadden 1987; Hammond 1985; Haynes 2006). People who self-identify as "religious" are fewer than in the past whereas those who identify themselves as "without religion" are less atheistic than in the past. The borders between the non-religious and religious have become diffuse. Believers became secularized whereas the "atheists" can be believers (Campiche 1997; Campiche et al. 1992, 2004; Martin 1978, 2005).

Lambert concludes his analysis by identifying three principal trends among Europeans in regards to religious matters: (a) a continuation of the decline of religion, (b) an internal recovery among the Christians, and (c) the development of a religiousness without belonging. Recovery of the religious occurs both through the internal process of the Christian community and through the "belief without belonging." The three trends exist in all the countries though in different proportions; no coherent explanation of the reasons for these differences is provided. (Lambert 2004: 319). Lambert (1994: 241–271) also had established previously that indicators of religious vitality, especially the rates of religious practice, do not have any correlation with the different types of relationship between the Churches and the State. Said differently, the fact that different kinds of associations survive between a State and a religion or among school and religion does not appear to have particular effects on the religious attitudes of the individuals.

Though this fact does not question necessarily the association established between functional differentiation and secularization, it is interesting that a direct relationship does not exist between the level of the religious practices and what can seem a minor differentiation between religious institutions and other activities such as education. At the European scale, differentiation between politics and religion, though it exists, does not correspond to the model of the French laicism (where the State is separated legally from the Church), which demonstrates that this differentiation, as well as that of religious and educational, does not lead to a marginalization of religious institutions (Davie & Hervieu-Léger 1996; Hervieu-Léger 2001, 2003; Hervieu-Léger & Willaime 2001).

In a work titled *Theorising Religion: Classical and Contemporary Debates*, Willaime (2006a, b) considers the ultra-modernity as a secularization of modernity. Ultramodernity is for the author the end of what Touraine called "triumphant modernity," that far from meaning a failure of the modernity and secularization represents a liberating force that becomes exhausted as it triumphs (Touraine 1992: 111). Ultramodernity continues being modernity but is disillusioned, problematized, and auto-relativized. The secularization of the modernity is for Willaime a process that destroys the sacred aura of the modernity and removes the mythological character of the central institutions of modernity, such as work, politics, the family, and education. These institutions, though they seemed to be modern, have become emancipated from references or from religious guardianships. The certain thing is that they were continuing being traditional under secularized forms (Armogathe and Willaime 2003). With ultramodernity, the secular sanctification of values such as work, family, education or even political matters is reached in turn by a process of secularization in which the sacred character that had been granted to these secular questions is in turn questioned and they lose a good part of their relevancy Willaime (1977, 1996, 2004a, c).

114.4 The Secularization in Mexico

The political situation of the world and the complicated relation of the West with Islamic countries have provided evidence that the most significant mistake in the theories of secularization is the conviction that modernization inevitably leads to the loss of the importance of religion. Thus the proposition of Berger (2001) replacing this paradigm with an analysis of the interaction between the forces of the secularization and a counter secularization is pertinent.

In this sense, Haynes (2006) notes that the secularization is a double-route process that is accompanied of a renovation of the religious. To demonstrate this, Haynes uses the examples of Poland for the role of the Catholic Church, Spain's democratic Spanish transition, and of the actions of the fundamentalist Protestants and traditional Catholics in the United States, who can be considered a "religious right wing."

In a similar way, it can be said that globalization reminds us that the relationship between religion and politics as established in the Western world, that is to say the passage of the religious to the sphere of the private life, is rather an exception. It is not necessarily exportable and for the same reason, its expansion all over the planet is not inevitable (Blancarte 1992, 2003).

In the particular case of Mexico, we cannot forget that it is a preponderantly Catholic country where the problematic tied to Islam is known only to a limited group of persons who are associated with international matters or those employed in the academy. It is not surprising that Catholicism, and especially its institutional trajectory, is the most studied aspect of religious phenomena, to such a degree that, it could be said that Mexico has had a political science orientated towards the study of the Catholic Church as an institution rather than a Mexican sociology of religions (Martínez-Assad 1992: 9–10).

Religion in the history of Mexico is a key element in understanding the cultural reality of the country (Reilly and De la Rosa 1985; Revista Mexicana de Sociología (1981). In fact we have to say that in Mexico not only does there exist a collective memory or a reservoir of symbolism fund linked to the Catholic Church, but this institution continues to be a central social actor. An example of this is found in research conducted in 1993 of students in the Universidad Iberoamericana network[2] by Enrique Luengo. He discovered that the majority of students believed in the existence of the Catholic God and recognized God as occupying an important place in their lives. However, the first years of this century were an undeniable setback for Catholicism (Blancarte and Casillas 1999; Legorreta 2003). We can talk about changes in the nature of the religious that opened new perspectives for study. In that way studying the study of the process of secularization, its modalities and its particularities, has a good future, since it has been barely studied to date in Mexico.

In this context, there are particularly useful categories, including external and internal secularization. In this sense García-Alba (2011: 339) notes that the first is also called laicism or laicization because it refers to the separation between the State and religion, with the corresponding transfer of responsibilities. Here matters such as public education, public health, or the civil register that belonged to the ecclesiastical realm relate to the political field. Related to this point, non-confessional, modern and secularized governments govern in name of the people, not in the God's name.

On the other hand, the concept of internal secularization refers to the lesser religious institutional influences on the conscience and mentality of the individual, and on his/her behavior or religious practice. This lesser influence, having expanded, strikes at the whole culture, affecting families' education, the treatment of diseases, the relation with the body, and the treatment of the deceased.

[2] The Latin-American University is one of the most important non-public universities in the country. It was founded by the Jesuits.

For Blancarte (2003) the concept nearest to that of laicism is that of popular sovereignty or constitutional legitimacy. In this point of view, the State is non-confessional when it does not need religion as an element of social integration. For this author, laicism can be defined as a social regime of conviviality in which political institutions are legitimized principally by popular sovereignty and not by religious elements. Likewise, secularization can be understood as the manifestation of religious change itself (2001: 847).

In Mexico churches have legal problems transmitting their religious messages on public television. In analyses of surveys, internal secularization has affected the level of the conscience concerning the acceptance of laicism, especially among young people and at university levels. The people who answered the questionnaires for the World Survey of Values in Mexico (women and men 15 years old and older) did not agree with the participation of the churches in the structure of political power, although they agree and support the assistance the poor, the solidarity that they promote, and the promotion of the human rights. In these spheres, great confidence exists toward religious institutions.

In analyzing national surveys, there are more secularization indicators in the urban zones that in the rural or indigenous areas. In these latter areas, it is very difficult to speak about modernization, since the material marginalization is high, they are more vulnerable and, therefore, have been more receptive to the new religious messages, especially evangelical.

Though, the educational level influences the orientation towards modernity, especially concerning ways of life, it is interesting that, according to the International Survey of values, 98 % of the Mexicans affirm a belief in God, but there is an observed decline in the religious practice. 87 % affirmed that they pray quite often and 80 % said that they have "important" or "enough" confidence in the Church. Most Mexicans value marriage and the family, but 51 % stated that it was not necessary for a woman to have children. 89 % thought that they owe unconditional respect to their parents. On the other hand, they agreed to a limit of two children per family. There is a high percentage of belief in the life after the death, the soul and heaven. They reject abortion (67 %), euthanasia (53 %) and suicide (79 %).

For García-Alba (2011), secularization exists in Mexico in the sense that it is a non-confessional society differentiated in the structures concerning the religious institution and demonstrating an internalized acceptance of laicism. On the other hand, a syncretic and racially mixed religiousness persists among the population. Pluralism has been taking place gradually and conversions to new religious movements have augmented. Likewise, in the census information there is registered a gradual increase in the number of persons "without religion." But what is not clear is if we are facing cases of atheism or a withdrawal from the conventional religion. The available information is not sufficient to establish an interpretation based on the facts.

The processes of modernization have been selective. They concern different ways of life and different social sectors as recipients of the modernization. But in all the processes the underprivileged persons and excluded subjects have continued to be those who live in poverty in the peripheries of large cities and in rural or indigenous areas. A significant number of religious conversions take place in these spaces.

According to Suárez (2010), the studies of the National Institute of Statistics and Geography – INEGI – (2005) and the findings from the university sphere (De la Torre y Gutiérrez Zúñiga 2007a, b; De la Torre et al. 2006; De la Peña 2004; De la Torre 2002; Rivera 2005) show that religious plurality is one of the characteristics of the Mexican contemporary society. In this regard Masferrer-Kan in his work religious *Pluralidad religiosa en México* (2011) believe that the number of Catholics has been overestimated intentionally, using data from the General Census of Population and Housing. He contrasts the official numbers with those of baptisms, first communion and confirmations, which give a percentage of Catholics in Mexico of 46.37 of the total national population.

Masferrer-Kan uses the population census and historical data to create the category of religious dissenters, who he identifies as the persons that, for one reason or another, do not express, mention, or claim, or who reject, a Catholic identity (Masferrer-Kan 2001: 12). He identifies six levels of religious plurality:

1. Minimal Plurality – conditions of nearly Catholic unanimity, when the Catholic presence is larger than 97 %;
2. Incipient plurality – where the Catholic unanimity is already broken; Catholic presence that ranges between 92 and 96.99 %;
3. Low Plurality – conditions of rupture of the Catholic hegemony, between 88 and 91.99 %;
4. Medium Plurality – from 80 to 87.99 % Catholic;
5. Consolidated Plurality – when the catholic population is among 63–79.99 %;
6. High Plurality – where the Catholic population is below 63 %

According to his classification, Mexico would be considered as a country of Medium Plurality.

114.5 The Secularization in Mexico, Distrito Federal

In a country like Mexico, it is evident that modernity has been unequal in regards to the distribution of the wealth generated in the country. It is possible to speak about a peripheral modernity in relation with other societies or in relation with previous stages of the country's development. It can also be said that Mexico has risen to modernity through different processes of modernization and by incorporating liberal ideological elements such as the separation between church and state or the loss of religious centrality in daily life.

In this sense it is possible to affirm that a sector of the Mexican population, in which the elites are included but also the middle classes, has followed the process of secularization of modernity that Touraine called "triumphant modernity" and that Willaime identifies as a process that destroys the sacred aura of modernity and removes the mythological character of the central institutions of modernity such as work, politics,, the family, and education. This situation is easily identifiable in the large cities but especially in Mexico, Distrito Federal, where in a gradual but clear form we are in transition towards religious pluralism.

According to Masferrer-Kan (2011: 149), in 1970 the Federal District was in the category of "incipient plurality" in which the Catholic dominance was already broken. This same situation remained through the 1980s and 1990s (Instituto Nacional de Estadística Geografía e Informática 2000, 2010). In 2000 it changed to the category of a "low plurality" where the Catholic unanimity is already broken; in 2010 it passed into "middle plurality" and according to that author's forecast, by 2040 it will become a "consolidated plurality."

In this context, a fact that must be underlined is that external secularization has faced a decade of implicit and explicit questions, which began with the defeat of the Partido Revolucionario Institucional (PRI) in 2000. This development opened the door for a narrower relationship between the State and the Catholic Church in the sense that a political party with secular, and even to a point anti-clerical. These precedents were replaced with another institution, the Partido Acción Nacional (PAN) that did not have these prejudices and ideological roots related to Social Catholicism.

During the government of Vicente Fox, actions like kissing of the papal ring by a president of the Republic or the utilization of religious symbols in public ceremonies provoked an important outcry among those in the political and academic sectors. Nevertheless, this outcry was not accompanied by significant legal changes such as could be accomplished by the modification of a national constitutional article or amendment to any of the 31 constitutions of the Mexican federal states.

A different situation emerged in 2006 during the government of Felipe Calderón when the second president from the PAN promoted constitutional changes in the relationship between the State and the religious institutions. The basic issues were Article 24 in the constitution, which relates to religious freedom and Article 40, which defines Mexico as a non-confessional Republic, which can be considered to be favorable for the Catholic Church.[3]

Apart from these changes, it must be emphasized that the historical confinement of religion to the private sphere has changed dramatically in the last 12 years. This observation can be illustrated by changes in religious practices, especially the beliefs of Catholics, by changes in the federal Mexican bureaucracy, and in overturning of previous taboos about religions manifestations in public places. Among noticeable changes were placing Christian crosses in visible places of federal secretariats and government bureaus, the attendance at mass by the president of the republic, and the fact that he publicly takes communion.

These facts question the validity of the concept of "internal secularization" that alludes to a lesser religious institutional influence in the conscience and mentality of the individual, and one's behavior or religious practice that characterizes persons

[3]The articles of the Mexican Constitution that have direct relation with the religious question are 1, 24, 40 and 130. Article 1 recognizes the rights of all the Mexicans and prohibits discrimination; Article 24 recognizes the freedom of religious worship; Article 40 defines Mexico as a non-confessional republic (república laica), and Article 130 regulates the relations between the State and the churches.

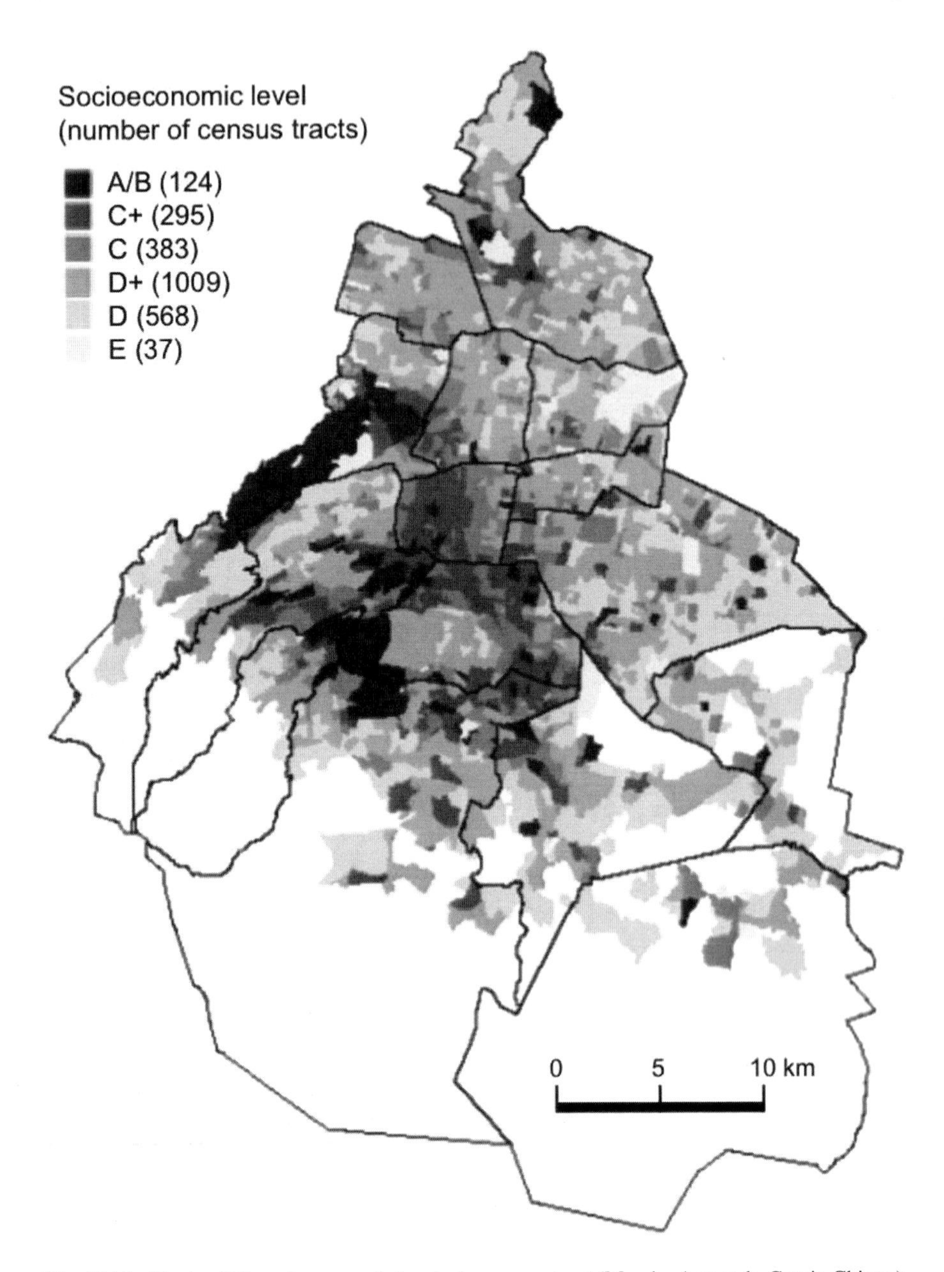

Fig. 114.1 Mexico DF, socioeconomic levels, by census tract (Map by Armando Garcia Chiang)

immersed in the ultramodernity. Using information from surveys, Blancarte (2001: 854) shows that people with a higher educational level education and greater income do not consider religion as being extremely important in their lives (Fig. 114.1).

It is evident that certain differences regarding religious questions exist between the different social strata. In fact it is possible to speak of constants that appear basically among the popular and middle low sectors. The first difference is the presence of a syncretic and racially mixed religiousness that identified with the popular Catholicism. An example of this is the numerous groups that meet on the 28th of every month to celebrate San Judas Tadeo (the saint of the poor) in the Temple of San Hipólito, situated in the central part of the Distrito Federal.

The second is a phenomenon of conversions to different expressions of Protestantism such as the so-called historical Protestants, the Pentecostals and Neo-Pentecostals, and the groups known as Biblical non-evangelical (Mormon Church, Seventh-Day Adventist Church) groups.

Directly related to conversion is religious plurality. A good example of this situation is the work by Hugo Jose Suárez (2010) on religious pluralism in the Ajusco neighborhood (Mexico, D.F.). In this paper, he notes that in a relatively small territory there are: two Catholic parishes, two chapels, four Protestant churches, four Pentecostals churches, two Biblical non-evangelical churches, two shops of holiness, a shop devoted to the Holy Death and numerous expressions of popular religiousness. Every Sunday in this neighborhood there are 14 Catholic masses and 13 celebrations of our groups (Fig. 114.2).

In contrast to the Ajusco neighborhood are the three different middle class neighborhoods as the Condesa, Roma Norte or Roma Sur have together 12 religious buildings (Fig. 114.3).

Suárez also indicates that, as has been observed in other empirical studies on the beliefs in Mexico, D.F, religion plays a very important role: 86 % belong to some

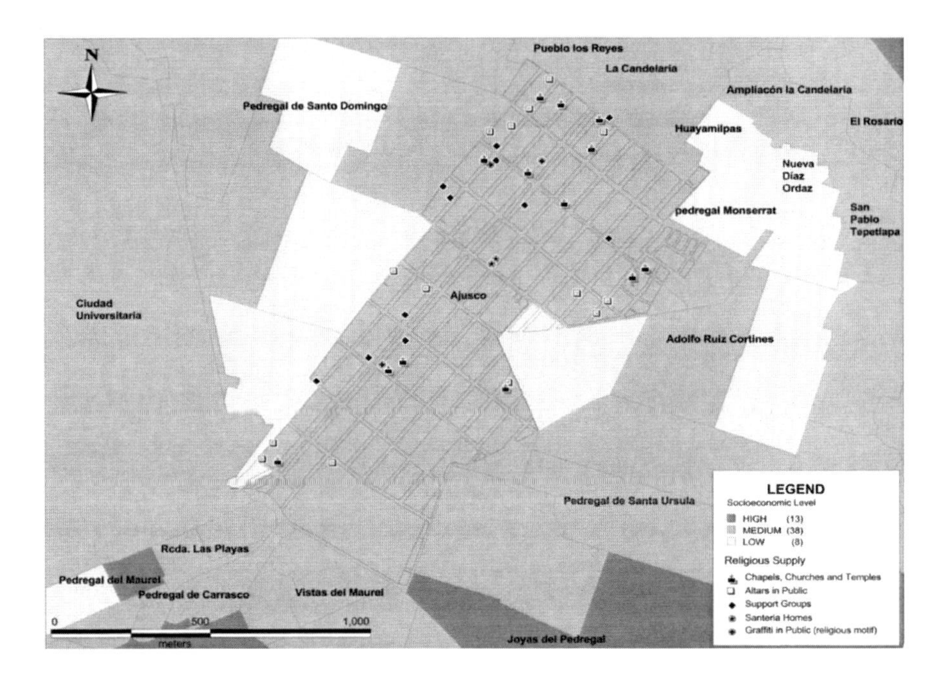

Fig. 114.2 The Ajusco neighborhood (Map by Armando Garcia Chiang)

Fig. 114.3 Condesa, Roma Norte, Roma Sur 1 (Map by Armando Garcia Chiang)

religion, 81 % have an altar or religious object in the home, 21 % said that they are very interested in the religion and the sacred questions and 60 % show themselves moderately interested. Only 16 % hold that religion does not appeal to them. The Catholic Church remains the institution that has the strongest ties: 78 of every 100 persons.

In regards to beliefs, the traditional references of Catholic theology have high agreement: 9 of every 10 persons interviewed believe in God, 7.5 in the paradise, 7 in miracles, 8 in the Virgin of Guadalupe and 7.5 in the Holy Spirit. The negative expression of similar concepts has less impact: only 46 % believe in the devil and 55 % in hell.

Likewise, from a geographical point of view, we are in front of a dispute for the territory, in front of a differentiated process of sacralization of the space where exists different forms to create a sacred place.

114.6 Towards a Conclusion

Different policies regarding religious space can operate at the same time. Thus, we can have a policy of positioning based on the conquest of the *denominational space*, *a policy of possession*, in which a sacred place is assigned or possessed by a group, or a *policy of exclusion* where local holiness is preserved by the establishment and a support of borders that differentiate the "interior" and the "exterior" (García-Chiang 2011),[4] but in no case does it lead us to support the inroads of secularization in Mexican society.

[4] In this case we can identify the different groups that seek to recover the culture and also the Aztec religion. These groups are generally integrated by dancers best known as "concheros."

The process of sacralization of the space provides a new meaning to ordinary elements in the urban landscape. In this situation a chapel can born in any place, someone installs an image, other persons give sense to a cross; and anyone can fill the space with forms of the divinity. In this process popular religiousness plays an influential role, the majority of its iconographic supports, places and practices have an amazing and expanded vitality. Also its practices, holidays, contents and specialists [can be separate?] from the institution, since that relationship has been always ambiguous and elastic. This fact allows for popular religiousness to be able to adapt itself to a current context.

On the other hand, it is important to indicate the situation in which religious differences are tied to the belonging to a specific social class which is also a differentiation that goes back several decades. At present, everything indicates that age is the principal factor of the differentiation of the contemporary religious attitudes.

In conclusion, based on interviews with civil servants of the financial sector, small and medium businessmen, independent professionals and university students, it is possible to conclude that the Mexican reality corresponds to Berger's indication that the secularized population of the world is represented by a sector carrying an international subculture which has high levels of education, generally in humanities and social sciences and which we might add, that also holds liberal political ideas.

All this is to say that in Mexico we are turning into a rational world in which the importance of religion has been in a state of constant decline has turned out to be false. Rather we are facing a reality where religion is changing, but not disappearing.

References

Armogathe, J.-R., & Willaime, J.-P. (Eds.). (2003). *Les mutations contemporaines du religieux*. Turnhout: Brepols.

Barker, E., Beckford, J. A., & Dobbelaere, K. (Eds.). (1993). *Secularization, rationalism and sectarianism. Essays in honour of Bryan Wilson*. Oxford: Clarendon Press.

Baubérot, J. (1994a). Laïcité, laïcisation, sécularisation. In A. Dierkens (Ed.), *Pluralisme religieux et laïcité dans l'Union européenne* (pp. 9–20). Bruxelles: Éditions de l'Université de Bruxelles.

Baubérot, J. (Ed.). (1994b). *Religions et laïcité dans l'Europe des Douze*. Paris: Syros.

Berger, P. L. (1967). *The sacred canopy: Elements of a sociological theory of religion*. Garden City: Doubleday.

Berger, P. L. (1971). *La religion dans la conscience moderne. Essai d'analyse culturelle*. Paris: Éditions du Centurion.

Berger, P. L. (Ed.). (1999). *The desecularization of the world*. Grand Rapids: Eerdmans.

Berger, P. L. (2001). La désécularisation du monde: un point de vue global. In P. Berger, G. Weigel, D. Martin, et al. (Eds.), *Le réenchantement du monde* (pp. 13–36). Paris: Bayard Éditions.

Bhargava, R. (1998). What is secularism for? In R. Bhargava (Ed.), *Secularism and its critics* (pp. 486–542). New Delhi: Oxford University Press.

Blancarte, R. (1992). *Historia de la Iglesia Católica en México*. México: FCE, El Colegio Mexiquense.

Blancarte, R. (2001). Laicidad y secularización en México. *Estudios Sociológicos, 19*(3), 843–855.

Blancarte, R. (2003). Religión, política y libertades en los albores del tercer milenio. *Metapolítica, En el nombre de Dios: Política y Religión, 23*, 39–45.

Blancarte, R., & Casillas, R. (Eds.). (1999). *Perspectivas del fenómeno religioso*. México: Secretaría de Gobernación: Facultad Latinoamericana de Ciencias Sociales.
Bruce, S. (Ed.). (1992). *Religion and modernization. Sociologists and historians debate the secularization thesis*. Oxford: Clarendon.
Bruce, S. (1995). *Religion in modern Britain*. Oxford: Oxford University Press.
Bruce, S. (1997). *Religion in the modern world from cathedrals to cults*. Oxford: Oxford University Press.
Bruce, S. (1999). *Choice & religion. A critique of rational choice*. Oxford: Oxford University Press.
Bruce, S. (2001). The curious case of the unnecessary recantation: Berger and secularization. In L. Woodhead, P. Heelas, & D. Martin (Eds.), *Peter Berger and the study of religion* (pp. 87–100). London: Routledge.
Bruce, S. (2002). *God is dead. Secularization in the West*. Oxford: Blackwell Publishers.
Bruce, S. (2006). Secularization. In R. A. Segal (Ed.), *The Blackwell companion to the study of religion* (pp. 413–430). Oxford: Blackwell.
Campiche, R. J. (Ed.). (1997). *Cultures jeunes et religions en Europe*. Paris: Le Cerf.
Campiche, R. J., Dubach, A., & Bovay, C. (Eds.). (1992). *Croire en Suisse(s)*. Lausanne: L'Âge d'Homme.
Campiche, R. J., Broquet, R., Dubach, A., & Stolz, J. (Eds.). (2004). *Les deux visages de la religion. Fascination et désenchantement*. Genève: Labor et Fides.
Casillas, R. (1996). La pluralidad religiosa en México. In G. Giménez (Ed.), *Identidades religiosas y sociales en México* (pp. 103–144). México: IFAL, Instituto de Investigaciones Sociales UNAM.
Chaves, M. A. (1991). *Secularization in the twentieth-century United States*. Unpublished Ph.D. dissertation, Harvard, Harvard University, Department of Sociology, Cambridge.
Chaves, M. A. (1994). Secularization as declining religious authority. *Social Forces, 72*(3), 749–774.
Chaves, M. A. (2007). Religious pluralism and religious participation. *Annual Review of Sociology, 27*, 261–281.
Davie, G. (1990). Believing without belonging: Is this the future of religion in Britain? *Social Compass, 37*(4), 455–469.
Davie, G. (1994). *Religion in Britain since 1945. Believing without belonging*. Oxford: Blackwell.
Davie, G. (2001). The persistence of institutional religion in modern Europe. In L. Woodhead (Ed.), *Peter Berger and the study of religion* (pp. 101–111). London: Routledge.
Davie, G. (2002). *Europe: The exceptional case. Parameters of faith in the modern world*. London: Darton, Longman and Todd Ltd.
Davie, G. (2004). Rituels royaux en Angleterre: Deux enterrements et un jubilé. In E. Dianteill, D. Hervieu-Léger, & I. Saint-Martin (Dirs.), *La modernité rituelle. Rites politiques et religieux des sociétés modernes* (pp. 23–37). Paris: L'Harmattan.
Davie, G., & Hervieu-Léger, D. (1996). *Identités religieuses en Europe*. Paris: La Découverte.
De la Peña, Guillermo. (2004). El campo religioso, la diversidad regional y la identidad nacional en México. In *en Relaciones, Estudios de Historia y Sociedad* (Vol. 25, no. 100, pp. 21–71). Colegio de Michoacán.
De la Torre, R (2002). Guadalajara y su región. cambios y permanencias en la relación religión-cultura-territorio. In Covarrubias, K (comp), *Cambios religiosos globales y reacomodos locales* (pp. 41–76). Colima: Altexto.
De la Torre, R., & Gutiérrez Zúñiga, C. (Eds.). (2007a). *Atlas de la diversidad religiosa en México*. México: El Colegio de Jalisco, El Colegio de la Frontera Norte A.C., Centro de Investigaciones y Estudios Superiores en Antropología Social, El Colegio de Michoacán, A.C., Secretaría de Gobernación; Subsecretaría de Población, Migración y Asuntos Religiosos.
De la Torre, R., & Gutiérrez Zúñiga, C. (2007b). Territorios de la diversidad religiosa hoy. In R. De la Torre & C. Gutiérrez Zúñiga (Eds.), *Atlas de la diversidad religiosa en México* (pp. 35–37). México: El Colegio de Jalisco, El Colegio de la Frontera Norte A.C., Centro de Investigaciones y Estudios Superiores en Antropología Social, El Colegio de Michoacán, A.C., Secretaría de Gobernación; Subsecretaría de Población, Migración y Asuntos Religiosos.

De la Torre, R., et al. (2006). Perfiles socio-demográficos del cambio religioso en México. In *Enlace. Revista Digital de la Unidad de Atención de las Organizaciones Sociales*, Nueva Época Año 4, número 4, Abril-Junio. www.organizacionessociales.segob.gob.mx/UAOS-Rev4/perfiles.html,consultadoel10dejuniodel2008

Diotallevi, L. (2000). Alternativas a la laïcité. Para un conocimiento menos augusto de la cultura europea. *Ciencias Sociales y Religión/Ciências Sociais e Religião, Porto Alegre, 12*(13), 15–36.

Dobbelaere, K. (1981). Secularization: A multi-dimensional concept. *Current Sociology, 29*(2), 3–213.

Dobbelaere, K. (2002). *Secularization: An analysis at three levels*. Bruxelles: P.I.E. Peter Lang.

Einsenstadt, S. N. (2000). Multiple modernities. *Daedalus, 129*(1), 1–29.

Fukuyama, Y. (1963). Groupes religieux et sociologie aux États Unis. *Christianisme Sociale, 71*(9–12), 739–746.

García-Alba, P. E. (2011). *Modernidad, secularización y religión: El caso de México*. Tesis doctoral, Universidad Complutense de Madrid, Madrid.

García-Chiang, A. (2004). De la década de los ochentas al primer lustro del siglo XXI. Panorama de los estudios sobre lo religioso en México. Biblio 3v. *Revista bibliográfica de geografía y ciencias sociales Any, 12*(2007). www.raco.cat/index.php/biblio3w/article/view/73044/83401

García-Chiang, A. (2007). Los estudios sobre lo religioso en México. Hacia un estado de la cuestión. *Scripta Nova: revista electrónica de geografía y ciencias sociales, Any, 8*(2004), 8.

García-Chiang, A. (2011). *Territorialidad de lo sagrado. Geografía de lo religioso*. Madrid: Editorial Académica Española.

Glasner, P. (1977). *The sociology of secularisation. A critique of a concept*. London: Routledge & Kegan Paul.

Hadden, J.-K. (1987). Towards desacralizing secularization theory. *Social Forces, 65*(3), 587–611.

Hammond, P. E. (1985). *The sacred in a secular age. Toward revision in the scientific study of religion*. Berkeley: University of California Press.

Haynes, J. (2006). Introduction. In J. Haynes (Ed.), *The politics of religion: A survey* (pp. 1–9). London/New York: Routledge.

Hervieu-Léger, D. (2001). Le christianisme en Grande-Bretagne: débats et controverses autour d'une mort annoncée. *Archives de sciences sociales des religions, 116*, 31–40.

Hervieu-Léger, D. (2003). *Catholicisme, La fin d'un monde*. Paris: Bayard.

Hervieu-Léger, D., & Willaime, J.-P. (2001). *Sociologies et religion. Approches classiques*. Paris: Presses Universitaires de France.

Instituto Nacional de Estadística Geografía e Informática. (2000). *XII Censo General de Población y Vivienda 2000: Tabulados Básicos Veracruz- Llave* (t. III). México: Instituto Nacional de Estadística Geografía e Informática (INEGI), Aguascalientes.

Instituto Nacional de Estadística Geografía e Informática. (2010). *Censo de Población y Vivienda 2010: Tabulados del Cuestionario Básico*. Instituto Nacional de Estadística Geografía e Informática (INEGI). www.inegi.org.mx. 23 de febrero de 2012.

Isambert, F.-A. (1976). La sécularisation interne du christianisme. *Revue Française de Sociologie, 17*(4), 573–589.

Lambert, Y. (1994). Les régimes confessionnels et l'état du sentiment religieux. In J. Baubérot (Ed.), *Religions et laïcité dans l'Europe des douze* (pp. 241–258). Paris: Fayard.

Lambert, Y. (2004). Des changements dans l'évolution religieuse de l'Europe et de la Russie. *Revue Française de Sociologie, 45*(2), 307–338.

Legorreta, Z. J. (2003). *Cambio religioso y modernidad en México*. México: Universidad Iberoamericana.

Luengo, E. (1993). *La religión y los jóvenes de México ¿El desgaste de una relación?* México: Universidad Iberoamericana.

Martin, D. (1965). *Towards eliminating the concept of secularization*. London: Penguin.

Martin, D. (1978). *A general theory of secularization*. New York: Harper & Row.

Martin, D. (2005). *On secularization. Towards a revised general theory*. Aldershot: Ashgate.

Martínez-Assad, C. (1992). Religión y desarrollo en América Latina. In M. Assad & C. Asaad (Eds.), *Religiosidad y política en México*. México: Cuadernos de Cultura y Religión, núm 2, Universidad Iberoamericana.
Masferrer-Kan, E. (2001). La configuración del campo religioso. In Masferrer-Kan, E. (Comp.), *en Sectas o iglesias. Viejos o nuevos movimientos religiosos*. México: Plaza y Valdez.
Masferrer-Kan, E. (2011). *Pluralidad religiosa en México. Cifras y proyecciones*. Buenos Aires: Libros de la Araucaria.
Monod, J.-C. (2002). *La querelle de la sécularisation de Hegel à Blumenberg*. Paris: Vrin.
Parker Gumucio, C. (1994). La sociología de la religión y la modernidad: por una revisión crítica de las categorías durkheimianas desde América latina. *Revista Mexicana de Sociología, 4*, 229–254.
Proctor, J. D. (2006). Introduction: Theorizing and studying religion. *Annals of the Association of American Geographers, 96*(1), 165–168.
Racine, J. B., & Wachter, O. (2006). Geografía de las religiones. In en Hiernaux, Daniel y Alicia Lindón. (eds.). *Tratado de Geografía Humana* (pp. 481–505). Barcelona/México: Editorial Anthropos/UAM Iztapalapa.
Reilly, C., & De la Rosa, M. (Eds.). (1985). *Religión y política en México*. México: Siglo XXI Editores.
Revista Mexicana de Sociología, Vol. 43 núm. Extraordinario, 1981.
Rivera, C. et al. (2005). *Diversidad religiosa y conflicto en Chiapas. Intereses, utopías y realidades*. México: UNAM-IIFL/PROIMMSE/CIESAS.
Robertson, R. (Ed.). (1980). *Sociología de la religión*. México: Fondo de Cultura Económica.
Suárez, H. J. (2010). El pluralismo religioso en la colonia Ajusco. In *Estudios Sociales* (6, pp. 286–309). México: Universidad de Guadalajara, Guadalajara.
Touraine, A. (1992). *Critique de la modernité*. Paris: Fayard.
Tschannen, O. (1992). *Les théories de la sécularisation*. Genève: Droz.
Wallis, R., & Bruce, S. (1992). Secularization: The orthodox model. In S. Bruce (Ed.), *Religion and modernization. Sociologists and historians debate the secularization thesis* (pp. 8–30). Oxford: Clarendon Press.
Willaime, J.-P. (1977). La relégation superstructurelle des références culturelles. Essai sur le champ religieux dans les sociétés capitalistes postindustrielles. *Social Compass, 24*, 323–338.
Willaime, J.-P. (1996). Surmodernité et religion duale. In L. Voyé (Ed.), *Figures de dieux* (pp. 235–246). Bruxelles: De Boeck Université.
Willaime, J.-P. (2004a). The cultural turn in the sociology of religion in France. *Sociology of Religion, 65*(4), 373–389.
Willaime, J.-P. (2004b). *Europe et religions. Les enjeux du XXI e siècle*. Paris: Fayard.
Willaime, J.-P. (2004c). *Sociologie des religions*. Paris: Presses Universitaires de France.
Willaime, J.-P. (2006a). Religion in ultramodernity. In J. A. Beckford & J. Wallis (Eds.), *Theorising religion: Classical and contemporary debates* (pp. 73–85). Aldershot: Ashgate.
Willaime, J.-P. (2006b). La sécularisation: Une exception européenne ? Retour sur un concept et sa discussion en sociologie des religions. *Revue Française de Sociologie, 47*, 755–783.
Wilson, B. (1966). *Religion and secular society. A sociological comment*. London: CA Watts.
Wilson, B. (1985). Secularization: The inherited mode. In P. E. Hammond (Ed.), *The sacred in a secular age* (pp. 9–20). Berkeley: University of California Press.

Chapter 115
Secularization and Church Property: The Case of Czechia

Martina Hupková, Tomáš Havlíček, and Daniel Reeves

115.1 Introduction

The mutual interaction of political and religious (church) power plays a significant role in the development of every society along with the territories associated with each given society. Church-state relationships have undergone numerous changes in recent years, depending on the type of political power being wielded and the various types of religion involved (Geyer 2004; Madeley 2003). Madeley (2003) explores church-state relationships in Europe and distinguishes three broad belts: the historic mono-confessional culture belt, the historic Northwest-Southeast multi-confessional culture belt and the historic Northeast-Southeast multi-confessional culture belt. He places Czechia in the NW-SE multi-confessional belt, which is characterized by a very low number of inhabitants without any church affiliation (that is, non-denominationalists). Current figures on the portion of non-believers in Czechia, which is nearly 60 %, demonstrate that, in comparison with its neighboring states, Czechia exhibits a very unique position in terms of religiosity. The role of religion in society is significantly less than that found in other European states. When considering church-state relations, it is clear that the state controls a much more dominant position of influence over society and space than other countries (Havlíček 2006).

Currently no official declaration mandating the separation of church and state has been made in Czechia, as has been done in other countries of Europe. Nonetheless, the Charter of Fundamental Rights and Basic Freedoms in the Constitution of the Czech Republic proclaims that the state is founded on democratic principles and may not be bound to exclusive ideologies or religious denominations. Consequently, the salaries of the clergy of most of the churches are paid from the

M. Hupková (✉) • T. Havlíček • D. Reeves
Faculty of Science, Department of Social Geography and Regional Development, Charles University, Albertov 6, 128 43 Prague 2, Czechia
e-mail: martina.hupkova@natur.cuni.cz; tomas.havlicek@natur.cuni.cz; danvreeves@gmail.com

S.D. Brunn (ed.), *The Changing World Religion Map*,
DOI 10.1007/978-94-017-9376-6_115

national budget, as set forth in the Act on Financial Provisions by the State for Churches and Religious Societies (Karlová 2011).

Since the beginning of the millennium, an evident trend is towards greater secularization and a more dominant role of profane culture, both which have become increasingly evident in Czech society. From 1991 to 2001, the portion of inhabitants declaring no religious affiliation increased from 39.9 % (1991) to 58.2 % (2001). The lower numbers in 1991 can, in part, be attributed to a short-lived period of societal openness to spiritual phenomena, immediately following the fall of the communist regime. Generally speaking, Czechia's religious landscape is increasingly diversified and secularization is on the rise (Havlíček et al. 2009). The larger churches are losing adherents, while smaller groups, particularly small Christian communities, are gaining members (Havlíček 2005). Czech society's tendencies towards secularization, over the past 20 years, are clearly reflected in the issue of the restitution of church properties seized during the communist era.

Over the course of their existence, religious institutions acquired property – for the most part real estate – that was used for worship services and related purposes. Incidentally, by managing and utilizing this property, churches were self-sufficient in day-to-day, practical matters. With increasing secularization of European, or Czech society, the role of religious institutions has decreased, along with the extent of their property and, consequently, the impact of said properties as symbols in the religious landscape. During the past 50 years, the uses of many religious structures and church-owned real estate changed from religious to non-religious purposes (Knippenberg 2005). The uses of some church properties also changed as a result of political changes, particularly due to the emergence of authoritative (primarily communist) regimes. These dictatorships viewed religious institutions as an obstacle impeding the process of controlling and manipulating a society. They tried to limit freedoms of speech and assembly, in part, by seeking to control or, more directly, seize church-owned property, thereby weakening churches' role in society.

This situation arose during the Cold War (1948–1989) in the communist states of Central and Eastern Europe. A large portion of church properties was either entirely or partially seized/controlled by the centralized communist governments. Churches became economically and proprietarily dependent on the state (Havlíček 2006). In addition, because of property losses, churches were no longer able to effectively participate in educational, health or social-welfare systems.

After the fall of the communist dictatorships, attempts to return formerly church-owned properties seized by the state to their previous owners began in the hopes of achieving a full restitution of properties and resolving disrupted church-state relations. And yet in Czechia, more than 20 years after the fall of the communist regime, the majority of previously church-owned properties remain unresolved. Most churches continue to be dependent on the state, because legislative conditions for the proprietary and financial separation of church and state have not been agreed upon or enforced. At present (May 2012), a new draft of the Act on Property Restitution has been prepared and the issue continues to be a subject of frequent discussion and media coverage. Optimistic views regarding this draft being approved and entering into force are, however, disrupted by the current political crisis.

The opening paragraph of the drafted legislation on property restitution presents a number of reasons, from the state's perspective, that property restitution should be carried out. The language reads:

> Parliament, remembering the painful experiences of periods when human rights and basic freedoms were suppressed in the territory of the present-day Czech Republic; determined to protect and develop our inherited cultural and spiritual wealth; led by attempts to mitigate the consequences of certain property-related and other grievances, perpetrated by the communist regime during the period from 1948 to 1989, to settle property relations between the state and churches or religious societies, as a prerequisite to complete religious freedom, thereby enabling the restoration of the proprietary holdings/resources of churches and religious societies, the free and independent status of churches and religious societies, whose existence and works it considers to be an essential element of a democratic society… (Governmental draft of a bill 2012: 1)

The objective of this chapter is to evaluate the process of returning church properties seized during Czechia's communist era as a reflection of continually developing church-state relations in post-1989, secularized Czech society. We will analyze mutual discussion on the topic and the priorities of the issues among various stakeholders, including the general public. Based on the generally low degree of religiosity in Czechia, we expect that the secularized public will exhibit insufficient interest in returning church properties to the churches which will, in turn, impact political decisions on the issue. Considering the block (limits to utilization) placed on formerly church-owned properties, we expect the effects of unresolved church-state relations to express themselves, even in local and regional development, and that the impacted stakeholders (local and regional governments who are key participants in regional development and business people) will play a significant role in the process.

115.2 Church-State Relations in Czechia in the Context of Central Europe

115.2.1 Restitution of Church Properties in Central Europe

Although the Czech examples presented herein are somewhat extreme in nature, the restitution of church-owned property seized by communist regimes is not unique to Czechia. It is a difficult political issue that has been forced upon all of Central and Eastern Europe's post-communist governments. There seem to be at least as many approaches to resolving church-owned properties as there are political states involved. According to a report issued by the Commission on Security and Cooperation in Europe (CSCE, also known as the U.S. Helsinki Commission), as of 16 July 2002, the majority of communal (church-owned) properties had been returned in most of the countries of Central and Eastern Europe. Estonia, Hungary, Latvia and Slovakia were presented as examples of states that had enacted or were in the process of enacting effective laws to restore previously confiscated properties (CSCE 2002). On the other hand, this hearing gave "a particular focus on claims in Poland, the Czech Republic [sic], and Romania," in effect, proclaiming that these three countries left much to be desired in terms of property restitution (Walsh 2002).

The official transcript from this 2002 hearing makes it clear that Poland's shortcomings in property restitution dealt primarily with private property, particularly the losses of Holocaust victims (CSCE 2002). In Romania, problems with property restitution stemmed from the confusing and excessive administrative requirements necessary for filing claims as well as from allegations of discrimination in the return of communal property, mainly against the Greek Catholic Church. Nonetheless, significant progress towards the restitution of seized communal properties, both in Poland and Romania, had been made by the time of this hearing. While Czechia also exhibited shortcomings concerning the restitution of private property, of more significance to the topic at hand, even 10 years ago (in 2002), was the absence of legal mechanisms to govern the return of communal properties.

A follow-up briefing held by the U.S. Helsinki Commission in 2003 praised the efforts of Slovakia, Slovenia and Bulgaria, countries that were nearing complete restitution of seized properties. Once again, however, the commission criticized a continuing lack of progress in Czechia, Poland and Romania, this time adding Serbia to their blacklist. We should reiterate here that, as an American political advisory body, this committee's focus was on personal claims and the majority of previously church-owned properties in Poland were resolved relatively smoothly, by 1995, with the exception of a number of Jewish claims (Pogany 1997).

Pogany (1997) points out a number of issues surrounding the return of Church-owned properties in Central and Eastern Europe, focusing in particular on his native Hungary. He explains parts of the complicated process claimants must navigate to request the restitution of property and explores some of the reasons today's society is, in many cases, not supportive of a complete return to former ownership conditions. For example, the city of Sopron near Hungary's western border has a secondary school that, in the 1940s, was owned and operated by a German Protestant church. At that time, nearly half of the town's population was comprised of ethnic Germans, many of whom were Protestants. As in Czechia, Slovakia and Poland, a large portion of Hungary's *Volksdeutsche* were forced to emigrate following World War II, leaving Sopron with a very different social and cultural composition today. In spite of local government support for its restitution, a Protestant "Gimnázium in Sopron could no longer be justified by reference to the current needs, or even wishes, of a substantial portion of the local population" (Pogany 1997: 194).

Generally speaking, it is clear that the restitution of previously church-owned properties in Central Europe has been and continues to be, at least for Czechia, a very tricky issue. In spite of the correctness and fairness of a complete return of these properties, even coupled with the good will and favorable intentions of political leaders, it is a difficult sell to a public that has changed drastically over 40 years of communist rule and – as the process drags on – more than 20 years of post-socialist globalization (Broun 1996).

115.2.2 Historical Aspects of Church-State Relations in Czechia

Throughout Czech history sacred and profane, or worldly, powers have been closely connected. This mutual connectivity and influence has led to many conflicts. Religion, more specifically, Christianity, played a decisive role in the initial process of creating a political state on the territory of present-day Czechia. For some time, beginning in the tenth century, Czechia found itself situated along the ideological divide between Eastern Orthodox and western, Roman-Catholic influences. Eventually, the Western influence of the Mainz Bishopric increased, giving the Catholic Church decisive power in the area. A very early phase of the reformation of the Roman Catholic Church began here with Jan Hus, who was both a preacher in Prague and rector of Charles University. After Hus' execution by fire, this phase culminated, during the first half of the fifteenth century, in the Hussite Wars and other local reformation efforts. This conflict led to increasing separation of the church from state powers. During the period of broader European reformation, the Czech lands fell increasingly under the influence of Protestant movements and churches that were connected with regional political forces. Protestant churches' attempts to take control of political power in the Czech lands were, in the end, suppressed by the Hapsburgs' strong anti-Reformation movement, following the Battle of Bílá Hora (White Mountain; in 1620). Religions other than Catholicism were forbidden Hapsburg rulers and, eventually, non-Catholic religious communities were made illegal by the state. This step led to the emigration of Protestants or to their Catholization during the baroque (counter Reformation) period, which, incidentally, also involved the construction of numerous sacred symbols in the landscape (wayside shrines, chapels and crosses).

The creation of Czechoslovakia in 1918 brought about the separation of state and church (primarily Roman Catholic) power. At the same time, the "state" Czechoslovak Hussite Church was established after the pattern of the Anglican Church in Great Britain. By the beginning of World War II, members of this church comprised more than 10 % of Czechia's population. After this point, there was a gradual decrease in the number of adherents of the Czechoslovak Hussite Church. At present only 1 % of the population claims membership in this "state" church (Havlíček 2006).

During the twentieth century, two separate political dictatorships (Nazism and communism) led to significant state restrictions limiting the operations and freedoms of church institutions. State power even assumed certain aspects of religious thinking and behavior in its ideologically charged discourses. So-called political religion (Mayr 1995) emerged, founded on ideologies taken from existing religious structures, which were, at the same time, strictly suppressed. Practically all religious orders were disbanded and many monks and priests were incarcerated. Until 1989 the Czechoslovak constitution included a provision, proclaiming the need to eliminate religion – an anachronism from an exploitive society – from people's consciousness.

During the period of communist rule, the state limited and regulated church operations. The fall of the communism in Europe and in Czechia, Czechoslovakia, brought a renewal of religious freedom and renewed separation of church and state powers. Under the communist regime, the state confiscated a large amount of church-owned properties (3 % of Czechia's land area) and this property is yet to be returned to the churches. The state continues to partially finance the operations of the various churches (contributing, for example, to clergy salaries). However, it no longer interferes in internal church structures. It merely formulates the basic framework conditions that shape church-state relations.

115.3 Negotiations for the Restitution of Church Properties After 1989

115.3.1 Negotiations After 1989

After the Velvet Revolution, in the context of newly developing legal systems, restitution began to made. In post-communist countries this meant the return of nationalized or confiscated property to its rightful owners. While in most cases, restitution was made relatively quickly, the restitution of formerly church-owned properties in Czechia was encumbered by complicated negotiations that continue to the present time. On several occasions, it looked as if ownership rights would finally be successfully resolved and debts from the past would be paid. Each time, however, the sensitivity of the topic, related to the decreasing religiosity of Czechia's inhabitants, resulted in public disagreement with the return of extensive property to churches. The lack of will among political representatives to negotiate the return of said properties is, naturally, tied to the general public's rejection of church restitution.

In connection with the return of church properties to the churches, people tend to use the term property "adjustment" rather than property "settlement." It appears that no matter what form of property return is eventually made, the damages caused by the seizure of church properties will never be completely compensated. This reality is only increased by the absence of any exact database of church properties (that is, no thorough enumeration of their value) and the fact that many properties simply cannot be returned, due to technicalities. Damages to property caused by various natural and human factors or by failure to invest into necessary repairs are not accounted for. Nor is any reckoning made for lost profits from the properties in question or other similar losses. In light of the realities of the situation, it is not possible to ensure a completely objective restitution, due to the fact that many of the properties have undergone irreversible changes. Thus, restitution must be viewed more as a renewal of former ownership rights under existing constitutional and legal conditions (Jäger 2012).

In 1990, the so-called Enumeration[1] Acts (Acts No. 298/1990 Coll. and No. 338/1991 Coll.) were enacted, which resulted in the transfer of certain monastery buildings (altogether, roughly 170 properties). The "blocking paragraph" (Section 29 of Act No. 229/1991 Coll., on the modification of ownership rights to land and other agricultural property) entered into force in 1991, in effect making the transfer of properties owned by a church to other owners impossible. Consequently, the state and its subordinate organizational units, which are responsible for the interim management of the properties in question, are not motivated to properly care for the property, a fact that only serves to increase internal property debt (Czech Ministry of Culture 2008). Between 1996 and 1998, executive mandates (in the form of government resolutions) ensured the return of additional structures used for spiritual, pastoral, social, educational or other similar purposes. This did not include economically managed properties. Over the following years, attempts were made to institute the restitution of church properties through the Enumeration Act. Both the churches as well as state institutions opposed and eventually put a halt to these attempts. The existence of so many outstanding properties awaiting transfer (approximately 100,000) meant a high likelihood of error and overly complicated legal processes (Palas 2004; Večeře 2001). Attempts to work things out through the Enumeration Act continued until 2004. Various forms of settlement were analyzed: the Enumeration Act, a general restitution order, rent, etc. In 2007, a government commission for the settlement of relations between the state and churches and religious communities concluded that it would be best to seek restitution in a combined form, that is, both properties and financial compensation (return property that can be returned and pay a negotiated amount for any remaining properties). The return of the actual properties is a better outcome for monastic orders and congregations, for whom working the land is part of their spiritual traditions. In 2008, the government (executive branch) approved and submitted to the Chamber of Deputies (legislature) a draft for an act on the settlement of property with churches and religious organizations (Government Resolution No. 333, dated 2 April 2008). At the end of the legislative session, this draft had still not been debated. On 1 July 2010 the Constitutional Court issued a decision (Judgment of the Constitutional Court Pl. CC 9/07), proclaiming the Parliament of the Czech Republic's failure to take action on the issue of the restitution of church properties to be unconstitutional. "The current state of affairs, that is, the absence of a reasonable settlement of historically church-owned property, in which the state – due to its own inactivity – continues to be the dominant source of income for the effected churches and religious communities, furthermore with no apparent connections to the profits accruing from the retained historical church properties, in effect violates Article 16 of Charter 1, in terms of the freedom to express faith within society, through generally public and traditional forms of religiously motivated activities that utilize the historically formulated

[1] The term "enumeration" indicated that the appendices to the act included an exact list of real estate properties that were subject to transfer.

economic resources in question and, in particular, Article 16 of Charter 2, the economic component of church autonomy" (Jäger 2012).[2]

In response to the court's decision, the "quickest possible" solution for property settlement was integrated into the government's program proclamation. In 2011, the Commission defined a new model for property settlement that had been approved by both a church commission and the government. Beginning with this act, negotiations for a new law have been ongoing. The new (proposed) act on the settlement of property with churches includes not only mitigation for wrongs committed by the earlier regime. It also sets forth new property relationships between the state and churches or religious communities. It is, therefore, more than simply a law governing restitution. Instead, the restitution of previously church-owned properties is a condition for the eventual separation of church and state.

115.3.2 The Extent of Church Properties

Former church properties are unevenly distributed throughout Czechia, depending on the site and situation of various sacred structures, economic (agricultural) buildings and surrounding properties. In light of the fact that three-fifths of the blocked parcels of land is comprised of forest (Fig. 115.1), one could expect the distribution of these properties to correspond with the spatial distribution of forests in Czechia. The remaining land parcels are currently controlled by the Land Fund of the Czech Republic and their distribution is shown in Fig. 115.2. Figure 115.1 also depicts the extent of buildings, seized from churches, along with the type of original owner. Nearly half of the buildings belong(ed) to dioceses of the Roman Catholic Church.

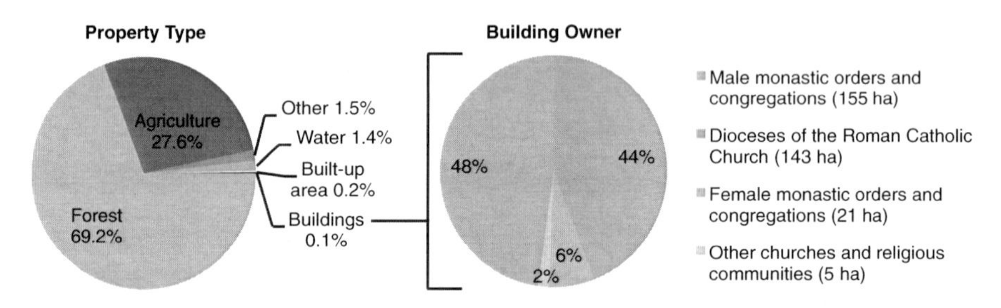

Fig. 115.1 Extent of formerly church-owned properties (Source: Martina Hupková, Tomáš Havlíček and Daniel Reeves, data from Government draft of a bill 2012)

[2]"It is also possible to point out that the failure to return (as actual real estate properties) any of a certain group of items from among the historical properties of churches and religious communities would constitute a violation of Article 15 (1), Article 16 (1) and (2) of the Charter of Fundamental Rights and Basic Freedoms" (Jäger 2012).

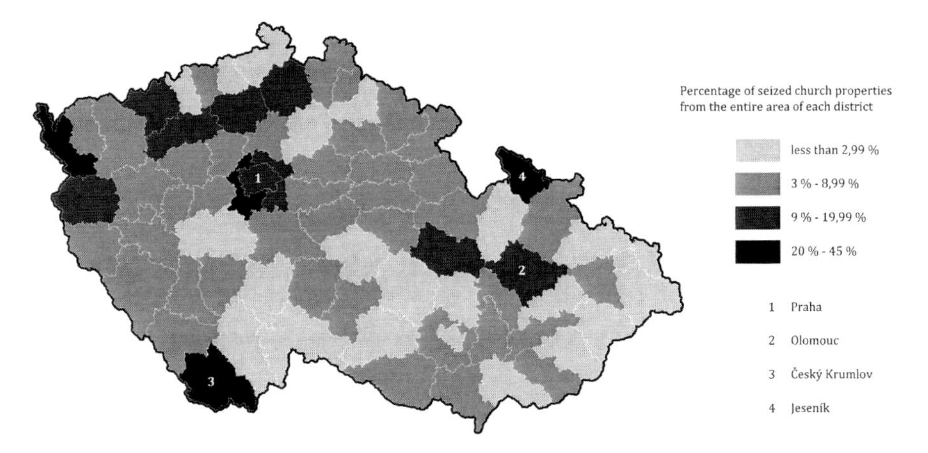

Fig. 115.2 Seized church properties in the Czech Republic, by district (Map by Martina Hupková, Tomáš Havlíček and Daniel Reeves, data from Land Fund of the Czech Republic 2011)

Regional differences in seized church properties (excluding forests) exhibit significant variability at the district level. Higher concentrations of seized church property are dependent, primarily, on the presence of important centers of the Roman Catholic Church (archbishoprics, bishoprics) as well as the presence of important monasteries which, in the past, controlled vast real estate holdings. The highest portions of seized church property from the overall area of districts (more than 9 %) are found in districts next to the archbishoprics of Prague and Olomouc. Above-average values are also found in the Český Krumlov and Jeseník Districts, both of which are home to important monasteries affiliated with extensive tracts of real estate. In contrast, low portions (less than 3 %) of seized church property (primarily agricultural land) are found in regions with a higher concentration of Protestant or non-Catholic inhabitants, particularly in Vysočina, northern Moravia and Silesia (Havlíček and Hupková 2008).

Church property evolved over several centuries. In the fourteenth century, one-third of Czech lands were controlled by the (Catholic) Church. During the fifteenth century, the time of the Hussite reform movement, a large amount of church property was destroyed and land was seized from the Church. This seized property was never returned and the Church acquired new properties from other resources (in part, from taxes on salt). Towards the end of the eighteenth century, as part of the reforms of Emperor Joseph II, a third of Czechia's monasteries were shut down[3]; their properties were placed under the administration of religious funds. Land reforms in the twentieth century had a great impact on church properties. The first land reform (beginning in 1919) included the division of large estates (larger than

[3]Those that were not involved in education, scientific research, healthcare or charity work.

150 ha (370.7 acres) and the reallocation of land to small-scale farmers. This reform did not apply to land owned by the state, the districts or the municipalities; nonetheless, it did apply to church-owned land to the tune of 232,000 ha (more than 537,000 acres). Contributions to clergy salaries, in accordance with Act No. 329/1920 Coll., were provided as compensation for the seized properties. A revision of the first land reform began to be applied in 1948. It called for the confiscation of estates larger than 50 ha (123.6 acres). This reform applied to church property and no form of legally-mandated compensation was ever made for the seized properties. A new land reform, also instigated in 1948, seized all properties larger than 1 ha (2.47 acres), that is, virtually all forest and agricultural lands. This left the churches without any economic resource base and robbed them of any chance at self-sufficiency. Because no compensation had been offered, Act No. 218/1949 Coll. on Financial Provisions by the State for Churches and Religious Societies was prepared and enacted and, from that time forward, the impacted churches have been forced to rely on aid from the state. The acquisition of church properties was carried out, not only through legal means; in some cases, it even resulted from the illegal, forced seizure of property.

In 2007, a church committee submitted a list of all seized church properties to the government (Štícha et al. 2008). As the objective of the settlement is to mitigate the losses caused by the communist regime after 1948, a critical period,[4] from 25 February 1948 to 1 January 1990, was determined. The list was compiled from land registers which, while they do include information about ownership, omit any information on the size of the real estate properties in question. Due to changes in divisions of land parcels, extending over nearly 60 years' time, specifications concerning the size of former church holdings were generated from a comparative assemblage of parcels or extrapolations. In this way, the extent of previously church-owned properties was determined to be 261,000 ha (645,000 acres). The list of properties was audited by an independent counseling firm, compared with data from the current administrators of the properties (state enterprises and institutions) and data from the various nationalized, large estates in the National Archive. Nearly 225,000 ha (556,000 acres) of formerly church-owned property has been documented and, therefore, the list of properties created by the church commission was designated as the starting point for property settlement between the state and the effected churches (Ministry of Culture of the Czech Republic 2012).

Only property currently owned by the state – not properties owned by municipalities, regions or private parties – shall be designated for return. Essentially, this means property controlled by the Land Fund of the Czech Republic or the Forests of the Czech Republic. Table 115.1 presents an overview of all the properties that churches claim in the list submitted by the church committee in 2007 along with the total area of formerly church-owned properties presently owned by the state. Forests of the Czech Republic records show an additional 23,128 ha (57,150 acres) of forest land that was previously owned by the Catholic Church. By 25 February

[4] Only losses that occurred during this time period are to be compensated.

Table 115.1 Overview of the extent of former church property

Property		Hectares		Perspective
Former church property		261,000	List of properties subject to restitution, according to the list submitted by the church committee in 2007	Church
Former church property currently owned by the state	Land Fund	48,412	Former church property currently administered by the Land Fund of the Czech Republic, according to Land Fund records	State
	Forests of the Czech Republic (Lesy ČR)	151,801	Former church property currently administered by the Forests of the Czech Republic, according to Forest CR records; only 24 ha of which was not previously owned by the Roman Catholic Church	

Source: Ministry of Culture (2008) and Government draft of a bill (2012)

1948, however, this particular land no longer belonged to any church and, as such, restitution does not apply to it.

115.3.3 Parameters of Settlement

The church committee, working in cooperation with Czechia's Ministry of Culture, completed an assessment of the value of former church-owned property,[5] in 2007. The overall amount[6] was calculated at 134 billion CZK (approx. 6.7 billion USD) and represents the state's financial debt to the churches. While 98 % of the property was taken from the Roman Catholic Church, the drafted bill on property settlement allocates 80 % of the overall amount to the Roman Catholic Church and 20 % to the other churches, in an attempt to encourage the self-sufficient financial management of all churches and religious organizations (Grulich 2012).

The current parameters of settlement come from a government-approved proposal prepared in 2008. In contrast with the proposal described in the preceding paragraph, the most substantial change is in the ratio of actual property returned to the amount of financial restitution provided. In light of the necessity to stabilize the financial management of Czechia, the volume of physical properties to be returned increased. Expressed monetarily, this increase was from 51 to 75 billion CZK (2.55 to 3.75 billion USD). On the other hand, financial restitution decreased from 83 to 59 billion CZK (4.15 to 2.95 billion USD). As the objective of restitution is to enable churches to operate financially independent of the state, economically useful

[5] Based on the average market price of agricultural land.

[6] The amount is the current value of the property that was unlawfully taken from the churches in the past.

property (fields, forests, fish ponds and usable (for example, agricultural) structures) currently administered by state institutions and enterprises – the Land Fund and Forests of the Czech Republic – will be preferentially transferred to the churches. This model is preferred, in part, because of the possibility of quickly transferring property from the Land Fund and Forests of the Czech Republic to churches.

Pastoral property currently owned by the state that is functionally tied to the properties of churches and religious communities shall also be subject to restitution. Additional adjustments to the settlement parameters focus on the payment of the financial settlement, including the time period of the payments (shortened from 60 to 30 years to ensure the quicker separation of church and state), the appreciation of financial amounts and implementation of a transitional period of financing churches (a 17-year period of decreasing church financing before it ceases altogether, which is intended to give churches time to adjust to the new conditions).

Smaller churches will use the newly acquired property and finances to expand and improve their current activities in the knowledge that their financial resources will not suffice for larger projects or investments (Grulich 2012). In general, churches will extend their activities beyond worship, education and healthcare to include social work, youth outreach programs and the prevention of socially pathologic problems in society. Finances will likewise be used to promote and present the religious communities themselves. Larger churches will be able to invest larger sums of money and use assets to co-finance projects from European funds.

115.4 Červená Řečice: An Example of the Effects of the Church Property Block on the Development of a Municipality

The municipality of Červená Řečice in the Vysočina Region (southeast of Prague, between Prague and Brno) has become a symbol of the negative repercussions of the unresolved property issues between the state and churches, at the municipal level. The municipality, home to roughly 1,000 inhabitants and with an area of 2,646 ha (6,538 acres), can only utilize about 20 % of its land area. More than 80 % of the municipality's land is subject to the block on formerly church-owned property and is, therefore, administered by the Land Fund. As Fig. 115.3 affirms, several of these blocked properties are located in the immediate vicinity of the center of the municipality, severely limiting its territorial development in all directions. It is as if the municipality were locked in a glass cage that limits development, both in terms of construction and infrastructure as well as society. Clearly, the greatest problem is the block of properties within built-up areas of the municipality.

The situation that now exists in Červená Řečice is a reflection of the cultural development of the region itself. The renaissance palace in the center of town served as the summer headquarters of the Prague Archbishop. This is the primary reason for their being such a large concentration of properties near the palace and even its broader surrounding area that belong(ed) to the Catholic Church. However, the unfortunate territorial configuration of blocked land parcels is not the only hindrance to the

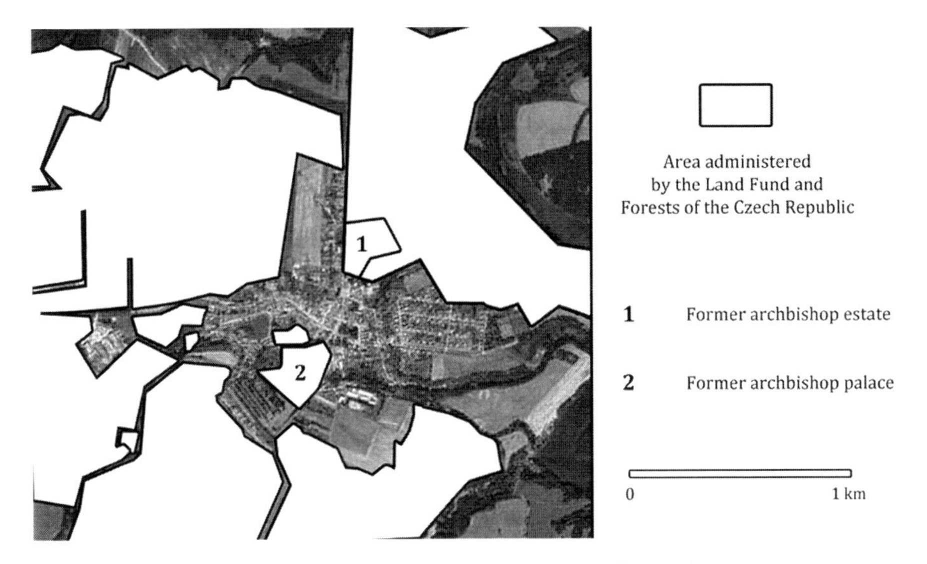

Fig. 115.3 Blocked properties (*white colour*) in the municipality of Červená Řečice (Map by Martina Hupková, Tomáš Havlíček and Daniel Reeves, data from Government draft of a bill 2012)

municipality's development. A second cause can be found in the failure to reach a working consensus between the municipality, the Catholic Church and the state. While purchasing or selling blocked former church-owned properties is strictly forbidden, it would be possible to realize certain objectives by alternative means. In practice, so-called three-way agreements do occur between municipalities, churches and the state. If a municipality wishes to create a residential zone on properties administered by the Land Fund or the Authority for Representing the State in Issues of Property, it makes a deal with the rightful former owner (that is, the Catholic Church) that, after a property settlement between the state and the churches (regardless of how it is carried out in the end), said rightful owner shall give the property to the municipality. A similar agreement is also made between the municipality and the state (in the event that the conclusion of property settlement would result in the state owning the property in question). These agreements are concluded as contracts for a future contract. The alternative described above requires the mutual consensus of three sides, each of which is seeking its own interests. Naturally, this is not an easy arrangement. In Červená Řečice, the negotiations described were not successfully completed due to an unclear situation concerning responsibility and the appropriate administering body of the Catholic Church. Červená Řečice is located within the České Budějovice Diocese, but its local palace served as headquarters for the Prague Archbishop. There is an ongoing internal dispute within the church as to which of these institutions (the Prague or České Budějovice Diocese) should be the one to decide in matters of the properties in question. Nonetheless, agreements with the Catholic Church have previously been made in Červená Řečice, concerning the construction of a roadway that transected part of a blocked parcel of land. This situation demonstrates that resolution of various situations is on a case-by-case basis and depends very much on the specific properties in question and on the specific stakeholders that are entitled to decide on the matter.

Not only does the block on church properties result in a stagnation in development, that is, infrastructure and new construction, it also brings into play a series of associated effects which prevent or hinder the natural socio-economic development of a given territory. Because of an inability to build new residential structures, a municipality misses out, in large degree, on the possibility of new residents moving in. According to its mayor, the municipality of Červená Řečice has experienced changes in its demographic structure due to its inability to provide housing for young families with children that could otherwise settle in the town. After 1989, the municipality noted interest from potential new inhabitants, to which it was unable to sufficiently react, in the form of new construction. Yet it failed to meet existing demand. Consequently, the influx of young families with children was limited, which in turn led to increases in the portion of older residents. In connection with the low number of children in the municipality (compared to expected numbers in the presence of normal development), the local elementary school was closed down. Beyond residential construction, the block on church properties has also impacted Červená Řečice in terms of business-related construction which, due to the arrangement of Czechia's system for financing municipalities, can play a significant role. Municipalities receive 30 % of the income taxes of individuals employed or conducting a business, on the basis of their place of residence, and 1.5 % of the statewide earnings from an income tax acquired from individuals from employment or functional benefits, based on the ratio of the number of employees in a given municipality, as of 1 December of the previous year, to the total number of employees in all of Czechia's municipalities (Act No. 243/2000 Coll.). Yet another unfavorable effect of the blocked properties is the inability to receive financial aid from European Union subsidy programs, which require that property ownership be clearly defined. Projects often completed in Czech municipalities with support from the EU include improvements to the heating and insulation of schools and nursery schools, construction of playgrounds in parks or schools (blocked in Vyskytná Municipality), the renovation/modernization of municipal buildings, for example, fire stations or cultural halls (blocked in Vyskytná Municipality), municipal roads, improvements to infrastructure (blocked in Nová Cerekev Municipality). Additional grant-awarding schemes at national and regional levels also require either clearly defined ownership of property or that the property owner must apply directly for any funding.

A block impacting even very small land parcels can cause problems to the implementation of linear transportation or technical infrastructure projects. As stated above, in Červená Řečice, the municipality, the Catholic Church and the state managed to come to an agreement over blocked properties in a location where the municipality intended to construct a municipal road. Czechia's extensive network of cycling paths constitutes another significant linear element of infrastructure development. Technical linear infrastructure, primarily water delivery and sewage systems, is a key ingredient in the potential development of a municipality and is, therefore, an item of interest to mayors. While this problem does not impact Červená Řečice, the municipality has an effective wastewater disposal system in place, it can and does affect municipalities of many sizes, both smaller and larger. On the basis of a directive on water (2000/60/EC), Czechia committed to make improvements in purifying wastewater in municipalities with more than 2,000 inhabitants. By 2012, and in all smaller municipalities by 2015, this is a significant current issue. According to Czech legislation (Act No. 254/2001), technical norms (CSN 75 6101) and general standards for the construction of wastewater treatment systems, implementation of linear sewer lines coupled with a classical water treatment plant is the most appropriate solution for treating wastewater, even in the smallest municipalities and settlements (Moldan et al. 2011). Failure to reach a consensus in the three-way negotiations, described above, concerning parcels of land designated for sewer line construction can be considered very risky to the communities involved. In the municipality of Sedlec-Prčice, for example, the necessary three-way agreement leading to sewer line construction was only reached after long and complicated negotiations.

This blockage of properties also results in the devastation of the cultural and spiritual values of an effected municipality. Architecturally, culturally and historically valuable buildings, situated on blocked properties, are frequently left to fall into disrepair and eventual ruin. The state which, in the present interim period, controls such properties with their buildings does not have sufficient capability or will (naturally, as it is merely a temporary steward) to invest into such buildings. Further postponement of an eventual solution will lead to irreparable damages to a number of valuable structures and, as such, to the loss of part of the local cultural-historical heritage.

The mayor of Červená Řečice expects to see renewed interest in small scale residential construction in the municipality, after the anticipated settlement between the state and the effected churches is complete (Bečková 2011). Such expectations are in line with general developments occurring in other small municipalities in the hinterland of larger regional centers. At the present time, however, she does not express much hope in new business-related construction nor does she expect such development in the future. The block on church properties, which are tied up in the ongoing settlement negotiations (or lack thereof), has interrupted the continuity of development in Červená Řečice. As to whether natural development has merely been interrupted or if it has been systematically altered to take on a new trajectory, at present, we can only speculate. In the case of Červená Řečice, it seems very unlikely that the municipality will be able to pick up where "natural" development left off in 1989, before the block on church properties took effect. Not only has the territorial development of the municipality been impeded, or temporarily halted, by this process. But it has been fundamentally changed with a direct impact on any future development of the town and its surrounding area. Červená Řečice is actively doing what it can (it has initiated discussions on property settlement with a letter to the president of Czechia as well as through the Union of Towns and Municipalities of the Czech Republic) to bring about a solution – any type of solution – to property ownership issues between the state and the churches.[7] The primary concern for the municipality is that contested properties be no longer blocked and that the new owners of said properties have complete control over them. In other words, one and only one entity will be able to decide on the fate of a given property, making it much easier for the municipality to make arrangements concerning its own current and future development. In connection with this, local leaders have come up

[7] There is a justifiable assumption that, in the case of certain parcels becoming property of the state, such property would be transferred for free or, at least, under favorable conditions, to the municipality, in which it is located.

with a slogan: "any solution is better than no solution." At present, Červená Řečice has an approved master plan (a fundamental, long-term development document, prepared by municipalities in Czechia, focused on land-use and zoning for development) in place that designates the blocked properties as developable and indicates possible future uses. In this way, the master plan anticipates settlement of the property issues and plans for the development of a post-restitution municipality.

115.5 Property Restitution in Czech Society

The long and drawn out nature of talks concerning the eventual form of property settlement between the state and the churches indicates just how complicated the topic is. Within Czechia's secularized society, a large majority of the general public is opposed to the idea of returning property to churches and religious societies. According to one church representative (Grulich 2012), 70 % of Czechia's inhabitants do not agree with the return of church properties to the churches. During the initial, post-communist transition period, efforts were made to implement a systematic settlement of the issue. In light of the political underpinnings of the problem, however, progress towards such systematic settlement has been heavily influenced by the policy of each of the subsequent government coalitions. Grulich (2012) called the property settlement of the churches and the state a "hot political issue." Is it realistic to expect any government coalition in Czechia to make actually returning property to the churches (not merely declaring good intentions) a priority? In increasingly secularized Czechia, postponement of this already overdue settlement is exacerbated by growing opposition from the public. While the residents of Červená Řečice are likely better informed, as compared to the inhabitants of other municipalities, about the issue of property settlement between the state and the churches – due to the fact that it impacts them directly – they generally do not share the opinion of municipal leaders who want to see any type of settlement implemented (as long as it frees up the blocked properties). Instead, Červená Řečice's residents tend to share the view of the majority, secularized population, which does not support the full-value return of properties to the churches (Bečková 2011).

As the discussion above implies, property settlement between the state and the effected churches is not a simple issue confined only to the subjects involved. We affirm that the consequences of the as yet unresolved church-state relations manifest themselves in local and regional development. Local self-governing authorities, that is, municipalities, and to a lesser degree regions and private entities are important stakeholders, in that they either use – or would like to use – blocked properties. Their primary interest is to see ownership of the properties resolved in any way that would make it possible for the land to be used or sold. Pressure to resolve this property settlement issue is coming from municipalities, but only from those municipalities impacted by blocked properties. This creates an interesting situation, in which certain municipalities and the churches are allies working to achieve property settlement with the state. The churches can be viewed as passive stakeholders due to the fact that they have never made an official proposal or request for settlement of the outstanding property issues. Instead, the churches have waited for the state to initiate negotiations. In contrast, impacted municipalities can be designated active stakeholders, which initiate and actively promote actions leading to eventual settlement of the matter. According to Bečková (2011), the mayor of Červená Řečice, the three-way agreements described above actually work against any systematic resolution of the problem, because any instance of a problem-free, three-way agreement – municipality-church-state – results in less pressure from municipalities to expedite property settlement. This indicates that, from a long-term perspective, three-way agreements are not in the best interest of stakeholders seeking property settlement.

We frequently use the term "churches" in this chapter to describe all the churches and religious societies involved in property settlement in Czechia. The churches speak together and negotiate collectively through a church committee, established for the sole purpose of negotiating the ultimate settlement of disputed properties with the state. According to Grulich (2012), leader of an ecumenical group in Czechia, the opinions and attitudes of all the effected religious societies are united and cooperation and communication within the committee are problem-free. Consequently, we can consider the "churches" to be an internally cohesive unit, seeking one common goal. This is a remarkable feat in and of itself, considering the differences among the various churches. Because the state determined the financial amount to be paid out to the churches, the church committee needed only to finalize the rates by which the amount would be divided (Table 115.2). An agreement was met without any significant disputes. The various individual churches were cordial

Table 115.2 Division of the lump-sum financial restitution among churches and religious societies after 30 years

Church	Sum
Roman Catholic Church	47,200,000,000
Czechoslovak Hussite Church	3,085,312,000
Evangelic Church of Czech Brethren	2,266,593,186
Orthodox Church in the Czech Lands	1,146,511,242
Apostolic Church	1,056,336,374
Brethren Evangelical Free Church	761,051,303
Silesian Evangelical Lutheran Church	654,093,059
Moravian Church	601,707,065
Seventh-day Adventist Church	520,827,586
United Methodist Church	367,634,208
Greek Catholic Church	298,933,257
Old Catholic Church in the Czech Republic	272,739,910
Federation of Jewish Communities in the Czech Republic	272,064,153
Czech Baptist Union	227,862,069
Evangelic Church Augsburg Confession in the Czech Republic	118,506,407
Lutheran Evangelical Church – Augsburg Confession	113,828,334
Religious Society of Czech Unitarians	35,999,847

Source: Government draft of a bill (2012)

and, in the end, the initial amount was divided according to a number of criteria, the two most important of which are the number of clergy and the number of adherents recorded in the Czech Population and Housing Census.

In terms of the issue at hand, the state should be understood to mean both legislative and executive powers, that is, parliament and the ruling government coalition. This power is obligated with both the fulfillment of generally social-moral commitments of a state (that is, implementing restitution for the wrongs of the former regime) as well as with responsibilities towards citizens/the voters and their public opinion. Considering public disagreement with the return of church properties, it is clear not only that a conflict exists, but that it is the primary cause for the long period of inactivity, on the part of the state, regarding the settlement of church properties. There is a confirmed connection between the rising secularization of Czech society and the tendency to object to the restitution of church property, which can be viewed as a consequence of political inactivity in implementing (not merely in declaring) a settlement of former church estates.

Internet discussions on the topic[8] enable us to tease out the primary reasons behind the public's negative attitude regarding the return of properties to the churches. The first subset of reasons arises out of Czech society's generally negative relationship towards churches (church institutions) and results in an aversion towards the return of church properties to churches and, in particular, to the implementation of any type of financial restitution, in light of the state's unfavorable economic situation. The second group of reasons results from ignorance or superficial knowledge concerning the issue. For instance, many people express the conviction that ownership (acquisition) of the properties in question was unauthorized to begin with. Such convictions are usually based on one of two faulty conclusions: (a) that church property equals state property and, consequently, nothing needs to be returned or (b) that after returning property to churches, it shall be owned/controlled by a given church, as a whole, and not by its local divisions (that is, the Roman Catholic Church with headquarters in the Vatican as opposed to a local parish or diocese). In the process, a discussion concerning the legitimacy of church ownership has become a significant component of property settlement. Four expert opinions have been prepared (Hrdina and Kindl 2007; Mikule and Kindl 1998; Šimáčková 2007; Zachariáš 2007). They show that former (as of 1948) church-owned properties were owned by the churches, at least, in the sense of the present-day term "ownership." The expert opinions also agree that the rightful subjects, to whom properties should be returned are the various individual parishes, parish beneficia, chapters and foundations, but not the larger church organizations.

It is interesting to note how developments on the topic of property restitution are closely followed by the media and through the media by the majority of population, at least, as long as "something interesting is happening" (for example, a new settlement proposal is brought forward). Most of the time, however, property

[8] Discussions with Pavla Bendová, director of the Churches Division in the Ministry of Culture, are particularly interesting as is the first reading of negotiations on the drafted legislation for an act on property settlement with churches and religious societies in the Chamber of Deputies of the Czech Parliament.

settlement between the state and the churches, along with its impacts on regional development, remains outside of the interest of mainstream media and, consequently, outside of the public interest as well. This is the situation in spite of the fact that it is one of the more significant processes taking place as part of the post-totalitarian transformation of Czech society that its position is considered a visible indicator of Czechia's ability to come to terms with its own past. Resolving this sticky situation is only becoming more complicated with the passage of time, particularly considering the spiritual development of Czech society.

Acknowledgment This article was prepared within the grant project Nr. P410/12/G113: Historical Geography Research Centre and Nr. 13-35680S: Development, transformation and differentiation of religion in Czechia in the context of global and European shifts (Grant Agency of the Czech Republic).

References

Act No. 218/1949 Coll., on Financial provisions by the State for Churches and Religious Societies.

Act No. 229/1991 Coll., on Adjustments to the ownership of land and other agricultural property.

Act No. 243/2000 Coll., on the Budgetary allocation of revenue from certain taxes to territorial self-governing units and certain state funds (the Budgetary Allocation Act).

Act No. 254/2001 Coll., on Water (the Water Act) and related regulations.

Act No. 298/1990 Coll., on Adjustments to the property rights of holy orders, congregations and the Olomouc Archbishopric.

Act No. 329/1920 Coll., on the Transfer and restitution for seized properties (the Restitution Act).

Act No. 338/1991 Coll., which alters and amends Act No. 298/1990 Coll., on Adjustments to the property rights of holy orders, congregations and the Olomouc Archbishopric.

Bečková, Z. (2011, November 11). *Authors' interview with Zdeňka Bečková.* Mayor of the municipality Červená Řečice, Červená Řečice.

Broun, J. (1996). The Catholic Church after communism. *East–West Church and Ministry Report, 4*(3), 2–4.

Commission on Security and Cooperation in Europe. (2002, July 16). *Hearing: Property restitution in Central and Eastern Europe: The state of affairs of American claimants.* Official transcript. Retrieved October 11, 2011, from www.csce.gov/index.cfm?FuseAction=ContentRecords.ViewDetail&ContentRecord_id=220&Region_id=0&Issue_id=0&ContentType=H,B&ContentRecordType=H&CFID=617521&CFTOKEN=25900967

CSN 75 6101. (2004). Czech technical norm. *Stokové sítě a kanalizační přípojky* [Sewer networks and connections]. ČNI.

Directive 2000/60/EC, dated 23 October 2000, establishing a framework for community action in the field of water policy.

Geyer, M., & Geyer, M. (2004). Religion und Nation – Eine unbewältigte Geschichte. In M. Geyer & J. Lehmann (Eds.), *Religion und nation. Nation und religion* (pp. 11–32). Göttingen: Wallstein.

Government Resolution No. 333, dated 2 April 2008: Návrh zákona o zmírnění některých majetkových křivd způsobených církvím a náboženským společnostem v době nesvobody, o vypořádání majetkových vztahů mezi státem a církvemi a náboženskými společnostmi a o změně některých zákonů.

Governmental draft of a bill. (2012). o majetkovém vyrovnání s církvemi a náboženskými společnostmi a o změně některých zákonů (zákon o majetkovém vyrovnání s církvemi a náboženskými společnostmi).

Grulich, P. (2012, April 18). *Authors' interview with Petr Grulich.* Chancellor of the Ecumenical Council of Churches in the Czech Republic, Prague.

Havlíček, T. (2005). Czechia: Secularisation of the religious landscape. In H. Knippenberg (Ed.), *The changing religious landscape of Europe* (pp. 189–200). Amsterdam: Het Spinhuis.
Havlíček, T. (2006). Church-state relations in Czechia. *GeoJournal, 67*(4), 331–340.
Havlíček, T., & Hupková, M. (2008). Religious landscape in Czechia: New structures and trends. *Geografie, 113*(3), 302–319.
Havlíček, T., Hupková, M., & Smržová, K. (2009). Changes of geographical distribution of religious heterogeneity in Czechia during the period of transformation. *Acta Universitatis Carolinae Geographica, 44*(1–2), 31–47.
Hrdina, A., & Kindl, M. (2007). *Posudek – č.j. DFPr 101/07*. Plzeň: Západočeská univerzita v Plzni.
Jäger, P. (2012). *Ústavněprávní aspekty finanční náhrady jako součásti vypořádání historického majetku církví a náboženských společností*. Brno: Ministerstvo Kultury ČR.
Judgment of the Constitutional Court Pl. CC 9/07: K restitucím církevního majetku ("blokační" ustanovení § 29 zákona o půdě).
Karlová, I. (2011). *Analýza problematiky majetkového vyrovnání státu a církví v ČR, postoj aktérů a vládní návrhy zákonů z roku 2007 a 2011*. Praha: Univerzita Karlova.
Knippenberg, H. (Ed.). (2005). *The changing religious landscape of Europe*. Amsterdam: Het Spinhuis.
Land fund of the Czech Republic. (2011). Pozemky ve správě Pozemkového fondu ČR dotčené ustanovením § 29 zákona č. 229/1991 Sb., ve znění pozdějších předpisů, evidované v CRN k datu 1. 5. 2011. Retrieved April 11, 2012, from www.mkcr.cz/cirkve-a-nabozenske-spolecnosti/majetkove-narovnani/default.htm
Madeley, J. T. S. (2003). A framework for the comparative analysis of church-state relations in Europe. *West European Politics, 26*(1), 23–50.
Mayr, H. (1995). *Politische Religionen*. Freiburg im Breisgau: Herder.
Mikule, V., & Kindl, V. (1998). *Právně historická expertiza Univerzity Karlovy v Praze právního postavení tzv. katolického církevního majetku v druhé polovině 19. a ve 20. století na území dnešní ČR*. Praha: Univerzita Karlova v Praze.
Ministry of Culture. (2008). *Zákon o majetkovém vyrovnání s církvemi a náboženskými společnostmi: Základní fakta*. Praha: Ministerstvo kultury ČR.
Ministry of Culture. (2012). *Rozsah původního církevního majetku*. Retrieved April 11, 2012, from www.mkcr.cz/cirkve-a-nabozenske-spolecnosti/majetkove-narovnani/default.htm
Moldan, B., Hupková, M., Kouba, B., & Hamák, L. (2011). *Studie variant koncepce zneškodňování odpadních vod z malých komunálních zdrojů znečištění do 500 EO*. Praha: Univerzita Karlova v Praze, Povodí Vltavy s.p.
Palas, J. (2004). *Připomínky k materiálu Zpráva o postupu řešení úkolu "Návrh zákona o uspořádání majetkových vztahů mezi státem na jedné straně a církvemi a náboženskými společnostmi na straně druhé" z výhledu legislativních prací vlády na rok 2005*. Praha: Ministerstvo zemědělství ČR.
Pogany, I. S. (1997). *Europe in change: Righting wrongs in Eastern Europe*. Manchester: Manchester University Press.
Šimáčková, K. (2007). *K problematice vlastnictví katolické církve a restitucí církevního majetku*. Brno: Masarykova Univerzita v Brně.
Štícha, K., Slavíčková, I., & Fojtík, K. (2008). Majetek církví a náboženských společností. Evidence původního církevního majetku provedená církvemi a náboženskými společnostmi v období od roku 1990 do roku 2007.
Večeře, K. (2001). *Stanovisko k materiálu. Zpráva o přípravě návrhu zákona o úpravě některých majetkových vztahů některých církví a náboženských společností*. Praha: Český úřad zeměměřický a katastrální.
Walsh, M. T. (2002). Property restitution efforts examined. Retrieved May 7, 2012, from www.csce.gov/index.cfm?FuseAction=ContentRecords.ViewDetail&ContentRecord_id=39&Region_id=77&Issue_id=64&ContentType=G&ContentRecordType=G&CFID=616522&CFTOKEN=16209038
Zachariáš, J. (2007). *Odborný posudek – Čj. 19/43/07/ST*. Praha: Ústav státu a práva, Akademie věd České republiky.

Chapter 116
Indian Secular Nationalism Versus Hindu Nationalism in the 2004 General Elections

Igor Kotin

116.1 Introduction

In this article an attempt is made to analyze the most dramatic event of recent politics, the struggle for power of the secular nationalist Indian National Congress and the Hindu nationalist Bharatiya Janata Party, the struggle that led to the victory of secular nationalists and the coming back to power of the Indian National Congress. It article deals with the history and nature of the nationalism in India, its relevance to classical models of nationalism, its recent development into secular and religious versions and the most recent political struggle of the two nationalisms.

116.2 The Phenomenon of Indian Nationalism

Multicultural, multi-ethnic and multi-religious India in an anthropological sense has never been an example of ethno-national unity. In political terms, however, for 100 years it is the country of Indians and it is the nation. Indian nationalism as a political anticolonial movement emerged as a mixture of intellectual anti-British struggles and a middle class patriotic movement. The strongest Indian nationalists at the start of movement were devout Bengali and Marathi patriots as well; they were mostly Hindus, Muslims and Parsees by origin. They shared a religious loyalty and a language loyalty with pan-Indian patriotism as far as they did not contradict each other. The main factor in this initial patriotism was anti-British sentiment. It is interesting that it is the English language that united Indian patriots against the English rule.

I. Kotin (✉)
Museum of Anthropology and Ethnography, Russian Academy of Sciences,
St. Petersburg, Russia
e-mail: igorkotin@mail.ru

S.D. Brunn (ed.), *The Changing World Religion Map*,
DOI 10.1007/978-94-017-9376-6_116

I want to include here a long quotation from the seminal book of Sunil Khilnani:

> The puzzle of India's unity and of Indianness raised a variety of contending responses within the nationalist movement that brought India to independence. Nehru's was only one among these and it was in no sense typical of nationalism as a whole.... Indian nationalism is a somewhat misleading shorthand phrase to describe a remarkable era of intellectual and cultural ferment and experimentation inaugurated in the late 19th century. The various, often oblique, currents that constituted this phase extended well beyond the confines of a political movement such as the Congress, with its high political, bilingual discourse. The possible basis for a common community was argued with ingenuity and imagination in the vernacular languages, especially in regions like Bengal and Maharashtra that had been exposed longest to the British... (Khilnani 1999: 153)

When the movement matured, it received, however, some characteristics of classical nationalism. The nationalism as the ideology was developed by Rousseau, Herder and Mazzini. As the ideology of young nations it replaced the idea of states being united by a monarch; in the nineteenth century many European states became republics. The phenomenon of nationalism became the subject of academic research later due to the contributions of Ernest Gellner (1983) and especially in the post-war England and Central Europe where nationalism studies were recognized as a specific discipline. Cubitt admits that 'The concept of the nation is central to the dominant understanding both of political community and of personal identity. Perceptions of politics are framed in terms of national interest and of international relations' (Cubitt 1998:1).

In some way nationalism theory became linked with the theory of "imagined communities" (not limited to nations) of Benedict Andersen. He observed,

> There is no disagreement that nationalism has been "around" on the face of the globe for, at very least, two centuries. Long enough, one might think, for it to be reliably and generally understood. But it is hard to think of any political phenomenon which remains so puzzling and about which there is less analytic consensus. No widely accepted definition exists. No one has been able to demonstrate decisively either its modernity or its antiquity. Disagreement over its origins is matched by uncertainty about its future.... (Andersen 1996: 1)

Gellner (1996: 98) speaks of nationalism as "the linking of state and of "nationally" defined culture." It is both the feeling of affiliation to a single nation (often "imagined") and the movement for the emergence of this nation that applies to the political entity.

> As Hutchinson and Smith conclude from reading pioneers of nationalist thought, Nationalism was, first of all, a doctrine of popular freedom and sovereignty. The people must be liberated – that is, free from any external constraint; they must determine their own destiny and be masters in their own house; they must control their own resources; they must obey only their 'inner' voice. But that entailed fraternity. The people must be united; they must dissolve all internal divisions; they must be gathered together in a single historic territory, a homeland; and they must have legal equality and share a single public culture. But which culture and what territory? Only a homeland that was "theirs" by historic right, the land of their forbears; only a culture that was "theirs" as a heritage, passed down the generations, and therefore an expression of their authentic identity. (Hutchinson and Smith 1984: 4)

Indian patriots "imagined" Indian nation, but certain groups of Muslim Indians (Muhammad Iqbal, M.A. Jinnah) favored for the special identity and "imagined" the nation of Indian Muslims. At a certain point Hindu and Muslim Indian nationalists started considering each other as "alien." Further development of this trend led to the rise of Muslim separatism, Sikh Separatism and Hindu Nationalism.

The breakup of India in 1947 led to the creation of secular but Hindu-dominated India and Pakistan, the state of Indian Muslims.

In secular India authorities introduced a sort of official nationalism, centrally sponsored patriotism based on the history of the independence movement. Partly as the reaction to it the Hindu Nationalism grew. In both cases of Indian Nationalism and Hindu Nationalism the terms are used slightly differently from the tradition of the western study of nationalism. Yet, both terms in the Indian context became the tradition themselves. Currently, the use of the term is widespread and often the meaning is different in different writings. As Jalal (2001) notes in her article on nationalism in South Asia, Indian nationalism is viewed as an idea and a historical force and a challenge to Western colonialism. It has both similarities and differences from the Western model of nationalism developed on European material.

Indian nationalism generally refers to the movement for independence of India. Also it is considered as the basis for the secular state ideology of the present-day Republic of India. One of the problems with this movement, however, lies in the absence of the single Indian nation as ethnological reality. It is rather a complex mixture of dozens of peoples, thousands of ethno-religious communities, castes plus 636 officially recognized tribal groups ('scheduled tribes' or 'backward tribes' according to the Anthropological Survey of India). Though religious and cultural links and linguistic proximity bind many regions together, there are significant centrifugal movements in western, southern and northeastern parts of the Indian subcontinent. Several huge empires, the Maurya Empire (3–2 c. B.C.), the Gupta Empire (5 c. A.D.), the Kanauj Empire under Harshavardhana (7 c. A.D.), the Delhi Sultanate at its heyday (12–14 c.) and the Mughal Empire (15–18 c. A.D.) formally controlled most of the Indian subcontinent but failed to take into their grips the small units like principalities and the jagirs (fiefs). The eighteenth century conglomerate of principalities known as the Maratha Confederation instead of reuniting imperial bonds ruined the old links by introducing a series of taxes ("Chauth" etc.) and periodical plunder of the lands of India. As the result, the British found it easy to win a divided and weakened India.

The British not only managed to unite hundreds of Indian principalities, but were able to expel their French opponents and neutralize the Dutch and the Portuguese and other Europeans who also had a presence on Indian soil. The Mutiny of 1857–58, known in the Indian historiography as the First Indian War for Independence, united Hindus and Muslims against the common enemy, the British. Ironically, the British more clearly recognized the Muslim danger for them and ignored Hindu mobilization by preferring the latter as the core of the colonial administration at basic level. Thus the Hindus formed the core of the new educated class of Indians who were dreaming of getting rid of their masters.

Though it was the British who supported the formation of the new Indian intellectual class, they became the main subject of hatred by the South Asian population. The British were hated as both "suppressors" and the "foreigners." The rise of the anti-British sentiments led to claims for self-administration of British India. Though local patriotism of the Bengalis and the Marathas provoked them of praising their "Golden Bengal" and "The Maha Rashtra," their influence and ambitions grew over their ethnic territories which led them to speak in pan-Indian terms. Local anti-British organizations like the Bengal Association and the newspapers in English and vernacular languages, like the "Amrita Bazaar Patrika," "Mahratta" and "Kesari." helped mold public opinion about the existence of the Indian nation. Their organizational efforts resulted in the formation in 1885 of the Indian National Congress, the first Indian national political party. In the first years of its existence (1885–1905) the loyalty was to the British crown, but the party turned into anti-British position during the anti-Partition (of Bengal) movement in 1905–1906. With Gandhi as its leader the Indian National Congress in 1919 became the largest pan-Indian organization of major opposition to British colonial administration. Though Mahatma Gandhi was in favor of the use of vernacular languages, the Indian National Congress accepted English as the language understandable for the majority of its members. Thus the English provided Indians with the idea and the language of unity. At the same time, two rival models of mobilization emerged in India in early twentieth century, that is. Hindu and Muslim communal movements. In early twentieth century two rival organizations emerged known as the Hindu Mahasabha (Great Council of Hindus) and the Muslim League. Both had nationalist slogans as well as communal ones, but they spoke of the two different Indias and two different nationalisms.

Hindu Nationalism grew along with Indian secular nationalism. Hindu nationalists of the Arya Samaj and of the Hindu Mahasabha, particularly, M. M. Malaviya, and Rajendra Prasad were also influential members of the Indian National Congress. Muslim nationalism developed into a separate force due to dissatisfaction of Indian Muslims over the poor economic conditions of their co-religionists and also their anger with what they considered "the Hindu (even Brahmin) dominance" in the Indian National Congress. It has both anti-Hindu and anti-colonial sides. Depending on the situation Indian Muslims sought either Indian Nationalist support or the British protection. The decision of the British Labour government in 1946 to grant India independence led Muslim nationalists to their struggle for their independent state called Pakistan ("the country of the pure" or abbreviation from words Punjab, Kashmir and Pushtunistan).

As Jalal (2001: 737) rightfully observed,

> A challenge to Western colonialism reflecting the aspirations of a subjugated and diverse populace, nationalism as the idea and a historical force in South Asia has remained fiercely contested terrain.

The problem of Indian nationalists is that they speak on behalf of the united nation in the country where nation is divided. They cannot speak even on behalf of the majority. As Anton Pelinka (2003: 212) observes,

> The nature of Indian democracy is characterized not simply by the rule of a "natural" majority of Hindus. Instead, it is the fact that in India a majority that includes minorities.... Above all, the reality of the castes destroys the notion of a given Hindu majority. The specificity of the castes makes the concept "Hindus" a multitude of partial concepts, such that one could even argue there is no majority in India ....

In this situation it is not surprising that there is plenty of political movement with ethnic, caste and religious bases and ideologies in India.

116.3 Hindu Nationalism vs. Secular Nationalism in India

Hindu Nationalism has features of both the nationalism and the communal movement. It also grasped the banner of the opposition ideology when Indian secular nationalism became the ruling ideology of the independent Indian state. Proponents of the Hindu Nationalism claimed that it is not "Indian," the English term, but "the Hindu" is right term for the members of the Indian nation. Having expressed strong anti-Muslim and anti-Christian feelings, Hindu nationalists stress the foreign character of Islam and Christianity in India. At the same time a safe place is allocated in their ideology for the "protected" local religions of Buddhism, Jainism and Sikhism. Linguistic, caste and class differences among the Hindus are downplayed by the Hindu nationalists, though they failed to overcome hostility of the Hindus in the south towards northern Indian Brahmans. In many ways Hindu Nationalism is more a nationalist than a communalist movement. The founder of the movement Vinayak Damodar Savarkar, a Marathi Chitpavan Brahman, was educated in London and the admirer of Mazzini; he wrote a seminal history of the 1857–58 Indian Uprising (known also as the Mutiny and the Sepoy Revolt) in which he claimed the main aim of Indians was their fight for independence. In the book Savarkar called for Hindu-Muslim unity. He later changed his mind, maybe because of his personal sufferings from the hands of the Pushtoon guards in the prison where he was sent for revolutionary activity. In 1923 Savarkar wrote the book "Hindutva" (Savarkar 1923, 1989), claiming that Hindutva or "Hindu"-ness is the real identity of Indians. Savarkar wrote that Hinduism was only part of Hindutva, and Hindutva was a history, a historical heritage of people in India. I wish to insert here another quote from Khilnani:

> In Savarkar's genealogical equation between the Hindu and the Indian, members of the Indian political community were united by geographical origin, racial connection (rather ambiguously specified), and a shared culture based on Sanskritic languages and "common laws and rites." Those who shared these traits formed the core, "majority" community. Those who did not – Muslims, who constituted a quarter of pre-Partition India's population, "tribals," Christians – were relegated to awkward, secondary positions... (Khilnani 1999: 161)

Khilnanin continues, arguing that Savarkar translated

> Brahmanical culture into the terms of an ethnic nationalism drawn from his reading of Western history. This created an evocative, exclusivist and recognizably modern definition of Indianness, with rich potentials to sustain future political projects and to induce direct political effects. It was contact with these ideas that in 1925 led another Brahmin, K. B. Hedgewar, to found the Rashtriya Swayamsevak Sangh (RSS-Association of National Volunteers), to this day the backbone of Hindu nationalist organization. It also inspired the Hindu Mahasabha, until 1930 the main party of Hindu nationalism. The Gandhian Congress adroitly marginalized the Savarkarite conception of Indian history and Indianness, but its presuppositions were never erased; many nationalists outside Congress, and even some within it, shared them. (Khilnani 1999: 161)

The assassination of Mahatma Gandhi on 30th January 1948 by Hindu nationalists led to the ban of some Hindu nationalist organizations and limitations on the activities of others. Yet, in 1951 Hindu communalist party, Jan Sangh, participated in the General Elections in India. Later it was transformed into the Bharatiya Jan Sangh, the Janata Party, and in 1990s – into the Bharatiya Janata Party or the India's Nation (or People's) Party. In 1996 for 13 days it was the ruling party. In 1998 it again became the ruling party, due to the failure of the Indian National Congress to win the majority in the hung Parliament. In 1999 the BJP became the ruling party thanks to its victory in the elections as the major partner in the multicultural alliance. In 2004 it called for earlier elections being sure of its new victory. Yet, it lost the to the Indian National Congress and its allies. Secular Nationalism appeared as the more appropriate ruling party for multi-religious and multi-ethnic India.

One of the major points of Hindu nationalists was that Muslims and Christians were enemies of Hindus and are foreigners in India. It is Muslims, they said, who demolished hundreds of Hindu temples including shrines in key Hindu sacred cities of Varanasi, Mathura and Ayodhya. It was the Ayodhya "Ramjanmbhumi-BabriMasjid" issue that mobilized Hindu masses in RSS and BJP led campaign for the demolition of the mosque (Babri Masjid) and construction of Hindu temple on the alleged place of Hindu Lord Rama's birth (Ramjamnbhumi). In 1992 Hindu nationalists demolished the Babri Masjid which led to their rise in popularity, but also to political crisis in India. So, when they achieved power in 1998 and remained in power until May 2004, Hindu nationalists did not dare construct a Hindu temple on the disputed spot both due to the respect to the Highest Court decision and because they understood such a construction could bring all India into turmoil.

Another issue was more important for Hindu nationalists while they were in power, rather than those in the opposition to the government. New history textbooks were introduced by the National Council of Educational Research and Training (NCERT). Having been critical of previous textbooks written by 'leftists' and having made a series of government-ordained deletions from the earlier generation of NCERT textbooks, the government of Hindu nationalists inspired the publication of its own vision of Indian history. It is noteworthy what was deleted previously. As Jha observes

> In Ancient Indian History, the deletions included (i) references to beef-eating and cattle sacrifice; (ii) a critical evaluation of the Puranic and epic traditions in the light of archeological and epigraphic testimony with reference to the antiquity of Ayodhya and the origin of Krishna worship in Mathura; the exposition of Brahmanical hostility towards Ashoka. (Jha 2003: 87)

The remain in power in 1998–2004 made the Bharatiya Janata Party more careful about its relations towards the 130 million of Indian Muslims. This made the party's ideology even more Indian nationalist than Hindu nationalist one, or rather Hindu Nationalist than Hindu communalist. As Ex-Vice-President of BJP K. R. Malkani observes on softening of the position of BJP (and RSS) on Muslims:

> It has nothing against Muslim Indians – as distinguished from Muslim invaders. Its position on this issue has all along been Justice for all and appeasement of none. But it has no doubt that we were and are a Hindu nation; that change of faith cannot mean change of nationality. (Malkani 2003: 1)

116.4 General Elections of 2004 in India: Observations

In the December 2003 three out of every four states of India elections resulted in the victories of the BJP candidates for the Legislative Assemblies. This success made BJP's leaders Atal Bihari Vajpeyee and Lal Krishna Advani believe that in coming elections to the Lok Sabha (Lower Chamber of the Indian Parliament) BJP would also be successful. A. B. Vajpeyee even asked the president to call parliamentary elections 8 months earlier than it was scheduled based on existing pro-BJP support among the electorate. Elections were announced to be held in April–May 2004. Pre-election polls showed that the middle class was in favor of the BJP. "India is Shining" and "Feel Good Factor" slogans of the BJP showed its belief in the future electoral victory.

The electorate has decisively rejected the Atal Bihari-led National Democratic Alliance and voted in a Indian National Congress-led coalition spearheaded by Sonia Gandhi. On its own Congress has emerged as the largest single party, with 145 seats under the belt. Along with the left parties (62 seats), the Congress coalition (with 216 seats) secured majority. Fears over left parties' influence over the new government today led to a historic crash in the stock markets. The Congress –led government managed, however, to settle things and has successfully ruled for the last 5 years under the Prime Minister Manmohan Singh.

116.5 Secular Nationalists vs. Hindu Nationalists in 2004 General Elections

The 2004 parliamentary elections were particularly heated. The most loved person of the elections, Atal Bihari Vajpayee, received much of the critique. Even if his personal virtue is not in doubt, his name was so intensively used by the Bharatiya Janata Party and its Parivar (family) communalist parties that he became a subject of ridicule. Other party leaders like Mr. Naidu and L. K. Advani remained in the shadow of A. B. Vajpayee. Vajpayee himself was mocked for his illusions over what was really happening in India.

The greatest political contest of the 2004 general elections was between the Bharatiya Janata Party and its National Democratic Alliance on one side and the Indian National Congress and its alliance of parties on the other side. Both parties had significant business and media backing. It is interesting that in general the incumbent parties received more criticism. If political analysts would have consulted political cartoons rather than public opinion polls they could have predicted the INC-led coalition victory and an anti-incumbency factor working against the BJP and its allies.

All the opinion polls predicted Vajpayee's victory. Vajpayee was praised for his successes in the economy and peace-making. The Congress had been experiencing the split of its state organizations. Yet, the winner was "Sonia Ji." Sonia Gandhi managed to unite allies. The big story of the 2004 elections personified was, of course, "the dynasty." Other issues were the "foreign" origin of Sonia Gandhi and difficulties to deal with Congress. The allies of Congress often were "polluted" by scandals or divided over different issues, even "foreign origin" question itself. Yet, the BJP and its allies got much more criticism. Characters of the Hindu epic well suited the theme of mocking the Hindu communalist party. AIDMK's leader Jayalalita was ridiculed as the despot among her parrot-like ministers and assistants. Many Bollywood stars and the rest of Indian filmdom were ready to showcase their talent for political theater and make-believe. Virtually every day the Bharatiya Janata Party, the Congress and some other political parties were springing their catches on the media and a presumably eager electorate. That big cinema people turned politicians is the traditional thing in the South. With cinema were associated C. N. Annadurai, M. Karunanidhi, M. G. Ramachandran, M. T. Rama Rao, Jayalalitha, Rajnikanth and others. Cinema appearances have become the norm in Indian politics and elections.

The mixture of modern political propaganda and the nostalgia for the "glorious past" found its expression in the Bharatiya Janata Party's "*yatras*." In ancient India *yatras* were movements of the pilgrims to sacred places. Similarly kings traveled to other places with their armies for "*digvijaya*" or conquering the world. Now Hindu nationalist politicians re-introduced *yatras* as the conquering of votes.

The "nationalist" element of the Bharatiya Janata Party's critique of the Indian National Congress's leadership was aimed at Congress's President Sonia Gandhi, Italian by origin. The BJP's anti-Sonia slogan "Ram Rajya, not Rome Raj" was aimed at stressing her foreign origin.

Both major parties, Indian National Congress and Bharatiya Janata Party were unable to win the necessary majority of votes and seats in Lok Sabha without external support, which they sought from the regional parties.

116.6 Regional Parties and Their Attitudes to the Two Main Nationalisms

Regional parties in India mostly developed mostly after 1967 when the Indian National Congress lost its monopoly on power in the southern state of Madras (now Tamilnadu). However, some old parties with regional support claim their origin in the 1920s. The Shiromany Akali Dal and its fractions in Sikh dominated state of Punjab claim origins in 1920s when the movement emerged to take control of Sikh temples and places of congregation (gurdvaras) from priests (Mahants) to the special people committees (Shiromani Gurdvra Prabandhak Committees). Other early southern Indian movements like the Dravidian Progressive Association are considered the predecessor of modern Tamil parties, such as the All India Dravida Munnetra Kazhagam or Dravida Munnetra Kazhagam. All regional parties had to develop their "attitude" towards main nationalist parties. This attitude partly was based on a tactical position towards the Indian capital of Delhi and the central government of the Republic of India, but also had its origins in ethnic, religious and caste compositions of leadership as well as voters in regional parties.

The Sikh component of the Akali Dal ideology could suggest that Sikhs would be very careful about Hindu nationalists who claim that there is no such nation as the Sikhs, but the Sikhs are Hindus as well as Hindus are Buddhists and Jains.

The paradox, however, is that for several reasons major fractions of the Akali Dal were more afraid of the Indian National Congress because of their support in its regional Punjabi branch for the BJP with whom they joined hands in several elections including the 2004 one. In some way it was also due to the traditional Sikh religious history that associated Delhi with evil rule since the time of the atrocities of eighteenth century during the rule of the Mogul Padishah Aurangzeb.

It was also thought that the anti-Brahman All-India Dravida Munnetra Kazhagam may would be more likely to support a secularist government in the center rather than the "rule of the Brahmans" of the BJP. Yet, its alliance with the BJP was based mostly on the tactical expectations of gains from the party then ruling in Delhi. It is also the result of the antagonism with the rival faction, Dravida Munnetra Kazhagam which joined hands with the INC. Dravida Munnetra Kazhagam was not always in good terms with the Indian National Congress, but such alliance was possible due to economic offers for Tamilnadu from the future INC–led government and tactical agreements of INC and DMK over the future ministerial portfolios.

Thus we observe that both nationalist parties during the 2004 elections managed to downplay their differences with their regional allies.

116.7 The Victory of Secular Nationalists in the 2004 Elections and Further Development of Secular India

Indian Secular Nationalism won these elections with the help of the regional parties that were afraid of the Hindu and Hindi-oriented BJP. This election resulted in the re-introduction of the official ideology of Indian Nationalism based on the INC's major contribution in the struggle of India for Independence.

The next parliamentary elections in 2009 were rather predictable and they did not change political situation in the country. The Indian National Congress easily won the elections with the help of its allies because it has allies in secularism or secularist nationalism. In May 2014, however, Hindu Nationalist Bharatiya Janata Party won the Parliamentary elections and its leader Narendra Modi became new Prime-Minister of India. Thus the pendulum moved again in favour of Hindu Nationalism.

116.8 Conclusion

Indian struggle for independence coined the main political party of India – Indian National Congress (INC). It remains the party of Indian secular nationalists. It has, however, includes today regional secular forces with which it joins hands to defeat Hindu nationalists. Hindu nationalists combine in their ideology elements of both nationalist and communal rhetoric. Despite tactical alliances with regional forces the Hindu nationalists have failed after several years in power to regain the ruling status in 2004.

Acknowledgement Research for this article was sponsored by the Special Projects Office, Special and Extension Programs of the Central European University Foundation (CEUBF). The thesis explained here represents ideas of the author and not necessarily those of CEUBF.

References

Andersen, B. (1996). Introduction. In G. Balakrisnan (Ed.), *Mapping the nation* (pp. 1–16). New York: Verso.
Cubitt, G. (1998). Introduction. In G. Cubitt (Ed.), *Imaging nations* (pp. 1–8). Manchester: Manchester University Press.
Gellner, E. (1983). *Nations and nationalism*. Oxford: Basil Blackwell.
Gellner, E. (1996). The coming of nationalism and its interpretation: The myths of nation and class. In G. Balkrishnan (Ed.), *Mapping the nation* (pp. 98–145). New York: Verso.
Jalal, A. (2001). South Asia. In A. I. Motyl (Ed.), *Encyclopedia of nationalism* (pp. 737–757). San Diego: Academic.
Jha, V. M. (2003, February 28). A new brand of history. *Frontline*, pp. 87–89.
Hutchinson, J., & Smith, A. D. (1984). Introduction. In J. Hutchinson & A. D. Smith (Eds.), *Nationalism* (pp. 3–13). Oxford: Oxford University Press.
Khilnani, S. (1999). *The idea of India*. New York: Farrar Straus Giroux.
Malkani, K. R. (2003). *BJP history: Its birth, growth & onward march*. New Delhi. BJP booklet.
Pelinka, A. (2003). *Democracy Indian style. Subhas Chandra Bose and the creation of India's political culture*. New Brunswick: Transaction.
Savarkar, V. D. (1989). *Hindutva. Leiden, 1923* (New ed.). Delhi: Bharati Sahitya Sadan.

Chapter 117
Atheist Geographies and Geographies of Atheism

Barney Warf

Faith is believing what you know ain't so.

(Mark Twain)

117.1 Introduction

Geographers have a long and rich tradition exploring the spatiality of religion, including religious landscapes, religious diversity, and the innumerable ways in which religion as a set of social practices and discourses shapes, and is in turn shaped by, broader social, political, and cultural processes (see Kong 2010 for a review). This literature need not be recapitulated here. Suffice it to say that the spatiality of those who reject religion, including atheists and agnostics, has been remarkably overlooked (but see Wilford 2010). The very absence of works on the geographies of atheism is perhaps testimony to the degree to which religion has become normalized, for nonbelievers typically comprise a small (but growing) share of the population.

The neglect of atheism in human geography is all the more surprising given that the discipline is itself decidedly secular in outlook. This is obviously not to say that all geographers lack religious beliefs, or even spirituality. But the volume of work in geography from an explicitly religious point of view – projects that take faith as their point of departure in their analytical interpretations of the world – is noticeable for its absence. From the positivist "quantitative revolution" of the mid-twentieth century to various poststructuralist accounts today, human geography is overwhelmingly characterized by an implicit, if not explicit, secularism. Other than a minuscule minority such as the Bible Specialty Group of the Association of American Geographers, the discipline analyzes space and place from a perspective that gives little credit to metaphysical or supernatural forces. (The Geography of Religions and Belief Systems also examines religions, but primarily from a secular point of view.) Academic geography, like academics everywhere (Ecklund and Scheitle 2007), is largely practiced by people for whom religion serves a minor intellectual purpose,

B. Warf (✉)
Department of Geography, University of Kansas, Lawrence, KS 66045, USA
e-mail: bwarf@ku.edu

S.D. Brunn (ed.), *The Changing World Religion Map*,
DOI 10.1007/978-94-017-9376-6_117

even if some participate in religious communities for the social and recreational ties they offer.

The analysis of atheism is important in several respects, for it brings to the fore issues such as whether or not secularization is a long-term, apparently inevitable trend. *Time* magazine, perhaps echoing Nietzsche, asked in its April 8, 1966 cover "Is God Dead?," a notion echoed by the influential book by Altizer and Hamilton (1966), which challenged the ability of conventional wisdom to satisfy Americans' spiritual needs. Long marginalized in the West, atheism has recently witnessed new vigor and vitality (Goodstein 2009). A slew of best-selling books on atheism has served as a call to arms for the non-religious and the anti-religious (Dawkins 2006; Harris 2005, 2006; Dennett 2007; Hitchens 2007). And, as will be seen, atheism and related belief systems have exhibited slow but steady growth in their number and proportion of practitioners in the world's economically advanced countries.

This chapter explicates the geography of atheism through several analytical approaches. First, it addresses the question as to whether religion, and metaphysics in general, has undergone a gradual decline in popularity in power; this Weberian interpretation has been hotly contested by those who point to a recent desecularization of social life (Berger 1999). Second, it summarizes the atheist critique of religion, including its logical inconsistencies, anti-scientific worldview, and the frequent social ills for which organized religion has been responsible. Third, the paper offers a brief historical overview of the emergence of secular and atheist thought under varying historical conditions, including Communist attempts to eradicate religion in the twentieth century. Fourth, it turns to global geographies of atheism and secularism as a means of examining how various secular forces have left an indelible geographical impact that altered the religious life of billions of people. The fifth section focuses on Europe, long held to be the apex of Weberian secularism. Sixth, it examines the United States, a country long judged to be "exceptional" in terms of its religiosity, but in fact one much more similar to the world norm than is Europe. The conclusion ties these themes together and points to the promise of a secular future.

117.2 Religion in Decline? Secularization in Historical Perspective

Ever since the Enlightenment, religion – at least in the West – has suffered a long, slow, and steady retreat. Since the sixteenth century, on the heels of ascendant capitalism, European culture witnessed the separation of church and state and the rise of an alternative, secular set of institutions (for example, universities). Religion, which held a monopoly over cultural, legal, and political power in Europe for a millennium, as well as much of the Muslim world, has seen a gradual decline in the number of believers, church attendance, its public visibility, and the depth of religiosity in everyday life. As Christianity has incrementally lost much of its ideological power and political clout, it has opened a space for secular and scientific thought. From the Copernican revolution to Darwin to the flourishing of secular social sciences and humanities in the twentieth century and today, alternatives to religious worldviews abound in power and popularity.

Of course, during the long millennium of the medieval era, the very idea of secular thought was beyond the horizons of the thinkable. Renaissance humanism, including figures such as Petrarch, Dante, Erasmus, and Montaigne, while still highly religious in inspiration, nonetheless made great strides in putting humans, rather than god, at the center of intellectual and much popular thought (Johnson 2002). The Copernican revolution assisted greatly in revealing the vastness of the universe and the lack of Earth's centrality in it. The printing press, rising literacy levels, and the scientific breakthroughs of the Enlightenment all contributed mightily to undermining the legitimacy of religion, the intellectual monopoly of the church, and the rise of secular alternatives such as universities. Reason and verification gradually came to rival blind faith as criteria of what counts as important and reliable knowledge. The empiricism of the British Enlightenment, including Francis Bacon, John Locke, and David Hume, was also an important milestone in the triumph of evidence over faith as the primary criterion for defining valid knowledge. In the seventeenth century, Baruch Spinoza's pantheism, which held that god was a social creation rather than a divine being, bordered on atheism, and he was excommunicated by both the Jewish and Catholic faiths. Socially, the ferocious anti-clericalism of the French Revolution marked the ascendency of militant secular thought, which included the closing of churches, execution of clergy, and prohibition of the Mass, anticipating in some respects the virulent and horrifying state-enforced atheism of the Soviet Union and Communist China in the twentieth century. In the nineteenth century, Darwinism and socialist thought, particularly Marxism historical materialism, posed severe challenges to the credibility of religious thought.

The orthodox, dominant explanation of secularism originates in the works of the famed sociologist Max Weber (Hughey 1979; Weber 1922/1991). Weber's argument began with his observations that the so-called Protestant ethic, which he held up as the motivating cause of capitalism, enticed believers into working hard, delaying gratification, and accumulating savings (i.e., capital) as a sign of god's grace, that is, the likelihood of entry through the pearly gates. To succeed as incipient capitalists, adherents had to become increasingly "rational" in outlook and behavior, adopting logical, even scientific norms. Rational secularization was accelerated by the growth of the market, with its unforgiving laws of profit maximization, and through the impersonal bureaucratization of politics and collective decision-making. Moreover, in a secular capitalist culture the commodity displaces god as the object of holy attention, a process Weber called the "disenchantment of the world." The very rise of capitalism, he correctly maintained, gradually lowered an "iron cage" of rationalism over the very culture that created it, squeezing religion into the domain of the irrational. The commodification of daily life and the adoption of explicit, impersonal rules of bureaucratic behavior entailed the steady decline in the appeal of religious authority. "Victorious capitalism, since it rests on mechanical foundations, needs its support no longer" (Weber 1904/1930: 181). Capitalism thus originated in religion but ended up cannibalizing that which gave birth to it, leading to an increasingly secularized society.

Secularization is a complex process fraught with ambiguity, multiple meanings, setbacks, and contradictions (Martin 1979; Bruce 2002; Taylor 2007). It may refer to the rise of non-religious institutions (including ones formerly given over to religion, such as Harvard University), the gradual displacement of religious authority through the growth of state services, declining church attendance, a less visible role for religion in public life, and changes in the outlook and behavior of residents that emphasize rational, non-metaphysical interpretations. Its aura of inevitability owes much to Parsonian structural-functionalism (Swatos and Christiano 2000), which firmly sutured secularization to the historical process of modernization. Religion, in this reading, was a hidebound relic of a premodern world, condemned to die a gradual, agonizing death in the inevitable march to a secular, modernized world.

While Weberian secularization continues to be the dominant lens through which secularization is viewed, it suffers from several empirical and conceptual flaws. Today, the argument of inevitable secularism faces stiff intellectual resistance. First, it is misleading to assume that secularism is some sort of unstoppable, teleological process. Quite the contrary: religion has stubbornly refused to disappear. Even as organized religious authority has suffered a slow, ignominious decline, religious belief has not (Chaves 1994; Berger 1999). In much of the world, especially since the end of the Cold War, a widespread revival of religious fundamentalism has occurred, largely as a backlash against globalization and its disruptive, largely secular impulses (Barber 1995; Stump 2000). Others maintain that religiosity has not decreased and that a broader resurgence of desecularization has even reversed the historical decline of religion (Berger 1999). Weberian secularism also suffers from intellectual and factual problems: it fails to explain the different trajectories of economically advanced societies in Europe and the United States (about which more later). Modernity, rather than fostering secularism, instead creates pluralism, a vast heterogeneity of life-worlds and outlooks that stands in contrast to the ideologically homogeneous universes in which the bulk of humanity has lived and continues to live. In modern societies, religions lose their ideological (and often political) monopolies and must compete with one another for adherents (Finke and Stark 1988, 1998; Stark & Finke 2000; Berger et al. 2008). In short, unlike Weberianism and its Parsonian offshoot, secularization should be seen as contingent, reversible, socially and geographically uneven, and intimately tied to changing global, national, and local economic, political, and ideological forces.

117.3 Faith Is an Excuse Not to Think: The Atheist Critique of Religion

Atheists are those who deny the existence of god(s). In this respect they differ from agnostics, who hold open the possibility of a deity, and other non-believers such as "freethinkers," a nebulous group that largely skirts the issue of god's existence. As Harris (2006: 51) argues, "Atheism is not a philosophy; it is not even a view of the world; it is simply an admission of the obvious." Some erroneously label atheism another religion, which is like calling abstinence a sexual position.

For atheists, Judaism and Christianity amount to little more than a set of Bronze Age fairy tales about a big ghost in the sky, a tale of the supernatural grounded in one magic book written by a nomadic people over a millennium and another book written several generations after the death of Jesus. It is difficult for atheists to take seriously faiths that include outlandish claims such as: talking snakes; a man born of a virgin was executed, came back to life, and can be eaten as a cracker; blessed wine can miraculously become blood (Catholicism); that a book written in the Neolithic is a reliable guide to the present; that handling poisonous snakes reveals divine protection; that the faithful will magically fly through the air during the Rapture (various Protestant cults); that god lives on the planet Kolob and wants people to wear sacred underclothes (Mormonism); that sins can be transferred to animals (Judaism); that there are 330 million gods; that one's karma is trapped in a social caste and cannot escape except through death; that the recycling of souls can transform one's dead grandmother into a cow (Hinduism); that paradise awaits martyrs with 72 virgins and 80,000 servants (Islam); or that a galactic warlord named Xenu brought billions of people to earth and killed them with hydrogen bombs (Scientology). Choosing among different gods on rational grounds strikes atheists as ludicrous: what makes the Christian or Muslim god preferable to the Flying Spaghetti Monster? For most people, the religions of others are silly, and they tend to believe that their specific theological interpretation alone has a monopoly over divine truth: atheists simply extend this line of thought to include all metaphysical beliefs. As Dawkins (2006: 150) puts it, "We are all atheists about most of the gods that humanity has ever believed in. Some of us just go one god further."

Atheism asserts that god's omnipotence is a logical impossibility (can god build a stone so heavy that even god cannot lift it? Who created god?), and a moral dead end: "If God exists, either He can do nothing to stop the most egregious calamities, or He does not care to. God, therefore, is either impotent or evil" (Harris 2006: 55). Similarly, "any God who would grant prayers for football championships, while doling out cancer and car accidents to little boys and girls, is unworthy of our devotion" (Harris 2006: 114). Atheists range in their attitudes from those who quietly assert that religion need play no role in their lives, but respect the rights of the religious to their beliefs, to those who militantly assert their antipathy to all religions. Atheists should not be confused with misotheists, who actively hate god (Schweizer 2010); to hate god, one must believe that he/she/it exists. Moreover, to view the world as an atheist is not to dismiss religion as unimportant, for atheists know full well the tremendous power that religion plays in the lives of most people.

From the atheist perspective, religion would be comical if not for its enormous destructive consequences. Hitchens (2007) thus argued that religion poisons everything that it touches. For example, ascendant Christianity, during the waning days of the Roman Empire, carefully and systematically murdered the Greek tradition of scientific inquiry (Freeman 2002). Religion has long fostered, and depended upon, willful ignorance and superstition, cultivating a hatred of those who believe differently (often even those with minute differences of theological interpretation), or those who do not believe in religion at all. Most religions regard women as second-class people and legitimate patriarchal and misogynistic forms of domination. For a millennium, feudal Europe suffered under the yoke of a Christian theocracy, during which time it lagged technologically and intellectually well behind its counterparts in the Middle East, India, and China.

Historically, religion has been at the heart of innumerable cruel, violent conflicts, including the Crusades, the mass murders of the counter-Reformation, the Inquisition, the legitimation of colonial conquests and genocide, and contemporary struggles in Israel/Palestine, Northern Ireland, Lebanon, Yugoslavia, Cyprus, Nigeria, Sri Lanka, Gujarat, and Kashmir, among others. Muslim fundamentalism was vital to the September 11, 2001, attacks and the rise of the Taliban. Without religion, there would be no religious wars, jihads, or pogroms against Jews, suicide bombings or fatwas, honor killings or female genital mutilation, apocalyptic cults, witch and widow burnings, stonings of adulterers, bombings of abortion centers, or priestly pedophilia. As Blaise Pascal famously said in 1670, "Men never do evil so completely and cheerfully as when they do it from religious conviction."

More broadly and pervasively, religion almost always reifies a social order, presenting humanly constructed hierarchies (particularly poverty) as natural and immutable, thwarting possibilities for effective social change. All too often religious authorities have ratified claims by political elites that their power is divinely mandated, such as the feudal "divine right of kings." For a millennium, Christianity greatly retarded European scientific thought. Religious believers have long waged a war against scientific inquiry, ranging from the Catholic Church's attempts to stop the rise of the Copernican perspective on the universe in the sixteenth century to contemporary efforts to deny evolution, astronomy's big bang, or embryonic stem cell research. Religions throughout history have sought to control way people eat and drink, what they believe, their health and dress, and repress their sexuality, particularly that of women and gays/lesbians. At times, various religions have celebrated slavery and child sacrifice. Christian fundamentalists have frequently and vehemently opposed contraception, birth control, and women's reproductive rights, which facilitates the spread of sexually transmitted diseases and unwanted pregnancies, while Muslims in Nigeria and Afghanistan have sought to prevent polio vaccines. At various times, religious extremists have opposed the introduction of new cultural forms, including jazz, rock and roll, and novel types of dance, policing the pleasures of others in order to impose their own narrow moral vision. Many apostates, agnostics, and atheists view religion as a wet blanket of conservative conformity laced with guilt and self-righteousness. As Hitchens (2007: 17) argued, "It may speak about the bliss of the next world, but it wants power in this one."

Of course, this is not to say that all religious people are ignorant or anti-scientific; indeed, many people of faith are tolerant and well-intentioned, and a few manage to reconcile religion and science. Occasionally, religion has been responsible for acts of enormous charity, rigorous educational programs (for example, the Jesuits), and the construction of beautiful and powerful artistic and architectural masterpieces (for example, cathedrals and mosques). For many people, religion is a comforting source of solace in the face of a bewildering and often terrifying world, offering an off-the-shelf explanation of suffering and death. But these contributions are far outweighed by religion's devastating social and psychological costs.

As an institutional force, often backed by the power of the state, religion to varying degrees has more often than not served reactionary purposes that minimize the power of human beings to make their histories, geographies, and lives in self-conscious and emancipatory ways, encouraging them to live in fatalistic fear of and submission to a petty and vengeful god, or at best, an imaginary friend, who oversees their every move. Marx (1844) famously wrote in *Contribution to the Critique of Hegel's Philosophy of Right* that "Religion is the sigh of the oppressed creature, the heart of a heartless world, and the soul of soulless conditions. It is the opium of the people." Bluntly, religion is the triumph of faith over reason, the denial of fact-based, logical, and rational inquiry. As one anonymous wag put it, "religion is philosophy with the questions left out" (Hitchens 2007: 278).

117.4 Historical Geographies of Atheism

Atheism has a long, rich, but largely unknown historical geography (Martin 2006; Hyman 2010). Several "free thinkers" in classical Greece openly expressed their doubts about the existence of god, including Diagoras (known as "the Atheist") of Melos, Democritus, and Epicuras, who said

> Either God wants to get rid of evil, but he can't; or God can, but doesn't want to; or God neither wants to nor can, or he wants to and can. If God wants to, but can't then he's not all-powerful. If he can, but doesn't want to, he's neither all-loving. If he neither can nor wants to, he's neither all-powerful nor all-loving. And if he wants to and can – then why doesn't he remove the evils? (Ranke-Heinemann 1992: 60)

Although some argue that atheism in the modern sense did not exist prior to the seventeenth century (Davidson 1992), others point to the rise of the humanist movement during the Renaissance. For example, the term "atheist" first came into being during sixteenth century France, although its original meaning was limited to an insult. The German critic of religion Matthias Knutzen was apparently history's first self-described atheist, in 1674. A very partial list of some famous atheists in history includes: Voltaire; Thomas Paine; Denis Diderot; Jeremy Bentham; Friedrich Nietzsche; Karl Marx; Sigmund Freud; Auguste Comte; Ludwig von Feuerbach; Charles Darwin; Edgar Allen Poe; Walt Whitman; Thomas Edison; H.G. Wells; Peter Kropotkin; Bertrand Russell; W.C. Fields; Diego Rivera; Pearl S. Buck; George Orwell; Ernest Hemingway; Ayn Rand; Jean Paul Sartre; Katherine Hepburn; John Lennon; Isaac Asimov; Carl Sagan; Arthur C. Clarke; and Michel Foucault. Some contemporary examples include: Billy Joel; Noam Chomsky; Oliver Sacks; Ted Turner; Gore Vidal; Woody Allen; Brad Pitt; and Steven Pinker. A large share, if not the majority, of the world's scientists are agnostic or atheist.

In nineteenth century Europe several currents propelled atheist thought forward. Friedrich Nietzsche announced that "god is dead," proclaimed himself the anti-Christ, and set the stage for later generations of relativist forms of moral and social thought. Auguste Comte paved the way for a secular science of sociology.

Marxism likewise emerged as the most serious, organized body of secular social thought to challenge the legitimacy of religion, a social counterpart to the natural sciences. Historical materialism, which arose in part in reaction to hegemonic German idealism, sought to lodge the dynamics of human affairs firmly in this world, focusing on real-world economic, political, and cultural processes rather than metaphysical ones. In the twentieth century the Vienna Circle of positivism, and related intellectuals such as Bertrand Russell, firmly rejected religion as metaphysics unsubstantiated by empirical evidence.

Twentieth century atheism is frequently discussed in terms of the triumph of Communism in the Soviet Union, China, Eastern Europe, Mongolia, North Korea, Vietnam, and Cuba. State-enforced atheism, rather than that which purchases its legitimacy from intellectual debate, is quite a different animal from its intellectual counterpart. Viewing religion as bourgeois decadence and a challenge to the power and authority of the secular state, Communists in the USSR and China attempted the most wholesale liquidation of religious faith in world history. For seven decades, atheism reigned as state ideology in the Soviet Union and client states, imposing incalculable human suffering and loss of cultural heritage. Under Stalinism, the atrocities against people of faith are too numerous to recall: religious instruction was prohibited; numerous places of worship were closed; tens of thousands of priests and nuns were imprisoned, tortured, or executed; entire religious communities were terrorized, moved to distant regions and/or annihilated; schools spewed forth anti-religious propaganda; and the populace was virtually forced to give up its religious beliefs under the threat of severe sanctions (Majeska et al. 1976; Pospielovsky 1987; Husband 1998; Simon 2009). By 1940, 97 % of Orthodox churches open in 1916 had been closed (Froese 2008: 9). In Central Asia, a nearly identical strategy was deployed against the region's Muslims. Similarly, the Soviet-backed regime in Mongolia destroyed thousands of Buddhist temples and executed monks.

Likewise, in the People's Republic of China, state-sponsored atheism was manifested in the suppression of its Buddhist heritage (Zuo 1991; Potter 2003), particularly in Tibet, including the murder of more than one million people and the ransacking or obliteration of nearly all of its 6,000 monasteries during the Great Leap Forward and the Cultural Revolution (Lopez 1999). The Chinese state has also harshly repressed Muslims in Xinjiang and the Buddhist movement Falun Gong. In Cambodia under the Khmer Rouge in the 1970s, Buddhists and other people of religious faith faced wholesale extermination.

Communist atheism imposed by force amounts to tyranny of the worst kind, and its legacy undermines atheism as a contemporary intellectual and social movement. From a progressive political stance, to deny the existence of god is not the same as denying freedom of religion, which rarely works in any case. The failure of state-sponsored secularism in the communist world, as reflected in the recent resurgence of the Orthodox Church in Russia and Buddhism in China, points to the cruelty of Communism's rulers more than it does to the intellectual legitimacy (or lack thereof) of atheism as a worldview. It is as inaccurate to tar atheism with the crimes of the USSR and China as it is to paint all Christians as anti-Semites or bloodthirsty Crusaders: as Baggini (2002) notes, "If the Soviet Union provides some kind of refutation of atheism, then atrocities such as the crusades or inquisitions would likewise refute Christianity."

In contrast to the dominance of religion in the lives of most people and most societies, atheism has long been confined to the social and political fringes of most societies. Atheists have long suffered, and continue to suffer, discrimination, imprisonment, exile, and even execution at the hands of religious authorities. Plato argued that atheism constituted a social danger and should be legally punishable (Gey 2007). Under the Inquisition, atheism was a crime similar to being a Jew or Muslim, and some accused of the sin of not believing in god were executed in gruesome ways, such as Italian priest-turned-agnostic Lucilio Vanini and Polish nobleman Kazimierz Łyszczyński (Davidson 1992). In the nineteenth century, atheist Charles Bradlaugh was elected to the British Parliament but was prohibited from taking office. Even today, atheists often suffer discrimination at work, are shunned in some communities, and have been harassed, threatened, and publicly insulted for their beliefs, which have been equated with devil-worship, communism, and immorality (Edgell et al. 2006).

Atheists and agnostics tend to be relatively well educated, and the growth of public school systems and mass literacy since the nineteenth century is undoubtedly one driver behind the gradual increase in the number of non-believers. As Zuckerman (2012: 11) notes,

> High educational attainment probably makes it more difficult to maintain a strong belief in one's religion when learning about the social construction of religion, the history of religious development, the diversity of religious claims, and the inability of many religious claims to be supported by scientific verification.

Indeed, atheists are often even more knowledgeable about religion than many of the self-professed religious (Goodstein 2010)! Secularists tend to be progressive politically, including: open-mindedness to cultural diversity; support for the rights of women and ethnic and sexual minorities; a concern for power, inequality, and social justice; compassion and empathy for the socially marginalized; an interest in science and scientific issues; and skepticism of received authority, including churches and governments.

Atheists and apostates are a heterogeneous bunch. Some were raised as children of atheists, while others left their faiths, sometimes at an early age (typically late adolescence or during college) or sometimes in mid-life. Some are induced to give up god because they find religious myths to be unconvincing (if everyone is descended from Adam and Eve, aren't we all products of incest?) or because religious teachings contradict the evidence from astronomy, geology, biology, history, and archaeology. Leaving one's faith often involves a wrenching emotional decision that isolates apostates from their communities and families; some understandably maintain social and cultural links to their religious traditions even if they no longer believe in them (Zuckerman 2012). Occasionally people of faith, including preachers, become radical advocates of atheism (for example, Loftus 2008; Barker 2008). Whereas much received opinion holds that religion is essential to the development of a moral worldview, in fact the vast majority of atheists do indeed have a strong sense of morality without god (Hunsberger and Altemeyer 2006; Aronson 2008; Maisel 2009; Smith 2010). As Epstein (2009: x) notes, "To suggest that one can't be good without belief in God is not just opinion, a mere curious musing – it is a prejudice."

117.5 Contemporary Global Geographies of Secularism and Atheism

Roughly a billion people, or 15 % of the planet, may accurately be described as nonreligious (Epstein 2009), although not all of them are atheists. Using data on self-identified followers of religions in 2010 from the World Christian Database (www.worldchristiandatabase.org/wcd/), it is possible to ascertain the distribution of this population across the world (Fig. 117.1). The geography of this population reflects several important world historical forces, including the differential distributions of wealth and poverty and the lasting influence of Communist state-enforced atheism (Warf and Vincent 2007). Most obviously are the substantial proportions of the population of countries such as North Korea (56 %), Russia (26 %), and China (40 %), where secularism is prominent largely due to the hideous enforced atheism of politically repressive regimes.

The Muslim world, typified as it is by widespread poverty, relatively low levels of education, frequently high levels of religious uniformity, and the lack of an historical uncoupling between church and state (Lewis 2003), not surprisingly exhibits low degrees of secularism. Exceptions include states in Central Asia, where the legacy of Soviet atheism endures. Even in the regions dominated by Islam, however, pockets of secularism exist (Marranci 2010) and even some apostates can be found, although rarely publicly (Warraq 2003).

Notably, global geographies of wealth and poverty are also important here. For individual atheists, personal tragedy or unanswered prayers are often key to

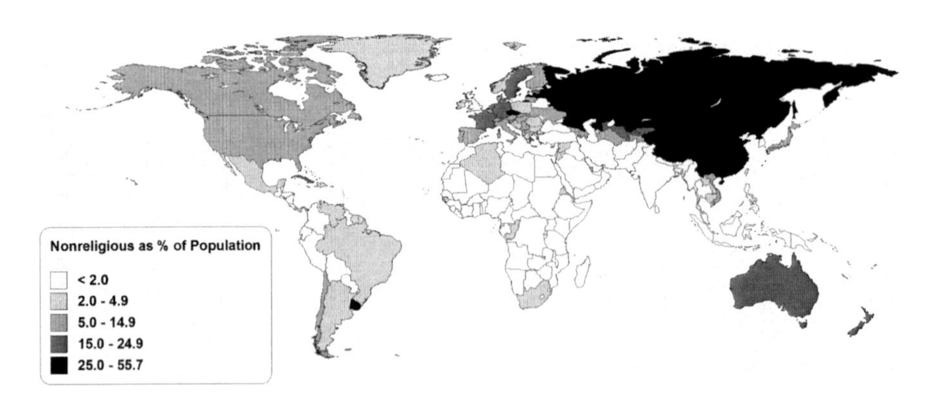

Fig. 117.1 The nonreligious as percent of population, 2010 (Map by Barney Warf with data from World Christian Database, www.worldchristiandatabase.org/wcd/)

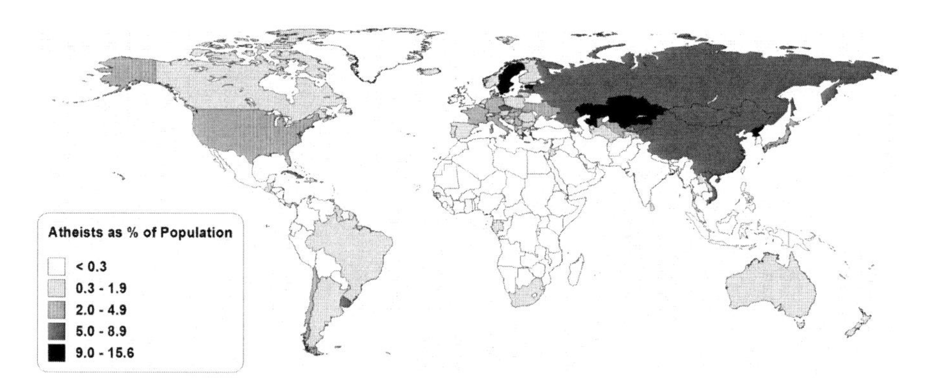

Fig. 117.2 Atheists as percent of population, 2010 (Map by Barney Warf with data from World Christian Database, www.worldchristiandatabase.org/wcd/)

their road to apostasy; however, at the scale of societies as a whole, severe social problems are closely linked to high degrees of religiosity. Thus secularism is least present in most of Africa, South Asia, Indonesia, and the poorest parts of South America. As Norris and Inglehart (2004) demonstrate, it is societies plagued by chronic insecurity, war, hunger, unemployment, poverty, and disease that are the most religious on earth. Zuckerman (2012: 54) contrasts the differing causes of secularism at the social and individual scales, noting that they comprise "a potential paradox. On the micro, individual level, personal misfortune can serve as a direct cause of apostasy. Yet on the macro, societal level, an abundance of misfortune seems to strengthen religiosity, not weaken it."

Self-described atheists comprise a small share of the world's population, and are unevenly distributed (Fig. 117.2). In Stalinist North Korea, 16 % of the population describes itself this way; other notable countries include Sweden (12 %), Estonia and Kazakhstan (10 % each), and China (8 %). Throughout the Muslim world, India, and most of Latin America atheists are virtually invisible, almost always forming <1 % of the population. Among India's 1.2 billion people, for example, atheists comprise only about 1.2 %: thus, "In terms of religion, India and Sweden can serve to mark the antipodes of religiousness and secularity. The American situation can be described as a large population of 'Indians' sat upon by a cultural elite of 'Europeans'" (Berger et al. 2008: 12).

117.6 Europe: Culmination of Weberian Secularism or World Exception?

Europe, with an historical legacy of social democracy and the enduring legacy of traumatic religious strife, has essentially largely become a secular continent, in marked contrast to the United States (Davie 2002). The proportion of Europeans

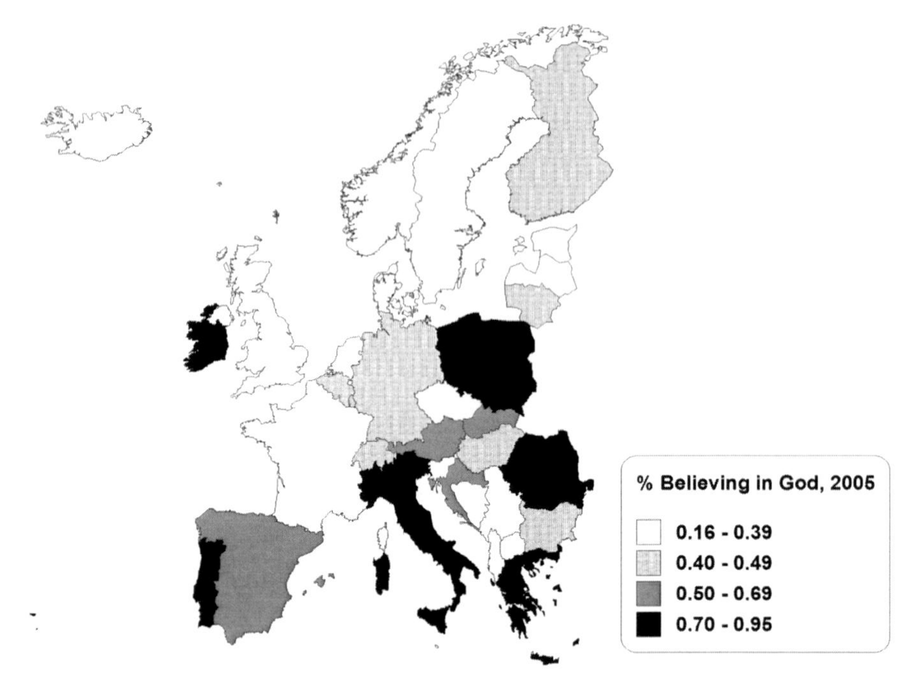

Fig. 117.3 Percent of Europeans believing in God, 2005 (Map by Barney Warf, adapted from Eurobarometer 2005)

who say they believe in god, 53 % in 2005, is among the lowest in the world (Eurobarometer 2005). Obviously there are widespread differences across the continent in terms of the prevalence and severity of religiosity and secularism, with stronger degrees of secularism in northern Europe than predominantly Catholic Southern Europe (Fig. 117.3). In Western Europe, Christianity has faced a steady diminution of its power, popularity, and respectability since the Enlightenment (McLeod and Ustorf 2003). The decline in faith has been uneven across the continent. Scandinavia, for example, has become a series of essentially secular societies (Zuckerman 2009); to a lesser extent, so has Germany (Shand 1998). Christianity in Britain has faced a long, slow decline (Brown 2001; Bruce 2001; Crockett and Voas 2006), including falling church attendance and ever-fewer numbers of people who say that religion makes an important difference in their lives. The highest rates of religiosity tend to be found in Romania, Greece, and heavily Catholic countries such as Ireland, Poland, and Italy. Eastern Europe, of course, suffered under decades of state-enforced atheism during the Soviet occupation following World War II (Froese 2008). In Estonia, only 16 % of the population believes in god. Despite official attempts to eradicate religion, churches continued to exist, and following the fall of the Iron Curtin in 1989, have often thrived, as seen, for example, in the enduring power of the Catholic Church in Poland.

Many religions in Europe find it increasingly difficult to recruit adherents. Rather than a lack of demand for religious services, Stark and Iannaccone (1994) offer a "supply side" interpretation of the continent's secularism, arguing that European churches, coddled by their governments and facing little competition, have become lazy, dysfunctional, and incapable of attracting a widespread audience. From the Catholic monopoly in Southern Europe to state funding for the Lutheran and Anglican churches in northwestern parts of the continent, churches have received government subsidies and support to an extent unfathomable in the United States. In Spain, for example, laws allowing Protestants to hold political office and to organize schools only were passed in the post-Franco era of the late 1970s. In Sweden, state taxes pay the salaries of Lutheran clergy, and in Denmark the Parliament still retains official administrative power over the Lutheran church. Thus, Stark and Iannaccone (1994: 241) conclude "that where regulation stifles competition among religious firms, religious participation also is stifled."

Eurosecularity has in many respects become the Parsonian model for the rest of the world "to follow." European politics generally lacks the religious dimensions found in the U.S., including an aggressive evangelical movement (Norris and Inglehart 2004). The constitution of the European Union does not mention god (for that matter, neither does the American). Berger et al. (2008) attribute Europe's divergence from the United States, both of which are economically advanced societies, to a series of factors, including: two interpretations of the Enlightenment legacy (for example, French radical secularism following the Revolution of 1789); different state policies toward religious organizations (including subsidies and tax policies); the relative absence or prevalence of religiously-oriented educational institutions (for example, nationalized European systems versus local control of schools in the U.S.); varying attitudes towards intellectuals and their roles in public life (particularly widespread American anti-intellectualism); and contrasting ways in which religion is closely associated with class in Europe, but much less so in the U.S. Thus, rather than the outcome of some universal process of secularization, the lack of religiosity and religious power in Europe results from its unique political and cultural circumstances: from this perspective, Europe is secular not because it is modern but because it is European, that is, the path-dependent result of centuries of historical change. Of course, Europe's religious pluralism and tolerance are being sorely tested given the growing, and often radicalized, Muslim minorities in its midst.

117.7 American Exceptionalism? Religion and Atheism in the United States

The United States is often regarded as the prominent exception to the Weberian secularism thesis, that is, a wealthy, post-industrial country that is still highly religious. Religion is deeply woven into the DNA of American culture and politics, and plays a role in public life unparalleled in the rest of the economically advanced world.

The founding myths of Euro-American culture typically point to the Pilgrims and Puritans, the notion of America as the New Zion or "city on the hill" (Hertzke 1988). While religious fundamentalists claim the country's Founding Fathers as fellow co-religionists, in actuality most were Deists, with a vague and typically unspecified set of religious beliefs that minimized the importance of organized faith: Thomas Jefferson, John Adams, James Madison, Thomas Paine, and Benjamin Franklin are examples, and their influence was critical in establishing the new country's legal boundary between church and state (notably the First Amendment). The Constitution does not mention god. As Alexis de Tocqueville observed in the 1830s, religion flourished in the United States in a way that it did not in Europe, which he attributed to the legal separation of church and state.

The U.S. is perhaps the most religiously diverse society in the world (Eck 2001; Warf and Winsberg 2008). Without state backing, a multitude of different faiths compete and jostle with one another for adherents and resources, including Catholics, a wide variety of Protestant sects, Mormons, Pentacostalists, and small minorities of Jews, Muslims, and Buddhists.

The country is not only religious diverse, it exhibits high levels of religiosity (the latter in no small part due to the former), in sharp contrast to all other industrialized countries. Roughly 90 % of Americans say they believe in god (Newport 2011). About 44 % of Americans attend church weekly, a higher proportion than almost all of Europe with the exceptions of Ireland, Poland, and Italy (Table 117.1). Nine out of ten Americans say they pray at least some of the time, 58 % claim they

Table 117.1 Percent attending church weekly or more often, 2008

Country	%
Ireland	84
Poland	55
Italy	45
United States	44
Canada	38
Austria	30
United Kingdom	27
Spain	25
France	21
Hungary	21
Romania	20
Switzerland	16
Bulgaria	10
Norway	5
Denmark	5
Sweden	4
Finland	4
Russia	2

Data source: World Values Survey www.wvsevsdb.com/wvs/WVSData.jsp

pray daily (Zuckerman 2012: 45), and 42 % explicitly reject evolution. Zuckerman (2012: 21) details the degree to which metaphysics and superstition permeate American life to a degree that resembles medieval Europe:

> 75 % of Americans believe in miracles, 75 % believe in the existence of heaven, 71 % believe that Jesus was born of a virgin, 58 % believe in the existence of the Devil, 31 % believe in the existence of witches, and 71 % believe in the existence of angels.

There are several historical forces that explain the uniquely high level of American religiosity. Immigrant groups, carrying their faiths from their countries of origin, often founded churches as central points of social support, particularly given the intense individualism and spotty nature of the public safety net in the country. In contrast to Europe, in which many churches still benefit from public subsidies, the constitutionally enshrined separation of church and state in the U.S. created a relatively free arena in which different faiths compete and flourish. Sociologists of religion (Finke and Stark 1988, 1998; Bruce 2002) liken the competition among faiths to that of firms seeking to maximize market share; adherents are comparable to consumers choosing among different brands of products. In short, the First Amendment obliges different religions to compete with one another for followers, a process that elevates religiosity in its political and social significance. As Tiryakian (1993: 45) noted, "religion is usually better off, in terms of its social vitality, in societies where it is not a state-regulated monopoly." However, given how widespread religion is throughout most of the world, the U.S. is not so exceptional after all; indeed, Europe, and perhaps Japan, are the secular exceptions.

More recently, fundamentalist religion has surfaced as a potent political force given the influential role of conservative Christian activists within the Republican Party (Wilcox 1991). Ironically, conservative religious activism has helped to fuel the growth of American secularism: Campbell and Putnam (2012: 34) state

> As religion and politics have become entangled, many Americans, especially younger ones, have pulled away from religion. And that correlation turns out to be causal, not coincidental.

Given the deep historical roots of religion in the U.S., the high degrees of public religiosity and the ferocity of politically conservative Christians, it is no surprise that American atheists are a reviled and highly distrusted group (Gervais et al. 2011). Half of Americans say they would never vote for a qualified atheist for president (Epstein 2009) or would want their daughter to marry one, and atheists were less popular or likely to be accepted than other marginalized groups such as gays or Muslims (Edgell et al. 2006). Atheists are regularly portrayed and regarded as angry, loveless, purposeless, and even sociopathic (Zuckerman 2012: 135). Atheism is so unpopular in the U.S. that is often considered legitimate to openly discriminate against them, or practice atheophobia. The constitutions of several states bar atheists from holding elected offices, including those of Maryland, North Carolina, South Carolina, Mississippi, Tennessee, and Texas, although these provisions have been largely overturned through legal challenges. In 1987, running for president, George H.W. Bush reportedly said "I don't know that atheists should be considered as citizens, nor should they be considered patriots" (Dawkins 2006: 43). The Boy Scouts of America does not allow atheists as members. Atheism has also been used to disqualify parents in divorce custody hearings. No other group suffers discrimination in such quantities: apparently, increasing acceptance of religious diversity does not extend to the nonreligious. Even worse, the political timidity of many secularists, in contrast to the loud activism of religious conservatives, amplifies the problem.

However, even in the godly U.S. organized religion is on the wane. Weekly church attendance, for example, has declined gradually (Chaves and Stephens 2003). The data on this issue yield different estimates, but point to a common trend. The World Values Survey, for example, holds that weekly church attendance dropped from 49 % in 1990 to 44 % in 2008, whereas American National Election Studies (2009) maintains that the proportion declined by 38 % in 1970 to 23 % in 2008. The proportion of Americans who self-identify as non-religious, agnostic, or atheist varies over time, place, and according to the source. In 2000, 14 % of the population indicated they had no religious preference, although this includes some who also say they believe in god. Only 3 % affirmed "they do not believe in god" (Hout and Fischer 2002), only 1 % explicitly called themselves atheists, and another 4.1 % maintained they did not know whether or not god existed. However, these numbers are changing rapidly: since 1990, the percentage of Americans who do not identify with any religion has doubled, to 15 %, and the proportion self-identifying as atheists rose from 4 % in 2003 to 10 % in 2008 (Zuckerman 2012). Only 53 % of Americans born after 1981 say they believe in god.

American secularism has a short but interesting history (Jacoby 2004). Thomas Paine, technically a freethinker who argued that Christianity, like all religions, was a product of people rather than god, was the first reviled as an atheist. In the nineteenth century, various Freethinkers, often influenced by socialism and Darwinism, mounted sporadic challenges to organized religion, including the famous orator and atheist activist Robert Green Ingersoll, Samuel Porter Putnam, and Charles Lee Smith. In the 1930s, the American Humanist Association began. In 1963, American Atheists was founded by Madalyn Murray O'Hair, an outspoken and often counterproductive advocate of secularism; she and her son were kidnapped and murdered by an employee in 1995.

There are no precise data on the locations of American atheists. The Census does not collect data on religious affiliation, and thus other sources must be used. This discussion utilizes the census on the nation's religions in 2000 published by the Glenmary Research Center (2000), which lists numbers of adherents by denomination by county. The 45 million Americans who state that they do not identify with any formal religion (not all of whom are atheists) are unevenly distributed across the country (Fig. 117.4). Several major clusters are evident, including parts of the Pacific Northwest, Colorado, Michigan, Florida, and Alaska. In contrast, monotheistic regions such as the Baptist Bible Belt in the South and the Lutheran-dominated upper Midwest tend to exhibit few proportions of the non-religious. As noted, however, American atheists are growing rapidly. Between 1990 and 2008, their numbers rose from approximately nine million to 31 million and their proportion more than doubled to 10 %; these increases were unevenly distributed across the country, with the highest rates of growth found in New England (not coincidentally the country's

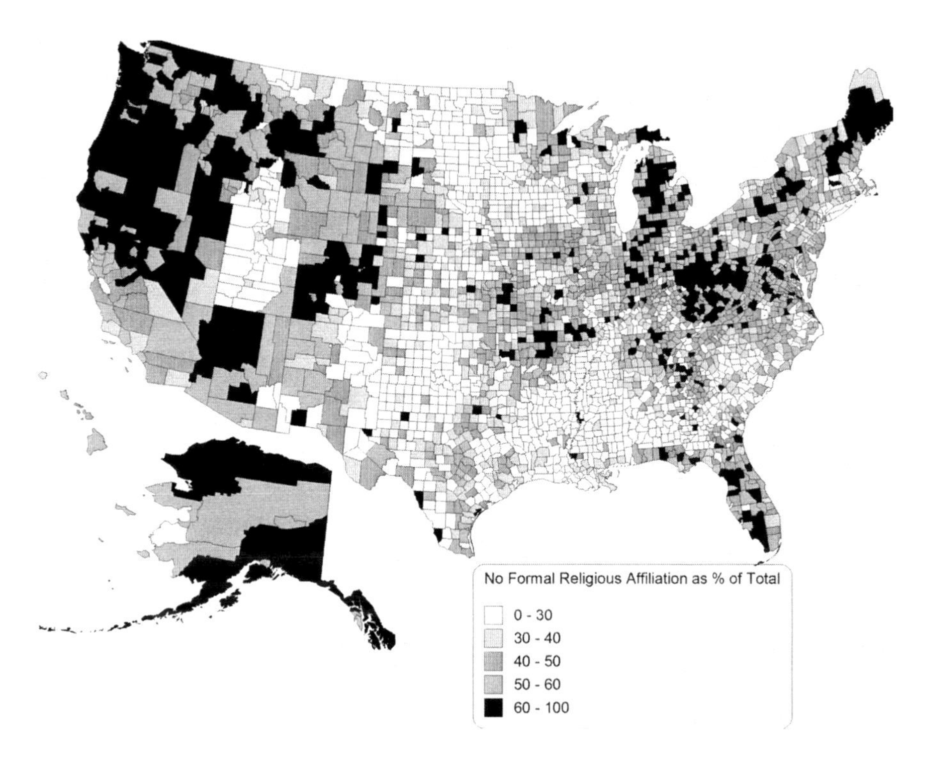

Fig. 117.4 Americans with no formal religious affiliation as percent of total, 2000 (Map by Barney Warf with data from Glenmary Research Center 2000)

best educated region) and parts of the West while the lowest rates of increases were concentrated in the South (Fig. 117.5).

There are, of course, few to no explicitly atheist places. Some locations, such as universities, museums, libraries, laboratories, astronomical observatories, and zoos serve as centers of secular knowledge, although religious learning and teaching can occur in these locales as well. Such places constitute centers for the continuing diffusion of rational secularism that began with Renaissance humanism and the Enlightenment and continues today, gradually and unevenly insinuating itself into the fabric of modern societies and everyday life (Livingstone 2003; Withers 2007; Meusburger et al. 2010).

More broadly, secularists in the U.S. have waged a long series of legal struggles to keep religion out of public life and space as an enforcement of the First Amendment's separation of church and state. American atheists have become increasingly militant and outspoken in regard to their beliefs (Goodstein 2009). Led by groups such as the Freedom From Religion Foundation (FFRF), the American Humanist Association, Americans United for the Separation of Church and State, People for the American Way, and the ACLU, secularists have opposed the display

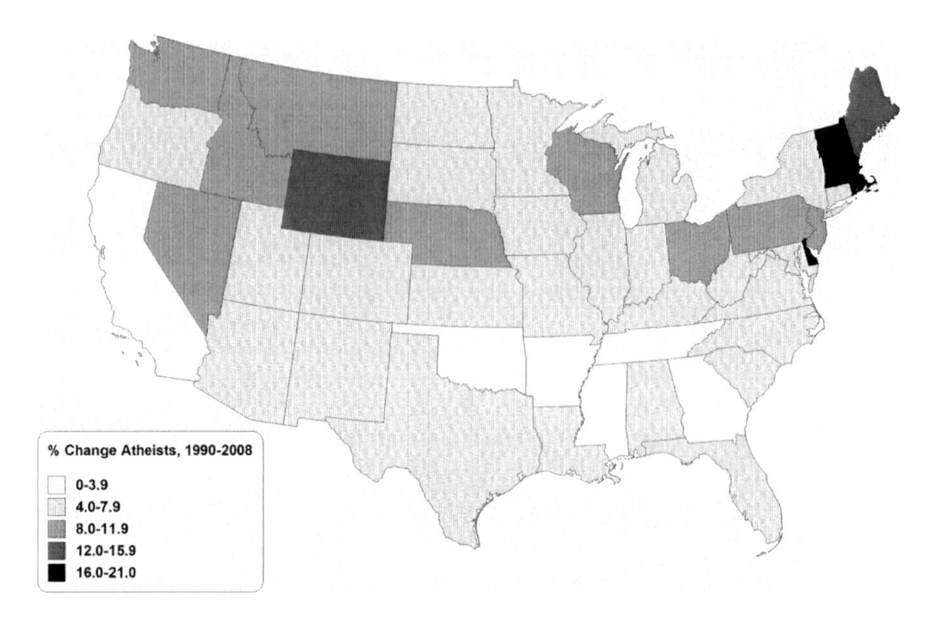

Fig. 117.5 Growth of American atheists, 1990–2008 (Map by Barney Warf, adapted from Goodstein 2009)

of Christmas crèches in public places; Christian crosses along public highways and public parks; prayers in public school classrooms and graduations, military functions, and athletic events; government vouchers for religious schools; and the posting of the Ten Commandments in public buildings (Provenzo 1990; Irons 2007). Frequently, such actions have been upheld by the Supreme Court, such as the 1980 case *Stone v. Graham* in Kentucky, which outlawed the display of the Ten Commandments in public school classrooms, and the 2005 case *McCreary County v. ACLU of Kentucky* forbidding their display in public courtrooms. In 2001, Chief Justice of the Alabama State Supreme Court, Roy Moore refused to remove a 5,000 pound granite statue of the Ten Commandments from the state judicial building; in 2003 he was removed from the bench by a state ethics panel following a federal Supreme Court ruling. In 2004, parents in Dover, Pennsylvania, successfully overturned the local school board requirement that "intelligent design" be taught as an alternative to evolution. In 2011, in Cranston High School in Rhode Island, atheists, led by 16-year old student Jessica Ahlquist, successfully petitioned to remove a plaque on the wall that called upon the "Heavenly Father" to bless the student body. Typically, religious groups in these cases are represented by conservative political organizations such as Jerry Falwell's Liberty Counsel, the American Center for Law and Justice, the Alliance Defense Fund, the Rutherford Institute, and the Thomas More Law Center. Other protests by atheists, such as requests to remove the words "under god" from the Pledge of Allegiance (which was inserted during the McCarthyite 1950s) or "In God We Trust" from coins, have been unsuccessful.

All of these cases, and countless others, reveal a theme that cultural and urban geographers have long studied (Mitchell 2003), that is, contests and debates over who gets to control and shape public space and whose ideology is represented there, only this time the conflict is couched in secular-religious terms.

117.8 Concluding Thoughts

Religion has enjoyed a long period of hegemony throughout human history and has imposed an indelible stamp on all societies and their landscapes. For many, religion offers hope and solace, especially relief from suffering by escaping to the afterlife. For atheists, religion of all sorts consists of fairy tales, often rooted in the Bronze Age, that are contradictory, nonsensical, and lacking empirical evidence. Other than the devastating intellectual critique of religion, atheists object to its reactionary social impacts, particularly the naturalization of inequality, subjugation of women, and repression of sexuality. Yet atheism for the most part has been confined to the social margins of almost all societies.

Nonetheless, the steady, inexorable commodification of everyday life and consciousness unleashed by capitalism has gradually eroded the social foundations of most religions, the process Max Weber likened to an iron cage of secularism (Weber 1922/1991). Since the Enlightenment, secular institutions, places, and bodies of knowledge have grown exponentially in size, number, and influence. As sociologists of religion have pointed out, there is nothing inevitable about this process, and much of the world has witnessed a recent desecularization and revival of religious fundamentalism. Simultaneously, there has also occurred a recent bloom in atheist thought and activism.

Secular societies tend to be wealthy ones. In vast swaths of the world, where poverty, disease, and lack of education are the norm, religion continues to hold sway to the same degree as it did in the West for a millennium. Nowhere in the world is secularism more advanced than in Europe. Yet in addition to a well-educated public, Europe's state support of religions may be as much to blame for the continent's lack of religiosity; in religiously pluralist societies such as the U.S., with a separation of church and state, religions have been far more aggressive in their competition with one another, and atheism tends to be less common. Thus, one must not accept Weberian secularism as some teleological turn away from the sacred to the profane. In this reading, a secular continent such as Europe is the world's exception, not a model that other regions will necessarily emulate.

As a political project, atheism has a distinctively dichotomous history. State-imposed atheism under Communist regimes in the Soviet Union, China, and elsewhere in the twentieth century led to incalculable human suffering for hundreds of millions of people, and is a moral and ethical abomination. Yet as an intellectual force, contemporary atheism has contributed significantly to holding back the tide of religious fanaticism that threatens to engulf the world during an era of unprecedented globalization. Atheists have long worked in support of abolitionism, women's rights, human rights, and similar causes; some support freedom of religion even as they advocate freedom from religion.

"Imagine no religion" sang John Lennon. This vision still holds promise. A humanity freed from the shackles of metaphysics, using reason and evidence to ground its views in this world and not some mythical afterlife, is a humanity empowered to unleash its formidable creative capacities to their maximum extent, to remake the world without being blinded by ancient fallacies about a big ghost (or ghosts) in the sky and to construct itself in emancipatory and self-liberating ways.

References

Altizer, T., & Hamilton, W. (1966). *Radical theology and the death of god.* New York: Bobbs Merrill.

American National Election Studies. (2009). Church attendance, 5 categories 1970–2008. www.electionstudies.org/nesguide/toptable/tab1b_5b.htm

Aronson, R. (2008). *Living without god: New directions for atheists, agnostics, secularists, and the undecided.* Berkeley: Counterpoint.

Baggini, J. (2002). *Atheism: A very short introduction.* Oxford: Oxford University Press.

Barber, B. (1995). *Jihad vs. McWorld: How globalism and tribalism are reshaping the world.* New York: Ballantine.

Barker, D. (2008). *Godless: How an evangelical preacher became one of America's leading atheists.* Berkeley: Ulysses Press.

Berger, P. (Ed.). (1999). *The desecularization of the world: Resurgent religion and world politics.* Grand Rapids: Eerdmans Publishing.

Berger, P., Davie, G., & Fokas, E. (2008). *Religious America, secular Europe? A theme and variations.* London: Ashgate.

Brown, C. (2001). *The death of Christian Britain.* London: Routledge.

Bruce, S. (2001). Christianity in Britain: R.I.P. *Sociology of Religion, 62*(2), 191–203.

Bruce, S. (2002). *God is dead: Secularization in the west.* Oxford: Blackwell.

Campbell, D., & Putnam, R. (2012). God and Caesar in America: Why mixing religion and politics is bad for both. *Foreign Affairs, 91*(2), 34–43.

Chaves, M. (1994). Secularization as declining religious authority. *Social Forces, 72*(3), 749–774.

Chaves, M., & Stephens, L. (2003). Church attendance in the United States. In M. Dillon (Ed.), *Handbook of the sociology of religion* (pp. 85–95). New York: Cambridge University Press.

Crockett, A., & Voas, D. (2006). Generations of decline: Religious change in 20th-century Britain. *Journal for the Scientific Study of Religion, 45*(5), 567–584.

Davidson, N. (1992). Unbelief and atheism in Italy. In M. Hunter & D. Wootton (Eds.), *Atheism from the Reformation to the Enlightment* (pp. 55–86). Oxford: Oxford University Press.

Davie, G. (2002). *Europe: The exceptional case. Parameters of faith in the modern world.* Maryknoll: Orbis Books.

Dawkins, R. (2006). *The god delusion.* New York: Houghton Mifflin.

Dennett, D. (2007). *Breaking the spell: Religion as a natural phenomenon.* New York: Penguin.

Eck, D. (2001). *A new religious America: The world's most religiously diverse nation.* San Francisco: HarperCollins.

Ecklund, E., & Scheitle, C. (2007). Religion among academic scientists: Distinctions, disciplines, and demographics. *Social Problems, 54*(2), 289–307.

Edgell, P., Gerteis, J., & Hartmann, D. (2006). Atheists as other moral boundaries and cultural membership in American society. *American Sociological Review, 71*(2), 211–234.

Epstein, G. (2009). *Good without god: What a billion nonreligious people do believe*. New York: HarperCollins.

Eurobarometer. (2005). Social values, science and technology. http://ec.europa.eu/public_opinion/archives/ebs/ebs_225_report_en.pdf

Finke, R., & Stark, R. (1988). Religious economies and sacred canopies: Religious mobilization in American cities. *American Sociological Review, 53*, 41–49.

Finke, R., & Stark, R. (1998). Religious choice and competition. *American Sociological Review, 63*(3), 761–766.

Freeman, C. (2002). *The closing of the Western mind: The rise of faith and the fall of reason*. New York: Vintage Books.

Froese, P. (2008). *The plot to kill god: Findings from the Soviet experiment in secularization*. Berkeley: University of California Press.

Gervais, W., Shariff, A., & Norenzayan, A. (2011). Do you believe in atheists? Distrust is central to anti-atheist prejudice. *Journal of Personality and Social Psychology, 101*(6), 1189–1206.

Gey, S. (2007). Atheism and the freedom of religion. In M. Martin (Ed.), *The Cambridge companion to atheism* (pp. 250–262). Cambridge: Cambridge University Press.

Glenmary Research Center. (2000). *Religious congregations & membership in the United States*. Atlanta: Glenmary Research Center.

Goodstein, L. (2009, April 26). More atheists shout it from the rooftops. *New York Times*, p. 1.

Goodstein, L. (2010, September 28). Basic religion test stumps many Americans. *New York Times*, p. 1. www.nytimes.com/2010/09/28/us/28religion.html

Harris, S. (2005). *The end of faith: Religion, terror, and the future of reason*. New York: Norton.

Harris, S. (2006). *Letter to a Christian nation*. New York: Vintage Books.

Hertzke, A. (1988). American religion and politics: A review essay. *Western Political Quarterly, 41*, 825–838.

Hitchens, C. (2007). *God is not great: How religion poisons everything*. New York: Hatchette Book Group.

Hout, M., & Fischer, C. (2002). Why more Americans have no religious preference: Politics and generations. *American Sociological Review, 67*(2), 165–190.

Hughey, M. (1979). The idea of secularization in the works of Max Weber: A theoretical outline. *Qualitative Sociology, 2*(1), 85–111.

Hunsberger, B., & Altemeyer, B. (2006). *Atheists: A groundbreaking study of America's nonbelievers*. Amherst: Prometheus.

Husband, W. (1998). Soviet atheism and Russian Orthodox strategies of resistance, 1917–1932. *Journal of Modern History, 70*(1), 74–107.

Hyman, G. (2010). *A short history of atheism*. New York: I.B. Taurus.

Irons, P. (2007). *God on trial: Dispatches from America's religious battlefields*. New York: Viking.

Jacoby, S. (2004). *Freethinkers: A history of American secularism*. New York: Metropolitan.

Johnson, P. (2002). *The Renaissance: A short history*. New York: Modern Library.

Kong, L. (2010). Global shifts, theoretical shifts: Changing geographies of religion. *Progress in Human Geography, 34*(6), 755–776.

Lewis, B. (2003). *What went wrong? The clash between Islam and modernity in the Middle East*. New York: Harper Perennial.

Livingstone, D. (2003). *Putting science in its place: Geographies of scientific knowledge*. Chicago: University of Chicago Press.

Loftus, J. (2008). *Why I became an atheist*. Amherst: Prometheus.

Lopez, D. (1999). *Prisoners of Shangri-La: Tibetan Buddhism and the West*. Chicago: University of Chicago Press.

Maisel, E. (2009). *The atheist's way: Living well without gods*. Novato: New World Library.

Majeska, G., Bociurkiw, B., & Strong, J. (1976). Religion and atheism in the U.S.S.R. and Eastern Europe, review. *Slavic and East European Journal, 20*(2), 204–206.

Marranci, G. (Ed.). (2010). *Muslim societies and the challenge of secularization: An interdisciplinary approach*. New York: Springer.

Martin, D. (1979). *A general theory of secularization*. New York: Harper & Row.
Martin, M. (2006). *The Cambridge companion to atheism*. Cambridge: Cambridge University Press.
Marx, K. (1844/1977). *Contribution to the critique of Hegel's philosophy of right*. Cambridge: Cambridge University Press.
McLeod, H., & Ustorf, W. (2003). *The decline of Christendom in Western Europe, 1750–2000*. New York: Cambridge University Press.
Meusburger, P., Livingstone, D., & Jöns, H. (Eds.). (2010). *Geographies of science*. Dordrecht: Springer.
Mitchell, D. (2003). *The right to the city: Social justice and the fight for public space*. New York: Guilford.
Newport, F. (2011). More than 9 in 10 Americans continue to believe in god. Gallup Polls. www.gallup.com/poll/147887/americans-continue-believe-god.aspx
Norris, P., & Inglehart, R. (2004). *Sacred and secular: Religion and politics worldwide*. New York: Cambridge University Press.
Pospielovsky, D. (1987). *A history of Marxist-Leninist atheism and Soviet antireligious policies*. New York: St. Martin's Press.
Potter, P. (2003). Belief in control: Regulation of religion in China. *China Quarterly, 174*, 317–337.
Provenzo, E. (1990). *Religious fundamentalism and American education: The battle for the public schools*. Albany: State University of New York Press.
Ranke-Heinemann, U. (1992). *Putting away childish things*. San Francisco: HarperCollins.
Schweizer, B. (2010). *Hating god: The untold story of misotheism*. Oxford: Oxford University Press.
Shand, J. (1998). The decline of traditional religious beliefs in Germany. *Sociology of Religion, 59*(2), 179–184.
Simon, G. (2009). *Church, state, and opposition in the U.S.S.R*. Berkeley: University of California Press.
Smith, J. (2010). Becoming an atheist in America: Constructing identity and meaning from the rejection of theism. *Sociology of Religion, 72*(1), 1–23.
Stark, R., & Finke, R. (2000). *Acts of faith: Explaining the human side of religion*. Berkeley: University of California Press.
Stark, R., & Iannaccone, L. (1994). A supply-side reinterpretation of the "secularization" of Europe. *Journal for the Scientific Study of Religion, 33*(3), 230–252.
Stump, R. (2000). *Boundaries of faith: Geographical perspectives on religious fundamentalism*. Boulder: Rowman & Littlefield.
Swatos, W., & Christiano, K. (2000). Secularization theory: The course of a concept. In W. Swatos & D. Olson (Eds.), *The secularization debate* (pp. 1–19). Lanham: Rowman and Littlefield.
Taylor, C. (2007). *A secular age*. Cambridge, MA: Harvard University Press.
Tiryakian, E. (1993). American religious exceptionalism: A re-consideration. *Annals of the American Academy of Political and Social Science, 527*, 40–54.
Warf, B., & Vincent, R. (2007). Religious diversity across the globe: A geographic exploration. *Social and Cultural Geography, 8*(4), 597–613.
Warf, B., & Winsberg, M. (2008). The geography of religious diversity in the United States. *Professional Geographer, 60*(1), 413–424.
Warraq, I. (Ed.). (2003). *Leaving Islam: Apostates speak out*. Amherst: Prometheus.
Weber, M. (1904/1930). *The Protestant ethic and the spirit of capitalism*. London: Allen and Unwin.
Weber, M. (1922/1991). *The sociology of religion*. Boston: Beacon.
Wilcox, C. (1991). *God's warriors: The Christian right in 20th century America*. Baltimore: Johns Hopkins University Press.
Wilford, J. (2010). Sacred archipelagos: Geographies of secularization. *Progress in Human Geography, 34*(3), 328–348.

Withers, C. (2007). *Placing the enlightenment: Thinking geographically about the age of reason*. Chicago: University of Chicago Press.

Zuckerman, P. (2009). Why are Danes and Swedes so irreligious? *Nordic Journal of Religion and Society, 22*(1), 55–69.

Zuckerman, P. (2012). *Faith no more: Why people reject religion*. Oxford/New York: Oxford University Press.

Zuo, J. (1991). Political religion: The case of the Cultural Revolution in China. *Sociology of Religion, 52*(1), 99–110.

Chapter 118
Representing the Unrepresentable: Towards a Strong Cultural Geography of Spirituality

Justin Wilford

Over the past decade, many human geographers have engaged in provocative critiques of representation, signification and culture. These provocations, gathered under the sign of non-representational theory, have grown out of a desire to account for the aspects of human and non-human life that are "more-than-representational," that are, to use a theological term, ineffable. Clearly, this interest in representing the unrepresentable has deep correlates in many religious traditions, Christianity especially.

It is, therefore, unsurprising that geographers of religion have taken up non-representational lines of inquiry in their own work. Most often the non-representational, or what Ethan Yorgason and Veronica Della Dora call "the more-than-rational" (2009: 630) is represented under the rubric of spirituality. The latter is often left undefined but some outlines have been sketched. Spirituality is "beyond rationality," "other-worldly," "immanent but not yet manifest," "mysterious, elusive and ethereal;" it produces "real and specific feelings" and gives us a "way to access experience directly" (Dewsbury and Cloke 2009: 696, 699, 707). It is a "calling," a "silent but nagging presence" (Rose 2010: 519), a "transcending [of] the problem of language" (Slater 2004), and it jars us out of our "forgottenness of origins" (Yorgason and Della Dora 2009: 635). In most every sketch, a central aspect of "the spiritual" is its experiential surplus, that it is "something more" (Dewsbury and Cloke 2009: 708).

For geographers of religion, a non-representational perspective on spirituality opens up quite fascinating ethical, political and phenomenological questions. But as a mode of socio-spatial research, it presents some rather thorny epistemological problems that cannot be overcome from within a non-representational framework. In this paper, I will argue that non-representational approaches to spirituality are ultimately based on modes of realism that obscure the historical, cultural and social relationships that undergird the meaning of any experience or practice of "the spiritual."

J. Wilford (✉)
Department of Geography, University of California, Los Angeles, CA 90024, USA
e-mail: jwilford@ucla.edu

S.D. Brunn (ed.), *The Changing World Religion Map*,
DOI 10.1007/978-94-017-9376-6_118

I do not want to defend, in any way, an approach that reduces 'the spiritual' to the historical, cultural or social, let alone the economic or political. Instead, I want to reclaim space for geographers of religion to work through the ways individuals and communities transform experiential surplus into something meaningfully represented as spirituality. In this regard, meaning might be (or perhaps always is) affected by a surplus of experience—what I will later introduce as Theodor Adorno's (1973) "preponderance of the object"—but it can never be reduced to it. We are bound, for better or for worse, in a thick web of signification that assures such surplus becomes conceptualized and incorporated into our collective worlds.

I will begin by arguing that researching spirituality in a non-representational mode (as opposed to a more social-constructivist perspective like that in the work of Julian Holloway 2003, 2006) turns us away from the historical, cultural and social. The ahistoricism of the "more-than-representational" does not run parallel with the "more-than-rational." In fact, in socio-spatial research, they run in opposite directions. Affect, embodiment, and daily, lived experience—all "more-than-rational" aspects of the spiritual—are transformed into meaning by all individual and collective subjects. This transformation is inherently historical, cultural, social, geographical, and thus, representational. I will conclude by showing how a strong cultural interpretivist approach to spirituality—one that sees culture, signification, and representation as fundamental and irreducible aspects of human life[1]—can incorporate experiential surplus while recognizing its historical and cultural constructedness.

118.1 Turning Away from the Social

"The spiritual" and spirituality, for many geographers of religion, lie beyond representation. They are found in "bodily existence [and] felt practice" (Dewsbury and Cloke 2009: 696) and in "exteriorities [that are] wholly infinite and eternally beyond our perceptible horizons" (Rose 2010: 508). In non-representational terms, they are 'irreducible, nonthematisable … dimensions of life' (Harrison 2007: 591). And while many argue that they are fundamental to daily life and human experience in general (Yorgason and della Dora 2009: 635), it is almost always recognized that they are somehow deeper and more real than this life.

There is, of course, an intellectual history for thinking through this deeper 'something more'. Many geographers of religion draw on Martin Heidegger, Marcel Merleau-Ponty, Emmanuel Levinas, Gilles Deleuze and Gianni Vattimo among others in order to gain purchase on this lived experience that outstrips

[1] By using the term "strong," I am following current debates in American and British cultural sociology that revolve around Jeffrey Alexander's push for a cultural pragmatics in which culture is viewed as an autonomous force in social life. In contrast, a "weak" cultural sociology sees culture as "epiphenomenal" and overdetermined by other social forces, such as material production or power politics (see Alexander and Smith 2001; Reed and Alexander 2009; Alexander 2008). I will elaborate on this in the final section of this paper.

conceptualization. The innovative philosophical tools these thinkers provide are attractive to geographers of religion because they make it easier to speak about spirituality as "something more." But the focus on this phenomenological objective surplus, often leads to a valorization and overemphasis of individual experience and action. This is because in the process of excavating the 'personal' experience of "something more" (see Harrison 2007), history, webs of culture and socio-political forces are bracketed. And thus, "human geographers should no longer ignore the individual religious dimensions" (Slater 2004: 251) in which individual identity is not "sociologically responsive . . . But ontologically responsive" (Rose 2010: 508). Such responsiveness is in essence a relationship between the experiencing individual and 'existential sensation, performative faith and immanence' (Dewsbury and Cloke 2009: 709).

History, culture and the social are necessarily bracketed in non-representational approaches because these draw us away from the surplus of experience and directly into representation—into themes, concepts, scripts, narratives, and meaning. At a more general theoretical level such bracketing leads to brilliant insights that can push us to reconsider our own subjective positionality, our conscious taken-for-granted worlds, our ethical responsibilities to others and the non-human world. But in socio-spatial research, the results are very different. One way to take stock of the difference in results between a philosophical and a socio-spatial approach is by taking a closer look at the philosophical nature of the surplus of experience.

One thinker who can shed light on this is Theodor Adorno, who strangely is missing from the non-representational canon in human geography. Adorno is largely known for his stinging critiques of the cultural production of mid-twentieth century capitalism. But a later work, *Negative Dialectics*, is perhaps his greatest philosophical contribution (or at least, as Adorno himself put it: It "represents the quintessence of my thought" [quoted in Wiggershaus 1995, 600]). In this and in two other works—an essay entitled "On Subject and Object" (1998) and his posthumously published book *Aesthetic Theory* (1997)—Adorno works through the epistemological and ethical problems of the surplus of experience, or what he calls the "preponderance" or "primacy of the object" (1973, 1998).

Prefiguring non-representational theory's antipathy towards the transcendental cogito, Adorno argues, "The prevailing trend in epistemological reflection [has been] to reduce objectivity more and more to the subject. This tendency needs to be reversed" (1973: 176). The subject (human consciousness) holds the object (the world that is experienced) to be equal to its cognition, but this is, of course, an illusion, says Adorno. Following this basic Kantian insight, we know the object is never equal to the subject; the former stands always outside and against the latter. But Adorno develops an immanent critique of Kant through this and begins to argue that the subject intuits this and thus, "exalts [itself as a] reaction to the experience of its impotence" (180). This exaltation is a mode of cognition Adorno describes as identitarian thought, in which the subject holds the object world to be identical to its concepts of the world. "The subject," writes Adorno, "reduces the object to itself; subject swallows object, forgetting how much it is object itself."

Identitarian, though, in itself is not the problem. It is the basis of reason and practical action. We need to act *as if* our concepts were identical to the world. The problem is that it "prevents self-reflection" (Adorno 1973: 180). It hides from the subject the fact that the subject's substance is entirely dependent on its relationship to the object. This dependency "feels to the subject like an absolute threat." Even the slightest realization of the subject's non-identity with the world "spoils" the subject's illusions that it is "whole" (183).

Adorno argues that we need to be conscious of our non-identity with the world for at least two reasons. First, the subject's unreflective identitarian thinking is a necessary condition for domination. The subject dominates the object world with its conceptual, thematizing cognition. And then it uses these identitarian categories to suppress and annihilate difference. Insofar as the subject longs for identity between its concepts and the object world, it "cannot love what is alien and different" (172). However, we can be reflective and self-critical about our conceptual framing of the world by "giving the object its due," allowing "the alien [to] remain what is distant and different, beyond the heterogenous and beyond that which is one's own" (191).

But the second reason to be aware of non-identity strikes to the heart of geographers' concerns with spirituality. The subject must be ready to recognize the inherent inadequacy of its concepts to mirror the object world, for the object always outstrips the subject. In Adornian terms, the subject must be ready for the non-identity of the "primacy of the object." The important ethical question is what do we do when confronted with this surplus of experience? Adorno argues that we can either regress back into identitarian thought, dominating and repressing this surplus with our old conceptual frameworks, or we can accept our inherent epistemological inadequacy and allow for an experience of non-identity.

This latter strategy is what many geographers of religion have in mind when writing of spirituality. This overflow of the surplus of the object appears to be just another description of the "more-than-rational," the "mysterious and ethereal," the inadequacy of language that geographers of religion have used to characterize spirituality. In harmony with this, Adorno writes,

> Knowledge of the object is brought closer by the act of the subject rending the veil it weaves about the object. It can do this only when, passive, without anxiety, it entrusts itself to its own experience. In the place where subjective reason senses subjective contingency, the primacy of the object shimmers through: that in the object which is not a subjective addition. (1998: 254)

But Adorno does not allow this experience of non-identity to remain simply a non-representational mystery. There would be no criticality, no dialectical movement in this. It would be a withdrawal and a refusal. And further, it would allow for the unmediated domination of the subject by the object. For, "only the strong and developed subject, the product of all control over [the material world] and its injustice, has the power both to step back from the object and revoke its self-positing" (1997: 266). Identitarian thought's opposite is as undesirable as it is improbable. Our access to non-representational experience is always and necessarily partial because it "is not completely and adequately open to knowledge, and . . . never beyond question" (Adorno 1997: 266). So, while ineffable mystery might be experiential surplus's true ontological status, its relationship to the subject is dialectical: "subjectivity mediates objectivity" (Adorno 1997: 266). It is thus historical, cultural and social. Insofar as we have any awareness of it, it must be represented.

118.2 The Reality of Realism

If we want to represent spirituality while attempting to preserve its non-representationality— spirituality as something-more—we must ultimately rely on an epistemology that allows us some access to ontology. In other words, we need a realist epistemology that purports to give us access to ontology. Although those of us interested in post-positivist geography generally eschew realism, even in its watered down, critical form (although compare Ferber 2006), we nevertheless find ourselves struggling against the realization that we only have interpretive, socially constructed access to reality. If we are comfortable with this fact in the context of the social world of, say, political discourse, local community action, or economic policy, then we seem less so in the face of affect, embodiment and pre-cognitive practices.[2]

In both cases—the socially constructed world of history, cultural and society in one case, and the experiential world of the non-relational and non-representational in the other—we are seeking to employ "depth epistemologies" that reject the 'positivist and postmodern contentions that interpretation beyond the "surface" of social life is metaphysical nonsense' (Reed 2008: 124). But those interested in non-representational approaches are seeking, however obliquely, a ground to this "depth." An ontology, a ground of being, is ultimately what is referred to by those interested in spirituality as "more-than." Even in the language of "performative presencing," "excess," (Dewsbury and Cloke 2009), and a forceful "outside" (Rose 2010), what is sought is some ontological reality that *can not only be accessed but also recovered*. To link this surplus of experience to the category of "spirituality," let alone "religion," is to seek to represent the non-representational while somehow sidestepping representation (that is, the historical, cultural and social work of meaning-making).

But, to return to Adorno's terminology, the surplus of the object world (an ontological breakthrough) has no content, no substance for the subject outside of the subject's concepts. 'The primacy of the object shimmers through,' but in itself it does no more than shock and humble the subject. To take account of this shimmer, to even think of it or remember it, is to give up pretensions of "direct experience" (Adorno 1997: 244). It is to give up the implied wish for a realist epistemology. But

[2] For examples, see the recent interest in the "ontological turn" (cf. Escobar 2007), "speculative realism" (Woodward 2011), and other types of "new empiricism" (McFarlane 2011; Anderson and McFarlane 2011).

what if a non-representational approach to spirituality is limited to a humbler sort of realism[3] that wishes to leave experiential surplus as a "mystery," a "calling," the "ineffable?"

118.3 Representing the Non-representational

'It is not the purpose of critical thought,' writes Adorno (1973), "to place the object on the orphaned royal throne once occupied by the subject. On that throne the object would be nothing but an idol" (181). To leave the "shimmer of the object" as a "more-than-rational" "mystery" of spirituality would be to halt the dialectic of critical thought just at its most important pivot. The skepticism towards dialectical thought in much non-representational work is largely tied to a rightful mistrust of positive dialectics that "construe contradictions from above and to progress by resolving them." For Adorno, true dialectical thought is negative dialectical thought; it "pursue[s] the inadequacy of thought and thing" to no end (Adorno 1973: 153). Pursuing this essential inadequacy leads us not to non-representation—for this is where we start—but rather to the historically, culturally, and socially contingent relationship between thought and thing. That is, it leads us back into the world of representation.

What happens if, in the course of our analysis of any spiritual experience, we stay in a non-representational mode? Must we "foreclose (transcendent) interpretation of what something means in favour of an immanent intuition of or participation in the process whereby something is produced" (see Dewsbury and Cloke 2009: 705)? Insofar as a non-representational approach leads us to devalue acts of meaning-making—to see interpretation as "distance," "codification," and a restricted sensory experience (Thrift 2003)—then we are stuck, perhaps in a moment of object-surplus, but certainly in a moment of non-relation (Harrison 2007). But as Harrison points out in his work on the non-relational surplus of life (grieving, in his particular study), as social researchers we cannot remain in a non-relational mode, even as we respect the surplus's non-relationality.

"Social analysis cannot but translate [the surplus of experience] and thereby put it to use," writes Harrison (2007: 597). And as we put the surplus to use, we give it meaning, but more, as socio-spatial researchers, we give meaning to others' meaning (Alexander 2011; Reed and Alexander 2009). And it is because this layered hermeneutic doubly effaces the original surplus that non-representational concerns and critiques have such force. But it is also because we cannot remove

[3] Critical realism has been presented as one such form of a humbler realism (in geography, see Ferber 2006; for some foundational texts, see Bhaskar 1975, 1979). Nevertheless, CR is still committed to an epistemology that allows for access to a non-cultural social ontology (that is, that we social scientists have access to ontologically secure social structures in the same way natural scientists have access to natural structures). An excellent cultural-interpretivist critique of CR can be found in Reed 2008.

ourselves from this hermeneutic circle that we must take representation and meaning-making even more seriously.

If we are, then, interested in allowing spirituality to "speak back" (Yorgason and della Dora 2009) we must be prepared to make meaning out of it. Still, if we wish to capture spirituality as direct experience, we must be prepared to accept that our very conceptualizations of it are representations. To categorize the embodied, felt force of some experiential surplus as "spirituality," let alone as some specific type of spirituality (for example, Christian), is to plunge headfirst into the ocean of historical, cultural and social representation. As geographers of religion and socio-spatial researchers, we should pay particular attention to the ways this plunge is taken. And we should keep in mind that we are always already tossing about in this ocean.

This is not to say, however, that there are not other reasons to stay with the non-representational. Clearly, there is a normative force and political potential in the non-representational (Thrift 1996, 2003; Hinchliffe 2001, 2003) as well as more intimate, ethical justifications (Harrison 2007).[4] But as a strategy or mode of thought for gaining understanding and constructing social knowledge, it is misleading. And it is even more misleading in the case of religion and spirituality because it is at the heart of many religious practices to efface the historical, cultural and social work that turns experience into representations.

And so while the theory and methodology of non-representational approaches in the study of geographies of spirituality are problematic, spirituality as an object of study surely is not. How then might we go about understanding geographies of spirituality, respecting their "other-worldiness" while interpreting them within this world?

118.4 A 'Strong' Cultural Approach to Spirituality

Respecting the 'more-than' of spiritual experience does not necessarily leave us in a non-representational mode. With Adorno, we can acknowledge the ontology of this experience while rejecting its "self-positing" (Adorno 1997: 346). In other words, its ontology does nothing in itself until it is subsumed in a specific historical-geographical context. As Clifford Geertz argued,

> The thing to ask about a burlesqued wink or a mock sheep raid is not what their ontological status is. It is the same as that of rocks on the one hand and dreams on the other—they are things of this world. The thing to ask is what their import is: what it is, ridicule or challenge, irony or anger, snobbery or pride, that, in their occurrence and through their agency, is getting said. (2000: 10)

What, then, is being said in and through the geographical expressions of spirituality? Yes, this takes us back to questions of that old 'new cultural geography.' And,

[4] Even as there are political pitfalls to non-representational approaches (see Barnett 2008).

yes, its 'novelty has worn off' (Castree and MacMillan 2004: 470). And yes, its chief concerns—meaning, signification, and representation—are 'in ruins, carpet bombed by the formidable arsenals of contemporary critical theory' (Prendergast 2000: ix, quoted in Castree and MacMillan 2004: 470). As Nigel Thrift (2003) has pointed out, meaning, signification and representation enlarge the gap between observer and observed. The research narratives we produce are dim reiterations of the undetermined representations given in our research fields. And through these layers of representation, Thrift argues, we seek to stabilize the identities of our research subjects. In this way, we impose the apparent solidity of our representational frameworks on the very fluid dynamism of the world. And our privileging of our representational practices as socio-spatial researchers leads us to attend only to similar representational practices we find in the field.

So then are there ways to answer the question of what spirituality *means* rather than what *it is* without succumbing to these pitfalls of representation? I believe we can, but not by turning away from culture but rather by taking it even more seriously. We can do this by staying with the paradoxes of representation: how does the "irreducibility of representation" (Barnett 1999: 268), the groundless hermeneutical circle, produce a meaning with force? How, in other words, can floating signifiers weigh a ton?

A "strong" cultural approach that sees culture as a force—not because, as in a 'weak' cultural approach (see Alexander 2011), it is ideology or mystification but because it structures the limits of our experience in the world—can provide us with the conceptual tools to give spirituality its due while recognizing its cultural irreducibility. One way to think about a strong cultural approach in critical human geography is suggested by Patricia L. Price (2010), who draws our attention back to narrative.[5] In a recent journal forum on the future of cultural geography, she summarized the way "stories" and their components have real and powerful effects, again, not as ideology or mystification, but as cultural structures that allow us to make sense of the world and each other. Narrative orders both time and space by sequencing actions in the former and positioning them in the latter. Through 'metaphor, similes, dualisms, allegory, personification, symbolism, [and] foreshadowing' boundaries are drawn and maintained between heroes and villains, the virtuous and evil, the sacred and the profane. Narrative, writes Price, "creat[es] value through the coherency of unity" (205).

But, non-representational critics might argue, what happens when, as we find so often in our research, coherence and unity breakdown? And more specifically, how are we to account for the experiences that produce narratives that are just forming, contradictory or partial? Are they without value? Price begins to address this by suggesting that "narrative itself expand" to incorporate the non-representational, that we enlarge our stories to take into account the 'haptic, aural, olfactory, and gustatory,' emotions as well as motion (1010: 208).

[5] Narrative has long been a concern of cultural geographers (for example, Entrikin 1991) but has been effectively diminished by the rise of non-representational approaches.

But the key criticism of non-representational theorists stands. If we are chained to the coherence of representation—and especially narrative—then how can we account for liminality that produces what appears to be a cultural breakdown? To account for this while remaining true to the irreducibility of representation we need a more encompassing conceptual framework than narrative, one that can incorporate narrative while accounting for its potential disintegration. For this we can draw on a touchstone of non-representationalism: performance. But here we will draw on a fully representational performance, one formulated in social theorist Jeffrey Alexander's cultural pragmatics (2004).[6]

To be sure, Alexander's cultural pragmatics relies heavily on narrative. But the latter is both too broad and too narrow to capture the production, negotiation, maintenance, and breakdown of meaning in contemporary societies. So instead, Alexander draws on dramaturgical theory (especially Erving Goffman) to outline a social theory of cultural performance in which any social action, to achieve its intended effect, must bind together all elements of performance into a 'fused' whole. Such performative fusion "display[s] for others the meaning of their social situation" (Alexander 2004: 529).

In this model, seven distinct elements make up performance:(1) background symbols; (2) foreground scripts; (3) actors (individual and collective); (4) observers/audience; (5) means of symbolic reproduction; (6) *mise-en-scéne*; and (7) social power (Alexander 2004: 530–533). Background symbols are related to Price's literary techniques' ('metaphors, similes, dualisms, allegory, personification, symbolism, foreshadowing'); but rather than thinking of them as techniques or actions, we can consider them as cultural resources. They are deep cultural symbols, not unlike Durkheimian binary codes.[7] And although they are not some fundamental ground of narrative and performance—as in the work Claude Levi-Strauss (1966)—neither can narrative and performance be constructed without them. The second element of Alexander's cultural performance, foreground scripts, are more straightforwardly connected to Price's narratives. They are more articulated stories that combine, juxtapose, negate, and invert deeper background cultural elements. In other words, narratives and scripts are not determined by deep cultural symbols, but have a "relative freedom," in Alexander's (2004: 550) words, to create and contest meaning, albeit within limits.

The advantage of the concept of performance over that of narrative is that the former puts us firmly into the world of action, spatiality, materiality, and history. The third and fourth elements of Alexander's model—actors and audience—make room not just for story*tellers*, but active, spatialized doers. Cultural performance is as much about cultural structure as it is cultural action. The deep background symbols and foreground narratives do nothing in themselves. Far from superorganic, they are the tools and parameters through which actors cohere and perform as

[6] I am not the first geographer to draw on Alexander's cultural pragmatics. See Nicolas Howe's (2008) excellent appropriation of it.

[7] Sacred/profane is a deep cultural binary code, but Durkheim (1912/2001: 40) draws attention to 'secondary species' of this fundamental binary as well.

intelligible social beings. And while we can analytically separate actors and their codes and narratives, they must appear fused to the actors' audience, otherwise the performance fails. In order to achieve performative fusion, actors must have a deep, visceral understanding of the codes and narratives being deployed, as well as access to the remaining elements of performance: the means of symbolic reproduction, *mise-en-scéne* (the skill of literally putting a performance in place) and adequate social-positional power.

What can this model offer to the socio-spatial study of spirituality? First, it allows us to recognize our epistemological limits as social researchers without giving into a relativist solipsism or materialist cynicism. Seeing spirituality as an experience and cultural performance allows us to interpret and explain it as having real effects without it also being false consciousness. As actors experience something "more-than-representational," they can only move within the cultural structures of their and their audience's histories and geographies. The meaning of this 'more than representational' experience is not derived referentially, from its objective reality. Rather, it derives from these cultural structures: the language, symbols, rituals, myths and so forth. These experiences are tested against what is already "known" and felt to be sacred and profane (see Alexander 2011, p. 90). Non-representational sensitivity is clearly important insofar as it forces researchers to consider the embodied, material and affective dimensions of these experiences. But cultural pragmatics alerts us to the ways these experiences are always already bound to worlds of representational meaning.

Second, we can account for what these "more-than" experiences mean while acknowledging their meaning is underdetermined by any specific cultural structure. The codes and narratives that actors draw on are not simply given by their personal history, but they are also drawn on and enacted in relation to a specific socio-spatial environment (Alexander's "audience" and "scene"), and are dependent on the actor's performative repertoire, skills and social power. In this sense, non-representational materiality has a double effect on the performance of spirituality: it can be a source (an "objective surplus" in Adorno's terms) of a spiritual experience, and it also affects in myriad ways the interpretation, translation and performance of this experience.

Finally, we can account for those more-than-representational experiences that remain unrepresented. Alexander's cultural pragmatics opens up cultural-theoretical room to talk about meaninglessness, disintegration, mistranslation, or what he calls "de-fusion." If an effective cultural performance is a fusion of all the performative elements, then a failed performance—that is, a failure of meaning—is their disassociation. Actors can misinterpret the codes and narratives they use to represent their experience. They can miscommunicate their representational self-understanding. Their audience may be distanced or pre-disposed to skepticism. Or an audience may see an actor, however skilled and well-positioned, as falsely or inauthentically related to the codes and narratives deployed. In the case of "spirituality" or "spiritual" experiences in open, complex, differentiated societies, such performative failure is at least as likely as success.

118.5 Meanings of Spirituality in American Evangelicalism

One way to demonstrate a cultural pragmatic approach to spirituality is through a brief examination of the performance of spirituality in American postsuburban megachurches. Many sociologists of religion have noted the way 'spirituality' is opposed to 'religion' in American Christianity, and Protestant evangelicalism in particular (Roof 1999). Some have even used this discursive binary as an analytical tool for tracking historical changes in secularisation and religiosity (Heelas and Woodhead 2005). In most of these studies, spirituality is seen as the product of a first-hand experience of 'God's presence'. It typically arises alongside a binary other, though the latter is widely varied. In one instance, the 'spiritual' might be contrasted with institutional religion, while in another it might be contrasted to mundane daily life. And yet, in another it might signify New Age heresy as opposed to 'true religion'.

At a large American postsuburban megachurch I studied for 18 months (see Wilford 2012), I observed these and many other uses of the terms "spiritual" and "spirituality." In all cases, however, these terms were not used to describe an intellectual state or soteriological status, but rather were performative. In interviews with lay members—as opposed to pastoral or administrative staff—the terms were used to signify an inner, personal transformation. Most often 'spiritual' was a descriptive qualifier, as in "spiritual growth," "spiritual maturity," "spiritual blindness," "spiritually hungry" or "spiritually minded." The terms came up most often in the context of conversion or finding a congregation that felt like "home." My interview subjects rarely referred to the "mysterious, elusive and ethereal" in the descriptions of "spiritual journeys," but nevertheless were signifying an experience, or a desire to experience, something "more-than."

For one subject, her time spent in her 20s and 30s in an American mainline Protestant congregation was "spiritually deadening." That time represented to her institutional religion, an overly-intellectual theology, and mundane daily life. She had little access to "the spirit," she said. But her finding and becoming committed to this new evangelical megachurch was not an ecstatic, effervescent experience. It was, as with many other members I interviewed, a series of experiences that were interpreted and re-interpreted within many different and fluid social and material spaces[8] within the megachurch. The small and variegated experiences of surplus came while singing in worship services, praying and sharing intimate stories with new friends in small groups in members' homes, or in quietly reading the Bible in the morning while alone (a classic evangelical practice that this church strongly advocated). These moments were sometimes described in non-representational terms ("I can't explain the feeling ..."), but insofar as she saw these experiences as affecting her, they were interpreted through the cultural structures of American

[8] See Wilford (2012) on the "flexible sacralization" fostered by U.S. megachurches' dependency on small groups, campus venues and short-term mission trips.

evangelicalism. She described a particularly powerful and emotionally overwhelming spiritual experience that occurred on a short-term mission trip to the suburbs of Kigali, Rwanda:

> We're [divided] up and we each go to different houses of the congregation because this is a very poor congregation. I ended up in a woman's house who … they brought in all these widows. Either their husbands died of genocide or they died of AIDS. And there's a lot of AIDS in this church. So they asked me to tell my story and I told my whole story like I told you but from a real female perspective. And they begin weeping. And then I begin weeping. And we're all just weeping, you know? It was just overwhelming. The spiritual bond was amazing.

She said that after that experience it took her the rest of the night to understand what had happened. "God laid it on my heart that I was going to find an orphanage" in the country of her mission trip and work from back in the U.S. to fund it. At the time of our interview, this interview subject had suffered a series of setbacks in her quest to form her own non-profit and fund this orphanage. Far from discouraged, she saw every obstacle (one of the largest ones was her church which had its own orphans program) as the binary other to her "spiritual" journey.

The audience for this performance was her small group, her family, and also herself "back home." Drawing on these deeper cultural structures she performed new interpretations of her Rwandan mission experience in which that moment of surplus was made to stand against an array of non-spiritual forces. She saw her actions as being implicit decisions between spiritual inspiration or religious bureaucracy; spiritual intuition or intellectual theology; spiritual selflessness or material consumerism; spiritual purpose or mundane daily life. And each of these cultural binaries was inextricably caught up with others. In her and others' narratives of spiritual experiences on short-term mission trips, global cultural binaries (America/Third World; Western industriousness/Southern laziness; modern individual autonomy/traditional dependency) were interwoven with local, daily ones (cosmopolitan/provincial; authentic/fake). Further underscoring the contingent nature of such socio-spatial hermeneutics, this church member was beginning to find that many of her new "spiritual" decisions were leading her away from the megachurch. Not only was her performance leading her to reinterpret her relationship with the megachurch, but, according to her, some church pastors were distancing themselves from her as well.

Another churchgoer who had attended several short-term mission trips recounted for me a visit to an orphanage in Pattaya, Thailand. After a few hours of helping to prepare and serve food to the children, she said she became overwhelmed with despair. "There was nothing I could do," she told me. "I felt like nothing good could happen there and I started crying. I was trembling. It was too much." Later in the trip she was able to talk and pray about it with her mission group. She began to think of the orphanage experience as a moment of spiritual growth in which her "heart was softened" and "opened." She came to perceive that back in her postsuburban neighborhood, parts of her life were shallow and superficial. If she could hold on to that moment, she believed, then her heart could remain more sensitive to those in need back at home. No important lifestyle changes were "laid on her heart." Instead, this experiential catalyst for "spiritual growth" led to delicate shifts in her perception. "Since the trip," she said, 'I think I'm more aware of hurting ... what's going on behind the scenes [in people's lives].'"

Her experience of surplus was reincorporated in very subtle ways. She related how difficult it was for her to watch television or engage with shallow gossip at work after her mission trip. These acts felt "silly" and "just not right" after what she had experienced. In other words, these actions were inauthentic to her new performative interpretations. In the instance of work, she performed her experience of surplus to her co-worker audience in such a way that reinforced her new self-understanding. And in the instance of abstaining from television at home, she performatively emplaced this new self-understanding within her daily, intimate, material environment. The living room became a stage for a continuous hermeneutical re-enacting of her spiritual experience.

While a sensitivity to non-representational approaches allows us to take seriously this moment of experiential surplus, it is clear that such an experience does not speak on its own. To allow it to "speak back" would be to listen carefully to how actors speak of *it*. And most importantly, it would be to understand that when actors speak (and of course not just through language but through action as well) they are also *speaking to*. They are, in essence, performing not just for others but for themselves as well. And in this process they are creatively drawing on cultural structures, bringing them to life and combining them in novel ways with the spatiality and materiality at hand. The "more-than" of spiritual experience is put in place as those who have experienced it seek to make it meaningful to themselves and others.

118.6 Conclusion

The goal of this paper is certainly not to disparage the study of spirituality in the geography of religion. Far from it. Instead its purpose is to argue against a desire that seems increasingly prevalent in the geography of religion—a desire to incorporate the 'more-than' of spiritual experience into our work through non-representational means. My use of Adorno's later philosophical work was intended to show a way of recognizing this "more-than" while always keeping in view its dialectical relation with signification, history, self-identity, and culture. This dialectic is always in motion, and this motion, I argue, is best explained through Alexander's cultural pragmatics.

I hope I have shown that to ignore the signifying side of this dialectic would be to ignore the most crucial and geographical aspect of spirituality: the process of turning an experience into embodied and emplaced meaning. Instead of reducing spirituality to economic rationality, power-political chicanery or ideological delusion, a thoroughly non-representational approach would reduce it to the mysteriously ineffable. However, by allowing spirituality to enter the endlessly referential web of cultural meaning, we can see it as truly irreducible. We can see that any efforts to reduce spirituality and the spiritual to some "really real" foundation, is itself a performance of cultural representation that is bound up in a series of cultural binaries, narratives, materialities, and spatialities. And these latter cultural elements only refer to still other cultural elements. There is no ontological ground to which these experiences can be reduced.

"A child counts on his fingers before he counts "in his head;" he feels love on his skin before he feels it "in his heart." Not only ideas, but emotions too, are cultural artifacts in man" (Geertz 2000: 81). As geographers, we understand, against notions of its immateriality, that culture is indistinguishable from the spatiality and materiality of "fingers" and "skin." And insofar as non-representational thought returns us to this awareness, it is vital and useful. But when this insight leads us to undialectically replace one idol (representation) with another (non-representation), we lose sight of how thoroughly performative socio-spatial life is. The question, then, for geographers studying religion is not only how best to think through the way people make meaning out of their spiritual experiences, but also, how we can make new meaning out of their meaning.

References

Adorno, T. W. (1973). *Negative dialectics*. New York: Continuum.

Adorno, T. W. (1997). *Aesthetic theory* (G. Adorno & R. Tiedeman, Eds., R. Hullot-Kentor, Trans.). Minneapolis: University of Minnesota Press. (Original work published 1970)

Adorno, T. W. (1998). *Critical models: Interventions and catchwords*. New York: Columbia University.

Alexander, J. C. (2004). Cultural pragmatics: Social performance between ritual and strategy. *Sociological Theory, 22*, 527–573.

Alexander, J. C. (2008). Clifford Geertz and the strong program: The human sciences and cultural sociology. *Cultural Sociology, 2*, 157.

Alexander, J. C. (2011). Fact-Signs and cultural sociology: How meaning-making liberates the social imagination. *Thesis Eleven, 104*, 87–93.

Alexander, J. C., & Smith, P. (2001). The strong program in cultural theory: Elements of a structural hermeneutics. In J. Turner (Ed.), *Handbook of sociological theory* (pp. 135–150). New York: Springer.

Anderson, B., & McFarlane, C. (2011). Assemblage and geography. *Area, 43*, 124–127.

Barnett, C. (1999). *Culture and democracy: Media, space and representation*. Edinburgh: Edinburgh University Press.

Barnett, C. (2008). Political affects in public space: Normative blind-spots in non-representational ontologies. *Transactions of the Institute of British Geographers, 33*, 186–200.

Bhaskar, R. A. (1975). *A realist theory of science*. London: Verso.

Bhaskar, R. A. (1979). *The possibility of naturalism* (3rd ed.). London: Routledge.

Castree, N., & Macmillan, T. (2004). Old news: Representation and academic novelty. *Environment and Planning A, 36*, 469–480.

Dewsbury, J. D., & Cloke, P. (2009). Spiritual landscapes: Existence, performance and immanence. *Social and Cultural Geography, 10*, 695–711.

Durkheim, E. (2001). *The elementary forms of religious life* (C. C. Cosman, Trans.). New York: Oxford University Press.

Entrikin, J. N. (1991). *The betweenness of place: Towards a geography of modernity*. Baltimore: Johns Hopkins University Press.

Escobar, A. (2007). The 'ontological turn' in social theory. A Commentary on 'Human geography without scale', by Sallie Marston, John Paul Jones II and Keith Woodward. *Transactions of the Institute of British Geographers, 32*, 106–111.
Ferber, M. (2006). Critical realism and religion: Objectivity and the insider/outsider problem. *Annals of the Association of American Geographers, 96*, 176–181.
Geertz, C. (2000). *The interpretation of cultures: Selected essays*. New York: Basic Books.
Harrison, P. (2007). How shall I say it…? Relating the nonrelational. *Environment and Planning A, 39*, 590–608.
Heelas, P., & Woodhead, L. (2005). *The spiritual revolution: Why religion is giving way to spirituality*. New York: Wiley-Blackwell.
Hinchliffe, S. (2001). Indeterminacy in-decisions—Science, policy and the politics of BSE. *Transactions of the Institute of British Geographers, 26*, 182–204.
Hinchliffe, S. (2003). Inhabiting landscapes and natures. In K. Anderson, M. Domosh, S. Pile, & N. Thrift (Eds.), *The handbook of cultural geography* (pp. 207–226). London: Sage.
Holloway, J. (2003). Make-believe: Spiritual practice, embodiment, and sacred space. *Environment and Planning A, 35*, 1961–1974.
Holloway, J. (2006). Enchanted spaces: The séance, affect, and geographies of religion. *Annals of the Association of American Geographers, 96*, 182–187.
Howe, N. (2008). Thou shalt not misinterpret: Landscape as legal performance. *Annals of the Association of American Geographers, 98*, 435–460.
Levi-Strauss, C. (1966). *The savage mind*. Chicago: University of Chicago.
McFarlane, C. (2011). Assemblage and critical urbanism. *City, 15*, 204–224.
Prendergast, C. (2000). *The triangle of representation*. New York: Columbia University Press.
Price, P. L. (2010). Cultural geography and the stories we tell ourselves. *Cultural Geographies, 17*, 203–210.
Reed, I. (2008). Justifying sociological knowledge: From realism to interpretation. *Sociological Theory, 26*, 101–129.
Reed, I., & Alexander, J. (2009). Social science as reading and performance: A cultural-sociological understanding of epistemology. *European Journal of Social Theory, 12*, 21–41.
Roof, W. C. (1999). *Spiritual marketplace: Baby boomers and the remaking of American religion*. Princeton: Princeton University Press.
Rose, M. (2010). Pilgrims: An ethnography of sacredness. *Cultural Geographies, 17*, 507–524.
Slater, T. R. (2004). Encountering god: Personal reflections on 'geographer as pilgrim'. *Area, 36*, 245–253.
Thrift, N. (1996). *Spatial formations*. London: Sage.
Thrift, N. (2003). Summoning life. In P. Cloke, P. Crang, & M. Goodwin (Eds.), *Envisioning human geography* (pp. 65–77). London: Arnold.
Wiggershaus, R. (1995). *The Frankfurt School: Its history, theories, and political significance*. Cambridge, MA: MIT Press.
Wilford, J. (2012). *Sacred Subdivisions: The postsuburban transformation of American evangelicalism*. New York: New York University Press.
Woodward, K. (2011). *The difference of things*. Paper session organized for the annual meeting of the Association of American Geographers, Seattle.
Yorgason, E., & della Dora, V. (2009). Geography, religion, and emerging paradigms: Problematizing the dialogue. *Social and Cultural Geography, 10*, 629–637.

Chapter 119
Postsecular Stirrings? Geographies of Rapprochement and Crossover Narratives in the Contemporary City

Paul Cloke

119.1 The Recency and Distinctiveness of Postsecular Rapprochement

Over recent years the phenomenon of "postsecularism" has received very considerable attention from the academic world, attracting some fascinatingly different interpretative postures around a broad consciousness of changing relations between religion and secular society. Dismissed by Zizek (1999, in Beckford 2012) as "postsecular crap," the use of the term postsecular has, for example, been implicated in the broad political desire to reify religion as a core motivator and player in particular forms of privatized activity within public governance (see Dalferth 2010). As such it can be read as an exploitative device for conservative and individualist politics, or perhaps more generously (for example according to Bretherton 2011) as a catalyst for political involvement that focuses on the pursuit of public and common good. Other interventions have lamented the recognition of postsecularity as some kind of grossly oversimplified stage in an historical progression from presecular through secular and beyond. Not only does such a progression misrepresent the historicity and specificity of the varying relations between secularity and religion, but it also, as Beckford (2010) has argued, fails to illuminate significant issues concerning religion and public life in contemporary society and geography. Kong (2010: 764) has emphasised that the "engagement of the sacred and secular is not re-emerging but continuing." Postsecularism is thus viewed as taking a "short-sighted view of history" (Beckford 2012: 16–17) and its use as a concept can be "like waving a magic wand over the intricacies, contradictions and problems of what counts as religion to reduce them to a single, bland category."

P. Cloke (✉)
Department of Geography, Exeter University, Exeter, UK
e-mail: p.cloke@exeter.ac.uk

S.D. Brunn (ed.), *The Changing World Religion Map*,
DOI 10.1007/978-94-017-9376-6_119

Yet despite these apparently overwhelming objections, recent research in human geography points to some significant and perhaps fresh expressions of partnership between faith-motivated and other people in the emerging landscapes of third sector activity, often occurring in spaces that arise from the neoliberal shrinkage of the formal state and subsequent forms of engagement with, or resistance to different welfare regimes. My own work, in collaboration with many others (see, for example, Cloke et al. 2010, 2012, 2013; Cloke 2010, 2011a, b; Cloke and Beaumont 2012; Beaumont and Cloke 2012; Williams et al. 2012), has sought to tread a very particular pathway of interpreting these expressions of partnership in terms of the concept of the postsecular. Research on emergency services for homeless people, and on the broader activities of faith-based organisations (FBOs) both in providing welfare and caring services and in representing and protesting against issues of injustice and exclusion, have confirmed a significant presence and in some cases emergence of activity, some of which includes partnership between faith-motivated and secular parties. It is clear that some of this faith-based/secular partnership is longstanding as well as recent. For example, the work of the Salvation Army with homeless people has a long and well documented history, but participation in contracted-out service provision (for example in programs to counter addiction) has led to a greater degree of co-working with specialist staff and volunteers with no religious motivation and connection. Equally, faith-based/secular partnership occurs both in and through FBO activity, and in and through the activities of faith-motivated individuals in secular roles, for example, in state welfare professions.

However, there is also research evidence from which to raise the question of whether "new forms" of partnership have emerged over recent decades due to the emerging contexts, critiques and opportunities arising from the playing out of neoliberal governance. As we have discussed elsewhere (see Williams et al. 2012), third sector activity is typically imagined as being enrolled directly into the ideologies and subject-formation of neoliberalism; FBOs and other partnership groups are assumed simply to be the "little platoons" of neoliberalism's army (see, for example, Hackworth 2010). However, it can be argued that this thesis of automatic co-option obscures at least two forms of resistance to neoliberalism. First, we have suggested that even if organizations are contracted into formal provision of neoliberal welfare, they can exercise a performative resistance to repressive subjectification of socially excluded people by providing spaces of care and relational possibilities that run contrary to the expectations of co-option. Secondly, significant third sector activity occurs beyond the formal welfare state, and often reflects a desire by concerned organisations and citizens to act in response to the needs of those who fall through the contemporary welfare net. Thus the activities of, for example, night shelters, drop-ins and soup runs are usually prompted by the unmet needs of homeless people, and indeed can run contrary to the neoliberal wisdom of not supporting people "on the street."

"New forms" of postsecular activity, then, can refer either to new organizations – such as the myriad service organizations for homeless people (but also undocumented migrants and trafficked people) that have sprung up over the last three decades or so, and now represent the bulk of services in many urban settings – or to ways in which existing organizations are being received differently in the public sphere (see also, Beaumont and Baker 2011). For example, longstanding organizations are perhaps being tolerated differently by people who are not their natural supporters, are finding different kinds of public voice or public acceptability, and are showing signs of being willing on occasions to loosen the strictures of faith-badging in order to work more collaboratively with those not motivated by faith. In amongst these partnerships, I have been particularly interested in processes of "rapprochement" and in the manner in which new ways of negotiating, agreeing and working together are emerging within the current dynamics of the third sector and the neoliberal state. The distinctiveness of these new forms of partnership is not, therefore, linked to grand claims about a "postsecular era," or the emergence of a "postsecular city," but rather to questions about the formation of fresh characteristics of partnership and rapprochement in evidence in both longstanding and newer organizational contexts.

In the remainder of this chapter, I discuss three avenues for discerning some of the key factors underlying these emerging rapprochements, before outlining what is for me an exciting agenda for continuing empirical research in this area. Before doing so, however, there are some obvious complexities and caveats that require clear articulation. First, the research that I have been involved with has usually taken as its focus a Christian perspective on religion, not because of disinterest in other religious inputs to rapprochement (which are clearly significant in a range of partnerships, for example in the development of city-based citizenry groups such as London Citizens – see Jamoul and Wills 2008), but due to the empirical predominance of Christian faith-motivation in UK services for homeless and other socially excluded people. This Christian focus needs to widened out to include recognition of how different religious ideas seep into the public realm in order to gain a fuller picture of potential rapprochement. Secondly, the Western world clearly shapes the context of the research, and obviously the relations between religion, state and society will vary significantly in other contexts. As a consequence the claims made about postsecular rapprochement should be seen in this context, and certainly not regarded as geographically universal. Thirdly, the underlying conditions for potential reterritorialization resulting from rapprochement will largely reflect emergent local spaces of care, tolerance, reconciliation and ethical agreement in urban settings. While there is no reason why such spaces cannot arise in rural areas, the scale and availability of interconnecting networks make cities the most likely context for significant rapprochement. Cities can induce an affective response to obvious and often visible socio-economic needs; they contain places and people that cry out for "something to be done about something." They are the locations for centralized services, which in turn offer potential spaces for subversion and resistance, and although these services are often located in the marginal spaces of the city, they are usually upheld by voluntaristic and charitable cross-subsidy from the more affluent suburbs from whence church networks and other differently –motivated individuals and groups provide flows of economic, social and even spiritual (Baker and Skinner 2006) capital into centralised spaces of care. In this way, urban socio-spatial connections are developed that can unsettle geographical boundaries as well as religious/

secular ones. Cities also attract particular campaigns to sponsor city-wide ethical tropes – such as City of Sanctuary and Fair Trade City (Darling 2010; Malpass et al. 2007) – reflecting religious and a range of other interests performing an ethical convergence of ideological, theological and humanitarian concerns. These place-related lines of flight are replicated at a much more local level in the ethical branding of schools and cafes and in spaces of reconciliation and tolerance amongst groups working across divides involving inter-religious, anti-religious or anti-secular sentiment.

119.2 Discerning the Postsecular: Discursive Technologies

Here I draw on the work of Jurgen Habermas (2002, 2005, 2006a, b, 2010; Habermas and Ratzinger 2006) to think through some key discursive technologies by which postsecular rapprochement might be discerned. Habermas' pronouncements on what he terms postsecularity have been received with much scepticism (Joas 2008). Part of the problem, as recorded by Beckford (2012), is a lack of clarity about whether Habermas' observations of the return of/to religion in the public sphere signify "an increase in religious thought, feeling and action" or simply "an increase in the salience of questions about religion in politics and the media"(p. 8). Habermas' arguments about the *impression* of religious resurgence are perhaps clarified in Eder's (2002, 2006) account of how religion did not disappear under secularization, but merely disappeared from the public sphere. In these terms, secularization "*hushed up*" religion, relegating it away from public societal debates. Unlike religion in the public sphere of the US, he argues, in Europe people gathered to practice their private faith in communal church settings rather than displaying that faith in the public sphere or out on the streets of the city. However, the impression of religious resurgence reflects that in Europe religion has *found its public voice* again, and has begun to frequent the public sphere making confident, if multifaceted, contributions to public affairs.

These contributions chime with the starting point of this chapter – that new forms of partnership involving Christian religious interests are appearing in the urban public arena, and in so doing the previously hushed up voice of religion is being released back into, and in some cases accepted back into the public sphere. Habermas argues that religion is gaining influence both as a community of interpretation – contributing to public opinion on moral and ethical issues – and as a community of service and care; in both ways public consciousness is being changed, and the question of how to participate in a more postsecular society is being worked out. He asks:

> How should we see ourselves as members of a post-secular society and what must we reciprocally expect from one another in order to ensure that in firmly entrenched nation states, social relations remain civil despite a growth of a plurality of cultures and different world-views? (Habermas 2008: 7)

Habermas' (2010) response lies in a learning process in which secular and religious mentalities can be reflexively transformed, rather than being maintained in dominant and subaltern positions respectively, and in which conditions are presupposed onto both the secular and the religious. In both cases, he argues against fundamentalist positioning. Secularist polemic of negative response to religious doctrine and action is seen as incompatible with shared citizenship and respect for cultural difference that lies at the heart of the postsecular balance. Secular citizens need to be able to discover, even in religious utterances, semantic meanings and personal intuitions that cross over into their discourses. Religious discourse equally needs to respect the authority of "natural reason" as the fallible outcome of science and of egalitarian law and morality. This latter insistence has been interpreted to suggest that "Habermas actually makes very few concessions to religion" (Beckford 2012), and indeed Habermas does admit that the divide between secular knowledge and revealed religious knowledge cannot be bridged. Nevertheless, not only does Habermas establish a framework of mutual tolerance as the foundation for postsecular rapprochement, but he also argues for the possibility of distinct crossover narratives between the religious and the secular – where different knowledges may not be bridged, mutually translating *narratives* can be deployed on which to found this rapprochement.

In *An Awareness of What is Missing* (2010), Habermas explores the discursive technologies of these crossover narratives, suggesting that democratic strength rests on broad moral stances that arise from prepolitical source such as from religious life. Rather than normative guidelines, these sources are significant motivational forces for political and ethical practices. If religious and secular utterances are mixed together in the prepolitical soup, they might also be liable to a process of "mutual translation" (Reder and Schmidt 2010: 7). As Habermas (2010: 16–17) puts it:

> The philosophically enlightened self-understanding of modernity stands in a peculiar dialectical relationship to the theological self-understanding of the major world religions, which intrude into this modernity as the most awkward element from its past.... modern reason will learn to understand itself only when it clarifies its relation to a contemporary religious consciousness which has become reflexive by grasping the shared origin of the two complementary intellectual formations....

Habermas suggests here that the role of religion in the contemporary liberal state cannot be confined to legalistic considerations of freedom of expression, but requires a legitimation of convictions that can be accepted by nonreligious and religious citizens alike. In order to replace "what is missing," therefore, postsecularity requires a technology of complementary learning in which the secular and the religious involve one another.

Critics (for example, Barbato and Kratochwil 2008; Jedan 2010; Malik 2007) worry that Habermas' analysis of what is missing is primarily concerned with "the health of public politics and morality" (Beckford 2012: 10) than about any respect for the integral value of religion in public life. In one sense, this turn towards religion is being shaped by high profile politics of multiculturalism and religious diversity which demand a more open recognition of religious elements of political difference. Alternatively, Habermas' attraction to what religion can offer has been seen as limited to that of a vague repository of moral values. It is certainly clear that his desire for complementary learning is embedded within a concern over the takeover by markets of an increasing slate of regulatory functions (Habermas and Ratzinger 2006) which he sees as leading to a depoliticizing of citizen action and a dwindling hope that creative political forces will emerge simply from secular critical reasoning. However, Habermas is clear in his call for a "conversion of reason by reason" (Habermas and Ratzinger 2006: 40) as reason reflects on its origins, discovers alternative directions, undergoes an exercise of philosophical repentance and transcends itself into an openness towards other frameworks of reason, including religious tradition. In particular, this transcending openness involves the *assimilation* by philosophy of ideas form Christian religion:

> His work of assimilation has left its mark in normative conceptual clusters with a heavy weight of meaning, such as responsibility, autonomy, and justification; or history and remembering, new beginning, innovation, and return; or emancipation and fulfilment; or expropriation, internalization, and embodiment, individuality and fellowship. Philosophy has indeed transformed the original meaning of these terms, but without emptying them through a process of deflation and exhaustion. (Habermas and Ratzinger 2006: 44–45)

It is this process of assimilation, then, that seems to offer a basis for rapprochement, whether longstanding or in new form. Assimilation can lead to different outcomes: it could lead to a dilution and deradicalization of religious concepts as they assimilate with the process of secularization (this was the underlying fear that has in the past fuelled opposition by evangelical Christians to the so-called social gospel – see Cloke et al. 2013); or it could re-connect religious concepts to their societal roots as part of a postsecular technology in which crossover narratives form the basis of new partnerships across the religious/nonreligious divide. I want to argue that these crossover narratives are perhaps more significant than some critics would credit as devices of assimilation and reflexive transformation on which emerging rapprochement might be built upon.

If the acid test of Habermas' ideas about assimilation and reflexive translation is whether religious legitimation for policy and action is overtly accepted in nonreligious arenas (see Wilford 2010), then the test is really only passed in the Western world in terms of the influence of the religious right in the U.S. (Berger et al. 2008; Dionne 2008). If however, assimilation and reflexive transformation are open to discernment within crossover narratives relating to ethical issues, for example, as seen in campaigns such as those relating to fair trade, anti-trafficking and fair wages which have held together the combined discourses and praxis of nonreligious and religious citizens, then the ideas may have more purchase in understanding emerging partnerships in the city. Accepting criticisms about the limited geographical reach and potential ethnocentricity of the concept, it remains possible that in the context of Western Europe Habermas' ideas about postsecularity and its attendant reflexive and discursive technologies can inform our understanding of the conditions in which crossover mutualities between religious and secular discourses can be enabled. This requires us to investigate the possibility that crossover narratives are occurring in particular arenas of discourse and praxis, and that they are permitting broad-based alliances to develop around a willingness to focus on ethical sympathies and actions even if that means setting aside other moral differences. Here we might expect to find some evidence, then, of a drawing back from different forms of secular and religious fundamentalism in order to forge alliances and partnerships on key ethical and political issues.

119.3 Emerging Rapprochement Beyond Secularist Fundamentalism?

In any discussion on the drawing back from secularist fundamentalism, it is clearly important to recognize that such fundamentalism is alive and well, not least in the public trial of religion by bestsellers in which the demonizing of religion as a fictitious and dangerous elusion has become almost a public sport (see, for example, the discussion in Cloke 2011b). There remains, then a continuing tendency towards building and maintaining distinct barriers between the secular and the religious both in public philosophy and in the academy (McLennan 2007). Such divisions, are not, however, the only relational current in contemporary Western society, and it can be argued that some secular (if not secularist) thinking has been attracted into exploring social and political options that involve different forms of crossover narratives with religion.

There have been at least two different flavors of discursive critique of secularism that have pointed to postsecular possibilities. The first stems from a philosophical and theological assault on the outcomes of secularist politics and society from within the religious and political establishment. Philip Blond's (1998) introduction to *Post-secular Philosophy* suggests three detrimental impacts of these secularist politics:

(a) That secularism has helped to drive religion into the hands of fundamentalist extremists with condemnatory attitudes towards others – a direction that Hedges (2006) has identified as a dangerous form of fascism.
(b) That secularism has resulted in scientific and economic modes of valuation and evaluation being transferred across into political and ethical fields, championing self-centered individualism and emphasising the appropriateness of force and counter-force. Milbank (2006) argues that these moves have results in a debased and sometimes tyrannical democratic politics that struggle to maintain civic ethics and moralities amongst the fragmented and relativistic identity politics of the postmodern.
(c) That secularism has sponsored a kind of hopeless vacuity – an almost fatalistic sense of life being dictated by the market-state's capacity to shape how we govern ourselves – that disavows any sense of transformative or transfiguring change.

Blond advocates a response that fuses cultural conservatism and radical communitarianism that is understood according to theological principles. As befits an Anglo-Catholic theologian and Conservative Party think-tanker, his *Red Tory*

(2010a, b) manifesto leans heavily on Radical Orthodox theology to endorse a sense of the common good based on transcendent religious values rather than secular norms, and to bring these values together with a belief in the power of citizenry self-help as part of the so-called "Big Society." This connection by Radical Orthodox thinkers between a critique of the secular and a re-emphasis of religious values has been represented by some as nothing less than religious empire-building (see, for example, Smith and Whistler 2010) that misappropriates postsecular ideas. Habermas' mutual translation technology here might be seen to have resulted in religious discourse being used in crossover narratives for distinctly conservative political ends.

However, the sort of critique advanced by Blond need not result in a conservative alliance between religion and politics. Connolly (1999), for example, sees the critique of the politics arising from secularism as paving the way for an "overt meta-physical/religious pluralism in public life" (p. 185) and Dalferth (2010) suggests that such a critique results in need to identify postsecular society as neither secular nor religious. These engagements seem to suggest an acceptance of the plurality of religious and metaphysical perspectives in the public arena that infer some kind of shift in public secularist self-understanding (de Vries 2006) rather than any paradigmatic shift in the public role of religion or in the nature of the secularist state. Moreover, some of the key contemporary architects of materialist socialism have leaned on concepts from religion in their discussions of hope and justice in the current context. Milbank (2005: 389) fleshes out this invocation of theology:

> Derrida sustains the openness of signs and the absoluteness of the ethical command by recourse to .... negative theology; Deleuze sustains the possibility of a deterritorialisation of matter and meaning in terms of a Spinozistic virtual absolute; Badiou sustains the possibility of a revolutionary event in terms of the one historical event of the arrival of the very logic of the event as such, which is none other than Pauline grace; Zizek sustains the possibility of a revolutionary love beyond desire by reference to the historical emergence of the ultimate sublime object, which reconciles us to the void constituted only through a rift in the void. The sublime object is Christ.

These theological connections should certainly not be used to suggest that left-leaning neo-Marxist philosophers are converting *en bloc* to a religious perspective. Indeed there has been considerable theological disquiet about how philosophical use of theological connections has exploited religion for philosophical resource and in so doing has tended to empty out the resources concerned of their distinctly religious connotations (see Sigurdson 2010). However, it should still be recognized that significant strands of the philosophy that underpin contemporary material socialism are interconnecting, albeit in partial and fragmentary manner, with an implicit horizon of faith and belief and on the *theo-ethics* (Cloke 2010) of otherness, grace, love and hope that carry an excess beyond material logic and rationale.

These *discursive* responses to the critique of secular fundamentalism have coincided with evidence of new arenas of postsecular *praxis,* although little is known how praxis precisely shapes or is shaped by discourse in this area. It is certainly over-simplistic to suggest that postsecular partnerships arise directly from the discursive moves discussed above, and it may well be that postsecular discourses serve to legitimate already existing postsecular practice. What we can say is that new forms of postsecular practice have been commonly associated with a particular niche within neoliberal landscapes. In some cases, third sector groups, including FBOs have become co-governing service providers, relying on state funds for some of their activities (Bretherton 2010). In this way (see Beaumont and Cloke 2012; Dinham and Lowndes 2009) the activity of FBO's appear to chime with Blond's conservative communitarianism. However such groups also engage in various forms of resistance to neoliberal governance (Williams et al. 2012).

The evidence of FBO activity is significant, but should not be overemphasised, not least because other – nonreligious – organizations in the third sector have also been shaping the conduct and subjectivities of the city overt the same period. What I want to emphasize here is the potential for postsecular forms of practice in these niches. As faith-based and other more secular organizations provide opportunities for both religious and nonreligious people to become staff and volunteers, so the boundaries between faith-based and other praxis become more porous, and reflect postsecular ideas of translation and crossover. Some of this partnership may well be rather pragmatic in nature, but to various degrees they demonstrate willingness for rapprochement that creates conditions for postsecular possibility. In turn, these conditions open up new spaces of opportunity for faith-groups to break out from their previously "hushed up" position in the public sphere, and for the performance of Habermas' preconditions for new postsecular consciousness – shared citizenship, mutual tolerance, reflexive transformation and crossover ethical narratives.

119.4 Emerging Rapprochement Beyond Fundamentalist Religion

If the mutual translation of postsecularity is partly fuelled by the critique of fundamentalist secularism, then it must also be associated with a move away from fundamentalist religion that accepts faith-action as a form of "no-strings" service and caritas rather than as a vehicle for evangelism. As with secularist thought, religious fundamentalism in Christianity remains a strong element in the practice of that faith, but the possibilities for postsecular rapprochement have increasingly been opened up as Christian people have become increasingly willing to eschew faith-by-dogma in order to practice the potential of faith-by –praxis. In this way, faith-based activists have increasingly found common ground with others fighting for a similar ethical cause, even if they do not share the same moral frameworks. As Brewin (2010) has discussed, faith communities engage with otherness both *internally*, within the self and within faith communities, and *externally* in their wider communities as opportunities to care and to serve arise in semi-autonomous zones (Bey 2003) where action spaces provoke ethical responses of responsibility and trust in favour of others. These external engagements reflect potential contributions to postsecular rapprochement and its critique of fundamentalism.

Two elements of this faith-by-praxis stand out as a response to the critique of the politics shaped by secularism. First, in contrast to the rational, material, non-transformable, *visible,* nature of the secularization thesis, the continuing emergence of Christian faith-praxis in the public realm has pointed to the possibilities of what is *invisible* as well as visible in contemporary society. Part of this focus derives from the prophetic expectation carried by many evangelical believers of a Jesus-centered kingdom that is contemporary as well as eschatological. However, this focus also arises in the poststructural reformulation of religion. For example, Caputo (2001) sees faith articulating a hyper-real vision of reality beyond the visible; a vision that rests not on the certainties of evangelical faith, but on a more uncertain translatability between the divine and the human played out in love, beauty, truth and justice. In this way, an always-becoming love of God is evident in the actions of faithful lives towards the transformative possibilities of the future. Faith, then is literally an enactment that bears witness to the love of God – a leap of love into the hyper-real of unhinging human control and power, and reliance on invisible powers for seemingly impossible hope. Such leaps of love are capable of unsettling the settled secular order (Caputo 2006), and are precisely illustrated by the kind of involvement of faith-motivated people in serving marginalized people that forms the basis of the arguments in this chapter about emerging postsecular spaces in the city. They seem impossible – how can providing advice, hospitality and sanctuary to a small number of immigrants and asylum seekers solve the deep-seated problems underlying multi-cultural tension, and how can providing food and shelter for a few homeless people solve the problems of homelessness? Yet a willingness to countenance the impossible results in faith-by-praxis spilling out agape and caritas into situations of need, which in turn can become attractive to others who, motivated by other than religion, will accept the risk that doing something about something is preferable to doing nothing.

This seeking after the possibilities of the invisible is aided and abetted by a concomitant willingness within the faith-networks concerned to embrace the public (rather than private) praxis of faith. Both the intrigue and experience of faith through ethical practice are attractive to those Christian faith-communities where theological dogma is inaccessible without practical arenas in which to explore, discover and constitute its meaning (Hutter 1997; Oakley 2005). This move towards being doers rather than just hearers of the Word has coincided with the rise of a form of Christian virtue ethics, that according to Wright (2010), blends tradition and immanence. In this way, the contribution of Christian faith-ethics to the possibility and hope of postsecular rapprochement lies in the cultivation of virtues based on the expectation that charity can be reproduced as love, and friendship as a creative gift of hospitality, and that both can be practiced in a counter-ethics of relational and empowering service rather than in proselytising or heavy evangelism (Coles 1997).

I am certainly not suggesting here any kind of totalizing transformation of religion that is somehow resulting in a homogeneous positioning of virtuous faith groups in the praxis of postsecular partnership. Many Christian practices of social action remain doggedly resistant to partnership with other faiths or secular interests, seeking instead to "badge" their work as Christian and to retain a theological purity of expression through restrictive membership and activity. However, there is a multi-dimensional current of faith-motivated activity that is open to the partnership involved in postsecular rapprochement. Large mainstream denominational churches are becoming increasingly involved in the performance of faith-ethics in the public sphere, including the establishing and sustaining of large-scale projects that can exhibit postsecular characteristics. In addition, the growth of new expressions of church (Gibbs and Bolger 2006) reflects intentional faith-communities that deliberately seek to embody a radically transformative theology of social engagement based on habitual caring and empowering praxis involving marginalized others, and are therefore making a strong contribution to the innovative possibilities of postsecular spaces. In these and other forms, faith-by-praxis is leading to participation in the public sphere that is capable of complementing more secular forms of political participation – claiming similar ethical concerns, and perhaps deploying mutual crossover narratives that in turn provoke crossover practices that can blur some of the boundaries between religious and secular action.

119.5 Conclusion

The argument expressed in this chapter uses preliminary evidence from research on caring responses to homelessness and on the welfare and caring services provided by FBOs to lay the foundation for an understanding of particular postsecular possibilities in the contemporary city. Taking note of those critics who regard postsecularity as conceptually inept in its apparently overgeneralized historic and organisational treatise of going beyond the secular, I nevertheless argue that some of the *technologies* of reflexive translation and crossover narratives proposed by Jurgen Habermas do seem to be apparent in the emergence of partnerships and rapprochements between different religious and nonreligious parties in the welfare sector. Indeed, the context of neoliberal governance has seemed to open out new spaces for such partnership, which I argue should not be simply dismissed as the enrollment and co-option of third sector organizations into neo-liberal welfare and subject formation. Instead, there is initial evidence that such partnerships can serve to resist neoliberal ideologies, and represent alternative ethically-based platforms for action on behalf of marginalized people. This more compact and modest conception of emerging spaces of postsecularity helps to explain how the hushed up voice of religion is being released back into the public sphere and is helping to reterritorialize public life in areas of care, welfare and justice.

Along with initial empirical research evidence, I have discussed by discursive and praxis-focused trends that seems to be associated with the critical (if only partial) rejection of fundamentalist discourse and practice in both secular and religious circles. These trends help to explain the potential for postsecular rapprochement between parties that may not be able to agree on the rational or immanent bases of knowledge, but that are able to come together around common ethical tropes in order to "do something about something." These postsecular stirrings seem to me to present an exciting agenda for new research into the actual organizational practices that are developed and negotiated in order to initiate and sustain such rapprochement. We need to know far more detail about how exactly crossover narratives are proposed, introduced, explained and acted upon, and about how secular/religious boundaries can be rendered porous or reinforced during these processes. We need to know if rapprochement is somehow attracted to particular issues or arenas, and in what circumstances it breaks down. When we know more about these issues, we can perhaps present more authoritative arguments about what currently are (for me at least!) stirrings of something fascinating in the social geography of the contemporary city.

References

Baker, C., & Skinner, H. (2006). *Faith in action: The dynamic connection between spiritual capital and religious capital*. Manchester: William Temple Foundation.

Barbato, M., & Kratochwil, F. (2008). *Habermas' notion of a post-secular society. A perspective from international relations* (EUI Working Papers. MWP 2008/25). Florence: European University Institute.

Beaumont, J., & Baker, C. (Eds.). (2011). *Postsecular cities: Religious space, theory and practice*. London: Continuum.

Beaumont, J., & Cloke, P. (Eds.). (2012). *Faith-based organisations and exclusion in European Cities*. Bristol: Policy Press.

Beckford, J. (2010). The return of public religion? A critical assessment of a popular claim. *Nordic Journal of Religion and Society, 23*, 121–136.

Beckford, J. (2012). Public religions and the postsecular: Critical reflections. *Journal for the Scientific Study of Religion, 51*, 1–19.

Berger, P., Davier, G., & Fokas, E. (2008). *Religious America, secular Europe*. Aldershot: Ashgate.

Bey, H. (2003). *T.A.Z.: The temporary autonomous zone, ontological anarchy, poetic terrorism*. New York: Autonomedia.

Blond, P. (1998). Introduction: Theology before philosophy. In P. Blond (Ed.), *Post-secular philosophy: Between philosophy and theology* (pp. 1–66). London: Routledge.

Blond, P. (2010a). *Red Tory*. London: Faber and Faber.

Blond, P. (2010b, September 21). Reclaiming a liberal legacy. *The Guardian*.

Bretherton, L. (2010). *Christianity and contemporary politics*. Chichester: Wiley-Blackwell.

Bretherton, L. (2011). A postsecular politics? Inter-faith relations as a civic practice. *Journal of the American Academy of Religion, 79*, 346–377.

Brewin, K. (2010). *Other: Loving self, God and neighbour in a fractured world*. London: Hodder and Stoughton.

Caputo, J. (2001). *On religion*. London: Routledge.

Caputo, J. (2006). *The weakness of God: A theology of the event*. Bloomington: Indiana University Press.

Cloke, P. (2010). Theo-ethics and radical faith-based praxis in the postsecular city. In A. Molendijk, J. Beaumont, & C. Jedan (Eds.), *Exploring the postsecular: The religious, the political, the urban* (pp. 223–243). Leiden: Brill.

Cloke, P. (2011a). Emerging postsecular rapprochement in the contemporary city. In J. Beaumont & C. Baker (Eds.), *Postsecular cities: Religious space, theory and practice* (pp. 237–254). London: Continuum.

Cloke, P. (2011b). Geography and invisible powers: Philosophy, social action and prophetic potential. In C. Brace, A. Bailey, S. Carter, D. Harvey, & N. Thomas (Eds.), *Emerging geographies of belief* (pp. 9–29). London: Cambridge Scholars.

Cloke, P., & Beaumont, J. (2012). Geographies of postsecular rapprochement in the city. *Progress in Human Geography, 37*, 27–51, (Online First, 18 April 2012).

Cloke, P., May, J., & Johnsen, S. (2010). *Swept up lives: Re-envisioning the homeless city*. Chichester: Wiley-Blackwell.

Cloke, P., Thomas, S., & Williams, A. (2012). Radical faith praxis? Exploring the changing theological landscape of faith motivation. In J. Beaumont & P. Cloke (Eds.), *Faith-based organisations and exclusion in European cities* (pp. 105–126). Bristol: Policy Press.

Cloke, P., Beaumont, J., & Williams, A. (2013). *Working faith*. Milton Keynes: Paternoster Press.

Coles, R. (1997). *Rethinking generosity: Critical theory and the politics of caritas*. Ithaca: Cornell University Press.

Connolly, W. (1999). *Why I am not a secularist*. Minneapolis: University of Minnesota Press.

Dalferth, I. (2010). Post-secular society: Christianity and the dialectics of the secular. *Journal of the American Academy of Religion, 78*, 317–345.

Darling, J. (2010). A city of sanctuary: The relational re-imagining of Sheffield's asylum politics. *Transactions. Institute of British Geographers, 35*, 125–140.

De Vries, H. (2006). Introduction: Before, around and beyond the theologico-political. In H. De Vries & L. Sullivan (Eds.), *Political theologies: Public religions in a post-secular world* (pp. 1–90). New York: University of Fordham Press.

Dinham, A., & Lowndes, V. (2009). Faith and the public realm. In A. Dinham, R. Furbey, & V. Lowndes (Eds.), *Faith in the public realm: Controversies, policies and practices* (pp. 1–20). Bristol: Policy Press.

Dionne, E., Jr. (2008). *Souled out: Reclaiming faith and politics after the religious right*. Princeton: Princeton University Press.

Eder, K. (2002). Europaische sakularisierung- ein sonderweg in die postsakulare gesellschaft. *Berliner Journal fur Soziologie, 3*, 331–343.

Eder, K. (2006, August 17). Post-secularism: A return to the public sphere. *Eurozine*.

Gibbs, E., & Bolger, R. (2006). *Emergent churches: Creating Christian community in postmodern cultures*. London: SPCK.

Habermas, J. (2002). *Religion and rationality; Essays on reason, God and modernity*. Cambridge, MA: MIT Press.

Habermas, J. (2005). Equal treatment of cultures and the limits of postmodern liberalism. *Journal of Political Philosophy, 13*, 1–28.

Habermas, J. (2006a). Religion in the public sphere. *European Journal of Philosophy, 14*, 1–25.

Habermas, J. (2006b). *Time of transitions* (G. Schott, Trans.). Cambridge: Polity Press.

Habermas, J. (2008, June 18). Notes on a postsecular society. *Sign and sight.com*, pp. 1–23.

Habermas, J. (2010). An awareness of what is missing. In J. Habermas et al. (Eds.), *An awareness of what is missing* (pp. 15–23). Cambridge: Polity Press.

Habermas, J., & Ratzinger, J. (2006). *The dialectics of secularization: On reason and religion*. San Francisco: Ignatius Press.

Hackworth, J. (2010). Neoliberalism for God's sake: Sectarian justifications for secular policy transformation in the United States. In A. Molendijk, J. Beaumont, & C. Jedan (Eds.), *Exploring the postsecular: The religious, the political, the urban* (pp. 357–379). Leiden: Brill.

Hedges, C. (2006). *American fascists*. New York: Free Press.

Hutter, R. (1997). *Suffering divine things: Theology as church practice*. Grand Rapids: Eerdmans.

Jamoul, L., & Wills, J. (2008). Faith in politics. *Urban Studies, 45*, 2035–2066.

Jedan, C. (2010). Beyond the secular? Public reason and the search for a concept of postsecular legitimacy. In A. Molendijk, J. Beaumont, & C. Jedan (Eds.), *Exploring the postsecular: The religious, the political, the urban* (pp. 317–327). Leiden: Brill.

Joas, H. (2008). *Do we need religion?* (A. Skinner, Trans.), Boulder: Paradigm.

Kong, L. (2010). Global shifts, theoretical shifts: Changing geographies of religion. *Progress in Human Geography, 34*, 755–776.

Malik, A. (2007, April 23). Take me to your leader: Post-secular society and the Islam industry. *Eurozine*.

Malpass, A., Cloke, P., Barnett, C., & Clarke, N. (2007). Fairtrade urbanism? The politics of place beyond place in the Bristol Fairtrade City campaign. *International Journal of Urban and Regional Research, 31*, 633–645.

McLennan, G. (2007). Towards postsecular Sociology? *Sociology, 41*, 857–870.

Milbank, J. (2005). Materialism and transcendence. In C. Davis, J. Milbank, & S. Zizek (Eds.), *Theology and the political: New debates* (pp. 393–426). Durham: Duke University Press.

Milbank, J. (2006). *Theology and social theory: Beyond secular reason*. Oxford: Blackwell.

Oakley, M. (2005). Reclaiming faith. In A. Walker (Ed.), *Spirituality in the city* (pp. 1–14). London: SPCK.

Reder, R., & Schmidt, J. (2010). Habermas and religion. In J. Habermas et al. (Eds.), *An awareness of what is missing* (pp. 1–14). Cambridge: Polity Press.

Sigurdson, O. (2010). Beyond secularism? Towards a post-secular political theology. *Modern Theology, 26*, 177–196.

Smith, A., & Whistler, D. (Eds.). (2010). *After the postsecular and the postmodern: New essays in continental philosophy of religion*. London: Cambridge Scholars.

Wilford, J. (2010). Sacred archipelagos: Geographies of secularization. *Progress in Human Geography, 34*, 328–348.

Williams, A., Cloke, P., & Thomas, S. (2012). Co-constituting neoliberalism: Faith-based organisations, co-option and resistance in the UK. *Environment and Planning A, 44*, 1479–1501.

Wright, T. (2010). *Virtue Reborn*. London: SPCK.

Chapter 120
Faith Islands in Hedonopolis: Ambivalent Adaptation in Las Vegas

Rex J. Rowley

And there shall be a tabernacle for a shadow in the daytime from the heat, and for a place of refuge, and for a covert from storm and from rain. (Isaiah 4:6)

I pray not that thou shouldest take them out of the world, but that thou shouldest keep them from the evil. (John 17:15)

Where great evil is there can also be great righteousness, not hand in hand, but side by side. (Tad Seargent, Las Vegas local)

120.1 Introduction

Las Vegas is a uniquely instructive place to explore religious experience. Religious and nonreligious locals alike often claim that local life in this tourist town is "separate" from the Strip and what it represents. This insider/outsider binary is, indeed, one of the city's defining characteristics (Rothman 2002; Rowley 2013). Locals sometimes describe the dichotomy invoking an "island" metaphor. Theresa Murakami, for example, explained that her path to work at a local library goes around (not through) the tourist and casino core of the valley containing the Strip and downtown. "That's how it is," she said. "There's the island of the Strip and downtown in the larger ocean of Las Vegas. We call it [the tourist core] 'Oz,' and I will go around and avoid it except when we take our family down there when they come to visit."

The experience of Las Vegas faith tradition often follows this same dichotomous pattern. A number of institutional religions in the city (including Islam, Judaism, and most evangelical Christian denominations) espouse principles that stand in opposition to the temptation and allure promoted by Sin City, and many local believers strive to avoid the vices in their city (Rowley 2012). The metaphor of the island can be extended to understand the experiential interaction between faith and the city. Physiographic islands are a refuge from the ocean for birds, plants, animals, and

R.J. Rowley (✉)
Department of Geography-Geology, Illinois State University, Normal, IL 61790, USA
e-mail: rjrowls@gmail.com

S.D. Brunn (ed.), *The Changing World Religion Map*,
DOI 10.1007/978-94-017-9376-6_120

humans. Faith islands—religious edifices or an institutional religion itself—exist as places of refuge for believers, an escape from "the world" around them. At the same time, an island is not totally isolated from or impervious to the influence of the surrounding sea. Waves lap at the shore, eroding and depositing sediments. Occasional storms rage, inundating large portions of coastline, decimating vegetation and villages, or disrupting an inland supply of fresh water. Furthermore, flotsam and jetsam can add to or detract from the cultural or natural life on the island. Simply put, surrounding water constantly interacts with the island, shaping it, changing it, and making it what it is. Similarly, followers of a faith tradition that stands in opposition to a surrounding context are often shaped and influenced by exposure to its influence (Stump 2008).

In this essay, I will elaborate on the idea of "faith islands" though the lens of practicing members of The Church of Jesus Christ of Latter-day Saints (often called Mormons or simply Latter-day Saints) in Las Vegas. The interactive relationship between place and religion has received some attention in geographies of religion (Watling 2001; Stump 2008; Mazumdar and Mazumdar 2009; Rowley 2012), but is an area in need of further empirical work (Kong 2001; Yorgason and della Dora 2009). Here I use this interaction as a stepping-off point, but go beyond it to ponder deeper questions of personal faith amidst the temptation and tension that exists outside the walls of a church or religious home. I base my discussion on interviews with 12 practicing Latter-day Saints who grew up or currently live in Las Vegas. I will also draw on my own life growing up Mormon in the city, reflexively using some personal experiences (Slater 2004), and many other observations and informal conversations I have had with local church members over the years. Interviewees quoted here have been given pseudonyms.

Following a short background on the Latter-day Saint experience in Las Vegas, I describe how the faith of Mormons acts as a spiritual sanctuary from the sin and vice of the city. Then I discuss how direct and indirect secular exposure to and interaction with the surrounding city affects their faith. I conclude by exploring how the experience of Mormons in this hedonopolis may be applicable in other places and circumstances around the world.

120.2 Mormons in Las Vegas

Latter-day Saints are not the only locals who share a relationship with the sin in Sin City. That "Sin City" moniker itself is, in fact, broadly applied by the media, critics of the city, and locals of all stripes to describe the overt acceptance of gambling and sexual mores that may be shunned elsewhere (Smith and Smith 2003). It is used in a negative light by those who oppose such acceptance, but carries more positive connotation for others who embrace or market this side of the city's personality. Such discourse with "sin" has been a longstanding concern for many Las Vegas residents, regardless of their religious affiliation. Its beginnings likely date to the city's 1905 origins as a whistle-stop frontier town that embraced prostitution and gambling (Paher 1971). Historians point to further development and promotion of industries of sin during the decades following World War II when locals grappled more intensely with dilemmas that came from living in a renowned gambling and adult-oriented resort city (Moehring 2000; Rothman 2002; Schwartz 2003).

The perspective of Mormons in such a place, however, provides an illustrative case for understanding the concept of faith islands. After all, the contrast between teachings of The Church of Jesus Christ of Latter-day Saints (hereafter, "the Church") and what is "taught" (and practiced) on the Strip is clearly demarcated. Church doctrine on the ills of gambling, drinking, pornography, and of abstinence before and fidelity after marriage are commonly and deliberately taught in family and church settings. In fact, Latter-day Saints are one of two religious groups in the city (Islam, which hosts five local congregations and more than 1,000 adherents, is the other) whose doctrine and leadership firmly oppose gambling (Hinckley 2005; ASARB 2010; Rowley 2012). Such teachings stand opposite of the *raison d'etre* depicted in Las Vegas marketing campaigns and in experiential realities of both locals and tourists.

Despite such opposition, Latter-day Saints have long played an important part in the city's history. Mormons are often given credit for creating Las Vegas (Paher 1971; Roske 1986). It is true that Mormon settlers were the first permanent, European settlers to take up residence near the life-giving springs of the valley in 1855, but they largely abandoned what is now called the "Mormon Fort" 18 months later. In reality, the city was founded as a way station on William Clark's San Pedro, Los Angeles, & Salt Lake Railroad during a land auction 50 years later. Still, the city's proximity to Utah has always brought large numbers of Latter-day Saints to Las Vegas and the influence this faith tradition has had on the area is undeniable (Fig. 120.1).

In 2010, more than 124,000 Latter-day Saints called southern Nevada home. According to a recent Association of Statisticians of American Religious Bodies survey, 64 out of every 1,000 people (the adherence rate) in Clark County—where Las Vegas contributes 97 % of the county's total population—is Mormon. By another measure, 17.8 % of local religious adherents are Latter-day Saints. This makes the Church the city's second largest single religious group after Catholics, who make up around 50 % of local worshipers (ASARB 2010). Based on numbers from the 2010 ASARB study and Metropolitan Statistical Area (MSA) definitions from the United States Census Bureau, Las Vegas is the most "Mormon" of any major metropolitan area in the country, after Salt Lake City, Utah, where the Church is headquartered (Fig. 120.2). In other words, no other American city with over one million people (outside of Utah) has an adherence rate higher than the Las Vegas-Paradise, Nevada, MSA (ASARB 2010).

Beyond raw numbers, Latter-day Saints also play an important role in local business and politics. Several non-Mormon interviewees discussed the influence church members have in local politics, business, and casino management.

Fig. 120.1 The Reed Whipple Cultural Center on North Las Vegas Boulevard was the first Latter-day Saint stake center in Las Vegas before the Church sold it to the city. (A stake is a multi-congregational geographic unit in the Church similar to a Catholic diocese). Located just north of downtown, this building symbolizes the legacy of the many Mormons who have lived in the city, even during its early days when most of the population lived near the historic city center (Photo by Rex J. Rowley 2012)

Anecdotally, in 2007, when I last lived in Las Vegas, I observed that the mayors of Henderson and North Las Vegas, and the chair of the Clark County Commission (which has jurisdiction over most of the unincorporated urbanized area in Las Vegas) were members of the Church. Such influence extends to national government. Another Mormon, Harry Reid, represents Nevada as majority leader in the United States Senate. In short, Latter-day Saints have a deep connection to Las Vegas in history, demography, and influence.

120.3 Spiritual Sanctuary

Faith is an important, protecting influence for Mormons in Las Vegas. Throughout the world, wards of the Church—geographical divisions of congregations, similar to parishes for Catholics—are a source of community and provide a sense of belonging for members. Indeed, the congregational relationship among Latter-day Saints is often compared to a family; the term "Ward Family" is common in church vernacular. It is not surprising then, that many Las Vegas Mormons consider the relative high density of members within the city as a source of courage and comfort. Savanna Frey, who grew up in Las Vegas but now lives in the upper Midwest, put it this way: "For me, I think it was easier than it would have been growing up in the Midwest. Our values

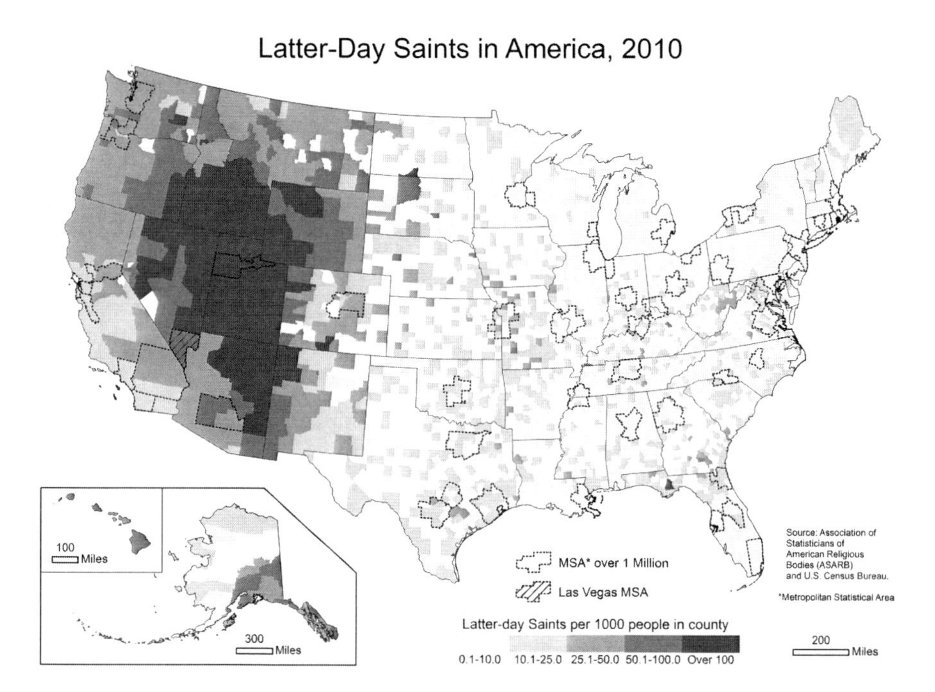

Fig. 120.2 Prevalence of Latter-day Saints in 2010, by county. Las Vegas has the highest adherence rate (number per 1,000 population) of any major MSA in the country after Salt Lake City (Map by Rex J. Rowley)

may not have been the same as those of the city, but we weren't out-numbered. I knew lots of people that shared the same beliefs I did." In fact, she noted that her "favorite part [about being Mormon in Las Vegas] was that you didn't have to drive very far to go to church and most of the kids you went to school with were also Mormon." Sam Marvin also spoke extremely positively about his experience growing up in the city and pointed to that same strength-in-numbers quality as one reason he has chosen to stay in his hometown with his three children: "I've chosen to raise my kids in Las Vegas because of the strength in the Church … and the strength of its members."

As Sam noted, the role of a large membership goes beyond just numbers and includes the support members can be to each other amid temptations and challenges to faith in Las Vegas (Fig. 120.3). Savanna likes her current home in the Midwest, but recognizes that many Latter-day Saint congregations there cover a much wider area. She said: "I miss that my support system (the other members) were walking distance (not hours) away." She went on relating how she felt protected among her church member friends:

> [In high school], we had parties where no one drank, everyone was invited, and everyone had a good time … and we didn't have to care about what everyone else was doing because there were so many of us that shared the same values.

Fig. 120.3 The full church parking lot on a Sunday morning illustrates the high density of Latter-day Saints that live and worship in Las Vegas. The density is a source of strength for many local members (Photo by Rex J. Rowley 2012)

Underscoring the significance of shared belief among friends, Savanna compared her past and present circumstances:

> My husband is not a member [of the Church] and all of our friends out here [in the Midwest] drink. There's no getting around it, really.... I'd love to say I'd never be seen at a bar, but out here if you want food or something to do the bars are the only places to go.... I miss that there were a lot of people who understood that you don't have to have any form of alcohol or substance to have a good time.

Marsha Lamb Leason, who grew up in Alamo, Nevada, and moved 90 mi (145 km) south to Las Vegas when she was a young adult, extended this notion of strength in the Church. Marsha put her comments in the context of expectations among non-Mormons about how Mormons should behave:

> If you're a member of the Church and start to do something wrong, non-member friends will tell you, 'You can't do that.' ... You have to stand up for what you believe in. (Hill 2008)

Ted Burke, a master plumber born in Las Vegas in the 1940s, has lived his entire life in the city. I interviewed him during his service as the bishop of his ward (all leadership in the Church is made up of lay members, and bishops are called and oversee a ward congregation often while keeping a full-time career). When asked about the challenges his six children faced in the city, he responded: "There's so much more positive here than any other place. How can I say this?" He stopped himself midsentence, paused, and continued: "Positive values can be taught in a negative environment only if you have an equally strong positive environment." He told of opportunities he had to start an arm of the family business in Phoenix, Reno, or Sacramento. In the end, after he and his wife had seriously considered the options, they decided against the move because, as Ted explained, "the Church was not as strong in any of those cities as it was in Vegas. And with the Church not being as strong, what kind of friends would [our children] have?" (Rowley 2013).

Such a bond between friends who share values connected to their shared membership in a church can make a difference in a place like Las Vegas. Kale Ismay, who moved to Las Vegas with his family at age 11, said that "sneaking into strip clubs was a common occurrence that was described in the halls of local high schools." He explained that his ability to avoid this element of the city came from his peer associations: "The temptation was always there, but I gravitated to friends with similar values and found little difficulty resisting the temptation for that reason."

For Latter-day Saint youth, the Church's seminary program is one place where such friendships are made. Seminary is a 1-h scripture study class for high-school-aged Mormons held at a church building near local high schools, usually early in the morning before school begins. Because seminary classes are typically made up of members that attend various wards in the area, it is where hundreds of teenage Mormons meet and make close friends with high school peers who share similar values. Bishop Burke stated how such a program, centered on spiritual teaching, encourages Mormon youth to surround themselves with positive influences. Seminary is where I found much of the strength I would draw on to avoid the spiritual pitfalls (gambling and drinking, predominantly) my parents and church leaders had warned me about. In fact, looking back on it, it feels almost as if I was isolated, secured from the nonspiritual influences that existed, because of my association with fellow students from seminary.

Underlying the strength that exists within Latter-day Saint membership are principles taught by parents, leaders, and seminary or Sunday school instructors. Helping to lift up one another is, in fact, one such tenant. When asked how his faith impacted his experience growing up in Las Vegas, Sam Marvin replied:

> Religion played a vital role. There are plenty of opportunities to get into trouble in Las Vegas, but the Gospel set the boundaries to follow and kept me away from most of the difficulties that many people in Vegas face. … Most people are surprised that you can be a religious person and live in Las Vegas. They wonder how it's possible and are pleasantly surprised when they see the results that can come from living the Gospel.

Savanna Frey gave a similar response to the query, again relating it to her entertainment landscape: "My religion taught me that having fun didn't require alcohol or other substances regardless of what the rest of the world thinks. In a town like Vegas, so much of the entertainment is centered around gambling and 'ladies' or 'gentlemen's clubs.' My religion taught me that those things were not going to make a person happy and that any entertainment found there (if any) was not worth the damage it would do to your spirit."

The temple is the pinnacle of worship for Latter-day Saints and is another source of spiritual sanctuary for Las Vegas Mormons. This edifice, also called "The House of the Lord," is where church members who meet certain standards participate in ordinances and instruction separate from regular Sunday meeting schedules. Temples are found in 137 communities around the world (LDS 2012). Members are taught to look to the temple as a beacon in their lives, a symbol of spiritual worthiness and attainment. Such is the case for Latter-day Saints in Las Vegas and their temple,

Fig. 120.4 A view from Frenchman Mountain, which stands as the eastern boundary of the Las Vegas Valley. Visible in the foreground is the Las Vegas Temple of The Church of Jesus Christ of Latter-day Saints, and an adjacent meetinghouse used for regular Sunday worship. The downtown area and north end of the Las Vegas Strip are visible in the background and left, respectively (Photo by Rex J. Rowley, 2007)

which was completed in 1989 and sits on the flanks of the eastern edge of the Las Vegas Valley (Fig. 120.4).

Mark Lewin, another Las Vegas bishop, pointed to the temple as symbol of the contrast between the teachings of Vegas and the teachings of God. He spoke regarding the youth of the Church, a group in his congregation for which he felt an important stewardship, but his comments represent the meaning of the temple for Las Vegas Mormons. Bishop Lewin acknowledged the challenge of accessibility of sin in the city, but also acknowledged the beacon of the temple: "There is such a contrast that you can delineate more easily between good and evil." Members don't necessarily need to be in the temple all the time or stay away from nongambling entertainment or dining in the casinos all the time, he continued, but the temple is a "symbol and a standard to help them choose" virtue over vice (Rowley 2012, 2013).

I have seen the reverence Las Vegas Mormons give the temple and the symbol it can be for their faith. As a teenager, I remember several trips to the temple with another dozen or so young people. I recall those in attendance were excited and personally impacted by the beauty of the building and the spiritual connection made with each other and God during these excursions. I have known several individuals and families who regularly make the drive from near or far locations across the metropolitan area to simply enjoy the temple grounds. One such family makes a special trip each Christmas Day to sing hymns outside the temple as a form of musical worship.

Such interactions with the temple are certainly not unique to Las Vegas Mormons, but are examples of how the representation of one building can stand as an oppositional force to the secular world that surrounds it. Tad Seargent, New Zealander by birth, Las Vegan by transplant, related a memory that brought this contrast into sharp focus. He was in attendance when Gordon B. Hinckley, then president of the Church, spoke at a multicongregational meeting held at University of Nevada, Las Vegas's Thomas and Mack Center. Tad recalled how President

Hinckley marveled: "Isn't it ironic that all of us are here in Sin City? That we have a temple [here]?"

An experience stuck in my memory further underscores this oppositional dichotomy. After one youth temple excursion, our group was walking around the grounds when we came to a point that overlooks the Las Vegas Valley. The sun was falling and the characteristic blanket of lights illuminated the desert metropolis. My bishop and I conversed about our experience and then he lamented, "Out of the temple, now back to Sodom and Gommorah." Indeed, from a purely spatial standpoint, it is not surprising that the temple is built on a location at almost the farthest possible point (in the eastward direction) from the Strip/downtown geographic and tourist center of Sin City.

The temple is indicative of the intense contrast between Mormon faith and the city, which is apparent in many of the above comments. Such is the nature of the faith island as refuge. In fact, I have spoken with many local members who see that contrast as an important way their faith is strengthened. Tad Seargent noted how such a contrast in this city can benefit one's faith: "I think that Las Vegas Mormons, if they don't get dragged down, have a more balanced view of the world. They have the influence of the Gospel, and they have the evil that's just out there, in your face." Sam Marvin put his thoughts on raising children here in a similar context: "I feel that few members take the Church for granted and my perception is that this happens frequently in other cities because there isn't [the] big contrast between right and wrong like there is in Las Vegas." And, Marsha Leason explained: "There is an opportunity to be strongly committed to what you believe when you have both sides pulling at you" (Hill 2008).

Bishop Ted Burke saw the opposition between church teachings and gambling as one of the benefits of living and raising children in Las Vegas. He remembered how his father used such opposition as a teaching opportunity. While Ted's mother prepared for a surprise party, she wanted the kids out of the house. Ted's father drove him and his brother down Las Vegas Boulevard, past some of the projects under construction in 1949. Having done some contracting work at the Desert Inn, which was nearly complete and was the biggest hotel in town at the time, Ted's father told his boys: "These big hotels are not built with the money that people win. They are built with the money that people lose." As a very young boy of around five years of age, this experience stuck in the mind of Ted and he has never had a desire to participate in that side of the city. He then explained the choice for local Latter-day Saints in a place where temptation stares you in the face: "If you live here, it's kind of hard to have a run of the mill lifestyle. You're either in the gambling, sex, and drugs side of things, or you're not. There's no middle ground" (Rowley 2013).

Sam Marvin saw that same contrast in the choices he made growing up in Las Vegas. "If I didn't have the boundaries and guidance provided by the Church, my life would have turned out much different," he said.

> It would have been very easy to participate in all the terrible things that Vegas prides itself on in marketing itself as a tourist destination. One thing about Las Vegas is there isn't much grey area in the middle. Most people [in the Church] either keep the commandments and rules of the Church or they don't want anything to do with it. They may vacillate in the middle for a short time, but it typically doesn't take long before they choose one course or the other.

A similar black-white-grey analogy is commonly expressed by other Las Vegas Mormons. Whereas the choices for people in some communities may fall in a nebulous grey area between what they are taught and what the community accepts, in Las Vegas that same choice is the starker one between black and white.

120.4 Secular Exposure

Even though many see their spiritual life as separate or opposite that of the city's sinful image, Las Vegas Mormons still face exposure to waves of temptation and tension that lap at their spiritual shorelines every day. One element of such interaction is spatial in nature. Whereas many observers consider "Sin City" to be only the Strip and downtown tourist corridor, the vice yielding such a nickname is not quarantined to any one part of the metropolitan area (Gottdiener et al. 1999; Rowley 2013). Indeed, no Las Vegan can remain separate from the "temptationscapes" of gambling and adult-oriented entertainment.

Gambling is especially pervasive in the activity and worship space of Las Vegas Mormons. Latter-day Saint congregations no longer meet in buildings near the tourist core. The historic church buildings near downtown have been repurposed or sold since the members they had served largely vacated the city center for the suburbs during the high-growth decades of the 1980s and 1990s (see Fig. 120.1). Beyond the city's gambling hub, however, banks of video poker machines are found in neighborhood grocery and convenient stores, mini-casinos exist in every corner at McCarran Airport, and massive casino-entertainment complexes (called locals or neighborhood casinos by residents) catering to local tastes permeate the city's suburban periphery where most Mormons live and worship today (Fig. 120.5). In short, simple proximity dictates that the worship world of Las Vegas Mormons exists inseparable from a gambling spatial context (Fig. 120.6).

Most Latter-day Saints live in neighborhoods at a distance from the strip clubs and sexually oriented businesses typically located near the tourist core. Still, everyday movement to and from work, school, or play often brings Mormons in contact with the city's landscape of sex. Two experiences from my life illustrate such contact. Friends, fun, and religion often took me from my childhood home, 4 mi (6.4 km) west of Las Vegas Boulevard, to attractions east of the Strip. After graduating from high school, for example, I regularly attended religion classes at the Church's Institute of Religion adjacent to the UNLV campus east of the Strip. (The Institute is a university student-oriented community center for Latter-day Saints seeking social interaction and religious education, similar to a Catholic Newman Center.) My route via Flamingo Road meant an eventual crossing at Las Vegas Boulevard. Beside the snarls of vehicle and pedestrian traffic on the Strip, at this point in my drive I often met an inundation of sexual imagery on hotel marquees, taxicab advertisements, and roadside newspaper racks (a common place for strip clubs and escort services to advertise).

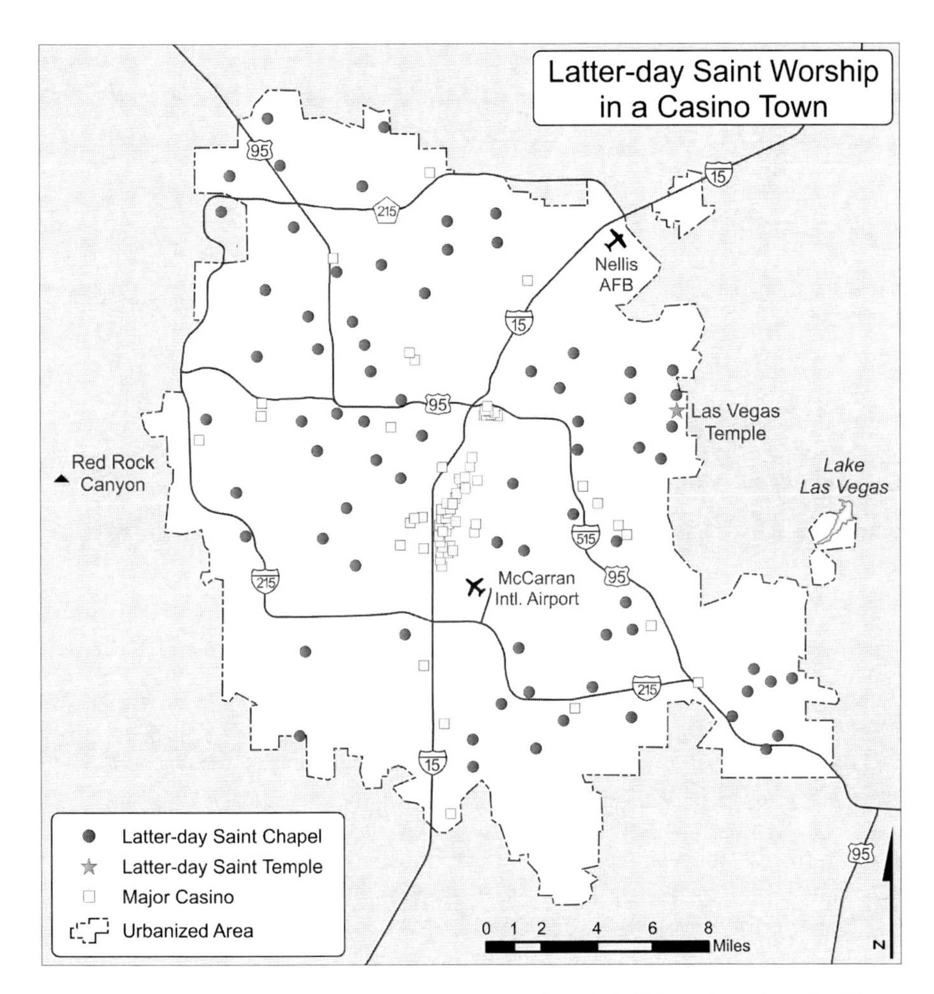

Fig. 120.5 Map of major casinos and Latter-day Saint church buildings throughout Las Vegas. Whereas congregations no longer meet near the city's tourist core, local Mormons hold worship and social functions in meetinghouses in the same neighborhood space as suburban, locals casinos located throughout the city (Map by Rex J. Rowley; data from www.maps.lds.org, 2012)

Several years later, the shortest route on my twice-weekly trip from my home on the valley's east side to my 5-year-old son's prekindergarten sports class to the west, was Desert Inn Road. In the mid-1990s local transportation authorities reengineered this road as a "superarterial bypass" that goes under the Strip instead of crossing it. This allowed me to avoid the gridlock I experienced on Flamingo, and some of the accompanying imagery as well. But the section of Desert Inn on the west side adjacent the Strip is notorious for risqué billboards. As my son and I passed through that particular stretch, I made a habit of asking Caleb random questions to draw his eyes and attention away from signage that revealed more of a woman's body than I would wish him to see.

Fig. 120.6 A view from a Latter-day Saint meetinghouse near Bermuda Road and Silverado Ranch Boulevard in Las Vegas. The tower of South Point, a casino resort catering to locals, in the background symbolizes the fusion of landscapes of sin and the spirit for churchgoing Mormons in the city (Photo by Rex J. Rowley 2012)

Beyond such matter-of-fact spatial interaction, many local Mormons choose to associate more directly with the vices of Sin City. The Church prohibits any gambling or patronizing of sexually oriented businesses, but, of course, some do it anyway. Many Latter-day Saints, however, participate in casino-related activities in other ways. Employment is one. I know several members of the Church, and my interviewees pointed to others, who have worked in one capacity or another in the gaming industry. Whether as an information systems director at one major Strip resort, a box office host at another, or a regulator for the Nevada Gaming Commission, employment in the city's most prominent industry is part of life for many Mormons and one can often work within it and still remain in good standing with the Church.

It is in the employment experience that a fusion between the spiritual world of Mormons and the secular, profane world of Las Vegas becomes apparent. Kale Ismay's experience is representative. As an 18 year-old preparing to become a full-time missionary for the Church, Kale worked as a casino "hard count," someone who collects coins from slot machines to be counted, bagged, and taken to the vault. (Incidentally, another Mormon in a management position within that casino's establishment recommended Kale for that job.) Latter-day Saints, in Las Vegas and everywhere, live around and associate with non-Mormons on a daily basis. But, as Kale explained, a job such as his brings the "experience [of] being directly involved with people who embraced the Las Vegas lifestyle." Through "frank discussions about what was acceptable to them [his coworkers] versus what was acceptable to me" Kale found it somewhat "surprising that so many people could embrace what Las Vegas stands for as such commonplace behavior."

Fig. 120.7 An aerial view of the intersection of Sunset Road and Stephanie Street in the southeast corner of the Las Vegas Valley. This typical suburban area illustrates the fusion of gambling and neighborhood landscapes in one space. Note the Sunset Station neighborhood casino (*A*), typical American shopping mall (*B*) and strip malls (*C*, *D*) on adjacent corners, with housing subdivisions all around (Photo by Rex J. Rowley, 2006)

A similar fusing of spiritual and profane can also be seen in the entertainment choices of local Latter-day Saints. On one occasion, I observed a grandmother escorting four children to a birthday party at the Strike Zone bowling alley in the Sunset Station (a locals-oriented casino-resort; Fig. 120.7). Whereas holding a child's birthday party at a casino may seem a bit out of place to anyone not from Las Vegas, it is largely acceptable to locals as part of life in a gambling town. What was surprising to me in this instance, however, was the Brigham Young University logo on the t-shirt of one child. Rarely will someone wear (or dress their child in) the BYU logo and not be a member of the Church, which sponsors the Utah school.

Indeed, local Latter-day Saints often interact with the gambling landscape by participating in a wide variety of leisure activities at casino-resorts on and off the Strip. Marsha Lamb Leason echoed the comments of many residents when she explained how she and other locals avoid the Strip unless they go to "see the lights" with visiting family and friends (Hill 2008). Flint Salvador and Savanna Frey agreed. Savanna explained: "The Strip was a place that we took visitors from out of town, but not necessarily somewhere we hung out regularly." Flint added another exception. He explained that he doesn't go to the Strip, "unless there is a specific reason, like a show or something. [We] go to the Bellagio for the flower exhibit." Ted Burke chronicled the handful of Strip casinos he has entered, each time for a specific attraction: a Kingston Trio concert at the Dunes, a good barbershop in the Tropicana, or a plumbing convention in Caesar's Palace. And, Sam Marvin said that, as a teenager, he went into Strip casinos on occasion, but mostly spent his time outside on rollerblades, weaving in and out of cars and pedestrians on Las Vegas Boulevard.

Activity in gambling space becomes more frequent and more commonplace in the neighborhood casinos-entertainment complexes located throughout the suburbs. Growing up in the city, and when I return to visit family and friends, I often go to movies in the multiplex theaters attached to most locals casinos. Six other local Latter-day Saint interviewees mentioned movies as one reason for visits to a casino property. Entering a casino is almost unavoidable for moviegoers; of the 23 theaters (including one remaining drive-in) in the city, only seven are located away from casinos. Dining is another locals-casino amenity enjoyed by several interviewees. Similarly, other entertainment amenities such as bowling, ice-skating, and most concerts only exist within the city's gaming properties. Ironically, the Mormon Tabernacle Choir held a 2004 concert in the Orleans Arena, which is part of the Orleans Hotel & Casino complex (see Osborne 2004). Ted Burke's other favorite barber (after the one in the Trop) was located in the Gold Coast, another off-Strip casino. His comment on this point, I think, underscores the amenity-based relationship between local Mormons and casinos in the city: "The only good barbers are in the casinos."

With any interaction between oppositional forces come benefits and challenges. On the positive side are the opportunities for employment and entertainment, much of which is available on the city's 24/7 schedule. Savanna Frey misses the fact that Las Vegas sidewalks don't "roll up at 9:00 at night like they do in a lot of towns in the Midwest. … Everything you needed was right at your fingertips anytime of the day."

Even if a resident does not work or play at casino properties, everyone in the city benefits from the city's core industry. Both Flint Salvador and Sam Marvin pointed to taxes as one specific, indirect benefit enjoyed by all. Sam explained: "We also don't mind that they [the tourists] come here and help pay for the room taxes and gaming taxes that keep the rest of our state taxes low." Furthermore, the readily apparent contrast between sin and virtue highlighted earlier is one of the most commonly praised elements of being Mormon in Las Vegas.

Even though such a benefit is often touted, the simple fact that "sin" is part of the sin/virtue dichotomy is bound to create tension and distraction for many practicing Mormons. A story Ted Burke told about his son's wayward friend, who had chosen a life of drugs and addiction, is one example of temptation's power in Las Vegas. Another is Flint's admitting that he "played around" with gambling when he was younger. Shawn Newman also acknowledged to her occasional gambling adventures since moving to the city in the early 2000s. Countless other church members, regardless of their knowledge of doctrine, undoubtedly have participated in gambling, alcohol use, and sexual entertainment. Even those Latter-day Saints who try to avoid the worst of the gambling and sex landscape by, for example, going to a movie or watching the fountains at the Bellagio, can't avoid being inundated by secondhand smoke in casinos or the flyers for strip clubs and escort services that litter the sidewalks of Las Vegas Boulevard.

That sexual imagery so prevalent in the city and its influence beyond the Strip is, in fact, an intense source of offense for some local Latter-day Saints. When I asked about his least favorite part of growing up and raising kids in Las Vegas, Sam Marvin responded: "All the strip club billboards you have to avoid." When we are in Las Vegas, my wife and I often discuss the dilemma of how to teach our children, not only about our values regarding sexually oriented businesses, but also about the objectification of the human body. We don't often see such revealing imagery blatantly portrayed in the Midwest, and the constant visibility of such things when we visit Las Vegas result in interesting and sometimes difficult conversations derived from questions such as, "Mommy, why isn't the woman on that sign wearing any clothing?"

The objectified image of the "perfect" female body was a source of consternation in the mind and faith of Savanna Frey. Her lamentation is representative of how other local women feel in the sexually charged environment of Las Vegas:

> As a woman it was hard living in a place like that because you have to have either the flat stomach or the voluptuous breasts in order to be considered attractive (even unfortunately among the men that were [church] members). . . . I was not considered attractive by Vegas standards. Even under 130 pounds, I was still considered too fat to be attractive. That image of what a woman should look like was everywhere and no matter how you grew up, most of the men that lived there [desired] 'that' body type.

Perhaps the biggest challenge facing Las Vegas Mormons is the overabundance of temptation. To be clear, the difficulties faced by the people I spoke to are present elsewhere. Several interviewees frankly noted that fact, but both also added that sin and temptation are more prevalent in Las Vegas. "They're just more available [and have] a higher profile here," said Ted Burke. Mark Lewin added: "You have [in Las Vegas] easy access to things that cause a degenerative process to occur within the family." Lewin then pointed to a unique and related trial for local Mormon families exposed to Sin City's omnipresent influence: "Another [problem] is the fact that it's a 24-hour town." He said that such an environment causes a "change in the family dynamics" and has seen the struggles of raising children and keeping a family close when mother and father work different shifts (Rowley 2013).

Flint Salvador put a more personal face on the 24-h town, there's-just-more-of-it-here issue. "Other towns are on a schedule," he said.

> That influences what you do. … At night, when teenagers are looking for something to do, they have more choices in Las Vegas. In Alamo [Nevada] they might just be able to try and get some beer or go to Ash Springs [a local spring-fed pool] or steal farmer Bob's cows, but in Las Vegas they might try to sneak into a cathouse or to a beach at the casino or to gamble or something. It's not that one place is better for kids than another. There is more to get into in Las Vegas.

When I asked whether or not more temptation exists for kids growing up in Las Vegas, he responded:

> Well the religious answer is no. Everyone is tested equally; there are just as many choices to be made no matter where you are. The non-religious answer is yes. How could there not be? But, really it's the same, just different choices. (Rowley 2013)

120.5 Ambivalent Adaptation

It is in the grappling with choices and challenges to faith that the outcome of the interaction between the "island of faith and the sea of sin" can best be seen. As one might expect, even faithful members of the Church reach a level of tolerance for Sin City as they become more accustomed to experiencing daily life in it. Such implicit acceptance of some influence and benefit from a source of vice they otherwise staunchly oppose is what I refer to as "ambivalent adaptation."

The willingness to accept that a faithful member of the Church can work in a casino is one example of such adaptability. In speaking about his job as a hard count and his interactions with coworkers, Kale Ismay explained that it was his "choice to decide how much of Las Vegas [to] let in." Working in such an environment is an acceptable amount to let in, it seems, as long as it does not involve direct participation in the activity.

It is unlikely that Latter-day Saints would find such an attitude elsewhere. One interviewee's father left his job in security at a Strip resort for a similar position at a riverboat casino in Kansas City. When members of his ward there found out about his career, he was looked at with shame and scorn that he didn't feel back home. Attending church services in any Las Vegas ward, one is likely to find a number of people employed directly or indirectly in the casino industry. This is understood and even predictable in a place like Las Vegas.

The commonly voiced view that sexual imagery and gambling are just part of your everyday viewshed is another example of ambivalent adaptation. After mentioning his dislike for the omnipresent strip-club billboards, Sam Marvin continued: "There are some pretty racy pictures shown in public that for us seem normal because they are commonplace, but other people that come to visit are shocked at what they see and even more shocked at how we think it's normal and not such a big deal."

It seems that ambivalent adaptive behavior develops with time and choice. Having dinner with a longtime friend and his wife shortly after they moved from Utah to Las Vegas, we discussed the unique situation of Latter-day Saints in this city. Both expressed some trepidation (bordering on reprehension) at the acceptability in patronizing casino movie theaters or bowling alleys. Similarly, Shawn Newman said that just after she moved to Southern Nevada from Salt Lake City, she was shocked at the movie theaters in casinos and slot machines in grocery stores. But, she added: "You kind of get used to it." In time, I suspect my friends will have a similar experience. Regarding the integration of movie entertainment in casinos, lifelong Las Vegan Flint Salvador said: "Never really thought anything of it. It's just normal. It's just there."

What leads Mormons in Las Vegas to see what would be shunned elsewhere as more acceptable or tolerable in this city? Certainly time, human nature, and exposure provide a partial answer. Perhaps the situation in Las Vegas simply reflects the notion that faith traditions often exhibit a certain level of "mutability" or "adaptability" when interacting with a different, sometimes oppositional cultural environment (Warner 1993; Stump 2008). I wonder, however, if such explanations do not overstate the issue. Knowing first-hand the experience of adhering to a faith both inside and outside of Las Vegas, I think theology itself plays a role. I have heard in countless Sunday school lessons the common Christian admonition that we must live in the world, but not of the world. Almost as frequently, I have heard the admonition, "bloom where you are planted." Both aphorisms connote how a person can live in a place where sin is present, but does not need to indulge in it to maintain a full life. Most Las Vegas Mormons do not condone behavior or billboards they confront or even the economic model of their employer, but instead they strive for faithful membership in a church while existing in such environs. They simply separate faith from the reality around them. And perhaps it is through such separation that they can, in turn, affirm faith while maintaining gainful employment in a gambling town or enjoying wholesome entertainment that is often connected to casinos.

120.6 Faith in Hedonopolis

The Mormon perspective in Las Vegas is unique. The idea that a faith-based group exists in a context of oppositional cultural forces, however, is not singular to practicing members of The Church of Jesus Christ of Latter-day Saints in this place. To elaborate on lessons that can be gleaned form the Las Vegas experience, I turn to the concept of "hedonopolis," introduced by the geographer John Sommer (1975). Sommer envisioned hedonopolis, and the associated "fat city," to be the future of urban America, and certain evidence points to at least some truth in his prediction (Sui 2003). But the concept itself, and how it applies to contemporary religious urban life, can help us understand the applicability of faith islands to other places.

Hedonopolis, Sommer explained, "is an urban society in the process of becoming, a society in which there is high investment in elaborate and complex personal services" (1975: 138). It is a city that revolves around ego-satisfaction, a service economy, and a population that spends much discretionary time in recreation. Such a characterization certainly fits Las Vegas. After all, the desert metropolis is the poster child for a post-industrial city (Rothman 2002). More than any other city in the United States, Las Vegas relies on leisure and hospitality revenue for its economic lifeblood (LVCVA 2010). Furthermore, what could be more ego-satisfying and self-serving than a game of blackjack or a lap dance? In such a light, faith islands may be seen in faith communities throughout the world that exist in tourist cities where sex, gambling, and drinking are a main attraction, such as Reno, Lake Tahoe, New Orleans, Monte Carlo, Amsterdam, Macau, or Bangkok.

If we understand Sommer's broader points, what he called "elemental principles" behind hedonopolis (146), the pattern of faith islands can be more generally applied beyond pleasuring cities. In such a world, Sommer wrote, "people now spend more time trying to satisfy themselves rather than a Supreme Being, [where] Hedonism has displaced Puritanism" (136). In this view, many cities may take on the hedonopolis label, even if it may not be a "sin city," per se (at which point it would perhaps make more sense to call Las Vegas and its cohorts "viceopolis"). In other words, many places in the world have become increasingly anti-spiritual, hedonistic, or, simply put, secular. After all, the "nones," or Americans claiming no religion, are the fastest growing "faith" tradition in the country (Kosmin and Keysar 2009). Spiritually driven groups—those that desire to satisfy a Supreme Being—still exist, however, and with an increase in secularism throughout the world, it follows that "believers" may feel at times like "islands in a secular sea."

At an even broader level, religious scholars should also consider the places around the world where minority faith traditions exist within another majority religious group. In Las Vegas, this may be Mormonism or any other faith tradition that pales in size to the "religion" of vice so pervasive in this city. Nearest to Las Vegas on the map, an inverse example might be the "gentile," or non-Mormon population in Salt Lake City, Utah, which operates as an agglomeration of minority faith (or agnostic) traditions separate from but invariably influenced by a deeply entrenched Mormon culture (Mitchell 1997; Garza 2012). We might also view the Muslim population in the mostly Judeo-Christian United States in a similar light. Further afield one might consider Christian communities in Indonesia or Japan.

At whatever level, further exploration of faith islands and their implications can yield rich fruit for scholars of religion. In doing so, we will more deeply understand the broadest generalization of the concept, which is found in the connections between people and place, culture and context.

References

ASARB [Association of Statisticians of American Religious Bodies]. (2010). *2010 U.S. religion census: Religious congregations and membership in the United States*. The Association of Religion Data Archives. www.thearda.com. Accessed 12 June 2012.

Garza, X. (2012, April 8). Growing up catholic in Mormon territory. *Las Vegas Review-Journal*. www.lvrj.com/living/growing-up-catholic-in-mormon-territory-146576555.html. Accessed 6 June 2012.

Gottdiener, M., Collins, C. C., & Dickens, D. R. (1999). *Las Vegas: The social production of an all-American city*. Malden: Blackwell Publishers.

Hill, G. (2008, July 12). Watching the Church grow in Las Vegas. *Church News*. www.ldschurchnews.com/articles/52721/Watching-the-Church-grow-in-Las-Vegas.html. Accessed 6 June 2012.

Hinckley, G. B. (2005, May). Gambling. *Ensign*, pp. 58–61.

Kong, L. (2001). Mapping "new" geographies of religion: Politics and poetics in modernity. *Progress in Human Geography, 25*(2), 211–233.

Kosmin, B. A., & Keysar, A. (2009). *American religious identification survey (ARIS 2008): Summary report*. Hartford: Institute for the Study of Secularism in Society & Culture.

LDS [The Church of Jesus Christ of Latter-day Saints]. (2012). *Temples*. www.lds.org/church/temples?lang=eng. Accessed 13 June 2012.

LVCVA [Las Vegas Convention and Visitors Authority]. (2010). *Economic impact series: The relative dependence on tourism of major U.S. economies*. Las Vegas: Applied Analysis.

Mazumdar, S., & Mazumdar, S. (2009). Religious placemaking and community building in diaspora. *Environment and Behavior, 41*(3), 307–337.

Mitchell, M. (1997). Gentile perceptions of Salt Lake City: 1849–1870. *The Geographical Review, 87*(3), 334–352.

Moehring, E. P. (2000). *Resort city in the sunbelt: Las Vegas 1930–2000* (2nd ed.). Reno: University of Nevada Press.

Osborne, J. (2004, April 22). Choir's secular music surprising. *Las Vegas Review-Journal*. www.reviewjournal.com/lvrj_home/2004/Apr-22-Thu-2004/living/23710569.html. Accessed 4 Mar 2009.

Paher, S. W. (1971). *Las Vegas, as it began-as it grew*. Las Vegas: Nevada Publications.

Roske, R. J. (1986). *Las Vegas: A desert paradise*. Tulsa: Continental Heritage Press.

Rothman, H. (2002). *Neon metropolis: How Las Vegas started the twenty-first century*. New York: Routledge.

Rowley, R. J. (2012). Religion in sin city. *The Geographical Review, 102*(1), 76–92.

Rowley, R. J. (2013). *Everyday Las Vegas: Local life in a tourist town*. Reno: University of Nevada Press.

Schwartz, D. G. (2003). *Suburban Xanadu: The Casino Resort on the Las Vegas strip and beyond*. New York: Routledge.

Slater, T. R. (2004). Encountering God: Personal reflections on "geographer as pilgrim". *Area, 36*(3), 245–253.

Smith, J. L., & Smith, P. (2003). *Moving to Las Vegas* (3rd ed.). Fort Lee: Barricade Books.

Sommer, J. W. (1975). Fat city and hedonopolis: The American urban future? In R. Abler, D. Janelle, A. Philbrick, & J. Sommer (Eds.), *Human geography in a shrinking world* (pp. 132–148). North Scituate: Duxbury Press.

Stump, R. W. (2008). *The geography of religion: Faith, place, and space*. Lanham: Rowan & Littlefield.

Sui, D. Z. (2003). Musings on the fat city: Are obesity and urban forms linked? *Urban Geography, 24*(1), 75–84.

Warner, R. S. (1993). Work in progress toward a new paradigm for the sociological study of religion in the United States. *American Journal of Sociology, 98*(5), 1044–1093.

Watling, T. (2001). "Official" doctrine and "unofficial" practices: The negotiation of Catholicism in a Netherlands community. *Journal for the Scientific Study of Religion, 40*(4), 573–590.

Yorgason, E., & della Dora, V. (2009). Geography, religion, and emerging paradigms: Problematizing the dialogue. *Social and Cultural Geography, 10*(6), 629–637.

Chapter 121
Marketing Religion and Church Shopping: Does One Size Fit All?

Stanley D. Brunn, Wesley Jetton, and Barbara Palmquist

> Choosing a church is now like exploring the stores and boutiques of the new malls and arcades in our town and city centres. Increasing numbers of those who go out "churching" on Sundays have loyalty cards which they use according to the needs of their spiritual shopping list for that particular week. (Scotland 1984: 136)

> Choosing a church (or mosque or synagogue or temple) isn't just a matter of theology for many Americans. They might decide where to worship because they adhere to a broad tradition ... or they might choose based on location or children's activities or preaching or music or potluck offering. The concept of church-shopping is uniquely American. (Sullivan 2009)

121.1 Introduction

Three words that describe the contemporary religions landscape in the United States are fluidity, identification and choices. All are related and evident in the identity of those who consider themselves religious and spiritual (not the same thing) or associate with religious institutions. The fluidity concept is evident, those who drop out of religious institutions, disassociate themselves with views and practices associated with childhood, and adopt views and practices in adult years that reflect changes in their worldviews of earlier religious teachings and the contemporary diversity that reflects much of contemporary society. Peter Ward (2002) describes these developments of fluidity and networking as characteristics of a "liquid church" and sociologist Kees de Groot (2006) uses the term "liquid modernity." Explorations into the meanings of participation, involvement and commitment are ventures by both those attending churches as well as church leaders themselves who try (often struggle) to make sense of what is going on around them. Identification is also associated with fluidity in that it relates to both changes in how and what one was taught about religions doctrines and creeds versus the potential adult emergence of new belief

S.D. Brunn (✉) • B. Palmquist
Department of Geography, University of Kentucky, Lexington, KY 40506, USA
e-mail: brunn@uky.edu; bjpalm0@email.uky.edu

W. Jetton
Student, University of Kentucky, Lexington, KY, USA
e-mail: wesley.jetton@uky.edu

S.D. Brunn (ed.), *The Changing World Religion Map*,
DOI 10.1007/978-94-017-9376-6_121

systems or the labeling oneself as agnostic or even an atheistic. Lewis Lugo (Sullivan 2009), director of the Pew Forum on Religion and Public Life, noted that "The U.S. has an unmatched religious dynamism. It's an open religious marketplace as well as a very competitive one. This is the supermarket cereal aisle." Sullivan (2009) writing about this noted that: "Without an established state religion all faiths can freely exist in the U.S. but must compete for adherents in order to survive." Santella (2009) notes that the "American faith comes in lots of flavors, but doesn't necessarily mean that today's church shoppers are buying into a superficial, strip-mall faith." The Barna (1988) group noted that this "rampant spiritual consumerism" may not altogether be a bad idea as is provides for a more informed and ecumenical culture.

The choice concept is also related to the previous two themes, in that there are many acceptable choices available to those in youth or adult years that are considered acceptable by a wider society. Kimberly Winston (2012) discussed a recent poll that showed that the percentage of those who declared themselves "religious" dropped from 73 % in 2005 to 60 %. Sullivan (2009) noted that "For the most part the unaffiliated report deep dissatisfaction with organized religion, believing that focuses too much on rules and religious leaders are too concerned with acquiring power and wealth." Rather than choosing to practice religion as a Catholic or Protestant, or one of many Protestant denominations, there are now almost unlimited religious institutions and belief systems one can associate with and faithful religions duties one might practice in public or in private.

A number of recent surveys about the religious views, beliefs and practices about the U.S. population illustrate the shifts in religious views, multiple labels, and practices. Santella (2009) reports that about one in seven adults changes churches every year and about one in six do so on a rotating basis; these results are from the Barna Group which analyzes marketing data for churches. In his regard Hendricks (2004) noted that one in four church members started coming to their current congregation in the last 5 years; of those 7 % to an entirely new to church. People switch and shop for many reasons, dissatisfaction with the liturgy or the music or sermon content or conflicts within a church, theological positions of the larger denomination, inclusiveness or exclusiveness. In this calculus could also be non spiritual issues such as timing of a service, lack of available parking, lack of children programs or community outreach.

Declines in the membership in mainline Protestant denominations, the rise in Hispanic Catholic membership, the rise of evangelical churches in the social and political arena, the megachurch phenomenon, the small but steady increases in the numbers labeling themselves as agnostic and atheists, the new minorities of Islam and Hindus in selected states and cities and the appearance of new groups including Wiccans, Bahai, neopagans, and others attest to this shifting in the religions landscapes of urban American especially. The Pew Forum acknowledges the "fluidity" in the American religious landscape. Sullivan (2009) summarized the key findings:

> Former Catholics who either switched to another tradition or became unaffiliated cited unhappiness with church teachings on abortion and homosexuality and disagreements over the role of women in the church. Protestants were more likely to switch because they marked someone from another tradition. And if they eventually left religion altogether, they were most like of all formerly religious adherents to have tried several different traditions before giving up – 38 % of unaffiliated former Protestants had switched traditions twice, and 32 % had switched three or more times.

The Forum noted, with some surprise, that whereas 7 % of Americans were raised unaffiliated, only half that number retained that category as adults. They recognized the positive aspects of being affiliated with a religious institution. When they were queried about joining an institution, one-third stated the benefits of being connected to a spiritual community and one-fifth acknowledge that something spiritual was missing in their life.

Maintaining an observant eye on the above dynamic religious landscape are the religious institutions themselves. Those that have seen sharp membership declines in the past several decades as well as church closures are ever mindful of what is going on at the grassroots, viz., fewer people are attending weekly worships. These observers are well aware of the visible and emergent secular society that they can observe in society at large. Some religions institutions and denominations have sought creative ways to try and stem the tides (or possibly tsunamis) of declines in membership and numbers worshipping weekly, and sought some innovative initiatives to attract new members or believers or at least reduce the number that become associated with a secular rather than a religious America. A number of these initiatives are devising and implementing some clever marketing strategies to bolster church numbers in membership or weekly services. These same efforts recognize the fluidity, identity, and choice concepts introduced above and may use many of the same marketing tools and techniques used by the private sector to entice, seduce and attract believers. These may include support for electronic church initiatives, including "distance religious services" where one can watch from a website or iPad, ad hoc religious/spiritual therapy sessions and support groups conducted on line, the observance of religious practices in one's home (without traveling to a formal church institution), and virtual weddings, conversions and funerals. In this regard, churches are viewing potential members as consumers or clients who need to be presented with creative, appealing and attractive packages about their product (that is, church programs, the worship experience, the advantages over competitors, etc.), and the benefits of supporting or consuming one church's product over another.

A significant component of this fluid, dynamic and consumer-oriented religious landscape is the amount of "shopping" that goes on. That is, what kind of "shopping behavior" is evident among those who move into a new area, those who want to change (or shop) for churches in an area where they currently reside, and also those who simply want to withdraw from any formal religions institution. All three varieties of "church shopping" are likely to occur in a given area. That is, there are newcomers to a city or suburb who wish to find a new home, so they "shop" until they discover one to their liking. Second, there are permanent residents in these cities or communities who wish to withdraw or "drop out" of one church and find another to their liking. These individuals or families also are engaged in church shopping. And third are those who for theological, ideological or personal reasons decide to withdraw from formal religious institutions; these "drop outs" may withdraw from a faith or denomination permanently or withdraw for a period of time during which they may eventually not be formally associated with any official church or become loosely associated with another group. This third group, as fascinating as it is in the American religious landscape, is not the focus of this study.

121.2 Marketing Religion and the Church Shopping Phenomenon

As important and visible as the church shopping phenomenon is in contemporary America, it is not a topic which has generated much scholarly research. The reasons for this lacuna in the literature by sociologists, historians and geographers who study religion are threefold. First, there are no readily available datasets one can access, analyze and map to describe, illustrate and examine these "shopping" dynamics at rural and urban scales. Second, the major headquarters of churches and related religious institutions seem largely unaware of the importance of gathering and analyzing such data in order to understand shifts (growths or declines) in membership and better understand what exactly is going on where and why. Third, and this we discovered in undertaking the study described below, is that local churches seem unwilling or unable to gather "shopping" information from potential members or those who eventually end up joining their fold. Again, ignorance of the importance of the "shopping experiences," even if anecdotal, are weighed against whether to share their stories with competitors (same denomination in the city or those in close proximity).

A search for church shopping entries in the Google Search Engine and Google Scholar yielded less than a dozen useful references; these include a few book references and websites by pastors and other religious writers. What literature does exist can be divided into two broad categories. First, it is treated as a phenomenon of postmodern America, that is, a strong consumer culture in which "church shopping" is analogous to other decisions one makes from a market place that promote multiple versions of the same product. The contributors to this literature are mostly sociologists and religious writers taking critical looks at contemporary societies (Shawchuck et al. 1992; Hoover 1998). A second cluster of studies looks at church shopping more from religious or theological viewpoints; many of these websites and articles are by church leaders, most who discuss what the church shopping experience means in a religions/spiritual context. These pastors also provide advice to prospective shoppers. Together both groups illustrate the importance of social scientists studying this topic, not only the dynamics of membership growth, decline and switching, but also the need for detailed case studies, even if only preliminary in its scope and findings.

A very insightful discussion of church shopping in a consumer-oriented society is provided by Nigel Scotland (1984). It is no surprise to him that the "western world has become totally consumption oriented with an appetite that devours not merely the products of capitalism, but just about everything else, including the church" (p. 135). He discusses both the church leaders, including their training for a consumer-oriented society, and also those individuals who engage in shopping. One of the challenges leaders face is that they "need to bring people into their churches, but at the same time those individuals are shopping around for what fulfills their expectations and meets their needs" (p. 137). It is understandable in the free-market approach to church competition that there will be definite winners and losers. "In the end it will be the most aggressive and powerful who will survive and prosper, while the weakest and smallest will go to the wall" (p. 142). Many of the competing churches are engaged in what Scotland calls "the cloning of success," where there are many similarities, but they engaged in "marginal differentiality." (p. 140) or identifying and promoting some small differences.

In regards to pastors, Scotland (1984: 143) notes that "church leaders are no longer pastors but marketers and managers." Their training also is changing; he writes that

> Theology has been largely hijacked by the academics who see it as their preserve, with the result that many congregational members now regard it as irrelevant to everyday needs. Indeed many church leaders have only a minimal grounding in academic theology, but who instead have expertise as psychologists, trained counselors, therapists or organisers. (2000: 144)

Potential members are those caught up in the rampant consumerism they see around them. Scotland (1984: 144) writes that:

> Because the churches are marketing their product in such an aggressive fashion, people are minded to keep looking around in case they missed out and there are better bargains and services to be had at St. Develictus-in-the-Marsh on the other side of town. Consumerism has created a generation of church shoppers who move from one fellowship to another in the same way that grocery shoppers changed from Tesco's to Safeway's to Sainsbury's to Waitrose to Gateway and back. In the same way, church goers move as the ads and the grapevine prompt them.

"Consumerism in contemporary Christianity, according to Andrew Walter (1998), has resulted in rampant individualism." To him "ascetic individualism" (doing good) has given way to "hedonistic individualism" (feeling good)."Sometimes this hedonism is translated into a "prosperity theology" (Starkey 1989: 222–227). All of this it is argued produces a jaundiced church, which, according to Scotland, is far removed from the deeply committed fellowships that formed the backbone of the early church" (p. 144).

There is, it should be mentioned, disagreement among scholars of religion and religious writers about the marketing of religion. According to George Barna (1988), "Jesus Christ was a marketing specialist" (p. 13). He continues: "Like it or not, it has a "product to sell – a relationship with Jesus and others, its "core product" is the message of salvation and each local church is a franchise." Scotland (1984) find that some parts of creedal faith do not market well, "notably the themes of judgment, punishment, suffering for one's faith, the cost of commitment and the demands on taking a public stand on Christian values in a culture that does not acknowledge them" (p. 140). Kenneson and Street (1997) note that when "marketing a personal religion the church is in fact dehumanizing it" (p. 61). Wells (1994: 82) acknowledges that while it is possible to market the church, the gospel, Christ and Christian character cannot be marketed.

James Twitchell's (2007: 1) book on *Shopping for God: Christianity Went from in Your Heart to in Your Face* is an insightful commentary into the "whys and wherefores" of church shopping, but also what it says about contemporary religion in a mass consumer culture. His focus is stated succinctly:

> This book is about how some humans-modern day Christians to be exact – go about the process of consuming – of buy and selling, if you will – the religion experience. The book is about what happens when there is a free market in religious products, more commonly called beliefs. Essentially, how are the sensations of those beliefs, generated, marketed and consumed? Why pays? How much?

About leaders he writes:

> Are the religious dealers, if I may call them that, in any way, like the car dealers on the edge of town? And what of the denominations do they represent? Do they compete?

An entirely new business model called the "Church Growth Movement" is looking at transferred or lapsed members as consumers. Twitchell's book is essentially "about how religion and sensation is currently being manufactured, branded, packaged, shipped out and consumed" (p. 3). The new megachurches, are being run by a "very market-savvy class of speculators whom I call *pastorpreneurs*. By clever marketing, they a have been able to create what are essentially, city states of believers" (p. 3).

One could also list in the church shopping literature Reverend Billy and the Church of the Life After Shopping performance group in New York City that challenges consumers to become involved in issues about economic justice, environmental protection, anti-war activities and the corrupt practices of multinational corporations and the manipulative initiatives of mass media (see: http://en.wikipedia.org/wiki/Reverend_Billy_and_the_Church_of_Life_after_shopping)

Chris Rule (2009), an editorial writer for Relevant, considers church hunting or church shopping as basically a search for a community where one can feel at home. He writes:

> When shopping for most products, most people have some idea of what they want. Just like some people will always buy Chuck Taylors, because that's what they've always bought. Some people settle into a church because it is a familiar denomination, or style. That is not necessarily wrong, just like Chucks aren't a bad thing, but choosing a community to enter into is a more serious decision than how you look on Saturday night.
>
> Think of how people would search for a church; instead of the Yellow Pages, they're much more likely to consult Google or other online tools. Websites have become more important than ever for church.
>
> As churches get more creative in attempting to attract new members with catchy websites and more programming, what are the essentials someone should look for? Sunday services are changing as fast as culture changes, offering a dizzying blend of church forms, music – everything from the latest praise music to ancient forms of taize –developing programming for kids, and many, many more options, all of which will affect how someone chooses a church. And that not just on Sunday (Relevant Magazine 2009).

Santella (2009) acknowledges that the church shopping, with the latest technologies used by pastors and marketers, are changing the nature of the religious service and also community and commitment.

> Web sites stream audio and video of sermons and music to let prospective members shop from home, and consultants help congregation market themselves to the "untouched" and the merely unsatisfied by deploying focus groups, surveys, product giveaways …and other tactics borrowed from the commercial realm. The Wall Street Journal recently reported on churches employing mystery worshippers, "a new breed of church consultant," who covertly attend services and evaluate them (were the bathrooms clean? Was the vibe friendly" as if they were first timers looking for a new church).

The extremes of consumerism have also been noted by some writers. Some churches have removed crosses, crucifixes, statues, pulpits and choir stalls; others have clergy who dress down to promote casual worship experiences (Scotland 1984: 148). Churches are starting to look like shopping malls. Some churches install the latest high video and audio systems, thinking and hoping these will attract church shoppers. Others conduct surveys about perceived and genuine needs to structure their worship experiences. Kennerson and Street (1997: 68) note that sometimes surveys serve "the needs of marketing rather than God's kingdom." Niche marketing or "ecclesiastical marketing" (Scotland 1984: 149) can have both upsides as well as downsides. Megachurches are among those churches that are singled out as using a variety of advertising and other promotional techniques to lure the unchurched or those already affiliated with existing churches. They have large staffs which can serve the needs of multiple niche groups, including youth, single adults, single parents, the elderly, and various sports, music and education programs for all ages. Their appeal, according to Scotland (1984: 136) is that "large numbers come because they are comfortable in an unthreatening atmosphere www.thecripplegate.com." Martyn Percy labeled the charismatic experience associated with Toronto Blessing, an evangelical congregation, as "quick, easy and consumer-oriented, a sort of McDonaldization of mysticism" (Scotland 1984: 142). He notes that

> it is possible to market the gospel using consumerist models without necessarily changing the product. It is also possible to embrace some aspects of postmodern culture without allowing the world to squeeze us into its mould. (2000: 150)

It is not only sociologists who are weighing in on the issues of church shopping, but so are pastors, some with websites. Rev. Eric Davis, pastor of the Cornerstone Church in Jackson Hole, Wyoming describes these mobile religious folks as "pew hoppers." He writes that a common thread of many Christians today who seek commitment is to "attend everywhere but committed nowhere (www.thecripplegate.com)." A concern expressed by more than one pastor was what the "shopping phenomenon" says about the church itself. Gordon Atkinson (2014), a former pastor, expresses this point in his website:

> Of course, a church can't spend all of its time trying to make visitors happy. If a church tries to become what it thinks people want, than it stops being the Body of Christ and starts becoming something more like Walmart or Disney World. The danger for those of us who are looking for a spiritual community is that we might slip into a consumer mentality. You can tell this has happened when you sound like a movie critic at lunch on a Sunday afternoon.

Some pastors have also provided specific advice for the church shoppers. Atkinson's tips are:

1. Think of your visit to a church as someone's home. Be gracious. Rather than concentrating on things you don't like, make an effort to notice and affirm things the church does well.
2. Seek to worship more than you seek to evaluate. Enjoy discovering new ways of experiencing and serving God.
3. Be thankful for friendly churches, but do not be overly impressed with superficial greetings. The real test of community is investing in each other's lives.
4. Pray for the church you visit each Sunday at lunch.

5. Do not put too much importance on the sermon. A good preacher is nice, but ministers come and go. The people of the congregation are the ones who sustain each other over the long haul.
6. Look for a church that will provide you with opportunities to serve others.
7. Try some small, out-of-the-way congregation. You might have to ask around to find them. Pay particular attention to churches that are new or are meeting in interesting locations.
8. Count your season of wandering as a time of spiritual growth and discernment.
9. Finally this: you will not find your perfect church. Your perfect churches does not exist. You should have a few things in your mind that are most important to you. Look for those things and don't about the rest. Part of the joy of being in a community is learning to live with the fault and frailties of others, just as they learn to live with you and your idiosyncrasies.

Rev. Cwirla (2005), pastor of Holy Trinity Lutheran Church in Hacienda Heights, California, has a lengthy blog that provides twelve things to look for when church shopping. These include: is the church Christ-centered, sacramental, liturgical, confessional or creedal and also does it confess the Triune God, teach that sinners are justified by God's grace, and make disciplines by baptism? He then adds "Twelve dos and don'ts for Christian shoppers." This list includes attend more than one service, lead with head, not heart, don't feel compelled to leave an offering, don't get sucked into music (which can be manipulative) and worship as entertainment, speak with the pastors and pray that God will lead you to a Christ-centered congregation.

Rev. Barry Howard (2012) of the First Baptist Church in Pensacola, Florida in his website expresses his dislike for the term "church shopping" because it "just sounds too commercial." Searching requires a deeper level of introspection and guidance. He prefers the term "mission driven" because it is "much less entertaining or performance-based than you might experience on a big screen church." Howard also considers that many of the church marketing schemes are "veiled attempts at proselytizing (encouraging believers to leave their church and come to join your church) and I unapologetically believe the proselytizing is a sin." In discussing the church shopping behavior with another pastor once, it was suggested that perhaps churches should issue a "church passport" for those frequent church swappers who "would not have to go the trouble of to keep transferring their membership every time they become disgruntled." His advice on choosing a church home is to:

> choose a place where your spiritual gifts are needed. Consider a church that offers diverse styles of worship, expressions that span the generations and the ages. Think about joining a church where your beliefs are going to be stretched and challenged by the preachers and teachers, not simply validated. Choose a church based on the opportunities you will have to serve, not just to be served, opportunities you will have to minister and not just be ministered too. And cultivate a sense of belonging by getting involved in the work of the church.

Santella (2009) notes that because "church goers have so many options (and) should keep pastors and preachers on their toes. And in this regard church shopping transfers a bit of power from the pulpit to the pews," which he acknowledges is never a bad idea.

121.3 A Case Study: Lexington, Kentucky

The Lexington, Kentucky metropolitan area has approximately 350,000 population. The city is in the heart of the Bluegrass and is identified with thoroughbred horse raising and racing, the University of Kentucky, major regional hospitals serving central and eastern Kentucky, and assorted services concerned with high tech ICTs. Lexington is about 80 miles south of Cincinnati and the same distance from Louisville, the state's largest city. Historical and cultural geographers would place Lexington in the Upper South, that is, in a similar category one would place the other cities in Kentucky and perhaps Tennessee. While the city is only 90 min away from the Ohio, which is one of the major cultural fault lines in North America, and a feature that separated free from slave states in the Civil war, the city is decidedly Southern in culture and heritage, more than southern Middle West. Ample evidence of this southern culture can be presented using several criteria. We use the dominance of Baptist churches as a key variable that describes this moderate to conservative religious city in the Bluegrass.

The Yellow Pages (business pages) of the Lexington Metropolitan Area telephone directory lists churches in Lexington and five surrounding communities, all within a half hour of Lexington: Midway, Versailles, Nicholasville, Georgetown and Wilmore. Many of these communities have residents who commute to Lexington daily and who probably worship in Lexington on weekends.

The aforementioned directory lists 42 church headings and nearly 250 individual churches in the Lexington metro area. The largest number, which would surprise no one who lives in the Inner Bluegrass, are Baptists churches (74). The next leading churches, which are far down the list in numbers, are: Methodist (18), Church of Christ (11), Presbyterian (UPUSA) (10), Catholic (9), Christian Church, Disciples of Christ (8), Assemblies of God (7), Episcopal (7), Christian (8), Southern Baptist Convention (7), Lutheran (6), and Non-Denominational (5) and Interdenominational (5). There are less than a handful of Nazarene, Church of God, Apostolic, Eastern Orthodox, Church of God, Cleveland, Tennessee, Seventh Day Adventists, and Pentecostal. Single churches include Quakers, Muslims, Hispanic, Jewish temples. There are also two Hindu temples.

Our first task was to learn how different denominations in Lexington promoted themselves through phrases and words. This information one could consider as an example of church marketing or the kinds of "catch words and phrases" one might use to appeal to and welcome new church shoppers. Examples of the wording and phrasing used include:

- Quest Community church www.questcommunity.com which is a megachurch. Its promotion includes these phrases: "What life are you waiting for? Become a wholehearted follower of Jesus. 10 Week Transformation Adventure."
- Second Presbyterian church www.2preslex.org describes itself as: "A living faith. Become parent God wants you to be. Book groups, music groups, courses for adult disciplines and more!"

- Faith Lutheran church www.faithlutheranchurch.com promotion is: "Exploring Christian faith or a lifelong Christian, there is a place for you in the faith family."
- St. Luke's United Methodist Church's website is: stluke@umc.org is "Committed to being a vibrant church today. Offers three services, blended, traditional and contemporary. Multicultural worship service @ 3:00."
- St. Andrew's Episcopal Church's website is www.standrewslex.com "Warmly welcomed to join us for our Sunday worship @ 10 and followed by coffee hour fellowship."
- Central Baptist Church's website is www.lexcentral.com uses these words: "All are welcome here, no exceptions." Find out more tabs: sermons, centered thoughts, what I love about Central Baptist Church.

121.4 Church Websites

Many church shoppers begin their search for a church congregation by using the Internet. The growth of online media coupled with the fact that the majority of Americans have access to the World Wide Web on their laptop, smart phone, tablet or other devices has created a need for churches and religious bodies to increase their presence on the Internet. For certain, many have already done so. After analyzing the Internet presence of nearly 20 churches in the Lexington area, including a variety of different denominations such as Baptist, Methodist, Presbyterian, Catholic, and Unitarian to name a few, we discovered in undertaking this project that most church congregations have some sort of presence online whether it is through a web site, blog or social media account. Even some smaller churches have a presence, although congregations with less than 50 regular attendees are much less likely to be represented online.

Websites are an important way that churches reach new and potential members. These sites mostly exist to present the Internet user with general information about the church. The subject matter content that is fairly commonplace on these sites usually includes information concerning when services are held, who are the leaders of the church, and what are some specific beliefs of the church among others. There are also a significant amount of information regarding youth services and adult ministries that have connotative names and slogans. For example, the high school ministry at Southland Christian church features a midweek service called "Amplify" and a middle school service called "Lift." This page features text such as, "Our mission is to prepare students to unleash a revolution of love!" Many other sites, such as Central Baptist's, simply feature a text listing of the times and descriptions on a page titled "Youth Programs." Unitarian Universalist Church of Lexington's youth page is similar to this but features the title "Religious Exploration (Sunday School)." It should be noted that the vast majority of websites did not contain information in a language other than English. Foreign language content was difficult or nearly impossible to find even on megachurches' websites.

An overall theme that we noticed when comparing the sites of multiple congregations was the impact that the size of the congregation had on the website design. First, it appears as though the larger Lexington congregations have much more modern websites. They feature moving graphics (questcommunity.com), embedded videos (southlandchristian.org), and sometimes, even links to where purchases can be made of Christian music, mostly contemporary, that is used in services (southlandchistian.org). This could presumably be because they have higher budgets and can afford to hire a company to design a website for them. Or they may have members within their ranks who voluntarily contribute time and expertise to promote "marketing" the church. These websites are also much more intuitive with fewer places to click and a much broader heading bar that may include the terms "New Here?" "Connect," and "About Us." Shoppers may find this approach less intimidating than having twenty or more buttons to choose from on the home page. Old Redeemer Lutheran Church (www.orlutheran.com) has 17 links in the main navigation bar while Northeast Christian Church (http://ncclex.org), a larger church, has only five. By providing much broader heading links these churches are allowing those who have very limited knowledge of church vernacular to still have the ability to navigate efficiently on their sites.

From our perspective there seems to be an extreme emphasis on the presentation of images rather than text. Examples of this include southlandchristian.org (Southland Christian Church), questcommunity.com (Quest Community Church) and ibc-lex.org (Immanuel Baptist Church). Most pages of Southland Christian Church's website feature a heading bar with a very large image beneath it with some sort of text overlay. Below this, there is generally a brief description of the content that was mentioned. These images, an example of which is shown in Fig. 121.1, can feature anything from colorful graphic designs that do not display people, to images

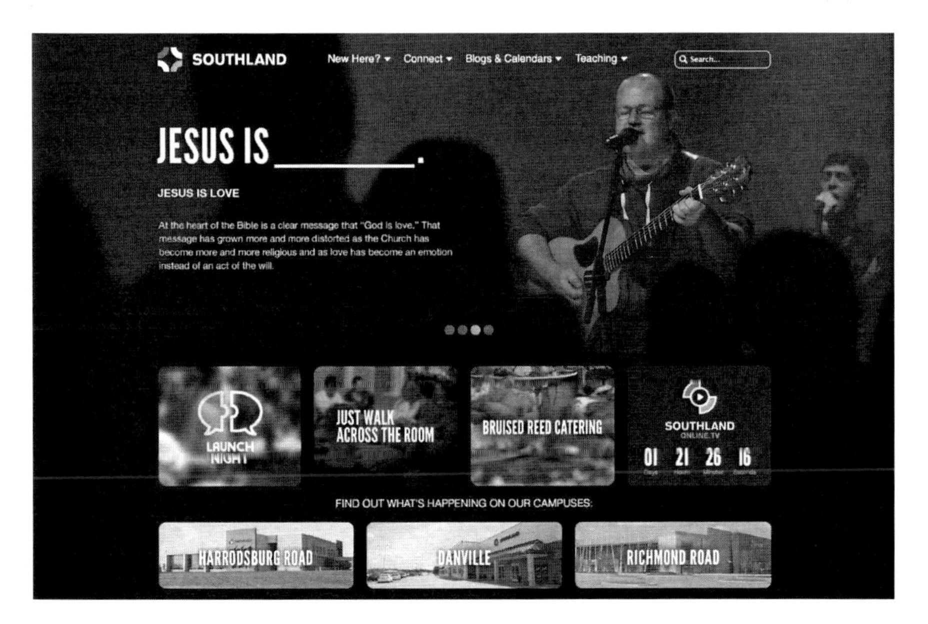

Fig. 121.1 Southland Christian Church web page (Image from SouthlandChristian.org, used with permission)

of the church itself or members who attend. The visual image is the interior of Southland Christian's sanctuary; it provides potential shoppers or consumers accessing the website of what attending a given service would be like. It is meant to provide comfort, a good size attendance of worshipers and, above all, a pleasant feeling about the church and church setting. According to Justin Meeker, Communications Director at Southland Christian Church:

> One of the interesting things we've found with our analytical data is that the overwhelming majority of use for our website is returning users looking for ways to grow (devotionals, study resources, etc.) and get connected to serving and community. We have about an 70/30 split of traffic with the majority coming from repeat visitors and existing congregants and a smaller percentage from "shoppers" -so the intent of our site is to be setup the same way with the majority of focus being on resources for people wanting to study more or get connected.

Most church websites feature images of members of the church participating in church planned activities. These may include mission trips, hand raising in service, or soloists performing on stage. Quest Community has several dozen photo galleries available to view on their website (Fig. 121.2). Some of the galleries contain photos taken during weekend services or baptismal ceremonies, others feature photos of the church having some sort of fellowship concerning a basketball game in the spring of 2012. Also, an image of the exterior of some churches was provided, perhaps for a sense of recognition, which is likely the case with the website of The Cathedral of Christ the King, which has a recognizable building. The heavy emphasis placed on the visual aspect of web pages among churches may be due to the fact

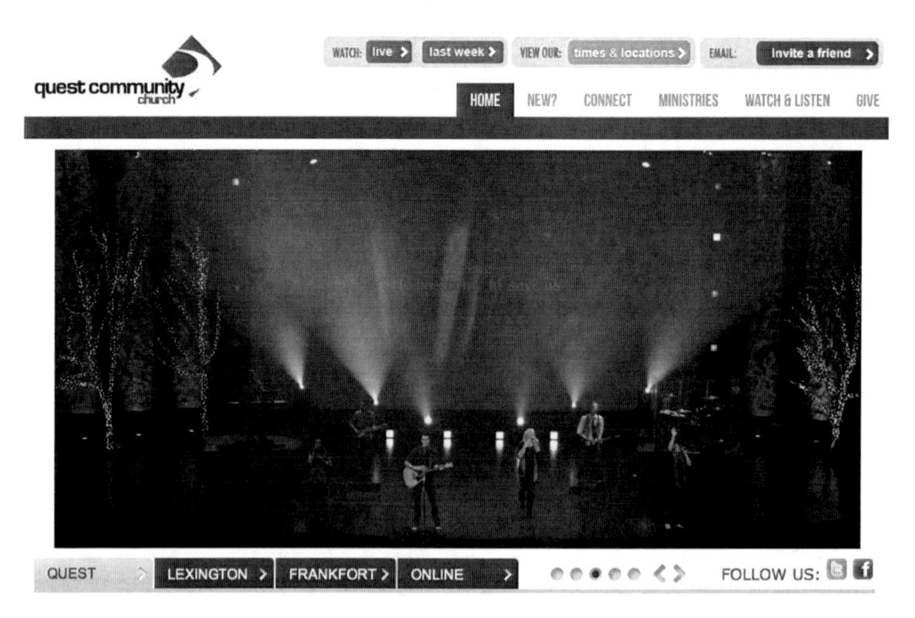

Fig. 121.2 Quest Community Church web page (Source: QuestCommunity.com, used with permission)

Fig. 121.3 Central Baptist Church web page (Source: Lexcentral.com, used with permission)

that they are trying to create a positive feeling in the shopper and an image is a relatively simple way to convey emotion or an appropriate message.

Smaller churches use their websites in some very different ways. First, there is a drastic difference in the web page design. As stated previously larger churches, overall, have very simple designs with few links and a significant number of images and multimedia presentations. Smaller congregations appear to have more text and more links. We can see this on the Central Baptist Church's website (Fig. 121.3) that features 466 words on the home page compared to just 63 words for Quest Community Church's home page. St. John's Lutheran Church is also an example of an overemphasis on text. They also appear to not be updated as frequently as evidenced by outdated calendar of program events and also sermon information. The last sermon that is displayed on Central Baptist's website is for September 30, 2012, 3 months prior to this writing. The sites for most large congregations are updated almost daily and certainly weekly. Featuring outdated or inaccurate information could be seen as a turn off to those who are in tune with modern technology and social media and could certainly determine whether they would even visit the congregation. Those who use the Internet often and are familiar with the ways modern sites operate and appear many find the sites of some smaller churches are both outdated and not visually appealing.

There is also heavy emphasis on social media on a number of sites that were accessed. There are links to the individual Facebook and Twitter accounts of the congregation and sometimes the personal accounts of ministers themselves. Tweets such as, "There are many opportunities to unleash love on local and global levels. Find out how you can get involved: http://bit.ly/unleashlove#E4." This quote is from

Southland Christian Church's twitter account (@southlandcc) with 3,457 followers as of mid-December 2012. By using these links you can observe updates, sometimes daily, of events that are happening in the church; all contain overall positive messages. If a church shopper is a contemporary, or even a casual, Internet user and has access to these social media sites, this information could be very influential on the first impression one has about a church.

It is clear that there are some very obvious differences in the visual content in many sites of Lexington churches. It is difficult so say what kind of exact impact these sites, and the differences among them, might have on the growth of the churches or the influence that they have on shoppers' decisions to join or attend. But it would not be difficult to argue that accessing these webpages in itself is in some way an exercise in religious discovery, exploration and shopping. And in a society that values the web and social media, these pages with images, color, movement and symbolism likely play an important role in religious consumerism.

We also examined the webpages of four additional Lexington churches: Immanuel Baptist Church (Fig. 121.4), Calvary Baptist Church (Fig. 121.5), St. Luke United Methodist Church (Fig. 121.6), and First Presbyterian Church (Fig. 121.7). All are considered large downtown and suburban churches in Lexington with sizable memberships, many who will pass by many other churches of the same denomination to worship at their place of membership. The content of these pages were similar, in that they provided information about upcoming Christmas services in December 2012 and links to church ministries, staff, programs and general "welcoming to worship" information. In the analysis of these sites it became evident that while the major goal

Fig. 121.4 Immanuel Baptist Church web page (Source: IBC-Lex.org, used with permission)

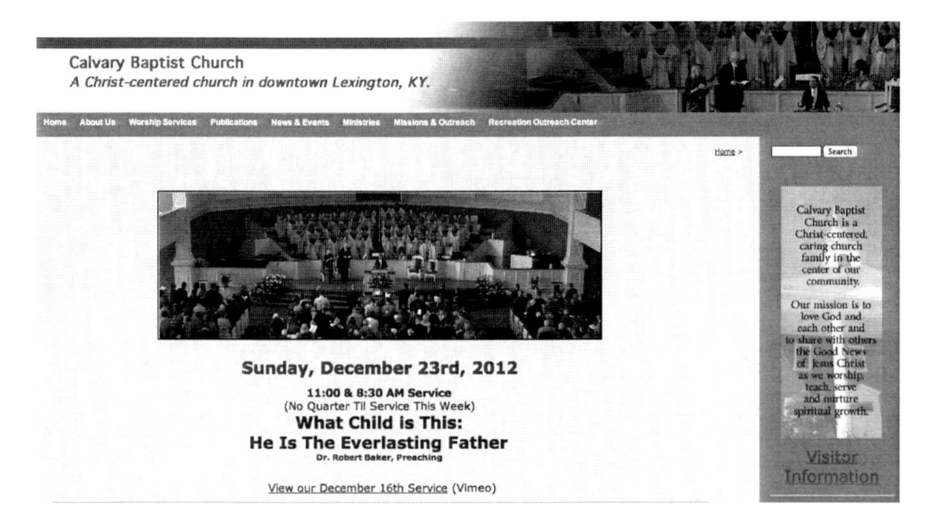

Fig. 121.5 Calvary Baptist Church web page (Source: CalvaryBaptistChurch.com, used with permission)

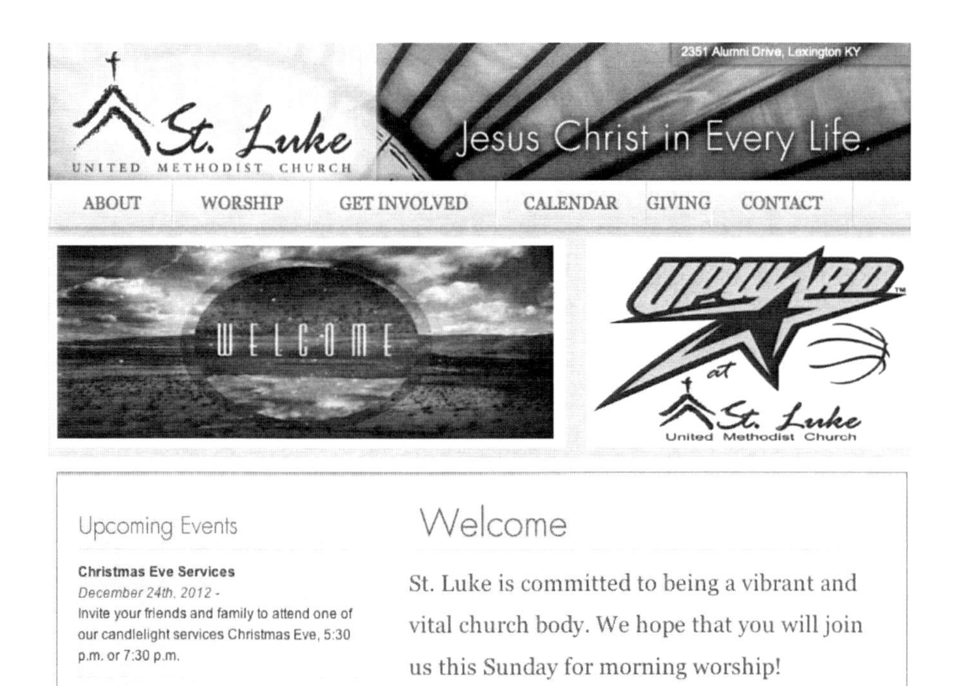

Fig. 121.6 St. Luke United Methodist Church web page (Source: www.stlukeumc.org, used with permission)

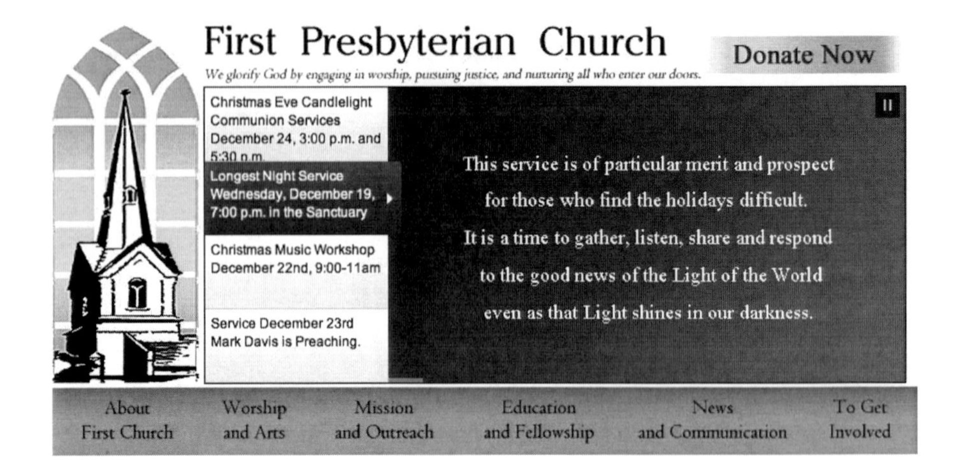

Fig. 121.7 First Presbyterian Church web page (Source: fpclex.org, used with permission)

of these sites is to provide general information about the church; another overriding objective is to create a desire to physically attend among those in the church shopping community.

121.5 Survey of Church Shoppers

Our next task involved constructing a two -page survey in summer 2012 of a sample (not statistically derived) of recent residents in Lexington. We asked six major questions:

1. How long have you lived in Lexington?
2. How long after arrival did you begin exploring places to worship?
3. How many churches did you visit? (name denominations)
4. What were the reasons why you decided to join a given church? (multiple items to check)
5. How far is the church you selected from your home?
6. How many churches of the same denominations are within a shorter distance of home?

Our initial plan was to give the survey to new members in a dozen different churches in Lexington. Our major objective from the start was to identify at least eight or ten different churches, that is, different faiths and denominations. We wanted to include a number of Protestant churches (mainline, traditional, liberal), Baptist churches (of which there are many in Lexington, as we discuss below), and several mainline churches (Lutheran, Methodist, and Presbyterian). We also wanted to include one or more Jewish temples, one or two Catholic churches, and several that are interfaith and interdenominational, that is, not affiliated with any formal denomination.

Contacts were made with pastors of nearly 20 churches thinking they would be interested in these results for their outreach and appeal for a new adherents. Some denominational leaders were enthusiastic about the survey and were willing to help us gather the results. We also promised those who helped us would be provided the generic results of the samples. The initial excitement that the topic rendered with pastoral leaders or chairs of new membership committees eventually dissipated for a variety of reasons, including some reluctance to share the names (even anonymously) with us. Some expressed that administering the survey to 20–25 new members was disruptive and counterproductive to their own recruiting efforts.

In spite of the lack of cooperation we proceeded to contact other leaders in late spring and early summer 2012. Behind all our thinking was that we wanted a good sample (thirty would have been ideal from a half dozen different denominations) so we could perform some statistical analysis. Realizing eventually that this objective would not be met, we decided to collect stories from eight or ten different individuals. We specifically wanted to try and include stories of new members from these groups and denominations in Lexington: mainline churches (Presbyterian, Methodist, Lutheran), a megachurch, two Southern Baptist churches, two African American Baptist Churches, and a Unitarian Universalist Church. Their stories, it was believed, even if not statistically would be useful in making some sense, albeit rudimentary, about the church shopping phenomenon in a small Middle West metropolitan area. These stories, which are presented below, are based on the nine individuals completing the survey we initially designed for a larger sample.

Below we summarize the stories of nine Lexington residents who have recently church shopped.

A megachurch. A single and never married mother of three who had not attended church in the past has now a found a home in a megachurch. She has lived in Lexington all her life, never having moved anywhere else. Family, friends and co-workers attend this church and invited her to come to a service and see if she would be interested in attending on a regular basis. This was her first visit to any church. She felt that the megachurch welcomed her and was friendly to strangers. Child care was offered to those who are attending so she can drop off her children on arrival and attending the worship service knowing they are being looked after in a safe and nurturing environment. New member classes are offered and she is currently attending those to enrich her spiritual experience.

A Lutheran experience. Job opportunities brought this young man to Lexington 2 years ago. He has attended and been a member of the same denomination in other cities in which he has lived. Spirituality and a feeling of holiness are very important to him and he felt that he found that in this church. One other church which he visited offered a contemporary service which he did not find appealing. Traditional music and use of a traditional liturgy make him feel closer to God. He has friends who belong to this church and they encouraged him to visit and become part of the congregation. The theological beliefs and all inclusive language follow those which he adheres. Adult education programs and times of fellowship which may be a meal or just a cup of coffee in the narthex make him feel welcomed as part of the church. He has been attending church services on a regular basis and has made the commitment to becoming a member of the congregation.

A Catholic. Traditional churches were first visited by this married young woman who has lived here for 8 years before she attended a megachurch. She attended services at the megachurch for several months and then felt that the services left something out which she was seeking. The messages that were sent were more social than based on religious beliefs. This was not what she was ultimately looking for in a spiritual home so she again attended a traditional Catholic church. She was not raised in this faith, but felt that the quality of the sermons and the Eucharist and communion services brought her closer to God.

Another Catholic. A current member of a Catholic church that is on the campus of the University of Kentucky moved to Lexington in summer of 2009. It took this individual 2 months before he began looking for churches in the area. Residing in the home are six people, all in the 20–30 year range. This person only visited one church before making the decision to join. It is the church he still attends today. According to the survey results he does not have any family or co-workers who attend and were not invited by a friend or relative prior to their visit. He was also not contacted by the minister after the first visit. Being a Catholic his entire life, he listed several reasons for joining including proximity to residence, he felt welcomed, the theological beliefs, and the fact that they were friendly to strangers.

An African American Baptist. This individual moved to Lexington in August 2010. She waited more than 4 months before beginning her search for a church. There are two to three people in her household. She listed two churches, both are African American churches with one being Baptist and one being non-denominational, that they visited before making the decision to attend one of these. Her decision to attend may have been influenced by that fact that they have family, friends, or co-workers who attend there currently. She was also invited by a friend or relative and was contacted by a minister after the visit to the church. She attended new members' classes and say that they visited more than 10 times before joining. She was not raised in this denomination but have been a member in the past. The reasons checked for joining were: like the meals and fellowship, felt welcomed, quality of sermons, baptism services, local mission projects, international mission programs, friendly to strangers, spiritually inspiring, and spiritual growth.

Another African American Baptist. This individual is currently a member of an African American Baptist church in the Lexington area. She moved here in the fall of 2010 and waited more than 4 months before visiting her first church. She has two to three people in her house of varying age and visited three churches before making a decision. She currently has a family member, friend, or co-worker who attends the church now and was invited by a friend or relative. The church pastor did not contact her after the first visit. She did not partake in any new members' classes and visited the church three times before joining. She has been a member of the denomination in the past. Some of the reasons she joined were proximity to residence, like the meals and fellowship, friends who belong, felt welcomed, and friendly to strangers.

Unitarian. This young woman in her late twenties attends one of the few Unitarian churches in the region. She is married with one child under the age of six. Prior to her joining this congregation she visited several more prominent churches in the city including a megachurch. She had never been a member of a Unitarian church before. She cited the welcoming atmosphere, as well as a more accepting tone in the sermons that drew her to join with her family. Her friends and co-worker invited her to a weekend service. She also cited fellowship, use of inclusive language, and children's education as other reasons for her joining. She stated that she lives within three miles of the church location.

Baptist student. This male student waited no more than a few weeks after starting college before he ventured out to look for churches to attend during his stay in the city. He is unmarried with no children and lives in a college dorm. Having been a member of another Baptist congregation in his hometown, he stated he focused his search on finding a similar one. None of his friends or co-workers invited him to this congregation but he instead found it on his own. Meals and fellowship were cited as reasons for joining, as well as the church's emphasis on college age members. The most important aspect for him was proximity to his residence, as he would be forced to walk to the church.

Presbyterian adult. This male church shopper moved to the Lexington area in September 2007 and waited approximately 4 months before beginning his search in earnest. There are between four and six people in his household, two of them are between 20 and 30 and the other two between 61 and 75 years of age. Before making a final decision of which church to join, he visited a single church, a Presbyterian church. He has no family members who attend this church with him, but at least one of his friends attends the same congregation; and he has co-workers who attend the same church. One of the pastors contacted him after his first visit and he also attended classes for new members. He visited the church at least twenty times prior to joining. Among the reasons that had an impact on his decision to join were: friends who belong, the quality of sermons, local mission projects, adult music programs, a welcoming of religious diversity, and also a spiritually inspiring congregation.

121.6 Discussion

What we learned from our inquiry into the webpages of selected Lexington churches and a small sample of church shoppers are four distinctive features. First, "the visual" is considered a very important and powerful way to attract potential church members. The colorful, interactive and even seductive webpages, especially of megachurches, is not that dissimilar from the kinds of visual marketing conducted by big box stores, trendy upscale shops or major mall developers. These are marketing "tools" used in written texts (advertisements in magazines and newspapers) as well as television commercials. Familiar symbols and icons, the depiction of families or selected demographics (the young or the elderly), and individuals using the latest technologies for information access or to contact with others are commonplace. Churches that provide interactive websites, images of church programs and groups, and even contemporary music are thought to appeal to the potential church shopper. It would be hard to underestimate the power and potential impacts of visual information in the "selling" and "buying" of religion and religious products.

Second, we have learned that there is vast unevenness in the quality and quantity of information for the potential church shopper. While there are high tech, interactive and even glitzy webpages for a few churches (megachurches seem to have the upper hand in this category), there are other churches that have little or very little information, including some that are not updating their websites regularly. The lag or delay in both presenting information or updating it regularly may be because the church leaders, staff and members do not consider it worth the time, money and effort to enter and compete in the "interactive church shopping game." Their reluctance also may be theological, that is, churches should not be engaged in the highly competitive "game" of seducing or luring souls, believing that old fashioned ways and methods of attracting members are preferred and successful, even in a society bombarded with high tech marketing of all kinds of consumer products.. Those in the latter group may reject the gimmicks and slick marketing strategies used by other churches, believing that the truly committed members are not lured by high powered religious marketing initiatives.

Third, from those newcomers to Lexington who shopped for churches, we learned that their experiences offered a variety of insights into both the "shopping" process and the selection itself. Not unexpectedly, some shoppers' experiences and "church searches" were limited, in that they preferred to stay with a familiar denomination. That is, they were just looking for the right Presbyterian, Lutheran or Catholic church in Lexington and did not explore other denominations in their search. These shoppers we could label as "denominational transplants." For others, their search meant exploring different denominations; some of these we could label as "membership transplants" in that they switched to a different church or denomination than in their previous residence. What we don't know is whether those "membership transplants" from the South had the same "shopping" experiences as those from the Middle West, Northeast or California.

Fourth, and finally, what we learned from the surveys is that there is a fair amount of fluidity in the church shopping world. This fluidity is evident at several levels. There are newcomers moving into Lexington each week; some are affiliated with the University of Kentucky, others with the major health institutions, the international horse economy, regional tourism, the housing, law and financial enterprises, and state and local government. Many, and it is difficult to discern how many, will visit places of worship on a given worship day to explore a place where they wish to establish religious roots. For some, this process may be short-lived, that is, a decision on a place to join will be made within 2 or 3 months; for others, it may be 5 or 6 months. A good analogy might be looking at newcomers as pieces on a gigantic moving chessboard, the board identifying shoppers and potential places of worship. There will be some pieces (new residents) who will move short distances, others will move long distances, others moving close to their homes and places of employment, and still visiting to potential churches on another side of the city. Some churches will attract many possible newcomers, others perhaps very few. Some will gain new members, others perhaps very few. And there may also be newcomers who shop for 3 or 6 months and eventually not to decide to join any church. These "non-attending church shoppers" are also part of the religious landscape of Lexington and many other contemporary U.S. cities. There are individuals, couples and families who "drop out" of formal religious institutions, perhaps for theological reasons or preferring to spend their time, money and energies in other ways to improve the human spiritual condition. In short, the religious "shopping" landscape of any large city is one of constant change: shopping and experimenting, denominational-switching and withdrawing.

121.7 Conclusion

Our results suggest that the church shopping phenomenon is "alive and well" in consumer –oriented America. There are youthful, middle aged and elderly "shoppers" of many different denominations who "shop" for various reasons: a comforting theology, alternative worship experiences, a good religious atmosphere to raise children, programs for youth and adults (sports, music, drama, dance, study, etc.), good community outreach programs, church suppers and other learning experiences. We agree with Twitchell (2007) in his calls for more research on what he terms the "scramble competition" in American religion, now the work of a few economists and marketing scholars. He notes that "what makes it unique is it allows new suppliers into the market with little more than a prayer. In 1900 there were 330 different religious groups, now there are over 2000" (p. 24).

Based on our study we identify three areas that would seem ripe for continued investigation. The first is to examine in detail the demographics and religious/spiritual lives of those who "church shop" and also those who are "church withdrawals." For those who fit the "church consumerist" mode, their "time/space" paths would be worth further scrutiny, especially their travels to prospective places (does distance play a role?), the influences of friends and neighbors, and even the theological positions of church leaders (if important). Perhaps collecting "church shopping diaries" would offer some further insights into the sorting out processes. Equally as important to study would be those who "drop out" or withdraw from formal religious experiences. We have uncovered no studies dealing with this group, yet, but with a growing number of Americans who identify themselves as secular and "not associated formally" with any religious institution (Winston 2012), it would be useful to know the "time/space" paths of institutional withdrawing and formal religious-disaffiliation. Was the process as long as that for newcomers into a community? Did the process include visiting alternative churches before formally withdrawing? Did distance to worship play a role? And, finally, were there elements about the worship experience that turned people off, for example, religious instruction for children, the stances of national denominations on issues such as abortion, women clergy or same-sex marriage, the quality of sermons and music program (language of hymns, unfamiliar tunes) and, incomprehensible theological views for the twenty-first century? Learning what "appeals" to potential shoppers is just as useful as what "detracts" from continued religious and worship experiences.

A second theme would look at the role of current information/communication technologies in marketing religion and especially efforts to attract or lure newcomers to a place of worship. How important are church websites to potential attendees and especially the images and words about the church? In an increasingly visual world are longstanding images about a church's history, holidays and festivals appealing to newcomers or not? Are images of stained glass windows, children playing and adult community service more important in a church's websites than crosses, crucifixes, and icons? Are biblical passages and biblical language appealing or less useful in marketing? Is inclusive language a plus or a minus? A detailed study looking at a cross-section of churches in a community would likely lend valuable insights into the uses and successes of "image vs. text" marketing messages.

A third theme would look at the extent of "shopping" by other religions and in other countries. One of the opening quotes acknowledged that "church shopping is uniquely American." While this phenomenon may be more popular in the U.S. than elsewhere, we suspect that one would also discover "mosque shopping" among new Muslims in London and Paris and possibly "synagogue shopping" among highly mobile Jewish populations in these and other European cities. We would also envision some degree of Christian "church shopping" among newcomers (the poor especially) into megacities in Africa and Latin America where, for example, there are shifts from formally-structured Catholic churches to rapidly growing Pentecostal communities (Jenkins 2011). The extent of "shopping" among the poor and marginalized in the Global South would likely be for many different reasons than the shifts toward secularized churches in the Global North. Another dimension of this shopping would be to examine the process in places where there is a predominance of a single church. For example, in the Bible Belt South where Southern Baptists are dominant, how much "shopping" or switching is there among Southern Baptist members moving to places where the denomination is predominant? The same question might be asked about cities in the U.S. and elsewhere the Catholic church is the dominant church.

In conclusion, we see the topic of "church shopping" as one that would, could and should appeal to scholars in the social sciences in many different world settings. Such studies would help us understand both the changes in identity that many religious followers are experiencing and also those who choose not to change their places of worship, religious worldviews, and personal behaviors.

Acknowledgements We want to thank the following individuals, including Lexington clergy, who helped us in various stages of this project: Lynn Phillips, Woody Berry, Cynthia Caine, Brian Cole, Rob Docherty, Chester Grundy, Sara Herbener, David Howard and Ruth Lundborg.

References

Atkinson, G. (2014). *8 church shopping tips from an Ex-Paster*. http://www.churchleaders.com/pastors/pastor-articles/153086-gordon-atkinson-8-shopping-tips. Accessed 8 Oct 2014.

Barna, G. (1988). *Marketing the church: What they never taught you about church growth*. Colorado Springs: Navpress.

Cwirla, W. M. (2005, September 11). *Rev. Cwirla's Blogosphere. A guide to church shopping. Higher Things, Inc.* http://blog.higherthingsorg/wcwirla/Articles/Shopping.html

De Groot, K. (2006). The church in liquid modernity: A sociological and theological exploration of a liquid church. *International Journal for the Study of the Christian Church, 6*(10), 91–103.

Hendricks, K. D. (2004). *Church shopping*. www.churchmarketingsucks.com/2004/09/church-shppping/

Hoover, S. M. (1998). *Mass media religion*. London: Sage.

Howard, B. (2012). *A pastor's take on church shopping*. www.abpnews.com.blog/ministry/a-pastors-take-on-church-shopping-2012-10-04

Jenkins, P. (2011). *The next Christendom. The coming of global Christianity*. New York: Oxford University Press.

Kennesson, P. D., & Sreet, J. L. (1997). *Selling out the church: The dangers of church marketing*. Nashville: Abingdon.

Relevant Magazine. (2009, September 11). *How to do church shopping. Finding out what really matters when searching for a faith community*. www.relevantmagazine.com/god/church/features/18261-church-shopping

Rule, C. (2009). *How to go Church shopping. Figuring out what really matters while searching for a faith community*. http://www.relevantmagazine.com/god/church/features/18261-church-shopping. Accessed 17 Sept 2009.

Santella, A. (2009). *The church search*. www.slate.com/articles/life/faithbased/2009/02/the_church_search.html.

Scotland, N. (1984). Shopping for a church: Consumerism and the churches. In C. Bartholomew & T. Moritz (Eds.), *Christ and consumerism: Critical reflections on the spirit of our age* (pp. 135–151). Cumbria: Paternoster Publishing.

Shawchuck, M., Kotler, P., Wrenn, B., & Rath, G. (1992). *Marketing for congregations: Choosing to serve people more effectively*. Nashville: Abingdon.

Starkey, M. (1989). *Born to shop*. Eastborne: Monarch.

Sullivan, A. (2009, April 28). Church-shopping: Why Americans change faiths? *Time*. www.time.com/time/printout/0.8816.1894361.00.html

Twitchell, J. B. (2007). *Shopping for God: How Christianity went from in your heart to in your face*. New York: Simon and Schuster.

Walker, A. (1998). *Consumerism, personhood and the future of Christian mission*. Presentation at conference on seduction or evangelism. Cheltenham and Gloucester College of Higher Education.

Ward, P. (2002). *Liquid church*. Peabody/Edinburgh: Hendrickson Publisher.

Wells, D. (1994). *God in the wasteland: The reality of truth in a world of fading dreams*. Leicester: IVP.

Winston, K. (2012, October 9). Lousing our religion: One in five Americans are now 'nones.' *Religious News Service*.

Part X
Megachurches and Architecture

Chapter 122
Sacred Ambitions, Global Dreams: Inside the Korean Megachurch Phenomenon

Michael Bégin and Caleb Kwang-Eun Shin

122.1 Introduction

In 1993 nearly half of the world's fifty largest churches were in Korea. The largest church in the world in terms of membership is Yoido Full Gospel Church, now with approximately 760,000 members throughout Seoul and a weekly attendance of over 150,000 (Thumma and Leppman 2011). Korea is presently home to fourteen churches with more than 10,000 members, and more than 1,000 churches with approximately 1,000 members each (Lee 2000). Although these numbers constitute only 2 % of all Korean churches, one can safely infer that Korean Christianity is under significant influence of the contemporary *megachurch*, having become a unique socioreligious phenomenon in its own right. The term itself has been typically used to indicate a church with a Sunday attendance of 1,000–3,000 members; however, we prefer to define it as an institution which recognizes no limits to growth (Shin 2009). Whereas traditional churches encountered impediments which precluded accelerated growth, such prohibitions have since disappeared, resulting in "super-sized" churches brought about by a specific series of real-world circumstances that constitute their social contexts (Table 122.1).

This chapter will examine the various mechanisms at work in giving rise to the global megachurch phenomenon, with a particular focus on the Korean example. These mechanisms are understood to greatly influence the rise of Korean megachurches, although in the Korean case there are many unique particularities that merit closer examination. The first section will offer some contextual background on the rise of the megachurch in the West, outlining the important precedents that

M. Bégin (✉)
Department of Global Studies, Pusan National University, Pusan, Republic of Korea
e-mail: mikebegin9@hotmail.com

C.K.-E. Shin
ABD, Korea Baptist Theological Seminary, Daejeon, Republic of Korea
e-mail: calebkshin@daum.net

S.D. Brunn (ed.), *The Changing World Religion Map*,
DOI 10.1007/978-94-017-9376-6_122

Table 122.1 Locations and rankings of Korea's ten largest churches, by membership size, 1993 (Map by Caleb K.E. Shin; data source: Chosun Ilbo News Service)

#	Church	Location	Membership	World ranking
1	Yoido Full Gospel Church	Seoul	600,000	1
2	Anyang Full Gospel Church	Anyang	105,000	2
3	Keumnan Methodist Church	Seoul	56,000	7
4	Soong-Eui Church	Incheon	48,000	9
5	Juan Presbyterian Church	Incheon	42,000	10
6	Seoul Sungrak Church	Seoul	30,000	11
7	Kwanglim Methodist Church	Seoul	30,000	12
8	Youngnak Presbyterian Church	Seoul	28,000	13
9	Hyesung Presbyterian Church	Seoul	23,000	15
10	Somang Presbyterian Church	Seoul	22,000	16

Data source: Chosun News Service Online, South Korea

eventually fueled the phenomenon in Korea. In the second section, we explore the Korean context, highlighting the important sociocultural, historical, and geographical particularities. The third section examines the four generations of the Korean megachurch, revealing their distinctive characteristics and providing some important examples. In the final section, we offer an evaluation of the Korean megachurch phenomenon, detailing its beneficial and less beneficial aspects, as well as some prognostications.

122.2 The Mechanisms of the Global Megachurch Phenomenon

122.2.1 The Theological Basis for Growth

Megachurches owe their phenomenal growth and development primarily to the passion for evangelism and missionary activity, which the traditional church had never emphasized. In the Protestant Church, this initiative found its origins in the seventeenth century, accelerating during the Great Awakening Movements of the eighteenth and nineteenth centuries. Biblical precedent for this was found in "The Great Commission" of Matthew 28:18–20, which had been considered a command only to the first century apostles until William Carey put a new interpretation on the same verse at the end of the eighteenth century (Bosch 1991). Consequently, the saving of souls and evangelizing unto the ends of the earth became the church's greatest commission, eventually becoming the supreme mandate for every church and every Christian. Bound with the principle of voluntarism, this passion compelled a great number of church pastors and lay persons to make great sacrifices for the sake of missionary activity. Evangelists who had embraced Dispensational Premillenialist eschatology proclaimed *the evangelization of the world in this generation* as their operating mantras, coming to believe that global evangelization would guarantee the second coming of Jesus Christ, the defining teleology of Christianity. As the church regarded evangelism and global missionary activity as its most urgent tasks, we recognize the megachurch phenomenon as a product of this sense of urgency undergirding church eschatology.

As European evangelists gradually shifted focus from pagan conversion to saving the souls of the newly autonomous, Enlightenment-era individuals, churches came to define religious "success" in terms of evangelization (Bosch 1991). In the New World, the church became a veritable arena of competition for success enhanced by the freedom from state control, as was experienced in Europe. A great shift in evangelical focus came with the influential mid-twentieth century *church growth* theories of evangelicals such as Donald McGavran (1955, 1970), who argued that churches should focus on growth in addition to evangelization. This mandate harmonized with the assertions of Drucker (1990), who argued that non-profit organizations must strive toward a certain kind of quantifiable success. Today, growth itself is seen as the most valuable and praiseworthy goal of the contemporary church.

122.2.2 Transcending Limitations

The modern megachurch did not create itself merely through internal transformation; the removal of important external barriers to church growth also played a role. Of course, the significant twentieth century boom in world population allowed for substantially greater opportunities for the church to bring souls to Christ. Taken together with the very rapid rates of urbanization in the developing world, and with nearly 80 % of the developed world now concentrated in urban areas, it is apparent that the geographic distribution of population is also of primary concern to the growth of churches. The sociopolitical shift associated with such dramatic demographic changes was identified in 1930 by Ortega y Gasset (1994) in his writing on mass movements, where mass society itself asserts its interests in a growing public sphere of democratic debate and discourse, and where modern democratic values represent influential ideas of shared destiny and mutual interest, building the foundations of mass culture that are easily transmitted to the congregation. Consequently, the techniques and technologies of public relations and mass communication became all the more central, especially in settings where many thousands of people congregate. Audiovisual technologies and other types of "hardware" systems that constitute the usual fare in an age of information and mass communication go hand-in-hand with the organizational techniques that render propaganda, marketing, and management essential aspects for the successful transmission of organizational messages. As McLuhan (1964) referred to these types of "extensions" that permit the transcendence of physical limitation, likewise these techniques are clearly extensions of the modern megachurch, sharpening the skills of persuasion in an age where church individualism and free-market competition for adherents thrive in the absence of the traditional strictures of catholicity.

122.3 The Context of Korean Protestant Christianity

As a socioreligious phenomenon that experienced a rapid global ascent in the 1960s, megachurch communities derived their vitality from a powerful theological justification for church growth and an eradication of the external impediments to growth. The Korean megachurch is no exception to this. However, it is necessary to consider the unique sociocultural, historical, and geographical contexts that have given rise to the contemporary Korean megachurch.

122.3.1 A Successful Introduction of Christianity in Korea

Korea's eventual abandonment of its isolationism opened up important opportunities for evangelism on the peninsula, and the late nineteenth century witnessed the ascent of the Protestant church in Korea. By this time, the Regent Heungseon Daewon-Gun, the last isolationist sovereign of the Chosun Dynasty, had fallen from power, and the succeeding Chosun leadership had signed treaties with Japan and the United States, dispatching envoys to many European countries in an effort to affect a rapprochement between Korea and the outside world. By this time, the Chosun leadership had developed a much more favorable attitude toward Christianity, and the Protestant Church in Korea succeeded in developing a more positive reputation there than did the Catholic Church. With voluntary requests for foreign missionaries and the continued dissemination of the translated Bible, Christianity enjoyed a somewhat elevated status in Korea, supported in part for political reasons (Lee 1985). Korean intellectuals, for example, saw Christianity as a means of extending political power through Westernization and the introduction of Enlightenment values, as had taken place in Japan. Korean independence fighters such as Kim Gu, An Chang-Ho, and Lee Seung-Hoon saw Christianity as a route toward national liberation (Lee 1980). Anxieties related to foreign invasion compelled many Koreans of the Chosun dynastic period to seek solace and hope in the Holy Gospel. Christianity in Korea subsequently enjoyed more remarkable success in Korea than ever before, becoming the "indigenous" brand of the faith that persists to this day. This phenomenal growth was part and parcel of the impressive influence attained by Korean megachurches.

122.3.2 The Internal Aspects of Church Growth in Korea

The Korean Protestant church was evangelical in nature from its very inception, owing to the precedents set by missionaries to Korea inspired by the Great Awakening and World Evangelization Movements. These missionaries were devoted to the establishment of an evangelical Korean church starting in the Chosun period, setting an influential and lasting precedent in Korean church history that came to resemble a form of revival movement. From the Pyeongyang Great Revival Movement of 1907 to the influential evangelical conferences known as EXPLO '74 and EXPLO '80, such movements persisted with great influence, with the new Korean church investing prodigious efforts to evangelize and Christianize Korea.

Church growth was also fueled internally by cultural factors. The pervasive *familism* in Korean society derives from an enduring Confucianism, which places significant emphasis and importance on the family (especially the family patriarch) as the fundamental social unit which expands to include nation, organization, and enterprise. With these guiding archetypal principles of Korean culture, the church is viewed as a broad extension of one's family, where the pastor is "father" to his followers. As cities rapidly expanded in Korea from the 1960s, a largely village-oriented culture organized around extended families coexisting in local communities transformed into fractured groups of urban migrants seeking new community belonging (Kim 1995). This was inspired by the Korean sense of *'we'-ness*, a sense of collective well-being and mutuality of destiny, a very important element of Korean familism. The seminal *Book of Filial Duty* by Confucius decrees the obligation of sons to bring success and prestige to their respective parents. A mandate still very prevalent in Korea, it certainly offers some explanatory weight for the success-oriented culture of Korean society as well as the goal-oriented, growth-seeking element of the Korean megachurch. The pervasive *face culture* of Korean society is closely related to these foundational Confucian ideas, providing definition to the Korean penchant for hierarchical organization in both families and institutions. The organizational ethics of these established hierarchies incentivize the upward mobility not only of individuals, but also of institutions such as churches.

With this, it is unsurprising that the Korean church in the 1960s and 1970s upheld the primacy of development and growth as its organizing principle. That both nation and church adopted the same *growth first* mantra can be acknowledged with the observation of the corresponding peak growth rates of Korea's national economy and that of the Korean church. More importantly, the Korean church accepted this mantra as its very theology, a point exemplified by Cho Yong-Ki, the founder of Yoido Full Gospel Church, who espoused the theology of the *Five-Fold Gospel and the Three-Fold Blessing*. This theology upholds the Gospels of Regeneration, the Fullness of the Holy Spirit, Divine Healing, Blessing, and the Advent of the Risen Christ, while emphasizing the spiritual, physical, and circumstantial blessings bestowed by God (Kim 2011b). Also at the core of this theology is economic growth and prosperity for all Christian adherents, ideas finding great political purchase in the early 1970s with former president Park Chung-Hee's *Saemaeul* campaign, which urged Koreans to improve living standards through diligent individual effort and cooperative communities. Meaning "new village," Saemaeul promoted a multi-faceted approach to the modernization of rural villages nationwide, initiating road improvement projects, a reorganization of agricultural production, and a modernization of home construction.

122.3.3 The Korean Megachurch in the Context of Social Change

Once again, the Korean megachurch phenomenon would not have taken place without the abolition of the impediments to church growth. Beset by a series of major changes and upheavals over a 150-year period (namely the end of Korean isolationism, the Japanese occupation and colonization in the early twentieth century, the Korean War, and the rapid industrialization beginning in the 1960s), the stubbornly traditional and reclusive Hermit Kingdom found itself unable to resist the profound changes that would ultimately facilitate this abolition. The Korean population boom of the 1960s and the corresponding dramatic rates of urbanization constituted intricate elements of the profound social changes taking place, resulting in a primarily urban population. The impressive population densities and the development of new districts of Seoul and other cities set the stage for the advent of megachurches, such as Yoido Full Gospel Church (Fig. 122.1), that owe their existence in part to urban development. In the 1970s, even the diminutive Baptist community in Korea expanded dramatically by simply redirecting evangelical activities from rural to urban areas.

The Korean example is unique in the sense that traditional districting and localization of church membership were freed from erstwhile geographic restrictions, resulting in a *translocal* church membership and a complex of overlapping districts. Consequently, megachurches often competed for membership in the same areas. For example, megachurches planning to build branch churches in neighboring

Fig. 122.1 Yoido Full Gospel Church in Seoul, Korea (Photo by Caleb Shin)

districts were faced with vigorous opposition by competing churches, resulting in territorial battles for church influence. Also, the brazen recruitment activities of some church promotional campaigns entailed the positioning of volunteers in front of other churches to "steal" members away, occasionally eliciting violent scuffles. These recruitment efforts were sometimes aided, sometimes impeded by other developments, including public transportation, shuttle bus service, and the boom in personal vehicle ownership. Communications technologies furthered this competition with the incorporation of cable TV and satellite communication service, and finally online broadcasting that relieved members of the necessity to physically attend church services.

The intense and protracted weakening of Protestant universality (that is, ecclesial unity) furthered the rise of Korean megachurches. Starting in 1952, the Korean Presbyterian Church divided time and again through 1979. Similar divisions also took place among the Methodist, Holiness, Baptist, and Pentecostal denominations (Cho 2008). Accordingly, Korean megachurches are largely separate from their parent denominations (for example, Onnuri Church in Seoul is no longer subject to the auspices of the Presbyterian Church of Korea), and can thus be considered autonomous *post-denominational* churches. During this seemingly infinite division, voices from the parent denominations themselves barely intervened, furthering the erosion of the binding power of the denominations, which had rather feebly lobbied for church union and cooperation. As church communities became individual entities fully responsible for their own survival, expansionism became the means by which these communities would persist. The next section provides some important Korean examples of this.

122.4 The Four Generations of Korean Megachurches

Since about 1960, the Korean megachurch phenomenon developed over the span of four discernible generations, offering a general narrative of its history and development. As the nature of these church communities gradually changed with time, the essence of church membership in Korea clearly has become *the extent of an individual's participation in church worship services*, where members acquire, and frequently transition through, important status distinctions (newcomers, declared believers, ordinary believers, delegates, volunteers, ministers, staff, group leaders). Although training for pastoral leadership in Korea comes mostly from seminaries (for example, the Seoul-based Presbyterian College and Theological Seminary, Methodist Theological University), many megachurch pastors have taken the charismatic route, laying the foundations for new post-denominational or non-denominational churches with occasional financial support or training from American schools and churches.

Architecturally speaking, early-generation Korean church buildings were typically fashioned after European Gothic-style churches, frequently with high, steep pinnacles displaying the Christian cross in anachronistic red neon, making such churches easily locatable especially at night. However, later-generation Korean churches feature a range of architectural styles including Romanesque and stadium-like designs, with some hybridized concepts. Contemporary Korean megachurches are often built in modern style, with sleek features, plate glass, and polished steel, much like stylized office buildings, borrowing occasionally from earlier styles. In the following subsections, we offer a description of the primary characteristics of each megachurch generation, including the key themes, figures and events that provide meaning and definition to these communities, as well as some important case studies.

122.4.1 The First Generation

The first generation of megachurches in Korea is the group of traditional churches centered in the old downtown area of Seoul, consisting of Saemoonan Church, Chungdong First Church, Youndong Church, and Yongnak Presbyterian Church (Fig. 122.2). These churches had expanded at a very early stage (for example, Yongnak had 3,000 regular members by 1950), made possible by their traditions, their locational accessibility, and the rapid post-war growth of the Seoul metropolitan area (Lee and Lee 2002). These rather exceptional churches expanded to unprecedented sizes for the era, demonstrating the incredible potential of Korean churches at the time. That these churches had evolved into what we now define as megachurches

Fig. 122.2 The former main chapel of Yongnak Church (Photo by Caleb Shin)

was quite unintentional, yet neither did they actively intervene to prevent growth. This more passive approach toward church expansion is characteristic of the first generation of Korean megachurches.

Yongnak Presbyterian Church, now the world's largest Presbyterian church, provides an excellent case in point. Situated in the ancient citadel of the Chosun dynasty, the church was constructed in 1945 by the pastor Han Kyung-Jik and 27 North Korean defectors, rapidly becoming a refuge for North Koreans and expanding to 1,400 members within one year. Staunchly democratic, Yongnak Church became an active center of the Korean anti-communist movement, expanding eventually to 60,000 members. Known for his ethical and intellectual sensibilities and his folksy, populist approach, Rev. Han's remarkable leadership struck a chord with many South Koreans as well, earning him the Templeton Prize in 1992. Yongnak is known for its active, Christian-based volunteerism, its theological research, the *Borin-won* orphanage, and a host of other social welfare projects promoting education and social responsibility. A prominent leader in the national evangelical movement and a common benchmark for many other churches, Yongnak had focused not only on its internal activities, but also on the larger goal of reorganizing Korean society through Christianity. This focus expressed the larger goal of establishing Christianity as the official state religion of Korea, bringing the nation closer to the American model.

122.4.2 The Second Generation

The Korean megachurch phenomenon gained great momentum as the drive for aggressive church growth became a common aspiration. The second generation of megachurches emerged in the 1950s, concurrent with the rapid growth of the national capital and its changing urban morphology, which saw the old downtown area extended to the outskirts of Seoul where new churches were founded at an equally rapid rate. *Kangbuk* megachurches, named in reference to their location north of the Han River, are now found throughout Korea (Fig. 122.3). Examples include the Kwanglim Methodist Church and Yoido Full Gospel Church. Kangbuk churches forged comfortable political alignments with the Korean government, encouraging the mobilization of all Koreans for economic growth while also drawing suspicion from other mainstream churches. Virtually despotic leadership, centripetal organizational structure, and aggressive and coercive evangelical programs are all commonly employed to justify the promise of material well being for its members (Ku 2008). Because the enormity of these communities makes monitoring rather difficult, many churches employ small "watchdog" groups assigned to scrutinize and control the activities of church members. Chaired by church-appointed group leaders better positioned to monitor attendance and control dissent or criticism, such groups are also sometimes used to send envoys to the homes of members who fail to attend important meetings or functions.

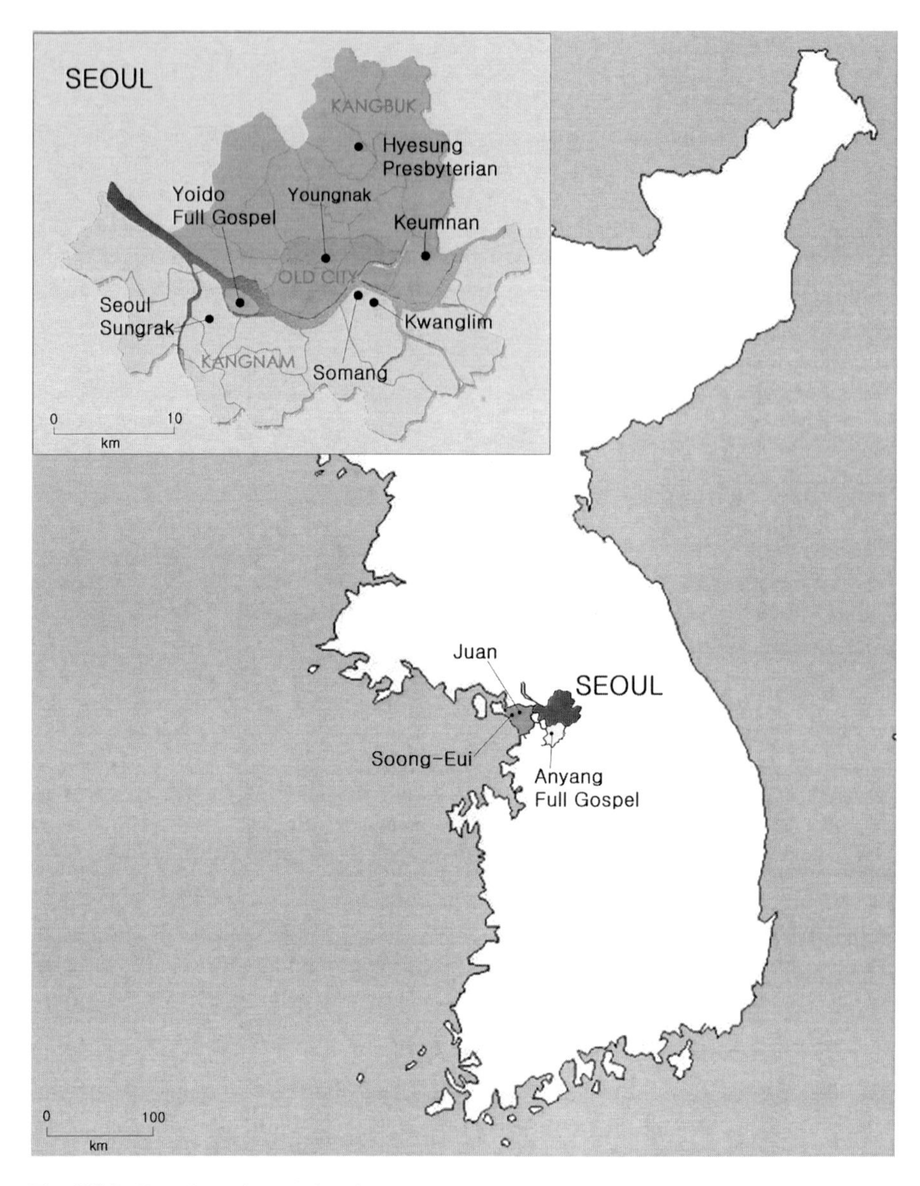

Fig. 122.3 Korea's ten largest churches

Founded in 1958, Yoido Full Gospel Church is an exemplar of the second generation. With an original congregation composed of the city's impoverished, labor-weary, and war-stricken cast into a harsh, new urban life, Yoido promptly became a place of sanctuary and salvation as pastor Cho Yong-Ki fervently preached acceptance of Jesus Christ for atonement, healing, wealth, and a guarantee of eternal

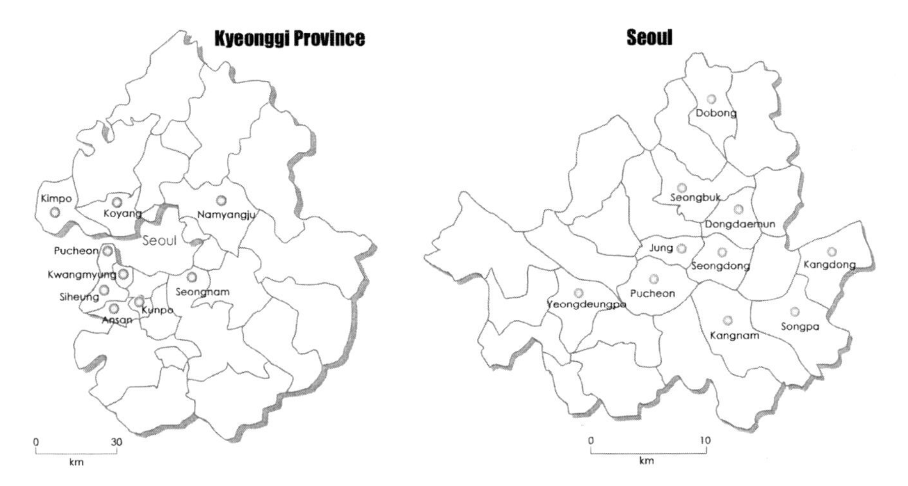

Fig. 122.4 Branch locations of Yoido Full Gospel Church. The multitude of locations is telling of Yoido's pervasive influence in the region (Map by Caleb K.E. Shin and M. Bégin, 2012)

salvation in heaven. With over 10,000 members by 1972, Yoido Full Gospel became an influential player in the global church growth movement, achieving a membership of 763,301 in 2007. Yoido Full Gospel boasts not only a cavernous main chapel and numerous worship facilities (Fig. 122.4), but also its very own Osanri Prayer Mountain on the outskirts of Seoul, all of which are devoted to earnest prayer, psalms, and divine healing ceremonies performed in a relatively spontaneous, informal, dynamic, and participatory atmosphere.

The sophisticated organizational technique of Yoido, which accommodates hundreds of thousands of members during services each week, is impressive. In 2008, Yoido Full Gospel consisted of 35 large parishes and 312 small parishes organized in tandem with the administrative divisions of Seoul city government. At their heyday, these parishes consisted of 43,330 small working groups forming micro-communities wherein fellowship, faith healing, and education were enabled. Challenged with the weekly accommodation of an astonishing number of parishioners, Yoido utilizes its main building and adjacent buildings for Sunday worship, admitting up to 200,000 members at a time. Numerous branch churches and satellite-based broadcasting bring worship services to an even greater number of devotees. Yoido further extends its influence through its affiliated publishing companies, university, broadcasting company, mission centers, schools, and social welfare centers, employing a great number of pastors, missionaries, and staff members.

The great significance of Yoido Full Gospel is that it signaled the advent of the global megachurch phenomenon while transcending itself, manifesting a phenomenon that is experienced by many as sublime, indeed sacred, by virtue of its extraordinary growth. As such, Yoido Full Gospel is a unique religious phenomenon, serving as proof that virtually limitless growth is indeed possible once obstacles to growth are circumvented.

122.4.3 The Third Generation

Korea's third generation of megachurch is known as the *Kangnam* megachurch. Situated to the south of the Han River, the Kangnam area is synonymous with the national elite and some of the most expensive real estate in Korea. Prominent Kangnam megachurches include Sarang, Somang, and Onnuri Community churches, all of which were established in the early 1980s. Following closely in sequence, the Kangdong and Bundang areas of Seoul saw the construction of similar churches, which also gained the classification of Kangnam-type megachurches. Typically, the members of Kangnam churches are of higher class standing than Kangbuk-type church members (Ku 2008). Incorporating post-Fordist organizational methods allowing for a higher quality and variety of programs, Kangnam worship services are characteristically calm, intellectual, and well-polished, focusing on the spiritual nurturance of the inner self as opposed to hands-on, faith-based healing. Although these churches do feature charismatic leadership, they are usually critical of the more narrow, growth-driven policies of Kangbuk-type churches and are reputed for balancing size and quality more effectively, thus seeing theirs as more "healthy" communities. Kangnam churches are also characterized by the *Four Big Figures,* or the four most influential megachurch pastors. Since 1980, their ministries have been recognized as role models of Korean church organization and practice. The precedents they set not only encouraged the incredible growth of these churches, but also vastly expanded the breadth of community practices in Korea, making the Korean megachurch a unique sociological phenomenon in its own right.

One of the *Four Big Figures,* Rev. Ha Yong-Jo held his first worship service with his eighty-member bible study group in 1985, heralding the inception of Onnuri Community Church (Fig. 122.5). Since then, Onnuri (Korean for "global") has become one of the fastest-growing churches in Korea, expanding to 1,194 members in three years. Rev. Ha was reputed for his ambitious and expansive projects, such as his *2,000/10,000 Vision* project that would dispatch 2,000 missionaries and 10,000 ministers for overseas evangelical work. By virtue of this ambitious vision, Onnuri expanded to 10,000 members by late 1996. In 2003, Onnuri proclaimed yet another ambitious goal: the *Acts 29 Vision*–another innovative project to bring the church itself to the mission field. By 2010, the Acts 29 project resulted in the establishment of eight new churches in Korea and 25 overseas churches, unified via its internal CGNTV network, nine satellites, and an online network.

Onnuri was relatively successful in distinguishing itself from traditional churches by reforming its own image. This entailed that Onnuri establish religious authenticity by establishing itself as *the very church*, in the spirit of the first Christian church of ancient Jerusalem, balancing expansion with the quality of church services. Another of Onnuri Church's distinctive features is its *parachurch*, a second core element of the Onnuri community known as Duranno Ministry Center. Preceding

Fig. 122.5 The main building of Onnuri Community Church (Photo by M. Bégin)

Onnuri itself, Duranno administers various programs, institutions, and operations including a publishing house, a bible college, a parenting school, a one-on-one ministry program, and an international mission project. Onnuri also developed many additional programs addressing children's education, leadership, inner healing, family ministry, and foster parenting. Far removed from the German choral hymnals of old, Onnuri is particularly well known among Korean Christians for its *Worship and Praise* musical program, complete with live, contemporary music in the American Christian Pop vein. This approach effectively attracted scores of youth to the church, at once inspiring many local churches throughout Korea to adopt similar musical approaches to worship, essentially revolutionizing church service formats in Korea (Fig. 122.6). Onnuri Church is relatively systematic in its administrative approach, placing emphasis on effective planning and organization of worship, mentoring, and ministry systems (Lee 2005). This approach has been recognized as profoundly effective in encouraging church growth, allowing the church's various elements to function independently of pastoral supervision. The death of Ha Yong-Jo in 2011 brought new leadership to Onnuri in the form of Rev. Lee Jae-Hoon, bringing some uncertainty to the church's future.

Fig. 122.6 Korean ministry group performing a worship service featuring contemporary music at Yoido Full Gospel Church (Photo by M. Bégin)

122.4.4 The Fourth Generation

Although the Korean church experienced nearly constant growth for 130 years, this growth was not fated to continue in perpetuity. The general decline in church growth from 1995 signaled the beginning of a new era where the growth of individual churches gradually gave way to growth in the number of converted Christians, effectively ramping up institutional competition for devoted souls in a kind of Hobbesian "war of all against all," an essentially anarchic condition. With this, we can argue that the unfolding scenario will eventually result in a fourth generation of megachurches that constitutes the victors of this incessant competition. Thus, the usefulness or practicality in identifying a fourth generation of Korean megachurch remains arguable. Nevertheless, it is worthwhile here to focus our attention on Sarang Community Church, actually a more fitting representative of the third generation of megachurches under the previous leadership of Rev. Ok Han-Eum (also known as Rev. John Oak).

Sarang's humble debut started with nine worshipers in 1978. Like Rev. Ha of Onnuri, Rev. Ok's stated intention was to build *the* church, emphasizing active involvement among the laity (Park 1998). Desiring nothing less than to awaken his congregants and transform them into active ministers, Ok introduced the discipleship training method he had gleaned from organizations like *Campus Crusade for Christ* and *Navigator*. His efforts bore fruit in the form of massive church growth, with 17,490 members in 1998 (Park 1998). Sarang Church became a model of the harmonious coexistence of qualitative growth and quantitative growth, inspiring thousands of pastors to further their leadership training via Ok's *Called to Awaken the Layman* (CAL) seminar. Consequently, Sarang Church was vastly influential among numerous pastors of different denominations while maintaining its status as a third-generation megachurch.

After 26 years of ministry at Sarang, Rev. Ok retired in 2004, to be replaced by Rev. Oh Jeong-Hyun, the former pastor of the Sarang Community Church of Los Angeles. Under the leadership of Rev. Oh, Sarang Church changed rapidly, due primarily to his efforts to combine the existing emphasis on discipleship training with the mandates of the Holy Spirit movement, a move reminiscent of Rev. Ha's pastoring philosophy at Onnuri Church. This effort had the effect of doubling the church's 2002 membership in 3 years and marked a great departure from the original intentions of Rev. Ok, who often commented that he never intended to build a megachurch, and feared such a reputation.

Sarang's ambitions for church growth became very apparent upon examination of its policies and decisions, providing further evidence that Sarang had indeed entered a new era. In 2009, Rev. Oh announced a plan for the construction of a new chapel a short distance from the former one, better suited to accommodate large congregations (Fig. 122.7). The problem was not only with the project's exorbitant price tag (approximately US$ 200 million), but also with a questionable contract agreement Sarang signed with Seoul City government to close off certain subway entrances and open others to facilitate access to the new chapel (Kim 2011a). The threat this posed to the many other churches in the area resulted in numerous objections, disappointing many who viewed this move as a radical departure in church values and a reflection of identity loss on the part of the church. Still, construction is now underway.

122.5 Prospects: Evaluating the Korean Megachurch Phenomenon

Evaluations of Korean megachurches are often conflicted. Koreans once shunned large churches, an attitude that changed considerably after the megachurch gained a foothold in Korea. Now recognized as models of healthy church communities, the advent of the third generation of megachurch played a significant role in eliciting this change, garnering attention and admiration from all denominations and communities, to the point where even the more progressive Christian denominations in Korea came to view size as a measure of success (Jeong 2007). However, with megachurches now being the end game for nearly every denomination and church, theological principle in Korea more frequently takes a backseat to sheer power and influence, extending into virtually every sphere of religious and public

Fig. 122.7 The new chapel of Sarang Church, 2013 (Photo by Caleb Shin)

life, for good or ill. The paradoxical nature of the megachurch phenomenon in Korea is further elaborated in this section, followed by a brief evaluation of its future prospects.

122.5.1 The Ups and Downs of Megachurches

One reason Koreans view the megachurch as a positive development is its enthusiasm for evangelical activity, justified by Matthew 28: 18–20, with the church as the means by which this is accomplished. This view positions the megachurch as an enthusiastic and active church which extends the kingdom of God via overseas missions. Guinness (1993) praises the megachurch for its ability to renew itself, claiming that an effective church can adjust to modern society by revising its message as appropriate to rapid cultural changes. Similarly, Williams (2009) argued that the megachurch is very relevant to a post-Christian era, providing sensitive interpretations of the timely technological and cultural changes that otherwise challenge church traditions. This point becomes evident with our examination of the third generation of Korean megachurches, which typically make innovation a working mantra. Significant departures from Christian traditionalism, these experimental churches freely embrace non-Korean ideas, programs and methods, purchasing a certain flexibility that permits innovation, such as the systematic methods of disciple training practiced by the influential Onnuri and Sarang churches. The Korean megachurch is also an amazingly dynamic institution, characterized by a membership that is rarely passive, uninvolved, or apathetic. With newcomer orientation meetings, small group activities, volunteer initiatives, and missionary training programs, megachurches like Yoido, Sarang, and Onnuri offer highly developed and systematic community functions in addition to innovative ministry.

Still another strength of the Korean megachurch is its concentrated power, rendering possible a variety of ambitious projects that traditional churches can barely fathom, such as overseas mission projects and social welfare programs which require prodigious human and financial resources. This is also the power to build schools, establish media networks, and assemble large groups of people to disseminate innovative ideas. The Korean megachurch demonstrates an efficiency and service quality that smaller church communities are less able to provide, creating a highly organized, vibrant, and enjoyable religious culture from which members can derive inspiration (Fig. 122.8).

Fig. 122.8 A cheerful Yoido youth group enjoying their fellowship outside the chapel's main entrance (Photo by M. Bégin)

At the same time, the Korean megachurch is not above legitimate criticism. From a theological standpoint, the traditionalism that megachurches have so effectively transcended may actually carry the more socially supportive and spiritually nourishing messages focused on discipleship rather than command-driven evangelism. Bosch (1991) suggested that a more thorough and erudite understanding of scripture permits the formation of genuine discipleship communities focused on Christian teachings of forgiveness, peace, and love rather than a crass "conversion" ethic that often fuels the evangelical zeal of modern megachurches. Another critique posits that the megachurch weakens more radical (and perhaps more relevant) Christian teachings. Here, the *cheap grace* concept Bonhoeffer described in 1937 challenges the often materialistic, bourgeois themes prevalent among Korean megachurches, begging the question as to whether the push for sheer growth tends to weaken genuine discipleship (Bonhoeffer 1995). This resonates with ecclesiological critiques offered by some Christian theologians (Ellul 1963; Jethani 2009) who question whether the modern public relations techniques embraced by megachurches may lead to cynicism toward Christianity.

This cynicism may, in turn, be fueled by perceptions of the corrupting influence of megachurch power such as those presented by Dawn (2001). In addition to these theological critiques, the Korean megachurch, and the Korean Protestant Church in general, have experienced a great deal of censure in recent times. According to a 2010 survey conducted by the Christian Ethics Movement of Korea, public credence of Korean Protestantism hovered at about 20 %, compared with 41 % for Catholicism and 33.5 % for Buddhism. The abounding criticisms of the Korean church, often very closely related to megachurches themselves, cite corrupt activities and scandals such as financial crimes, adultery, and exorbitant personal expenditures as some of the primary problems. With promises of pioneering reform failing to materialize, public disappointment gradually gives way to more cynicism in the wake of imperialistic church expansions, costly excesses, privatization of church resources or programs (as in the case of Yoido's publishing operations), and the luxurious lifestyles and indulgences of authoritarian pastors. Indeed, the Protestant church is now considered the most anti-democratic, most authoritarian institution in Korea, replete with instances of abuse, nepotism, fraud, and even sexual exploitation perpetrated by patriarchal pastors making ill use of the concentrated power afforded them by familistic church members (Shin 2009). Moreover, the Korean church's conservative leanings often ironically contradict more genuine Christian teachings and democratic principles, resulting in a powerful collective political force for privatization, neo-liberalism, anti-reunification, and self-interest/special interest lobbying. More recent examples of collusion between the Lee Myung-Bak administration and a handful of powerful megachurches such as Somang Church, of which the former Korean president is a member, point to the increasingly problematic distinction between church and state in Korea, a situation that comes much to the consternation of those Koreans who distinctly remember (and perhaps would rather forget) the nation's previous struggles with authoritarianism.

122.5.2 Some Prognostications

As for the prospects for Korean megachurches, we can make a few inferences. First, they will likely experience a downward turn for some time to come. As discussed, the Korean church has been on the decline since 1995, with many leaving the Protestant Church for Catholicism, Buddhism, or atheism (Cho 2007). This trend is likely to continue. Although the Korean church has been experiencing negative growth, many more pastors have emerged to build many more churches, ramping up the already prodigious competition among them. At the same time, social and organizational polarization in Korea continues. As churches increasingly compete for more existing Christians than new ones, horizontal movements become more likely. In this new game, advantage will go to those churches capable of providing a greater variety of social and religious services, amplifying competition for market share among the finite number of devotees who now consider "shopping" for the best church deals perfectly acceptable (Jethani 2009). For this new market, the Korean church is essentially transforming itself into one-stop welfare and cultural centers to serve the socio-spiritual needs of a demanding public. Eventually, Korean Christians will likely choose from a small number of giant megachurches or a larger number of *microchurches* comprising small congregations with tenuous futures (for example, *Yeoleumteo House Church* of Daejeon; *Beautiful Village Community* of Seoul; *Jesus Village Anabaptist Church* of Chuncheon). At any rate, the Korean church is not likely to peak again.

As these trends continue, critics of the church will attempt to offer legitimate alternatives. This is already happening to a lesser degree in the form of the microchurch movement, and in the emergence of associated groups. The focus of this fledgling movement is not size, but rather reformation of the Korean church, presenting some possible alternatives to megachurch membership (Kim 2011c). Alternatively, the Korean Protestant church may evolve into a kind of oligopoly, with megachurches and small churches seeking mutually beneficial arrangements through financial cooperation and/or program sharing. Regardless of the outcome, the Korean megachurch, at once a culmination and a continually unfolding process brought about by a complex array of cultural, social, and historical precedents, remains a fascinating, dynamic, and salient phenomenon that merits continued scholarly attention.

Acknowledgement The authors would like to thank Mrs. Sarah Shin for her invaluable translation assistance.

References

Bonhoeffer, D. (1995, originally published in 1937). *The cost of discipleship*. New York: Touchstone.
Bosch, D. (1991). *Transforming mission: Paradigm shift in theology of mission*. Maryknoll: Orbis.

Cho, S. (2007). *Why did they go to Catholic church? (in Korean)*. Seoul: Jeyoung Communication.
Cho, C. (2008). *The consciousness of Korean Protestant pastors and the adaptation of Korean churches to secularization (in Korean)*. Seoul: Korean Studies Information.
Church, Y. P. (1995). *The 50-year history of Yongnak church (in Korean)*. Seoul: Yongnak Presbyterian Church.
Church, Y. F. G. (2008). *The 50th Anniversary of Yoido Full Gospel Church (in Korean)*. Seoul: Yoido Full Gospel Church.
Church, O. (2010). *Talk & talk: 25 years of Onnuri church history (in Korean)*. Seoul: Onnuri Church.
Dawn, M. (2001). *Powers, weakness, and the tabernacling of God*. Grand Rapids: Eerdmans.
Drucker, P. (1990). *Managing the non-profit organization*. Woburn: Linacre House.
Ellul, J. (1963). *Fausse présence au monde moderne*. Paris: Les Bergers et Les Mages.
Guinness, O. (1993). *Dining with the devil: The megachurch movement flirts with modernity*. Grand Rapids: Baker.
Jeong, J. I. (2007). *Theology and church growth (in Korean)*. Seoul: The Bread of Life.
Jethani, S. (2009). *The divine commodity*. Grand Rapids: Zondervan.
Kim, B. (1995). *Korean society and the Korean Christian church (in Korean)*. Seoul: Hanul Books.
Kim, E. (2011a). *Will the controversy of preferential treatment for Sarang Church's new chapel building go to court? (in Korean)*. Retrieved March 3, 2012, from www.newsnjoy.or.kr/news/articleView.html?idxno=35103
Kim, J. (2011b). *The megachurches were built by absorbing the suffering of the era of development (in Korean)*. Retrieved March 3, 2012, from http://h21.hani.co.kr/arti/COLUMN/132/28812.html
Kim, J. (2011c). *The revolt of the small and the birth of the small religion (in Korean)*. Retrieved March 3, 2012, from http://h21.hani.co.kr/arti/society/society_general/30542.html
Ku, M. (2008, April). *A theo-ethical reflection about the faith-form of the women of Kangnam-type megachurches (in Korean)*. A paper presented at the annual symposium of the Korean Association of Feminist Philosophy in Seoul.
Lee, M. (1980). *The Christianity of the late Chosun Dynasty and the national movement (in Korean)*. Seoul: Pyungmin.
Lee, M. (1985). *The 100 years of Korean church history (in Korean)*. Seoul: Bible Reading.
Lee, J. (2000). The Korean Protestant Church and the ideology of 'growthism'. *Contemporary Criticism (in Korean), 12*, 225–240.
Lee, S. (2005). *Healthy church growth is now a system: An analysis of Onnuri Church (in Korean)*. Seoul: Logos & Life.
Lee, H., & Lee, K. (2002). The change of urban structure and megachurch growth. *Mission and Theology (in Korean), 10*, 41–72.
McGavran, D. (1955). *The bridges of God: A study in the strategy of mission*. New York: Friendship Press.
McGavran, D. (1970). *Understanding church growth*. Grand Rapids: Wm. B. Eerdmans Publishing Co.
McLuhan, M. (1964). *Understanding media: The extensions of man*. New York: Signet Books.
Ortega y Gasset, J. (1994). *Revolt of the masses* (Reissue ed.). New York: W.W. Norton & Company.
Park, Y. (1998). *Called to awaken the Korean church (in Korean)*. Seoul: Life Books.
Shin, K. E. (2009). *Against megachurches*. Seoul: Jeongyeon.
Thumma, S., & Leppman, E. (2011). Creating a new heaven and a new earth: Megachurches and the reengineering of America's spiritual soil. In S. D. Brunn (Ed.), *Engineering Earth: The impacts of megaengineering projects* (pp. 903–931). Dordrecht: Springer Science+Business Media B.V.
Williams, J. (2009). The new ecclesiology and the post-modern age. *Reviews and Expositor, 107*, 33–40.

Chapter 123
Megafaith for the Megacity: The Global Megachurch Phenomenon

Scott L. Thumma and Warren Bird

123.1 Introduction

Since 2010 more than half the world's population resides in urban areas. Along with this ever-increasing population density, a distinctive model of religious organization – the megachurch – has arisen. Urban population concentration and massive religious congregations have gone hand-in-hand for many centuries. Over the past half-century, both the United States and the rest of the Christian world began to see a rapid proliferation of these massive faith communities. Approximately 25 years ago, these very large churches collectively came to be perceived as a distinctive religious reality. An exploration of this religious reality globally provides insight into the function of religion in today's urbanized world and the role and power of religious organizations for individuals in a mass society.

This chapter describes the phenomenon, maps the megachurches known to the authors, traces the patterns of organizational change around the world as it relates to urbanization and migration, and highlights a number of specific examples from different countries throughout the globe (see Thumma 2012). The chapter uses the widely accepted definition of a megachurch as any Protestant Christian congregation with 2,000 or more average weekly attendees (both adults and children) at all services and physical locations. This form of religious organization combines mass gatherings of thousands of participants, often in stadium-like venues, coupled with intimate, small-community groups or service teams, extensive use of technology, a vast array of ministries and programs, authoritative charismatic leadership and

S.L. Thumma (✉)
Professor, Department of Sociology, Hartford Seminary, Hartford, CT 06105, USA
e-mail: sthumma@hartsem.edu

W. Bird
Research Division, Leadership Network, Dallas, TX 75204, USA
e-mail: warren.bird@leadnet.org

S.D. Brunn (ed.), *The Changing World Religion Map*,
DOI 10.1007/978-94-017-9376-6_123

increasingly multiple affiliate locations united into a single distinctive congregational identity. In addition, the success of these megachurches throughout the globe has created international networks of like-minded churches apart from traditional denominational models. Their rise has developed new power bases of religious authority and often new political players throughout the globe.

The originality of the megachurch phenomenon stems not just from it being a model of church that is evolving and adapting to these new contexts, but also from what it offers in a contemporary cultural climate, with the social needs of an urban context globally. The megachurch adaptation has become highly successful. In other words, this new religious form demonstrates a stronger resonance with contemporary cultural forces than traditional models of organizing religiously do.

123.2 The Paradigmatic Case of Yoido Full Gospel Church

Currently the world's largest-attendance church, Yoido Full Gospel Church sits on one end of Yeouido (formerly Yoido) Island in the Han River in downtown Seoul, Korea, and shares its choice location with major power brokers, including City Hall, banks, and broadcasting companies (Fig. 123.1). This urban context is one of the globe's largest metropolitan areas and is home to over half of South Korea's population. A brief look at this congregation, from its founding in 1958 to its world dominance with 250,000 attendees by the 1990s, sketches the dynamics that drive this global megachurch phenomenon.

Fig. 123.1 The world's largest attendance congregation, Yoido Full Gospel Church, Seoul, Korea (Photo courtesy of 3WM and Yoido Full Gospel Church, used with permission)

Following many decades of oppression during the Japanese occupation and 3 years of brutal warfare, South Korea began in 1953 a period of rapid urbanization, modernization and westernization. This growth took place first under the authoritarian rule of a military general turned president and then in the 1980s under a more democratic form of government, both of which were supported by the Christian minority in the country. Prior to this period, Christian missionaries had relatively little success; however, following the war Christianity appeared linked to the country's rapid development. Presbyterians, Methodists and the Assemblies of God all developed thriving churches. Many of the country's earliest and most influential megachurches were urban and upper-middle class Presbyterian and Methodist churches, such as Young Nak Presbyterian Church built at the heart of Seoul's metropolitan development. (See the Begin and Shin chapter for a more detailed description of the megachurch phenomenon in Korea.) However, it was a Pentecostal church belonging to the Korean Assemblies of God, founded by David (formerly Paul) Yonggi Cho and built at the city's then fringe among the poorest urban migrants, that soon dwarfed all of its competition.

Cho was born Buddhist but converted to Christianity in 1955 at age 19. Subsequently, he worked as an interpreter for an American missionary and later a Pentecostal healing evangelist. Following this he attended the Korean Assemblies of God Bible School and was further exposed to Pentecostal teachings. In 1958, David Yonggi Cho, with his future mother-in-law Jashil Choi, founded a small church that grew to 300 in 3 years and a year later expanded into a new 1,500-seat sanctuary.

By 1964, this church was attracting 3,000 weekly, and within 4 years had more than 10,000. Struggling with the management challenge, Cho devised an approach to member organization using lay leadership and small home fellowships. This hierarchical network of small groups simultaneously offered a way to provide social intimacy and pastoral care to the thousands, while also employing free volunteer labor from the wives of poor laborers who had recently arrived in the rapidly growing urban area. These women were organized, trained and empowered to be leaders at a time when Korean females had very little power in society. This structure also provided a semblance of sociability, personal connection and, in part, an urban replication of the small village system they had recently left to come to this unsettled and rapidly growing urban social context. This recreated Christian community structure offered both a sense of community and activities to engage the displaced women who could no longer practice their daily rural tasks.[1]

And indeed this new context was an entirely unfamiliar reality. Throughout the twentieth century, the Korean peninsula was predominantly rural and under Japanese occupation. In the early 1950s Seoul had a population of less than 1.5 million. However, by 1960, as a free and independent nation, the capital grew to 5 million, by 1970 to 9 million, and by 1980 to 15 million. In 1990, the city proper peaked at nearly 11 million inhabitants, with the metro area at 19 million. This urban metro area has continued to expand to nearly 24 million currently, encompassing roughly half the country's population, making it the third largest urban area in the world.

---

[1] Some of this analysis follows an interesting argument by Seongwoo "Linus" Lee (2007).

This population expansion was matched by a corresponding economic boom. South Korea had a gross domestic product per capita of less than $1,300 (2,010$) in 1950 and, even in 1965, only roughly $2,000 per person. However, the GDP then expanded dramatically to $30,200 per person by 2010. Among high-income world urban areas, Seoul's population growth has been greater than that of any other since 1950 except for Tokyo. With population density of 27,000 people per square mile (10,400/km^2), Seoul's density is second highest among the world's affluent urban areas, trailing only Hong Kong (Cox 2011).

From the time that Cho founded his church until 1973 when he opened the 12,000-seat church present sanctuary on Yeouido Island, the city's population grew from 2.4 to over 5.3 million. Many of those migrants were poor laborers from rural areas who settled in close proximity to the church's facility. Nevertheless, its new location symbolically signaled an upwardly mobile ideology. An earlier bridge project linked the island to the city and skyrocketed its development. Yeouido became home to the National Assembly Building, the massive skyscraper 63 Building and the center of Korea's investment and insurance industry, many of its broadcasting stations – and the world's most populated megachurch.

Currently, the worship services at Yoido Full Gospel Church are professional and polished, with contemporary music and excellent choirs. Each Sunday there are seven worship services with multiple projection screens, state of the art sound system, dramatic lighting and, it is claimed, simultaneous translation into any language if given 24 h notice. Cho's theology is Pentecostal, stressing ecstatic spiritual gifts including divine healing, prophecy, blessings of wealth and physical wholeness along with a literal reading of the Bible. In addition, the church emphasizes extended fervent prayer sessions, sacrificial giving of offering and a disciplined daily life of faith. The church has invested heavily in television, newspapers, and educational facilities, as well as retreat and prayer centers. Since the early 1990s, it has also adopted a multisite approach to growth, spinning off additional campuses throughout the metropolitan area in addition to planting new independent congregations, many of which have grown to megachurch status as well.[2]

This combination of a spiritual message of hope, healing and worldly success provided a powerful motivation to an underprivileged, struggling working class. The structure of cell groups led by women focused on both spiritual development and social connectedness. When combined with an emphasis on education, the media and programs of leadership instruction for these dislocated migrants, the church offered adaptation channels for members dealing with the dramatic sociopolitical, economic, and cultural upheaval taking place in their lives in urban Seoul. This church, in many ways, forms the paradigmatic model that has been replicated throughout the global urban expansion of the past several decades.

[2] The details for the description of the history of Yoido Full Gospel Church come from Dr. H. Vinson Synan (2012), the church history section of the church's website in English from http://english.fgtv.com/yoido/history.htm and from www.davidcho.com/NewEng/bd-1.asp. For details of the present activities, both authors have attended and studied the church.

123.3 The Mega Religious Phenomenon

The number of very large churches has increased dramatically globally in the last 50 years. It is estimated that throughout much of the twentieth century there were no more than 10–20 very large churches at any given time anywhere in the world. In the U.S. following World War II, the GI bill sparked an education and housing boom along with a baby boom. Additionally, the rise in automobile sales, road quality and a greater acceptance of everyday travel contributed to intense urbanization and suburbanization patterns. In the 1950s and 1960s churches vied for having the largest Sunday school programs, with several dozen U.S. churches having over 2,000 weekly students (Towns 1969). In 1958, Rex Humbard, among others, paved the way in part for the modern megachurch movement with his 5,000-seat domed Cuyahoga Falls, Ohio, church, prophetically named "The Cathedral of Tomorrow." By 1970 approximately 50 megachurches existed in the U.S., growing to 150 by 1980, 300 in 1990, over 600 by 2000 and then doubling to roughly 1,200 by 2005. As of 2012, an estimated 1,650 megachurches exist in the United States, while outside the U.S. there may be as many as 500 (Thumma and Travis 2007)[3] (Fig. 123.2).

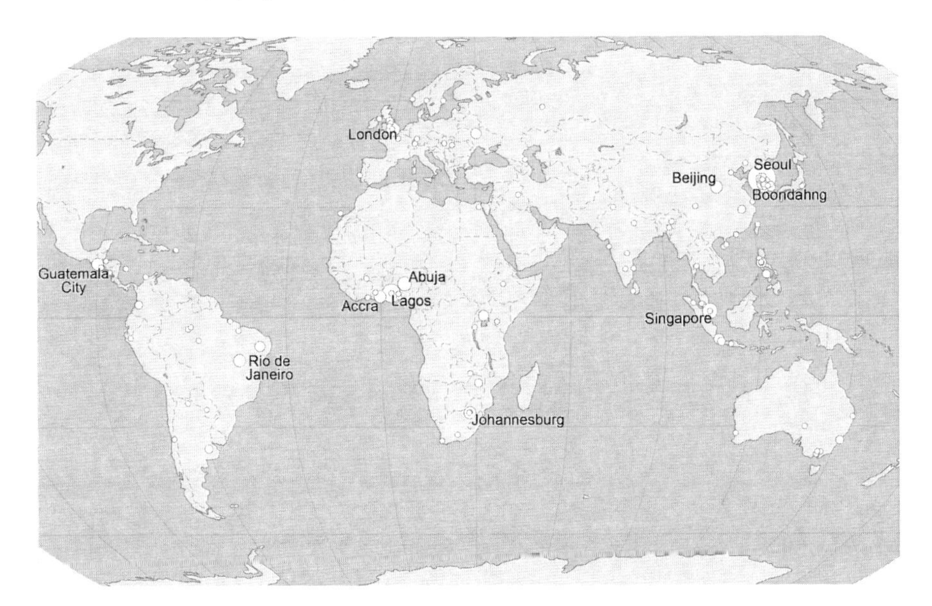

Fig. 123.2 Global megachurch locations cover urban areas on all habitable continents (Map by Jeff Levy, University of Kentucky Gyula Pauer Center for Cartography and GIS; commissioned by the editor)

[3] Unless otherwise noted, all the facts and data about megachurches referenced in this chapter come from the authors' national studies of megachurches in 2000, 2005, 2008 and 2011 that are available at www.hartfordinstitute.org, along with a database of U.S. megachurches. Information about the locations of those outside the U.S. is available from Warren Bird and Leadership Network at http://leadnet.org/world (Bird 2012).

Although the megachurch phenomenon is often defined exclusively in terms of size, it is better understood as a distinctive congregational variation. No doubt bigness is the most immediately apparent characteristic; however, megachurches throughout the globe generally share other traits. Virtually all megachurches have a conservative Protestant theology, even those within mainline Christian denominations. A large number are independent or nondenominational or, if affiliated with a denomination, have very loose ties with that national body. Megachurches combine distinct practical, programmatic and organizational characteristics while doing worship on a grand scale in an urban or suburban, market-driven, technologically rich context. This distinctive style and organizational complexity, not just size alone, make them different from their equally large counterparts in Catholicism, Islam, or other faiths. While considerable diversity exists among megachurches worldwide in terms of architecture, ministries, message and theology, nevertheless the appeal and structural characteristics across the phenomenon are remarkably similar.

Success is a powerful marketing tool, and the megachurch phenomenon is successful in cultures where bigger is better. In fact, it is in part this dominance of the religious marketplace that signifies economic, political and spiritual significance to members and the larger social world. The largest U.S. megachurch, Lakewood Church of Houston, Texas, gathers in a converted sports arena that seats 16,000 and draws approximately 45,000 weekly participants (Fig. 123.3). Three dozen U.S. megachurches have sanctuaries that hold 5,000 or more persons; several have a million square feet (92,900 m^2) under roof and a number own more than 300 acres

Fig. 123.3 Joel Osteen's Lakewood Church in Houston, Texas (Photo by Warren Bird)

Fig. 123.4 The Philippines Arena is a 50,000-seat mega church slated to open in Manila in 2014 (Photo from Populous, used with permission)

(121 ha) of property. While the median U.S. megachurch sanctuary seats 1,500 (often used many times a weekend), the websites and publicity from those churches typically emphasizes their bigness via images of an excited, very large crowd.

Globally, the numbers of attendees in some of these churches are staggering with Yoido Church, described above, claiming over 250,000 weekly attenders and 800,000 members. Other congregations in several African countries, Singapore and Brazil are also nearing similar enormous estimates of regular attenders. At least three different Nigerian churches contain seating for 30,000 or more people, as do churches in India, the Philippines (Fig. 123.4), and Indonesia Bird (2012).

The exact social shifts that created these changes and facilitated this rise of global megachurches are complex and multifaceted, and can be introduced only in a cursory way in this chapter. They are rooted in changing population patterns such as the U.S. baby boom and its effect on organizational size throughout America. Likewise, migration patterns and the transition of populations from rural to urban to suburban contexts as well as rapid immigration and the pluralism within the culture, play a part in the U.S. context, Brunn et al. (2011a). Globally, similar dramatic patterns of migration and urbanization, along with the intense social and personal adaptive pressures these changes bring, play a major role in the phenomenon. Additionally, the introduction of new religious options and charismatic religious entrepreneurs who stand out in a religious marketplace further drive the appeal of these large congregations.

The growing appeal of megachurches might also be attributed to a waning of the significance of denominational identities, distrust of national bureaucracies, increased cost of running an effective medium-sized church, or the rise in individualistic spiritual seeking. Additionally, transportation patterns, economic and market trends, as well as technological inventions from large screen video projection to the Internet, all contribute to the spread of this church model worldwide.

There are both push and pull factors at work in making this religious form appealing in many different cultures. This model of church excels in attracting religious participants through its technologically savvy, media rich, highly produced, and expertly marketed services. Additionally, the theological message is often one of hope, economic growth, success and individual advancement. The explicit sermons of megachurch pastors are quite often biblically based practical applications of the faith relevant to everyday situations. Pastors frequently add seemingly authentic and personally revealing illustrations of how the Biblical message has transforming their own human weaknesses and frailty into faithful, victorious living. The implicit message in many of the teachings is that faith in God produces some level of success and power in the world. Often the theology and values espoused are traditional, but the necessities of the organization and need for volunteers expands and in some sense liberalizes the boundaries of tradition, as can be seen with Reverend Cho's necessary use of women as leaders of their cell groups. Similarly Rick Warren, the well known U.S. pastor of Saddleback Church in Lake Forest California, has embraced both right and left-wing political figures and engaged in efforts to address the AIDS epidemic.

Societal changes in musical tastes, an expectation of professional quality, the attraction to technologically-enhanced productions and the desire for personal spiritual choice also contribute as much to the push away from smaller traditional churches as they do the pull into a megachurch model. Socio-psychologically, the need for adaptive, intentional communities that assist in re-creating community, developing networks of strangers and learning coping skills in a new social context further contribute to megachurch appeal. The leadership of these churches contributes to its social adaptability. Most megachurches globally have strong-personality authoritative senior pastors, some of whom are even described as authoritarian in their use of spiritual authority, with an accompanying top-down leadership structure. At the same time, nearly all of them must generate a very high level of both lay involvement in the functioning of the church and lay ownership of the vision and a strong sense of commitment. In some upper-middle-class global churches, especially in Western countries like the U.S., the move toward a team-leadership model has created a more collaborative environment. Megachurch leaders tend to adapt best practices from the corporate world for structuring and managing their staff. The megachurch system requires significant lay involvement, which empowers volunteers to function as small group or ministry leaders. This dynamic in turn trains them and gives them opportunities for gaining leadership experience that can also be applied to their lives outside the church.

Whatever the exact configuration of factors, it is undoubtedly a combination of these and others that have produced a reality where megachurches are now almost ubiquitous and play an increasingly significant role in shaping religious and social life. The urban and suburban context is rapidly outgrowing a smaller localized model of church. As such, the megachurch phenomenon can be seen as a reshaping of congregational forms to fit a new geographic, cultural and religious context.

123.4 The Megachurch in the United States

The megachurch phenomenon has blossomed only in the past 50 years. Since the 1990s, research on U. S. megachurches has begun to explore the phenomenon. Thus much more is known about North American megachurches than those in the rest of the world. It is clear that there are significant differences across countries and that local contexts reshape the phenomenon, but much more needs to be learned about the megachurches outside the U.S. to derive an accurate and complete picture.

The typical megachurch in the United States, however, appears considerably different physically from those in other global contexts, although quite similar in their function and appeal. They are generally smaller, with a median of 2,750 weekly attendees and median seating for 1,800, but have much more property, averaging around 40 acres (16.2 ha) per church. The phenomenon began slowly in the U.S. as an urban, inner-city reality at the turn of the twentieth century and continued to develop through the 1960s. With the dramatic suburban population movement post-World War II, the growth of megachurches shifted to new suburban contexts where more acreage was available, affordable and less regulated. The smallest U.S. megachurch is still a mammoth compared to the vast majority of American churches. The median U.S. church has 75 regular worshippers, while 94 % of U.S. churches have attendance of 500 or less people (Chaves et al. 1999). Megachurches, which account for barely 0.5 % of all U.S. congregations, are home to over six million, roughly 10 % of weekly religious attendees. These congregations are present in nearly all American metropolitan areas and most large cities (Fig. 123.5). This remarkable

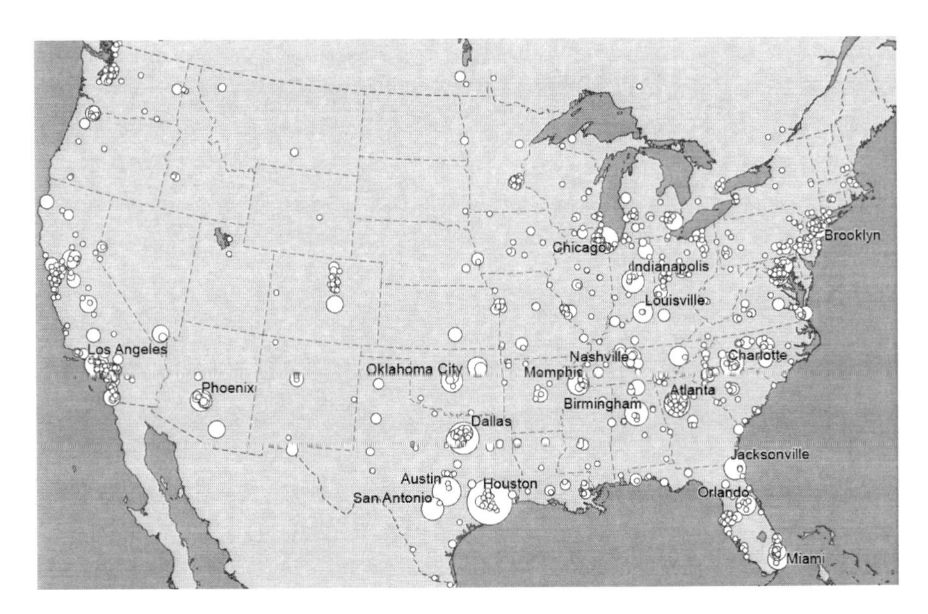

Fig. 123.5 U.S. megachurch locations include most states and most larger cities (Map by Jeff Levy, University of Kentucky Gyula Pauer Center for Cartography and GIS; commissioned by the editor)

concentration of people and resources in the country's largest churches is as dramatic a change in the nation's religious landscape as it is culturally significant.

123.5 Megachurches Globally

At present our knowledge of the global megachurch universe is sketchy at best, but even this initial investigation sheds light on the patterns and dynamics of the growing trend. Globally, the megachurch phenomenon is centered in the relatively recent dense urban areas, often with a new and/or rapidly growing Christian population. While there are suburban megachurches evident globally in the Seoul metro area (see the Shin and Begin chapter), in Australia and elsewhere, urban megachurches are more the norm outside the U.S. This is in part due to the urban configuration of these cities but may also be related to the presence of a large middle class. Of the currently known 229 non-U.S. megachurches, over 40 % are in Asia, 20 % in Africa, 15 % in Central and South America, 11 % in Europe and 6 % in Australia.[4] Korea, and its megacity Seoul, have the highest concentration of megachurches of any country and city outside the US (Tables 123.1 and 123.2). These non-U.S. megachurches (see Fig. 123.2) have a median weekly attendance of 6,000 – roughly twice as large as American megachurches. They are also relatively new, with an average founding date of 1975 (median 1982) compared to 1960 and 1972 respectively for the U.S.

Global megachurches cluster into two dominant models. The first is characterized by a very large single site on a small plot of land. This approach might be described as an urban vertical profile; the church and property are densely designed to fit on a small footprint with several stories above ground and often

Table 123.1 Top countries with megachurches

Country	Percent of megachurches
South Korea	17
Nigeria	8
China	6
Brazil	5
UK	4
Australia	4
Philippines	4
Indonesia	3
Singapore	3
South Africa	3

Source: S. Thumma and W. Bird

[4] Even after a diligent search we are certain that our listing is incomplete, however, it is the most complete list that presently exists and can be found at www.leadnet.org/world/ (Bird 2012).

Table 123.2 Top cities with megachurches

City	Number of megachurches
Seoul	17
Lagos	8
Abuja	5
Boondahng	5
London	5
Accra	4
Beijing	4
Johannesburg	4
Rio de Janeiro	4
Buenos Aires	3
Inchon	3
Kampala	3
Kiev	3
Sao Paulo	3

Source: S. Thumma and W. Bird

with parking underground. The second model is that of many smaller churches linked together as a single multisite congregation with one senior pastor, a common budget and uniform identity. This multisite church might have monthly or occasional massive gatherings to bring together members from all the locations for worship, but it is essentially a distributed church network of many smaller congregations. Based on information from roughly a third of all the global megachurches, three-quarters had multiple site locations but thought of themselves as a very large single church.

123.6 Societal Patterns that Contribute to the Phenomenon

No single explanation can account for the appearance, development and functional reality of megachurches in the United States or globally. As much as we would like a single explanatory model, the social and religious reality across the globe is just too complex and contextually diverse. Nevertheless, a number of possible factors drive this phenomenon worldwide, while also recognizing that certain of these reasons are more or less dominant in different parts of the world.

Huge shifts are taking place in the global population in terms of where people reside, how they interact and the possibilities available to them within the variety of religious options. In various ways, all these global modifications are contributing to the appearance and appeal of the global megachurch movement. Very large congregations offer a set of characteristics that make them more conducive to meeting the needs of contemporary urban (globally) and suburban (United States) residents. The population changes and corresponding megachurch characteristics begin to offer an explanation for why this religious form has become so prevalent in the latter part of the twentieth, and now the twenty-first, century around the world. In part, this religious form offers multiple attributes that resonate with the larger institutional and societal shifts taking place. Multiple dynamics contribute to the congruence of these realities.

The dramatic changes of population migration, megacity growth, industrialization, development of new economic classes and other factors have created new and unanswered personal and interpersonal needs. Many of these changes have also been accompanied by the growth of alternative religious meaning systems and new options in the religious marketplace, some of which are seen as symbolic of, and highly congruent to, the developing political and cultural future. These new religious meaning systems are often entrepreneurial and indigenous grass-roots enterprises of rising lower classes. The purveyors of these faiths are evangelically driven, politically savvy and innovative risk-takers within the new cultural reality. Finally, globalization trends tied to the world economy, a system of higher education, inexpensive rapid transportation and especially Internet technologies have all created a condition of international institutional isomorphism which leads to the increased likelihood not just of the spread of successful organizational forms but also of the power and scope of both cultural models and theological ideas of global charismatic personalities.

123.6.1 Population Growth, Migration and Concentration

Truly staggering changes in global population figures, concentrations of these populations in urban areas and the ever-expanding sizes of these urban megacities have taken place in a very short span of human history. The global population has skyrocketed from under two billion in 1900 to just over seven billion in 2012. In 1900, 20 % of humans lived in urban areas, whereas in 2010 more than half did and this percentage is expected to nearly double by 2050. In 1900, only 13 % of the world's population lived in cities, but by the end of the century almost half did. In 1950, there were 83 cities with populations exceeding one million; by 2007 this had risen to 468, with 27 of these megacity "agglomerations" having over ten million residents.[5] Much of this urban expansion was initially confined to the developed and industrialized countries; however, now nearly all this urban expansion is taking place in the developing world.

In the United States, the pattern of population growth was equally astonishing, but with a slightly different concentration and migration pattern. The population more than tripled in the twentieth century from 75 to 281 million Americans.

[5] See the excellent summary article of urban trends see Brunn et al. (2012) as well as www.un.org/News/Press/docs/2005/pop918.doc.htm, www.unfpa.org/swp/, and www.un.org/popin/ for details.

The decade of the 1990s, the period of greatest increase in the number of U.S. megachurches, also had the largest numerical population increase of any decade in the country's history. This population concentration also involved rural and small town migration to the cities. Prior to 1940, most Americans lived outside an urban area; but, by the end of the century, 80 % of the nation resided in a metropolitan area. This growth did not swell the size or density of the central city, but rather predominantly grew the suburban portion of metropolitan areas, so that by 2000 half the U.S. population lived in suburban areas.

This rapid population expansion and concentration has begun to shift the social reality throughout the world. The "scale, geography, form and institutions make [immense urban areas] entirely new in the history of human experience" (Sudjic 2012) and, as such, require "new ways of living" and congregating in this setting. Many of these areas of 15–30 million people look more like independent nation-states than what is traditionally thought of as a city. Singapore is the most obvious example but likewise Tokyo, Seoul, Mumbai, Shanghai, New York, Hong Kong and Los Angeles function as world players semi-autonomous in relation to their state or national political entities. Many of these massive global population centers are the most likely homes for the planet's megachurches. They are currently found in 22 of the 34 largest global megacities.

123.6.2 Urban Social Dislocation

The scale of these megacities does not come without a cost. For those rural poor trekking into the megacities of São Paulo or Lagos or Seoul in hopes of a better future, they face near total social, and even physical, uncertainty. The global megacity phenomenon initially meant millions of impoverished people pouring into urban areas to reside in squalor and derelict shanty sprawl around these developing cities. This pattern continues, but has also begun to change in recent decades. A host of newly developed urban areas throughout the globe, predominantly centered in Asia, show dramatic growth driven by equally dramatic economic production (Global Health Observatory 2012). Likewise, the growth of the suburban reality in the U.S. and other nations calls for a different configuration of social interaction and community organization, although not on the same scale as the megacity. Not only does this migration and concentration cause the disruption of their previous social reality with the unsettledness and social dislocation it brings, but it also causes a fragmentation of the social order that governed their former life in rural, small town or village locales. Andrew Kim notes that in South Korea, and especially Seoul, the intensive urbanization, industrialization and rapid social change of the past 40 years have resulted in the highest population density in the developed world with a correspondingly highest suicide rate as a result of the country's social anomie and dislocation (Kim 2002).

123.6.3 Megacities with Megasize Congregations

Within the landscapes of these cities, certain distinctive social landmarks are ubiquitous, such as their airports, high-rise apartment complexes, shopping malls, museums, universities, massive hotel complexes as well as massive impoverished areas. Therefore, it is not surprising that with these population concentrations have come the rise of mega-retail outlets like Walmart, IKEA, and the ubiquitous suburban megamall. Likewise, mega-religious structures are an integral part of these landscapes. In these massive concentrations of population, it is not just large-scale transportation, industrial or economic institutional forms that get created to support this geo-political reality but also mega-scaled social, cultural, commercial, sports and even spiritual institutional forms.

Recent research has shown that nearly every major religious denomination in the United States has considerably more large churches now than it did at the turn of the twentieth century (Chaves 2006). Although many of these are not of megachurch proportions, nevertheless the move toward increasingly larger congregations, whether due to an economy of scale or a desire for large institutional forms, is a dramatic change across the religious landscape. In fact, it could simply be a factor of the dramatic increase in population and its concentration in metropolitan areas. A 2010 census of U.S. religious congregations and their adherents shows a strong correlation between the average size of religious congregations and the county's population in which it resides. In the U.S. the average size of a congregation, no matter what the faith tradition, is increasingly larger as the population of the county becomes denser, as Table 123.3 demonstrates.

Unfortunately, we do not have comparable information for all nations; nevertheless, there does seem to be a clear correlation between the size of the religious form and an area's population concentration. As the size of the community increases, the scale of all its institutions grows. This is not only a matter of needing increased space and service systems to handle the population and a greater demand for limited real estate; it also reflects an organization's to stand out in a congested competitive marketplace. This is especially true in America (and increasingly throughout the modern world) where one of the cultural values is that "bigger is better." The larger the organization, the more cost-effective it is,

Table 123.3 Relationship between population and average congregation size, by faith tradition

	Average adherents, by county population				
Religious group	5,000,000 or Larger	1,000,000–4,999,999	250,000–999,999	Under 250,000	Smaller
Catholic	6,087	3,682	2,928	1,477	551
Black Protestant	420	380	306	203	163
Evangelical Protestant	368	349	305	248	170
Mainline Protestant	416	383	343	244	151
Other Christian	598	435	264	115	93

Data source: U.S. Religious Census www.religioncensus.org

and the more likely it is to offer a wider array of professional and efficient services – whether this is the delivery of market goods, consumer electronics or spiritually vital worship services.

123.6.4 Social Hubs for Celebrities, Politicians and Others

The increased size of these very large churches means that they are able to command the attention of politicians, government leaders, business executives and even sports and media celebrities. It is not surprising that many professional athletes and entertainers can be found in the largest churches worldwide or that, for instance, the president of South Korea is an active member of one of Seoul's megachurches and Barack Obama attended a Chicago megachurch prior to his presidency. Nor is it unusual for would-be politicians to visit in order to court the vote of the thousands of attenders. Likewise, the names of these churches and their sometimes infamous senior pastors often make newspaper headlines for either doing good or being embroiled in scandal.

Most of all these megachurches offer a ready-made community for individuals faced with social dislocation, disruption of traditional norms and folkways, and a new urban context that requires new skills and patterns of behavior. These megachurches, and especially their networks of small groups and intentional programs, teach adaptation to the urban or suburban environment and provide a sense of community that functions as a replacement to the world they left behind. In a sense these massive congregations are the functional village in a postmodern highly urbanized world of the megacity. As Philip Jenkins notes,

> In the last thirty or forty years, urbanization has been one of the most significant factors in spreading not just Pentecostal Christianity, but also fundamentalist Islam, because it is very much these religious traditions that have organized networks of human services, education, welfare, community, self-help, and which go so far to explaining why these religions have boomed so dramatically…. It is likewise more valuable to be a member of the church than to be a citizen of Nigeria or Peru, because in these vast cities, where you have deracinated masses, it tends to be churches that provide the human community. (Myers 2012)

123.6.5 Rise of New Religious Options

There is more to the rise of megachurches than just societal and cultural factors; after all they are *religious* entities. The faith traditions and worship expressions of these churches offer a distinctive alternative in the religious marketplace. Within the U.S., it is the simultaneous growth of a mainstream, middle-class evangelicalism as well as a neo-Pentecostalism within the charismatic movement, that has shaped the distinctive theological and worship style that characterize megachurch worship.

Outside the U.S., the rise of megachurches must be partly attributed to the dramatic growth of Christianity, specifically Pentecostalism, in the Southern hemisphere.

In America, the 1960s through 1990s witnessed the evangelization of the country. A marginalized Southern evangelicalism became an increasingly acceptable national option with the presidency of Jimmy Carter and the rise of the religious right simultaneously with political right. This influence spread as the U.S. population migration moved former Southerners throughout the country and the South became the new suburban home to large portions of the population previously unfamiliar with the evangelistic fervor of the Bible belt (Brunn et al. 2011b). Compounding this theologically conservative religious expansion was a concurrent Charismatic movement that introduced a domesticated Neo-Pentecostalism into middle-class Mainline congregations throughout the country. As a result, the mainline denominations have decreased by 50 % since 1965, while the market share of the more conservative denominations and nondenominational churches has increased. In addition, nearly 90 % of all U.S. megachurches claim an evangelical or Pentecostal/Charismatic theological label and are filled with solidly middle-class adherents.

Together, this religious reorientation interjected distinctive alternative options into the religious marketplace. These movements "reinvented American religion," (Miller 1997) offering something that broke with traditional models of church. A life of faith included expressive emotionalism, an evangelistic fervor to reach those outside the faith, a seriousness of orthodox beliefs and a holistic vision of Christianity as a daily lifestyle. Likewise, the corresponding worship styles shifted to a come-as-you-are informality, non-hymnal-based guitar-led contemporary praise music, enhanced by the use of projection technology and an outreach orientation with a church growth imperative. In all, the leaders of these religious initiatives were distinctive in their willingness to break with traditional forms and to innovate entrepreneurially. This sparked the megachurch movement to grow with baby boomer youth through great Sunday school or Youth programs. It wasn't long, however, before these influences also reshaped worship around a new vision of church that sustained them as adults while also attracting other growing families.

A considerably different story took place outside the U.S., where the catalyst was either the successful influx of Christianity into a formerly non-Christian area (such as Islam and indigenous faiths of Nigeria, Shaman and Buddhist Korea and more recently secular China), or the impassioned incursion of Protestantism in the form of Pentecostalism into a region of Catholic dominance (including many countries throughout Central and South America). These developments happened in these countries during economic development with a growing middle class, while Christianity, specifically of the Pentecostal variety, spread dramatically southward, to the point that "in 1900 over 80 % of all Christians lived in Europe and Northern America; however, by 2005 this proportion had fallen to under 40 % and will likely fall below 30 % before 2050. The Christian population in Africa alone grew from roughly 10 million in 1900 to nearly 360 million by 2000" (Johnson and Kim, p. 24).

The Pentecostal movement, which began officially in the United States in 1900, directly sparked this shift. This theological tradition incorporates an eschatological urgency and fervent evangelism, with ecstatic, expressive worship forms, supernatural spiritual gifts and an appeal to poorer and downtrodden persons with the message that God cares about them and can act supernaturally to better their lot, both in this world and the life to come. This movement has become the most rapidly growing Christian religious form worldwide and shapes the complexion of global megachurches, since at least 65 % of them for which we have a denominational affiliation are related to the Pentecostal or Charismatic traditions.

Global Pentecostalism does not represent Western colonial dominance, nor does it look to America for spiritual inspiration. Nevertheless, it does contain an impetus for education, personal betterment and social progress in this life, not just spiritual escapism to a future paradise. Likewise, the implicit egalitarianism inherent in the belief that everyone can directly encounter and individually experience the Divine can be a powerful message of self-worth and personal motivation.

Pentecostalism combines active evangelism with the demonstration of the gifts and fruits of the Holy Spirit. This creates a faith that is spiritually expressive and dynamically engaged in the world. A recent study showed that over 65 % of Pentecostals from 10 nations indicated they were highly evangelistic, speaking to others about their faith at least once a week (Lugo et al. 2006). The study's author, Luis Lugo, said the Pentecostal success stories in Latin America and Africa are understandable. African converts "don't have to leave behind their world of spirit . . . and Pentecostalism is second to none in providing a sense of community," especially in countries affected by massive displacement and migration. "I don't think it's too farfetched at this point to seriously consider whether Christianity is well on its way to being Pentecostalized," Lugo said, "certainly in the developing world" (Dart 2006: 12).

For example, the largest-attendance African church at present is a Baptist congregation named Eglise Protestante Baptiste Oeuvres et Mission Internationale (The Works and Mission Baptist Church Int'l). Based in Abidjan, Ivory Coast, it is led by founding pastor Dion Robert, who was brought to the Christian faith by a Southern Baptist missionary from America. Yet when this congregation of cell churches holds its frequent all-church gatherings in stadium-like venues, a common part of the program involves the casting out of demons and other dealings with the supernatural associated with the Pentecostal movement.

Additionally, Singapore's largest-attendance church at present is the nondenominational City Harvest Church, founded and pastored by Kong Hee. The average age of the congregation is under 30. The enthusiastic singing, accompanied by the widespread waving of arms, emphasizes a worship that is both emotional and cognitively challenging and stands clearly within the Pentecostal tradition.

When the dynamism within Pentecostalism is coupled with the organizational structures of large churches, whose programs offer training in leadership and organization, skills in technology and coping with mass society and activities for literacy and education, the result is essentially a school for modern urban life embedded in a faith that emphasizes hope and trust in God's spiritual, emotional blessing. As David Martin (1998: 126, 130) has suggested,

> The poor, caught up in the maelstrom of the transition to global capitalism and postmodernism respond with more enthusiasm to the Pentecostal option because they see with their own eyes its capacity to transform lives, here and now, for the immediate better….and to transform familiar elements of ethnic, familial and other habits of collective solidarity within a movement which also inaugurates new experiences of Postmodern individualism, autonomy, mobility and self-determination.

Considerable evidence has shown that Pentecostalism produces a "competitive economic advantage" for those believers who give up drinking, drugs, womanizing and gambling (Miller and Yamamori 2007: 33). When this implicit upward mobility advantage is combined with a "prosperity gospel" theme within some Pentecostal preaching, the push toward social betterment is evident. Prosperity preaching suggests that the blessings of a godly life are not just spiritual but also material health and wealth. This can be a powerful message for the under classes and for a rising middle class, especially as they also begin to live as Christians and turn from the vices that consumed much of their income.

The Pentecostal faith also means that charisma – the personal expression of spiritual gifts of preaching, healing, prophecy, etc. – carries power and authority. This diminishes the authority of external authorities: if one can directly encounter the divine spirit, the civil authorities pale in comparison and can therefore be challenged and supplanted. Donald Miller strongly suggests that, "contrary to widespread perceptions, Pentecostals are anything but apolitical." Ninety percent of Pentecostals and 85 % of charismatics in the U.S. agreed that Christians have a responsibility "to work for justice for the poor." Between 72 and 93 % agreed in Brazil, Chile and Guatemala while 97 % of Kenyan Pentecostals agreed (Dart 2006: 12). Nowhere is this attitude better seen than in Pentecostal movements in Central and South America and the political shifts taking place there, as many of these countries move from Catholic dominance to a Pentecostal faith often embedded in their megachurches (Miller and Yamamori; 2007: 125–26).

123.6.6 Holistic Programming Leads to Community and Personal Integration

Not only does the evangelical and Pentecostal theology with this-worldly middle-class asceticism develop more successful attenders of megachurches globally, but so too do the programs of these churches themselves reinforce this effort. As a congregation grows larger, it is more able to a mass both the financial and human resources to provide programs that assist the members and the larger community. As such, a comparison of programs offered by U.S. churches shows a distinct advantage for the congregation that grows larger (Table 123.4).[6]

Megachurches use this advantage to provide not only for participants' spiritual needs but also their everyday social needs. A larger percentage of megachurches in

[6] Data analysis of Faith Communities Today 2010 survey, www.faithcommunitiestoday.org, done by authors.

Table 123.4 Relationship between church size and special services offered

Percent of U.S. congregations claiming a lot of emphasis or a program specialty	All congregations (%)	Megachurches (%)
Community service activities	45	75
Support groups (bereavement, job loss, 12-step)	15	60
Parenting or marriage enrichment activities	18	56
Young adult activities or programs	32	61
Youth [teen] activities or programs	58	91
Team sports, fitness activities, exercise classes	11	27

Source: Author's "Data Analysis of Faith Communities Today 2010 Survey," www.faithcommunitiestoday.org

Table 123.5 Relationship between church size and upward mobility programs

Percent of U.S. congregations that directly provide the following for church and/or community members	All congregations (%)	Megachurches (%)
Financial counseling or education	30	85
Day care, pre-school, before or after-school programs	18	55
Job placement, job training, employment counseling	13	42
Tutoring or literacy programs	15	36
Health education, clinics	18	24
Community organizing, organized social issue advocacy	16	24
Programs for migrants or immigrants	7	16

Source: Author's "Data Analysis of Faith Communities Today 2010 Survey," www.faithcommunitiestoday.org

the U.S. offered training in marketable skills, job training, and educational efforts than churches of other sizes Table 123.5.[7]

What this indicates is a social responsiveness of the highly successful megachurch leadership to the needs of the community, whether in suburban America or urban global cities. This entrepreneurial willingness to risk and adapt, along with an ability to experiment and innovate, keeps the megachurch model congruent with the changes in culture and society.

123.6.7 Globalization and an Interconnected Planet

Nowhere is this innovation in evidence as much as the networks of interconnections across these global mega-congregations. Scholars such as Philip Jenkins and Mark Juergensmeyer describe a "global Christianity" with a "fluid process of cultural interaction, expansion, synthesis, borrowing and change (Johnson and Kim 2012: 26; Juergensmeyer 2003: 4–5; Jenkins 2002b). Increased immigration,

[7] Data analysis of Faith Communities Today 2010 survey, www.faithcommunitiestoday.org, done by authors.

cross-continental travel and advancements in communication and the Internet created opportunities for many megachurches to participate in global networks, sharing best practices and mission efforts. These transnational religious connections mean that David Yonggi Cho's strategy of cell-groups rapidly migrated to American megachurches through his books and oversees speaking engagements. Former Nigerian megachurch pastor Benson Idahosa influenced countless megachurches as an Archbishop of the global International Communion of Charismatic Congregations. Likewise, U.S. megachurch Pastor Rick Warren exerts profound international influence through his website Pastors.com.

Many megachurches also exercise direct cross-country sway in their branch locations planted throughout the world. Over 40 % of non-US megachurches, for which we have information, have a branch of their congregation in the U.S., while quite a few American megachurches have church plants, television ministries and conference tours to spread their influence internationally.

123.6.8 Global Isomorphism

The global migration of populations has also influenced the megachurch phenomenon, especially in the urban European context. In Europe, the majority of the largest megachurches are led and populated by immigrants. The largest European churches-Kingsway International Christian Center, London, UK; Church of the Embassy of the Blessed Kingdom of God for all Nations in Kiev, Ukraine; and Redeemer Christian Church of God in London, UK, are all led by Africans. Interestingly, the latter church actually started in Nigeria and then branched out in a European Diaspora. Recently, a major academic study by the Church of England (Goodhew 2012) found that, contrary to widespread assumption, Christianity in Britain is not suffering terminal decline. While the majority of traditional parishes are shrinking, church planting and growth among immigrants has offset that trend. Indeed, even the cover of the book symbolizes a new reality as it depicts a baptism by a pastor of African descent. This pattern is also evident in the U.S. where increasingly megachurches can be found where members speak only Korean, Spanish or Russian.

123.7 Conclusion

The global appearance of these massive megachurches throughout the past half-century can be linked to patterns of intensive urbanization and suburbanization, unprecedented population migration and concentration, along with technological advances within an increasingly modernized and interconnected global reality. Simultaneously, the growth of new varieties of Protestantism has further contributed to the megachurch movement, with contemporary religious worship and ideologies that provide a sense of community, meaning and economic advantages during times of instability.

In addition, a dramatic shift in the geographic areas where Christianity is concentrated and growing has also taken place globally during this same time. None of these variables directly caused the phenomenon, but together they all contributed to its rise and continued success.

The megachurch phenomenon can be seen as representing a new model of faith community for a contemporary mega-metropolitan social reality. This model of church provides the necessary characteristics, structures, programs and adaptability to address the challenges of a highly urbanized area with a population of dislocated and under-equipped migrants. Not only do these massive congregations offer an enthusiastic, modern and indigenously customized version of Christianity, but they also provide a community of support, networks of training and assimilation into the chaotic life of the intense urban context in ways that smaller congregations cannot offer.
It is likely that this form of global faith will be around for many decades to come. Much scholarly attention has focused on megacities and urbanization, with their problems and challenges into the future. However, far too little focus has been paid to the religious and spiritual networks within these cities that provide a moral foundation in the midst of an uncertain time. This chapter offers some initial evidence of interesting and distinctively urban religious innovation taking place. Much further study needs to be undertaken within specific megacities (as the Shin and Begin chapter does) to assess the development of religious institutions in relation to the growth of the city. In addition, a concerted effort needs to be undertaken to identify and map, not just the largest Protestant and Catholic Christian congregations, but also parallel trends within Hinduism and Islam. Finally, not all the religious vitality in the megacity happens on a grand scale. A rapidly growing small intentional cell-church and house church movement is evident, but understudied, throughout the developing urban world, and fits the megacity context perhaps as well as the megachurch (Thumma 2006). The global megacities are far from being "secular cities;" Therefore, to fully understand the moral glue and strategies for coping with large scale social disorder, much greater attention should be paid to the religious innovation happening in these contexts.

Acknowledgement We want to thank Mr. Jeff Levy, Gyula Pauer Cartography Laboratory, Department of Geography, University of Kentucky, Lexington, KY for constructing the maps.

References

2006 Global Pentecostalism Study by the Pew Research Institute. Retrieved August 1, 2012, from www.pewforum.org/Christian/Evangelical-Protestant-Churches/Spirit-and-Power.aspx

Bird, W. (2012). *The world's largest churches: a country-by-country list of global megachurches maintained by Warren Bird*. Retrieved August 1, 2012, from www.leadnet.org/world

Brunn, S. D., Ghose, R., & Graham, M. (2011a). Cities of the future. In S. D. Brunn, M. Hays-Mitchell, & D. J. Zeigler (Eds.), *Cities of the world: World regional urban development* (5th ed., pp. 556–559). Lanham: Rowan and Littlefield.

Brunn, S. D., Webster, G., & Archer, J. C. (2011b). The Bible Belt in a Changing South: Shrinking, Relocating, and Multiple Buckles. *Southeastern Geographer, 51*, 513–549.

Chaves, M. (2006). All creatures great and small: Megachurches in context. *Review of Religious Research, 47*, 329–346.

Chaves, M., Konieczny, M. E., Beyerlein, K., & Barman, E. (1999). The national congregations study: Background, methods, and selected results. *Journal for the Scientific Study of Religion, 38*, 458–476.

Cox, W. (2011). *The evolving urban form: Seoul 2-17-2011*. Retrieved August 1, 2012, from www.newgeography.com/content/002060-the-evolving-urban-form-seoul

Dart, J. (2006, October 31). Debunking some Pentecostal stereotypes, *The Christian Century*.

Global health, observatory, situations and trends in key indicators. Retrieved August 1, 2012, from www.who.int/gho/urban_health/situation_trends/en/index.html

Goodhew, D. (2012). *Church growth in Britain: 1980 to the present*. Farnham: Ashgate.

Jenkins, P. (2002a). *The next Christendom: The coming of global Christianity*, interview by Joanne J. Myers, Carnegie Council. Retrieved August 1, 2012, from www.carnegiecouncil.org/resources/transcripts/136.html

Jenkins, P. (2002b). *The next Christendom: The coming of global Christianity*. Oxford/New York: Oxford University Press.

Johnson, T., & Kim, S. (2012). *The changing face of global Christianity*. Retrieved August 1, 2012, from www.bostontheological.org/assets/files/02tjohnson.pdf

Juergensmeyer, M. (2003). Thinking globally about religion. In *Global Religions: An Introduction*. Oxford: Oxford University Press.

Kim, A. E. (2002). Characteristics of religious life in South Korea: A sociological survey. *Review of Religious Research, 43*, 291–310.

Lee, S. (2007). *Three Korean megachurches: The growth and decline of three Korean megachurches in their cultural and generational backgrounds*. Hartford: Hartford Seminary. MA thesis.

Lugo, L., et al. (2006). *Spirit and power: A 10-country survey of Pentecostals*, Pew Research Center. Retrieved August 1, 2012, from www.pewforum.org/Christian/Evangelical-Protestant-Churches/Spirit-and-Power.aspx

Martin, B. (1998). From pre- to postmodernity in Latin America: The case of Pentecostalism. In P. Heelas (Ed.), *Religion, modernity and postmodernity* (pp. 102–146). Oxford: Blackwell.

Miller, D. E. (1997). *Reinventing American Protestantism: Christianity in the new millennium*. Berkeley: University of California Press.

Miller, D. E., & Yamamori, T. (2007). *Global Pentecostalism: The new face of Christian social engagement*. Berkeley: University of California Press.

Myers, J. J. *The next Christendom: The coming of global Christianity*. Interview with Philip Jenkins. Retrieved August 1, 2012, from www.carnegiecouncil.org/resources/transcripts/136.html

Sudjic, D. *Identity in the city. Third Megacities Lecture*. Retrieved August 1, 2012, from www.megacities.nl/lecture_3/lecture.html

Synan, H. V. (n.d.). The Yoido Full Gospel Church. *Cyberjournal for Pentecostal-Charismatic Research*. Retrieved August 1, 2012, from www.pctii.org/cyberj/cyberj2/synan.html

Thumma, S. (2006). The shape of things to come: Megachurches, emerging churches and other new religious structures supporting an individualized spiritual identity. In L. Charles (Ed.), *Faith in America: Changes, Challenges, New Directions* (pp. 185–206). Westport: Praeger Press.

Thumma, S. (2012). *Database of U.S. megachurches and national megachurch reports*. Retrieved August 1, 2012, from www.hartfordinstitute.org

Thumma, S., & Travis, D. (2007). *Beyond megachurch myths: What we can learn from America's largest churches*. San Francisco: Jossey-Bass Leadership Network Series.

Towns, E. (1969). *The ten largest Sunday schools and what makes them grow*. Grand Rapids: Baker Books.

Chapter 124
Houston Mosques: Space, Place and Religious Meaning

Akel Ismail Kahera and Bakama BakamaNume

Thinking about and organizing [urban] space is one of the pre-occupations of power.

(Jean-Michel Brabant 2007: 25)

124.1 Introduction

The subject of this chapter is twofold. First, we want to re-examine Foucault's *Heterotopias,* which has opened up a radical new possibility for the reading of urban mosques in the city of Houston, Texas. This thesis seeks to reconcile "the 'heterotopias' that we encounter … the 'erosions' that occur there are our prejudices, delusions, ideals but also our disillusionments and negativities … and they are specific [urban] spaces" (King 1996: 220). Second, *Heterotopias* seeks to reconcile the philosophical tensions implicit in Foucault's theory of space/knowledge/power with the contentious political discourse and the widespread misunderstanding of the urban mosque. Above all our theoretical assumptions will serve to examine the urban space, public knowledge and religious power—using the city of Houston as a case study—to critique the urban mosque. To situate Foucault, it is important to note that "The twentieth century is perhaps the era when such kinships were undone … And if [urban] space is, in today's language, the most obsessive of [trans-national and global] metaphors, it is not that it henceforth offers the only recourse; but it is in [urban] space that, from the outset language unfurls" (Crampton and Elden 2007: 163). We argue that the urban mosque is a valid, but contentious assumption, because it serves to define Foucault's *Heterotopias*; it is an obsessive trans-national and global metaphor. "It is in [urban] space that it transports itself, that its very being "metaphorizes" itself" (2007: 163).

A.I. Kahera (✉)
College of Architecture, Art and Humanities, Clemson University, Clemson, SC 29634, USA
e-mail: akelk@clemson.edu

B. BakamaNume
Division of Social Work, Behavioral and Political Science,
Prairie View A&M University, Prairie View, TX 77466, USA
e-mail: bbbakamanume@pvamu.edu

S.D. Brunn (ed.), *The Changing World Religion Map*,
DOI 10.1007/978-94-017-9376-6_124

The first assumption is that "there can be no emancipation without unmasking all the linkages and kinships between space, knowledge and power … in which spatial relations can be represented" (King 1996: 220). Another assumption is that at the core of Western paternalism is the obsessive impulse to marginalize, exclude or eliminate non-Christian "minority" communities (Hindu, Buddhist, Muslim et al.) and their religious spaces; this is so because the understanding of "freedom of religion" has been undeniably tainted by widespread political propaganda. We may compare such prejudices with the negativities of the German political thinker Carl Schmitt (Stanford Encyclopedia of Philosophy). Schmitt advanced the notion that the term "friend" is a homogenous category and anyone who diverges from that category–in political, cultural and religious terms–is an enemy. Given the universal forms of terror exhibited by the betrayal of the faith of Islam by a minority of Muslim radicals, the displaced anxieties of the media and their political allies may be readily understood; in general such anxieties recall Schmitt's definition of "friend" and "enemy." We have no doubt that Schmitt himself would today label all non-Western non-Christian religious practices as unfriendly and would view the urban mosque as inimical to any city in a Western society.

Returning to Foucault's *Heterotopias,* we wish to highlight the censorship of religious space in one well-known example from the city of Córdoba, where the urban mosque was built during the ninth century C.E. However, after the Spanish Reconquesta (1492)–five centuries later–a decision was made by the reigning Spanish sovereign to condemn the mosque, and, therefore, to subjugate the edifice to a hybrid identity; notwithstanding the motivation to emancipate the city of Córdoba from the grip of Islamic rule, religious power and public knowledge. Subsequently a cathedral was constructed within the sacred space of the mosque. Barthes, has identified this as "the denaturing" of the Córdoba mosque, as noted by Barnes and Duncan in *Writing Worlds* (1992a, b: 21). A more recent example is the destruction of the Babri mosque in 1992 in the city of Ayodhya, India; Hindu fundamentalists claimed that a previous temple was removed to create the mosque in 1527 and that the mosque was built on the fort (hill) of the Hindu god Rama. In yet another example, Bevan's *The Destruction of Memory* (2006), points to the ethnic cleansing of Bosnian Muslims by the Serbs and Croats and the violent destruction of mosques in the Balkans; he vividly illustrates another way in which contested identities are painfully silenced. The aforementioned examples illustrate the manner in which Foucault's *Heterotopias* are commonly invoked; each example describes xenophobic prejudices, the condemnation of urban space, religious power and public knowledge.

Foucault's thesis space/knowledge/power raises a key question. Has the Patriot Act (Public Law 107-56) been instrumental in composing and decomposing expressions of uncertain religious beliefs putting all Muslims—and indeed, all minority religious faiths—in the West off-balance? Almost two decades after 9/11 we are in the midst of another contentious debate recently sparked by the so-called "Ground Zero" mosque (Park 51) controversy. When taken to its extreme conclusion the so-called "Ground Zero" mosque is considered by some critics to be an illegitimate religious edifice, which accounts for and is supported by many instances of Islamophobia. If the First Amendment implicitly supports religious space for all Americans and all religious faiths, then the rhetorical stance of contentious politics advanced by Schmitt and others may force us to ask these questions:

- How should the First Amendment guarantees apply to current controversies and contentions in the political discourse about religion, public space and public life?
- How should we interpret afresh the separation of church and state while allowing the First Amendment and free speech to also define, excoriate and exclude others (Jews, Hindus, Muslims, Buddhists et al.) who do not share Christian beliefs?
- While there is enough evidence to support the freedom of religion, in what way has the contentious political discourse placed the Qur'an and the religion of Islam on a collision course with the U.S. Constitution?

It is not possible to address all these issues in this chapter, because the problem is much more complex and intractable. But in thinking about space/power/knowledge as King has noted in *Emancipating Space* (1996), our analysis seeks to focus on the urban mosque in the city of Houston, Texas. Thinking about space/knowledge/power we offer a rich and descriptive account of the urban mosque, which has far too long been neglected. Our account reveals a descriptive language as a basis for aesthetic syncretism and the creation of an American Muslim *leitmotif*. The *leitmotif* can be identified as "a thing of space;" therefore, the community "determines its [aesthetic] choices, draws its figures [plans], and [informal] translations. It is in [urban] space that it transports itself" (Crampton and Elden 2007: 163). On the other hand Foucault's *Heterotopias* helps us to deconstruct the xenophobic commodification of political propaganda, but to address the urban mosque as an American *leitmotif* we want to draw the attention of the reader to the characteristics of its physical adaption within the city of Houston—and indeed in any American city, Boston, New York, Detroit, et al. A few observations derive from a close examination of Houston's demographics, Diaspora and the successive variations in style of the building which together ask us to restate one of Foucault's key concerns: "What are the relations between knowledge (*savoir*), war and power?" (2007: 3). Another aspect of this critique can also be found throughout many of Foucault's writings, viz., Jean-Michel Brabant tells us that "every strategy of power {including war] has a spatial dimension" (2007: 25).

Because Craig and Thrift have also noted that, "thinking about space occurs through the medium of language" (2000: 4), we are aware that the concept of language, public worship and cultural aesthetics is always related to particular qualities of urban life and practices (Fig. 124.1). One component of the thought process occurs through the particular liturgical practice of five daily prayers, which are performed in a common language: Arabic. A second aspect deals with time as a regulating element bringing together time and space, "All of which still tends to leave a sense of time and space in language where the formal relationship is a static pattern of places" (Crang and Thrift 2000: 23). A final aspect of language deals particularly with the communal space of public worship. In this regard we may conclude that the

Fig. 124.1 Moharram Celebration (Shi'ite community) Houston, Texas. This is a depiction of the Shrine of Abu Fadhl Al-Abbas—the half brother of Imam Al-Hussain (as), which is located in present day Karbala, Iraq (Photo © Afroz David Okhowat, 2008, used with permission)

urban mosque has a spatial language, starting with the orientation of the worshipper and the physical edifice, which is met by a static pattern: facing Makkah. The extent to which this practice can be considered as a determinant of urban space in the city of Houston is very relevant to the thesis of space/knowledge/power.

Our first contention is as follows: since the urban mosque is independent of a historical chronology, each enigmatic feature is directly related to the problem of human consciousness or the communal need for cultural identity. The second contention recognizes the emotional value placed on the use of an extant aesthetic precedent, which is a burdensome consideration for an architect and his/her client. When an attempt is made to copy an extant structure from the past, viz., a mosque, the result will invariably end up as an anomaly and the building's aesthetic image will be severely compromised. Nevertheless, together the ordered space and the practice of communal worship constitute the embodiment of the urban mosque. Because the character of urban spaces varies, these changing conditions offer scope for further observation and critique, as the urban public experiences the edifice. Likewise, the permanence of a place enables it to belong to the city and to a given urban context, which forms the basis for our qualitative interpretation.

124.2 The Diaspora Community and the Urban Mosque

The putative term "urban mosque" which first appeared in the text *Deconstructing the American Mosque: Space, Gender and Aesthetics* (Kahera 2002a), is again adopted here to refer to a religious edifice primarily constructed by Muslims who reside in an urban locale. The urban mosque, which is also referred to as an Islamic center, is the heart of the Muslim community; it is where the faithful gather to engage in daily communal worship, spiritual retreat, matrimony, education and social activities. The key challenge is how to interpret the broad range of aesthetic, liturgical requirements and site planning considerations. In other words how does this particular building type arrive at a synthesis of recurrent elements such as a minaret, a prayer-niche (*mihrab)*, or a dome? Because Islam is often imbedded in a transnational identity, the collective activity of worship treats the mosque as a reflection of the Diaspora community (Arabs, Africans, Asians, Turks et al.). Yet, the cultural customs cannot be ignored and as such the type of aesthetics that we find in Houston mosques exhibit a range of cultural nuances, modern schemes, traditional styles or a hybrid appearance.

To situate the problem we will later attempt to interpret and bring critical analysis to bear on space, form, symbol and order. At the heart of the problem of cultural customs is a key question: how does one come to know and identify the urban mosque in the city of Houston? In the absence of zoning regulations that determine the scale, size, typology and character of a religious edifice, we will focus on location analysis, aesthetic visualization, site context and the spatial syntax that informs the nuances of cultural customs. Because the urban mosque remains an emerging concept, it illustrates the way the edifice has responded to epistemic urban configurations, the production of public space and communal place and a unique design problem for architects and planners. In the context of architecture and urban planning these thematic elements may advocate a mode of cultural expression related to the sentiments of a patron/community or the expertise of builders. However, the many urban images, forms and spatial elements of each mode of cultural expression are also controlled by the widespread use of the mosque as a place of congregational worship, with its uniform observance of worship tied to common belief within the

Fig. 124.2 Masjid Hamza Mission Bend Houston, Texas. Surrounding the Masjid are residential neighborhoods and apartment complexes, making the Masjid (mosque) easily accessible (Photo © Michele Scurry, 2008, used with permission)

urban Muslim community (Fig. 124.2). This characteristic has maintained an urban congregational prototype, but has also allowed for a creative process of aesthetic development to occur. A cursory observation of mosques in the United States reveals an urban prototype that is tied to the diversity of aesthetic language, widely influenced by the sentiments of a Diaspora community.

The appropriation of urban space must also consider religious values to adequately explain any design strategy. Conversely, it could be argued that while architectural theory differs from exegesis per se, architectural theory is nonetheless a by-product of religious belief and practice and, therefore, carries the same rigor. Following this line of reasoning and the multifaceted problems of urban mosques, we offer here some thoughts about religious values, expanding on the earlier discussion about space/power/knowledge. First, aniconism, or the absence of iconography in any place of Muslim worship is a most fundamental principle of *tauheed* or monotheism. *Tauheed* has no associated symbolic form, only the primary act of individual submission to Allah (God), and as such the reluctance of directly investing an edifice with any human deity or anthropomorphic expression. Human or anthropomorphic expressions are considered sacrilegious and against the principle of monotheism, and in this regard the rejection of idolatry is commonly understood. The individual act of submission to Allah is paired with the physical experience, thus the cognitive rule of facing toward the *Ka'bah* in Makkah (Mecca) can be explained as the ontological axis of prayer (*qiblah*). Makkah is the universal omphalos for Muslims anywhere on the planet. A difficulty arises when we try to identify the themes associated with sentiment and syntax, viz., how can we identify the origin or the attributes of each theme? In this sense the urban mosque provides an opportunity to purse this inquiry because it demonstrates both 'sentiment' and 'syntax', both in a syncretic language, which can be defined in terms of space/knowledge/power. It could also be argued that 'sentiment' can be taken to mean feelings of expression and cultural identity that are invariably linked to syntactical nuances, aesthetics and belief.

124.3 Urban Space, Religious Power and Public Knowledge

Since feelings are first, who cares about the syntax of things. –E.E. Cummings

By recalling the past, one is no longer in a foreign and unfamiliar environment, but in an environment where one's sentiments are nourished by familiar aesthetic features. Many Houston examples demonstrate how the past is recalled and how image is represented in the structure of an urban mosque. The urban space of the mosque is a localized condition and the physical organization of the edifice in the city of Houston is a way of seeing the world, viz., a cultural image and a pictorial way of representation (King 1996: 221). Against the secular vision of the city the plan of the edifice is shaped by language, meanings and cultural ideas that often conform to religious beliefs or customs. Consider, for example, the segregation or non-segregation of women, representing the cultural reproduction of a Diaspora community with masculine-oriented roles. Three descriptive markers—cultural and social norms, religious belief and religious practice—further characterize the institution of congregational worship and in many crucial respects give reason to dispute various statements about the physical component of women's prayer space. Some urban mosques support an emancipatory vision of female congregants, as it were to the collective act of communal worship, thus allowing women unfettered and equal access to mosque. Most important is the fellowship hall where men and women gather to pray on a daily basis, read the Qur'an and engage in a host of pious activities. Aside from cultural rivalries over women's unfettered access to the mosque, notwithstanding conflicted feelings common to an overabundance of legal opinion, which carry authority, and in extreme cases censorship—such cases are hard to assess without a background in Islamic legal theory. Firstly, religious practice defines the etiquette of women and men who share the same right of entry to the mosque, for the performance of the five conventional daily prayers, the *jummah* or Friday prayer and the *eid* or feast prayers (Fig. 124.3). Secondly, while both premodern and present-day commentators point to the merits of public worship without gender distinction, scholarly discourses and legal consensus differ due to the interpretation of the *hadith* (prophetic injunctions). In summary, the view that supports the merits of public worship that includes women is derived in part from the following *hadith*: "…the congregational prayer is twenty seven times greater" [than that of the prayer offered alone] (Bukhari vol. 1 book 11: 621).

Fig. 124.3 Masjid At-Taqwa SW Houston (Sugarland), Texas. The Masjid (mosque) provides community health care (al-Shifa clinic) and an elementary school (Darul Arqam) in addition to the fellowship hall (musalla) (Photo © Akel Kahera and Bakama Bakamanume 2012)

Likewise, religious power is always at work in signs, symbols, the daily observance of prescribed rituals and the common understanding of the world. Public knowledge on the other hand is dialectically related to power and the effective power of the community draws from effective knowledge, relies on it, and more than often, public knowledge shapes communal and social relations. These three categories of urban space, religious power and public knowledge deserve further comment because various aspects of material reproduction link the secular nature of the city with the symbolic reproduction. One aspect forces a reformulation of the argument; the other would compel the rejection. In this sense Foucault has argued that subjects are the relations of power, the ethics of conduct and the transmissions of knowledge (Belsy 2002: 53; King 1996).

Perhaps the reformulation of Foucault's space/knowledge/power finds focus in the question of aesthetics and the specific instance of the role of the architect. For example, in the absence of design standards American architects who have been commissioned to design a mosque have exercised absolute freedom in interpreting planning prerequisites to meet American code requirements. Architects wield tremendous power, and in some instances they have exercised "free interpretation" of common aesthetic themes. In *Deconstructing the American Mosque: Space, Gender & Aesthetics* (Kahera 2002a) and *Design Criteria for Mosques & Islamic Centers: Art, Architecture & Worship* (Kahera et al. 2009) these issues were addressed at length.

Fig. 124.4 Muhammad Mosque No. 45, Nation of Islam, Houston, Texas (Photo © Akel Kahera and Bakama Bakamanume 2012)

Our focus in the present study is to examine the urban fabric of the city of Houston to provide the reader with a better collective understanding of the discourse. For example, a practical problem resides with the accuracy and the method of determining the *qiblah* axis, the Makkah orientation of the edifice. There is a goal to achieve some given aesthetic and functional purpose, which the architect and his/her client must carefully consider and agree upon at the start of the design process. On the one hand there is the urge to reject modernity, given the ethnic diversity that exists in the Muslim community. The confusion that can form with such a vast array of ideas and images along with the general opposition towards traditional influences within modernist thinking can also lead to purely reductive views and processes which create building forms completely detached from the fundamentals that form the quintessential Muslim edifice and its embodied, global identity (Kahera 2002a). The forces that shape the urban mosque are complex, but they are equally dynamic (Fig. 124.4). On the other hand many types of design issues related to ethnicity have resulted in an architectural discourse that have provided the cultural codes regardless of the context; a modest but growing body of literature has dealt with the problem of cultural interpretation and ideas of cultural identity. Foucault argues that the forms of resistance, that is, integration, assimilation and differentiation—that accompany or belong together involve two elements: codes of behavior and modes of subjectivity (Belsy 2002: 54).

The debate about the relevance of aesthetics and interpretation has taken many turns over the last decade and more recently the *International Journal of Islamic Architecture* (2012) offers fresh insights. Here, too, we attempt an uncoupling of the material, religious and symbolic aspects of the debate. The rules that prescribe aesthetics for an urban mosque are complex and bear a direct relation to codes of behavior that show signs of integration, assimilation and differentiation. So firstly we should ask what is Muslim art? Why is Muslim art not naturalistic? Briefly, Muslim art does not seek to embody as a priori the idea of *man as the measure of all things* and nature in portraiture and stone. In Islam monotheism (*Tauheed*), Divinity is not an *idee-fixe,* which is not conceived as an image, an *apotheosis*, or the transfiguration of a "being" into Allah (God). Islam is free of idolatry and polytheism since only Allah is divine, transcendent, omnipotent, and non-representational, Allah is *Al-Ahad* (The One), *Al-Khaliq* (The Creator) etc. In this context, the intuitive senses of the Muslims are necessarily without, and do not need the image as a sensory aid to enhance his/her faith or practice because this would constitute a condition of *shirk* (polytheism) or the abrogation of monotheism. The highest form of esthetic needs of the Muslim is, therefore, achieved through the words of the Qur'an and its meaning which emanate when read or heard or transcribed on buildings. Thus, Muslim art and the features of mosque architecture are not an imitation of created form, or a representation of nature, but an aesthetic principle whose objects, patterns, motifs, epigraphy are articulated in light of *tawheed*–the essence of monotheism.

Our problem, however, is not one of defining the theological methodology tied to a creative impulse, which makes and remakes *Muslim* aesthetics. If the methodology is properly applied, it may demonstrate that the form of reasoning and the operational structure of aesthetics of the mosque are linked to the understanding of the nature of being and the reality of the intellect (*'aql)* and the realization that all man/woman made creations (objects) are transient. Above all, the design conceptualization of faith, spirituality and aesthetics can support two primary tropes: first, to preserve the identity of the various forms that constitute the elements of a religious edifice for men and women, and, second, the relationship between spiritual repose, spatial equity and aesthetics (Kahera 2002a, b). This rule applies in America as well and the practice of Islam in America takes its religious precedents and aesthetic traditions quite seriously drawing heavily from doctrinal understanding of iconography and architectural knowledge largely derived from extant styles in the Muslim world. The aesthetic traditions of Muslim religious edifices, specifically, the decorative and spatial styles of mosque architecture have evolved since the first mosque (*masjid*) was built by the Prophet Muhammad in seventh century CE Arabia; it shares a common aesthetic value with similar structures recently built in North America. Five aesthetic principles have shaped the formative aesthetic principles of *masjid* since the seventh century; they are the structure of belief, order, space, materials and symbols. These formative principles can be found in mosques built in the United States since the 1950s, but in America they are largely responsive to the religious and cultural sentiments of the immigrant community.

To situate the problem it is important to realize that both architect and client face two related design choices: first an approach, which attempts to interpret and to bring critical analysis to bear on space, form, symbol and order. This approach makes it possible to avoid an aesthetic anomaly. Secondly, because the mosque is a building type endowed with a 1,500-year history it may be very difficult for any client or architect to suspend the temptation to randomly borrow ideas from the corpus of examples that exist throughout the Muslim world. Finally, many communities lack decisive power over cultural style and imagery largely because of the inability to reach a consensus. The literature on this topic is now more frequently available as a resource, but a key question remains for the architect, viz., how to interpret the specific symbolic and aesthetic associations for Houston, Texas and the larger American context? Is it at all possible to conceive of an American Mosque as an authentic representation?

Authenticity means that the architecture of the mosque has a two-fold space conception: spiritual and physical, epistemological and aesthetic. These are fundamental areas of concern, which an architect will have to decipher. It is within these areas that the architect will also have to learn to become a mediator or negotiator of experiences, beliefs, and ways of knowing, while also bringing to bear on the design the basics of building codes, local zoning ordinances and socio-cultural dogmas (Kahera 2002a, b). Above all it is important for us to remember that the Promethean myth does not exist in Islam; religion and the sacred have traditionally been the major factors organizing the human space of the mosque. In Muslim aesthetics the hegemony of the discipline called *Shari'ah* (sacred law) has dominated the written and the spoken word often altering the relationship of faith to aesthetics. However, everyone will agree that Muslim religious aesthetics is a Theo-centered epistemology.

The examples featured here of Houston mosques illustrate the way they have responded to this very epistemic configuration and unique design problem linking the sacred texts to the foundations and power of belief. Second, it is important to pin down precisely the semantic components of the mosque. We may begin simply by asking the question: what is a mosque? A mosque is primarily a place of spiritual repose, a spiritual sanctuary. It is very important to the faithful, but mosques are not built according to divine patterns; the two main religious texts for Muslims, the Qur'an and the *hadith* provide no clear prescriptive rules as to what a mosque should look like. However, the Qur'an does stress the value of the edifice as a place for the remembrance of Allah and the *hadith* prescribes a list of profane actions that are not allowed to take place in a mosque, for example, the absence of human or animal imagery (iconography) that we find in a mosque could be understood as follows. The essence of sacred art remains always reflective, contemplative and Theo-centric; the acceptance of revealed truths requires a keen intellect [*al-aql*], the purity of heart [*qalb*] and the piety of one's soul [*ruh*].

There are more than two dozen mosques in the city of Houston, Texas that have been established over that last five decades. Houston, mosques display a wide variety of styles based on this broad interpretation of aesthetic vocabulary and the need to meet the liturgical requirements. However the plan of a mosque's fellowship hall for men and women is a fundamental criteria; it is primarily governed by the liturgical axis towards Makkah [Mecca]. The indication of this axis is a niche (*mihrab*) in the wall facing Makkah. Historically three kinds of visual patterns have evolved in sacred art: (1) Designs derived from plant life often called arabesque in the West, (2) Arabic calligraphy which is the most revered art form in Islam because it conveys the word of God, and (3) Tessellation or the repetitive "ordering" of a geometric pattern. In general these three are not common to mosques in Houston, Texas, although isolated examples of their occurrence do exist. Of particular importance to the aesthetics of the Houston mosque is the realm of meaning, in other words the religious power of sign and symbol. One can describe the process and properties and elements employed in the characteristics of spatial treatments of sign and symbol, but at the level of construction we find an exhaustive category of examples. There have been many interpretations in *Making Muslim Space in North America and Europe* (1996). Another interpretation, *Deconstructing the American Mosque* (Kahera 2002a) treats Islamic practice itself as prescriptive in terms of behavior. Undoubtedly, the desire for communal worship is rooted in a long tradition of public gathering especially on Friday and important religious occasions. While this is an important point, we must also acknowledge the reality of urban life in cities such as Houston.

124.4 Spatiality of Mosques in the Houston Metropolitan Region

> Our epoch is one in which space takes for us the form of relations among sites. (Foucault 1984: 47)

124.4.1 The Spatial Location of Urban Mosques: Sites, Spaces, and the Home

The literature reminds us that we should pay "special attention to the construct of identity for the individual and the community" (Schopflin 2000; Bourdieu 1993). It is well known that identities are anchored/founded around a set regulated values and behavior. The church, mosque, synagogues and temple do contribute to anchoring identity and have been the integral components of the urban layout (Ayhan and Cubukcu 2010). We are further advised to be aware of how we present questions about the commitment to place and space; they must be viewed within the context of current processes of globalization and their impact on local and regional communities as well as geographical profiles. We may ask two related questions. *First*, what processes (local and global) have influenced the production of urban mosques/space in Houston? Why should we study the urban mosque? There are several good reasons for the study. First, the number of mosques (sites) in Houston area has increased rapidly in the last 10 years. Yet, the last 10 years have also been turbulent years for the Muslim communities in the area and the country (Pew Research Center 2011). The Pew Report indicates that 25 % of Muslim Americans report that mosques or Islamic centers in their communities have been target of controversy or outright hostility (2011: 3). Despite the increase in the urban mosques, their numbers are far less than that of other churches in the Houston region. There are 4,378 churches within the region and just over 100 mosques (Survey 2011; Siddiquiz 2009). *Second*, the urban mosque warrants academic research because of the increasing diversity of the city and country (globalization) and the cultural implications associated with such spaces (commitment to place and space). This section examines the urban mosques using GIS maps. A descriptive approach is used with the help of maps – visual images. Foucault (1983) provides us an insightful comment about maps. He stated that "Maps can be statements if they are used as representation of geographical order" (1983: 45). We will attempt to find the statement(s) of the urban mosques in the ordering of the Houston landscape.

Since September 11, 2001, Muslim Americans have experienced increasing discrimination (Pew Research Center 2011). Discrimination or the segregation of a group of people often leads to increasing search for a place of comfort for the discriminated population. This may be the tendency for the Houston Muslim community and other communities in the U.S. They seek a place of comfort. To most people, the place of comfort is/can be home, an isolated location in park, library, or place of worship (Tuan 1991). A place of worship is truly important when it is viewed also as a place of socialization (matrimony, social activities), education, a home away from the residential home, and a place that reminds us of our ancestral homes and communities. This last situation refers mostly to immigrant communities, who often make up the majority population in the urban mosques. They worship and get together for socialization, outside their homes, with other fellow Muslims and without any fear of discrimination.

The relationship between sites (home and mosques) is examined using GIS maps. The home is spatially identified as the zip code of residence. The percentage of ethnic population living in the zip code was used as the measure. Mosques are plotted against the specific selected immigrant populations within a zip code area. Then, using a descriptive method we provide an explanation for the relationship between population and location of the urban mosque. Houston can be divided into sub-regions (vernacular regions): Northeast, Northwest, Southeast, and Southwest. Highway Interstate 10 divides the metropolitan area into the North and South regions. Highway 45 North provides the boundary for East and West in the area north of Interstate 10. Highway 59 South divides the boundary line for the east and west regions south of Interstate 10. Figure 124.5 shows the distribution or location of major mosques in the Houston area. The mosques are shown by zip codes where they are located. The highest concentration of mosques is in the Southwest area. This is an area that extends from Interstate 10 West to Highway 59 South. The Northeast and Southeast have the fewest.

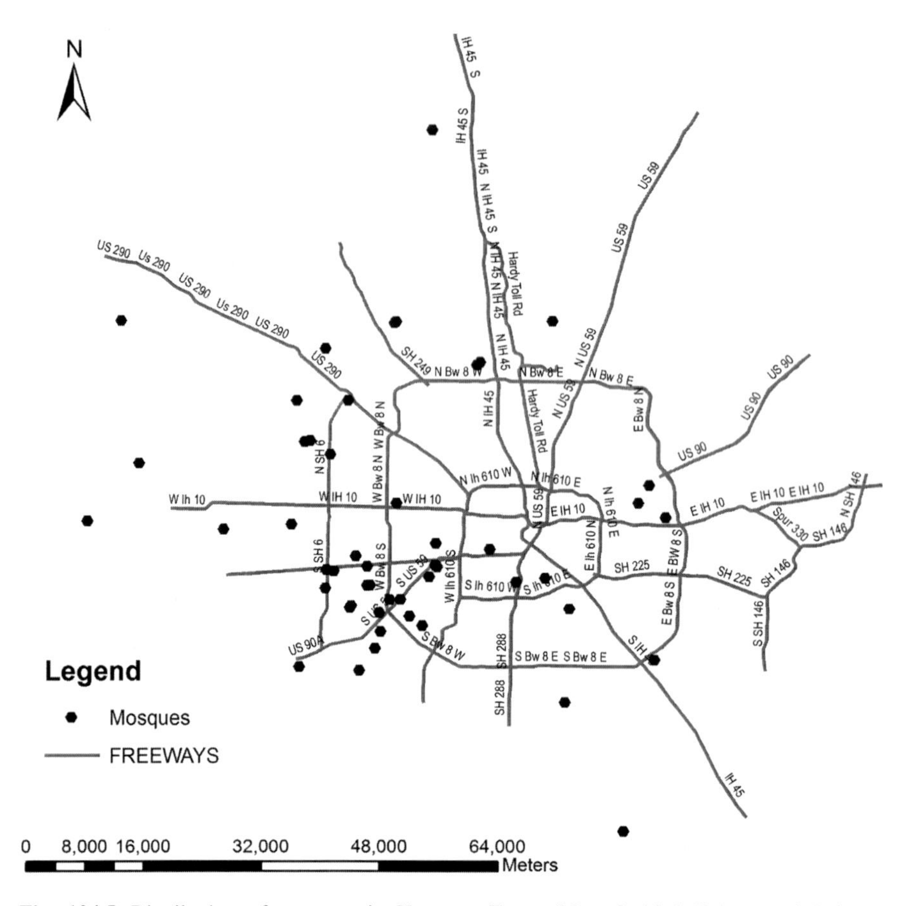

Fig. 124.5 Distribution of mosques in Houston, Texas (Map © Akel Kahera and Bakama Bakamanume 2012)

The location of urban mosques in Houston is far from a random pattern and the cluster in the Southwest is very noticeable. The area is one of the most diverse in terms of population (U.S.A. Bureau of the Census 2010). Most zip codes in the Southwest have the largest percentages of foreign born population (U.S.A. Bureau of the Census 2010). It is mostly this culturally diverse population, but with a spiritual faith, that frequents, finances and builds (has built) these mosques (Survey 2011). The urban mosque or Islamic center is not just a place for worship. There are hall rooms for entertainment such as weddings; they have basketball courts and football pitches which are also used for the game of cricket and a playground for children. These amenities have turned the urban mosques into a lived space – representation (Crang and Thrift 2000). Some of these mosques have schools as well. It can be said that urban mosque has provided this particular (Muslim) population with another "home" or place of comfort besides their residences.

Part of our study also included fieldwork. We visited several mosques in Houston. Our observations of the congregations revealed that that the majority of those who frequented the urban mosque were a foreign born population. The exception was one mosque in the Third Ward of the Nation of Islam (Muhammad Mosque # 45) which is frequented by an African American population. Mosque # 45 was also different from other urban mosques in that it was originally a bank, which has been converted into a mosque. It has a community hall and a school on the same site. Our analysis of locations of the urban mosque and population is primarily based on immigrant population. This is because 63 % of Muslim Americans are first generation immigrants in the U.S (Pew Research Center 2011: 8). The Pew report also reports that 45 % arrived in the U.S. after 1990 and one-quarter of the U.S. Muslims adults since 2000. The majority of the foreign born Muslim Americans are from Middle East and North Africa.

Distance is a variable that is also used widely in geographical analyses (Allen 2000). However, the distance between mosques is not considered for analysis here. Consideration was given to the distance between the home and mosques. We sought to address the question of where the mosque is in relation to the home. Previous studies have looked at the distances between mosques in their attempt to understand city neighborhoods. Raymond (1984) examined the location of mosques built in the fourteenth century to explain the expansion of Cairo, Egypt. Ayhan and Cubukcu (2010) used GIS approach to explain historical urban development in Turkey. The Houston mosques besides being places of worship, serve as education, social and civic centers. Most of the urban mosques have their own websites and have multi-language programs (see www.isgh.org/). This interaction between people located at different spaces reinforces the idea of a mosque as a representative space with spatial practices. Representational space is a lived space (Merriffield 2000: 217).

124.4.2 The Urban Mosque and Arab Population in Houston

Figure 124.6 shows the relationship between the Arab population in Houston zip codes and mosques in the region. It shows a visual spatial correlation between the two variables the area with highest concentration of mosques. The Arab population is concentrated in the Southwest and Northwest sections of Houston. These areas also have the highest concentrations of mosques. As noted earlier, the Northeast and Southeast have the least number of mosques, the zip codes in these areas also have the lowest or no Arab population and least foreign born population. It should be pointed out that the Arab population in Houston is well grounded. "Houston has one of the more settled Arab communities" (Klineberg 2006, 2008). The rootedness of the Arab population is described by demographer Stephen Klineberg (2005): "Arab immigrants arriving in Houston in recent decades tend to be highly educated, with incomes higher than the average American – good predictors of future political involvement." This statement is true of other ethnic groups examined here.

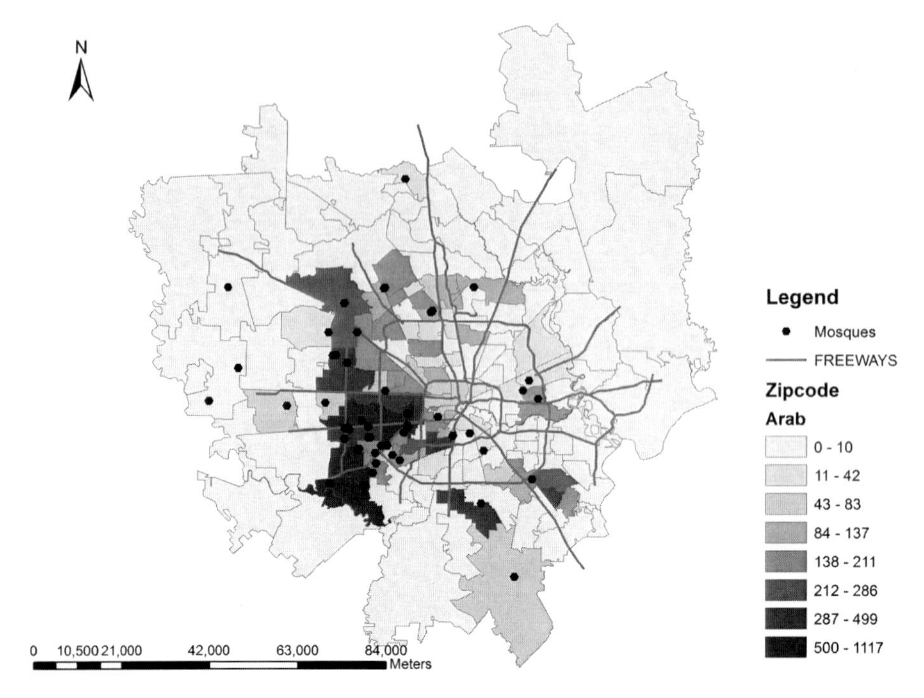

Fig. 124.6 Distribution of the Arab population and the location of mosques in Houston, Texas (Map © Akel Kahera and Bakama Bakamanume 2012)

124.4.3 The Mosques and Population Born in Asia

The relationship between mosques and Asian population shown in Fig. 124.7, probably best illustrates the connection, but must be interpreted with care. Not all Asians are Muslims. Unfortunately, the U.S. Census data used did not have a category for Bengalis, Pakistanis, Malays and Indonesians. So Fig. 124.7, which clearly shows a very strong spatial correlation between Houston's Asian population and location of mosques must be viewed and interpreted with some reservation. It is, nonetheless, a good illustration of the positive relationship between the foreign born populations and mosque sites. The visual image presented on the map points to some geographical agreement between zip codes which are home to the foreign born Asians population cluster of the location of urban mosque. Within the Asian born population we have Pakistanis, Bengalis, Indonesians, and Malays. These four countries are predominantly Islamic; and Indonesia has the highest Moslem population of any country in the world.

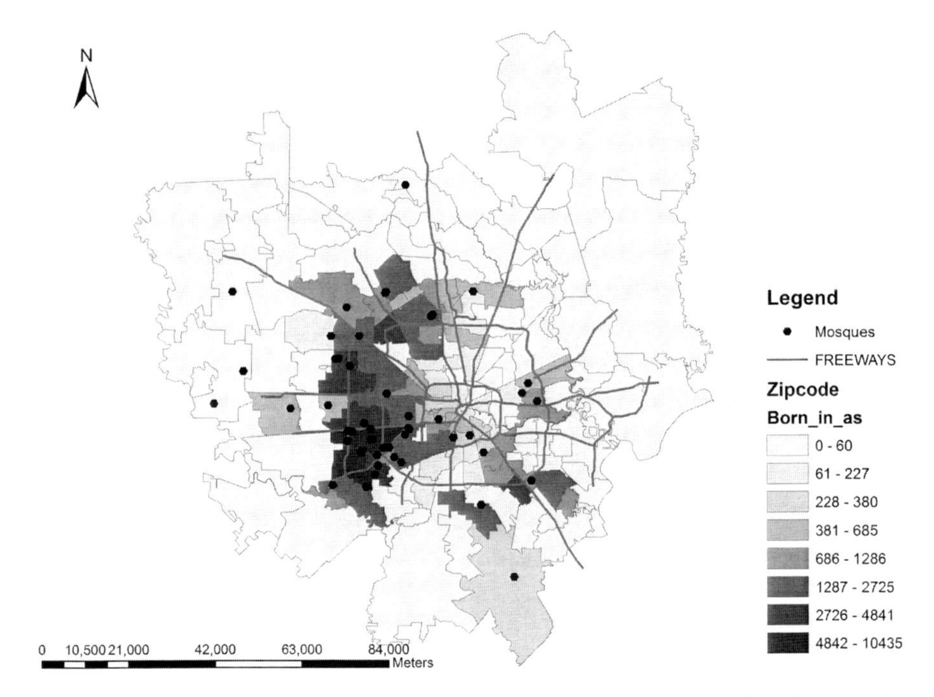

Fig. 124.7 Distribution of the immigrant population born in Asia and the location of mosques in Houston, Texas (Map © Akel Kahera and Bakama Bakamanume 2012)

124.4.4 Turkish Population and Mosques in Houston

An examination of mosque location and Turkish population suggests that the mosques in Houston are concentrated in those zip codes with high percentage of Turkish population. This association was anticipated because the Turkish population is predominantly Moslem; it has also established several schools, one being the Harmony Independent School districts in the country. The Harmony independent school district is unique. Even though, Turkey is a secular country, our observation of the Harmony schools leads to a conclusion that the school allows for what we will call a "better Muslim environment," which is very inviting surrounding for the expression the Muslim religious belief. The school system is secular and enrolls non-Muslims and Muslims.

124.4.5 Mosques and Family Median Incomes in Houston

Is there a relationship between the urban mosque and the income of individual? The study sought to examine this relationship using average median income in zip codes where the mosques are located (Fig. 124.8). Mosques tend to be located in upper

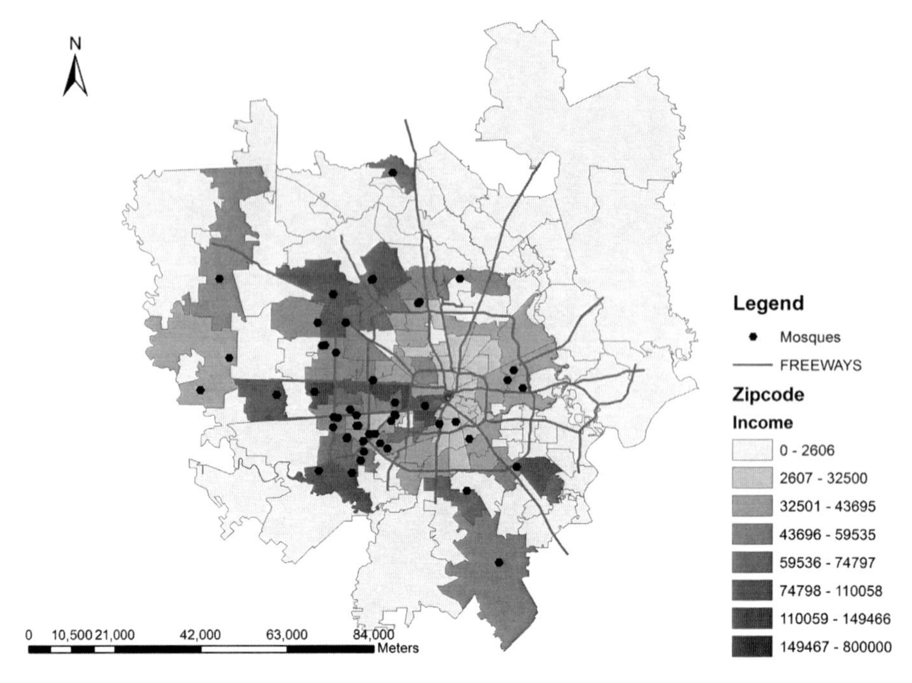

Fig. 124.8 Median family income and the location of mosques in Houston, Texas (Map © Akel Kahera and Bakama Bakamanume 2012)

low, middle, and low upper income zip codes. There is clear absence of the urban mosques in the upper income, and the low income zip codes. In the upper incomes zip codes (River Oaks etc.), the population is predominantly white and probably Christian or Jewish.

In low-income zip codes, the population is mostly Hispanic, African American and some whites. These three groups are predominantly Christian (Survey 2011). The zip codes with the lowest family median incomes tend to be prominently Christian, viz., Catholic. There is one mosque, which serves the predominantly African American zip codes in the Southeast. The Northeast has three mosques; this area extends from Highway 45 South to Highway 45 North. It includes Highway 10 East, Highway 59 North, and the Hardy Toll Road. The zip codes are predominantly Latino in terms of population. The location of mosques may be explained by the fact that the foreign born population which tends to be slightly more Islamic than the general population, also tends to have higher levels of education than the general population (U.S.A. Bureau of the Census 2010). The foreign immigrant population is more qualified and has relatively higher incomes than the general population. It is those with high incomes who financed the building of the urban mosque. The urban mosques in Houston reflect the growing number of Muslim Americans in the region, increasing number of structures in the last decade to what is now is a more complex establishment of the mosque space. Houston has seen a great influx of South Asians, Southwest Asians and Northern Africans (from the Arab countries). The new immigrants are responsible for the mushrooming of the mosque spaces. The mosques are operated mostly by and independent Society (ies). For example, the Islamic Society of Greater Houston operates several mosques in the region. Several of the urban mosques operate facilities dedicated to spiritual and educational outreach both in the Islamic community and the Houston non-denominational population at large. These programs are focused on assisting youngsters with the study of Islamic languages and texts, as well as regular academic courses (Kleinberg, 2006). The urban mosques are "places of knowledge … where things are rendered transparent, intelligible and visible" (Crang 2000: 174).

124.5 Conclusion

Edward Said has reminded us that Foucault was neither a historian, nor a philosopher, but all of those things together (1991: 3). "In short Foucault was a hybrid writer, dependent on—but in his writings going beyond—the genres of fiction, history, sociology, political science and philosophy" (1991: 3). In thinking about space we are drawn to history and the autochthonous conditions of urban and rural Blacks many of whom were the descendants of Muslim Slaves. In the 1940s Malcolm Bell Jr. (1913–2001) photographed and interviewed many coastal Georgia blacks as part of the Works Progress Administration program. The results of this Georgia Writer's Project, headed by Margaret Granger, were published in 1940 and entitled *Drums and Shadows: Survival Studies among the Coastal Georgia Negroes* (Bell et al. 1986; Joyner 2010). The stories are based on interviews with a number of coastal Georgia blacks who were descendants of former slaves who were also Muslims. A few described the daily ritual prayer that was still being performed at the time of the interview. One interview mentions a building that a slave master ordered to be torn down; from the description it could have been intended to be a gathering place (*mus-allah*) for Muslims rather than a domicile.

Likewise the presence of African Muslims in antebellum America has been well documented by Allan Austin, *African Muslims in Ante-bellum America* (1997) and in other texts. The preservation of the unique events of African Muslims in antebellum America, and the identity and the forms of space/power/knowledge is an ongoing project in the areas of history, sociology and plantation archeology. But today one of the most visual expressions of Muslim identity in North American cities is the existence of the urban mosque; examples can found in Washington DC, Boston, Atlanta, New York City, and Houston. These are some of the most populous cities where Muslims reside. The urban mosque in these cities also serves as an Islamic center; it is the heart of the Muslim community, where the faithful gather to engage in communal worship, spiritual retreat, matrimony, education and social activities. The urban mosque has thus far demonstrated its social efficacy in myriad ways. Undoubtedly, the urban mosque has forged various types of favorable alliances between the Muslim community and public life surrounding the mosque. Hence most Americans would be quite surprised to find out that there are an estimated 2,000 mosques that have been built in the United States. "The U.S. Mosque Survey 2011 counted a total of 2,106 mosques; as compared to the year 2000 when 1,209 mosques were counted—representing a 74 % increase from 2000" (Baghby 2012: 4). Similarly in a 2001 study The Mosque in America: A National Portrait, conducted by Bagby, Perl, and Froehle, the study estimated that the total number of Muslims living in the U.S. was between six and seven million and in the breakdown, African-Americans were reported to comprise 42 % of the total; 24.4 % are Indo-Pakistani; 12.4 % are Arabs; 5.2 % are Africans; 3.6 % are Iranian; 2.4 % are Turks; 2 % are from Southeast Asia; 1.6 % are white Americans; 3.2 % are Albanians; and all other groups comprise 5.6 % (Baghby et al. 2001).

As we have noted in the introduction Foucault's *Heterotopias* have opened up a radical new possibility for architectural and social, cultural and political theory. We have attempted to critique the political and philosophical tensions, which we have highlighted, because they have been summarily glossed over by critics who failed to recognize the issue from an entirely different perspective. Hayden's *The Power of Place: Urban Landscapes and Public History* (1997) outlines the elements of space and how they connect people's lives but McRoberts sums it up best, "In urban America religious institutions are woven deeply into the physical and social fabric of the city. In nearly every neighborhood we find temples, churches, synagogues and mosques. These places of worship are perhaps the oldest and most ubiquitous forms of the urban community—the religious congregation" (2005: 1). Drawing on the Muslim, Hindu and Jewish communities in Montreal, Gagnon and Germain discuss "Espace urbain et religion: esquisse d'une géographie des lieux de cultemi-noritaires de la région de Montréal" (2002). Likewise we will also examine some elements of the spatial strategies used by ethno-religious groups in the choice of location of their places of worship in the U.S. and the regulatory framework for the siting and construction of places of worship at the municipal level. Finally, Isin and Siemiatycki *Making Space for Mosques* (2002) argue, "this specific struggle was one of many for Toronto's growing Islamic population seeking appropriate places of worship" (189).

The most salient feature of "Immigrant Islam" is a search for ways to accommodate tradition and modernity while reinventing religious identity, reinterpreting religious practices, and confronting the idiosyncrasies of a secular and politically charged society. In this regard, we embrace Jackson's argument: "At the bottom their depictions point to the fact that … traditionally, this has entailed at least two interrelated challenges. First how are religious communities to relate to the dominant culture? Second how is religion to operate under a secular democratic state?" (2005: 136). Our aim in this chapter has been to document ubiquitous forms of the Muslim Diaspora in the urban community by using GIS mapping and field surveys of the urban mosque in the city of Houston. Examples that we discussed exhibit diverse land-use requirements, functional requirements and unique typological features of urban mosques. The study graphically illustrates the way each urban mosque has responded to an epistemic urban configuration and the unique design problem of facing Makkah (Mecca): 45.362642° from North clockwise. This is a fundamental orientation requirement of all mosques.

In the course of this chapter we extended the range of meanings and interpretation when we quantified a number of activities accommodated by the urban mosque (congregational worship, education and social), the area required and a range of complex aesthetics features. We also provide for further study an analysis of planning and design criteria that include the following: demographics, economic factors, land use, aesthetic and cultural diversity of the Houston Muslim community. The specific study area and site of each example including five existing examples in built up urban areas, and any neighboring areas made an assessment of housing, commercial etc. At the root of the framework of the analysis are the ascribed values of aesthetics that cannot be ignored. Evidence for this comes from the fact that looking for aesthetic satisfaction is tied to Diaspora, demographic and socio/economic factors. This mix is undeniably relevant to Foucault's *Heterotopias*, thus, we have brought to the analysis, the human disposition, the problem of aesthetic interpretation and, in part, matters of language and communication. What does the building communicate to the city observer or the community? Multiple interpretations are presented here of the urban mosque, creating an overall picture of the urban mosque and an analysis of the aesthetic and architectural features. The typical approach for this aspect of the study was based on narrowly defined criteria recorded and cross-referenced as an aesthetic audit of exterior/interior spaces and the environmental aspects of each site. First is to understand the aesthetic character of the urban mosque and the way it operates as a necessary point for verifying our claims; second is to identify urban issues and planning concerns relevant to the land-use in the city of Houston one attends to urban space and the edifice as an object. But to be much more inclusive one must know something about the subject (the community) about their intentions, life and socio-religious conventions. Likewise the West African Muslim Parades in Harlem, New York and religious public ceremonies in Houston, Texas are similar cultural exchanges that are occurring. Today, the principles of belief, order, space and form can, therefore, be perceived as a synthesis, because they are the determinants of practice and the production of space. In this view the function of an edifice alone is static and devoid of a heuristic understanding, which is why we have sought to explain the development of the aforementioned thematic elements and to unmask the underlying formative and generative aesthetic principles that inform the production of space.

One implication of Foucault's *Archeology of Knowledge* (1972) is where he seems to suggest that social life be viewed as being played out in the "spaces of dispersion" (King 1996: 219). It has been our intent in this chapter to capture the spaces of dispersion, fully aware that the process of reporting is far from being value-free. Our objectives are to be seen as an opportunity to study the urban mosque. Furthermore, it will assist future research to identify sites and zones, to map the distribution of mosques and to provide readers with a field survey reports about surrounding land use. The urban character and the planning of urban mosques is important, yet it is impossible within the scope of this chapter to unmask all the links that depend on urban space, religious power and public knowledge. What is being argued is perhaps the difficulty of finding an appropriate design language to explain the urban mosque in the city of Houston and the identity of the forms to be constituted and the function, which they serve, in real spaces engaging both the city and the community. That is why we find the elements employed in the spatial treatment of the urban mosque an exhaustive category of styles; yet, in its simplest function, the mosque is a space for contemplation, repose and communal worship.

Because religious traditions persist, space matters and in thinking about urban space American Muslims have been constructing mosques (*masajid*) and Islamic centers (*marakiz*) for several decades now in towns, cities and neighborhoods where they reside. Because space matters, it is for this reason that mosques in the city of Houston—and in North America—have a two-tiered identity, which changes according to the cultural interaction of the émigré. In this case one might say that the term urban mosque, in its architectural framework, is as associated with memory and as such any style adapted from far away foreign places and cultures. One way of summing up these important cultural tropes would be to suggest that stylistically the design of an urban mosque falls within two common genres: first, a strict adherence to an aesthetic tradition influenced by sign, symbol and building convention, and, second, an attempt at design interpretation employing experimental and popular ideas and resulting in a hybrid image. And finally, a faithful attempt is made to understand its place in modernity, tradition and urbanism. In Houston, Texas this formula holds true with the added proviso that the principles of belief, order, space and form can be perceived as a synthesis of composition and various cultural productions with their forms of oppositions, all which make it possible to study the recurrent elements of space, knowledge and power.

References

Allen, J. (2000). Georg Simmer: Proximity, distance and movement. In M. Crang & N. Thrift (Eds.), *Thinking space* (pp. 80–99). London: Routledge.

Austin, A. (1997). *African Muslims in Ante-bellum America*. London: Routledge.

Ayhan, I., & Cubukcu, K. M. (2010). Explaining historical development using the locations of mosques: A GIS/spatial statistics-based approach. *Applied Geography, 30*, 229–238.

Baghby, I. (2012). *The American mosque 2011*. www.cair.com/Portals/0/pdf/The-American-Mosque-2011-web.pdf

Baghby, I., Perl, P. M., & Froehle, B. T. (Eds.). (2001). *The Mosque in America: A national portrait*. Washington, DC: Council on Islamic Relations.

Barnes, T. J., & Duncan, J. S. (Eds.). (1992a). *Writing worlds: Discourse, text and metaphor in the representation of landscape*. Oxon: Routledge.

Barnes, T. J., & Duncan, J. S. (Eds.). (1992b). Ideology and bliss: Roland Barthes and the secret histories of landscape. In *Writing worlds: Discourse, text and metaphor in the representation of landscape* (pp. 18–37). Oxon: Routledge.

Bell, M., Bell, M., & Joyner, C. (1986). *Drums and shadows: Survival studies among the coastal Georgia Negroes*. Georgia: University of Georgia Press.

Belsy, C. (2002). *Poststructuralism: A very short introduction*. Oxford: Oxford University Press.

Bevan, R. (2006). *The destruction of memory: Architecture at war*. London: Reaktion Books.

Bourdieu, P. (1993). *The field of cultural production*. Cambridge: Polity Press.

Brabant, J.-M. (2007). *Space, Knowledge and power: Foucault and geography* (pp. 25–28). Hampshire: Ashgate.
Crampton, J., & Elden, S. (Eds.). (2007). *Space, knowledge and power: Foucault and Geography*. Hampshire: Ashgate.
Crang, M. (2000). Relics, places and unwritten geographies in the work of Michel De Certeau (1925–86). In M. Crang & N. Thrift (Eds.), *Thinking space* (pp. 170–190). London: Routledge.
Crang, M., & Thrift, N. (Eds.). (2000). *Thinking space*. London: Routledge.
Foucault, M. (1972). *The archeology of knowledge & the discourse on language*. (A. M. Sheridan Smith, Trans.). New York: Pantheon Books.
Foucault, M. (1983). *Beyond structuralism and hermeneutics*. Chicago: The University of Chicago Press.
Gagnon, J. E., & Germain, A. (2002). Espace urbain et religion: Esquisse d'une géographie des lieux de culte minoritaires de la région de Montréal. *Cahiers de Géographie du Québec, 46*(128), 143–163. www.cgq.ulaval.ca/textes/vol_46/no128/02-Gagnon.pdf
Hayden, D. (1997). *The power of place: Urban landscapes and public history*. Cambridge: MIT Press. http://dspace.maag.ysu.edu:8080/dspace/bitstream/1989/1096/2/SCI3-2.pdf.
Isin, E. F., & Siemiatycki, M. (2002). Making space for mosques. In S. H. Razack (Ed.), *Race, space and society: Unmapping a white settler society* (pp. 185–210). Toronto: Between the Lines.
Jackson, S. (2005). *Islam and the Blackamerican: Looking toward the third resurrection*. Oxford: Oxford University Press.
Joyner, C. (2010). *Drums and shadows: Survival studies among the coastal Georgia Negroes*. Los Angeles: Indo European Publishing.
Kahera, A. (2002a). *Deconstructing the American mosque: Space, gender and aesthetics*. Austin: University of Texas Press.
Kahera, A. (2002b). Urban enclaves, Muslim identity and the urban mosque in America. *Journal of Muslim Minority Affairs, 22*(2), 369–380. www.taylorandfrancis.metapress.com/index/8F80UR2015RFAJ2Q.pdf
Kahera, A., Abdulmalik, L., & Anz, C. (2009). *Design criteria for mosques and Islamic centers*. Oxford: Architectural Press.
King, R. (1996). *Emancipating space: Geography, architecture and urban design*. London: Guilford Press.
Klineberg, S. (2005). *Public perceptions in remarkable times: Tracking change through 24 years of Houston surveys*. Houston: Rice University, printing by the Houston Chronicle, November 2005.
Klineberg, S. (2006). The 2006 Houston area survey: The latest findings in the context of a quarter century of Houston survey, April 22, 2006. http://report.rice.edu/sir/faculty.detail?p=78127229FABCE456
Klineberg, S. (2008). Demographic and related economic transformations of Texas: Implications for early childhood education and development. In A. R. Tarlov, & M. P. Debbink (Eds.), *Investing in early childhood development: Evidence to support a movement for educational change* (pp. 159–176). New York: Palgrave-Macmillan. With Steve Murdock.
McRoberts, O. M. (2005). *Streets of glory: Church and community in a black urban neighborhood*. Chicago: University of Chicago Press.
Merrifield, A. (2000). Henri Lefebvre: A socialist in space. In M. Crang & N. Thrift (Eds.), *Thinking space* (pp. 208–228). London: Routledge.
Pew Research Center. (2011). Mainstream and moderate attitudes Muslim Americans: No signs of growth or alienation or support for extremism. *PewResearchCenter*. www.pewresearch.org
Raymond, A. (1984). Cairo's area and population in the early fifteenth century. *Muqarnas, 2*, 21–31.
Said, E. (1991). Michel Foucault 1926–1984. In J. Arac (Ed.), *After Foucault: Humanistic knowledge, postmodern challenges* (pp. 1–11). New Brunswick: Rutgers, The State University.
Schopflin, G. (2000). *Nations, identity, power*. New York: NYU Press.

Siddiquiz, A. (2009). *President, Islamic Society of Greater Houston*. The interview was conducted by Houstonian Corner and published on September 17, 2009.
Survey. (2011). The authors conducted field survey during the research. Mosques in Houston area were identified, geo-coded, and a total 25 were visited by one of the authors or both. The information obtained included mathematical location (latitude and longitude), elevation and pictures. In some cases we talked to the official who oversaw the mosque.
Tuan, Y.-F. (1991). Language and the making of place: A narrative-descriptive approach. *Annals of the American Association of Geographers, 81*(4), 684–696.
U.S.A. Bureau of the Census (2010) Census 2010.

Chapter 125
Sacred Place-Making: The Presence and Quality of Archetypal Design Principles in Sacred Place

Arsenio Rodrigues

125.1 Introduction

What makes place sacred? How does this process unfold within the built environment? What are the contributing factors? Can they be objectified? In order to identify how architecture can embody transcendental qualities, one has to explore multiple ways in which the sacred finds inclusion in *place-making*. To that effect, there should be some discussion on the nature of the sacred and its relationship to place, if we are to create healing and sustainability within the spaces we inhabit. However, according to Meurant (1989), defining the sacred is improper, because definitions imply limits and are not all-inclusive. According to Critchlow (1980), that which is transcendental and all-inclusive is difficult to understand; it requires integral, unified thinking and experience. Concepts of *totality* and *absolute reality* become difficult to comprehend, since our entire education is based upon contention, polarities, and the nature of categorization (Lawlor 1994). It follows that the sacred cannot be sensibly named, expressed, or imaged. Its presence, however, can be apprehended through relationships involving cosmological, metaphysical, ontological and spiritual contexts (Tabb 2006). That which is sacred can be sensed and its presence can be known in place.

According to Tabb (1996), sacred place could express higher intensions, exemplifications, and important cultural values via its physical and spatial characteristics. These characteristics, when uplifted in everyday place, hold the potential of transforming secular place into sacred. Desacralization of place, however, has made it increasingly difficult for modern societies to rediscover existential dimensions of the sacred that were once immediately recognizable and readily accessible to

A. Rodrigues (✉)
School of Architecture, Prairie View A&M University, Prairie View, TX 77466, USA
e-mail: arodrigues@pvamu.edu

S.D. Brunn (ed.), *The Changing World Religion Map*,
DOI 10.1007/978-94-017-9376-6_125

humans of the archaic societies in their everyday places (Eliade 1959). Further, in a time dominated by vapid architectural styles and trends, the re-discovery of sacred design characteristics, which have guided and informed place-making from time immemorial, is often overlooked or forgotten. The motivation for this chapter comes from the need to re-connect with timeless universal themes – *design principles* that seem to contribute to sacred place-making. To that extent, the material presented herein focuses on the secular and spiritual dimensions of architecture and is intended to foster a renewed interest in the phenomenology of sacred place. The chapter is a process of clarification, yet what remains untouched are the elements of complete mystery, something "wholly other" as remarked by Eliade.

125.2 Relevance of Sacred Place to Health and Well-Being

Does the built environment contribute to health? The World Health Organization, a growing number of health-care professionals, and the public today equate good health not merely to the absence of disease, but also to the presence of positive well-being (Brannon and Feist 2002). George Stone (1987) classified available definitions of health into two broad categories: (1) health as an ideal (static) state of positive well-being; and (2) health as a (dynamic) state of consistently moving towards positive well-being. The dynamic definition of health implies a greater holistic approach; it acknowledges health, not merely as a static state of being, but more as a direction on a continuum towards positive well-being. In this context, the traditional twentieth century biomedical model falls short in defining health as a holistic dimension, as it addresses health exclusively in terms of a single condition – the absence of disease. The biomedical model of health has, therefore, given way to the biopsychosocial model – a model that advocates a greater holistic approach towards health, considering not only our biological, but also our social, psychological, physiological, as well as our *spiritual health*. For the purpose of our current investigation, we need only focus on physical-spatial factors (place-related characteristics) that contribute to spiritual health.

Through time, sacred places have remained special and essential to humans. Profanation of place, however, remains pervasive in our current era of secular dominance. This phenomenon is observed through the widespread construction of structures that provide habitation, but fail to address the spiritual health of their inhabitants (Lawlor 1994). According to Venolia (1988), disharmonious environments can contribute to psychological stress and lead to a decline in physiological health. Among various place-types, sacred places are speculated to support healing, help us feel energized, bring about a unique transformation of consciousness, and contribute to transcendental experiences (Swan 1990; Brill 1986). Further, sacred places may also function as fertility, prophecy, and astronomical sites (Steele 1988). According to Eliade (1959), archaic humans settled and founded their world around sacred space, thereby giving meaning to sacred place as the center of their world. By living in close proximity to consecrated space and by merely entering it, humans were ensured of transcending the profane world, thereby, sharing in the world of the sacred.

Several authors have expounded upon the unique nature of sacred place and its subsequent effect on our experience of place. According to Brill (1986), specific emotions triggered and awakened at sacred places include ecstasy, ancient stirrings within the self, feelings of repose, feelings of sensory unification, and a sense of dissolution of the self – emotions that are not only powerful, but also intensely real, human and similar to those experienced by archaic people in their sacred places. Pioneering psychological studies have shown that transcendental experiences are characterized by a sense of ego surrender (James 1902), feelings of unification and harmony with all things, feelings of blessedness or joy, a sense of timelessness and spacelessness, a sense of connecting with some sort of objectivity or ultimate reality, and a sense of divine presence or sacredness (Stace 1960). Research involving neuroscience suggests that sacred experiences relate to elevated states of awareness and feelings of awe (Eberhard 2005). According to Venolia (1988), such experiences associated with feelings of calm, relaxation, and balance are unique characteristics of healing environments. Sacred places, on account of their inherent purpose and functions, therefore, have the potential of contributing significantly to our spiritual health and well-being.

125.3 Theory of Place-Making

According to Eliade (1959: 11), the sacred is the "opposite of the profane," something that reveals and manifests itself as a wholly different order from the profane, thereby making us aware of its reality. He affirms that sacred place by its very nature, implies a superabundance of reality. It is a significant break in the physical and spatial plane of homogenous mundane space whereby, that which is special and extraordinary become elevated. Likewise, Brill (1986) assimilates sacred place to the triumph of order over chaos and formlessness. Both authors emphasize that sacred place-making in its entirety is founded upon the Cosmogony – the paradigmatic act of the creation of the Universe. Being patterned around the cosmic model or primordial myth of creation, it is suggested that sacred places could possibly share a common origin. Sacred place-making, therefore, remains a consecrative act – one that seeks to manifest the hierophany or irruption of the transcendental in the physical terrestrial world.

Being qualitatively different from our other inventory of places, sacred place remains our most special type of place; it symbolizes a break in the homogeneity of mundane space. It is suggested that this qualitative difference is observed at sacred place via the presence of an exclusive set of design principles. Further, it is speculated that these recurring design principles are what make sacred place unique and distinct from our everyday ordinary/mundane place (Brill 1985; Tabb 1996). According to Rodrigues (2008), however, it is not merely the presence of the design principles, but also their quality of expression in space that significantly contributes

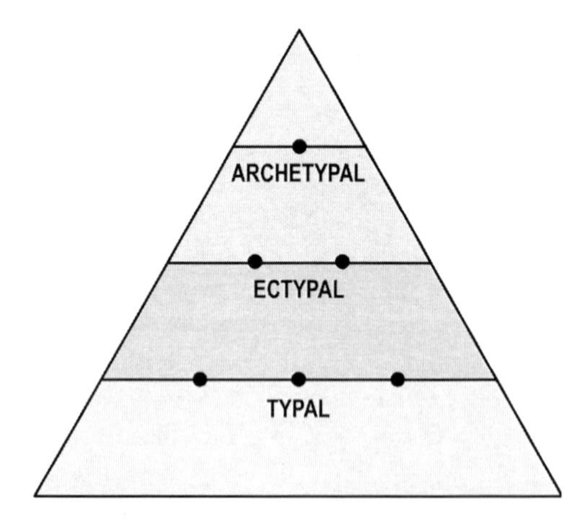

Fig. 125.1 Hierarchical levels of the Tectractys (Source: Arsenio Rodrigues)

to place being experienced as sacred. And since the quality of expression of the principles is synonymous with their evocative qualities, underlying universal themes that govern these principles can have deep impact on how and why we experience sacred place as being different and unique as compared to secular place.

125.3.1 Hierarchical Levels of the Tectractys

There is a correspondence between the internal structure of our being, in Jungian terms, a "collective unconscious" which resonates with the design principles (and their underlying substantive and philosophical meanings) at sacred place. Early Greek philosophy teaches of a multi-layered model or system known as the *Tectractys* (Fig. 125.1), through which the underlying meanings of the design principles may be apprehended. In addition to expressing mathematical and geometrical concepts, the Tetractys embodies the canon of hierarchy. The three horizontal divisions relate to the realms of *archetype*, *ectype* and *type*. Robert Lawlor (1982: 22) provides the following description of how these three levels operate:

> Let us take an example of a tangible thing, such as the bridle of a horse. This bridle can have a number of forms, materials, sizes, colours, uses, all of which are bridles. The bridle considered in this way, is typal; it is existing, diverse and variable. But on another level there is the idea or form of the bridle, the guiding model of all bridles. This is an unmanifest, pure, formal idea and its level is ectypal. But yet above this there is the archetypal level which is that of the principle or power-activity that is a process which the ectypal form and typal example of the bridle only represent.

Within the context of our investigation, the lowermost level *type* can be likened to the physical, sensual and sensible attributes of place – that which can be seen, felt, measured, and directly experienced in place. The intermediate level *ectype* could represent plans, sections, elevations, or other architectural drawings that seek to guide material physicality at the typal level. The highest level *archetype* seeks to exemplify the perfect idea, underlying process, which the typal and ectypal level only seek to represent. The archetype, therefore, occupies both transcendent and ideal orders; it exemplifies the energy and power and is concerned with universal processes or dynamic patterns that have no material carrier. In this context, archetypes manifest expression through their substantive counterparts in the material/typal realm. The Tectractys when viewed through the three hierarchical levels becomes a valuable tool for extracting deeper phenomenological meanings that underlie the processes and sensitivities of sacred place-making. That which remains "unseen" at sacred place could thus be given meaning via the hierarchical levels of the Tectractys.

125.3.2 Place-Making Design Principles

Several authors have theorized the presence of principles/characteristics/patterns/themes in their respective analytical work on place-making. Such material has subsequently contributed to the development of the theory of place-making. More prominent among these include: (1) *The Ten Books on Architecture* (by Marcus Vitruvius Pollio in 1st BC); (2) *The Four Books on Architecture* (by Andrea Palladio in 1570); (3) *The Sacred and the Profane: The Nature of Religion* (by Mircea Eliade in 1959); (4) *A Pattern Language: Towns, Buildings, Construction* (by Christopher Alexander, Sara Ishikawa and Murray Silverstein in 1977); (5) *Using the Place-Creation Myth to Develop Design Guidelines for Sacred Space* (by Michael Brill in 1985); (6) *Chambers of a Memory Palace* (by Charles Moore and Donlyn Lyndon in 1994); (7) *Sacred Place: The Presence of Archetypal Patterns in Place Creation* (by Phillip Tabb in 1996); and (8) *The Nature of Order: The Phenomenon of Life* (by Christopher Alexander in 2002).

What follows, is a brief description of design principles and ways in which they seem to be embodied in place (Fig. 125.2). This is a work in progress and may not be exclusive in its entirety or definitions. The design principles have been derived through analysis of literature, sourced to Eliade (1959), Brill (1985), Lyndon and Moore (1994), Tabb (1996), and Alexander (2002), including personal observations conducted via several case studies. Each of the design principles may be conceived within the hierarchical realms of the Tetractys – archetype, ectype, and type. Within the context of our current investigation, we need only focus on archetypes and their typal counterparts in place. Archetypes associated with individual design principles are introduced as action verbs – 'ing.'

Center (centering, beginning, focusing): Center is expressed via features such as sand paintings, fireplaces, altars, monuments, landmarks, temples, shrines, buildings, rocks, trees, gardens, fountains, streets, markets, town squares. It is a very special place and typically articulates the focal point or geometrical center of the place – physical location associated with intense activity and meaning. To that extent, it may be the focal point of ceremonial experience and express the conceptual essence of place, it forms an expression of the whole and could represent the spatial point

Fig. 125.2 Place-making design principles (Source: Arsenio Rodrigues)

where a connection between the aspirant and the Sacred is most likely to occur. Creating and acknowledging a Center, articulates a reality that *first*, seeks to counter the non-reality of uninterrupted, homogenous, and formless space; and *second*, in doing so, establishes a substantial and fixed location. Sacred place, therefore, embodies an articulated Center that is distinct from its surroundings, has a fixed location, and signifies an absolute beginning. It is likened to the fundamental unit that embodies the coding for manifestation into the world of space and form. Its attributes are, therefore, revealed in the overall character and expression of the place, and its embodiment symbolizes victory over chaos.

Boundary (bounding, containing, enclosing): Boundary is expressed via enclosures – floors, walls, roofs, property/village/naturally occurring geological edges. These enclosures are *first*, substantial, clearly defined, and physically fixed; *second*, possess a fixed relationship with the center (thereby creating a comprehensible surrounding edge); and *third*, define extent (thus resulting in an ordered and distinct domain). This Center-Boundary-Domain relationship remains formless and triangular – it comprises of unity at the center where all parts are unified into one, diversity at the periphery where everything remains distinct, and homogeneity within the domain where parts mediate between unity and diversity. Boundaries may reveal differentiation depending on their orientation with the cardinal directions or other contextual features in the surrounding landscape. In this context, they are responsible for differentiating the coherence of design and providing containment to place. Further, they could be solid or have openings (windows, doors) at specific locations to provide views, and to enable physical movement. The sense of enclosure, therefore, may be substantially complete – oriented openings and implied directions may be present. Of the specified enclosure-types, walls and roofs are typically expressed as distinct Boundaries, where roofs could signify the heavens – roofs are most expressive of our desire to reach that which is transcendental. To that extent, at sacred place, roofs may be absent, open or the most permeable of all specified Boundaries. Floors and walls, on the contrary, are expressed as staunch Boundaries and serve to isolate us from the chaos of the underworld, including the world that extends beyond in the four cardinal directions. Hence, floors and walls are generally absolute, fixed, and impenetrable, except for specifically oriented openings that imply orientation with the cardinal directions or with geologically important features in the surrounding landscape.

Orientation (acknowledging, aligning, responding): Orientation is expressed via significant building alignment, acknowledgement or response on the site with features such as the cardinal directions (north, south, east, west), positions of the sun in the sky, natural contours of the site, nature-related views (seas, mountains, plains, plateaus, etc.), or other important geological features in the surrounding vicinity. It is likened to information or the in-form – the evolution and growth of the *center* into the manifest world. It relates with materialization of the coding contained within the center, expressed physically in the terrestrial world and manifested by means of the cardinal and vertical directions. The impulse of the center to manifest itself outward into the physical world, and the embodiment of our bodies in space, generate the three axes of the body – front-back, left-right, and up-down. The three axes subsequently define the four cardinal directions or north, south, east, west, the upward direction (heavens), and the downward direction (underworld). At sacred place, each of these axes is equally important in generating wholeness to place. Nonetheless these axes are qualitatively different from each other – each of these axes relate with the center, yet encompass different meanings (sunrise-sunset, forward-backward movement, left-right movement, moving upward-downward). In addition, the cardinal axes also acknowledge distinct philosophical teachings related with the four elemental qualities (fire, water, air, earth). By means of these directions (aligned with the center), axes could be pivotal in *first*, articulating the creation of symmetry; *second*, providing initial organization to place; and *third*, generating overall form to place. Orientation, therefore, may provide differentiation and diversity to place, while articulating axes, hierarchy, order, visual symmetry, balance to overall organization, and visual wholeness to place.

Descent (descending, penetrating, grounding): Descent is expressed via features that occupy the under-realm – features that signify a connection and downward gesture with the earth. When acknowledged, these features allow visual or physical descent such as ground and lower floors, foundations, footings, and other features that are controlled and bordered, including wells, water fountains, still pools, and cisterns. These features provide grounding to place and are symbolic of the gravitational energy of the place – they carry with them a sense of grounding that is exchanged and transferred directly into the earth at the lowermost point of the built-form. The vertical axis in the upward direction generates the counter property of reaching downward into the watery chaos of the underworld, a process that is conquered through the place-making process. Descent, therefore, forms an important relationship and transition with the earth, created on account of vertical elements of the building penetrating the earth's horizontal plane. The process of Descent involves a resolve of forceful action, associated with the deep psyche – the instinctual and primitive. Descent therefore, relates with the principle of fecundity – certain ancient sacred sites comprise of earth-wombs with penetrating solar rays. Such places are associated with healing, cosmological or ceremonial principles and functions. To that extent, Descent could mark a place of communal gathering.

Ascent (ascending, levitating, uplifting): Ascent is expressed via verticality or vertical features that allow visual or physical ascent – features that express an upward gesture or connection with the sky. It is physically articulated by providing tall columns, soaring walls, high ceilings, towers, vertically ascending roof lines, pierced roofs or canopies with celestial references, shafts of light from above, or by providing openings toward the sky. Other examples of Ascent include pyramids and ziggurats. It is also observed naturally in the landscape through mountains and ridges. Ascent signifies a path or movement in the upward direction, symbolizing the uplifting force or aspirational source related with the will or energy of life – it acknowledges our need to reach higher, subdue chaos, connect us with the greater cosmos, and subsequently come closer to the sacred. To that extent, Ascent is a part of the vertical axis, connecting the downward direction and middle plane to the upward direction – it articulates a connection or breakthrough that occurs between the realms. In doing so, Ascent provides uplift or levity to place. Hierarchy of space and important public and sacred sites may be identified by means of Ascent – the highest point or peak element delineates hierarchy to the overall form of the structure or organization of place.

Passage (initiating, separating, mediating): Passage is expressed via thresholds of continuity or distinct transitional spaces (doorways, foyers, entrances) between two differing realms. To that extent, it forms the neutral space between two opposing domains – sacred and profane. It is experienced as an actual space with distinct features – it acknowledges the point and place of actual entry into a domain, and demarcates a realm that is distinct from sacred and mundane space. To that extent, the function of Passage is likened to the role of mediation – it functions as the middle realm, between the upward-downward or inside-outside domains. It is physically articulated such that, one can enter and leave sacred space, while partaking in both the sacred and mundane domains. At sacred place, Passage is generated by the dematerialization of boundary walls that separate the inner precinct from the outside world. It is significantly marked to differentiate it from the pragmatic function of mundane doorways. It reinforces the inner process of transition between the sacred and mundane realm by functioning as a preparatory space for profound spatial experience when entering or exiting the realms – it provides a scaled relationship and transition, and appropriate cleansing for initial entry and subsequent penetration into the sacralized realm from mundane space. Further, it is typically large in size to accommodate the divine and godly enhancement that occurs upon exit from sacred place. In this sense, Passage forms a symbolic gesture of a welcome space, while simultaneously providing a gradual and comfortable transition back into the mundane realm. It seeks to bring about transformation and harmonious resonance with movement, by providing a sense of entry, distance, and means of communication between the two modes of being – sacred and profane.

Numeric Order (quantifying, enumerating, symbolizing): Numeric Order is expressed via numerical identity, revealed as pattern in place – acknowledging the recurrence of significant sets of numbers such as the singularity or duality of forms, number of towers, doors, windows, columns, walls, steps. It relates to the Pythagorean school of thought – the belief that numbers are evocative of hidden meanings. Since numbers serve to describe and distill qualitative and quantitative characteristics associated with themselves in actual form and physical detail. Numeric Order when acknowledged, results in the creation of ceremonial architecture. To that extent, it engenders quantitative as well as qualitative character to place. Numeric Order is expressed by acknowledging numbers' one through nine (primary numerical identities), numbers' one through ten (the Tectractys – an ancient Greek study tool), numbers' 11, 12, 16, 19, 22, 360 or other specific sets of numbers that are considered significant in various esoteric and philosophical traditions.

Geometric Order (shaping, proportioning, harmonizing) Geometric Order is expressed by means of shapes that generate the physical form of the structure or built environment. It signifies number expressed as volume in space and pervades all physical entities – it exists across all natural elements and its ordering principle governs the structure of all physical manifestations in space. Geometric Order is suggestive of significant relationship between the measurable and immeasurable numbers. It embodies the transcendental root powers (square root of two, three, and five) – immeasurable numbers that function as geometric metaphors and transformational agents. This transformation occurs in space through three processes – the formative, generative, and regenerative. To that extent, it generates harmonic proportion and progression to physical built-form.

Spatial Order (organizing, articulating, corresponding): Spatial Order is expressed by means of spatial organizations that are successive and rhythmic – circular, linear, radial, triangular, orthogonal, or spiral. The creation of center with subsequent orientation generates the first ordering principle of visual symmetry – Spatial Order, articulated through the fixed relationship of the center with the boundaries. It defines the wholeness of a pattern or the process of cutting-in-half – it articulates the understanding of parts of a pattern by simple division of the whole. To that extent, it transforms chaotic-undeveloped landscape, generates corresponding relationships between successive spaces, and reveals rhythmic order by means of visual symmetry. In this sense, sacred place reveals Spatial Order that suggests victory over chaotic space. Spatial Order can be based upon celestial references such as locations and cycles of the sun, moon, stars, and wind. This play of celestial rhythms within sacred place suggests our very need for spatial order. In the organization of the domain, the three transcendental orders or sacred geometries (square root of two, three and five), can be used to yield expanding or collapsing proportional relationships. Such corresponding relationships express ethereal force lines, directed in the outward direction (as seen in the lotus or mandala). It generates a field-like-effect of multiple centers or events and articulates appropriate meaning, hierarchy, and proportion to the overall pattern of spatial organization.

Anthropomorphic Order (scaling, humanizing, resonating): Anthropomorphic Order is expressed via human references/proportions/behavior that is projected onto inanimate objects, including architecture – the articulation of built-forms and details based upon anthropomorphic attributes or measurements of the human body. It can be articulated through plans, elevations, details or other architectural characteristics. It can also be expressed by means of certain geometric proportions and relationships such as the Golden Mean proportion. The cardinal directions and vertical axis are expressed in the human body – front-back, left-right, and up-down. Architecture, therefore, could be proportioned with reference to various attributes of the human form, scale, or features – the building design can be expressive of body height, facades that represent facial features, curvilinear forms, including details that engender human-like attributes to place. To that extent, Anthropomorphic Order may provide for resonance and harmony between built-form and the human body.

Ordered Nature (nurturing, revitalizing, nourishing): Ordered Nature is expressed via natural features – special plants, trees, gardens, groomed natural ground cover, geological formations, and other landscape features that are controlled and tamed (continually taken care of, by humans). At sacred place, such natural features are bordered and subdued versus the boundless, expansive, unknown, wild, chaotic, unruly, and disordered environment that exists in the surrounding landscape. Nonetheless, in an effort to maintain natural processes and sustain the natural qualities of nature, at times, natural wild areas at sacred place are left undisturbed and unspoiled at specific locations within the precinct. To that extent, nature maintains its natural characteristics, but remains bounded and ordered. In this sense, nature is constantly cared for, and controlled, displaying an image of balance and chaos-under-control, thereby, yielding potency to sacred place. Sacred place therefore, expresses balance between nature and cosmic order, symbolizing the earthly paradise. Ordered Nature is therefore, expressed through the spirit of natural features at the place and could articulate seasonal changes – the natural features at a place can contribute toward the ceremonial ordering of the year at that place.

Celestial Order (sky-referencing, temporal-marking, earth-centered viewing): Celestial Order is expressed via openings or markers that articulate the movement of the sun, moon, celestial objects, constellations, or by means of formal orientations that articulate solstices or equinoxes – orientation of built-form that acknowledges temporal changes or the changing of light. The cosmos expresses an order that is harmonious and whole in context with the Earth. Celestial Order therefore, signifies an Earth-centered perspective. It expresses a connection with, and understanding of the greater cosmos – it relates to the visual experience and comprehension of the greater cosmos or celestial wallpaper. To that extent, it can also be articulated via building ceiling forms, such as domes, vaults, and open-to-sky roofs.

Void (emptying, dematerializing, diffusing): Void is expressed via emptiness – a spatial quality that is infinite in depth and contrasted by buzz and detail in the surroundings. Voids connect with the infinite emptiness of space, and also relate with our inner sense of center. To that extent, Voids are typically embodied at the center of sacred buildings (the altar in a church, mosque or temple), and relate to the idea of deep silence and nothingness. It is likened to the diffusion of the buzz and detail, and destruction of the materialized self-structure, thereby, providing calm. It is functional and experiential in nature, and signifies silence, concentration of simplicity, unclutteredness, clarity, stillness, profound emptiness, and centrality. Further, Voids contrast with the materiality of the surrounding structure and provide myriad opportunities for social and emotional experiences. Centers require the emptiness of the Void, because it collects the energy of the center together within itself and forms the foundational strength of the center. Voids therefore, balance the materialization of delirious detail in the surroundings, by instilling profound emptiness and calm, thereby generating intense life to the boundary.

Ordered Views (marking, connecting, reinforcing): Ordered Views are expressed via openings (doorways, windows, etc.) that enable visual connection between corresponding sacred or unique/significant places. Since sacred places share in each other's space and time, Ordered Views can be expressed by providing specific views to gardens, trees, geologic rock formations, mountains, water, etc., that sustain or enhance the integrity of experience within the place. Openings that provide direct visual interaction, between corresponding sacred places or between sacred places and other special features are, therefore, intentionally provided. At times, openings are avoided to restrict views that would otherwise negate the experience within the place – direct visual connection between sacred and mundane space is avoided. The absence of openings is, therefore, also suggestive of Ordered Views. Openings are therefore, restricted, limited, or strategically positioned along the boundaries to enable sacred place to maintain and reinforce its sacrality, thus keeping it distinct from the mundane world.

Materiality (generating, manifesting, realizing): Materiality is expressed via construction components – brick, stone, wood, cement, steel, ceramic tile, plaster, glass, etc., that generate the actual physical built-form. Materiality relates to the physical state or the quality of being material – the materialization of substance in space that result in the expression of physical built-form. Materials are revealed by light via their textures and formal characteristics. At sacred place, they are typically rare (not easily obtained), cumbersome to work with (difficult to move), and distinct from everyday building materials (different from everyday common-place materials). In addition, these materials are carefully crafted, far from their natural state, and resistant to erosion brought about by natural forces. They maintain their formal integrity and physical order – the making of the whole from parts, including their joinery is suggestive of order. The selection and placement of building materials at sacred place is therefore suggestive of struggle and sacrifice. To that extent, Materiality at sacred place signifies the cosmic struggle or victory over chaos – the triumph of order over formlessness and chaos.

Water (permeating, conforming, cleansing): Water is expressed via features such as pools, ponds, fountains, cascades, cisterns, brooks, streams, lakes, rivulets, and rivers. Water is given definition by the manner in which it is contained. It has no fixed form itself, but is given meaning and shape by materials and physical characteristics that make up its surroundings. Significant qualities related with water include clarity, purity, freshness, and power – attributes that evoke specific emotional responses in place, which contribute to the overall experience of the place. Water's clarity can be expressed by means of its depth and its power to reflect. Its purity is employed to literally cleanse oneself physically, mentally, and spiritually, prior to, or upon entering a place. Its kinesthetic quality of freshness could provide a cooling effect in hot weather or warmth when it is cold. Its power is observed through its presence (hurricanes and floods) and through its absence (droughts). In addition, water lends essential qualities such as movement, stillness, and sound when embodied in place. Its movement could take the form of fountains or meandering streams running along paths that lead people visually and physically from one point to another in place, while its stillness can be observed by means of pools and cisterns that provide for calm and contemplation. Different types of sounds, ranging from active and vigorous to reflective and silence can be generated with water features, ranging from energetic waterfalls to still pools.

Light (illuminating, revealing, contrasting): Light is expressed via luminance that reveals form in space. At sacred place, light is typically provided from above and may provide significant contrast with surrounding darkness, enable orientation with the cardinal directions, and/or demarcate the passage of time (with movement of the sun across the sky). With the rising of the sun each day, light enables us to experience the changing world or passage of time, until darkness sets in. In this sense, the daily cycles of day and night signify the unending cosmic struggle. To that extent, the absence of light (darkness) also becomes an essential quality of light.

Ceremonial Order (embodying, ritualizing, celebrating): Ceremonial Order is expressed via spaces that enable ceremonies, meditation, prayer, and other temporal and seasonal celebrations, including ritual and consecrative acts in place. It facilitates connection between the self and physical place, resulting in the creation of boundary and orientation with the geometry of the place. To that extent, Ceremonial Order functions as a trigger for inducing transcendental states of consciousness, such that the wholeness of the place and its centrality are experienced. In doing so, it engenders communal presence to place, while unifying our consciousness with the spatial wholeness of place. The completion of construction of sacred place is typically marked by ceremonial events that signify an absolute beginning, reinforce us and remind us of the beginning of all things. Such consecrative acts are likened to the divine repetition of victory over chaotic space. The finish of construction of sacred place, therefore, signifies the reality and enduringness of our efforts in establishing place for habitation.

125.4 Presence and Quality of Design Principles at Sacred and Secular Place

In a doctoral dissertation conducted by the author (Rodrigues 2008), it was found that it is not merely the presence of the design principles, but also their quality of expression that contributes to place being experienced as sacred. Rodrigues hypothesized that built environments which possess a higher presence and higher quality of expression of the design principles are more likely to be experienced as sacred, than built environments with a lower presence and lower quality of expression of the design principles. To test this hypothesis, qualitative and quantitative research methods were utilized at a selected sacred and secular building – Rothko Chapel and Contemporary Arts Museum (Fig. 125.3). Both settings serve as museums and are located in Houston, Texas. Rothko Chapel is an acknowledged sacred building and houses a group of 14 paintings by Mark Rothko. The paintings are exhibited along the periphery of the interior octagonal-shaped plan of the Chapel. Besides exhibiting Rothko's work, the Chapel functions as a place for private meditation, common worship, and hosting colloquia related with philosophical and religious themes. In contrast, Contemporary Arts Museum is a secular building, dedicated to exhibiting contemporary art to the public.

Fig. 125.3 View of sacred Rothko Chapel (*left*) and secular Contemporary Arts Museum (*right*) (Photos by Arsenio Rodrigues)

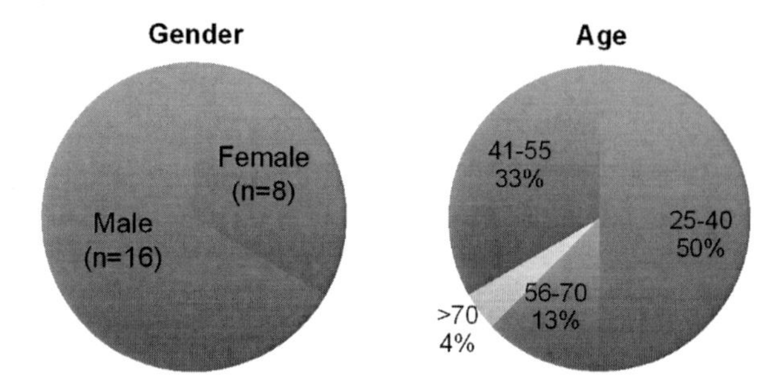

Fig. 125.4 Age and gender distribution of questionnaire participants (Source: Arsenio Rodrigues)

The purpose of the study was to examine differences in the presence and quality of expression of design principles between the two locations. A total of 48 questionnaires (24 at each setting) were administered to 24 participants at the sacred and secular place. Questionnaire items were specific in their usage of architectural language – completing the questionnaire required participants having an architectural background. The sample population for the questionnaire was comprised of architects from firms in Houston, providing basic architectural services and specializing in the design of both, religious and secular facilities. A total of 90 architecture firms in Houston were contacted. Of these, 24 firms (architects) agreed to participate. Of the 24 Houston architects, 16 participants were male, while 8 participants were female (Fig. 125.4). A total of 12 participants were in the age group of 25–40 years, 8 participants were in the age group of 41–55 years, 3 participants were in the age group of 56–70 years, and 1 participant was in the age group of 71 years or above (see Fig. 125.1).

The questionnaire was comprised of open-ended as well as multiple choice questions. The presence of the design principles at the sacred and secular building was scored on a 5-point Likert scale – very low, low, uncertain, high, and very high. Similarly, to assess the quality of expression of the design principles at the sacred and secular building, questionnaire responses allowed for scoring also on a 5-point Likert scale – very low, low, intermediate, high, and very high. During data analysis, relative frequencies were calculated for multiple-choice answers, while open ended questionnaire items were subjected to inductive content analysis, *first*, reading responses to identify emerging categories and; *second*, coding for category inclusion.

125.4.1 *Difference in Presence of Place-Making Design Principles*

Based on the opinion of questionnaire participants, a comparison of the presence of design principles between Rothko Chapel and Contemporary Arts Museum is shown in Fig. 125.5.

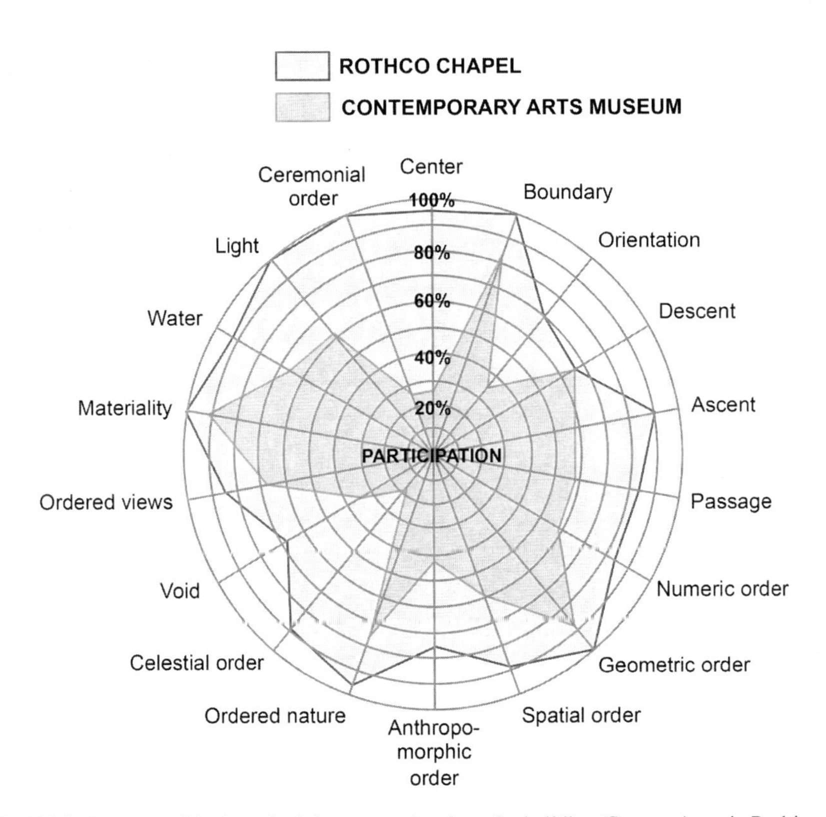

Fig. 125.5 Presence of design principles at sacred and secular building (Source: Arsenio Rodrigues)

At Rothko Chapel, the presence of 14 design principles – Center, Boundary, Ascent, Passage, Numeric Order, Geometric Order, Spatial Order, Ordered Nature, Celestial Order, Ordered Views, Materiality, Water, Light, and Ceremonial Order was found to be *very high*, while the presence of the remaining four design principles – Orientation, Descent, Anthropomorphic Order, and Void was found to be *high*. At Contemporary Arts Museum, the presence of three design principles – Boundary, Geometric Order, and Materiality was found to be *very high*, while the presence of five design principles – Descent, Ordered Nature, Ordered Views, Water, and Light was found to be *high*. The presence of four design principles – Center, Orientation, Void, and Ceremonial Order was found to be *low*, while the presence of one design principle – Celestial Order was found to be *very low* at Contemporary Arts Museum. The presence of five design principles – Ascent, Passage, Numeric Order, Spatial Order, and Anthropomorphic Order remained *uncertain* at Contemporary Arts Museum.

The questionnaire results indicated that the presence of all 18 design principles at Rothko Chapel was higher than the presence of their counterparts at the Contemporary Arts Museum. The difference in percentage values of the presence of design principles at Rothko Chapel over Contemporary Arts Museum are shown in Fig. 125.6. In the figure, the design principles are arranged in descending order,

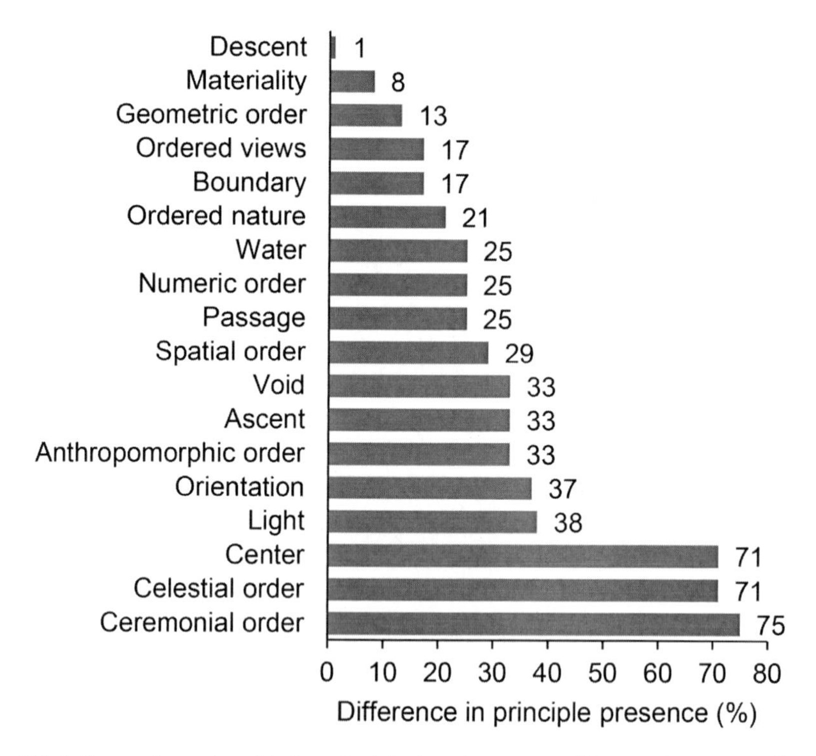

Fig. 125.6 Descending order of design principles based on percent difference in presence (Source: Arsenio Rodrigues)

based on difference in percentage values (associated with presence). The differences in the % age values of the presence of *Ceremonial Order* was highest, while *Descent* displayed no difference in percentage values of presence.

125.4.2 Difference in Quality of Expression of Place-Making Design Principles

Based on the opinion of questionnaire participants, a comparison of the quality of expression of design principles between Rothko Chapel and Contemporary Arts Museum is shown in Fig. 125.7. At Rothko Chapel, the quality of expression of 12 design principles – Center, Boundary, Descent, Ascent, Numeric Order, Geometric Order, Spatial Order, Ordered Nature, Ordered Views, Materiality, Light, and Ceremonial Order was found to be *very high*, while the quality of expression of six design principles – Orientation, Passage, Anthropomorphic Order, Celestial Order, Void,

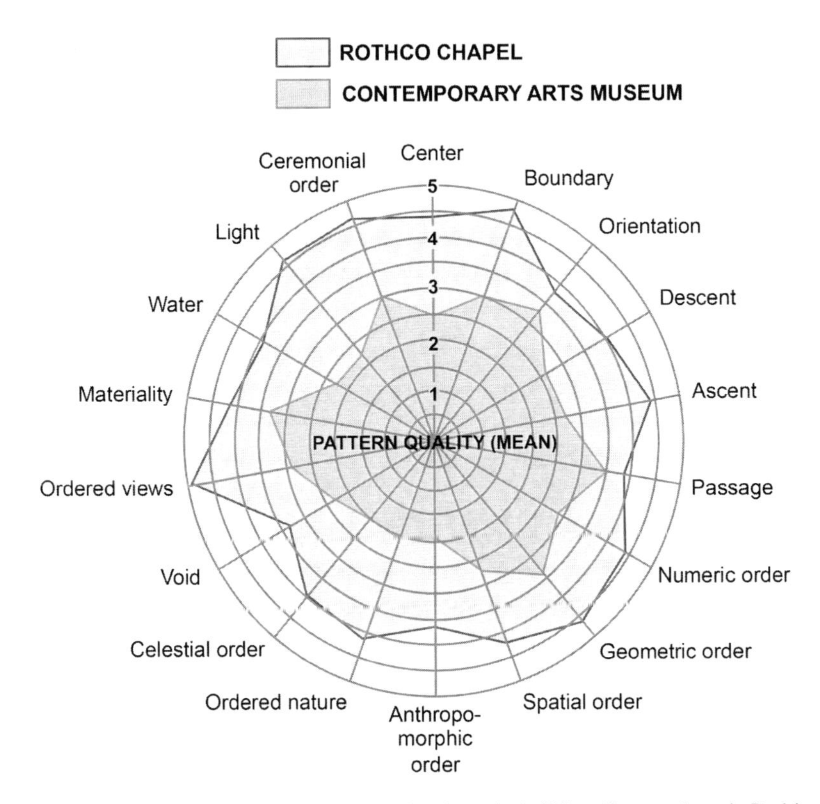

Fig. 125.7 Quality of design principles at sacred and secular building (Source: Arsenio Rodrigues)

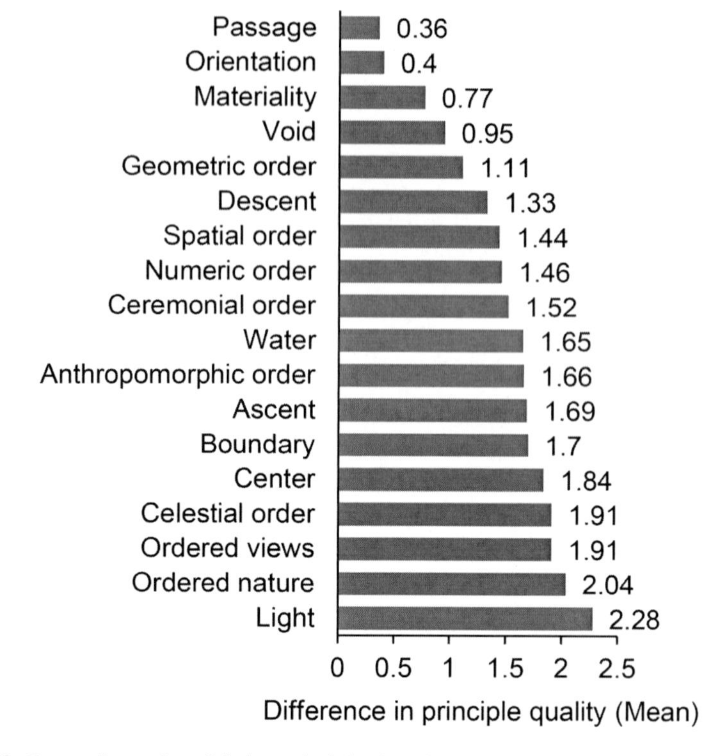

Fig. 125.8 Descending order of design principles based on mean difference in quality (Source: Arsenio Rodrigues)

and Water was found to be *high*. At Contemporary Arts Museum, the quality of expression of six design principles – Boundary, Orientation, Passage, Geometric Order, Materiality, and Ceremonial Order was found to be high. The quality of expression of 11 design principles – Center, Descent, Ascent, Numeric Order, Spatial Order, Ordered Nature, Celestial Order, Void, Ordered Views, Water, and Light was found to be *intermediate*, while the quality of expression of one design principle – Anthropomorphic Order was found to be *low* at Contemporary Arts Museum.

The questionnaire results indicated that the quality of expression of all 18 design principles at Rothko Chapel was higher than the quality of expression of their counterparts at Contemporary Arts Museum. The difference in mean values of the quality of expression of design principles at Rothko Chapel over Contemporary Arts Museum is shown in Fig. 125.8. In the figure, the design principles are arranged in descending order, based on difference in mean values (associated with quality). The difference in mean values of the quality of expression of *Light* was highest, while the difference in the quality of expression of *Passage* was lowest.

125.5 Development of Place-Making Design Principle Matrix

Research findings produced through Rodrigues' doctoral dissertation were utilized to develop a *Place-Making Design Principle Matrix* (Fig. 125.9), a tool meant to serve as a guide for creating place that is meaningful and extraordinary. The Matrix is composed of 18 design principles arranged in two concentric layers around *Unity* at the center. The hierarchical ordering of place-making design principles within the Matrix was based on differences in their presence and quality of expression at the sacred and secular buildings. A scoring system of values from 1 to 18 was used to determine the hierarchical ordering of place-making design principles within the Matrix. Each of the 18 place-making design principles was assigned two scores, ranging from 1 to 18 (one score for its presence and one score for its quality of expression) based on its order of listing in Figs. 125.6 and 125.8. The two scores

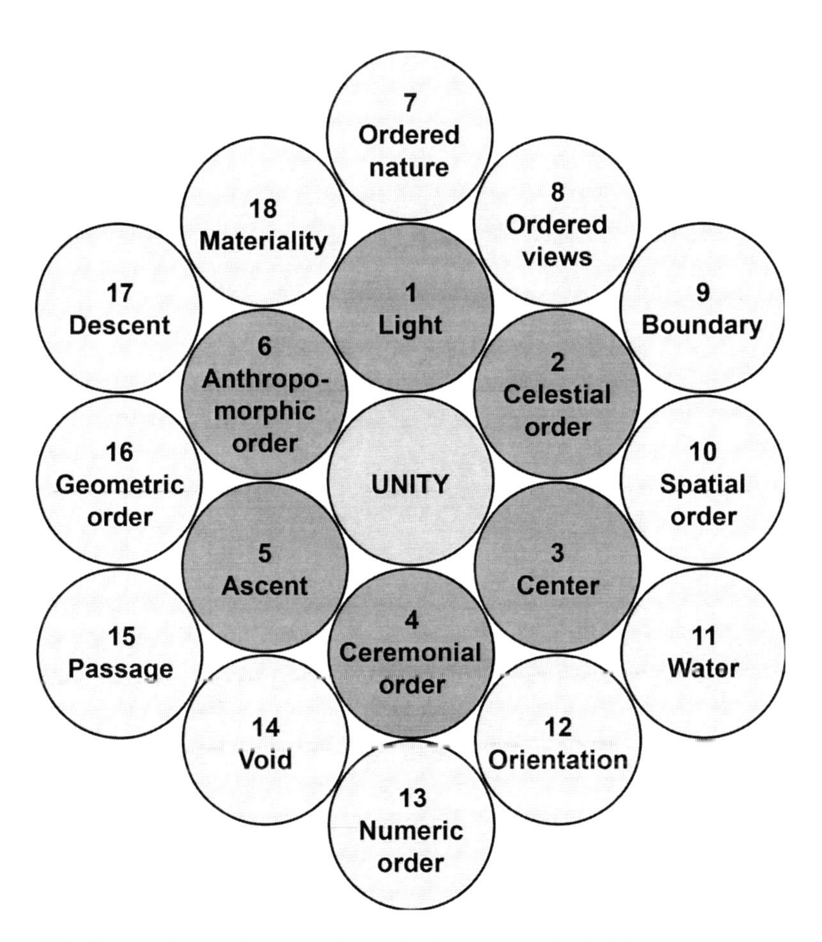

Fig. 125.9 Place-making design principle matrix (Source: Arsenio Rodrigues)

Table 125.1 Presence scores, quality scores, and total scores for place-making design principles

Principle	Presence value	Quality of expression value	Total score	Hierarchical order in matrix
Light	15	18	33	1
Celestial order	17	15	32	2
Center	16	14	30	3
Ceremonial order	18	9	27	4
Ascent	12	12	24	5
Anthropomorphic order	13	11	24	6
Ordered nature	6	17	23	7
Ordered views	4	16	20	8
Boundary	5	13	18	9
Spatial order	10	7	17	10
Water	7	10	17	11
Orientation	14	2	16	12
Numeric order	8	8	16	13
Void	11	4	15	14
Passage	9	1	10	15
Geometric order	3	5	8	16
Descent	1	6	7	17
Materiality	2	3	5	18

Source: Arsenio Rodrigues

(presence score and quality of expression score) were then added to determine a total score for each place-making design principle as shown in Table 125.1. The total score determined the hierarchical order of each of the eighteen place-making design principles within the Matrix.

The hierarchical ordering of place-making design principles in the matrix is a work in progress and could function as an indicator of the importance of individual design principles in contributing to sacredness of place. It is likely that the presence and quality of expression of place-making design principles in the middle layer of the matrix (in descending hierarchical order – Light, Celestial Order, Center, Ceremonial Order, Ascent, and Anthropomorphic Order) have a greater impact in contributing to sacredness of place, than the presence and quality of expression of the remaining 12 place-making design principles in the outermost layer of the matrix. In this sense, it is speculated that the presence and high quality of expression of *Light* may have greater impact in contributing to sacredness of place as compared to the presence and quality of expression of *Materiality. Unity* signifies the experience of "wholeness" or "oneness" between all place-making design principles in the Matrix. It is the integral and meaningful unification of all 18 design principles that yield wholeness to place. It is important to note that research on additional case studies in other parts of the world might help to validate universal applications of the findings of Rodrigues' work. Examples of additional case studies might include Notre Dame du Haut Chapel (Ronchamp, France), Bahai Lotus Temple (New Delhi, India) and Jubilee Church (Rome, Italy).

125.6 Relevance of Design Principles in Place-Making

The contents of this chapter are intended to reinvigorate a dialogue on sacred place-making. It is a small step in re-evaluating the process, goals, and status of using specific principles in place-design as an activity. The design principles, via their presence and quality of expression, seem to contribute tremendously to creating a sense of place. Via their presence, they seem to provide fundamental qualities to place, while their quality of expression engenders distinguishing attributes to form and space. It is important to note, however, that other design principles may be important, which may not be addressed in this study – design principles examined in this chapter may not form an exclusive list. In focus group discussions, conducted by Rodrigues (2008), it was revealed that seven additional characteristics contributed to sacredness experienced at Rothko Chapel. These include: (1) Quality of Sound; (2) Presence of Exhibits/Objects; (3) Presence of Path/Procession; (4) Intentions of Architect/Designer; (5) Sense of Timelessness; (6) Presence of Deity; and (7) Sense of Unity. The speculation of the embodiment of these and other design principles is intended to form a bridge between the contemporary place-making process and evolving esoteric wisdom. It is not intended to de-mystify the sacred, but rather intended to indicate the formal design characteristics which can help enable our connection and experience of place.

At times, spirit and matter are dislocated from one another giving distance or separation between the sacred and profane; while other times they seem to infuse into one another are giving rise to a dynamic and vital sacred design. The design principles, in this regard, could embody the coding for sacral manifestation. They seem to resonate with the ethereal recesses of the human mind. To this extent, the design principles seem to be inextricably related not only to architectural creativity but also to archaic and vernacular ideologies. Each of the design principles, therefore, is intended to serve as a guide in creating a more meaningful and memorable place, to the extent where, a sense of *Unity* may be achieved via the containment and agglomeration of all the design principles into a memorable whole. It is through the lens of the design principles that the qualitative nature of place may be understood. When meaningfully embodied and uplifted in place, the design principles seem to function as sentient and mnemonic devices in architecture – archetypes that help us re-remember our quest for the most exemplary model of place – place that is sacred and place that heals. To discover and experience all the design principles while they dissolve into the wholeness of place is truly a wonderful experience, for it is the purpose of sacred place to bring us back into the living universe – meaning "all taken together."

References

Alexander, C. (2002). *The nature of order: The phenomenon of life*. Berkeley: Center for Environmental Structure.

Brannon, L., & Feist, J. (2002). *Health psychology: An introduction to behavior and health*. Belmont: Wadsworth/Thomson Learning.

Brill, M. (1985). Using the place-creation myth to develop design guidelines for sacred space. Urbana: University of Illinois. *Proceedings of the Council of Educators in Landscape Architecture's Annual Conference. September,* pp. 17–27.

Brill, M. (1986). The mythic consciousness as the eternal mother of place-making. Lawrence, KS: University of Kansas, *Proceedings of the Built-form and Culture Research's Conference, November,* pp. 39–48.

Critchlow, K. (1980). *Lindisfarne Letter 10, Geometry and Architecture: What is sacred in architecture?* West Stockbridge: Lindisfarne Association.

Eberhard, J. (2005). Sacred spaces: A neuroscience perspective of sacred spaces. *American Institute of Architecture Journal, 1*(5), 6–7.

Eliade, M. (1959). *The sacred and the profane: The nature of religion*. New York: Harcourt.

James, W. (1902). *Varieties of religious experiences*. New York: Modern Library.

Lawlor, R. (1982). *Sacred geometry: Philosophy and practice*. New York: Thames and Hudson.

Lawlor, A. (1994). *The temple in the house: Finding the sacred in everyday architecture*. New York: G. P. Putnam's Sons.

Lyndon, D., & Moore, C. (1994). *Chambers for a memory palace*. Cambridge: MIT Press.

Meurant, R. (1989). *The aesthetics of the sacred: A harmonic geometry of consciousness and philosophy of sacred architecture*. Whangamata: The Opoutere Press.

Rodrigues, A. (2008). *The sacred in architecture: A study of the presence and quality of place-making patterns in sacred and secular buildings*. College Station: Texas A&M University, Department of Architecture. Unpublished Ph.D. dissertation.

Stace, W. (1960). *Mysticism and philosophy*. New York: Lippincott.

Steele, J., V. Brenman, & F. Gerceker. (1988). *Geomancy of consciousness and sacred sites* (Film). (Available from Trigon Communications, Inc., P.O. Box 1713, Ansonia Station, New York, NY 10023).

Stone, G. (1987). *Health psychology: A discipline and a profession*. Chicago: University of Chicago Press.

Swan, J. (1990). *Sacred places: How the living earth seeks our friendship*. Santa Fe: Bear and Company.

Tabb, P. (1996). *Sacred place: The presence of archetypal patterns in place creation*. Denver: Academy for Sacred Architectural Studies.

Tabb, P. (2006). *First principles: Architecture of the unseen*. Denver: Academy for Sacred Architectural Studies.

Venolia, C. (1988). *Healing environments: Your guide to indoor well-being*. Berkeley: Celestial Arts.

Chapter 126
Reinventing Muslim Space in Suburbia: The Salaam Centre in Harrow, North London

Claire Dwyer

126.1 Introduction

In January 2011 members of a Shia Ithna'ashari Muslim community in the suburb of Harrow in North London were finally given planning permission for the realization of a new building, the Salaam Centre, which will replace the converted NAAFI[1] hut which they have used for the last twenty years. Designed by award winning architects, Mangera Yvars, the Salaam Centre has been conceived as a "shared space for the community" (The Salaam Centre) and the planned building invites visitors to experience a modern and contemporary space of worship which also offers leisure and social facilities. The innovative design of the building incorporates influences from Islamic heritage with modern design drawing particularly on ideas about light and space. Woven into the façade of the building are patterns which connect the migratory journeys of the Shia Ithna'ahsaris and the suburban vernacular landscape of the northwest London (Fig. 126.1).

The Salaam Centre is perhaps the most innovative in a range of spectacular new religious buildings which have emerged in London's suburbs in the last two or three decades. Celebrated examples in north and west London include the Swaminaryan Hindu Temple in the suburb of Neasden, the Mohammedi Park Mosque in Northolt, the Shri Guru Singh Sabha Sikh Gurdwara in Southall, the Shree Sanatan Hindu Temple in Wembley and further out towards the green belt in Potters Bar the Shikharbandhi Jain Temple (Fig. 126.2). New religious buildings are often described as contested spaces challenging established narratives of urban planning or normative framings of appropriate architecture. Geographers of religion have explored the processes by which new religious buildings have been constructed and their presence

[1] The Navy, Army and Air Force Institutes (NAAFI) was set up by the British government in 1921 to provide recreational establishments for service personnel and their families.

C. Dwyer (✉)
Department of Geography, University College London, London WC1H OAP, UK
e-mail: claire.dwyer@ucl.ac.uk

S.D. Brunn (ed.), *The Changing World Religion Map*,
DOI 10.1007/978-94-017-9376-6_126

Fig. 126.1 The Salaam Centre, Harrow (Image courtesy of Mangara Yvars Architects, www.myaa.eu, used with permission)

in urban and suburban landscapes negotiated. In this chapter I use the case study of the Salaam Centre in Harrow in northwest London to extend and develop these debates. Within the context of my ongoing interest in the theorizing of religious landscapes in the suburbs, I suggest that the vision and architectural design of the Salaam Centre heralds the emergence of new forms of hybrid religious space.

126.2 New Religious Buildings and Suburban Geographies

For geographers of religion the study of the establishment, built form and purpose of formal and congregational spaces of worship has remained a significant topic of interest despite important calls for attention both to more informal, domestic and improvised places of worship (Kong 2010; Gale 2012) and to everyday, embodied practices of religion and spirituality (Dewsbury and Cloke 2009; Holloway 2006, 2011). Study of formal, public worship spaces remain interesting because of the insights they offer into how faith communities in different places are actively involved in contesting their presence in public space. While geographers have traced disputes surrounding places of worship for different faith communities (Germain and Gagnon 2003; Hackworth and Stein 2011; de Witte 2011; Naylor and Ryan 2003), it is the building of mosques, particularly by migrant and post-migrant communities in contexts such as Australia, North America and Europe, which has attracted the most attention (Dunn 2005; Ehrkamp 2012; Gale 2004, 2005, 2008; Isin and Siemiatycki 2002). Unsurprisingly, research has emphasised the specific opposition faced by Muslim groups in building mosques within a wider context of Islamophobia. Opposition to new religious buildings is usually expressed indirectly

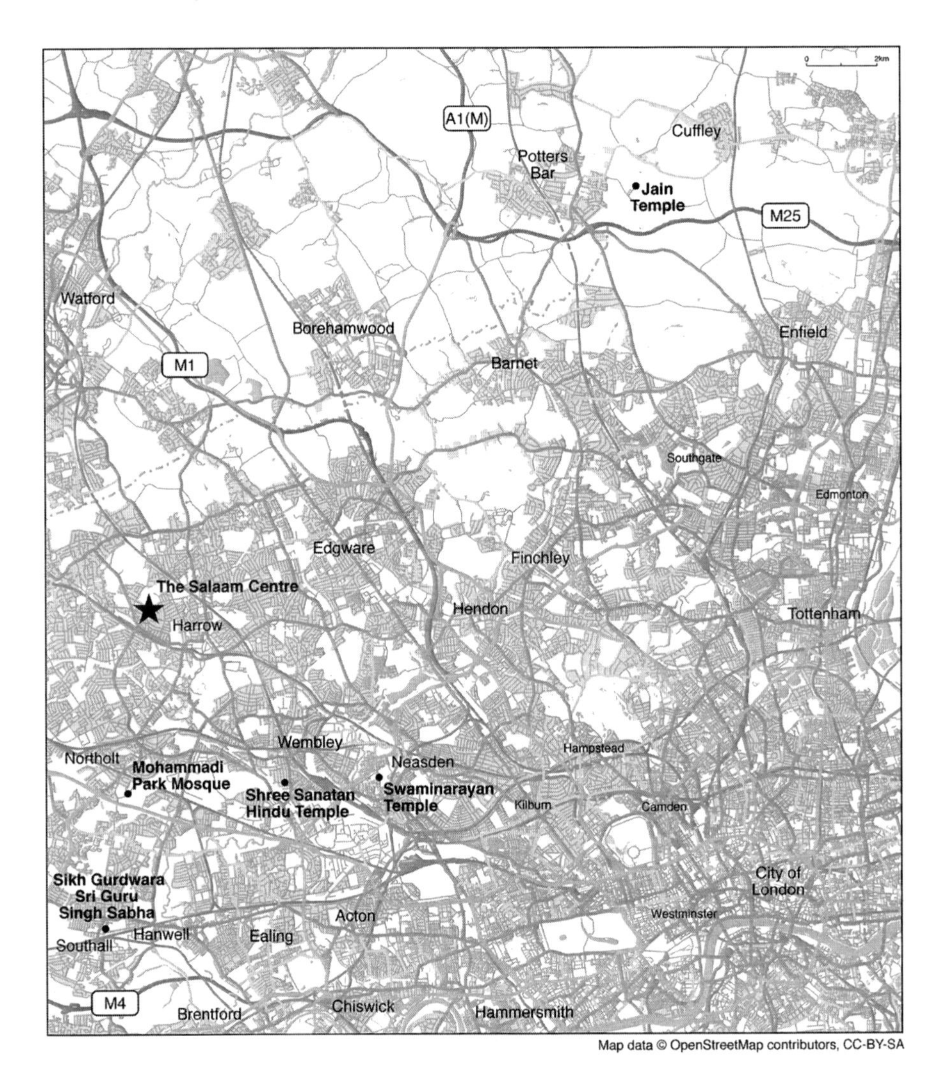

Fig. 126.2 Sites of new religious buildings in London's suburbs and the site for the Salaam Centre (Map prepared for the author by Miles Irving, UCL; data © OpenStreetMap contributors, CC-BY-SA)

through what Gale and Naylor (2002) refer to as "the subjective problematic of amenity" a focus on issues such as noise and disturbance to local communities and or concerns about increased traffic or car parking. However sometimes objections will be raised about the distinctive (and "alien") architecture of the building and its unsuitability for a specific urban or suburban location (Naylor and Ryan 2002). In the most extended analysis of mosque building in the UK, Richard Gale's work in Birmingham (2004, 2005, 2009) emphasises the institutional contexts within which planning for religious worship is negotiated and how support from local political actors has been crucial in enabling mosques to be built or in the procurement of a suitable site (see also McLoughlin (2005) on mosque building in Bradford). In my discussion in this chapter of the Salaam Centre in northwest London, I extend these debates focusing particularly on two issues concerning the architectural style of new religious buildings, a topic which has arguably garnered less attention from geographers of religion.

The first is the interesting typology of "new religious landscapes" in England first developed by Peach and Gale (2003) which postulates a "four-stage development cycle" (p485) for minority faith groups developing places of worship which they summarise as "denial, confrontation, accommodation, celebration." As Peach (2012) elaborates this begins with the initial tacit or illegal use of domestic buildings as places of worship before expanding to larger scale conversion of existing buildings (sometimes older religious buildings but often industrial premises). While the third and fourth stages of the "development cycle" involve the development of purpose built premises, Peach and Gale (2003) distinguish between those which are "hidden" and those which are "embraced and celebrated" (p.485). For the former they include elaborate and ornate new buildings like the Shri Swaminaryan Temple in north-West London whose eventual location in the "downmarket site of inner suburban" (p. 482). Neasden followed difficulties in gaining planning permission in more prominent and affluent locations. Similarly the Mohammedi Park Mosque in nearby Northolt is located on an industrial estate next to the canal screened by town houses which are also built and owned by the Dawoodi Bohras community (Eade 1993, 2011; Crinson 2002). In contrast the prominence of some new buildings, such as Birmingham's Dar ul Uloom Islamia Masjid built on a busy arterial entrance to the city, is seen to reflect Birmingham city council's enthusiasm for multiculturalism (Gale 2004) although here too Gale (2012) traces the ways in which other Birmingham mosques, such as the Jame Masjid (formerly the Saddam Hussain Mosque) in Handsworth only gained planning permission because it was "screened" by a flyover.

In their discussion of new religious landscapes in Britain, Peach and Gale (2003) reflect on the aesthetics of Islamic, Sikh and Hindu architectural styles which are seen to challenge the "normalising language of planning discourse" in their "architectural forms, building materials and decorative colours" (2003: 286). As such new religious buildings, even those which are admired or celebrated as emblems of British multiculturalism, are cast firmly outside existing frameworks of vernacular architecture. The idea of new religious buildings as "exotic" or incongruous is often particularly marked in relation to their suburban location. So the Neasden Temple was variously described as "Neasden's Taj Mahal" set in contrast to the "cut price, crinkly tin shopping shed" and "grim-grey, pebble-dashed suburban streets" (The Guardian 28th July, 1995, cited in Dwyer et al. 2012). As we have recently argued (Dwyer et al. 2012) such representations draw on familiar tropes of the suburbs as "sites of modernisation, materialism and secularism" but fail to engage with a more complex story of suburbs both as significant sites of creativity and innovation and as extroverted sites of transnational connectivity. In their account of the history of one of the earliest mosques in Britain, the

London Fazl Mosque in the suburb of Putney in South West London built by the Ahmadiyya UK Mission in 1926, Naylor and Ryan (2002) argue that the building which was variously described as "exotic" and "Orientalist picturesque" fitted within a longer trajectory of relocation of different architectural forms to the suburbs, such as bungalows and villas (King 2004), but also was echoed in other forms of suburban "exotic" such as the "Orientalist style" cinemas of the 1930s. We might therefore rethink suburbia recognising the multiple ways in which imperial transnational connections have shaped suburban space (Gilbert and Preston 2003).

In contemporary times, suburbs are increasingly important for new religious buildings, as migrant communities suburbanize (Li 2009; Nasser 2004; Friesen et al. 2005; Ehrkamp and Nagel 2012) and as real estate prices and restrictive planning laws prohibit the construction of new religious buildings in more central areas (Dwyer et al. 2011; Hackworth and Stein 2011). These processes have stimulated my own recent collaborative work about the distinctiveness of the geographies of suburban faith spaces. In a case study of the Shikharbandhi Jain Temple (Shah et al. 2012; Fig. 126.3), in the "edge-city," commuter-belt location of Potters Bar, we suggest that representations of the temple as a faithful replica of a temple in India transplanted to an incongruous English pastoral landscape ignores more complex hybridities. Not only were the architects of the building required to make compromises to the building style and materials to accommodate local planning regulations, but crucially they had to rethink the ways in which the sanctity of the space could be performed. Instead of a

Fig. 126.3 Shikharbandhi Jain Temple, Potters Bar (Photo by Claire Dwyer)

traditional elevated location the temple was sunken into a garden whose geometry forms the Jain symbol for "Triloka" or Cosmos, but also draws on the setting and traditions of the English country house, Hook House, in which it is situated. This setting becomes a signifier for the peace and spiritual retreat which the temple hopes to recreate in its aspiration to be a "tirth" or official Jain pilgrimage site (Shah et al. 2012).

The case of the Shikbarbandhi Jain Temple provides a compelling example of how new faith spaces in British suburban landscapes might be reinterpreted. While its architects and founders narrate their temple as an authentic replica of other Jain temples in India, handcarved by master craftsmen in India and carefully reassembled in its outer London location, we suggest that a more hybrid form has emerged where the Jain Temple has been reimagined and reanimated in relation to its semi-pastoral setting in the commuter town of Potters Bar. We argue that this example provides continuity with other forms of religious creativity in the suburbs (Dwyer et al. 2012). The case study of the Salaam Centre which I discuss below presents a further dimension to this process of the creation of suburban religious space. What is particularly interesting about the Salaam Centre is the vision of its architects, Mangera Yvars, in the creation of a distinctively new form of worship space in Britain which not only reinterprets an Islamic heritage but also engages creatively with its local suburban setting as well as the transnational narratives of its founders.

126.3 The Salaam Centre: The Vision and the Planning

The London borough of Harrow is one of the most ethnically and religiously diverse boroughs in the UK with a third of the borough's population describing themselves as "Asian" or "Asian British" in the 2001 census.[2] The Ithan'ashari Shia Muslims in Harrow are a Khoja community who trace their ancestral linkages to Gujarat in India, but are migrants from East Africa (Tanzania, Uganda and Kenya) who first came to Britain in the 1960s and 1970s as students and professionals. The expulsion of Asians from Uganda by Idi Amin in 1974 prompted more long term settlement in the UK and the community now includes second and third generation British Asians who are highly educated and professional. Many settled in the northwest London suburbs of Wembley, Harrow and Edgware and over time established sites of community and worship. The community first rented their current premises on Station Road in North Harrow in the 1980s for religious worship and finally bought it from the local council in 1990. Subsequently named the "North Harrow Assembly Halls," the buildings are used for worship, yoga classes, children's activities and education

[2] More than 30 % of the borough's population defined themselves as "Asian or Asian British" in the 2001 census. While 47 % of the borough defined themselves as "Christian," 19.7 % defined themselves as Hindu, 7.2 % as Muslim and 6.3 % as Jewish. (Harrow's Diverse Communities, June 2006, Harrow Borough Council).

and social activities. However, the premises were originally built as a temporary restaurant or NAAFI in the post-war years and by the 1990s the pre-fabricated buildings were already showing signs of dilapidation.

The plans to construct a purpose built centre on the site involved the purchase of four neighbouring houses to create a site of 0.86 acres (2.07 ha). Dr Nizar Merali, the project's director, explained that ideas for the new center were generated through intensive debate with the community. His own generation, he suggests "wanted to make some significant and positive contribution to the community that has nurtured us, our children and grandchildren."[3] From the younger, second generation members of his community emerged an idea to create a "radical" and "evolutionary" building. Thus from the outset the ambition was to create a significant architectural building, but also a civic building which was open to the wider community. For Merali it was important that the building offered a range of leisure facilities for all age groups, particularly for young people, so that different generations might share the space. Thus the plans emerged for a center which would include a sports hall, library, a café, and children's play centre as well as halls and meeting rooms. The openness and inclusivity of the space was also seen as a means by which anti-Muslim prejudice could be overcome:

> Negative publicity is a problem for all Muslim communities. How can we address this? The idea is to open up the space so that the community at large can come and meet us. (Dr Merali)

This ambition was echoed in the final proposal which was submitted to Harrow Council (MYAA Planning Submission, 20 July 2010): "it is hoped that [the project] will act as a catalyst for intercultural and inter faith cohesion, with emphasis on family values, health and well-being and countering stereotyping" (p. 1). This emphasis on a community centre which would be open to all was seen as something "very different" from the traditional mosque projects developed by other Muslim communities anywhere else in the UK. Central to the vision for the Salaam Centre is that it is not a mosque but is instead an inter-faith centre. Even within the wider Shia community the plans were seen as controversial raising questions about how appropriate safeguards to the sanctity of a place of worship would be maintained if it was used by non-believers who might not respect Muslim codes of conduct.

In 2006 the trustees of The Battlers Wells Foundation, the charity set up to develop the centre, appointed architects for the new center via a competitive tender. The architects appointed were Mangera Yvars, a London-Barcelona partnership founded by Ali Mangera and Ada Yvars. Mangera Yvars, established in 2001, have designed buildings in Qatar and Kuwait, including the new Faculty of Islamic Studies for Qatar University in Doha as well as in the UK where they are best known for their plans for the Abbey Mills Mosque in the London Borough of Newham. Plans for the Abbey Mills Mosque were commissioned by a Deobandi Muslim group Tablighi Jamaat in Newham, East London, who had bought the large site, a former chemical works, in 1996 with plans to establish a large purpose built

[3] Quotes taken from interview with Nizar Merali, (2 April 2012).

mosque (DeHanas and Pieri 2011). For Mangera Yvars the commission was an opportunity to set a new standard in the creation of the design of contemporary mosques in Europe. Mangera points out that most mosques in the UK have little architectural merit, adapted from existing buildings or incorporating "postmodern pastiche … they have that cartoon look, all plastic domes and minarets" (Mangera cited in Glancey 2006). Instead their design for the new mosque complex was imagined as "a kind of tented city," set into the "dunes of East London" (Mangera, op.cit.) evoking the earliest nomadic mosques in Arabia. The designs included traditional Islamic geometric patterns and Islamic gardens but also incorporated wind turbine minarets, solar and water-generated power. Mangera Yvars plans for Abbey Mills Mosque gained widespread media coverage in 2005 when they appeared in *The Sunday Times* beneath a headline warning "A Giant Mosque for 40,000 may be built at London Olympics" (DeHanas and Pieri 2011: 806). Controversy quickly mounted against the scale of the new mosque, dubbed a "mega-mosque" and "Olympic mosque" by its opponents which included a campaign led by a local councillor associated with the Christian People's Alliance Party. While the scale of the mosque on an 18 acre (8 ha) site in the same borough as the Olympic site was the main focus of opposition, subsequent coverage was marked by Islamophobia centering on the sources of funding for the proposed mosque and the associations of Tablighi Jamaat with terrorist threats.

The Abbey Mills Mosque remains unbuilt. In 2010 Newham Council sought unsuccessfully to evict Tablighi Jamaat from their site on the grounds that their permit to develop the site had expired. For Ali Mangera the publicity generated was disproportionate and ill-informed. He describes how "our not-for-publication render … an imaginary building still undesigned" was made public and "became a polemic" (Mangera 2011: 13). He argued that his former clients, the Tablighi Jamaat were unable to engage in the public and political debate needed to secure support for his innovative design, but nonetheless space had been opened up to "redefine what a mosque might be in the context of modern-day Britain" (Mangera 2011: 14). As I discuss below, Mangera's plans for The Salaam Centre develop some of the concepts suggested for Abbey Mills, although worked through on a different scale, to a different brief and for a very different site. What is most different in the case of The Salaam Centre, however, is that the group commissioning Mangera have been much more effective in working with the local council to secure planning permission for their innovative design and to counter negative publicity.

For the trustees of the Salaam Centre the commissioning of an innovative architectural design has been accompanied by a coordinated approach towards the local council in Harrow to ensure that their plans are supported. However, it has still been a considerable challenge to achieve planning permission. In 2008 the first plans for the Salaam Centre were submitted to Harrow Council. There was support from councillors for the plans, as Mangera comments, "they saw this as an opportunity to create a landmark building for Harrow, to put Harrow on the map."[4] However, the plans to redevelop the site generated opposition particularly from local residents, including the Headstone Residents' Association, who mobilized a campaign against

[4]Quotes taken from discussion with Ali Mangera (7 March 2012, 19 March 2012).

the project. Letters of opposition focused particularly on issues of local amenity such as noise, disturbance, increased numbers of people visiting the site and increased car parking as well as more specific opposition which suggested that there were sufficient mosques in Harrow already and no further provision was necessary. Such opponents did not appreciate that the building was not a mosque and were sceptical that non-Muslims would be welcome at the new center. An additional context to this opposition might have been the virulent anti-Muslim activities of the English Defence League (EDL) which had targeted the central Harrow Mosque in 2009 (*The Independent*, 11th September 2009) and fears from residents that another center associated with a Muslim group might attract further protests. The planning application was initially deferred for further consultation and was then rejected in July 2009. The reasons given for the rejection were its "obtrusive and overbearing size," concerns about flood risk, loss of green space and increase in traffic flow.

A second application was lodged in July 2010. The scale of the project had been considerably revised from the previous application, with a reduction in the overall footprint of the building of 30 % and decrease in its elevation from four stories to two. Plans to include residential spaces within the complex were abandoned. The revised plans ensured that the scale of the building was more consistent with the neighboring semi-detached suburban homes, which were also shielded by green space. However the architects argued that the "urban development of Station Road is in transition" with "no overall architectural vocabulary" (MYAA Planning Submission 2010: 15). In particular a large new housing complex had recently been constructed immediately opposite the site of the proposed center and the elevation of the Salaam Centre was in line with this permitted development. The trustees were very active in public consultation inviting local residents to inspect the revised plans before they were submitted. They placed particular emphasis that provision had been made in relation to parking and reducing the visual impact on neighbouring properties. They also sought to combat criticisms that the site was "only for Muslims" reiterating that this was a genuine initiative to develop a shared and open civic space. Again the support they had from some local councillors, particularly the leader of the Council, herself from a Jewish background, was seen as significant in challenging opposition. In January 2011 Harrow Council approved the plans for the Salaam Centre. Since then the trustees have embarked upon a concerted round of transnational fundraising, particularly reaching out to other Shia Muslim groups across the diaspora to donate to support the cost of the building. Building on the site commenced in January 2014.

126.4 The Salaam Centre: Reimaging Islamic Morphologies in the Suburbs

For Mangera Yvars, the architects who have designed The Salaam Centre, their brief has been to shape a building which not only provides a range of facilities for social and religious activities for the existing Muslim users of the site, but also opens up the use of the site to the wider community. The ambition and aim of the building is stated in the plans they submitted for the building which stated: "the new building will provide a purpose built venue for interfaith and intercommunity dialogue" (MYAA Planning Submission 2010: 1). For Mangera Yvars the commission has provided an opportunity to develop architectural concepts which translate ideas and draw upon Islamic heritage and culture into a British setting. Mangera explains that "the community's East African ancestory has provided us with a conceptual basis for the scheme." This connection is narrated in the overall design of the building which is described as "less of an 'edifice'" and more of a "fluid landscape reminiscent of the landscapes of the Levant and the Indus Valley" (MYAA Planning Submission 2010, see Fig. 126.1). A key feature of the buildings is what Mangera describes as "fractal geometric patterning" which is incorporated into the façade of the building. The aim of this distinctive patterning is to allow daylight to filter into the building which will create "distinctive geometric shadows." The use of light in the building to open up the space is also complemented by the public spaces such as the entrance "agora piazza" with open air seating alongside the café, and two gardens; "the garden of contemplation" and the "garden of discovery" adjacent to the children's play centre (Fig. 126.4). The piazza draws upon notions of "an iconic Islamic courtyard" while the water streams in the gardens also evokes traditions of Islamic gardens.

In many ways the fluid lines and soft silhouettes of the drawings for the Salaam Centre echo some of the concepts which Mangera Yvars have developed elsewhere whether in the "tented city" mosque concept of Abbey Mills or in the Islamic courtyards, gardens and classrooms of the Qatar Faculty of Islamic studies in Doha. In the latter the use of light in the building is connected to ideas about knowledge and "enlightenment" from the Qu'ran, while verses from the Qu'ran provide a "ribbon of calligraphy" linking the buildings (MYAA 2012). Thus the Salaam Centre can be

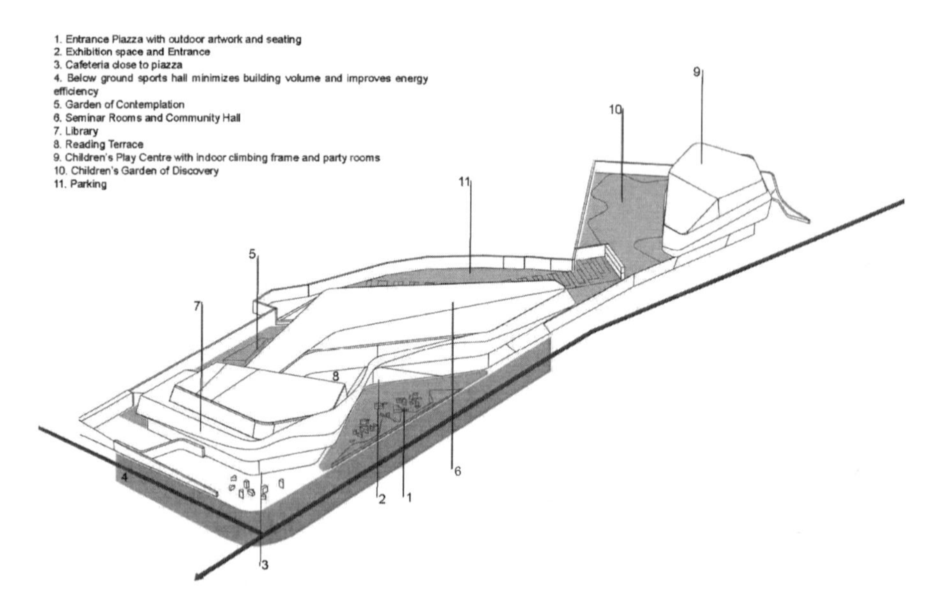

Fig. 126.4 Plans for the Salaam Centre, Harrow (Image courtesy of Mangara Yvars Architects, www.myaa.eu, used with permission)

seen as the realisation by Mangera Yvars of a long term project to "reinvent" and or challenge existing "Islamic Urban Morphologies" (Mangera 2012). Mangera offers a parallel but more critical trajectory of Islamic architecture in the UK to the typology proposed by Peach and Gale (2003). He suggests that Muslim communities have moved from "adaption," the conversion of domestic and commercial premises to mosques, to "assertion-pastiche;" purpose built mosques which announce an Islamic presence but do little to engage with their immediate surroundings and are often poorly built and jarring within the urban landscape. He sees new buildings such as the Islamic Studies centre in Oxford as well as non-Islamic buildings like the Jain Temple in Potters Bar as "postmodern crafted" buildings which similarly fail to engage with their British context. What the plans for the Salaam Centre offer, he suggests, is an opportunity to reinvent Islamic morphologies in the UK through architectural design which enables both education (about Islamic heritage for example) and integration.

That the Salaam Centre is being developed in a suburban location is testament to the ethnic and religious diversity of London's suburbs and evidence that, contra to dominant perceptions, outer suburbs, as much as inner cities, are shaped by processes of migration and transnationalism (Dwyer et al. 2012). For the community behind the Salaam Centre, civic support for their venture reflects enthusiasm for an innovative building which will "provide a landmark for Harrow" or "put Harrow on the map," echoes of familiar narratives about unspectacular suburban landscapes. However what is particularly intriguing about the Salaam Centre are the ways in which Mangera Yvars has sought to engage directly with Harrow's vernacular suburban architecture and the processes of transnationalism which shape the contemporary suburban landscape. In a presentation of the concepts underpinning their designs for the Salaam Centre, Mangera presents images of archetypal 1930s suburban landscape including the square tower of Wealdstone Motors at the corner of Harrow's main shopping street and the brick built tower of St Alban's Anglican Church in North Harrow (Mangera 2012). His starting point for the site for the Salaam Centre is Jon Betjamin's (1973) BBC documentary "Metroland" which paints a nostalgic and affectionate portrait of how the networks of the Metropolitan railway created suburbia, drawing on Betjamin's own poems which included "Harrow-on-the-Hill" (1954). Mangera juxtaposes the original "Metroland" map of railway networks and connections produced by the Metropolitan Railway with another set of maps which illustrate the migratory history of the Shia Ithan'ashari community from Iran, to India, to Tanzania to north-West London. A third map illustrates the web of transnational networks between the US, Canada, Australia, East Africa and India which characterise the contemporary Shia Ithan'ashari diaspora and underpin the transnational fundraising campaign for the Salaam Centre. Mangera explains how these networks and connections have shaped the architects' response to the building:

> We are inspired by John Betjamin's "Metroland". Our North Harrow Community Centre resides within it and were it not for the electrification of the London Metropolitan Railway, suburbia and the context for our project would not have been the same. North Harrow still retains vestiges of its former urban grain, modernity and deco building forms, and an emphasis on landscape. But North Harrow is no longer suburban – it is now a multicultural

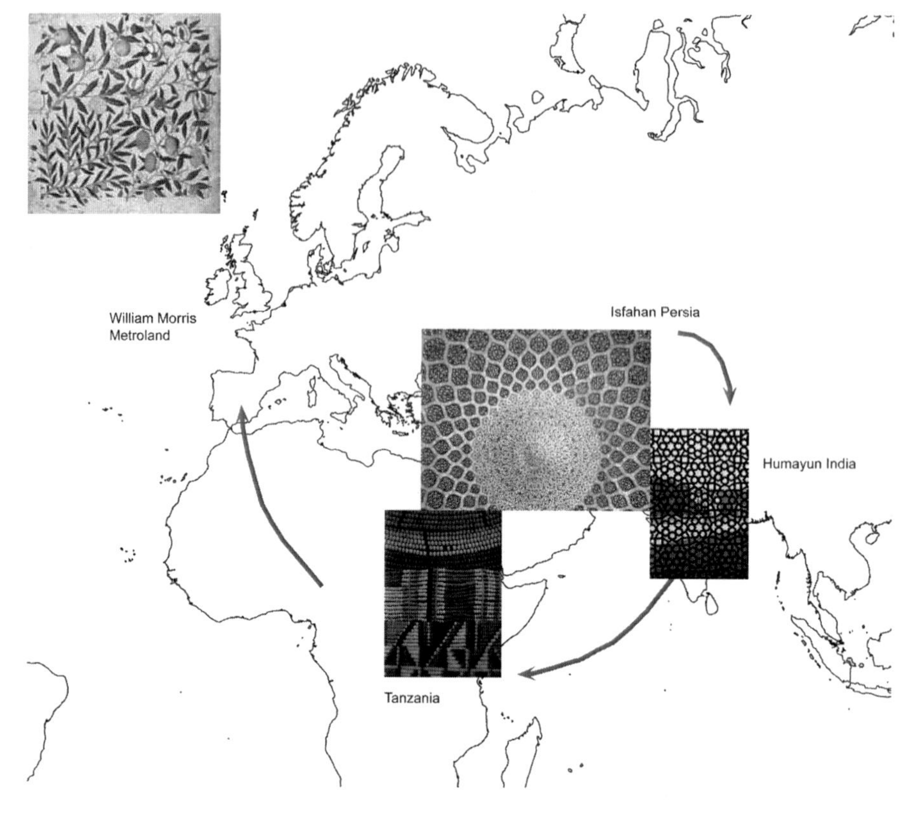

Fig. 126.5 "Reinvention: Patterns" (Map courtesy of Mangara Yvars Architects, www.myaa.eu, used with permission)

> urban centre, an ethnoburbia of sorts. MYAA is investigating the North Harrow suburban context to make explicit migratory movement using architectural pattern making as a conduit for storytelling. (MYAA 2012)

Mangera Yvars ideas for the façade of the building use patterns drawn from Isfahan in Persia, Humayun in India, textiles from Tanzania, and William Morris wallpaper designs from London (Fig. 126.5). Different parts of the building will be fronted by different patterns, narrating a migratory history and a changing suburbia. These themes are echoed in the gardens and "green wall" of planting within the complex which are planned to suggest both an "English Metroland garden" as well as Islamic gardens in India and Persia. The patterning on the façade of the building is thus a creative engagement by the architects with the transnational geographies which shape contemporary Harrow. The multiple webs of these transnationalities (Crang et al. 2003) and translocations are extended by reflections on how William Morris and the wider British Arts and Crafts movement was shaped by engagement with pattern and design from the Empire (Driver 2010; Parry 2005). As Navid Akhtar (2009) illustrates, Islamic designs were a particularly important source for William Morris, inspired by the Persian textiles and ceramics he encountered in contemporary exhibitions in London. The Salaam Centre thus emerges as a dynamic site of engagement with Harrow's suburban history and contemporary multicultural geographies. It is a building whose architecture opens an invitation for inter-faith and community engagement and whose façade narrates the complex transnational geographies of contemporary London suburbia.

126.5 Conclusion

In this chapter I have suggested that the construction of religious buildings, and particularly the negotiations and contestations around the emergence of *new* religious buildings, remain significant topics of interest for geographers of religion. New religious buildings are powerful symbols of how different communities define themselves and how they seek to project their identities to others. Of course, there is considerable variation amongst different faiths about how important a building is for religious practice with Muslims differing from Jews and Christians, for example, in the significance of a dedicated consecrated space for worship. Nonetheless the growth of new religious spaces for minority faith communities within multicultural societies has been a significant trend in recent decades. Such buildings might be read as important signs of integration and multiculturalism but often emerge as contested spaces of national identity and belonging, particularly in the case of mosques. While the opposition to the Abbey Mills Mosque discussed above is one example, nowhere is this more evident than in the furore surrounding the so-called "Ground Zero Mosque," the Islamic Cultural Centre at 51 Park Place in Lower Manhatten (Kilde 2011; Takim 2011). Yet, like the Salaam Centre, plans for Park51 outlined by its architects SOMA, reveal the ways in which building design can open up possibilities for reflection and engagement. In this case through the use of open latticework exterior (echoing traditional "mushrabiya" lattice work windows) "with the clear message that the building has nothing to hide" (Abboud 2011: 35). Park51 and the Salaam Centre, as well as more modest buildings such as the unrealised "Polder Mosque" in the Netherlands (Buğdaci and Erkoçu 2011), are architectural interventions which open up spaces for worship and interaction often through their creative engagement with the existing urban or suburban landscape. The Salaam Centre is an exciting and innovative example whose cutting edge modern architecture draws upon Islamic heritage and self-consciously explores the dynamic transnational geographies of suburban space. Of course, the story does not end here and it is only when the building has been realized that there will be scope for exploring how both members of the Shia Ithna'ashari Muslim community and the wider suburban community of Harrow engage with the new civic space.

Returning to the wider debates set out at the outset of this chapter about the growth of a range of different new religious buildings in many different post-migration contexts in Europe, North America and Australia my discussion here, and some of the wider work cited, suggests a broader research agenda for geographers of religion. There is much to be gained from more comparative studies of emerging new religious architecture in different contexts and associated with different religious groups. Such research might focus on the different planning and regulatory contexts which facilitate new religious buildings or the different ways in which religious communities are accepted and integrated with wider settled populations. There might be scope here to extend or apply the typology suggested by Peach and Gale (2003) to other contexts. Another route for comparison might be in relation to the different architectural forms which emerge and the transnational geographies underpinning religious communities as well as a critical engagement with the different urban and suburban contexts within which new forms of religious architecture are emerging. Within my own work there is certainly scope to compare the changing geographies of religion in suburban London (Dwyer et al. 2012) with the transnational suburban religious landscapes I have been exploring in Canada (Dwyer et al. 2011).

Acknowledgements I am very grateful to Dr. Merali and Shenaz Gulamhusein, of The Salaam Centre, and Ali Mangera, from Mangera Yvars architects, for providing so much information and discussion about the Salaam Centre Project and to Louis Kettle for additional documentary research.

References

Abboud, M. (2011). The architecture of Park51: From controversy to resolution. In J. Jaeckle & F. Türetken (Eds.), *Faith in the City: The mosque in the contemporary urban west* (pp. 32–35). London: The Architecture Foundation.

Akhtar, N. (2009). *William Morris and the Muslims*, BBC World Service, first broadcast 10th Aug 2009. www.bbc.co.uk/programmes/p003vdc5

Buğdaci, C., & Erkoçu, E. (2011). The architect as social interpreter. In J. Jaeckle & F. Türetken (Eds.), *Faith in the City: The mosque in the contemporary urban west* (pp. 28–29). London: The Architecture Foundation.

Crang, P., Dwyer, C., & Jackson, P. (2003). Transnationalism and the spaces of commodity culture. *Progress in Human Geography, 27*(4), 438–456.

Crinson, M. (2002). The mosque and the metropolis. In J. Beaulieu & M. Roberts (Eds.), *Orientalism's interlocutors: Painting, architecture, photography* (pp. 79–102). Durham: Duke University Press.

De Witte, N. (2011). Exploring the postsecular state: The case of Amsterdam. In J. Beaumont & C. Baker (Eds.), *Postsecular cities* (pp. 203–223). London: Continuum.

DeHanas, D., & Pieri, Z. (2011). Olympic proportions: The expanding scalar politics of the London 'Olympics mega-mosque' controversy. *Sociology, 45*(5), 798–814.

Dewsbury, J. D., & Cloke, P. (2009). Spiritual landscapes: Existence, performance and immanence. *Social and Cultural Geography, 10*(6), 695–712.

Driver, F. (2010). Exhibiting South Asian textiles'. In C. Breward, P. Crang, & R. Brill (Eds.), *British Asian Style* (pp. 160–173). London: Victoria and Albert Museum.

Dunn, K. M. (2005). Repetitive and troubling discourses of nationalism in the local politics of mosque development in Sydney, Australia. *Environment and Planning D: Society and Space, 23*(1), 29–50.

Dwyer D., Tse, J., & Ley, D. (2011, April). Highway to heaven: The making of a transnational suburban religious landscape. *Paper presented at the Annual Conference of the Association of American Geographers*, Seattle.

Dwyer, D., Gilbert, D., & Shah, B. (2012). Faith and suburbia: Secularisation, modernity and the changing geographies of religion in London's suburbs. *Transactions of the Institute of British Geographers, 38*(3), 403–419.

Eade, J. (1993). The political articulation of community and the Islamisation of space in London. In R. Barot (Ed.), *Religion and ethnicity: Minorities and social change in the Metropolis* (pp. 27–42). Kampen: Kok Pharos.

Eade, J. (2011). From race to religion: Multiculturalism and contested urban space. In J. Beaumount & C. Baker (Eds.), *Postsecular Cities* (pp. 154–168). London: Continuum.

Ehrkamp, P. (2012). Migrants, mosques and minarets: Reworking the boundaries of liberal democratic citizenship in Switzerland and Germany. In M. Silberman, K. Till & J. Ward (Eds.), *Wall, borders, boundaries: Strategies of surveillance and survival*. New York: Berghahn Books.

Ehrkamp, P, & Nagel, C. (2012). Immigration, places of worship and the politics of citizenship in the U.S. South. *Transactions of the Institute of British Geographers, 37*(4), 624–638.

Friesen, W., Murphy, L., & Kearns, R. (2005). Spiced-up Sandringham: Indian transnationalism and new suburban spaces in Auckland. *New Zealand Journal of Ethnic and Migration Studies, 31*(2), 385–401.

Gale, R. (2004). The multicultural city and the politics of religious architecture: Urban planning, mosques and meaning making in Birmingham, UK. *Built Environment, 30*(1), 18–32.

Gale, R. (2005). Representing the city: Mosques and the planning process in Birmingham. *Journal of Ethnic and Migration Studies, 31*(6), 1161–1179.

Gale, R. (2008). Locating religion in urban planning: Beyond 'race' and 'ethnicity'. *Planning Practice and Research, 23*(1), 19–39.

Gale, R. (2012, March 12). '...And make your Dwellings into Places of Worship': Mosque Development and the Politics of Residence' Paper given at Negotiating Religion in Urban Space Workshop, Department of Geography, University College London.

Gale, R., & Naylor, S. (2002). Religion, planning and the city: The spatial politics of ethnic minority expression in British cities and towns. *Ethnicities, 2*(3), 387–409.

Germain, A., & Gagnon, J. E. (2003). Minority places of worship and zoning dilemmas in Montreal'. *Planning Theory and Practice, 4*, 295–318.

Gilbert, D., & Preston, R. (2003). Stop being so English. Suburbia and national identity. In D. Gilbert, D. Matless, & B. Short (Eds.), *Geographies of British modernity: Space and society in the twentieth century* (pp. 187–203). Oxford: Blackwells.

Glancey, J. (2006, October 30). Dome sweet dome. *The Guardian*. Retrieved June 12, 2012, from www.guardian.co.uk/artanddesign/2006/oct/30/architecture.religion

Hackworth, J., & Stein, K. (2011). The collision of faith and economic development in Toronto's inner suburban industrial districts. *Urban Affairs Review, 20*(10), 1–27.

Holloway, J. (2006). Enchanted spaces: The séance, affect and the geographies of religion. *Annals of the Association of American Geographers, 96*(1), 182–187.

Holloway, J. (2011). Tracing the emergent in geographies of religion and belief. In C. Brace, A. Bailey, S. Carter, D. Harvey, & N. Thomas (Eds.), *Emerging geographies of belief* (pp. 30–52). Newcastle: Cambridge Scholars Publishing.

Isin, E. F., & Siemiatycki, M. S. (2002). Making space for mosques: Struggles for urban citizenship in diasporic Toronto. In S. H. Razack (Ed.), *Race, space and law, Unmapping a White settler society* (pp. 185–209). Toronto: Between the Lines Press.

Kilde, J. (2011). The Park51/Ground Zero Controversy and sacred sites as contested space. *Religion, 2*, 297–311.

King, A. (2004). Suburb/Ethnoburb/Globurb: The making of contemporary modernities. In *Spaces of global cultures* (pp. 97–111). London: Routledge.

Kong, L. (2010). Global shifts, theoretical shifts: Changing geographies of religion. *Progress in Human Geography, 34*(6), 755–776.

Li, W. (2009). *Ethnoburb: The new ethnic community in urban America*. Honolulu: University of Hawai'i Press.

Mangera, A. (2011). The mega mosque. In J. Jaeckle & F. Türetken (Eds.), *Faith in the City: The mosque in the contemporary urban west* (pp. 12–14). London: The Architecture Foundation.

Mangera, A. (2012, March 7). Negotiating religion in urban space. Presentation at the workshop *Negotiating Religion in Urban Space*, Department of Geography, University College London.

McLoughlin, S. (2005). Mosques and the public sphere: Conflict and cooperation in Bradford. *Journal of Ethnic and Migration Studies, 31*(6), 1045–1066.

MYAA Planning Submission, North Harrow Community Centre. (2010). Retrieved June 3, 2012, from www.thesalaamcentre.com/application

MYAA website. Retrieved June 12, 2012, from www.myaa.eu/

Nasser, N. (2004). Southall's kaleido-scape: A study in the changing morphology of a west London suburb. *Built Environment, 30*(1), 76–103.

Naylor, S., & Ryan, J. (2002). The mosque in the suburbs: Negotiating religion and ethnicity in South London. *Social and Cultural Geography, 3*(1), 39–59.

Naylor, S., & Ryan, J. (2003). Mosques, temples and Gurdwaras: New sites of religion in twentieth-century Britain. In D. Gilbert, D. Matless, & J. Short (Eds.), *Historical geographies of twentieth century Britain* (pp. 168–183). Oxford: Blackwell.

Parry, L. (2005). *Textiles of the arts and crafts movement*. London: Thames & Hudson.

Peach, C. (2012). Islam and the art of mosque construction in Western Europe. Annual Meeting of the Association of American Geographers. Fourth annual GORABS Lecture, New York.

Peach, C., & Gale, R. (2003). Muslims, Hindus and Sikhs in the new religious landscape of England. *The Geographical Review, 93*(4), 469–490.

Shah, B., Dwyer, C., & Gilbert, D. (2012). Landscapes of diasporic belonging in the edge-city: The Jain temple at Potters Bar. *Outer London South Asian Diaspora, 4*(1), 77–94.

Takim, L. (2011). The Ground Zero mosque controversy: Implications for American Islam'. *Religion, 2*, 132–144.

The Independent. (2009, September 11). Retrieved June 20, 2012, from www.independent.co.uk/news/uk/home-news/riot-police-called-in-as-demonstrators-clash-at-antimuslim-protest-1785797.html

The Salaam Centre. Promotional Brochure. Retrieved June 8, 2012, from www.thesalaamcentre.com/accessed

Chapter 127
Islam and Urbanism in Indonesia: The Mosque as Urban Identity in Javanese Cities

Hafid Setiadi

127.1 Introduction

Many scholars from various disciplines have studied Islam in Indonesia in relation to issues such as local tradition, social class and cultural change (Bruinessen 1995; Abdullah 1966; Geertz 1976; Bowen 2003), interreligious-relations (Jayamedho and Ekoputro 2010); feminine Muslim identity and women's right (Dewi 2012; Hull 1976), the role of political power (Alfian 1989; Azra; 2004; Burhanudin 2012), Chinese Muslims (Dickson 2008; Maulana 2010), and globalization (Lukens-Bull 2000). Some scholars also focused their interests on reviewing developments and the institutionalization of Islam in Indonesia. The existence of prominent Islamic organizations, especially *Muhammadiyah* and *Nadhlatul Ulama* (NU), have been frequently used as a main focus to elaborate on contemporary Islamic thought in Indonesia. However, several recent studies have also begun to examine some "peripheral organizations" such as *Hizbut Thahir Indonesia* (HTI), *Jaringan Islam Liberal* (JIL), or *Front Pembela Islam* (FPI) which provide alternative views on contemporary Islam in Indonesia (see Fox 2004). As is shown in these studies, Islam in Indonesia is performed in various forms, beliefs and practices among people and in different places due to its interaction with local, national, and global factors (Hefner 2002). It means that, as in other countries, religious practice in Indonesia is basically a cultural phenomena (Maarif 2009).

There is no doubt but that there is a strong relationship between religion and social transformation. Every religion has several familiar identities and practices which are performed by its followers in various ways. The realization of these identities may be symbolized in the form of rituals, life style, fashion, or buildings

H. Setiadi (✉)
Department of Geography, University of Indonesia, Depok, West Java 16424, Indonesia
e-mail: hafid.setiadi@ui.ac.id

S.D. Brunn (ed.), *The Changing World Religion Map*,
DOI 10.1007/978-94-017-9376-6_127

Essentially, these symbols are the important means through which followers maintain and communicate worldviews, norms, values, and ideology, all which are part of their religion (Arkoun 1990). All provide an ethical dimension as well as a collective motivation for the community to understand, organize, and progress through life. In this context, we can see religion not only as a course or source of faithful messages, but also as a foundation for a civilization. According to Weber (1930), a major change in the nature of a religion in a culture will be followed by a dramatical transformation in its economic and social order.

As a geographical field, religion study is a branch of cultural geography. As identified by Park (2004), there are two geographical approaches to studying religion. The first is "religious geography" which focuses on the role of religion in shaping and developing human perception. The second is a "geography of religion" which tries to understand religion as a human institution and its manifestation in human life. Before Park, Levine (1986) proposed some philosphical views on the nature of religion as a foundation to the study religion in geographical field, including phenomenology, structuralism, functionalism, historicism, and ecologism. Each view has a distinctive definition as well as a distinctive methodology for studying religion. Meanwhile, Stoddard and Prorok (2003) introduced the branch of geographical field called the "geography of religion and belief system" to show a critical diferentiation between "religion" and "belief system." They wrote:

> Even though beliefs may not necessarily be organized systematically, the popular acceptance of national mythology, revered symbols, and hallowed places closely resemble characteristics of formal religions. As with other belief systems containing culturally perceived ultimate priorities, the motives for honoring patriotic icons, preserving hallowed grounds, and visiting sanctified sites originate from a sense of "shouldness." (Stoddard and Prorok 2003: 259)

As a part of culture, religion plays an important role in the foundation of human existence. It shapes moral code, ethic, perception, spirit, and institutions for human life. Such elements assist people in designing their society and environment. For example, a built environment, such as city can be defined as the incarnation of relationship patterns and belief systems (Hiller and Hanson 1984). In other words, city is a product of civilization (Polanyi 1968). It represents the progress of cultural innovation, social communication, and economic production (Castells 1983). Without a doubt, every city requires a suitable physical environment in terms of land, water, climate, and accessibility for sustaining its growth. However, its existence is not dependent only on such environmental factors. It is more dependent upon the specification of basic ideas or assumptions which form the background of city development itself (Coulanges 2001). It is also constrained by social and political structures and because of these one particular city will be different with other cities despite, having the same causal factors.

In relation to the Islamic city, some earlier scholars tried to elaborate the above issues. One notable elaboration was conducted by Paul Wheatley (2001). He focused on the changes of urban design and structure of old and new Islamic cities in the Middle East and Africa from the 7th to 10th centuries. He concluded that there was not a trend to unite Islamic city and region by common urban design. In context of urban planning, Al-Hathloul (2004) showed this mode of planning differencing between the Islamic city of modernists and traditionalists. While the modernist mode considered Islamic tradition as an obstacle to urban planning, the traditionalist made restrictions for involving modern technology into the planning process. Subsequently, Muazaz Y. Soud et al. (2010) examined the quality of traditional Islamic planning in order to identify aspects of decline in contemporary Islamic cities. In a Java setting, Rutz (1987) described the changes of traditional nucleus of Javanese cities starting from the Hindu period to the Islam period. More recently, Santoso (2008) explained the role of belief systems – including animism, Hindu, and Islam – in creating urban space of pre-colonial cities in Java. Lombard (2005b) and Tjandrasasmita (2009) also identified some important points about the impact of Islamic penetration on the spatial order of Javanese cities.

This chapter focuses on the interplay between Islam and urbanism in Java. It uses the urbanism as a main foundation to examine the interplay between the role of Islam and Islamic urban landscape creation. The definition of urbanism as used here relates to the role of urban place-making under a certain culture. It also associated with urban symbolization or urban identity (Hakim 1990). As used here, urbanism contains three mutual concepts, they are behavioral, structural, and processual. According to Wheatley (1983), while the *behavioral* refers to transformation attitude and value; the *structural* focuses on attention on the patterned activities; the *processual* is concerned with the progressive concentration of population. Wheatley argued that the very nature of urbanism is always ideological.

In his research, Brass (2003) considered ideological factors together with topographical factors as the main elements important in understanding urbanism in Ancient Egypt. In conclusion, he stated that there was a plurality of hierarchical settlements which were constrained and reinforced by ecological, mobility, logistical and ideological considerations. In another study, Fletcher (2009) discussed agrarian low-density urbanism in lowland Mesoamerica, in Sri Lanka and interior Southeast Asia. Also Evers and Korff (2002) investigated urbanism in Southeast Asia in relation to the meaning and power of social space. They found that political systems as well as globalization forces had stimulated the emergence of various actors and that social networks which played a significant role in creating urban identity. When studying urban identity, we have to take into account Mohammed Arkoun's about urbanism in Islamic cities. In his short chapter entitled *Islam, Urbanism, and Human Existence Today*, Arkoun (1983: 38) wrote:

> ... if we wish to identify a typical urban tradition, we should not look at those great monu ments or mosque which engage the attention of the specialist and even amaze us. I emphasise that our approach must be pragmatic. When considering the history of Islam, Muslim as well as Islamicists usually take into account a fictitious historical continuity represented by architectural monuments or great philosophical works. In so doing the real social fabric is totally omitted .... It celebrates works of civilization which we continue to protect at the expense of local invention and creativity.

Regarding Arkoun's statements, this chapter attempts to apply the concept of centrality. As revealed by Bouchair and Dupagne (2003), this concept in many Islamic cities is related to the role of the mosque as a dominant point. In this sense, the mosque symbolizes the existence of a "super power agent" which centralizes authority over religious, cultural, social, and political lives. They argued that in the past it was important for every Islamic settlement to have only one mosque in order to avoid manifesting a divisive impression. In some cases, the market place along with the mosque played central positions.

The central question is the role of the mosque in Indonesian cities. Does the mosque still hold its position as a dominant symbol in creating urban identity? This chapter addresses this question by discussing several Javanese cities. For that purpose I discuss architectural durability for representing, the sturdiness of the mosque dealing with external forces. In an Islamic context, Janavese architecture has a distinctive form which represents a genre of local tradition (Wahby 2007; Santoso 2008; Ashadi 2002). This architectural conversion can be seen as a way to examine changes in the cultural relations between Islamic values and local Javanese traditions.

127.2 An Overview of Islam in Java

Currently there are various opinions about the arrival of Islam into Java. When interviewed by Anthony Shih, Professor Mark Mancall stated that Islam in Indonesia, including Java, originated from India that led to the emergence of *Sufism* (Shih 2002). This view is confirmed by Woodward (1989). *Sufism* indicates that there is a dominant Hindu-Buddhist culture that lay just below the Javanese Islam surface. Meanwhile, Wahby (2007) stated that the arrival of Islam to the region is attributed to many groups, mainly merchants, *ulama*, and Sufis. Other scholars such as Reid (1992), Lombard (2005a), and Tjandrasasmita (2009) believe that Muslim merchants' domination in the golden age of maritime-trade in Southeast Asia was a key factor in the introduction of Islamization in Java. And more specifically, Mulyana (2005) and Graaf and Pigeaud (2003) have argued the decisive role of Chinese travellers in spreading Islam into Java. According to this debate, Azra (2006) revealed that one of the problems of this historical complexity is the disagreement among scholars on the meaning of "Islam." While certain scholars define "Islam" as the statement of faith or *shahadah*, others describe it from sociological point of view. The later definition refers to the society which Islam provides the actual working principles for cultural, social, and political life.

Therefore, the debates provide strong confirmation that to the Javanese Islam was an alien tradition (Muhaimin 2006). This alien tradition was obviously introduced first on the north coast of Java as noted by the finding of Fatimah Maemun's tombstone at Gresik with the marking 475 Hijriah (1,082 M). Prior to the fifteenth century, the coming of Islam was more noticeable. Then the Hindu kingdoms still held cultural and political domination over the Javanese (Munoz 2009). In order to build economic and political relations with supporters overseas, those kingdoms had by necessity to maintain some ports in the north coast of Java such as Banten, Kalapa, Tuban, and Gresik.

When the trading activities in Southeast Asia experienced a remarkable improvement in the fifteenth century, these ports became more crowded by the presence of Muslim merchants (Reid 1999). Then, the long peaceful assimilation persuaded the Javanese to embrace the Islam tradition. This assimilation ran gradually from coastal areas to the interior; from the merchant to the farmer communities; and from the heterogenous to a homogenous society. As a result of these developments, this overall process generated various types of Islamic societies in Java. From a dichotomic standpoint, for instance, there were the Santri versus Abangan tradition (Geertz 1976), the coastal versus interior Islam (Syam 2005), the modernist versus traditionalist (Lukens-Bull 2006), and the aristocracy versus and orthodox Islam (Jay 1963). These dichotomics represent the complexities of cultural exposure between Islamic and Javanese values.

However, when Islam came into Java, the region was still full of indigenous mystical life that was enriched by a Hindu-Buddha civilization mainly in a statecraft constitution developed for more than thousand years (Ricklefs 2005). With regard to this circumstance, there are two interrelated themes in explaining Javanese Islam. The first theme is related to the Wali Songo as a group of nine outstanding figures who had been widely recognized as key actors in the dissemination of the Islamic faith. They employed local tradition including myths and arts to preach about Islam among the Javanese. One of them, Sunan Kalijaga, for example, inserted Islamic elements to the Wayang – the traditional Javanese puppet show – in his preaching into Hindu-Buddha societies in an interior region (Moenthadim 2010). In many cases, the preaching activities of Wali Songo obtained political support from the ruler. Moreover, the ruler himself was often involved directly in disseminating Islam (Setiadi 2012).

In overall, Wali Songo was responsible for the transformation of Javanese religion and culture. This transformation is related to the syncretism as the second theme. It refers to the mixture of animism, Hindusm, and Islam as a distinctive feature of Islamization in Java. Syncretism did not mark the end of the cultural movement of Islamic ideas among Javanese, but rather the beginning. Islam which evolved initially in Java having an affinity with sufism (Yusuf 2006; Azra 2006). In simple definition, sufism can be described as a mystical dimension of Islam which involves a spiritual journey for obtaining intimacy with God (Azra 2006). It is "a devotional and mystical current within the Islamic tradition" (Howell and Bruinesse 2007). Because of the nature of its rituals, sufism matched with the cultural setting of indigenous people who were more colored by animism and Hindu asceticism. This was indicated, for example, by semedi, a ritual to gain an intimate contact with God and finally merge with God. In Javanese tradition, this ritual also represented the integration between the human sphere (microcosmos) and supra-human sphere (macrocosmos). This condition allowed the Javanese people, either at the highest or at the lowest status of society, to receive Islam as their new religion. It is in line with the term "received Islam" which refers to the Islam that is taught to a particular community (Lukens-Bull 2006)

127.3 Islam and Urbanism in Java

127.3.1 Basic Pathways of Urbanism

It is not easy to define "Islamic city" in Java. Almost all cities in Java did not employ the principles of Islam in planning and regulating the built environment. Most scholars more often used the term of "Javanese city," even in their investigation focused on the capital cities of Islamic kingdoms. To examine the influence of Islamic elements on the Javanese cities, they used the term of "the city during the Islamic period" (see Wertheim 1956; Keyfitz 1976; Rutz 1987; Rahardjo 2007). In relation to urban studies, this tendency indicates that the Javanese features was more salient than the Islamic features. The Javanese features were a product of socio-cultural values of local traditions which were enriched by Indianization, Islamization, and Westernization and also globalization today.

The above mechanism of urbanism can be distinguished between an agrarian and trade pathways (Nas 1986). The first pathway dominantly emerged in interior regions which were situated at mountainous terrain. It was influenced either by the swidden or paddy cultures. Even though there were many differences between the two cultures in terms of production system, life style, and social organization (see Harris 1972; Johnson 1974; Sarmela 1987; Nakashima and Rouè 2002; Sumardjo 2002; Rahardjo 2007), both shared the same vision such as dependence of God (Thompson 2010).

Primarily, this vision appeared as a tribal religion called animism, which believes in the role of spiritual power coming from some particular objects. The Hindu-Buddha civilization then modified this tribal religion through inserting the conception of Gods, caste system and elite domination (Ricklefs 2005). Indeed, it prompted a deep impression in the matters of sanctity relationship and amalgamation between the human and God's sphere (Lombard 2005b; Heine-Geldern 1942). These Hindu-Buddha elements provided a major contribution to social as well as a regional integration process in the form of the kingdom (Lombard 2005c). And the king was a symbol of God. So, the establishment of a kingdom came as a new social and political tradition in Java. In line with this new tradition, some particular places in the upland terrain were pointed out as capital cities where the king's palaces were located (Ekadjati 2009). It represented the flowing down of God's power continuously from the "upper" to the "below" (Sumardjo 2002). By such rule, the king domiciled in the upland and the common people lived in the lowland. In the context of urban tradition, the spatial distribution of cities as showed by Fig. 127.1 reflected that rule.

Islam penetration infiltrated a new culture into Java; not only Islamic values but also maritime-trade tradition. During Islamization the coastal cities developed and advanced mainly as international or regional hubs. They replaced the interior cities as focal points in the social and political life in Java as a whole. As noted by Lombard (2005a) and Vickers (2009), the way of life of the people, especially in coastal region, became more open, more cosmopolitan, and more outward-oriented. The sacred relationship between humans and God no longer became as a determinant factor of social identity (Lombard 2005b). Hindu traditions such as the

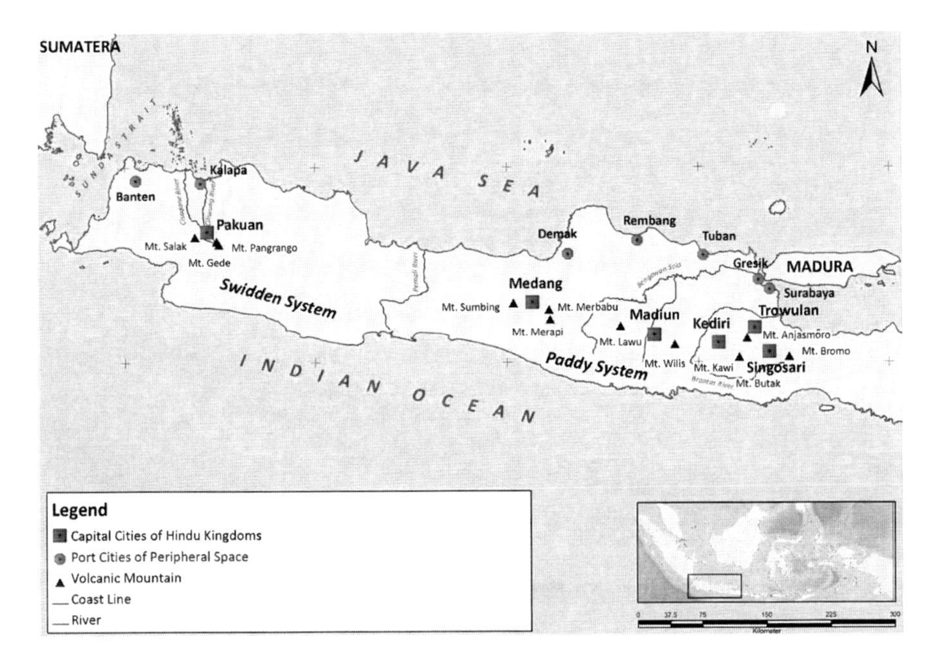

Fig. 127.1 Distribution of cities in Java under the rule of Hindu-Buddha traditions (Map by Hafid Setiadi)

implementation of caste system increasingly disappeared (Mulyana 2005). Social as well as economic competition were more noticeable. This second pathway also regarded water routes and equal-distance transit points as important elements (Wastherdahl 1992). Thus, the port and commercial cities replaced the king's palace as focal points.

However, in examining either the agrarian or trade-based cities, it is important to consider carefully the conception of mancapat in the urban tradition in Java as explained by Santoso (2008). This concept basically divides urban space into four sections with the intersection of those sections is a point of major interest. After examining the view of Roufflaet Van Ossenbruggen, Santoso concluded that mancapat is a Javanese spatial framework in order to shape and ensure social stability. This concept contained both sacred and social values. The sacred value refers to the existence and integration of five basic elements of the universe: water, fire, land, air, and ether. In the context of community life, the intersection point of mancapat symbolizes a center of social, economic, and political dimension for the city. Actually, this concept was also used as a foundation to describe the "city" in the Hindu period. When Islam came into Java, the conception of mancapat was still maintained as a main feature in defining urban identity. Overall, this concept articulates the preeminence of the "center" against the "periphery" (Lombard 2005b). Spatially, this concept seem to have denoted a four-square arrangement of a village at each cardinal point with the biggest village at the center (Martono 1974; Santoso 2008).

Fig. 127.2 A pair of banyan trees at alun-alun of Yogyakarta (Photo from http://wikimapia.org/16085557/Alun-alun-Kidul-Selatan#/photo/3297880)

Fig. 127.3 The royal symbols as a dominant ornament at the main gate of the Great Mosque of Yogyakarta (Source: http://en.wikipedia.org/wiki/File:GrandMosqueYogya.JPG)

At a micro scale, it would also be manifested by a distinctive form of a city's town square or alun-alun which represents the existence of religious space as well as social space in the Javanese life (Ashadi 2002; Santoso 2008). In terms of religious space, alun-alun is characterized by the presence of the mosque on the west side for symbolizing the power of God (Handinoto 2010). On the south side, there is the king's palace or kraton representing the king as the God's representative on the earth. On the east and north sides, there are social activities such as a daily market and public settlement. At the center of alun-alun there is a banyan tree to symbolize the integration of religious and social space (Figs. 127.2 and 127.3). It means that the conception of Javanese space not only refers to physical object, but also to things that have a spiritual aspect.

127.3.2 The Javanese Mosque as Urban Identity

This section provides a short description about the position of mosque in the urban landscape in Javanese cities. To obtain a good comparison, four different cities will be discussed, they are: Tuban, Demak, Yogyakarta, Jakarta. The city of Tuban represents the small town in Java which is rooted in maritime-trade tradition. In its history, Tuban never appeared as a center of Islamic power. Meanwhile, the city of Demak is characterized as a coastal city that had an important role as Islamic power and religious center. Similar to Tuban, this city is also based on maritime-trade tradition. The third city, Yogyakarta, is a large city in which situated at interior of Java. It was a center of the largest and longest Islamic kingdoms in Java, the Sultanate of Mataram. Even today, Yogyakarta is still known as a core of Javanese civilization with its main foundation in an agrarian tradition. Jakarta's history and Islamic landscape is sharply different than the other three.

Tuban Tuban is a small town in the north coast of Java. In the Hindu period, it played an important role as a main port together with Gresik and Surabaya for the Kingdom of Majapahit. At that time, many overseas merchants from China, India, Malay, and Middle East came to the city. Because of this Tuba experienced a remarkable progress in terms of economic and social life as well. It is not surprising that several Muslim communities were there at that time. However, the progress of Islam became more obvious in the fifteenth century after the north coast of Java was occupied entirely by the first Islamic power of Java, namely, the Sultanate of Demak. During the reign of Demak, Tuban turned into the center of Islam dissemination, which was led by a prominent clergyman namely Sunan Bonang, one of Wali Songo. Currently, Sunan Bonang is remembered by the name of a narrow street in Tuban's periphery area (Fig. 127.4).

As with other Javanese cities, urban identity at Tuban was also mainly marked by the existence of the *alun-alun* or town square (Fig. 127.5). There were many interesting places and buildings around the square such as the Great Mosque of Tuban, local government offices, Chinese temples, and the Sunan Bonang's grave (Hartono and Handinoto 2005). The Great Mosque was established in the colonial period and designed by the Dutchman, H. M. Toxopeus. In regard to this mosque's initial status as a pioneer of Islam dissemination in Java, this fact is very interesting. As a colonial product, it appears that the mosque never became a main feature noted as a cultural and political symbol for disseminating Islam.

The losing of local tradition should be a reason why in today's identity Tuban's mosque has changed continuously into its Middle Eastern style (Fig. 127.6). Moving further away from local tradition. It confirmed that there was a cultural penetration by foreign forces to impose Islamic symbols as a part of urban imposition process. According to Wheatley (1983), such penetration implanted and created new symbols for retaining the existence of foreigner then misplacing local identity. In the case of Tuban, after the introduction of foreign culture, the Islamic identity of Tuban became stronger. Now, the mosque is known as the largest mosque in east Java. And the city of Tuban has received a nickname as "the country of 1,001 nights," just like the city of Baghdad in Iraq.

Fig. 127.4 Immortalizing Sunan Bonang as the name of a narrow street at Tuban, East Java (Photo by Hafid Setiadi)

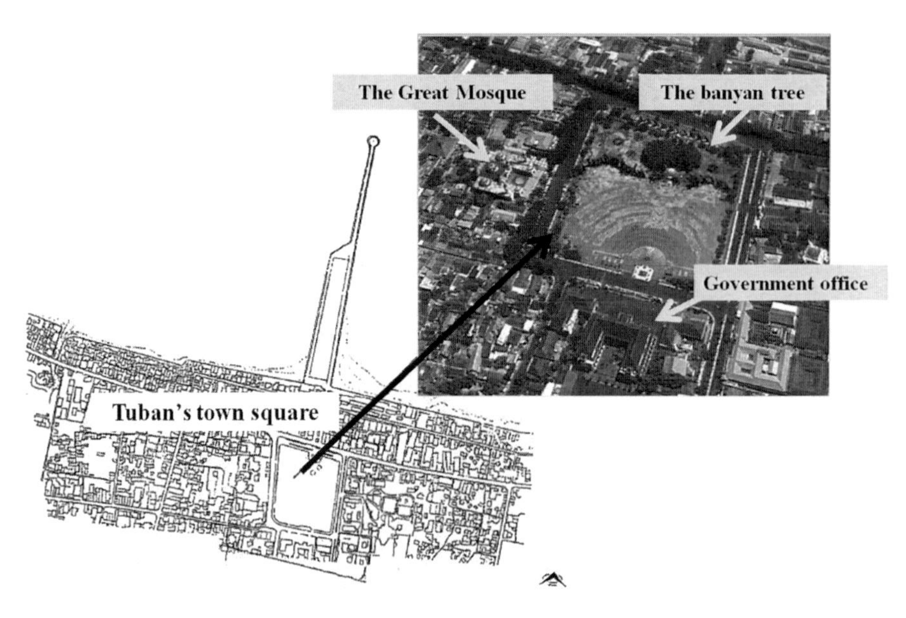

Fig. 127.5 Today's situation at Tuban's town square (Map by Hafid Setiadi, modified from Pratomo (2001: 178), image from Google Earth, November 30, 2012)

Fig. 127.6 Architectural change on the Great Mosque at Tuban from the Colonial style in 1894 (**a**) to the Middle Eastern's style in 2004 (**b**) (Source: Hartono and Handinoto (2005: 137), used with permission)

Demak Tuban's experience was totally different from that of Demak. Similar with the mosque at Kudus, Banten, and Cirebon; the Great Mosque of Demak was established by *Wali Songo* during the Islamization era. Demak itself was a capital of the first Islamic kingdoms in Java, namely the Sultanate of Demak. In the early period of Islamization, Demak provided the political and economic foundation for spreading Islam the entirely of Java. Thus, Islam spread progressively by means of military attacks, clergymen missions, and trading activities. Furthermore, the ruler of Demak – Raden Patah – proclaimed himself as "the King of Java." He had an obsession to lead Demak as a hegemonic power in order to control the main routes of maritime trade throughout Southeast Asia.

To symbolize his power, the ruler built a mosque near his palace as a representation of the amalgamation of political and religious values. In this case, the ruler required the presence of the mosque not only to accommodate religious-social concerns, but also to guarantee the stability of political power. Although the ruler still existed as the dominant actor, the determining factor of regional integration was the mosque. The mosque of Demak then became as the center of Islamization in Java (Graaf and Pigaued 2003; Purwadi and Maharsi 2005). It is unfortunate that when the Sultanate of Demak collapsed in the sixteenth century, the palace (*kraton*) was shattered.

However, the mosque still exists on its original style. There were many myths behind the mosque construction that were generally connected to the divine power of *Wali Songo* (Graff and Pigeaud, 2003; Ashadi 2002). One of the main features of Demak's mosque is on the philosophical values that are embedded in its architectural style. There is a pyramidical roof with three layers being shaped like a isosceles triangle (Fig. 127.7). This kind of roof is quite different from the Middle Eastern's style is more in the form of a dome. The design originated from the Javanese Hindu tradition to represent the Mount Meru in India as a center of the universe (Lombard 2005c; Ashadi 2002; Graff and Pigeaud, 2003). Meanwhile, the three layers of roof represent the three dimension of Islam, they are the "faith" (*iman*), "submission" (*Islam*) and "doing the beautiful" (*ihsan*) (Ashadi 2002; Wahby 2007;

Fig. 127.7 The Great Mosque of Demak, with three layers on its pyramidal roof, is similar to The Great Mosque of Yogyakarta shown in Fig. 127.3 (Photo from Ashadi (2002), used with permission)

Graff and Pigeaud 2003; Chittick 2000). However, currently there is still debate on the core idea of such an architectural feature, specifically, whether it derived from India, China, or is fully Javanese.

Even there were many similarities between Demak and Tuban, especially if we refer to them as coastal cities. The most important difference between the two cities was related to their political status in the past. Today, because of its status as the former center of Islamic power, Demak has a stronger urban identity than Tuban. That status led to differences in the role of the mosque between two cities in historical development of Islam in Java. The mosque at Demak was situated at the center of political power and was a basis for a "king-oriented Islam." The main character of such kind of Islam was integrated into the king's power. As a result, political power symbols such as the palace appeared as a dominant identity and then gave a major contribution in strengthening the existence of the mosque as a religious symbol. This circumstance implanted the mosque of Demak deeply in collective memory not only among residents of Demak, but also the Muslim population at a whole of Java. Thus, the mosque of Demak became as an essential part of Islamic tradition and history in Java. Most Javanese Muslims consider it as *the* most sacred historical Islamic place in the country. Any major and fundamental change to the style or physical features of the mosque would bring disappointment to some Muslims.

Yogyakarta It is interesting to make a comparison between Demak's and Yogyakarta's experiences. Similar to Demak, Yogyakarta was a center of Islamic power in Java. Its origin came from the largest Islamic kingdom in Java namely the Sultanate of Mataram, established in the seventeenth century. Even today, the

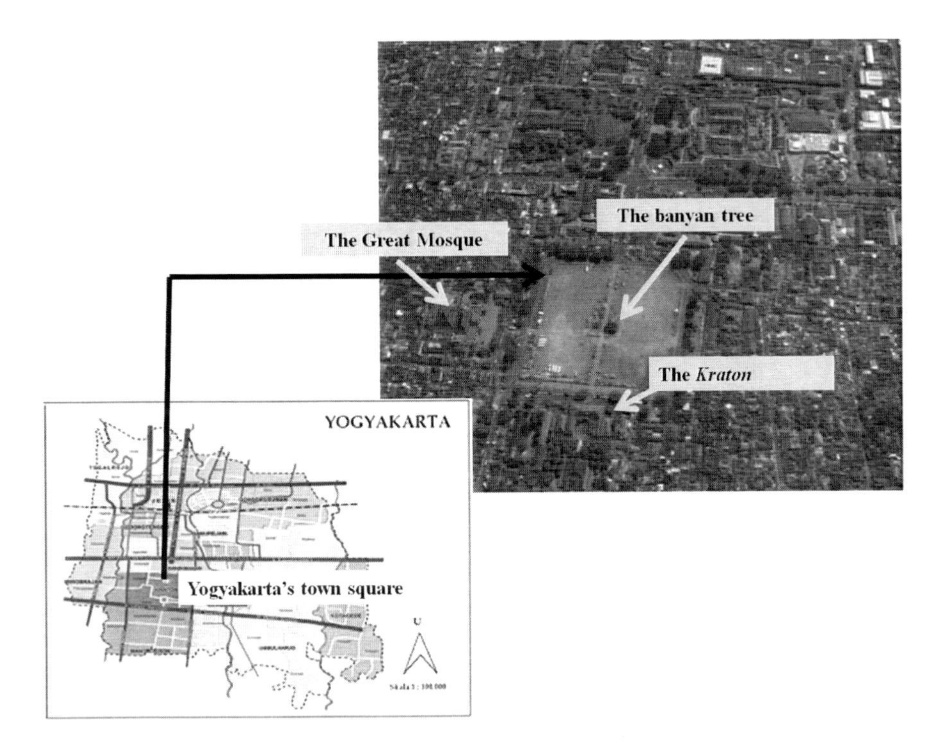

Fig. 127.8 Today's situation at Yogyakarta's town square (Map by Hafid Setiadi, modified from www.citrainfo.com, December 4, 2012, image from Google Earth)

Sultan, commonly known as "Sultan Yogyakarta," still plays an important role in political and cultural life of Javanese people, whether Muslim or non-Muslim. He has a formal political status as the Governor of Special Province of Yogyakarta. The term of "special province" is a tribute to the success of the Sultan and the people of Yogyakarta in maintaining Javanese culture.

Many scholars such as Santoso (2008), Rahardjo (2011), and Mulder (2001) acknowledge that the Ancient-Java tradition, called as *kejawen,* embraced tightly by Muslim people in Yogyakarta. This tradition has both an agrarian and Hindu culture as well. According to Gertz (1976), this kind of Muslim might be grouped as *Islam Abangan*. This group stressed Islamic culture by syncretism. Thus, the Javanese culture was more salient than the Islamic one. Because of that, Yogyakarta was called as "the real Javanese city." In addition, Yogyakarta also know as an aristocratic city due to a hegemonic role of the Sultan. The Sultan was positioned as a center of the universe. Figure 127.8 shows today's situation at the *alun-alun* of Yogyakarta.

In Yogyakarta the mosque had never been emphasized as a basis of urban symbolization. The mosque is just a part of the Kraton. Thus, as a Islamic symbol, it was incorporated into political symbols as a consequence of the conception of centralized power which was defined as the amalgamation of God and king into a single entity. The king symbolized God's existence in the world (Adiyanto 2011). In line with that, there were two important symbols in representing this centralized power: the palace (Kraton) and banyan tree. Santoso (2008) stated that in Mataram's tradition the Kraton represented the figure and personality of the Sultan who was able to unify honor, justice, sovereignty and prosperity. The banyan trees represented the sanctity, solidity, sturdiness, and guardianship of power (Herusatoto 2008). Because of these interpretations, either the Kraton or banyan trees was more sacred than the mosque. Nevertheless, as a result of the strong symbolic interactions between the Kraton, banyan tree, and mosque; changes against the mosque itself were not possible (see Fig. 127.7). Any alteration, however minor, might affect those interactions.

The three above cities all have a strong connection with the Islamisation of Java. Therefore, the mosques have a special meaning not only for local residents, but also for the Muslim population in Java, and also Indonesia. On certain days, pilgrims from various regions visit the mosques. Not all cities in Java have such a special meeting, not even Jakarta. Jakarta is considered an outstanding example that combines nationalism, modernism, and capitalism.

Jakarta Currently Jakarta is the largest city in Indonesia in terms of its spatial scale as well as population and economic size. This city originated from a port called Kalapa which was subject to the Hindu Kingdom of Sunda. In the fifteenth century, the Sultanate of Demak conquered Kalapa and renamed it as "Jayakarta" to symbolize its victory against the Portuguese who had always tried to seize the port for years. However, in the sixteenth century the Dutch took over the port and changed the name into the new one "Batavia." Then it became a center of Dutch colonialism in Asia for centuries. After the Indonesia's declaration of independence in 1945, the name "Jakarta" officially replacing Batavia; this name was inspired by "Jayakarta" which it means victory. Then, the city was designated as the capital of Indonesia. Thus, throughout its history, Jakarta had never been as center of Islamic power nor Islam dissemination. Its substantial development was due to its status as the main port and administration center during Dutch colonialism. Not surprisingly, Jakarta is better known as a product of the colonial culture, not the Javanese culture.

Islam came into Jakarta in various ways. Beside the cultural exposure between the local people and Muslim merchants, the existence of the Sultanate of Banten in the westernmost part of Java was also an significant factor. The sultanate was situated at about 40 km (24 mi) to the west of Kalapa. As a vassal-state of Demak, the ruler of Banten established Islam in Kalapa as a part of political and economic efforts to strengthen Demak's position dealing with competition against the Portuguese (Guillot 2008). One of top figures in this campaign was Fatahillah. He was a commander of Demak's troop and also the son-in-law of Sunan Gunung Jati; a member of *Wali Songo*. Some believe that in 1527 he built the first mosque in Jakarta near the harbor (Zein 1999). Currently, that mosque is known as the Al Alam Mosque. Its roof type is similar to the Demak's mosque.

Despite having the high historical value, Al Alam Mosque has never appeared or been recognized as a great mosque in Jakarta. Recently, the people of Jakarta are more familiar with the Al Azhar Mosque and the Istiqlal Mosque as the symbols of

Fig. 127.9 The Al Azhar Mosque with Indian-Middle Eastern style and the Istiqlal Mosque (shown in Fig. 127.10) with the modern style are the most famous mosques in Jakarta (Photo courtesy of Tropenmuseum of the Royal Tropical Institute (KIT), http://id.wikipedia.org/wiki/Berkas:COLLECTIE_TROPENMUSEUM_De_Al-Azhar_moskee_in_Kebajoran_Baru_TMnr_20018508.jpg)

Fig. 127.10 The modern style Istiqlal Mosque is the largest mosque in Southeast Asia (Photo by Gunawan Kartapranata, http://en.wikipedia.org/wiki/File:Istiqlal_Mosque.jpg)

Islam (Fig. 127.9). The Istiqlal Mosque is the largest mosque in Southeast Asia (Fig. 127.10). It was designed with the modern style by a Christian architect, Frederich Silaban, and built in 1963. It is located at the Jakarta's downtown close to the National Monument and the Presidential Palace and next to the most prominent church in Jakarta, the Cathedral of Saint Maria Assumption (De Kerk van Onze

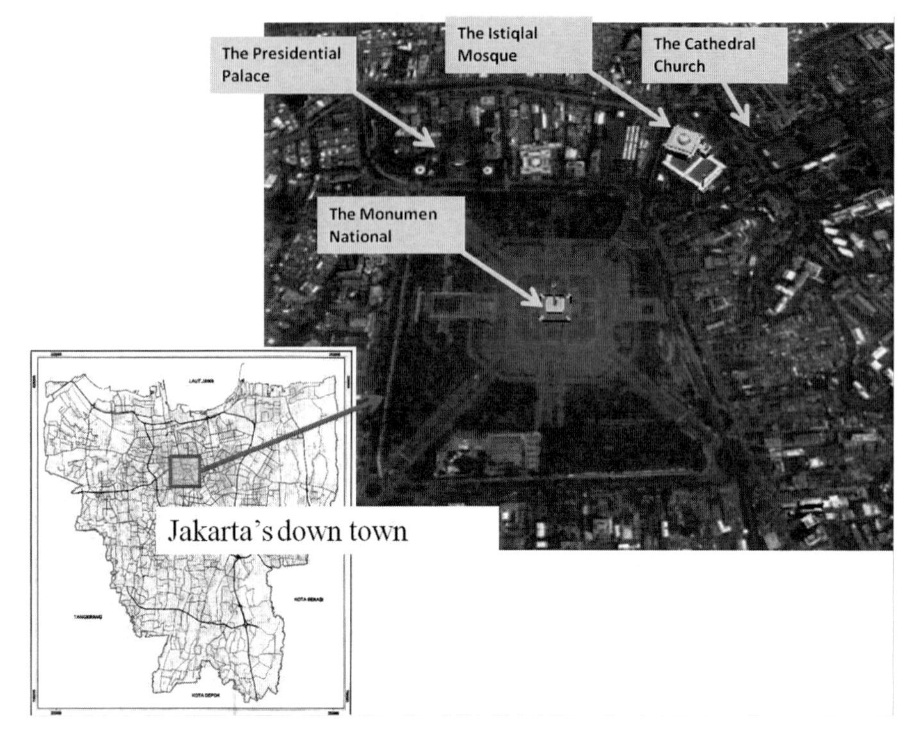

Fig. 127.11 Spatial setting of the Istiqlal Mosque in Jakarta (Map by Hafid Setiadi, modified from Setiadi (2007: 4), image from Google Earth, December 26, 2012)

Lieve Vroume ten Hemelopneming) built by the Dutch in 1810 (Figs. 127.11 and 127.12). "Istiqlal" is an Arabian word for "independence." Meanwhile, the Al Azhar Mosque is situated in one of the prominent commercial districts of Jakarta. The style of its architecture represents a fusion between the Indian and the Middle Eastern's styles.

127.3.3 Reflections on Contemporary Issues

This far I have discussed the differences in Islamic urbanism in four different cities in Java, Tuban, Demak, Yogyakarta, and Jakarta and especially the role of mosque. Based on such discussion, we can see that as a an Islamic symbol, the mosque contributes a major feature in the urban identity of each city. However, the ways and forms of identity creation were different among the three cities. Demak was the only city which showed the important durability of the mosque as a prominent factor in creating urban identity. In Tuban, initially the mosque seemed not played an important role. But, in line with the changes of its style as well as architecture, the mosque suddenly increased its role significantly as a dominant factor in urban landscape.

Fig. 127.12 The Istiqlal Mosque, the Cathedral Church, and the National Monument harmonizing diversity (Photo by Michael J. Lowe, http://en.wikipedia.org/wiki/File:Istiqlal_Mosque_Monas.jpg)

Meanwhile, in Yogyakarta the role of the mosque was a faint symbol of political power. Jakarta's history is quite different inasmuch as it is the political/administrative center of the country.

All examples that in the cases of Javanese cities, the foundation of city, whether agrarian or trade, also contributes in determining the strength of the mosque and the creation of urban identity. The agrarian tradition in the interior region of Java helped found aristocratic cities like Yogyakarta which was more important symbolically and politically than for tis religious meaning. In the coastal region, there was a maritime-trade tradition that stimulated a plural society with various political interest and Islamic currents. In such society, the existence of the mosque was a crucial feature meant to achieve social stability. But, that role increased significantly only in the coastal city (Demak) that had ever served as the center of political power. For this reason the mosque in Demak is more sacred than the mosque in Tuban.

Based on the this analysis, I conclude that the mosque is not always positioned as the city center, even if it is located at the town square. The meaning of religious symbols attached to the mosque may be hidden by the meaning of other symbols such as political or economic symbols in the landscape. When religious symbols are very prominent and strongly connected to local tradition, external incluences, even globalization in today's world, will not undermine the sturdiness of the mosque as a main factor in determining urban identity. In fact, in cities where the mosque has a weak connection to local tradition, the current of globalization may raise the symbolic status of the mosque. But, that status may not be fully fused with the history of the city itself, so that it may turn or influence the urban identity in some new direction.

This last point leads me to apply the term "territorialization" when exploring urbanism and identity. According to Delueze and Guattari (1983), there are two kinds of territorialization, namely de-territorialization and re-territorialization. While the first explains the weakening ties between identity and place, the later focuses on landscaping cultural space with a new identity. They are twofold movements that always emerge as the opposite faces of a single process. From this perspective we may introduce the term of "nomadic" which has a contemporer meaning as "a never ending process of becoming" (Piliang 2005). It describes a gradual and fundamental changes of identity, symbolism and other features of the human landscape. Such changes move constantly in cyclical patterns between territorialization and deterritorialization (Cupers 2005).

If we put the terms of "territorialization" and "nomadic" in the context of this chapter and then we combine them with the concept of centrality, we may provide a short extension with this reflections. We understand that the existence of the mosque as the center often appears together with other structures in the city centers. Therefore, since the beginning, the mosque does not necessarily act as the main feature or as a "big center." In other words, the position of the mosque may be just in a "center in periphery" location, where it may receive only limited attention. To appear as the big center or large central feature, the mosque has to discharge its original identity and also create the new one. This new identity may be far from some of the locational attributes which were important in providing public support for its original existence. But, if there it is in a "big center" location that reflects absolute power, a new identity may also be hampered. Under these conditions both the symptoms of "territorialization" as well as "nomadic" do not occur.

127.4 Concluding Remarks

Without doubt, Islamic values have contributed significantly to urbanism in Java. However, Javanese cities have responded differently to urbanism in accordance with their basic pathway and socio-political mechanism respectively. Urbanism in Java has exposed some conformities about the conception of centrality as it is associated with the position of clergy in political power circles in the past. In line with that, the manifestation of the conception appeared in various forms in Indonesian cities, variations that reflected the history of individual cities. In a coastal city that had never served as political center such as Tuban, the existence of the mosque did not play an important role in creating urban identity until it experienced a dramatic change in contemporary Islamic style and architecture. Otherwise, in the coastal city which had an experience mainly as a political power, the mosque in its original style contributed significantly to urban identity creation, even with the political status of the city have been reduced. In the interior region where there was an aristocratic city, the role of the mosque served as an identity creation and was seen as a political power symbol along with the palace or *kraton*. The above findings illustrate that in the case of Java, the process of urban imposition over Islamic symbols would be challenged both by Islamic tradition and the collective memory of the people at local level, especially in regards to their socio-religious values.

References

Abdullah, T. (1966). Adat and Islam: An examination of conflict in Minangkabau. *Indonesia, 2*, 1–24.

Adiyanto, J. (2011). *Konsekuensi Filsafati Manunggaling Kawula Gusti pada Arsitektur Jawa*. Surabaya: Doctoral Program in Architecture Graduate Program, Institut Teknologi Sepuluh Nopember (ITS). Ph.D. dissertation.

Alfian. (1989). *Muhammadiyah: The politicalbehavior of a Muslim modernist organization under Dutch colonialism*. Yogyakarta: UGM Press.

Al-Hathloul, S. (2004). Planning in the Middle East, moving toward the future. *Habitat International, 28*(4), 64.

Arkoun, M. (1983). Islam, urbanism, and human existence today. In H. Renata & D. Rastorfer (Eds.), *Architecture and Community* (pp. 38–39). New York: Aperture.

Arkoun, M. (1990). The meaning of cultural conservation in Muslim societies. In A. J. Imamuddin & K. R. Longeteig (Eds.), *Architectural and urban conservation in the Islamic world* (pp. 25–33). Geneva: The Aga Khan Trust for Culture.

Ashadi. (2002). Masjid Agung Demak sebagai Prototipe Masjid Nusantara: Filosofi Arsitektur *Journal Arsitektur NALARs, 1*(1), 1–10.

Azra, A. (2004). *The origins of Islamic reformism in Southeast Asia: Networks of Malay-Indonesian and Middle Eastern 'Ulama' in the seventeenth and eighteenth centuries*. Australia/ Honolulu: Allen & Unwin and University of Hawaii Press.

Azra, A. (2006). *Islam in the Indonesian world: An account of institutional formation*. Bandung: Penerbit Mizan.

Bouchair, A., & Dupagne, A. (2003). Building traditions of Mzab facing the challenges of re-shaping of its built form and society. *Building and Environment, 38*, 1345–1364.

Bowen, J. R. (2003). *Islam, law, and equality in Indonesia: An anthropologic of public reasoning*. Cambridge: Cambridge University Press.

Brass, M. (2003). *The nature of urbanism in ancient Egypt*. Essay for Degree MA in Archaeology, University College London.

Bruinessen, M. (1995). Shari'a court, tarekat, and pesantren: Religioun institution in the Banten Sultanate. *Archipel, 50*, 165–200.

Burhanudin, J. (2012). *Ulama dan Kekuasaan, Pergumulan Elite Muslim dalam Sejarah Indonesia*. Bandung: Penerbit Mizan.

Castell, M. (1983). *The city and its grassroots*. London: Arnold.

Chittick, W. C. (2000). *Sufism, a beginner's guide*. Oxford: Oneworld.

Cupers, K. (2005). Towards a nomadic geography: Rethinking space and identity for the potential of progressive politics in the contemporary city. *International Journal of Urban and Regional Research, 29*, 729–739.

De Coulanges, N. D. F. (2001). *The ancient city: A study on religion, laws and institutions of Greece and Rome*. Kitchener: Batoche Book.

Delueze, G., & Guattari, F. (1983). *Anti-Oedipus, capitalism and schizophrenia*. Minneapolis: University of Minnesota Press.

Dewi, K. H. (2012). Javanese women and Islam: Identity formation since the twentieth century. *Southeast Asian Studies, 1*(1), 109–140.
Dickson, A. (2008, July 1–3). *A Chinese Indonesian mosque's outreach in the reformasi era*. Melbourne: Paper presented to the 17th Biennial Conference of the Asian Studies Association of Australia.
Ekadjati, E. S. (2009). *Kebudayaan Sunda Jilid 2 Zaman Pajajaran*. Jakarta: Pustaka Jaya.
Evers, H., & dan Korff, R. (2002). *Urbanisasi di Asia Tenggara, Makna dan Kekuasaan dalam Ruang-Ruang Sosial*. Jakarta: Yayasan Obor Indonesia.
Fletcher, R. (2009). Low-density, agrarian-based urbanism: A comparative view. *Insights, 2*(4), 1–19.
Fox, J. J. (2004). *Currents in contemporary Islam in Indonesia*. Cambridge, MA: Harvard Asia Vision 21.
Geertz, C. (1976). *Religion in Java*. Chicago: University of Chicago Press.
Graaf, H. J. dan, & Pigeaud T. H. (2003). *Kerajaan Islam Pertama di Jawa. Tinjauan Sejarah Politik Abad XV dan XVI*. Jakarta: PT Utama Pustaka Graafiti.
Guillot, C. (2008). *Banten, Sejarah dan Peradaban Abad X-XVII*. Jakarta: Kepustakaan Populer Gramedia.
Hakim, B. S. (1990). The Islamic city and its architecture. *Third World Planning Review, 12*(1), 75–89.
Handinoto. (2010). *Arsitektur dan Kota-Kota di Jawa pada Masa Kolonial*. Yogyakarta: Graha Ilmu.
Harris, D. R. (1972). Swidden system and settlement. In P. J. Ucko, R. Triangham, & G. W. Dimlebly (Eds.), *Man, settlement, and urbanism* (pp. 245–262). London: Gerald Duckworth.
Hartono, S., & Handinoto. (2005). Alun-alun dan Revitalisasi Identitas Kota Tuban. *Dinamika Teknik Arsitektur, 33*(1), 131–142.
Hefner, R. W. (2002, March 7–9). *Globalization, governance, and the crisis of Indonesian Islam*. University of California, Santa Cruz: Center for Global, International, and Regional Studies, Conference on Globalization, State Capacity, and Muslim Self Determination.
Heine-Geldern, R. (1942). Conceptions of state and kingship in Southeast Asia. *The Journal of Asian Studies, 2*(1), 15–30.
Herusatoto, B. (2008). *Simbolisme Jawa*. Yogyakarta: Penerbit Ombak.
Hiller, B., & Hanson, J. (1984). *The social logic of space*. Cambridge: Cambridge University Press.
Hull, V. J. (1976). *Women in Java's rural middle class: Progress or regress?* Yogyakarta: Lembaga Kependudukan, Universitas Gadjah Mada. Working Paper Series no. 3.
Jay, R. R. (1963). *Religion and politics in rural central Java*. New Haven: Yale University, Southeast Asia Studies. Cultural Report Series.
Jayamedho, V., & Ekoputro, Z. A. (2010). *The hinterland harmony: Buddhist and Muslim relations in Indonesian rural areas*. Paper presented at the ASIA-AFRICA Diversity in Globalised Society; The Role of Asia and Africa for a Sustainable World 55 Years after Bandung Asian-African Conference 1955. Yogyakarta: Gadjah Mada University Graduate School.
Johnson, A. (1974). Ethnoecology and planting practices in a swidden agricultural system. *American Ethnologist, 1*(1), 87–101.
Keyfitz, N. (1976). The ecology of Indonesian cities. In Y. M. Yeung & C. P. Lo (Eds.), *Changing South-East Asian cities: Readings on urbanization* (pp. 125–130). Singapore: Oxford University Press.
Levine, G. J. (1986). On the geography of religion. *Transactions of the Institute of British Geographers, New Series, 11*(4), 428–440.
Lombard, D. (2005a). *Nusa Jawa: Silang Budaya, Kajian Sejarah Terpadu. Jilid 1. Batas-Batas Pembaratan*. Jakarta: Gramedia Pustaka Utama & Forum Jakarta-Paris.
Lombard, D. (2005b). *Nusa Jawa: Silang Budaya, Kajian Sejarah Terpadu. Jilid 2. Jaringan Asia*. Jakarta: Gramedia Pustaka Utama & Forum Jakarta-Paris.
Lombard, D. (2005c). *Nusa Jawa: Silang Budaya, Kajian Sejarah Terpadu. Jilid 3. Warisan Kerajaan-Kerajaan Konsentris*. Jakarta: Gramedia Pustaka Utama & Forum Jakarta-Paris.

Lukens-Bull, A. (2000). Teaching morality: Javanese Islamic education in a globalizing era. *Journal of Arabic and Islamic Studies, 3*(2000), 27–47.

Lukens-Bull, R. (2006). Pesantren education and religious harmony: Background, visits, and impressions. In M. Pye, E. Franke, A. T. Wasim, & A. Mas'ud (Eds.), *Religious harmony: Problems, practice and education* (pp. 315–321). Berlin: de Gruyter.

Maarif, A. S. (2009). *Islam dalam Bingkai Indonesia dan Kebhinekaan: Sebuah Refleski Sejarah.* Bandung: Penerbit Mizan.

Martono, S. (1974). *State and statecraft in old Java: A study of the later Mataram period, 16th to 19th Century.* Monograph Series. Ithaca: Cornell University, Southeast Asia Program Department of Asian Studies.

Maulana, R. (2010, November 1–4). *Pergulatan Identitas Muslim Tionghoa: Sebuah Cerita dari Yogyakarta* Banjarmasin: Annual Conference on Islamic Studies (ACIS).

Moenthadim, M. (2010). *Pajang: Pergolakan Spiritual, Politik, dan Budaya.* Jakarta: Genta Pustaka & Yayasan Kertagama.

Muhaimin, A. G. (2006). *The Islamic traditions of Cirebon, Ibadat and Adat among Javanese Muslims.* Canberra: Australian National University Press.

Mulder, N. (2001). *Mistisme Jawa, Ideologi di Indonesia.* Yogyakarta: LKIS.

Mulyana, S. (2005). *Runtuhnya Kerajaan Hindu-Jawa dan Timbulnya Negara-Negara Islam di Nusantara.* Yogyakarta: LKIS.

Munoz, P. M. (2009). *Kerajaan-Kerajaan Awal Kepulauan Indonesia dan Semenanjung Malaysia: Perkembangan Sejarah dan Budaya Asia Tenggara (Jaman Pra Sejarah – Abad XVI).* Yogyakarta: Penerbit Mitra Abadi.

Nakashima, D., & Rouè, M. (2002). Indigenous knowledge, peoples and sustainable practice. In P. Timmerman (Ed.), *Social and economic dimensions of global environmental change. Encyclopedia of global environmental change* (Vol. 5, pp. 314–324). Chichester: Wiley.

Nas, P. J. M. (1986). The early Indonesian town: Rise and decline of the city state and its capital. In P. J. M. Nas (Ed.), *The Indonesian city, Studies in urban development and planning* (pp. 18–36). Dordrecht/Cinnaminson: Foris Publication.

Park, C. (2004). Religion and geography. In J. Hinnells (Ed.), *Routledge companion to the study of religion* (pp. 439–455). London: Routledge. Chapter 17.

Piliang, Y. A. (2005). *Transpolitika, Dinamika Politik dalam Era Virtualitas.* Yogyakarta/Bandung: Jalasutra.

Polanyi, K. (1968). *The great transformation: The political and economic origins of our time.* Boston: Beacon Press.

Pratomo, S. (2001). Makna Struktur dan Unsur Pembentuk Pusat Kota Pelabuhan Tuban, Kajian Morfologi dan Silang Budaya Pusat Kota Pesisir. Semarang: Graduate Program University of Diponegoro. MA Thesis in Architecture.

Purwadi, & Maharsi. (2005). *Babad Denak: Sejarah Perkembangan Islam di Tanah Jawa.* Yogyakarta: Tunas Harapan.

Rahardjo, S. (2007). *Kemunculan dan Keruntuhan Kota-Kota Pra-Kolonial di Indonesia.* Depok: Universitas Indonesia. Fakultas Ilmu Budaya.

Rahardjo, S. (2011). *Peradaban Jawa: Dari Mataram Kuno Hingga Majapahit Akhir.* Depok: Komunitas Bambu.

Reid, A. (1992). *Asia Tenggara Dalam Kurun Niaga 1450–1680: Tanah di Bawah Angin.* Jakarta: Yayasan Obor Indonesia.

Reid, A. (1999). *Dari Ekspansi Hingga Krisis: Jaringan Perdagangan Global Asia Tenggara 1450–1680.* Jakarta: Yayasan Obor Indonesia.

Ricklefs, M. C. (2005). *Sejarah Indonesia Modern 1200–2004.* Jakarta: PT Serambi Ilmu Semesta.

Rutz, W. (1987). *Cities and towns in Indonesia.* Berlin/Stuttgart: Gebruder-Borntraeger.

Santoso, J. (2008). *Arsitektur Kota-Kota Jawa: Kosmos, Kultur, dan Kuasa.* Jakarta: Centropolis-Magsiter Teknik Perencanaan Universitas Tarumanegara.

Sarmela, M. (1987). Swidden cultivation in Finland as a cultural system. *Suomen antropologi – Finnish Anthropologist, 4*(1987), 1–36.

Setiadi, H. (2007). *Urbanization and Urban Environment Quality in Jakarta*. Paper presented at Joint Seminar University of Indonesia (UI) – University Kebangsaan Malaysia (UKM) Current Research in Natural and Mathematical Sciences: Collaboration Opportunities at UI and UKM, 19–20th June 2007. Depok: Faculty of Science University of Indonesia.

Setiadi, H. (2012). *Islamisation and urban growth in Java, Indonesia: A geopolitical economic perspective*. Cologne: Paper presented to the 32th International Geographical Congress.

Shih, A. (2002). The roots and societal impact of Islam in Southeast Asia. Interview with Professor Mark Mancall. *Stanford Journal of East Asian Affairs, 2*, 114–117.

Soud, M. Y., Hasshim Bin Mat, & Kausar Binti Hj Ali. (2010). The aspects of decline in contemporary Islamic cities. *European Journal of Scientific Research, 47*(1), 43–63.

Stoddard, R. H., & Prorok, C. V. (2003). Geography of religion and belief systems. In C. J. Willmott & G. Gaile (Eds.), *Geography in America at the Dawn of the 21st Century* (pp. 759–767). New York: Oxford University Press.

Sumardjo, J. (2002). *Arkeologi Budaya Indonesia*. Yogyakarta: Qalam.

Syam, N. (2005). *Islam Pesisir*. Yogyakarta: LKIS.

Thompson, P. B. (2010). *The agrarian vision: Sustainability and environmental ethics*. Lexington: University Press of Kentucky.

Tjandrasasmita, U. (2009). *Arkeologi Islam Nusantara*. Jakarta: Kepustakaan Populer Gramedia.

Vickers, A. (2009). *Peradaban Pesisir, Menuju Sejarah Budaya Asia Tenggara*. Denpasar: Pustaka Larasati and Udayana University Press.

Wahby, A. E. I. (2007). The architecture of the early mosques and shrines of Java: Influences of the Arab merchants in the 15th and 16th centuries. Bamberg: University of Bamberg. Ph.D. Dissertation in der Fakultät Geistes und Kulturwissenschaften (GuK).

Wastherdahl, C. (1992). The maritime cultural landscape. *The International Journal of Nautical Archaeology, 21*(1), 5–14.

Weber, M. (1930). *The Protestant ethic and the spirit of capitalism*. London: Allen and Unwin.

Wertheim, W. F. (1956). *Indonesian society in transition. A study of social change* (2nd ed.). Bandung: Sumur Bandung.

Wheatley, P. (1983). *Nagara and Commandery: Origins of the Southeast Asian urban tradition*. Chicago: Department of Geography University of Chicago.

Wheatley, P. (2001). *The places where men pray together: Cities in Islamic lands, 7th through the 10th centuries*. Chicago: University of Chicago Press.

Woodward, M. R. (1989). *Islam in Java: Normative piety and mysticism in the Sultanate of Yogyakarta*. Tucson: University of Arizona Press.

Yusuf, M. (2006). *Sejarah Peradaban Islam di Indonesia*. Yogyakarta: Pustaka.

Zein, A. B. (1999). *Masjid-Masjid Bersejarah di Indonesia*. Jakarta: Gema Insani Press.

Chapter 128
The Catholic Church and Neo-Gothic Architecture in Latin America: Scales for Their Analysis

Martín M. Checa-Artasu

128.1 Introduction

The construction of churches, temples and cathedrals in the neo-Gothic style in Latin America was a constant during the final quarter of the nineteenth century and the first three decades of the twentieth century. The construction of these temples, beyond their architecture, became part of a solution serving the political and social needs of the Church. Through this idea we can understand these buildings as being symbols of the balance, sometimes conflicting, sometimes fully collaborative, between the ecclesiastical hierarchy and the national governments which arose at that historical moment. In addition, those buildings are a reflection of the Catholic revival which occurred in the Western hemisphere starting in the last third of the nineteenth century, as they express attempts at the positioning, both social and territorial, of a Catholic hierarchy that was attempting to emerge after years of wars, conflicts, transfers of property and expulsions.

In this paper, I formalize the early stages of research on the extent and forms of neo-Gothic religious architecture in Latin America. This is an analysis which seeks to understand how the Catholic Church has taken this style and used it, directly or indirectly, as an additional element in a complex policy of integrating itself into societies of the then-new (nineteenth century) Latin American countries. Thus, we seek to go beyond both the simple listing of works and architects and the mere architectural and stylistic description of these places of worship.

To understand this phenomenon, we use the geographical concept of scale as the modulating element. It allows us to structure the role of the Church on the continent in the late nineteenth century, using the neo-Gothic architectural style as a pretext

M.M. Checa-Artasu (✉)
Department of Sociology, Universidad Autónoma Metropolitana,
Unidad Iztapalapa, Mexico, DF, Mexico
e-mail: martinchecaartasu@gmail.com

S.D. Brunn (ed.), *The Changing World Religion Map*,
DOI 10.1007/978-94-017-9376-6_128

through three scales attached to geographical and political notions: the nation or the state, the territory, and the local. This analysis will also allow us to offer an outline of the spatial impact of the Catholic Church throughout the continent in the late nineteenth century.

128.2 The Concept of Scale as Related to the Catholic Church and Neo-Gothic Style in Latin America

Scale is a concept that has been used to explain various concepts emanating from geography that investigate the relationships between humans and space. This is not just a technical issue associated with a geometricized view of the world, but a concept with a more complex understanding with which we can associate humans with the series of objects or elements that exist or occur in space. The dynamics and links between humans, objects and scale and are also part of an ongoing geographical discourse (Batlori 2002:7).

In light of this thinking we can consider the Catholic Church to be an object in geographical space which is able to generate a certain territoriality that is mediated by a series of scales in which it interacts with humankind and from which different characteristics are derived which are susceptible to analysis in a study. This perspective explains why recent works that relate geography and religion consider the role of scale as important. It is also used as an explanatory concept that is useful in studying religion in a spatial context (Rosendahl 2005:12933; Stump 2008; Sack 1986:92–127).

Thus, for example, Brazilian geographer Zeny Rosendahl (1996:32) classifies this linkage between religion and geographic space into three main groups: *faith, space and time*, which incorporate the dissemination and area of coverage of a faith or belief in a given space; *the centers of convergence and irradiation* of the same, and *the sacred space and the perception* which considers the experience and symbolism which are derived from this relationship American geographer Roger Stump (2008:221) also suggests the existence of three types of scales, which he calls: *communal, narrower and wider*. These define religious territoriality in a secular space. These scales, in turn, interact among themselves with the objective of defining a specific territoriality. In my view, the *communal* scale would be that which is determined by territoriality in the community spaces for practice of a particular religion. For the Catholic world, these include churches, parishes, shrines, chapels, convents and monasteries. The *narrower* scale would link the territoriality of the religious act with the person; that is, it would be confined to the body, the home, the family. The *wider* scale would be that which expresses territoriality from institutions with a religious bias, but that coexist with social and political structures, which may be secular or of other religions (Stump 2008:224). For the Catholic religion, we would be referring to categories such as the diocese or archdiocese. The scales can also be categorized within themselves in terms of actions that believers of one religion or another perform. For example, the action of prayer, worship or meditation confers a specific territoriality that coexists with different actions performed in the same scale or in a different one. For example, in one Catholic church, a person may be praying, while another may be walking around, varying the sense of scale if the person walking around is a tourist who arrives in this building with the intention of seeing it or if the traveller is someone who completes a pilgrimage to make a request to a particular saint.

In this regard, the geographer Robert Sack (1986:92–125) tells us about territorialization, which has evolved over time in the Christian church in various forms since its origins, all which are attached to the church hierarchy and with a clear economic component mitigated by the processes of reform, the emergence of industrialization, capitalism and liberalism. For Sack, this reasoning aids in our considering the Catholic Church to be a traditional institution, since it has clung to a territorial distribution which, while it has been diluted by the development of other economic powers, continues to be a notion creating contingent spaces which contain the sacred: the temple, the church and the cathedral, all which also support it in an exclusive manner with other spaces that do not contain it. From this perspective, Richard Kieckhefer (2004:15) limits the scale analysis to the sacred space of a temple because it incorporates concepts such as liturgical use, which consists of spatial dynamics or the central focus and a desired response and where both the ascetic impact and symbolic resonance are considered. This interpretation considers the spatiality of the temple and how humans unfold in it, not only from a mobility standpoint, but also how that space stimulates and recreates perceptions, feelings and events relating to connectivity with the sacred or turning that space into a socio-cultural production of the first magnitude.

128.3 Scales of Action

The brief theoretical framework outlined above helps us, using the context of the neo-Gothic architecture of different Catholic churches, to determine the role of the Catholic Church in Latin America between the last third of the nineteenth century and the first three decades of the 20th. These were years when this type of architectural style had a high degree of significance on the continent. Neo-Gothicism was one of many emerging historicisms which arose at that time, it was used by architects both because of personal taste or their formative influences, as well as due to the wishes of their clients, that is, members of the Catholic hierarchy, who had a fondness for this style based on its symbolic and ecclesiological nature. This school of architecture can be traced back to previous times of grandeur in the Church in socioeconomic terms. But it also had a versatile mystical component, since Gothic architecture was associated with the idea of a heavenly Jerusalem and a transcendent approach to the sacred.

Nevertheless, Gothic architecture was a foreign and imported style into Latin America, which looked at the American universe through different lens. This fact alone places it in an ambivalent logic. For certain Latin American political constituencies, it was synonymous with a necessary grandeur attached to a concrete modernity that linked it to processes justifying the construction and consolidation of the State, and which, in addition, were also blessed by the Church. For this view to be realized, especially for the Church's hierarchy, it represented the style that suggests a return to a glorious past where the Church had an axial role in society. Neo-Gothic, perhaps like no other architectural style, hid the desire for a return to the past. It was associated with nation-building and at the same time, it proposed a different access to modernity (Gil 1999:24–25). Thus history and modernity met in local environments; cities and towns and were viewed within the construction or restoration of Catholic churches in the neo-Gothic style.

These architectural features, which reflect diverse and complex decisions, can be analyzed considering different scales where space, humans and objects interact and generate different processes with and among each other. This diversity scale, for the case in question, can be grouped into three broad categories.

128.3.1 First Scale: Church vs. State (Nation)

The role of the Catholic Church in Latin America throughout the nineteenth century can be considered to be as complex as it is multifaceted (Krebs 2002:315–320). This paper does not seek to analyze that occurrence, but it is useful to address some issues that connect this period with approaches used in this research. Specifically, during the second half of the nineteenth century, new trends arose beyond the already familiar defensive attitude of the Church. These were translated into diverse developments, including the review of approaches regarding religious pluralism and the spread of Protestantism on the continent, the development of positivism among the intellectual and political elites or state royalism, Romanization with the establishment of a formal agreement between the Holy See and the nations in the Americas and the reliable dissemination of papal encyclicals as well as other directives. Also, the consolidation of an Ultramontane church (it asserts the superiority of Papal authority over the authority of local temporal or spiritual hierarchies) in various nations, which was then fighting not only against liberalism, but also with other phenomena such as Protestantism or secularism and a notable extension of the action of the religious orders associated to the development of social Catholicism. Some orders like the Jesuits were reinstated in Latin America after the expulsions in the eighteenth century. Other, newly created orders were the Salesians and the Vincentians and more ancient including Discalced Carmelites and the Franciscans.

It should be noted that the Jesuits were expelled from the Spanish and Portuguese colonial dominions between 1754 and 1767. After that their role in structural social and economic terms through the Jesuit traditions vanished. The Society of Jesus would not return to Latin American until the last third of the nineteenth century.

This development explains the nineteenth century religious revival which covered the entire Catholic world, and especially Latin America. This revival was not without violence, exiles, the disposal of properties and even war; for example, the well-known conflicts between Church and state or between liberals and conservatives throughout the nineteenth century.

It is precisely this confrontation that led the Church to the renewal of education of the clergy in new seminaries, some of which were coteries of Ultramontane positions, forming most learned priests. It also involved improving the internal management and organization of the church, and especially the expansion of its geographical distribution. Thus, new dioceses and parishes as well as synods and episcopal conferences were created, with the most prominent example being the Latin American Plenary Council held in Rome in 1899. These conferences provided structure to and revitalized the social and evangelizing mission of the Church, which in some cases, evolved into active roles in politics, forming parties of Catholic inspiration that were blessed by the hierarchy in order to combat ongoing secularism.

It is in this context that we must understand the extension of neo-Gothic revival, its role beyond architecture and its nature as a milestone on the scalar action of the Church in the Americas. The sense of that milestone was evident in the many neo-Gothic religious buildings constructed and their role as shapers of new landscapes. Most of these structures reflected a dialogue between the political and economic elites and also the Catholic Church, all who played important roles in national building in the American republics. When we examine these dialogues, we discover new and refined invocations that justify general policies made by governments, always in key of national construction. The power of justifying these invocations would be defined by the sacred attributes that gave the Church, through its first voice, the Pope, much respect on such matters.

None of these developments ignored the diverse and complex relationships between church and State which occurred throughout Latin America. These issues were often core political issues throughout the nineteenth century in many states were experiencing, including being subjected to riots, revolutions and coups.

An example of the aforementioned is the use of the invocation of the Sacred Heart to consecrate cities, regions and even entire nations, all during the same period of emergence and distribution of neo-Gothic architecture. It should be recalled that in 1856, Pope Pius IX established the liturgical feast of the Sacred Heart and in 1899, Leo XIII consecrated the entire human race to the Heart of Jesus, explaining this action in the encyclical *Annum Sacrum Pos* (1899). The dedication, however, had its origin in 1675 with the unveiling of a statue of Jesus with an open heart which speaks of the ingratitude of humans before the love emanating from Jesus Christ to the religious of the Order of the Visitation of St. Mary: Saint Margarita María Alacoque (1674–1690). It should be noted that this invocation arrived in the Americas in the early decades of the eighteenth century by the Jesuits. After their expulsion in 1767 from Spanish America, the dedication remained, but was taken up with renewed vigor with the return of the Society of Jesus and other religious orders (Diaz Patiño 2010:97). It is from this source that the idea of a blessing or dedication should focus on the spirituality formalized by an institution and one which proposes a relationship good faith. It was also a blessing that included the construction of methods of persuasion, domination and submission, through which members of a community believed they were really represented (Diaz Patiño 2010:99). This explains why the consecration of a territory or a nation is understood as an act of collective efforts, where the Catholic faithful, the community formalized in a

Fig. 128.1 Basilica del Voto Nacional in Quito, Ecuador, the result of the consecration of Ecuador to the Sacred Heart of Jesus. It was designed by French architect Emile Tarlier in 1883 (Photo by Martín Checa-Artasu)

territory, and in some cases, a nation, recognized the harmful effects against the Heart of Jesus and by extension, the Church, and made prayers both to mitigate this grief and to seek atonement (Capelluti 2007:240). To undertake this change would necessitate the construction of large Catholic churches, which were not exclusively neo-Gothic in style, although most were in that style.

The most emblematic example of Neo-Gothic faces is certainly that of the *Basilica del Voto Nacional* in Quito, Ecuador, the result of the consecration of Ecuador to the Sacred Heart of Jesus (Kingman, 2003:101–105) (Fig. 128.1). This structure was instigated by the pro-Catholic government of General Gabriel García Moreno, president of the country in two periods from 1861 to 1865 and from 1869 to 1875. He saw it as a justification both of his policy and of the nation-building process in Ecuador throughout the nineteenth century (Ayala 1981:145–151). Suffice it to say that Garcia Moreno had established excellent relations with the Holy See through an arrangement in 1863 which gave considerable privileges to the clergy, the Jesuits and other orders who were included as an integral part of the power structures. His government, ruled by a conservative Messianism, was also an example of shifting movements, which at times was the extremes between political ideologies of conservatives and liberals in Latin American countries in the nineteenth century.

The Quito basilica was promoted by a group of Quito citizens, headed by the representative and priest José Julio María Matovelle Maldonado (1852–1929), who proposed the construction of a large National Temple to the Sacred Heart with the Andean nation taking advantage of the devotion conferred by Pope Pius IX by a

Conciliar Decree from August 31, 1873. Ten years later, on June 23, 1883, the government decreed the building of a National Basilica to the Sacred Heart of Jesus. The 1884 Convention approved it in the February 29 meeting, granting 12,000 sucres (the same equivalent as the U.S. dollars at that time) to construct it (Larrea 1976:117–119; Egas et al. 1994:57; Matovelle 1934). This example illustrates perfectly the process described, viz., a dedication to which the papacy granted a number of attributes to those who adhered to its policies and results in support for a large cathedral or temple, the conciliator and protective role attributed to it in a nation. In turn, the structure becomes an urban landmark in the nation's capital, associated with political power and granting the Catholic Church exceptional visibility for endorsing the power which continued to maintain its role as a moral protector and beacon for the faithful in society. Today the Catholic Church serves as an element of political reconciliation in the interest of nation building.

128.3.2 The Second Scale: Church vs. Territory

The second scale, Church versus territory, considers the fact that many countries in the Americas, at the same time as the spread of neo-Gothic style, developed strategies for the expansion, domination and control of territories, referred to as "deserts" or "empty" space. These territories were not yet under effective state control, since the state had scarcely exercised its presence and, in many cases, even though it had acquired this contact since colonial times. For those expansionist strategies, the religious orders that were again allowed in Latin America, beginning in the last third of the nineteenth century, would play a major role. The religious orders introduced their evangelizing efforts and socioeconomic strategies in large areas where there was much modernization and development. These efforts helped integrate these territories into the nations to which they belonged. All of these initiatives were associated with the construction of hospices, hospitals, schools, craft and trade centers, agricultural institutes, universities and, of course, churches and chapels, where neo-Gothic became a predominant style to help to shape symbols on the landscape. They would also provide evidence of the construction and modernizing actions for the state from the Church.

There are several examples of the above church-state relationship which occurred in all types of spaces, whether urban or rural. One of the most striking examples was in Colombia. The architectural model focused on the civic work of a Spanish and Discalced Carmelite friar: Andrés Huarte Arbeloa who made some Gothic temples in Medellín, notably the church of *Señor de las Misericordias* (1921–1931) (Vélez White 2003:51–53; Osorio Gómez 2008:108) and also in other villages in the Antioquia department. In Frontino, in northwestern Antioquia, in the jungle region of Urabá, Arbeloa designed the *Basílica Menor de Nuestra Señora del Carmen* (1914–1929) with the help of Father Daniel García Puelles (Elejalde Arbeláez 2003:105). This basilica is the centerpiece, due to its symbolism, of the Carmelite mission of Urabá de los Katíos, the name being associated with the Discalced Carmelites in the region (Miranda Arraiza 2003:47). That mission was a major component of a "colonization" operation of the northwest Antioquia indigenous subregion promoted by the Colombian government given the intense activity of gold mining that was linked to the elites' interests the area (Piazzini 2009:195). The Carmelites, many of them coming from Spain, proved to be the vehicle which the Catholic religion extended its mantle of civilization. In 1916 they began promoting the construction of Frontino as the capital of the Apostolic Prefecture of Urabá. It was an ecclesiastical jurisdiction governed by a prefect appointed directly by the Vatican; it was granted independence from the diocesan system. The jurisdictional hierarchy worked diligently in line with the territorial scheme of the Colombian state. This arrangement explains that while the Prefecture existed (1916–1941), there were tensions between the order of the Carmelites and the clergy of Santafé de Antioquia, the nearest diocese (Gálvez Abadía 2006:83 and 90). The two Carmelite Fathers, García Puelles, who was a member of the mission of Uraba and builder of the temple, and the other, Andrés Huarte, a Medellín resident who made the plans, designed the neo-Gothic temple. Their thinking was that the "modernizing" impact of this architecture style in a jungle context and in the framework of the prevailing socioeconomic organization, which the monks sought to develop in that territory, was protected by the Colombian state. Puelles and Huarte also required the "pacification" of that territory within the context of the emergence of gold mining and the attraction of national and foreign capital that it required.

128.3.3 The Third Scale: Church vs. Local

The third scale, which I call *Church versus local* analyzes the role that is played in urban spaces, specifically where there is the construction of new churches and parishes, many in neo-Gothic styles. The role of the Church goes beyond the construction itself and considers the position of these new places of worship within the context of the growth of cities and towns. Here again, we need to consider a multi-scale relationship between the Catholic Church with the city, but also with humankind. There are several scalar arrays which illustrate this relationship. The first architecture is inside the church itself, which will be determined by the series of sections and architectural components of the building, all which encourage the believer, man or woman, to provide the full range of functions associated with a connection to the sacred. A second scale array will be in the external environment closest to the temple, where the interaction between humankind and the temple is marked with functions inherent to the relationship itself: belief, faith or prayer including processions, novenas, coronations or Via Crucis. The third scale brings us close to the relationship between the church and the city, that is, through building elements that contribute to what the architecture represents to the public. The atrium, the square in front of the church, is the main element of this scalar dialectic. This feature invites the access of both the believer and the citizen. It can also be extended and integrated into the city and acquiring, depending on the size of the plaza, a new centrality. Thus, civic spaces were created, which are susceptible to being used in many different ways and with diverse functions by residents

Fig. 128.2 Templo del Sagrado Corazón de Jesus in Leon, Guanajuato, started in 1921 and designed by Mexican architect: Luis G. Olvera (Photo by Martín Checa-Artasu)

(leisure, tourism, belief, etc.). The towers, steeples and the volumetric structure of the temple also enhance the connection between the sacred and the city. The monumentality derived from those elements makes the cathedral or temple a symbol of the city, an icon of it and of its economic, social and cultural development.

An example of the above symbolism is the series of monumental neo-Gothic churches started in the late nineteenth early twentieth century in western Mexico: the *Templo Expiatorio del Santísimo Sacramento* in Guadalajara, designed by Italian Adamo Boari in 1901 (Moya Pérez 1998:207), the Church of *San José Obrero* in Arandas, Jalisco, started in 1902 (Checa-Artasu 2012), the *Santuario de la Vírgen de Guadalupe* in Zamora, Michoacán from 1898, and the *Templo del Sagrado Corazón de Jesus* in Leon, Guanajuato, started in 1921 and designed by Mexican architect: Luis G. Olvera (Fig. 128.2) (Checa-Artasu 2011a, b). These are large churches which have maintained this internal and external scalar set, reinforcing it after long periods of construction and enduring all types of economic and political shocks and which today, have revived it through tourism and city marketing. Similar examples are seen in other large neo-Gothic churches located in cities on the continent, such as the Basilica de Nuestra Señora de Luján in Luján, Argentina, where the aforementioned scale set is enhanced with the annual pilgrimage, which reinforces the symbolic value for a broad community with a national nature (Hadad and Venturiello 2007:34–36). Another example of this symbolism which is both urban centrality and evidence of the city's growth, is of a specific historical monument, represented by the Igreja de la Sé, the neo-Gothic cathedral of Sâo Paulo (Gonçalves et al. 2006:84–86) (Fig. 128.3).

Fig. 128.3 Igreja de la Sé, the neo-Gothic cathedral of São Paulo in Brazil built from 1913 by German engineer Maximilian Emil Hehl (1861–1916) (Photo by Martín Checa-Artasu)

It was through this set of scales that a number of agents interacted to promote, build and develop the church and surrounding community. Among these agents were the priests and pastors appointed by the hierarchy to meet the needs of worship, in many cases of newly established or growing populations. There were also members of religious orders with functions similar to the above, but with a greater capacity to generate civic and social infrastructure that are mediated within the urban territory. They played an important role in building temples of neo-Gothic features, town halls and accompanying buildings and also decisions about the construction of colleges, technical schools, hospitals, infirmaries and universities. These decisions were made by many of the same religious officials who often came from the same European countries as the major religious orders; in short, they would bring medieval-art styles with them. There are many examples of the above structures throughout the continent. In Bolivia there is the prominent work of neo-Gothic facies of Spanish Jesuit Eulalio Morales (1837–1907) in La Paz (Menacho 2001:2738). In Quito and also in various other provincial cities of Ecuador, there is a notable Central Europeal Gothicism of Pedro Humberto Brüning (1869–1938), a German priest of the Congregation of the Mission (Cevallos Romero 1994).

In Argentina, the Italian architect, Ernesto Vespignani (1861–1925), member of Salesians of Don Bosco, is associated with important projects in medieval style, between Gothic and Romanesque, throughout the country and even in Uruguay, Peru and Bolivia (Cufré Pedro 2009:309; Gil and Wichepol 2004; and Petriella and Sosa Miatello 1976; Bruno 1984). In the Colombian piedmont there is the Salesian, Giovanni Buscaglione (1874–1941), disciple of Vespignani, with works related to that order throughout the country (APRAARQ 2004:45–60; Rozo Montaña 2000; Carrasco 2004:137–168; Del Real 1942; Patiño, Patiño de Borda 1995:114). Among the most outstanding works of this Salesian is the Santuario Nacional de Nuestra Señora del Carmen in Medellín. This temple was designed mainly by Buscaglione, an imitation of the Buenos Aires Church of San Carlos Borromeo, designed by Vespagni a few years before. Buscaglione built it between 1927 and 1938. Its facade is a Gothic style with traces of Florentine influence and Arabic tones and has exuberant external chromatic decoration, which place it in the category of a relevant urban structure in the La Candelaria Barrio, where it is located. The decoration and architecture reinforce its character as a civic and religious symbol. Suffice it to note that the Colombian Congress approved it in 1926 as national temple. This fact justified the continuation of very important following to the Virgin del Carmen in the city and also was the excuse for channeling the river San Agustin that allowed for the urban development of the area (Vélez White 2003:39–40; Saldarriaga Roa 2007).

Residents in these church communities are also agents based on their religiousness. It is not uncommon to find examples where the owner of an undeveloped plot of land transferred it to the Church so that it can build a temple considered necessary for that community. Also from an ecclesiastical perspective the transfer was justified with a miracle or a mystical sign. Finally, the architect or secular master of works would build the temple, adhering to the desires of the main client, whether the parish priest, the hierarchy or the parishioners.

All of these constructions coincide with the migration patterns from European countries that reached Latin America at the end of the nineteenth century and the beginning of the next. Among them were architects and engineers, who brought their knowledge of styles such as Gothic and Romantic, as well as masons, carpenters, cabinetmakers, blacksmiths and artists, all who participated in the design and construction of many types of buildings, including churches. Numerous examples tying immigrant architects and neo-Gothic architectural achievements abound. We cite three examples: Maximilian Emil Hehl (1861–1916), a German engineer who moved to Brazil to work as a mining technician, later developing an extensive work, which includes three Catholic temples with neo-Gothic features: the Metropolitan Cathedral and the Church of Consolation, both in Sâo Paulo, and the Cathedral in Santos (Malta Campos and Simões Júnior 2006:32 and 77; Francia Cerasoli 2010;

Niccoli Ramirez and Lindenberg 2010). Belgian engineer Agustín Goovaerts (1885–1939), who arrived to Medellín after World War I and developed extensive construction projects, notably some churches resembling Gothic styles, even though this technique included numerous influences of Art Noveau. Among these are the following: in Medellín: the church of Sagrado Corazón de Jesús began 1923 and the construction in the Antioquía Department of several churches: Nuestra Señora del Rosario in Donmatias, designed in 1926; the Inmaculada Concepción in Caramanta, both done with the architect Tomas Uribe and Nuestra Señora de las Mercedes, in Montebello (Vélez White 1994; Molina Londoño 1998). Another example of a professional who fled old Europe due to WWI was Italian architect, painter and photographer, Augusto Cesar Ferrari (1871–1970), who after arriving in Buenos Aires in 1914, did several religious works, inspired by and imitating the French Gothic style, such as the Church of Sagrado Corazón in Córdoba, which began construction in 1928, and the parish of Nuestra Señora del Carmen in Villa Allende (Aliata 2003).

In addition to the specialized migration or church architects, we must mention local professionals educated in their countries or elsewhere. One example of many is the Costa Rican Lesmes Jiménez Bonnefil, a mining engineer educated in Louvain, Belgium, and designer of the most notable neo-Gothic examples in that Central American nation, the Church of La Merced in San José, San Isidro in Coronado, and San Rafael in Heredia (Fig. 128.4). These integrate the use of structures and cast-iron frames with brick; stylistically they have a clear resemblance of

Fig. 128.4 Costa Rican Lesmes Jiménez Bonnefil, a mining engineer designed the neogothic Church of La Merced in San José in 1894 (Photo by Martín Checa-Artasu)

German Gothic style (Vargas Cambronero and Zamora Hernández 1999; 123; Sanou and M. Ofelia 2001:140–143). These include constructions, all done in a governmental context marked by a hard-lined liberalism that would strain Catholics' political action and which was seen in those temples as a stronghold of the Church-State confrontation. It was not resolved in Costa Rica until well into the twentieth century (Troyo Calderón 2002:13–15).

128.4 Conclusion

As stated above, this paper analyzes the multifaceted role of the Catholic Church in Latin America at the end of the nineteenth century, using as an example the construction of churches in neo-Gothic styles. This task justified the use of a modular element, viz., geographic scale, in order to understand these religions and political processes.. In this way, we are able to provide an analysis through three scales: the nation or the state, the territory, and the local context. Each offers a number of possible future contributions. This study has provided us a perspective on the social impact the Catholic Church at the end of the nineteenth century across Latin America; it goes beyond traditional specific disciplinary boundaries of geography, history and architectural history.

In this study, architecture and art, as George Kubler stated well (1988:79), becomes a socio-cultural construct, viz., a signal that explains actions and processes proper to each period of time that go beyond form and style. We observe that it also includes religious, political and social elements as being integral parts in understanding the geographical distributions of such features. We also believe that this study opens the doors for others to explore the historical, religious and architectural interfaces.

References

Aliata, F. (2003). Eclecticismo y experimentalismo en la obra arquitectónica de Augusto César Ferrari. In L. Ferrari, A. C. Ferrari, & L. F. Noé (Eds.), *Augusto C. Ferrari (1871–1970): Cuadros, panoramas, iglesias y fotografías*. Buenos Aires: Ediciones Licopodio.

Asociación Pro Rescate de Archivos de Arquitectura (APRAARQ). (2004). El Fondo Buscaglione. *Boletín Cultural y Bibliográfico*, *66*(41), 45–60.

Ayala Mora, E. (1981). Gabriel García Moreno y la gestación del estado nacional en Ecuador. *Revista Cultura, 10*(4), 141–174.

Batlori, R. (2002). La escala de análisis: un tema central en didáctica de la geografía. *Íber: Didáctica de las Ciencias Sociales, Geografía e Historia, 32*(8), 6–18.

Bruno, C. (1984). Los Salesianos y las Hijas de María Auxiliadora en la Argentina: 1911–1922. In C. Bruno (Ed.), *Los Salesianos y las Hijas de María Auxiliadora en la Argentina* (Vol. 3). Buenos Aires: Instituto Salesiano de Artes Gráficas.

Capelluti, L. (2007). La devoción al Sagrado Corazón de Jesús. *Consideración al libro Amó con corazón de hombre. Revista Teología, 93*(44), 239–252.

Carrasco, F. (2004). Breves semblanzas de ocho arquitectos del siglo XX en Colombia. *Ensayos. Historia y teoría del Arte, 9*(9), 137–168.
Cevallos Romero, A. (1994). *Arte, diseño y arquitectura en el Ecuador. La obra del Padre Brüning (1899–1938).* Quito: Ediciones del Banco Central del Ecuador, Ediciones Abya Yala y Conferencia Episcopal Ecuatoriana.
Checa-Artasu, M. (2011a). Revisitando el papel del templo en la ciudad: los grandes templos neogóticos del Occidente de México. *Religião e Sociedade, 31*(2), 179–206.
Checa-Artasu, M. (2011b). Monumentalidad, Símbolo y Arquitectura Neogótica. El Santuario Guadalupano de Zamora, Michoacán. In O. Montes Vega & O. González Santana (Eds.), *Estudios Michoacanos* (Vol. 14, pp. 143–194). Zamora: El Colegio de Michoacán.
Checa-Artasu, M. (2012). El templo de San José en Arandas, Jalisco: un ejemplo del neogótico mexicano inconcluso y monumental (1879–2011). *Revista Academia XXII, 2*(3), 84–95.
Cufré Pedro, D. (2009). Vespignani Ernesto/Augusto Ferrari. Aportes a la arquitectura religiosa argentina. In VV.AA (Ed.), *Temas de patrimonio cultural, 25: Buenos Aires italiana* (pp. 309–321). Buenos Aires: Comisión para la Preservación del Patrimonio Cultural de la Ciudad Autónoma de Buenos Aires.
Del Real, L. J. (1942). *Tríptico modelo: Rasgos biográficos de tres coadjutores salesianos*. Bogotá: Escuelas Gráficas Salesianas.
Díaz Patiño, G. (2010). Imagen y discurso de la representación religiosa del Sagrado Corazón de Jesús. *Plura, Revista de Estudos de Religião, 1*(1), 86–108.
Egas, V., Manuel, P. A., & Francesia, J. B. (1994). *Cuando el premio es el destierro Luis Calcagno, fundador de la obra Salesiana en el Ecuador*. Quito: Editorial Abya Yala.
Elejalde Arbeláez, R. (2003). *A la sombra del plateado: Monografía de frontino*. Medellín: Gobernación de Antioquia.
Francia Cerasoli, J. (2010, July 12–15). History and the configuration of the architectural repertoire from Brazilian architects travelling to Europe in the early XX century. In *14th International Planning History Society Conference*, Istanbul. Urban Transformation: Controversies, Contrasts and Challenges.
Gálvez Abadía, A. (2006). *Por obligación de conciencia. Los misioneros del Carmen Descalzo en Urabá (Colombia), 1918–1941*. Bogotá: Universidad de Antioquia & Universidad del Rosario & ICANH.
Gil, P. (1999). *El templo del siglo XX.* Barcelona: Editorial Serbal; Collegi d'arquitectes de Catalunya.
Gil, C. C., & Wichepol, S. N. (2004). Vespignani, Ernesto. In J. F. Liernur (Ed.), *Diccionario de arquitectura en la Argentina* (pp. 353–354). Tomo S/Z, Buenos Aires. Editorial Clarín (F. Aliata, Comp.).
Gonçalves, L. R., De Lemos, A. G., & Freire, C. (2006). *São Paulo imaginado*. Bogotá: Convenio Andrés Bello.
Hadad, M. G., & Venturiello, M. P. (2007). La Virgen de Luján como símbolo de identidad popular: Significaciones de una virgen peregrina. In R. R. Dri & D. O. Bocconi (Eds.), *Símbolos y fetiches religiosos: En la construcción de la identidad popular* (pp. 27–44). Buenos Aires: Editorial Biblos.
Kieckhefer, R. (2004). *Theology in stone: Church architecture from Byzantium to Berkeley*. New York: Oxford University Press.
Kingman Garcés Eduardo. (2003). *Discurso y relaciones de poder en el Quito de la primera mitad del siglo XX.* Tarragona: Universitat Rovira i Virgili. Programa de Doctorado en Antropología Urbana del Departamento de Antropología, Filosofía y Trabajo Social. Tesis para optar al título de Doctor en Antropología Social y Cultural, Director, Dr. Joan Josep Pujadasersitat.
Krebs, R. (2002). *La Iglesia de América Latina en el siglo XIX*. Santiago de Chile: Ediciones Universidad Católica de Chile.
Kubler, G. (1988). *La configuración del tiempo: observaciones sobre la historia de las cosas*. San Sebastián: Editorial Nerea.
Larrea, C. M. (1976). *Breve historia de la iglesia catedral de Quito durante cuatro siglos*. Quito: Corporación de Estudios y Publicaciones.

Malta Campos, C., & Simões Júnior, J. G. (2006). *Palacete Santa Helena: um pioneiro da modernidade em São Paulo*. Sao Paulo: Senac.

Matovelle, J. (1934). La de Sdo. Corazón de Jesus: voto nacional ecuatoriano: Apuntes históricos, tomados de una obra inédita por el Rmo. Padre Dr. D. Julio Matovelle, fundador y superior de la congregación de sacerdotes oblatos de los Corazones SS. de Jesús y María. Quito: Editorial Ecuatoriana.

Menacho, A., & S. J. (2001). Eulalio Morales. In Piatkiewicz O'neil, E. L. Charles & Domínguez Joaquín María (Eds.), *Diccionario histórico de la Compañía de Jesús: biográfico-temático* (p. 2.738), Volumen 3, Salamanca: Universidad Pontifica de Comillas.

Miranda Arraiza, J. M. (2003). *Misioneros carmelitas en Urabá de los Katíos*. Vitoria: Ediciones El Carmen. Biblioteca Carmelitano-Teresiana de Misiones, Tomo XII.

Molina Londoño, F. (1998). *Agustín Goovaerts y la arquitectura colombiana en los años veintes*. Bogotá: El Áncora Editores; Banco de la Republica.

Moya Pérez, A. (1998). *Arquitectura religiosa en Jalisco: Cinco ensayos*. Zapopan: Amate Editorial.

Niccoli Ramirez, K., & Lindenberg, H. (2010, January 04–08). The structural behavior of the cathedral of Sé in São Paulo. In *11th Pan-American Congress of Applied Mechanics*, Foz do Iguaçu.

Osorio Gómez, J. (2008). *Patrimonio arquitectónico del valle del Aburrá: final del siglo XIX y principio del siglo XX*. Medellín: ITM.

Patiño de Borda, M. (1995). *Monumentos nacionales de Colombia: siglo XX*. Bogotá: Instituto Colombiano de Cultura.

Petriella, D., & Sosa Miatello, S. (1976). Ernesto Vespignani. In *Diccionario biográfico Ítalo-Argentino*. Buenos Aires: Asociación Dante Alighieri. Retrieved January 20, 2012, from www.dante.edu.ar/web/dic/diccionario.pdf

Piazzini, C. F. (2009). Planeación y procesos espaciales: configuración territorial del municipio de Frontino en el noroccidente de Antioquia (Colombia). *Boletín de Antropología, 40*(23), 186–228.

Rosendahl, Z. (1996). *Espaço e religião. Uma abordagem geográfica*. Río de Janeiro: EdUERJ.

Rosendahl, Z. (2005). Território e territoralidade: Uma perspectiva geográfica para o estudo da religião. In *Anais do X encontro de geógrafos da América Latina* (pp. 12928–12941). 20–26 de março de 2005, Universidade de Sâo Paulo.

Rozo Montaña, N. (2000). *Giovanni Buscaglione, 1920–1940. Arquitectura religiosa en Colombia*. Tesis de Maestría en teoría e historia, Universidad Nacional de Colombia.

Sack, R. D. (1986). *Human Territoriality. Its theory and history*. Cambridge: Cambridge University Press.

Saldarriaga Roa, A. (2007). La imagen de la iglesia y del estado en la arquitectura republicana: gótico, clasicismo y eclecticismo fueron los estilos definitorios. *Revista Credencial Historia, 86*, 34–45.

Sanou, A., & M. Ofelia. (2001). *Arquitectura e historia en Costa Rica: templos parroquiales en el Valle Central, Grecia, San Ramón y Palmares 1860–1914*. San José: Editorial Universidad de Costa Rica. Colección Nueva historia.

Stump, R. W. (2008). *The geography of religion. Faith, place and space*. Lanham: Rowman Littlefield Publishers.

Troyo Calderón, A. (2002). *El Catolicismo en el contexto religioso de Costa Rica*. San José: EUNED. Vol. 14 de Colección Ideario XXI.

Vargas Cambronero, G. A., & Zamora Hernández, C. M. (1999). *El patrimonio histórico-arquitectónico y el desarrollo urbano del Distrito Carmen de la ciudad de San José: 1850–1930*. San José: Ministerio de Cultura, Juventud y Deportes.

Vélez White, M. L. (1994). *Agustín Goovaerts y la arquitectura en Medellín*. Medellín: Editorial El Propio Bolsillo.

Vélez White, M. L. (2003). *Arquitectura contemporánea en Medellín*. Medellín: ITM. Biblioteca básica de Medellín, Vol. 5.

Chapter 129
Changing Russian Orthodox Landscapes in Post-Soviet Moscow

Dmitrii Sidorov

129.1 Introduction

Orthodoxy, or in Russian *Pravoslavie*, means "right worshipping" and proper service is not conceivable without special, physically extant Orthodox church buildings (hereafter, churches). They have become the *most important identifier symbols of the traditional Russian landscape*, the earliest manifestations of national identity, as well as essential elements of societal organization. In the late Russian Empire Orthodoxy was the only religion represented in all regions: in 1914, on the eve of the Communist period there were about 55,000 of these churches (plus 23,000 chapels) (Vsepoddanneishii Otchet 1916: 6–7; Sidorov 2000a: 211, 283–4) (Fig. 129.1).

Immediately after the 1917 revolution the new Bolshevik authorities started targeting the Russian Orthodox Church (hereafter, the Church), and, finally, by a decree of January 23, 1918 separated the state and the Church, and nationalized all its property (Mitrofanov 1995: 5). The assault on churches was unprecedented: thousands of them were closed, reused for secular purposes, or simply ruined (Sidorov 2000a: 208). As Fig. 129.1 shows, the year 1939 was the worst in the history of the Church; in the whole of the USSR, the number of open churches was only 200–300 (Davis 1995: 13) or even 100; in most regions of the USSR, only one church was functioning, 25 regions were designated as "church-less" (Chumachenko 2002: 4). That was one of the most drastic secularization attempts in history; the Church was literally on the eve of annihilation as the new authorities promoted an atheist landscape. The end of Communist rule resulted in yet another dramatic change of the trend; religion was again allowed, and a religious renaissance occurred, in Russia more than in other post-Communist countries (Dzis'-Voynarovsky 2012).

D. Sidorov (✉)
Department of Geography, California State University, Long Beach,
Long Beach, CA 90840, USA
e-mail: Dmitrii.Sidorov@csulb.edu

S.D. Brunn (ed.), *The Changing World Religion Map*,
DOI 10.1007/978-94-017-9376-6_129

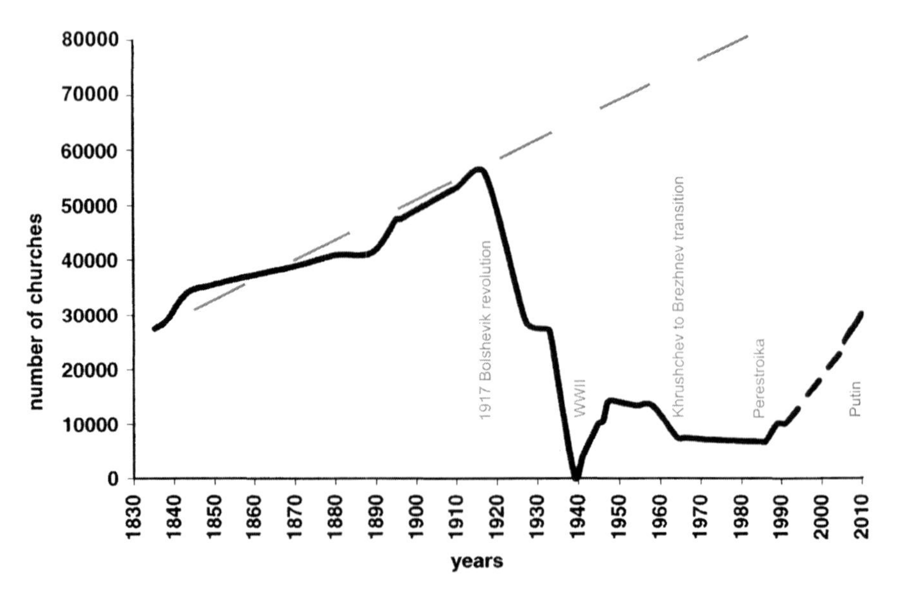

Fig. 129.1 Number of Orthodox churches, the Russian Empire, the USSR, and the Russian Federation, 1834–2010 (Graph by Dmitrii Sidorov)

Moscow is an especially exemplary case as the religious landscape of the city has experienced some of the most dramatic changes in recent history. In 1917, there were in the city close to 800 churches (Sidorov 2001: 156), yet due to the harsh Soviet anti-religious policies, in 1985 Moscow had only 54 functioning churches (Sidorov 2001: 151). The end of the Communist rule reversed the trend and in 1995 the city had quadrupled this number to more than 200 (Sidorov 2001: 160). As of 2011, the total number of churches and chapels in Moscow was 849 (of them, 316 with functioning parish communities) (Kirill 2011) equaling or exceeding the pre-Soviet level. Yet it would be superficial to conclude that in less than 100 years Moscow has managed to entirely recreate its traditional Orthodox landscape because in the same period the city itself has become dramatically different.

As a result of rapid Soviet urbanization, Moscow's population and territory in the twentieth century have increased more than 10 times. The Church was not allowed to grow equally to adjust to these changes; thus a *territorial mismatch* has been created between the numbers of residents and churches in the city. It also has a geographical dimension; most of the churches nowadays are in the pre-1917 downtown area of the city while the majority of the increased population resides at the outskirts added during the Soviet period. (Hereafter, for convenience, they often will be labeled as inner suburbs or just suburbs even if strictly speaking they are not; they are within the city limit and do not possess the qualities of low-rise sprawling suburbs of the west).

This discussion of the changing post-Soviet Orthodox landscape will pay particular attention to two trends in two recent periods. First, in the first post-Soviet decade (1990s), initial church reconstructions were often politically motivated and shaped by new political forces, often for the purpose of their legitimization. The effect of such a nexus of the political and the religious was certain *corporatization* of church reconstruction projects. Second, I would like to highlight a new trend in the 2000s, *suburbanization* and *mass-construction* of new churches at the outskirts of the city. Privately funded through an umbrella foundation, these new constructions are driven by more pragmatic considerations yet may cause conflict with residents if one takes the form of in-fill additions to already densely settled neighborhoods.

129.2 The Church and the City in the Past

When French marquis Astolph de Custine in 1839 visited Moscow, he observed a unique cityscape with numerous churches:

> ...[b]right chains of gilded or plated metal unite the crosses of the inferior steeples to the principal tower; and this metallic net, spread over an entire city, produces an effect that it would be impossible to convey, even in a picture. ... a phalanx of phantoms hovering over the city. ... in 1730, Weber counted at Moscow 1500 churches. Coxe, in 1778, fixes the number at 484. As for myself, I am content with endeavouring to describe the aspect of things, I admire without counting; I must, therefore, refer the lovers of catalogues to books made up entirely of numerals. (de Custine 1989: 394–395)

Thus, churches at that time were relatively evenly distributed over the territory of the city. The density of churches may have declined towards the margins, yet no significant part of the city was without churches. Even suburbs had churches at cemeteries as well as in rural settlements around Moscow (territories of expansion in the twentieth century, Fig. 129.2).

This territorial match between population and churches was dramatically broken during the Soviet period, due to three processes. First, Soviet Moscow, once again the capital city since 1918, was reshaped by the Cultural Revolution with its intense *politico-architectural reconstruction* of the historical downtown, including the ruthless demolition of many churches (see the story of the Cathedral of Christ the Savior below). Several churches also were demolished under Khrushchev in the late 1950 and early 1960s for purely propagandistic purposes.

Second, Moscow's cityscape has been affected by *urban modernization*. This profound restructuring of the city resulted in the demolition, closure, and re-utilization of many churches for secular purposes. For example, the construction of Moscow's first underground railway in the late 1920s by the cut-and-cover method resulted in the demolition of many churches and other architectural landmarks, despite the fact that many of them had been restored only a few years earlier (Palamarchuk 1992: 14). In the late 1920s, 23 churches were demolished to clear places for new school construction, because at that time 80 % of pre-revolutionary school buildings in Moscow were occupied by other new organizations, while the city population was rapidly growing. Another 80 downtown churches were demolished in favor of residential and office construction (Palamarchuk 1992: 14–15). As a result, by 1933, the city had only 112 functioning churches; in 1985, only 42 churches remained open on the territory of 1917 Moscow (Sidorov 2001: 160). By that time, half of all 1917 churches had been demolished and the rest were simply closed or reutilized. Several churches with tall bell-towers were dismantled during WWII, for fear that they could

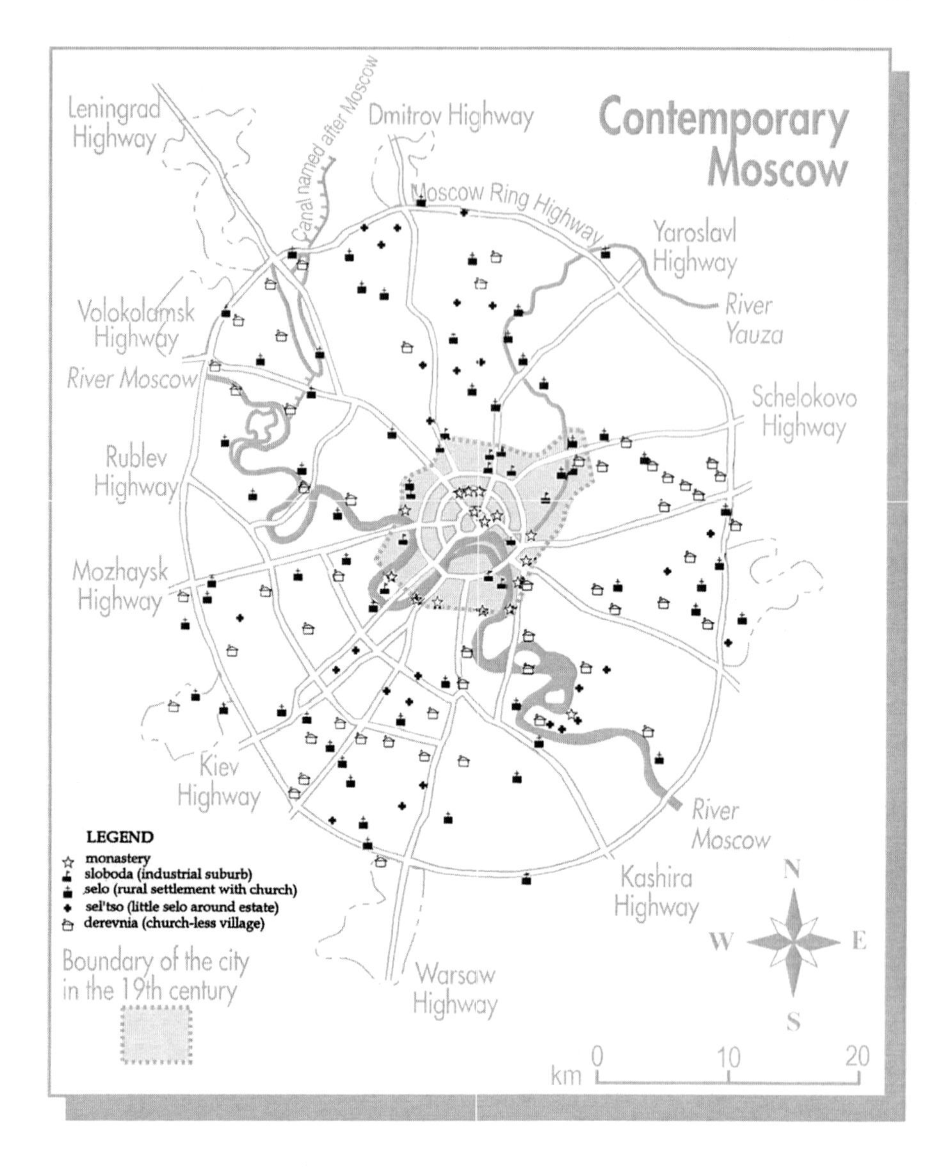

Fig. 129.2 Types of seventeenth century settlements on the territory of contemporary Moscow. The core area is the nineteenth century Moscow extent. Within the limits of the ring road (the city boundary of 1960) there were 157 settlements with their own places of worship (Map by Dmitrii Sidorov, after Sidorov, 2001: 156, based on unpublished map by V. S. Kusov)

serve as landmarks for German artillery. It is difficult to cite an example of another major world city with a similar dramatic secularization of its landscape.

A third force altering Moscow's Orthodox cityscape during the Soviet era was *urban expansion*. In 1897 Moscow was a city of slightly more than one million inhabitants, but by the 1990s its total population was close to nine million and the daytime population (including commuters and tourists) exceeded ten million. The territorial extent of the city had also expanded well beyond the original pre-Soviet 1917 city boundary. With Moscow's population quintupled and its territory quadrupled, the city is certainly different from its pre-revolutionary predecessor. The Church was not allowed to respond to these changes, however, so the typical residential suburbs of modern Moscow are either churchless, or include small, originally rural churches, which geographically and architecturally seem out of place in their new urban setting. A profound territorial mismatch thus has emerged between the church-packed downtown (the city of the nineteenth century) and the rest of the city (Fig. 129.3).

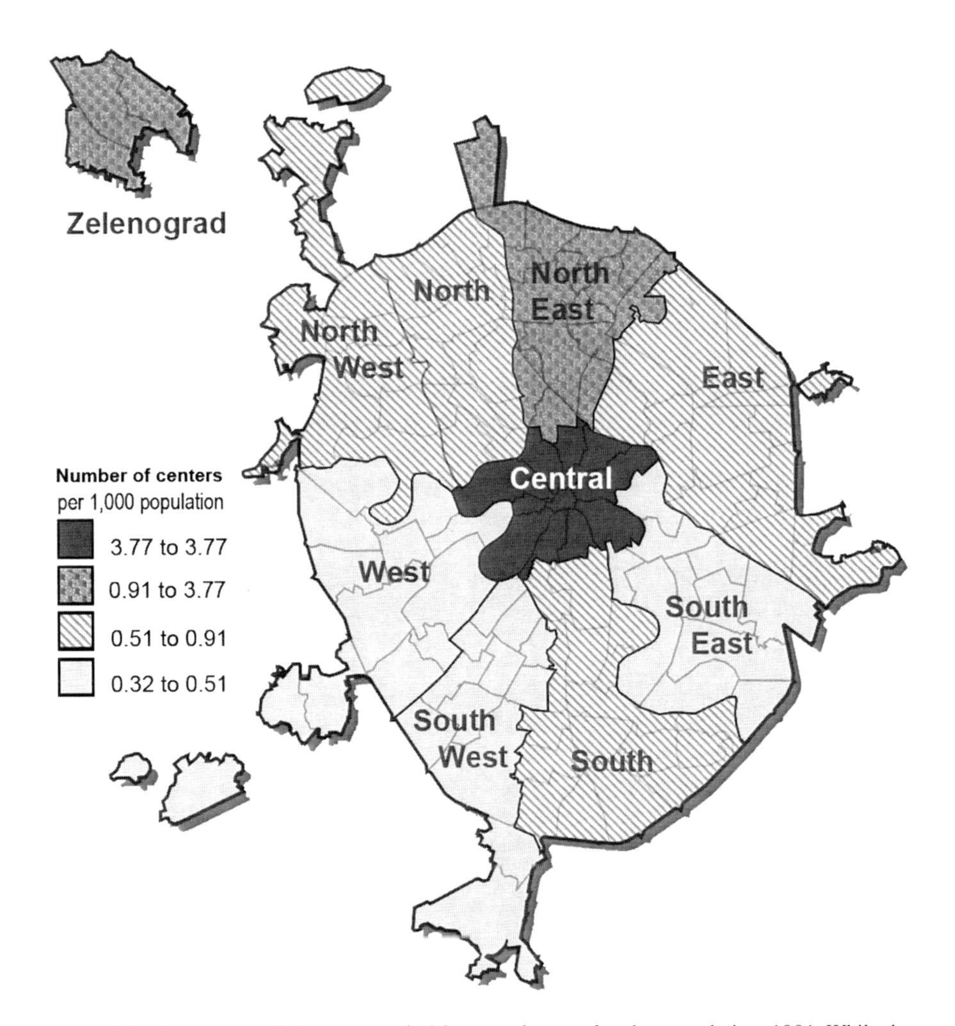

Fig. 129.3 Permanent religious centers in Moscow okrugs related to population, 1991. While the average density of the centers for the city as a whole was 0.83 per 1,000 people, the same parameter for the Central okrug (3.77) was 4–5 times higher. Overall, the southern half of the city had lower densities of church centers than the northern part. In particular, the north-eastern part of the city had a relatively high density of centers (Map by Dmitrii Sidorov)

129.3 The 1990s: First Church Reconstruction and Politics

In the first post-Soviet decade (1990s) there have been in Moscow perhaps three main dimensions of the church property transfers. Leaving aside *restitution of church property rights* (de-nationalization), this section discusses two of them: *restorations* of ruined Orthodox buildings as well as *new church constructions* trying to show that these early post-Soviet processes were often driven by political motivations and, therefore, had predominantly focused on the center of the city, thus paradoxically contributing to *continuation the Soviet territorial mismatch* between where the churches are (downtown) and where the majority of people live (residential churchless outskirts with the standardized high rise residences). Of ten new church (re)constructions in Moscow in the mid-1990s (Pravoslavnaia Moskva 1995: 8) this section highlights several most emblematic of them (Fig. 129.4) focusing on the impact of the political forces that backed them.

129.3.1 Cathedral of Christ the Savior, Volkhonka

The tragic fortune of the Church could be best exemplified by the fate the Cathedral of Christ the Savior (Fig. 129.5). Built to commemorate people's sacrifice to the victory over Napoleon in 1812, this was the most popular church in the Russian collective national imagination. The Communist authorities wanted to strike into the very heart of Orthodox religiosity by demolishing what the people regarded as the main cathedral and dynamited it in 1931. The emptied site was to be occupied by a new "cathedral," Stalin's proposed Palace of the Soviets: as a symbol of coming Communist global hegemony, it should become the highest building in the world with Lenin's statue on top. WWII changed priorities and saved Moscow from this gargantuan structure. After the war and Stalin's death, the project was eventually abandoned. In 1960–1993 its pit was recycled as an outdoor steam-heated swimming pool, one of the world's largest. Finally, in 2000, Moscow authorities under controversial Mayor Yu. Luzhkov reconstructed the Cathedral in its original site, as a symbol of both the break with the Soviet past and their relative power in transitional Russia (see details of the story in Sidorov 2000b).

The desperate condition of many active churches in Moscow, however, made the $US 650 million reconstruction of the Cathedral with 10,000 people capacity geographically questionable. This politically important reconstruction paradoxically reinforced the mismatch created during the Soviet time; most of the old churches are in downtown while most of people reside on the margins of cities. The nouveau riche that started moving into the gentrified area (Volkhonka is now known as Golden Mile, one of the most expensive neighborhoods in the world) are less likely to attend services.

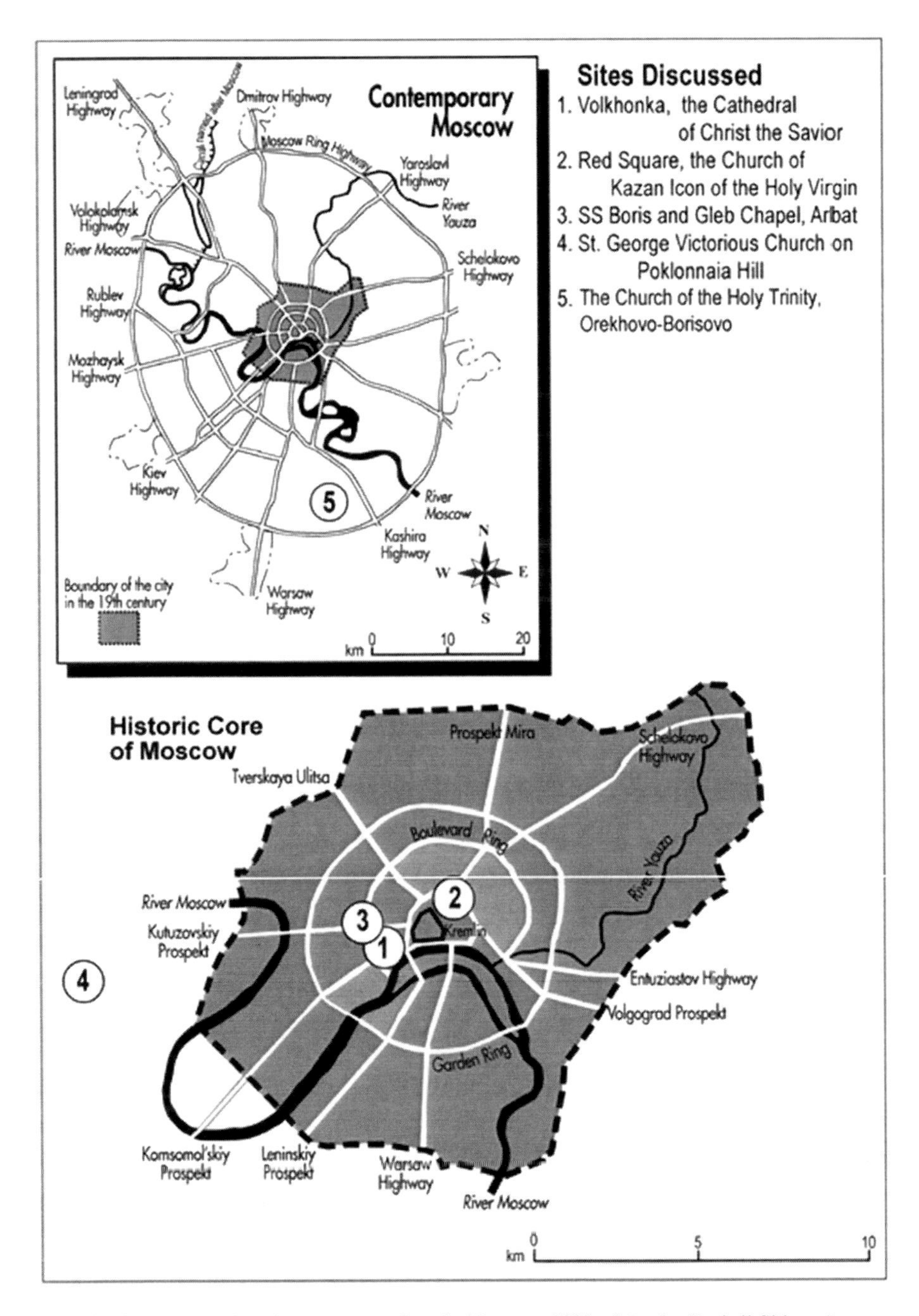

Fig. 129.4 Key new church (re)constructions in Moscow, 1990s (Map by Dmitrii Sidorov)

Fig. 129.5 The Cathedral of Christ the Savior, 2009 (Photo by Dmitrii Sidorov)

129.3.2 Church of Kazan Icon of the Holy Virgin, Red Square

The Church of Kazan' Icon of the Holy Virgin on Red Square (Kazanskii Church) (Fig. 129.6, see Fig. 129.4) was first built in 1626 by Tsar Mikhail Fedorovich Romanov and Count Pozharskii, the liberator of Moscow from Polish-Lithuanian invaders. Here, in 1812 the Commander-in-Chief of the army, Mikhail Kutuzov, received a blessing for his victory over Napoleon's troops. In 1927–1932 Petr Baranovskii undertook the heroic task of measuring the building and preserving the church's original design for future restoration (Beliaev and Pavlovich 1993). Soon after, in 1936, the church was demolished, a victim of campaign to "clear historical trash." Instead, the site was humiliatingly reused for public toilet (as two dozen other former church sites) (Palamarchuk 1992: 15). Finally, the city government under post-Soviet Mayor Luzhkov undertook the reconstruction and after a year and a half of work, the church was officially reopened in 1993. Kazanskii Church was the first church restored in Moscow during the 1990s and served as a trial test for the restoration of the Cathedral of Christ the Savior (see above). In both cases, the Church itself helped little in the restoration; expenses have been paid mostly by the Moscow government, not the federal one or the public that was offered only negligible participation in the reconstructions. The restored church on Red Square also openly symbolizes power: in its entrance hall a golden plaque states, among other things, that "this church was restored by the order of the first Russian President Boris Yel'tsin … and the efforts of Moscow Mayor Yury Luzhkov and his deputy A. S. Matrosov."

Fig. 129.6 The restored Church of Kazan' Icon of the Holy Virgin on Red Square, 2013 (Photo by Leonid Sidorov for Dmitrii Sidorov)

129.3.3 SS Boris and Gleb Chapel, Arbat Square

SS Boris and Gleb Chapel (Fig. 129.7, see Fig. 129.4), one of the oldest churches in the area of Arbat Gates (yet in its western European Baroque style of today it has existed only since 1764), was ruined in 1930 (K. 1997). Resurrection of the chapel in 1997 was carried out with remarkable speed, just 3 months since blessing of its foundation. It is noteworthy that the foundation ceremony was attended by the highest political figures, Prime-Minister Viktor Chernomyrdin, Moscow Mayor Yury Luzhkov, and the President of the Russian Federation Boris Yel'tsin who expressed his hope that the reconstruction would contribute to reconciliation and accord in the country. Indeed, this was an important goal at the time of greatest challenge during Yel'tsin's presidency, the Chechen war. Rumors suggested that the flawless reconstruction was personally and discretely backed by Yel'tsin whose name is Boris and his only grandson's name is Gleb. A plaque inside informs that this chapel was restored "with support of the President of the Russian Federation B. Yel'tsin and efforts of Moscow Mayor Yu. Luzhkov." SS Boris and Gleb are traditionally protectors of the men with arms; the chapel re-emerges now not as a major church of the area, but rather as a church of a powerful ministry behind it (the headquarters of the Ministry of Defense). The restoration in front of the major military complex of the chapel devoted to the military seemed for the authorities highly appropriate, even if it now occupies the site of another church: it is not well-publicized that SS Boris

Fig. 129.7 The restored SS Boris and Gleb Chapel near Arbat Gates in front of the Ministry of Defense, 2012 (Photo by Dmitrii Sidorov)

and Gleb Chapel has been restored on the site of another seventeenth century church ruined by the Soviet authorities in 1933 (the Church of Holy Hierarch Tikhon, the Bishop Amafutinskii). Being initially situated next to one another, both churches witnessed the same historical events and, therefore, could be equally revered.

129.3.4 St. George Victorious Church, Poklonnaia Hill

The Memorial on Poklonnaia Hill (Figs. 129.8 and 129.9, see Fig. 129.4) was built in 1995 to commemorate 50th anniversary of the Victory in Russia's second Patriotic War of 1941–1945. It took the country half a century to design the complex, mainly a result of a general ideological uncertainty about the Memorial's purpose and design (the war had a disastrous casualty list). The hill is historically significant primarily in the context of a different patriotic war (with Napoleon), creating some ethical doubts about the construction of a war memorial on the historical site of

Fig. 129.8 The Church of St. George Victorious, Poklonnaia Hill, 2013 (Photo by Leonid Sidorov for Dmitrii Sidorov)

another war; this geographical juxtaposition perhaps elevated the Memorial's image, yet the actual hill as a landform ceased to exist, being flattened by the construction (Kusov 1995). The erected 109 m (358 ft) obelisk, tallest of its kind, could not substitute for the loss of this landmark hill. The final late-Soviet design of the Memorial is characteristically totalitarian in its grandiose style, yet mass-cultural in its eclecticism. Mayor Yu. Luzhkov has acknowledged publicly the aesthetic shortcomings of the project, yet the local Moscow authorities again subsidized the actual construction (through a loan to the federal government). In the last stages of design, the project was upgraded by adding a modern church named after St. George the Victorious. Initially designed as a small church (15 × 15 m, 40 m (49 × 49 ft, 130 ft) to the cross) for 30–50 parish believers, or as a place for special events, it has

Fig. 129.9 The Church of St. George Victorious, Poklonnaia Hill, 2008 (Photo by Dmitrii Sidorov)

become popular, constantly packed with visitors, showing the public power of even a "touch of Orthodoxy." The idea of using church construction as efficient legitimizing addition to the risky public projects of local Moscow officials has become a common practice. The authorities' domination of the process, however, compromised the design. Against the canons of Orthodoxy the church has been decorated not by icons, but by bas-reliefs, both inside and outside (Rokhlin 1995). The Memorial represents one of the first manifestations of what has become a unifying "Luzhkov style" for Moscow's newest generation of projects. This "grand-style" is characterized by the dominance of monumental decorations made by Mayor Luzhkov's favorite artist, Zurab Tsereteli (Shimansky 1996).

129.3.5 The Church of the Holy Trinity, Orekhovo-Borisovo

The cases discussed above share the dominant if not monopolizing role of the local Moscow authorities and Mayor Yu. Luzhkov. The Church itself often had somewhat different ideals. For instance, it was interested in resurrection of the Cathedral of Christ the Savior (see above), yet was cautious about simply replicating the lost monument. Instead, the Church originally planned to indirectly revive the idea of the Cathedral elsewhere. To celebrate the 1988 millennium of the Russian conversion to Christianity, the state allowed construction of a new cathedral. Designed as a reminder of the lost Cathedral, the project was, nevertheless, to signify the new policy of separation of Church and state, and, therefore, to be located not in the center, but rather in the residential outskirts of the city, a residential district in the southern part of Moscow (Orekhovo-Borisovo) as a rare alternative to the

Fig. 129.10 The Church of the Holy Trinity, Orekhovo-Borisovo, 2012 (Photo by Marina Lystseva, http://en.wikipedia.org/wiki/Orekhovo-Borisovo_Metochion)

downtown-focused church reconstructions in the 1990s (Sidorov 2000b). This project was not realized at that time since the Church characteristically lacked sufficient resources.

This project has eventually transformed into the Church of the Holy Trinity (Fig. 129.10, see Fig. 129.4) built in 2001–2004 in the same Byzantine revival style as the Cathedral. It is now a metochion of the Church' Patriarch, and includes the 70 m (227 ft) tall main church, the compound with a chapel, a free-standing prothesis, a belfry, and a school. Originally it was planned to be erected in 1988, in commemoration of the millennium of the Orthodox Baptism of Rus, but those plans, however, did not materialize for 15 years until, characteristically, the second post-Soviet period of the 2000s.

129.4 The 2000s: Suburbanization of Churches

In the 2000s, the political-economic environment has become different both in Russia as a whole and the city of Moscow. The Moscow power group has lost some of its power with ousting of its controversial Mayor Yu. Luzhkov, while the transition from Soviet to post-Soviet system has resulted in the emergence of new influential business groups that, very much like the political groups of the 1990s, want legitimization and seek social association with still the most respected institution of the Church, yet to do that in a more business fashion somewhat different from the previous period. As a result, the focus of new church construction projects is on the suburbs; the resultant *suburbanization of churches* may further profoundly transform the religious landscape of Moscow.

Moscow still lacks churches. Nowadays it has about 40,000 parishioners per one church (while in the region of Mordovia only 3,000) (the Church itself estimates 60,000–65,000 parishioners per church) (Leont'ev 2012). This number is based on the following estimation. While there are 836 churches in Moscow, the city of 10.5 million people, only 263 of them are open to everyone with the rest being the partially accessible monastery churches, chapels, home churches and alike. To reach Russia's average church density of 11.2 urbanites per church, there should be built 800 new churches in the city (Mironov 2010). In some parishes clerics have to use speakers to serve the people outside their churches unable to enter the packed interior during major holiday services like the Easter (Leont'ev 2012).

A program to build 200 new churches in the church-less suburbs of Moscow is highly emblematic for the new period. It was initiated in the summer of 2010 by Kirill, the Patriarch of Moscow and All Russia, and supported by ex-Mayor Yu. Luzhkov and his successor, the current Mayor S. Sobianin (Ogil'ko 2011; Smolitskiy 2012). The Patriarch originally requested 600 churches, yet Mayor Luzhkov noted a lack of vacant spots for construction and the Church agreed to lower the number to 200 (Mironov 2010). The same year the first 35 construction sites were defined by the authorities (Mironov 2010). As of March 2012, construction has begun on about 20 churches Leont'ev 2012), Fig. 129.11); the program is expected to be accomplished in 8–10 years (Leont'ev 2012). It is expected that the modular design would allow relatively short construction time: 8 months for small churches and about a year for large churches. During the construction process, temporal wooden chapels will be erected (Ogil'ko 2011).

Unlike in the previous period, these new churches are "modular" (or partially standardized) relatively inexpensive buildings due to use of several standardized designs ("modules," "types") with one or five cupolas. These are meant to be "convenience churches" "within walking distance." The architects of three leading architectural companies have prepared 24 versions of facades (more than 30, according to Mironov (2010)) based on the ancient Vladimir-Suzdal style. One-cupola churches are designed for 246–369 attendants and have six variations in their windows, cupola shape, and wall decoration (Ogil'ko 2011). Total church area is to be up to 830 sq. m (8,934 sq. ft.) (Mironov 2010). There will be also "upscaled" variants. A somewhat separate type is represented by large one cupola churches for 300–500 attendants designed in the style of the thirteenth century; they will include additional sections, individual decisions of brick layout, and six color schemes (ivory, yellow, red, blue, purple, and green) (Ogil'ko 2011). The Moscow Architectural Committee published designs on their website (http://mka.mos.ru/mka/mka.nsf/va_WebResources/File/$File/2011051102.pdf) as well as location sites of the first 57 constructions (www.mka.mos.ru/mka/mka.nsf/va_WebResources/File/$File/201110250101.pdf). To allow construction of both the

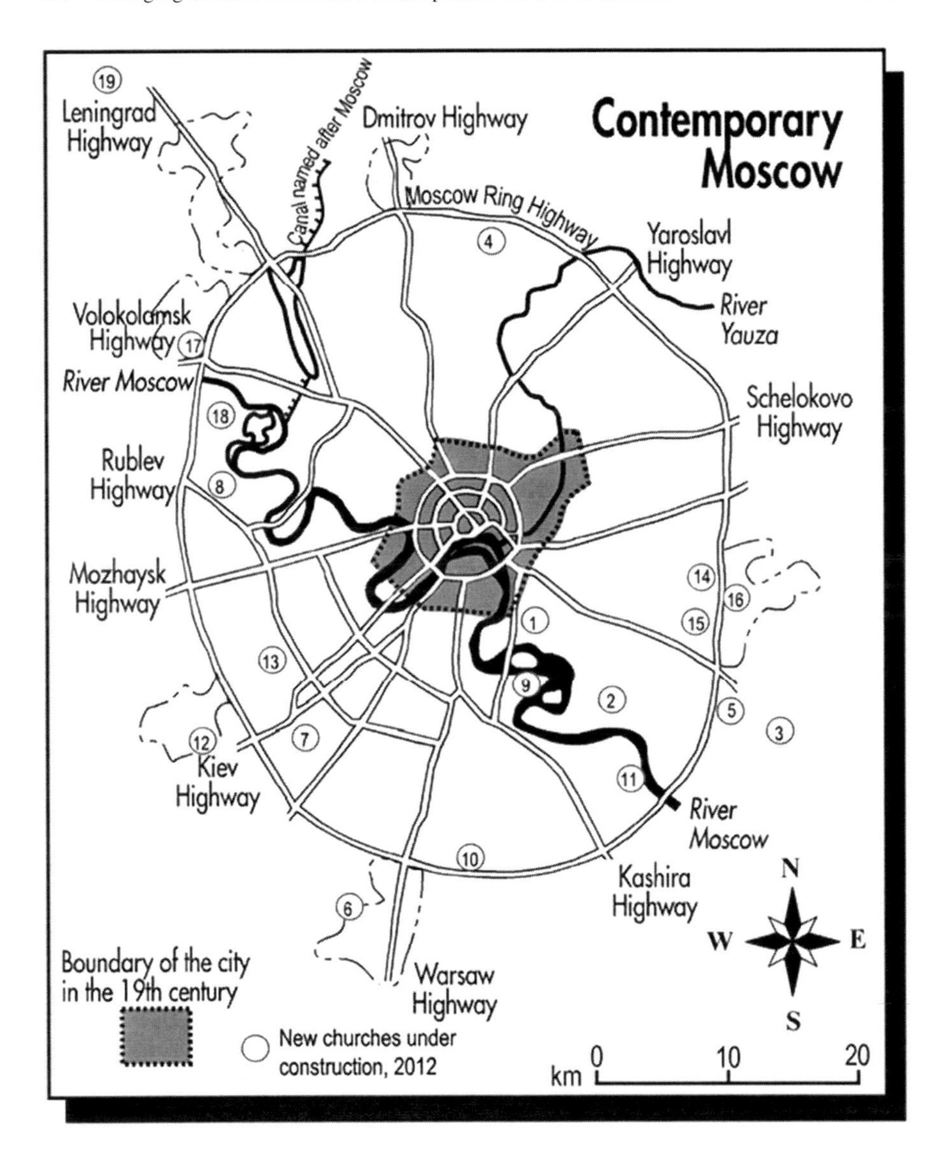

Fig. 129.11 Program "200 churches": new Orthodox churches to be constructed in 2012. Churches discussed in the text include #1 SS Cyril and Methodius church, Dubrovka; #6 St. Stephen of Perm Church, Southern Butovo; #13 FSB church (Map by Dmitrii Sidorov)

church and the house of the clergy, the Church requested sites up to 2 ha (4.9 acres), yet a typical site provided is of 0.6. ha (1.48 acres) or 0.3 ha (.74 acres), according to Mironov (2010).

Estimated cost of a typical modular church of 49 sq. m (527 sq. ft.) construction is 2–2.5 mln rubles (Smolitskiy 2012), or US$ 62,000–78,000. However, there are significant differences in estimations of the cost: Mironov (2010) cited 50–60 mln rubles (US$ 1.6–1.9 mln); Leont'ev (2012) cites 220 mln rubles (US$ 6.9 mln) for a 300 attendants church (incl. all costs) and 250–270 mln rubles (US$ 7.8–8.4 mln) for a 500 attendees church. Unlike in the 1990s, no municipal taxpayers' money is to be used; all funds come from private and investors' sources (Smolitskiy 2012) via an established foundation to support the program. As the foundation reported, most of donations are lower than 5,000–10,000 rubles (US$ 156–312); that was enough to start 20 projects yet for their completion the foundation had to apply to business (Golubeva 2012). Especially active sponsors now are Moscow construction companies who cite their "values" yet may be interested in both positive publicity related to charity work and good relations with local authorities (Golubeva 2012).

The period is also different in that the newly emergent civil society is especially active at the municipal level. There have been conflicts of NIMBY (not-in-my-backyard) kind: 70 cases of the program have been considered and local residents have succeeded in requesting relocation of 35 original construction sites (Smolitskiy 2012). Conflicts occurred in such districts as Pechatniki, Kurkino, Khoroshevo-Mnevniki, Zelenograd (Smolitskiy 2012). For example, residents of a neighborhood in Voykovo petitioned for relocating a church construction half a year after it started citing its close proximity to their residences as the reason; the 40-m (131 ft.) tall building is planned to be built just 4–5 m (13–16 ft.) away from existing buildings (Chibisova 2012).

According to the Church, the amount of donated funds has been sufficient to launch pilot construction of the first 19–20 churches that, after public approval, are to be erected already in 2012 (Gavycheva 2011; Golubeva 2012; Lokotkova 2012; see Fig. 129.11). Several noteworthy of them are highlighted below.

129.4.1 Saints Cyril and Methodius Church, Dubrovka

This church (Fig. 129.12) is the very first project started under the program (Ogil'ko 2011), perhaps because the site is located in front of the theater on Dubrovka that became globally known during the brutal terrorist siege in 2002 with its enormous casualties list. The church is expected to be opened by the 10th anniversary of the tragedy (October 23–26, 2012) (Leont'ev 2012). The 23 m (75 ft.) tall 1,404 sq. m (15,112.5 sq. ft.) tented-roof 9 cupola white-stone church is to be for 500 attendants, includes a Sunday school house and is designed to symbolize a universal church building (Moscow Committee 2011a; Ogil'ko 2011).

129.4.2 St. Stephen of Perm Church, Southern Butovo

This project in Southern Butovo is especially emblematic of the program. The area located beyond the ring road (the 1960 boundary of the city (see Fig. 129.11) is both geographically and politically "on the edge." A few years ago it became a household toponym in Russia when some local owners of private village-like houses resisted forced police relocation to allow new high-rise expansion of the city. This distant

Fig. 129.12 SS Architectural rendering of the SS Cyril and Methodius Church, Dubrovka, 2012 (Source: Leontiev 2012)

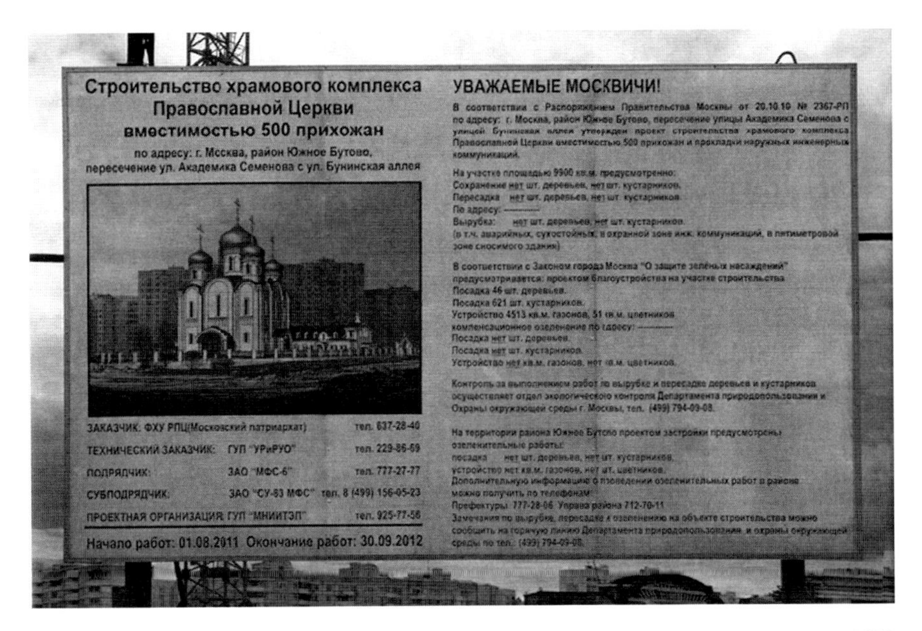

Fig. 129.13 Architectural rendering of St. Stephen of Perm Church, Southern Butovo, 2012 (Public poster outside the church construction site; photo by Dmitrii Sidorov)

area is expected to get several churches through the program; St. Stephen of Perm Church is the first of them (the foundation ceremony was in Sept. 2011). The project (Fig. 129.13) includes a five cupola church for 500 attendants (826.5 sq. m; 8,896.3 sq. ft.), a clergy house (1,129.6 sq. m; 12,158.9 sq. ft.) for administration, dining,

Fig. 129.14 Construction of Iverskaia Icon of God's Mother Church, Michurinsky Prospekt, 2012 (Photo by Dmitrii Sidorov)

Sunday school, baptizing, and resting; and a technical support house (260.2 sq. m; 2,800.8 sq. ft.) (Moscow Committee 2011a; for further design details see Moscow Committee (2011b)).

129.4.3 Iverskaia Icon of God's Mother Church, Michurinsky Prospekt

This church may become the largest in the program and accommodate up to 1,000 attendants (Smorodinova 2012). Located on the territory of the Academy of Russia's Federal Security Services (Fig. 129.14, see Fig. 129.11), this church is designed in a more individual style (Leont'ev 2012) and is constructed by Krost Company, which is also investor of the project (Foundation 2012; Golubeva 2012).

129.5 Conclusion

The religious landscape of post-Soviet Moscow is changing as are changing the changes themselves reflecting the dynamics of new Russian society. What remains constant is the fact that the Church and state in the country are deeply intertwined through history – and geography. This match is illustrated here through analysis of the mismatch created by the atheist Communist authorities. Even if it finally to be reduced, geographical perspectives on the changing religious landscape deserve to be continued. This study is an opportunity to contribute to recent developments in the studies of religion in geography in particular and in social sciences in general. Since the early 1990s, western geography of religion has become more interested in the political and the constructivist dimensions of spatial religious phenomena, new spaces and scales (see Kong 1990, 2001, 2010). At the same time, social sciences in general and Russian studies in particular have experienced what some observers labeled "spatial turn" with its interest in the regional and local dimensions of what previously have been studies primarily at the universally Russian national (state-wide) level (see review in Rolf 2010). Both of these trends may benefit from this study with its focus on the political construction of the religious landscape in post-Soviet Moscow and on the dynamic geography of the Church and society relationship (the center-periphery mismatch). Hopefully, in the future research attention will be paid not only to the traditional social categories "state" and "society" but also to the regional as well as the corporatist dimensions of religions. This is relevant not only for the geographic studies of Orthodoxy, but also of Islam and Buddhism.

References

Beliaev, L., & Pavlovich, G. (1993). *Kazanskii sobor na Krasnoi ploschadi* [Kazanskii Cathedral on Red Square]. Moscow: Biznes MN.

Chibisova, D. (2012, April 24). *Muscovites revolt against church in the courtyard* [Moskvichi vzbuntovalis' protiv stroitel'stva khrama vo dvore] *Moskovskii Komsomolets*. Retrieved August 18, 2012, from http://www.mk.ru/moscow/article/2012/04/24/696705-moskvichi-vzbuntovalis-protiv-stroitelstva-hrama-vo-dvore.html

Chumachenko, T. (2002). *Church and state in Soviet Russia: Russian Orthodoxy from World War II to the Khrushchev years*. Armonk: M.E. Sharpe.

Custine, A. 1989 (1839). *Empire of the czar: A journey through eternal Russia*. Foreword by D. J. Boorstin. Intr. by G. F. Kennan. New York: Doubleday.

Davis, N. (1995). *A long walk to church: A contemporary history of Russian Orthodoxy*. Boulder/San Francisco/Oxford: Westview Press.

Dzis'-Voynarovsky, N. (2012, June 8). *Atheists are disappearing in Russia* [Ateisty v Rossii vymiraiut], *slon.ru*. Retrieved June 9, 2012, from http://slon.ru/world/gde_vera_silnee-796227.xhtml

Foundation for Support of Church Construction in Moscow. (2012). *Iverskaia icon of god's mother church*. Retrieved August 1, 2012, from http://www.200hramov.ru/4/temple/71/

Gavycheva, A. (2011, August 29). Investors agreed to build churches "for publicity" [Investory soglasilis' stroit' khramy "dlia imidzha"] *Izvestia*. Retrieved on July 23, 2012, from http://izvestia.ru/news/498839

Golubeva, A. (2012, April 10). Funds for new churches have been collected from Moscow construction companies [Den'gi na novye khramy sobrali s moskovskikh stroitel'nykh kompaniy] *RBC-Daily*. Retrieved August 18, 2012, from www.rbcdaily.ru/2012/04/10/focus/562949983520700

K., O. (1997). Foundation ceremony for a new chapel in Moscow [Zakladka novoy chasovni v Moskve]. *Zhurnal Moskovskoi Patriarkhii* 6.

Kirill, the Holy Patriarch of Moscow and All Russia. (2011, December 23). *Report to the episcopal meeting of the city of Moscow*. Retrieved June 10, 2012, from www.moseparh.ru/publications/reports/2011/doklad-svyatejshego-patriarxa-kirilla-na-eparxialnom-sobranii-moskvyi-23-dekabrya-2011.html

Kong, L. (1990). Geography and religion: Trends and prospects. *Progress in Human Geography, 14*(3), 355–371.

Kong, L. (2001). Mapping 'new' geographies of religion: Politics and poetics in modernity. *Progress in Human Geography, 25*(2), 211–233.

Kong, L. (2010). Global shifts, theoretical shifts: Changing geographies of religion. *Progress in Human Geography, 34*(6), 755–776.

Kusov, V. (1995). Poklonnaia Hill [Gora Poklonnaia]. *Moskovskii Zhurnal, 5*, 15–16.

Leont'ev, A. (2012, March 5). 200 churches – new history of Moscow [200 khramov – novaia istoriia Moskvy]. Retrieved on July 25, 2012, from www.pravoslavie.ru/put/51977.htm

Lokotkova, Zh. (2012, March 25). Nineteen more churches will be erected [Postroiat esche 19 modulnykh khramov]. *Vecherniaia Moskva*. Retrieved June 10th, 2012, from www.vmdaily.ru/news/postroyat-eshe-19-modylnih-hramov1332700397.html

Mironov, N. (2010, October 6). 35 new Orthodox churches will appear at Moscow outskirts [Na okrainakh Moskvy poiavitsia 35 novykh pravoslavnykh khramov] *Komsomol'skaia Pravda*. Retrieved July 25, 2012, www.kp.ru/daily/24570/742375.

Mitrofanov, G. (1995). *Russkaia Pravoslavnaia Tserkov' v Rossii i v emigratsii v 1920-e gody: k voprosu o vzaimootnosheniiakh Moskovskoi Patriarkhii i russkoi tserkovnoi emigratsii v period 1920–1927 gg*. [The Russian Orthodox Church in Russian and in Emigration in the 1920s: On the Question of Relationship between the Moscow Patriarchate and Russian Church Emigration in 1920–1927]. St. Petersburg: Noakh.

Moscow Committee for Architecture and Urban Development [MosKomArchitektura]. (2011a). All architectural solutions for the modular Orthodox complexes have found their place in Moscow [Vse arkhitekturnye resheniia tipovykh pravoslavnykh kompleksov nashli svoe mesto v Moskve]. Retrieved July 23, 2012, from www.pravoslavie.ru/put/51977.htm

Moscow Committee for Architecture and Urban Development [MosKomArchitektura]. (2011b). Retrieved July 23, 2012, from www.mka.mos.ru/mka/mka.nsf/va_WebResources/File/$file/201105170210.pdf

Ogil'ko, I. (2011, September 8). Here is cross [Vot vam krest]. *Rossiyskaia Gazeta*. Retrieved July 23, 2012, from www.rg.ru/2011/09/08/hram.html

Palamarchuk, P. (1992). *Sorok sorokov: kratkaia illiustrirovannaia istoriia vsekh moskovskikh khramov* [Forty by forty: a brief illustrated history of all Moscow churches], vol. 1. Moscow: Kniga i biznes.

Pravoslavnaia Moskva. (1995). Moscow: Izdatel'stvo bratstva Sviatitelia Tikhona.

Rokhlin, A. (1995, April 26). St. George the Victorious put into bronze: church on Poklonnaya Hill is decorated not in accordance with Orthodox Church's canons [Georgiia Pobedonostsa zakovali v bronyu: khram na Poklonnoy Gore raspisan vopreki kanonam Pravoslavnoy Tserkvi]. *Moskovskii Komsomolets*.

Rolf, M. (2010). Importing the "spatial turn" to Russia recent studies on the spatialization of Russian history. *Kritika: Explorations in Russian and Eurasian History, 11*(2), 359–380.

Shimansky, D. (1996, January 24). Tretii Rim epokhi raspada: chto ob'ediniaet moskovskie "stroiki veka"? [The Third Rome of decay epoch: what units Moscow's "constructions of the century"?]. *Nezavisimaia Gazeta*.

Sidorov, D. (2000a). Playing chess with churches: Russian Orthodoxy as re(li)gion. *Historical Geography, 28*, 208–233.

Sidorov, D. (2000b). National monumentalization and the politics of scale: The resurrections of the Cathedral of Christ the Savior in Moscow. *Annals of the Association of American Geographers, 90*, 548–572.

Sidorov, D. (2001). *Orthodoxy and difference: Essays on the geography of Russian Orthodox church(es) in the 20th century* (Princeton Theological Monograph Series 46). San Jose: Pickwick Publications.

Smolitskiy, G. (2012, July 19) Muscovites slow down the 200 churches construction program [Moskvichi tormozit programmu stroitel'stva 200 khramov] *Izvestia*. Retrieved July 25, 2012, from http://izvestia.ru/news/530682

Smorodinova, E. (2012, July 10). The largest church will be able to accommodate up to 1000 attendants [Samyy bol'shoy khram smozhet vmestit' do tysiachi prikhozhan] *Vecherniaia Moskva*. Retrieved August 1, 2012, from www.vmdaily.ru/news/samii-bolshoi-hram-smojet-vmestit-do-tisyachi-prihojan1341923160.html

Vsepoddanneishii otchet ober-prokurora Sviateishego Sinoda po vedomstvu Pravoslavnogo ispovedaniia za 1914 god. [Report of the Holy Synod's Over-Procurator on Orthodox Faith Department in 1914]. (1916). Petrograd.

Part XI
Culture: Museums, Drama, Fashion, Food, Music, Sports and Science Fiction

Chapter 130
The Nature Theatres of the Occult Revival: Performance and Modern Esoteric Religions

Edmund B. Lingan

130.1 Introduction

In his 1918 book, *The Open-Air Theatre*, Sheldon Cheney discusses the emergence of an "open-air" movement in U.S. and European theatre that strove to rescue theatrical productions from the interior confines of closed architectural structures. This abandonment of interior spaces was inspired by a desire to free the art of theatre from the "trickeries and artificialities of the modern stage" and to once again "learn the simplicity, the directness, and the joyousness of dramatic production under the sun and stars" (Cheney 1918: 8). In addition to its admirable artistic and aesthetic characteristics, Cheney credited the open-air theatre movement of the early twentieth century with meritorious "social aspects." Cheney asserts that open-air productions produce "hygienic and economic effects" that revivify "mind and body," "solidify" the communities that make up the audience by bringing them together to share a "common artistic purpose," and achieve a tangible "communal spirit." Cheney argued that the theatres of the open-air theatre movement bore a resemblance to the effect of the tragedies that were performed to honor the god Dionysus in fifth-century Athens and also the Christian biblical plays of medieval Europe (Cheney 1918: 9).

While discussing the merits of open-air theatre productions, Cheney notes a peculiar effect that he experienced as an audience member. He describes:

> [A]n intangible spiritual aspect, a subtle, almost religious effect on each individual, which collectively must make for social betterment. For man is never else so near God as when certain sorts of dramatic beauty are revealed to him under the open sky. (Cheney 1918: 10)

E.B. Lingan (✉)
Department of Theater, University of Toledo, Toledo, OH 43606, USA
e-mail: edmund.lingan@utoledo.edu

S.D. Brunn (ed.), *The Changing World Religion Map*,
DOI 10.1007/978-94-017-9376-6_130

One of the theatre artists that Cheney credits with producing this spiritual, religious effect on audiences is Katherine Tingley. Tingley was the head of a spiritual organization called the Universal Brotherhood and Theosophical Society (UBTS), which had its headquarters in a residential settlement in Point Loma, California. Cheney notes that Tingley constructed an outdoor theatre inspired by the ancient Greek model on the grounds of the UBTS headquarters in 1901 and he notes that she consciously seeks to create productions that will "prove in some measure a spiritual revelation" (Cheney 1918: 36).

The spiritual purpose of Tingley's theatrical productions were enhanced by the placement of the theatre within a strikingly beautiful landscape. Cheney writes:

> Certainly no theatre in ancient Greece ever had a greater loveliness, or a more idyllic background. As one comes to it on its precipice above the sea, it seems to glisten like some gleaming white jewel in a setting fashioned with perfect artistry. (Cheney 1918: 37)

At the time when Cheney wrote this description, Tingley was one of the best known advocates of Theosophy, which was one of the most successful and influential spiritual movements to arise during the "Occult Revival." The Occult Revival manifested as a surge of interest in esoteric beliefs and practices (that is, astral projection, astrology, alchemy, ceremonial magic, invocation, evocation, Hermetic philosophy, Gnosticism, cabbala, etc.) that flourished between the late nineteenth and mid-twentieth century in Europe and the U.S. A wave of alternative spiritual movements developed during the Occult Revival, and Theosophy was one of the most well-known of these movements.

Theosophy proposed the existence of a primordial and secret doctrine that was the mother of all modern religions and which shared the principles of karma and reincarnation that were central to the Buddhist and Hindu traditions. For Tingley, the art of theatre was essential to the propagation and practice of Theosophy and she also believed, like Cheney, that theatre performed in natural settings can reveal a divine aspect of life that, according to Tingley, is a spiritual reality. Tingley was not alone in her attitudes toward the blending of theatre and nature to the end of teaching occult principles. In fact, other leaders of the Occult Revival staged theatre and dramatic rituals in natural settings to embody their teachings and practice their beliefs.

This essay will explore how two of the leading figures of the Occult Revival – Katherine Tingley and Gerald Gardner (who established Wicca, the modern witchcraft religion) – did just this. This exploration of Tingley and Gardner's work will provide insight into a legitimately religious current of nature-focused theatre that seems to have been a part of the broader open-air theatre movement that Cheney describes in his book. It will also demonstrate that the sacred, natural theatre that Tingley and Gardner practiced anticipated contemporary forms of alternative spiritual and religious performance that are rooted in Western "ecospirituality." Ecospirituality is a contemporary religious perspective that expands the conventional Western "sense of sacrality" beyond "its overly human center" by including within its scope everything that occurs in the natural world, such as soil, water, plants, animals, and the atmosphere (Mazis 2007: 125). Today ecospirituality is promoted by adherents of various esoteric movements, including Neopagans, Goddess worshippers and modern witches.

The theatrical and performance forms to be discussed hereinafter represent two kinds of performance spaces that Cheney discussed in *The Open-Air Theatre*. The first of these Cheney is referred to as "architectural theatre," by which Cheney referred to the "classic type" of theatre, such as large stone or concrete structures with "massive stone backgrounds" (Cheney 1918: 10). The ancient stone outdoor theatres of Greece and Rome are examples of Cheney's conception of the architectural outdoor theatre. Cheney also referred to what he called "nature theatres" or "forest theatres." These kinds of theatre were often little more than natural settings whose geographic qualities were particularly suitable to theatrical entertainments and performances. Cheney explains that the stage of a nature theatre might be as simple as an "open place in a forest or on a mountainside" and that the auditorium might be "nothing more than an open hillside or sloping meadow" with no artificial seating to mar the scenery (Cheney 1918: 11). Tingley staged work in both an architectural theatre and in nature theatre environments while Gardner's Wiccan rituals often took place in nature theatres, as well.

130.2 The Architectural Theatre of Katherine Tingley and the UBTS

Between 1897 and 1929, Katherine Tingley became a well-known propagator of Theosophy when she became the leader of the Theosophical Society in America, which she renamed the Universal Brotherhood and Theosophical Society (UBTS) in 1897. The UBTS originally had its headquarters in New York City, but Tingley relocated the organization to Point Loma, California shortly into her tenure as the official leader. Raising funds and support from those who valued her teachings, Tingley managed to make the Point Loma headquarters, which were located a few miles west of San Diego, into a wonderland of fantastical architecture that rested within a landscape of striking natural beauty. Photos of the UBTS headquarters, which the UBTS residents referred to as "Lomaland," reveal whimsical circular buildings topped with illuminable, stained glass domes, such as the Temple of Peace, the Raja Yoga Academy, and the home of Albert Spalding, who founded Spalding Sporting Goods. These photos of Lomaland reveal the careful landscaping and the beautiful, well-groomed foliage that surrounded these unique domed structures.

The monochromatic glass domes that sat atop the round buildings of Lomaland were often illuminated at night and they could be seen by travelers on board ships in the Pacific Ocean, which the cliffs of Lomaland overlooked. According to Greenwalt, these glass domes were said to have "occult significance" (Greenwalt 1955: 48). Although, Greenwalt does not mention the specific meaning of the domes, it is safe to assume that they relate in some way to the heightened spiritual knowledge that Tingley and her associates felt was achievable through the study and practice of Theosophy.

Modern Theosophy began in 1875, when Helena Petrovna Blavatsky co-founded the American Theosophical Society in New York City with Henry Steel Olcott and William Quan Judge (White 2006: 77). The Theosophical Society quickly became one of the most well-known spiritual movements of the Occult Revival and Blavatsky gained celebrity status as a figurehead of Theosophy and occultism. In her 1889 book, *The Secret Doctrine: The Synthesis of Science, Religion, and Philosophy*, which became one of the most influential works of the Occult Revival, Blavatsky asserted the existence of an Infinite Principle that was the source of all life, delineated a process of reincarnation by which human beings were said to evolve through countless physical incarnations, and identified Karma as law that is linked to the spiritual development of humanity (Santucci 2005: 183). By blending ideas from a variety of scientific, esoteric, and philosophical sources, Blavatsky formulated a "belief system" that attracted thousands of followers from all over the world (Carlson 1993: 29, 35). Tingley adhered to Blavatsky's belief system, and, in particular, she valued Blavastsky's conceptualization of "Universal Brotherhood," that is, the belief that all human beings receive life from a single Infinite Principle and are bound by this relationship to care for one another (Blavatsky 1889: 34–5). Tingley built upon the principle of Universal Brotherhood by proposing that the performance of altruistic acts constituted the path toward spiritual enlightenment and she spoke of a coming era in which Universal Brotherhood would be realized throughout the world. All of the productions that she staged in the Greek Theater that Cheney described in *The Open-Air Theatre* were created to promote the concept of Universal Brotherhood, as well as Tingley's other Theosophical teachings.

The construction of the Lomaland Greek Theater was completed in 1901 and the photographs of this structure show that it was very carefully placed within a circle of trees overlooking the Lomaland cliffs and the Pacific Ocean. With a seating capacity of about 2,500, the outdoor Greek Theater was accessible to the total population of Lomaland (which exceeded 500 by 1911) as well as many more visitors (Waterstone 1995: 295; Greenwalt 1955: 80). Within this theatre, Tingley staged three kinds of dramas: Greek tragedies, plays by Shakespeare, and original Theosophical dramas that she referred to as "symposia." As will be seen, all of the dramas that Tingley directed were staged in visual harmony with the natural surroundings in which they took place, and, in fact, the Greek Theatre itself had been carefully positioned to make this harmony between production and environment possible. Tingley attempted to preserve the aesthetically-pleasing relationship between the theatre and the natural features that surrounded it by having the members of the UBTS Photograph Department (Greenwalt 142) take photos of the productions that took place on its stage. Through an arrangement with the Theosophical Society Archive located in Pasadena, California, I have been allowed to reproduce several of those photographs in this article, and they will reveal not only the intersection of architectural and natural beauty in the Greek Theater, but also the manner in which Tingley's productions revealed her beliefs about the religious function of theatre and the existence of a divine and spiritual aspect of life.

The collection of photographs at the Theosophical Archives include a variety of photographs of the Greek Theater. Some of these photos are direct views of the front

Fig. 130.1 Front of the Greek Theater at sunset (Photo courtesy of The Theosophical Society Archives, Pasadena, CA, used with permission)

of the theatre; others are taken from various side angles. What all of the photographs have in common, however, is that they carefully frame the theatre within a larger landscape. One of the most striking of these photographs displays the front of the Greek Theater at sunset. The photographer who took the photograph stood directly in front of the stage, and behind the pillared pavilion, beams of bright sunlight pierce billowing clouds and dance upon the ocean's waves. Between the glistening water and the Greek Theater stand the dark shapes of the rugged cliffs that overhang the sea (Fig. 130.1).

Another photograph taken from the auditorium shows a northwest view of the Greek Theater. In this photograph we see a small portion of the circular stage and part of the pavilion is visible as well. What is most striking about this photograph is the composition, for the visibility of the Greek Theater is quite intentionally diminished by plants and trees that stand between the photographer and the subject (Fig. 130.2). The composition of this photograph gives the impression of the theatre suddenly appearing in an opening in the foliage to someone who happens to be walking along a path. It is almost as if the photographer has discovered a secret theatre hidden in a natural clearing.

The careful framing of the Greek Theater within the beautiful plant life of Lomaland constitutes an intentional combination of geography and architecture for the purpose of promoting two ideas that were essential to Tingley's worldview. The first of these ideas was that everything in nature is a direct expression of the Divine Principle from which all life comes, and the second idea was that art, if created in the proper frame of mind and with the proper intentions, could also constitute an expression of the divine. Many of Tingley's ideas about interconnections between nature, art, and the divine were published in the pages of several Theosophical periodical publications of the UBTS that she edited. In an issue of Universal Brotherhood, for instance, an author argues that the key to living an "inward . . . life of spirituality"

Fig. 130.2 A northwest view of the Greek Theater, showing a small portion of the circular stage and part of the pavilion (Photo courtesy of The Theosophical Society Archives, Pasadena, CA, used with permission)

is aligning one's "highest ideals along the lines of nature and divine law" (Adhiratha 1899: 567). In another issue, Hattie A. Browne writes that "every flower of the field is the embodiment of divine thought" and asserts just staring at a landscape causes the presence of this Divine Fire to "mingle with our Soul" (Browne 1899: 655). For Tingley and her followers, nature was more than an expression of biological and botanical life: it was a message from a divine intelligence.

What is more, Tingley and the other members of her organization felt that theatrical productions could, if created in just the right way, become "true mystery dramas," that is, operate as a "religious symbolic representation" that could lead the audience member to spiritual enlightenment. In a 1902 article from another UBTS publication, *Universal Brotherhood Path*, an anonymous author writes:

> May we too resort to a symbol by comparing the true mystery drama to a clear, starlit sky? Hardly a mind is so unformed as not to be altogether filled by its splendor and to be lifted a little above its normal, but to the true student come visions of cosmic heights and depths, and a mystic imagination adds itself to the knowledge of the mind. From the starry sky, as from the sacred drama, every mind receives that which it can retain and just a little more. It is food for babes and wisdom for the wise. (Anonymous 1902: 113)

In an earlier issue of *Universal Brotherhood Path*, Tingley included an article that declared the "sacred dramatic work" of the UBTS to be of "the most valuable nature" and credited the productions that Tingley directed with a "saving power" that was causing the "minds of men" to be "permeated by the fundamental principles of Brotherhood." The article goes on to say that the results of Tingley's dramas would "be seen in the attitude of the masses when opportunities for striking out new lines of action occur in social and political life" (Anonymous 1901: 110). For Tingley and her fellow Theosophists, every aspect of the theatre experience – the play, the performance, the architecture of the theatre building, and the landscape surrounding the theatre – shared with nature the capacity to communicate spiritual truths concerning a connection between the souls of human beings and a Divine Principle from whence those souls originated. This symbolic connection that Tingley made between nature and theatre arts is vividly revealed in the previously-discussed photographs of the Lomaland Greek Theater.

The nature-theatre connection in Tingley's philosophy was also in the performances that she staged in the Greek Theater. Among the most lavish of these was a 1911 production of an original Theosophical drama, or "symposium," entitled *Aroma of Athens*. Theatre critic Bertha Hofflund wrote a lengthy review of this production in a 1911 issue of *The Theatre*. Hofflund, who lauds the residents of Lomaland for making the principles of "art-loving, nature-loving Athens a part of their lives" (Hofflund 1911: 219), vividly describes how natural and theatrical scenery blended in *Aroma of Athens*:

> From the level, sanded floor of the amphitheatre, softly lighted with encircling torches, rises the sweet, pungent scent of incense, mingling with the fresh odors of the night. . . . In a neighboring temple are seated a group of musicians. High on a cliff above stands another temple, with groups of young warriors on the outlook, leaning easily on their long bows. At a nearby fountain, classically draped women, bearing on their heads graceful water urns, leisurely fill and depart with their burdens. Near and more remotely, statuary, the woodland deities, Apollo, Diane, Venus, and Pallas Athene, glimmer among the shrubbery.
>
> Who, in the witchery of the hour and the scene, lighted by the changing glow of the signal fires of red and green and gold, rising at intervals from the silent hilltops and canyons, would not say, in these days of the twentieth century, it is the unsubstantial fabric of a dream?

For critics such as Hofflund and Cheney, the manner in which Tingley fused the natural setting of Lomaland with her theatrical productions was central to the overall effect of her work as a director. For Tingley, however, the fusion of landscape and theatre was a means of illuminating her theosophical ideas about the relationship between human beings, the world and the divine.

130.3 Tingley's Procession of the Seven Kings as "Nature Theatre"

Tingley not only staged theatrical events in the Greek Theater, but she also staged public presentations and processions that took place in what Cheney described in *The Open-Air Theatre* as "nature theatres." These nature theatres were often unaltered natural environments whose particular qualities were suitable to a specific theatrical event or performance. In the case of the procession of the Seven Kings, which Tingley directed and presented in Point Loma, a nature theatre served as the final stage of a perambulatory performance that advocated for peace, opposed war, and promoted the humanitarian principles of Theosophy. The procession of the Seven Kings took place on June 23, 1915, and it consisted of a parade of demonstrators and performers who marched through the streets of Point Loma to the grounds of Lomaland in 1915. This performance was presented as part of a conference associated with a UBTS sub-organization called The Parliament of Peace and Universal Brotherhood, which took place June 22–25, 1915. Tingley founded the Parliament of Peace on March 3, 1913 and it was intended to be an "international permanent organization for the promotion of peace and Universal Brotherhood" (Parliament of Peace 1915). Many prominent people attended or sent representatives to this conference, including: Prof. Osvald Sirén, professor of History of Art at the University of Stockholm; Mrs. Josephine Page Wright, President of the San Diego Woman's Press Club; Hon. Geo. W.P. Hunt, Governor of Arizona (represented); Hon. James E. Ferguson, Governor of Texas (represented); and Rev. Howard B. Bard of the First Unitarian Church of San Diego. Representatives of the Swiss International Bureau of Peace, the Swedish Peace and Defense Society, the United Daughters of the Confederacy, the American Humane Association, and the New York Anti-Vivisection Society presented addresses at this function (Parliament of Peace 1915).

The 1915 congress featured diverse entertainments, including a Grand International Pageant that included music, the procession of international flags and anti-war banners, a Peace Symposium performed by students of Raja Yoga Academy, and performances of *A Midsummer Night's Dream* and *The Aroma of Athens.* Out of these various entertainments at the conference, the procession of the Seven Kings was one of the most photographed. The story behind the procession was a Swedish legend of world peace, the essence of which was displayed on the primary banner of the procession. Carried by students of the Raja Yoga College, this banner read:

> A prophecy of permanent peace – Vadstena's legend of the seven kings: seven beech trees will grow from one root; seven kings will come from seven kingdoms. Under the trees they will establish permanent peace. This will be at the end of the present age. (Parliament of Peace 1915)

Other banners in this procession bore slogans such as "Universal Peace—A Protest Against War," "Win Peace By the Sword of Knowledge," "Universal Brotherhood is the Lost Chord in Human Life," and "Helping and Sharing is what Brotherhood Means" (Parliament of Peace 1915). The performers who portrayed the Seven Kings marched through the streets of San Diego and they were followed by the Raja Yoga students, who carried their own banners. The seven kings entered the Lomaland grounds on horseback through a large entry gate from an adjoining boulevard, where they came together under several trees and signed a peace treaty.

Natural elements were again represented as the manifestation of divine elements in Tingley's procession of the Seven Kings. In this procession, the manifestation of a higher, divine force through nature was suggested in the correspondence between the "seven beech trees that grow from one root" and the meeting of the "seven kings from seven kingdoms," which was described on the banner carried by the Raja Yoga students. As the photographs of this process show, Tingley made a concerted effort to find a tree on the grounds of Lomaland that bore a striking resemblance to the one described in the legend. In one photograph, the seven kings sit on their horses behind a multi-trunk tree, in front of which sits a small table holding the peace treaty that they ceremoniously signed at the end of the procession (Fig. 130.3).

Fig. 130.3 Lomaland tree resembling the "seven beech trees that grow from one root" (Photo courtesy of The Theosophical Society Archives, Pasadena, CA, used with permission)

The goal of the Pageant of the Seven Kings was to contemplate and dramatize the possibility of permanent world peace, which Tingley (1923: np) believed could not be achieved without the understanding that, as human beings,

> ...our responsibilities are not for ourselves alone, not for our own countries alone: but for the whole human family. Territory and trade may be much, national honor may be much, but the general salvation of human society here in this world—that is all.

In other worlds, Tingley believed that world peace could only be achieved when human beings unanimously agreed upon the Theosophical principle of Universal Brotherhood. With the procession of the Seven Kings, Tingley urged spectators to re-assess their cultural differences and national identities by embracing a larger kinship that transcends such boundaries and extends into the realm of the divine. Tingley believed that when all human beings recognized this affinity, a new cycle of peace would begin.

130.4 Wiccan Ritual as "Nature Theatre"

Tingley was not alone in her belief that the careful blending of natural settings and theatrical events might contribute to the spiritual enlightenment of human beings, for this idea was shared by other occultists and mystics who came after her. Although Tingley's own beliefs and practices differ greatly from those of many of the teachers of occultism and esoteric spirituality who appeared after she died in 1929, she shared with many of them her faith in the spiritual efficacy of theatre. Perhaps the most well-known of such occultists who came to prominence after Tingley's death was Gerald Gardner, who first publicized the Wiccan religion in the early 1950s and described Wiccan rituals as "sacred dramas" (Gardner 1954: 82).

Tingley had absolutely no interest in promoting witchcraft, but the practice of witchcraft was central to Gardner's religion, which he began to publicize in England after 1951, which is the year in which the UK government repealed the Witchcraft and Vagrancy Acts of 1736 (Hutton 1999: 242). Gerald's performance work is distinguishable from Tingley's because it did not take the form of public performances for large audiences. On the contrary, Wiccan performance generally takes the shape of rituals performed by covens, which are made up of much smaller numbers of members than Tingley's UBTS. Most Wiccan covens do not have more than twelve members (Hutton 1999: 403).

Gardner publicly revealed the details of Wiccan beliefs and practices in 1954, with the publication of his still-influential book, *Witchcraft Today*, and since the time of that publication many different versions of Wicca and modern witchcraft have emerged. Gardner's explications of his beliefs were not very specific. Like Theosophists, Gardner asserted that Wiccans believed in reincarnation and a divine force that gives life to all; however, Gardner professed ideas that he did not share with Theosophists and other occultists. Perhaps most unique to the first known school of Wicca, as it Gardner revealed it, was the worship a feminine deity and a masculine deity. The masculine deity was described as a "horned god" who controls the powers of death and reincarnation, and the feminine deity is an unnamed goddess who is associated with the moon and is credited with conveying to human beings the secrets of reincarnation and ritual unification with the divine (Gardner 1954: 40–41). Gardner and his followers attempted to magically unite with the horned god and the unnamed goddess through the performance of magical rituals, which were often performed out of doors and in natural settings.

Gardner and the members of his first coven, which was based in South England's New Forest, performed their rites in forest clearings, which they believed was a continuation of the practices of a "Stone Age cult" that had been carried on in secret by pagan priests and priestesses for countless centuries (Gardner 1954: 43). Gardner did not make up this idea; in fact, it came from a 1921 anthropological text entitled *Witch-Cult in Western Europe: A Study in Anthropology*, which was written by anthropologist, Margaret Alice Murray (Murray 1921). Although Murray's argument that many of the victims of the medieval and early-modern witch hunts of Europe had actually been practitioners of a prehistoric religion was later debunked and is no longer regarded as accurate by modern anthropologists (Hutton 1999: 206–07), Gardner and many advocates of modern witchcraft accept Murray's theories, and have found inspiration in them for the development of new religious rituals and practices. In fact, Murray wrote an introduction for the original version of *Witchcraft Today* and in that essay she does not attempts to debunk Gardner's assertions to have discovered "in various parts of England groups of people who still practice the same rites as the so-called 'witches' of the Middle Ages." Nor does Murray discount Gardner's assertion that the rituals he describes are "a true survival" of a Stone Age religion (Gardner 1954: 15).

In her introduction to *Witchcraft Today*, Murray outlines the recurring patterns of the prehistoric witch cult that she believes once existed and suggests that Gardner's Wiccan rituals bear an affinity to them. Murray (1954: 15–16) writes:

> Personal worship may take any form, but a group of persons worshipping together always devise some form of ritual, especially when the worship takes the form of a dance. . . . All the movements are rhythmic, and the accompaniment is a chant or performed by percussion instruments by which the rhythm is strongly marked. The rhythmic movements, the rhythmic sounds, and the sympathy of numbers all engaged in the same actions, induce a feeling of exhilaration, which can increase to a form of intoxication. This stage is often regarded by the worshippers as a special divine favour, denoting the actual advent of the Deity into the body of worshippers. The Bacchantes of ancient Greece induced intoxication by drinking wine, and so making themselves one with their God.

Murray's description of the Bacchantes brings to mind Euripides' play, *The Bacchae*, which depicts the revels of the Bacchantes. One detail from that play which is relevant here is that Euripides represented the revels of the Bacchantes as a form of ecstatic worship that took place in outdoor, natural settings. Gardner and other Wiccans adhere to this practice.

In *Witchcraft Today*, Gardner describes a Wiccan ritual known as the "Drawing Down of the Moon." Gardner describes an outdoor event at which participants throw "various leaves" onto a fire, dance themselves into trance-like states, and hold torches as they stand in a magic circle (Gardner 1954: 24–5). The magical circle is central to Wiccan ritual practice and it is viewed as a boundary between two worlds: the physical world and "the dominion of the gods" (Gardner 1954: 26). Although Wiccans often perform their rituals outside, the nature of the magical circle is to create a connection, or a unity, that enables passage between the natural world and a realm populated by divine beings. It is this link between the natural world and the spiritual world that has identified Wicca as a "religion of nature" (Bracelin 1960: 200). By establishing a religion rooted in the practice of magical rituals in natural settings, Gardner and his students anticipated a wave of nature religion, goddess worship, and ecospirituality that has continued into the present.

130.5 Conclusion

Presently there are a wide array of alternative religions and spiritual movements that have continued the strain of occult nature theatre that was evident in the theatrical and performance work of Katherine Tingley and Gerald Gardner. A few examples of this work includes the ecologically-focused rituals and demonstrations of Starhawk, a highly-influential witch and leader of nature-focused religion who has linked the creative practice of spiritual ritual to ecological activism. On more than one occasion, Starhawk's pro-environment demonstrations have placed her in confrontation with law enforcement officials and led to her arrest. In the 1980s, for instance, Starhawk's reverence for the Goddess as nature led her being "arrested and imprisoned for non-violent direct actions undertaken in protest against installations where nuclear weapons were made" (Hutton 1999: 348). Other pagans, occultists, and witches who have attempted to reveal spiritual truths by staging rituals and theatrical performances in natural environments include, include Las Vegas magician, Jeff McBride, who operates a group ritual known as the Alchemical Fire Circle, in which dancers enter trance by dancing around a large flame from dusk until dawn (*Vegas Vortex*, www.vegasvortex.com), and Australian occultist, Orryelle Defenestrate-Bascule's Metamorphic Ritual Theatre, who stages occult dramas at sacred locations, such as Stonehenge and Amesbury (Mutation Parlour, www.crossroads.wild.net.au). In fact, the range of ecospiritualitities – spiritual world views that equate love and respect for the natural environment with love and respect for the divine – continues to flourish and expand in the present time.

One of the significant aspects of the relationship between nature and the performance of theatre and dramatic rituals that was established by spiritual teachers such as Tingley and Gardner during the Occult Revival is that it seems to mark an early stage of a relationship between ecological interest, religion, and art that is quite perceivable today. As we have seen, the link between these elements was noted by theatre critics such as Cheney and Hofflund in the early years of the twentieth century. Later, Gardner strengthened the links between ecology, new spirituality, alternative religion, and performance when he introduced Wicca to the general public as a "religion of nature" (Bracelin 1960: 201). Many of today's esotericists – goddess worshippers, witches, neopagans, nature worshippers, and others – continue to create dramatic ritual and sacred theatre as a form of dramatic expression. The annual Starwood Festival, for instance, which takes place in Pomerey, Ohio, is the "largest Pagan/Magickal/Consciousness" gathering in North American" and perhaps, according to the festival's website, "perhaps the world" (*Association for Consciousness Exploration*, www.rosencomet.com/starwood). The 32nd annual Starwood Festival will take place June 10–16, 2012 and the festival participants will "celebrate … diversity and alternatives of belief systems, lifestyles, and spirituality" with "music, drumming, dance, and theatre" (*Starwood Festival XXXII*. www.rosencomet.com/starwood/2012/). Pagan theatre and occult dramas are listed in a wide range of pagan newsletters, blogs, and websites, such as The Wild Hunt (The Wild Hunt, www.patheos.com/blogs/wildhunt/2010/06/spiritual-theatre-and-ritual-performance.html), Terra Mysterium (Terra Mysterium, www.terramysterium.com/), and Capital Witch (Capital Witch, www.capitalwitch.com/2012/03/theatre-company-preps-for-pagan.html), all of which include listings and articles that discuss pagan, magical, and occult theatre. The range of web resources that discuss Wiccan, pagan, magical, and occult theatre suggest that the link between the divine, natural settings, and sacred religion and performance that was established in the nature theatres of the Occult Revival is growing more diverse in the present and will continue to flourish and expand for years to come.

References

Adhiratha. (1899). Fragment – Omniscience. *Universal Brotherhood, 13*(10), 566–567.

Anonymous. (1901). New dramatic symposium, "The Conquest of Death." *Universal Brotherhood Path, 16*(2), 110.

Anonymous. (1902). The Easter festival in Isis theater at San Diego. *Universal Brotherhood Path 17*(2),112–114.

Blavatsky, H. P. (1889). *The key to theosophy*. Reprint; Adyar: The Theosophical Publishing House.

Blavatsky, H, P. (1888/1999). *The secret doctrine: The synthesis of science, religion, and philosophy.* Reprint; Pasadena: Theosophical University Press. 2 vols.

Bracelin, J. L. (1960). *Gerald Gardner: Witch*. London: Octagon Press.

Browne, H. A. (1899). Divine fire. *Universal Brotherhood, 13*(12), 655–657.

Carlson, M. (1993). *"No religion higher than truth:" A history of the theosophical movement in Russia, 1875–1922*. Princeton: Princeton University Press.

Cheney, S. (1918). *The open-air theatre*. New York: Mitchell Kennerly.

Gardner, G. (1954). *Witchcraft today*. London: Rider and Company.

Greenwalt, E. A. (1955). *The Point Loma community in California 1897–1955: A theosophical experiment*. Berkeley: University of California Press.

Hofflund, B. (1911). The revival of ancient Greek drama in America. *The Theatre Magazine* 18–19.

Hutton, R. (1999). *Triumph of the moon: A history of modern pagan witchcraft*. Oxford: Oxford University Press.

Mazis, G. A. (2007). Ecospirituality and the blurred boundaries of humans, animals and machines. In L. Kerns & C. Keller (Eds.), *Ecospirit: Religions and philosophies for the earth* (pp. 125–155). New York: Fordham University Press.

Murray, M. (1954). Introduction. In G. Gardner (Ed.), *Witchcraft today*. Reprint. New York: Citadel Press, 2004.

Murray, M. A. (1921). *The witch-cult in western Europe: A study in anthropology*. Oxford: Clarendon Press.

Parliament of Peace and Universal Brotherhood. Program from a congress that took place at Point Loma, CA, 22–25 June 1915. No pagination. Pasadena: Theosophical Society Archives.

Santucci, J. A. (2005). Helena Petrovna Blavatsky. In W. J. Hanegraaff, et al. (Eds.), *Dictionary of Gnosis and Western Esotericism* (pp. 177–185) Leiden: Brill.

Tingley, K. (1923). Towards permanent peace. Reprint *Sunrise, 47* (April/May 1998), 209–211.

Vegas Vortex. www.vegasvortex.com. Accessed 30 Apr 2012.

Waterstone, P. B. (1995). *Domesticating universal brotherhood: Feminine values and the construction of utopia, Point Loma Homestead, 1897–1920.* Dissertation. (Department of History. University of Arizona.) Ann Arbor: UMI Microform.

White, R. A. (2006). Stanislavsky and Ramacharaka: The influence of turn-of-the-century occultism on the system. *Theatre Survey, 47*, 73–92.

Chapter 131
Affect, Medievalism and Temporal Drag: Oberammergau's Passion Play Event

Jill Stevenson

131.1 Introduction

Every 10 years, thousands of people from across the globe take a pilgrimage to Oberammergau. The reason for journeying to this small Bavarian town is a Passion Play, one the villagers have performed nearly every 10 years since 1634. According to the play's official history, when plague struck Central Europe in 1630 and killed 80 Oberammergauers, the remaining villagers took an oath that if their lives were spared "they would perform the 'Play of the Suffering, Death and Resurrection of Our Lord Jesus Christ' every 10 years. At Pentecost 1634, they fulfilled their pledge for the first time" (Passion Play Oberammergau 2010, "Chronicle"). Except in a few instances due to war, the villagers have maintained this promise up until the present.

The five-and-a-half hour long play is a unique example of religious community theatre. Its cast of 2,400 actors, drawn entirely from the village's 5,200 residents, includes 21 principal roles (each shared by two actors who rotate performances), 120 additional speaking roles, and 100 choristers, as well as a 55-person orchestra. Only people who were born in the village, have lived in Oberammergau for at least 20 years, or who marry villagers (making them eligible after only 10 years) can perform in the Passionsspiele (Shapiro 2000: 3–4). As a result, the play is inextricably bound to the villagers' identity; as a former director of the play remarked, "No native in the village can think about his life without thinking about the play. The play is a part of us. It controls the rhythms of our life, like it or not" (Quoted in Shapiro 2000: 5–6)

Claire Sponsler has analyzed how Oberammergau's play is marked by medievalism. Medievalism constitutes the way the Middle Ages have been used, recycled, and remade in order to serve the purposes of later periods. Although we certainly

J. Stevenson (✉)
Department of Theater Arts, Marymount Manhattan College,
New York, NY 10021, USA
e-mail: jstevenson@mmm.edu

S.D. Brunn (ed.), *The Changing World Religion Map*,
DOI 10.1007/978-94-017-9376-6_131

undertake such work with respect to other periods, the Middle Ages are unique in that, as Catherine Brown notes, they "were invented to be a foreign country" (Brown 2000: 547). Likewise, Sponsler (2004: 185) explains:

> Created by early modern humanists to define what they thought they were not, but which in many ways they still were, the Middle Ages have, since their inception as a cultural construct, been thoroughly ideological: the Middle Ages are a historical period tied to a social myth. That myth has to do with concerns about the present that have positioned the past as a place of refuge and reassurance. This positioning requires a simultaneous distance from and nearness to the past; the past has to be different enough from the present to provide solace and yet similar enough to be meaningful.

It is precisely this odd mixture of self and other in Oberammergau that prompted the British tourist Ethel B. T. Tweedie to write in 1890 that "sitting among those simple surroundings, it seemed almost impossible to realize that one was living in the nineteenth-century… [I]t is strange to be suddenly transported back, as it were, to the Middle Ages" (Quoted in Shapiro 2000: 116). Tweedie is of that "medieval" world—she inhabits it comfortably and seamlessly—yet she also remains separate enough to appreciate its alterity.

As Tweedie's comments suggest, Oberammergau's medievalism does not function solely through cultural references; crucially, it creates meaning by supplying affective encounters with an idealized medieval past. Considering medievalism as an affective force can therefore help us to better understand why Oberammergau's Passion Play event "works" for many contemporary spectators, particularly the many who travel to the village from the U.S., and, thus, to recognize its cross-geographic power. For this reason, my analysis is related to work by Rebecca Schneider and Jonathan Gil Harris, both of whom analyze how material objects and historical events can challenge linear chronology. Significantly, Schneider argues that this occurs, in part, through the "relational aspects of affect," which allow us to build bridges across periods and geographies (Schneider 2011: 36). She notes that many scholars "argue for the value of crossing disparate and multiple historical moments to explore the ways that past, present, and future occur and recur *out of sequence* in a complex crosshatch not only of reference but of affective assemblage and investment" (Schneider 2011: 35, original emphasis). In this essay, I argue that Oberammergau's affective medievalism cultivates exactly this kind of temporal and geographic crosshatching in order to satisfy certain cultural longings.

For many visitors, then, the experience in/of Oberammergau functions, to use Harris' term, as a polychronic event that compresses different times and geographies within a single surface (Harris 2009: 16). Doing so provides visitors with "felt" access to a past that, although largely constructed through medievalist fantasy, works for them. Indeed, Harris could be describing medievalism when he explains how, within a polychronic event, "the past is always potentially alive. And in its untimely life, that past speaks with and through us in the accents of the present and, in ways we can never quite predict, the future" (Harris 2009: 25). The Middle Ages has always spoken with and through the present, dragging present into past and past into present, and thereby creating affinities and distinctions between different geographies and temporalities. I, therefore, propose that Oberammergau's "felt" affective medievalism not only reveals something about this particular Bavarian Passion Play event, but that, in some respect, it also allows us to capture traces of the "real" Middle Ages.

In her analysis of medieval praying, Rachel Fulton shifts her readers "from thinking about prayers as things made (or crafted) to thinking about prayers as themselves tools for making" (Fulton 2006: 717). In doing so, she acknowledges that "as the case is with all such tools…it is difficult to appreciate exactly how they are to be most effectively used without trying them oneself." Therefore, although she recognizes that "some readers may find this unsettling, not what they are used to in reading an academic article," as part of her research Fulton allows herself to experience something of these prayers' "intended effect" in order to gain access to the medieval experience of praying (Fulton 2006: 708, 707). Similarly, I will suggest that opening ourselves up to the affective trajectory supplied by certain medievalist events can give us insight into the valuable "felt" work—the intended effects—of performances in the Middle Ages. Affect can, therefore, play an important role in our methods of historical inquiry.

131.2 Making Oberammergau Matter

Many theorists working on affect distinguish between affect and emotion. However, in her critique of the recent scholarly "turn to affect," Ruth Leys encourages us to be skeptical of such models, since they typically propose "that affect is independent of signification and meaning" (Leys 2011: 443).[1] For this reason I find Sianne Ngai's work useful. Ngai (2005: 27) acknowledges a distinction between affect and emotion that does not empty the former of meaning:

> the difference between affect and emotion is taken as a modal difference of intensity or degree, rather than a formal difference of quality or kind. My assumption is that affects are less formed and structured than emotions, but not lacking form or structure altogether; less "sociolinguistically fixed," but by no means code-free or meaningless; less "organized in response to our interpretations of situations," but by no means entirely devoid of organization or diagnostic powers…. What the switch from formal to modal difference enables is an analysis of the transitions from one pole to the other: the passages whereby affects acquire the semantic density and narrative complexity of emotions, and emotions conversely denature into affects.

[1] Leys notes that many affect theorists, especially those using evidence from the neurosciences, share "a commitment to the idea that there is a disjunction or gap between the subject's affective processes and his to her cognition or knowledge of the objects that caused them. The result is that the body not only 'senses' and performs a kind of 'thinking' below the threshold of conscious recognition and meaning but … because of the speed with which the autonomic, affective processes are said to occur, it does all this before the mind has time to intervene" (2011: 450, original emphasis).

According to Ngai, the meaning affect carries is less "fixed" and, thus, more potential than certain. Due to this dynamism, affect can be manipulated in different ways and for different purposes than emotion.

Ngai's model puts affect into productive dialogue with notions of embodied experience as a site for meaning construction. It is especially important to recognize the meaningful "work" of affect when studying religious genres since, throughout history, various devotional cultures have understood affect to be a valued bearer of ideological meaning. This is especially true of the affective piety that developed during the European Middle Ages. This devotional tradition, which emerged in the late eleventh and twelfth centuries, and revolved around meditation on vivid visual, textual, and mental images of Christ's Passion, resulted in "a shift in theological perspective and a radical reschooling of religious emotion" (Kupfer 2008: 8).[2] Sarah McNamer demonstrates that affective piety supplied medieval devotees with ways to establish and maintain beneficial spiritual identities. For example, she argues that some of the earliest affective texts on Christ's Passion were designed for religious women and that the compassion these texts generated was intended to help women verify their status as true brides of Christ, a status that, among other things, secured them a degree of agency over their earthly and spiritual lives. In this respect, affectively oriented genres engendered "'intimate scripts' … quite literally scripts for the performance of feeling—scripts that often explicitly aspire to performative efficacy" (McNamer 2010: 12). In other words, these scripts cultivated emotional production that resolved very real problems for medieval devotees.

For some people, visiting Oberammergau during a Passionsspiele year may hold the same affective, performative efficacy by cultivating an intimate script that resolves cultural longing. However, rather than piety, I contend that the affective force of these scripts is medievalism. Melissa Gregg and Gregory J. Seigworth suggest that "affect is in many ways synonymous with *force* or *forces of encounter*" (Gregg and Seigworth 2010: 2, original emphasis). With respect to a performance, we might then interpret dramaturgical elements as affective forces that can, by directing "those intensities that pass body to body…those resonances that circulate about, between, and sometimes stick to bodies and worlds" (Gregg and Seigworth 2010: 1), influence how spectators make meaning from their experience at/in a performance event. As Andy Lavender explains, the function of staged elements is determined by how they impact spectators, "this impact is likely to be in relation to meaning-effects – but it may also have a *felt* charge that structures our experience of the event" (Lavender 2006: 64, original emphasis). Affect theory can help us to examine how this "felt" charge guides performance spectators toward/into certain emotional responses, responses that can promote particular ideologies or beliefs. Affective medievalism is, therefore, a force that extends out, resonating across geography and history, sticking to bodies and worlds, in order to create moments of "felt" contact across time and space. Such contact not only engenders beneficial

[2] For more complete discussions of these trends in Passion imagery in relation to affective piety, see: Fulton (2002), Binski (2004), MacDonald et al. (1998), Beckwith (1993).

emotional production, but it also remakes the spectator into a living link to the medieval(ism) tradition.

Furthermore, Sponsler suggests that plays like Oberammergau's Passionsspiele function as "performative historiographies that enact a set of relations with history." Such historiographies do not offer strictly linear chronologies, but instead follow their "own historical logic based on assertion of tradition," narrating a past that is "fluid" and "shaped by the needs of the present" (Sponsler 2004: 8, 187). I will use affect theory to pursue this idea further in hopes of providing new insights into why Passion Plays, and perhaps religious performances more generally, matter today. Linda Kintz explains that "matter" refers to "the very nature of materiality in general and the body in particular," as well as to "the way things come to matter, the way emotions are learned and taught" (Kintz 1997: 55).[3] "Mattering" is, therefore, an affective construction that fosters powerful moments of emotional stability (Kintz 1997: 61), a process directly related to the intimate scripts McNamer describes.[4] Consequently, studying why a particular cultural phenomenon remains significant over a period of history—why Oberammergau's Passion Play still matters to so many different types of spectators today—requires us to consider how it functions affectively.

131.3 Adaptation as Tradition: The Oberammergau Passion Play

In his exceptional study of Oberammergau's 2000 production, James Shapiro reviews the Passion Play's history, including its troubling connections with anti-Semitism and Nazism; Hitler attended the 1930 and 1934 productions, and later praised the play's negative depiction of Jews.[5] Shapiro also describes the different revisions to the script and changes to production elements that have occurred over the centuries. The script has undergone at least four significant known revisions since the seventeenth century, with minor changes made each season. The oldest surviving manuscript is dated to 1662 and it retains several features characteristic of medieval religious drama, such as elaborate spectacle, a mixture of high and low elements, and characters such as Satan, devils, and allegorical figures appearing onstage. Moreover, the climax of this version was not the Crucifixion or Resurrection,

[3] Here Kintz is referencing Grossberg (1992).

[4] Kintz explains: "The intensity of mattering, while ideologically constructed, is nevertheless 'always beyond ideological challenge because it is called into existence affectively.'" Here she is quoting Grossberg (1992: 86).

[5] At a dinner on July 5, 1942, Hitler is recorded as saying: "it is vital that the Passion play be continued at Oberammergau; for never has the menace of Jewry been so convincingly portrayed as in this presentation of what happened in the times of the Romans. There one sees in Pontius Pilate a Roman racially and intellectually so superior, that he stands out like a firm, clean rock in the middle of the whole muck and mire of Jewry" (Quoted in Shapiro 2000: 168).

but instead the Harrowing of Hell, an episode recounted in the Apocryphal Gospel of Nicodemus that describes Jesus' time in Hell between his death and resurrection. This dramaturgical choice effectively shaped the play into a revenge drama with Jesus as the hero and Satan and his devils as villains (Shapiro 2000: 59–60).

A Benedictine monk named Ferdinand Rosner undertook the next major script revision in 1750. Rosner's changes included adapting the play into verse, adding elements like the laments of the Virgin Mary and Mary Magdalene, and formalizing the prefiguration tableaux vivant that, despite subsequent changes in specific content, still open each act of the play. These and other revisions nearly doubled the play's length. Shapiro describes Rosner's text as having a "Baroque, Jesuit-influenced, operatic style" that allowed historical and allegorical characters to mingle together onstage (Shapiro 2000: 61).

In 1811 and in the midst of controversy over Passion playing in Germany, another monk, Othmar Weis, significantly revised Rosner's text. According to Shapiro, Weis adapted the play into realistic prose and cut many elements that lacked biblical authority, including the devils and other allegorical figures. Removing the devils also radically altered the play's dramaturgy since it eliminated the villains. But, as Shapiro explains, "Weis's answer was the Jewish priests and merchants, who now became Jesus' main persecutors and were given biblical names and stereotypical attributes" (Shapiro 2000: 70). The 1811 text also had a new musical score written by Rochus Dedler, a schoolteacher and church organist in Oberammergau.

In addition, the early 1800s witnessed a major change in staging practice. The play, which had previously been performed in the village churchyard, moved into a new, more permanent stage in 1830; the current Passionsspielhaus stands on that same site. This shift also supported Weis's script changes: "the play was moving away from [a] hierarchical arrangement at roughly the same time that it became free of the cramped churchyard. The new wide stage was ideal for the horizontal staging of the increasingly realistic production" (Shapiro 2000: 71). Thus, as Shapiro notes, by 1830 all the ingredients for commercial success were in place: a permanent stage, new music, and a revised text that contained little to offend Protestant spectators, who began attending the production in larger numbers (Shapiro 2000: 71).

Joseph Alois Daisenberger, a student of Weis's, completed the last major overhaul of the Oberammergau script around 1860. He also served as the production's director for multiple seasons. Daisenberger shortened many of the play's speeches, updated its language, and focused more attention on characters' psychological motivation. Moreover, as Shapiro explains, Daisenberger formalized the play's typology, especially with respect to the tableaux vivant (Shapiro 2000: 72). This 1860 version continues to serve as the production's base text, but certain changes (some small, some more extensive) have been made to the production nearly every season. Consequently, for both the 2000 and 2010 productions the director Christian Stückl and dramaturg/deputy director Otto Huber revised and added elements to Daisenberger's text. For example, the 2010 play devoted much more time to debate among the Jewish priests over how to respond to Jesus than had previous scripts.

The same is true of adapting the set and costumes; as Stefan Hageneier, set and costume designer for the 2010 production explained:

> My intention for the 2010 Passion Play is to bestow a more intimate and artistic appearance to the wide proscenium and the wings, i.e. the open-air sections of the Oberammergau stage. The proscenium floor, hitherto made up of large square flagstones, has been covered with screed and fashioned as a homogeneous blue space. The previously bare walls of the side alley – now adorned with olive trees – are reminiscent of the walls of Jerusalem. (Hageneier 2010: 6)

Furthermore, in 2010, musical director Markus Zwink added new compositions to Dedler's base score. Thus, although Oberammergau's Passion Play maintains a powerful connection to a centuries-old tradition, ultimately it has always been a dynamic cultural product designed to serve the needs and inclinations of the present.

131.4 Medieval Atmosphere in 2010

Nestled in the Ammergau Alps, Oberammergau's location is somewhat remote, approximately 56 miles (91 km) southwest of Munich and 50 miles (81 km) north of Innsbruck. Nevertheless, beginning in the nineteenth century, foreign visitors—especially those from the U.S.—have comprised a large percentage of the audience and now constitute the majority. Shapiro explains how by 1860, the uniqueness of the village's ancient Passion Play began to attract intense foreign attention, which directly impacted ticket sales; 13,000 spectators attended the 1830 production, while that number rose to 100,000 in 1860. This international fame was due, in part, to how the village itself supplied spectators with the experience of visiting an idealized past. As Shapiro demonstrates, mid-nineteenth-century accounts reveal foreign travelers who feel they "have gone back in time, a sense reinforced by the startling appearance of the villagers, with their long hair, serious demeanor, [and] biblical aura." Oberammergau represented a community that had "somehow managed to avoid the rupture between the spiritual and material that increasingly defined life in the West" (Shapiro 2000: 115).

Thus, for many visitors, Oberammergau's Passion Play served as a reassuring living link to a nostalgic pre-modern, pre-industrial, pre-Reformation medieval past. An 1883 piece in Century magazine illustrates this view: "The antagonism and enlightenment of the Reformation did not reach the Bavarian peasant; did not so much as disturb his reverence for the tangible tokens and presentations of his religion. He did not so much as know when Miracle Plays were cast out and forbidden in other countries" (The Passion Play at Oberammergau 1883: 914). Moreover, accounts like this also demonstrate that medievalism was not confined to the play, but permeated the larger Oberammergau experience. For example, the village's secluded location reinforced a "medieval" aura; in an 1870 essay in *Putnam's*

Monthly Magazine, Lucy Fountain exclaimed, "when one reflects that this marvel of beauty … is the production of untutored peasants in a remote village of Bavaria … the *Passionspiel* [sic] of Ober-Ammergau becomes indeed a Miracle-play" (Quoted in Sponsler 2004: 129). Here, an experience in the present folds back upon the past, thereby remaking the play into a remnant of the Middle Ages. This is performative historiography in action.

Medievalism remains an important part of the Oberammergau experience today. In the summer of 2010, I was one of approximately a half-million people who travelled to Oberammergau in order to see the village's Passion Play. The 2010 production ran from May 15th through October 3rd, with 102 performances total. Although spectators in the 4,700-seat Passion Play Theatre are covered, the play has always been presented on an open-air stage with the Bavarian sky and landscape serving as backdrop. For some spectators, this feature enhances the play's spiritual dimensions: "This closeness of nature is an accessory of illimitable effect; the visible presence of the sky seems a witness to invisible presences beyond it, and a direct bond with them" (The Passion Play at Oberammergau 1883: 920). Traditionally, the Passionsspiele started in the morning, but in 2010 for the first time the play began at 2:30 pm, with a 3-h dinner break mid-performance. The second half ended at approximately 11:00 pm. The night sky added considerable dramatic impact to the play's final scenes, particularly the Crucifixion (Fig. 131.1). However, since the play is performed in German, the evening darkness also meant that non-German speakers who wanted to read along in their translated textbooks had to do so using small flashlights, a somewhat distracting necessity.

The Passion Play, which most visitors see at the end of their 1- or 2-day stay in Oberammergau, is obviously the reason pilgrims descend upon the village each decade. And it is a truly remarkable event. The 2010 play was composed of 11 acts,

Fig. 131.1 Crucifixion. Passionplay Oberammergau, 2010 (Photo by Brigitte Maria Mayer and Oberammergau Production, used with permission)

Fig. 131.2 Jesus clears the Temple. Passionplay Oberammergau, 2010 (Photo by Brigitte Maria Mayer and Oberammergau Production, used with permission)

beginning with Christ's Entry into Jerusalem and ending with a brief scene that depicted Mary Magdalene's encounter with the risen Jesus at the tomb. Choral interludes, accompanied by the tableaux vivant, opened each act. The crowd scenes, some containing as many as 1,000 actors, often featured live animals (Fig. 131.2). The costumes and properties were exceptionally designed and the attention to "authentic" detail, reflected in elements like the real beards and long hair that actors begin growing out a year in advance, helped create a sensually rich performance experience. Yet, I consider the Passionsspiele as only the final (although, arguably, climactic) act of a larger, equally sensual and equally staged performance event. My focus is therefore not the play itself, but rather the Oberammergau event experience as a whole.[6] More specifically, I am interested in how affective medievalism helps this event experience to matter to spectators, particularly U.S. spectators who may not know German and, therefore, upon whom this "felt" medievalism may work as a particularly powerful force.

For many visitors, Oberammergau's medievalism begins the moment they book their travel. Despite the cost, packaged tours are an extremely popular and convenient way for visitors, especially foreign visitors, to arrange their trips. These tours are usually one- or two-night packages that include accommodation and most meals, admission to the show, a textbook with the complete play text in German and in translation, and free transportation to local attractions. According to one U.S. agent, the price of two-night packages that include the best hotel lodging options can be as high as $1,130 and those with more modest hotel lodging usually cost no less than

[6] For a more complete description of the 2010 play, see my review: Stevenson (2011a), or Montgomery (2011).

$740 (Browne 2009).[7] Furthermore, the tour companies who offer these packages purchase nearly all of the seats in the Passion Play Theatre (Browne 2009).[8] This tour-package system also creates small, temporary communities; I sat with the same couple for all the meals at my hotel, and alongside fellow hotel guests at the Passion performance.

For some visitors, this tour system may help to situate Oberammergau as a modern-day pilgrimage site. Eamon Duffy explains that the purpose of medieval pilgrimage was "to seek the holy, concretely embodied in a sacred place, a relic, or a specially privileged image" and thus to experience "a temporary release from the constrictions and norms of ordinary living" (Duffy 1992: 191). This release offered time to reflect upon one's life, while the hardship of the journey served certain penitential functions. Travel to Oberammergau may not be particularly arduous, but it requires more planning and effort than many excursions, a fact that may increase the trip's personal value and encourage spectators to perceive their visit as extraordinary. Moreover, although not necessarily approached as a sacred site, nineteenth-century accounts demonstrate that the village provides visitors with an experience that feels outside of ordinary time and space.

The remarkable nature of this journey is underscored by the fact that those who wish to see Oberammergau's Passionsspiele must travel to the village; although you can buy official production photos or a CD of the choral music, no video or audio recordings of the performance have ever been made available for sale, and a strict no camera policy is enforced in the theatre. In fact, throughout the twentieth century Oberammergauers have rejected lucrative offers in order to keep the play firmly situated within the village. For example, Sponsler reports that "in 1900, an American producer tried to offer the main performers at Oberammergau 5,000 dollars each to perform in a passion play in New York, but they turned him down, apparently out of fear of lessening the value of Oberammergau's production" (Sponsler 2004: 133). Similarly, Shapiro recounts how in 1922 a Hollywood studio offered the village one million dollars to film the production; this bid was also rejected. Decades later, in 1966, the British press announced that a production of the village's play was touring various British cities. However, it was soon revealed that this production was not the Oberammergau play, but another "Bavarian" Passion play (Shapiro 2000: 126). A representative from the village publicly announced to the British press that the

---

[7] The price list available on one of the main Bavarian tour company's websites includes more lodging options, ranging from hotels to rooms in villagers' homes. The prices for these two-night packages run from €839 to €275 ($1,031–$338, using the current exchange rate), with one-night packages ranging between €575–€199 ($706–$245) (Oberammergau Passion Play 2010).

[8] "The Passion Play season is known throughout the Christian world and demand for tickets and tours has usually been high. Tickets for the performance are only available in advance as part of a package with one or two nights local accommodation in Oberammergau, the neighboring villages of Ettal, and Unterammergau and other villages in the region. As Oberammergau is a big draw for incoming church groups, tour operators have had to scramble for allocations to fulfill demand" (Browne 2009). Although the point of this article is the initially slow ticket sales for the 2010 show, later reports from travel agencies and on the Oberammergau Passion Play website suggest that sales improved and tickets were difficult to obtain once the production opened.

Oberammergau production would never be performed outside of the village. As Shapiro notes, "If you wanted to see the play you had to make the pilgrimage" (Shapiro 2000: 127).[9] The same holds true today.

Different displays and events around the village, all listed in a "Supporting Programme" that accompanies the play textbook, further remind visitors of the extraordinary nature of their trip to Oberammergau. For instance, in 2010 an Oberammergau Museum exhibit entitled "A Village Performs the Redemption," contained different scripts, pictures, musical recordings, videos, costumes, and props in order to showcase the play's venerable legacy. This and other exhibits emphasized the play's historically unique characteristics, and, thus, the remarkable nature of the visitor's own excursion. Other events underscored the play's religious dimensions. For instance, each evening the Catholic Church of Saints Peter and Paul presented "Son et Lumière: Making faith shine," a Passion-themed light-show with musical accompaniment staged in the sanctuary. "Passion of Life," a juried art exhibit outside the Passion Play theatre, included sculptures by nine artists, each a meditation on the Passion theme.

The "Supporting Programme" also included information about worship services at local churches and about the church-sponsored "Ecumenical introduction to the Passion Play." This introduction, a presentation that included images, text, and music from the play, was offered on Mondays, Wednesdays, and Saturdays in both German and English. The "Supporting Programme" states:

> In this introduction, you hear about the development of the Passion Play from its early beginnings and its arrival at its present form. The structure of the play and its main messages are looked at in detail and explained.

This event was jointly sponsored by the Catholic Church of Saints Peter and Paul and the Lutheran Church of the Holy Cross. The "invitation by the churches" printed in the "Supporting Programme" reads: "The Passion Play deals with the most important messages of our Christian faith. We would like to play our part in unlocking and sharing their meaning."

All of these exhibits and events occur within a distinctive material context. For example, most of the restaurants, houses, and stores that line the narrow village streets are two- or three-story timber-framed buildings, an architectural style that first arose during the later Middle Ages and that was especially popular in Bavaria (Liebold 1874: 97) (Fig. 131.3). In addition, gift shops sell an assortment of Passion Play souvenirs—postcards, CDs of the play's choral music, a coffee table book filled with color production photos, t-shirts, mugs, and mouse pads. These objects function as relics, material evidence of the pilgrim's participation in this remarkable event. Oberammergau is also renowned for its tradition of wood carving, and most shops have shelves filled with wooden Nativity sets, Christmas ornaments, and religious statues. These carved objects have the aura of hand-work and are therefore

[9] There are certainly financial reasons for this. According to Bayerischer Rundfunk, the state broadcast and news agency of Bavaria, the village made a net profit of €25,700,000 ($31,318,547) from the 2010 production season. Das war die Passion 2010 (2011).

Fig. 131.3 Oberammergau Street (Photo by Jill Stevenson, 2010)

even more effective at concentrating the Oberammergau event into a single, material relic. Finally, visitors are surrounded by the magnificent Bavarian Alps, a beautiful setting that may promote positive feelings about their excursion.[10] (Fig. 131.4).

The combination of these elements (events, exhibits, souvenirs, architecture, and landscape) generates an atmosphere that not only reinforces the experience's extraordinary nature, but that may also encourage spectators to perceive the Passionsspiele through that same lens, as a play impressive enough to sustain such a long-standing tradition by drawing people from across the globe toward it. As Erika Fischer-Lichte explains, atmosphere results from the general impression

[10] As Deana Weibel explains: "Research in the field of evolutionary biology suggests that a significant component of pilgrims' positive responses to particular shrines … may in fact be positive responses to particular qualities of landscapes." This response "may certainly be part of the reason that pilgrims to particularly beautiful pilgrimage centers report being cured of illness or transformed in some way" (Weibel 2012: 195).

Fig. 131.4 Oberammergau village (Photo by Jill Stevenson, 2010)

spectators have of a performance space and then continues to impact audience perception throughout the entire performance (Fischer-Lichte 2008: 115). Atmospheres "pour out into, and thus shape, the space. They neither belong just to the objects or people who appear to radiate them nor to the people who enter a space and physically sense them.... [A]tmospheres exist in the interplay of elements and usually form a carefully calculated part of a theatre production." Atmospheres, therefore, have a felt, sensual quality and, according to Fischer-Lichte, represent "spheres of presence" that "intrude on and penetrate the perceiving subject's body" (Fischer-Lichte 2008: 115, 116). A performance's narrative, material, sonic, and visual elements together generate an atmosphere—a sphere of presence—that penetrates the spectator's body, thereby influencing perception and emotional production. Consequently, atmosphere can function as a powerful "felt" affective force.

In the case of Oberammergau, the performance space is not limited to the Passion Play amphitheatre, but encompasses the entire village. The moment visitors arrive in the town they are, in fact, engaging a performance event whose atmosphere is purposefully crafted to support a nostalgic experience of community and tradition. And, if they are taking a packaged tour, they are also staged as part of that event: the established mealtimes and evening events generate communal processions throughout the day; the village's small size, to some extent, limits pilgrim activities before and after the play; Ettal monastery, virtually the only major local attraction, becomes an almost mandatory day-trip.

In addition, as many scholars note, the villagers themselves make a concerted effort to foster a particular atmosphere; before the 1870 production, Daisenberger cautioned villagers: "Let nothing go on either within or without the theatre, in the streets, in your houses, or in the church, which can give occasion for offence. The eyes of many strangers will be fixed not only on the Play, but on ourselves" (Quoted in Shapiro 2000: 134). Indeed, the *Century* review confirms that, for some spectators, the villagers are a key element of Oberammergau's atmosphere: "every one of these [performers] acts with an enthusiasm and absorption only to be explained by the mingling of a certain element of religious fervor with native and long-trained dramatic instinct"; "There is not another village in the world whose members would work so hard, and at so great personal sacrifice, for the good of their community and their Church" (The Passion Play at Oberammergau 1883: 917, 916). And, due to the village's size, visitors regularly bump into actors during their stay, many of them recognizable because of their long hair and beards; a cast member was eating lunch at a table near me a mere hour after I arrived in Oberammergau. Consequently, the actors and pilgrims, the narrow streets and architecture, the shops filled with "handmade" woodcarvings, all set against an idyllic landscape, together generate a "felt" atmosphere whose affective force allows visitors to make contact with a nostalgic, "medieval" past. Such contact influences not only how spectators will finally perceive the play, but also what meaningful intimate scripts it will engender within them.

131.5 Pious Intensities in Arkansas

Acts of medievalism do not simply reveal similarities and differences between the past and present; rather, in Catherine Brown's words, they challenge us to *do* something with that difference (Brown 2000: 548).[11] Because, as Ngai suggests, affect is more potential than certain, each spectator will do something different with Oberammergau's affective medievalism. Elsewhere I have argued that performances of medieval plays that reconstruct the original performance context and practices supply enchantment (Stevenson 2011b: 91–112). Jane Bennett describes how

[11] Brown writes: "The question isn't whether medieval people did things differently than we do now; the question is what we as putative nonmedievalists are going to do with the difference. What stories do we tell ourselves about it? What do they do to and for us?" (2000: 548).

enchantment is an "acute sensory activity. To be simultaneously transfixed in wonder and transported by sense, to be both caught up and carried away" (Bennett 2001: 5). Significantly, Bennett explains how in Western society disenchantment is often recounted as the acceptance of modern science, reason, and religion that necessitates a simultaneous rejection of the cosmology, theology, and superstition typically associated with the Middle Ages. The medieval past is, therefore, "positioned as an object of desire," a golden age of enchantment now lost (Bennett 2001: 58–9, 67, 63). For some pilgrims—be they from the nineteenth century or today—Oberammergau's affective medievalism may supply cross-geographic, cross-temporal moments of contact whose felt alterity generates intimate scripts that re-enchant the modern world and, thus, satisfy a (perhaps only subconscious) cultural longing.

Medievalism may therefore enable some spectators to experience the Oberammergau Passion Play as a meaningful secular event that functions through the kind of affectivity usually associated with religious encounters. This may particularly be the case for certain U.S. spectators given the way American culture typically positions affective experience. Ann Pellegrini argues that because U.S. secular discourse has largely ceded the language of affect to the religious sphere, it has become difficult for secularism to appreciate affect as a legitimate and valuable mode of world-creation and meaning-making in non-religious contexts (Pellegrini 2009: 211–2). And yet, she recognizes a genuine desire for exactly this kind of meaning construction, and suggests that the recent turn to affect in scholarship may, in fact, signal a desire within secular discourse "for the forms of world-making and plenitude and excess that have otherwise been set down as religion's provenance" (Pellegrini 2009: 215). For certain spectators, then, Oberammergau's medievalism may satisfy cultural longing by enabling them to experience this religious Passion Play's affectivity as a reassuringly secular mode of plenitude and excess.

Yet, it is also important to recognize that many people, including many from the U.S., attend the Oberammergau Passion Play expressly for religious reasons.[12] This is a demographic that the village actively cultivates; as Nicole Estvanik Taylor reported, at the beginning of 2010, two of the actors playing Jesus and Mary "completed a major North American media tour, including a stop at an L.A. megachurch" (Taylor 2010: 58). For these spectators, the difference between the medieval past and present may enhance Oberammergau's religious value by giving them an unparalleled "felt" experience of authentic devotion. Affective medievalism can therefore

[12] For example, many churches and religious organizations in the U.S., representing a variety of Catholic and Protestant denominations, arrange group tours to the village during Passion Play season. These tours are often described in marketing materials as pilgrimages. A few examples include: Pilgrim Tours, "Oberammergau, Germany – Passion Play 2010," http://christian-tour.org/christian_tour/oberammergau.htm; 206 Tours, "The Oberammergau Passion Play Pilgrimages," www.206tours.com/oberammergau/; Nawas International, www.nawas.com/; Fiat Holidays, "Oberammergau 2010, www.fiatpilgrimages.net/pilgrimages/Oberammergau_2010.html; Casterbridge Church Tours, "Oberammergau 2010: An Unforgettable Portrait of Christ's Love," www.casterbridgetours.com/index.php/church_tours/church_tour_specialists/special_touring_events/oberammergau_2010. All Retrieved June 15, 2012.

be a powerful devotional force that pulls people toward Oberammergau and into a religious experience that is emotionally productive (in both senses of the word). However, in other instances, the devotional force of Oberammergau's affective medievalism may travel in the opposite direction, radiating out across space and cultivating meaningful scripts for people in markedly different contexts.

As Sponsler notes, for certain U.S. Passion Plays, Oberammergau has always "lurked in the background" (Sponsler 2004: 144). Here, Sponsler is specifically referring to the Black Hills Passion Play, which was mounted annually in Spearfish, South Dakota from 1939 until 2008.[13] This production, which was performed by a cast of 250 local actors three times a week between early June and late August, was initiated by Josef Meier, an immigrant from Germany. The script was an English adaptation of a German Passion Play dating to 1242 and Meier originally toured the production throughout the U.S. during the 1930s before finally settling in Spearfish. As with Oberammergau, from that point forward the play was tightly linked to geography; even when on tour, publicity materials referred to the play as the "Black Hills Passion Play" (Sponsler 2004: 143–4).

References to Oberammergau helped this Passion Play draw upon the power of medievalism. For example, Sponsler shows how Meier allayed anxiety about representing the divine in a live performance by "cannily invoking the play's medieval origins" (Sponsler 2004: 147). This was accomplished, in part, by situating both the play and Meier himself as "inheritors of a centuries-old performance tradition," but one that had found a safe "haven in the remote reaches of the American West" (Sponsler 2004: 144, 147). According to Sponsler, by referencing a medieval tradition Meier was able "to turn an 'Old World' tradition into an 'American institution.'" Medievalism thereby allowed this play to claim "historical stability for itself while, not coincidentally, also removing itself from history and especially from that part of modern history that enfolds current controversies over the role of religion in public life" (Sponsler 2004: 148, 147). Although the play's publicity proclaimed its "living link" to the medieval dramatic tradition and invoked references to Oberammergau, medievalism ultimately functioned to dehistoricize the Black Hills Passion Play in ways that this production's U.S. spectators found reassuring (Sponsler 2004: 154).[14]

As Sponsler notes, Oberammergau's influence is also evident with regard to other annual American Passion Plays, such as one that has been presented in Union City, New Jersey since 1915,[15] or the Great Passion Play of Eureka Springs, Arkansas, first performed in 1968. However, medievalism serves these productions in very different ways than it did the Black Hills Passion. For instance, elsewhere I have discussed how the Great Passion Play in Eureka Springs was established as an

[13] Based upon the articles and press reports I have read, it is unclear exactly why the production closed, but it was likely a combination of financial and logistical reasons. See Garrigan (2008).

[14] Sponsler devotes a section of her chapter on American Passion playing to the Black Hills Passion Play. See Sponsler (2004: 142–55).

[15] The production bills itself as "a modernized version of the Passion Play performed in Oberammergau, Germany." Park Performing Arts Center (2012).

evangelical event, a model of communal devotional activity designed to combat perceived assaults from an increasingly secularized and fragmented U.S. culture (Stevenson 2013). Here I approach this play from a slightly different angle by analyzing how this production draws upon Oberammergau's affective medievalism. I would argue that, for this production, affective medievalism serves to direct devotional intensities so that they will circulate around and stick to the Great Passion Play and those spectators who attend it. Importantly, the Eureka Springs production is not appropriating the affective force of medieval theology; in fact, many aspects of medieval Catholic theology are at odds with the evangelical Protestant tenets promoted by the Great Passion Play project, most specifically the laity's direct access to God and the salvic importance of spiritual rebirth.[16] Therefore, like all instances of medievalism, the Great Passion Play event takes from the Middle Ages only what is valuable to its specific context. In this case, the production employs the affective intensity of Oberammergau's dedication to a common religious purpose, a dedication that is often associated with romanticized notions of medieval piety. Thus, by passing through Oberammergau, the alterity of the European Middle Ages becomes useful to an evangelical U.S. audience.

Every year from May through October, local, amateur actors perform the Great Passion Play five nights a week. The play debuted on July 14, 1968 and by 2008 over seven million visitors had attended the production (Kovalcik 2008: 78). The play was established by Gerald L. K. Smith, a religious entrepreneur, ordained minister, and political orator infamous for his anti-Semitic views, most openly expressed through his magazine, *The Cross and the Flag*.[17] The Great Passion Play was the second of several Sacred Projects that Smith founded, the first being the 70-foot high "Christ of the Ozarks" statue. All of the projects are located on 600 acres (242.82 hectares) of land deep in the Ozark mountains of Arkansas. I visited the Sacred Projects and attended a performance of the Great Passion Play on Saturday, June 19th, 2010.

As the program and other marketing materials explain, "The Eureka Springs production is modeled after the Oberammergau, Germany play in form and function to involve the local community to glorify God" (Program for *The New Great Passion Play* 2010: 1). The 2-h play is performed on an open-air, three-story tall

[16] As I will explain, the Great Passion Play is one of several Sacred Projects and these other projects are even more explicit about promoting evangelical tenets. For instance, staff members at the Sacred Arts Museum or along the Living Bible Tour regularly testify to visitors about their own spiritual rebirth, ask visitors about their personal salvation, profess biblical literalism, and refer to pre-tribulation End Times theology. I discuss this in detail in Chapter Four of my book, *Sensational Devotion*.

[17] Given his personal opinions, Smith may have originally intended to draw upon the affective force of Oberammergau's anti-Semitic features: "With stubborn insensitivity to public feeling at the time, the Smith Foundation named the project 'Mount Oberammergau Passion Play' in honor of its Bavarian counter, which was then being widely criticized for perpetuating anti-Jewish stereotypes" (Long 2003: 75). Currently, marketing materials for the production downplay its connection to Smith and the play itself has been stripped of many anti-Semitic features often found in Passion imagery.

permanent set built into a hillside; Smith named the hill Mount Oberammergau (Jeansonne 1988: 5). The production replicates Oberammergau's attention to "authentic" costuming and props, and its use of live animals. Moreover, the cast of approximately 250 people has always been comprised entirely of local amateur performers, a choice that evokes a nostalgic image of devotional community theatre, something the actors help perpetuate; as one performer said, the cast is "like a family" (Quoted in Simon 2005).[18] Referencing Oberammergau as inspiration and reproducing its "medieval" community-theatre production model therefore allows the Great Passion Play to evoke a similar aura of authentic piety. However, as in Oberammergau, a larger extra-performance atmosphere is also critical to how the Great Passion Play cultivates valuable intimate scripts and, I would argue, it is vital to how this production draws upon Oberammergau's affective medievalism in order to construct the image of a community united by a longstanding and genuine religious commitment.

The atmosphere in Eureka Springs is constructed in many ways that are similar to Oberammergau. For example, the Great Passion Play amphitheatre is not only located in a somewhat isolated area, but it is also surrounded by a circumscribed set of activities and exhibits designed to keep spectators occupied the morning and afternoon before the play. These include the "Christ of the Ozarks" statue, a Sacred Arts Museum, a Bible Exhibit that includes 6,000 Bibles, the 100-year-old one-room Church in the Grove, and an original section of the Berlin Wall (Fig. 131.5). There is also a Living Bible Tour that lasts over 2 h; it is clear that spectators are meant to take this tour the day they see the play since characters repeatedly make comments such as "tonight you'll see this" or "remember this bible verse tonight." The actors along this tour are dressed in "historically accurate" attire and their monologues are often typologically oriented. For example, at the "Passover House" station, the female character recounted the story of the Jewish exodus out of Egypt and then explained how the sacrifice of the lamb in the Passover account had prefigured, and then been supplanted by, Jesus' sacrifice on the cross (Fig. 131.6). The episodes along this tour aim to situate the Passion Play within a particular salvation narrative.

In the evening before the play, visitors can attend the "Top of the Mountain Dinner Theater" for a sit-down home-style meal and hour-long gospel music show. The gospel performers describe their show as the "worship" that prepares spectators for the "sermon" later that evening. Two short plays are also staged on the grounds each evening before the Passion Play: "Parables of The Potter" and "David the Shepherd Boy." Together these attractions help generate an atmosphere that encourages spectators to perceive the Great Passion Play through "felt" notions of national nostalgia, ideal community, and authentic piety.

Finally, as in Oberammergau, the theatre's remote location—near a small picturesque village and set against a striking mountain backdrop—is critical to the pro-

[18] Kovalcik writes: "The Great Passion Play is a large production. But the employees of the foundation are a family. This idea was fostered from the beginning" (2008: 111).

Fig. 131.5 Berlin Wall section. Great Passion Play grounds, Eureka Springs, Arkansas (Photo by Jill Stevenson, 2010)

duction's "felt" atmosphere.[19] It also helps to shape an excursion to see the Sacred Projects into a purposeful religious pilgrimage, for some spectators, even an annual one. One couple from Oklahoma returns every year on their anniversary: "We use it to recenter our marriage and our lives…. Living in the world we live in, it's easy to get away from your values" (Quoted in Simon 2005). A nurse from Tulsa, who by 2005 had attended the play five or six times, said her of her visits, "It keeps me humble." Moreover, as is the case with most religious pilgrimages, some spectators

[19] In other work I discuss how Eureka Springs' reputation as a gay-friendly town fosters a sense of cultural embattlement that, for some visitors, helps strengthen their evangelical Christian identity (Stevenson 2013).

Fig. 131.6 "Passover House" station. Living Bible Tour, Great Passion Play grounds, Eureka Springs, Arkansas (Photo by Jill Stevenson, 2010)

imply that a trip to see the Great Passion Play may hold divine power; one woman brought her 18-year-old son to see the play the day he graduated from Army boot camp. She explained, "My biggest fear is that my baby will go to war.… If anything should happen to him, I need to know I did everything I could to prepare him" (Quoted in Simon 2005).

While the Black Hills Passion Play demonstrates how hard it can be "to make explicitly religious performances acceptable in American contexts" (Sponsler 2004: 154), the Great Passion Play reveals something different—how explicitly religious performance events still play a vital role in constructing the identities and legacies of certain American communities. The Great Passion Play matters because it offers visitors "felt" contact with Oberammergau's longstanding religious communal tradition, one that functions through medievalism. Therefore, the Great Passion Play uses Oberammergau's affective medievalism to situate itself within a "felt" legacy of authentic, devotional community and, in doing so, it makes the difference of the Middle Ages work for its own specific context.

Both Oberammergau and Eureka Springs use ideas about the Middle Ages to create "felt" experiences that bridge temporal and geographic divides by generating intimate scripts that matter. Therefore, while the two Passion Plays are separated by spatial and theological geography, it is through affective medievalism that they meet in a temporal conjunction that pleats past and present against one another. The past gives the present value, even as the present (its needs, desires, and preoccupations) remakes the past into something whose difference matters to spectators. I will briefly conclude by suggesting how affective medievalism can also be a mode of historical inquiry that offers scholars new insight into how past religious experiences mattered.

131.6 Affective Historiography

As I noted in my introduction, as part of her research into medieval praying, Rachel Fulton engages the affective forces of prayers, in particular those preserved in the Admont Manuscript (MS) number 289. Although she admits it is "very difficult phenomenologically to reproduce another's experience, particularly an experience both temporally and spatially distinct from our own" (Fulton 2006: 722), in the end her conclusions emerge directly from her difficult experience with/of that difference:

> I would suggest, if we find books like Admont MS 289 filled with nothing but "vain exercises of words" and "mere repetition of sacred formulas," it is not the artifacts (books, exercises, formulas) that are at fault but, rather, ourselves. We are frustrated when we try to use them—and I challenge the reader not to have felt a certain degree of frustration working through the exercise above—because we do not know how, and so we blame the tools for being unwieldy or, ironically, too "mechanical" when it is we ourselves whom we should blame for being unskilled. (Fulton 2006: 732)

Opening herself up to medieval prayer's affectivity was critical to Fulton's methodology and it enabled her to interrogate medieval religious experiences that scholars before her had failed to recognize or appreciate.

If the cultural forms we study are not merely finished products, but tools for cultivating intimate scripts, it is important for us to, in Stephen di Benedetto's words, "be bold and break from passive modes of critical inquiry" (Di Benedetto 2007: 128). Doing so can, as Fulton demonstrates, reveal new questions and perspectives that will enrich our arguments. This seems to be what Catherine Brown is seeking in her attempt to find "a model for relation to the past that allows for and encourages the very particular and concrete liveliness of two historical moments, of past and present, attempting to avoid what [Frederic] Jameson called the 'ideological double bind between antiquarianism and modernizing "relevance" or projection'" (Brown 2000: 550). The solution Brown proposes is that we learn to read and think "performatively, *per artem*—in the middle of them, from the inside out"

(Brown 2000: 566). This approach requires us not to simply imagine how artifacts worked for people in the Middle Ages or to try putting ourselves into a "medieval" frame of mind. It also necessitates entering into experiences with these artifacts, allowing their affective forces to work on us, circulating through and sticking to our modern bodies, in order to try to recapture "felt" traces of past experiences.

In their articles, both Fulton and Brown use medieval texts. True, texts that work performatively on bodies, but still, texts that one can read and practice repeatedly in different ways. For scholars of medieval performance, the text (when we have one) is insufficient—it is a shell that was originally enlivened by bodies, gestures, sounds, smells, and other synaesthetic elements, but that now feels somewhat empty and lifeless even when we imaginatively "fill" it with evidence of past performance practices. It is, therefore, especially difficult to engage medieval performances *per artem* unless we find ways to move beyond texts and records. Events such as the York Mystery Plays (www.yorkmysteryplays.co.uk/) or the Poculi Ludique Societas productions (http://homes.chass.utoronto.ca/~plspls/pastprod.html), which use historical research to reconstruct some of the original medieval performance tradition, certainly provide opportunities to try thinking *per artem*. Yet, oftentimes these and other productions of medieval plays get stuck in the very ideological double bind that Brown aims to avoid; Sponsler acknowledges something similar in her analysis of academic reconstructions of medieval plays (Sponsler 2004: 156–83).

Admittedly, the Oberammergau and Eureka Springs Passion Play events are not artifacts from the "real" Middle Ages. Therefore, it can be dangerous to look to such events for evidence of the medieval past. However, building upon Leslie Workman's claim that "the study of the successive recreation of the Middle Ages by different generations *is* the Middle Ages," William Diebold reminds us that medievalism is not simply a "realm of passive reception," but a "sphere of active creation" (Diebold 2012: 250, original emphasis).[20] Consequently, I would argue that contemporary events like those I have analyzed here offer us opportunities to employ medievalism as a mode of creation and, by doing so, to engage with and think about medieval performances *per artem*. As I suggested at the beginning of this essay, such productions allow us to think about temporal and geographic crosshatchings that are constructed not only through references, but also through affective assemblage and investment. The affective medievalism of these events does not just reveal our emotions *about* the Middle Ages, but it can also be a way to experience emotions *from* the Middle Ages: the pleasure of community; the awkwardness of small talk with strangers over meals; the satisfaction of finally completing a highly anticipated journey in order to see an extraordinary, occasional performance that moves you outside ordinary time and space; the thrill of realizing that the mundane person you ate lunch beside is now onstage before you; the desire to purchase a material relic of this unique experience; the uncanniness of walking through a village that all year round carries the "felt" traces of a centuries-old religious performance in its streets,

[20] Diebold is quoting from Leslie Workman, "Preface," *Studies in Medievalism* 8 (1996), 1.

buildings, inhabitants, and narrative, and that compresses many different pasts and presents into a single surface. Most of all, the feeling of having a large scale, communal religious performance experience prompt emotions that satisfy certain desires. Those emotions and desires are almost certainly different than those medieval spectators experienced, but the affective trajectory itself can be an important artifact from the medieval past.

As I stated near the beginning of this essay, I am inspired by scholars like Schneider and Harris who, like medievalism itself, challenge chronological and geometric trajectories, and who see the value of asking "what if time (re) turns? What does it *drag* along with it?" (Schneider 2011: 2, original emphasis). For Schneider,

> The act of revolving, or turning, or pivoting off a linear track, may not be "merely" nostalgic, if nostalgia implies a melancholic attachment to loss and an assumed impossibility of return. Rather, the turn to the past as a gestic, affective journey through the past's possible alternative futures … bears a political purpose for a critical approach to futurity. (Schneider 2011: 182–3)[21]

Medievalism—often dismissed as "merely" nostalgic—operates through a similar "temporality of conjunction" that "combine[s] active elements from different times." Its purpose has always been to hold the past and present together in a rhizome whose "connections are horizontal rather than vertical, and presume a dispersed heterogeneity" (Harris 2009: 145, 144).[22] The performance events I have analyzed here give scholars a unique opportunity to not simply observe temporal drag, but to feel it work on their bodies, to make physical contact with "a heterodox temporality … that disregards the entrenched partitions and distances informing the geometric lines of chronological time" (Harris 2009: 171), and thus to understand why such temporal connections matter to people. And, it is important to remember that religious performances in the Middle Ages also functioned through temporal drag as they presented biblical and extra-biblical stories anachronistically, through a lens blurry with medieval desires, needs, and preoccupations; as Marcus Bull reminds us, medievalism was a medieval invention: "running different parts of the past together and getting things mixed up are not recent inventions: one comes across something similar, for example, in a twelfth-center epic loosely based on events in the eighth century, or in a thirteenth-century Arthurian romance set in a

[21] Harris concludes his study by asserting: "But Shakespeare's untimely stage materials … are far more than mute fossils of dead civilizations. They enter into clamorous, global networks of agency in their pasts—and in our presents. These global networks, moreover, are not necessarily those of capitalism or superpower imperialism. If Eastern Standard Time (or EST) is the time of Washington, D.C., and hence the time of American imperial power, Shakespeare's stage materials …allow us to produce something we might call Eastern Nonstandard Time: that is, they help confound the fantasy that insists on treating the orient and the past as synonyms partitioned from the west. And in our war-addled time, such untimely dis-orientations couldn't be timelier" (2009: 194).

[22] Harris is not referring to medievalism specifically, but as I have already asserted, I interpret his ideas as relevant to the ways in which medievalism functions.

distant past that we would nowadays locate sometime around the sixth century" (Bull 2005: 9).

Oberammergau and Eureka Springs offer us opportunities to live inside an historical artifact, to rub up against multiple histories at once and then to see what residue—what intimate scripts—are left behind. I am suggesting that we place our felt experiences of those intimate scripts alongside texts, images, objects, and other material remains, as valuable historical evidence that, if approached critically, can inform our historical inquiries into medieval performance culture.

References

Beckwith, S. (1993). *Christ's body: Identity, culture and society in late medieval writings*. London/New York: Routledge.

Bennett, J. (2001). *The enchantment of modern life: Attachments, crossing, and ethics*. Princeton: Princeton University Press.

Binski, P. (2004). *Becket's crown: Art and imagination in gothic England, 1170–1350*. New Haven: Yale University Press.

Brown, C. (2000). In the middle. *Journal of Medieval and Early Modern Studies, 30*(3), 547–574.

Browne, D. (2009, May 27). Oberammergau Sales Start Slowly in Core U.S. Market. *Global Travel Industry News*. Retrieved June 15, 2012, from www.eturbonews.com/9477/oberammergau-sales-start-slowly-core-us-market

Bull, M. (2005). *Thinking medieval*. New York: Palgrave.

Di Benedetto, S. (2007). Guiding somatic responses within performative structures: Contemporary live art and sensorial perception. In S. Banes & A. Lepecki (Eds.), *The senses in performance* (pp. 124–134). New York: Routledge.

Diebold, W. J. (2012). Medievalism. *Studies in Iconography, 33*, 247–256.

Duffy, E. (1992). *The stripping of the altars: Traditional religion in England 1400–1580*. New Haven: Yale University Press.

Fischer-Lichte, E. (2008). *The transformative power of performance: A new aesthetics*. London/New York: Routledge. Trans. S. L. Jain. Originally published as *Ästhetik des Performativen*. Frankfurt: Suhrkamp, 2004.

Fulton, R. (2002). *From judgment to passion: Devotion to Christ and the virgin Mary, 800–1200*. New York: Columbia University Press.

Fulton, R. (2006). Praying with Anselm at Admont: A meditation on practice. *Speculum, 81*(3), 700–733.

Garrigan, M. (2008, August 27). Passion Play's 70-year run to come to an end. *Rapid City Journal*. Retrieved July 10, 2012, from http://rapidcityjournal.com/news/local/article_4d334b7e-3dec-5d7a-abc5-1b06cb043485.html

Gregg, M., & Seigworth, G. J. (2010). An inventory of Shimmers. In M. Gregg & G. J. Seigworth (Eds.), *The affect theory reader* (pp. 1–25). Durham: Duke University Press.

Grossberg, L. (1992). *We gotta get out of this place: Popular conservatism and postmodern culture*. New York: Routledge.

Hageneier, S. (2010). The stage sets of the 2010 Passion Play 2010. In *Oberammergau Passion Play 2010 press kit* (pp. 6–7). Oberammergau: Gemeinde Oberammergau. Retrieved July 11, 2012, from www.passionplay-oberammergau.com/uploads/media/english_Press_kit_30_November_01.pdf

Harris, J. G. (2009). *Untimely matter in the time of Shakespeare*. Philadelphia: University of Pennsylvania Press.

Jeansonne, G. (1988). *Gerald L. K. Smith: Minister of hate*. New Haven: Yale University Press.

Kintz, L. (1997). *Between Jesus and the market: The emotions that matter in right-wing America*. Durham: Duke University Press.
Kovalcik, T. (2008). *Images of America: The great passion play*. Charleston: Arcadia Publishing.
Kupfer, M. (Ed.). (2008). *The passion story: From visual representation to social drama*. University Park: Pennsylvania State University Press.
Lavender, A. (2006). Mise en scène, hypermediacy, and the sensorium. In F. Chapple & C. Kattenbelt (Eds.), *Intermediality in theory and performance* (pp. 55–67). Amsterdam/New York: Rodopi.
Leys, R. (2011). The turn to affect: A critique. *Critical Inquiry, 37*, 434–472.
Liebold, B. (1874). Mediæval wood architecture in Germany. *The Workshop, 7*(7), 97–101.
Long, B. (2003). *Imagining the Holy Land: Maps, models and fantasy travels*. Bloomington: Indiana University Press.
MacDonald, A., Ridderbos, B., & Schlusemann, R. M. (Eds.). (1998). *The broken body: Passion devotion in late medieval culture*. Groningen: Egbert Forsten.
McNamer, S. (2010). *Affective meditation and the invention of medieval compassion*. Philadelphia: University of Pennsylvania Press.
Montgomery, E. J. (2011). Oberammergau passion play 2010. *Theatre Journal, 63*(2), 260–262.
n.a. (1883). The passion play at Oberammergau. *Century, 25* (6), 913–920.
n.a. (2010). Program for *The new get passion play.* Eureka Springs: The Elna M. Smith Foundation.
n.a. (2011, November 9). Das war die Passion 2010. *Bayerischer Rundfunk*. Retrieved July 11, 2012, from www.br.de/themen/bayern/inhalt/kult-und-brauch/passionsspiele-2010-bilanz100.html
Ngai, S. (2005). *Ugly feelings*. Cambridge: Harvard University Press.
Oberammergau Passion Play 2010. *How to book*. Retrieved July 20, 2012, from www.oberammergau-passion.com/en-us/how-to-book/experience-the-passion-play.html
Park Performing Arts Center. *The passion play: America's longest running play*. Retrieved June 15, 2012, from www.parkpac.org/NewHTM/newpassion.htm
Passion Play Oberammergau. 2010. *Chronicle*. Retrieved July 10, 2012, from www.passionplay-oberammergau.com/index.php?id=126
Pellegrini, A. (2009). Feeling secular. *Women and Performance: A Journal of Feminist Theory, 19*(2), 205–218.
Schneider, R. (2011). *Performing remains: Art and war in times of theatrical reenactment*. London: Routledge.
Shapiro, J. (2000). *Oberammergau: The troubling story of the world's most famous passion play*. New York: Vintage.
Simon, S. (2005, September 22). Acting on faith in Arkansas. *Los Angeles Times*. Retrieved June 15, 2012, from http://articles.latimes.com/2005/sep/22/nation/na-passion22
Sponsler, C. (2004). *Ritual imports: Performing medieval drama in America*. Ithaca: Cornell University Press.
Stevenson, J. (2011a). Oberammergau's Passion play 2010: Performance and context. *Material Religion: The Journal of Objects, Art and Belief, 7*(2), 304–307.
Stevenson, J. (2011b). Embodied enchantments: Cognitive theory and the York mystery plays. In M. Rogerson (Ed.), *The York mystery plays: Performance in the city* (pp. 91–112). Rochester: York Medieval Press.
Stevenson, J. (2013). *Sensational devotion: Evangelical performance in 21st-century America*. Ann Arbor: University of Michigan Press.
Taylor, N. E. (2010). Global spotlight. *American Theatre, 27*(4), 58–59
Weibel, D. L. (2012). Magnetism and microwaves: Religion as radiation. In D. Cave & R. S. Norris (Eds.), *Religion and the body: Modern science and the construction of religious meaning* (pp. 171–198). Leiden/Boston: Brill.

Chapter 132
Windows on the Eternal: Spirituality, Heritage and Interpretation in Faith Museums

Margaret M. Gold

What is the purpose of an object of faith? Is it a physical embodiment of a divinity? An evocation of religious narrative? An illumination of a spiritual tenet? What structure does it document? Is it a tool, or a container? Does it bear an essential message? (Carr 2009: 100)

132.1 Introduction

Religious museums defy easy definition. Religious belief so pervades the material culture, art and everyday life of humanity that it is hardly surprising that many objects in museum collections reflect, consciously or unconsciously, the faith of the communities that produced them. Western audiences are accustomed to seeing religious objects in museums—whether in collections of art, archaeology, ethnography or social history. Indeed many major museums and galleries have permanent galleries devoted to religious objects. These include "Faiths and Empires, 300–1250" and "Devotion and Display, 300–1500" in London's Victoria and Albert Museum (V&A London) and the Islamic galleries at the Metropolitan Museum of Art in New York. Temporary exhibitions on religious themes also appear frequently, with examples including the National Gallery's exhibition "Seeing Salvation the Image of Christ" in London (which marked the Second Millennium) and "Rituals and funerary traditions in Bahrain" at the Hermitage Museum in St Petersburg (Russia) in 2012.

However, processes of social and economic change, migration, globalization and growing diversity have provided the backdrop to the growth of a new genre of faith museums, rooted in the desire of specific faith communities either to understand their own experience more fully or to allow greater public understanding of that

M.M. Gold (✉)
London Guildhall Faculty of Business and Law, London Metropolitan University,
London EC2M 6SQ, UK
e-mail: m.gold@londonmet.ac.uk

S.D. Brunn (ed.), *The Changing World Religion Map*,
DOI 10.1007/978-94-017-9376-6_132

experience. Within that framework, specific motives can be as varied as the desire to document the traditions and history of particular faiths, to preserve the diaspora experience of migrant communities, to safeguard the memory of people or events (particularly in the case of Holocaust memory of Jewish communities), to celebrate the achievements of particular faith groups in art or science, to reinforce the faith among adherents, or to engage and even convert non-believers.

This chapter defines the faith museum, provides a typology based on the purpose and content of their collections, and examines the practice of interpretation, exhibition design and visitor engagement at contrasting examples of religious museums. Three case studies are taken to develop these themes in more detail. The first, St Mungo Museum of Religious Life and Art in Glasgow (Scotland), was launched in 1993 to promote mutual understanding and respect for different faiths. The second, the Jewish Museum (London), opened in 1932 to explore Jewish heritage and identity as part of the wider story of Britain. The final example, The Holy Land Experience in Orlando (Florida), opened in 2001 with the claim to be the "world's first living, historical, biblical museum."

132.2 Defining Museums: Definitions, Role, Mission

The museum as we know it in contemporary society is a relatively recent phenomenon. Its emergence in the eighteenth century challenged earlier patterns of princely and aristocratic collecting based on taste, fashion, patronage and diplomatic networks, embracing instead the western enlightenment tradition of collecting for scholarly, scientific, aesthetic and public understanding. The academy has normalized that tradition in its approach to knowledge in museums in terms of the practice of collecting, classifying and interpreting people's cultures into systems that made sense in terms of western art, culture and knowledge. Museums, by definition, remove objects from their original context and in so doing change their status and meaning even before attempts are made to interpret and display them. The International Committee of Museums (ICOM) defines a museum as:

> A non-profit, permanent institution in the service of society and its development, open to the public, which acquires, conserves, researches, communicates and exhibits the tangible and intangible heritage of humanity and its environment for the purposes of education, study and enjoyment. (ICOM 2013)

This definition finally acknowledged the importance of intangible heritage—the cultural practices and beliefs which animate and give objects meaning. This is particularly important for faith collections where the values, beliefs, use, ritual meaning and symbolism are integral to the understanding of objects. Nevertheless, there remains concern that such a definition tends to reflect western museum practice and fails to accommodate the variety of practice found in the non-western world where, for example, local and indigenous practices fall outside traditional definitions (Krepps 2005: 5). Virtual museums also lie outside the remit of this definition.

132.3 Defining Religious Museums

Given the way in which objects related to faith are an integral part of so many museum collections, it may be asked as to how it possible to define faith museums as a specific category without ignoring large collections of objects in museums that do not see themselves as relating to religion *per se*. In order to be as inclusive as possible, therefore, two dimensions are addressed here, namely: first, the degree of emphasis a museum places upon religion, ranging from those for which religion is the core focus and those in which it is incidental; and, second, the number of faiths which a museum encapsulates, ranging from museums that focus on a single faith to those that that encompass more than one religion. This supplies four broad categories of faith museums.

The first category embraces museums of material culture that focus on a single region or faith group though their artifacts in terms of fine art, applied or decorative arts, archaeology or ethnography rather than religion itself. The focus is on the object, its production, materials, aesthetics or history without the desire to relate the objects to a fundamental understanding of the religious tenets represented by those objects. Examples include the Museum of Islamic Art (Berlin) which 'preserves, studies, restores and communicates the cultural memory of Muslim societies from the Mediterranean to the Pacific, and from Antiquity to the Modern Age' (Freunde Museum Islamische Kunst 2013), the National Gallery in London with its collection of West European art, or the Rubin Museum of Art in New York which exhibits the art of the Himalayas so as to offer the public opportunities to "explore the artistic legacy of the region…and to appreciate its place in the context of world cultures" (Rubin Museum of Art 2013). The mission statements of such museums are littered with phrases connected with public understanding and enlightenment, the desire to unlock or bring the past alive, and to inspire, engage or encourage learning. Of particular note recently has been the growth in interest in Islamic art with major renovations taking place in established museums, such as the Benaki Museum (Athens), the David Collection of Islamic Art (Copenhagen), and the Museum of Islamic Art (Cairo). The development of new museums in the Persian Gulf, many in signature buildings by international architects, is creating a cluster of world class museums. These include the Museum of Islamic Art which opened in Doha (Qatar) in 2008 designed by L. M. Pei, and the Sharjah Museum of Islamic Civilization which also opened in 2008 (Ouroussoff 2010).

The second category are so called 'universal' museums that present the material culture of many, diverse cultures. They include the British Museum with its strap line 'a museum of the world for the world' (2008: 3) and the Metropolitan Museum of Art with its mission: *"to collect, preserve, study, exhibit, and stimulate appreciation for and advance knowledge of works of art that collectively represent the broadest spectrum of human achievement"* (MMA 2000). The latter has over 400 galleries covering archaeology, ethnography (Americas, Africa, and Oceania), Islamic Art, Asian Art, European mediaeval and Byzantine art. These museums see themselves as having shaped the "universal admiration for ancient civilizations" through their diverse collections which allow comparative study (DIVUM 2006: 248). None of these museums sees their role as explaining religious belief, endorsing religious values or providing interpretation for the faithful.

The third category comprises museums of Comparative Religion. Their main interest lies in religion itself, aiming to interpret a number of faiths through the material culture that they produce. There are surprisingly few such museums. Paine (2000: xvi) identified just six, with the earliest being the Marburg Museum of Religions (*Religionskundliche Sammlung*) established in 1927 by the theologian Rudolf Otto (see Gooch 2000). This was an academic collection that offered particular strengths in its Asian, African and Oceanian collections alongside the monotheisms of Judaism, Christianity and Islam (Bräunlein 2005: 179). It aims to "acquaint" the university and general public with the diversity of the world's religions (Marburg Museum of Religions 2013). Paine's list also includes Le Musée des Religions du Monde (Nicolet, Canada) which opened in 1986; the Eternal Heritage Museum (Puttaparthi, India), dating from 1990; the St. Mungo Museum of Religious Art and Life (Glasgow) which opened in 1993; and the Museum of World Religions (Taipei, Taiwan) launched in 2001. The largest museum in this category, however, is found in St Petersburg (Russian Museums 2013). Dating back to 1932, it originated as part of the same Soviet anti-religion campaign that established Museums of Atheism for the purpose of convincing the population of the falsity and futility of faith (Paine 2009: 2). Although it collected objects from outside Russia, during the Soviet period it actively collected objects by sponsoring archaeological and ethnographic expeditions, accumulating a collection that reflected the diversity of beliefs within the USSR (notably with respect to Siberia). The museum survived the collapse of communism. Now housed in its own building, having vacated its original home in Kazan Cathedral which has been returned to religious use (Koutchinsky 2005: 172), it was transformed first into the Museum of the History of Religion and Atheism and currently into the State Museum of the History of Religion.

Faith museums, the largest and the most diverse of the four categories, focus on a single religion with the aim of explaining and interpreting it to its believers or the wider world. Their mission ranges from those that aiming at public understanding in a traditionally neutral museum space to those that actively promote the religion concerned. Faith museums can be further divided according to content with five of the key sub-categories being: fine and decorative arts; ethnography and anthropology; social history; historic sites and buildings; and faith organizations (religious movements, churches, denominations, local religious activity).

The first sub-category includes the Museum of Biblical Art in New York (established by the American Bible Society although now independent). In the words of Ena Heller (2013), its founding director, it aims to "put scripture back into culture," seeking to present the public with "an interpretation of art through the lens of biblical religions and an understanding of religion through its artistic manifestations." The Museum of Christian Art (Goa, India), a museum of Catholic art, aims to "create a love for such art objects that reveals the rich heritage of our ancestors; motivate people to treasure such art objects in their Churches and homes" (Museum of Christian Art 2013). This is one of many Catholic museums that house religious art and provide a documentary record of their communities in order to remember "those human activities which have accompanied the history of the Church" as the "cultural embodiment" of God's message (Marchisano 1995: 48). Within this sub-category are also the Museums of Judaica (Jewish ceremonial art). These grew out of private nineteenth century collections and essentially seek to challenge negative perceptions of the artistic culture of the Jewish community (Grossman 2008: 16).

While much of nineteenth century ethnographic collecting was for scholarly or scientific purposes, some was overtly for religious purposes. The Missionary Museum of the London Missionary Society, dating from 1814, was one of a number of such institutions established to house objects collected by missionaries—covering the artifacts of daily life and the discarded gods of the converted. It was used to publicize the society's work, raise funds for its ministry, and to train aspiring missionaries (Huang and Chen 2001: 87). Much of the material collected by such bodies subsequently found their way into the major museums but some, like the Mission Museum (Stavanger, Norway), still remain. Nevertheless, ethnographic material now features in a new genre of museums that treats indigenous material in religious terms. The most influential of these is perhaps the National Museum of the American Indian in Washington DC, which explains and supports the religious beliefs of native cultures in the Western Hemisphere.

Of the museums recording the social history and way of life of communities in which religion is a defining characteristic of identity, one of the most important forms is the Jewish museum. These chart the experience of established communities (as in Italy, Greece or Germany) and of more recent diasporas (such as the U.S., Australia, and South Africa). Some European examples record an almost vanished community in areas affected by the Holocaust (Clark 2004: 99). Initiatives such as the Islamic Heritage Museum in Washington DC (which opened in 2011) and the Islamic Museum of Australia project in Melbourne (established in 2010) aim to record the Muslim experience of migration to, settlement in and experience of their respective new countries. Social history museums are a particular feature of the U.S., where communities as varied as Shakers, Baptists, Mennonites, Moravians, Presbyterians, Quakers and Lutherans are all represented in museums.

Moving on, a diverse group of museums are associated with historic sites and buildings. These include places of worship such as temples, synagogues, mosques, churches and Cathedrals; historic sites associated with particular events including miracles, shrines, sites of persecution, and the Holocaust; and historic house museums associated with religious figures or leaders. The Beehive House in Salt Lake City, home of Brigham Young (leader of the Church of the Latter Day Saints) is a representative example of this sub-category.

Finally, there are the museums that chart the history of a particular religious faith or organization as opposed to the social history of the faithful. These cover a wide range of the world's religions and denominations, with examples including the Central Sikh Museum (Amritsar, India), the Canadian Museum of Hindu Civilization (Toronto), the World Methodist Museum in North Carolina (USA), and Salvation Army museums such as those in Wellington (New Zealand), Melbourne (Australia), London, New York and Atlanta.

132.4 The Geography of Religious Museums

Given the difficulty that most countries have identifying the precise size of their museum sector, any global assessment of their presence and distribution is fraught with difficulty. As an approximate guide, however, the directory *Museums of the World*—a source that presents data from 222 countries and lists over 55,000 museums—supplies details of their continental distribution and key concentrations (MOW 2009). These are shown, respectively, in Tables 132.1 and 132.2.

This directory distinguishes between museums that concentrate on religious art (788) and on religious history and traditions (705). However, it identifies other categories of museum with religious objects: namely, treasuries (mainly linked to Cathedrals), Icons and Monasteries. All are principally in Europe. Only one category of religious community, Judaica, is picked out by the directory as being more broadly distributed—only 75 of 122 such museums are in Europe. Some of the largest collections of religious art and artifacts are omitted from these categories as they are not classified as religious museums. So, for example, the Museum of Islamic Art in Berlin and the V&A in London are both classified as decorative arts museums, and the Metropolitan Museum of Art in New York is defined as a fine arts museum.

What Table 132.1 reflects is the western notion of the museum, with Europe and European diaspora nations featuring strongly compared to Africa and Asia. The former are countries with established and often publicly-funded, highly-professionalized and networked museum sectors, with systems of accreditation that make their identification more straightforward. While this model of museum-making was exported through colonial networks at the end of the nineteenth century, the number of "formal" museums in Africa and Asia is smaller by comparison and many traditional collecting and curatorial practices fail to be included as museums in official directories (see Krepps 2003).

Table 132.1 Continental distribution of religious museums

Continent	Arts and symbolism	History and traditions	Treasuries[a]	Icons	Monasteries	Judaica	Total
Australasia	–	9	–	–	–	3	12 (0.6 %)
Africa	4	11	–	1	1	1	18 (0.9 %)
Asia	65	60	3	1	5	17	151 (7.8 %)
North America and the Caribbean	47	128	1	3	1	23	203 (10.5 %)
Central and South America	98	56	–	1	2	3	160 (8.2 %)
Europe	574	441	73	90	145	75	1,398 (72.0 %)
Total	788	705	77	96	154	122	1,942

Data source: MOW (2009)
[a]Mainly cathedral and abbey collections

Table 132.2 Countries with ten or more religious museums

10–19 museums	20–49 museums	50–99 museums	>100 museums
Argentina	Belgium	Austria	Germany
Norway	Brazil	Canada	Italy
	Colombia	France	Spain
	Croatia	Greece	United States of America
	Hungary	Japan	
	Israel	Portugal	
	Mexico	Romania	
	Netherlands	Russia	
	Peru		
	Philippines		
	Poland		
	Switzerland		
	United Kingdom		

Data source: MOW (2009)

Table 132.2 lists the countries with the largest numbers of religious museums. This shows considerable concentration, with 27 countries having 10 or more religious museums and accounting for 88 % of the total. The top four states are Germany with 227 museums, Italy 191, Spain 169 and the U.S. 141; these together account for 37 % of the total. Yet besides the significance of Europe, Table 132.2 also indicates the importance of South America with its embrace of the Catholic tradition of museum-making. Not surprisingly perhaps, the U.S. has the most diverse set of religious museums, reflecting the strong tradition of using museums to preserve the memory of its various faith communities and record the existence of historic sites.

132.5 Windows on the Eternal

Presenting and interpreting objects with religious associations raises complex issues for the museum community. The status of objects always changes when they enter a collection, but questions of context, intangibility and meaning have particular resonance when objects leave the world of living religion and enter a museum space. Traditionally, this meant that objects once regarded as holy, sacred or divine could be treated as art and valued for their beauty, craftsmanship, or history. Western audiences were familiar with this trope and museums in any case generated their own secular rituals of reverential gaze and reflection. However, this has been increasingly challenged by what has become known as the new museology, which argues for museums being "people-centered and action-oriented" rather than collection-centered and is concerned with democratization and participatory practices (Krepps 2003). This means engagement not just with audiences but with source communities, reflecting the shifting power relationships in the modern world as groups challenge the authority of the museum and demand a say in how they are represented in the museum (Witcomb 2003:103). This is exemplified by the Glenbow Museum in Calgary Canada, which spent a 10-year period of collaboration in developing its new gallery "Niitsitapiisinni: Our way of Life" to represent the history and culture of the Blackfoot people (Conaty 2003; Glenbow Museum 2013). The new National Museum of New Zealand (NMNZ), which opened in 1998, is similarly a partnership between the indigenous Maori and the settler communities (NMNZ 2013). Consequently, museums have to consider four issues: their policy towards the "religious" in the museum context; the treatment and care of objects with religious associations; the practice of interpretation; and the renegotiation of the relationship with the visitor.

While not all objects in museums with religious associations would be regarded as sacred or holy, those that are have additional layers of significance. Paine (2013: 15) identifies three such layers: the intrinsic, extrinsic and ascribed. *Intrinsic* information is that "carried" by the object which can easily be observed and interpreted including its shape, composition, design, style, materials and condition. *Extrinsic* information is external to the object such as its place of origin, who owned it and how it was used. *Ascribed* information comprises its religious and spiritual meaning, and its social value as an object. While intrinsic and extrinsic information is the province of the museum expert, the ascribed involves the detailed knowledge of religious practice over time which is more likely found within faith communities themselves. It is in the realm of the "ascribed information" that difficulties for museums arise—the tension between the neutral space of the museum and the faith world of the believer. Yet these objects on one level are the windows on the eternal, or as Rudolf Otto expressed it, "the numinous," supplying a sense of the scared that is at the core of every religion (Arthur 2000: 7). In displaying these objects, one engages with the religious truths—something with which the modern museum is not always comfortable. Art museums may feel it is not their place to delve beyond symbolism and style in order to interpret their collections. Publicly funded museums, with their diverse audiences, may consider it inappropriate to explain religion. Museums of comparative religion take care not to privilege one faith over another. Single faith museums run by 'insiders' may feel no compunction to assert the 'truths' embodied in their sacred objects or to use their collections to underpin religious belief.

132.6 The Treatment and Care of Religious Collections

It is part of the professional practice that has developed in western museums that all objects are treated with care, but it is frequently argued that an extra layer of management—encapsulated in the word 'respect'—is also required for religious objects. According to Paine (2013: 57), 'respect' implies 'paying attention' to objects 'in a culturally appropriate way'. This includes how they are handled, stored, and displayed. Icons, saints' relics, and crucifixes are all objects which their faith communities would expect to be treated thoughtfully, but in some contexts the demands may be very specific. The treatment of the most sacred items of Judaica—such as Torah scrolls and associated objects—dictate who should repair the objects, when, with what type of materials, and how any discarded material should be dealt with (Morris and Brooks 2007: 244). Similarly, principles governing the handling of the Qur'an or artifacts with Qur'anic inscriptions aim to maintain the boundaries between the sacred and profane (Paine 2013: 59).

In many religious traditions there may be practices which dictate who may see an item, who may handle or conserve it, how it should be stored, and how it should be treated when not in use. This is particularly the case where objects are deemed to be living entities. Allowing objects to breathe goes against traditional museums practice such as the use of plastic bags, freezing, or low oxygen atmospheres (Rosoff 2003: 76). Questions of storage take on new dimensions when feelings attributed to objects are considered. The Natural History Museum in Washington, for instance, has received requests to orient a Dream Dance Society drum with its red painted stripe facing south, the 'spirit line' on a Navaho wedding basket facing east, and a Sun Dance buffalo skull to be stored facing east but also upside down to signify it was not 'active'. Issues of use include ceremonial feeding, ritual cleansing (smudging or smoking objects) and wearing objects for ceremonies (Flynn 2001: 3–6).

Changes in museum practice regarding tribal sacred objects has been greatly influenced by representations from the indigenous peoples of the U.S., Canada, Australia and New Zealand where the 'determination to preserve and regain control of their cultural heritage' and a 'reawakening of cultural identity' have led to campaigns to address the injustices committed in the collecting of materials and treatment of indigenous culture by European settlers and experts (Simpson 2001: 191). While repatriation of sacred objects—one demand of such campaigns—lies outside the scope of this paper, they also seek the right to consultation over the treatment of objects by museums. The passing of the Native American Graves Protection and Repatriation Act (NAGPRA) in 1990 and policies and ethical codes adopted in other countries have introduced new ways of engaging with source communities. This has been accompanied by a re-evaluation of 'indigenous curation'—the traditional practices of object care in many non-western societies—as culturally appropriate for the communities involved. This includes traditional buildings or Keeping Places to house objects, the methods that protect the spiritual integrity of objects, and the involvement of local 'caretakers' such as priests, shamans, ritual specialists, or elders. Krepps (2009: 193, 5–6) argues that these practices are themselves 'unique cultural expressions' and should be protected as intangible cultural heritage.

This issue of restricting access to objects by preventing them being displayed or even 'seen' in storage by certain groups of people is one that has generated debate. Tiffany Jenkins (2006), who supports notion of the enlightenment museum and opposes the restricted access to museum objects on the basis of ethnic or tribal background, sex, or religious belief, argues that it goes against the role of museums to limit 'what can be seen and what can be known' and 'undermines the secular task of exhibiting artifacts as evidence of the history of human civilizations'. She feels this is particularly pertinent in publicly-funded museums whose role is public education. The counter-argument is that the role of museums is about 'preservation and understanding' and that displaying objects that should not be seen is 'unwarranted contempt' (O'Neill 2006: 366)

Yet it cannot be denied that religions and secular museums do differ in the nature of the engagement between the faithful and the objects of their faith. Visitors in museums are not expected to touch, kiss, or leave offerings for museum objects or to perform the bodily movements associated with worship, such as circling, kneeling, genuflecting or removing footwear. Some museums have embraced change by creating spaces for ceremonies and others allow staging of ceremonies in the galleries. An example of this is the annual celebration of Wesak Day (Buddha Day) that has been taking place since 2006 in Birmingham Museum and Art Gallery (UK). Here Buddhist monks celebrate in the gallery before the Sultanganj Buddha—a 2.3 m bronze Buddha (c 700 CE)—with chanting and flowers (Paine 2013: 38).

132.7 Case Studies

Discussion to this point has indicated the complex range of issues that interact in the formation of religious museums and their collections and in shaping the display and interpretation offered to the visitor. In seeing how these issues pan out in practice, it is valuable to consider case studies of contrasting types of museum. The first of the three case studies to be considered here is St Mungo Museum of Religious Life and Art in (Glasgow)—a rare example of a museum of comparative religion that aims to promote a greater understanding of faith. The second is the Jewish Museum (London), which interprets the Jewish experience in the United Kingdom. The third is the Holy Land Experience (Florida), which uses museum objects to illustrate biblical narratives.

132.7.1 The St Mungo Museum of Religious Life and Art

Located next to St Mungo's Anglican Cathedral in Glasgow, this opened in 1993 and is operated by Glasgow Museums, the city's museum service. Offering free admission, it seeks to 'promote understanding and respect among people of different faiths, and of none' and to explore 'the importance of religions in people's lives across the world and across time' (Glasgow Museums, n.d.). As such, it combines three approaches to religion: an aesthetic approach to religious artifacts, an ethnographic approach to religious practice, and a social history approach to religious life in Scotland. Spatially, it combines a global perspective with a concern for localism (Spalding 1995: 6).

Unusually, this museum was not the result of long-term strategy or planning, but rather originated in a stalled project to build a visitor center for Glasgow Cathedral. Although the building had been constructed, problems over funding in 1989 caused the City Council to intervene. After discussion, a new multi-faith museum concept developed in place of the visitor center (Spalding 1995: 6), encompassing some elements specific to the City of Glasgow and others part of the wider *Zeitgeist*.

To elaborate, by the 1980s Glasgow had developed a city regeneration strategy that placed culture in a central role and creating a museum that would attract tourists fitted this strategy. However, museum staff were keen to develop a more instrumental approach within the museum service. Glasgow, it was argued, had no museum that reflected the city's multicultural character; a place where ethnic communities could find reference to their own culture or where difference could be celebrated. In addition, there was a desire to promote understanding between faith groups in order to combat racism and, particularly, the Catholic-Protestant sectarianism that had traditionally bedeviled community relations (Sandal 2007: 47; O'Neill 2011: 229).

The final shape of the museum was partly dictated by the strengths of collections within the city in terms of Catholic, Hindu and Buddhist items and partly by the need for objects of 'sufficient visual interest to communicate something of importance to the visitor'. Loans or commissions were primarily used to fill gaps in areas where existing collections were weak or because certain religions have richer visual art traditions than others—as with the painting and sculpture associated with Catholicism or Buddhism compared to the emphasis on text and items used in Judaic or Islamic worship (O'Neill 2013).

A number of tensions had to be resolved in creating a municipally-run museum that addressed questions of faith. First, the team wanted to go beyond presenting objects in purely aesthetic or cultural terms (O'Neill (2011: 211), but could not appear to offer official endorsement to any specific claim that different faiths were making to religious truth. Secondly, while wishing to celebrate religious diversity, it was important to address the negative side of religion. To create this balance, topics such as war and persecution, the oppression of women, and the prevalence of sectarianism were included (O'Neill 1996: 197, 2011: 230). Thirdly, a dialogue was developed not just with faith leaders but also with voices that might express the faith experience of ordinary Glaswegians. To that end, an oral history project was carried out (Carnegie 2009: 162). This raised questions about whether the religious leader or the practitioner was the authentic voice on religious matters, an issue exemplified by the case of a loaned painting of a Sikh family where the father had cut his hair. While this was an authentic representation of the migrant experience, in which the individual had cut his hair to assimilate more easily, it did not represent the Sikh ideal. The painting, therefore, was not shown (*ibid*, 163).

Care was taken when designing and installing the exhibitions not to privilege one faith above another. For example, the organization of the galleries dealing with the six faith traditions sees Buddhism, Christianity, Hinduism, Islam, Judaism, and Sikhism treated in alphabetical order. The dating system used was not BC/AD (Before Christ; Anno Domini) but BEC/CE (Before the Common Era; Common Era). Staff actively consulted about the positioning of objects, what should or should not be placed alongside each other, and how far it was possible to accommodate religious requirements. Thus a blessing ceremony was held for the installation of the god Ganesha, but a request to put a Sikh holy book to bed each night was not accepted (Arthur 2000: 18; Carnegie 2009: 163).

Fig. 132.1 St Mungo Museum of Religious Life and Art: the Gallery of Religious Art showing the bronze sculpture of Shiva as Nataraja (Lord of the Dance) (Photo by Margaret Gold)

The outcome is a museum covering 120 faiths and over 5,000 years of history. It begins with an audio-visual presentation in which adherents of different faiths relate what their religion means to them in everyday life and provides glimpses of the religion's importance in the working of their community. This multicultural discourse sets the scene for the interpretation of the artifacts in the three permanent galleries that follow. The first of these is the Gallery of Religious Art. Containing works intended 'to reflect the meaning of the religious traditions which inspired their creation' (Glasgow Museums, n.d.), it contains objects from classical Greece, Buddhas, stained glass, and a statue of Shiva as Nataraja, 'Lord of the Dance' (Fig. 132.1). The ensuing Gallery of Religious Life aims to show what the museum's information board describes as 'the separate dignity of each faith while also demonstrating the human themes common to many religions'. It explains the religious beliefs of the six world religions, the ways in which religion more generally is 'interwoven' with daily life through the rituals that mark the life cycle from birth to death, and themes such as health, hope and happiness, the afterlife, spreading the word, persecution, war and peace. The third room, the Scottish Gallery, tackles such themes as education, charity, sectarianism, and religion in secular Scotland, and provides activities for children relating the six world religions. In addition to these three galleries are a small exhibition outlining the Cathedral's history and an exterior Zen Buddhist garden.

The museum's impact when it first opened can, perhaps, be seen on the strong reaction of stakeholders and visitors. Both the Moderator of the Church of

Scotland and the Archbishop of Canterbury are recorded as expressing disappointment that the museum failed to represent adequately Scotland's heritage as a Christian country and gave 'undue prominence' to non-Christian faiths (O'Neill 2011: 230). Hindu groups objected to the presentation of the god Shiva and so a plinth was introduced to demonstrate physically the respect due to the Hindu god. A photograph of female circumcision caused a strong reaction, prompting a demonstration by a local feminist group. Rather than remove the photograph, the explanatory text was expanded to meet their objections. However, images of men and women in the Islamic displays were removed after protests from community leaders. (O'Neill 2011: 230; Arthur 2000: 18, Carnegie 2009: 163; Sandal 2007: 68). Display boards were added to the museum galleries some weeks after opening as a way of giving visitors a way of recording their reactions. A complaint that this feedback system pinpointed, concerned the lack of pagan traditions in the museum, which was addressed by placing a clooty tree near the Zen Garden. This pre-Christian Celtic tradition encourages people tie strips of cloth representing hopes or prayers to the branches (Dunlop 2013). The most violent reaction to the museum came from a young man who, offended by seeing a statue of Shiva in a gallery with Christian objects, attacked the statue causing considerable damage. The picture of the broken statue is now used in citizenship workshops with schools to initiate discussion on peoples' use of religion to divide (Singh 2006).

Since opening, events in the wider world have continued to place religious conflict high on the political agenda. The museum attracts some 200,000 visitors per year including 7,000 school visitors. Reaction is generally positive (Sandal 2007: 80), but an analysis of the feedback boards also identifies two main negative reactions: from those that argue that there is one true faith and that the museum should make clear which one it is; and those that believe that the religious diversity portrayed in the museum demonstrates the falsity of religion and that the museum should make this point clear (O'Neill 2011: 230–1)

132.7.2 The Jewish Museum

Located in Camden Town in north-central London, the Jewish Museum is one of only two Jewish museums in Britain and the only one in London devoted to a single minority group. It is an independent museum with charitable status, meaning that it does not receive regular public funding but relies on admission charges, a membership scheme, a patrons' scheme, trusts, foundations, business sponsorship, and venue hire to supply its income stream. Indeed, its history has been shaped by the patronage of supporters and philanthropists who have donated money and objects and contributed to the purchase of buildings and the cost of refurbishments.

Effectively, this is a museum that tells a community's story 'from the inside' (Burman 2000: 136), with its mission is to 'explore and preserve Jewish heritage, celebrate diversity and challenge prejudice.' In order to fulfill that mission, it:

> …collects, preserves, interprets and exhibits material relating to Jewish history, culture and religious life. It draws on the Jewish experience as a focus for the exploration of identity in a multicultural society, actively engaging with the shared experiences represented in the diverse cultural heritage of London, Britain and the wider world. As a forum for education, learning and interfaith dialogue, the museum encourages understanding and respect by challenging stereotypes and combating prejudice in all its forms. (Jewish Museum 2013)

As such, it exemplifies the practice of a museum aiming to record the historic experience of a community in its particular geographical context (London and Britain), demonstrating its deep roots in British society and the contribution made by the community to British life and identity over the centuries. At the same time, it seeks to preserve collective memories through objects, prints, drawings, paintings and photographs, ceremonial art and oral histories, actively working to fill the gaps in the collection by concentrating on contemporary Judaica, migration since 1945, and migration to the suburbs in order to keep pace with the changing socio-economic patterns within the community (Jewish Museum 2005: 4–5). The museum's activities also address a wider social and political agenda of engaging with multicultural London by finding common themes with other migrant groups, inter-faith dialogue and anti-racism programs. Beyond that, it enjoys historic connections with Jewish communities around Europe and is thereby embedded in a series of networks, cooperative arrangements and digital projects that seek to make the collections available to a wider audience. These initiatives include the European Union-funded Judaica Europeana project that aims to create a fully searchable central database (Jewish Museum 2013) and its own digitization program.

The Jewish Museum stemmed from two separate initiatives—the Jewish Museum founded in 1932 and the Museum of the Jewish East End dating from 1983. These represented the two main strands of Jewish collections, with the former reflecting the fine-art collecting that emerged in Britain at the end of the nineteenth century and the latter recording of the Jewish social and historical experience that emerged later in the twentieth century. As such, the Jewish Museum concentrated on Judaica, textiles, prints, drawings, and manuscripts in the fine art and decorative arts, but did not cover social history or artifacts less than 100 years old (Stephens 2008: 41). For its part, the Museum of the Jewish East End was founded to complement the Jewish Museum's remit, with collections centering on more recent social history, particularly aiming to record the rich cultural experience of an area that, by the 1980s, had lost much of its original Jewish community. It quite quickly broadened its scope, changing its name to the London Museum of Jewish Life in 1988. It recorded the religious, social and political life of the community, the businesses, Jewish welfare and community organizations, Yiddish theatre, immigration experiences and individual life stories recorded in an oral history archive and Holocaust education (Grossman 2008: 214, 216). In 1995, the Jewish Museum moved to new premises in Camden Town—at which point it amalgamated with the London Museum of Jewish Life. The two collections continued to be housed in their separate locations until the purchase of an adjacent building in Camden allowed for the expansion and refurbishment of the site to accommodate both collections. The museum re-opened in 2010 providing four permanent galleries, a temporary exhibition space, education center, auditorium, café and shop.

Rickie Burman, Curator at the London Museum of Jewish Life (1984–95) and then Director of the Jewish Museum (1995–2012) offers an overview of the challenges of the narrative that the Jewish Museum wishes to tell (Burman 1998, 2000, 2006, 2012; also Stephens 2008). These include: the contrast between an ancient religion and its ceremonial, and modern contemporary life; its rootedness in the United Kingdom with a history dating back to the Norman Conquest alongside its connections with Central and Eastern Europe, Aden, North Africa and India; the characteristic of being an integral part of British life while maintaining a distinct identity; analyzing the destruction of the Holocaust while avoiding the representation of Jews as victims; the need to preserve memory for the community while explaining the Jewish culture to the wider community and combatting stereotypes; and finally recording the Jewish history of migration while making connections with other immigrant experiences.

The museum's entry section combines the present and the distant past—a multimedia display showing members of London's Jewish community effectively welcoming the visitor to the museum and a thirteenth century Mikveh ritual bath excavated in 2001 that represents the presence of a Jewish community in medieval London. On the first floor, three galleries take the three key themes of the museum—the Jewish faith, the Jewish community in Britain and the Holocaust. Each adopts a very different atmosphere and approach in an attempt to illicit a response from the visitor that is reflective, vibrant and somber by turn.

'Judaism: a living faith' displays Judaica arranged by ritual setting—the synagogue, the home, and the personal (life cycle events such as circumcision, marriage and so on). This gallery aims to evoke an atmosphere of 'spirituality' and 'reverence' (Burman 2000: 134), using interactivity and film to illustrate religious belief and contemporary practice. Figure 132.2, for example, shows the Torah interactive display in the center of an otherwise more formal-looking space.

The atmosphere of the 'History: a British story' gallery is more celebratory, documenting the Jewish presence in Britain from 1066 onwards. Notably, it uses sound, smells, film, reconstructions, family-friendly interactive displays and oral history in evoking immigrant life in the East End. Figure 132.3 shows 'The Immigrant Home', illustrating life in 'in rented rooms in the crowded streets and alleys of the East End.' This display explains the difficult conditions, with the doubling-up of living accommodation as workspace symbolized by the sewing machine.

By contrast, the Holocaust Gallery uses objects, documents and the powerful personal testimonies of Holocaust survivors who made their homes in Britain. The most prominent of these is Londoner Leon Greenman, who was living in Rotterdam with his wife and baby son when Germany invaded Holland in 1940. Unable to prove their British nationality they were deported to Auschwitz where his wife and baby son were killed. Leon survived the war finally being released from Buchenwald in 1945. His story is complemented by poignant objects—clothing, photographs,

Fig. 132.2 The Jewish Museum London: the permanent gallery Judaism: a Living Faith, displaying Jewish ceremonial art (Photo by Margaret Gold)

a toy and documents. Greenman was involved in Holocaust education at the museum up until his death in 2008 at the age of 97.

In the first year after its re-opening, visitor numbers doubled from the pre-refurbishment total of 30,000 per annum, although remained short of the 85,000 that Burman had hoped for (Joseph 2012; Stephens 2008: 41). The difficult financial climate for an independent museum reliant on its own resources resulted in a controversial experiment with Saturday opening from May 2011, ostensibly to attract non-Jewish visitors and contribute to interfaith understanding (Rocker 2011). However, this was abandoned at the end of 2011 as visitor numbers had not justified the decision and costs of replacing volunteers with paid staff outweighed any gains from income (Elgot 2012).

Fig. 132.3 The Jewish Museum London: The immigrant home. A recreation of a room in the Jewish East End of London (Photo by Margaret Gold)

132.7.3 Holy Land Experience

The Holy Land Experience refers to itself as the world's first 'living, historical, biblical museum and park' and illustrates the use of a themed environment, biblical exhibits, drama, music and interactivity to transport the visitor 6,000 miles and 2,000 years back in time 'to the land of the Bible.' Its purpose is to 'educate believers and to bring the Gospel of Christ to the World' in 'a thrilling swirl of characters, costumes, and color' (HLE 2012). It was opened in 2001 on a 15-acre site in Orlando (Florida) by Zion's Hope, a body run by the Baptist minister Marvin Rosenthal. In 2007, ownership passed to Trinity Broadcasting Network, who run it as a not-for-profit attraction. It has enjoyed tax-free status as a religious museum since 2006 (Wilson 2009) and charges admission—initially $17 per adult but having risen to $40 by 2013.

Rosenthal's vision focused very much on the Old Testament Holy Land and its relationship to the passion and resurrection of Christ, with his primary interest in setting up Zion's Hope being to evangelize the Jewish Community. While making much of his own Jewish origins and conversion experience in this regard, his belief was also that the modern church needed a better understanding of its Jewish roots. In particular, he highlighted the importance of Israelite sacrificial practices, Jerusalem and the Jewish Temple because of the role all three would play in the 'end times'—the restoration of Jerusalem to Israel, the rebuilding of the Temple and the reintroduction of sacrificial practices including Yom Kippur (Beale 2005: 67). This belief partly explains the idiosyncratic mix of Old Testament and Christian elements selected for the Park.

Before moving to Florida, Rosenthal had led teaching tours of the Holy Land in the belief that there was no substitute for 'seeing, touching, feeling the landscape' in which biblical events happened. The Holy Land Experience attempted to bring that experience closer to home by creating an environment that resembled the real thing as much as possible. To that end, he employed the theme park specialists, ITEC Entertainment Corporation, to design the park, even taking the lead designer to the Holy Land to get 'a sense of the color and touch of the place' (Wharton 2006:195). In taking this approach, Rosenthal, perhaps unwittingly created an attraction that resonated with a long representational tradition, extending back into the nineteenth century, where groups attempted to recreate first century Jerusalem free from later accretions (*ibid*: 197)—often incorporating a particular view about the location of Calvary and the Holy Sepulcher (e.g. see Kark and Frantzman 2010).

When it opened, the park took visitors through recreations of first century Jerusalem's narrow streets (Fig. 132.4), beyond which they would find Calvary's Garden tomb (Fig. 132.5), the Qumran Caves where the Dead Sea scrolls were discovered, the Wilderness Tabernacle (a representation of the home of the Ark of the Covenant during the Israelites 40-year journey through the Sinai Desert), and Herod's Temple (built half-scale but even so rising to six stores). Within the Temple, the Theatre of Life provided a 25-min film showing God's purpose for humanity. Visitors could then receive a 30-min guided tour of the world's largest model of

Fig. 132.4 Holy Land Experience: The Jerusalem City Gate which acts as an entrance to the site with the Jerusalem Street Market beyond (Photo by Margaret Gold)

Fig. 132.5 Holy Land Experience: Calvary's Garden Tomb (Photo by Margaret Gold)

Biblical Jerusalem, beyond which was the Shofar Auditorium with its lecture program on aspects of the Biblical scholarship. Dramatic presentations were given at the Garden Tomb and Temple Plaza throughout the day. In 2002 the Scriptorium (Center for Biblical Antiquities) opened in a Byzantine-themed building to house the substantial collection of bibles, books, artifacts and manuscripts. The Scriptorium was designed as both a museum and an archive for scholarship. Rosenthal put considerable emphasis on this museum element as it lent 'authenticity and veracity' to the rest of the site and provided 'the biblical anchor' for the whole project (Beale 2005: 60). It was organized around a series of 13 themed spaces that put the key objects of the collection in context and told a narrative of the development of the bible starting with ancient Mesopotamia and ending with a twenty-first century American home taking in *en route* Babylon, Alexandria, a Byzantine bindery in Constantinople, a mediaeval monastic scriptorium, John Wycliffe's study, Gutenberg's print shop in Mainz in 1455, Tyndale's print shop in Cologne, John Bunyan's cell, the Metropolitan Tabernacle in London, the Mayflower, and an American Prairies church. The finale was located on the slopes of Mount Sinai surrounded by representations of Biblical figures and the Ten Commandments. Visitors were shepherded through the collection in groups of 15 every 7 min (Samworth et al. 2003).

Given the activities of Zion's Hope, it is not surprising that the park attracted adverse criticism and even a demonstration at its opening. The Jewish Defense League claimed that its aim was to convert Jews and its leader Irv Rubin described Rosenthal as a 'soul snatcher' bent on the 'spiritual destruction of the Jewish people' (Brabant 2001). This, according to Rosenthal, led to an unexpected level of national and international press coverage to which he attributed good initial visitor figures. Despite claiming 4 million visitors in the first 4 years of operation, the lack of any advertising budget and reliance on church groups, word of mouth and local tourist information meant that the park struggled to generate income (Reed 2007). Reactions in the press and internet suggest that the visitors found it 'a humble and earnest depiction' (Wallace 2001, 18) and the lack of activities for children compared unfavorably with Orlando's reputation as a family-centered theme-park paradise.

The park's purchase in 2007 by the Trinity Broadcasting Network, a Christian broadcasting operation headed by Paul and Jan Crouch, radically changed matters. They invested heavily in the attraction to increase its appeal to a wider age spectrum. Activities for children were added and the range of attractions for adults was expanded to include a model of the town of Bethlehem and replicas of the bell tower of its Church of the Holy Nativity and of the birthplace of Jesus, a Jesus Boat (based on those found on the Sea of Galilee), a devotional Garden containing wax figures illustrating the life of Christ, Pilot's Judgment Hall, and the Garden of Gethsemane. Outside in the car park area were installed the Garden of Eden, the Shepherd's Field (where the angels announced the birth of Jesus) and a Face of Jesus statue. In 2012, the 2,000-seater auditorium opened. Doubling as a performance space and recording studio, this offers a continual program of worship and theatrical performances that include a daily passion play. Film screenings have been increased and presentations at the various themed sites run throughout the day including Holy Communion with Jesus in the Qumran Cave and the trial of Jesus in Pilot's Judgment Hall.

This final case-study shows a single-faith museum, presenting themed environments and authentic artifacts to engage visitors with a particular interpretation of the sacred. It has attracted charges of edutainment—that mixture of education and entertainment that has bedeviled attractions trying to make their offer more accessible to a wider audience. Parallels are often drawn with the Magic Kingdom at nearby Disney World or, less convincingly, with Colonial Williamsburg (Beale 2005: 53, 56). It divides opinion between those for whom it reinforces belief and those uneasy at its portrayal of Christ and commodification of the Gospel. The earlier internal logic, although not always readily apparent, has been overwritten by the addition of the later attractions. The whole operation has shifted theologically from being fundamentalist and anti-charismatic to charismatic (Religious Tolerance 2001). Evangelic in tone, it offers an essentially Protestant interpretation of the Holy Land's landscape and sacred sites, ignoring Catholic and Orthodox traditions. Judaism is presented as a prequel to Christianity and the Islamic presence in the region is ignored altogether. Its educational role is purely internal.

132.8 Conclusion

This survey of religious museums demonstrates the diversity of the sector and how the material culture of religion can be appropriated for very different ends. From this basis, all three case studies look simultaneously inwards to the needs of faith communities and outwards to the wider community, albeit in contrasting ways. On the one hand, they are all repositories of memory, and serve to reaffirm faith, reassert identity, or demonstrate the richness of their heritage. On the other hand, they seek to create a dialogue with those of other faiths and, indeed, with those of no faith. It is here that key differences emerge. In the case of St Mungo and the Jewish Museum the aim is to promote understanding and respect, dispel misconceptions and challenge stereotypes. The Holy Land Experience also wishes to engage with 'outsiders' but with a view to convincing them of the truth of the message on offer.

Ultimately all the collections discussed in this chapter give audiences the pleasure of looking at some of the world's greatest treasures. Museum visitors can experience something of the inspiration to be had from engaging with objects that try to express themes that transcend the material world—that 'window on the eternal'. What the collections also do is to provide a window back into the soul of society, revealing some of the worst aspects of human behavior as well as the best. It is the job of these museums to resolve these tensions.

These are issues that continue to exercise the minds of museum professionals. There have been a number of international conferences and collaborations to discuss, for example, issues connected with the display of Islamic collections (see Guzy et al. 2010; Junod et al. 2013), or related to the handling of sensitive material from indigenous collections (see Sullivan and Edwards 2004). Some of the most thoughtful work carried out by museums in the area of religion takes the form of temporary exhibitions. Examples include: the 2011 exhibition "Treasures of Heaven" which transferred from the Cleveland Museum of Art to the Walters Art Gallery (Baltimore) and the British Museum, London; and the 2012 exhibition 'Hajj: journey to the heart of Islam' at the British Museum. Initiatives that take religious collections beyond the walls of the museum have included the "Museum with no Frontiers," which creates exhibition trails and virtual exhibitions. This began with topics such as Gothic art and Baroque art, but since 1995 has moved into projects to promote Islamic art: notably, an exhibition trails project "Islamic Art in the Mediterranean" and the "Discover Islamic Art" (MNF 2006). The creative collision of cross-cultural sensitivities, curatorial practices and advances in information technology, *inter alia*, guarantee that this process of innovation will continue in diverse and unpredictable ways for the foreseeable future.

References

Arthur, C. (2000). Exhibiting the sacred. In C. Paine (Ed.), *Godly things: Museums, objects and religion* (pp. 1–27). London: Leicester University Press.

Beale, T. K. (2005). *Roadside religion: In search of the sacred, the strange, and the substance of faith*. Boston: Beacon Press.

Brabant, M. (2001, February 5). Jewish fury at Holy Land Park. *BBC News*. Retrieved: 23 February, 2013, from http://news.bbc.co.uk/1/hi/world/americas/1154643.stm

Bräunlein, P. J. (2005). The Marburg museum of religions. *Material Religion, 1*(1), 177–180.

British Museum. (2008). *The British Museum strategy to 2012*. London: British Museum. Retrieved 25 February, 2013, from www.britishmuseum.org/pdf/Strategy%20to%202012%20web%20version.pdf

Burman, R. (1998). Holocaust work at the Jewish Museum, London: Preservation, exhibitions and education. *Journal of Holocaust Education, 7*(2), 44–50.

Burman, R. (2000). Presenting Judaism: Jewish museums in Britain. In C. Paine (Ed.), *Godly things: Museums, objects and religion* (pp. 132–150). London: Leicester University Press.

Burman, R. (2006). The Jewish Museum, London introduction and history. In R. Burman, J. Marin, & L. Steadman (Eds.), *Treasures of Jewish heritage: The Jewish Museum London* (pp. 10–19). London: Scala Publications.

Burman, R. (2012). Jewish museums help the wider Community. *Museums Journal, 112*(2), 17.

Carnegie, E. (2009). Catalysts for change? Museums of religion in a pluralist society. *Journal of Management, Spirituality and Religion, 6*(2), 157–169.

Carr, D. (2009). Our conversations among faith objects: A statement of the museum problem. In A. M. Hughes & C. H. Wood (Eds.), *A place for meaning: Art, faith and museum culture: Learning from the Five Faiths project at the Ackland Art Museum* (pp. 100–105). Chapel Hill: Ackland Art Museum.

Clark, D. (2004). Managing Jewish museums in a multi-faith society: Notes from Italy. *Journal of Management, Spirituality and Religion, 1*(1), 92–112.

Conaty, G. T. (2003). Glenbow's Blackfoot gallery: Working towards co-existence. In L. Peers & A. K. Brown (Eds.), *Museums and source communities* (pp. 227–241). London: Routledge.

DIVUM (Declaration on the Importance and Value of Universal Museums). (2006). Declaration on the importance and value of universal Museums: 'Museums serve every nation'. In I. Karp, C. A. Kratz, L. Szwaja, & T. Ybarra-Frausto (Eds.), *Museum frictions: Public cultures, global transformations* (pp. 247–249). Durham: Duke University Press.

Dunlop, H. (2013) Final transcript AD317 VC1. *Religion Today*, Band One, St Mungo Museum of Religious Art and Life. Retrieved February 24, 2013, from https://moodle.fp.tul.cz/file.php/393/AD317_1_tra.pdf

Elgot, J. (2012, January 5). Museum suspends Shabbat opening. *Jewish Chronicle Online*. Retrieved March 28, 2013, from www.thejc.com/community/community-life/61468/museum-suspends-shabbat-opening

Flynn, G. A. (2001). *Merging traditional indigenous curation methods with modern museum standards of care*. Paper submitted for the Marie Malaro Award Competition Museum Studies program. Retrieved February 28, 2013, from www.gwu.edu/~mstd/Publications/2001/gillian%20flynn.pdf

Freunde Museum Islamische Kunst. (2013). Perspectives and aims. Accessed 16 Feb 2013, from http://freunde-islamische-kunst-pergamonmuseum.de/ziele/

Glasgow Museums. (n.d.). *Welcome to St Mungo Museum of Religious Life and Art!* Glasgow: Glasgow Museums.

Glenbow Museum. (2013). *Niitsitapiisinni*: Our way of life. Retrieved April 6, 2013, from http://www.glenbow.org/exhibitions/permanent.cfm

Gooch, T. A. (2000). *The numinous and modernity: An interpretation of Rudolf Otto's philosophy of religion*. Berlin: Walter de Gruyter.

Grossman, G. C. (2008). *Jewish museums of the world*. New York: Universe Publishing.

Guzy, L., Hatoun, R., & Kamel, S. (Eds.). (2010). *From imperial museum to communication centre? On the new role of museums as mediators between science and non-western societies*. Würzburg: Königshausen and Neumann.

Heller, E . (2013). Founding Director's note, Museum of Biblical Art. Retrieved February 28, 2013, from http://mobia.org/about/directors-note/

HLE (Holy Land Experience). (2012). History. Retrieved: December 20, 2012, from www.holylandexperience.com/education/history.html

Huang, P. H. J., & Chen, J.-H. (2001). Of the sacred and the secular: Missionary collections in university museums. *University Museums and Collections Journal, 3*, 85–90.

ICOM (International Council of Museums). (2013). Museum definition. Retrieved: February 25, 2013, from http://icom.museum/the-vision/museum-definition/

Jenkins, T. (2006). Comment. *Material Religion, 2*(3), 353–358.

Jewish Museum. (2005). Collections management policy. Retrieved January 5, 2013, from www.jewishmuseum.org.uk/domains/jewishmuseum.org.uk/local/media/audio/Collections%20Management%20Policy.pdf

Jewish Museum. (2013). The museum's mission and mandate. Retrieved January 1, 2013, from www.jewishmuseum.org.uk/jobs-new

Joseph, A. (2012, 11 April). Behind the scenes at the museum of ourselves. *Jewish Chronicle Online*. Retrieved February 28, 2013, from www.thejc.com/arts/arts-features/66380/behind-scenes-museum-ourselves

Junod, B., Khalil, G., Weber, S., & Wolf, G. (Eds.). (2013). *Islamic art and the museum: approaches to art and archaeology of the Muslim world in the twentieth century*. London: Saqi Books.

Kark, R., & Frantzman, S. (2010). The Protestant Garden Tomb in Jerusalem, Englishwomen, and a land transaction in late Ottoman Palestine. *Palestine Exploration Quarterly, 142*(2), 199–216.

Koutchinsky, S. (2005). St Petersburg's museum of the history of religion in the new millennium. *Material Religion, 1*(1), 168–172.

Krepps, C. (2003). *Liberating culture. Cross-cultural perspectives on museums, curation and heritage preservation*. London: Routledge.

Krepps, C. (2005). Indigenous curation as intangible cultural heritage: Thoughts on the relevance of the 2003 UNESCO Convention. *Theorizing Cultural Heritage, 1*(2), 1–8. Retrieved January 3, 2013, from www.folklife.si.edu/resources/center/cultural_policy/pdf/ChristinaKrepsfellow.pdf

Krepps, C. (2009). Indigenous curation, museums, and intangible cultural heritage. In L. Smith & N. Akagawa (Eds.), *Intangible heritage* (pp. 193–208). Abingdon: Routledge.

Marburg Museum of Religions. (2013). Welcome to the Museum of Religions. Retrieved March 3, 2013, from www.uni-marburg.de/relsamm/

Marchisano, Archbishop F. (1995). Cathedral and diocesan museums: Crossroads of faith and culture. In Cathedral Museum (Ed.), *Cathedral and diocesan museums: Crossroads of faith and culture: Proceedings of the International Symposium held in Malta 27–29 January 1994 on the occasion of the 25th anniversary of the Mdina Cathedral Museum* (pp. 46–56). Malta: Cathedral Museum Publication.

MMA (Metropolitan Museum of Art). (2000). Museum mission statement. Retrieved March 1, 2013, from www.metmuseum.org/about-the-museum/mission-statement

MNF (Museum with No Frontiers). (2006). *Museum with no frontiers 10 years 1996–2006*. Retrieved 25 February, 2013, from http://www.museumwnf.org/doc/MWNF_10years_en.pdf

Morris, B., & Brooks, M. M. (2007). Jewish ceremonial textiles and the Torah: Exploring conservation practices in relation to ritual textiles associated with holy texts. In M. Hayward & E. Kramer (Eds.), *Textiles and text: Re-establishing the links between archival and object-based research* (pp. 244–248). London: Architype.

MOW (Museums of the World). (2009). *Museums of the world* (16th ed.). München: De Gruyer.

Museum of Christian Art. (2013). Museum of Christian art, Old Goa. Retrieved March 1, 2013, from www.museumofchristianart.com/index.php?flag=PR&sflag=K

NMNZ. (2013). Work with iwi & museums. Retrieved April 6, 20–13, from http://www.tepapa.govt.nz/AboutUs/Pages/Workwithiwiandmuseums.aspx#national

O'Neill, M. (1996). Making histories of religion. In G. Kavenagh (Ed.), *Making histories in museums* (pp. 188–199). London: Leicester University Press.
O'Neill, M. (2006). Enlightenment traditions, sacred objects and sacred cows in museums response to Tiffany Jenkins. *Material Religion, 2*(3), 359–368.
O'Neill, M. (2011). Religion and cultural policy: Two museum case studies. *International Journal of Cultural Policy, 17*(2), 225–243.
O'Neill, M. (2013). Final transcript AD317 VC1. *Religion Today*, Band One, St Mungo Museum of Religious Art and Life. Retrieved February 24, 2013, from https://moodle.fp.tul.cz/file.php/393/AD317_1_tra.pdf
Ouroussoff, N. (2010, November 26). Building museums, and a fresh Arab identity. *New York Times*. Retrieved April 6, 2013, from http://www.nytimes.com/2010/11/27/arts/design/27museums.html?pagewanted=1&_r=0&ref=design
Paine, C. (2000). Preface. In C. Paine (Ed.), *Godly things: Museums, objects and religion*. London: Leicester University Press.
Paine, C. (2009). Militant atheist objects: Anti-religion museums in the Soviet Union. *Present Pasts, 1*, 1–8. Retrieved: December 12, 2012, from www.presentpasts.info/rt/printerFriendly/pp13/18
Paine, C. (2013). *Religious objects in museums: Private lives and public duties*. London: Bloomsbury.
Reed, T. (2007, July 29). Christian theme park a holy land experience, *USA Today*. Retrieved: February 7, 2013 from http://usatoday30.usatoday.com/news/religion/2007-07-29-holy-land-experience_N.htm
Religious Tolerance. (2001, March 11). 2001-Mar-11 Florida: Christian theme park criticized for religious intolerance. *News of Religious Intolerance and Conflict*. Retrieved: February 23, 2013 from www.religioustolerance.org/news_01mar.htm
Rocker, S. (2011, May 5). Jewish Museum to open Saturdays. *Jewish Chronicle Online*. Retrieved March 28, 2013 from www.thejc.com/news/uk-news/48495/jewish-museum-open-saturdays
Rosoff, N. B. (2003). Integrating native views into museum procedures: Hope and practice at the National Museum of the American Indian. In L. Peers & A. K. Brown (Eds.), *Museums and source communities: A Routledge reader* (pp. 72–79). London: Routledge.
Rubin Museum of Art. (2013). About the Rubin Museum of Art: Mission and values. Retrieved February 25, 2013, from www.rmanyc.org/about
Russian Museums. (2013). The State Museum of the History of Religion. Retrieved March 2, 2013 from www.russianmuseums.info/M113
Samworth, H., Holmgren, S., & Kinniburgh, S. (2003). *A guide to the scriptorium*. Orlando: Sola Scriptura. Retrieved February 2, 2013 from www.answersingenesis.org/assets/pdf/am/v4/n2/scriptorium-guide.pdf
Sandal, R. (2007). *Museums, prejudice and the reframing of difference*. London: Routledge.
Simpson, M. G. (2001). *Making representations: Museums in the post-colonial era*. London: Routledge.
Singh, K. (2006). Bigot buster: Tackling sectarianism. *Interpret Scotland, Issue 14*. Retrieved December 18, 2012 from www.interpretscotland.org.uk/website/interpretscotland.nsf/byunique/issue14.html/$FILE/bigot14.pdf
Spalding, J. (1995). Preface. In Glasgow Museums (Ed.), *The St. Mungo Museum of Religious Life and Art* (pp. 5–6). Edinburgh: Chambers.
Stephens, S. (2008). United we stand. *Museums Journal, 108*(4), 38–41.
Sullivan, L. E., & Edwards, A. (Eds.). (2004). *Stewards of the sacred*. Washington, DC: American Association of Museums Press.
Wallace, M. (2001). Land of our Father. *Financial Times Weekend Magazine, 14*(04), 2001.
Wharton, A. J. (2006). *Selling Jerusalem: Relics, replicas, theme parks*. Chicago: University of Chicago Press.
Wilson, B. (2009, October 29). Holy Land experience: Religious theme park bucks ominous trend. *Daily Finance*. Retrieved: December 20, 2012 from www.dailyfinance.com/2009/10/29/holy-land-experience/
Witcomb, A. (2003). *Re-imagining the museum: Beyond the mausoleum*. London: Rou

Chapter 133
The Creationist Tales: Understanding a Postmodern Museum Pilgrimage

Jeffrey Steller

Then do folk long to go on pilgrimage,
And palmers to go seeking out strange strands,
To distant shrines well known in sundry lands. (Chaucer)

133.1 Introduction

The word pilgrimage evokes thoughts of weary travelers on a long journey to some distant shrine, some seeking salvation and others answers. A pilgrim's journey may take weeks marked with innumerable pitfalls. The journey may take pilgrims from inn to inn along worn countryside paths or across a vast expanse of desert. People have been undertaking pilgrimages to an assortment of sites for untold centuries.

Yet, in an ever-changing world where blank portions of the world map no longer exist pilgrimages have changed. Pilgrimage sites that took weeks to reach by foot are accessed in a day using modern transportation. Society increasingly encroaches on sites that were once situated on the edges of civilization. In addition to these changes, modern conceptions of what constitutes holy places are changing. Gone are the days of miracles establishing a site as a shrine. Religious sites are now manufactured and available on-demand in the confines of one's house.

These changes necessitate new understandings of what constitutes a pilgrimage. The purpose of this essay is to develop an understanding of a *postmodern pilgrimage*. In order to do so, one must first evaluate older meanings of the term and wed these to new interpretations. Using the Creation Museum, located 20.5 miles (33 km) southwest of Cincinnati, Ohio, in Petersburg, Kentucky, as a case study, it is argued that a postmodern pilgrimage should not be seen only as arduous journey to a traditional holy place or shrine. Instead, a postmodern pilgrimage should be understood as a religiously motivated journey from the local to a religious site. The words *local* and *religious site* are left intentionally ambiguous to be more inclusive of different understandings and interpretations of the terms. The concept of local must not be bound up

J. Steller (✉)
Public History, Northern Kentucky University, Highland Heights, KY 41009, USA
e-mail: jeff.steller@gmail.com

S.D. Brunn (ed.), *The Changing World Religion Map*,
DOI 10.1007/978-94-017-9376-6_133

in mere measures of distance, but needs to be expanded to include cultural understandings as well. The religious motivation is largely in the eye of the beholder, but should be the express reason for the journey. Finally, what constitutes a religious place has undergone a dramatic transformation in recent years. The manufacture and production of religious sites create new spaces, such as the Creation Museum and ever increasing megachurch campuses, for religious journeys. When conceived of in this light, it becomes evident that places like the Creation Museum are at the forefront of a new understanding of religious pilgrimage.

133.2 Understanding a Postmodern Pilgrimage

In order to determine the validity of the supposition that the Creation Museum is in fact a postmodern pilgrimage site, one must first clearly elucidate the operational definitions of terms involved. This is no easy feat as the literature abounds with different meanings of the terms and criteria involved, as discussed below. There is no more telling statement than Morris's (2001: 1): "It is notoriously difficult to define a pilgrim." In addition, the rise in the secular uses of terms such as "pilgrimage," in the literature, media and common vernacular further muddle the issue. Scholars from a wide array of nearly all the social science disciplines proffer a host of definitions of "pilgrim" and "pilgrimage." Quite often the theoretical framework of a scholar's chosen discipline underpins the definition, lending slight variation and nuance to each iteration of the term. A brief survey of some of the more prominent definitions allows for the teasing out of commonalities and provides the foundation for conceiving a postmodern pilgrimage.

Every scholar agrees that a pilgrimage must involve a journey to a religious site. Some of the definitions of pilgrimage are simple and straightforward, offering very few conditions. Take the following definition of pilgrimage as an example: "A journey to a sacred place as an act of religious devotion" (Sykes 1982: 776). Some scholars seem content with such vague generalities, offering little in the way of specifics. Other scholars expand their definitions to include a range of different criteria such as more specificity on what occurs during the experience as well as more detail about the journey and destination. For example, McKevitt (1991) argues that the sacred site to which the pilgrim travels must lie outside the scope of the everyday and commonplace experience. In his work, Morris (2001) gives a lengthy, detailed analysis of how the term varies across space and time, evolving with new understandings and interpretations.

In other instances scholars posit that a pilgrimage is more than the physical travel and visit, arguing it includes an internal journey as well (Wiederkehr 2001:11). This internal journey may be marked by deeper understandings and new meanings, or it may only include a meditative element, with no mention of a transformation occurring after the experience (Barber 1991: 1; Blackwell 2007: 46). Other scholars make the argument that the spiritual journey is much deeper and more intense. According to Morinis (1984), a pilgrimage is the allegorical journey an individual makes to find the very soul of God. While not all scholars expressly add the internal journey element to their definition, tangential references to an internal or spiritual journey as an element of the pilgrimage are common.

Still other scholars continue adding more stipulations to their definition. Some argue pilgrimage is a group exercise, a notion consistent with many historical understandings of the term (Stoddard 1997). In some instances this is meant as a form of group travel, while others focus on the experience at the site. The site itself is also a point of divergence among scholars. Very often the site is referred to as a "sacred place" or "shrine," but little or no mention is made concerning the physical location. Others are more explicit about the site, suggesting that it should be on the periphery of society, far from the major centers of population (Turner and Turner 1978).

What becomes clear from this cursory survey is that considerable difference exists among scholars even when using common themes across disciplines. As stated previously, the idea of a journey is always consistent. There is little dissension that a pilgrimage involves the physical movement from one location to another. However, as Stoddard (1997) astutely points up, there is limited agreement on the nature of travel in order for a trip to qualify as a pilgrimage.

One issue is the benchmark for a "minimum distance" that a person must travel in order for the journey to be considered a pilgrimage. Such a discussion indirectly brings into the discussion a pilgrim's choice of transportation. This is not simply a peripheral issue, but a fundamental element in developing an understanding of a postmodern pilgrimage. The incredible progression of transportation technology and rapid globalization over the last century has significantly altered conceptions regarding length of travel. Journeys that once took weeks are completed in day's travel. The popular image of pilgrims walking for days has given way to a world where high speed trains run to Canterbury and the *hajj* can be made in first class comfort.

Globalization has compressed humanity's understanding of time and space, creating a convergence unparalleled in human history. It has also transformed another understanding of pilgrimage: leaving the local. As discussed previously, some scholars argue that pilgrimage sites were well removed from the local and located on the periphery of society (Turner and Turner 1978). Yet, in the ever evolving "global village" what constitutes the "periphery?" How does one define a site that is not near any major population center, as the Turners argued pilgrimage sites were, when almost our concept of distance and periphery have evolved?

This compression of space-time does not render the argument about leaving the local completely irrelevant. While globalization certainly does much to transform understandings of "local," the exclusion of local movements is still important to delineate a pilgrimage from other religious travel. In this sense, local may be better thought of as an area of familiarity and commonality. Using this understanding, regular trips to church would not be a pilgrimage, a point widely accepted in the literature (Stoddard 1997). At the same time, someone who does not travel a considerable distance, but must experience a dramatic cultural shift or navigate unfamiliar territory, might still be understood as a pilgrim. A pilgrim also makes his or her journey infrequently (Pavicic et al. 2007). Once again, criteria other than mere distance help inform conceptions of postmodern pilgrimages.

This discussion does not completely discount the issue of distance. While distance is no longer a defining feature, it is still indicative of other elements of a pilgrimage. For example, although air travel shortens the time and physical burdens of travel, distance traveled offers insight into a pilgrim's desire to reach a site. Longer distances tend to cost more money, resulting in a greater financial undertaking for the journey. Furthermore, if a site draws visitors from the world over, such as the shrine at Lourdes, it offers some insight into the importance of the site. However, on the whole, distance still remains an indicator of other issues rather than defining a pilgrimage directly.

The above discussion transforms the postmodern understanding of a pilgrimage because it reshapes the understanding of how important is the "journey." It also shifts the definitional emphasis to other criteria. Chief among these is a traveler's motivation. As Griffin (2007) observes, the commonality among most definitions is that pilgrimage is motivated by a spiritual and/or religious desire within. This definition returns to the idea of the internal journey. A religious tourist may seek out a shrine or other scared space as part of a larger journey. He or she might be driven by curiosity or perhaps an interest in the architecture or history of a location. The pilgrim seeks the destination to build a closer relationship to his or her God, searching for deeper meaning and understanding (Murray and Graham 1997). In fact, many pilgrims are seeking "conversion and salvation in their everyday lives" (Giuriati and Arzenton 1992: 9 in Ambrósio 2007).

It is not simply the nature of the motivation that determines classification of a journey as pilgrimage. The destination itself it is also important in differentiating types of religious tourists. Most of the definitions cited previously, as well as many not included in this chapter, refer to "sacred places" or "shrines" as the destination for pilgrims. Shrines are generally viewed as "repositories for a revered body or venerated relic. In its broader meaning, a shrine refers to a sacred site that houses holy artefacts, promotes ritual practice and attracts religious travelers" (Pavicic et al. 2007: 51). These authors go on to discuss scared places as "mediating spaces or transitional zones" that bring the spiritual and human realms closer together (Pavicic et al. 2007: 51).

Other definitions are not as constricting. Guerra (1989) views a shrine as a religious place that delivers a sense of pleasure as well as lingering absence. It also involves memory, presence and prophecy; that is to say, it venerates the origin of God, is an expression of the Christian community's values and the hope of what God may deliver in the future (Guerra 1989). There is little mention of the artefacts, but implicit in the definition is the idea of the mediation zones and promotion of ritual. This is best understood when taking into account Griffin's (2007) point about nature, in the way of tree and rocks have been accepted as pilgrimage sites.

However, this place focus only clarifies the understanding of pilgrimage sites to some degree. The issue is complicated further by growing trends of manufacturing sacred spaces. One of the most prominent examples of this is the Holy Land Experience in Orlando, Florida. Designed as what many term a religious theme park, the park creators state that "above all, beyond the fun and excitement, we hope that you will see God and His Word exalted and that you will be encouraged in your search for enduring truth and the ultimate meaning of life" (Blackwell 2007; Holy Land Experience 2012). It seems clear the park's encouragement of the search for truth and meaning are very consistent with aforementioned conceptions of sacred places and pilgrimages. Despite the park being purpose-built and non-sacred, as pointed out by the experience, it is nonetheless an attempt to provide a medium for people to revere their God and participate in rituals while simultaneously providing the memory, presence and prophecy elements (Blackwell 2007). In that sense then, even though the attraction was not born of miracle or martyr, nor does it provide a home for ancient relics, it is certainly a manufactured sacred space in a more holistic postmodern interpretation.

Only recently have scholars started to investigate whether journeys to burial places such as Graceland might be considered pilgrimage sites (Griffin 2007). While these are well outside the scope of this chapter, they raise an interesting point. If a site holds immense spiritual value for an individual and their express purpose for undertaking a journey is an inner desire for growth, healing or reverence, must the site by "sacred" or "holy" in the religious sense of the word? In a century where megachurches repurpose warehouses and sports arenas, televangelism continues to grow and groups manufacture their own shrines, the question arises: what constitutes sacred space?

Part of the answer to this question exists outside this research, and the research of countless other scholars, because sacred space varies across, time, space, religion, culture and individuals. Even within denominations of the same faith the idea of shrines and pilgrimages vary. Where Catholicism venerates Lourdes for its Marian mysticism, other denominations place little stock in its role as a pilgrimage center.

Perhaps the most inclusive definition proffered in the literature is that "pilgrims go to shrines to invoke and to welcome the Holy Spirit, transferring it, later on, in terms of everyday actions" (Ambrósio 2007: 80). Taken in this way, a shrine then is any site that allows for the invocation, welcoming and transfer of the Holy Spirit. It encompasses a much wider array of religious sites, expanding the possible places of pilgrimage around the world.

Bringing all this together, we arrive conceptually at a much different understanding of pilgrimage. It obscures any hard and fast criteria that might be used to mark a pilgrimage. Given the rise of the global village, the compression of space-time and an ever-evolving understanding of manufactured sacred spaces, a postmodern pilgrim looks very different from Chaucer's miller and knight. A postmodern pilgrimage then fits Sykes's definition offered at the beginning of this section, viz., "a journey to a sacred place as an act of religious devotion." As demonstrated above, in the postmodern sense, this journey need only leave the local, as understood to be a place uncommon to the traveler, regardless of physical distance or means of transport. The motivation remains unchanged for the modern pilgrim, viz., some personal understanding of undertaking the trip for strictly spiritual and/or religious reasons. Finally, the postmodern pilgrimage may be to a site that offers the ability to experience ritual and a connection with God. The destination need no longer be a shrine in the traditional understanding of the term, but may fit new conceptions of scared spaces.

133.3 The Creation Museum as a Postmodern Pilgrimage Site

With the definitional issues sorted through, it becomes possible to examine the Creation Museum as a case study of a postmodern pilgrimage site. This section gives a brief history of the museum and its mission before delving into an examination of its potential role as a postmodern pilgrimage. The section employs primary data and informal interviews collected by the author as well as data and information disclosed by the Creation Museum.

133.3.1 A Brief History of the Creation Museum

The foundation for the Creation Museum harkens back to 1970s when Ken Ham and other creationist Christians in Australia formed the Creation Science Foundation, which, at the time, was headquartered in Brisbane, Australia (Answers in Genesis 2012). Ham came to the United States "on loan" as a speaker to the Institute for Creation Research. In 1993, Ham and other members of the ICR left to form their own creationist group, which ultimately evolved into Answers in Genesis.

By this time, the idea of a creation museum was already in the planning stages. The inclusion of the museum in future plans strongly influenced the decision to relocate the headquarters of Answers in Genesis to Northern Kentucky (LiveScience 2005). If a museum was indeed to be built, the leaders want to maximize visitors be situating themselves in a readily accessible location. Seven years after the move to Kentucky ground was broken for the museum. On May 28, 2007 the Creation Museum officially opened in Petersburg, Kentucky at a price of $27 million (Stevenson 2012). By November of the same year, 250,000 people had visited the museum. At the end of the second year of operation, 719,206 visitors had passed through the museums gate and by 2010 over one million guests had visited (Answers in Genesis 2012). Data for the fourth and fifth years of operation are not publically available.

The Ark Adventure represents the next phase of development for Answers in Genesis, but it is not officially a product of the group. The development will have nine themed attractions with a full-scale replica ark as the centerpiece. Ark Encounter LLC, a for-profit venture out of Springfield, Missouri, will head up development while Answers in Genesis will be the managing partner. As is discussed in more detail below the new development is very much in line with the mission of Answers in Genesis (Interview with Mark Looy).

As of August 2012, tickets were priced to the Creation Museum were as follows (Table 133.1).

In addition, the Museum offers free admission to all on Christmas Eve and to certain groups on select days. There are also options to purchase season passes to the Museum. Several different levels exist, with a variety of options available for a range of prices.

The museum has not been without its share of controversy and media attention since its earliest days. Eugenie C. Scott, director of the National Center for Science

Table 133.1 Entrance fees to the Creation Museum (Information as of Jan. 1, 2014, from http://creationmuseum.org/tickets/; http://creationmuseum.org/plan-your-visit/groups/)

	Standard rate	Group rate
Adult (age 13–59)	$29.95	$19.95
Senior (age 60+)	$23.95	$15.95
Children (age 5–12)	~~$15.95~~ FREE in 2014	$11.95
Children (<5)	FREE	FREE

Education, a vocal critic of young Earth creationism, once called the museum "the creationist Disneyland" (Slevin 2007). Protestors have staged several events including a gathering on opening day (Slevin 2007). PZ Meyers, a professor of biology at the University of Minnesota, Morris, led a group of 300 scientists through the museum for a critical examination (Powell 2009). Detractors argue that the version of history and scientific evidence is misleading at best and false at worst, a charge the museum steadily refutes by pointing to the work of its in house scientists. The Museum and its staff are quick to remind detractors that these scientists have obtained their PhDs from respected institutions, including Ivy League universities. This essay does not mean to address or unpack the issue of controversy other than to note it is simply a matter of history that controversy seems is associated with the museum.

133.3.2 A Walk Through the Museum

A full-scale casting of a dinosaur standing in front of the modern glass façade welcomes visitors to the Creation Museum and offers the first glimpse at the experience before them (Fig. 133.1).

Once the visitor has chosen from the wide array of ticketing plans, he or she enters the Main Hall. Towering above the visitor is a mastodon skeleton which lords over the primary departure point for the Museum. From the Main Hall, a visitor reaches the Planetarium, the bookstore, special effects theater, café and main exhibits (Fig. 133.2).

The main exhibits begin with the Dig Site, a recreation of an archeological dig that beings to explore natural history. Following on is the Starting Points gallery which uses many panels to examine how the same facts lead to different points of view. The panels in the gallery, which focus on events such as the Big Bang, juxtapose God's word and Man's word to explain the different interpretations (Fig. 133.3). The Biblical Authority gallery uses two recreated writing scenes and more panels to explain how God's word came to supersede the word of man before the visitor enters the Biblical Relevance gallery.

The visitor then enters the Culture in Crisis gallery. The gallery uses darkened light, graffiti and images from popular magazines such as Time to portray an image of crisis. Included in the gallery are video screens giving visitors a peek into various common situations in which people are deviating from God's word and contributing to the crisis. Leaving this gallery, the visitor walks through the Time Tunnel into the next gallery. The Wonders of Creation gallery bathes the visitor in bright light with colorful video boards discussing the wonder of life.

Fig. 133.1 The front entrance to the Creation Museum (Photo by Jeffrey Steller)

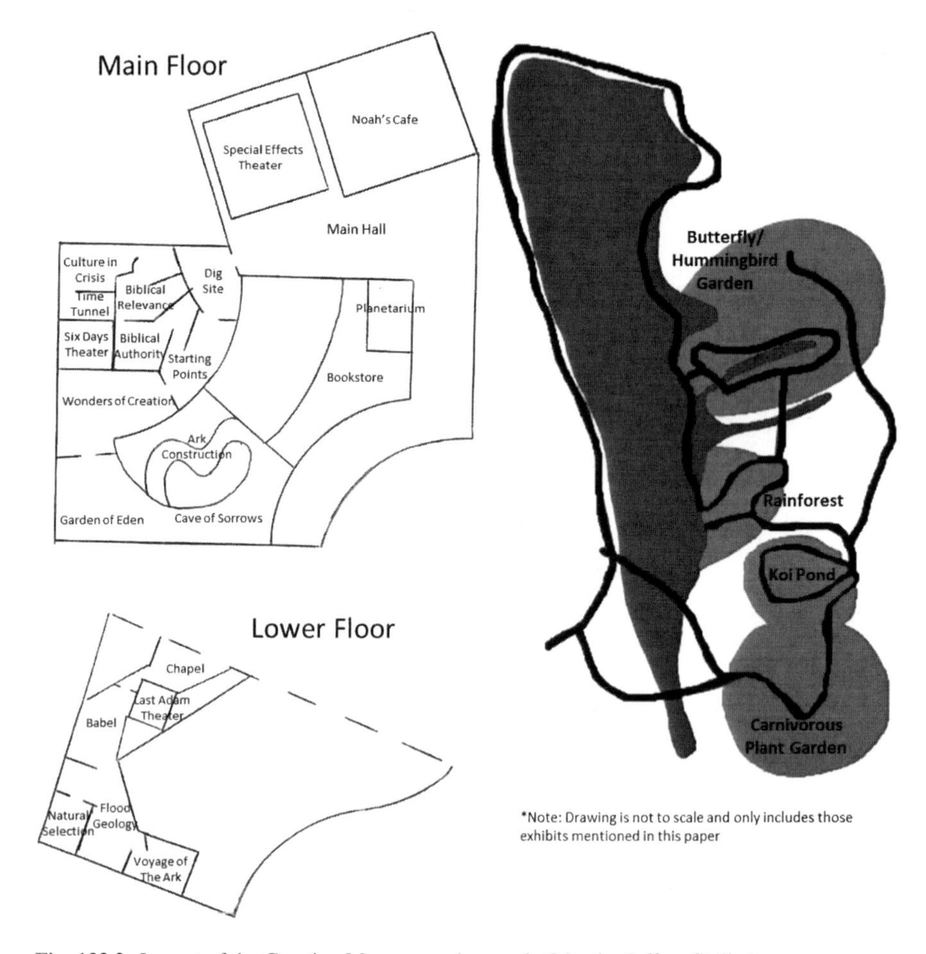

Fig. 133.2 Layout of the Creation Museum and grounds (Map by Jeffrey Steller)

Fig. 133.3 The Starting Points Gallery explains how facts are interpreted differently (Photo by Jeffrey Steller)

Fig. 133.4 Adam naming animals in the Garden of Eden (Photo by Jeffrey Steller)

Next, the visitor finds him or herself transported to the beginning of time as they step into the Garden of Eden. Wax figures and figurines of Adam and Eve walk the visitor through the Creation story in a veritable cornucopia of recognizable wildlife (Fig. 133.4). Interspersed throughout are Bible verses that help tell the Genesis story.

The visitor leaves the lush Garden for the Cave of Sorrows, a dark gallery with moving images depicting the results of the fall of man.

This is followed by the Ark Construction gallery, considered by many to be the most popular in the Museum. The gallery includes a partial, full-scale replication of the Ark in construction with a host of panels to augment it. The Voyage of the Ark gallery that comes next has a Noah figure that interacts with a visitor through a touch screen panel, one of the high tech features in the Museum (Fig. 133.5). The Natural Selection and Flood Geology galleries help explain the science behind the two topics through panels and common examples. The Babel Gallery helps to account for differences among humans.

This ends the major galleries. There is one theater located on each level, each playing a short educational video. The visitor can also purchase a ticket to the Planetarium to learn more about God's work in the universe.

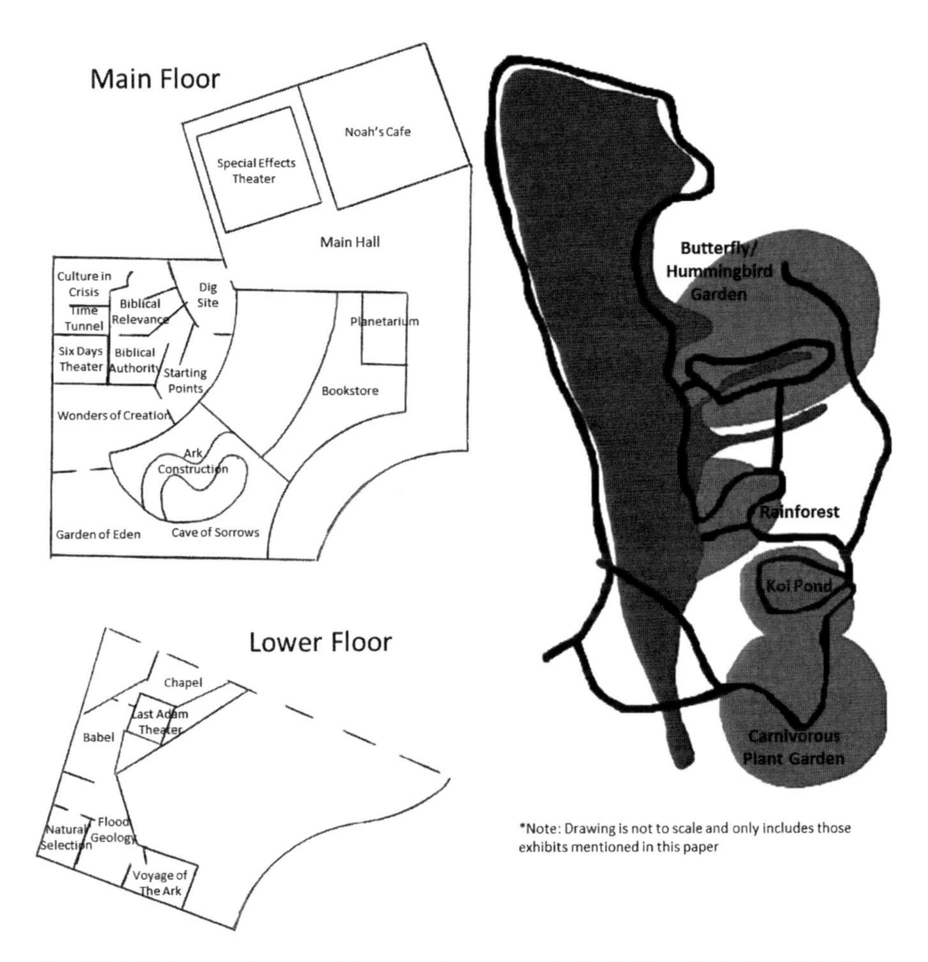

Fig. 133.5 Visitors can interact with a dynamic animatronic Noah (Photo by Jeffrey Steller)

133.3.3 A Postmodern Shrine to Creationism

During the discussion of what constitutes a postmodern pilgrimage, it became evident that the traditional concept of a shrine or sacred space as a pilgrimage site is largely irrelevant. New forms of scared spaces are being constructed in a multitude of manners. The first step in determining if the Creation Museum is a postmodern pilgrimage site is to determine whether the site is a religious space in the postmodern understanding. According to its website, the Creation Museum "brings the pages of the Bible to life, casting its characters and animals in dynamic form and placing them in familiar settings" (Creation Museum 2012). As a natural history the museum takes the visitor through the history of the Bible. In doing so it certainly fits the argument that a pilgrimage site must venerate religious memory.

Yet, there is little in the way of suggestion that the visitor might connect with God though the experience or partake in traditional religious ritual. Promotional materials such as the website and brochures emphasize the exhibits and other offerings such as the audio/visual programming. There is little mention of church services or chapels for prayerful reverence. The Museum does not promote the possibility to speak directly with God or experience a miracle on the grounds. It does not venerate a fallen saint or religious figure. Through and through the Museum promotes the telling of Biblical history. Even many of the workshops provided by the museum (http://creationmuseum.org/events/workshops) focus on biblical education and science as opposed to directly offering a powerful religious experience.

However, while the museum website (www.creationmuseum.org) and promotional materials may not openly suggest such religious experiences, it is in fact one of, if not the, primary purpose of the museum. According to Ham, "people will get saved here… If nothing else, it's going to get them to question their own position of what they believe" (LiveScience 2005). This is certainly in line with many of the shrines and religious places presented earlier. It suggests that a pilgrim to the museum will undergo a religious and/or spiritual journey at the site, complete with a total transformation for some. At the very least the developers of the museum expect that each visitor will experience some degree of religious and/or spiritual growth during a visit. There is also a strong attempt to "inspire change" and have a lasting impact on the visitors once they leave (Stevenson 2012: 110). On these points, informal interview data suggests there is marked success, with many visitors lauding the experience as giving them the confidence to no longer willingly subject themselves to criticism without offering counterarguments.

Furthermore, for believers, the museum represents a space where they can celebrate their faith. According to Aucourt (1990: 19 in Ambrósio 2007), some of the hallmarks of pilgrimage include the ability to "see, hear, remind, testify, celebrate and pray." Certainly the Creation Museum affords such a space, especially through the indulgence of the senses. The exhibits in the museum are replete with extensive artwork as well as a variety of models and audio/visual demonstrations. The very nature of the museum, with its historical presentation of the Bible is a powerful reminder of the teachings religious pilgrims adhere to. It also offers considerable testimony through the presentation of scientific evidence for one to reinforce his or her beliefs. Amidst all of this excitement, there also remains a serene garden area that provides a quiet space for prayerful reverence (Fig. 133.6). This then remains in line with a space that fosters the personal development pilgrims seek at their destination.

133.3.4 Motivations of a Creationist Pilgrim

Supposing then that in the postmodern conception of holy place the Creation Museum meets the criteria, one moves to examine the visitors themselves. Specifically, their motivations provide considerable insight as to whether they might

Fig. 133.6 Garden scene for reverence (Photo by Jeffrey Steller)

be considered a postmodern pilgrim. As noted previously, for a person to be a pilgrim it is essential that his/her journey must be religiously/spiritually motivated. It is important to realize that these religious motivations are extremely complex, bound up closely with personal and cultural beliefs (Blackwell 2007: 39). This adds a level of difficulty in differentiating the religious tourist from the pilgrim, as the two are intimately related and not completely distinct. However, one simple distinction begins with the nature of the journey. Blackwell (2007: 39) is quick to point out that the simple act of traveling to a religious site, no matter how sacred it may be, does not automatically qualify a person as a pilgrim. The motivation for the journey should be the visit to the sacred place, not as an addition to a previously planned trip or as a stop on an otherwise secular trip.

For a moment, it is instructive to eschew the postmodern understanding of a shrine and pilgrimage, testing the idea of pilgrimage against a derivation of Catholic teachings. Pilgrimages are intended to be celebrations of faith with specific stages. Each pilgrim experiences the beginning of the journey, marking a concerted effort to realize certain religious ideals. The journey itself prepares the pilgrim for his/her meeting with God while the actual visit to the shrine is an invitation for interfacing with the sacred. Lastly, the return is a delivery back into the world as evangelists to spread their beliefs (Ambrósio 2007). To those ends, the Creation Museum certainly fits as a postmodern pilgrimage site because it meets such qualifications and a traveler who creates his/her journey in this fashion is a full-fledged pilgrim. A person who makes the Museum a stopover, regardless of how pious he/she might be, is not a pilgrim, because the journey failed to follow such a pattern.

Undoubtedly, not every visitor to the Creation Museum is a pilgrim, nor do all of them even identify as religious tourists. The groups of protesters alluded to in the history section signify a group of visitors who would certainly not consider the museum a pilgrimage site. Similarly, there are those visitors drawn by simple curiosity. With the extensive coverage in the press, including a number of international profiles by such well-known media sources as *The Guardian* and *Vanity Fair* (Europe), interest among the general population is palpable. In fact, during an informal interview, a Finnish individual said that the aforementioned *Vanity Fair* article, inspired him to experience the museum for himself. Of course, this was part of a lengthier trip through the United States, one in which he wanted to "experience the most interesting aspects of America". Although this individual is by no means representative of foreign tourists to America, his case highlights that some individuals come to the museum for nonreligious purposes whereas a true pilgrim makes his/her trip only to visit the shrine.

Even Christians who visit the museum are not necessarily doing so as pilgrims. During several informal interviews, individuals self-identifying as Christians said they came to the museum out of curiosity. They viewed the trip as a learning experience rather than a religious one. When asked what they meant by a learning experience, most responded that they were interested in learning more about what Creationist Christians believe. A religious awakening or religious journey, as the majority of those who fell into this category said they did not subscribe to the creationist mantra. Instead, their goal was to understand how two groups reading the same Bible could develop such divergent interpretations.

Of course, these examples do not invalidate the hypothesis that the Creation Museum is a postmodern pilgrimage site. To operate under the assumption that all visitors to well-known and widely accepted pilgrimage sites such as the Wailing Wall are in fact pilgrims would be a gross mischaracterization. In fact, Poira et al. (2003) found that a primary motivation for Christian visitors to the Wailing Wall was the historical nature of the wall rather than its religious nature. It is the Jewish pilgrims who seek an emotional experience at the site. Similarly, many tourists visit Canterbury as much for its architectural and historical significance as they do for its role as a shrine to Thomas Becket. Does this mixed population of visitors detract from any of these sites' position as a pilgrimage site? Indeed it does not; all are widely accepted as pilgrimage destinations.

Their role as a religious pilgrimage site depends primarily on the motivations of specific segments of the visitor population. For the Wailing Wall, pilgrims are composed largely of Jews whereas at Lourdes pilgrims are primarily Catholics in search of a miracle. In both cases, it is specific groups that comprise the pilgrim population. The situation is no different at the Creation Museum. Those visitors who fall into the category of a postmodern pilgrim follow a specific brand of Christianity. While not all who would be classified as pilgrims fall into this category, nor would all Creationists necessarily make the pilgrimage, the pilgrim population at the museum largely subscribes to the Creationist doctrine. More specifically, many are young earth creationists, those Christians who believe the Earth to be no more than 6,000 years old based on a literal interpretation of the Bible.

During informal interviews, a recurrent theme was celebration of faith. Many of the visitors came expressly to experience a site dedicated to sharing their beliefs. This was especially common among members of groups. Interviewees remarked that they enjoyed the ability to strengthen their sense of religious community and share with others outside of their home regions. Others commented on the power of leaving the local. Two individuals in particular emphasized "breaking a cycle of complacency." By removing themselves from their daily routines, both secular and non-secular, the two respondents felt the message was more powerful for members of their group.

Another recurrent them in line with pilgrim motivations was the defense of faith. Several interviewees felt the visit would provide them the strength to fight against what they perceive as an encroaching society threatening their Creationist beliefs. These visitors came because they heard or read about the voluminous data provided throughout the Museum. Their intention was to arm themselves with their own scientific evidence to "fight their battle" against those who denounce the Creationist belief. Furthermore, they aimed to take their knowledge back to others who did not make the journey, in turn strengthening the faith and defense of their entire community. In her own research at the Museum, Stevenson (2012) put forth a similar argument about the museum experience, arguing that the setup of exhibits such as *Culture in Crisis* adds to the ability to defend one's beliefs, in turn strengthening one's faith. Some respondents indicated a total change in attitude concerning its defense. Where before they did not feel confident arguing against anti-creationists, they believed the experience at the Museum would instill in them an ability to engage in defense of their beliefs.

Other visitors were on a personal journey to renew their faith. Echoing other statements to some degree, they felt society had worn them down and they needed the visit to regain an internal fire which no longer burned brightly. Still others brought new believers and recent converts to experience the gifts of the Lord and celebrate newfound faith. One individual remarked that in the early stages of his own born-again experience he struggled against the onslaught of information against his new beliefs. He stated that he only wished something as "spiritually empowering" as the Creation Museum would have existed at the time.

Returning then to the concept of religious motivations, it is evident that nearly every response by someone identifying him/herself as a Creationist Christian paralleled the motivations of pilgrims. Whether interviewees came alone or in groups, for spiritual growth or strengthening their defenses against persecution, the motivations were all religious. It is also important to note that vast majority of these same individuals said they made their trip strictly to visit the Museum. With these two facts in mind, it is easy to understand that many of the visitors are indeed pilgrims on account of their motivations.

133.3.5 Pilgrims Leaving the Local

One of the most confounding issues in determining a site's status as a postmodern pilgrimage is the issue of the journey. Globalization invited a radical departure from traditional understanding of travel and journeys as the world shrank. New forms of transportation opened up once inaccessible locations and compressed our understandings of space and time. As argued previously, the result of this new understanding of travel is that a person need only leave his or her locality to embark on a pilgrimage.

From the outset, the length of this journey was mitigated by the decision to locate the Creation Museum in Northern Kentucky. According to Ken Ham 69 % of the U.S. population is within 1 h air travel. By locating the museum less than 10 miles (16 km) from the Greater Cincinnati/Northern Kentucky International Airport leaders of Answers in Genesis hoped to make the museum extremely accessible (Sheehan 2005). Furthermore, this same swath of population, which includes the whole of the Bible Belt, is within 1 day's drive from the museum as well. It is no wonder then that a brochure for the Creation Museum boasts that it is "conveniently located." (Creation Museum Brochure)

Such a location virtually eliminates any discussion of the pilgrims facing a long, let alone perilous, journey. It would be folly to argue that the Museum is on the peripheries of society when it is well within reach of so much of the American population. This fact does not diminish the status of the museum as a pilgrimage site. As discussed previously, a pilgrim need only leave his/her local surroundings to "journey" on a pilgrimage. Using this understanding, even a short plane ride dislocates the pilgrim from his or familiar environment. A short car trip accomplishes the same. The location on the outer edges of Greater Cincinnati, situated at the confluence of the South and the Midwest, in a former border state neighboring one of the most important political battlegrounds means that even those driving a short distance are likely to experience cultural differences. Even within Kentucky, the short drive from Lexington or Louisville involves a certain degree of cultural shifts, where accents, attitudes and political beliefs change.

Of course, such a nuanced argument does not even apply if the majority of visitors are coming from further away. In an analysis of the top 100 visitor origins, the average distance traveled turned out to be significant (Fig. 133.7). According to internal survey data provided by the Creation Museum and using distances provided by Google Maps, a visitor to the museum traveled an average of 273 miles (439 km) (Google Maps). To put this in perspective, that means the average visitor made a trip from a city as far away as Chicago, Cleveland or Nashville. Additionally, visitors within 150 miles (241 km) only account for 43 % of visits. When the distance is expanded to 300 miles (483 km), 67 % of visitors are accounted for; at 450 miles (724 km), 82 % of visitors are included (Fig. 133.8).

These travel distance are by no means insignificant, nor could they reasonably be argued as a "local" visit. In fact, the closest thing to a local travel is from Cincinnati and it accounts for a mere 16 % of visits. Even if one were to make an argument that

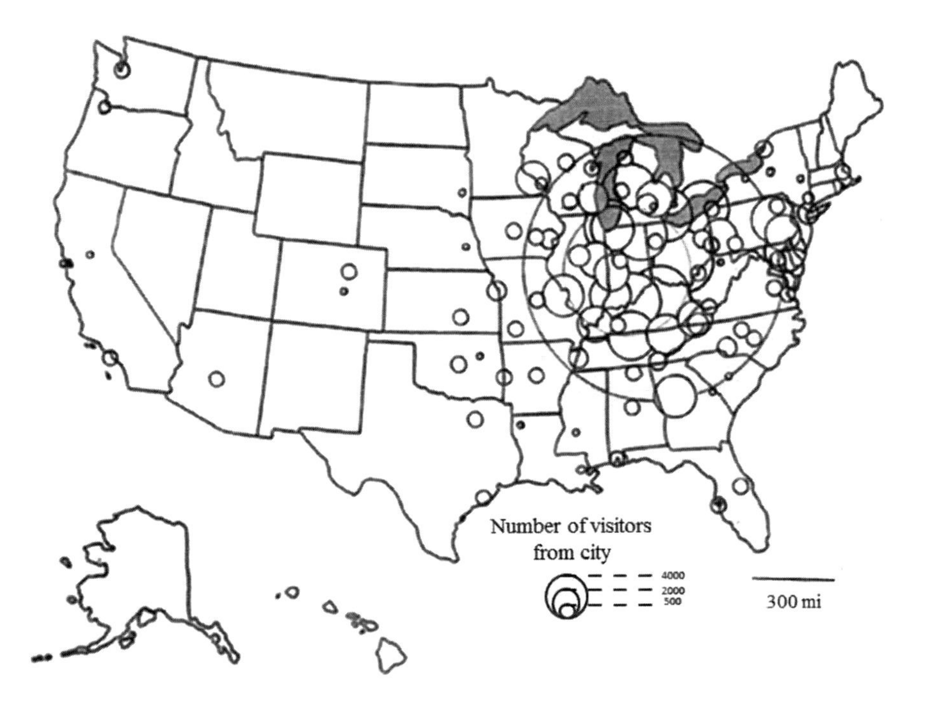

Fig. 133.7 The top 100 visitor origins from March 2007 (opening) to January 2012 according to Creation Museum data. The *red* (Cincinnati) and *blue* (Indianapolis) are highlighted specifically because they account for such high numbers of visitors: Cincinnati had 23,886 and Indianapolis had 13,191 (Map by Jeffrey Steller) (Color figure online)

Fig. 133.8 Distance traveled by Creation Museum visitors. The *blue circle* is a 150 miles (2,421 km) radius that accounts for 43 % of group visits to the museum. The *red circle*, 300 miles (483 km) represents 67 % of visitors and the *black circle*, at a distance of 450 miles (724 km) includes 82 % of all visits (Map by Jeffrey Steller) (Color figure online)

Fig. 133.9 The traditional Bible Belt (Map by Jeffrey Steller)

anyone traveling from within the Bible Belt was still traveling "locally" in the sense of not going to a culturally distinct area, the argument is tenuous. The Bible Belt, as traditionally accepted, accounts for only 40 % of the visitors (Fig. 133.9). Coupled with that fact that that religion is but one facet of culture and that the Bible Belt is by no means culturally homogenous, it is hard to use this line of argument that most visitors are local.

Another way to see how far people are traveling to visit the museum is to investigate the vehicles in the parking lot on a typical day. Over the course of 12 Saturdays from October 2011 to April 2012 license plates in the parking lot were categorized by state. Many states do not issue license plates with county identification, making an accurate measurement of the distance traveled difficult. However, if for this experiment one were to include only the tri-state region (Ohio, Kentucky and Indiana) only a slim majority of the visitors are local. Furthermore, given that most charter busses and 12 passenger vans originate outside the tri-state, it is plausible to argue that local visitors do not actually account for the majority of museum patrons, but only for the majority of vehicles.

It must also be taken into account that while travel by car or airline eases the burden of travel, it also offers more time for reflection. Only one person in a car needs to drive, while the others are free to spend their time preparing for the experience. This is especially the case for some large groups that take charter busses. Informal interviews revealed that the journey to the museum was one of silent prayer and reflection. Another group said the bus ride was a time for sharing Bible verses and talk among the group about their daily tribulations. These findings certainly lend credence to the idea that the journey should be a spiritual one, preparing the pilgrim for the experience at the site.

This also marks the opportunity to address an interesting point about the Museum itself. The promotional tagline of the Museum is "Prepare to Believe." While Mark Looy stated the museum was not intended as a pilgrimage site, nor had he ever thought of the museum as one (Interview 2012), the phrase nonetheless links closely with the understanding of pilgrimage. As discussed previously, historical as well as contemporary interpretations of pilgrimage include the idea of prayer and reflection. Historical pilgrims used the journey as a time to prepare for the wonders to come. However the phrase is intended, the inclusion of the verb "prepare" invites the museum goer to experience at least one part of the pilgrimage during his/her journey to the museum.

When taken as a whole, these interview responses and the distances traveled build a compelling case that people traveling to the museum are doing so for the expressed purpose of celebrating their faith outside of a local setting. While the journey may no longer be as arduous as pilgrimages once were, it is imperative to reformulate the understanding of pilgrimage sites. With very few exceptions, all the major pilgrimage sites are now easily accessible. If people are still willing to undertake a religiously motivated journey to a holy place that dislocates them from the local, they are in fact pilgrims. We must alter of perceptions of holy places to match with new understandings, just as scholars have adapted their understandings of space and time in the face of modern changes.

133.4 Conclusion

The popular image of pilgrimage linked with the *Canterbury Tales* must be reimagined to fit our postmodern understanding of religious pilgrimages. As this chapter has demonstrated, the very nature of the criteria used to determine if something is a pilgrimage has changed. Modern transport has fundamentally changed travel by all but rendering distance irrelevant. Travel is far less arduous or time consuming as in the past. In order to understand the journey aspect of a pilgrimage, it is now necessary to conceive of the journey as leaving the local, in a cultural sense of the word. The motivations for undertaking a pilgrimage remain unchanged: the pilgrim must be religiously or spiritually motivated.

The nature of the destination has not changed either, but conceptions of shrines and sacred space must. In an era of manufacturing all manners of religious place it is important to not discount these as pilgrimage sites simply because they do not fit a traditional image of the shrine. As long as a destination is revered by the devout and fulfills the proper criteria presented in this essay, it is indeed a postmodern shrine.

In these ways, the Creation Museum is a clear example of a postmodern pilgrimage destination. It fulfills the purpose of providing a location that allows a complete religious experience, one that celebrates the past, present and future of religion. It affords a space for an internal journey and prepares the visitor for a future evangelistic endeavor. Many of the visitors undertake their trips for the sole purpose of visiting the museum. And as the evidence shows, the museum draws not only from culturally distant places, but physically distant as well. As Chaucer said, "folk long to go on pilgrimage" and for young earth creationists, that longing takes them into a land where Adam and Eve walk one with the dinosaurs.

References

Ambrósio, V. (2007). Sacred pilgrimage and tourism in secular pilgrimage. In R. Raj & N. D. Morpeth (Eds.), *Religious tourism and pilgrimage management: An international perspective* (pp. 78–88). Cambridge: CAB International.

Answers in Genesis. (2012). About. www.answersingenesis.org/about/history Accessed 17 Sept 2012.

Aucourt, R. (1990). Pelerins, Touristes ou tourists religieux. *Espaces, 102*, 19–21.

Barber, R. (1991). *Pilgrimages*. London: The Boydell Press.

Blackwell, R. (2007). Motivations for religious tourism, pilgrimage, festivals and events. In R. Raj & N. D. Morpeeth (Eds.), *Religious tourism and pilgrimage management: An international perspective* (pp. 35–47). Cambridge: CAB International.

Creation Museum. http://creationmuseum.org/. Accessed 17 Sept 2012.

Giuriati, P., & Arzenton, G. (1992). Le tourisme religieux. In Le Tourisme: Le Fait, les Virtualites Chretiennes, Ebauche d'une Pastorale, La Pastorale du Tourisme de la Conference Catholique Candienne, Montreal, Canada.

Griffin, K. A. (2007). The globalization of pilgrimage tourism? Some thoughts from Ireland. In R. Raj & N. D. Morpeth (Eds.), *Religious tourism and pilgrimage management: An international perspective* (pp. 15–34). Cambridge: CAB International.

Guerra, L. (1989). O turismo Religioso no mundo de amanhâ. In *Tourism education for the 21st century*. Lisbon: NP.

Holy Land Experience. (2012). About. www.holylandexperience.com/about/about.html. Accessed 17 Sept 2012.

LiveScience. (2005). Museum claims earth is 6,000 years old. Retrieved August 15, 2012, from http://web.archive.org/web/20071212003726/www.livescience.com/strangenews/ap_050523_creation_museum.html

McKevitt, C. (1991). San Giovannni Rotondo and the shrine of Padre Pio. In J. Eade & M. J. Sallnow (Eds.), *Contesting the sacred: The anthropology of Christian pilgrimage* (pp. 77–97). New York: Routledge.

Morinis, E. A. (1984). *Pilgrimage in the Hindu tradition: A case study of West Bengal*. Delhi: Oxford University Press.

Morris, C. (2001). Introduction. In C. Morris & P. Roberts (Eds.), *Pilgrimage: The English experience from Becket to Bunyan* (pp. 1–11). New York: Cambridge University Press.

Murray, M., & Graham, B. (1997). Exploring the dialectics of route-based tourism: The Camino de Santiago. *Tourism Management, 18*(8), 513–524.

Pavicic, J., Alfirevic, N., & Batarelo, V. J. (2007). The management and marketing of religious sites, pilgrimage and religious events: Challenges for Roman Catholic pilgrimages in Croatia. In R. Raj & N. D. Morpeth (Eds.), *Religious tourism and pilgrimage management: An international perspective* (pp. 48–63). Cambridge: CAB International.

Poira, Y., Butles, R., & Airey, D. (2003). Tourism, religion and religiosity: A holy mess. *Current Issues in Tourism, 6*(4), 340–363.

Powell, D. (2009, August 14). A firebrand visits creation museum. *Minnesota Star Tribune*.
Sheehan, P. (2005, January 17). Onward the new Christian soldier. *Sydney Morning Herald*.
Slevin, P. (2007, May 27). A monument to creation. *The Washington Post*.
Stevenson, J. (2012). Embodying sacred history performing creationism for believers. *TDR: The drama review, 56*(1), 93–113.
Stoddard, R. H. (1997). Defining and classifying pilgrimages. In R. H. Stoddard & A. Morinis (Eds.), *Sacred places, sacred spaces: The geography of pilgrimages* (pp. 41–60). Baton Rouge: Geoscience Publications, Louisiana State University.
Sykes, J. B. (1982). *The concise Oxford dictionary of current English* (7th ed.). Oxford: Clarendon.
Turner, V., & Turner, E. (1978). *Image and pilgrimage in Christian culture: Anthropological perspectives*. New York: Columbia University Press.
Wiederkehr, M. (2001). *Behold your life: A pilgrimage through your memories*. Paris: Ave Maria Press.

Chapter 134
The Religious Exhibition at the Capital Museum in Beijing: What It Tells Us and Does Not Tell Us

Shangyi Zhou and Stanley D. Brunn

134.1 Introduction

One of the emerging research themes for cultural, historical and political geographers in the past decade has been the importance of memory in human landscapes (Forest et al. 2004; Hoelscher and Alderman 2004; Johnson 1995, 2004, 2005; Till 2012). These studies focus on the importance of "the visual" as evident in contemporary and heritage settings, including memorials and monuments (Kearns 1993; Foote 2003; Dwyer 2004; Johnson 1995, 2004, 2006, 2012; Till 2005; Dwyer and Alderman 2008; Alderman and Dwyer 2009), but also museums (Karp et al. 1992; Bennett 1995; Bann 1998; MacDonald 2003; Modlin 2008; Hooper-Greenhill et al. 2009; Kaplan 1994). These studies, in a conceptual and theoretical context, examine several overriding themes, including discourses behind the constructed places, monuments and museums, the state or power issues that are always at play in visual objects and displays, and the actual "visual" nature of the exhibition, festival, or monument, including the place/power interfaces in the architecture of buildings themselves (Foucault 1977; Tagg 1988; Haraway 1989; Bal 1991; Bourdieu et al. 1991; Reynolds 1995; Barthes 1977; Rose 2001: 170–186; Christians and Bracci 2002). Many of these constructed place images, whether a war memorial or public commemorative event or a museum exhibit, would and could bring about controversies, not only on the part of the designer and curator, but also to the viewer who seeks to identify with what is visually displayed. The extant literature on these topics suggests there are numerous additional research studies worthy of scrutiny, including those that look at religion and faith themes in local, national or global contexts.

S. Zhou (✉)
School of Geography, Beijing Normal University, Beijing 100875, China
e-mail: shangyizhou@bnu.edu.cn

S.D. Brunn
Department of Geography, University of Kentucky, Lexington, KY 40506, USA
e-mail: brunn@uky.edu

S.D. Brunn (ed.), *The Changing World Religion Map*,
DOI 10.1007/978-94-017-9376-6_134

No matter country or region, the purpose of a religious exhibition is a propagation of religious thoughts. There might be a large difference between countries having a national religion and having no national religion in regards to a government's policy regarding religious exhibitions. The countries which declare Christianity as the state religion in their constitutions or other include the United Kingdom, Norway, Denmark, Finland and others. The countries which declare Islam as their state religion in their constitutions and laws include Iran, Syria, Saudi Arabia, and many other Arab countries. The countries which define Buddhism as the state religion are mostly in Southeast Asia; they include Thailand and Cambodia and others. To a certain extent, the United States can be regarded as a country characterized by its recessive state religion because of the emphasis on "one nation, under God" which is part of the Pledge of Allegiance that all office-holders take and new citizens acknowledge as part of their swearing in ceremony. The pledge is verbalized by many in public events. In regards to China, some foreign scholars believe that China has no real religions faith (Fellmann et al. 2005: 167). Some Chinese scholars also think that Chinese people lack a sense of true religion. Nevertheless, the majority of Chinese scholars and most people believe that China is a multifaith country (National People's Congress of China 2004). Thus China has no national religion in a country where the governing authorities authorized and opposed any formal religious beliefs of a particular faith. In a country where there is a wide variety of different religions expressions, that the government sponsored and approved a religions exhibition is noteworthy. The exhibition itself will reveal what the government wanted viewers and visitors to see, but also there is another side, what it does not reveal about religion in China.

This paper describes and summarizes the exhibits of religious objects in the Capital Museum. The museum is the largest museum owned by the Municipality of Beijing and is located on Chang'an Street which is so-called "first street" and a political symbolized street in China (Fig. 134.1). The exhibition was entitled "Religious Art and Culture from the Museum of World Religions in Taipei." The exhibition opened 29 December 2012 and closed 10 March 2013. The Capital Museum and the Museum of World Religions in Taipei reached an agreement to include an exhibition of collections from both museums. The year 2011 marked the tenth anniversary of the Museum of World Religions in Taipei. In November of 2011 the first exhibition of Avatamsaka World Wisdom: Buddhist Art from the Capital Museum was held at the Museum of World Religions in Taipei. As an exchange, the exhibition of "Religious Art and Cultural from the Museum of World Religions in Taipei" launched an exhibition at the Capital Museum in Beijing.

The exhibition was held in the best temporary gallery on the first floor of the museum which has four galleries for temporary exhibitions; it had a distinctive title (Fig. 134.2). It consisted of two parts. The first part was called "An Overview of World Religions (Fig. 134.3)." It showed videos and had interactive maps. Some multimedia technologies introduced basic information on the place origins of major

Fig. 134.1 The Capital Museum, Beijing (Photo by Shangyi Zhou)

Fig. 134.2 The *arrow* marks the entrance to the exhibition of Religious Art and Culture from the Museum of World Religions in Taipei (Photo by Shangyi Zhou)

Fig. 134.3 The first part of the exhibition is titled "An Overview of World Religions" (Photo by Shangyi Zhou)

religions. The second part of this first floor exhibition introduced the religions with most believers in the world. They are Christianity, Buddhism, Islam, Judaism and Hinduism. It also included Taoism (the local religion in China); Shintoism comes from Taoism.

134.2 The Major Objectives of the Exhibition

Important questions about the display of any subject are what does a museum represent, what does it include, and what exhibits are suitable, or not suitable, in this case, to help understand regional cultures in China? These are specific questions for the director of the museum and the curator of an exhibition. The curators from Taipei and Beijing considered this exhibition as a medium to promote the cultural exchange of cross-region relations and believed this exhibition will enhance the communication and cultural identify of people living across the strait.

People living on both sides of the strait share a common starting point. Religion emerged, as historians and anthropologists inform us with first dawn light of human civilization. It profoundly affected the development of human societies in the world even in ancient times. It still plays an important role in high tech societies in the contemporary world. Religion has an enormous impact on the human spirit. In tracing human history and also projecting the future of humanity, we are unable to turn a blind eye to religion. Secular culture, including science and technology, cannot answer various important questions about people's life and cannot meet the demands of psychological needs. People have to find or seek to find the answers to these questions from a supernatural power, as well as pinning their hopes on it. People imagine all kinds of gods, create intricate religious interpretation system, and build a grand and magnificent temples. The problem we are facing is not the reasonableness

Fig. 134.4 Flyer for the exhibit on Religion and Culture from the Taipei Museum of World Religions in Beijing, 2013 (Photo by Shangyi Zhou)

of religion existence in our life. It is how we live together peacefully in a varied and complex religious world. The history of the world has documented countless painful lessons that result from religious conflicts. The main cause of many conflicts is arrogance, self-centeredness and thinking other religions are wrong.

The Exhibition Religious Art and Culture provides an example that matches the cultural value of a contemporary modern society and applies it to different religious cultures. This exhibition aims to inform the visitors about the popularity of religious culture and knowledge, widening citizens' in sights, promoting mutual respect and understanding the differences among religions, and increasing harmony and peace. Beijing is a multi-religious city with a long history and rich cultural legacy. The exhibition also tries to show the spirits of this megacity, viz., the inclusiveness of different religions and respect for its religious tradition (see www.capitalmuseum.org.cn/zlxx/content/2012-12/28/content_47260.htm). The exhibition's objective is clearly stated in the last sentence of the brief introduction of the exhibition flyer (Fig. 134.4).

The exhibition curators from both sides of the strait accepted the overall objective of the exhibit and obtained official permission from their respected higher authorities. Their hope was that all of the exhibits about the origin and current status of the religions will lead to harmony and respect for each religious group depicted in the exhibits. This idea comes from the many painful lessons in the history of religions as basically being self-serving events.

The Museum of World Religions in Taipei is a private museum. Its founder is Master Shi Hsin Tao who lives at Lingjiu Peak (or Vulture Peak). His purpose was to set up the museum as a public statement in support of brotherhood and tolerance. Vulture Peak is one of famous Buddhist branches in Taiwan. Its scale is not as large as that in Foguangshan Mont, but it is very influential in Taipei and has many faithful followers. The Museum of World Religions is a foundation museum. It claims it belongs to all faiths. Its founding is inspiring and encouraging interfaith dialogue so that all can work together to create peace and understanding in the world we share (see www.mwr.org.tw/content_en/introduction/origin-concept.aspx). That matches the aim of the Capital Museum's exhibition.

134.3 State Laws and the Principles Governing the Exhibition

An exhibition is a kind of public communication in a public space financed fully by the municipal government, which must be determined by the institutional factors and decisions in a given country. These institutional forms emerge within specific historically and culturally situated contexts (Woodward 2009). China has thousands of years of history and diversified regional cultures. Each exhibition in a cultural museum cannot ignore this basic historical point. Searching for a universal and common value is a trend in the world (Christians and Cooper 2009). Many scholars realized that the mass media have to face audiences and readers from different cultural groups (Christians and Cooper 2009:50). Expressing a common value or a universal value could be a fair way for mass media to convey information, including exhibitions.

In regards to religious exhibitions, the freedom of religious belief is the universal value in most modern countries. It was guaranteed by the *Déclaration des Droits de l'Homme et du Citoyen* in 1789 (see www.syti.net/DDH.html) and confirmed also in the U. S. Constitution. Since then many countries have added similar language about religious freedom into their written constitutions. P. R. China is one of them. In its thousands of years of history, Taoism, Buddhism and Confucianism have interacted and integrated. It is an historical base for the freedom of faiths.

The official state law related to this exhibition is entitled the "Regulations on Religious Affairs." It was approved by the 57th executive meeting of the State Council on July 2, 2004. It was implemented on March 1, 2005. It includes seven aspects: general principles, religious societies, sites for religious activities, religious personnel, religious property, and liability, supplementary articles (The State

Council 2010). This regulation makes reference to the Constitution (of China) and relevant laws as its legal basis. Chapter I lists General Provisions that form the guidelines for the exhibition.

According to Articles 1 and 2 of *Regulations on Religious Affairs*, this exhibition has to show something while not showing something. Firstly, it must protect citizens' freedom of religious belief, the maintenance of religious harmony and social harmony, and standardizing the management of religious affairs. The exhibition does not take a position in support of any specific religious group. It does not compel citizens to believe or not believe in religion. And the specific exhibition does not discriminate against religious citizens or non-religious citizens. It seeks to enhance religious citizens and non-religious citizens, and encourages citizens of different religions to respect each other and live in harmony.

Some of the local governments in China have issued their own local regulations on what could be included in religions exhibitions. For example, religious exhibitions, film and video production of religious sites must be approved by governments in Hangzhou, the capital of Zhejiang Province. The approval criteria by the Ethnic and Religious Affairs Bureau in Hangzhou include the following. Firstly, any exhibitions, films or television shows must be approved by authorities. Secondly, the contents of exhibition, films or TV shows shall not violate state religious laws, policies and religious rituals. Thirdly, permission must be approved by a management committee of religious sites. Fourthly, it does not forbid believers from carrying out everyday religious activities. Fifthly, the local law is not contrary to the law of Cultural Relics Protection. These are based on Article 25 of *Regulations on Religious Affairs* by central government (The State Council 2010).

In regards to the Beijing exhibition, the local officials of the Bureau of Religious Affairs in Beijing supported this exhibition enthusiastically. They went to Taipei to participate in the opening ceremony of *Avatamsaka World Wisdom: Buddhist Art from the Capital Museum*. The representatives were well aware that a religious cultural exchange is an important way to promote cross-strait cultural exchanges. Therefore, the approval of this exhibition with very smooth (Interview record of Prof. Xiao ling Guo). A report of this exhibition is also on the website of the central government (Fig. 134.5).

134.4 Displays of Religions Items

The Beijing curators decided what could be included in an exhibition after consulting with the director of the Capital Museum. One hundred and four relics in this exhibition are fine collections from the Museum of World Religions in Taipei. These relics were collected by the museum or donated by people from many places. The exhibits label the notes to each elegant religious item. Each item included very detailed notes regarding the specific religious item. From these descriptions viewers would obtain some sense about what the curators wished to emphasize about the items' artistic values.

中文简体 | 中文繁体 | 邮箱 | 搜索 本网站搜索 搜索

中华人民共和国中央人民政府

The Central People's Government of the People's Republic of China

www.GOV.cn

网站首页 | 今日中国 | 中国概况 | 法律法规 | 公文公报 | 政务互动 | 政府建设 | 工作动态 | 人事任免 | 新闻发布

当前位置：首页>> 今日中国>> 图片新闻

台北世界宗教博物馆宗教艺术文化展首博开展

中央政府门户网站 www.gov.cn 2012年12月28日 21时49分 来源：新华社

【字体：大 中 小】【E-mail推荐 发送 】 打印本页 关闭窗口

12月28日，观众在首都博物馆观看展览。当日，台北世界宗教博物馆宗教艺术文化展在首都博物馆开展。展览分为两部分，第一部分以影片、互动地图等多媒体技术介绍了宗教的起源、世界宗教的分布等基本信息；第二部分则通过宗教艺术品、文字、多媒体等介绍了佛教、基督教、伊斯兰教、道教等宗教的特点、内涵和核心价值。展览展出

Fig. 134.5 Web news of the exhibit on Religion and Culture from the Taipei Museum of World Religions in Beijing, 2013 (Photo by Shangyi Zhou)

The display of religious art is a form of the cultural expression of religions groups in the world. When searching "religious exhibition" using the Google database, mention the earliest was the *Great Exhibition of Religious Paintings*. This is a manual of an exhibition which had a detailed description of the subject, a Bible story, a history of each major painting, even the figures surrounding Pilate and Jesus (Pennsylvania Academy of the Fine Arts 1843). Exhibiting of religious items seems to be a tradition in Christian societies. There are many books about religious exhibitions. For instance, Amon Carter Museum published a book to introduce an exhibition in 1964 about Christian arts (Kubler 1964). The exhibition in that museum showed Christian churches, sculptures, paintings and other art forms in Central America, South America (for examples of Argentina and Chile). The exhibits

Fig. 134.6 A plate with Arabic calligraphy (Photo by Shangyi Zhou)

mentioned in the book have their origins in 17th-18th-century European colonial religious art. Some are Southwest Spanish prototypes. Those exhibits also represented religious understandings of rustic makers and their religious expression of folk arts. The author of the book implied that a religious art exhibition is undoubtedly a popular artistic geography atlas.

In Mainland China, many exhibitions with religious objects prefer to use the label "religious art." The artistic value of religious exhibits is stressed in the Beijing exhibition while the religious meaning is ignored. Most exhibit descriptions in any sector introduce various art forms, for instance, the description used for an Islamic plate (Fig. 134.6). It tells visitors how beautiful the decorative plants on the edge and the calligraphy in the center are. In fact, the statement provides no reason for idolatry about Islam or Islamic architecture and the images that have provided a basis for Muhammad's religious revolution. He developed a unique God (Allah) in the universe to unify the spiritual world with many gods in Mecca and the surrounding regions, and then he transformed the tribes into a unified state. Eliminating tribal idolatries is a way to promote harmony. There is another example of recent research by the Cambridge Studies in Islamic Civilization group that noted that Islam arose out of conflict with other monotheists whose beliefs and practices were judged to fall short of true monotheism and were, in consequence, attacked polemically as idolatry (Hawting 2006: 150). Emphasizing religious art simply is one effective way to avoid arguments about religious history.

In order not to display any idolatry towards Islam, Muslim artistic talent is more reflected in the art of calligraphy which has a very high cultural value and aesthetic value. In the Koran, it is recorded that Allah taught people writing by pen. Arabic calligraphy is not only used for copying of scriptures, but also used in the decoration of a Muslim's daily life and in mosques. Because of opposing idolatry, exquisite calligraphy, geometric patterns, plants and flowers become important decorative art forms on the plate.

Emphasizing the aesthetic value of this exhibition is an implied exhibition principle in mainland China. It could also be seen in other religious exhibitions in China. Another religious and cultural art exhibition was held in Shanghai Library exhibition hall Oct 12–15, 2012. Comparing the two exhibitions, the Beijing exhibition used fewer words to introduce the art value. It included calligraphy, painting, photography, sculpture, and other. The Shanghai exhibition was jointly organized by the major religious groups in the city: Buddhism, Taoism, Islam, Catholicism, and Protestantism. The exhibition was supported by the Ethnic and Religious Affairs Committee of Shanghai and the Shanghai CPPCC (Ethnic and Religious Committee). One item was a large wood carving which was entitled the Stories of Jesus. It made by Mr. Zhang Wan; long the work is 3.7 m long, 1.64 m wide, engraved with 75 stories from the Bible, and more than 1,000 biblical figures. The wood carving is made by a more than 100 year old camphor wood (lignum cinnamomi camphorae). It has six layers. The artist used the technique of high relief carving on the pavilions, rocks, and spring. The figures are as natural as living. It is a valuable treasure that combines traditional Chinese crafts with biblical themes. It took ten years to complete. Another Shanghai exhibit was about the "King's Edition Bible ". It was published in 1894 which was the 60th birthday of Empress Dowager Cixi. Some of her female followers presented it to her. Considering her age, they designed this Bible with enlarged characters, with a bordered gold font on its cover. It is particularly stylish and is also known as the "King's Edition Bible" (see www.chinatibetnews.com/zongjiao/2012-10/16/content_1079173.htm).

Here is another example of a religious exhibition. The Nanjing Museum held an exhibition "Getting Close to Buddha" in December 2008. It showed public religious relics from Forbidden City. Dozens of large and small Shrine temples in the Forbidden City were filled with Buddhist statues, pagodas, Buddhist sutras, and Offering multipliers. Some of the items displayed were from the museum in the Forbidden City. The most valuable one is the Buddhist sutras which was transcribed by Emperor Qianlong. Another attractive item is a Gebura drum which was made from the skull of an eminent monk. The news report of the Nanjing exhibition emphasized the artistic value of each exhibit. From an artistic point of view, the exhibition shows the beautiful and ornate palace creation associated with Tibetan Buddhism. The court life in the Forbidden City is always mysterious for common people. Besides glory and poetry, mysterious religious items show you the deities in the world with you (see www.njmuseum.com/zh/zl/content/content_1012.html).

The exhibition principle of emphasizing the aesthetic value of religious spirits is not only practiced in China, but also used in the exhibitions outside of China, which are organized by Chinese government. For example, a "China-Singapore 2009 Religious Cultural Exhibition" was held in Suntec City, Singapore from December 15–22, 2009. A delegation of over 200 religious personnel from Mainland China led by Mr. Wang Zuoan, Director of State Administration of

Religious Affairs attended this grand event together with representatives from Hong Kong, Macao and Taiwan. The theme of this cultural exhibition was to "Join Together to Build a Beautiful World: Peace, Friendship, Cooperation and Progress". In this exhibition, the Chinese Christian delegation displayed different versions of Bible from China and artwork related to Christianity as Chinese calligraphy and painting, bamboo carving and paper-cutting. The Chinese Christian delegation also attended inter-religious seminars, took part in local social service programs, Sunday worship and workshops co-organized with United Bible Societies. An ethnic minority praising group took part in the series of Christmas evening activities organized by the National Council of Churches of Singapore at Orchid Road from December 18–25, presenting the best songs and dances to God (see www.ccctspm.org/english/enews/2010/120/10120991.html). This exhibition emphasized religious arts more than religious spirit, because Singapore is a religious freedom country without a state religion.

134.5 Hiding the Doctrine Behind the Exhibits

Hiding the doctrine of religions is a principle of the Beijing exhibition. One of the materials stated:

> The purpose of the exhibition is education and popularization of culture and arts, enhance understand of the truth, goodness, beauty of the world, and the objective existence in the world. We do not refer to the religious policies of both sides cross the strait and do not want to lead religious controversy. All the descriptions of the exhibits are decided by the experts of religious studies in Beijing, especially in Chinese Academy of Social Sciences, and are identified by the Museum of World Religions in Taipei. (Xiaoling Guo)

This exhibition seems to let visitors recall their own religious knowledge from the information displayed in the art works, rather than telling detailed information. For example, a series of paintings of hell in Taoist world shows the cruel punishment to bad persons by the king of the hell. Figure 134.7 is one of the series. The visitors could not understand it from only the short note. Only some Chinese can figure out what are in the picture. It is said that hell has ten temples. This picture is the scene of the third temple on the birthday of the temple lord. On this day, the person who likes to incite conflicts will be punished by Rock Pressing and then sent to the fourth temple. People will get the sense of the painting if she/he knows the Taoist doctrine.

All exhibition items contain descriptions, but the sources of the items are not always provided. It is stated on the web page of the Museum of World Religions in Taipei that each relic in the exhibition has its own story (The flyer of the exhibition). But it is hard to know the real story of the exhibit in this exhibition, such as its designer, its former owner, and its handover process. The information on the website of the Museum of World Religions in Taipei sparks visitors' interests to know the stories behind the exhibits. The tantalized expects of visitors drop down after their web visit. Supposedly, designers would like to leave room interpretation for further.

Fig. 134.7 A Taoism picture in the exhibition (Photo by Shangyi Zhou)

Illustrations with historical stories are very common in museum displays. It is just like the following introduction to the relics of Parthenon on display in the British Museum.

The Parthenon was built as a temple dedicated to the goddess Athena. It was the centre piece of an ambitious building program on the Acropolis of Athens.

Fig. 134.8 The Sefer Torah on parchment (Photo by Shangyi Zhou)

The temple's great size and lavish use of white marble were intended to show off the city's power and wealth at the height of its empire. Room 18 exhibits sculptures that once decorated the outside of the building. The pediments and metopes illustrate episodes from Greek mythology, while the frieze represent the people of contemporary Athens in religious procession (see www.britishmuseum.org/explore/galleries/ancient_greece_and_rome/room_18_greece_parthenon_scu.aspx). The visitors receive no background information to some objects on display in the exhibition. Taking the example of a parchment with Sefer Torah in the sector of Christian arts (Fig. 134.8). The translation for the note in the photo is:

> "Torah" refers to the first five volumes of the "Tanakh" ("Old Testament"). They are Genesis, Exodus, Leviticus, Numbers and Deuteronomy. They are also known as Moses Pentateuch. The full text of the Torah is written on parchment or sheepskin scrolls. Torah is an essential sacred object in synagogues as part of a worship ritual. The text of Torah is without of punctuation. Its vowel letters of the Hebrew are written in cube letters. The text starts from right to left. The 248 pieces of parchment are fixed at both ends in wooden stick which is known as the "Life Tree", and loaded by volume sets, placed in the ark of the synagogue.

The visitors might like to know how this more than 100 year old *Torah* became a collection in The Museum of World Religions and where it came from. It would be a vivid story in the Jewish world.

Comparatively, the following paragraph on the webpage of the Museum of World Religions in Taipei shows us a real Buddhist in Taiwan and what his view is.

If one sees Buddhist monks and nuns spreading their religion or raising funds among the masses in Taiwan, the most usual explanations are that they wish to erect a magnificent and reverent temple where monks and lay people can offer their worship to Buddha or to build a hospital or refugee center. In other words, their desire is to relieve people's physical suffering. Master Hsin Tao does not share this point of view however. What he wishes is to relieve the suffering endured in people's hearts and minds (see www.mwr.org.tw/content_en/introduction/origin-concept.aspx).

134.6 Showing Religious Wedding Rites

The curators of this exhibition have in their wisdom shown how religion plays a role in our lives, even under restricted exhibition principles. This religious art exhibition not only shows the artistic charm of religious articles, but also shows the practice of religion in daily life. It is a really difficult task to show religious rites in this exhibition within the complicated religious situation in mainland China. A subtle issue might pose problems to religious believers, governments, even foreign religious groups. An event which happened on 6 July 2012 may be a reason for no religious rites at the exhibition. It was one of the top ten religious events in 2012[1] in China. Fusheng Yue, the bishop of the Heilongjiang Catholic Parish, held a Consecrated Ceremony. The Vatican declared that Bishop Fusheng Yue had not been nominated by Vatican, so the ceremony was "illegal." On July 11, the spokesman of Chinese Catholic spokesman responded to the Vatican statement. He said that Bishop Fusheng Yue held a successful ceremony. It was not only a great activity for Chinese Catholic society, but also for ecumenical societies. The ceremony received congratulations and supports from these societies. The Vatican's statement was not conducive to the unity of the Chinese Catholic society. China's State Bureau of Religious Affairs spokesman also said that the Vatican's accusations and interference in the normal religious activities in China was is a restriction on freedom and intolerance performance. The last sentence of Article 36 in the Constitution reads "Religious bodies and religious affairs are not subject to any foreign domination." The principle of independence and self organizing has become one of the basic principles of China's religious circles. It is because China had suffered from imperialist aggression and plunder in the past. Catholicism had been controlled by imperialism. The bishops in China should

[1] The remaining nine of the top ten events (Zhou 2012) are: (1) A debate of an official in Kunming (Yunnan Province), who drove a government car to a temple to worship for his own purpose. (2) A Buddhist practice class that cheated people on money. (3) Two false monks who stayed in a hotel with their girl friends. (4) Chinese Buddhist delegation that quit from the 26th World Buddhist Association Conference. (5) Some of the famous religious mountains in China are listed on stock market. (6) A temple in Shenzhen invested in a new hospital. (7) The Chinese Islam Association strongly condemned an American film which smears Islam. (8) Sichuan polices detected that it is Dalai who incited people for self-burning immolation. (9) Cult of the "Almighty God" which rumored "doomsday" for gathering money from people.

enjoy equal theocracy, status and respect to all the bishops of universal church or other religions (Zhou 2012). So the curators did not allow religious people to perform any rites at the exhibition, or to distribute religious leaflets.

The curators of this exhibition selected the wedding ceremony as the point of religions practice in daily life. Like many museums, this exhibition also uses the video. It was one of variety of interesting multimedia presentations. These hi-tech media highlight the "audience experience" atmosphere, guiding the visitors to explore the knowledge, explore themselves. In a variety of video shown in the museum, the most attractive was a series of wedding animations.

Although religious forms in weddings are very different around the world, it is the same people that express their best wishes for the future life of the new couple. That may be the universal point of religious practices in weddings, even in everyday life. The Hindus wedding date is chosen by the priest according to an almanac. The wedding is presided over by a Hindu priest. In some places, the wedding ceremony is sophisticated while it is simple in other places. But three ceremonies are essential in any Hindu regions. The first is a handshake ceremony in which the father of the bride takes his daughter's hand on the hand of the groom, and sprinkles a little water by saying his daughter's life is in the hand of the groom. The second is a sacred thread ceremony. The third is going around seven fires which are on the ground and burned by seven woods with religious meanings. The priest is chanting through every ceremony. A Christian wedding is also presided over by a religious leader. Each step shows God with the couple: the officiating priest is speaking, reading the Bible, praying, offering a poem and singing to the new couple. In Han Chinese' wedding, the first step of newlyweds is to worship God of Heaven and God of Land, and then giving thanks to the parents; the last step is bowing to each other.

134.7 Conclusion

What the exhibition tells us and does not tell us is understandable in China's institutions. The representation of religious culture or arts in a state owned museum is quite different from a private museum. In a multifaith country, the exhibition contents must be consistent with the constitution, which emphasizes the equality of religions. A state owned museum stresses equality of religions charm, the pursuit of religions art, the art behind the day-to-day religious life, and the harmonious coexistence of grassroots religious social practice. All stress the importance of the central government and national identity.

The creative way of the exhibition is giving a space for visitors to think about the function of religions in the world through an animation of the wedding. The role of a museum is to lead visitors 'thinking' and 'doing' (MacDonald 2003). Some scholars discuss the poetics and politics of museum display. There is a shared point that sees the role of museums as intermediaries between the makers of art or artifacts and the eventual viewers (Rockefeller Foundation 1991). The religious art exhibition in the Capital Museum seeks to be an intermediary. Everyone knows that the selection criteria of knowledge, the presentation of ideas and images of an exhibition are enacted within a power system. Whatever this exhibition tells us or does no tells us, its power of representation, which is from the institution or is limited by the institution, that has reproduced the structures of beliefs and faiths in the country. The studies of exhibiting cultures show that the representation power of a museum does not work in the same way for all types of museums (Karp et al. 1992: 2). Emiko Ohnuki Tierney, a native of Japan and a researcher in the National Museum of Ethnology (in Osaka) states that the government of Japan takes the position to unify the diverse peoples in the history, as well as in the national museums (Tierney 1998). This paper serves as one example how the Capital Museum reconstructs the religious world in China as an agent, even within strict regulations. Kaplan (1994) has suggested that museums themselves should be analyzed as important social institutions.

Acknowledgements I am honored to contribute a short chapter to *The Changing World Religion Map* volume. Prof. Stanley D. Brunn not only encouraged me, but also gave me many detailed comments. Thanks also to Prof. Xiaoling Guo, the director of the Capital Museum for his cooperation and interview.

References

Alderman, D. H., & Dwyer, O. J. (2009). Monuments and memorials. In R. Kitchin & N. Thrift (Eds.), *International encyclopedia of human geography* (Vol. 7, pp. 51–58). Oxford: Elsevier.

Bal, M. (1991). *Double exposure: The subject of cultural analysis*. London: Routledge.

Bann, S. (1998). Art history and museums. In M. A. Cheetham, M. A. Holly, & K. Moxey (Eds.), *The subjects of art history: Historical objects in contemporary perspective* (pp. 230–249). Cambridge: Cambridge University Press.

Barthes, R. (1977). *Image-music-text* (Ed. & Trans., S. Heath.). London: Fontana.

Bennett, T. (1995). *The birth of the modern museum: History, theory, politics*. London: Routledge.

Bourdieu, P., Darbel, A., & Schnapper, D. (1991). *The love of art: European art museums and their public*. Cambridge: Polity Press.

Christians, C. G., & Bracci, S. L. (2002). Moral engagement in public life. In C. McCarthy & A. N. Valdivia (Eds.), *Intersections in communications and culture: Global approaches and transdisciplinary perspectives* (Vol. 3, p. p50). Wien: New York/Bern/Berlin/Bruxelles/Frankfurt/Oxford.

Christians. C. G., & Cooper, T. W. (2009) The searching for universals. In L. Wilkins & C. G. Christians (Eds.), *The handbook of mass media ethics* (pp. 55–70). New York: Routledge.

Dwyer, O. J. (2004). Symbolic accretion and communication. *Social and Cultural Geography, 5*, 419–4335.

Dwyer, O. J., & Alderman, D. H. (2008). Memorial landscapes: Analytic questions and metaphors. *GeoJournal, 73*, 165–178.

Fellmann, J. D., Getis, A., & Getis, J. (2005). *Human geography* (p. 167). Boston: McGraw-Hill College.

Foote, K. E. (2003). *Shadowed ground: America's landscapes of violence and tragedy*. Austin: University of Texas Press.

Forest, R., Johnson, J., & Till, K. E. (2004). Post-totalitarian national identity: Public memory in Germany and Russia. *Social and Cultural Geography, 5*, 357–380.

Foucault, M. (1977). *Discipline and punish: The birth of the prison* (A. Sheridan, Trans.). London: Allen Lane.

Haraway, D. (1989). *Primate visions: Gender, race and nature in the World of modern science*. London: Routledge.

Hawting, G. R. (2006). *The Idea of idolatry and the emergence of Islam: From polemic to history* (p. 150). Cambridge: Cambridge University Press.

Hoelscher, S., & Alderman, D. H. (2004). Memory and place: Geographies of a critical relationship. *Social and Cultural Geography, 5*, 347–355.

Hooper-Greenhill, E., Phillips, M., & Woodham, A. (2009). Museums, schools and geographies of cultural value. *Culture Trends, 18*, 149–183.

Johnson, N. C. (1995). Cast in stone: Monuments, geography and nationalism. *Environment and Planning D. Society and Space, 13*, 51–65 (Reprinted in N. Thrift & S. Whatmore (Eds.), *Cultural Geography, critical concepts in the social sciences*. London: Sage, 2004).

Johnson, N. C. (2004). Heritage landscapes: Geographical imaginations of material culture tracing Ulster's past. In T. Mels (Ed.), *Reanimating places: A geography of rhythms* (pp. 227–239). London: Ashgate.

Johnson, N. C. (2005). Memory and heritage. In P. Cloke, P. Crang, & M. Goodwin (Eds.), *Introducing human geographies* (2nd ed., pp. 314–325). London: Edward Arnold.

Johnson, N. C. (2006). Memorializing and marking the Great War: Belfast remembers. In F. Boal & S. Royle (Eds.), *Enduring city: Belfast in the twentieth century* (pp. 207–220). Belfast: Blackstaff.

Johnson, N. C. (2012). The contours of memory in post-conflict societies: Enacting public remembrance of the bomb in Omagh, Northern Ireland. *Cultural Geographies, 19*, 2237–2258.

Kaplan, F. E. S. (1994). *Museums and the making of "ourselves": The role of objects in national identity*. London: Leicester University Press.

Karp, I., Kreamer, C. M., & Levine, S. (1992). *Museums and communities: The politics of public culture* (p. 2). Washington, D.C.: Smithsonian Books.

Kearns, G. (1993). The city as spectacle: Paris and the celebration of the bicentenary of the French revolution. In G. Kearns & C. Philo (Eds.), *Selling places: The city as cultural capital, past and present* (pp. 49–102). London: Pergamon.

Kubler, G. (1964). *Santos: An exhibition of the religious folk art of New Mexico*. Fort Worth: Amon Carter Museum of Western Art.

MacDonald, S. J. (2003). Museums, national, postnational and transcultural identities. *Museum and Society, 1*(1), 1–16.

Modlin, E. A. (2008). Tales told on the tour. Mythic representations of slavery by docents at North Carolina plantation museums. *Southeastern Geographer, 48*, 265–287.

National People's Congress of China. (2004). *Constitution of P.R. China*. Beijing: Law Press.

Pennsylvania Academy of the Fine Arts. (1843). *Great exhibition of religious paintings*. Philadelphia.

Reynolds, A. (1995). Visual culture. In L. Cooke & P. Wollen (Eds.), *Visual display: Culture beyond appearances* (pp. 82–108). Seattle: Bay Press.

Rockefeller Foundation. (1991). *Exhibiting cultures: The poetics and politics of museum display*. Washington, D.C.: Smithsonian Institution Press.

Rose, G. (2001). *Visual methodologies*. Thousand Oaks: Sage.

Tagg, J. (1988). *The burden of representation: Essays on photographies and histories*. London: Macmillan.

The State Council. (2010). *Regulations on religious affairs*. Beijing: Religion and Culture Publishing House.

Tierney, E. O. (1998). A conceptual model for the historical relationship between the self and the internal and external others: Conceptual model for the historical relationship between the self and the internal and external others: The agrarian Japanese, the Ainu, and the special-status people. In D. C. Gladney (Ed.), *Making majorities: Constituting the nation in Japan, Korea, China, Malaysia, Fiji, Turkey, and the United States* (pp. 31–54). Stanford: Stanford University Press.

Till, K. E. (2005). *The New Berlin: Memory, politics. Place*. Minneapolis: University of Minnesota Press.

Till, K. (2012). Wounded cities: Memory-work and a place-based ethics of care. *Political Geography, 31*, 3–14.

Woodward, W. (2009). A philosophically based inquiry into the nature of communicating humans. In L. Wilkins & C. G. Christians (Eds.), *The handbook of mass media ethics* (pp. 3–14). New York: Routledge.

Zhou, F. (2012, December 25). Top ten domestic religions events in 2012. *China EthnicNews.*, p. 7.

Chapter 135
Islam on the Catwalk: Marketing Veiling-Fashion in Turkey

Banu Gökarıksel and Anna J. Secor

135.1 Introduction: "Prayer at a Fashion Show!"

Tekbir Inc.'s fashion show in Istanbul on 30 June 2011 was as scandalous as ever. On the catwalk professional models displayed Tekbir's 2011–2012 winter collection of what we call "veiling-fashion:" striking combinations of headscarves with skirts, pants, and coats in the season's trendy colors (leopard prints, deep red, and neutrals), fabrics, and designs (Fig. 135.1). In attendance were more than 1,000 people, a mix of men, women, and some children. For the first time, male models were on stage with female models showcasing the company's new men's collection. While the women's styles on display were overtly indexed to Islamic modes of modest dress for women, the men's clothes were quite indistinguishable from any other men's fashion in Turkey. The only thing that signified the 'Islamicness' of the men's dress was the stagecraft of the fashion show, and indeed this staging would become the focus of the frenetic media coverage that began the following morning and continued to make waves several days after the event.

Several major newspapers covered the fashion show on their front page, with the headline: "Prayer at a fashion show!" Below this headline was a photo of three male models lined up on the stage, heads turned to the side and hands clasped at mid-body (Fig. 135.2). In the daily newspaper *Milliyet,* the description right below the headline elaborated: "At the fashion show of Tekbir Giyim, a company known for *tesettür* clothing [veiling-fashion], three male models prayed on the catwalk to the sounds of *ilahis* [religious music and hymns]. This is the first time prayer has been

B. Gökarıksel
Department of Geography, University of North Carolina, Chapel Hill, NC 27599, USA
e-mail: banug@email.unc.edu

A.J. Secor (✉)
Department of Geography, University of Kentucky, 817 Patterson Office Tower, Lexington, KY 40506, USA
e-mail: ajseco2@uky.edu

S.D. Brunn (ed.), *The Changing World Religion Map*,
DOI 10.1007/978-94-017-9376-6_135

Fig. 135.1 A scene from Tekbir's 2011-2012 winter collection fashion show, June 30, 2011 (Photo by Banu Gökarıksel and Anna J. Secor, used with permission of Tekbir Giyim, Inc.)

Fig. 135.2 Controversial scene that led to the headline "Prayer at a fashion show!" (Photo by Banu Gökarıksel and Anna J. Secor, used with permission of Tekbir Giyim, Inc.)

instrumentalized at a fashion show in Turkey! The main female model of the fashion show was Ece Gürsel who is famous for her nude poses and vibrant nightlife" (Milliyet, 02 July 2011; see also Söylemez 2011). Describing the controversial scene, the article mentioned the images of knit skullcaps (traditionally worn for prayer), men praying in a mosque, and prayer beads projected on the screen right behind the models. The sound of *ilahis*, the religious images, and the posture of the models all suggested an act of prayer, and particularly the type of prayer that is performed at funerals by men (*cenaze namazı*).

Tekbir's CEO Mustafa Karaduman denied the resemblance of the scene to Islamic prayer, instead suggesting that it was all part of the show (personal communication, July 2011). The fashion show was supposed to be a public spectacle and he had hired professionals to put on the most memorable show ever. In an announcement published on the Tekbir website on 18 July 2011, he explained the scene as an artistic take on an Islamic motif and defended the scene citing both the Kuran and the Hadith (Karaduman 2011). Rejecting the accusation that Tekbir was, once again, "selling Islam," he accused the journalists of actively sensationalizing the event in order to artificially create "news" and sell newspapers.

This was not the first time Tekbir fashion shows had come under scrutiny. In fact, ever since its first fashion show in 1992 (in which Tekbir put a black *çarşaf*, an all-enveloping clothing similar to abaya, on the catwalk), the company has been criticized by secular and conservative Islamic communities alike for its production, marketing, and sale of veiling-fashion. The combination of veiling and fashion seems scandalous to many. Veiling is commonly accepted as an obligation for Muslim women. This obligation is usually explained with reference to three verses of the Kuran (24:30, 24:31, 33:53), yet these verses are quite ambiguous in their references to covering, and exactly what and how to cover is far from agreed upon. While veiling is thus linked to the Islamic moral code, fashion inserts women into the circuits of consumer capitalism and its imperatives of taste and distinction (Bourdieu 1984). Criticisms of veiling-fashion often call into question whether such fashionable styles of clothing and headscarves can be considered Islamically appropriate. Perceived as showy and extravagant, fashion shows featuring "Islamic" themes become the flashpoint of controversy because they seem to contradict standard tenets of Islamic morality, especially those that call for modesty and the avoidance of excess.

Despite the harsh criticisms targeting Tekbir, the company has continued to expand exponentially and taken a leadership position in the growing and diversifying veiling-fashion industry. Today, over 200 firms specialize in the production of veiling fashion. The largest of these companies market their products through fashion shows, glossy catalogues, and advertisements. And they find a vibrant consumer base as women continue to wear veiling-fashion styles despite accusations of hypocrisy and inconsistency. How do companies market their products and employ visual and discursive strategies to seamlessly combine veiling and fashion? In this chapter we examine the emergence and growth of veiling-fashion with a focus on its visual representations in fashion shows and catalogues. This analysis traces the aesthetic and functional expansion of the field of veiling-fashion. We argue that in catalogues and on the catwalk, prominent veiling-fashion companies construct an ideal image of the Muslim woman as cosmopolitan, trendy, beautiful, and of a certain class status. This multiple signification of veiling-fashion moves beyond the Islamic moral code as its only referent and brings into play a plurality of cultural, historical, and geographical references from the Ottoman past to contemporary destinations of global tourism in Europe. Veiling-fashion, thus, shifts the accepted boundaries of both religion and fashion.

135.2 Veiling-Fashion in Turkey

When Tekbir Inc., whose name literally means "God is great," put on its first fashion show in Istanbul in 1992, veiling-fashion could not have been considered an industry yet. Along with a number of other small scale tailors and apparel stores located mostly in conservative neighborhoods of Istanbul, Ankara, Konya, and other large cities of Turkey, Tekbir's CEO Mustafa Karaduman had been sewing special order ankle-length overcoats (*pardesü*) and long skirts since the early 1980s. However, the styles were relatively fixed, and the colors were often muted with preference given to beige (especially for the summer), grey, navy, and black. The scope of production was quite small. Women combined the coats and skirts with oversized headscarves usually draping over the shoulders to cover the bosom. The advertisements of the 1980s also reflected the pre-fashion industry status of veiling. Advertisements mainly appeared in highly specialized Islamic women's magazines and journals and they were generally black and white silhouette drawings of veiled women (rather than photos of models), often with no facial features (Kılıçbay and Binark 2002). Yet, the distinctive look of what would become veiling-fashion a decade later developed from these early styles.

It was also in the 1980s that the headscarf became increasingly politicized, turning into the prime "political issue" intensely debated by the public (Secor 2005) for decades to come. Following a military coup in 1980, there was a sea change in the political, economic, and social landscape of Turkey with the rise of political Islam and Islamic capitalism and the emergence of new conservative Muslim middle classes. Women's dress, particularly the headscarf, had been symbolically important at least since the early decades of the Turkish Republic (established in 1923). The republican policies focused on women's emancipation as crucial to Turkish secularization and modernization, measured visibly by women's un- or non-veiling (Göle 1996). While women were discouraged from veiling, many women, especially those outside of urban centers, continued to cover their hair, and there was no national regulation of the headscarf. In the 1980s the headscarf entered a new era of contention. A "headscarf ban" in the form of dress codes at universities, courts, parliament, and other state spaces was instituted. Although the application of this ban varied (tightening at certain political moments and sites, and loosening at others), it effectively brought the headscarf into the spotlight as a politically coded object subject to state regulation. Particularly controversial was the new style of wearing the headscarf tightly around the face, covering completely the hairline, neck, and shoulders, and with an overcoat that covered the shape of the body. This style – called *türban* by its critics in distinction to non-threatening *başörtüsü* – was worn by many, especially in urban centers, including some young university students who organized mass protests to oppose the headscarf ban. The images of these protests and some key events, such as the "Merve Kavakçı affair" (the elected parliamentarian was prevented from taking oath because of her headscarf and later stripped of her Turkish citizenship for her failure to disclose her dual citizenship as legally required (see Kavakçı Islam 2010 and Shively 2005), contributed to the association of this kind of veiling with political activism aiming to "Islamicize" Turkish society. This association of veiling with political Islam continues to haunt women as they constantly face the accusations of being agitators.

Veiling-fashion became a lucrative and sensationalized industry within this political context. Its growth into a segment of Turkey's highly profitable textiles and apparel sector also benefited from and contributed to a flourishing "Islamic consumptionscape" (Sandıkçı and Ger 2007) composed of a variety of products (from television channels and clothing to spas and resorts) targeting the new conservative Muslim middle classes. The commodification of veiling and its insertion into circuits of fashion also brought about its aesthetic and functional expansion and diversification (Navaro-Yashin 2002). Today, in contrast to the 1980s, veiling-fashion is not limited to large headscarves that drape over the shoulders and loose ankle-length overcoats. These styles can still be found, but they are seen as outmoded and possibly only appropriate for older women. For example, women in our focus groups (conducted in 2009 in Istanbul and Konya)[1] called such styles the "grandma" style. Instead, veiling-fashion since the 1990s refers to a complete outfit marked by the diversification and expansion of clothing items. Although the headscarf remains the essential element of veiling in Turkey, it is by no means outside of fashion. Indeed, the headscarf itself is quite a fashion item: its size, shape, color, design, accessories, change dramatically from one fashion season to another, and according to individual taste, location, occasion, etc. Fashion trends affect the extent of coverage (shoulders, necks, or foreheads) and the ways in which headscarves are tied, fastened with pins or broaches, or affixed with bows. Women often invest considerable amount of money and time, acquiring scarves by dozens. Moreover, not only have previously questionable clothing items (such as pants, which are thought reveal more of the shape of the body than a skirt) and colors (such as bright red, which is thought to draw attention) been incorporated into veiling-fashion assemblages, but the pieces that build it constantly change (Gökarıksel and Secor 2010a).

In 2008, we conducted a survey of firms within the veiling-fashion sub-sector of the apparel industry in Turkey.[2] Our survey population was defined as those companies that were engaged to a significant degree in the design and production (whether for wholesale or retail) of veiling-fashion. Our survey sample consisted of 174 veiling-fashion firms, out of an estimated 200–225 such firms (wholesalers and

[1] In the summer of 2009, we conducted 11 focus groups with women consumers of Islamic dress, salespeople, and industry workers in Istanbul and Konya. This research was supported by the National Science Foundation, Geography and Regional Science, Proposal No: 0722825. The title is "Collaborative Research: The veiling-fashion industry: transnational geographies of Islamism, capitalism, and identity." Research for this study was conducted with the assistance of Sosyal Araştırmalar Merkezi (SAM), Levent, Istanbul. Only the authors are responsible for the content of this article.

[2] The survey was also part of the NSF project cited above.

retailers) operating in Turkey at that time. These firms were identified through multiple channels, including the membership lists of textile associations, industry fairs, advertisements, and networking. The survey took place in the offices of the companies mostly with CEOs, company partners, or senior managers. The survey included questions about production, financing, marketing and sales.

The results of our survey paint a picture of a sub-sector that in many ways reflects broader trends within the Turkish apparel industry,[3] but at the same time can also be seen as part of the relatively recent rise in Islamic consumption (White 1999; Saktanber 2002; Sandıkçı and Ger 2007). The median year in which these companies were founded is 1996, with 80 % of them founded after 1984. Similarly to other apparel companies in Turkey, the firms in our study are mostly small to medium sized operations, employing on average 42 people, with 35 % employing fewer than 10 people and half of all firms reporting 18 or fewer employees. One company stands apart from the others in terms of size, and that is Tekbir, which employs 1,450 people, has 33 branches, and reported earnings of over 20 million YTL in 2008 (over 11 million $US). For comparison, the next largest veiling-fashion firm in terms of earnings (SetrMS) employs 212 people and has 9 branches. As we will demonstrate below, the attention that Tekbir has garnered in national and international media and scholarship (see Navaro-Yashin 2002) is no doubt due in part to the strategies of CEO Karaduman, who has used flashy fashion shows and celebrity-style interviews to position himself, his firm, and his products as both cutting-edge and pious. However, before turning to Tekbir's fashion shows, the next section presents an analysis of print and online catalogues from a range of firms. As this discussion will show, even smaller and less publicly visible companies than Tekbir market their products by navigating the polarities of fashion and piety.

135.3 Marketing Veiling-Fashion

When companies began to market overcoats and large headscarves to urban women in the 1980s, veiling was neither fashionable nor an industry. Yet a few early companies began to advertise in magazines using simple pen-and-ink illustrations that left women's faces undrawn (Kılıçbay and Binark 2002; Sandıkçı and Ger 2007). Today, veiling-fashion companies advertise their products on massive urban billboards, in magazines, on television, in catalogues and on-line (Fig. 135.3). Indeed, the industry has developed its own advertising conventions with recognizable visual tropes, which include how women are posed and the background scenes into which they are inserted. In this section, we will discuss how firms think about their target market and how they present their products in catalogues and promotional videos.

When asked about the characteristics that best describe their customers, veiling-fashion firms paint a picture that weaves together many of the seeming contradictions of the industry. Indeed, firms see their customers as being chic, but modest;

[3] See Gökarıksel and Secor (2010b) for more detailed analysis of the veiling-fashion industry in relation to the Turkish apparel industry more broadly.

Fig. 135.3 Armine store and billboard in Fatih, Istanbul (Photo by Banu Gökarıksel and Anna J. Secor, 2009)

conservative, but desirous of change (Table 135.1). The expectation that customers are likely to be married women fits with the general sense that the market for these clothes is not teenagers but rather anchored more in middle age (Table 135.2). And despite the high elegance of the fashion shows, catalogues, and promotional videos, the products of veiling-fashion are, for the most part, middle class (Table 135.3).

If we understand that veiling-fashion is being marketed to pious, middle class, married women, the aspirational elements of the advertising images are thrown into relief. In the online catalogues of firms such as Miss Yağmur, Tuğba & Venn, Rabia Yalçın, Zühre, and Kayra, models are photographed posing at beachside villas, on

Table 135.1 Characteristics of customers, according to veiling-fashion firm survey (n = 174)

Customer characteristic	Number of firms ranking high or very high	Percentage of firms ranking high or very high
Chic	117	67
Conservative	111	64
Married	111	64
Religious	106	61
Likes change	105	60
Urban	105	60
Modest	104	60
Modern	97	56
Active	81	47
Cultured	80	46
Showy	79	46
Classy	62	36
Professional	58	33

Source: Banu Gökarıksel and Anna J. Secor

Table 135.2 Target age groups of consumers, according to veiling-fashion firm survey (n = 174)

Target age groups	Number of firms	Percentage of firms
12 and under	10	6
13–18	60	35
18–25	156	90
26–40	170	98
41–60	142	82
60 and up	85	49

Source: Banu Gökarıksel and Anna J. Secor

Table 135.3 Socio-economic status of consumers, according to veiling-fashion firm survey (n = 174)

Socio-economic status of customers	Number of firms	Percentage of firms
Low	13	8
Middle	145	83
High	16	9

Source: Banu Gökarıksel and Anna J. Secor

Aegean islands, and in airports, art galleries, malls, and other chic locales.[4] Interior spaces are often white and minimalist, a motif emphasized in SetrMS's 2012 spring/summer catalog themed "White Look," and featuring models in a white room with doves (Fig. 135.4). Models are often posed at the edges of posh homes, standing by

[4] Many of these catalogues are posted on line at the site www.tesettur.gen.tr/. All of the catalogues referenced in this paragraph were available at this site as of June, 2012.

Fig. 135.4 Sample from SetrMS 2012 spring/summer online catalog (Source: www.e-setrms.com)

Fig. 135.5 Sample from Kayra's 2012 spring/summer collection video (Source: YouTube video published on www.tembi.net on Mar 21, 2012 by tesetturdunyasi)

glassy pools, or visiting sea-side resorts. In Kayra's 2012 spring/summer collection, a model is posed wearing reading glasses and looking at an iPad. The Kayra video for the same collection presents a vignette, in which a woman who wears the headscarf and one who does not (but who also wears Kayra styles) are shown having a sea-side vacation, relaxing on a hammock, reading a book, gazing out at the sea, looking at an iPad (Fig. 135.5). Tekbir's 2011 summer catalogue features models posing in recognizable Istanbul locations, such as Arnavutköy, Ortaköy, and Taksim (with a blurred Hilton in the background). In short, like fashion advertisement more broadly, these catalogues aim to associate their products with a certain lifestyle, one that is characterized by leisure and elegance. For the average older, married, middle class consumer, this lifestyle is quite likely to be part of the aspirational appeal of veiling-fashion.

The marketing of veiling-fashion works across the polarities of Islam and fashion, transforming them both in the process. On the one hand, some of the visual tropes of advertising images are clearly influenced by Islamic ideals. While the models in Tekbir's fashion shows do strike the usual provocative poses at the end of the catwalk, the catalogue images are generally more conservative in this regard. For example, the prim poses of the models – standing in the middle of the scene, clutching a handbag, staring off to the side rather than directly at the camera – are consistent with ideals of modest female comportment. Catalogue models do not wear colored nail polish or bright lipstick, and they rarely sit or lounge (with some exceptions, notably Kayra). On the other hand, the scenes within which models pose almost never reference religion. These scenes are, as discussed above, indexed to notions of cosmopolitanism, leisure, and modernity. Interesting, one shot of a model in the courtyard of Istanbul's famous Süleymaniye mosque does feature in amongst all of the malls, airports, and art galleries in the Miss Yağmur summer collection, but probably signifying tourism rather than prayer. In short, veiling-fashion is marketed in such a way that both accommodates Islamic ideals and positions the styles within a materialistic, aspirational lifestyle of travel, leisure, and affluence. This is the chic-and-modest, conservative-and-trendy market niche of veiling-fashion.

135.4 Islam on the Catwalk: Veiling Fashion Shows

Just as veiling-fashion catalogues do much more than merely display clothes, fashion shows also work to associate the showcased products with a certain lifestyle and identity: one that is chic, pious, modern, and upwardly socially mobile. Fashion shows are carefully choreographed events that *stage* the clothes to create an image for the company and the brand name and an aura around the new season's styles. In fact, Tekbir's CEO Mustafa Karaduman considers the fashion shows it has organized almost every year since 1992 very important for the rise of his firm and the establishment of veiling-fashion as an industry. He often describes the attention his company received in Turkey and abroad because of its fashion shows. Much of the media coverage of Tekbir fashion shows was (and continues to be) sensational, and criticisms abound of Tekbir's strategic use of Islamic symbolism and references. For example, in a widely cited commentary, M. Şevket Eygi of the conservative daily *Milli Gazete*, called Tekbir's fashion show: "… the greatest disgrace of over 1,400 years of Islamic history!" (Eygi 2008). Eygi's critique targeted the use of Islam in the fashion show and what he saw as the violation of the "sacredness" of veiling. Yet, for Karaduman, organizing a Tekbir fashion show year after year was essential to the formation of the Tekbir brand name. It was through these fashion shows that Tekbir came to be recognized as *the* brand of veiling-fashion in Turkey; today it is recognized as such even among those women who never shop there. Several other large veiling-fashion companies (such as SetrMS based in Ankara) organize fashion shows, but they rarely get the media attention that Tekbir shows receive. Furthermore, not all companies can make the investment in resources required for a fashion show. For example, we asked the up and coming company Armine's representatives whether they had plans to organize a fashion show in 2009. The chief designer and the marketing director pointed out that fashion shows are artistic events and that the best ones exhibit the stylistic and thematic coordination of clothes with the show. They did not consider Armine ready to take on such an endeavor, at least not yet.

The question of Islamic ethics also seems to enter a company's decision making about putting together a fashion show that will exhibit women's bodies. For several companies, even catalogues seem to pose a challenge (see also Jones 2010 and Lewis 2010). For example, Selam's catalogues do not show the models' faces, in compliance with the generally known Islamic prohibition on realistic depictions of human beings (Gökarıksel and Secor 2010a). The strategy of cropping out faces from photos works in a print catalogue, but would not work in a fashion show. Even though many companies do not understand the Islamic code as strictly as Selam and do publish models' faces, fashion shows still bring to the fore a series of questions about the combination of veiling and fashion. We turn to Tekbir fashion shows to examine some of these questions and how they have shifted since the 1990s.

Tekbir's fashion shows stage the company's collection just like any other fashion show would. They use lighting, music, décor, and other visual and sound effects to create an ambience. Tekbir employs professional models because the company seeks to locate itself within the established norms of fashion and to associate its products with the accepted aesthetics of contemporary fashion. Professional models know how to perform feminine beauty on stage. In Tekbir's fashion shows the models walk down the catwalk with hips swaying, strike hip-jutting poses, stare sultrily at the cameras, and sometimes act out mise en scenes (Fig. 135.6). Thus, Tekbir fashion shows are spectacles that emphasize visuality, aim to capture and animate the gaze of the spectators and the public at large, and participate in the production of a Tekbir aesthetic and distinction. Yet, Tekbir fashion shows are different from mainstream fashion shows in several ways, and each of these differences becomes a point of controversy.

First, Tekbir fashion shows are marked by the absence of exposed skin and hair; Tekbir styles completely cover the body and hair, although they may be quite slim-fitting and reveal the shape of the body. None of the professional models hired by Tekbir ordinarily practices veiling, and most are known for daring poses that show a lot of skin. Yet, in Tekbir fashion shows, they appear on stage in headscarves with no skin showing. However, the way they walk and strike poses on the catwalk is the same as in any other fashion show. This display of Tekbir's "Islamic" clothing on the bodies of women who are not known for their modesty and piety and whose postures practically define 'sexy' becomes a controversial issue debated by the audience and in the media. While some of the women in our focus groups see no

Fig. 135.6 A model showing off the generous sleeves of her dress (Photo by Banu Gökarıksel and Anna J. Secor, 2011, used with permission of Tekbir Giyim, Inc)

problem with the hiring of professional models, others in the same focus groups see a rupture emerging between the veil and the veiled body, between what appears on the surface and what is underneath. The news coverage of veiling fashion shows insistently writes about this rupture, often publishing models' photos from other fashion shows (in bikinis, etc.) right next to the ones from the show. The same point is strikingly made in one scene of the feature film *Büşra* (released in 2010) when the stage collapses at a veiling-fashion show to expose the naked bodies of the models behind it. This juxtaposition of veiling and nudity highlights the perceived hypocrisy implicit in veiling-fashion and becomes a sticking criticism of Tekbir, as well as of the women who wear veiling-fashion.

Second, unlike the catalogue images discussed above, Tekbir fashion shows make explicit references to Islam. Indeed, Tekbir as a company strives to present itself and its products as "Islamic." This Islamic self-identification is prominent in the company's name, logo, and in the many interviews Karaduman gives (Gökarıksel and Secor 2010b). The public relations manager of Tekbir similarly emphasizes that Tekbir customers trust the company to know what is Islamically appropriate and to guide them in both Islam and fashion. Fashion shows become crucial for the company to actively produce its identity and products as simultaneously Islamic and cutting edge. The Islamicness of the Tekbir brand name and fashions is created not only by the styles on display, but also in the many other elements that come into play during the fashion show. The music often consists of the latest high beat acoustic techno sounds interspersed with Islamic- and Middle Eastern-sounding tunes. The soundscape of Tekbir fashion shows is dominated by artists like Mercan Dede, whose music fuses Sufi-inspired, traditional rhythms with electronic sounds by intermixing a traditional wind instrument, the *ney*, with computer-generated beats.

Islamicness is also visually constructed at these shows. Images projected onto the screen placed on the back of the stage make Islamic references, as in the controversial images of prayers beads, prayer skull caps, and men praying in the 2011 fashion show. The show itself occasionally includes scenes in which models' gestures suggest praying. As discussed in the introduction, 2011 became controversial because of the male models' suggestion of prayer on stage. However, just a few years ago female models had kneeled on stage, lifted their hands up, and gazed as if praying as well (*Milliyet* 2008). Tekbir's 2007 fashion show entitled "Sultans' Luminosity" started with a performance by the Sufi whirling dervishes. The host of the event was Yaşar Alptekin, a famous actor/model who ended this career to embrace a pious lifestyle and has become a spokesperson for an Islamic way of life since then. Tekbir aims to construct an Islamic identity for itself and its products through these choreographed performances on stage. In addition, Tekbir makes sure there is a separate space within the building for prayer. The early media coverage of Tekbir fashion shows pointed out the presence of these make shift *mesjid*s and the perceived irony in praying first and then participating in the fashion show's this-worldliness, in its visual economy of desire, commodities, and consumerism afterwards. While the practice of ritual prayers ideally orients one to the spiritual path to piety, the fashion show draws the person into the material world, stoking the very desires that one has to learn to control and rid oneself of to become more pious.

Tekbir's signification of Islamicness also shifted since the 1990s. The diamond studded *çarşaf* that symbolized a newly pronounced Islamic sensibility in Tekbir's early fashion shows rarely has made an appearance again on the catwalk. Instead, there has been an expansion of the scope of veiling-fashion, as well as a diversification of its styles. Tekbir's products have expanded over the years to include specialty clothing such as maternity-wear and active-wear. In 2004 models took to the stage with tennis rackets in hand to present Tekbir's new collection of veiled active-wear. The stylistic diversification of Tekbir's veiling-fashion included a move away from traditional decorative elements such as embroidery to include more innovative and constantly shifting design details that focus on patterned stitching, buttons, ruffles, and the like. Today it is common to see more daring veiling-fashion styles. Tekbir also seems to have shifted its own approach to what is Islamically appropriate at its fashion shows. For example, following the Islamic code of modesty about not mixing unrelated men and women, the audience was gender segregated at Tekbir's first fashion shows. In 2004, this practice was abandoned and men and women started to sit together during Tekbir fashion shows (*Tercüman* 2004; Tezel 2004). The company defended this shift by simply stating that there was nothing strange about men and women sitting together. In 2007 Tekbir hired the German fashion designer Heidi Beck to design part of its collection. The irony of having a Christian designer produce "Islamic" clothing was highlighted in the media coverage of the event (*Milliyet* 2008). Beck's collection referred back to Ottoman opulence and brought designs inspired by the Ottoman elite onto the catwalk as suggested by the title of the show, "Sultans' Luminosity."

This analysis of Tekbir fashion shows reveals how this company, the leader of the veiling-fashion industry in Turkey, carefully orchestrates the image of veiling-fashion as Islamic and modest but also chic and up to date. In trying to present veiling and fashion seamlessly coming together under the Tekbir brand name, Tekbir fashion shows mobilize multiple references and strategies. Aiming to associate the Tekbir brand name unmistakably with Islamic piety, the fashion shows use music, images, and performances on stage that explicitly reference Islam. At the same time, these Islamic references are mediated through advanced technology and are often juxtaposed to or intermixed with distinctively cutting edge sounds, visuals, or designs. The Ottoman past is often invoked to claim an Islamic cultural and religious heritage that is also glamorous, signaling upward social mobility. In addition to historical and geographical references, Tekbir fashion shows bring tennis rackets to the catwalk to establish Tekbir as the brand name of the newly rising Islamic elite. On the bodies of professional models, veiling-fashion also becomes linked to mainstream conceptions of beauty and sexiness and enable the production of veiling-fashion as aesthetically distinct but still within the world of fashion. Through these strategies, veiling-fashion shows help redefine both veiling and fashion.

135.5 Conclusion

The first veiling fashion show in Turkey was over three decades ago. However, the juxtaposition of Islam (as the moral referent of veiling) and the catwalk (signifying everything this worldly, including consumerism, class, taste, and beauty) continues to stir up fiery debates about morality, Islamic piety, consumerism, and gender. This chapter has focused on the flashpoints of controversy by analyzing the representation of veiling-fashion in catalogues and in fashion shows. This analysis also has documented the striking aesthetic and functional diversification and expansion of veiling-fashion in Turkey. The ideal image of the Muslim women produced in catalogues and on the catwalk is cosmopolitan, trendy, visible, and beautiful. These images are aspirational, insofar as veiling-fashion is primarily marketed to middle-class women. In the marketing of veiling-fashion, images play with a range of desires – from modesty to affluence, from fashion to piety. The marketing of veiling-fashion in catalogues and on the catwalk thus at once maintains the Islamicness of the dress *and* reinvents it in relation to a multiplicity of cultural references. In this way, veiling-fashion works to transform both practices of veiling and the boundaries of fashion.

References

Bourdieu, P. (1984). *Distinction: A social critique of the judgment of taste*. Cambridge, MA: Harvard University Press.

Eygi, M. Ş. (2008, 25 April). Evlere şenlik tesettür defilesi (A puzzling veiling fashion show). *Milli Gazete*. www.milligazete.com.tr/makale/evlere-senlik-tesettur-defilesi-102085.htm. Accessed 27 Aug 2012.

Gökarıksel, B., & Secor, A. (2010a). Between fashion and *tesettür*: Marketing and consuming women's Islamic dress. *Journal of Middle East Women's Studies, 66*(3), 118–148.

Gökarıksel, B., & Secor, A. (2010b). Islamic-ness in the life of a commodity: Veiling-fashion in Turkey. *Transactions of the Institute of British Geographers, 35*(3), 313–333.

Göle, N. (1996). *The forbidden modern: Civilization and veiling*. Ann Arbor: University of Michigan Press.

Jones, C. (2010). Images of desire: Creating virtue and value in an Indonesian Islamic lifestyle magazine. *Journal of Middle East Women's Studies, 6*(3), 91–117.

Karaduman, M. (2011). Important announcement. www.tekbir.com.tr/basindadetay.php?id=2. Accessed 27 Aug 2012.

Kavakçı Islam, M. (2010). *Headscarf politics in Turkey: A postcolonial reading*. New York: Palgrave Macmillan.

Kılıçbay, B., & Binark, M. (2002). Consumer culture, Islam and the politics of lifestyle: Fashion for veiling in contemporary Turkey. *European Journal of Communication, 17*, 495–511.

Lewis, R. (2010). Marketing Muslim lifestyle: A new media genre. *Journal of Middle East Women's Studies, 6*(3), 58–90.

Milliyet. (2008). Önce Namaz, Sonra Defile (First Prayer, Then the Catwalk). 21 April.

Milliyet. (2011). Defilede Namaz! (Prayer at a fashion show!) 02 July http://gundem.milliyet.com.tr/defilede-namaz-/gundem/gundemdetay/02.07.2011/1409425/default.htm. Accessed 18 June 2012.

Navaro-Yashin, Y. (2002). *Faces of the state: Secularism and public life in Turkey*. Princeton: Princeton University Press.

Saktanber, A. (2002). *Living Islam*. London: I.B. Tauris.

Sandıkçı, Ö., & Ger, G. (2007). Constructing and representing the Islamic consumer in Turkey. *Fashion Theory, 11*, 189–210.

Secor, A. (2005). Islamism, democracy and the political production of the headscarf issue in Turkey. In G. Falah & C. Nagel (Eds.), *Geographies of Muslim women: Gender, religion and space* (pp. 203–225). New York: Guilford Press.

Shively, K. (2005). Religious bodies and the secular state: The Merve Kavakçı affair. *Journal of Middle East Women's Studies, 1*(3), 46–72.

Söylemez, E. (2011, 2 July). Defilede Namaz! (Prayer at a fashion show!), *Posta*, p. 1.

Tercüman. (2004, 18 May). Harem-Selam Karıştı (Women's and men's spaces mixed up). p. 1..

Tezel, C. (2004, 18 May). Tesettür Defilesinde Gelinlik Krizi (Crisis about wedding dress at a veiling-fashion show). *Hürriyet*, p. 1.

White, J. (1999). Islamic chic. In C. Keyder (Ed.), *Istanbul: Between the global and the local* (pp. 77–91). Lanham: Rowman & Littlefield.

Chapter 136
Fashion, Shame and Pride: Constructing the Modest Fashion Industry in Three Faiths

Reina Lewis

136.1 Introduction

This chapter was initially prompted by a request to contribute something on the topic of modest dressing to a special issue on shame for the European fashion journal *Vestoj*.[1] The piece you are reading here has been augmented somewhat for its present purpose but retains the focus and tone of its original iteration. I wanted to keep traces of its initial location because the crossover address serves as a reminder of how unusual it is to find modest dressing included in the framework of fashion, even though there have now been several academic publications in this field (many of which are incorporated as references below). Sometimes it feels as if modest dressing is all over the media, but on the news pages, not the fashion pages. Muslim women in headscarves or face veils are constantly used to illustrate security scare stories in the European press about young Muslim men becoming jihadised, whilst more recently pictures of ultra-orthodox Jewish women have been featured in reports of how ultra orthodox Jewish communities in Israel and the U.S. are seeking to regulate the dress of all women in public spaces and transportation through districts in which they seek spatial control. In both types of stories dress that is understood to be religiously defined is presumed to be the opposite of fashion, to be spatially and temporally apart from the fast changing individuating global trends that characterize modernity.

This situation is full of paradox. In Britain the sartorial displays of cool young hijabis continue to remain under the radar in celebrations of British street style whilst around the world marketers are honing in on Islamic branding as the latest emergent market for neoliberal consumer capitalism. This chapter looks at the

[1] See Lewis (2012). I am grateful to the editor for permission to reproduce. See www.vestoj.com.

R. Lewis (✉)
London College of Fashion, University of the Arts, London W1G OBJ, UK
e-mail: reina.lewis@fashion.arts.ac.uk

S.D. Brunn (ed.), *The Changing World Religion Map*,
DOI 10.1007/978-94-017-9376-6_136

development of a new niche market for modest fashion that crosses the three Abrahamic faiths and that works internationally (research was primarily in Britain, Canada and the United States), paying particular attention to the creation of new taste communities that are simultaneously localized and transnational. The stretchy geographies of this research are a natural consequence of its focus on how the internet and digital communications have facilitated the development of new forms of and discussions about modest fashion that mix commerce and commentary. It was interesting to be asked to write about shame in a volume devoted to fashion rather than religion because it un-hooked the presumption that religious dress alone must be motivated by shame, de-exceptionalising religious clothing cultures and placing them within the wider consideration of how shame impacts on fashion as an embodied practice. For this reason, the expanded article below does not seek to provide an overview of literature on religious debates about shame, the body, and dress. Rather, it aims to reposition shame and as one among other considerations that factor into the current explosion of creativity in modest fashion.

136.2 Fashion, Shame and Pride

Shame surrounds practices of modest dressing, not least because it suffuses the minds of those who look at modestly dressed women, even if those women themselves do not incorporate it consciously into their motivation. Distinctive clothing and appearance has long been a core component in the public expression and recognition of religious identity, but dress has also often served to stigmatize (sometimes through sumptuary legislation) minority communities. Why now do we see more and diverse forms of religiously affiliated fashion in territories that are not governed by state regulation of women's dress?

The last four decades have seen an increase in religious revivalism across major world faiths, and clothing is often one the most conspicuous indicators of increased religious participation (whereas eating halal or kosher food, or attending church can more easily remain imperceptible to outside observers). In much of Western Europe and North America, modest dressing presumed to be religiously motivated is most often associated with Muslim women wearing some form of head covering, hijab, or face covering, niqab/burqa. Yet women from Christian and orthodox Jewish communities in Western Europe and North America are also dressing modestly in greater numbers (as is the case elsewhere, such as the increase in the modest fashion market catering to Pentecostal Christians in Brazil for example, outside the scope of this article). The change in modest dress practices is not simply numerical. Modesty is now being achieved and asserted through an overt engagement with fashion.

The modest body that external observers often presume to be an understood within religious communities as an object of shame to be hidden away is ever more spectacularly visible on the streets of the "secularized" west and is seen ever more fashionably styled in Muslim majority states (such as Indonesia or secular Turkey). The image of the sartorially chic Muslim woman appears in print on the pages of

Fig. 136.1 Dresses available from Shabby Apple e-retail web page (Image from www.shabbyapple.com, March 2012, used with permission)

specialist Muslim lifestyle magazines and is highly visible on the internet, seen in myriad forms on modest fashion blogs, discussed in fora, and marketed to on commercial websites serving the rapidly expanding niche market for modest fashion and associated lifestyle services (Fig. 136.1).

The research project *Modest Dressing: faith-based fashion and internet retail*[2] has been looking at dress practices and related commercial activities that accompany and foster the expansion and diversification of modest self-presentation among women from the three Abrahamic faiths.[3] This is not to presume that only women from these religions dress modestly, or that all women in these faith communities are even remotely concerned with dressing modestly. There are enormous variations within each faith. Similarly, there are plenty of women who do not define themselves in religious terms who also dress in ways that might be regarded as modest. Like all dress practices it is impossible to discern from the outside the motivations

[2] Modest Dressing: Faith-based fashion and internet retail is part of the Religion and Society Programme funded by the UK Arts and Humanities Research Board and the Economic and Social Science Research Board. The project is based at the London College of Fashion, and included as co-investigator Emma Tarlo and as postdoctoral researcher Jane Cameron. See: www.fashion.arts.ac.uk/modest-dressing.htm

[3] This research did not focus on the dress cultures of religious groups like the Amish whose forms of clothing (and its production) are intended to signal separation from contemporary fashion cultures and the mainstream society with which they are associated.

behind any woman's chosen clothing. But in cases of modest dressing that are sufficiently distinctive to be visible to the external observer the stakes in marking oneself out in these ways can be high.

The assertive wearing of modest clothing within a framework of fashion that has developed in recent years can be seen as a practice aiming to reclaim the positive associations of dress cultures that have been or often are still used as a form of stigmatisation. Like the 'black is beautiful' slogan of the Civil Rights movement and the Black Power aesthetic that successfully recoded as hip previously disparaged "natural," untreated, Afro hairstyles, or the slogans and public assertions of sexuality associated with all strands of LGBTQ Pride, the creation and parade of, for example, trendy hijabi fashion is also a riposte to mechanisms of shame and shaming. This is the flip side of shame, an affect that is foundational in identity formation. Whilst different cultures might classify as shameful different things, to experience shame, Eve Kosofosky Sedgewick argues, is to have a "a bad feeling attached to what one is: one therefore *is something* in experiencing shame" (Sedgewick 1993).[4] As Butler (1993) demonstrated in relation to gay drag balls in the United States, the socially situated spectacle of drag allows marginalized subjects to counteract the experience of being shamed by revealing the constructedness of the very categories of gender identity in relation to which as gay and transsexual men they are judged to have failed. Like all reversals, the ability to reclassify a stigmatized form of identity or behaviour is likely to be temporary and partial, and can risk further strengthening the stereotype it sets out to undermine. In making this link I am thinking more about how new performances of modesty might intervene in the perceptions of non-participating majority cultures (that form one of the modest dresser's key audiences) and less about how the positioning of modesty as a positive or empowering practice might relate to concepts of shame internal to cultures invested in modesty. The ways in which different religions and religious cultures conceptualize sexuality and seek to control female sexuality in particular often form the backdrop against which women's practices of modest dressing are constructed, with strictures about which parts of the female body are to be seen when and by whom, how the body is to be covered, and how it should be comported. Modest dressers and designers can experience themselves transgressing such conventions not only in what they show of themselves, but in the colors that they wear, the materials that they use (bulky or body hugging), the ways that they picture their designs. Many of the women who engage in contemporary re-fashionings of modest dressing feel themselves to be engaged in a finely tuned process of negotiation and accommodation with existing religious and community practices, often adjusting their style for different locations and audiences, as they seek to assert the validity of their interpretations of the faith against potentially shaming codes of gender containment. However, for cultures that operate on codes of honor and shame (regional rather than religious in origin and span) simply wearing religiously recognizable clothing, like the hijab, can interpellate women into prevailing forms of shame-based

[4] I have previously considered shame in relation to the formation of imperial and postcolonial ethnic and religious identities, Lewis (2004: ch4).

regulation and surveillance (Werbner 2007). This, as is well documented in the case of young Muslim hijabis in Britain (Dwyer 1999) can mean that wearing hijab operates as an alibi for movements and activities outside the home away from community regulated spaces, or can be used by young women to guarantee respectable behaviour from young men within the community. Modest dressers, therefore, have several audiences, within and without their communities, but it is important to remember that for religiously motivated modest dressers the presentation of the body is also, or for some primarily, about the construction and comportment of a spiritually appropriate self.

Anthropologist Saba Mahmood (2005) urges us not to focus only on political or social gains that might be made by assertions of modest dressing, arguing that the spiritual dimensions of modest dressing must be taken into account. For religiously motivated women modesty can be experienced as simultaneously a requirement of their faith, a testament to faith, and a means to facilitate faithfulness. In her fieldwork with Islamic revivalist women in Cairo in the 1970s, Mahmood found that for many women the initial and repeated experience of dressing in hijab was fundamental to the construction and maintenance of the pious self. It is not that the subject is faithful a priori to donning the clothing, which might then merely clad the physical self. Rather, it is through the act of wearing, being seen in, and comporting appropriately the veiled body that the pious disposition is cultivated and exercised.

In other contexts and other faiths, women are electing to form themselves as pious subjects through a discourse of obedience to divine will that finds accommodation with the individuating validation of the fashion system. For Muslims this is seen most especially among a young cohort of modest dressers who are quite categorically choosing their modest ensembles from mainstream fashion rather than from conventional community or "ethnic" clothing. This turn away from tradition is not simply sartorial, nor is it total – in the UK for example, many young hijabis of South Asian descent will still wear traditional clothing for family parties, events, festivals, and to the mosque – but it is ideological. The practices of fashionable modesty among western young Muslim women are often rooted in a revivalist study of the holy texts and an affiliation to postmodern forms of what Olivier Roy (2004) calls "global Islam" that prioritise international bonds with other Muslims over familial or community bonds with countries of parental origin (see also Gole 2011). Many of the young women I spoke to come from families where their mothers and grandmothers did not veil, or did so in a habitual and less stringent manner (the loosely wrapped *dupatta* of a *salvar kameez* ensemble would be a case in point). For these young women their practices of covering are distinct from their mothers in style (generally covering the head and neck/chest more completely and carefully) and in purpose. Whilst women in Muslim majority locations and in previous generations might cover in ways to do with distinctions of class, caste, ethnicity and region as much as gender, revivalist practices often seek to resituate the veil as inherently Islamic, premised on an affiliation to the pure precepts of the transnational *umma* but often played out in very localized family and community relations.

Based on religious study, young hijabis can assert to internal audiences a doctrinal validation for their practices – often using this to challenge previously accepted community norms (be they about dress, movement outside the home, inter-ethnic marriage) as originating in culture not religion. The use of mainstream clothing in the fashioning and presentation of the modest self therefore plays to internal audiences as part of disputation within the religion. Creating alternative modes of religious interpretation and authority, new practices can be validated as more properly in accordance with Islam than inherited parental or local "cultural" practices. Whereas Roy is concerned primary with committed and active Islamists, especially jihadi young men, the practice of validating new behaviors as in keeping with true Islam has trickled down to many of today's under thirties. Growing up with global consumer culture and internet access, the desire to achieve modesty by adapting items from the high street rather than by wearing what is regarded as traditional can lead to new areas of potential conflict and disputation that are not the conflicts that majority observers might expect, using forms of fashion capital to create forms of distinction from other types of religious dressers as much as from the secularized majority (Sandıkcı and Ger 2010).

Yet for the fashion industry as a whole, as still for many commentators, religious dress suffers the same fate as non-western or "ethnic" clothing and is regarded as outside fashion. Fashion and faith are often seen as mutually exclusive by the fashion industry, just as previous generations of religious communities have often regarded fashion and consumer cultures as oppositional to the practice of faith. But now, with younger generations accomplished in the modes and forms of contemporary mass culture and mass fashion, many parents have come to regard modest fashions as a way to keep youngsters within the faith. As Shellie Slade (2010) of Utah-based Mormon brand ModBod explained, her decision to start the company, initially offering long line "shells" or camisoles and T-shirts, was prompted by,

> watching my girls [then aged ten and seven] just struggle with going into the teen, tween years and all of a sudden going from cute little lovely girls' things to very revealing short-cut, low-cut clothing… in our religion [the Mormon Church of the Latter Day Saints] we don't believe in showing certain parts of your body and so… we teach our children from a very young age to dress modestly… I just saw these types of clothing [low rise jeans, cropped tops] coming out from my girls and I'm thinking I just don't want to have to have my girls feel alienated by not having some of these fashion pieces that they can wear … Yes, it was really the Britney Spears era. I mean she'd just come out of Disney and gone crazy…

Jewish companies also often began with the design and distribution of "basics" that could adapt regular products for modest requirements. Recently, across the faiths, companies are bringing out fashion forward clothing that is assertively on-trend. In this, designers and manufacturers share with journalists and bloggers the challenge of pushing the visual language of the field away from its initial deference to literal modes of reading in which fashion and lifestyle campaigns and editorial were likely to face vociferous criticism over any depiction of the female form that conservative viewers might find insufficiently modest (Lewis 2010). Now, newer websites are beginning to feature products in edgily styled shoots confident that

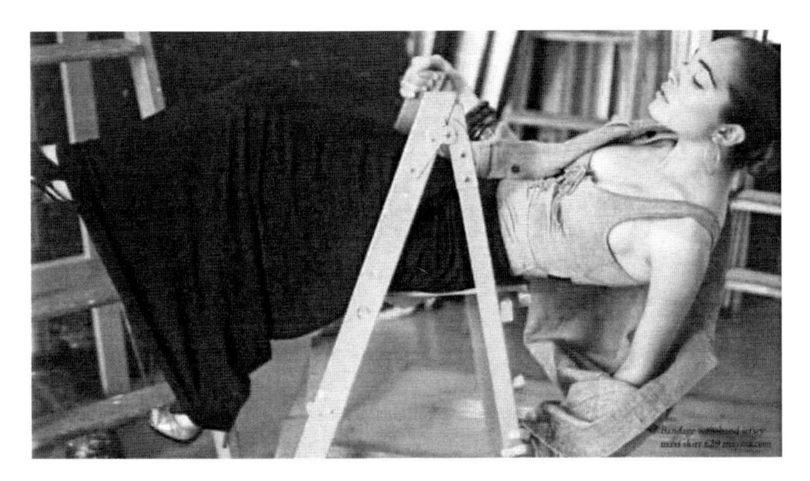

Fig. 136.2 Maysaa digital magazine used this promotional photo when the company web site launched (Image from www.maysaa.com, September 2010, Issue 2, used with permission)

they can address their consumer as sufficiently fashion literate to appreciate the aesthetic and as sufficiently modesty literate to understand how to combine items in situations requiring modesty. British convert to Islam designer Hana Tajima-Simpson (2010) who garnered a strong international following for her blog *Stylecovered.com* before she launched her company Maysaa in 2010, was from the first distinctive for her strong visuals that (like the long skirt styled with only a bustier in Fig. 136.2) did not defer to conservative visual regimes:

> ... obviously we were aware that other people had a really strict policy on, can't show the head, can't show anything... it's kind of ridiculous. [You] have to give the consumers some credit. Just because you show it a certain way doesn't mean that they're going to wear it that way. [T]hese clothes aren't just meant for Muslims, also they're not just meant for when you go out wearing hijab....

For Christian, Muslim, and Jewish companies and bloggers alike, concerns with how to picture the body are constantly negotiated with their changing sense of their various audiences and consumer groups.

When it comes to asserting preferred versions of modesty, and to judging the self-presentation of other women, style comes up against shame time and again. I see in blogs and social media (both "independent" and corporate) a delicate set of manoeuvres as women try to assert their belief in correct forms of covering without appearing to criticize other women or to be controlled by comments of co-religionists. Discussing how their decisions as young adults to wear hijab made them more visible as Muslims to other Muslims, Onjali Bodrul and Nathasha Ali (personal communication in 2007), interns at Muslim lifestyle magazine *emel* in London, noted:

> NA: I get more acknowledgement, you know, you do get a lot more smiles. That's one thing that I noticed when I started wearing hijab, it was like, wow, people know I'm Muslim all of a sudden, it's like I'm part of a bigger community. ... I remember the first time I walked on a bus and a lady turned round to me and said, 'Assalamu Alaikum.' I was like, oh, you know I am Muslim.

But the glance of recognition can also bring judgement:

> NA: They will look twice at you, more so than maybe if you weren't wearing a hijab because they feel like they have the right almost to judge you … You do get more surveillance… I don't really enjoy that when people kind of like give you that second look.
> OB: No, no, because it's the same kind of criticism. Just as non-Muslims are judging you on what you wear, the Muslims are doing it too.

These modes of informal internal community regulation point to an inherent conflict for all the religiously motivated modest dressers I spoke to, in that modesty is understood as both a divine requirement to which a believer should submit and as a personal choice. Similarly, young hijabis come up against a progressivist model that holds that the hijab is just the first step on the path to "full," read correct, covering in the floor length robe, or *jilbab*, and face veil, *niqab* (see Tarlo 2010). Several young women said that once they presented in the hijab there were more likely to be urged to take the "next step." These and other oft-reported similar experiences can be seen as instances of *da'wa*, the obligation to guide other Muslims into virtuous behaviour that, though once the preserve of the ulema and the territory of the mosque, were expanded by Islamic revivalists in the early twentieth century to enjoin all Muslims in the practice of steering others to greater faith. Although many of the women I spoke to expressed great reluctance to judge other women's modest body management, encouragement to increased veiling is in many parts of the world a key plank in revivalist and Islamist political practice (Ahmed 2011).

Within this discourse of choice, modest style leaders struggle to demarcate the boundaries of acceptable modesty without disrespecting the different choices of other women. Modest fashion blogger Jana Kossaibati (2010) of *hijabstyle.co.uk*, one of the early and very British successful Muslim style blogs, invites photos for her reader style features, but treads carefully when she feels compelled to turn down a reader whose understanding of modesty does not cohere with her own (see also Lewis 2013).

One might be tempted to argue that the activity around modest fashion is an attempt to reclaim from shame practices that are intrinsically and irredeemably rooted in the distaste for the female body, and that are, therefore, inherently shaming. This would attribute false consciousness to women – regarding them as only and ever unwittingly contributing to their own oppression. Or one might emphasize the socially liberating potential of modest dressing as a mechanism that allows women (in minority and majority contexts) to exercise greater public autonomy – and there are indeed across the world examples of how wearing hijab acts as an alibi to allow girls and women greater social mobility, able to work or study outside the home because their modest dress acts as a presumed guarantor of their behavior and reputation. This is sometimes the case, but it is never the only effect of hijab. Many of the key proponents of new modes of modest dressing are young women (including converts) with considerable social, economic and cultural capital – who have always had the "freedom" to move about the modern

Fig. 136.3 Jacqueline Nicholls, "The Shame Kittel" (Image from www.jacquelinenicholls.com/shame-kittel.html, April 2013, used with permission)

city, and the globe, as and when they liked (Kariapper 2009). For them the hijab does other work, creating forms of spiritual capital and social distinction. The same is true in the other Abrahamic faiths with modest fashion providing often a starting point for far reaching discussions about gender roles, rather than their reinforcement. Women are engaging in modes of critical evaluation that can be contestatory of as well as conformist to prevailing religious opinion and practice in discussions that operate across as well as within denominational divisions, spanning the boundaries between faiths and between religion and the secular. The online modest fashion sector renders porous the boundaries between commerce and commentary in the development of a network of websites, blogs and social media in which women discuss modest dressed embodiment (Tarlo 2013; Cameron 2013). This operates in conjunction with the growing body of literature and visual art (Fig. 136.3) that covers related themes to create a critical cultural (and commercial) mass that validates as worthy of serious attention the ways in which embodiment can be central to women's articulation of their spiritual and religious selves.

The creation of a religious disposition does not in itself require a fashion system – but in this day and age for women, especially young women, from different religions it is hardly surprising that these exercises of the self are conceptualised through the commodities and visual regimes of global consumer culture. Contemporary global Islam, along with forms of Jewish revivalism, is now understood by many young adherents in terms of personal spiritual fulfilment, a quest for self development that has more in common with Christian confessional traditions and new age spirituality than with the habitual and community embedded practices of their parents and grandparents. Among the under thirties who have grown up with neo-liberal consumer culture this interest in religious practices of self-improvement are developed, expressed, and communicated through consumption practices (Grewal 2005). The attendant development of specialist commodities and services – modest fashion, halal food, Christian rock music – has attracted attention from the mainstream companies and marketers, who increasingly regard faith-based consumers as significant emerging markets (Temporal 2011). The growing interest in Muslim branding and Muslim marketing extends to Islam and Muslims the longer established Christian capitalism of the U.S. – attempts by early department stores to routinise marketing strategies around the Christian festivals were not without criticism at the time (Leach 1980) – and similar techniques are now being used for Muslim consumers. The Jewish market in North America is a more well established demographic, with global and regional brands regularly posting greetings in the minority and mainstream press on high holy days as is characteristic of the strategies of global and regional brands in the U.S., though less so the UK or Europe. The odor of shame hovers over marketing initiatives, however, in the desire not to be too closely associated with the marked, non-universal, minoritized consumer segment that they target. In seeking to reach (constitute) newly recognized faith-based markets, brand specialists operate similar tactics as those deployed in the first round of queer literate adverts in the lifestyle sector that were able in the early 1990s to garner queer consumers (delighted to be recognized in ads at all) through subtle gay coding without making the brand or product too overtly gay and, thereby, risk turning off heterosexuals not wishing to be tainted by a potentially shaming gay association (Clarke 1993).

Staging their public presentation of religious modesty within the frame of global fashion was important to many women that I spoke to. For young Muslims, facing rising Islamophia and constant press coverage linking Islam with terrorism, to be visibly Muslim but visibly fashionable was a way to promote positive understandings of their community (itself another form of *da'wa*). These women are developing new covered fashion that deploy the colors, shapes and styles of global fashion industry in order to present faith-based subjectivities as one among other forms of social distinction achieved by fashionable presentation. As well as their own pleasure in fashion and style, these hijabis hope that by looking like other women on the high street they will move away from stigmatisation and towards social inclusion. For ultra-orthodox Jews, to position oneself outside fashion by wearing clothes that were spectacularly frumpy was to make oneself, and, therefore, one's community, negatively conspicuous to the majority culture. But, it was also ill-advised to render oneself overly conspicuous to co-religionists by being too fashionably dressed. Avant-garde fashionability was likely to read as attention-seeking rather than modest behaviour. Brooklyn based ultra orthodox company H2O Pink Label keeps up with the trends of the global fashion industry, but cannot simply adopt every style that provides sufficient cover: it is just as bad to be conspicuously avant-garde as frumpily old-fashioned as company director Mindy Schlafrig (2010) explains:

> We'll be starting a style sometimes two years after everybody ... we want to look nice and modern and not really old-fashioned. ... we don't want to stick out either way. We like to look nice but we're not going to be the first.

The niceties of getting it right but not too right, and certainly not too wrong, do not only apply to the orthodox or the religious. All women face potentially shaming forms of public scrutiny of their bodies and their dress. But modest dressers who are perceived to be religiously motivated face particular challenges.

The affect of shame is not activated only in those that are the shamed object. One may feel shame also at witnessing, let alone contributing to, the shaming of others. One response to this may be to project blame onto the shamed subject, another to feel motivated to intervene. The mania in Europe at the moment to control through legislation the dress of Muslim women (for summaries see Scott 2007 and Werbner 2007), extended recently to focus on the tiny percentage of the European Muslim population that wears a face veil, activates mechanisms associated with shame. It is clear anecdotally that the experience of having co-religionists shamed is a motivator for many young women to up-veil (and has been the case in Europe since the first attempts in France to prevent school students wearing the hijab in 1989). It is also a motivator for non-Muslims to express solidarity on the grounds of discrimination (and the racism and xenophobia that lies behind these policies), even though for many, especially feminists, it is challenging to support a women's right to dress as she pleases when modest dressing is understood as part of a wider cultural frame that seeks to regulate female sexuality. Apologias that honor codes (not to be equated with religion) and religious doctrines of gender segregating behaviour impinge on men as well as women do not take away from the fact that in practice it is women's bodies and behaviors that are most often more closely regulated. But then, so goes the counter-argument, the same is true for "secular" cultures and societies (Duits and van Zoonen 2006). Muslims who do not believe that Islam mandates a particular, or for some any, form of head and face covering risk being shamed through association, struggling to argue for rights to freedom of religious expression and the multiplicity of Muslim interpretations and practices.

Perhaps shame, and its obverse pride, can be put to use in the formation of new attitudes to the body in society. The development, diversification, and segmentation of a niche market in modest fashion stimulate and demonstrate the emergence of taste communities within faith groups, between faiths, and between the religious and the secular. Awareness of convergence and divergence not only creates a larger, generic modest market for brands, but also fosters social, political, and spiritual dialogue between women and between communities, whose boundaries like the modest styles of the moment can be seen as mutable and mutually created.

136.3 Conclusion

The growing presence of modestly motivated fashion online and on the streets demonstrates that wardrobe decisions can be of world significance when viewed through the appropriate cross-disciplinary frame. Poised at the intersections of commerce and commentary, of faith and fashion, and of the politics and the personal, the issues raised by the design, distribution, and display on the body of contemporary modest fashion immediately bring to mind several areas for future research within and across related fields of intellectual and creative endeavour. The new style trends introduced in this chapter have so far had little serious and aesthetic attention within the frame of fashion studies, this they deserve. For the sociology of religion, the ways in which women use dress and discussions about dress in the formation of religious and spiritual subjectivities provides fruitful avenues for the study of everyday and embodied religion. The (flexible and shifting) mixture of and relationship between the monetised and non-monetised in online business and dialogue for and about modest fashion points to new ways of thinking about entrepreneurship in the creation of religious authority for women and indicates future areas of investigation about the gender specific uses of information and communication technologies. At the same time, the ability of neoliberal capitalism to incorporate (and create) religious consumer demographics cannot simply be celebrated. Yet, the new markets for modest fashion and the commentary that accompanies and supports it also indicate new spatialities that might otherwise be hard to detect. The taste communities and markets that I have indicated in this study can be situated in relation to work on other religions and other secularities in order to map out new constellations of alliance based on concepts of modesty (plural) that not only trace new transnationalisms (mapping relations from periphery to periphery not just between the imperialised centre and its margins), but also connect detailed localisms so that intra-faith disputations can be seen as determining factors as much as relations between faiths and secularities.

References

Ahmed, L. (2011). *A quiet revolution: The veil's resurgence from the Middle East to America*. New Haven: Yale University Press.

Ali, N., & Bodrul, O. (2007, 23 October). Personal interview, London.

Butler, J. (1993). *Gender trouble: Feminism and the subversion of identity*. New York: Routledge.

Cameron, J. (2013). Modest motivations: Religious/secular contestation in the fashion field (pp. 137–157). In R. Lewis (Ed.), Modest fashion: Styling bodies, mediating faith. London: IB Tauris.

Clarke, D. (1993). Commodity lesbianism. In H. Abelove, M. A. Barale, & D. M. Halperin (Eds.), *The lesbian and gay studies reader* (pp. 186–201). London: Routledge.

Duits, L., & van Zoonen, L. (2006). Headscarves and porno-chic: Disciplining girls' bodies in the European multicultural society. *European Journal of Women's Studies, 13*(2), 103–117.

Dwyer, C. (1999). Veiled meanings: Young British Muslim women and the negotiation of differences. *Gender, Place and Culture, 6*, 5–26.

Gole, N. (2011). *Islam in Europe: The lure of fundamentalism and the allure of cosmopolitanism*. Princeton: Markus Wiener.
Grewal, I. (2005). *Transnational America feminisms, diasporas, neoliberalisms*. Durham: Duke University Press.
Kariapper, A. S. (2009). *Walking a tightrope: Women and veiling in the United Kingdom*. London: Women Living Under Muslim Laws.
Kossaibati, J. (2010, October 14). Personal interview, London.
Leach, W. (1980). *True love and perfect union: The feminist reform of sex and society*. New York: Basic Books.
Lewis, R. (2004). *Rethinking orientalism: Women, travel and the Ottoman harem*. London: I.B. Tauris.
Lewis, R. (2010). Marketing Muslim lifestyle: A new media genre. *Journal of Middle East Gender Studies, 6*(3), 58–90.
Lewis, R. (2012). Styling modesty. *Vestoj* (3), 210–224.
Lewis, R. (2013). Establishing reputation, maintaining independence: The modest fashion blogosphere (Chapter 14). In D. Bartlett, S. Cole, & A. Rocamora (Eds.), Fashion media. New York: Bloomsbury.
Mahmood, S. (2005). *Politics of piety: The Islamic revival and the feminist subject*. Princeton: Princeton University Press.
Roy, O. (2004). *Globalised Islam: The search for a new Ummah*. London: Hurst.
Sandıkcı, Ö., & Ger, G. (2010). Veiling in style: How does a stigmatized practice become fashionable? *Journal of Consumer Research, 37*, 15–36.
Schlafrig, M. (2010, July 13). Personal interview, New York.
Scott, J. W. (2007). *The politics of the veil*. Princeton: Princeton University Press.
Sedgewick, E. K. (1993). Queer performativity: Henry James's The art of the Novel. *GLQ, 1*(1), 1–16.
Slade, S. (2010, July 8). Personal interview, Orem, Utah.
Tajima-Simpson, H. (2010, 3 November). Personal interview, London.
Tarlo, E. (2010). *Visibly Muslim: Fashion, politics, faith*. Oxford: Berg.
Tarlo, E. (2013). Meeting through modesty: Jewish-Muslim encounters on the internet. In R. Lewis (Ed.), *Modest fashion: Styling bodies, mediating faith* (pp. 67–90). London: IB Tauris.
Temporal, P. (2011). The future of Islamic branding and marketing. In Ö. Sandıkı & G. Rice (Eds.), *Handbook of Islamic marketing* (pp. 465–483). Cheltenham: Edward Elgar.
Werbner, P. (2007). Veiled interpretations in pure space: Honour, shame and embodied struggles among Muslims in Britain and France. *Theory, Culture and Society, 24*(2), 161–186.

Chapter 137
Putting Christian Congregational Song on the Geographer's Map

Andrew M. McCoy and John D. Witvliet

137.1 Introduction

All over the world each week, hundreds of millions of Christians gather for public worship, in cathedrals and huts, in shopping malls and village greens. Congregational song is a regular and normative feature of this activity, common in almost all settings and locations, regardless of theological, political, economic, or social differences. Singing is widely viewed as both expressive of human experience and formative of human identity (Welch 2005). Christian practice, in turn, is an example of an even larger phenomenon of religious singing practices across a wide range of the world's religions (Hoffman and Walton 1992; Beck 2006).

Worship through congregational song has been studied—with attention to both texts and music, both practices of composition and reception—using a variety of methods, and by practitioners of several academic disciplines including musicology, theology, sociology, ethnography, and social and intellectual history—with relatively minimal engagement with geography. Meanwhile, cultural geographers have cultivated a rich body of studies of geography of music (Carney 1990, 1994, 1998; Byklum 1994; Kong 1995; Nash and Carney 1996; Smith 1997; Krims 2007; Leyshon et al. 1995; Revill 2000; Waterman 2006; Duffy 2009) with little engagement with Christian congregational song. Yet geographers are uniquely positioned to understand key aspects of communal singing, including regional differences, the spatial distribution patterns of singing practices, the interplay of immigration and singing practices, the relationship between congregational singing and spatial

A.M. McCoy (✉)
Center for Ministry Studies, Hope College, Holland, MI 49423, USA
e-mail: mccoya@hope.edu

J.D. Witvliet
Calvin Institute of Christian Worship, Calvin College and Calvin Theological Seminary, Grand Rapids, MI 49546, USA
e-mail: jwitvlie@calvin.edu

S.D. Brunn (ed.), *The Changing World Religion Map*,
DOI 10.1007/978-94-017-9376-6_137

dimensions of human experience and identity. Congregational song, meanwhile, offers geographers access to a potent site for the formation of worldview and identity of millions of people. When Farhadian (2007) and other scholars of sociology and religion call attention to the contemporary rise of Christianity in the global South and East and allude to a "new geography of global worship," their description constitutes a summons to geographers to unpack the significance of the geographical dimensions of worship.

Here, more specifically, are *six* reasons why congregational singing is ripe for geographical study.

137.1.1 Because Congregational Songs Reflect and Shape Perceptions of the Physical World and Geopolitics

Descriptions of the physical world often feature significantly in collections of congregational song. Faith traditions based in the scriptures of the ancient Near East typically affirm time, space and matter as the creative work of the divine. Their songs regularly offer praise or thanksgiving to a divine creator and give thanks for the physical world understood as gift. Examples range from the ancient Essene context of Qumran (Gordley 2008) to the contemporary "God the Sculptor of the Mountains;"

God the sculptor of the mountains,
God the miller of the sand,
God the jeweler of the heavens,
God the potter of the land:
You are spark of all creation
We are formless; shape us now.

Congregational song can also identify those gathered for worship in terms of location. Worshipping communities come to know *who* they are, at least in part, by singing about *where they are* and *where they are not*. Claims over space and location are the focus of many nineteenth century missionary hymns such as "From Greenland's Icy Mountains," which calls for the spread of Christian worship in Western nations throughout the exotic and "heathen" environs of "earth's remotest nation[s]"(Richards 2001: 366–410). Certain collections of congregational songs also have been used to identify one particular group as God's true people located within the world over and against all others. Some hymns also express the church as only finally located completely beyond the world which in its very physicality may not be redeemable or reconcilable in faith practice or belief (Brown 2006). Contemporary scholars regularly acknowledge the problematic dynamics of power and oppression implicit in each example above; "[t]here certainly are enough heathen hymns and battle hymns to justify any charges of cultural imperialism" (Schneider 2004: 82).

Recent movements towards global perspectives on hymnody aim to incorporate critical reflection on various aspects of Western hegemony alongside increasing appreciation for cultural diversity and plurality within the self-understanding of all worship communities and their liturgies. One example, "Siyahamba," a South African freedom song, has become well known *both* as a hymn in many current North American congregations *and* as an expression of solidarity with the new-found freedom and equality for all peoples of that nation. This need not be dismissed as a convenient multiculturalism or "liturgical ethnotourism" (Hawn 2011: 428) and "[c]ongregations should not be allowed to think that showing solidarity with others by singing global songs is the same as 'having fun in the sun in Mexico'" (Marshall 1995: 162 as cited by Hawn 2007: 210). Just as Paul Simon's *Graceland* (1986) was an example of a relatively new commodity of "world music," so, too, Connell and Gibson (2004: 342) argue, "[t]he expansion of world music exemplifies the deterritorialization of cultures and emphasizes how the rise of a particular cultural commodity (world music) is primarily a commercial phenomenon, but could not have occurred without the construction and contestation of discourses of place and otherness." The rise of global worship songs offers a religious counterpart to this phenomenon.

Recent trends in congregational song also locate the church amidst ongoing ecological concerns for the global environment. A series of articles in *The Hymn* (Buley 2009a, b; Wallace 2009a, b, c) examines the "Greening of Hymnody" and musical examples such as "Touch the Earth Lightly," and "Come Join the Cosmic Family." Buley considers how worship might include soundscapes from the natural world, a possibility celebrated by the third verse of the Troeger hymn, "Learn from All the Songs of Earth:"

> The creation God conceives brims with melody and beat.
> From the wind among the leaves to the thunder storm's retreat,
> from the whispering of snow to the water fall that sings—
> psalms and anthems rise and flow from the plainest, simplest things. (Troeger 2009).

Creation theology in hymnody no longer means an uncritical celebration of human dominion over nature, but increasingly includes examples of hymns used by various Christian denominations which confront the ecological impact of modern human societies and cultures (Seeger 1999; Morgan 2009).

137.1.2 Because So Many Practices of Congregational Singing Are So Closely Identified with Geographic Location

Congregational songs are often deeply connected to the locations where they are composed and practiced. Patterns of practice take shape and particular bodies of literature take root, not only along theological or sociological lines, but also in terms of place. Congregational song may become explicitly identified by or linked with a certain region (for example, Appalachian folk hymns), nation (for example, Sibelius' tune FINDLANDIA), or type of geographic location (for example, urban hymns). This identification may happen (1) because of register, voice, and dialect of the texts of song, (2) because of musical aspects of melody, harmony, or rhythm, or (3) because of practices associated with the composition, diffusion, or leadership of song.

Examples abound. In the 1950s, Catholic worshipers in Kentucky sang songs of thanks and blessing about plentiful tobacco crops (Woods 2012), while in Wisconsin Polka Masses evoke associations with nostalgia for Central Europe folk dance. Vaughan William's folk tunes, many taken up into the *English Hymnal* (1904), evoke associations with English countryside landscapes. Some hymns foreground particular landscapes ("There's a Church in the Valley by the Wildwood" or "The City is Alive, O God"). Frequently, church song is adapted from other forms of folk or popular music with strong regional associations (for example, Reggae worship music, Cajun hymns, Celtic hymns, country and western hymns).

These regional associations can be complex. The contrast between two traditions associated with the southern United States—Southern Gospel and the shape note singing related to the Sacred Harp hymnbook—provides an instructive example. Goff (2002: 7) suggests that pervasive interest in Southern gospel transcends its regional orientation because it reflects "something about which many Americans feel deeply," while Harrison (2008: 31) argues that the tradition more precisely demonstrates a "geographical modifier conscripted in service of a stylistic—and cultural—distinction." Understood in this light, "Southern" reflects a particular cultural or religious identity shared beyond the boundaries of the American South, and thus a Southern gospel singer in Massachusetts may in fact claim to find more in common with a fellow singer from Texas (in terms of theology, politics, or social concerns) than, say, with a fellow citizen of New England.

In contrast, pronounced differences persist between practitioners in traditional settings of rural southern communities and the recent renewal of Sacred Harp practice in the more theologically and sociologically progressive environments of northern states. Because northern singers do not share the evangelical worldview so prevalent in southern singers, Marini (2003: 93) observes that the rituals of the music itself provide the only real overlap between the religious experiences of the two groups. Whereas a closer look at Southern gospel music reveals the intersection of a common identity across geographic boundaries, the renewal of Sacred Harp hymnody across the same geographic boundaries reveals regional differences.

Regional associations may also be contentious in terms of power and discrimination. Southern gospel, for example, draws from many different streams of musical practice including African-American spirituals and bluegrass music associated with the rural white south. Yet, the label "Southern" has also been a source of controversy (even within the Gospel Music Association itself) as it regularly functions to differentiate this music and its Caucasian practitioners from the traditions and practice of "black" gospel music (Price 1996). Chapman (2011) calls for practitioners of both black and white gospel music to examine further the historical contexts in which these regional associations are embedded.

Further complexity emerges when religious communities explicitly resist the use of forms or examples of music closely associated with their local context, perhaps due to a strong distinction between what is permissible or advisable in public worship vs. other contexts. Christians in some European contexts—but not in Texas—sing religious lyrics to the tune "The Yellow Rose of Texas." Meanwhile, Christians in Texas may be more likely to sing religious lyrics to the tune "Danny Boy" than do some Irish Christians.

These linkages between songs and places not only affect how people perceive music, but also how they perceive the potential and meaning of specific places or landscapes. Singing the hymntune "Jerusalem" in England or singing Jerusalem temple psalms in puritan New England, not-so-subtly reinforces the theological claim that these respective locations are a place of God's own choosing (Carroll 2011: 235). A hymn like "I Come to the Garden Alone" affirms gardens as places for divine encounter, but might also unwittingly suggest that urban streetspaces are not, a tendency that is similar to, and possibly reinforced, by country and western music's tendency to picture cities as places of exile, isolation, and even godlessness (Wood and Gritzner 1990). While Christian theologians might argue about the relative merits of agrarian or urban images of the kingdom of God, the vast majority of Christians' conceptions are powerfully shaped by the hymns they sing.

137.1.3 Because Hymns Frequently Draw Upon the Elements of the Physical World as Metaphors and Mental Models by Which People Construe Other Realities

The form and language of hymns often draw upon the physical world for the purposes of lyric imagery, simile and metaphor. Through such poetic conventions, mental images of the physical world evoke or become representative of other realities. Hebrew psalms of ancient Israel—sung and adapted throughout the history of the Christian church—feature the pervasive use of mountains and streams, sunlight and darkness, trees and animals as metaphors for describing emotion, interior experience, and perceptions of the sacred (Brown 2002).

These natural images can function in quite different ways. Congregations might sing of mountains as a place of order and protection or disorder and terror. The former theme pervades the final verse of "Where the mountains rise to open skies," a contemporary hymn from the New Zealand indigenous collection *Alleluia Aotearoa*:

> Where mountains rise to open skies
> Your way of peace distil the air,
> Your spirit bind all humankind
> One covenant of life to share!

Alternatively, an early nineteenth century Universalist hymn speaks of "the bowels of the mountains flame," associating the power and might of God with the destructive force of a volcano (Ballou and Turner 1828: 354). Mountains can also be places for public proclamation ("Go Tell It On the Mountain"), isolation ("Out on the Mountain, Sad and Forsaken"), places of mystical encounter and/or political power ("Jesus, Lead Me Up to the Mountain").

Bodies of water, likewise, may provide images of danger, or abundance, or even both simultaneously, such as in the famous Charles Wesley hymn, "Jesus, Lover of my Soul." The first verse reads:

> Jesus, lover of my soul, let me to thy bosom fly,
> While the nearer waters toll, while the tempest still is high.

> Hide me, O my Savior, hide, till the storm of life is past:
> Safe into the heaven guide; O receive my soul at last.

Images of toll, tempest and storm become transformed by "healing" waters in the fourth and final verse:

> Plenteous grace with thee is found, grace to cover all my sin;
> Let the healing streams abound,
> make and keep me pure within.
> Thou of life the fountain art, freely let me take of thee;
> Spring thou up within my heart, rise to all eternity.

The "waters" are thus described as both life-giving and death-dealing, either a simile for the pervasive "love of Jesus" or the source of death. Likewise, rivers can be understood as barriers ("There's a Land Beyond the River"), meeting places ("Shall We Gather at the River"), and icons of redemption ("I've Got Peace Like a River"; "There's a River of Life").

The inherent danger of the sea further functions as a powerful metaphor for the perils of life from which Christ redeems. Mouw (2004: 235) describes "the connection between imagery and the sense of mission" which pervades the "rescue" motif of nautical and maritime hymns, even when they are sung in an urban environment far away from any sea or coastline.

Frequently, melodies, rhythms, and harmonies reinforce these themes either through explicit "word-painting" or through the placement of musical emphasis or affective texture of the music—a good reminder that not just texts, but also the sonic aspects of music, are important for geographers.

137.1.4 Because Patterns of Use and Reception Confirm, Shape and Sometimes Resist Patterns of Immigration, Diffusion, Colonization and Colonialism, and Globalization

Congregational song and associated practices often become central to the identity of immigrant communities, helping them negotiate their sense of displacement. Marini (2003: 210) calls sacred song "a profoundly stabilizing force in Protestant religious culture." Numerous studies examine the role of hymns in the lives of nineteenth century European immigrants to the United States (Schalk 1995; Granquist 1996; Grimminger 2001; Ericson 2004; Leaver 2004; Holter 2008) and the continuing strength of these traditions in places such as Minnesota where Norwegian Lutheran congregations yearly invoke their heritage by singing "Jeg er sa glad hver julekveld" ("I am So Glad Each Christmas Eve"). Others focus on more recent non-European immigrant communities such as Korean (Min 2005) and Latino (Ramírez 2009) Protestants. The latter provides an especially potent context for examining the interface between music and geographical borders. Ramirez (2009: 169) examines the migration of popular Latino Pentecostal hymnody into both mainline Protestantism in the United States and prevailing Catholicism in Mexico, and proposes "the study of American Protestantism must transcend geographical boundaries to include heretofore ignored influences from regions south of the border and at the periphery." Such musical practices can reveal otherwise difficult to observe loci of power at work both within and without the boundaries of incorporated borders.

Geographers might especially note potent metaphorical use of the themes of peregrination and pilgrimage in Christian hymnody, with some similarities with secular road songs already studied by geographers (Krimm 2005). Echoing the ancient psalms of ascent in the Hebrew Bible, many Christian hymns conceptualize life as journey or pilgrimage—all the more so poignant for immigrant people groups. Thus "Guide Me, O Thou Great Jehovah" or "Somos pueblo que camina/We are People on a Journey."

Equally interesting are patterns of geographic diffusion of styles or approaches to singing. Just as geographers have studied the diffusion of blues or rock and roll from their various places or origin (Ford 1971), so too students of hymnody might study how particular patterns of congregational song travel. For example, when North American congregations today sing songs from Africa, is it very likely that their choice of musical idiom is more heavily influenced by Tom Colvin, John Bell, Michael Hawn—European and North American song leaders who have traveled extensively in Africa—than by leading African Christian musicians.

Frequently discussions of immigration and diffusion lead to morally and politically charged discussions of power, money, globalization, and colonialism. On a global scale, the power of European-derived congregational song—through overtly militaristic presentations of such hymns as "Onward Christian Soldiers"—has contributed to Western colonialism throughout many parts of the world (Glover 1990: 441; Doughty 2005; Jagessar and Burns 2008). This critique may not be limited to lyrical or theological content alone; even the tonal nature of much hymnody has been called into post-colonial question (Akrofi 2003). Yet Groody (2007: 216) also describes congregational song as a possible response of resistance to problematic aspects of both colonialism and globalization: "Music is one of the primary ways through which the liturgy is inculturated, and it is often one of the ways in which oppressed groups in particular resist the cultures that dominate them." On this score, Krabill (2008:76) cites Warnock's observations about hymnody in African missionary movements:

> Once committed to the way of life proposed by the missionary, Africans made the necessary adjustments to increase the enjoyment they could derive from the new music. They made conscious and unconscious adjustments of tunes and rhythms. They allowed hymns which they disliked to drop out of circulation. They started their own churches where they hoped to enjoy the benefits of Christianity without the unwanted trappings. In subtle ways the [Africans were] showing that [they were] in control of [their] destiny. The history of African church music is fascinating not so much from the standpoint of what the missionaries did, but how the Africans responded to the missionary impetus. Even when severe restrictions on musical expression were in operation, African Christians were able to filter the music through the African personality to create something new.

Here again, the examination of congregational song practices allows issues of identity and power in colonized spaces to be brought into sharper relief and critique. This critical examination also provides impetus for growing movements of hymnody which reincorporate indigenous instruments and musical practices back into Christian liturgical practice in non-Western contexts. A trenchant example is I-to Loh's focus on Asian music in the collection *Sound the Bamboo* (2011) and his discussion of that collection's namesake hymn written to address Asian issues of migration, marginalization, and power.

> Over four hundred thousand migrants work in Taiwan; they come from the Philippines and other Asian countries. Some of them have taken jobs that formerly would have been held by Taiwanese aboriginal people. This causes resentment and tension between Taiwanese tribes and migrant workers. When I discovered the New Zealand poet Bill Wallace's hymn "Sound a Mystic Bamboo Song" which vividly depicts Christ in Asian ways of life—wearing tribal cloth, living in a squatter's shed, bending while planting rice—I was deeply moved. I decided to use this text to create a contextualized hymn bringing Philippine and Taiwanese musical cultures into unity. My immediate idea was to use non-lexical syllables to establish the link between the Kalinga people from the Northern Philippines and Taiwanese aboriginal people, both of whom are fond of singing in non-lexical syllables. …I created a new context by putting non-lexical phrases from each culture into the same song. Thus, the Kalinga motive begins the first half of the song, and the Taiwanese tribal motive completes the second half. I hoped that Kalinga and Taiwanese singers and listeners would experience a feeling of unity and reconciliation through bringing the familiar and unfamiliar sounds together. (Loh 2005)

Recent hymnody from Western cultures has further responded to the harmful effects of globalism through the theme of confession of sin. Examples include "Children from Your Vast Creation:"

> We have grasped for more possessions,
> wanting things we do not need;
> help us, Lord, lest our obsessions
> soon consume us in our greed.
> Cure our tendency to plunder
> scarring forest, wasting ore.
> Come and turn our schemes asunder;
> take away our lust for more.
> We are learning how much damage
> spreads throughout the world from greed;
> though you made us in your image,
> we are less than you decreed.
> Wanting ease and pleasure strongly,
> craving things your love deplores,
> asking not, or asking wrongly,
> we resort to waging wars.

Thus, Christian congregational song demonstrates both the capacity to promote and to resist colonialist attitudes and perspectives.

137.1.5 Patterns of Use and Reception Confirm Perceptions of What Locations Are "Central" and What Is "Peripheral" in Religious Experience

Just as moviegoers perceive Hollywood to be the center of the film industry, so too those who engage in shaping congregational song have a sense of being oriented elsewhere. This orientation can be reinforced by theology, politics, church polity or the economics of the "worship industry."

Many Hebrew Psalms evoke Mt. Zion, the temple space in Jerusalem, as a place of orientation (Maier 2008; Miller 2010). Later, many traditional Christian worship spaces were built with an East-ward orientation, so that worshipers would pray and sing in the direction of humanity's supposed origin in the garden of Eden or towards Jerusalem as the anticipated place of Christ's second coming (Kieckhefer 2004: 154–155). Much more recently, the influence of Pentecostal "praise and worship" on contemporary congregational song emulates the structure of Israel's temple through a pattern of worship that presents a figurative progression from the outer gate into the inner sanctum of the Holy of Holies (Music 2001: 7).

Whether sung in Jerusalem or Rome, early Christian hymns which laud the lordship of Jesus were sung in the face of Roman authority (Kraybill 2010; Walsh and Keesmaat 2004). In a different way, spirituals associated with nineteenth century American slavery have been used differently over time to describe, in terms of location, the struggle for and movement towards freedom. A song like "If You Don't Go, Don't Hinder Me" proclaims "I'm on my way to Canaan land" both evoking the ancient Israelite exodus from Egypt as well as encoding present intentions to travel north to freedom in Canada following the U.S. ratification of the 1850 Fugitive State Law. A century later, the biblical referent is dropped and the lyric rendered, "I'm on my way to freedom land," for the purpose of direct protest *from within* the context of the U.S. civil rights movement (Reagon 2001: 2).

In congregational song as in other forms of music, musicians and music industry leaders are concentrated in relatively few regional centers (Florida et al. 2010; Scott 1999). This can lead North American congregations to locate their musical practices in relationship to places like Nashville (known for its substantial Christian recording and publishing industry), Detroit or Memphis (cities with long histories of association with blues and black gospel music), to particular megachurch congregations such as Willow Creek near Chicago or Saddleback Church in Southern California (Witvliet 2003). Liturgical practices derived from worship at these locations influences not only congregational song trends but also how participants perceive their local community in a wider national and cultural context (Ellingson 2007: 76). Likewise, in England and throughout the Anglican diaspora there is an ongoing, dynamic relationship between cathedrals and parish churches in which either one may at times be perceived as the center or the periphery of liturgical innovation and excellence (Perham 1995).

137.1.6 Because Hymns Reveal Patterns of Response to Catastrophes

For any geographer interested in the interaction between catastrophes and cultural development, congregational singing practices offer a direct and immediate source of material. Geographers who are engaged with hazards research will be able to explore new data in the form of congregational songs that provide insight into the observations and experiences along the continuum of various natural to technological to social hazards (Montz et al. 2003).

Congregations sing in good times and in bad, and collections of hymns often include localized responses to events perceived as tragic and overwhelming. Hymns do not simply memorialize specific events, but evoke the experiences suffered at the locations of catastrophes. Hymns can reinforce a sense of vulnerability for those who live on faultlines or in the frequent pathway of hurricanes or tornadoes, but also may reinforce resolve to face these recurrent dangers. Frequently at issue are perceptions of God or the divine character arising in relation to environmental events. Just as some psalms of ancient Israel draw parallels between the terror of natural forces and divine power, so also does congregational song, as exemplified by this early nineteenth century Universalist hymn:

How great is our Creator God,
In wisdom, majesty and might;
When he displays his power abroad,
And brings his wonders forth to light.
Behold what cloudy columns rise,
Terrific as the shades of night;
What peals of thunder rend the skies,
The light'ning, how sublimely bright.
How dreadful is the threatening hail,
Th' approaching tempest, O how grand!
What terror doth the mind assail,
When deep convulsions shake the land.
The seas with hollow murmurs groan,
The bowels of the mountain flame;
The elements affrighted own
The awful greatness of thy name.
Almighty God! thy chariot wheels
In solemn pomp and grandeur roll;
Thy presence trembling nature feels,
And humble rev'rence fills our soul. (Ballou and Turner 1828: 353–354)

By contrast, some contemporary hymnody may disavow any direct correlation between God's might and natural disasters in order to emphasize divine care and concern, as in this hymn written in the aftermath of Hurricane Katrina:

Katrina, oh Katrina,
where did you get your pow'r,

when you came busting in here
like some dark vengeance hour?
An act of God they call you,
who last week said God's dead.
No way would my God send you,
He sends His love instead. (B. Prigge 2005)

Other hymnody may incorporate both themes above by including affirmations of divine sovereignty *and* caring concern alongside admissions of human limitations in the context of environmental extremities:

God of nature, at whose voice the waves and winds must now obey,
Give your people words of comfort, acts of grace to share today.
Yours the pow'r of devastation, yours to gather, help and heal;
We know not your ways of wisdom; let your light our paths reveal. (Lawton 2005).

Hymnody also addresses localized experiences at sites of poverty and repression, warfare and terrorism. This tradition is inspired by the gripping and prophetic Psalms of protest and lament in the Hebrew Bible (Witvliet 2007). Philipino hymn writer I-Toh Loh's tune "Smokey Mountain" is named after an infamous garbage pit near Manila, a poignant choice to set Shirley Murray's prophetic Christmas carol "The Hunger Carol," which prophetically challenges the blight and inhumanity of consumerist Christmas celebrations (Loh 2011: 370). Thornburg (2002:18) surveys the literature written in direct response to September 11, 2001, including a tune given the name "PENTAGON" and a separate text titled "See the Newfound Terror," from a New York pastor who lyrically confronts the aftermath of attacks on the World Trade Center: "smell devolving terror/stink and smoke and death." In the immediate aftermath of those events, many congregations in both the U.S. and beyond sung Gillette's hymn, "O God, Our Words Cannot Express" (Gillette 2011). Gillette (notably before 9/11) responded to the circumstances of Afghan refugees with, "God, How Can We Comprehend," a hymn often featured in recent liturgies for Refugee Sunday.

God, how can we comprehend— though we've seen them times before—
Lines of people without end fleeing danger, want, and war?
They seek safety anywhere, hoping for a welcome hand!
Can we know the pain they bear? Help us, Lord, to understand! (Gillette 1999)

In locations of sustained violence congregational song also responds with texts of thanksgiving and praise as forms of protest and defiance. Byasse (2010) points to recent music from the Episcopal Church of Sudan, including this English translation of a Jieng hymn penned by Mary Alueel Garang:

Let us give thanks
Let us give thanks to the Lord in the day of devastation,
and in the day of contentment.
Jesus has bound the world round with the pure light of the
word of his Father
When we beseech the Lord and unite our hearts and have hope,
then the demons have no power
God has not forgotten us
Evil is departing and holiness is advancing,
these are the things that shake the earth.

A closer study of these hymns could provide those who research disasters or catastrophes with ample opportunity to examine the relationship between trauma and human efficacy.

137.2 Conclusion

In sum, Christian congregational singing is ripe for geographical study. We have offered six illustrative trajectories for future exploration, which fit well with typical thematic tracks in geography as a whole (Pattison 1964; Taaffe 1974) as well as with recent work in geography of music (Carney 1998; Nash and Carney 1996; Leyshon et al. 1998; Duffy 2009). Geographers with interests in the nature and society tradition will find in Christian congregational song a rich repository of descriptions of the physical landscape, as well as potent interactions between physical landscapes and cultural and theological metaphors. Geographers who find their work situated in phenomenological-based exposition will discover in Christian congregational song a rich body of material regarding perceptions of nature, geopolitical categories, central and peripheral places, as well as sacred and secular places. Those whose work focuses more specifically on immigration, diffusion, natural catastrophes, power, coercion, colonialism, or environmentalism, among other themes, will find examples that are both potent and formative, both problematic and redemptive.

References[1]

Akrofi, E. (2003, Sept 18–19). *Tonal harmony as colonizing force on the music of South Africa*. Unpublished paper presented to the Swedish South African Research Network (SSARN) seminar on the impact of political systems on African music and music education, Pretoria, South Africa.

Ballou, H., & Turner, E. (1828). *The Universalist hymn-book: A new collection of psalms and hymns for the use of Universalist societies* (4th ed.). Boston: Munroe & Francis.

Beck, G. L. (2006). *Sacred sound: Experiencing music in world religions*. Waterloo: Wilfred Laurier Press.

Brown, W. (2002). *Seeing the psalms: A theology of metaphor*. Louisville: John Knox Press.

Brown, C. G. (2006) Singing pilgrims: Hymn narratives of a pilgrim community's progress from this world to that which is to come, 1830–1890. In M. A. Noll & E. L. Blumhofer. (2002) *Seeing the psalms: A theology of metaphor* (pp. 194–213). Louisville: John Knox Press.

Buley, D. M. (2009a). The greening of hymnody? Touching the earth lightly through congregational song. *The Hymn, 60*(1), 27–28.

Buley, D. M. (2009b). The greening of hymnody, part 3: "Come join the cosmic family.". *The Hymn, 60*(3), 46–47.

Byasse, J. (2010). The muse of church revival in Sudan. Retrieved May 14, 2012, from www.faithandleadership.com/blog/09-22-2010/jason-byassee-the-muse-church-revival-sudan

[1] Unless otherwise noted, references for all hymns mentioned in the chapter may be found at www.hymnary.org (search by hymn title).

Byklum, D. (1994). Geography and music: Making the connection. *Journal of Geography, 93*(6), 274–278.
Carney, G. O. (1990). Geography of music: Inventory and prospect. *Journal of Cultural Geography, 10*(2), 35–48.
Carney, G. O. (1994). *The sounds of people and places: A geography of American folk and popular music*. Lanham, Maryland: Rowman & Littlefield.
Carney, G. O. (1998). Music geography. *Journal of Cultural Geography, 18*(1), 1–10.
Carroll, J. (2011). *Jerusalem, Jerusalem: How the ancient city ignited our modern world*. New York: Houghton Mifflin Harcourt.
Chapman, R. (2011). Worship in black and white: Racial reconciliation happens when we not only sing each other's songs but learn the stories embedded in those songs. *Christianity Today, 55*(3), 26–28.
Connell, J., & Gibson, C. (2004). World music: Deterritorializing place and identity. *Progress in Human Geography, 28*(3), 342–361.
Doughty, S. (2005). Archbishop: "It was a sin to teach the world our hymns." Retrieved April 23, 2012, from www.dailymail.co.uk/news/article-367333/Archbishop-It-sin-teach-world-hymns.html
Duffy, M. (2009). Sound and music. In R. Kitchin & N. Thrift (Eds.), *The international encyclopaedia of human geography* (pp. 230–235). Oxford: Elsevier.
Ellingson, S. (2007). *The megachurch and the mainline: Remaking religious tradition in the twenty-first century*. Chicago: University of Chicago Press.
Ericson, S. E. (2004). The anatomy of immigrant hymnody: Faith communicated in the Swedish Covenant Church. In E. L. Blumhofer & M. A. Noll (Eds.), *Singing the Lord's song in a strange land: Hymnody in the history of North American Protestantism* (pp. 122–139). Tuscaloosa: University of Alabama Press.
Farhadian, C. E. (2007). Introduction: Beyond lambs and logos: Christianity, cultures, and worship worldwide. In C. E. Farhadian (Ed.), *Christian worship worldwide: Expanding horizons, deepening practices* (pp. 1–24). Grand Rapids: Eerdmans.
Florida, R., Mellander, C., & Stolarick, K. (2010). Music scenes to music clusters: The economic geography of music in the US, 1970–2000. *Environment and Planning A, 42*(4), 785–804.
Ford, L. (1971). Geographic factors in the origin, evolution, and diffusion of rock and roll music. *Journal of Geography, 70*(8), 455–464.
Gillette, C. W. (1999). God, how can we comprehend? Retrieved May 14, 2012, from http://gbgm-umc.org/umcor/worship/refugeeshymn.stm
Gillette, C. W. (2011). Hymns for September 11. Retrieved May 14, 2012, from http://sojo.net/blogs/2011/08/15/hymns-september-11
Glover, R. F. (1990). *The hymnal 1982 companion. Vol. 1: Essays on church music*. New York: The Church Hymnal Corp.
Goff, J. R., Jr. (2002). *Close harmony: A history of Southern gospel*. Chapel Hill: University of North Carolina Press.
Gordley, M. E. (2008). Creation imagery in Qumran hymns and prayers. *Journal of Jewish Studies, 59*(2), 252–272.
Granquist, M. A. (1996). American hymns and Swedish immigrants. *Lutheran Quarterly, 20*(4), 409–428.
Grimminger, D. J. (2001). *Das Neu Eingerichtete Gesangbuch* of 1821: A forgotten voice of Ohio Lutheranism. *The Hymn, 52*(1), 7–15.
Groody, D. G. (2007). *Globalization, spirituality, and justice: Navigating the path to peace*. Maryknoll: Orbis Books.
Harrison, D. (2008). Why Southern gospel music matters. *Religion and American Culture, 18*(1), 27–58.
Hawn, C. M. (2007). Praying globally—Pitfalls and possibilities of cross-cultural liturgical appropriation. In C. E. Farhadian (Ed.), *Christian worship worldwide: Expanding horizons, deepening practices* (pp. 205–229). Grand Rapids: Eerdmans.
Hawn, C. M. (2011). The truth shall set you free: Song, struggle and solidarity in South Africa. In J. S. Begbie & S. R. Guthrie (Eds.), *Resonant witness: Conversations between music and theology* (pp. 408–433). Grand Rapids: Eerdmans.

Hoffman, L., & Walton, J. (1992). *Sacred sound and social change*. Notre Dame: University of Notre Dame Press.
Holter, S. W. (2008). Singing with the past: An account of Norwegian hymnody in an American context. *Cross Accent, 16*(3), 26–38.
Jagessar, M. N., & Burns, S. (2008). Hymns old and new: Towards a postcolonial gaze. In S. Burns, N. Slee, & M. N. Jagessar (Eds.), *The edge of God: New liturgical texts and contexts in conversation* (pp. 50–66). London: Epworth.
Kieckhefer, R. (2004). *Theology in stone: Church architecture from Byzantium to Berkeley*. Oxford: Oxford University Press.
Kong, L. (1995). Popular music in geographical analysis. *Progress in Human Geography, 19*(2), 183–198.
Krabill, J. R. (2008). Encounters: What happens to music when people meet. In R. King, J. N. Kidula, J. R. Krabill, & T. A. Oduro (Eds.), *Music in the life of the African church* (pp. 57–80). Waco: Baylor University Press.
Kraybill, J. N. (2010). *Apocalypse and allegiance: Worship, politics, and devotion in the Book of Revelation*. Grand Rapids: Brazos.
Krimm, A. (2005). *Route 66: Iconography of the American highway*. Santa Fe: Center for American Places.
Krims, A. (2007). *Music and urban geography*. New York: Taylor & Francis.
Lawton, S. (2005). Song written in response to Hurricane Katrina: God of nature, at whose voice – Stephen Lawton. Retrieved May 14, 2012, from http://worship.calvin.edu/resources/resource-library/song-written-in-response-to-hurrican-katrina-god-of-nature-at-whose-voic--tephen-lawton
Leaver, R. A. (2004). Hymnody in English and Dutch exile congregations ca. 1552–1561. *Jahrbuch für Liturgik und Hymnologie, 43*, 152–179.
Leyshon, A., Matless, D., & Revill, G. (1995). The place of music. *Transactions of the Institute of British Geographers, 20*(4), 423–433.
Leyshon, A., Matless, D., & Revill, G. (1998). The place of music. *Transactions. Institute of British Geographers, 20*, 423–433.
Loh, I. (2005). Contextualization versus globalization: A glimpse of sounds and symbols in Asian worship. Retrieved May 14, 2012, from www.yale.edu/ism/colloq_journal/vol2/loh4.html
Loh, I. (2011). *Hymnal companion to Sound the Bamboo: Asian hymns in their cultural and liturgical contexts*. Chicago: GIA Publications.
Maier, C. M. (2008). *Daughter Zion, Mother Zion: Gender, space and the sacred in ancient Israel* (p. 2008). Minneapolis: Fortress Press.
Marini, S. A. (2003). *Sacred song in America: Religion, music, and public culture*. Urbana: University of Illinois Press.
Marshall, M. F. (1995). *Common hymn sense*. Chicago: GIA Publications.
Miller, R. D. (2010). The Zion hymns as instruments of power. *Ancient Near Eastern Studies, 47*, 218–240.
Min, P. G. (2005). Religion and the maintenance of ethnicity among immigrants: A comparison between Indian Hindus and Korean Protestants. In K. I. Leonard, A. Stepick, M. A. Vasquez, & J. Holdaway (Eds.), *Immigrant faiths: Transforming religious life in America* (pp. 99–122). Walnut Creek: AltaMira Press.
Montz, B., Cross, J., & Cutter, S. (2003). Hazards. In G. Gaile & C. Willmott (Eds.), *Geography in America at the dawn of the 21st century* (pp. 481–491). New York: Oxford University Press.
Morgan, M. (2009). Caring for creation: Hymns for worship in a challenged world. *Call to Worship, 42*(4), 16–21.
Mouw, R. J. (2004). "Some poor sailor, tempest tossed": Nautical rescue themes in evangelical hymnody. In R. J. Mouw & M. A. Noll (Eds.), *Wonderful words of life: Hymns in American protestant history and theology* (pp. 234–250). Grand Rapids: Eerdmans.
Music, D. W. (2001). *Christian hymnody in twentieth-century Britain and America: An annotated bibliography*. Westport: Greenwood Press.
Nash, P., & Carney, G. (1996). The seven themes of music geography. *Canadian Geographer* (Le Géographe canadien), *40*(1): 69–74.

Pattison, W. D. (1964). The four traditions of geography. *Journal of Geography, 43*, 211–216.
Perham, M. (1995). Liturgical laboratories of the Church: the role of English cathedrals in Anglican worship today. In M. Dudley (Ed.), *Like a two-edged sword: The word of God in liturgy and history* (pp. 179–194). Norwich: Canterbury Press.
Price, D. E. (1996). Opinion divided on new category names for Doves. *Billboard, 108*(10), 5.
Prigge, B. (2005). Katrina. Retrieved May 14, 2012, from www.newhymn.com/034Katrina.htm
Ramírez, D. (2009). Alabaré a mi Señor: Culture and ideology in Latino Protestant hymnody. In J. F. Martínez & L. Scott (Eds.), *Los Evangélicos* (pp. 149–170). Eugene: Wipf & Stock.
Reagon, B. J. (2001). *If you don't go, don't hinder me: The African American sacred song tradition*. Lincoln: University of Nebraska Press.
Revill, G. (2000). Music and the politics of sound: Nationalism, citizenship and auditory space. *Environment and Planning A, 18*, 597–613.
Richards, J. (2001). *Imperialism and music: Britain 1876–1953*. Manchester: Manchester University Press.
Schalk, C. F. (1995). *God's song in a new land: Lutheran hymnals in America*. St Louis: Concordia Publishing.
Schneider, R. A. (2004). Jesus shall reign: Hymns and foreign missions, 1800–1870. In R. J. Mouw & M. A. Noll (Eds.), *Wonderful words of life: Hymns in American protestant history and theology* (pp. 69–95). Grand Rapids: Eerdmans.
Scott, A. J. (1999). The US recorded music industry: On the relations between organization, location, and creativity in the cultural economy. *Environment and Planning A, 31*(11), 1965–1984.
Seeger, P. (1999). Ecological hymnody: A survey of denominational hymnals for ecologically themed hymn texts. *Arts, 11*(1), 30–35.
Smith, S. J. (1997). Beyond geography's visible worlds: A cultural politics of music. *Progress in Human Geography, 21*, 502–529.
Taaffe, E. J. (1974). The spatial view in context. *Annals of the Association of American Geographers, 64*(1), 1–16.
Thornburg, J. (2002). From rubble to hope: Texts and tunes after September 11, 2001. *The Hymn, 53*(3), 18–21.
Troeger, T. H. (2009). You are a woodwind and a drum. *The Hymn, 60*(2), 10–21.
Wallace, W. L. (2009a). The greening of hymnody? "Touch the earth lightly.". *The Hymn, 60*(1), 24–26.
Wallace, W. L. (2009b). The greening of hymnody, part 2: "Charge our hearts with wonder.". *The Hymn, 60*(2), 43–45.
Wallace, W. L. (2009c). The greening of hymnody, part 3: "Come join the cosmic family.". *The Hymn, 60*(3), 44–45.
Walsh, B. J., & Keesmaat, S. C. (2004). *Colossians remixed.*. Downer's Grove: IVP Academic.
Waterman, S. (2006). Geography and music: some introductory remarks. *GeoJournal, 65*, 1–2.
Welch, G. F. (2005). Singing as communication. In D. Miell, R. MacDonald, & D. J. Hargreaves (Eds.), *Musical communication* (pp. 239–259). Oxford: Oxford University Press.
Witvliet, J. D. (2003). The blessing and bane of the North American evangelical megachurch. In *Faith seeking understanding: windows into Christian practice* (pp. 251–268). Grand Rapids: Baker Academic.
Witvliet, J. D. (2007). *The biblical psalms in Christian worship*. Grand Rapids: Eerdmans.
Wood, L. A., & Gritzner, C. F. (1990). A million miles to the city: County music's sacred and profane images of place. In L. Zonn (Ed.), *Place Images in media: Portrayal, experience, and meaning* (pp. 231–254). Savage: Rowan and Littlefield.
Woods, M. (2012). Cultivating soil and soul: The intersection of the National Catholic Rural Life Conference and the American liturgical movement, 1920–60. In J. Bratt (Ed.), *By the vision of another world: Worship in American history* (pp. 105–131). Grand Rapids: Eerdmans.

Chapter 138
Tracing the Migration of a Sacred/Secular Tune Across Tunebooks and Its Traditions in Evangelical Nineteenth Century America: "The Peacock" Variations

Nikos Pappas

138.1 Introduction

During the eighteenth and nineteenth centuries, early settlers of the southern, northern, and western hinterlands of the United States brought with them items of material culture, as well as the knowledge to replicate and manufacture them. Typical for rural and urban frontier settlement patterns, these individuals relocated from older, more-established areas of the Eastern Seaboard and other colonial centers in the Western hemisphere as well as Europe and Africa. Settlers migrated either individually or in larger, cultural and religious groups or family units. Surviving items of material culture, including music, reflect these patterns of settlement and migration, allowing scholars to trace the geographic dispersal of people and culture.

However, delineating the geographical history of music before recorded sound poses certain challenges, in that a composition only survives in physical items of print and manuscript culture. Tune repertories and insights into their application to specific cultural groups remain locked within the physical object itself. Paralleling other items of material folk culture in the U.S., such as chair types, foodways, or plow types, the documentation of musical items pertaining to specific cultural groups or individuals representative of these larger groups requires locating surviving examples of print and manuscript culture. From these pieces, observations can be made to denote their provenance based upon the characteristics of the individual or individuals who produced them. In this way, the migration of the tune as object reveals its dissemination from the Eastern Seaboard to the interior of the country.

Conversely, music can also deviate from other types of material culture. For instance, a piece of music can reflect both popular and folk origin simultaneously, transcending the historical limitations of its medium. Certain melodies

N. Pappas (✉)
Musicology, University of Alabama, Tuscaloosa, AL 35405, USA
e-mail: napappas1@music.ua.edu

S.D. Brunn (ed.), *The Changing World Religion Map*,
DOI 10.1007/978-94-017-9376-6_138

remain a product of popular culture, others of specific cultural or regional practice representing folkloric treatment. Outside of its physical containment, versions of a melody were often transmitted orally from one individual to another. Paralleling the concept of textual negative tropism, deviations to the original form of the melody (if a piece of popular music) can reveal regionally or culturally specific variant forms based upon the identification of the sources and the individuals who copied or printed these tunes.

Thus, a surviving setting or variant of a tune reflects both its dissemination as a piece of oral culture (that is, a tune variant) and also as a physical object (a tunebook). Tune families, or all of the known variants of a particular piece of music, document not only general trends in musical practice among Americans in the nineteenth century, but also those aspects of music that reveal practices among specific cultural groups. Regarding religious music, it also pertains to theologically like-minded denominations. U.S. culture prior to the American Civil War followed mostly regional trends, with numerous localized practices carried out under an overarching national expression. An investigation into the interconnectivity of region, culture, and denomination as it relates to music reveals dissemination patterns that parallel settlement into the interior of the U.S. Exploring these indicators of object, tune, and variant allows for the mapping of musical culture, which encompasses a number of factors: the physical object in which the tune survives, the nature of the melody itself, and the specific folkloric aspect of a melody variant within its tune family.

This study uses two primary methods for documenting antebellum sacred music geography. First, it applies eighteenth-and-nineteenth century identifications of place for the delineation of region and culture. Besides antebellum conceptions of region, transportation networks and routes of cultural exchange serve as the means for the dissemination of material folk culture, including music. Second, it establishes the identification of melodies and their variants within their physical containment as the means to chronicle this dissemination over a 90-year time span. Through these processes, music serves as an indicator of cultural and denominational settlement patterns.

138.2 The Folk Hymn

A folk hymn is a type of religious song that circulated throughout the U.S. by various denominations, beginning in the eighteenth century with tunes such as the melody known as "Amazing Grace." The folk hymn occurs in three basic forms. In its most widespread use, it is a sacred appropriation of a secular song, termed a contrafactum, appearing in a number of variant melodic settings from the popular original. Though it is most often associated historically with Baptist, Presbyterian, and Methodist congregations, the folk hymn in fact appears in every denomination of every religion from every culture in the United States before 1860.

Among the more enthusiastic and evangelical groups, folk hymns were just as commonly a piece of music with no apparently popular secular source for the melody, existing in tune families. Individual melody variants followed the same sense of variation as the first type. These pieces apparently originated as sacred parallels to their secular cousins. More rarely, a final form of folk hymn constitutes a parody tune that seems directly based on a pre-existing tune, either through indirect or direct caricature.

Because folk hymns appear among all U.S. denominations, it becomes necessary to determine their folk or popular modes of expression. Many examples illustrate the cultural and denominational boundaries that mirrored those cultural groups migrating West and South, functioning much in the same way as the dissemination of a type of notch used in log house construction, a floor plan to a type of dwelling, or Fraktur motifs. In contrast, the musical style favored by liturgical-based musicians (that is, Episcopalians, Lutherans, Catholics, Jews) remains popular-based because the style of harmony, accompaniment, and compositional genre employs the popular mannerisms of its time period. Conversely, evangelical congregations often adapted this popular style initially to folk aesthetic or rejected its use in favor of their own folk traditions. Paralleling popular artistic trends that were disseminated to all regions of the country, music can also undergo regional or folkloric adaptation, such as Greek Revival architecture, the Hancock chair, or the folk use of stencils to imitate popular French wallpaper. Thus, a folk hymn contrafactum can demonstrate its popular origin through the tune selection, but the variants of the melody and their style and harmony its folk identity.

Together, the dissemination not only of a tune or tunebook, but also variants of folk hymns reveal patterns of print-based and manuscript material culture, as well as the specific cultural groups and/or denominations that produced this item. Placing it within a four-dimensional time frame over a 90 year period reveals the migration patterns of tune repertories. Linking the identification of melodies and their variants to the place of origin for the physical item in which they are found, and placed within the time parameters of their appearance, demonstrates the dissemination of music as an indicator of cultural migrations and settlement patterns.

138.2.1 Earlier Scholarship on American Sacred Music and the Folk Hymn

Within American music historiography, the study of early American sacred music comprises the oldest and most established area of scholarship and descends from two major sources: *Church Music in America* by Nathaniel Duren Gould (1853) and *A History of Music in New England* by George Hood (1846). This history was given a strong New England bias until German scholar George Pullen Jackson wrote his pioneering work, *White Spirituals in the Southern Uplands* (1933). In this volume Jackson discovered modern performers of eighteenth and nineteenth century American sacred music living throughout the American South.

Jackson's findings revealed that in addition to the New England repertory, the singers also performed pieces composed by nineteenth-and-twentieth century Southern musicians. He came to term these pieces spiritual folksongs, or folk hymns because of melodic characteristics shared between secular oral vocal repertories (that is, ballads) and those sacred pieces found in printed collections used by Southern singers. All of his major subsequent work centered on folk hymns. Much of the early scholarship on Southern American sacred music, including that by Jackson and Annabel Morris Buchanan in her *Folk-Hymns of America* (1938), followed the subjective aesthetic of white nativism concurrent to the academic thinking of the 1920s. For Jackson (1926: 6), the performance of these pieces represented "a sort of lost tonal tribe which was plying its musical art in pure pre-revolutionary form," paralleling the quest for modern ancestors in the Appalachian Mountains by early ethnomusicologists such as Cecil Sharp in his *English Folk Songs from the Southern Appalachians, collected by Cecil J. Sharp* (Karpeles 1932).

Other scholars saw some overriding purpose in the use of harmony and general compositional procedure in these folk hymns. Because these works often defied the rules of harmony as practiced by European classical composers, folk hymns became a symbol of rugged American individualism. Scholars then began to consider them a type of folk classicism, beginning with Charles Seeger in his article on contrapuntal style (1940). This self-styled American folk classicism view has continued to the present through such studies as David Brock's dissertation, *A Foundation for Defining Southern Shape-note Folk Hymnody from 1800 to 1859* (1996).

Earlier scholarly trends of white nativism and folk classicism spawned a renewed interest in the folk hymn during the post-WWII era. In an attempt to place this music within its greater cultural sphere, scholars started to apply notions of geography and region to understanding folk hymns, beginning with Irving Lowens' article on John Wyeth's *Repository of Sacred Music* (1813) as a northern precursor of Southern folk hymnody (1952). All subsequent scholars accepted uncritically the link between the Mason-Dixon line and the boundaries of North and South created by formal Southern nationalism and the American Civil War. They applied this concept to all music in the U.S., regardless of the date of a particular source. However, by selecting a body of repertory that predates the American Civil War, the appellation of bellum and post-bellum political and cultural identifiers on an antebellum repertory appears somewhat inapplicable. The Reconstruction-era identification of North and South simply did not exist during the time period of the development and maturation of the folk hymn (c. 1750–1850). Further, the first tunebook with any self-proclaimed Southern identity did not appear until 1835 with *The Southern Harmony* by William Walker of Spartanburg, South Carolina; this text contains no notions of Southern nationalism, only Southern regionalism. Even as recent as 2005, scholar David W. Music stated that "a 'southern' tune book is considered to be one compiled by a resident of one of the states that later formed the Confederate States of America, plus the border states of Kentucky and Missouri" (2005: xiii).

Finally, all of the above scholars have studied folk hymns primarily from one specific type of item of material culture: a published tunebook, printed in a musical innovation called shape-note notation that was patented in 1798 by William

Little and William Smith. Unlike traditional notation, shape-note note heads are printed with four different shapes corresponding to the English form of the solfege (fa-sol-la-fa-sol-la-mi), instead of the Italian form (do-re-mi-fa-sol-la-ti). Jackson viewed the four-shape books as representing a type of musical purity because these collections generally eschewed the compositions of nineteenth-century Northern reform musicians in favor of pieces by late eighteenth-century New England composers and early nineteenth-century composers of the post-bellum American South. Later scholars perpetuated this ideology largely because of the unbroken tradition of performance of this music from tunebooks such as *The Sacred Harp* by B. F. White and E. J. King (1844), despite the fact that books employing the seven-shape system using the Italian solfege also appear in unbroken tradition. However, these books contained more pieces by reform musicians, presenting what was perceived as a tainted selection of a pure repertory found in the four-shape books. Indeed, many folk hymns appear in both forms of shape-note notation as well as in books published in standard notation and in a form of numeral notation used in the Ohio River Valley.

As a result of these developments, studies of folk hymnody have primarily focused on older conceptions of purity. This pure repertory is found in four-shape shape-note tunebooks emanating from accepted regions of shape-note activity, the American South of formal Southern nationalism. Since the 1960s, major contributions to the scholarly literature include the following: Harry L. Eskew's dissertation, *Shape-note Hymnody in the Shenandoah Valley, 1816–1860* (1966), Dorothy D. Horn's *Sing to Me of Heaven: A Study of Folk and Early American Materials in Three Old Harp books* (1970), Ellen Jane Lorenz's *Glory, Hallelujah! The story of the Campmeeting Spiritual* (1978), David W. Music's (2005) article, "Seven 'new' tunes in Amos Pilsbury's *United States Sacred Harmony* (1799) and their use in four-shape shape-note tunebooks of the Southern U.S. before 1860" (1995), John Bealle's *Public Worship, Private Faith: Sacred Harp and American Folksong* (1997) and Kay Norton's *Baptist Offspring, Southern Midwife – Jesse Mercer's 'Cluster of Spiritual Songs' (1810): A Study in American Hymnody* (2002). With the exception of Horn and Norton, all sources follow the conceptions of repertory, region, and purity as delineated by George Pullen Jackson.

These earlier studies discussed above followed three basic approaches to understanding repertory: the music itself, a source's place of origin (provided that it emanated from accepted areas of musical activity), and notation type. Few scholars have examined manuscript sources, nor devoted much research to non-English-language source material. This study applies antebellum concepts and definitions of place and society to music. It also draws upon all existing types of surviving source material to document as many instances of occurrence as possible. The results present a different picture of regional identity, geographic dissemination, and their ties to culture and denomination. Applying the concepts of region, culture, geography, and their place within American society – that is, contemporaneous to the individuals that produced folk hymns – allows for a more accurate and nuanced understanding of the dissemination of sacred music culture and its artifacts throughout the United States.

138.3 Method

Among the many surviving folk hymns, the contrafacta variants of the popular song "The Peacock," reveal the dissemination of a tune family throughout all main regions of the U.S. as it existed in the antebellum period. Sacred versions of "The Peacock" are found in all types of sacred musical material culture, including printed tunebooks, manuscript copybooks and commonplace books. For the purposes of this study, all of the musical settings consulted were printed or copied between 1780 and 1870, coinciding with the tune's earliest date of known appearance and concluding with the institution of the Reconstruction Era.

Because academic scholarship on this topic has focused almost exclusively on English-language, Protestant Christian sources, no index exists to document: (1) sacred settings in American source material after 1820, (2) American musical imprints for German, Latin, Hebrew, and Native American source material, and (3) manuscript sources for sacred music, outside of insular communities such as the *Unitas Fratum* or Moravians (Gambosi 1970; Steelman 1981; Cumnock 1980). As a result, participating source material for this research was drawn from all of these categories to document as many sources as possible. Selection was determined by consulting databases of source materials found in libraries around the world using the following websites: WorldCat, Early American Secular Music and Its European Sources, 1589–1839: An Index, and The Hymn Tune Index. Several older print-based sources were also consulted, including James J. Fuld and Mary Wallace Davidson's *18th Century American Secular Music Manuscripts* (1980), Richard J. Stanislaw's *A Checklist of Four-Shape Shape-Note Tunebooks* (1978) and *Resources of American music history* edited by D. W. Krummel et al. (1981). As a result, this study consists of two major inquiries: secular settings of "The Peacock" (1789–1830), and sacred appropriations of the secular original (c. 1800–1870).

The sample size constituted 511 printed and manuscript sacred music compilations mostly from the U.S., but also Great Britain and Germany. Creating a database of 55,000 pieces of music, the results of all of these compilations, all entries include detailed information about every tune or tune title listed in the source. These sources were selected because of several factors. The primary focus of the database centered on source material emanating from the American South and West from 1750–1870. Other sources featured in the database included influential material from the Eastern Seaboard as well as Europe. These influential sources were determined by (a) tune contents and their relationship to Western and Southern source material, (b) known copies of European or Eastern Seaboard imprints used by Western and Southern musicians, and (c) other surviving indicators such as tunebooks or tunes mentioned in diaries, household inventory lists, wills, letters, and broadsheets.

Using FileMaker Pro software, the database documents tune titles, melodic and lyric incipits as well as composers, arrangers, and lyricists. It also provides information to assess regional sub-styles within a larger national framework, including shared repertories, paleographical connections among sources, texts accompanying the tunes, theological content, harmonic style, tune types and their musical form.

Other data entry fields document instrumental accompaniment, chorus and refrain-coda structure, compositional structure, instances of repetition, shared or dissimilar devices found between verses and choruses, special compositional or stylistic features if relevant, and subject matter of secular songs. Documentation of folk hymns follows the three basic forms: contrafacta, tune families originally circulating through oral transmission, and parody tunes. Finally, documentation of form, structural features, compositional devices between verse and chorus, and repetition of melodic material document folk and popular techniques.

As a result, these characteristics or data features indicate popular use as well as regional adaptation and folkloric or ethnic identity associated with the distinct cultures found throughout the different geographic regions of the U.S. Through this method of documentation, patterns shared among Northern, Middle Atlantic, Western, and Southern sources reveal the way in which popular and folk-based cultural traits moved from region to region, and changed throughout the antebellum period.

138.4 British Origins of "The Peacock"

"The Peacock," like many contrafacta embraced among American evangelicals, originated as a popular tune. It survives in a number of folk variants both in its secular and sacred forms. However, tracing the origins of this particular tune family proves somewhat convoluted on account of problems relating to its provenance, beginning with the earliest surviving sources. Of the former, scholar David Music noted a connection of this tune to folk music in his *Selection of Shape-Note Folk Hymns* (2005), specifically linking it to Child ballad 56, a paraphrase of the parable of the rich man and Lazarus in Luke 16: 19–31 (xxv). However, this connection does not constitute a reference to traditional music, but rather a popular song. In fact, Lucy E. Broadwood and J. A. Fuller Maitland personally wedded the Lazarus ballad to a tune set by A. J. Hipkins, Esq., in their publication, *English Country Songs* (1893).

Subsequently, English composer Ralph Vaughn Williams canonized its supposed folk status in the early twentieth century through two works, "Five Variants of Dives and Lazarus" for harp and orchestra, and a sacred hymn setting titled KINGSFOLD. In contrast, William Chappell, in *The Ballad Literature* (1859) stated that the tune had in fact circulated in England as early as the 1820s, being sung by and associated with unemployed vagrants, including sailors and gardeners (pp. 747–748). As a result, association of the tune with James Child and the Lazarus legend remains a product of late Victorian fancy. These sources also suggest the melody's origin as a popular song.

In contrast, four common threads bind most of the secular settings together, and a few of the sacred too: Robert Burns, freemasonry, Scotland, and a theme of farewell. Indeed, many of these settings are linked to two particular poems composed by Burns in the summer of 1786 when he planned immigrate to

Jamaica. The first poem, "The Farewell," was written as a parting song for his fellow Masonic brothers at Tarbolton. In particular, portions of Burns's text include a reference to some of the titles given to subsequent, sacred settings of this melody. The first and second verses referred to his fellow Lodge brothers as a "social band." One of the more widespread sacred variants was entitled SOCIAL BAND. In this sense, the social band of Freemasons was later appropriated to a sacred Christian assembly. Furthermore, Burns conceived his poem as a secular contrafactum to be set to the tune "Guid night, and joy be wi' you a'." His version of this tune appeared in the final volume of the *Scots Musical Museum* (1803).

The second poem referenced is Burns' "Farewell Song to the Banks of Ayr." According to the poet: "I composed this song as I conveyed my chest so far on my road to Greenock, where I was to embark in a few days for Jamaica. I meant it as my farewell dirge to my native land" (Burns 1895: 18). This poem deals with Burns's sadness at having to leave his friends and native country behind, and his fear for the unknown dangers on his journey across the sea. As with the previous poem, Burns conceived his "Farewell Song to the Banks of Ayr" as a contrafactum, this time, specifying "Roslin Castle" as the accompanying tune. However, neither of these tunes match the one found in sacred source material.

Despite a seemingly dead end with Burns's Scottish sources, the earliest documented version of the tune family does in fact originate from Lowland Scotland, possibly Perthshire. A version is found in a manuscript compiled by John Fife, a sailor (Fife [1780–c. 1804]). Fife's version, "The Barley Strow. A Quick Step," indicated its use as a piece of functional military music to accompany the quick march. It also reinforces William Chappell's observation on the British use of the tune in the mid-nineteenth century, equating it with a song sung by vagrant sailors. Further, many secular instrumental settings titled this piece "The Peacock" or "The Peacock's Feather," a reference to military officers who wore feathers in their hats. From these origins, the tune entered the vocal repertory.

138.5 American Secular Variants of "The Peacock" (c. 1790–1830)

Although the earliest version of the melody originated from a European source, all subsequent settings and melody variants within the study's time parameter appeared in source material from the United States. Indeed, if the tune had not been printed in nineteenth-century English publications, it would seem that it emerged from the U.S. Between 1789 and 1825, at least 8 variants appeared in 11 secular sources, spanning a geographic area north from the Connecticut River Valley, west to the Great Lakes area of western New York, and south to Philadelphia in the Middle Atlantic. Even though it originated as a popular tune, "The Peacock" was disseminated in a more folk-based expression as evidenced by its number of melody variants. In particular, the most concentrated instances of secular variants lay within southern and western New England (Fig. 138.1).

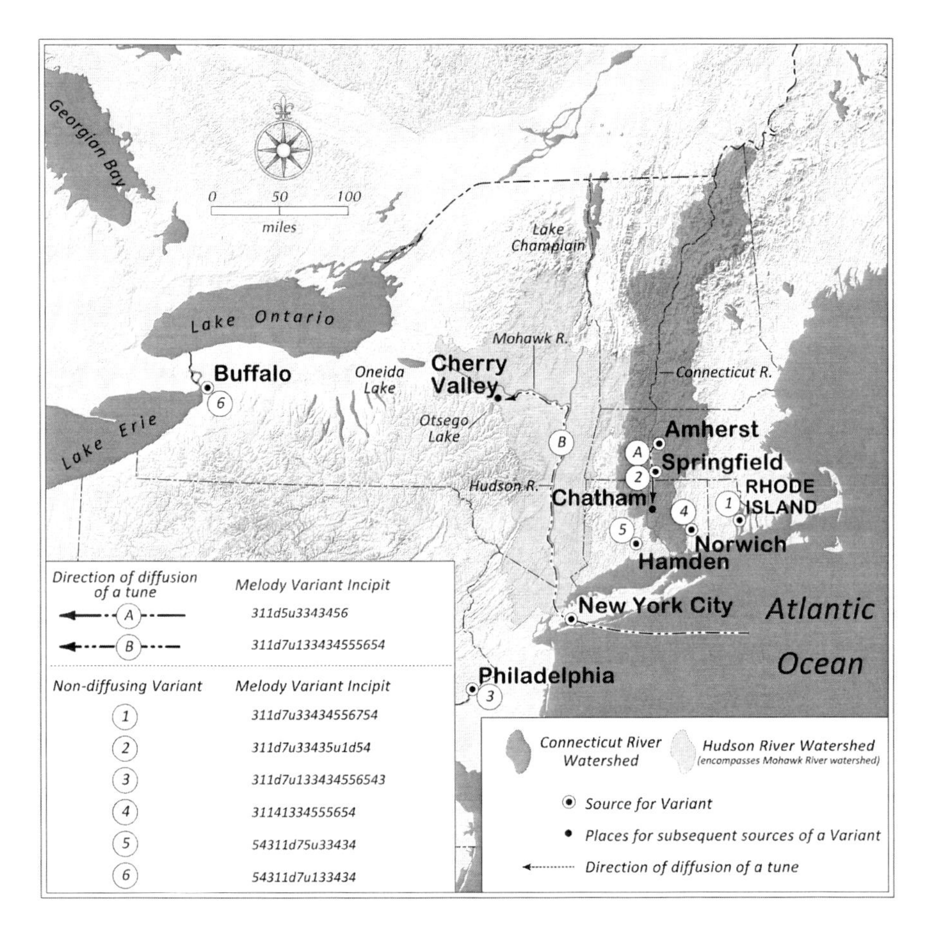

Fig. 138.1 Secular variants of "The Peacock" tune family in New England (Map by Dick Gilbreath, University of Kentucky Gyula Pauer Center for Cartography and GIS; commissioned by the editor)

Melody variant	Tune title	Source	Compiler location	Setting
311d5u3343456	1. Jemmy's Farewell	Silas Dickinson copybook (c. 1800)	Amherst, MA (?)	V
	2. The Farewel, or Jemmy's Farewel	Ishmael Spicer, "Spicer's Pocket Companion," manuscript commonplace book (1797–[1814])	Chatham, CT	V

(continued)

Melody variant	Tune title	Source	Compiler location	Setting
311d7u33434556754	The Peacock	"The American Volunteer," manuscript commonplace book (c. 1797)	New England, Rhode Island (?)	I
311d7u33435u1d54	The Masons Adieu	Bathsua Pynchon, "Bathshua Pyncon's Book" (1797–1805)	Springfield, MA	V
311d7u133434556543	Burns Adieu	A Professor, *An Introduction to the Art of Playing the Bassoon* (1826)	Philadelphia, PA	I
311d7u133434555654	1. The Barley Strow. A Quick Step	John Fife, manuscript copybook (c. 1780)	Perthshire (?), Scotland and at Sea	I
	2. The Peacock	*Gentleman's Pocket Companion for the German Flute or Violin* (c. 1802)	New York City	I
	3. Masonic Adieu, or Burns' Farewell Address	George White, "The Scale of the Gamut for the Violin" (c. 1790–1830)	Cherry Valley, NY	I
31141334555654	Peacocks Feather	Cushing Eells, "Cushing Eells's Music Book" (1789)	Norwich, CT	I
54311d75u33434	Freemasons Farewell	"Property of the Bellamy Band" (1799-c. 1805)	Hamden, CT	I
54311d7u133434	Burns' Farwel to Ayrshire	Manuscript copybook (c. 1820)	Buffalo (?), NY	I

Siglia: *V* vocal, *I* instrumental

The tune's appearance largely within New York and its border states follows the patterns of Scottish immigration to the British colonies in the 15 years preceding hostilities between England and the U.S. According to scholar Bernard Bailyn, 70.8 % of all immigrant Scots settled in two colonies, New York and North Carolina ([1986] 1988: 205–06). In contrast, only 5.5 % of English immigrants went to either of these colonies. Significantly, Massachusetts Bay and its largest urban center, Boston, remained a closed society, despite its status as a port city and one of the larger urban centers within the British North American colonies. Virtually no immigrants arrived in Boston, either English (0.9 %) or Scottish (0.8 %). The fact that the earliest setting of the tune appears in a Lowland Scottish manuscript confirms the pattern of dissemination presented by the secular instances of the tune's appearance in the U.S. However, "The Peacock" did circulate outside of its initial Scottish diaspora, becoming popular among sailors in England and Yankees of English descent in western and southern New England.

Secular variants of "The Peacock" circulated by the most common and efficient means of transportation possible in early nineteenth-century America: water navigation. All remained tied to important waterways in the northeastern United States. Further, the variants that survive in more than one setting follow precisely those paths of Scottish emigration as well as existing transportation patterns in the northeastern U.S. After its initial appearance in a Perthshire manuscript, the same variant of this tune was found in areas of strong Scottish influence: New York City and the Cherry Valley connected to the Mohawk River. This waterway constituted an important tributary of the Hudson River and served as the main transportation route connecting Lake Ontario to eastern New York before the completion of the Erie Canal (Bailyn [1986] 1988: 578). Likewise, the two vocal settings titled JEMMY'S FAREWELL appeared in sources originating along the Connecticut River Valley.

Secular settings of "The Peacock" also preserved their military association. Many versions of this tune family are found in instrumental settings with indicators of military use, such as "The American Volunteer" (c. 1797) manuscript, the John Fife manuscript and the settings for a military-style band from Hamden, Connecticut and the bassoon tutor from Philadelphia. Most surprisingly, no known secular variants are found in sources from the southern United States, becoming particularly revealing as to its sacred appropriation. No original southern version appeared in any source material before 1844.

138.6 Sacred Appropriation of "The Peacock" and Its Connection to Regionality

Shifting to the sacred appropriation of "The Peacock," certain trends parallel the secular dissemination patterns. Others differ significantly, mostly concerning the distribution of the melody over a greater geographic area. However, all explicate the basic regional character of the U.S. before the Civil War. Only towards the end of the period do instances of a variant's occurrence suggest a more national sphere for public display and publication. The following discussion will be structured according to the regions of the United States where various settings originally appeared. Migration patterns revealed that a few tune variants travelled outside of their place of origin. As a result, mapping their dissemination explicates how a melody variant, as well as the person or persons who performed it travelled throughout the nation.

Between 1805 and 1857, 11 sacred variants of the tune in 51 settings are found throughout the 4 basic regions of the U.S.: the *north* centered around New England, the *Middle Atlantic*, the *west* centered along the Ohio River Valley and northern Mississippi River, and the *south*. Surviving melody variants provide particularly neat parameters for regional dissemination. As such, focus will be given to geographic demarcation and migration patterns for these sacred versions of "The Peacock" tune family (Fig. 138.2).

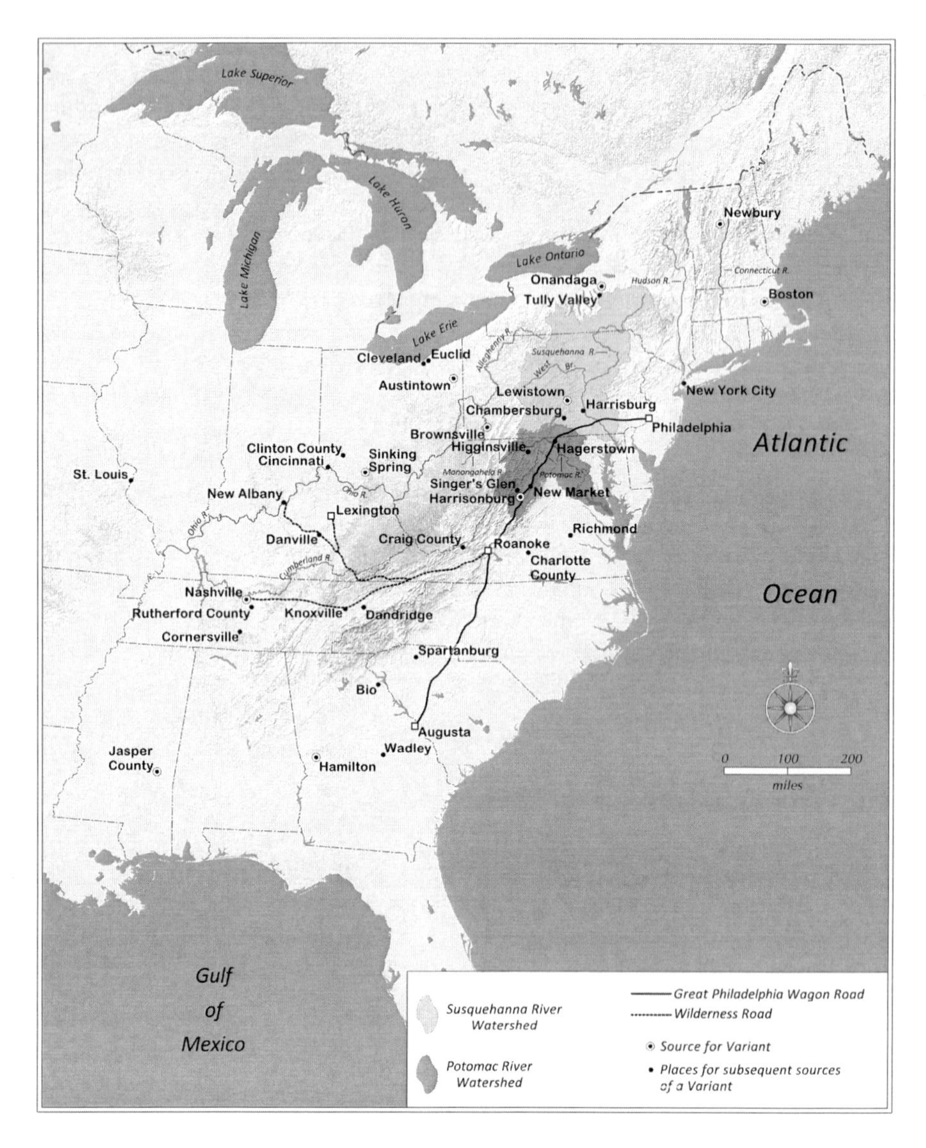

Fig. 138.2 Geographic locations and transportation networks for the sacred variants of "The Peacock" tune family in the eastern U.S. (Map by Dick Gilbreath, University of Kentucky Gyula Pauer Center for Cartography and GIS; commissioned by the editor)

Northern tune variants					
Melody variant incipit		Tune title	Source	Compiler location	Notation
5(4) 311d7u334	3(4)5575 (4)311	Mary's Dream	A. Churchal, manuscript copybook (c. 1810–30)	Onandaga C[ourt] H[ouse], NY	R

(continued)

Northern tune variants					
Melody variant incipit		Tune title	Source	Compiler location	Notation
5311d7u334	3(4)5565 (4)311	1. Shouting Hymn	Jeremiah Ingalls, *The Christian Harmony* (1805)	Newbury, VT	R
		2. Shouting Hymn	Deerin Farrer, *The Christian Melodist* (1828)	Tully Valley, Onandaga County, NY	R
5311d7u334	355753(2)1 (d7)u1	1. Resurrection	Rev. W. McDonald and S. Hubbard, Esq., *The Wesleyan Sacred Harp* (1857)	Boston, MA; New York City, NY; and Cleveland, OH	R
		2. Cross of the Lord	*The Jubilee Harp* (1866)	Boston, MA	R

Middle Atlantic tune variants					
Melody variant incipit		Tune title	Source	Compiler location	Notation
5311d5u334	3(4)5675 (4)311	1. Ayrshire	J[ohn] H[oyt] Hickock, *The Sacred Harp* (1832)	Lewiston, PA	4S
		2. Mary's Dream	J[ohn] H[oyt] Hickock, *The Social Lyrist* (1840)	Harrisburg, PA	4s
5311d7(u1)334	3(4)5575 (4)311	1. Clamanda, attr. Chapin	Ananias Davisson, *A Supplement to the Kentucky Harmony* (SKH), ed. 3 (1825)	Harrisonburg, VA	4S
		2. Clamanda	William R. Rhinehart, *The American or Union Harmonist* (1831)	Hagerstown, MD	4S
		3. Clamanda, attr. Chapin	William Caldwell, *Union Harmony: or Family Musician* ([1834] 1837)	Dandridge, Jefferson County, TN	4S
		4. Social Band	Joseph Funk, *A Compilation of Genuine Church Music* (CGCM) (1835)	Mountain Valley (Singer's Glen), VA	4S
		5. Clamanda	John W. Steffey, *The Valley Harmonist* (1836)	Newmarket, VA	4S
		6. Clamanda	John B. Jackson, *The Knoxville Harmony of Music Made Easy* (1838)	Knoxville, TN	4S

(continued)

Middle Atlantic tune variants					
Melody variant incipit		Tune title	Source	Compiler location	Notation
		7. Clamanda	Henry Smith, *The Church Harmony*, ed. 12 (1841)	Chambersburg, PA	4S
		8. Social Band	Joseph Funk, CGCM, ed. 3 (1842)	Mountain Valley, VA	4S
		9. Social Band	William Walker, *The Southern and Western Pocket Harmonist* (1846)	Spartanburg, SC	4S
		10. Social Band	Joseph Funk, CGCM, ed. 4 (1847)		
		11. Clamanda, attr. Chapin	William Hauser, *Hesperian Harp* (1848)	Wadley, GA	4S
		12. Social Band	George Hendrickson, *The Union Harmony* (UH) (1848)	Midway, Craig County, VA	4S
		13. Social Band	Eli Ball, *The Manual of the Sacred Choir* (1849)	Richmond, VA	4S
		14. Social Band	George Hendrickson, UH, ed. 2 (1850)	Midway, Craig County, VA	7S
		15. Social Band	Joseph Funk and Sons, *The Harmonia Sacra* (HS), ed. 5 (1851)	Singer's Glen, Virginia	7S
		16. Social Band	Levi C. Myers, *Manual of Sacred Music* (1853)	Higginsville, WV (then VA)	7S
		17. Social Band	William Walker, *The Southern Harmony*, new ed., rev. and imp. (1854)	Spartanburg, SC	4S
		18. Social Band	Joseph Funk and Sons, HS, ed. 6 (1854)	Singer's Glen, VA	7S
		19. Social Band	George Hendrickson, UH, ed. 3 (1855)	Midway, Craig County, VA	7S
		20. Social Band	Joseph Funk and Sons, HS, ed. 7 (1856)	Singer's Glen, VA	7S
		21. Social Band	Joseph Funk and Sons, HS, ed. 8 (1857)	Singer's Glen, VA	7S

(continued)

Middle Atlantic tune variants					
Melody variant incipit		Tune title	Source	Compiler location	Notation
		22. Social Band	Joseph Funk and Sons, HS, ed. 10 (1860)	Singer's Glen, VA	7S
		23. Social Band	William Walker, *The Christian Harmony*, ed. 2 (1873)	Spartanburg, SC	7S
5311d7u334	3(4)5575 (4)311	1. Clamanda, attr. Chapin	Ananias Davisson, SKH (1820)	Harrisonburg, VA	4S
		2. Clamanda, attr. Chapin	Ananias Davisson, SKH, ed. 2 (1820)	Harrisonburg, VA	4S
		3. Clamanda, attr. Chapin	Ananias Davisson, *Introduction to Sacred Music* (1821)	Harrisonburg, VA	4S
		4. Clamanda, attr. Chapin	Marshall family. Manuscript copybook (1809[-c. 1825])	Charlotte County, VA	4S
		5. Burns' Farewell	Anthony Joseph, *The Western Minstrel, or Ohio Melodist* (1831)	Clinton County, OH	4S

Western tune variants					
Melody variant incipit		Tune title	Source	Compiler location	Notation
5(4)311d7 (u1)33(5)4	3(4)55(6)75 (4)31(d7)u1	1. Shouting Song	Allen D. Carden, Samuel J. Rogers, F. Moore, and J. Green, *The Western Harmony, or, The Learner's Task Made Easy* (1824)	Nashville, TN	4S
		2. Shouting Song	Allen D. Carden, *United States Harmony* (1829)	Nashville, TN	4S
		3. Shouting Song	John B. Seat, *The St. Louis Harmony* (1831)	St. Louis, MO (?)	4S
		4. Shouting Song	Andrew W. Johnson, *The American Harmony* (1839)	Rutherford County, TN	4S
5(4)311d7 (u1)334	3(4)55 (6)75(4)311	1. Amboy	Amos Sutton Hayden, *Introduction to Sacred Music* (ISM) 1835	Austintown, OH	4S
		2. Amboy	Amos Sutton Hayden, ISM, ed. 2 (1838)	Austintown, OH	4S
		3. Amboy	Amos Sutton Hayden, *The Sacred Melodeon* (1848)	Euclid, OH	7S

(continued)

Western tune variants					
Melody variant incipit		Tune title	Source	Compiler location	Notation
		4. Amboy	Silas W[hite] Leonard and A[ugustus] D[amon] Fillimore, *The Christian Psalmist*, rev. and enl. ed. (1850)	New Albany, IN; Cincinnati, OH	N
		5. New Richmond	Silas W[hite] Leonard, *New Christian Psalmist* (1870)	New Albany, IN	N
5(4)311d7 (u1)334	3(4)567 (6)5(4)311	The Mason's Farewell	James Krepps, Manuscript copybook (1830)	Brownsville, Fayette County, PA	4S
531d77u334	3(4)5575311	1. Clamanda, attr. Chapin	J. Amen (?), Manuscript copybook (c. 1820)	Sinking Spring, Adams County, OH (?)	4S
		2. Clamanda, attr. Reed	Benjamin Shaw and Charles H. Spilman, *Columbian Harmony, or Pilgrim's Musical Companion* (1829)	Danville, KY	4S

Southern Tune Variants					
Melody variant incipit		Tune title	Source	Compiler location	Notation
5311d7 (u1)334	3(4)55 (6)75(4)311	1. The Christian Soldier	L[azurus] J. Jones, *Southern Minstrel* (1849)	Jasper County, MS	4S
		2. Amboy	Andrew W. Johnson, *The Western Psalmodist* (1853)	Cornersville, Giles County, TN	7S
5311d7u3 (4)54	3(4)55 (6)75(4)311	1. Clamanda	B[enjamin] F[ranklin] White and E[lisha] J. King, *The Sacred Harp* (1844)	Hamilton, GA	4S
		2. Clamanda	John G. McCurry, *The Social Harp* (1855)	Bio, GA	4S

Siglia: *R* standard (round-note) notation, *4S* four-shape shape-note notation, *7S* seven-shape shape-note notation, *N* numeral notation

138.6.1 Northern Variants

Paralleling trends in secular dissemination, the earliest sacred version originated from northern New England in the Connecticut River Valley town of Newbury, Vermont. Presumably set by Jeremiah Ingalls, a tavern keeper, musician, and composer, SHOUTING HYMN appeared in his only tunebook, *The Christian Harmony* (1805), a collection of hymns and spiritual songs. The text, harmonic setting, and musical notation illustrate northern trends in evangelical enthusiastic composition. From a theological and denominational perspective, Ingalls' tunebook demonstrates his alignment with Calvinist Nonconformists or the older Calvinist churches that adopted some of the trends of Nonconformists or proto-Methodists from mid-eighteenth-century initiatives.

Beginning in the 1740s, Calvinists along the Connecticut River began adopting the tenets of the First Great Awakening following the initiatives of Jonathan Edwards (1703–1758), a minister in Northampton, Massachusetts. Slightly later, Nonconformist Calvinist and Anglican evangelist, George Whitefield (1714–1770) conducted a series of preaching tours throughout Connecticut and western Massachusetts. These clergy created a split between a more traditional staid style of worship and one that personally excited the religiosity of a congregant (Benes 1982: 129). This split created a theological rift between conservative Calvinists and those that had felt a new spiritual calling.

Ingalls' setting featured several Calvinist Nonconformist traits. Many spiritual songs including SHOUTING HYMN appeared in three-part harmony mimicking Nonconformist British publications like the *Harmonia Sacra* by Thomas Butts (1754). Reflecting current and fashionable trends of popular and theatrical music, these works were composed of two unspecified melodic parts, with a bass line to be performed either by an instrument or voice (Temperley 1993: 63–64). Similarly, for his texts, Ingalls chose many verses from regional Calvinist Nonconformists and other enthusiastic denominations such as Baptist Joshua Smith (1760–1795), a lay minister from Canaan, New Hampshire. Not just a generic hymn of praise, the texts used by Ingalls bespoke a familiarity with God that was embraced by enthusiastic Calvinist and Calvinist-influenced Protestants. Finally, Ingalls did not embrace shape-note notation, but printed his compilation in its standard form. Though shape-note notation did exist somewhat throughout the North; most regional evangelicals did not embrace it.

Ingalls' setting only appeared in one other compilation, *The Christian Melodist* (1828) by Methodist Deerin Farrer of Tully Valley, New York. In this sense, Ingalls' setting in Farrer's tunebook followed the theological trends of the Connecticut River Valley established over the past 80 years. Significantly, Ingalls' SHOUTING HYMN also reflected migration patterns of New Englanders settling the Great Lakes area of the early Middle West, first in northern and western New York in the area known as the Burned-Over District (Cross [1950] 1965: 4–7) and later in the Connecticut Western Reserve of northern Ohio and other areas of the original Northwest Territory as established by the Northwest Ordinance of 1787. In this sense, Ingalls' setting reflected denominational, regional, and cultural migratory patterns.

Similarly, RESURRECTION also maintained its connection to enthusiastic Calvinist Nonconformist initiatives through its variants' original appearance in a Wesleyan collection that was later adopted by Seventh Day Adventists and influenced by William Miller and the Millerite movement in western New York. Printed in Boston,

The Wesleyan Sacred Harp was published by two other firms: one in New York City, the other in Cleveland, the urban center of the Connecticut Western Reserve. Also reflecting regional trends, this variant appeared in standard notation. In sum, northern sacred settings most closely replicated distribution patterns of its secular progenitor.

138.6.2 Middle Atlantic Variants

In contrast to northern settings, all of the Middle Atlantic variants appeared in a form of shape-note notation invented by John Connelly of Philadelphia and patented by William Little and William Smith, two Middle Atlantic singing masters, in 1798. Not coincidentally, tunebook compilers in this area produced many of the most influential shape-note tunebooks among all English and German-speaking citizens. Further, the two settings from Harrisonburg, Virginia proved the most influential on subsequent tunebook printings by compilers in the West and the South, disseminating over the largest geographic area of any of the other variants.

Regarding tune settings from the Middle Atlantic, three melody variants had origins in two of its major watersheds: the Susquehanna and the Potomac Rivers. The Susquehanna watershed encompasses an area spanning southern New York, Pennsylvania, and Maryland, the Potomac, Pennsylvania, Maryland, and Virginia. Both empty into the northern Chesapeake Bay. AYRSHIRE, from a central Pennsylvania tunebook by John Hoyt Hickok, appeared in two of his compilations. However, the most influential settings originated from one individual, Ananias Davisson (1780–1857), a Presbyterian musician living within the Shenandoah Valley of Virginia. Residents along the Shenandoah River, a major tributary of the Potomac, maintained closer ties to Philadelphia than other places in Virginia east of the Blue Ridge Mountains.

The valley featured a religious pluralism distinct from coastal areas of the South including the Low Country of South Carolina and the Tidewater area of Virginia within the Chesapeake Bay. According to scholar Stephen Longenecker (2000: 185) "no denomination dominated religious life [in the valley], making the Shenandoah a community of minorities, much more similar to Pennsylvania than to eastern Virginia." Musical practice remained fairly uniform among Calvinist, Nonconformist, and Anabaptist denominations, including Presbyterians, Methodists, and Mennonites.

Besides the watersheds, settlement of this area occurred along the Great Wagon Road, which followed the older Cherokee Warriors' Path, an important hunting and trading route before white settlement (Rouse [1973] (2004); Britton et al. 1990). It extended west from Philadelphia and south through Maryland, paralleling the Appalachian Mountains through the Upper Piedmont areas of Virginia, North Carolina, and South Carolina until its terminus in Augusta, Georgia. In southern Virginia, the Wilderness Road split off from the Great Wagon Road and extended southwest into Tennessee and north into Kentucky. Most of the early settlers travelled these roads in the settling of the Southern Backcountry and the early Southwest between 1740 and 1820.

Musical practice among Shenandoah Valley musicians, especially between 1800 and 1835, followed a different set of compositional characteristics than northern melody variants of "The Peacock." Besides the use of shape-note notation, musicians tended to favor a four-part harmonization of tunes with the melody given to the tenor voice. Besides establishing a paternal form of expression with the male voice as musical leader and hence authority, musical style descended (in terms of American influence) ultimately from English Presbyterian initiatives in Philadelphia, the largest city within the Middle Atlantic. This style of presentation and harmony first appeared in the 1760s through the efforts of New Jersey-born James Lyon (1735–1790) and later Andrew Adgate (1762–1793). Like many singing masters active in the Middle Atlantic in the late eighteenth and early nineteenth centuries, Adgate was born in Connecticut but made his livelihood outside of New England.

South of Harrisonburg within the Shenandoah Valley of Virginia, Ananias Davisson (1780–1857) published two variants of this tune, each titled CLAMANDA. Davisson ascribed both to Chapin, referring to the brothers Amzi and Lucius Chapin of Springfield, Massachusetts along the Connecticut River. Lucius (1760–1842) was active as a singing teacher in the Shenandoah Valley from 1789 to 1797 before relocating to Washington, Kentucky near Maysville on the Ohio River. Amzi Chapin (1768–1835) had lived in Hartford, Connecticut before joining his brother in Virginia in 1791 (Scholten 1972). Amzi travelled throughout Virginia and North Carolina along the Great Wagon Road, teaching singing to various churches and communities before moving to Kentucky in 1795.

Tunes ascribed to Amzi and Lucius Chapin entered into the shape-note repertory and became quite popular. However, the ascription of CLAMANDA to Chapin remained suspect since the tune did not appear in any holograph manuscripts of either Lucius or Amzi (Chapin c. 1798). Further, both variants were ascribed to Chapin, and Davisson did not obtain any of his settings of Chapin tunes directly from either of the Chapin brothers, instead taking them from *Wyeth's Repository of Sacred Music, Part Second* (1813), an evangelical Presbyterian collection from Harrisburg, Pennsylvania influenced by Calvinist Nonconformist initiatives. Together these factors cast doubt on the correctness of this attribution. Regardless of its accuracy, Davisson's ascription indicates the status accorded the Chapins.

The earliest version of CLAMANDA to appear in Davisson's publications was found in the first edition of *A Supplement to the Kentucky Harmony* (1820a). His inclusion of this tune follows established trends connected to other sacred adaptations of "The Peacock." Davisson first issued the tunebook's predecessor, *The Kentucky Harmony*, in 1816. As a compilation exemplifying Middle Atlantic Presbyterian taste, it featured a wide range of tunes intended for congregational performance, including a number of folk hymns. *A Supplement to the Kentucky Harmony* (1820a) had a different intent. In it, Davisson stated that his "principle design in offering his Supplement is, that his Methodist friends may be furnished with a suitable and proper arrangement of such tunes as may seem to him best calculated to animate and enliven them, and all other zealous Christians, in their acts of devotion" (Preface).

Reflecting its use among Methodists and "zealous Christians" (for example, Separate Baptists) many of the pieces in the collection appear in three-part harmony mirroring the presentation of tunes in Ingalls' tunebook. While the 1820 setting of CLAMANDA followed the four-part form characteristic more of Presbyterians, the combined presence of three-part tunes together with his stated intention of the collection indicated at least an intended use by the same type of denominations as the Northern variants. Appearing in different regions in different melody variants, CLAMANDA and SHOUTING HYMN expressed the same form of religiosity and catered to the same clientele.

Though the first version of CLAMANDA had a somewhat limited circulation around Virginia and along the Ohio River, the second melodic variant, found in the third edition of *A Supplement to the Kentucky Harmony* by Davisson (1825), enjoyed the widest distribution of any other surviving variants. After its initial appearance in Davisson's tunebook, the second variant of CLAMANDA circulated throughout Virginia, Maryland, eastern Tennessee, southern Pennsylvania, western South Carolina and Georgia. This particular variant explicated not only the distribution of Davisson's book south and west, but also the migration of evangelical Calvinist Nonconformists and their musical style and taste.

Paralleling the Great Wagon Road, Davisson's setting travelled north, appearing in later collections from Hagerstown, Maryland and Chambersburg, Pennsylvania. These collections betray a cross-cultural sphere of influence. Rhinehart's collection was intended for the Church of the United Brethren in Christ, an evangelical German denomination influenced by Arminian (that is, Free Grace) Methodism. Henry Smith's collection from Chambersburg included a supplement designed for members of the German Reformed Church, a Calvinist denomination. CLAMANDA also circulated throughout the Shenandoah Valley in the ensuing decade and travelled as far south as Wadley, Georgia, southeast of Augusta. Davisson's setting also followed the Wilderness Road into eastern Tennessee, appearing in a collection in Knoxville and another in nearby Dandridge. In this way, Davisson's second variant of CLAMANDA, though originating from within the Middle Atlantic, was disseminated outside of its region of origin to the south and the west.

Although first appearing in Davisson's tunebook, another harmonization of this melody variant enjoyed almost as wide a distribution as the earlier setting. Revised and renamed SOCIAL BAND, it was printed first in the tunebooks of Mennonite Joseph Funk and his sons (1835, 1842, 1847, 1854) of nearby Mountain Valley, later Singer's Glen, about 20 miles (32 km) north of Davisson's farm. Beginning in the 1840s, the Funk family also performed contractual publishing work for other compilers. Through these publications, SOCIAL BAND travelled northwest into Higginsville, (West) Virginia, and south to Midway in Craig County, Virginia. It also travelled into the south to Spartanburg, South Carolina to appear in a few tunebooks compiled by William Walker, author of *The Southern Harmony* (1835).

Middle Atlantic variants and settings demonstrated more than any other region the presence of a cultural and religious pluralism. Unique to other sections of the country, much cross-cultural interaction occurred over a wide geographic area, extending outward from its point of origin. Rather than limiting interaction and influence among denominations, many clergy and congregants embraced a more plastic conception of the role and place of religious music, preferring tunes with a similar theological sentiment above cultural background. In this sense, Davisson's publications functioned more within a popular framework because his collections were embraced across a wide spectrum of individuals, denominations, and cultural groups. All of these characteristics, combined with the geographic centrality of the Middle Atlantic region within the boundaries of the U.S. partly explain its popularity. Fortunately, the large number of surviving tunebooks containing the two Davisson tune variants allowed both for a more nuanced understanding of the circulation of a particular melody, and an insight into those persons or groups that purchased or copied them.

138.6.3 Western Variants

Western variants remain less uniform in terms of theological disposition and the nature of their dissemination. Some reflected trends similar to those found along the Eastern Seaboard while others arose because of uniquely western religious developments. Perhaps as a result, more variants survive from the west than any other region of the U.S. Paralleling northern settings, their distribution occurred mostly along waterways instead of roads. In particular, all are connected to the Ohio River, emanating from western Pennsylvania, Ohio, and Tennessee. Within this watershed, variants also circulated via important tributaries of the Ohio, including the Cumberland and Monongahela Rivers. One variant went into the Far West across the Mississippi River into Missouri. As with Middle Atlantic trends, tune circulation moved freely north, south, and west within the larger region.

Two western variants maintained close connections to Virginia and Ananias Davisson though through independent means. The first originated from Allen D. Carden (1792–1859), a Virginia-born singing master active throughout the Ohio River watershed and St. Louis, Missouri. In 1817 Carden had acted as one of Davisson's agents for *The Kentucky Harmony* in Wythe County, Virginia, but within a year he had a falling out with the older musician (Ananias Davisson to Solomon Henkel 1817). Shortly thereafter, Carden relocated to Cincinnati and St. Louis to teach singing schools. In 1820 he published his most popular and influential tunebook, *The Missouri Harmony*. However, Carden's variant of "The Peacock" first appeared in *The Western Harmony* (1824), and co-authored with S. J. Rogers, F. Moore, and J. Green after he had settled in Nashville, Tennessee.

In the book's preface, besides re-igniting his quarrels with Davisson, Carden iterated a common trope found throughout many western tunebooks: a strong sense of Western identity as a statement of regionalism. A typical fashion, he proclaimed that "[t]his collection of sacred music was made for the express purpose of accommodating and associating all the friends of psalmody in the Western country" (3). Northern and

Middle Atlantic compilers had emphasized more the theological connections among certain denominations to help promote their book. For many westerners, their region of residence often acted as a unifying agent to advance their product.

Besides its connections to Virginia, the setting of SHOUTING SONG featured in the *Western Harmony* mirrored other trends manifested throughout the Eastern Seaboard. First, it used the same title and text as Ingalls' northern setting. However, these similarities probably derive from the text's second stanza ("Then will I shout then will I sing,"), as opposed to Ingalls' setting exerting a direct influence on Carden and his associates. This tradition of naming a tune after a significant word or phrase in the text hearkens back to Calvinist spiritual songs among progressive English Presbyterians of the early eighteenth century.

Second, Carden intended its use by Calvinists and Calvinist Nonconformists including Baptists and Methodists, based upon poetic source attributions. Further, the compilation included a mixture of three-and-four-part settings following trends found in the north with Ingalls and the Middle Atlantic with Davisson. In this sense, the book followed established theological printing and compositional conventions of other enthusiastic evangelical denominations in the east.

The dissemination of the western variant of SHOUTING HYMN occurred mainly throughout central Tennessee, appearing in a subsequent Methodist-influenced compilation by Carden, as well as a later tunebook compiled in Rutherford County. However, the tune did travel northwest into Missouri along the Cumberland, Ohio, and Mississippi Rivers. All of these compilations betray the same religious sentiment and repertory mix as *The Western Harmony*, forming a unified expression independent of but closely resembling trends encountered in the Shenandoah Valley.

Another western compilation and its variant of "The Peacock" demonstrated a strong connection to the Middle Atlantic, and again to Davisson. CLAMANDA, the earliest western variant, found in a manuscript copybook used by a peripatetic minister travelling between central Kentucky and Ohio, is almost exactly contemporary to Davisson's first setting of CLAMANDA. This source includes an extensive listing of travel dates and locations over a number of months, beginning in Lexington, Kentucky and concluding in south central Ohio. Though smaller than *A Supplement to the Kentucky Harmony*, this manuscript shares many similarities to its Middle Atlantic cousin. It presents a similar repertory as Davisson's tunebook, with a number of pieces shared between them, in particular, the tunes by J. C. Lowry. Further, both also contain tunes by Lowry that are unique to each compilation. This phenomenon can tentatively be explained by two factors: (a) the itinerant nature of its compiler suggests a possible Methodist hand in its creation, and (b) Sinking Springs, Ohio lies within the older Virginia Military District, a section of the state apportioned to Revolutionary War veterans from Virginia. Thus, the similarities may have occurred via denominational or cultural/geographical connections.

Although it did not circulate over as broad an area in a large number of sources, the western CLAMANDA variant does appear in the *Columbian Harmony* (1829), a central Kentucky tunebook compiled by students at Centre College in Danville, Kentucky. Co-authors Benjamin Shaw and Charles H. Spilman (1829) compiled the book in the hopes of covering their tuition and boarding expenses through the sale of the collection. As before, this source maintains close ties not only to Davisson, but also the earlier manuscript copybook. It contained the same variant of "The Peacock" as the manuscript book and maintained a direct connection to central Kentucky.

The final variant that proved popular in the west maintained a strictly theological identity with members of the Primitive Christian Movement, a grouping of churches arising from the Second Great Awakening in Kentucky in 1801 that sought to restore the original church founded by Christ's disciples as described in the Book of Acts in the Bible. Also known as the Restoration Movement, the church recognized two main streams of influence: one under Barton Stone (1772–1844), a Maryland-born Kentuckian, the other under Alexander Campbell (1788–1866), a Northern Ireland-born Scottish immigrant living in Bethany, (West) Virginia near the Ohio River. Both men had served as ordained Presbyterian clergy before they independently instituted their own form of Primitive Christianity. As such, their congregants were drawn from the prevailing evangelical denominations in the early nineteenth century West: Presbyterians, Methodists, and various branches of Baptists. Campbell's Scottish training combined with the theological background of early church members may explain the presence of a variant associated with the Disciples of Christ.

AMBOY first appeared in shape-note publications by Amos Sutton Hayden (1813–1880), the first musician to publish a tunebook within the Restoration Movement. At the time, Hayden, born in Youngstown, Ohio but whose family had arrived from Pennsylvania, resided in Austintown in the Connecticut Western Reserve. He was a member of the Mahoning Association, the first formal organization of churches in the Campbellite Disciples of Christ. Thus, his cultural and religious heritage remains somewhat separated from the predominant cultural and geographical delineations of northern Ohio. Distinct from other northern compilers, he also published exclusively shape-note tunebooks until the 1870s, both in four-shape and seven-shape systems. His area of activity influenced musicians and trends in the Ohio River Valley much more than his area of residence.

After its occurrence in three of Hayden's tunebooks, AMBOY appeared in compilations by other Disciples of Christ musicians in southern Ohio and Indiana.

In particular, Silas White Leonard from New Albany, Indiana across the river from Louisville, and Augustus Damon Fillmore of Cincinnati re-harmonized Hayden's four-shape setting and printed it in a form of numeral notation invented and patented by Springfield, Ohio musician and schoolteacher Thomas Harrison (Harrison 1839). Numeral notation became the preferred form of musical notation among the Disciples of Christ in the Ohio River Valley beginning in the 1840s, largely through the efforts of Leonard, and Fillmore and his descendants. Because of these individuals, Hayden's variant enjoyed a wide distribution not only throughout the upper Ohio River Valley, but also among members of their church into the late nineteenth century.

Western sacred music practice among evangelicals, though unified often by a common theme of western identity, reflected a more disparate set of cultural and denominational interactions within the region, involving variant circulation, theological identity, and musical notation type. Rather than circulate over broad areas within region, or travel outside of their area of origin, surviving variants remain tied to their sub-region or denomination. The western variant of SHOUTING HYMN circulated only around Carden's area of professional employment in central Tennessee and Missouri. The western variant of CLAMANDA appeared only in a small area confined to south central Ohio and central Kentucky. Similarly, AMBOY only appeared in collections associated with Restoration Movement musicians. In this sense, western practice deviated most strongly from Middle Atlantic practice despite the strong connections among variants originating within the Ohio River watershed and the Shenandoah Valley. Also, notational type as an indicator of theological disposition does not follow a uniform practice. More variants in non-standard musical notation such as shape-note and numeral notation originated from the west than any other section of the country.

However, the opposite can be true in some regards. Although the variants to "The Peacock" do not demonstrate a cross-cultural relationship among sources, they do follow the same type of crossover evangelical denominational-theological connection. Carden's tunebook catered to the same clientele as Ingalls, Davisson, and Funk. Similarly, the denominational origins of many members of the Disciples of Christ reflected this same mixing of Calvinists, Calvinist Nonconformists, and other enthusiastic sects. That a tune should appear from within this mixing of individuals was neither surprising nor unexpected.

Because of the disparate nature of the west, its connections to the east were easier to trace. Together, these variants demonstrated that the U.S., rather than forging a single national practice, instead comprised a series of regional and localized subregional spheres of expression and practice during the Early Nationalist and Antebellum periods. All variants fit into two basic forms of expression, tied either to geography and culture, or theology and/or denomination. Some constituted a series of outward-arching geographical identifiers beginning with city or county, state, region, and finally nation. Others circulated within their region and beyond because of their denominational affiliation. Variants from the west most clearly explicated these patterns of identity and distribution.

138.6.4 Southern Variants

With the exception of the Wesleyan setting from Boston, the two southern variants were the last to appear. Further, both suggest for their origin a direct adaptation of Davisson's CLAMANDA variants, paralleling somewhat the two western variants with ties to the Shenandoah Valley. Because no secular sources existed documenting "The Peacock" in the American South, southern musicians may have learned the tune not through oral dissemination, but through print culture. Thus, the southern variants suggest a form of direct parody or adaptation, unlike other surviving versions.

As with the collections by Carden, and Shaw and Spilman, Southern sources including the *Southern Minstrel* by Lazarus Jones (1849) and *The Sacred Harp* by B. F. White and E. J. King (1844) took much of their repertory from Davisson's publications. Once again, these tunebooks indicated that Davisson's activities transcended his region of origin, influencing repertory and expression throughout much of the country. As such, his collection, though rooted in folk and regional practice, became an object of popular consumption. In this sense, the folk and popular strains of evangelical Christian song could occupy simultaneous but independent roles in the dissemination of repertory.

Perhaps because of their late appearance, both variants did not influence other publications much. As with their western counterparts, southern variants disseminated over a largely localized area, as seen with the southern variant of CLAMANDA found in *The Sacred Harp*. This particular variant, though found in separate parts of the state did not circulate outside of Georgia. Its appearance in *The Social Harp* by John G. McCurry owes to the fact that he reprinted a number of items from *The Sacred Harp*.

In spite of its limited circulation, Jones' variant from Mississippi manifested a cross-regional influence. Tennessee compiler, Andrew Johnson, who had earlier published Carden's SHOUTING HYMN in a four-shape tunebook in 1839, published Jones' Mississippi variant, THE CHRISTIAN SOLDIER, in his final tunebook, *The Western Psalmodist* (1853). Johnson retitled the setting AMBOY borrowed from Amos Sutton Hayden. Significantly, *The Sacred Melodeon* (1848) by Hayden heavily influenced Johnson's tunebook, extending also to the use of Jesse B. Aiken's seven-shape notation system. As a result, Johnson, then living in south-central Tennessee attempted to draw upon a relatively local southern variant setting and to present it following the conventions of Hayden's immensely popular western tunebook, *The Sacred Melodeon*. This converging of south and west in Tennessee illustrates the flexible nature of identity and borders, and cultural and regional influences found in regional border areas during the late Antebellum Period leading up to the American Civil War.

138.7 Summary and Conclusions

Despite its obscure origins and dubious ties to Freemasonry, Jamaica, and Scotland, "The Peacock" tune family demonstrates many facets of sacred music used and appropriated by Calvinists, Calvinist Nonconformists, Primitive Christians, and other enthusiastic denominations. Originating as a Scottish naval military quickstep and associated with two secular poems by Robert Burns, "The Peacock" became one of the most popular and widespread of America's folk hymns. As a sacred contrafactum, "The Peacock" variants do not replicate any one secular version of the tune. Rather, the wide variations that existed among versions demonstrated its popularity and broad dissemination among evangelical Christians throughout the Northern, Middle Atlantic, Western, and Southern states. Reflecting simultaneously folk process, parody, Calvinist and Nonconformist theological propensity, and geographical dissemination tied to cultural migration and settlement patterns, folk hymns revealed the subtle and multi-layered facets of material culture in the United States.

Given the nature of this study and its method, other inquiries might have expanded upon the information conveyed by "The Peacock" variants. This study chose a contrafactum to reveal geographic distribution and identity. Choosing a folk hymn that had no secular original popular source may have resulted in a different set of denominational parameters. Likewise, studying a folk hymn that appeared in an even broader range of source material would likely have revealed other patterns of migration. For instance, none of the variants of "The Peacock" appeared in Hawaiian missionary imprints, nor were they printed in collections associated with liturgy-based denominations and religions. Finally, choosing a folk hymn found in German-language source material might have presented a different set of cultural parameters for understanding the migration of non-English-speaking residents. However, though no surviving variant chronicles all of the cultural and geographical variations within the repertory, all conformed to the same established trends and patterns of migration throughout the U.S. during its formative period of development.

138.8 Figure and Caption Information Notes

All melody variants in the caption information table and maps appear in a number code that equates a number with its degree on the scale. The incipit uses numbers instead of letters because the same numbers apply to any key, following the principle of the moveable do. If the key is G major, then the note G would be 1, in E major the note E would be 1. Since there are seven notes in a scale, each melody variant only employs the numbers 1–7. For instance, the opening theme of "Twinkle, twinkle little star" would be rendered 1155665 4433221 for the first two lines of text. The symbols u and d indicate a shift in octave (u = up, d = down). Thus, a complete octave scale in this notation would be rendered 1234567u1 for its ascent, and 1d7654321 for its descent. Finally, for the incipits listed in Fig. 138.3, parenthetical marks denote melismas, or a single syllable sung to more than one note. Any number appearing in a parenthesis is joined to the number preceding the parenthetical mark. For instance, the opening melody to "Amazing grace" would be written thus: 5u13(1)321d65.

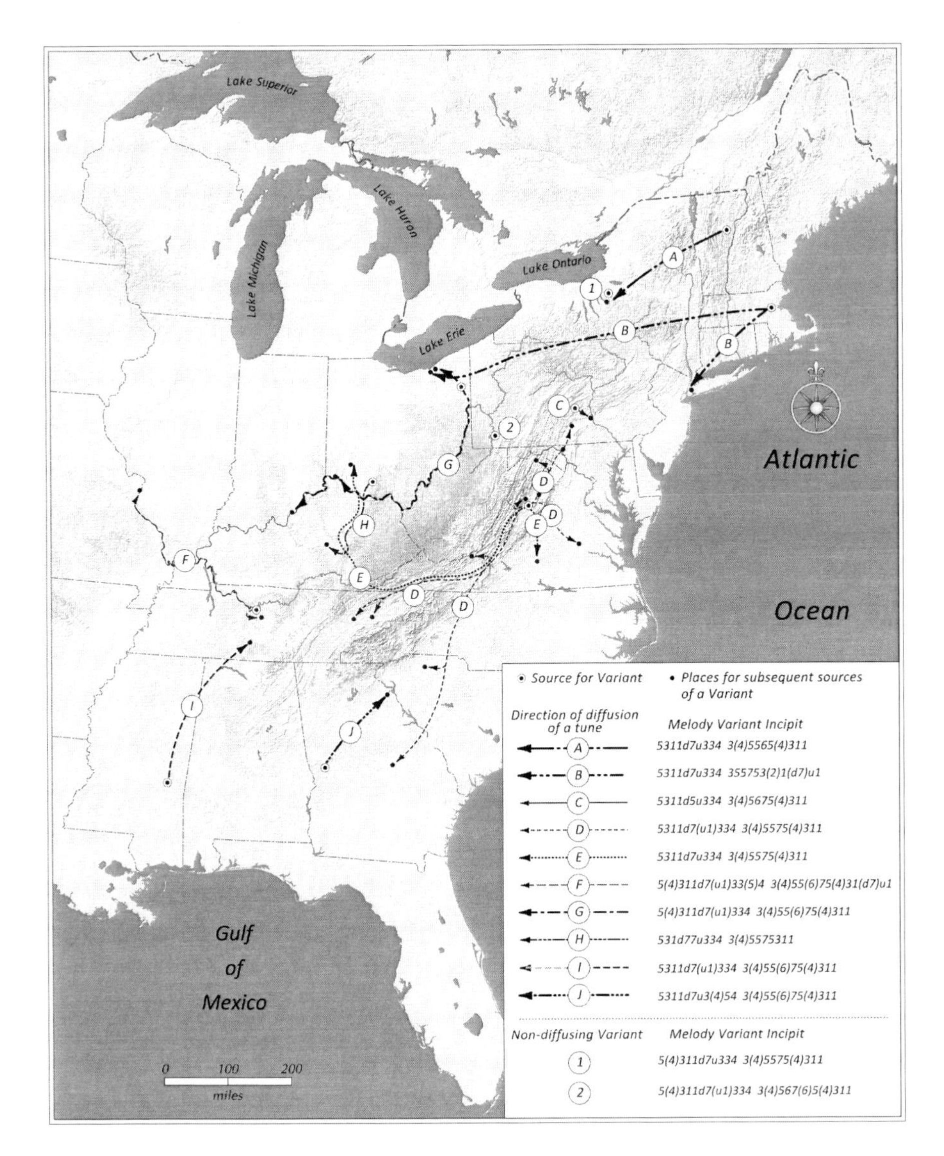

Fig. 138.3 Dissemination of sacred variants of "The Peacock" tune family in the eastern U.S. (Map by Dick Gilbreath, University of Kentucky Gyula Pauer Center for Cartography and GIS; commissioned by the editor)

References

Amen, J. (c. 1820). *Manuscript copybook of sacred music*. Sinking Spring (?) (Annabel Morris Buchanan Papers, # 4020, Folder 434). Chapel Hill: Manuscripts Department, Southern Historical Collection, University of North Carolina.

Anthony, J., Jr. (1831). *The Western minstrel, or Ohio melodist*. Cincinnati: E. H. Flint, printed at the Cincinnati Journal Office.

Bailyn, B. ([1986] 1988). *Voyagers to the west: A passage in the peopling of America on the eve of the revolution*. New York: Vintage Books.

Ball, E. (1849). *The manual of the sacred choir: A selection of tunes and hymns*. Richmond/Philadelphia: Harrold & Murray/Thomas Cowperthwait & Co.

Bealle, J. (1997). *Public worship, private faith: Sacred Harp and American folksong*. Athens: University of Georgia Press.

Benes, P. (Ed.). (1982). Psalmody in coastal Massachusetts and in the Connecticut River Valley. In *The bay and the river: 1600–1900*. Boston: Boston University.

Britton, A. P., Lowens, I., & Crawford, R. (1990). *American sacred music imprints 1698–1810: a bibliography*. Worcester: American Antiquarian Society.

Broadwood, L. E., & Maitland, J. A. F. (1893). *English country songs*. London: Leadenhall Press.

Brock, D. A. (1996). *A foundation for defining southern shape-note folk hymnody from 1800 to 1859 as a learned compositional style*. Ph.D. dissertation, Claremont Graduate School.

Buchanan, A. M. (1938). *Folk-hymns of America*. New York: J. Fischer and Brother.

Burns, R. (1803). *Scots musical museum*. Edinburgh: James Johnson.

Burns, R. (1895). *The Works of Robert Burns, 2*. Edinburgh: James Thin.

Caldwell, W. ([1834], 1837). *Union harmony: Or family musician. Being a choice selection of tunes*. Maryville: F. A. Parham.

Carden, A. D. (1829). *United States harmony, containing a plain and easy introduction to the grounds of music*. Nashville: John S. Simpson.

Carden, A. D., Rogers, S. J., Moore, F., & Green, J. (1824). *The Western harmony, or, the learner's task made easy*. Nashville: Allen D. Carden and Samuel J. Rogers.

Chapin, A. (c. 1798). *Manuscript copybook of sacred music*. Washington County (?): Sherrill Papers, private collection of David Thomas, Peninsula, OH.

Chapin, L. (c. 1780–1830). *Manuscript copybook and hymns*. Springfield, MA/Springfield, VA/Washington, KY/Cincinnati: Blinn Family Papers, Cincinnati Historical Society.

Chappell, W. (1859). *Popular music of the olden time: a collection of ancient songs, ballads, and dance tunes, illustrative of the national music of England*. London: Cramer, Beale & Chappell.

Churchal, A. (c. 1810–30). *Manuscript copybook of sacred and secular vocal music, and instrumental music*. Onondaga: Antiquarian Society, Worcester, Ma. (photocopy of original). Octavo Series, vol. 5. Irving Lowens Collection, American

Colonial Music Institute. Early American Secular Music and Its European Sources, 1589–1839: An Index. www.colonialdancing.org/Easmes/Index.htm

Cross, W. R. ([1950] 1965). *The burned-over district: the social and intellectual history of enthusiastic religion in western New York, 1800–1850*. New York: Harper & Row.

Cumnock, F. (Ed.). (1980). *Catalog of the Salem Congregation music*. Chapel Hill: University of North Carolina Press.

Cushing, E. (1789). *Cushing Eells's music book*. Norwich: The Connecticut Historical Society, Hartford.

Davisson, A. (1820a). *A supplement to the Kentucky harmony*. Harrisonburg: Printed and sold by the Author.

Davisson, A. ([c. 1821] 1820b). *A supplement to the Kentucky harmony* (2nd ed.). Harrisonburg: Printed and sold by the Author.

Davisson, A. (1821). *Introduction to sacred music, extracted from the Kentucky Harmony, and chiefly intended for the benefit of young scholars*. Harrisonburg: By the Publisher.

Davisson, A. (1825). *A supplement to the Kentucky harmony* (3rd ed.). Harrisonburg: Printed and sold by the Author.

Davisson, A. to Henkel, S. (December 12, 1817), Harrisonburg: University of Virginia, Alderman Library, MSS14434, Box Number 1, Folder date 1817–1820.

Dickinson, S. (c. 1800). *Silas Dickinson's book*. Amherst (?): New York Public Library, Performing Arts, *MNZ MUS RES.

Eskew, H. L. (1966). *Shape-note hymnody in the Shenandoah Valley, 1816–1860*. Ph.D. dissertation, Tulane University.

Farrer, D. (1828). *The Christian melodist; containing a selection of tunes in the different metres.* Utica: William Williams.
Fife, J. (1780–[c. 1804]). *Music copybook of instrumental music.* Perthshire (?) and at sea: Private Collection. Microfilm copy, National Library of Canada, Music Division, M2233.
Fuld, J. J., & Davidson, M. W. (1980). *18th-century American secular music manuscripts: an inventory.* Philadelphia: Music Library Association.
Funk, J. (1835). *A compilation of genuine church music, comprising a variety of metres, all harmonized for three voices* (2nd ed.). Winchester: J. W. Hollis.
Funk, J. (1842). *A compilation of genuine church music* (3rd ed., imp. and enl., besides an appendage of 40 pages of choice music). Harrisonburg: Henry T. Wartmann.
Funk, J. (1847). *A compilation of genuine church music* (4th ed., imp. and enl. with the appendage blended with the body of the work). Mountain Valley: J. Funk, S. Funk.
Funk, J. and Sons. (1851). *The harmonia sacra, being a compilation of genuine church music. Comprising a great variety of metres* (5th ed.). Singer's Glen: By the Authors.
Funk, J. and Sons. (1854). *The harmonia sacra* (6th ed.). Singer's Glen: By the Authors.
Funk, J. and Sons. (1856). *The harmonia sacra* (7th ed.). Singer's Glen: By the Authors.
Funk, J. and Sons. (1860). *The harmonia sacra* (10th ed.). Singer's Glen: By the Authors.
Gambosi, M. (Ed.). (1970). *Catalog of the Johannes Herbst Collection.* Chapel Hill: University of North Carolina Press.
Gentleman's pocket companion for the German flute or violin. (c. 1802). New York: G. Gilfert.
Gould, N. D. (1853). *Church music in America, comprising its history and its peculiarities at different periods, with cursory remarks.* Boston: A. N. Johnson.
Harrison, T. (1839). *Music simplified: or a new system of music.* Springfield: Office of "The Republic.
Hauser, W. (1848). *Hesperian harp: A collection of psalm and hymn tunes, odes and anthems; and Sunday-school.* Philadelphia: For the Author.
Hayden, A. S. (1835). *Introduction to sacred music; comprising the necessary rudiments, with a choice collection of tunes, original and selected.* Pittsburgh: Johnston & Stockton.
Hayden, A. S. (1838). *Introduction to sacred music* (2nd ed.). Pittsburgh: Johnston & Stockton.
Hayden, A. S. (1848). *The sacred melodeon, containing a great variety of the most approved church music.* Pittsburgh: Wm. Overend & Co.
Hendrickson, G. (1848). *The union harmony, or, a choice collection of psalm tunes, hymns, and anthems.* Mountain Valley: Joseph Funk and Sons.
Hendrickson, G. (1850). *The union harmony* (2nd ed.). Mountain Valley: Joseph Funk and Sons.
Hendrickson, G. (1855). *The union harmony* (3rd ed.). Mountain Valley: Joseph Funk and Sons.
Hickock, J.[ohn] H.[oyt] (1832). *The sacred harp, containing part first, a clear compendium of the rules and principles of vocal musick. Part second, a collection of the most approved church tunes . . . Part third, a collection of popular airs, and devotional hymns.* Lewistown: Shugert & Cummings.
Hickok, J.[ohn] H.[oyt] (1840). *The social lyrist: A collection of sentimental, patriotic, and pious songs, set to music, arranged for one, two, and three voices.* Harrisburg: W. Orville Hickok, ster. L. Johnson, Philadelphia.
Hood, G. (1846). *A history of music in New England with biographical sketches of reformers and psalmists.* Boston: Wilkins, Carter & Co.
Horn, D. D. (1970). *Sing to me of heaven: A study of folk and early American materials in three Old Harp books.* Gainesville: University of Florida Press.
Ingalls, J. (1805). *The Christian harmony; or, songster's companion.* Exeter: Henry Ranlet.
Jackson, J. B. (1838). *The Knoxville harmony of music made easy.* Madisonville: For D. & M. Shields & Co. and John B. Jackson, by A. W. Elder.
Jackson, J. B. (1840). *The Knoxville harmony of music made easy* (2nd ed.). Pumpkintown: D. and M. Shields and Co., and John B. Jackson, Proprietors: Johnson and Edwards, printers.
Jackson, G. P. (1926). The fa-sol-la folk. *Musical Courier, 93*(11).
Jackson, G. P. (1933). *White spirituals in the southern uplands.* Chapel Hill: The University of North Carolina Press.
Jackson, G. P. (1937). *Spiritual folk-songs of early America.* New York: J. J. Augustin.

Jackson, G. P. (1939). *Down-east spirituals and others*. New York: J. J. Augustin.
Jackson, G. P. (1943). *White and Negro spirituals, their life span and kinship*. New York: J. J. Augustin.
Jackson, G. P. (1952). *Another sheaf of white spirituals*. Gainesville: University of Florida Press.
Johnson, A. W. (1839). The American harmony: Containing a plain and easy introduction to the grounds of music (2nd ed.). Nashville: For the author by W. H. Dunn.
Johnson, A. W. (1853). *The western psalmodist: A new system of notation; a collection of church music, consisting of a great variety of psalms and hymns*. Nashville: Nashville Union Office for A. W. Johnson.
Jones, L.[azurus]. J. (1849). *Southern minstrel: A collection of psalm and hymn tunes, odes, and anthems: in three parts*. Philadelphia: Grigg, Elliot and Co.
Karpeles, M. (Ed.). (1932). *English folk songs from the southern Appalachians, collected by Cecil J. Sharp; comprising two hundred and seventy-four songs and ballads with nine hundred and sixty-eight tunes*. Oxford: Oxford University Press.
Krepps, J. (1830). *Manuscript copybook of sacred music*. Brownsville: Historical Society of Western Pennsylvania, Pittsburg.
Krummel, D. W., Geil, J., Dyen, D. J., & Root, D. L. (Eds.). (1981). *Resources of American music history: A directory of source materials from colonial times to World War II*. Urbana: University of Illinois Press.
Leonard, S. W. (1870). *New Christian psalmist: A collection of psalms, hymns, and spiritual songs, with appropriate music*. Cincinnati: R. W. Carroll & Co.
Leonard, S. W.[hite], & Fillmore, A.[ugustus] D.[amon] (1850). *The Christian psalmist, a collection of tunes and hymns, original and selected, for the use of worshiping assemblies* (Rev. and enl. by S. W. Leonard). Louisville: S. W. Leonard; Stereotyped and Printed by Morton & Griswold.
Library of Congress. Online Catalogue of the Library of Congress. www.worldcat.org
Longenecker, S. L. (2000). The narrow path: Antislavery, plainness, and the mainstream. In K. E. Koons & W. R. Hofstra (Eds.), *After the backcountry: rural life in the Great Valley of Virginia 1800–1900* (pp. 185–193). Knoxville: University of Tennessee Press.
Lorenz, E. J. (1978). *Glory, hallelujah! The story of the camp meeting spiritual*. Nashville: Abingdon.
Lowens, I. (1952). John Wyeth's repository of sacred music, part second: a northern precursor of southern folk hymnody. *Journal of the American Musicological Society, 5*(2), 114–131.
Manuscript copybook of sacred vocal and instrumental pieces (c. 1820). Western NY: Center for Popular Music; Murfreesboro: Middle Tennessee State University, [06–027] N5877.
Marshall Family. (1809–[c. 1825]). *Manuscript copybook of sacred music*. Charlotte County (?): The Library of Virginia, Richmond (Personal papers collection, Accession 22349).
McCurry, J. G. (1855). *The social harp, a collection of tunes, odes, anthems, and set pieces*. Philadelphia: T.K. Collins, Jr.
McDonald, Rev. W., & Hubbard, S., Esq (1857). *The Wesleyan sacred harp a collection of choice tunes and hymns for prayer class, and camp meetings, choirs, and congregational singing*. Boston: John P. Jewett & Company; Cleveland: H.P.B. Jewett; New York: Sheldon, Blakeman & Company.
Music, D. W. (1995). Seven 'new' tunes in Amos Pilsbury's United States' sacred harmony (1799) and their use in four-shape shape-note tunebooks of the southern United States before 1860. *American Music, 13*(4), 403–447.
Music, D. W. (Ed.). (2005). *A selection of shape-note folk hymns from southern United States tune books, 1816–61*. Middleton: A-R Editions, Inc.
Myers, L. C. (1853). *Manual of sacred music, or a choice collection of tunes, psalms, hymns, and spiritual songs, arranged for three voices*. Mountain Valley: Joseph Funk & Sons.
Norton, K. (2002). *Baptist offspring, southern midwife – Jesse Mercer's 'Cluster of spiritual songs' (1810): A study in American hymnody*. Warren: Harmonie Park Press.
Professor, A. (1826). *An introduction to the art of playing the bassoon*. Philadelphia: G. E. Blake.
Property of the Bellamy Band. (1799–[c. 1805]). *Manuscript copybook*. Hamden: Library of Congress, M1200. B45 Case.

Pynchon, B. (1797–1805). *Bathshua Pyncon's book*. Springfield. Connecticut Valley Historical Museum, Misc. Pynchon Family Material, Box 1.
Rhinehart, W. R. (1831). *The American, or union harmonist: or, a choice collection of psalm tunes, hymns and anthems*. Chambersburg: Henry Ruby.
Rouse, P., Jr. ([1973] 2004). *The great wagon road*. Richmond: Dietz Press.
Scholten, J. W. (1972). *The Chapins: A study of men and sacred music west of the Alleghenies, 1795–1842*. Ed. D. dissertation, University of Michigan, Ann Arbor.
Seat, J. B. (1831). *The St. Louis harmony: Containing the rudiments of music, made easy, and carefully arranged, to suit the capacity of the young learner*. Cincinnati: Lodge & L'Hommedieu, 1831.
Seeger, C. (1940). Contrapuntal style in three-voice shape-note hymns. *Musical Quarterly, 26*(4), 483–493.
Shaw, B., & Spilman, C. H. (1829). *Columbian harmony, or pilgrim's musical companion; being a choice selection of tunes, selected from the works of the most eminent authors, ancient and modern*. Cincinnati: Lodge, L'Hommedieu and Hammond.
Smith, H. (1841). *The church harmony. Containing a selection of approved psalm and hymn tunes* (12th ed., impr. and enl.). Chambersburg: Henry Ruby.
Spicer, I. (1797–[1814]). *Spicer's pocket companion*. Chatham: Connecticut Historical Society, MS Stack, Facs.
Stanislaw, R. J. (1978). *A checklist of four-shape shape-note tunebooks* (I.S.A.M. monographs: Number 10). Brooklyn: Institute for Studies in American Music, Brooklyn College of the City University of New York.
Steelman, R. (Ed.). (1981). *Catalog of the Lititz Congregation Collection*. Chapel Hill: University of North Carolina Press.
Steffey, J. W. (1836). *The valley harmonist, containing a collection of tunes from the most approved authors, adapted to a variety of metres. Also–a selection of set pieces and anthems*. Winchester: Robinson and Hollis.
Temperley, N. (1993). The Lock Hospital Chapel and its music. *Journal of the Royal Musical Association, 118*(44–72).
Temperley, N. (2000/2006). The Hymn Tune Index. http://hymntune.library.uiuc.edu.
Temperley, N., Manns, C. G., & Herl, J. (Ed.). (1998). *The hymn tune index: A census of English-language hymn tunes in printed sources from 1535 to 1820*. Oxford: Clarendon Press; New York: Oxford University Press.
The American volunteer. (c. 1797). New England. John Hay Library, Brown University, HARRIS, Music MA 638.2.
The jubilee harp, a choice selection of psalmody, ancient and modern, designed for use in public and social worship (1866). Boston: Advent Christian Publication Society.
Walker, W. (1846). *The southern and western pocket harmonist, intended as an appendix to The Southern Harmony*. Philadelphia: Cowperthwait & Co.
Walker, W. (1854). *The southern harmony, and musical companion: Containing a choice collection of tunes, hymns, psalms, odes, and anthems* (New ed., rev. and imp). Philadelphia: E.W. Miller.
Walker, W. ([1869] 1873). *The Christian harmony: In the seven-syllable character note system of music* (2nd ed.). Philadelphia: Miller's Bible and Publishing House.
White, G. (c. 1790–1830). *The scale of the gamut for the violin*. Cherry Valley: Private Collection, Ray Hauley, Valatie, NY
White, B. F., & King, E. J. (1844). *The sacred harp, a collection of psalm and hymn tunes, odes, and anthems, selected from the most eminent authors*. Philadelphia: B. F. White and Joel King.
Wyeth, J. (1813). *Wyeth's repository of sacred music. Part second. Original and selected from the most eminent and approved authors in that science*. Harrisburgh: John Wyeth.

Chapter 139
Landscapes and Soundscapes: How Place Shapes Christian Congregational Song

C. Michael Hawn

The Gospel is like a seed and you have to sow it. When you sow the seed of the Gospel in Palestine, a plant that can be called Palestinian Christianity grows. When you sow it in Rome, a plant of Roman Christianity grows. You sow the Gospel in Great Britain and you get British Christianity. The seed of the Gospel is later brought to America and a plant grows of American Christianity. Now when missionaries came to our lands they brought not only the seed of the Gospel, but their own plant of Christianity, flower pot included! So, what we have to do is to break the flowerpot, take out the seed of the Gospel, sow it in our own cultural soil, and let our own version of Christianity grow.

(D. T. Niles Chandler 1977:16)

139.1 Introduction

Daniel Thambyrajah Niles (1908–1970), a Ceylon native (now Sri Lanka) and ecumenist, was one of the first to employ indigenous musical idioms in Asian congregational singing. For Niles singing hymns in one's own musical idiom as well as in one's mother tongue was a way for the gospel to be planted into the local soil of the people—a way to experience the Incarnation of Christ in one's own cultural terms.

Before the 1960s most of the congregational song of the Christian church was generated by the colonial missionaries from the Western churches and carried to the rest of the world. A confluence of world events, ecclesial realignments, and cultural awakenings led to changes in perspective. National independence movements around the world, especially in Africa, coincided with the Second Vatican Council (1962–1965) whose *Sacrosanctum Concilium* (Constitution on the Sacred Liturgy 1963) encouraged the use of the vernacular language in the liturgy and recognized

C.M. Hawn
Sacred Music Program, Perkins School of Theology, Southern Methodist University, Dallas, TX 72752, USA
e-mail: mhawn@mail.smu.edu

S.D. Brunn (ed.), *The Changing World Religion Map*,
DOI 10.1007/978-94-017-9376-6_139

the "genius" of various ethnic traditions. The field of ethnomusicology came to maturity in the United States resulting in fieldwork in the musics of various world cultures, even influencing missionaries from the United States and Europe who began to encourage local musical traditions in worship.

More than a decade into the twenty-first century, we are now able to map the spread of global Christian song and foster an understanding of the world church that changes our ecclesiology from a Western church that sustains the rest of the world, to a world church whose identity is shaped by reciprocal relationships. Placed in this light, the classic paradigm of lex orandi and lex credendi[1] suggests that our faith, once shaped almost exclusively by the songs and prayers of our own cultural context, now has the possibility of being enriched by a global vision of the Incarnation—Jesus, born in one culture at a specific time and place, may now become the Christ for all cultures, times and places. Singing a global faith is indeed one way to experience a global vision of Christianity.

This essay will track the affect of this world-wide Christian diaspora by citing some examples of how soundscape intersects with landscape—how the sounds of congregational songs spring from the natural, social and political soil of specific places in the world. This is at best an anecdotal survey.[2] The specificity of the congregational songs cited here, however, is a nascent attempt to place our understanding of the Incarnation of Christ in a global perspective that may shift our viewpoint from one of cultural myopia to worldwide vision.

I have chosen three broad categories of congregational song to give some focus to the plethora of examples that are available. *Engaging the Landscape* explores songs that draw specifically upon the natural world and events within a given cultural setting. This is particularly important since Western hymnody often drew specifically on the seasons and images of the northern Euro-American landscape. *Mission of the Church* provides some insight into the social-political landscape of particular regions of the world and how this reality has inspired congregational song. *Around the Table* draws upon songs for the Eucharist—the central sacrament of the Christian liturgy—expressed in the language of several local cultures. Theology, like art, is not easily placed in pigeonholes. The songs cited here often overlap the three categories chosen.

139.2 Soundscape and Landscape: Beginning the Journey at Home

> From all that dwell below the skies,
> let the Creator's praise arise;

[1] Prosper of Aquitaine (fl. 435–442) proposed that prayer shapes belief. He states "…ut legem credendi lex statuat supplicandi" (…that the law of belief stands on or is founded upon the law of supplicating or praying).

[2] For a more comprehensive perspective, see Hawn 2013).

let the Redeemer's name be sung,
through every land, by every tongue.
Eternal are thy mercies, Lord;
eternal truth attends thy word.
Thy praise shall sound from shore to shore,
till suns shall rise and set no more. (Isaac Watts, 1719)

Isaac Watts's paraphrase of Psalm 117, the shortest of the psalms, provides a historical example from the European landscape. Though a classic hymn, it too reflects a cultural perspective that is for many a normative rather than an illustrative worldview. Just like all congregational song, it comes from the soil of a specific culture, place and time.

Watts draws his paraphrase from the Authorized Version (KJV) of the Bible (1611), barely a century old at that time:

1. O praise the Lord, all ye nations: praise him, all ye people. 2. For his merciful kindness is great toward us: and the truth of the Lord endureth for ever. Praise ye the Lord.

One cannot help but speculate about Watts's emphasis upon "all nations" and "all people." Mentioned briefly in the original first verse of the scripture, this theme receives expanded treatment in Watts's paraphrase, perhaps influenced by Great Britain's world dominance as major colonial power. By the time this hymn was published in 1719, the British naval forces during the reign Queen Elizabeth had defeated the Spanish Armada in 1588; the British East India Company was already over 100 years old; the American colonies were chaffing under British rule and barely 50 years away from declaring their independence.

Perhaps it is not stretching a point that belief in the divine right of kings, the theological and political foundation for the monarchy, also provided a basis for the colonial expansion of the British realm "from shore to shore, till suns shall rise and set no more." To modern ears we may hear echoes of that famous phrase, "The sun never sets on the British Empire." It is not unlikely that those who sang this hymn in Watts's day could not easily separate their vision for the spread of Christianity, from their celebration of the expansion of the British Empire.

This interpretation in no way denigrates the poetic excellence of Watts, nor his approach to scripture. All hymns are imbued with the spirit of their age, shaped in subtle ways by the language and thought processes of their time. Watts's approach to Psalm 117 may be seen as prophetic from the viewpoint of the twenty-first century. In the nearly 300 years since its composition, Watts's dream has in many ways come true. Christ's name is sung in virtually around the globe "from shore to shore." However, Watts would be shocked at the shift that has taken place in Christian demographics. The majority of Christians are no longer from Europe or even North America. Rather, at least two-thirds of Christians live in the southern hemisphere. He would also be surprised by the soundscape of global song springing from the richness of local dirt rather than transplanted directly from Euro-American soils.[3]

[3]For a discussion of the changing demographics of Christianity see Philip Jenkins, The Next Christendom: The Coming of Global Christianity (New York: Oxford University Press, 2002).

139.3 Engaging the Landscape

Elizabeth J. Smith (b. 1956), an Anglican vicar in Perth, has been a lover and writer of hymns from a young age. Nurtured on the classic hymns of Isaac Watts, Charles Wesley and the translations of John Mason Neale, she was encouraged by Brian Wren, a hymn writer from the England now residing in the United States, to find her own voice.

The Australian context also has informed her writing. "Where Wide Sky Rolls Down" is her only hymn that contains specifically Australian words that draw upon natural images from the continent (see www.youtube.com/watch?v=eFX7VBH_2ZQ for an interpretation of this hymn). In many ways, this hymn is a twentieth-century Australian equivalent of "All things bright and beautiful" from England by one of the great women hymn writers of the last half of the nineteenth century, Cecil Frances Alexander.

"Where wide sky rolls down"

(© *Elizabeth J. Smith, Australia)*

Where wide sky rolls down and touches red sand,
where sun turns to gold the grass of the land,
let spinifex, mulga and waterhole tell
their joy in the One who made everything well.

Where rain-forest calm meets reef, tide and storm,
where green things grow lush and oceans are warm,
let every sea-creature and tropical bird
exult in the light of the life-giving Word.

Where red gum and creek cross hillside and plain,
where cool tree-ferns rise to welcome the rain,
let bushland, farm, mountain-top, all of their days
delight in the Spirit who formed them for praise.

Now, people of faith, come gather around
with songs to be shared, for blessings abound!
Australians, whatever your culture or race,
come, lift up your hearts to the Giver of grace.

The hymn has the feel of an Australian psalm of praise for creation. Because the religious language is subtle, this hymn is often used for civic occasions in Australia. While over 60 % of Australians would ascribe to Christianity in some form, less than 10 % attend church on a given Sunday.[4] This hymn fits this demographic well—religious in feel but not explicitly. A more subtle feature is the Trinitarian structure of the first three stanzas: God: "the One who made everything well." Jesus Christ: "light of the life-giving Word." Holy Spirit: "delight in the Spirit… ." By choosing her theological language carefully, the author has provided a hymn that celebrates the uniqueness of the Australian landscape that may be sung by most Australians.

[4] See Bellamy and Castle (2004).

A subtler feature of the hymn is its use of direct, non-sentimental, monosyllabic, simple language that fits the Australian sensibility. The poet also celebrates the natural and ethnic diversity of the continent as well. Not only may an Australian identify with the breadth of the landscape and the unique flora and fauna of the continent mentioned in the hymn, they would also find a deeper authentic message in a hymn that names the diversity of its people, and carefully chooses language that brings together religious and nationalistic sentiments.

Francisco Feliciano (b. 1941–2014), noted composer and President of the Sambalikhaan (The Asian School of Music, Worship and the Arts), lived in Manila. His hymn, "Still I Search for My God" expresses the mysticism of one who seeks the source of life for the universe (see http://rockhay.tripod.com/worship/hymnlist.htm to hear the melody). The theme of mystery is prevalent in his compositions, both in the texts he writes or chooses and in the musical settings.

"Still, I search for my God"

(© *Francisco F. Feliciano, Philippines.*)

Still, I search for my God in silence, I marvel at the
Universe, the world it contains, its beauty, its harmony
Creator of such perfection, who else could it be?

Come, listen to the trees, and green fields, the rivers and the
Morning breeze, the birds of the air all singing their Maker praise!
Creator of countless wonders, who else could it be?

Yes, I am filled with joy and breathe in the presence of the
LORD, my God, your praise I will sing, your power declare in chant, and
When I am moved to worship, peace reigns in my heart.

One senses the awe of Psalm 19:1, "The heavens declare the glory of God; and the firmament shows His handiwork." (NKJV) Living in the crowded and frenetic existence of Manila, the poet created an oasis of several acres of tranquility and natural beauty in the heart of Manila for the study of music and liturgy. The scene depicted could describe many places on earth. For those living in a dense urban environment, however, one must seek peace and mystery intentionally.

Though Feliciano is a Roman Catholic, the music was inspired by a Buddhist meditation exercise.

> Imagine that you are standing in front of a calm lake. The lake is still and tranquil, and you can see the reflections of the further shore. You take a tiny pebble, and toss it into the midst of the reflections with respect, as if it were an offering to the ancient gods that live in the depths of the waters. The stone plops into the water, and disappears without trace, leaving behind waves of concentric ripples.
>
> Each ripple presents you with a slightly different perspective on the reflections of the other shore. You watch the ripples radiating from the place where the stone vanished, as they widen and fade and eventually disappear altogether. Then once the lake's surface is still once more, the reflections have returned to normal, and you toss in another offering.[5]

[5] This exercise was described to me by the composer. The quotation is from "Reflections in mindfulness" www.wildmind.org/mindfulness/four/reflections (accessed March 21, 2012).

The melodic shape was inspired by this exercise. The first phrase begins on a single and a half-step interval. Each subsequent phrase parallels the increasingly concentric circles described in the exercise, returning at the end of each stanza to a single pitch as the tranquility of the water returns.

In this way the composer has created a hymn of awe and mystery that can be embraced in a pluralistic context—Christian and Buddhist—or any seeker who stops to ponder the source of the universe and the mystery of life.

"Ulo Tixo" is the earliest Christian hymn composed by a black South African, Ntsikana son of Gaba (c. 1780–1821), usually considered to be the first Xhosa Christian. The hymn of the prophet Ntsikana is the prototype of church music in a traditional Xhosa style. David Dargie describes the prophet as an attractively mysterious figure in Xhosa history. A Cirha, and son of a councilor of the famous chief Ngqika, he was the first Xhosa Christian. It was probably as a herd-boy that he heard the preaching of the first missionary among the Xhosa, Dr. J. T. van der Kemp of the London Missionary Society, who worked in Ciskei from 1799 to 1801. The missionary, however, made no Xhosa converts; it was only years later, about 1815, that Ntsikana underwent a conversion experience without the presence of any missionary or white person (Dargie 1982).

"Ulo Tixo"/"The Great God"

(*Song by Ntsikana, Xhosa, South Africa;*

Translation by John Knox Bokwe.)
The Great God, He is in heaven.
Thou art thou, Shield of truth.
Thou art thou, Stronghold of truth.
Thou art thou, Thicket of truth.
Thou art thou, who dwelleth in the highest.
Who created life (below)
and created life (above).
The Creator who created, created heaven.
This Maker of the stars, and the Pleiades.
A star flashed forth, telling us.
The maker of the blind,
does He not make them on purpose?
The trumpet sounded, it has called us,
As for His hunting, He hunteth for souls,
Who draweth together flocks opposed to each other.
The Leader, he led us.
Whose great mantle, we put it on.
Those hands of Thine, they are wounded.
Those feet of Thine, they are wounded.
Thy blood, why is it streaming?
Thy blood, it was shed for us.
This great price, have we called for it?
This home of Thine, have we called for it?

Ntsikana's vision was dramatic. He was standing outside his hut when he saw a strange light strike his ox, called Hulushe, as the sun rose. A boy nearby could not see this light. Later in the day he went to a celebration with his family and miraculous occurrences continued. Janet Hodgson describes the vision as follows:

> Three times, as he started to dance, a raging wind arose out of clear blue sky forcing all dancers to stop. Tradition has it that he now became aware that the Holy Spirit had entered him, but the people thought him bewitched. He promptly took his family home and they were amazed when on the way he washed the red ochre from his body in the Gqora river, as a sign of his entry into a new life. (Hodgson 1980:4)

Eventually he gathered a group of disciples and settled into a new life, teaching them this hymn. The text includes the theology of the Creator and one true God, the coming Messiah who suffered for all people, and reconciliation among opposing groups. The latter theme was particularly timely due to the increasing tension between the Xhosas and the white settlers during this time.

The authenticity of Ntsikana's hymn may be discerned ultimately by its abiding influence on the spiritual life of the Xhosas, their persistent presence as a living artifact of Xhosa cultural heritage, and, therefore, the embodiment of hope for liberation for the Xhosa people.

Ntsikana's hymn (sometimes presented as four songs) is rooted in the traditional music culture of the Xhosa people. The music is in the style of the musical bow tradition, alternating around a progression of two major chords a step apart. The landscape is evident both in the conflict between cultures and in the incorporation of nature. For example, the Pleiades is a constellation seen only from the southern hemisphere. While never mentioning Jesus by name, images of Christ's suffering abound. The metaphor of Christ as the "hunter" who "draweth flocks together opposed to each other" is an image of reconciliation drawn from the Xhosa tribe as herders. The oldest example cited in this essay, this hymn is usually the first selection in Xhosa hymnals for over 100 years.

The hymns of Xiao-min Lü (小敏,吕) have become legendary in China and beyond. A rural young woman from a poor farming family in Fancheng, Nanyang, Henan, her hymns grow out of the persecution faced by the house-churches (Jiating Jiao-hui, 家庭教會) in China during the cultural revolution (1966–1976). After her conversion to Christianity in 1990, she has composed more than 1,400 songs, known as the Canaan Hymns, sometimes considered to be the "official hymnal" of the house-churches in China. Xiao-min conceives the melody and texts as a unit in the brief span of 5–10 min. They spread orally initially but have been notated by others, some even being arranged for choirs and orchestras.[6] Many Chinese consider these songs to be truly gifts from God because Xiao-min has little education and no formal musical training.[7]

---

[6] A brief biography of Xiao-min is available at http://womenofchristianity.com/hymn-writers/xaio-min/ (accessed on September 20, 2014).

[7] For a fuller introduction to the Canaan Hymns, see Irene Ai-Ling Sun," Songs of Canaan: Hymnody of the House-Church Christians in China," Studia Liturgica 37:1, 2007, 98–116.

The geographical topography of the Henan province, the largest in China, provides the source for the metaphors of the journey of the Christian. "Severe Times" was the result of her imprisonment because of her faith. A prominent image found in the mountainous area of her home province, the pine tree, becomes the metaphor for expressing her faith and the strength that God gives her strength to withstand difficult ("severe") times. In spite of the continual assault of the weather through the changing seasons, the pine tree remains erect, green, and endures for generations. The mountain itself is also a metaphor—an obstacle on the road of travail that must be traveled. Notwithstanding the difficulties endured by the composer and by many Christians, the hymn contains no self-pity or regret as life's struggles only serve to refine the traveler into a "useful vessel" for God, a reference to Romans 9:21.

"Severe Times"

(*Song by Xiao-min Lü, Canaan Hymns, China.*)

It is during severe times
that we are refined and train ourselves up.
It is during severe times that we grow.
Look at the pine tree in the mountain
that faces wind from all sides;
enduring long years of hardship, enduring extreme heat
and cold, passing summers and winters,
yet it remains erect
yet it retains its green, bowing to nothing, enduring to all ages.
It is during severe times that our will be refined and tested.
It is during severe times that we experience life.
It is true that the road is uneven and full of difficulties
yet let us take a deep breath and wipe off the sweat
from our face and mount the mountain.
This road that we must go through is in the hand of LORD.
We will be refined to become His useful vessel.

Geonyong Lee (b. 1947) was born in what is now North Korea or officially the Democratic People's Republic of Korea. His father was a Presbyterian minister. Following 40 years of occupation by Japan (1905–1945) and by war in 1950, the Korean people found themselves divided again after the signing of the Armistice Agreement of 1953. It was at this time when Lee left the north with his family and moved to South Korea, the Republic of Korea. In the late 1990s, he was among the first group to visit North Korea as a part of a cultural exchange (Interview with G. Lee 1999, Seoul, Korea).

Lee, President of the Korean National University of Arts (2002), is one of the leading composers of serious music in South Korea composing in a style that draws both upon the classical Western and Korean musical traditions, forming The Third Generation, a composer's forum devoted to "creating music that represents the unique identity of the third worlds and Korea."[8]

---

[8] See ""Korean-American Music Composition Competition-Judges" 2005" www.sejongsociety.org/2005scs_composition/2005scs_composition_judges.htm (accessed April 2, 2012).

Unlike the other hymns in this section, "O-so-so" does not mention the natural landscape of the Korean peninsula. The physical landscape of the border has everything to do with this song, however. This is a plea for political and spiritual unity on a peninsula divided by the strip of land 160 miles (250 km) long and 2.5 (4 km) miles wide called the Demilitarized Zone (DMZ) at the 38th parallel, one of the most heavily militarized borders on earth.

"O-so-so" / "Come Now, O Prince of Peace"

(*Song by Geonyong Lee, Korea.*)

Come now, O Prince of peace,
Make us one body, come, O Lord
Jesus, reconcile your people.

"Al despuntar en la loma el día" ("When o'er the hills"), by Cuban composer Heber Romero (b. 1954) is from his *La Liturgia Criolla*, published in the 1970s (see www.youtube.com/watch?v=ZutVLjosth0 and www.youtube.com/watch?v=badxiskvTWA for renditions of this song). Born in Santa Clara, Cuba, Romero has been part of the ecumenical renewal of Cuban worship composing in the local folk music styles of the people in the spirit of Vatican II. He was one of the "Nueva Trova Cubana," a movement that began around 1967 following the Cuban revolution in 1959. Known for composing in folk music idioms, composers combine them with progressive or even political lyrics (Rosales 1985:373–376). The hymn draws upon the *guajira*, an improvisatory Cuban folk style found especially in the Oriente (eastern) Province (Lockward 2002:21). Guajira music ("country music") evokes the music of the countryside in its rhythms and instruments. The style represented in this song developed in the 1930s as the *guajira-son*—a 4/4 m with guitars and African-influenced rhythms. The lyrics of the *guajira*, extolling the beauty of the countryside, make this the perfect vehicle musical vehicle for the hymn. The rising eastern sun brings the assurance of God's presence. Stanza two references the "green coffee fields"—a reference to the primary coffee growing region in Cuba, the Sierra Maestra Mountains which rise from the southeastern coast of the island. The view of the coffee fields along with their aroma, the sounds of the birds, and the feel of the morning breeze, provide a total sensory experience. Our voices join with the sounds of nature as we praise God.

"Al despuntar en la loma el día" / "When o'er the hills"

(*Words and melody © 1980 Heber Romero*)

When o'er the hills morning light is breaking,
once more your glory is born.
Filled with your joy all the fields are waking
and growing grass greets the morn.
A different day now is dawning
and yet, with fear, I am torn;
but you're the same God each morning;
as the day dawns, I'm reborn.

On the horizon the sun is blending
with hues of green coffee fields,
and from the bush there's a birdsong wending
till it a new life reveals.
The air with fragrance is swelling,
a sweet aroma I smell.
My song the heavens are telling;
"God indeed does all things well!"
I want to be like a brook that's flowing;
O how refreshing such grace!
Or like the sound of a soft wind blowing—
your thoughts through palm groves I'd trace.
I hear the rooster that's crowing,
the thrilling birdsong takes wing!
My voice ascends and its flowing.
How I sing, God, how I sing!

139.4 Mission of the Church

"When the church of Jesus" draws its social location from inner city London. Fred Pratt Green (1903–2000) was a Methodist minister who, though a poet much of his life, did not discover his hymn-writing voice until retirement. This hymn, his first, was based on James 2:14–17 and written in 1968 for the Stewardship Campaign of Trinity Methodist Church in London. Though the "noise of traffic" entices some congregations to close their doors to block out the sounds of the city, Green bids us open the doors to the pain, strife and suffering of the world.

"When the church of Jesus"

(*Fred Pratt Green, UK*)

When the church of Jesus
shuts its outer door,
lest the roar of traffic
drown the voice of prayer,
may our prayers, Lord, make us
ten times more aware
that the world we banish is our Christian care.

If our hearts are lifted
where devotion soars
high above this hungry,
suffering world of ours,
lest our hymns should drug us
to forget its needs,
forge our Christian worship
into Christian deeds.

Lest the gifts we offer,
money, talents, time
serve to salve our conscience,

to our secret shame,
Lord, reprove, inspire us
by the way you give;
Teach us, dying Savior,
how true Christians live.

One of the most unusual ironies of twentieth-century hymnody appears in the second stanza where we are chastised as we sing this hymn blithely while ignoring the needs of the world: "lest our hymns should drug us to forget [the world's] needs."

The author continues to challenge the hymn singer in the final stanza. Just because "we offer…money, talents [and] time" as a part of our gifts to the church, these offerings should not "serve to salve our conscience, to our secret shame," the shame of ignoring the suffering world around us. The final stanza ends with a petition to "reprove [and] inspire us" to follow Christ's example, and to "teach us, dying Savior, how true Christians live."

The stirring tune King's Weston by noted British composer Ralph Vaughan Williams (1872–1958) rides above majestic harmonies designed for the organ. While the music serves the text well, it is ironic that the very volume of a powerful organ can also "drown the voice of traffic" and "drug us to forget [the] needs" of the "hungry, suffering world of ours."

"Siyahamba" (Freedom Song, Xhosa/Zulu, South Africa) is song of protest and struggle (see www.youtube.com/watch?v=1RZqLwqRaKQ for one of many examples of this song). Imagine yourself present for one of the most infamous of massacres against Black South Africans that took place in the township of Sharpeville on March 21, 1960. When the police opened fire on a group demonstrating against pass laws, 69 people were killed and many others maimed. News reports from CNN during the 1980s captured the attention of the world. The anti-apartheid struggle in South Africa was given voice when black South Africans and their supporters were featured singing songs of freedom.

Usually translated as "We are marching in the light of God," the deceptively simple text contains layers of meaning: "We" may be seen as a word of community—the community of those living, and the community of the living dead who watch over the actions of the living as protectors and ethical guides. In spite of suffering, "we" are not alone. "Marching" is an action that unifies the community as they move physically and spiritually in the same direction. It is a physical, kinesthetic response to the Spirit, not a passive acquiescence. In spite of our powerless social position, we do not submit to those in power.

"The Light of God" has meaning on several levels. While it is a symbol of creation and of Jesus Christ, the light of the world, it is also a common refrain in songs of healing or *ngoma* throughout Southern and Central Africa. According to anthropologist John Janzen, "Let darkness be replaced with light" is coded language for "seeing clearly" (Janzen 1992:111–118). God is the source of clear sight in the midst of the struggle, the basis for discernment and truth. As we march we can see our way ahead. Our path is clear. Where there is light, there is hope.

When this message is amplified with engaging music, the words become embodied in the lives of the community that sing and dance it. The song accommodates and even facilitates a growing, evolving community of believers. "We are marching," knowing that the living dead are singing with us and thus gives us hope. When this song is taken into worship as a processional, it brings with it the struggle of the streets and sanctifies it.

"Bring Forth the Kingdom of God" (Marty Haugen, USA) is a hymn that reflects the social location of the United States during the tumultuous 1960s and the folk music and the protest songs of that era. Haugen (b. 1950) seizes on the direct and unassuming nature of the folk idiom to call for the coming of the reign of God now. He uses the "sing-a-long" quality of his folk-like melody and the call-response (solo-congregation) structure of this genre to give the feel of participating in a movement on the streets rather than singing a hymn in a sanctuary. The acoustic guitar is ideal—portable, flexible, easily learned, supporting the singing voice without dominating—a people's instrument. For many, this is part of the soundscape of authenticity and directness—emphasizing the message more than the messenger, inviting participation rather than a solo performance.

"Bring Forth the Kingdom of God"

(by Marty Haugen, U.S.A.)

Leader: Y ou are the salt for the earth, O people:
All: Salt for the Kingdom of God!
Leader: Share the flavor of life, O people:
All: Life in the Kingdom of God!

Refrain: Bring forth the Kingdom of mercy,
bring forth the Kingdom of peace;
bring forth the Kingdom of justice,
bring forth the City of God.

Leader: You are a light on a hill, O people:
All: Light for the City of God!
Leader: Shine so holy and bright, O people:
All: Shine for the Kingdom of God!

Leader: You are a seed of the Word, O people:
All: Bring forth the Kingdom of God!
Leader: Seeds of mercy and seeds of justice,
All: grow in the Kingdom of God!

Leader: We are a blest and a pilgrim people:
All: Bound for the Kingdom of God!
Leader: Love our journey and love our homeland:
All: Love is the Kingdom of God!

The song is replete with biblical images such as Mathew 5:13–16 from the Beatitudes. The first two stanzas draw directly upon the metaphors of "salt of the earth" and "Light of the world." Stanza four calls the "pilgrim people" on a "journey" towards our "homeland"—"the Kingdom of God." While Haugen composes congregational songs in a wide range of musical styles, this song is the successor of the anti-Viet Nam era protest song of Peter, Paul and Mary, Pete Seeger and the Kingston Trio. For those who lived in the 1960s, the soundscape evoked by the songs of that era brings the movement of the streets into the solemn sanctuary.

During the Vatican II years Federico Pagura (b. 1923) was a chaplain at the ecumenical seminary in Buenos Aires just starting to work with Homero Perera (b. 1939), a talented young musician who had come to the seminary from Uruguay in 1958 to study organ and composition. At this time, the church in Argentina was dependent almost totally upon Europe and North America for its hymns and musical idioms. Pagura offered a fresh challenge to the students and faculty of the theology school in November 1961:

> The times are ripe that the bosom of the church in our America produce a liturgical and musical renovation to make of worship an act of pleasurable and solemn adoration. The times are ripe for our young churches not only to familiarize themselves with the richest of the Christian musical heritage that has come to us, but also to encourage our new musicians and Christian poets to put wings to the gospel, in agreement with our own methods and talents. (Sosa 1995:73–74)

Drawing upon D. T. Niles's visit to Buenos Aires in 1960, Pagura and Perera wanted to sing songs that breathed the soil of Argentina. No other music expresses the Argentine spirit more than the tango. The tango is in many ways the essence of popular music and dance in Buenos Aires where the sounds and sights of the tango are ubiquitous. There are live tango shows, tango bars, tango movies and TV channels, tango books and recordings, and tango street musicians.

After an earlier unsuccessful effort in the 1960s, Pagura returned to the form in the 1970s when he and Perera wrote a trilogy of "Porque" ("Because") hymns. The most famous of these is the tango "Porque él entró en el mundo" (also known as "Tenemos Esperanza" or "We Have Hope").

"Tenemos Esperanza"/"We Have Hope"

(Federico Paguara, Argentina, and Homero Perera, Uruguay)

Because he came into the world and in history;
because he broke the silence and agony;
because he filled the earth with his glory;
because he brought light into our cold night;
because he was born in an obscure manger;
because he lived sowing love and life;
because he broke hard hearts
and raised hopeless souls.
This is why today we have hope,
this is why we struggle with persistence,
this is why we look with confidence
to the future in this my land.
This is why today we have hope,
this is why we struggle with persistence,
this is why we look with confidence
to the future.

In 1977 Pagura was elected Methodist bishop of Argentina. From the visibility of this position, he addressed *derechos humanos* (human rights) throughout Latin America.

For many within the church, "Tenemos Esperanza" became their "Ein feste Burg" during the epidemic of political turmoil and oppression of the 1970s throughout South America. After overthrowing Salvador Allende, the first elected Socialist president in Latin America, the military regime of Augusto Pinochet in Chile (1973–1989) was especially vicious. As a result of the struggles in Chile, people throughout Latin America were asking "¿Por qué?" or "Why?"—Why is this happening to us? Why has God abandoned us?

Pagura's hymns play on the Spanish interrogative "¿Por qué?" and respond by using anaphora, beginning many of the lines of each stanza with "Porque" ("Because"). In "Tenemos Esperanza" Pagura states that we can have hope because Christ came to live among us and suffered with us. By using the tango to embody Christ's ministry and hope in a difficult time of persecution and oppression, Pagura and Perera planted the gospel of the Incarnation deeply in Latin American soil.

139.5 Around the Table

Shirley Erena Murray (b. 1931) is a New Zealand Presbyterian minister's wife and one of the leading hymn writers of the late twentieth and early twenty-first centuries. While some consider her hymns radical, they in part grow out of a theology that is rooted in the landscape of New Zealand.

Though several hymn writers around the globe articulate a theology of radical inclusion, Shirley Murray pushes the envelope further than most. New Zealanders display their egalitarian spirit in many ways including use of the post-colonial Maori designation for their country, Aotearoa. Because New Zealand is an island nation and it is expensive to import goods, the people are very conscious of environmental and agricultural sustainability. In addition to those who are the descendants of the settlers, primarily from Great Britain, the Maori people have been honored with bilingual signs and the incorporation of Maori traditions throughout church and society. Though much remains to be done, two women have been recent prime ministers. Penny Jamison served as the Bishop of Dunedin in the Anglican Church from 1989–2004 and the Right Reverend Paul Reeves became the first Maori Archbishop and Primate of the Anglican Church in New Zealand in 1980. Jabez Leslie Bryce became the first Pacific Islander to become bishop of the Anglican Diocese of Polynesia and elected Co-Archbishop of New Zealand in 2006.

This sampling of egalitarian spirit in Aotearoa New Zealand provides the ideal soil for nurturing a theology of radical inclusion around the table—both the communion table and the table of life—proposed by Shirley Murray: women and men, young and old, just and unjust, abuser and abused—all deserve clean water, food, and shelter at the table of "justice and joy!" (see www.youtube.com/watch?v=t30-Lka5Feo for an interpretation of this hymn).

"A Place at the Table"

(*Shirley Erena Murray, New Zealand*)

For everyone born, a place at the table,
for everyone born, clean water and bread,
a shelter, a space, a safe place for growing,
for everyone born, a star overhead.

Refrain: And God will delight when we are creators of justice, justice and joy,
yes, God will delight when we are creators of justice, justice and joy!

For woman and man, a place at the table,
revising the roles, deciding the share,
with wisdom and grace, dividing the power,
for woman and man, a system that's fair.

For young and for old, a place at the table,
a voice to be heard, a part in the song,
the hands of a child in hands that are wrinkled,
for young and for old, the right to belong.

For just and unjust, a place at the table,
abuser, abused, with need to forgive,
in anger, in hurt, a mindset of mercy,
for just and unjust, the new way to live.

For everyone born, a place at the table,
to live without fear, and simply to be,
to work, to speak out, to witness and worship,
for everyone born, the right to be free.

The Russian communion hymn, "Do you know the stream that runs," captures the image of the suffering Christ (Sokolov 2000: Hymn 421). In a country with a long history of war, starvation, political upheaval, and hardships of all kinds, the Lord's Supper is a time to meditate on the Christ who also suffers. Hope may be found in a God who took on human form (Philippians 2:5–11) and experienced extreme agony. This hymn reflects the piety of Evangelicals in Russia, drawing upon several biblical passages and images.[9]

"Do You Know the Stream That Runs"

(*Anonymous, Russia*)

Do you know the stream that runs
From the cross where Christ died?
Do you know the one who gives
Redemption from suffering and tears?

[9]Two renditions of this hymn may be found at www.youtube.com/watch?v=ded3jIRsB0s&feature=related and www.youtube.com/watch?v=6icqkX1jCNI&feature=related (accessed May 2, 2012).

Chorus:
Though my fleshly sin is like purple
Though my guilt is a grievous mountain
The blood of Christ flows in a stream
In it, I am made whiter than snow.

Abandoned by the Eternal Father
Christ poured out his own blood;
On the cross, with a crown of thorns
He suffered, but not for nothing.

I have come on the call of Christ
He is the source of my life;
In Him I have salvation in its fullness
In Him I am made whiter than snow.

My spirit is weak and tired
I searched long for peace and happiness;
Like the shepherd comes to a lost sheep
My Lord came and gave me peace.

I love to gaze with my spirit
On the flowing stream from the cross;
Grace for all flows in it;
In it I am made whiter than snow.

The hymn opens with two rhetorical questions, inviting us into the drama. The refrain explores the contrast between "purple" sin and the forgiveness for those cleansed in Christ's blood making us "whiter than snow" (Psalm 51: 7). Stanza two recalls Christ's passion represented in the Gospels. Stanza three identifies Christ as the "source of my life" (John 1:4) and that we may experience "salvation in its fullest" (Psalm 51:12) and returns with the theme that we may be made "whiter than snow." Lines one and two of stanza four recall many of the psalms, e.g., Psalm 42; lines three and allude to the parable of the Lost Sheep (Luke 15:3–7). Stanza five returns again to the cleansing blood of Christ that makes us "whiter than snow." Hope is found in forgiveness offered from the one who has suffered himself and therefore understands our suffering.

Swedish Lutheran minister Per Harling (2002), an internationally known musician who participates in many ecumenical events, provides a joyful song, "You Are Holy," for the procession of the communion elements. Harling incorporates a samba feel to the music of the Sanctus from the mass resulting in a joyful eucharistic celebration in the spirit of the Second Vatican Council. The theme of wine and bread coming from the soil of the earth complements the praise of the entire cosmos. The Holy One is the bridge between the cosmos and, through the Incarnation, the church nourished by the wine and bread from the soil. Harling's influences include both the "harmonies and melodic inflections" of his Lutheran heritage as well as "the American/British rock music movement . . . with a deep appreciation for Latin American music" (Kimbrough 2002). Per Harling's *Sanctus* is rooted in his Lutheran tradition's appreciation for the sacrament of the Eucharist, but also reflects his global ecumenical experience and his vision of the table as a place where all cultures converge and the "cosmos" joins in the celebration.

"Du är helig"/"You are holy"

(Per Harling, Sweden)

You are holy, you are whole. You are always ever
more than we ever understand. You are
always at hand. Blessed are you coming near.
Blessed are you coming here to your church in
wine and bread, raised from soil, raised from dead.
You are holy. You are wholeness, you are present,
let the cosmos praise you, Lord! Halleluja, halleluja,
halleluja, hallelauja, our Lord.

Monseñor Cesáreo Gabaráin (1936–1991) was one of the best-known composers of Spanish liturgical music. He was inspired by the feelings and actions of the humble people he met during his ministry.

"Una espiga," composed in 1973, within a few years of the Second Vatican Council, beautifully combines the ecumenical spirit of Eucharist following Vatican II with agrarian imagery from the Spanish countryside (see www.youtube.com/watch?v=MmaZWWcDkFc for a recording) . Using a melodic folk-style of the region, Gabaráin provides a reflective, joyful and hopeful song that invites the congregation's participation in the Paschal mystery.

"Una espiga"/"Sheaves of summer"

(Cesáreo Gabaráin, Spain)

Sheaves of summer turned golden by the sun,
grapes in bunches cut down when ripe and red,
are converted in to the bread and wine of God's love
in the body and blood of our dear Lord.

We are sharing the same communion meal,
we are wheat by the same great Sower sown;
like a millstone life grinds us down with sorrow and pain,
but God makes us new people bound by love.

Like the grains which become one same whole loaf,
like the notes that are woven into song,
like the droplets of water that are blended in the sea,
we, as Christians, one body shall become.

At God's table together we shall sit.
As God's children, Christ's body we will share.
One same hope we will sing together as we walk along.
Brothers, sisters, in life, in love, we'll be.

The first stanza recognizes that the bread and wine are the result of the gifts of grain and grapes made available to us through the warming of the sun and by the hands of those who cultivate them for our consumption. This strong agricultural grounding adds a sense of authenticity and abundance to the elements of communion, an image drawn from those who can visualize the fields of wheat and vineyards of luscious grapes ready to be harvested.

The second stanza continues the agricultural theme and links it to Christ the great Sower of seed (Matthew 3:1–23). Just as the millstone grinds down the grain, so life grinds us down; just as the ground wheat is transformed into bread, Christ transforms us in our adversity into "pueblo nuevo en el amor" (new people through [God's] love).

The use of the poetic device of simile characterizes stanza three. Christians will become one body in Christ just as each note comes together to make a melody. The final stanza looks to the future when all Christians will sit together around one table with Christ at the *Gran Fiesta*. Our celebrations of communion offer us the opportunity to experience a foretaste of this fiesta on earth.

"Somos Pueblo que Camina" (*Misa Popular Nicaragüense*, Nicaragua) is the processional song for the Eucharist from one of several Central American folk masses composed in the first decade after the Second Vatican Council.[10] The opportunity for Central Americans to express themselves within Catholic Mass through their music was empowering. Years of political oppression for impoverished peoples gave way to hope for justice and liberation. Joyful folk music from the region expressed of unity more completely than earlier imported musical traditions from Europe. While most parts of the Catholic Mass must subscribe to approved texts, congregational song has more flexibility. This song describes those who will attend "la cena del Señor" and unites them as singers on a journey toward the table and therefore toward liberation.

"Somos pueblo que camina"/"We Are People On A Journey"

(Misa Popular Nicaragüense, Nicaragua)

For all the world and all its people
we address our prayers to God.

Refrain: Confidently, all can worship in the presence of the Lord.

All the powerless, all the hungry
are most precious to their God

For the poor, God has a purpose,
for the desperate, a word.

Christ is here and Christ is stronger
than the strength of sin or sword.

God will fill the earth with justice
when our will and his accord.

[10]These are discussed in José María Vigil and Angel Torrellas, eds. Misas Centro Americana (Managua: CAV-CEBES, 1998, with cassette). It contains the music of three Central American Masses: Misa Popular Nicaragüense (c. 1968), Misa Campesina Nicaragüense (c. 1975), and Misa Popular Salvadoreña (c. 1978–1980).

139.6 Conclusion

These examples reveal that cosmic truth finds it source in local communities. They are incarnational examples—that is, instances of how a given community views Christ—that add texture and depth to broad landscape of the Christian faith. Incarnational meaning from a local community may be transmitted to others through the complex rhetoric embedded in the original language and musical style. Musicologists, biblical scholars and church historians have interpreted the rhetorical symbol systems of past cultures for the present age, e.g., presentations of J. S. Bach's *Passion According to St. Matthew*. The fields of ethnomusicology and the social sciences available in the twenty-first century open the possibility of exploring the witness of living cultural rhetoric for the broader world community.

In this essay, an attempt has been made to identify the most salient rhetorical features of selected communities through a single song, admittedly anecdotal exercise. On the one hand, this may be a fascinating exercise for its own sake. On the other hand, such an examination reveals a global Christian consciousness as local perspectives inform collectively our understanding of our faith. Furthermore, if we are to understand Christianity in the twenty-first century, this awareness may lead us to a new kind of ecumenism not based on denominational unity, but on Christian cultural interdependence that reconstructs our understanding of the faith beyond the provincial bias of any single worldview.

Though the 50 years following the Second Vatican Council have given rise to Christian musics rising from the landscapes of various cultures throughout the world, Western media and popular music has also had an increasingly powerful influence. The soundscape of music is always changing and many indigenous musical styles around the world either give way to the sounds of the charismatic and Pentecostal churches such as Hillsong in Sydney, Australia, or produce a hybrid style combining indigenous music with Western popular sounds. Because music is an organic expression of culture, this is a natural development. Since the musical soundscape reflects an ever-changing cultural landscape, this is to be expected. Regardless of the assumed normative nature of Western musics, nuances drawn from the languages and musical traditions of local communities around the world will continue to leave their imprint on the people's song. This imprint is not only a witness to a diverse cultural landscape, but also sonic manifestation of Emmanuel—God with us—throughout the world.[11]

[11] Print music and recordings of many of the songs beyond North American and Europe are available through the Global Praise Program of the General Board of Global Ministries of the United Methodist Church under GBGMusik http://www.umcmission.org/Find-Resources/Global-Praise (accessed April 2, 2012). Of note are the following resources: Global Praise 1 (1996, 1997), Global Praise 2 (2000), Global Praise 3 (2004), Africa Praise Songbook (1998), Caribbean Praise (2000), Russian Praise (1999), Tenemos Esperanza (2002).

Acknowledgements I am grateful for the research and translations of the hymns of Xiao-min Lü provided to me by my graduate student Irene Dhing-Dhing Lai from Malaysia (February 2012). I am also grateful to Hoonseok Kang, a seminary student at Perkins School of Theology, SMU, for submitting the Korean and a literal translation of "O-so-so."

References

"A Place at the Table" by Shirley Erena Murray and Lori True. www.youtube.com/watch?v=t30-Lka5Feo. Accessed 2 May 2012.

"Al despuntar en la loma el día" ("When o'er the hills") by Heber Romero. www.youtube.com/watch?v=ZutVLjosth0 and www.youtube.com/watch?v=badxiskvTWA. Accessed 2 May 2012.

"An Interview with Per Harling". (2002). *Global song: A newsletter of the Global Praise Program* (S. T. Kimbrough, Jr., Ed.), Issue #1, 2.

Bellamy, J., & Castle, K. (2004). *2001 Church attendance estimates: NCLS Occasional Paper 4*. Sydney: NCLS Research. Available at www.ncls.org.au/download/doc2270/NCLSOccasionalPaper3.pdf. Accessed 21 Mar 2012.

"Bring forth the Kingdom of God" by Mary Haugen. www.youtube.com/watch?v=R7_v__5q9ws. Accessed 2 May 2012.

Chandler, P.-G. (1977). *God's global music: What we can learn from Christians around the world*. Downers Grove: InterVarsity Press.

Dargie, D. (1982). The music of Ntsikana. *South African Journal of Musicology, 2*, 7–26.

"Do you know the stream that runs" ("Знаешь ли, ручей, что бежит") found at www.youtube.com/watch?v=ded3jIRsB0s&feature=related and www.youtube.com/watch?v=6icqkX1jCNI&feature=related. Accessed 2 May 2012.

Hawn, C. M. (2013). Stream seven 'Through every land, by every tongue:' The rise of ecumenical-global song". In *New Songs of celebration render: Congregational song in the 21st century*. Chicago: GIA Publications, Inc.

Hodgson, J. (1980). Ntsikana's 'Great Hymn,' A Xhosa expression of Christianity in the early 19th century Eastern Cape. *Communications 4*. University of Cape Town, 4.

Janzen, J. M. (1992). *Ngoma: Discourses on healing in Central and Southern Africa*. Berkeley: University of California Press.

Jenkins, P. (2002). *The next Christendom: The coming of global Christianity*. New York: Oxford University Press.

Kimbrough, Jr., S. T. (2002). "An Interview with Per Harling," Global Song: A Newsletter of the Global Praise Program, Issue #1, p. 2.

"Korean-American Music Composition Competition-Judges". (2005). www.sejongsociety.org/2005scs_composition/2005scs_composition_judges.htm. Accessed 2 Apr 2012.

Lee, G. (1999). *Personal interview*. Korea: Seoul.

Lockward, J. A. (Ed.). (2002). *Tenemos esperanza/Temos esperança/We have hope*. New York: GBGMusik.

"Reflections in mindfulness". www.wildmind.org/mindfulness/four/reflections. Accessed 21 Mar 2012.

Rosales, D. M. (1985). The influence of the missionary heritage on liturgical forms. *International Review of Mission, 74*(295), 373–376.

Sacrosanctum Concilium. (1963). www.crossroadsinitiative.com/pics/Sacrosanctum_Concilium.pdf. Accessed 26 Mar 2012.

"Siyahamba". www.youtube.com/watch?v=1RZqLwqRaKQ. Accessed May 2012.

Sokolov, E. (Ed.). (2000). Песнь возрождения: сборник духовных гимнов и песен евангельских церквей/Pesn Vozrozhdeniia: Sbornik dukhovnykh gimnov i pesen evangelskikh tserkvei (Songs of renewal: Collection of spiritual hymns and songs of Evangelical Churches) (R. Harris, Trans.). Minsk: Belarus Publishers, Hymn 421.

Sosa, P. (1995). *"Pagura…El Cantor", Por Eso Es Que Tenemos Esperanza: Homenaje al Obispo Federico J. Pagura*. Quito: CLAI.

"Still, I search for my God" by Francisco Feliciano. http://rockhay.tripod.com/worship/music/stillisearch.htm. Accessed 2 May 2012.

Sun, I. A-L. (2007). Songs of Canaan: Hymnody of the house-church Christians in China. *Studia Liturgica, 37* (1):1, 98–116.

"Tenemos esperanza" by Federico Pagura and Homero Perera. www.youtube.com/watch?v=TEr-lAOLvLA. Accessed 2 May 2 2012.

"Una espiga" by Cesáreo Gabaráin. www.youtube.com/watch?v=MmaZWWcDkFc. Accessed 2 May 2012.

Vigil, J. M., & Rorrellas, A. (Eds.). (1998). *Misas Centro Americana*. Managua: CAV-CEBES.

"Where wide sky rolls down" by Elizabeth J. Smith. www.youtube.com/watch?v=eFX7VBH_2ZQ. Accessed 2 May 2012.

Chapter 140
Streams of Song: The Landscape of Christian Spirituality in North America

C. Michael Hawn

140.1 Introduction

Are hymns relevant to Christians today? Albert van den Heuvel (1966) of the World Council of Churches reflected on the problem of finding hymns that addressed the issues of his day in a preface written for a collection of new material:

> There was a minister in a European country not very long ago, who told his congregation on a Sunday morning that they would only sing one hymn: "What we should like to sing about," he said, "is not in the hymnal; what is in the hymnal about our subject is obsolete or heretical. So let us be silent and listen to the organ."
>
> This little story is, of course, irritating. I can already hear lots of people say: but there are beautiful hymns in our hymnal! Our fathers have sung them for many centuries! We have learned them from our mothers! What is wrong with Ambrosius' hymns, Luther's hymns, the Psalms, the Wesleyan treasury, and all the others? The man in our story would have shrugged his shoulders, I am afraid. His point is not that there are not good hymns, but that there are very few which support his preaching and that of his generation. I am with him on this. There are many things in the life of the denominations which are frustrating, but few are so difficult to live with as this one. Choosing the hymns for Sunday morning worship is an ever-recurring low ebb in my ministry.

The concerns raised by the European minister and echoed by van den Heuval suggest that it may be time to see what has happened in the people's song since the Second Vatican Council (1962–1965). In the nearly 50 years that have followed this quotation, the concerns voiced above have been answered many times over by an abundance of new congregational songs. Indeed, the mid-1960s signaled the beginning of an explosion of congregational song around the world. It is only now, at the beginning of the twenty-first century, that scholars are in a position to

C.M. Hawn (✉)
Sacred Music Program, Perkins School of Theology, Southern Methodist University, Dallas, TX 72752, USA
e-mail: mhawn@mail.smu.edu

S.D. Brunn (ed.), *The Changing World Religion Map*,
DOI 10.1007/978-94-017-9376-6_140

begin to understand the diversity and wealth of congregational music available to the church since these years of liturgical reform—a diversity and wealth of hymnody unprecedented in Christian history—and what this means for the shape of Christian spirituality.

140.2 Methodology

In the early 1990s I embarked on a project for the Hymn Society in the United States and Canada, examining hymnody from an ecumenical perspective. In the 1970s, soon after Vatican II, the Consultation on Ecumenical Hymnody (CEH) prepared a list of 227 hymns for ecumenical use. The first stage of my work was to survey the impact of the CEH in 38 North American English-language hymnals published between 1976 and 1995 (Hawn 1996).

I observed that the work of the CEH published as "Hymns and Tunes Recommended for Ecumenical Use" (1977) was limited in its recommendations, failing to include congregational songs from many voices in the North American Christian community, especially minority groups and songs widely known in free-church traditions. The participants in the CEH were aware of these shortcomings, but could not muster a fuller participation from these groups in their process. The primary purpose of the CEH was formative—that is, to provide a body of hymns that would influence the shape of the church's sung faith, balancing theological and liturgical concerns. The participants hoped that the hymns on this list would be chosen by future hymnal committees, creating a common body of sung faith at least in the church in the United States. I attempted to discern the impact of this noble effort in the article.

Following this article, I realized that the data I had collected revealed a much more complex picture of congregational song than the CEH process was able to demonstrate. What about newer hymns written since the CEH report in 1977? What about rich congregational resources that seemed to fall outside of earlier definitions of "hymn"?

It was at this point that I continued my research, but with a different goal than the CEH. Rather than developing a prescriptive list for guidance, I chose to develop a descriptive list that might inform future hymnal committees concerning the items that appear most often in North American hymnals, making no value judgments on the quality or ecumenical possibilities of a given item. The goal of this list was modest—simply to indicate what was being included in hymnals. Items written before 1960 comprised one list and items composed after 1960 a second list, the latter group needing to be separated out since they had not had the benefit of time and may not have been recognized by hymnal committees to the same degree. The results of this list produced from 40 English-language hymnals published in Canada and the United States between 1976 and 1996 were then published (Hawn 1997). This chapter is an attempt to discern meaningful patterns in recent congregational song that map the landscape of Christian spirituality in the United States.

140.3 Streams of Song[1]

The quantity and diversity of our generation's stanza does not lend itself easily to organization—the work of the Holy Spirit rarely manifests itself in ways that are easily discernable to human patterns of understanding. Yet, there are reasons why it may be valuable to attempt to recognize the gifts of the Spirit that have been given to the church in our age.

I have chosen "streams of song" as the overarching organizational metaphor. Streams add definition to a landscape and mold it. Because of flowing streams the landscape is always changing. The organic quality of the earth, shaped by rivers, is not unlike the continual changes in the spiritual landscape that respond to transformations in society due to political crosscurrents, current events, denominational shifts, and emerging cultural patterns. Streams have a source, and each of the proposed seven streams of song comes from particular sources of faith—a particular expression of piety. Streams come in various widths and depths. Not all streams are the same. Some of the song streams are rushing and seem to be overflowing their banks because of the musical outpouring being generated from their particular piety source. Others are steady in their flow, and yet others may be either drying up or merging with other streams. Streams meander; they do not flow in straight lines like canals. They occasionally crisscross each other. Such is the case with many songs in this overview. Some songs fit comfortably in two or more streams. Though seven streams are proposed here, most are braided streams—a stream containing complex currents within the same flow that run at various depths, widths, and speed. The spirituality indicated in each of the seven streams of song introduced in this model provides a panoramic view of what is a complex series of sub-currents upon closer examination.

This fluid model stands in contrast to a pigeonhole approach where everything is organized neatly. The fluidity of this model reflects how these songs usually appear in hymnals— a song from one tradition based on a particular season of the Christian year or theological theme may be placed in juxtaposition to songs from other streams. Hymns demonstrate flexibility in their liturgical possibilities. Many hymns embody a range of themes. An examination of the topical indexes of several recent hymnals reveals that the same hymn may appear in varying sections of different hymnals.

Finally, streams are vibrant parts of creation, carrying us along with them, offering constant changes in depth and rate of flow. Some parts of a stream are smooth with almost no sense of movement while others rush to a waterfall. Some songs still our souls while others raise us to an emotional apex. Streams are always changing. Every time we sing a song, it is a new experience. More familiar hymns still may surprise us with a new insight or provide security in a constantly changing world. Like an unforeseen turn in the bend of a river or an unanticipated crosscurrent, new songs often challenge us in unexpected ways, catching us off guard, delighting us as they provide words for feelings never before articulated, or confronting our previously held notions.

---

[1] Each of the streams introduced in this section receives extended treatment in C. Michael Hawn (2013) New songs of celebration render: Congregational song in the twenty-first century. Chicago: GIA Publications, Inc.

Few, if any, will navigate all of these streams with equal confidence. We all have our primary sources of piety and preferences for expressing this piety in song. Yet, the people of God who gather in the common assembly we call Christian worship may enrich their prayer by expanding the number of streams from which they draw.

140.4 Naming the Streams

My research and experience of singing in a wide variety of Christian traditions indicates seven streams of song, drawing on seven broad sources of piety, each with its own identity while overlapping in some cases with others in varying degrees. Table 140.1 provides an overview of each stream with examples.

Stream One—Roman Catholic Liturgical Renewal This stream reflects directly the reforms of Vatican II and the outpouring of song for the assembly that came and continues to come from this historic council. Virtually no hymnal is untouched by at least some congregational songs from this stream. At the center of this stream are songs for the sacraments, music for the lectionary, compositions for the Christian year, responsorial psalms and ritual music. Because of the breadth of the Catholic Church, these songs come to us from various parts of the world, but especially from Spanish-speaking locations as well as Euro-North American English speakers. This has become an increasingly braided stream since the early 1960s with folk, popular, western classical and global styles of music.

Stream Two—Classic Contemporary Protestant Hymnody This stream is a swelling stream originating in the "hymn explosion" of Great Britain in the 1960s and 1970s and joined by rivers in other English-speaking countries, especially Australia, Canada, New Zealand and the United States. While quite varied, the center of this stream includes paraphrases of scripture including fresh metrical paraphrases of the Psalms, hymns for the Christian year and sacraments, prophetic hymns on justice themes such as inclusion, peace and ecology, hymns on ministry, and, in some sections of the stream, a strong interest in inclusive language.

Stream Three—The African American Stream This stream finds a voice in virtually all confessional traditions. Here one will find a variety of musical expressions from spirituals and hymns to various styles of gospel music. This stream offers us songs born in the crucible of struggle, reflecting scripture and, often, expressing faith in the first-person. Since the middle of the twentieth century, virtually all hymnals include songs from this stream, even in predominately Anglo confessional traditions. Songs from this stream are seen by many as major, even unique, contribution from the United States to the larger church. This stream is increasingly becoming an anastomosing or braided stream with various genres of gospel, spirituals, hymns and Hip Hop styles.

Table 140.1 Seven streams of congregational song that have shaped hymnals since Vatican II

Stream	1. Roman Catholic liturgical renewal hymnody	2. Protestant contemporary classical hymnody	3. African American spirituals and Gospel songs	4. Revival/Gospel songs	5. Folk song influences	6. Pentecostal songs Azusa Street Revival (1906)	7. Global and ecumenical song forms
Themes	Sacraments	Justice/Environ	Personal experience	Salvation	Scriptural storytelling	Praise of God	Freedom
	Psalm settings	Lectionary	Salvation	Spreading the Gospel	Narratives	Adoration of Jesus	Justice
	Lectionary	Scriptural paraphrases	Refuge in times of trouble	Christological focus	Social concerns	Use of scriptural fragments	Liturgical music
	Community	Worship and the Arts	Praise of God	Triumphal Faith	Guitar	Personal, first person	Liturgical Inculturation
	Offices	Inclusive Lang.					Modern Missions
	Refrain forms	Sacraments					Sung prayer
		Christian Year					
Precursors	Of the Father's Love	God of Grace and God of Glory (Fosdick)	Lift every Voice and Sing (Johnson)	Nineteenth Century Gospel Songs	O Love, How Deep	Spirit of the Living God (Iverson)	Many and Great, O God (Renville)
	O Come, O Come Emmanuel	Hope of the World (Harkness)	Precious Lord	Early Twentieth Century Contributions:	Tomorrow Shall be My Dancing Day	His Name is Wonderful (Mieir)	Here, O Lord, Your Servants Gather (Yamaguchi-Stowe)

(continued)

Table 140.1 (continued)

Stream	1. Roman Catholic liturgical renewal hymnody	2. Protestant contemporary classical hymnody	3. African American spirituals and Gospel songs	4. Revival/Gospel songs	5. Folk song influences	6. Pentecostal songs Azusa Street Revival (1906)	7. Global and ecumenical song forms
	Where Charity and Love Prevail	Lift High the Cross (Kitchin-Newbolt)	We Shall Overcome	Great is Thy Faithfulness (Chisholm)	Let There Be Peace on Earth (Miller-Jackson)	There's Something About that Name (Gaither)	In Christ There is No East or West (Oxenham)
		Where Cross the Crowded Ways (North)	Sweet, Sweet Spirit (Akers)	How Great Thou Art (Boberg-Hine)	Blowing in the Wind (Dylan)	**Publishers:**	
			Stand by Me (Tindley)	Victory in Jesus (Bertlett)	To Everything Turn (Seeger)	Hosanna/ Integrity	
			Leave it There (Tindley)			Maranatha	
			Yes, God is Real (Morris)			CCLI	
Examples	Alleluia, Give Thanks (Fishel)	Jubilate Group (UK)	Appearance of Spirituals in mainline hymnals	Because He Lives (Gaither)	Bring Forth the Kingdom (Haugen)	Alleluia (Sinclair)	**Taizé chants:**
	Blest Are They (Haas)	All Who Love and Serve Your City (Routley)	Bless the Lord (Crouch)	Freely, Freely (Owens)	I Was There to hear Your Borning Cry (Ylvisaker)	As the Deer (Nystrom)	Bless the Lord
	Cantemos al Señor (Rosas)	Blessed be the God of Israel (Perry)	Come Sunday (Ellington)	He Touched Me (Gaither)	Lord of the Dance (Carter)	Awesome God (Mullins)	Ubi caritas

	Celtic Alleluia	Earth Prayer (Murray)	Give me a clean heart (Douroux)	People Need the Lord (Nelson & McHugh)	Pass It On (Kaiser)	El Shaddai (Card-Thompson)	**Africa:**
	Cuando el Pobre	God of Many Names (Wren)	Let It Breathe on Me (Lewis-Butts)	Share His Love (Reynolds)	She Comes Sailing (Light)	Emmanuel, Emmanuel (McGee)	Jesu, Jesu (Colvin)
	Gather Us In (Haugen)	God of the Sparrow (Vajda)	My Tribute (Crouch)	Shine, Jesus, Shine (Kendrick)	The First One Ever (Egan)	Great is the Lord (Smith)	Jesu Tawa Pano
	Here I Am, Lord (Schutte)	I Come with Joy (Wren)	Praise Ye the Lord (Cleveland)		Two Fisherman (Toolan)	How Majestic is Your Name (Smith)	Thuma Mina
	Make Me a Channel (Sebastian)	O Day of peace (Daw)	Someone Asked the Question (Franklin)		We Are the Church (Avery & Marsh)	I Love You Lord (L Klein)	Siyahamba (Nyberg: Freedom is Coming)
	My Soul Gives Glory (Winter)	Silence, Frenzied, Unclean Spirit (Troeger)	The Lamb (M McKay)		What Does the Lord Require (Strathdee)	Majesty, Worship His Majesty (Hayford)	**Asia:**
	On Eagle's Wings (Joncas)	Tell Out My Soul (Dudley-Smith)	The Lord is in His Holy Temple (G Burleigh)		When Jesus the Healer (P Smith)	My Life is in You (Gardner)	Saranam (DT Niles)
	One Bread, One Body (J Foley)	We Utter Our Cry (Kaan)	Through It All (Crouch)		Use of Southern Folk Tunes and Shaped-note melodies	Seek Ye First (Lafferty)	Come Now, O Prince of Peace (Lee)

(continued)

Table 140.1 (continued)

Stream	1. Roman Catholic liturgical renewal hymnody	2. Protestant contemporary classical hymnody	3. African American spirituals and Gospel songs	4. Revival/Gospel songs	5. Folk song influences	6. Pentecostal songs Azusa Street Revival (1906)	7. Global and ecumenical song forms
	Santo, Santo, Santo (Cuéllar)	What does the Lord Require (Bayly)	Total Praise (Smallwood)	In Christ Alone/ How Deep the Father's Love (Getty-Townend)	British, Irish and Scottish tunes	Shout to the Lord (Zshech)	**South America:**
	Taste and See	When in Our Music (Pratt Green)	Black Gospel Rap/HipHop			Spirit Song (Wimber)	Tenemos Esperanza (Pagura)
	Te Ofrecemos Padre Nuestro (Misa Popular Nicaragüense)	Womb of Life (Duck)				Thy Word is a Lamp (Grant)	**Iona:**
	Tú has venido (Gabaraín)					We Will Glorify (Paris)	Goodness Is Stronger Than Evil
	Una Espiga					God of Wonders/ God of this City (Tomlin)	Will You Come and Follow me
	You Satisfy the Hungry Heart (Westendorf)					PASSION (Giglio)	**Leaders:**
						Retuned hymns igracemusic. com	John Bell
							David Dargie
							I-to Loh
							P Matsikenyiri
							Pablo Sosa

Stream Four—Gospel and Revival Songs This stream is perhaps on the wane as a separate stream. It appears to be merging with others, especially with Streams Three and Six. These songs of praise, personal salvation and experience, and a triumphal faith continue, however, to find their way into a remarkable number of hymnals, even in some traditions where they have not been a dominant voice.

Stream Five—Folk Hymnody This stream draws from several sources of piety and has always been a part of the church's song. This stream experienced a revival in the Civil Rights Movement and the anti-Vietnam era of the 1960s, spreading into folk masses and continuing as an idiom in its own right today. The use of the acoustic guitar lends informality to songs of praise and protest as well as narrative ballads that are immediately accessible to groups.

Stream Six—Pentecostal Song This stream is often called "Praise and Worship" or "Contemporary Christian Music" finds its piety source in early twentieth century American Pentecostal traditions such as the Azusa Street Revival (1906), but has expanded into a world-wide expression of Christianity in many languages. Its electric sounds have influenced other streams, especially Stream Three, and those devoted to this stream have their spiritual roots in a wide variety of confessional traditions. Texts from these songs, often scriptural fragments, range from ecstatic praise to intense prayer, and frequently address God directly in the second person and petition Christ in the first person.

Stream Seven—Ecumenical and Global Stream This final stream attempts to bring into focus the contributions of two-thirds of the world's Christians, especially those that come from Africa, Asia, and Latin America and the Caribbean. European contributions come from two well-known ecumenical communities, Taizé and Iona, to global song. A direct result of the reforms of Vatican II and the pronouncement from *Sacrosanctum Concilium* (1963) to "respect… and foster… the genius and talents of the various races and peoples," this stream includes songs around the world from many confessional traditions that have been included in North American hymnals. Because of the cultural and geographical breadth of this stream, it is perhaps the most complex and braided stream of all. Due to the social location of a congregation, some may draw extensively from some of the sub-currents and not from others. Most congregations have not tapped into the rich variety and of this stream even though it is increasingly represented in many hymnals.

140.5 Limitations of This Model

This model focuses on the breadth rather than the depth of current congregational song practice. While this approach demonstrates the considerable breadth of the church's song in the last half of the twentieth and early years of the twenty-first century, taking a broad perspective sacrifices a deeper focus on any one particular aspect of the church's song. Just as many hymns may appear in different sections of a hymnal, many songs may be included in more than one stream. The purpose of this approach is not to select a slot for every song, but to suggest that the various pieties that give birth to song since Vatican II provide the church with unprecedented variety, a variety that reflects a diverse spiritual landscape.

This approach is limited by its social location. Songs from North American hymnals have determined the contents of this study. A similar study conducted in England, Argentina, Australia or Korea, for example, would reveal a significantly different reality. Some of these streams might be reduced to a rivulet in other cultural settings (if they exist at all), and others might be added.

This approach is based on what is published rather than on congregational song practice. The focus is on what songs appear in hymnals and hymnal supplements since the Second Vatican Council, not on what congregations are actually singing. Some congregations, especially those who draw heavily on Streams Three, Six and Seven, have often moved on to more recent material not available in the hymnals reviewed. By the time a hymnal comes out, some of its contents are for many churches out of date. Other congregations do not use hymnals at all, relying on the projection of texts on screens or transmitting of songs through oral/aural means. Most recently, leaders of Emerging Church worship make extensive use of "home grown" compositions. Though locally composed music is not sung exclusively in Emerging worship, many websites from these congregations contain MP3/MP4 files of their original compositions. By the congregational song choices made by pastors and church musicians each week, they serve as local "hymnal editors" or gatekeepers to the song of the people. In essence they control the range of spiritual expression offered to their people. Most congregations sing only a small percentage of their hymnal—a canon within a canon. Only exceptional congregations would sing more than 50 % of the contexts of their hymnbook. Those who do not use hymnals bear a greater responsibility for selecting a breadth of sources that they use to shape the congregation's sung faith. While the pastors and church musicians of every church are in effect the hymnal editors for their congregation, those who do not use hymnals bear a greater responsibility for selecting a breadth of sources that they use to shape the congregation's sung faith.

While I do not want to suggest that there is no correlation between congregational song collections and congregational practice, many congregations sing only a small percentage of their hymnal's contents, shaping in essence their own sung canon; others supplement their hymnals with additional resources by using one of several copyright licenses available to churches. Basing a study of the breadth on congregational song practice, while desirable, would be a very difficult, perhaps overwhelming, project to organize and report. While such a study based on actual practice would be valuable, I hope that this approach lends insight into the breadth and depth of our generation's diverse spiritual landscape.

Some groups may not be included. The ritual music and chants of various Orthodox Christian traditions are not included though a few examples appear in hymnals. Many hymnals include hymns originally in Greek and translated into English, usually in the nineteenth century. Since Orthodox congregations do not use hymnals and draw heavily on long-established traditional repertoire, however, it was difficult to include them in this study. Christian songs from the Middle East are rarely sung in North America. Translations of hymns emerging from Eastern Europe following the fall of Soviet Union have not found their way into many hymnals yet save the more recent *Evangelical Lutheran Worship* (2006), the hymnal of the Evangelical Lutheran Church in America. Relatively little Native American hymnody is available to the larger church. Though there are many singing traditions in Oceania in the South Pacific, they seldom find their way into North American hymnals. These are but a few of the many song traditions that are currently practiced, but not included in this study. These gifts of the Spirit may yet come to us in the future, however.

140.6 General Trends Observed

Various patterns emerge in this study. The following overview highlights some of these patterns with the hope that awareness of them may enhance the reader's understanding of the various streams.

140.6.1 Solo/Congregational Balance

Although congregational song is the focus of this study, several streams employ soloists or cantors in a variety of ways. In Stream One, cantors are essential to the performance of responsorial psalmody. Stream Three often draws upon soloists as catalysts for call/response singing—a standard feature of many African American styles. The gospel songs of Stream Four were often solos initially and later claimed by the congregation. The irregular meter of many folk songs in Stream Five is more conducive to solo singing than congregational participation in many instances. Contemporary Christian artists have extensive solo careers. Those who participate in the songs in Stream Six often learn them from CDs, DVDs and YouTube as solo or ensemble selections, and then sing them as congregational selections. Many songs of the world church, the focus of Stream Seven, draw upon solo singers in a variety of ways, especially call/response songs in Africa. Cantors enhance the sung prayer of the Taizé Community. Only Stream Two, contemporary classical hymns, does not regularly make use of solo singing to enhance the performance of congregational song. Yet, when one examines the average hymnal, the importance of cantors or soloists is not apparent. This calls upon church musicians to understand the wide variety of performance practices needed to bring alive various musical styles. Musical performance practice is indeed a part of the spirituality of a given song.

140.6.2 Written and Oral Traditions

A popular notion of the compositional process often imagines a creative and thoughtful hymn writer sitting at the computer composing a hymn text or a musician seated at the piano notating the melody of a hymn tune on staff paper as it emerges from her or his artistic imagination. While this may be true for many composers, songs throughout the spectrum of streams are often composed and sung without the benefit of or even the need for a written score. Notation on a physical piece of paper may be an afterthought in many African American styles. Musicians who perform from a written score in Stream Three often see musical notation as a general guide to be melodically and harmonically modified on the spot or as a basis for extensive improvisation. Choirs and congregations in Streams Three, Six and Seven often learn music through a process of oral transmission—totally without any written music. Written music is notoriously unrevealing in African music, providing only the barest outline that is enhanced through improvised solos, percussion parts, dance and improvised harmonies. What looks so simple on the page becomes complex in performance for those who know the style. Many songs from Asia depend upon heterophonic improvisation around the melody and minute melodic variations, all of which resist notation.

Regardless of the musical style, a written score should never be confused with the sound and experience of congregational singing. Much, perhaps most, of the world's Christian song is sung by people who do not read music and, as a result, is primarily an oral experience. Those congregations who participate primarily in Stream Two, the most consistently literate of the seven streams, may feel disoriented as they attempt to participate in oral or semi-oral musical styles. Church musicians trained only in classical western music may have difficulty in bringing orally conceived music to life.

140.6.3 Text/Music Independence

Stream Two, the English and American hymn tradition, has a heritage of tunes and texts composed by different people. Furthermore, the metrical nature of the texts allows numerous melodies to be paired with the same text. For example, depending upon which side of the Atlantic one resides, a different melody may be sung to the same text in England and the United States. Various faith traditions also sing different tunes to the same words. The majority of popular meters with the most options for tunes are common in English hymnody, for example, SM, SMD, CM, CMD, LM, LMD. While these and other commonly used meters are still used by current text writers in Stream Two, many poets explore new meters that demand new tunes. Only a few composers in Stream Two have the skill to successfully compose both texts and tunes, for example, Dan Damon, Marty Haugen, Michael Joncas, Jane

Marshall, Thomas Pavlechko in the United States, Pablo Sosa from Argentina, Per Harling in Sweden, and the late Spanish priest Cesáreo Gabaráin. Erik Routley, the eminent hymnologist, could also write both texts and tunes of lasting value.

By contrast, hymnals contain an increasing number of congregational songs with universally fixed text and tune pairings or where the tune and text are integrally linked and composed together because of the nature of the text or original language. Most selections listed as "irregular" in the metrical indexes at the back of hymnals have fixed text-tune pairings. Irregular meters include African American spirituals and gospel songs from Stream Three, gospel and revival songs from Stream Four, folksongs from Stream Five, contemporary worship music from Stream Six, global songs from Stream Seven, as well as hymns from Stream Two with stanzas of varying metrical length. In general, Stream Two demonstrates the most independence between texts and tunes while other streams require a more integral, fixed relationship between words and music.

140.6.4 Variety of Accompanying Instruments

The pipe organ once was the dominant instrument for leading congregational song. Recently, the organ has become one of many instrumental possibilities for supporting the people's song. Although the range of possible instruments varies widely within a stream, each stream has its normative instrumental sounds. Stream One may use organ extensively, but also piano and acoustic guitar are common in various Catholic masses. The pipe organ has traditionally been the domain of Stream Two, though piano is commonly used along with other instruments such as handbells. African American styles call upon a variety of instrumental sounds; thus one will find everything from pipe and electronic organs and electronic keyboards to electronic guitars and percussion in Stream Three. The gospel and revival songs of Stream Four have usually been noted for piano and organ (pipe or electronic) in combination. The acoustic guitar is the normative sound of the folk idiom of Stream Five, though piano is common along with light percussion of tambourines and congas. Stream Six is associated primarily with electronic guitars and keyboards and heavy use of percussion—both trap sets and congas. It is not uncommon in congregations with more resources, however, to have the bands of praise teams augmented with brass sections (including saxophones) and electrified strings and other wind instruments. With Streams Six and Seven the center of instrumental gravity switches from a melodic keyboard sound to a percussion-dominant sound, especially in various styles of world Christian music. Stream Seven may include a wide array of instruments associated with specific ethnic groups or regions of the world, ranging from particular kinds of drums and specialized percussion to string and wind instruments not common in western music. For example, the *sruti* box, producing open fifth drones, is a staple of music from south India and other countries in the Asian subcontinent. The guitar either supplants or enhances various keyboards (electronic, organ, or piano) in steams five, six and seven. The increasing role of church orchestras has made the sounds of woodwind, brass and string instruments more common, especially in streams Two, Four and Six.

A cappella singing also takes place in a number of the streams to varying degrees; indeed, singing without instruments is an option throughout the spectrum of musical styles. Unaccompanied plainsong is characteristic of Stream One. Several musical styles used to support the strophic hymns of Stream Two were essentially a cappella in their origins, for example, the music of the Sacred Harp or oblong tunebook tradition. The music of the African American Spirituals may be best experienced when unaccompanied and harmonized by ear. Streams Four, Five and Six employ unaccompanied singing more sparsely, but effectively, as points of variety. The unaccompanied voice is characteristic of various African and Asian music. Within broader parameters of artistic creativity, one can identify each stream by its instrumental soundscape alone without texts.

140.6.5 Many Songs Demonstrate Characteristics of Several Streams

The artistic imagination of poets and composers often bridges streams. For example, Fred Kaan's celebrative communion text, "Let Us Talents and Tongues Employ," is essentially from Stream Two, a strophic hymn with a refrain. When paired with Doreen Potter's Caribbean-based tune LINSTEAD performed effectively with guitar, claves, shakers and tambourines—sounds associated with the world church—it also drawn from Stream Seven. Many examples of text and musical exchanges take place between Streams Three and Four. For example, Andraé Crouch's "My Tribute" is one of the signature songs of African American gospel music, but alludes directly to Fanny Crosby's "To God be the Glory" and draws upon the metaphors of the redemptive power of the blood of Christ common in the gospel and revival music of Stream Four. Increasingly, a classic strophic hymn associated with Stream Two receives a musical treatment common to the charismatic music of Stream Six. The folk styles of Stream Five still influence music written out of the piety of Stream One. Some of the music composed by David Haas and Marty Haugen and others still easily fits an acoustic guitar and has echoes of earlier folk masses.

Bridging streams may be a sign of vitality in the life of the church and is increasingly a sign of the eclectic nature of congregational spirituality during the last 50 years. When a text from one stream is placed in counterpoint with a musical style from another stream, the result may be enlivening to both. A number of text and tune writers appear in more than one stream. These are signs of the Holy Spirit at work and fly in the face of those who take refuge in divisive camps that segregate musical styles from each other.

140.6.6 Variety of Song Structures

I have written extensively about the significance of various structures employed in congregational song—especially strophic (a form that I call sequential), cyclic, and refrain forms—and the relationship of these structures to worship (Hawn 2003a, b). Each structure has many variations that serve the text in a different way. Strophic hymns consist of several stanzas with many words that form a progression (sequence) of thought. Cyclic songs use fewer words that are repeated with musical variations—a theme and variation approach to structure. Refrain forms have attributes of both—sequential stanzas with cyclic refrains. Stream Two, the classic Protestant hymn, has characteristically though not exclusively used the strophic structures, a form that defines what a hymn "looks like" for many singers. This study indicates that other structures have gained in their prominence during the last half of the twentieth century. The refrain form has long been associated primarily with both Streams Three and Four as a characteristic of gospel songs. Refrain structures are also a primary feature of much Roman Catholic Renewal music (Stream One) since Vatican II. Cyclic structures appear in several streams, but especially in Streams Six, the music of the charismatic movement, and Seven, principally music from Africa and the Taizé Community. Stream Five, the folk stream, uses primarily refrain and cyclic structures, but may employ variations of all.

The importance of this observation is ontological in nature: What is a hymn? Some definitions recall the Greek *hymnos*, a term indicating a song praising "a god or gods, a hero, a nation, or some other entity or reality" (Schilling 1983:5). Augustine's classic definition of a hymn places God as the object of worship. Thus a hymn is "a song in praise of God." From a literary perspective, "a hymn is usually a lyric poem with a metrical and strophic text. Literally, a lyric poem suited for singing to the accompaniment of a lyre or a harp, but more broadly, it is simply a poem appropriate for singing" (Schilling 1983:5). Paul Schilling continues this train of thought indicating that a lyric poem "gives voice to the poet's feelings rather than to external events. Hymns are … lyrics [that] express the feelings, attitudes, needs, and commitments of their authors and those who use them" (Schilling 1983:5). It is this definition that may broaden our ontological response to the questions: What is a hymn and how does singing relate to the structure of liturgy?

While the structures of hymns have always demonstrated variety, historically strophic poetry (in stanzas) has shaped western consciousness about the nature of hymns and how they communicate. Other structures were thought of as alternative or supplementary at best and, perhaps, inferior at worst. While strophic hymnody remains vital, even a cursory look at the most recent hymnals reveals that refrain and cyclic structures are on the rise. Depending on the liturgical tradition represented in a hymnal, cyclic structures appear, in part, in a section of the hymnal labeled, "Service Music." Others include cyclic structures throughout on a thematic basis. Often both approaches are followed in the same book. Diversity of structure characterizes twenty-first century congregational song to such an extent that our notion of what constitutes a valid hymn has been challenged. Those who plan worship may benefit from an understanding of the liturgical possibilities inherent in the various structures so that they might integrate the congregation's song more purposefully and effectively into worship.

140.6.7 Physical Responses to Congregational Song

Since the middle of the twentieth century, Streams Three, Six and Seven insert another element into congregational singing—movement or dance. To those who sing in these streams extensively, part of the piety is expressed through kinesthetic involvement while singing. Movement is not optional, but is integral to the experience of singing. Specific musical styles and song structures, especially cyclic forms, lend themselves to a physical response. As these songs have become a part of our hymnals, they bring into our worship the possibility of congregational singing that is more fully embodied. Dance is a part of the diverse landscape of congregational singing in the twenty-first century.

The discussion in this section indicates the diversity of expression in the spiritual landscape today and some of the challenges in leading the breadth of congregational song available to the church.

140.7 Implications

Assuming that what we sing plays a significant role in shaping our faith, this study attempts to examine the breadth of our sung prayer. Prayers of praise, thanksgiving, adoration, invocation, confession, intercession, and blessing all come in sung forms. While not all congregational singing is a form of prayer, even in this expanded sense, learning to pray well is part of our liturgical responsibility. In an essay entitled "The Integrity of Sung Prayer," Don Saliers (1981: 291–292) notes:

> At the heart of our vocation as church musicians and liturgical leaders is the question of how we enable the Church to "pray well"—to sing and dance faithfully and with integrity … When we are engaged in sung prayer, we are not simply dressing out words in sound; rather, we are engaged in forming and expressing those emotions which constitute the very Christian life itself.

The Holy Spirit has provided today's church with a diverse spectrum of possibilities for praying well. This approach explores ways in which our sung prayer is changing.

When I examined the 40 hymnals as well as additional hymnal supplements that have shaped this study, a noteworthy pattern seemed to emerge: *I sensed, in broad terms, a correlation between musical style and theological emphasis.* This hypothesis challenges some widely held assumptions that musical style, as an artifact of culture, is neutral and conveys nothing in and of itself, i.e., a given content may fit into any style and communicate the same content. While it is true that some theological themes emerge across streams, specific differences appear from stream to stream; some theological themes are more prominent in some streams than in others.

Don Saliers (2007:28) acknowledges the link between musical style and theology:

> The musical idiom conveys a great deal about how the community conceives of God. Acoustic images reflect theological imagination at work. When the quality of the music is grandiose or pompous, the projected image of God may contain more of the self-image of the worshiping community than the community realizes. When the quality of music is pleasant and folksy, the projected image of God may be strong on intimacy and ease, but lacking in awe or mystery… [M]uch depends upon the language used to address God or to describe God's relation to the world and to human beings. Gregorian chant is simple in one sense, but not without mystery. This may be said for melodies from folk traditions…such as are found in Appalachian traditions or in the Spirituals. So we must attend to the wedding of text and tune, and to the way in which the assembly actually sings—what musicians refer to as the "performance practice" of the words set to music.

For Saliers and for us, music matters in worship not just as a conveyor of emotion (though music certainly has affective import), but as a window into the piety of a worshiping body and a partner the process of articulating the sung theology of a tradition.

The unprecedented musical eclecticism of congregational song styles since the Second Vatican Council replaces the assumptions of hundreds of years when a given tradition might be recognized more or less monolithically by a single musical style, for example, plainsong or Renaissance polyphony for the Roman Catholic Church, Gospel Songs for nineteenth and early twentieth century Revival traditions in the United States, or Victorian hymnody for nineteenth-century Anglicanism. To some, the eclecticism of our era connotes a fragmented church. From these who are perplexed by the sheer variety musical options, we hear the questions, "Where is all of this leading? What style will finally win out?" To others, the diversity of musical styles found in church music indicates a hopeful trend. Rather than asking where we are headed, these people relish a time when the musical fullness of the church, manifest through stylistic diversity and theological perspectives, seems to be at hand. Rather than a fragmented church, the myriad styles of music may be a theological indication that this is a time of unprecedented creativity in the Spirit, a time of unparalleled cross-cultural and ecumenical exchange, and a time to conceive God and God's actions among humanity in the broadest, least-restricted terms.

Congregational song styles are more restrictive than many other forms of music because they must be effective with large groups of people that rehearse very little (if at all) and who are as a whole untrained in music or singing. Effective congregational singing requires accessibility and, to varying degrees, immediacy. Because all congregational song has its fulfillment in worship, some music—albeit within a wide variety of musical styles and forms—is better than others for this purpose.

The assumption that music carries no inherent meaning and is, therefore, a neutral conveyer of content is largely an assumption of western classical aesthetics that values "absolute" music. Move into popular culture or traditional societies and this assumption breaks down very quickly. Based on this study, the poets in each stream tend to prefer a general musical style or a group of closely related styles to convey their theology. This study does not explain this assumption completely, but hopefully brings it to a greater awareness as an important factor.

If this is indeed true, one can assume a general relationship between musical style and theology. Congregations in North America have the possibility of singing many more musical styles than their forbearers in any other time in history. We may also deduce that singing out of only one stream, as varied as it might be and as comfortable as it may seem to the congregation, could limit the breadth of sung theology a congregation encounters. *This approach challenges all who lead worship to not limit their songs to a single stream, but to dip into several streams for an abundant sung faith with the hope of broadening the theological perspectives of their congregation and enjoying the variety of ways of praying that congregational singing offers*. Each congregation will have its own starting place or preference within this spectrum of current congregational song practice. Regardless of which stream is the center of a faith community's sung faith, vital singing congregation should broaden the range of its sung (and prayed) faith by incorporating songs from the depth of its particular confessional historical tradition and from the breadth of the current streams suggested in this article.

140.8 Future Streams—Changing Landscapes

This spectrum of congregational song is but a particular snapshot of how people are singing in a particular place (English-speaking North America) and time (the last two decades of the twentieth century and the first decade of the twenty-first century). How might such a survey look in 2050? As new hymnals are published and songs are increasingly disseminated through the media, Internet, and individual mechanisms such as iPod, I propose that we will have more streams than now. The landscape of congregational song will become increasingly varied as will the visual resources incorporate into worship. Given the exploding population of Latinos/as throughout the United States, at least one stream devoted to Spanish-language congregational song will be necessary. The many currents found in African American congregational song may divide and form separate streams. Hip Hop and Rap (Holder 2006), for example, already show signs of becoming a significant voice in the African American church. Stream Seven, Ecumenical and Global Song, is already potentially overflowing its banks. As communities who represent these songs continue to immigrate to North America, they may require separate and broader consideration. New voices are emerging, including Stuart Townend's Celtic sounds that are being embraced across many streams. The Emergent Church movement is producing a plethora of grassroots musicians, some of whom will find broader acceptance. Increasingly, texts and music will cross streams. For example, a Stream Two classical text may be set to Stream Six music. Regardless, this overview is but a point along the journey in the continuing expansion of congregational song and the growing diversity of the spiritual landscape.

At the beginning of this essay, I quoted Aalbert van den Heuval who boldly claims: "Tell me what you sing, and I'll tell you who you are!" Perhaps through singing more broadly we may also discover who we may become.

References

Hawn, C. M. (1996). The Consultation on Ecumenical Hymnody: An evaluation of its influence in selected English language hymnals published in the United States and Canada since 1976. *The Hymn, 47*(2) (April), 26–37.

Hawn, C. M. (1997). 'The Tie That Binds': A list of ecumenical hymns in English language hymnals published in Canada and the United States since 1976. *The Hymn, 48*(3) (July), 25–37.

Hawn, C. M. (2003a). Form and ritual: Sequential and cyclic musical structures and their use in liturgy (chapter 7). In *Gather into one: Praying and singing globally* (pp. 224–240). Grand Rapids: Wm. B. Eerdmans.

Hawn, C. M. (2003b). How can we keep from singing? The role of musicians and music in enabling multicultural worship (part III). In *One bread, one body: Exploring cultural diversity in worship* (pp. 126–139). Bethesda: The Alban Institute.

Holder, T. (Ed.). (2006). *The hip hop prayer book*. Harrisburg: Church Publishing.

Hymns and tunes recommended for ecumenical use. (1977). *The Hymn, 28*(4), 192–209.

Sacrosanctum Concilium. (1963). Par. 37. www.vatican.va/archive/hist_councils/ii_vatican_council/documents/vat-ii_const_19631204_sacrosanctum-concilium_en.html

Saliers, D. E. (1981). The integrity of sung prayer. *Worship, 55*(4), 290–303.

Saliers, D. E. (2007). *Music and theology*. Nashville: Abingdon.

Schilling, S. P. (1983). *The faith we sing*. Philadelphia: The Westminster Press.

van den Heuvel, A. (1966). *Risk: New hymns for a new day*. Geneva: World Council of Churches, Preface.

Chapter 141
Streams of Song: Developing a New Hymnal for the Presbyterian Church (USA)

Beverly A. Howard

141.1 Introduction

Congregational singing has been an integral part of Christian worship in America for 400 years, with the first British colonists bringing their psalters and religious singing practices to the New World. They sang from psalters, metrical settings of psalms, which they had used in England: *The Whole Book of Psalms* (Sternhold and Hopkins 1562), *The Whole Book of Psalms Englished Both in Prose and Metre* (Ainsworth 1612) and *The Whole Book of Psalms* (Ravenscroft 1621). They continued their singing practice of "lining out" a psalm: a cantor prompts a congregation with a line of text and tune, followed by the congregation singing that line. Thus, they sang their way through a psalm one line at a time. Congregational song was so essential in colonial congregations that the first English-language book published in North America was *The Whole Book of Psalms Faithfully Translated into English Metre* (1640), commonly called the Bay Psalter (Music and Richardson 2008: 80).

However, psalm singing in Britain and the colonies had a new rival: the hymn. Singing actual scripture (psalms) was the only acceptable practice among many English-speaking congregations. Isaac Watts (1674–1748) and others in England were challenging this notion by writing texts based on scripture and Christian doctrine, but of "human composure." Although Isaac Watts is popularly called "the father of the English hymn," other immigrant groups brought their own hymns to America. The first hymnal published in North America was John Wesley's (1703–1791) *Collection of Hymns and Songs* (1737) (Eskew and

B.A. Howard (✉)
School of Music, California Baptist University, Riverside, CA 92504, USA
e-mail: bhoward@calbaptist.edu

S.D. Brunn (ed.), *The Changing World Religion Map*,
DOI 10.1007/978-94-017-9376-6_141

McElrath 1995). Its content included hymns by Watts, John and Charles Wesley, and translations of German-language hymns he heard Moravian shipmates singing while sailing from England to colonial Georgia in 1735. As the hymn became firmly planted in North America, it thrived, evolved with a changing nation, and shaped the faith of future generations.

Christians view hymns as a unique type of poetry sung to God that may express praise, prayer, lament, confession, or petition. The text, featuring rhyme and poetic devices, is organized in verses or stanzas. The hymn is set to a tune, with each stanza repeating the same melody (strophic). A hymnal is a book containing the sung theology of a religious group. Its contents reveal a congregation's core beliefs, theological convictions and their practical application. New hymnals are usually produced once a generation (or 20–25 years), in response to a group's evolving theology and practice. Hymnal committees select hymns that both preserve a denomination's unique heritage and reflect that group's changing religious thought and practices.

Presbyterians have a long history of hymnal publication in America with the first hymnal appearing in 1831 (Psalms and Hymns Adapted to the Public Worship). Subsequent Presbyterian hymnals include:

The Presbyterian Hymnal (1874)
The Hymnal (1895)
The Hymnal (1933)
The Hymnbook (1955)
The Worshipbook (1972)
The Presbyterian Hymnal: Hymns, Psalms, and Spiritual Songs (1990).

In September 2008, 15 newly appointed members of the Presbyterian Hymnal Committee met for the first time in Louisville, Kentucky to begin work on a new Presbyterian hymnal. Over the next 3 years, this committee worked within its changing denominational landscape to develop *Glory to God: A Presbyterian Hymnal* (2013). This new resource includes materials reflecting the denomination's new demographics, new musical and liturgical styles, and new emphases. This article discusses the development of this new collection of congregational song focusing on the need for a new hymnal within the changing denominational landscape, its theological vision statement, and the selection of contents with special attention to contemporary Christian and global songs.

Cognizant of the denomination's changing landscape, the committee knew that this collection would include more than the standard strophic hymn and believed that the word "hymnal" in the committee's name conveyed a limited scope in its work. In fact, this collection encompasses texts and tunes streaming from many sources: chant, contemporary song idioms, historic and contemporary hymnody, global and ecumenical sources, praise choruses, and psalms. To convey the breadth of this project, the committee voted to change its name to the Presbyterian Committee on Congregational Song [PCOCS].

141.2 Methodology

The PCOCS met quarterly for 14 meetings during which the committee considered procedural processes, developed criteria for evaluating submissions, worked within subcommittees, and sang every psalm, hymn, or song under consideration. David Eicher, Editor, describes some aspects of the committee's process:

> We did all of the selection by a two-thirds majority vote. We had various task forces and groups working within the full committee that would do initial screening, and within those groups it took a two-thirds vote to move it on to the next level of consideration. Everything received several rounds of consideration by various groups before it reached the full body to make the final decision.
>
> Another important piece of the selection process is that we did it all anonymously. (Eicher 2012a: 58)

During the entire project, the PCOCS conferred with as wide a constituency as possible to create a collection that would respect the denomination's diversity. They familiarized themselves the PH 1990 committee's process, reviewed literature pertinent to the discussions, interviewed recognized experts in congregational song (by conference call and Skype), and consulted advisory groups and denominational leaders. Data from the following formal surveys informed the group:

- The Presbyterian Panel Survey: Hymns
- Feedback about Hymns in *The Presbyterian Hymnal*.

To realize the committee's intent for a transparent process, the Presbyterian Publishing Corporation launched a website that posted:

- News
- Project Blog
- FAQ
- Meeting archives
- Educational Resources
- Contents
- Ordering Information.

Working with thousands of psalm settings, hymns, and spiritual songs, the committee needed a means of categorizing and tracking materials. Hymnologist Michael Hawn's "Streams of Song," provided a useful metaphor and practical tool for organizing and discussing the texts.

This article draws on surveys, unpublished reports and minutes of the meetings, and the "streams of song" model to trace aspects of this collection's development.[1]

[1] As of May, 2012, survey results, articles, and general resources are available at www.presbyterianhymnal.org/resources.html. Meeting archives may be accessed at http://blog.presbyterianhymnalproject.com/. Pre-publication content list is available at www.presbyterianhymnal.org/sneakpeek.html.

141.3 Needs Assessment

141.3.1 Why a New hymnal?

The Presbyterian Church (U.S.A.) [PCUSA] last published a hymnal in 1990, *The Presbyterian Hymnal: Psalms, Hymns and Spiritual Songs* [PH 1990]. Prompted by the action of the 216th General Assembly (2004) to "authorize the Presbyterian Publishing Corporation, the Office of Theology and Worship, and the Presbyterian Association of Musicians to begin research into the feasibility of a new Presbyterian Hymnal," an ad hoc committee of representatives from the three entities requested a survey to determine need (Marcum 2007: i). The office of Research Services of the Presbyterian Church (U.S.A.) surveyed the Presbyterian Panel to gather opinions on the need for a new hymnal, how the current hymnal was being used, and musical styles used in worship.[2] From data collected in this survey, conducted in 2005, the Presbyterian Publishing Corporation concluded that there was sufficient need and interest to move forward with plans to develop a new hymnal.

Some results from the Presbyterian Panel survey encouraging the Presbyterian Publishing Corporation to embark on this endeavor included:

1. Almost all of the congregations use a hymnal, with two in three congregations using PH 1990. Most panelists reported the degree of satisfaction as either "very satisfied" or "satisfied."
2. A majority of ministers responded "very likely" or "generally likely" (57 %) that a new hymnal would be needed by 2013.
3. In contrast, only 44 % of members and elders responded "very likely" or "generally likely" that a new hymnal would be needed by 2013.
4. Approximately a third of laity (members, 30 %; elders, 33 %) and half of ministers (pastors, 47 %; specialized clergy, 49 %) are personally "very interested" or "generally interested" in the PCUSA "developing a new hymnal to be published in the year 2013. (Marcum 2007: 1)

The survey highlighted some arguments for why a new hymnal would be needed. From the subset responding that the need new hymnal will be "very likely" or "generally likely" these reasons emerged:

- Younger worshipers will want more contemporary hymns (members, 69 %; elders, 80 %; pastors, 79 %; specialized clergy, 62 %)
- Newer hymns will have been written and need to be included (members, 59 %; elders, 66 %; pastors, 75 %; specialized clergy, 70 %)

[2] The Presbyterian Panel consists of three representative groups from the PCUSA: members, elders currently serving on session (the governing body of a Presbyterian congregation, and ministers (ordained and specialized). Panelists serve for three years with surveys conducted quarterly. The Panel follows standards developed by the American Association of Public Opinion Research Guide.

- Congregational song preferences will have shifted such that many worshipers will want a different collection of hymns (members, 55 %; elders, 70 %; pastors, 67 %; specialized clergy, 63 %)
- Different hymns will be needed to reflect the changing cultural/racial/ethnic composition of the church (members, 45 %; elders, 63 %; pastors, 62 %; specialized clergy, 70 %) (Marcum 2007: 1).

The Presbyterian Publishing Corporation [PPC], in collaboration with the PCUSA Office of Theology and Worship, and the Presbyterian Association of Musicians announced the formation of a hymnal committee and welcomed applications. The 15-member committee, selected from 220 applicants, "represented all the geographical regions of the country, with racial ethnic diversity, men and women, older and younger members, small churches and large churches, musicians, pastors, and academic theologians" (Eicher 2012a: 57–58). In addition to the quarterly meetings at the Presbyterian Center in Louisville, KY, the PCOCS carried on extensive work between meetings.[3]

141.3.2 Hymnals as a Reflection of Changing Landscapes

As is the case with other mainline Protestant denominations, Presbyterians work in a changed and changing landscape. Declining membership, increased numbers of ethnicities, a diversity of theological groundings, and competing musical styles in worship create new and sometimes slippery terrain for Presbyterian congregations.

The committee spent considerable time discussing how the worship landscape in the early twenty-first century differs from that in 1990 and what that would mean for a new hymnal. Such a discussion was not unique to this committee or even this century; historically, other Presbyterian hymnal committees have wrestled with change. Studying a hymnal's preface offers clues to how earlier generations responded to their changing milieus because a preface includes the rationale for a hymnal's content. Comparison of the prefaces from four Presbyterian hymnals (1895, 1933, 1955, 1990) shows that the editors and committees valued two criteria in the selection process:

1. Preserving the heritage of congregational song
2. Expanding the selections to represent a current generation (Howard 2012a: 2).

For example, in the preface to *The Hymnal* (1895), editor Louis Benson wrote that hymnal was "a manual of the Church's praise, a treasury of things new and old, chosen for actual service, expressive in some degree of the culture of god's people"

[3] Mary Louise (Mel) Bringle, Chair; Chi Yi Chen, Adam Copeland, Alfred V. Fedak, Stephen H. Fey, Charles D. Frost, Karen Hastings-Flegel, Beverly Howard, Paul Junggap Huh, Mary Beth Jones, Eric T. Myers, Chelsea Roeder Stern, Edwin van Driel, Michael Waschevski, and Barbara G. Wheeler. Ex-officio: William Mc Connell, David Gambrell, David Maxwell, Mary Margaret Flannagan.

(Benson 1895). The contents preserving the culture were "those endeared to the Church by proved fitness" (Benson 1895: iv). The editors of *The Hymnal* (1933) sought to preserve the "rich treasure of the heritage hymns of the Church" (Dickinson 1933: iii). Working guidelines for *The Presbyterian Hymnal* (1990) included continuing "the diversity of our historical traditions" (McKim 1990: 9). An important part of Presbyterian heritage is singing the psalter; that is, a musical setting of psalms. Each generation of Presbyterians has preserved historic psalm settings, either as chanted texts or metrically (rhythmically) set texts.

Hymnal prefaces also reflect changes occurring within the Presbyterian church's theology and worship by offering a rationale for the new materials included. The preface from the 1933 hymnal states that Presbyterians were interested in congregational song that gave "expression to certain new emphases in religious thought today," which included "social service, brotherhood of man and world friendship" and the devotional life of the Christian (Dickinson 1933: iii). Its table of contents contained categories such as "Brotherhood," "World Friendship and Peace," and "The Inner Life." New emphases allow for the addition of new hymns to a collection.

The Presbyterian Hymnal (1990) reflected latter twentieth-century Presbyterian thought regarding images of God, diversity, and inclusivity. The preface alerted the reader that this collection would

1. Express a full range of biblical images for the Persons of the Trinity
2. Include of all God's people—sensitive to age, race, gender, physical limitations, and language (McKim 1990: 9).

Prefaces from these four hymnals indicate the editors' sensitivity to the changing demographics of Presbyterians in America. The first Presbyterians in America were largely of Scottish descent. As immigrants changed the face of America, the Presbyterian church also changed. *The Hymnal* (1933) expanded Presbyterian song to include hymns from "various nationalities so largely represented in the Presbyterian Church in this Country," namely Welsh, Irish, and Scandinavian traditions. *The Hymnbook* (1955) was a compilation of materials from five Presbyterian-Reformed denominations. While preserving the Presbyterian heritage by including more metrical psalms, *The Hymnbook* (1955) "secured the admission of a representative body of so-called 'gospel songs'" which were used by some of the groups (Jones 1955: 5). The PH 1990 committee made "conscious efforts to recognize various racial and ethnic musical traditions" (McKim 1990: 9). In fact, the PH 1990 may "have the best representation of global music in a denominational hymnal up to that time" and "made great strides in getting us to realize that many Christians around the world worship in languages other than English" (Eicher 2012a: 56).

As was the case with earlier Presbyterian hymnal committees, the PCOCS embraced the delight and challenge of selecting contents for a hymnal that both preserves and expands congregational song. Several studies helped the committee interpret the changing landscape in American religion and its impact on the PCUSA.

141.3.3 PCUSA: Mirror of the American Protestant Landscape

Over the past 25 years, American Protestantism has experienced a changing landscape with regard to church affiliations, attitudes toward musical styles, and liturgical practices, with the PCUSA mirroring many of these changes. *The Pew Forum on Religion & Public Life: U.S. Religious Landscape Survey* details data showing the declining percentage of Protestants (evangelical, mainline and historically black) in the United States from 60–65 % in the 1970s to 51 % in 2008 (Pew Forum 2008:18). The largest decline is among the mainline Protestants, the tradition in which Presbyterians are situated.

The Landscape Survey also notes that there is movement by Americans from one religious group to another: one in four Americans have changed their religion from that in which they were raised (Pew Forum 2008: 22). Of particular interest to this study is that if changing "within the ranks of Protestantism (for example, from Baptist to Methodist) and within the unaffiliated population (for example, from nothing in particular to atheist) is included, roughly 44 % of Americans now profess a religious affiliation that is different from the religion in which they were raised." (Pew Forum 2008: 22)

The PCUSA has experienced a parallel decline with a net loss of 31.6 % membership (Marcum 2009:1). Results of a 2010 survey of PCUSA congregations revealed that only 40 % of regular congregational participants 18 and older identified themselves as "life-long Presbyterians." (FACT 2010: 2) This would indicate that congregations are populated by people from other faith traditions, who bring with them their own hymn and liturgical traditions.

Although Presbyterians are experiencing a net loss of members, the membership is more ethnically diverse than in previous generations. From 1999–2008, whites dropped almost 2 % points from 93.5 to 91.7 % of the total PCUSA membership (Marcum 2009:7). These racial-ethnic groups increased:

Asian (2.3–3.3 %)
Hispanics (1.1–1.4 %)
Blacks (2.8–3.3 %)
Native Americans (.29–.30 %).

In this 9-year period, racial ethnics grew from 6.5 % of the total membership to 8.3 %, owing to both numerical growth among these groups and the sharp decline of whites (Marcum 2009:7).

141.3.4 Changing Landscape of Worship Music Styles

Among the many changes in American religious groups has been the rise of new musical styles labeled as Contemporary Christian or Praise and Worship. These styles, that began to emerge in the 1960s, use popular song idioms and instrumentation to attract a new generation of parishioners. In some churches, the standard strophic hymn has been abandoned for songs, organ has been supplanted by a praise band, and choirs replaced by a small vocal ensemble. Delivery of texts and tunes has also changed. Congregations may now sing in worship following a text projected on a screen rather than reading from a printed page in a hymnal. Churches enjoy unprecedented, instant access to congregational song resources downloaded from the Internet, frequently subscribing to various copyright licensing services such as Christian Copyright Licensing International [CCLI], Oregon Catholic Press [OCP], or onelicense.net. Before the advent of the Internet, many worship leaders illegally reproduced texts and tunes in orders of worship in an attempt to resource their congregations with the latest Christian songs. Subscription services, such as CCLI, began to emerge in the 1980s to provide churches a legal means to access and reproduce new materials. These subscriptions enable a church to download, print or project the text and tunes of numerous copyrighted hymns and songs for use in worship, if the copyright holder (usually a publisher) is covered by that license. One significant proviso is that the reproduction of this material is for congregational singing only. These licensing services specialize in contemporary Christian songs, although public domain and contemporary strophic hymns are available. Because of the wealth of materials now legally available online, some churches have abandoned hymnals altogether.

The PPC sought to determine the impact of this changing milieu on Presbyterian congregations. The 2005 Presbyterian Panel Survey stated:

- Half of pastors report congregational singing of "contemporary praise choruses" either "weekly" (39 %) or "every 2–3 weeks" (12 %). Only 19 % report "never" doing so.
- More than one in four pastors (28 %) report that their congregation has a "praise band/choir (Marcum 2007: 7).
- Around one in four pastors report that "every week" their worship service includes "projections on a screen or wall, using a computer."
- Almost three in four pastors (72 %) report that their congregations have a "copyright license to reproduce hymns." (Marcum 2007: 9)

141.3.5 Streams of Song

At the first PCOCS meeting in 2008, three committee members agreed to prepare needs assessment reports for the next meeting with goals of analyzing the contents of the current hymnal and projecting what types of hymns and songs might be needed for the new collection. They used data from the August 2005 Presbyterian Panel Summary: Hymnals and analyzed the contents of the PH 1990. Other resources examined included topical indices from several contemporaneous denominational hymnals, numerous single author collections, and literature on congregational song. Their reports included a statistical breakdown of the contents of the current collection, a spreadsheet indexing hymns within "streams of song," and discussion of underused or absent themes from the Revised Common Lectionary. Topical indices from other hymnals indicated what other denominational groups deemed important for their congregations to sing. Gleaned from several hymnals surveyed, these topics reveal postmodern thought and concerns: Alienation, Complacency, Care of the Earth, Ecology, Ecumenism, Expanded Images for God, Food and Hunger, Grief, Healing, Human Dignity, Lament, Multicultural and World Church Songs, and Stages of Life. Single-author collections proved a helpful source for recent texts authored by current hymn writers responding to today's world and its needs. Some perceived needs that emerged from these reports were:

- Continued inclusion of contemporary hymn writers
- Inclusion of songs from Contemporary Christian Music and Praise and Worship repertoire
- Inclusion of congregational song tracing God's works and acts in the lives of biblical witnesses, with attention to women
- Expansion of Holy Spirit imagery
- Hymns reflecting life in a global, post-modern, fractured world. (Howard et al. 2009)

An invaluable tool for categorizing congregational song emerged during this process: the "streams of congregational song." Hymnal committees grapple with finding ways to categorize hymn texts. With the vast quantity and diversity of texts available today, historic labels are not sufficient. Hymnologist Michael Hawn proposes seven "streams of song" as an overarching organizational metaphor for current congregational song. He writes:

> Streams have a source, and each of the proposed seven streams of song comes from particular sources of faith – a particular expression of piety. Streams come in various widths and depths. Not all streams are the same. Some of the song streams are rushing and seem to be overflowing their banks because of the musical outpouring being generated from their particular piety source. Others are steady in their flow, and yet others may be either drying up or merging with other streams. Streams meander; they do not flow in straight lines like canals. They occasionally crisscross each other…
>
> … streams are vibrant parts of creation, carrying us along with them, offering constant changes in depth, rate of flow and character. (Hawn 2010: 20)

Hawn names his seven streams:

1. Roman Catholic/Liturgical Renewal
2. Classical Contemporary Protestant Hymnody
3. African American Stream
4. Gospel and Revival Songs
5. Folk Hymnody
6. Pentecostal Song
7. Ecumenical and Global Stream (Hawn 2010: 20–21).

Although meant to describe the types of congregational song available today, each of the streams has historical precursors, making this a useful tool for itemizing the contents of a hymnal (Hawn 2010: 18–19) (Table 141.1).

Table 141.1 Streams of song as proposed by Michael Hawn

Stream 1. Roman Catholic Liturgical Renewal	Songs for sacraments and ritual, chant, responsorial psalms used in the roman church with renewed interest after Vatican II
Stream 2. Classical Contemporary Protestant Hymnody	Strophic hymns, metrical psalms, scriptural paraphrases, prophetic hymns; includes texts from the early church, Reformation, the hymn renaissance of 1960s, current writers
Stream 3. African American	Spirituals, hymns, gospel style
Stream 4. Gospel and Revival Songs	Songs of praise and personal salvation and experience; associated with 19th century Sunday School era songs and evangelistic music
Stream 5. Folk Hymnody	Folk, ballad style text, using acoustic guitar; rebirth during the 1960s
Stream 6. Pentecostal Song	Labeled Praise and Worship and Contemporary Christian; use scriptural fragments; range from ecstatic praise to intense prayer
Stream 7. Ecumenical and Global	Hymnic and song contributions of Christians worldwide, especially Africa, Asia, Latin America and the Caribbean; includes song from two ecumenical communities: Taize and Iona

Source: Hawn (2010:18–19)

These streams of song helped the PCOCS navigate the selection process and provided:

1. Fluid means to categorize texts under consideration
2. Perspective on the balance of the hymnal's contents.

141.4 Theological Vision Statement

While the "streams of song" gave structure to hymn categorization, the committee needed a theological framework against which to evaluate hymn submissions. A subcommittee developed such a statement to guide the committee's work throughout the process. This passage from the theological framework acknowledges the challenging milieu in which the committee developed *Glory to God.*

> The next Presbyterian collection of hymns and songs, however, will be published amid different conditions than those that molded previous ones. It will be offered in a world in which trust in human progress has been undermined and eclectic spiritualities often fail to satisfy deep spiritual hungers. It will be used by a church many of whose members have not had life-long formation by Scripture and basic Christian doctrine, much less Reformed theology. It is meant for a church marked by growing diversity in liturgical practice. Moreover, it addresses a church divided by conflicts but nonetheless, we believe, longing for healing and the peace that is beyond understanding. (Wheeler and van Driel 2009: 1)

This theological vision statement also addresses some of the needs expressed in the committee's Needs Assessment report, specifically offering a theological rationale for singing songs from other cultures and different musical styles.

141.5 Selection Processes

141.5.1 Material "Carried Forward"

With a preliminary understanding of what is needed for the new hymnal and a guiding theological framework, the PCOCS determined what content from the PH 1990 should be "carried forward" for inclusion in *Glory to God.* For historical perspective, a committee member examined *The Hymnbook* (1955) and reported that approximately one-third of its contents were included in PH 1990. Hymnal committees face difficult choices when deciding which hymns will be retained in a new hymnal. Deleted hymns usually feature archaic language or outdated ideas.

By the time the PCOCS began the winnowing process from PH 1990, the data from "Feedback about Hymns in *The Presbyterian Hymnal,*" was available. Of all the congregations using PH 1990, 292 congregations participated in the survey with each church having a minister, musician, and member rate every hymn. The responders noted how often it was sung during the last year and opinions on if the hymn should be retained. Using a five-point Likert scale, responders chose the frequency of a hymn's use ranging from "more often than once or twice" to "never" and whether to keep/drop a hymn in the new hymnal ranging from "definitely keep" to "definitely drop."

Not surprisingly, the list of "most sung" hymns shows a heavy concentration of Christmas carols and Easter hymns. Table 141.2 compares the two ranked lists which exhibit a strong correlation between hymns that are "most sung" with those to "definitely keep," but not necessarily with the same ranking (PPC 2009: 3–4).

The list of "least sung" hymns, in Table 141.3, includes psalm settings and some of the more difficult global songs. There is a notable difference between that list and the "definitely drop" list. Although some of the global songs appear on both lists, traditional hymns also appear on the "definitely drop" list (PPC 2009: 3–4).

Table 141.2 "Most sung" and "definitely keep" hymns from PH 1990

Rank	"Most sung" hymns	"Definitely keep" hymns
1	O Come, All Ye Faithful	O Come, All Ye Faithful
2	Come, O Come, Emmanuel	Joy to the World
3	Hark! The Herald Angles Sing	Angels We Have Heard on High
4	Joy to the World	Jesus Christ is Risen Today
5	Jesus Christ is Risen Today	O Come, O Come, Emmanuel
6	Angels We Have Heard on High	Hark! The Herald Angels Sing
7	Silent Night, Holy Night	Holy! Holy! Holy! Lord God Almighty
8	Holy! Holy! Holy! Lord God Almighty	Joyful, Joyful We Adore Thee
9	Amazing Grace, How Sweet the Sound	Silent Night, Holy Night
10	Be Thou My Vision	Were You There?
11	Joyful, Joyful We Adore Thee	

Source: Feedback from The Presbyterian Hymnal Research Services, Presbyterian Church, USA

Table 141.3 "Least sung" and "definitely drop" hymns from PH 1990

Rank	"Least sung" hymns	"Definitely drop" hymns
1	Earth's Scattered Isles and Contoured Hills	Holy Night, Blessed Night (*Shen Ye Qing*)
2	Our King and Our Sovereign, Lord Jesus (*Jesus – Jesus Es Mi Rey Soberano*)	When Twilight Comes (*Awit Sa*)
3	When Twilight Comes (*Awit Sa*)	Lonely the Boat (*Kahm Kahm hahn Bom Sanaoon*)
4	Holy Night, Blessed Night (*Shen Ye Qing Sheng Ye Jing*)	Sheep Fast Asleep (*Hitsuji Wa*)
5	Midnight Stars Make Bright the Sky	Creating God, Your Fingers Trace
6	Praise Is Your Right, Oh God, in Zion - Psalm 65	Walk On, O People of God (*Camina, Pueblo de Dios*)
7	Psalm 72	Christian Women, Christian Men
8	Psalm 118:19-29	God Created Heaven and Earth
9	With Joy I Heard My Friends Exclaim - Psalm 122	Psalm 84
10	God Folds the mountains Out of Rock	When I Had Not Yet Learned of Jesus (*Yee Jun Ae Joo Nim Eul Nae Ka Mol La*)
11	When I Had Not Yet Learned of Jesus (*Yee Jun Ae Joo Nim Eul Nae Ka Mol La*)	Psalm 146

Source: Feedback from The Presbyterian Hymnal, Research Services, Presbyterian Church, USA

This survey was valuable for committee deliberations not as a tool to dictate what was included or dropped, but rather to add perspective to discussions. The PCOCS retained approximately 60 % of the PH 1990 for inclusion in *Glory to God.*

141.5.2 Task Forces/Committees

With 60 % of the PH 1990 carried forward, there remained room for about 400 new selections. The committee considered around 10,000 options gleaned from open and invited submissions and subcommittee recommendations.

There was a 2-year open submission process for anyone to submit a previously published text/tune or an original text/tune. Review teams evaluated these submissions anonymously. Any submission receiving a 2/3 approval moved on to the large committee for a vote. All votes in the large committee carried with a 2/3 majority.

Subcommittees, or task forces, accomplished much of the work, bringing specific recommendations to the full committee. During the course of deliberations, these subcommittees functioned:

- Contemporary
- Global
- Other Hymnals
- Single-author Collections

- Psalms
- Review Teams
- Service Music
- Text
- Tune.

The Other Hymnals and Single-author Collections groups reviewed massive numbers of hymns, bringing their recommendations to the full committee for consideration. The Other Hymnals group examined more than 20 hymnals, many of which were published after 2000. Their work gave insight to what other denominations had recently added and, in some cases, restored to their collections. This committee noted that other Reformed groups were introducing more nineteenth century gospel hymns in their most recent hymnals. Perhaps the changing demographic of only 40 % of congregants 18 and older being life-long Presbyterians provides a partial reason for this. Single-author collections proved to be an outstanding source for new texts dealing with topics such as aging, healing, abusive relationships, ecology, disabilities, fractured families, and gender issues. In fact most of the hymns in *Glory to God* dealing with those special topics were gleaned from single-author collections.

The Psalms and Service Music committees reviewed items for metrical settings of psalms and music used in communion and other liturgies.

The Text committee examined every text approved by the PCOCS and recommended versions of a text, language changes, and number of stanzas to include, sometimes restoring stanzas the PH 1990 committee had removed. Because the 1990 PH committee paved the way for inclusive language, a generation of Presbyterians have become accustomed to language that does not stereotype persons according to "categories of gender, race, ethnicity, socio-economic class, sexual orientation, age, or disabilities." (Wheeler and van Driel 2009: 2) The PCOCS maintained that standard in *Glory to God.*

The Tune committee recommended tunes for texts submitted without tunes, alternate tunes, melodic and harmonic settings, keys, guitar chords, and accompaniments.

Of particular interest to an article exploring the shifting terrain in Presbyterian worship is the work of what became known as the Contemporary and Global groups.

141.6 Contemporary Group: Developing the Pentecostal Stream

The Contemporary group worked with texts and tunes belonging to the Pentecostal stream of song, a shallow stream with need for infusion. According to the needs assessment generated by the PCOCS, the PH 1990 contained only one item in that category, "Seek Ye First," a short praise and worship chorus. Within the Pentecostal stream, there are various styles of texts and songs identified as praise choruses, worship choruses, modern hymns, and adaptations of contemporary Christian music for congregations.

As music from the Pentecostal stream wends its way into the repertoire of congregational song, its musical style and performance are not without controversy; that is, non-hymnic literature accompanied by praise bands. For all the people who embrace the new styles and sounds, there are equal numbers who do not. Parishioners may be quick to voice their personal preferences, musicians may not be comfortable with the new performance demands, and some people simply resist change. Worship leaders are not always skillful incorporating the style into the liturgy. For Presbyterians, there may be language and doctrinal differences owing to much of this repertoire coming from the evangelical and Pentecostal traditions. However, this stream is vibrant and viable in the Presbyterian church and deserves thoughtful consideration.

The Contemporary group affirmed the portion of the Theological Vision Statement that addresses the need for this style:

> …the notion of salvation history invites us to bridge the divide between different musical styles and traditions. As scribes who have been trained for God's reign will bring out of their treasures "what is new and what is old" (Mt. 13:52), so musicians are invited to lead us in songs both old and new, in praise of a God who is the first and the last, the ancient of everlasting days and the Lord of the new creation. (Wheeler and van Driel 2009: 2)

This group developed an extensive reading list, interviewed church music directors actively involved with contemporary Christian music, and conferred with renowned leaders and academicians in this genre, including:

- Ron Rienstra, Western Theological Seminary and author on Contemporary Christian repertoire
- Greg Scheer, Associate at Calvin Institute of Christian Worship and author of *The Art of Worship: A Musician's Guide to Leading Modern Worship.*

Members met with focus groups at youth meetings held at Presbyterian conference centers. The group initiated a Facebook page to solicit suggestions for songs and gather input from people interested in this aspect of the hymnal project.

One of the challenges in choosing which contemporary Christian music goes in the hymnal is that churches committed to using this music on a regular basis use Internet resources that provide text and music for the latest songs. Hymnals, existing as print media, contain published materials that may be 5–15 years old. The Contemporary group's approach was to develop a core of songs for consideration; that is, a list of songs that are being sung regularly and gaining acceptance throughout the church.

Discussions with leaders in this field and the focus groups centered on establishing a core repertoire: what songs are becoming the canon. The group compiled a spreadsheet of 88 songs that enjoy widespread usage in churches. Each song carried annotations of scriptural references, performance style, liturgical function, and alignment with the theological framework.

Other challenges in selecting contemporary Christian music for a hymnal involve language issues and singability. Language in worship, whether spoken or sung, is powerful and shapes people's image of God. The PCOCS ratified a Statement on Language which reads:

> Scripture uses an abundantly rich array of prose and poetry to tell us about God's powerful acts of creation, redemption, and final transformation. Much biblical imagery is indeed masculine, but there is also a wide variety of other metaphors that are either feminine or gender-neutral. Most important, behind all biblical narrative lies the deep and prevailing sense that God is the one whose ways and thoughts are as beyond human speech as the heaven is higher than the earth. (Isa.55:8)
>
> The collection will draw from the full reservoir of biblical imagery for God and God's gracious acts. The final product will include both metaphors that are comfortable in their familiarity and those that are enriching in their newness.
>
> The collection will emphasize that the God who meets us so graciously and intimately in salvation history is at the same time one who is wholly other and beyond gender. Therefore, texts will reflect a strong preference for avoiding the use of male pronouns for God. In evaluating each hymn or song, issues of tradition, theological integrity, poetic quality, and copyright will all be considered. The goal is a collection in which traditional hymns and songs are balanced with others that are more gender-neutral or expansive in their reference to God. (PCOCS Language 2009: 2–3)

Much, but not all, of the repertoire from the Pentecostal stream employs limited and usually patriarchal imagery for God. The PCOCS took great care to recommend texts from this genre that display a wider array of images. In some cases, songs were referred to the Text committee for editing. For example, a text reading "He reigns" might be altered to "God reigns." Because most of these materials are protected by copyright, PPC must acquire permission for such changes to appear in *Glory to God.*

Many songs in this genre first gained popularity by contemporary Christian recording artists who wrote them for their own concerts or recordings. As such, some songs require adaptation for a congregation to sing. These songs can feature difficult melodic lines for the untrained singer or highly syncopated rhythms. The Contemporary group and the Tune committee searched for adaptations of songs that accommodate the average singer and are rhythmically accessible. The Tune committee also sought arrangements that are accessible for the average accompanist. Occasionally only the refrain (chorus) was selected if the verses and bridge were deemed too difficult. These refrains and some of the shorter songs from the Praise and Worship genre work well in the liturgy as responses or for communion.

In their own deliberations, and as recommended to the larger PCOCS body, the Contemporary group suggested the following guidelines when evaluating selections from the Pentecostal stream:

1. At the end of our selection process, do we find a good variety of songs and choruses from different sources?
2. Do our selections include a generally identified growing "core" of Contemporary Praise and Worship music being included in other mainline denominational hymns and songbooks?
3. Are the selections singable by congregations without professional musicians?

4. Are adequate performance notes/suggestions included in order to facilitate the accompanying and singing of these songs?
5. Are the texts acceptable for Reformed/Presbyterian congregations? (Stern et al. 2009)

After careful consideration, thoughtful discussion and deliberations, the PCOCS voted to include 74 songs from the Pentecostal stream. This increase from 1 % in PH 1990 to 9 % adds depth to the Pentecostal stream.

141.7 Global and Ecumenical Group: Expanding the Stream

The entire hymnal committee was committed to continuing and expanding the work of the 1990 PH committee to include songs from global Christian communities "as gathering songs, as short responses, as meditative music, or music during communion." (Eicher 2012a: 56) The Theological Vision statement underscored the need for including songs from Christian communities worldwide:

> The framework of the history of salvation offers a theological rationale for asking us to learn songs that come from cultures different than our own: Pentecost teaches us to speak and hear the gospel in many tongues and languages and only thus, "with all the saints," to comprehend the breadth and length and height and depth of the love of Christ (Eph. 3:18). We do not sing hymns and songs because they were birthed in our culture; we sing them because they teach us something about the richness that is in God. (Wheeler and van Driel 2009: 2)

Congregations who sing global songs, generally defined as texts and/or tunes from non-western countries, discover new, rich images of God, Christ, and the Spirit; worship with new sounds, rhythms and instruments; view the gospel through the lens of Christians worldwide. The numerous materials available make choosing what goes in a hymnal difficult.

Like the Contemporary group, the Global group carefully developed a core of songs for consideration. They reviewed numerous published resources and solicited recommendations for commonly-used songs from leaders in the field, such as Michael Hawn, author of *Gather into One: Praying and Singing Globally*. The entire PCOCS body participated in a conference call with Hawn and a Skype interview with Jorge Lockward, Director of the Global Praise Program of the General Board of Global Ministries in The United Methodist Church.

The Presbyterian Panel survey indicated that global songs were among some of the "lesser sung" materials. Pastors polled reported singing songs from other countries "less than once a month (38 %) or "once a month" (29 %) (Marcum 2007: 6). Some parishioners and church leadership share a perception that this music is too difficult. Indeed, churches with limited musical resources may have difficulty teaching and learning tunes from non-Western cultures. Another frustration expressed is the layout of these songs in the PH 1990 where the original language is placed immediately below the tune with the English printed further down the score. This forces English-language singers to find a note in the melody and then scan several lines of foreign language before finding the English text. However, both the Global group and the larger committee believe the advantages of resourcing Presbyterian congregations with these songs far outweigh the problems cited here.

The Global group began conversations that continued with the full body concerning the authenticity of the musical style. With the Tune committee, they located accompaniments that respect the culture's aesthetic. With the full committee, they dealt with the readability issue by proposing a consistent layout where English appears with the tune and the original language is printed following the hymn. Where existing English translations are awkward, new translations were located. The PCOCS recommended using tutorial videos as a means of assisting congregations in learning these non-Western styles of music. The Presbyterian Hymnal Project website features videos of 19 global songs appearing in *Glory to God*, with the promise of more to come. These videos, prepared by PPC and PCUSA staff, provide authentic performance styles and offer ways to teach the songs to a congregation.

The Global group produced an impressive spreadsheet of 196 items with annotations noting country of origin, original language, scriptural basis, liturgical function, and lectionary references (Jones et al. 2010). From this core list and other submissions, the PCOCS selected over 100 songs representing eight African countries, the Middle East. Spanish language hymns come from several South and Central American countries, Mexico and Spain. Asia is represented by songs from China, Japan, Philippines, and other areas of the Pacific region. Also represented are materials from the Taizé and Iona ecumenical communities. Global song in *Glory to God* constitutes 13 % of the contents, an increase from the 4 % in PH 1990.

141.8 Final Selection

The agenda for the last 3 day meeting of the PCOCS was devoted to finalizing the selections for *Glory to God*. Having already been through multiple layers of scrutiny, each proposed psalm, hymn, and spiritual song was brought once more to the full committee. The finalized collection includes more than 850 items from each of the 7 "streams of song," resourcing the next generation of Presbyterians with congregational song that preserves its Reformed heritage and yet is representative of a denomination with increased racial ethnic membership, and diverse musical styles (Fig. 141.1).

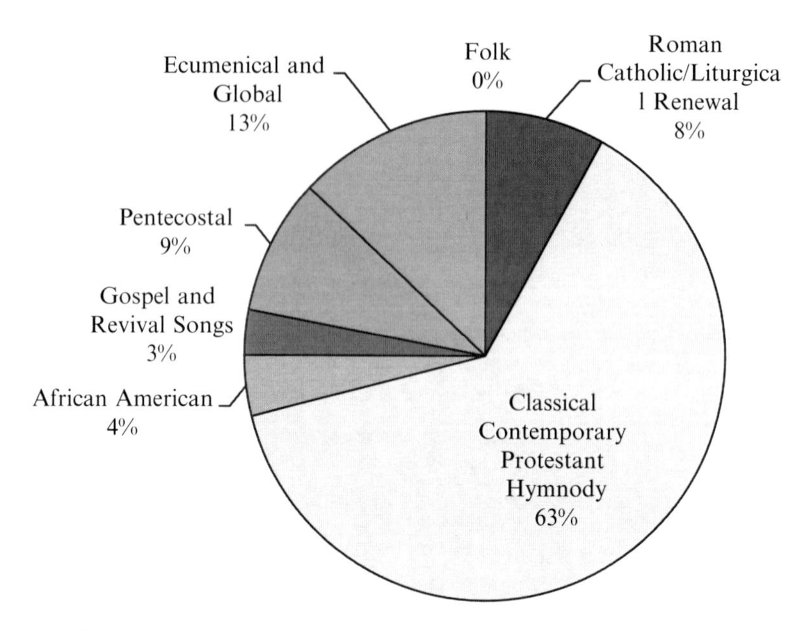

Fig. 141.1 Contents of Glory to God, by stream (Source: Beverly Howard)

141.9 In Production

In June 2012, the new hymnal was unanimously endorsed by committee and sent to the Presbyterian General Assembly, which commended it for use in congregations. As of this writing, *Glory to God* is in production with marketing underway. *Glory to God* will be the first Presbyterian hymnal available in print, projected, and e-book editions, yet another indicator of a changing landscape.

The PCOCS, PPC, and PCUSA have made good use of social media networking and technology to prepare the denomination for the publication. A series of articles, written by committee members, have been published for *thepresbyterianleader.com,* an online resource for Presbyterian leaders (Bringle 2010, 2012a, b, c; Copeland 2012, Eicher 2012b; Gambrell 2012a, b; Howard 2012b; Waschevski 2012). Topics include how to use a hymnal, hymnal contents, rationale for writing new hymns, global song, and praise bands. *Call to Worship,* PCUSA journal for music and liturgy, has featured articles on congregational song and spiritual formation. The Presbyterian Hymnal Project Blog has proven an excellent tool for disseminating information for the forthcoming hymnal.

Although the preface to *Glory to God* has yet to be written, it will, undoubtedly, assure readers that Presbyterian heritage has been preserved with a significant core of hymns, psalms, and spiritual songs from Reformed tradition. The preface will address that new materials reflect the denomination's current landscape by noting the expansion of global song, begun by the PH 1990 committee, and the significant addition of text/music from Contemporary Christian/Praise and Worship genres.

Glory to God will be the hymnal that Presbyterians use in worship and shapes their faith for the next generation.

References

Ainsworth, H. (1612). *The whole book of Psalms: Englished in both prose and metre*. Amsterdam: Giles Thorp.

Benson, L. (1895). *Preface to The Hymnal/published by the authority of the General Assembly of the Presbyterian Church in the U.S.A.* (p. iv). Philadelphia: Presbyterian Board of Publication and Sabbath School Work.

Bringle, M. (2010). Family albums for the people of God. *Call to Worship, 43*(4), 1–7.

Bringle, M. (2012a). *Making full use of your hymnal*. Retrieved May 1, 2012, from www.presbyterianhymnal.org/PDF/Making%20Full%20Use%20of%20Your%20Hymnal.pdf

Bringle, M. (2012b). *No more new hymns?* Retrieved May 1, 2012, from www.presbyterianhymnal.org/PDF/No%20More%20New%20Hymns.pdf

Bringle, M. (2012c). *Studying the Bible through hymns*. Retrieved May 1, 2012, from www.presbyterianhymnal.org/PDF/Studying%20the%20Bible%20through%20Hymns.pdf

Copeland, A. (2012). What goes in a hymnal. Retrieved May 1, 2012, from www.presbyterianhymnal.org/PDF/What%20Goes%20In%20a%20Hymnal.pdf

Dickinson, C. (1933). *Preface to the hymnal* (p. p. iii). Philadelphia: Presbyterian Board of Christian Education.

Eicher, D. (2012a). Interview with David Eicher, Editor of Glory to God: The Presbyterian Hymnal. *Call to Worship, 45*(4), 55–60.

Eicher, D. (2012b). Why don't hymns end with "amen"? Retrieved May 1, 2012, from www.presbyterianhymnal.org/PDF/Why%20Dont%20Hymns%20End%20with%20Amen.pdf

Eskew, H., & McElrath, H. (1995). *Sing with understanding* (2nd ed.). Nashville: Church Street Press.

FACT. (2010). *2010 Survey of Presbyterian church (U.S.A.) congregations*. Louisville: Research Services, Presbyterian Church (U.S.A.).

Gambrell, D. (2012a). The use of psalms in worship. Retrieved May 1, 2012, from www.presbyterianhymnal.org/PDF/The%20Use%20of%20Psalms%20in%20Worship.pdf

Gambrell, D. (2012b). Glory to God: The Presbyterian hymnal. *Call to Worship, 45*(4), 77–84.

Hawn, M. (2010). Streams of song: An overview of congregational song in the twenty-first century. *The Hymn: A Journal of Congregational Song, 61*(1), 16–26.

Howard, B. (2012a). Hymnals as denominational scrapbooks. Retrieved May 1, 2012, from www.presbyterianhymnal.org/PDF/Hymnals%20as%20a%20Denominational%20Scrapbooks.pdf

Howard, B. (2012b). Why sing global songs? Retrieved May 1, 2012, from www.presbyterianhymnal.org/PDF/Why%20Sing%20Global%20Songs.pdf

Howard, B., Stern, C., & Hastings-Flegel, K. (2009) *Unpublished Needs Assessment Report of the Presbyterian Committee on Congregational Song*.

Jones, D. H. (1955). Preface. In *The Hymnbook* (p. 5). Richmond: Presbyterian Church in the United States/The United Presbyterian Church in the U.S.A../Reformed Church in America.

Jones, M., Chen, C., Hastings-Flegel, K, Huh, P., Myers, E., & Waschevski, M. (2010) Unpublished report of Global group.

Marcum, J. (2007). *Hymnals: The report of the august 2005 Presbyterian panel survey*. Louisville: Research Services, Presbyterian Church (U.S.A.).

Marcum, J. (2009) *The Presbyterian church (U.S.A.) at 25: A statistical look at denominational change*. Louisville: Research Services, Presbyterian Church (U.S.A.).

McKim, L. (1990). Introduction. In *The Presbyterian Hymnal: Hymns, psalms, and spiritual songs* (p. 9). Louisville: Westminster/John Knox Press.

Music, D., & Richardson, P. (2008). *I will sing the wondrous story: A history of Baptist hymnody in North America*. Macon: Mercer University Press.

PCOCS. (2009). *A statement on language* www.presbyterianhymnal.org/PCOCS%20statements%20HO.pdf. Accessed May 2012.

Presbyterian Publishing Corporation [PPC]. (2009). *Feedback about hymns in the Presbyterian hymnal.* Louisville: Research Services, Presbyterian Church (U.S.A.).

Ravenscroft, T. (1621). *The whole book of Psalmes*. London: Company of Stationers.

Stern, C., Chen, C., Waschevski, M., & Wheeler, B. (2010) Unpublished report of contemporary group.

Sternhold, T., & Hopkins, J. (1562). *The whole book of Psalms*. London: John Daye.

The Pew Forum. (2008). *U.S. religious landscape survey religious affiliation: Diverse and dynamic*. Washington: Pew Research Center.

The Presbyterian Hymnal. (1874). Philadelphia: Presbyterian Board of Publication.

The worshipbook. (1972). Philadelphia: Westminster Press.

Waschevski, M. (2012). Will a praise band help our church grow? Retrieved May 1, 2012, from www.presbyterianhymnal.org/PDF/Will%20a%20Praise%20Band%20Help%20Our%20Church%20Grow.pdf

Wheeler, B., van Driel, E., & PCOCS. (2009). Theological vision statement. www.presbyterianhymnal.org/PCOCS%20statements%20HO.pdf. Accessed May 2012.

Chapter 142
"Tune Your Hearts with One Accord": Compiling *Celebrating Grace*, a Hymnal for Baptists in English-Speaking North America

David W. Music

142.1 Introduction

The hymnal has long been a staple of congregational worship in Christian churches. Hymnals have both embodied the beliefs of those who sing from them and helped shape those beliefs. Thus the publication of a new hymnal is important for giving voice to the expression of faith at a particular time and in forming future faith. This article traces the publication of a new hymnal designed for Baptists against the backdrop of the denomination's recent history and other contemporary hymnals that have been compiled in this tradition.

Baptists in the United States and Canada are divided into many organizational groupings, ranging from independent churches that adhere to no wider structure to bodies that are among the major Christian denominations in North America. For the 65 years between 1940 and 2005 churches of the Southern Baptist Convention—the largest group of Baptists in the United States and the one with the most consistent hymnal publication record during that period—relied principally upon congregational song books issued by the denominational publishing house, with new volumes being issued every 16–19 years: *The Broadman Hymnal* (1940), *Baptist Hymnal* (1956), *Baptist Hymnal* (1975), and *Baptist Hymnal* (1991). The publication of *Celebrating Grace* (2010) falls within this chronological framework, but its compilation was unlike that of any hymnal issued by and for Baptists in the South since before 1940.

D.W. Music (✉)
School of Music, Baylor University, Waco, TX 76798, USA
e-mail: david_music@baylor.edu

S.D. Brunn (ed.), *The Changing World Religion Map*,
DOI 10.1007/978-94-017-9376-6_142

142.2 Publication of the Previous Hymnal

The last Southern Baptist hymnal to be published in the twentieth century appeared in 1991. The book was compiled under the auspices of the Baptist Sunday School Board's Church Music Department and edited by Wesley L. Forbis, "secretary" (that is, director) of the department. This was an "official" hymnal—official in the sense that it was issued by the denominational publishing house, though the churches were under no obligation to use this or any other hymnal.

142.2.1 The Process of Compilation

Initial groundwork for *Baptist Hymnal* (1991) was laid in 1986, with the public announcement that the book was in preparation coming in the following year (1987). This early planning was done principally by Forbis and Terry W. York (who was named Hymnal Project Coordinator), and included moving the proposal for the book through the administrative channels of the Baptist Sunday School Board, selecting the release date for the volume, determining the structure of the hymnal committee, and choosing the persons who would be invited to serve on the committee. The date 1991 was selected for the premiere of the hymnal, this being the 50th anniversary of the formation of the Church Music Department, the 100th anniversary of the founding of the Baptist Sunday School Board, and the 300th anniversary of the English Baptist Benjamin Keach's *Spiritual Melody*, the first hymnal to be published by a Baptist compiler.

Five subcommittees were named: New Materials (which included an Ethnic Hymnody Council), Theology/Doctrine, Worship Aids, Music, and Design/Organization. Promotion was to be handled principally by Baptist Sunday School Board staff. The directors of the Baptist state convention music departments formed a special group for promotion and service on the other committees. The members of all these committees combined made up a plenary committee of nearly 100 people, which met twice during the process. An executive group, named the Hymns Recommendation Committee, was composed of the editor (Forbis), the chairs of the five subcommittees and the Ethnic Hymnody Council, plus six at-large members. While the plenary committee included a number of women and persons from different races, the Hymns Recommendation committee contained only one woman and no racial minority representatives. Suggestions from the subcommittees were sent to the Hymns Recommendation Committee, which either rejected them or passed them along to the plenary committee. In turn, decisions of the plenary committee went back to the Hymns Recommendation Committee for final disposition.

Scientific surveys were conducted by the statistics department of the Sunday School Board to determine which of the 512 hymns from the previous *Baptist Hymnal* (1975) were sung most often in Southern Baptist churches. Almost all of the "top 100" hymns from the earlier hymnal were more or less automatically retained; the exceptions were mostly pop-style pieces from the 1960s and early

1970s that were beginning to seem dated. Conversely, most of the "bottom 200" items were dropped except for a few that had historical or other significance. The most difficult choices involved the "middle 200," which had found some usefulness among Baptists, but had not achieved frequent use. All of these decisions were made by "yes" or "no" votes on each hymn, with a few being placed in a "hold" category in case they were needed for balance or pagination in the final product.

142.2.2 Special Considerations

Two important considerations impacted both the formation of the committee and the content of the 1991 hymnal. One was the revolution in Christian music that had taken place since the previous collection. While the 1975 hymnal had included incipient representation of choruses and Contemporary Christian Music (at that time called "Jesus Music"), by the mid-1980s these types of songs had left the realm of "youth music" to become almost the sole repertory sung by some churches, especially those following the church growth model. In fact, a number of churches had abandoned hymnals altogether in favor of choruses sung from song sheets or overhead projection, but even many of the congregations that retained hymns and hymnals were singing choruses. Though the words and music of choruses were seldom learned or sung from printed music, putting them in the hymnal did have a certain symbolic value. Thus, in addition to the usual representation from churches large and small, rural and urban, etc., the committee had to include persons whose churches were primarily chorus-singing and would, therefore, be unlikely to purchase the hymnal, in addition to those who would employ only the hymnal or some combination of hymns and choruses.

The second consideration was that since the late 1970s the Southern Baptist Convention had been embroiled in bitter controversy between fundamentalist and moderate factions, with the fundamentalists seeking to take control of the convention by capturing the office of president for at least ten consecutive years. The president of the convention appoints the committee on committees, which then appoints the other committees and ultimately the trustees of the various institutions; thus by electing presidents for an extended period the group could eventually take control of the convention's committee structure. The fundamentalists, proclaiming the "inerrancy" of the scriptures, the dominant role of the pastor, and opposing the ordination of women as deacons and pastors, claimed that the denomination had drifted from its historic roots. Moderates—though still generally conservative by the standards of other mainline denominations—held that the Bible was neither a history nor a science book but one of faith, stressed the freedom of the individual believer to interpret the scriptures for him- or herself, and supported (or at least did not oppose) ordination of women. By the time the hymnal was announced, the fundamentalists were close to achieving their 10-year goal, having won every presidential election since 1979. Thus the committee also had to reflect a political balance, and its members were chosen with care to involve pastors and church musicians from both parties, as well as those who sought to stay in the middle of the road.

It was—perhaps naively—hoped that the compilation of the hymnal with such a balanced group would help bring about a degree of unity in the convention, but the fundamentalists were determined to have their victory, and after a few more years of denominational struggle, the moderate elements in the convention formed a "fellowship" in August of 1990 to channel their energies and resources into Southern Baptist causes they felt they could in conscience continue to support (Vestal 1993) The hymnal was published about half a year later (March 1991). Because the Cooperative Baptist Fellowship had not made a clean break with the Southern Baptist Convention, and since many persons who were sympathetic to the Fellowship had served on the hymnal committee, the volume was readily accepted by churches with both fundamentalist and moderate leanings, though the denominational publishing house that issued it was now firmly in fundamentalist hands. Thus the hymnal sold well among churches of both parties, achieving sales of more than 4,500,000 copies by 2001 (*Glorious is thy name*, 2001: 23), fewer than the previous two hymnals—both of which went out of print when the 1991 book came off the press—but still a significant number for a denomination whose churches were under no obligation to purchase it and that was wracked by internal strife.

142.3 The Publication of *Celebrating Grace*

142.3.1 The Need for a New Hymnal

During the years after publication of *Baptist Hymnal* (1991) the leadership of the Southern Baptist Convention began rejecting the financial contributions (gifts) of the Cooperative Baptist Fellowship that were not sent through the normal channels (called the "Cooperative Program"). With the 1994 firing of Russell H. Dilday as president of Southwestern Baptist Theological Seminary in Fort Worth, Texas, by the fundamentalist-controlled board of trustees, the schism between the Cooperative Baptist Fellowship and the Southern Baptist Convention became complete, though some churches have maintained dual alignment. Once the separation was final, moderate churches were loath to continue purchasing materials (including hymnals) from the fundamentalist-controlled publishing house. On the other hand, no other organizations or groups were either willing or equipped to step into the breach and publish a hymnal for moderates. Churches that had purchased *Baptist Hymnal* (1991) generally continued to use it; some congregations that had taken a "wait and see" attitude bought hymnals from independent publishers or "trade" editions of books from other denominations. So, for example, some churches acquired *The Worshiping Church* (1990, edited by Southern Baptist Donald P. Hustad) from Hope Publishing Company, an independent evangelical firm, while others opted for the *Chalice Hymnal* (1995) of the Christian Church (Disciples of Christ).

By 2006 *Baptist Hymnal* (1991) was approaching the 16-year mark, the typical time that a new denominational hymnal would be published, but there was no indication from LifeWay Christian Resources (as the Baptist Sunday School Board had been renamed in 1998) that a new book was being contemplated. Even upon the release of *Baptist Hymnal* (1991) there had been widespread talk that this might be the last published Southern Baptist hymnal because of the increasing popularity in the churches of overhead projection of choruses and Contemporary Christian Music.

However, many Baptist churches in the South continued to use hymns and hymnals, sometimes instead of, and occasionally in combination with, overhead projection. The aging of *Baptist Hymnal* (1991), the lack of any announcement of a new hymnal project by LifeWay, plus the fact that few moderate churches would be likely to adopt any hymnal from this source, meant that the time seemed ripe for publication of a new collection designed particularly with moderate Baptists in mind. It was precisely these congregations that tended to maintain more formal worship patterns and continued to utilize hymns and hymnals instead of, or in addition to, choruses, Contemporary Christian Music, and overhead projection.

142.3.2 The Beginnings of the Project

The project that sought to fill this need, ultimately named *Celebrating Grace*, was begun independently of any formal denominational structure. The genesis of the book came in a conversation between John E. Simons, a professor of music at Mercer University in Macon, Georgia, and businessman J. Thomas McAfee, a graduate of Mercer. McAfee, recalling the important place church music had held in his life, indicated his desire to do something for the enhancement of this ministry and asked Simons to suggest a project. Simons responded that what was most needed was a new hymnal, an idea with which McAfee readily agreed. Together they set out to build a foundation for the publication of a new book of congregational song.

The first step was to develop a team of editors and a committee structure. The two originators of the project, McAfee and Simons, served as Project Chair and Coordinating Editor, respectively. Stanley L. Roberts, another music professor at Mercer, and Milburn Price, retired dean of the School of Music at Samford University in Birmingham, Alabama, were invited to round out the editorial team. With this (all-male) group in place the hymnal project was publicly announced in June of 2006 as a joint venture between the Townsend-McAfee Institute of Church Music at Mercer University and Mercer University Press. Release of the hymnal was projected for 2009, the 400th anniversary of the founding of the first Baptist church by John Smyth in Amsterdam (Cameron and Lunsford 2006).

Over the next few months, five subcommittees were planned and chairs were named for each: Texts and Tunes (David W. Music), Format and Organization (Paul A. Richardson), Extra-Musical Worship Materials (Alicia W. Walker), Supplemental Music Resources (Mark Edwards), and Product Promotions and Publicity (J. Thomas McAfee). The first meeting of the editors and committee chairs took place September 29–30, 2006, at which time a plenary committee of more than 50 church musicians, pastors, educators, composers, hymn writers, and laypersons was agreed upon. A significant component of this task was a determination to broaden the scope of the project beyond the Southern Baptist orbit by inviting Baptists from other traditions to serve on the committee, though it was still weighted heavily toward current or former Southern Baptists. Among persons from other streams of Baptist identity who were added to the process were an American (Northern) Baptist hymn writer and the editor of *The Hymnal* of the Canadian Baptist Federation (1973). A desire was also expressed that the hymnal might see ecumenical use by congregations from outside the ranks of Baptists. A further important decision was to prepare a survey of ministers of music at Baptist churches to determine what hymns were being used most often in their congregations.

Ironically, at the very moment the editors and chairs for *Celebrating Grace* were holding their first meeting, LifeWay Christian Resources issued a press release noting that a new hymnal would be forthcoming from the Southern Baptist denominational publishing house, with publication expected in 2008. Since the "Mercer" project had already been announced and it was presumed that the LifeWay book would be aimed primarily at a different constituency, the editors and chairs for *Celebrating Grace* agreed to continue with their project.

The first meeting of the plenary committee and subcommittees for *Celebrating Grace* occurred February 1–2, 2007, preceded by the second meeting of the editorial board and committee chairs. In the meantime, the survey that had been recommended at the September editors/chairs meeting had been taken. An inventory of 324 hymns was developed for the survey, derived principally from an article on ecumenical hymnody by C. Michael Hawn (Hawn 1997). To Hawn's list were added a number of items from recent Baptist hymnals that in the perception of the editors formed a sort of core repertory for the denomination. The respondents were not necessarily to choose the hymns that were most frequently sung by Baptists—though, of course, that was always a factor—but those that the book should not be without. Unfortunately, due to time constraints, the survey was received by ministers of music during one of the busiest times of the year for them (Advent/Christmas 2006), it was not based on a scientifically-selected sample, and it was therefore of limited value. Nevertheless, there were few surprises in the responses that were received, and at the editors/chairs meeting that preceded the plenary committee session it was recommended that the Texts and Tunes Committee simply accept the "top 125" hymns from the list as the core of the book.

The meeting proper began with a relatively brief plenary session of all the committees, the only time this would happen throughout the entire project. That evening and the next day the subcommittees began their work. The Texts and Tunes Committee adopted the recommendation of the editors to include the 125 standard hymns identified by the survey (actually adding one more to make 126), though changes in wording or other alterations were often recommended by the committee. After the subcommittee meetings the editors and chairs met again to receive reports and plan for the future. At this time the title of the hymnal was selected, not without some disagreement, principally because of its similarity to the title of *The Celebration Hymnal*, a non-denominational collection that had been published in 1997 and that was used by some Baptist churches; the principal concern was that potential constituents might think *Celebrating Grace* was a sequel to *The Celebration*

Hymnal, which was itself a sequel to *The Hymnal for Worship and Celebration* (1986). Nevertheless, because of the desire to communicate that worship leads to joyful acts of mission and discipleship ("Celebrating"), and since *sola gratia* is a common belief among evangelicals ("Grace"), *Celebrating Grace* was ultimately agreed upon.

142.3.3 Changes in Ownership and Scope

The hymnal had been undertaken in somewhat provincial fashion largely as an in-house project of Mercer University. Early on, it was thought that students in the university's School of Music would prepare the hymn pages using a digital music notation program and that the university's lawyers would handle copyright issues. However, the project was rapidly growing in size, scope, and complexity and none of the editors or chairs could devote full time to it. It quickly became apparent that the original conception was unsustainable. McAfee worked with the administration of Mercer University to fashion a mutually agreed upon separation from the school to form a not-for-profit company, Celebrating Grace, Inc., which had (and has) no formal or financial connection with the university.

In July 2007, Mark Edwards, who had been minister of music at the First Baptist Church in Nashville, Tennessee for 30 years and was serving as chair of the Supplemental Music Resources Committee, was employed as full-time Worship Resource Manager to assist in coordinating the entire undertaking (Vanderhoek 2007). This was a fortuitous step for a number of reasons: Edwards had many contacts with the music industry in Nashville, had been a member of the committee for the 1991 *Baptist Hymnal*, and provided an experienced minister of music's perspective for the heavily academic make-up of the editorial board. The new Worship Resource Manager put the editors in touch with typesetting and copyright professionals; these persons were employed in order to facilitate those areas of the work. The venture was further expanded in the fall of 2007 with the addition of David W. Music as a fifth member of the editorial team.

As the work progressed, and because of the expanded scope of the task, it became evident that the 2009 target date could not be met without seriously jeopardizing the quality of the final product. Thus the editors decided to move publication back to 2010, with the book to become available by Easter of that year.

142.3.4 The Work of the Committees

Meanwhile, the subcommittees were at work on their assigned tasks. Meetings of all subcommittees except the Texts and Tunes Committee were occasional, with most of the work being done through email, correspondence, and telephone. An important task was accomplished by the Format and Organization Committee in recommending to the editors a bipartite division of the book into the categories "I Will Be Your God" and "You Shall Be My People," with appropriate subcategories under each of these headings. The Extra-Musical Worship Materials Committee developed readings and other items to be included either in the hymnal itself or in online resources after publication of the book. The Supplemental Music Resources Committee worked to provide an array of materials to enhance congregational singing from the new hymnal, including orchestrations, handbell settings, "congregational anthems" (anthem-like arrangements that include congregational participation), and other music. The responsibilities of the Product Promotions and Publicity Committee were obviously heaviest as the hymnal neared publication, but plans had to be laid and resources gathered long before this.

A crucial element in making people aware of the hymnal was the establishment of "Hymnposia" led by Mark Edwards and Tom McAfee in various major cities, primarily in the southern states: Houston, Austin, and Dallas, Texas; Birmingham, Alabama; Nashville and Knoxville, Tennessee; Raleigh, North Carolina; Atlanta, Georgia; Richmond, Virginia; Annapolis, Maryland; Kansas City, Missouri; Louisville, Kentucky; and Jackson, Mississippi. Ministers of music from area churches were invited to meet in a central location for lunch, receive an introduction to the hymnal project, sing through a few hymns that might appear in the new book, and provide feedback for the editors.

Obviously, the Texts and Tunes Committee held a key role in assembling the book, since it is the actual hymnic content of a hymnal that will usually determine the success or failure of the enterprise. The committee was composed of twelve people, including three full-time music ministers, a professional theologian, a pastor, a former hymnal editor, two laypersons, and four university or seminary music professors. Two of the members were African-American and three were women. This committee met face to face much more frequently than the others, generally about once a quarter, with additional work done between meetings through correspondence.

One of the challenges in compiling a hymnal is not that there is too little material from which to choose, but that there is too much. The Texts and Tunes Committee formally considered over 2,300 hymns, not counting items from the survey and pieces that were screened before presentation to the full committee. This screening came into play when large numbers of hymns were received from a single source, as, for example, when a publisher sent an entire published collection for consideration or when an author or composer submitted dozens of pieces. A great deal of the Texts and Tunes Committee's time was spent wrestling with which fine hymns to forward to the editors. In general, the approach taken was that if a majority of the group believed a piece showed promise or could potentially fill a particular role in the finished product, it should be approved and passed on to the editors. There was a realization from the start that not everything that passed out of the Texts and Tunes Committee would ultimately go into the hymnal, but this approach would allow the editors some choices to help assure a balance of theological, liturgical, musical, and functional content.

Of course, a hymnal has to be both practical and prophetic. Publication of a hymnal is expensive and time-consuming, and for the effort to be worthwhile the end result must be widespread use. Thus the book must contain enough of the familiar and popular for churches to adopt it. At the same time, a new hymnal should make a distinctive contribution to the repertory of congregational song, and thus recent hymns and songs—even some that may be a bit "on the edge"—need to receive serious consideration.

142.3.5 Issues and Solutions

Two particularly contentious issues that had to be dealt with by both the Texts and Tunes Committee and the editors were the language of some hymns and certain theological perspectives. Generally speaking, the most serious language issue was the use of non-inclusive language for humans in older hymns. While this concern had been raised in the compiling of *Baptist Hymnal* (1991), the approach taken there was quite conservative: new texts were expected to be in inclusive language, and some less-familiar older hymns were altered, but most hymns that were considered to be in the congregation's "memory banks" were retained in their "received" versions. For example, the original first line of John Mason Neale's translation of a Medieval Latin text, "Good Christian Men, Rejoice," had been kept. In part, this approach reflected not only the generally conservative bent of Baptist churches in the South, but also the fundamentalist/moderate controversy that surrounded the compilation of that book.

Celebrating Grace took a more nuanced approach when it came time to vote on the individual hymns. Some on the committee objected to the "men" in "Good Christian Men, Rejoice" while others had a problem with altering such a familiar text (which, after all, would be heard in its non-inclusive version over loudspeakers all through the Christmas season at every shopping mall in the country). The hymn was passed along to the editors by the Texts and Tunes Committee with the alteration of the first line as "Good Christians All, Rejoice." After wrestling with this and other alternatives, the editors were not able to come up with what they deemed to be a satisfactory version and decided simply to drop the hymn, only to revisit it toward the end of the project and include it as "Good Christian Friends, Rejoice." "Friends" was chosen mainly because it has the same vowel sound as "men" and would be less obvious if the original and new versions were being sung simultaneously.

On the other hand, Reginald Heber's original line "Though the eye of sinful man thy glory may not see" from "Holy, Holy, Holy! Lord God Almighty" was retained as it stood. "Flesh" was put forward as a substitute for "man," but some felt that this suggested the "flesh/spirit" dichotomy of fourth-century Arianism. Another suggested alteration was "Though the eye made blind by sin," but this was deemed to be awkward. These are just a few examples of the situations encountered. The alterations (or non-alterations) had to be approached on a case-by-case basis.

One item that was of little concern in the *Celebrating Grace* project was the matter of inclusive language for God. Few efforts were made to alter traditionally masculine references to God ("Father," "King," "Lord," etc.), since these were largely biblical in origin. However, there was an attempt to provide at least some balance by including several hymns using female imagery for Deity ("Like a Mother with Her Children").

In general, the alteration of familiar hymn texts is a no-win situation. Some people will be offended by the use of non-inclusive language; others will be put off by changes in texts that have been committed to memory and are a significant part of their piety. The alteration of language in song—regardless of the reason it is done—is much more difficult than in prose or even in some other forms of verse because rhyme, poetic meter, syllable count, and the number of notes available must be dealt with, in addition to considerations of familiarity. Ultimately, the Texts and Tunes Committee and editors simply had to make the choices they felt were best and live with the consequences.

One theological issue that surfaced in the Texts and Tunes Committee related to differing views of the atonement, the Christian doctrine that refers to the redemption of humans through the death and resurrection of Jesus. Some committee members voiced particular objection to expressions of the "ransom theory" (humans have been captured by the Devil but God bought them back through the death of Christ, meanwhile tricking Satan by raising Jesus from the dead) and "penal substitutionary" atonement. The latter holds that (1) all humans have sinned, (2) God is perfect and cannot abide sin, therefore (3) all humanity is subject to his righteous wrath and must be punished, but (4) in God's love for humanity Christ was provided as a substitute so that the punishment was released on Jesus rather than on humans who put their trust in him. The objection to the ransom theory was that humans belong to God rather than to the Devil. While the penal substitution theory includes equal parts of wrath and love, some on the committee believed that the balance in hymnody was unequally weighted toward wrath.

The ransom theory debate centered mainly on the hymns of invitation (or altar call) that have been used to conclude many Baptist services since the last quarter of the nineteenth century, as can be seen in such phrases as "with his blood he purchased me;/on the cross he sealed my pardon,/paid the debt and made me free" ("I Will Sing of My Redeemer") and "Jesus paid it all,/all to him I owe" ("Jesus Paid It All"). The question over penal substitutionary atonement came to a head in the consideration of Keith Getty and Stuart Townend's recent hymns "In Christ Alone" and "The Power of the Cross," which include the lines "till on that cross, as Jesus died,/the wrath of God was satisfied" and "took the blame, bore the wrath," respectively. It was suggested that these be changed to "the love of God was magnified" and "took the blame, bore the shame." The wisdom of altering what were already becoming very familiar songs was debated, but in the end the committee and editors voted that the changes were worth any potential problems in these particular hymns.

142.3.6 Baptist Hymnal (2008)

In the meantime, the most recent version of *Baptist Hymnal*—which had been publicly announced 3 months after *Celebrating Grace*—was issued by LifeWay Christian Resources in August of 2008 (House 2008). Publication of *Celebrating*

Grace was still about a year and a half away. The quicker release of *Baptist Hymnal* (2008) was due largely to the fact that this project was done almost entirely in-house at LifeWay. There was a single meeting of interested parties in January 2007 but this was not a "hymnal committee" meeting in the usual sense. A blanket letter of invitation was sent to a variety of church music leaders inviting them to meet for information about the project and to provide some input into it. Afterward the attendees were sent lists of songs to approve, disapprove, or comment upon, but whether or exactly how these were tabulated or used in making the final selection is not evident. The acknowledgements page of the 2008 hymnal lists seventy-three names under the heading "Music Selection Committee" but this group appears to have functioned more as a survey sample than as a committee, and the compilation of the book appears to have been done primarily by employees of the publishing house. Because LifeWay already had a long history of hymnal publication, design, editing, printing, and copyright personnel and equipment were already in place, whereas the *Celebrating Grace* project had to build these from the ground up. Even so, the speed of publication of *Baptist Hymnal* (2008) contrasts starkly with the previous three denominational collections (1956, 1975, 1991), and is more in line with how *The Broadman Hymnal* (1940) was apparently produced.

142.4 Final Steps and Release of Celebrating Grace

Approximately 2 months after the release of *Baptist Hymnal* (2008) the editors and Worship Resource Manager for *Celebrating Grace* hosted a read-through of mostly new material that had tentatively been approved for the latter book. This event was held in Atlanta, Georgia, in October, and was designed principally for ministers of music in Baptist churches. Every person was given an evaluation sheet and asked to indicate whether or not the hymn should be considered for the book, either by its potential use in their church or by other criteria. The editors also commissioned a prominent pastor, a former hymnal editor, and a church music professor from outside the project to evaluate a provisional version of the manuscript. The principal concern expressed by this team of select evaluators was that the book not be "elitist," and the editors took this critique to heart, adding a number of popular-style hymns. While both of these forums caused a rethinking of some aspects of the book, in the main the editors were assured that they were on the right track for their primary intended target.

Regularly-scheduled editorial meetings continued into January of 2009, at which time the "final" contents of the hymnal were approved—pending copyright clearance—and the finished manuscript went to the copyright manager and typesetter, along with a few "extra" items that had been placed in a second tier; these additional pieces were to be used in case copyright permissions were not received or they were needed to make the formatting of the book come out right. While the copyright and typesetting tasks were in process other items were developed, including the indexes. In a very few instances, copyright permission was not forthcoming or was not received in timely fashion, and an item had to be changed. Chiefly this affected the harmonization of a few tunes. After the copyright permissions and typesetting were completed the editors met a final time to go over corrections. The book was then shepherded through the final stages by the Worship Resource Manager and Coordinating Editor.

Celebrating Grace was released at a 2-day event in Atlanta, Georgia, March 7–8, 2010, the culmination of nearly 4 years of work (Allen 2010). Two years after publication *Celebrating Grace* had sold over 73,000 copies, divided almost equally between those with the generic title *Celebrating Grace: Hymnal* and the more specific *Celebrating Grace: Hymnal for Baptist Worship*. While the number of units sold may seem modest in comparison to the gaudy figures put up by the Baptist hymnals published before 2005, it considerably exceeded the original expectations of the editors. It should also be noted that the previous hymnals could rely in large part upon denominational loyalty and promotion through the denominational structure, neither of which could be counted on for *Celebrating Grace*.

The largest numbers of churches purchasing 100 or more hymnals were located in Georgia (28), Texas (23), North Carolina (15), Tennessee (15), Virginia (11), and Kentucky (9), with congregations in Alabama, South Carolina, Louisiana, Mississippi, Maryland, Florida, Arkansas, Oklahoma, Kansas, Missouri, and as far outside the "Bible Belt" as Ohio, Wisconsin, Indiana, and Pennsylvania also acquiring at least 100 copies. The book is in use in at least five educational institutions. As might be expected, the majority of the churches and schools are Baptist in orientation, but a few Methodist, Episcopal, and non-denominational community churches have it in their pews as well.

It should be emphasized that the Cooperative Baptist Fellowship had no part in the compilation of *Celebrating Grace* and that it has never officially adopted or endorsed this or any other hymnal, though of course it is used in a number of churches affiliated with this organization. The earlier mention of the Fellowship simply serves as part of the background against which the book was published.

142.4.1 Two Baptist Hymnals: Comparisons and Contrasts

With two hymnals designed by and for Baptists published in such close chronological proximity, and both seemingly being widely accepted, a few comparisons may be in order. The differences in the method of compilation and time frame have already been mentioned. The books contain many of the same hymns that are standard to most Christian denominations, as well as songs that are part of the common stock among Baptists. Both contain texts and tunes by modern writers in traditional hymnic forms, as well as examples of Contemporary Christian Music.

It is in the differing proportions of the latter idioms that the differences between the two books are most pronounced, with *Baptist Hymnal* (2008) placing greatest emphasis on songs in popular style and *Celebrating Grace* stressing creative expressions in more historic forms. *Celebrating Grace* also addressed modern issues of language (albeit without complete consistency), whereas *Baptist Hymnal* (2008) showed little or no concern over this subject. The latter book contains optional last stanza free accompaniments and transitions between songs, while no material of this sort was included in *Celebrating Grace* (which offered such aids strictly on-line).

An examination of the topical indexes of the two books is also revealing. Both make provision for the Christian Year but under different terminology and in different proportions. The topical index of *Celebrating Grace* includes a specific category labeled "Christian Year" with the subtopics arranged in order of the calendar, while *Baptist Hymnal* (2008) puts most of the hymns under the general heading "Jesus" and lists the subtopics alphabetically.

On the whole, the provision of hymns for the Christian Year is much more important for *Celebrating Grace* than for *Baptist Hymnal* (2008). For example, the first three seasons of the Christian Year are provided for as follows: Advent—*Baptist Hymnal* (2008), 9 hymns/*Celebrating Grace*, 23; Christmas—35/41; Epiphany—1/13 (though some Epiphany items found in *Baptist Hymnal* [2008] are missing from this category in the topical index, most notably "We Three Kings of Orient Are"). While *Celebrating Grace* is larger than *Baptist Hymnal* (2008) by about thirty items this alone does not account for the significantly larger number of hymns provided for Advent, Epiphany, and certain other topics since in some areas (such as the more general category of "Commitment") *Baptist Hymnal* (2008) outnumbers *Celebrating Grace* by nearly two-to-one (56/29). Thus *Celebrating Grace* appears to have been designed in large part to serve as a liturgical book—in the sense of providing material for specific events and occasions in worship—whereas *Baptist Hymnal* (2008) is more general in its orientation.

Ironically, in its manner of compilation, content, and approach to language, *Baptist Hymnal* (2008) is more like the non-denominational *Celebration Hymnal* of 1997 than its three *Baptist Hymnal* predecessors, while the independently-produced *Celebrating Grace* seems to follow more closely in the steps of the 1956, 1975, and 1991 denominational collections. These differences are perhaps not surprising, since several persons who were intimately connected with *The Celebration Hymnal* played important roles in *Baptist Hymnal* (2008), while others who were part of the committee for *Baptist Hymnal* (1991) were significantly involved with *Celebrating Grace*. The dichotomies between the two volumes reflect the different Baptist constituencies at which each was directed and illustrate the vast number of songs, styles, and approaches that are available to Christian churches of all traditions in the early decades of the twenty-first century.

References

Allen, B. (2010). New Baptist hymnal favors singing traditions, innovations simultaneously. Retrieved April 16, 2012, from http://abpnews.com/content/view/4916/53/

Baptist Hymnal. (1956). W. H. Sims (Ed.). Nashville: Convention Press.

Baptist Hymnal. (1975 ed.). W. J. Reynolds (Ed.). Nashville: Convention Press.

Baptist Hymnal. (1991). W. L. Forbis (Ed.). Nashville: Convention Press.

Baptist Hymnal. (2008). M. Harland (Ed.). Nashville: LifeWay Worship.

Cameron, R. L., & Lunsford, J. (2006). Townsend-McAfee Institute to Produce New Hymnal. Retrieved April 16, 2012, from www2.mercer.edu/News/Articles/2006/060621Hymnal.htm.

Celebrating Grace. (2010). J. E. Simons (Coordinating Ed.). Macon: Celebrating Grace.

Chalice Hymnal. (1995). D. B. Merrick (Ed.). St. Louis: Chalice Press.

Glorious is thy name: A celebration of the 60th anniversary of the Music Ministries Department 1941–2001 (2001). Nashville: LifeWay Christian Resources.

Hawn, C. M. (1997). The tie that binds: A list of ecumenical hymns in English language hymnals published in Canada and the United States since 1976. *The Hymn, 48*(3), 25–37.

House, P. (2008). *2008 Baptist Hymnal makes official debut*. Retrieved April 16, 2012, from www.bpnews.net/bpnews.asp?id=28672&ref=BPNews-RSSFeed0812

The Broadman Hymnal. (1940). B. B. McKinney (Ed.). Nashville: Broadman Press.

The Celebration Hymnal. (1997). T. Fettke (Ed.). N.p.: Word Music/Integrity Music.

The Hymnal. (1973). C. M. Giesbrecht (Ed.). N.p.: Baptist Federation of Canada.

The Hymnal for Worship and Celebration. (1986). T. Fettke (Ed.). Waco: Word Music.

The Worshiping Church. (1990). D. P. Hustad (Ed.). Carol Stream: Hope Publishing Company.

Vanderhoek, M. (2007). *Edwards, distinguished Baptist music minister, named to staff of Hymnal.* Retrieved April 16, 2012, from www2.mercer.edu/News/Articles/2007/070614Edwards.htm

Vestal, D. (1993). The history of the Cooperative Baptist Fellowship. In W. B. Shurden (Ed.), *The struggle for the soul of the SBC: Moderate responses to the fundamentalist movement* (pp. 253–274). Macon: Mercer University Press.

Chapter 143
The Musical Shape of Cultural Assimilation in the Religious Practice of Pennsylvania-Dutch Lutherans

Daniel Jay Grimminger

143.1 Introduction

German-speaking Christians in the early American Republic were struggling to maintain continuity with their ethnic past. While the English hegemony pushed these early Americans to assimilate into the culture around them, many churches refused until they could not hold out any longer, leaving behind their cultural peculiarities and folk ways through a gradual process. Historic music sources of the Pennsylvania Germans (or "Pennsylvania Dutch") show us vividly this struggle between continuity and change, and at the same time they give us details concerning the performance practice of this culturally rich group. This essay will examine two main genres of historic music documents used by the German-speaking Lutherans, Amish, and Mennonites in the late eighteenth and early nineteenth centuries: tune books and broadsides. Both of these categories of printed music and texts give us clues into the ethnic change that took place among German-speaking Christians, especially Lutherans, on the North American continent between 1780 and 1840 in many Pennsylvania communities. Further, by looking at contemporary changes in religious practices from the standpoint of the stages of assimilation, outlined and applied in this essay, present-day scholars can shed light on topics such as the use of Contemporary Christian Music and the Black Gospel genre on the North American and European continents.

Grimminger, D. J. (2009). *Pennsylvania Dutch tune and chorale books in the early Republic: Music as a medium of cultural assimilation*. Ph.D. dissertation, The University of Pittsburgh, Pittsburgh, is the basis for this chapter.

D.J. Grimminger (✉)
Faith Lutheran Church, Kent State University, Millersburg, OH 44654, USA

School of Music, Kent State University, Kent, OH 44242, USA
e-mail: daniel.grimminger@rocketmail.com

S.D. Brunn (ed.), *The Changing World Religion Map*,
DOI 10.1007/978-94-017-9376-6_143

143.2 Lutheran Music in Early America

The Lutherans in America like the Lutherans of Europe were a singing people. They came to this country carrying text-only hymnbooks from their individual regions and cities, singing the four-part chorales as they had learned them in their homeland. Heinrich Melchior Muhlenberg noted in a journal entry dated July of 1751 that the people, in a congregation where he preached, were "not able to sing even the best-known hymns," and that "the miserable and lamentable noise" they made sounded "more like a confused quarrel than a melody" (Tappert and Doberstein 1942: 297). This cacophony was due to "many kinds of hymnbooks" in use among the German-speaking Lutherans (Schalk 1996: 13).[1] Muhlenberg vowed to remedy the situation by uniting all Lutherans in America under a common hymn book. After many years he kept his promise by publishing *Erbauliche Lieder Sammlung zum Gottesdienstlichen Gebrauch in den Vereinigten Evangelisch-Lutherischen Gemeinen* [*sic*] *in Nord-America* (Edifying Collection of Hymns for Congregational Use in the United Lutheran Congregations in North America), Germantown, 1786, which would remain the text-only hymn book used by German-speaking American Lutherans well into the twentieth century in many places.

While English-speaking evangelicals were turning to primitive forms of Gospel music in the early nineteenth century, German-speaking Lutherans stubbornly embraced their chorale tradition as a sign of ethnic vitality, and an embodiment of the Judeo-Christian scriptures. Unlike the Moravians who felt at liberty in their love feasts, daily devotions, and liturgies to put verses of different chorales together to form a homogenous whole or to pluck a single chorale verse out of context to be used as a daily text, many Lutherans insisted that chorale texts be kept intact. They believed that each chorale contained a chain of theological ideas, the basic dialectics of their Lutheran theology (including law and gospel, death and life, etc.), and the core sacramental teachings that nourish faith in the believer. Chorales were a way of praising God and teaching the faith and they sang them whenever they could, with organ accompaniment if possible.

Lutherans had some of the few organs in early America, most of which came from Germany. In 1751 three Pennsylvania churches (Philadelphia, Trappe, and New Hanover) imported organs from Germany under the supervision of the schoolmaster and organist Gottlieb Mittelberger (Tappert 1980: 67). In other instances American Lutheran organ builders supplied church organs to non-Lutheran churches. Philip Feyring (1730–1737), a Lutheran organ builder originally from Arfeld, Germany, produced organs for such churches as the German Reformed Church of Philadelphia and the High Dutch Reformed Church of Germantown (Market Square Church) (Ochse 1988: 49).

In other cases, Lutherans owned the best organs that money could buy, even if they were not built by Lutheran organ builders. The 1790 organ in Zion Lutheran

[1] The translation is Schalk's (from the introduction of Muhlenberg's Erbauliche Lieder Sammlung, 1786).

Church of Philadelphia, built by Moravian David Tannenberg (1728–1804), was declared "the finest [organ] in America" (Sachse 1901: 4): it was beautifully made

> with its multitude of pipes, and beautiful ornamentations, which portrayed the symbol of the Halle institution [where most of the early immigrant pastors attended seminary], together with the special hymns and services, all tended to further unify the German Lutheran congregation, and demonstrate its strength in the community as a bulwark against the growing infidelity of the age.

Instruments like this one made the edifices that housed them popular places for important public events. President George Washington's funeral, for example, was held in St. Michael's Lutheran Church in Philadelphia and drew a huge crowd because of the size of the church and the organ that was resident there.[2] But, Lutheran chorales and organ accompaniment were not the only modes of musical expression for these American Lutherans.

Many of the Lutherans who spoke Pennsylvania Dutch (*Deitsch*) had a folk music tradition, which ceased to exist during the early years of the twentieth century. This folk music "had no counterpart in European Protestantism," but was akin to black Gospel music or revivalist choruses in white evangelical circles (Yoder 1961: 1). Those German Lutherans who did not use this form of folk music made fun of it by creating musical parodies, sung to the original folk tunes. Any opposition that did exist to this folk genre was due to the fact that "Pennsylvania Spirituals" were seen as an Americanizing agent, something that will be taken up later in this essay (Yoder 1961: 434, footnote 33). These spirituals, regardless, were accompanied not by ranks of organ pipes, but by the strings of a folk instrument unique to the Pennsylvania German people.

The instrument of choice among the rural Pennsylvania Lutherans to accompany this folk music was the *Scheitholt*, a homemade six string dulcimer made of wood that could be held on the lap or placed on an empty chest while it was being played (Raichelson 1975: 42). More often than not this *Deitsch* dulcimer had a heart or a tulip carved into the body of the instrument, something akin to the decoration of Pennsylvania blanket chests and step back cupboards. There is some evidence to suggest that the *Scheitholt* was used at times for hymn singing accompaniment in the home, but because it produced a softer tone, it presumably was used for the most part at intimate social gatherings including dances and the many different kinds of "frolics" when neighbors would help each other complete their work tasks (Smith 2006: 32–33). Just as the organ accompanied the church chorales, so the *Scheitholt* played folk music (sacred and secular) and accompanied the singing of it. In short, the German-speaking Lutherans in Pennsylvania were a very musical people. Music assisted them in learning, working, worshiping, and in most cases it led them down the road of ethnic change.

---

[2] A copy of "Dead March and Monody" in the Library of Congress was composed and printed by J. Carr of Baltimore, 1799, for the occasion of Washington's funeral in the Lutheran Church of Philadelphia. See Hinke (1972: 283). Hinke reprints the first page of the music piece.

143.3 German Assimilation

Don Yoder, an important ethnographer of the Pennsylvania Dutch people, has identified several sub groups among the Pennsylvania Lutherans. On one hand, the "Germanizers" sought to retain their ethnic distinctiveness by maintaining German speaking schools and churches. For these German-speaking Lutherans, "The language bond was stronger than the confessional heritage. German-speaking Reformed might be preferred to English-speaking Lutherans [for the Germanizing Lutherans] … ." (Repp 1982: 111). Their staunch promotion of ethnic retention became most evident in the writings and works of Justus Heinrich Helmuth (1745–1825), a pastor in Philadelphia, whose congregation had conflict after conflict that resulted from the language issue.[3] On the other hand, "Americanizers" were progressive in their move away from ethnic autonomy. Many Lutherans who wanted an "American Lutheranism" were pleased with the use of English and the consolidation of German schools with the English system of public education. Americanizers were wrapped up in a process of ethnic assimilation to differing degrees (Yoder 1985: 41–65).

Stephanie Grauman Wolf has gone a step further than Yoder by documenting a pattern or process of ethnic assimilation among the Pennsylvania Dutch. She writes that when a smaller (less dominant) culture cannot resist the lures and weight of the dominant culture that presses in on it, *Adaptation* takes place. *Adaptation* is when the community "modifies its own patterns into new forms related to—but not identical with—those of its neighbors." The next step in the process of assimilation is visible when "significant new values and patterns of behavior [are introduced] into the system," changing the landscape of the ethnic group's behavior so that it begins to mirror that of the dominant culture. Scholars call this step *Acculturation*. The final stage in assimilation is *Amalgamation*, when a "group disappears into a surrounding, more dominant culture" (Wolf 1985: 67). This usually meant for Lutherans the taking over of printed materials published by English evangelicals or the publication of their own English publications. For Gettysburg Lutherans under the influence of Samuel Simon Schmucker (1799–1873), who called for an "American Lutheranism" that resembled evangelical revivalism, this meant that materials would be printed to totally replace the resources available from the Ministerium of Pennsylvania, the German synod that Muhlenberg founded that promoted an ethnic Lutheranism. Tune books and broadsides help to tell this story of assimilation as they embodied the social structure and changing ethnic landscape of the people who created and used them.

143.4 Tune Books

All German-American tune books contain evidence of their users' identity. By determining the books used in a given community or by a given church, even when other evidence lacks, we can understand where these people fell in the

[3] One such conflict ended in a court battle. See Carson (1817).

spectrum of retention to assimilation. Four main areas of these tune books serve as evidence in this respect:

1. Musical Notation,
2. Appearance of new English tunes or full retention of chorales from Europe,
3. Tune book introductions, and
4. Voicing (number of voices and print layout of these voices)

The first German tune book printed in America, *Sammlung Geistlicher Lieder nebst Melodien* (Collection of spiritual songs with melodies), Lancaster, 1789, was printed for the German Reformed school and congregation in Lancaster and was most likely used by some Lutherans as well. It and Helmuth's 1813 chorale book, *Choralbuch für die Erbauliche Lieder Sammlung* (*Chorale Book for the Erbauliche Lieder Sammlung*), printed to accompany Muhlenberg's hymn book mentioned above, stood for ethnic retention in the face of many competing forces. Both of these books were printed entirely in the German language. Their notation was, like their European predecessors, made of round note heads with mostly an even quarter note movement (Fig. 143.1). In the case of the 1813 chorale book, unrealized figured bass was printed, indicating that the people using this resource clung to the things that they saw as the last bastions of their European praxis and tradition. There were some new tunes in the 1813 chorale book written by the compiler (Helmuth), but in Doll's 1798 tune book all of the tunes were of Swiss origin. Neither book made use of American music genres or American texts. They were concerned with retaining ethnic culture.

Fig. 143.1 Unrealized "figured bass" as found in Heinrich Helmuth's Choralbuch für die Erbauliche Lieder Sammlung (Chorale Book for the Erbauliche Lieder Sammlung), 1813 (Source: Courtesy of the Samuel Putnam House Collection, Paris, Ohio)

Doll's 1798 tune book had a "rudiments of music" introduction as was the case in English-speaking singing-school book, probably for its practical didactic effectiveness. Helmuth's 1813 chorale book on the other hand only employed an introduction outlining the content of the book and it put forth the need for liturgical reform in matters of congregational singing. The emphasis was on giving German Lutherans a strong unifying bond through fine church music. In both of these music books of ethnic retention, there were never more than three voices notated on the printed page. Helmuth and Doll were maintaining most, if not all, of the peculiarities of the European sources they sought to emulate.

Tune books seeking to *Adapt* to English culture, like Joseph Doll's *Der Leichte Unterricht, von der vocal Musik*, Harrisburg, 1810, and its second volume with a slightly different title, *Leichter Unterricht in der vocal Musik*, Harrisburg, 1815, modified so that their layout and content looked like Anglo tune books, but at the same time maintained some distinction from their English-language counterparts. In fact the title of Joseph Doll's book was a direct translation of the first English singing school book, *The Easy Instructor*, Philadelphia, 1801, by William Little and William Smith, and its introduction was pedagogical to teach the rudiments of music to common people with little or no education.

Der leichte Unterricht made use of the German language, except for the English names assigned to each tune. In another adaptive tune book, *Choral Harmonie enthaltend Kirchen = Melodien* (Chorale harmony containing church melodies), compiled by Isaac Gerhart and Johann Eyer, Harrisburg, 1822, German and English were used in interlinear fashion. For example, the "Easter Anthem" tune is paired with the texts "The Lord is ris'n indeed! Hallelujah!" and "Der Herr ist erstanden! Hallelujah!" on the same page (Fig. 143.2). Each tune, instead of being scored for

Fig. 143.2 Interlinear text, German and English, accompanying the music for the Anthem, "Der Herr ist erstanden" (Source: Courtesy of the Samuel Putnam House Collection, Paris, Ohio)

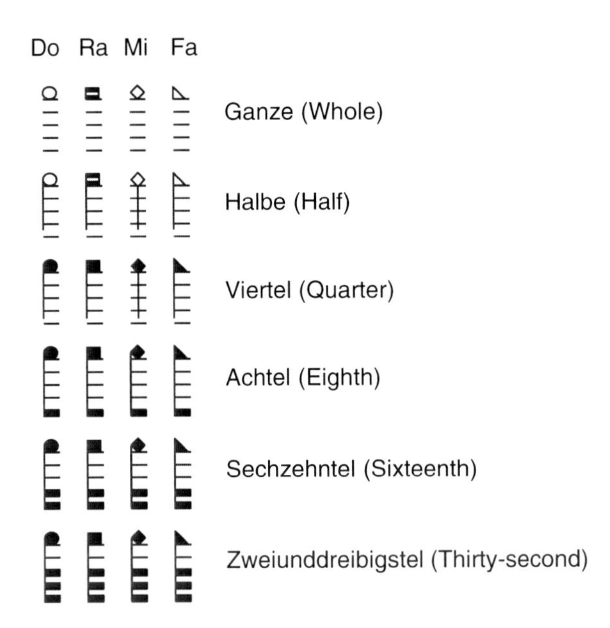

Fig. 143.3 Shape-note notation from Joseph Doll's tune books (Source: Joseph Doll's Leichter Unterricht In der Vocal Musik, 1810, courtesy of the Samuel Putnam House Collection, Paris, Ohio)

melody and figured bass, was put on three staves for three voice parts, an element that became a reality for all German American tune books as assimilation was advanced further. The notation itself also changed in Joseph Doll's tune book. He adopted the shaped note notation of the English-language singing school (Fig. 143.3). The various shapes of note heads assigned to different lines and spaces of the music staff were meant to teach common people how to read music. Further, musical forms foreign to German chorale books were incorporated into adaptive tune books. For example, "Passions Andacht" (Passion Prayer) set to the tune WORCESTER was canonic in form, set like a children's round instead of with a homorythmic chorale texture.

Tune books participating deeper in the assimilation process at the *Acculturation* level verged away from traditional German tune and chorale books in two ways. First, they were bilingual throughout (for example, John Rothbaust's 1821 *Franklin Harmony*). This is first apparent from their title pages. These liminal publications indicated an identity crisis of a people who were in between their German heritage and the predominant ways of American society. Second, there were few if any chorales to be found in these books. Instead, Anglo-American genres (for example, fuguing tunes, Anthems) dominated.

Tune books used by fully Americanized or *Amalgamated* Lutherans were identical to tune books used by English evangelicals. The General Synod's tune book *Carmina Ecclesiae*, Baltimore, 1860, compiled by William Diller Roedel, is the

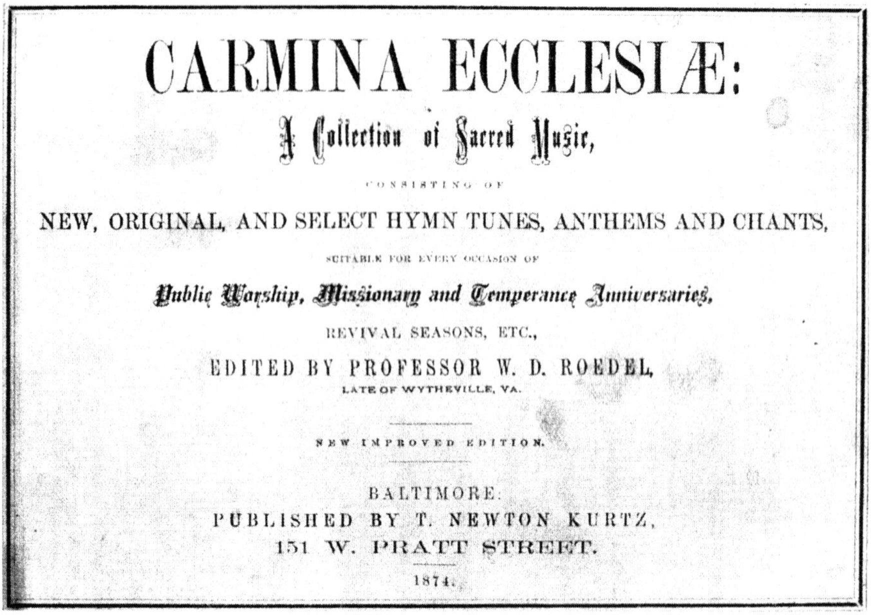

CARMINA ECCLESIÆ:

A Collection of Sacred Music,

CONSISTING OF

NEW, ORIGINAL, AND SELECT HYMN TUNES, ANTHEMS AND CHANTS,

SUITABLE FOR EVERY OCCASION OF

Public Worship, Missionary and Temperance Anniversaries,

REVIVAL SEASONS, ETC.,

EDITED BY PROFESSOR W. D. ROEDEL,

LATE OF WYTHEVILLE, VA.

NEW IMPROVED EDITION.

BALTIMORE:

PUBLISHED BY T. NEWTON KURTZ,

151 W. PRATT STREET.

1874.

Fig. 143.4 The title page from the General Synod's tune book, Carmina Ecclesiae, Baltimore, 1860, compiled by William Diller Roedel, the most visible example of Amalgamation (Source: Courtesy of the Samuel Putnam House Collection, Paris, Ohio)

most visible example of *Amalgamation* (Fig. 143.4). Its contents were entirely in English (except for 26 German chorales in the earliest edition of the book) and it utilized round note head notation, something that became standard in American hymn and tune books after the middle of the nineteenth century. Its repertory was made of mostly English tunes to match each hymn found in the then new General Synod hymn book *Hymns, Selected and Original* as well as those in the synod's Sunday school hymn book. Anthems, chants, as well as adapted and altered German tunes were included. Only about one-tenth of the hymns included made use of the homorhythmic chorale form (55). As was the case with tune books in earlier stages of assimilation, *Carmina Ecclesiae* had a didactic singing school introduction and the voicing for the tunes is identical to what would be found in an English-language singing-school tune book of the day (three or four voice parts with some counterpart).[4]

Tune books, whether they represent retention or assimilation, present for us what was going on in the German-speaking Lutheran community on a macro level. They point us to the overarching direction of their users at any given time. They also indicate that revivalist theology, which put great emphasis on sanctification, a conversion experience, and heaven, was at the heart of nineteenth century English-speaking

[4]For a more detailed discussion of English tune books among Amalgamated Lutherans, see Pannebaker (1998).

Lutheranism. Broadsides, on the other hand, allow us to analyze the particularities of German Lutheran music and culture.

143.5 Broadsides

Broadsides tell us different things about Lutheran music in the early Republic than tune books because they were meant to be ephemeral. They were most often directions for a given event or rehearsal. For many years broadsides were regarded by music scholars as trivial pieces of ephemera not worth keeping beyond their initial use. Fortunately this has changed. Pennsylvania German Broadsides in particular are becoming a popular genre of scholarly study, evident in the recent exhibition of Pennsylvania-German broadsides at the Library Company in Philadelphia curated by Don Yoder. Broadsides were printed most as a means for training choirs that were actively promoting ethnic retention. This gives them an important place in the study of assimilation and ethnic conflict. Broadsides played a similar role in Reformation and Counter-reformation Germany when there was a maelstrom of conflicting cultural and religious "ideas and new approaches to Christianity" (Yoder 2005: 1).

Music broadsides of the Pennsylvania Lutherans go beyond overarching trends in ethnic identity to the working out of particulars in specific places. Three key questions can be answered concerning music making among Lutherans in the early Republic when investigating broadsides. We will take each one of these questions in turn using two exemplar broadsides from Philadelphia to provide the answers.

1. **What were Germanizers doing musically to prevent Americanization?** A broadside discovered recently at the Lutheran Seminary Archives in Philadelphia (Mt. Airy) titled *Die Frohe Pfingst = Freude und Andacht der Evangelischen Lutherischen Singschulen der Gemeine in Zion, abgesungen unter der Anführung des Herrn David Otts* (The happy Pentecost joy and prayer of the Evangelical Lutheran singing school of the Zion Congregation, Performed under the direction of Mr. David Ott), 1785, tells us one of the major ways Germanizers in Philadelphia were resisting assimilation (Fig. 143.5). They started *Singschulen* or singing schools, much like the English-speaking Americans. However, in contrast to its Anglo equivalent, the German singing school did not serve the main purpose of teaching musical notation and rudimentary musicianship. Instead, its sole purpose was to promote German theological identity through music rehearsal and performance. The 1785 broadside states in its title that this singing school was from the congregation (der Gemeine) and David Ott, the organist of the Zion Church in Philadelphia, led this group. From the research gathered for this article, it is not for sure how many singing schools existed, but it is clear that the Philadelphia singing school was the strongest and produced copious broadsides, most likely because the Philadelphia Germans were in closer proximity to their

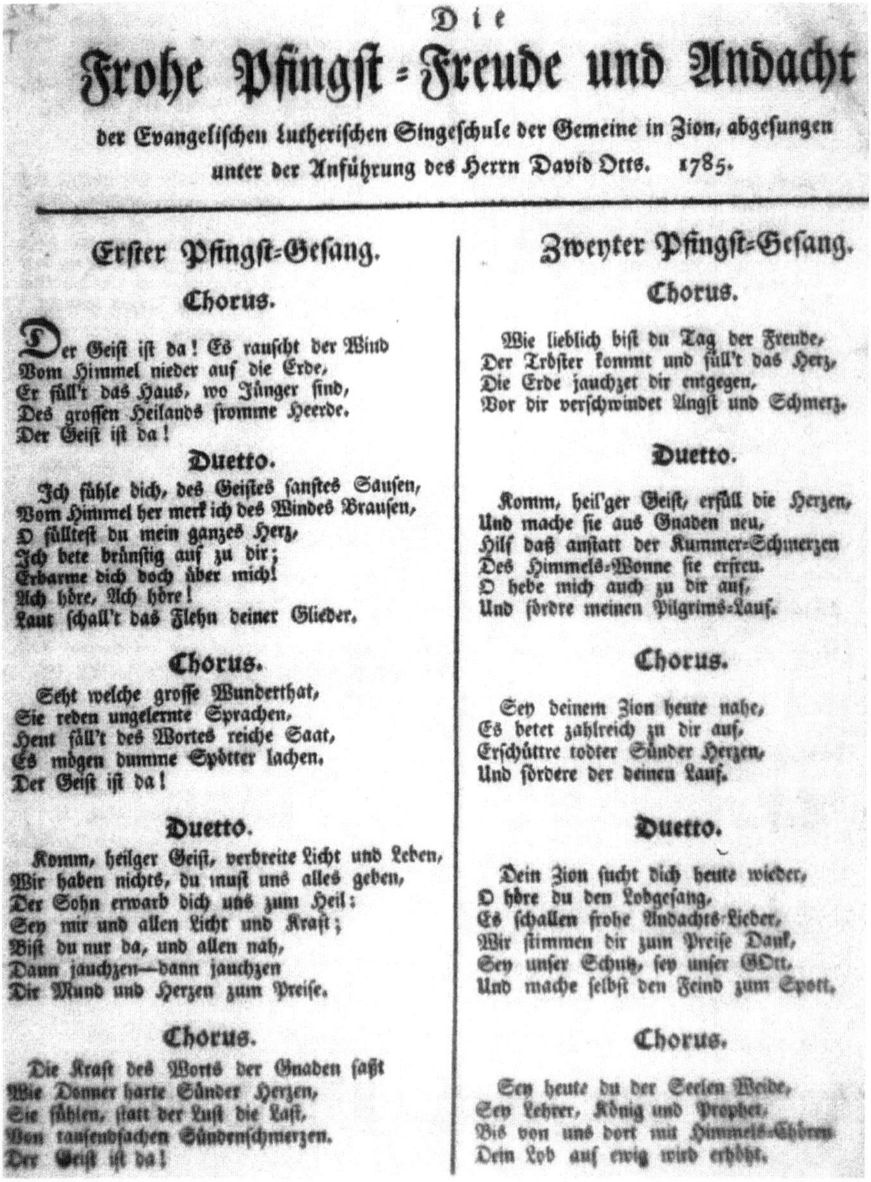

Die

Frohe Pfingst=Freude und Andacht

der Evangelischen lutherischen Singeschule der Gemeine in Zion, abgesungen unter der Anführung des Herrn David Otts. 1785.

Erster Pfingst=Gesang.

Chorus.

Der Geist ist da! Es rauscht der Wind
Vom Himmel nieder auf die Erde,
Er füll't das Haus, wo Jünger sind,
Des grossen Heilands fromme Heerde.
Der Geist ist da!

Duetto.

Ich fühle dich, des Geistes sanftes Sausen,
Vom Himmel her merk ich des Windes Brausen,
O füllest du mein ganzes Herz,
Ich bete brünstig auf zu dir;
Erbarme dich doch über mich!
Ach höre, Ach höre!
Laut schall't das Flehn deiner Glieder.

Chorus.

Seht welche grosse Wunderthat,
Sie reden ungelernte Sprachen,
Heut fäll't des Wortes reiche Saat,
Es mögen dumme Spötter lachen,
Der Geist ist da!

Duetto.

Komm, heilger Geist, verbreite Licht und Leben,
Wir haben nichts, du must uns alles geben,
Der Sohn erwarb dich uns zum Heil:
Sey mir und allen Licht und Kraft;
Bist du nur da, und allen nah,
Dann jauchzen—dann jauchzen
Dir Mund und Herzen zum Preise.

Chorus.

Die Kraft des Worts der Gnaden faßt
Wie Donner harte Sünder Herzen,
Sie fühlen, statt der Lust die Last,
Von tausendfachen Sündenschmerzen.
Der Geist ist da!

Zweyter Pfingst=Gesang.

Chorus.

Wie lieblich bist du Tag der Freude,
Der Tröster kommt und füll't das Herz,
Die Erde jauchzet dir entgegen,
Vor dir verschwindet Angst und Schmerz.

Duetto.

Komm, heil'ger Geist, erfüll die Herzen,
Und mache sie aus Gnaden neu,
Hilf daß anstatt der Kummer=Schmerzen
Des Himmels=Wonne sie erfreu.
O hebe mich auch zu dir auf,
Und fördre meinen Pilgrims=Lauf.

Chorus.

Sey deinem Zion heute nahe,
Es betet zahlreich zu dir auf,
Erschüttre todter Sünder Herzen,
Und fördere der deinen Lauf.

Duetto.

Dein Zion sucht dich heute wieder,
O höre du den Lobgesang,
Es schallen frohe Andachts=Lieder,
Wir stimmen dir zum Preise Dank,
Sey unser Schutz, sey unser GOtt,
Und mache selbst den Feind zum Spott.

Chorus.

Sey heute du der Seelen Weide,
Sey Lehrer, König und Prophet,
Bis von uns dort mit Himmels=Chören
Dein Lob auf ewig wird erhöht.

Fig. 143.5 Broadside titled, Die Frohe Pfingst=Freude und Andacht der Evangelischen Lutherischen Singschulen der Gemeine in Zion, abgesungen unter der Anführung des Herrn David Otts ("The happy Pentecost joy and prayer of the Evangelical Lutheran singing school of the Zion Congregation, Performed under the direction of Mr. David Ott"), 1785 (Source: Author's collection)

Anglo neighbors than the Dutch of Lancaster County. In the more rural German communities, it appears that the ethnic inhabitants created a closed *Gemeinschaft,* a society that only would be penetrated first by the enthusiasm of Anglo Revivalism, promoted by traveling evangelists speaking *Deitsch* (Pennsylvania Dutch) or accompanied by an interpreter. Music served an important role in reminding the people of their heritage and identity.

2. **How was music learned and practiced in a given location and how was it performed ("performance practice")?** A broadside dated October 12, 1786, *Folgende Ordnung* (The following order), divides the students of the singing schools into small groups to practice for a rehearsal that was held in the Zion Church (Fig. 143.6). This one broadside alone gives us a picture of intergenerational music activity. Both "Studenten" (Students) and "Schüler der hiesigen Deutschen Academie" (Pupils of the local German academy) sang together in small groups of two to eight singers. Some of them were young and others

Folgende Ordnung

Wird Heute, als am 12ten October, 1786, bey der Vocal Musik unserer Singeschule, und Rede=Uebung verschiedener Studenten und Schüler der hiesigen Deutschen Academie, in Zion gehalten werden:

1. Die Singeschule: Erhebe dich mein matter Sinn.
Marcus Kuhl;
Jacob Kitz;
Andreas Geyer, Christian Endreß, Joseph Stauß und Henrich Mühlenberg.

2. Die Singeschule: Wie schön ist doch das Band der Liebe.
Georg Wagner;
Philip Dirck;
Jacob Wack, Johann Justus und Johann Hägner.
Adam Seyberth.

3. Die Singeschule: Der Tag ist hin, mein GOtt, wie bald.
Johann Zerban;
Jacob Braun;
Johann Helmuth, Heinrich Zanzinger, Jacob Senn und Friederich Kuhl.
Georg Lochmann.

4. Die Singeschule: Hilf mir, HErr! aus dem Verderben.
Johann Nieß;
Wilhelm Hahn, Conrad Zentler, Carl Schäffer, Johann Zerban und Philip Dirck.
Wilhelm Hahn.

5. Die Singeschule: Ich sinke hin, du hebst mich auf.
Friedrich Schubart, Adam Seybert und Friedrich Schmidt;
Friedrich Schubart;
Marcus Kuhl, Georg Lochmann, Johann Hoffmann und Johann Rediger.

6. Die Singeschule: Auf, Seele! halte dich gefaßt.
Johann Steiner, Henrich Kämmerer und Wilhelm Stedekorn.
Joseph Stauß.

7. Die Singeschule: Ach, mein Heiland, ich Verirrter.
Carl Schäffer;
Johann Ott.

8. Die Singeschule: Das Echo: Stirbst Du, mein Heiland, rc.

Fig. 143.6 Broadside titled, Folgende Ordnung ("The Following Order"), 1786 (Source: Courtesy of the Lutheran Theological Seminary in Philadelphia)

were older. This is how young students learned the Lutheran faith: via the chorale tradition.[5]

Traditional "Alternatim Practice" is evident in the 1785 broadside. Two hymns listed on the document show an alternation between duet and larger choral sections for each hymn. This was the practice in Baroque Germany both in cantatas (including J.S. Bach's sacred cantatas) and in organ accompaniment of congregational singing. It allowed the liturgical leaders to elaborate on the key theological themes latent in the chorales through variation techniques, organ registration, modulations in key, and uses of differing timbres among other things. The singing school was obviously continuing this German performance practice. Further details of performance practice are missing until other broadsides assist us in this matter. The sketches of nineteenth century folk artist Lewis Miller (1796–1882) housed in the York Historical Society, York, Pennsylvania, may give more information about performing in the early Republic, for many of these sketches were iconic of situations when music was performed and religious connections to music made apparent, but such an investigation cannot be accomplished here.

3. **Which repertory was sung in congregations using broadsides?** Tune books tell us a lot, but they cannot tell us (without hand written marginalia and notation) exactly what repertoire was actually being sung in a given place like broadsides can. The 1786 broadside indicates that Zion Church in Philadelphia only made use of the German chorale repertory. In their October singing school rehearsal, singers practiced eight chorales, presumably an hour or an hour and a half of practice material depending on the skill of the singers. By referencing *Erbauliche Lieder Sammlung* it becomes evident that these texts were not in the Ministerium's official repertory and that they may have been written by Pastor Helmuth or David Ott. These two men collaborated on occasion to create new chorale texts for mostly preexistent chorale tunes. This may have seemed like a necessity to them as the newness of English-language hymnody in the Pennsylvania-Dutch community was appealing to so many. Helmuth's 1813 chorale book, of course, was not published until later and looking to it for a tune-text pairing would be useless. Beyond broadsides, an important way to view German Lutheran music in early America is through a comparison with other German American music.

143.6 Other German-American Music

Among the "Plain Dutch" (Anabaptist) groups in early America, sometimes called *Sektenleute*, there was a different kind of music culture. The Amish were less occupied with music than the Lutherans. In the eighteenth and early nineteenth centuries, the Amish printed a sole music resource, the *Ausbund*.[6] Its original core of 51 hymn texts

[5] For more information concerning the Lutheran education system, especially the "Academy," see Pardoe (2001: 190).

[6] The full title of this Amish hymn book is: Ausbund, Etlicher schöner Christlicher Geseng wie die in der Gefengnuss zu Passaw im Schloss von den Schweitzern und auch von rechtgläubigen Christen hin und her gedicht worden. The old low-German spelling has not been corrected here.

were written by imprisoned Anabaptists (who were not trained in composition or text writing as many Lutherans were) between 1537 and 1540, and it was enlarged and printed in several different American editions, including some by the Germantown printer, Christopher Saur (1693–1758). Saur's 1742 edition was the first German hymn book printed in North America. Later editions were printed in 1751, 1767, 1785, and 1815. As was the case with all Anabaptist groups in German-speaking America, there were no organs used in public worship. They may have used the *Scheitholt* in their home and social music making, but this has not been proven beyond a reasonable doubt. It was not until later in the nineteenth century that the Amish adopted other instruments for their amusement and to accompany their work (Raichelson 1975). Church hymns were passed from generation to generation by rote (Yoder et al. 1992: 7). The *Ausbund* contained no music notation. Scholars have concluded that most of the *Ausbund* hymns were *Contrafacta*, (Falck and Picker 2001: 367) showing a great deal of cross pollination of secular and sacred worlds. In comparison to the earliest Lutheran repertory passed down from Luther (despite long held misconceptions by hymnologists), the Amish tradition is quite unique in this way.[7] The Amish never assimilated or Americanized. Their way of life and religious beliefs in the sixteenth century and early Republic are basically what they have retained to the present day and their music reflects this fact. This is not true of all of the *Sektenleute*.

Mennonites in the Colonial and Republic periods also used the *Ausbund* and maintained an oral culture. But, unlike the Amish who refused to give up their ethnic autonomy, the Mennonites assimilated. Starting in the first years of the nineteenth century, they became dissatisfied with the slow monophony (unison singing) of their primary hymn book and started to print other books for worship, the most notable being *Ein Unpartheyisches Gesang-Buch* (An impartial hymn book), Lancaster, 1804. In their recent scholarship, Suzanne Gross and Wesley Berg have revealed that instead of broadsides, the main music resource Mennonites used on the local level was the *Notenbüchlein* (Little note book). This Pennsylvania-Dutch tune book was a small hand written manuscript book given to school children as an award. Printed tune books with shaped notes did not come into vogue with the Mennonites until later in the nineteenth century, something that many have retained to this day. While a few Lutheran *Notenbüchleine* do exist, they were not as common outside of the Mennonite community. Interesting enough, Gross and Berg have established that Johann Adam Eyer, a Lutheran School master who taught in Franconia Mennonite schools from 1779–1787, produced six of these manuscript books and may have "introduced the idea of compiling manuscript songbooks to Mennonite Communities" (Gross and Berg 2001: 193).

Unlike the 1813 Lutheran chorale book and the chorales printed in other Lutheran tune books discussed above, there was a considerable amount of variation in the ordering of tunes, number of tunes, and consistency of mode and notational style in the Mennonite *Notenbüchlein*. These tune books had several purposes: to impart

[7] Luther originally used a secular tune for "Vom Himmel hoch" (From Heaven Above), but by 1539 replaced it with a newly-composed tune. "There is in any case no justification for the argument that Luther attempted to promote congregational singing by catering to the tastes of the masses." Herl (2004: 22).

moral values, record a local repertoire and singing tradition, and to assist in the teaching of the rudiments of music (thus their "rudiments of music" introductions and charts prior to 1817) (Gross and Berg 2001: 194). The variation in white note heads (that resemble half or whole notes) and a more florid style (utilizing eighth notes) document a transformation of musical style among the Mennonites when the pressure to assimilate was the highest. This created diversity from book to book, but a unity remained in each community. The Lutheran chorale taken over by the Mennonites, *Allein Gött in der Höh' sei Ehr'* ("All glory be to God on high"), for example, is found in numerous Mennonite manuscript books, but "no version is an exact copy of any known preexisting printed source" (Gross and Berg 2001: 197). Each appearance of the tune in different Mennonite *Notenbüchleine* is different and this no doubt speaks as much to the issue of assimilation as it does to the variations of Mennonite beliefs from place to place.

The *Notenbüchleine* are like the German Lutheran broadsides in some ways because they describe hymn singing as it happened in a specific location. Gross and Berg note that these books did indeed play a role in the assimilation controversy. Those who inscribed them were "trying to engage their young people in their own German tradition, thereby fostering German Mennonite identity … ." Put another way, these books were a way that the "Franconia Mennonites defined and defended their identity at the turn of the 19th century" (Gross and Berg 2001: 205). Suffice it to make this brief mention of the music of the *Sektenleute* to bring the Germanic Lutheran tradition into sharper focus.

143.7 What Does This Mean?

What do these things mean for the study of Lutheran church music and religious practice, and why are these details important? First, Pennsylvania Dutch Lutherans like all ethnic Americans expressed their thoughts, beliefs, and feelings via music. "The functions and uses of music are as important as those of any other aspect of culture for understanding the workings of society. Music is interrelated with the rest of culture" (Merriam 1964: 15). This makes music not only important to the music scholar and practicing musician, it also raises it as a significant and credible field of inquiry for all scholars of religion and American spirituality. Second, studying particulars of music praxis and printing fills in the gap within our understanding of the everyday religious practice music of early American Lutherans, which has either been ignored or taken for granted in past scholarship. Third, viewing the ways that ethnic culture was maintained and changed may lead us to a better understanding of the parallel continuity and change in religious practices, even the "worship wars," within the last 50 years in the United States. How have Lutherans changed their liturgical culture and musical language in recent years to reflect the dominant culture around them?

143.8 Implications for Future Study

The study of assimilation in the Pennsylvania Dutch Lutheran community is valuable not only to help modern-day people understand the largest non-English-speaking ethnic group in early America, but this exercise can point us in a helpful direction for studying religious practice in our own day.[8] In contemporary scholarship, theologians could use the assimilation model to study how Lutherans (and other Christians) have changed their liturgy to reflect the pop culture surrounding the Church. Why has Contemporary Christian Music ("CCM") replaced traditional chorales and hymns in some American Lutheran churches today? How have theological language and ideas evolved, and which have played a role in this assimilation? How and why has music media gone from a traditional print format to overhead projection in many churches and why has rhythmic complexity become denser in this new church music? These are questions that can find answers in a deeper study of cultural change, using the current study as a starting point.

In the context of European Lutheranism, scholars can start with cultural change in order to understand the spread of the Black Gospel music style and repertoire. The prevalence of Gospel choirs in German churches is a phenomenon that scholars need to explore especially after a recent press release by the *Evangelische Kirche in Deutschland* that claims that there are over 3,000 gospel choirs in Germany containing over 100,000 singers (Röhmhild 2009). Scholars should investigate how and why this religious musical genre infiltrated Germany when this genre's theology and style are foreign to the German cultural heritage and its Reformation-era theology. The stages of assimilation hold the answers.

143.9 Conclusion

Lutherans in America heartfully made music in the home, school, and church to express their faith. This music either supported keeping the ethnic culture of the Lutheran community or it promoted ethnic change and a move away from key theological emphases in favor of revivalist religion. Through the study of tune books and broadsides, we can see how ethnic retention and assimilation took place. This modest body of music source material is a fertile ground which has produced a fuller picture of who the Pennsylvania-Dutch Lutherans were, how they lived, what they sang beyond the official hymnal, and what they believed in the early American context. With more investigation scholars today can understand contemporary changes in religious communities, their practices, and their theology using this study as a model for future work.

[8] For deeper exploration of the assimilation model in modern research and other suggestions for the modern scholar, see Grimminger (2012). The following sources would be a good starting point for the theologian exploring contemporary changes in liturgical practice: Dawn (1995), York (2003), Frankforter (2001), Day (1990), and Grimminger (2002). These authors explore music pertaining to this evolution at various levels.

References

Carson, J. (1817). *Trial of Frederick Eberle and Others at a Nisi Prius Court, held at Philadelphia, July, 1816, before the Honorable Jasper Yates, Justice, for illegally conspiring together by all means necessary and unlawful, with their bodies and lives to prevent the introduction of the English language into the service of St. Michael's and Zion's Churches, belonging to the German Lutheran congregation in the city of Philadelphia*. Philadelphia: James Carson.

Dawn, M. J. (1995). *Reaching out without dumbing down: A theology of worship for this urgent time*. Grand Rapids: Eerdmans.

Day, T. (1990). *Why Catholics can't sing: The culture of Catholicism and the triumph of bad taste*. New York: Crossroad.

Falck, R., & Picker, M. (2001). Contrafactum. In *The New Grove dictionary of music and musicians* (Vol. 6, 2nd Ed., p. 367). New York: Macmillan.

Frankforter, A. D. (2001). *Stones for bread: A critique of contemporary worship*. Louisville: Westminster John Knox Press.

Grimminger, D. J. (2002). *The concept of reform in sacred music: Historical figures as guides for the current crisis*. Doctor of Music dissertation, Claremont Graduate University.

Grimminger, D. J. (2012). *Sacred song and the Pennsylvania Dutch*. Rochester: The University of Rochester Press.

Gross, S., & Berg, W. (2001). Singing it 'our way': Pennsylvania-German Mennonite *Notenbüchlein* (1780–1835). *American Music, 19*, 190–209.

Herl, J. (2004). *Worship wars in early Lutheranism: Choir, congregation, and three centuries of conflict*. Oxford: Oxford University Press.

Hinke, W. J. (1972). Lutheran and Reformed church hymnody in early Pennsylvania. In *Church music and musical life in Pennsylvania in the eighteenth century* (Vol. 3, pp. 259–300). New York: AMS Press.

Merriam, A. P. (1964). *The anthropology of music*. Evanston: Northwestern University Press.

Ochse, O. (1988). *The history of the organ in the United States*. Bloomington: Indiana University Press.

Pannebaker, J. R. (1998). *Early Lutheran music in America: The hymnody of the General Synod*. Ph.D. dissertation, University of Pittsburgh, Pittsburgh.

Pardoe, E. L. (2001). Poor children and enlightened citizens: Lutheran education in America, 1748–1800. *Pennsylvania History, 68*, 162–201.

Raichelson, R. (1975). The social context of musical instruments within the Pennsylvania German culture. *Pennsylvania Folklife, 25*, 35–44.

Repp, A. C. (1982). *Luther's catechism comes to America: Theological effects on issues of the small catechism prepared for and for America prior to 1850*. Metuchen: The Scarecrow Press.

Röhmhild, S. (2009). *SPIRITed Gospel singing*. EKD Press Release, June 16, 2009. http://www.ekd.de/english/ekd_press_releases-pr_090616_gospel_singing.html

Sachse, J. F. (1901). *The religious and social conditions of Philadelphia, 1790–1800*. Philadelphia: No printer listed.

Schalk, C. (1996). *Source documents in American Lutheran hymnody*. St Louis: Concordia.

Smith, R. L. (2006). A great Scheitholt with some remarkable documentation. *Dulcimer Players News, 32*(3), 32–33.

Tappert, T. G. (1980). The church's infancy 1650–1790. In E. Clifford Nelson (Ed.), *The Lutherans in North America* (pp. 3–77). Philadelphia: Fortress Press.

Tappert, T. G., & Doberstein, J. W. (1942). *The journals of Henry Melchior Muhlenberg* (Vol. 1). Philadelphia: Muhlenberg Press.

Wolf, S. G. (1985). Hyphenated America: The creation of an eighteenth-century German-American culture. In F. Trommler & J. McVeigh (Eds.), *America and the Germans: An assessment of a three-hundred-year history* (pp. 66–84). Philadelphia: University of Pennsylvania Press.

Yoder, D. (1961). *Pennsylvania spirituals*. Lancaster: Pennsylvania Folklife Society.
Yoder, D. (1985). The Pennsylvania Germans: Three centuries of identity crisis. In F. Trommler & J. McVeigh (Eds.), *America and the Germans: An assessment of a three-hundred-year history* (pp. 41–65). Philadelphia: University of Pennsylvania Press.
Yoder, D. (2005). *The Pennsylvania German broadside: A history and guide*. University Park: The Pennsylvania State University Press.
Yoder, P. M., Bender, E., Graber, H., & Springer, N. P. (1992). *Four hundred years with the Ausbund*. Scottdale: Herald Press.
York, T. W. (2003). *America's worship wars*. Peabody: Hendrickson Publishers.

Chapter 144
Understanding Churchscapes: Theology, Geography and Music of the Closed Brethren in Germany

Friedlind Riedel and Simon Runkel

But ye, be not ye called Rabbi; for one is your instructor, and all ye are ***brethren****.*

Gospel of Matthew 23:8, Darby's translation

Do not wonder, ***brethren****, if the world hate you.*

1st John 3:13, Darby's translation

144.1 Introduction

In September 1854 John Nelson Darby[1] (1800–1882), a former Anglican priest and Anglo-Irish Christian met the German teacher Carl Brockhaus (1822–1899) in Elberfeld, Germany. It was the beginning of a long-standing friendship. As a result of their shared theological effort that was essentially concerned with the rejection of institutionalized, religious churches and hierarchical structures, a religious movement known as the Brethren (*Brüderbewegung*) was formed in Germany from 1843[2] on (Gerlach 1994; Eylenstein 1927). The Brethren movement established itself as a new congregational form alongside but in distinction to Baptists, Methodists and

[1] There is a vast amount of historical and biographical literature on John Nelson Darby available. Most of it was published by authors associated with the Brethren movement. However, a comprehensive account on Darby's theology is given by Schwarz (2008).

[2] On a few precursors formed in the 1840s, see Ouweneel (1977) and Gerlach (1994). In 1843 a Brethren meeting had been formed in Stuttgart by Georg Müller. Some years later, Peter Nippel, who also translated Darby's writings into German, formed another meeting in Tübingen (Ouweneel 1977: 93; Gerlach 1994: 21). Around the same time, a former Catholic, Julius Anton Eugen von Poseck, translated many of Darby's writings and began to meet with other Christians in the West Ruhr area. A Brethren assembly in Hilden was initiated in 1849 (Ouweneel 1977: 93).

F. Riedel
Department of Musicology, Georg-August-University of Göttingen, Göttingen, Germany
e-mail: friedlind.riedel@phil.uni-goettingen.de

S. Runkel (✉)
Department of Geography, University of Bonn, Bonn, Germany
e-mail: runkels@uni-bonn.de

S.D. Brunn (ed.), *The Changing World Religion Map*,
DOI 10.1007/978-94-017-9376-6_144

various other evangelical Free Churches in Germany that did not concur with the national Evangelical Church (*Evangelische Kirche in Deutschland*). Due to divisions and schisms resulting from personal disagreements, theological questions and political repression during the Third Reich,[3] the Brethren movement nowadays consists of at least four different main strands of Brethren meetings in Germany (Fig. 144.1), with central aggregations in four regions in western-central Germany (*Bergisches Land*, *Siegerland*, *Lahn-Dill-Kreis*, and *Hessisches Hinterland*). Steinmeister (2004: 225), at that time a Closed Brethren teacher, identifies four groups in the German Brethren network: *Bundesgemeinden* (congregations within the association EFG), *Bundesfreie Gemeinden* (Open Brethren), *Geschlossene Versammlungen* (Closed Brethren) and *Blockfreie Gemeinden* (Block-Free Brethren). Today the overall amount of members or affiliated individuals[4] (cf. Schwarz 2008: 8) is estimated by us to be 12,000–13,000 in the Closed Brethren assemblies and up to 17,000 in other Brethren assemblies[5] in Germany (cf. Henkel 2001: 199). The Closed Brethren meet at 212 locations in Germany in 2012.[6]

144.1.1 Methodology

Our analysis focuses on the Closed Brethren, although we intend to add to the knowledge about the Brethren movement in Germany in general. Some has been written on Brethren theology and history both by religious scholars and lay historians (Geldbach 1972; Boddenberg 1977; Ouweneel 1977; Orth 1977; Jordy 1979, 1981, 1986, 2001, 2003; Bister 1983; Heinrichs 1989: 341; Gerlach 1994;

[3] The movement was banned during the Third Reich, leading to a split between those who accepted a certain degree of accommodation with the Nazi regime (the implementation of the Führer-principle; see Liese 2002: 325), and those who resisted any intrusion of Nazi ideology into their church-doing by holding assemblies in private or even with other groups such as the pietist *Gemeinschaftsbewegung* (Kretzer 1987: 465). However, it would be euphemistic to describe the Brethren movement in general as opposed to Nazi ideology. For an extensive account on Nazi politics of religion and the Brethren movement: see Liese (2002), Kretzer (1987) and Menk (1980, 1986).

[4] At least theoretically there is no status of official membership. However, those who are approved at their local perish to take communion, are received by Closed Brethren communities worldwide. The procedure of approval is carried out by a group of locally recognized Brothers. Visitors who are not approved within one of the Closed Brethren churches as certified in a letter of recommondation of their home church are not allowed to take communion. In effect, the Lord's Supper is the place where membership is negotiated.

[5] The group of Block-Free Brethren is difficult to estimate, due to the fact that over the last years several newly emerged groups allocate themselves or are associated with as Block-Free Brethren.

[6] The lack of more precise statistics is due to the fact that a central organization of the Closed Brethren is rejected on theological grounds. An account of the global distribution of the Brethren movement is difficult to give. It is known that Brethren meetings exist in nearly all European countries. We estimate that the Closed Brethren for instance gather at circa more than 150 places in Egypt, a large number of meetings exist as well in Central Africa, India, the Caribbean and South America.

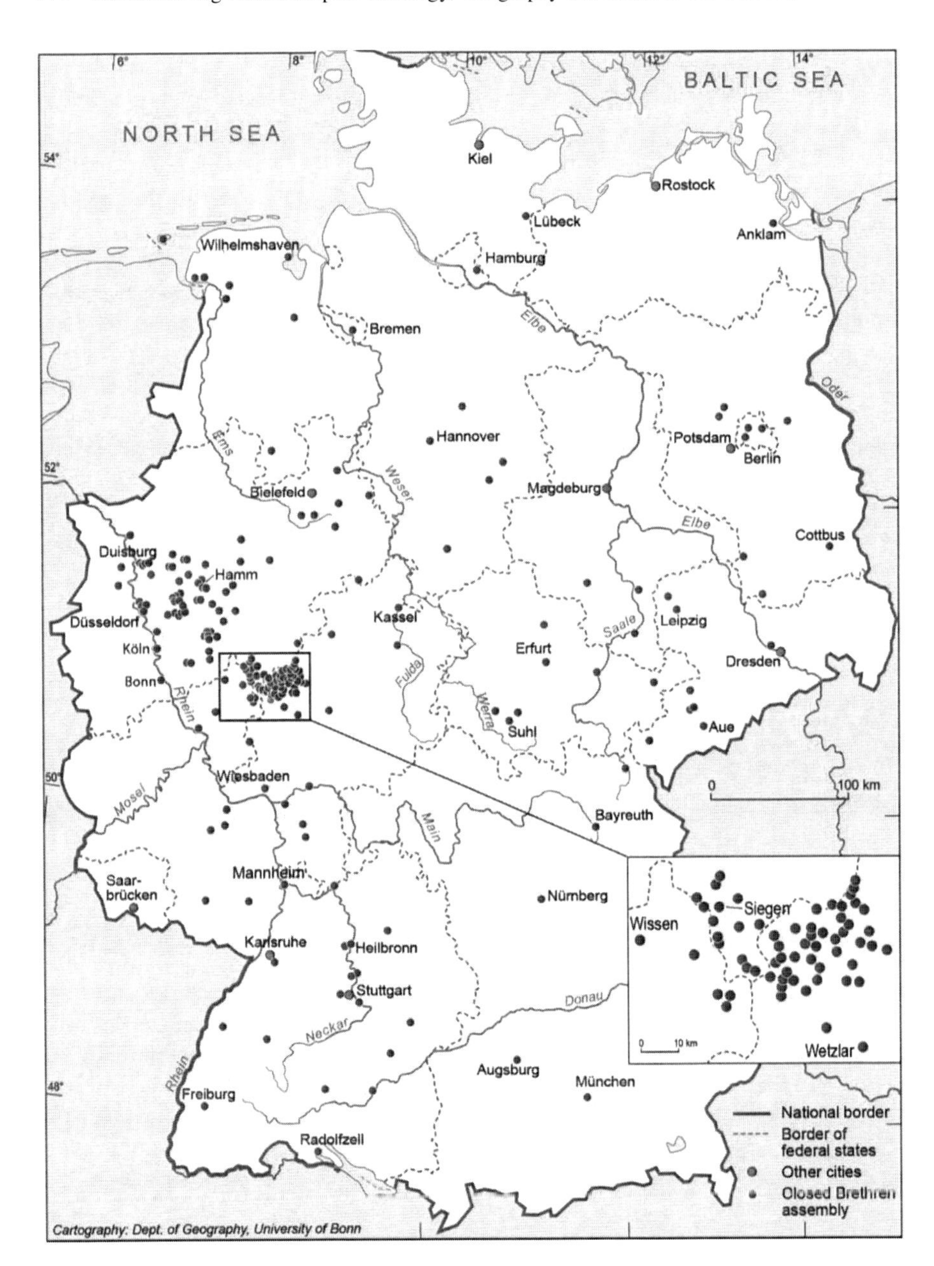

Fig. 144.1 The distribution of Brethren churches in Germany (Data provided by authors)

Arbeitskreis Geschichte der Brüderbewegung 2001; Steinmeister 2004; Schwarz 2008; Weremchuk 1988). Most of these scholars chose a historiographic approach by focusing on the origins of the movement in England as well as its dissemination in continental Europe. This approach is most often accompanied by a positivist reconstruction of Brethren theological convictions and their development over time. This perspective on Brethren theological identity that designates its forefathers' insight as being highly relevant for the movements present is analogous to Closed Brethren discourse about the "*ideal biblical way to assemble*" (cf. Briem 2004; Weremchuk 1989). Until today, however, the socio-cultural implications of these theological imaginations have rarely been touched upon. Choosing a rather ethnographic approach will therefore help to give insight into today's Brethren religious *practices* – rather than into theological ideals – and their cultural and spatial manifestations especially as they present themselves in congregational "musicking" (Small 1998) and music. Using Brethren theological conceptions of the church as main frame of reference in spiritual practices allows us, to interpret these cultural practices as manifestations of what we call *church-doing*. Thus we understand "church" not as a stable given in culture that is solely inhabited by its members, but as culturally ever new (re-)enacted and performed as in congregational musicking, hence *doing church*. On the other hand, we acknowledge the fact that theological imaginations and faith-based narrations are key elements to understand Brethren mind-set and their spiritual practices. As Brethren theological and ecclesiological discourses entail spatial imaginations of the intangible, Brethren church-doing can be understood as tangible performance of these places. Reading place in its religious constitution "*allow*[s] *religion to speak back*" (Yorgason and Della Dora 2009: 629).

This research draws on participatory observation in various Brethren meetings, conferences and other occasions, and is informed by our own biographical involvement in Brethren churches. The analysis of Brethren theological and edificatory literature was accompanied by three in-depth interviews and many informal conversations with various informants. One interviewee can be considered as a well-known key figure among the Closed Brethren. The second interviewee understands himself rather as critic and is at the same time proponent of and preacher among the Closed Brethren, whereas the third left Closed Brethren meetings and attends Block-Free Brethren meetings.

Two main challenges appeared during our study. Firstly, a consistent dichotomy between praxis and theoretical conceptions seemed to be immanent in the statements of the interviewees. This was challenging, because talking about personal experiences and practices were often covered by statements about the 'wish to do better' in regard to theological ideals. Secondly, the inclusion of theological discourses in research on religion is widely disputed among scholars (cf. Henkel 2011) and not unproblematic. This is due to the discursive construction of "religious convictions" and "scientific truth" as allegedly being dichotomous. However, in Brethren reasoning, spiritual as well as scientific arguments are used, but religious understandings establish a master frame from which the world and the Self are understood and science is comprehended. Our approach aims to meet these concerns by giving a thick description (Geertz 1973).

To advance the study of the spatiality of religious musicking (cf. Kong 2006), it seems necessary to deploy a rather specific theoretical approach in order to be able to conceptually grasp the questions of how Brethren assemblies are imagined and musically enacted. Thus, in the following, we will develop the concepts of "theological imagination" "churchscapes" and "religious movements" as well as contextualise the term "musicking" within the study of religion and geography.

144.1.2 Theological Imaginations

The object of faith of the Brethren is the Triune God. Jesus Christ as the son of God is believed to be the head of a worldwide church, to which all true believers belong. These convictions continuously constitute individual as well as group identity. However, the shared practices of believing, worshipping and musicking are ecclesiologically and thus spatially confined to those, who share the same theological conceptions. Using the broader term "imagination" (Appadurai 1996), we seek to link agency with the imaginary, a "*taken for granted spatial ordering of the world*" (Gregory 2009: 282) and – especially in theological terms – the suppositional matter-of-factness of truth and knowledge of the world in specific communities. To understand Brethren lifeworlds (cf. Appadurai 1996) it is crucial to not just identify the globally dispersed Brethren assemblies and individuals as an "*imagined community*" (Anderson 1991) but – with Appadurai extending Anderson – rather to reach beyond group identity and to acknowledge "community" merely as part of Brethren "*imagined worlds*" (Appadurai 1996: 33). These imagined worlds are constituted by manifold distinct but interconnected fields of imagination. We will identify *theological imagination* and *ecclesiological imagination* as imaginations that are concerned with what one might call "theology" or "ecclesiology" that constitute Brethren imagined worlds. Thus, practices of church-doing are not just embedded in and produced by imaginations of community, but of theology and ecclesiology. As these spheres are imagined to be hierarchically structured, theological imaginations are discursively authorized to be determinative for any other social practice respectively imagination, such as geographical or musical as well as musicological imaginations. It is particularly the theological imagination where Brethren agency emerges that fundamentally structures quotidian and dominical religious practices – such as congregational musicking.

144.1.3 Churchscapes and Religious Movements

Furthermore, we argue, that the theological imaginations and the spatiality of churchscapes are mutually constitutive. Brethren assemblies in Germany are embedded in global and local landscapes of Christian religiosity, which we refer to as "churchscapes." The term has been defined by various scholars. Zelinsky (2001: 565) uses it to describe an "*assemblage of objects*" in the visible landscape, while focusing on houses of worship and other buildings and enterprises as well as their signage in local settings. Leppman (2005: 84) extends the term as a description of "*a particular type of cultural landscape, created by and belonging to a group of adherents to specific religious beliefs.*" She stresses an iconographical perspective on religious landscapes in their spatio-temporal, communicative organization of meaning. The formation of churchscapes is explained in a process-related model through "belief," "attitude" and "intentionality" of religious communities, whereas the iconography is inscribed with meaning both by its designers and its users, whether they have religious or secular concerns. Thus, Leppman differentiates between churchscapes and their contexts, even if she links social group identity to the creative design of churchscapes. In addition, she acknowledges that the "*sacred space is both local and universal, as it attempts to link a particular place with wider truths*" (Leppman 2005: 85). In summary, the term churchscape appears deeply interwoven with ecclesiological, liturgical and eschatological imaginations[7] (cf. Ingalls 2011: 256), and their spatial expressions.

Taking this in consideration, churchscapes need to be understood as global configurations and concatenations of theological ideas, concepts, images, metaphors and especially scripts, which serve as cornerstones for the imaginative geographies of communal religious practices of individuals and groups around the world (cf. Appadurai 1996: 33). Understanding theological imaginations (ecclesiology, liturgy, etc.) as social practices and thus as building blocks of churchscapes offers a better insight of shifts and transformations in the local expressions of such religious landscapes.[8]

As they inform each other, shifting ecclesiological imaginations and transforming churchscapes do not seem to be locally bounded phenomena but rather a deterritorialized set of cultural phenomena. We propose to understand fundamental shifts in the global churchscape with the sociological term of "religious movements": "*A religious movement is a relatively organized attempt by a number of people to cause or prevent change in a religious organization or in religious aspects of life. Religious movements* [...] *are collective human attempts to create or to block change*" (Bainbridge 1997: 3). The argument here is that theology needs to be considered for the understanding of geographical diffusion of religious ideas and concepts, that is, "*the historically situated imaginations of persons and groups spread around the globe*" (Appadurai 1996: 33). Inasmuch as specific topologies are immanent to theological imaginations, churchscapes form specific geographies, for example, based on imagined, ecclesiological proximities. To grasp these proximities, spatial and cultural practices of exclusion/inclusion, for example present in musicking, gives a rather holistic insight into the emergence of Christian denomination, here, the Closed Brethren churchscape.

144.1.4 Musicking

Recently, Engelhardt (2012: 299) pointed to the profound and "*universal relation between music, sound and religion.*" The embodied practices of musical worship are framed by theological concepts of divine revelation, spiritual presence and the

[7] Nebeker (2001) speaks of "Darby's eschatological hope." Darby's eschatology is intimately connected with ecclesiology (cf. Geldbach 1972: 78; Schwarz 2008: 417).

[8] Leppman's concept of churchscapes tends to be ahistorical, since theological developments over time are not fully grasped by her model which is mainly a reflection on local settings of church-doing. Appadurai (1996: 64) argued for an ethnography that considers the "historical present."

expression of emotional attunements beyond speech. Accordingly, music is a constitutive part in processes of community formation. We suspect, that not just verbal discourse such as theology and ecclesiology initiates and sustains denominational difference, but that music is equally part of that process. Hence, we conceptualize music as discourse i.e. as a musical discourse intertwined with verbal discourse about music as it is theologically imagined (cf. Ingalls 2011: 256). As these levels of discourse are interdependent as well as conflicting they induce each other reciprocally, thus *"music makes religion and vice versa"* (Engelhardt 2012: 301). The practices of exclusion/inclusion on the various scales, crucial to Closed Brethren church-doing reveal a distinguishable Brethren musical discourse that constitutes and demarcates the Closed Brethren churchscape. This leads back to an understanding of musical discource as performative practice. Therefore, we use the term musicking (Small 1998). This allows us to depict individuals as well as groups as agents inhabiting agency and to grasp music as meaningful in singularly, situational performative acts.

Besides that, religious music seems to transcend local communities due to the latter's ecclesiological imaginations. This – among others – constitutes churchscapes. Shelemay (2011: 367) "*set[s] forth a tripartite framework*" for the study of musical communities and identifies "*processes of descent, dissent, and affinity as expressed through music*" that generate "*collectivities.*" Focusing on the Closed Brethren churchscape, all three processes that occur in a continuum can be outlined in congregational musicking, which mainly is multipart singing of hymns without instrumental accompaniment. However, processes of dissent are primary among Brethren communities since processes of descent[9] and affinity derive from these (cf. Shelemay 2011: 376).

As the outcome of the research suggests we identify the practice of separation and segregation as pivotal in Brethren church-doing. Focusing on the so-called Closed Brethren,[10] we argue that the movement emerges, diffuses and maintains through practices of delineation, dissent (Shelemay 2011) and exclusion. We understand "delineation" as strategies of making or displacing borderlines, "dissent" is the ground laying contradiction with the secular or worldly Other, whereas "exclusion" encompasses the practices of eliminating incoherencies and threats from within the community. We propose that these practices are due to Brethren theological imaginations constituting a Brethren "ideoscape" (Appadurai 1996) that appears to be what we call a *Brethren churchscape*. As the theological imaginations are in

[9] Shelemay mentions besides "biological kinship" the "religious practice" as decisive factor for "descent communities" (2011: 367).

[10] The Closed Brethren refuse to be labelled due to their ecclesiological understanding that there is only one Church and that they are not one group among others. However, in praxis various (exonymical and endonymical) labels are used to describe the group and their ecclesiological imagination. There is even a distinction between the labelling, for example, in English-speaking countries and in Germany, where the Closed Brethren are known as well as the "Exklusive Brüder." However, the community of the "Closed Brethren" should not be confused with the so-called "Exclusive Brethren," that is, the Raven-Taylor-Hales Brethren. This group gained repeatedly media attention because of various scandals. In Germany this splintered group of Brethren is not widely spread and the number of local groups is estimated by us to be less than 20.

their aspirations global and universal, they facilitate geographical diffusion of Brethren theology and church-practice, thus produce hubs, passage points and travel practices that generate exclusion and constitute the dissent community. Although religious movements like the Brethren need to be understood as deterritorialized in terms of their theological and especially ecclesiological imaginations, they constitute specific geographical localities that, among others, appear as spaces of – and through – musicking. As argued by Holloway and Valins (2002: 6): "*religious and spiritual matters form an important context through which* [people] *live their lives, forge a sense* [...] *of Self, and make and perform their different geographies.*" From a geographical perspective on religious musics, the communal spaces of musicking within the theological imaginations reveal themselves.

Identifying congregational musicking as the place where theological imaginations and "*musical ontology conflate*" (Engelhardt 2009: 33), we argue that music plays a crucial role in confining this Closed Brethren churchscape; hence the Churchscape comes into being as a musical, thus tangible performance of the Brethren "*place in the world*" (Dewsbury and Cloke 2009: 698). To understand these places, we will be concerned with Brethren musical strategies of performing the dissent, constituting demarcation, and establishing exclusion, while relating to an intelligible world (the Triune God) and fellow believers. Considering musicking as church-doing we will explore its impact in constituting the dissent on the various levels of Brethren spiritual practices.

144.2 The Geography of Brethren Churchscapes

144.2.1 Cultural Geographies and Theological Imaginations

It is argued among Brethren that what is a biblical and hence correct understanding of "Church" had been lost over centuries of church history. Asked for causes of the emergence of the Brethren movement, one interviewee replied that the reasons were:

> [...] firstly, the at that time emerging Bible criticism, already at the beginning of the 19th century in the French Revolution. This whole free-thinking was emerging at that time. And secondly, the tremendous demoralization with only dead Christendom, similar by Luther, who only saw dead, formal Christendom. And then the Bible criticism was added, which evoked a general awakening over whole Europe. Many, many communities came into being in the first half of the 19th century. Nearly everything what exists today outside the main church. Methodists evolved much earlier and then the whole Pietism – which has its roots already earlier – especially in the Siegerland, there had been a Pietism which has nothing to do with us directly. These are the main reasons, firstly the flattening of Christendom and secondly the Bible criticism. This had God affected through the Holy Spirit in many hearts. But there were only few, who said that we need to leave all these traditions and return to God, and many things became revealed, which did not become apparent in other circles of the awakening. [...] This was the matrix on which God led the Christians to make a step back to the Word of God.

Since the nineteenth century, the term "Free Churches" has been used in the UK and continental Europe for Christian groups that separated voluntarily from the main Churches (Anglican, established Protestant and Roman-Catholic churches) (cf. Henkel 2001: 131; Geldbach 1989: 29). However, Heinrichs (1989: 16) pointed out, that the development of "Free Churches" has to be understood as a "*Protestant response to modernity*" or, in other words as escapism during times of crisis (cf. Sweetnam 2010: 206). Basically, the social form of associations as civic organizations had been transmitted into religious lifeworlds. The argument of Free Churches and their theology being a response to modernity needs to be set in context with the rise of Dispensationalist models of Bible-based understandings of history, as systematically conceptualised by Darby. Dispensationalism and other Christian futurist worldviews are shaped by a prophetic teaching of a corrupted world and a strong emphasis on the expectation of the (pre-millenial and pre-tribulational) rapture.[11] This dystopic notion of future events leads argumentatively to cultural conservatism (concerning fashion, music, education, morality beyond biblical reference, etc.). For most of the Closed Brethren, Darby's dispensationalist theology is understood as biblical truth rather than as a hermeneutical approach among others such as covenant theology,[12] as the interviewee explains:

> The term "hermeneutic," is like many terms of theology a human term. I am a layman in these things, too. [But] to act on the assumption that there are different approaches which stand side by side – there is already a mistake in that. There can't be different interpretations of the Word of God, which have a right to exist in parallel, therefore it is a nebulous matter for me, because then one will only give explanations that one thinks different than someone else and this is camouflaged with a scientific term.

The writings of Darby are still held to be very important, which is mostly noticeable through the fact that new literature rarely contravenes him. Theologically Darby is considered authoritative, so that Brethren are often labelled as "Darbyists." Schwarz (2008: 148) therefore speaks of a "*nomistic normativity of Darby's 'opinion.'*"

The theology of the Brethren movement stresses the concept of the "Unity of the Church."[13] Strategies used to represent this unity vary from inclusive/ecumenical tendencies of the Open Brethren to exclusive strategies of the Closed Brethren, based on different hermeneutics of the bible. One general understanding of the Brethren has been, that the authority of the Church belongs to Jesus Christ and that it is necessary to be detached from the "*evil world*" (Harding 2006: 122). This is at best exemplified in a statement given in a letter of Darby himself, which is cited by Schwarz (2008: 483): "*I have no home – though countless mercies; on earth my home, for the home belongs to the heart, is the place of His* [God's] *will; for the rest,*

[11] See 1st Thes 4:17 (cf. Darby 1992: 5, 101). On the biblical hermeneutics of J.N. Darby, see Schwarz (2008).

[12] Cf. Sweetnam (2010), Schwarz (2008) Ouweneel (1977: 284).

[13] See Ephesians 4:4-6: "There is one body and one Spirit, as ye have been also called in one hope of your calling; one Lord, one faith, one baptism; one God and Father of all, who is over all, and through all, and in us all." Darby (1992: 4, 440) emphasizes that "we cannot exercise faith, nor enjoy hope, nor express Christian life in any form whatever, without having the same faith and the same hope as the rest, without giving expression to that which exists in the rest. Only we are called to maintain it practically."

it will really be in heaven; and Montpellier, Düsseldorf, or New Zealand – what is the difference? ... I wait for heaven." This notion of a "being-not-of-this-world" is commonly taught in Closed Brethren meetings and manifested in religious and cultural practices of what we call "self-othering". Thus, cultural forms become inverted to mark the theologically imagined distinction. This is especially visible in traditional clothing of Closed Brethren women and audible in congregational musicking. However, this exclusionary notion of the world is not only limited to the secular society, but to other denominations as well (Harding 2006: 167). Darby (as cited in Geldbach 1972: 128) wrote in a letter: "*Feeling that Christendom is corrupt, we are outside the church-world, by whatever name it is called.*" One interviewee reflected this notion of Christianity as follows:

> However, any congregation has its form, its liturgy, its singing, its doctrine, its speech. And I read the bible and there is ONE word of god, there is ONE god, who revealed what it is. And then suddenly there is Matthew 18, this verse [verse 20] had been so obvious – 'for where two or three are gathered together in my name, there am I in the midst of them'. Why is that we not just simply do it? 'Do this in remembrance of me.' These are simple laws, simple actions and thoughts. Why is this so complicated? And then I got to know these thoughts more and more and thus: away from all this human forms, out in the 'green field'! Simply: there is one body! And – as Darby said it – I don't know any other membership as the membership of the body of Christ. Either I am in or out. There is nothing in-between. And: a god-fearing way of life. Where these two things are, I am on the same par with them. [...] And this was a gift of God to them [Darby and his contemporaries], that this unity is possible. I was fascinated by this and I still agree with it today. More than ever!

The Brethren movement understands its theological imagination as being untouched by historically developed hermeneutics and based on the Bible, as found in pre-Constantine times or during the writings of the apostle Paul, although they relate themselves as well to the Reformation and movements or traditions in its aftermath. A literal interpretation of the Bible (cf. Sweetnam 2010: 199; Schwarz 2008: 147) is seen as opposed to academic and therefore "worldly" exegetical traditions.[14] Even more, it is regarded as an exposure of the truth.

Closed Brethren oppose any institutionalized forms of Christianity. They claim to not be a "church" but rather what we call a "non-church" (cf. Augé 1995), as they intend to not be a Christian denomination among other cultural and religious formations, believing that there is only "one, global Church." Although there is no association or central organization there is a practical need to know places where Christians, "*who walk the same way of separation,*"[15] hold their gatherings, due to the fact that the local churchscape is pluralist and fuzzy. The ambivalence is shown in the difficulty of opposing any forms of central organization while at the same time maintain-

---

[14] Even though Heinrichs (1989: 371) observation of a certain adaption of modern and rationalistic principles of bible study (e.g. the dichotomy of 'doctrine' and 'praxis' and eo ipso 'exegesis' and 'application') seems to be coherent. Heinrichs (1989: 371) calls this "modernity vs. modernity."

[15] For example, the Brethren author Graf (1990) uses this term in reference to 2nd Cor 6:14–41, Heb 13:13–15 and 2nd Tim 2:19–22. Darby (1992: 5, 214) writes: "One cannot walk with those who dishonor Him, and, at the same time, honor Him in one's own walk."

ing best practices. Thus, more or less informal lists (for example, an access controlled online database, whose password is issued on annual Bible conferences) of places and people exist as well as the practice of writing "letters of recommendation" for those local affiliates who have the intention of visiting other national or international assemblies.[16] Although the list is considered informal, it represents a quasi-denominational and exclusionary practice.

Besides that, various other strategies of organizing social life and religious practices in a community that considers itself as non-organized have developed throughout the history of the Brethren in Germany. The geographic distribution and diffusion are deeply interwoven with the theological imagination. Geographies of the Brethren movement are shaped by the ecclesiological imagination and vice-versa.

144.2.2 Diffusing Concepts: Translations and Publications

Although the Brethren movement in Germany was initiated by few, but more or less independent persons in different regions, it is worth noting that the impulse of the Brethren movement from national to a transnational phenomenon was mainly a cultural practice of translation. The fact that the English language was widespread on the European continent facilitated this process as well as Darby's high level of education and his apparently good knowledge of languages. The *passage points* of these transnational activities always included a translator. Thus, Carl Brockhaus can be identified as key figure due to his work as translator and his (co-)founding of the *Evangelischer Brüderverein*[17] with its central publication *Der Säemann (*"The Sower"*)*, which served as a central network for the influx of new theological ideas. Brockhaus separated from the *Evangelische Brüderverein*[18] because of theological dissonances and started to contact Darby in 1853 by post. Darby and Brockhaus met in 1854, probably through H. Thorens (Ouweneel 1977: 95). One of Brockhaus' key roles was his founding of a journal called *Botschafter des Heils in Christo* ("Messenger of Salvation in Christ") in 1854, which massively facilitated the distribution of Brethren exegetical ideas. The translation of the Bible by Brockhaus, Darby and von Poseck was published from 1855 (the New Testament) to 1871 (the whole edition) in Brockhaus' newly founded publishing company (Ouweneel 1977:

---

[16] Basically, this practice calls for an ecclesiological understanding of all local representations of the one Church being dependent in their decisions about authorization and excommunication of persons regarding the "breaking the bread," that is, taking part in the "Lord's supper." This has led to recent separations (from the 1990s until now) due to the fact that certain decisions were not accepted by other local groups. Thus, the "inter-dependence principle" has been cause for dispute (cf. Steinmeister 2004: 245).

[17] This union was led by Bouterweck and Grafe from Elberfeld (Ouweneel 1977: 94).

[18] A good background on these events is given by Jung (1999: 108).

95).[19] The literature itself was distributed by so-called "travelling brothers," who inter alia constituted the Brethren churchscape. The intensive distribution of literature as a dimension of the Brethren churchscape is definitely one of the main pillars of their successful geographical diffusion. As Jordy (1981: 32) states:

> The journals in their regularity of their publication could have influenced the awareness of gemeinschaft as well as the opinion formation. Thus, the strong solidarity of the German Brethren movement [...] is last but not least caused by the connecting function of their literature. [our translation]

It is even noticeable that a Brethren vernacular traced back to von Poseck's translation of Darby's writings and the *Elberfelder Übersetzung* of the Bible (Geldbach 1972: 51; Schneider 2004) has been maintained until today.

During the history of the Brethren movement many publishing companies were founded and closed. Today, the German strands of the Brethren movement can be identified by their main publishing houses for example in Dillenburg (Open Brethren) or Hückeswagen (Closed Brethren) and often tend to be deeply connected with the theological reputation of their owners. While not officially recognized as centers of the networks, they at least function as hubs for theological unification or cultural representation[20] (cf. Henkel 2001: 200).

> Insofar I will say that those publishing houses we know [...] that they are led by brothers and in which staff works, which really say "we do not want to simply publish Christian literature but literature which leads hearts to the Lord and which is based on the Word of God," and who want to interpret clearly, not to bring any fantasies or any special exegesis [...] And this, I would say, I would grant our publishing houses just like that. Because I know them, and I know their literature, which is published there [...] and I do not have any difficulties in recommending them publicly – and this I cannot say about any other publishing house in Germany. Starting with [name of a person] ... They always bear their doctrinal premises. And this is [name of a person]. An Open Brother! Through and through! Independent, autonomous! The literature he publishes is only this orientation, that he sells and promotes our publications as well is a dangerous thing.

Publishing houses have played a significant part in various separations and even the Brethren themselves acknowledge to a certain degree the centralist function of publishers in the theological discourses.

---

[19] The foundation of the Brethren movement in other countries such as the Netherlands, Switzerland, France, Denmark, Sweden and Norway pretty much occurred in similar patterns through a few personal contacts, family relations, and was massively supported by the translation of Darby's writings. In addition to his new English and German translation of the Bible, he was also responsible for a new French translation; also he supported the translation of the New Testament into Dutch, Italian and Swedish (Ouweneel 1977: 95). Furthermore, Darby visited the U.S., Canada, British Guyana, Barbados, Jamaica, New Zealand, Australia and Italy (Ouweneel 1977: 102).

[20] In 1891 the publishing house of Emil Dönges started to publish calendars and later on tear-off calendars for families and children (Jordy 1981: 31). This practice of nurturing the spiritual imagination is consistent until today. It must be added to the list of reasons why the Brethren tend to have their own terms and words for specific phenomena. For example, calendars, mostly read before or after family meals, facilitate spaces of communalization and belonging on the scale of the private home.

144.2.3 Travelling Brothers

The colportage and distribution of the literature had been primarily effected by the "travelling brothers" until the late 1930s, when common bookshops took over those offices (Jordy 1981: 26). New literature was greatly appreciated among the geographically dispersed Brethren groups, even though the literature and journals did not really serve informational or social functions but primarily theological ones (Jordy 1981: 34).[21] This has not much changed, although the journal *Ermunterung und Ermahnung* ("Encouragement and admonishment") (from 1947 until today) provides a small supplement with changes of addresses and small advertisements since 1980.[22] No official social platform has been initiated in recent years except for smaller mailing lists and online newsgroups. Thus, the main interregional spaces of social interaction are Bible conferences, holiday camps or evangelistic events, where the main social activities involve communal Bible study, musicking or sports.[23]

Basically, Brethren agency is morally and spatially confined to theological discourses. However, distance and proximity is constituted by theological topographies. Therefore, mapping Brethren churchscapes is not feasible in geographical terms but rather needs to be understood as deterritorialized theological landscape (cf. Appadurai 1996: 15) with exclusionary "territories," where "*everyone is invited to come but not everyone is welcomed.*"[24]

The growing network of Brethren meetings throughout Germany, ecclesiologically understood as "dependent" on each other, needed a relational technique to remain integrated and theologically unified. Smaller groups were especially challenged by the fact that not many men actually ministered. The establishing of "travelling brothers" as distinct from fixed pastoral districts of preachers was and still is a unique feature of the Brethren movement.[25] As mentioned before, Darby travelled

---

[21] The only exception is a journal which provides the Closed Brethren with information from transnational remote assemblies ("Mitteilungen aus dem Werk des Herrn in der Ferne"), which is published by the Gute Botschaft Verlag (GBV) in Dillenburg-Eschenburg.

[22] Another journal is "Folge mir nach" ("Follow Me"), which is mostly geared towards young people. Topics include Bible studies and practical guidance in challenges of everyday life such as clothing, musics, marriage, sexuality, etc. The journal produces protocols, which manifest themselves in "notions of etiquette and appropriate behavior, which have a geography that must not be overlooked" (Bailey et al. 2007: 144).

[23] Weddings seem to play a specific role in the sociology of the Brethren. Marriages mostly tend to be endogamous. Wedding ceremonies appear as well structured cultural events, where room is given to musical aptitudes beyond musicking for spiritual purposes. It is noticeable that other artistic practices than musicking seem to be not well-represented on social events in "Brethren culture."

[24] This anecdotal evidence goes back to a dispute in one local assembly of the Closed Brethren, in which the street sign was changed from "welcomed" to "invited." The story was tongue-in-cheek told by some informants in that area.

[25] Apart from the foundation of the "Verein für Reisepredigt" (Association of Travel Preaching) of the "Gemeinschaftsbewegung" in the Siegerland, which happened around the same time (1853). The idea behind this regional association was the spread of the Gospel and the encouragement of dispersed Christians in the area (cf. Jordy 1981: 15; Schmitt 1984 [1958]: 326).

throughout the world, in order to distribute his theological concepts and ideas. So did the German Brethren. Visiting the geographically dispersed Brethren was seen as a necessity and in accordance with the New Testament and the journeys of the Apostles. The purpose of each visit was to preach, to counsel, to encourage, to admonish, and last but not least to share theological insights (cf. Jordy 1981: 16). Carl Brockhaus and others quit their jobs and started to travel throughout Germany. It was common sense that only men in their 40s, who had "*proved themselves in family life, job and church*" (Jordy 1981: 17) could be called into "full-time service" by older Brothers or local Brethren communities. There was no payment by an institution or association of any kind, their income was completely dependent on donations by other Brethren (cf. Horn 1977). This risky sacrifice of a solid sustentation was considered to be a "*living in faith*" (Jordy 1981: 17). The "travelling brothers" were considered to be familiar with the bible and its exegesis. The coordination of travel routes and schedules was organized during their conferences held quarterly (Jordy 1981: 18). In 1932, 60–70 "travelling brothers" were in active service (Jordy 1981: 18), whereas today there are around 20–30 "travelling brothers" among the Closed Brethren in Germany.

144.3 Closed Brethren Musicking

As stated earlier, we understand congregational musicking as integral to the movement's dissemination. As the spatiality of the churchscape and the ecclesiological imaginations are mutually constitutive, so is congregational musicking not only an agent in articulating and confirming the delineation, but likewise constituent for the dissent.

Actions[26] of delineation that demarcate what is "Brethren" and what is not, can be outlined at three levels: Firstly, on the level of the Self, theologically imagined to be consistent of spirit, soul and body; secondly on the local level of the assembly, and thirdly, the practices and interdependency of the wider Brethren community, imagined within the context of a pluralist evangelical churchscape (cf. Kong 2006: 105). As dispensationalism is a fundamental dogma to Brethren churchdoing (Schwarz 2008), it also shapes the theological imagination of music and the cultural performance of Brethren spirituality. Therefore, the imagination of dispensational time-spaces can be read as a fourth level of Brethren delineation that intersects with the three others. In other words, the self-positioning within the "dispensation of grace" gives rise to cultural practices of musically imagining the Self, performing the worship service and musicking an imagined Brethren community within global evangelical churchscapes.

Theological imaginations are believed to be authoritative and primordial thus, theoretical concepts of music as well as the Self, the church and the secular are domi-

[26] Referring to Brian Alleyne (2002: 608) Shelemay points out that the term community "is best understood in action" (2011: 364).

nant in the discourse about congregational musicking. However, the ethnographic interviews as well as participatory observation give insight into the entanglement of concepts and practices in congregational musicking.

144.3.1 Moral Topologies of Self

In terms of Brethren theology, human consists of body, soul and spirit. This trichotomy serves as an epistemic baseline for the imagination of "sanctification" and gives rise to their understanding of music's modes of operation. As impressive consistently rendered by all interviewees as well as preached in gatherings and found in literature written or commonly adapted by Brethren on music (Bäumer 1984, 1988; Graul 2010; Heide 1986; Lessmann 2007; Liebi 1987), the spirit is associated with the text of the hymn, for the soul, it is the melody that moves it, while the body is imagined to be subject to the rhythm. One interviewee describes the succession of these three as top down. Thus, spirit, soul and body ought to be in perfect harmony, but hierarchically organized, in a way that the spirit (identified with rationality and comprehension) controls the soul (the bearer of feelings) and the body.[27] In the same way, music is discursively stratified with the text as pivotal and superior in the theological imagination. The individual ought to rationally understand the staves and consciously direct his/her singing towards God in prayer (cf. Kelly 1985: 148 pp; Briem 2004: 316).

> I can discern the text with my spirit [mind]. Especially spiritual contents. These are words – "words of the bible" – spiritually inspired words of God. They bring me in conjunction with God. But generally speaking, words are more. In contemplation – this is always difficult: does it belong to the soul or the spirit [mind]? – with my mind I can interpenetrate [these Words]. I can weigh them and draw inferences from them; consciously reflecting on them – words.

As in German language "spirit" and "mind" can be both described with the term "Geist," they become interchangeable and the emphasis on the spiritual appears as well as an emphasis on the mind. "Holy" thoughts should only derive from an understanding of the text that is supported and enunciated by the melodic movement. Thus, melody[28] ought only to lead to an anticipated intention of the poetic text and its imagined musical "character." Accordingly, one interviewee states that "*feelings are very dangerous for the human being. They can one tear apart from the Lord.*"

[27] As mentioned before, the theological imagination of the Brethren is fundamentally based on dichotomies. In the trichotomy of human, soul and spirit are excluded and identified with other dualisms. Thus, soul and body are linked and designated as the place of "feeling" whereas rationality is associated with the spirit (cf. footnote 18).

[28] When discussing melody, the interviewees seemed to understand the term as encompassing harmonic structure, as melodies always occur in the musical context of a four-part harmonic texture. Compare Lessmann (2007), in which the Brethren author identifies the soul with melody and harmony.

Ingalls (2011: 265) identifies this ontological divide of text and music as a dispositive in evangelical discourse about music, including its inference "*that music, in and of itself, is a morally neutral carrier of the Christian message.*" However, along these lines, Closed Brethren are not "evangelical" since music is not understood as neutral but rather essentially carnal and, therefore, spiritual. Hence, Closed Brethren discourse about music and musical discourse appears as a search for putative appropriate sounds.

According to the interviewees, rhythm, which is believed to move the body, is the most carnal and, therefore, most demonized musical category. If the body and its lusts are not spiritually and thus musically restrained, a "tension" comes into being that is imagined to be ecstatic, which hinders the ability to understand the texts of the hymns. Here, another argument is posed as music is moralized in accordance of what is believed to be "natural" and within the "order of creation," opposed to chaos, identified with the "devil" as taught in Brethren sermons. What is believed to be a "*natural expectation of rhythm perceived by the human body*" (Lessmann 2007 (2): 32) is a 4/4 beat with its repetition of "*relaxation*" on the second and fourth beat. In contrast, music that is rhythmically determined to be "*unnatural*" such as "*afterbeat,*" "*backbeat*" and "*offbeat*" is therefore believed to "*dope the nervous system*" and stimulate the compulsion to move.

> Depending on, how vigorous and permanent such rhythmic patterns are used, the balance of spirit, soul and body as a natural given in accordance to the creation, is violated and shifted to the disproportion of bodily stimulation (Lessmann 2007 (2): 33).

This is why syncopated and offbeat music is successfully rejected, at least in Sunday worship services and the discourse about music manifests itself in the musical discourse. Here, cultural abstinence of specific musical sounds serves as a temperance of the body (cf. Bailey et al. 2007: 143). Hence, the practice of singing without allegedly strong rhythmic patterns, can be read as actions of disembodiment, as the believer supposedly worships in "spirit and truth."[29] This musical practice causes there to be an imagined border to the "dispensation of law" as the Old Testament worship is believed to have been out of an "emotional drive" to offer God one's emotional content, whereas the New Testament believer of the "dispensation of grace" ought to worship in a spiritual manner with his "*heart*" (Gerlach 1994: 153). To "*worship in spirit,*" therefore, is within the Brethren discourse, contradicted with a "*carnal, external, ritual, ceremonial worship*" (Briem 1990: 227). The Brethren churchscape on the level of the Self is theologically constituted by the delineation of individual musical involvement, marking out a moral topology of the Self and of the musical sound.

Participatory observations in worship services, however, reveal a different story, where believers worship appears not just "rational" or spiritual but as well

[29] See John 4:24. Darby (1992: 3, 441) writes: "[…] the worship of their hearts must answer to the nature of God, to the grace of the Father who had sought them. Thus true worshippers should worship the Father in spirit and truth."

"emotional"[30] and somatic, thus rather holistic. Depending on the course of contributions given during the service, it occurs that a man, even in between sung verses, would raise his voice to invite the whole assembly to upraise and sing the remaining verses while standing. Hymns often sung that way are *Dem der uns liebt* (KSGL, 126), *Jesus Lamm Gottes, in Herrlichkeit droben gekrönet* (KSGL, 129) or *Die Ruh' auf immerdar* (KSGL, 166).[31] While singing, some individuals close their eyes or weep. Furthermore one interviewee phrases: "*when I had the feeling that the service was notably blessed or especially spiritual, by tendency more hymns were sung*."

Even though it would be misleading to argue that these emotional and somatic responses are triggered by music only, however we contend that they are entangled with music. Thus, music gives rise to emotional responses in times and spaces of worship.with music. This is supported by assertions of the informants who on the one hand argued that emotions are rather problematic, but on the other hand acknowledged their own affective involvement in singing: "*sometimes I need to weep sometimes I don't, even though it is the same song.*" As musical practices and theological imaginations are conflicting, the delineation on the level of Self appears as an ever contested borderline.

144.3.2 Musicking the Sacred

When it comes to religious music, what is relevant is the notion of the sacred (cf. Bohlman 2005: 3; Ingalls 2011; Engelhardt 2012: 299). According to Closed Brethren spiritual imagination, the sacred is not attached to the physical world, but located in transcendent "heavenly" spheres. Like other Christian denominations, the Brethren conventions claim to not have physical sacred places, persons, holidays or objects (cf. Ingalls 2011: 258). In fact, what is believed to be sanctified spiritually is the believer himself/herself[32] as well as the church, the community of all "saints".[33] But as everyday life and services are said and experienced to be imperfect, the notion of a "not-yet"[34] is stressed in Brethren discourses and the sacred is perceived as a spiritual truth. Consequently the assembly, as the local representation of the Church, produces the ecclesiastical site as a non-place, of constant pilgrimage, a place for passengers awaiting rapture in prolonged sanctification (cf. Augé 1995:

[30] We understand these categories as discursively constructed in Brethren communities. They parallel many other (secular) discourses where this dichotomy stands as an episteme.

[31] Whom, who loves us (KSGL, 126), Jesus Lamb of God crowned in glory aloft (KSGL, 129) or Peace forevermore (KSGL, 166).

[32] See Col 3:12 (cf. Darby 1992: 5, 52).

[33] See Eph 5:27. Darby (1992: 4, 466) writes: "[...] Christ sanctifies the assembly for which He gave Himself. [...] Thus in sanctifying the assembly He must needs to cleanse it. This is, therefore, the work of the love of Christ during present time, but for the eternal and essential happiness of the assembly."

[34] See 1. Cor 13:12.

103). Among Brethren, the world is mentally mapped as a "desert"[35] where there is nothing attractive but all is transient.

The meeting rooms of the Brethren do not have a stage or presbytery that separates church members between listeners and performers as in most other Free-Churches; rather the whole assembly performs and listens at the same time thus every person is equally engaged in musicking. One interviewee reflects this by phrasing: "*the attendees are the stage*." The only spatial division made is according to gender: men and women sit divided in blocks compliant with their agency in the services. While men lead the service, women are silent apart from joining the singing.

The focal point of the assembly room is a table on which bread and wine are placed. Since the tenet of the "*Lord's table*"[36] as an expression of the "Unity of the Church" is central to Brethren theology, it was repeatedly pointed out by informants that the very physical table with bread and wine, spatially central to the church, is not to be confused with the theological imagination of the "*Lord's table*." This possible concurrence of theological imagination and spatial practice finds a parallel in other cultural practices, for instance in clothing, since the Closed Brethren highlight the necessity to attire oneself appropriately considering the presence of the Lord during worship services. Even though any notion of the place being sacred is rejected, the physical space and the spiritually imagined sacred space appear to be nested together, thus a quasi-sacred place emerges.

The Sunday worship service plays a crucial role in the ecclesiological imagination where "brothers and sisters" (or "siblings" as they call themselves in German language) assemble. In doing so they constitute the place of the local church, accepting the promise "*for where two or three are gathered together unto my name, there am I in the midst of them*"[37] as true. The regular Sunday service consists of scripture reading, communal singing, saying of prayers, moments of silence ("*spiritual collectedness*" [*Sammlung*][38]) and Holy Communion. Any alleged liturgy is rejected "*as this is a sign of* [spiritual] *weakness of an assembly*" as an interviewee explained. A distinct feature of Darby's theological understanding was the emphasis of a "*general priesthood*" of all believers and a turning-away from any notion of clericalism (Jordy 1981: 15) or hierarchical structures. Until today the Brethren do not have trained pastors;

---

[35] This is based on the rendering of an exegetical landscape as a spiritual and moral landscape of the present world. Following this exegesis, the situations of the Israelites, God's people, serve as metaphors to understand the spiritual circumstances of today's Christians. The Israelites travel through the desert (the Christian life as pilgrimage) out of Egypt (leaving behind the evil and worldly system) facing Canaan, which is construed as heaven, the ultimate goal, finally reached in death. Yet, based on the interpretation of Eph. 1,3, Canaan (the heavenly place) is understood to be already "entered" spiritually, framing the mundane life of the pilgrim. This spiritual reality however, is not yet revealed for non-Christians to see.

[36] See 1. Cor 10,21.

[37] See Mat 18:20.

[38] The term "Sammlung," which is still in use in Brethren culture, is derived from scholastic teachings and has been described by theologians as "balance between the inside and the outside" (cf. Dettloff 1965).

instead every man is encouraged to serve as lay preacher. In Brethren spiritual imagination, the Holy Spirit guides and ministers the church service through the voices of it's male participants. This as well concerns the spontaneous selection of hymns.

While reading and praying are done by male individuals only, the singing of hymns from a hymnal[39] (and of course the silence) is a mutual act of the assembly as a whole. When asking interviewees about Brethren congregational singing, they first and foremost mentioned the practice of four-part singing, a capella, using the hymnal *Kleine Sammlung Geistlicher Lieder*.[40] In the beginning of the Brethren movement, one of the first hymnbooks compiled by Carl Brockhaus (1853) had only a noted melody. However, the second edition (1858) contained 119 hymns including a four-part texture (Karrenberg 1962: 5; Gerlach 1994: 144). Since then, the hymnal has evolved, as new hymns were added and became canonized sporadically. Today, the hymnbook consists of 250 hymns in four-part texture of which the last 75 were added in 2012.[41] Accordingly to an announcement by letter to all German Closed Brethren assemblies of the hymnals extension in February 2011, it was the aim of an all-male committee that the new hymns added were a continuation of the existing ones regarding their musical style. Yet textually the new hymns were intended to be thematically supplementary to the existing ones, but as well on the same "high spiritual level" as claimed by the committee themselves.

As individuals join in the singing by freely choosing their musical part, the theological imagination of being part of the "body of Christ" is sensually evident for the participants. Here, the perceived musical harmony apparent in the complementary polyphonic character of four-part harmony as well as in its movement, coherent in itself, contributes to a sought communal "harmony" and theologically imagined unity. Thus, "harmony" is understood as a musical principle of mathematical order identified with God in reference to nature as God's creation, whereas "disharmony" is demonized and identified with Satan and "*sin as disorder of harmony*" (Blankenburg 1975: 207). Here, music is theologically imagined and thus opens up a discourse, where musical sounds and moral values coincide (cf. Kong 2006). Furthermore, this is emphasized through aesthetic value-judgements as one Interviewee describes the singing in his assembly of origin: "*As there was no instrumental accompaniment and the hymns were multipart compositions, the assembly was capable of singing multipart. Multipart singing was cultivated in a way that even without Instruments,* [...] *high-valuable music developed.*" In Brethren theo-

[39] The haptic appearance of the hymnal is quite similar to the widely-used "Elberfelder" Bible, that is, commonly bound in leather and personalized with the name of the owner engraved on the outside cover.

[40] "Small Collection of Spiritual Songs," abbreviated as KSGL. For a more extensive account on the development of the hymnbook among the different strands of Brethren dominations, see Karrenberg (1962). In other accounts on the Brethren movement, much attention was paid to the text of the hymns, as in Gerlach (1994: 143), Jordy (1979: 106 pp, 1986: 357). Here, we focus on singing and musicking practices.

[41] At the time of this study the edition has not yet been published. Therefore, the 75 hymns added were not included in the study. The hymnal was published in September 2012.

logical and musicological imagination, harmony is in many ways entangled with moral beauty as well as aesthetic beauty. This is especially based on the biblical metaphor (cf. Rev. 21: 1–2) of the body of Christ being an adorned bride. As notions of unity and harmony are essential to Brethren ecclesiological imagination, multipart singing is its tangible performance, a "*collective effervescence*" to use Durkheim's (1976) term. This experience seems often to be labelled as a "spiritual experience" rather than the experience of communality (cf. Ingalls 2011).[42]

To Closed Brethren, the praxis of multipart singing without instrumentation[43] musically marks out the position of the local assembly in the dispensation of grace. This is due to the theological imagination of instrumentation being part of the "dispensation of law." Analogous to the differentiation of these two dispensations on the level of the Self, the employment of instruments is believed to represent an emotional steering towards God, as practiced in the Old Testament.[44] One interviewee pictures this:

> There were musical instruments. There was a shouting that did resound in the far distance because it was external. It was not worship in spirit and in truth. In truth, because it was conform to God's word, but it was not a spiritual worship.

As vocal music is based on the utterance of words believed to be rational and superior to the musical sound, congregational musicking in Closed Brethren churches is appropriated to the "dispensation of grace" in the lack of use of musical instruments.[45] However, instruments[46] are not morally banned or demonized in general. In fact they play a vital part in the everyday-lives of many Brethren families and even serve as markers of a Christian identity beyond denomination. Therefore the use and disuse of instruments demarcates the divide between the private and the congregational sphere in fact not on the level of morality, but theology and spirituality. Thus, on the local church level the musical segregation of the "*everyday geographies of civic and domestic space*" and "*formal spaces of worship*" (Bailey et al. 2007: 142) back the constitution of the Closed Brethren churchscape.

[42] This is especially noticeable in reports coming from mostly younger Brethren people attending larger bible conferences, which seem to be 'spiritual mass events'. Ingalls describes it as "the idea, that the conference gathering was an experience of heavenly community on earth" (2011: 260).

[43] Gerlach indicates that at some assemblies a pump organ was allowed (Gerlach 1994: 153). However, according to the informants, this instrument is not found in Brethren churches anymore.

[44] Gerlach (1994: 153) cites a letter from 1990 where one of the Closed Brethren gives a detailed description of the practice of not having instruments in worship services that is analogous to the statements we recorded in interviews and informal conversation.

[45] Not all informants agreed to this and gave a rebuttal based on their own understanding of the Bible. But still, Brethren assemblies do not have instrumental accompaniment during the time of worship service.

[46] This is true for instrumentation in general. Some instruments, however, are demonized, esp. electrified instruments of the modern age.

144.3.3 Among Others: The Moral Geographies of Music in the Closed Brethren Churchscape

The third level of the Closed Brethren community, imagined by Brethren themselves as a global assembly, is that of denomination. In the making of the Brethren movement, the musical practice of hymn singing established a tangible continuity between the churches the Brethren had left and whose theology was rejected. Even though some new hymns were composed,[47] most other melodies and their movements were adapted from the repertoire of Lutherans and other Protestant churches (Gerlach 1994: 153). Reverting to traditional sources while acting on the authority of the reformation correlated to the *zeitgeist* of the mid-nineteenth century (Stalman 2001: 98; Morath 2001: 114). As among the emerging Brethren assemblies, the ideal of four-part a capella singing (Stalman 2001: 105), the recycling of old melodies with new texts (Morath 2001: 115), and church specific adaptions of theological content (ibid. 2001: 116) were practiced in other Protestant churches and in other separatist communities (for example the *Gemeinschaftsbewegung*). Serving as a marker of denomination was the *canon* of hymns compiled by Carl Brockhaus in the hymn book *KSGL*. Here as in other churches, "*the 'right' songbook functioned as a weapon in quarrels over which denomination or church provided the true path towards righteousness*" (Holzapfel 2005: 177). By virtue of the travelling brothers, the loose collection of hymns was disseminated among the Brethren churches of the German-speaking world and soon canonized, marking out the Brethren churchscape. The various strands of Brethren domination that arose from separations over theological discrepancy are paralleled by the publication and use[48] of hymnbooks (Karrenberg 1962). Today, only the Closed Brethren sing from the hymnal *KSGL*. One Interviewee described the prominent role of the hymnal according to his experience a few years ago: "[...] *some clung this tight on to the hymnal as if it was the bible itself and as if the hymns were inspired and so on. No one would phrase it this way but the handling of the hymnal was suchlike.*"

While Closed Brethren churches adhere to their hymnal, the landscape of Christian congregational musicking that surrounded the Closed Brethren massively changed (cf. Morath 2001: 116). In the wake of global shifts and an "Americanization" (as expressed by one interviewee) of popular culture, congregational musicking in German Free Churches became highly influenced by contemporary Christian

[47] Wilhelm Brockhaus, brother of Carl, set 28 poems to music. Besides this many other texts were appropriated to the tenet (Gerlach 1994: 153).

[48] Different Brethren groups have used same hymnals but for different purposes. For example, the 2007 published hymn book "Loben" ("Praise") was a cooperation of Closed and Block-Free Brethren, used in both groups. While the Block-Free Brethren employ the hymn book in their Sunday services, Closed Brethren only use it outside of the official church meetings, such as for camps or in choirs, because the music is imagined as being too profane and "worldly".

worship music[49] that had its roots in the English-speaking world.[50] Today, most churches integrate these pop music oriented hymns in their services, thus bridging discrepancies between mundane and congregational musical spaces. In contrast to Brethren congregational musicking, most of these Free Churches in Germany have teams of lay musicians that lead Sunday worship services.

Closed Brethren assemblies do not join in this musical discourse of contemporary Christian worship music.[51] The exclusion of specific musics and musicking is precipitated with strategies of reaffirming the own position thus musically highlighting denominational distinctions. This is particularly apparent in dealing with the hymnal *KSGL*. During the lifetime of Carl Brockhaus, who himself wrote the words of 53 hymns, the hymnal was extended four times, growing from 83 to 137 hymns. After his death the ninth edition of the book was released in 1909 with 147 hymns (Gerlach 1994: 144).[52] Afterwards the practice of regular extension was not continued for almost 80 years. The 33 hymns that were newly added to the hymnal in 1986 pursued the musical discourse by confirming it, instead of using melodies coming from contemporary Christian churchscapes, as Brockhaus had done.[53] Old traditional hymns were added such as the popular "*Oh Haupt voll Blut und Wunden*"[54] thereby authenticating J. S. Bach as truly Christian, drawing on his alleged musical authority (cf. Shelemay 2011: 377). Others were appended from the Swiss German Closed Brethren hymnal thereby extending the churchscape of the Closed Brethren on the level of musical discourse.[55] The only known

[49] For a more detailed account on Worship music see: Ingalls (2011). Worship music can be categorized as Contemporary Christian Music (CCM), a genre label for a musical movement formerly known as "Jesus music," which basically subsumed popular music with Christian lyrics (cf. Livengood and Book 2004: 119). For an extensive account on CCM: Gormly (2003) and Howard (2004).

[50] In contrast to many other musical genres, music that is identified as Christian music, is "defined by its lyrical content instead of musical style" (Livengood and Book 2004: 119).

[51] This is only true for the Sunday worship service of the Closed Brethren. Many groups among the Open and Block-Free Brethren churches have teams of musicians that accompany the singing, as in other Free Churches, even though they mostly do not lead in choosing the hymns. Outside the worship service, many listen to 'profane' music, play in bands and even in Brethren hymnbooks such as "Loben" (2007), hymns are included that can be identified as being contemporary Christian worship music.

[52] Gerlach (1994: 144) gives the following dates: 1st ed. 1853: 83 hymns; 2nd ed. 1858: 115 h.; 3rd ed. 1863: 123 h.; 5th ed. 1870: 127 h.; 7th ed. 1891: 135 h.; 8th ed. 1898: 137 h.; 9th ed. 1909: 147 h.

[53] While Brockhaus wrote many new hymns, the extension of 1986 has only very few newly written hymns.

[54] Paul Gerhardt (1607–1676): "O Sacred Head, Now Wounded" that became popular through J.S. Bach's St Matthew's Passion.

[55] In terms of music, it is noticeable that globally all the hymnals of the Closed Brethren churchscape tend to be the same in terms of musical and theological content as well as outward appearance. This includes the French (also used throughout African countries), the Portuguese, the Romanian and the Arab hymnal.

hymn being a more recent composition was "*Herr zu deinen Füßen.*"[56] Even though melody and movement (and text) were newly written, it draws back on the musical style of the older hymns. By maintaining coherence the hymn confirms the Closed Brethren musical discourse. Just as the older hymns, the "new" one is strophic, has a four-part musical texture, is sung slowly and solemnly, and displays alterations in the melodic and harmonic structure that are reminiscent of ecclesiastical modes.

In doing so, musical history blurs and the hymns appear to be heterochronic rather than chronic. This is supported by the dominical vocal praxis, as the various hymns[57] are mostly sung in the same slow and solemn manner.[58] The heterochronic character of the hymnal is deliberately supported by the lack of information about author, composer or time of origin; hence the hymns appear to be beyond time.[59] On the other hand, the adherence and assertion of the musical heritage of the pioneers of the Brethren movement, can be understood as a search for belonging. Theological and spiritual identity are tied up to a specific time and place, persons and ideas and not least to the musical style of Brethren hymns (cf. Ingalls 2011: 264).

The stagnation of musical evolution can as well be understood in conjunction with the theological imagination of time. According to dispensationalism the present time is designated as the "*dispensation of grace*" or "*church age*" (Sweetnam 2010: 211, 201). It is the time in-between, characterized by God's assurance to not change in his "dispensational acting." According to the verses "*and they persevered in the teaching and fellowship of the apostles, in breaking of bread and prayers*"[60] persistence of the assembly in the tenet and in the Sunday services is a practice of church-doing and of placing one's self within the time-space of the "dispensation of grace." Therefore traditionalism, theologically imagined as atemporal, is not a corollary, but an intended religious act that is in itself spiritual and in its materialization cultural thus musical.

However, as the musics of the Closed Brethren churchscape are contested by newly emerging musics, strategies of exclusion take place in order to reaffirm the own cultural values and practices as being correct and biblical. Thus, as music is spiritually transcended, the musical discourse itself becomes a moral one. Hence, specific instruments, sounds, rhythms and musicians are being perceived as moral hazards (Kong 2006: 104). The reasoning runs along two different paths. On the one

[56] KSGL No. 170 by Christian Briem.

[57] www.bruederbewegung.de quantifies the KSGL hymns: 60 % originate in the nineteenth century, 20 % in the seventeenth century, 8 % in the sixteenth century and 3.5 % in the fifteenth century.

[58] The practice of singing is complained about by many members of the assemblies. As a reaction in some churches sessions were appointed to exercise the hymns.

[59] Today in many assemblies of the Open and Block-Free Brethren, the hymnal "Glaubenslieder" (1952) with 498 hymns is used that comprises the hymns of KSGL. Here the name of the author and composer is given and therefore the disguise of the time of origin is another praxis of self-othering.

[60] Acts 2:42.

hand one is concerned with the Self and the potential influence of music on and its possessive power over spirit (mind), soul and body, as outlined above. Thus, music is seen as intrusive for the topology of the Self.

On the other hand, moral hazards are historically and geographically located in the world of the Other: the imagined place of origin of today's rock musics sounds, rhythms, and instruments is linked to "Black Africa" as one interviewee asserts when elaborating on syncopated music,

> Rhythmic [music] or something with syncopation or the like, [...] this is not a good thing. As children, we would sing syncopated songs but we said this does not fit in the assembly. Till this day the Brothers retained this understanding, that this is just too strong – that the whole story – it is not deniable, that in the end this is all coming from Africa. This whole syncopated music with these strong beats. [...] The problems with black music are firstly these unbelievable strong rhythms by what means people entranced themselves. In Africa, even until today this is the same. They sit all night while drumming and after half a night they are all shaking. This sounds one kilometre through the forest. This is all common knowledge. And the slaves brought it with them to America and the whole modern music got started beginning with jazz, soul and rock. [...] It is not God behind all this but other powers.

The Self in Brethren discourse is understood to be deflected from spirituality with strong rhythmic pattern as outlined above. In order to support this contention, rhythmical instruments are repeatedly traced back in history as done by Heide (1986: 66) or Bäumer (1988). This imagined history of mainly "rock music" constitutes a spiritual geography of the world where "African music" is undifferentiated imagined to be (or to have been) thoroughly spiritually charged and in this demonized. This argumentation is engendered and paralleled by hegemonic colonial discourses of the "African Other." As "black people" are tendentially identified by Closed Brethren with demonized music, the verbal and musical discourse seem to be racialized and, in fact, profoundly racist.

Similarly but more subtle, a moral geography is constituted as the United States is in Closed Brethren discourse deemed to be the place of origin of music identified as "rock music," which is imagined to be born out of "African rhythms." As it is understood by many Brethren, the rebellious character of the 1950–1960s rock'n roll is grounded in the "Satanic forces" behind it. This is paralleled with the idea that American Christians are spiritually flat, since they admit these musical elements into their churches. As musical and cultural differentiation to a worldly lifestyle is no longer possible, American Christianity is imagined to be overwhelmed by its worldly Others. In this, American music is seen as syncretism of Christian and worldly elements. However, even as the cultural and musical context of "rock music" has changed, it is still understood as spiritually and morally "charged" and rhythmic instruments are still demonized among most of the Closed Brethren. Thus, music is essentialized and seen as statically connected to its cultural and spiritual place of origin.

As these rejected sounds can be found in other churches particularly in popular Worship music (Ingalls 2011) and its German adaptions, the segregation against other denominations and against a non-Christian world coincide; as one Interviewee states: "*if they* [Christians] *stand there in front on the stage and sway back and forth, where is there a difference to worldly performances?"* Whereas the marker of dissent originally was a theological one, now congregational hymn singing has become a musical marker of differentiation to religious systems that were and still are discursively rejected.

The Closed Brethren discourse about music can be interpreted in two ways. On the one hand it can be understood as trigger of the distinct musical tradition, as multipart singing without instrumental accompaniment or worship leader theologically imagined to be the only biblical way in the New Testament way of worshipping. On the other hand, one can argue that these discourses about music emerge from the musical discourse itself and the need, to authorise the own musicking as biblical.

As stated above, Brethren theological imagination implies the self-construction of a non-church and non-denomination. Thus cultural practices of church-doing can be understood as an act of self-othering, as music "*give*[s] *voice to dissent*" (Shelemay 2011: 370). The theological imagination of the Closed Brethren is deliberately constructed against the multiplicity of denominations (Jordy 1981: 59). Hence, Closed Brethren assemblies musically mark themselves as Other. The communal spaces of musicking shape the identity of imagined communities and thus involve strategies to expel deviant musics as "*moral hazards*" (Kong 2006: 104) from these shared imaginations, mostly through moral panicking. The key elements of such moral panic have been described by Goode and Nachman (1994) as being "*concern, hostility, consensus, disproportionality and volatility.*" Thus, moral geographies based on judgements of what belongs where are created. In the same way the Closed Brethren churchscape consists through moral geographies of "*dominative power of control and exclusion*" (Matless 1995: 396).

144.4 Conclusion

The practices of translation, the travelling brothers and the diffusion of theological concepts, congregational musicking and liturgical imaginations constitute the Brethren churchscapes. Yet, the diffusion of the particular Closed Brethren theology is driven by transformation especially on the level of Closed Brethren ecclesiology itself. On the one hand, the Brethren movement can be best described as a deterritorialized phenomenon in regard to its global and furthermore spiritual aspirations towards unity, both spiritually imagined and literally enacted in the Lord's Supper. As immanent to the particular Brethren concept of Church, the movement – from its early beginnings onwards – inevitably had to be more-than-local, in order to meet its own theological aspirations.

On the other hand, the diffusion of the movement is ensured by a strong inflection towards territorialized concepts of church-doing and a sustained solidification of liturgical practices. In effect, the cultural particularities of Closed Brethren liturgy – as it could be shown in regard to congregational musicking – and the practices of dissemination – as exemplified in the case of the travelling brothers -, are backed up theologically.

Today however, the growing fragmentation of Brethren churches reflects the effect of an increasing plurality and simultaneity of religious flows. The availability of diverse theologies through online sermons, the spread of new forms of church-doing that have adapted popular cultural flows, the local presence of these new movements often in the close neighborhood, a vast offer of theologically diverse bible schools and counselling facilities or the unmediated and instant availability of multi-denominational Christian music via music streaming platforms, allow an increasing cultural mobility of the individual Brethren congregants. Traditional constitutions of the Brethren communities are challenged and necessitate a continual justification of the own congregational form.

This cultural mobility, seems to further cultural conservatism among Closed Brethren communities for instance in respect to congregational musicking. It is due to this process of territorialisation that Closed Brethren congregational practices are increasingly culturally and not merely theologically distinct from their evangelical others.

It is emphasized, that what is considered sacred is intangible and by no means manifest in architectural structures, nor inherent to the liturgy itself or the musical texture, and certainly not embodied by a singular anointed person within the community. Yet, the cultural threshold between Closed Brethren everyday life and the Sunday worship service as well as between Closed Brethren and their 'Others', becomes charged with theological meaning and renders the allegedly intangible Sacred affective. The supposedly "not-sacred" space is performed as quasi-sacred. This is not to say that the theological becomes devoid of meaning. Rather, the affective and cultural substantially add to the intensity of divine encounters in times of worship and are not just its means.

In other words, the transition of theological and spiritual imaginations into cultural forms suggests, that Closed Brethren services and musicking are not culturally unmarked but are in fact heterotopic and – as they evoke both spiritual and mundane temporalities – also heterochronic. Closed Brethren musicking opens up a distinct tradition of musical discourse that in fact is connected to specific times and places and has become affective in reference to Closed Brethren spirituality. As the sacred is located in relation to a specific musical sound, the solemn four-part harmony without instrumentation and syncopation becomes a signifier for "true" worship. The spiritual experience is thus territorialized culturally and hence religiously as Closed Brethren distinct to other free churches in Germany.

A *close listening* to congregational musicking – that is in-depth music analysis based on auditory perception – would induce further insight in the struggles and strategies of musicking denomination or the intended discontinuation of denominational church-doing. Further research may explore the "*affective forces*" (Holloway 2011: 386) at play by focusing on aspects of sensations, felt attachment and "*somatic modes of attention*" (Holloway 2011: 387) in spiritual spaces. To understand these forces (either human or non-human) that may intersect in these spaces, an engagement with the qualities of atmospheres in church-doing and musicking may be fruitful. Furthermore, the fabrication and proliferation of spiritual atmospheres within global churchscapes may help to understand the contested zones of delineation, dissent and exclusion in religious practices – as in Closed Brethren assemblies through the conservation of specific atmospheres – as well as the intensities of affect in communal spirituality.

Acknowledgements We are grateful to Birgit Abels, John D. Boy, Stanley Brunn, Rashad Chichakly, Florian Neisser, Johannes Leder, Jürgen Pohl and Eva-Maria van Straaten for helpful comments on an earlier version of this article. We especially appreciated the confident collaboration of the interviewees who prefer to stay anonymous.

References

Primary Sources

Bäumer, U. (1984). *Wir wollen nur deine Seele*. Bielefeld: Christliche Literatur-Verbreitung.
Bäumer, U. (1988). *Rock – Musikrevolution des 20. Jahrhunderts*. Bielefeld: Christliche Literatur-Verbreitung.
Briem, C. (1990). *Da bin ich in ihrer Mitte – Die Kirche, nach dem Ratschluss Gottes und wie sie sich darstellt*. Hückeswagen: Christliche Schriftenverbreitung.
Briem, C. (2004). *Aus der Finsternis zum Licht* (5th ed.). Hückeswagen: Christliche Schriftenverbreitung.
Darby, J. N. (1992). *Synopsis of the books of the Bible* (Vol. 1–5). Kingston: Believers Bookshelf.
Graf, M. (1990). Der Weg der Absonderung. *Halte fest* (p. 294). Online available on www.halte-fest.ch/themen/die-versammlung/absonderung-der-weg-der-absonderung.html. Last accessed 14 July 2012.
Graul, A. (2010). *Rock-, Pop- und Technomusik und ihre Wirkungen. Eine wissenschaftliche und biblische Untersuchung* (2nd ed.). Bielefeld: Christliche Literatur-Verbreitung.
Heide, M. (1986). *Musik um jeden Preis?* Bielefeld: Christliche Literatur-Verbreitung.
Kelly, W. (1985). *Die Versammlung Gottes* (3rd ed.). Neustadt/Weinstraße: Ernst-Paulus-Verlag.
Kleine Sammlung Geistlicher Lieder (*KSGL*). (1986 [1853, 1858, 1863, 1870, 1891, 1898, 1909, 2012]). Wuppertal/Hückeswagen: R. Brockhaus Verlag/Christliche Schriftenverbreitung.
Lessmann, M. (2007). Musik – Geschenk oder Gefahr? Teil 1–3. *Folge mir nach, 6, 7 & 8.*
Liebi, R. (1987). *Rockmusik – Daten, Fakten, Hintergründe. Ausdruck einer Jugend in einem sterbenden Zeitalter*. Zürich: Beröa-Verlag.
Menk, F. (1980). "Brüder" unter dem Hakenkreuz. Das Verbot der "Christlichen Versammlung" 1937. Self-published. Reprint.
Menk, F. (1986). *Die Brüderbewegung im Dritten Reich*. Bielefeld: Christliche Literaturverbreitung.
Ouweneel, W. J. (1977). Die Geschichte der Brüder. 150 Jahre Versagen und Gnade. In *Uit het Woord der Waarheid*, Winschoten. (An undated reprint in German has been used)
Steinmeister, A. (2004). *...ihr alle aber seid Brüder. Eine geschichtliche Darstellung der "Brüderbewegung"*. Lychen: Daniel Verlag.
Weremchuk, M. S. (1988). *John Nelson Darby und die Anfänge einer Bewegung*. Bielefeld: Christliche Literatur-Verbreitung.
Weremchuk, M. S. (1989). *Ihr liefet gut... Nachgedanken zur Brüderbewegung*. Albsheim: Notting Hill Press.

Secondary Sources

Anderson, B. (1991). *Imagined communities: Reflections on the origin and spread of nationalism* (2nd ed.). London: Verso.

Alleyne, B. (2002). An idea of community and its discontents: Towards a more reflexive sense of belonging in multicultural Britain. *Ethnic and Racial Studies, 25*(4), 607–627. doi:10.1080/01419870220136655.

Appadurai, A. (1996). *Modernity at large. Cultural dimensions of globalization*. Minneapolis/London: University of Minnesota Press.

Arbeitskreis Geschichte der Brüderbewegung (Eds.). (2001). *200 Jahre John Nelson Darby* (Edition Wiedenest). Hammerbrücke: Jota Publikation.

Augé, M. (1995). *Non-places. Introduction to an anthropology of supermodernity*. London/New York: Verso.

Bailey, A. R., Harvey, D. C., & Brace, C. (2007). Disciplining youthful methodist bodies in nineteenth-century Cornwall. *Annals of the Association of American Geographers, 97*(1), 142–157.

Bainbridge, W. S. (1997). *The sociology of religious movements*. New York/London: Routledge.

Bister, U. (1983). Die Brüderbewegung in Deutschland von ihren Anfängen bis zum Verbot des Jahres 1937 – unter besonderer Berücksichtigung der Elberfelder Versammlungen. Inaugural dissertation, Philipps-Universität, Marburg.

Blankenburg, W. (1975). Der Harmonie-Begriff in der lutherisch-barocken Musikanschauung. In E. Hübner & R. Steiger (Eds.), *Kirche und Musik – Gesammelte Aufsätze zur Geschichte der Gottesdienstlichen Musik* (pp. 204–217). Göttingen: Vandenhoeck & Ruprecht.

Boddenberg, D. (Ed.). (1977). *Versammlungen der "Brüder". Bibelverständnis und Lehre, mit einer Dokumentation der Geschichte von 1937–1950*. Dillenburg: Christliche Verlagsgesellschaft.

Bohlman, P. V. (2005). Introduction: Music in American religious experience. In P. V. Bohlman, E. Blumhover, & M. M. Chow (Eds.), *Music in American religious experience* (pp. 3–22). Oxford: Oxford University Press.

Dettloff, W. (1965). Heideggers Analyse des alltäglichen Daseins und mittelalterliche Lebenslehre über die Sammlung. *Sein und Sendung: Zweimonatsschrift für Priester und Laien, 30*, 354–361.

Dewsbury, J. D., & Cloke, P. (2009). Spiritual landscapes: Existence, performance and immanence. *Social & Cultural Geography, 10*(6), 695–711.

Durkheim, É. (1976). *The elementary forms of the religious life* (Vol. 2). London: George Alien & Unwin.

Engelhardt, J. (2009). Right singing in Estonian Orthodox Christianity: A study of music, theology and religious ideology. *Ethnomusicology, 53*(1), 33–57.

Engelhardt, J. (2012). Music, sound and religion. In M. Clayton, T. Martin, & R. Middleton (Eds.), *The cultural study of music. A critical introduction* (2nd ed., pp. 299–307). New York: Routledge.

Eylenstein, E. (1927). Carl Brockhaus. Ein Beitrag zur Geschichte der Entstehung des Darbysmus. *Zeitschrift für Kirchengeschichte, 46*(9), 275–312.

Geldbach, E. (1972). *Christliche Versammlung und Heilsgeschichte bei John Nelson Darby*. Wuppertal: Theologischer Verlag Rolf Brockhaus.

Geldbach, E. (1989). *Freikirchen – Erbe, Gestalt und Wirkung* (Bensheimer Hefte 70). Göttingen: Vandenhoeck & Ruprecht.

Gerlach, R.-E. (1994). *Carl Brockhaus – ein Leben für Gott und die Brüder*. Wuppertal/Zürich: R. Brockhaus Verlag.

Goode, E., & Nachman, B.-Y. (1994). Moral panics: Culture, politics and social construction. *Annual Review of Sociology, 20*, 149–171.

Gormly, E. (2003). Evangelizing through appropriation: Toward a cultural theory on the growth of contemporary Christian music. *Journal of Media and Religion, 2*(4), 251–265.

Gregory, D. (2009). Geographical imaginary. In D. Gregory, R. Johnston, G. Pratt, M. J. Watts, & S. Whatmore (Eds.), *The Dictionary of human geography* (5th ed., p. 282). Malden/Oxford/Chichester: Wiley-Blackwell.
Geertz, C. (1973). *The interpretation of cultures: Selected essays*. New York: Basic Books.
Harding, J. (2006). "Come out of her my people" (Rev. 18:4). *The use and influence of the Whore of Babylon motif in the Christian Brethren movement, 1829–1900*. Dissertation. Liverpool: Liverpool Hope University.
Heinrichs, W. E. (1989). *Freikirchen – eine moderne Kirchenform. Entstehung und Entwicklung von fünf Freikirchen im Wuppertal*. Giessen: Brunnen Verlag.
Henkel, R. (2001). *Atlas der Kirchen und der anderen Religionsgemeinschaften in Deutschland. Eine Religionsgeographie*. Stuttgart: W. Kohlhammer.
Henkel, R. (2011). Are geographers religiously unmusical? Positionalities in geographical research on religion. *Erdkunde, 65*(4), 389–399.
Holloway, J. (2011). Spiritual life. In V. J. Del Casino Jr., M. E. Thomas, P. Cloke, & R. Panelli (Eds.), *A companion to social geography* (pp. 385–400). Malden/Oxford/Chichester: Wiley-Blackwell.
Holloway, J., & Valins, O. (2002). Editorial: Placing religion and spirituality in geography. *Social & Cultural Geography, 3*(1), 5–9.
Holzapfel, O. (2005). Singing from the right songbook: Ethnic identity and language transformation in German American hymnals. In P. V. Bohlman, E. L. Blumhofer, & M. M. Chow (Eds.), *Music in American religious experience* (pp. 175–194). Oxford: Oxford University Press.
Horn, H. (1977). Wilhelm Alberts (1823–1865). Ein Vorkämpfer der sog. Brüderbewegung aus dem Oberbergischen. *Monatshefte für Evangelische Kirchengeschichte des Rheinlandes, 26*, 167–186.
Ingalls, M. (2011). Singing heaven down to earth: Spiritual journeys, eschatological sounds and community formation in evangelical conference worship. *Ethnomusicology, 55*(2), 255–279.
Jordy, G. (1979). *Die Brüderbewegung in Deutschland. Teil I*. Wuppertal: R. Brockhaus Verlag.
Jordy, G. (1981). *Die Brüderbewegung in Deutschland. Teil II*. Wuppertal: R. Brockhaus Verlag.
Jordy, G. (1986). *Die Brüderbewegung in Deutschland. Teil III*. Wuppertal: R. Brockhaus Verlag.
Jordy, G. (2001). Carl Brockhaus: Ein Vater der deutschen Brüderbewegung. In Arbeitskreis Geschichte der Brüderbewegung (Eds.), *200 Jahre John Nelson Darby* (Edition Wiedenest, pp. 32–56). Hammerbrücke: Jota Publikationen.
Jordy, G. (2003). *150 Jahre Brüderbewegung in Deutschland*. Dillenburg: Christliche Verlagsgesellschaft.
Jung, A. (1999). *Als die Väter noch Freunde waren. Aus der Geschichte der freikirchlichen Bewegung*. Wuppertal: R. Brockhaus Verlag.
Karrenberg, K. (1962). *Die Liedsammlungen der "Brüder."* Online available: www.bruederbewegung.de/pdf/karrenbergliedersammlungen.pdf. Last accessed on 17 July 2012.
Kong, L. (2006). Music and moral geographies: Constructions of "nation" and identity in Singapore. *GeoJournal, 65*, 103–111.
Kretzer, H. (Ed.). (1987). *Quellen zum Versammlungsverbot des Jahres 1937 und zur Gründung des BfC*. Neustadt/Weinstraße: Ernst-Paulus-Verlag.
Leppman, E. J. (2005). Appalachian churchscapes: The case of Menifee county, Kentucky. *Southeastern Geographer, 45*(1), 83–103.
Liese, A. (2002). *Verboten, geduldet, verfolgt. Die nationalsozialistische Religionspolitik gegenüber der Brüderbewegung* (Edition Wiedenest). Hammerbrücke: Jota Publikationen.
Livengood, M., & Book, C. L. (2004). Watering down Christianity? An examination of the use of theological words in Christian music. *Journal of Media and Religion, 3*(2), 119–129.
Matless, D. (1995). Culture run riot? Work in social and cultural geography, 1994. *Progress of Human Geography, 19*(3), 395–403.
Morath, R. (2001). Das evangelische Kirchenlied. In W. Opp (Ed.), *Handbuch Kirchenmusik – Der Gottesdienst und seine Musik, Band I* (pp. 91–127). Kassel: Merseburger.
Nebeker, G. L. (2001). The theme of hope in dispensationalism. *Bibliotheca Sacra, 158*, 3–20.
Orth, J. (1977). Die Entstehung der Brüderbewegung. *Die Botschaft – Monatsschrift bibelgläubiger Christen, 118*(11), 255–260.

Schmitt, J. (1984). *Die Gnade bricht durch. Aus der Geschichte der Erweckungsbewegung im Siegerland, in Wittgenstein und den angrenzenden Gebieten* (3rd ed.). Gießen: Brunnen-Verlag. Reprint.

Schneider, M. (2004). *Die Sprache der "geschlossenen Brüder". Kennzeichen und Probleme.* Online available: www.bruederbewegung.de/pdf/schneidersprache.pdf. Last accessed on 17 July 2012.

Schwarz, B. (2008). *Leben im Sieg Christi. Die Bedeutung von Gesetz und Gnade für das Leben des Christen bei John Nelson Darby*. Gießen: Brunnen-Verlag.

Shelemay, K. K. (2011). Musical communities: Rethinking the collective in music. *Journal of American Musicological Society, 64*(2), 349–390.

Small, C. (1998). *Musicking. The meanings of performing and listening*. Middletown: Wesleyan University Press.

Stalman, J. (2001). *Kompendium zur Kirchenmusik – Überblick über die Hauptepochen der evangelischen Kirchenmusik und ihrer Vorgeschichte*. Hannover: Lutherisches Verlagshaus.

Sweetnam, M. S. (2010). Defining dispensationalism: A cultural studies perspective. *Journal of Religious History, 34*(2), 191–212.

Yorgason, E., & Della Dora, V. (2009). Geography, religion, and emerging paradigms. Problematizing the dialogue. *Social & Cultural Geography, 10*, 629–637.

Zelinsky, W. (2001). The uniqueness of American religious landscape. *Geographical Review, 91*(3), 565–585.

Chapter 145
The Bible, the Hymns and Identity: The Prophet Isaiah Shembe and the Hymns of His Nazareth Baptist Church

Nkosinathi Sithole

145.1 Introduction

The Nazareth Baptist Church (Ibandla lamaNazaretha or Nazaretha Church) marked hundred years of existence in 2010 and it can be argued that those have been years of engagement with colonialism, cultural imperialism and the Bible. Isaiah Shembe, who founded the Nazaretha Church in 1910, was an avid reader of the Bible, and his tremendous knowledge of the Bible is amazing because he was not educated in missionary schools. A number of scholars have commented on Isaiah Shembe's command of the Bible. Gerald West states, "That Shembe was familiar with the Bible is plainly apparent to anyone who listens to or reads his hymns and teachings" (2006: 163). One of the earliest scholars to study Isaiah Shembe and his Nazaretha Church, Esther Roberts (1936), writes that Isaiah Shembe was reported "to have been able to cite biblical references by chapter and verse, outwitting most European missionaries" (Quoted in West 2006: 163). What made this even more remarkable, argues West, is that "there is no clarity on whether or to what extent Shembe was literate" (163). West's point that while Shembe "seized" and "reconstituted" the Bible, it "also [took] hold of him, drawing him and his female followers to its narrative [and here I think not just his female followers]" (2007: 498) is validated by the fact that the Bible still plays a significant role in the life of the Church today. In this chapter I look at how Shembe "seizes" the Bible and uses it in his hymns (*izihlabelelo*) to negotiate his own identity in relation to colonialism, precolonial African life and missionization. I argue that in the Bible Shembe found, in Duncan Brown's terms, "a mode of spiritual power and personal articulation" (2006: 40–41), which allowed him to imagine himself as a Black Messiah while at the same time celebrating the life of Jesus Christ and rejoicing in being "saved" by him.

N. Sithole (✉)
Department of English, University of Zululand, KwaZulu-Natal, South Africa
e-mail: nkosinathi.sithole1@gmail.com

S.D. Brunn (ed.), *The Changing World Religion Map*,
DOI 10.1007/978-94-017-9376-6_145

As West has pointed out, there is a dearth of scholarship on Africa's engagement with the Bible: "an important task awaiting an African biblical hermeneutics is a comprehensive account of the transaction that constitute the history of the encounters between Africa and the Bible. While the accounts we have of the encounters between Africa and Christianity are well documented, the encounters between Africa and the Bible are partial and fragmentary" (2003: 65). In the similar vein Roland Boer laments the absence of the Bible in colonial and postcolonial studies, even though there is "perpetual, if not overwhelming, presence of the Bible in colonialism and postcolonialism themselves" (2001: 7). And Duncan Brown, coming from the point of view of literary scholarship, rather than theology and religion, has pointed to the central role the Bible plays in the lives of Africans. Drawing on the assertion made by Terence Ranger, he argues:

> Terence Ranger says, "any scholar who aspires today to 'think black' about many of the people in eastern Zimbabwe has to learn also to think Methodist" (1994: 309). While I do explore the specific influences of Methodism below, I would extend that provocative statement further by saying that any scholar who aspires to 'think in black' about the African (sub) continent has to learn also to think biblically. (2008: 81–82)

With its focus on the postcolonial subjects (ordinary readers) as agents who, in their encounter with the Bible produce "readings often at odds with, or resistant to, the normative discourse of the missionaries" (Brown 2009: 12), this work leans towards postcolonial studies. It deals with one of the religious groups that, in the words of Robert Young, "have taken on the political identity of providing alternative value systems to those of the west" (2001: 337). Even though Young's religions that "provide alternative value systems to those of the west" are only Islam and Hinduism, Brown points to the "abundant evidence of other religions, including various forms of indigenized Christianity, expressing "subaltern concerns" (2008b: 3–4).

145.2 What Is Postcolonial Studies?

Postcolonial Studies is a complex interdisciplinary theoretical approach, which "involves texts from different times, places, and cultures and whose boundaries often blur. It propounds a myriad of methods and theories, all of which examine literature and its participation in the building, collaboration, or subversion of global imperial relationships" (Dube 2000: 52–53). A Postcolonial hermeneutical approach, writes Jeremy Punt this time,

> includes and gives voice to the voiceless, the muted voices of the colonised, the marginalised, and the oppressed. It investigates and addresses disproportionate power relationships at the geo-political as well as subsidiary levels, at the level of the empire and the relationship between the imperial and the colonial, but also at social and personal level of the powerful ruler and the subaltern, to the extent of investigating relationships and interactions between the centre and the periphery (2010: 6).

The amaNazaretha members, and their leader Isaiah Shembe especially, who are the subjects of this study are indeed "the colonised, the marginalised and the oppressed," but not "voiceless" and, therefore, this study does not profess to give them a voice. Instead it examines the voice of the oppressed and renders it audible to the "oppressor" (both western and native), who, thanks to the work that has been done on the Nazaretha Church (and other African Initiated Churches) is not completely unaware of the presence of this voice. I am also mindful of the significance of Brown's call for a "South (or periphery) centred" approach to postcolonial studies: "rather than subjecting inhabitants of the postcolony to scrutiny in terms of postcolonial theory/studies, how can we allow the theory and its assumptions also to be interrogated by the subjects and ideas it seeks to explain?" (2009: 9). While this call is laudable, it is unclear who the "subjects" and their ideas are. Is a biblical scholar in the postcolony the same as an ordinary reader of the Bible like Isaiah Shembe? This suggestion seems to imply that there is a certain known point where the "aims and deepest aspirations" (to use Chinua Achebe's words), of the academic and the ordinary person in the South meet (1989: 44). That said, while Isaiah Shembe's reading of and engagement with the Bible would definitely unnerve a number of people in the South (both academic and ordinary readers), it is, however, an example of the South centered approach Brown is calling for.

In the similar vein West proposes a refocus of biblical scholarship on the impact Africa has had on the Bible, rather than the impact the Bible has had on Africa. Drawing on Kwame Bediako's statement that, "Further developments in African Christianity will test the depth of the impact that the Bible has made upon Africa" (1994: 252), West comments that:

> Bediako's statement points to the significant role the Bible has played in the formation of African Christianity. Unfortunately, this formulation perhaps gives the impression that the encounter between the Bible and Africa is in one direction: from the Bible to Africa. The Bible, in this formulation, is the subject and Africa is the object... But, what if we make Africa the subject and the Bible the object? We would then have the following formulation: Further developments in African Christianity will test the depth of the impact that Africa has made upon the Bible. (2000: 29)

The engagement with and transaction between Isaiah Shembe and the Bible that is explored here is epitomized in the "conversation" between Gerald West and a member of the Church in which West had spoken in the launch of *The Man of Heaven and the Beautiful Ones of God* (Gunner 2002) about how Isaiah Shembe was "a remarkable re-memberer of the Bible" (West 2006: 179). The member of the Church rebuked Gerald for his statement, stating that, "We do not interpret the Bible, the Bible interprets us" (179). This interpretation and counter-interpretation, the church's action upon the Bible and their being acted upon by it, or, this "mutual engagement" (Peel 2003: 1) constitute the relationship between Isaiah Shembe (and his Church) and the Bible which is the focus of this chapter.

145.3 A Brief History of Ibandla lamaNazaretha

Ibandla lamaNazaretha is one of the largest and rapidly growing African Initiated Churches (AICs) in South Africa. It was founded by Isaiah Shembe around 1910 in what is now KwaZulu-Natal (Fig. 145.1). Although it started as a local and

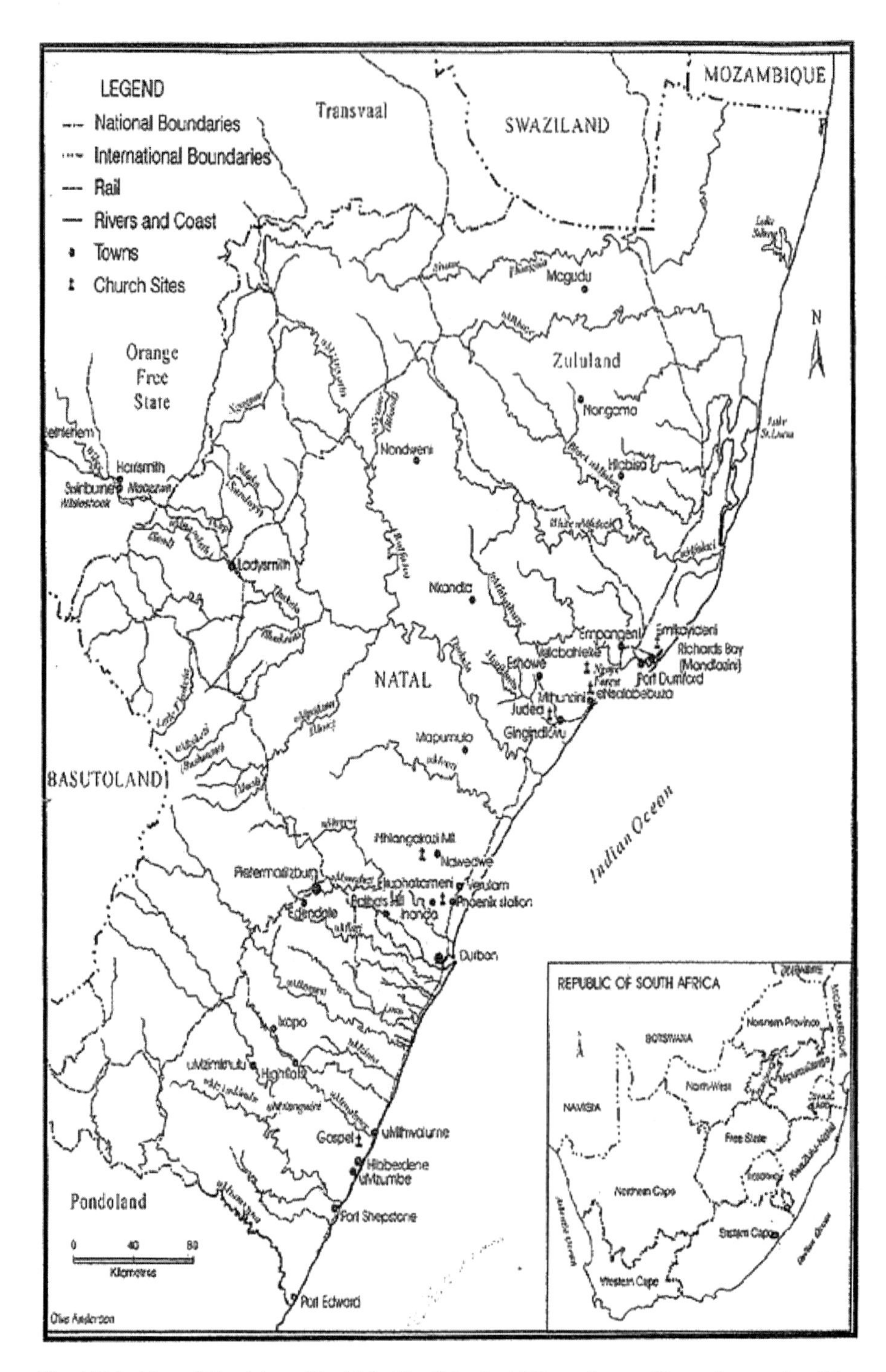

Fig. 145.1 Map of Natal (now KwaZulu-Natal) in the 1930s and some Nazaretha temples (Map by Nkosinathi Sithole)

ethnically specific church, today Ibandla lamaNazaretha has a strong national following; attracting members from all walks of life across ethnic groups, and has members as far afield as Mozambique and Zimbabwe. The church is rapidly growing in numbers. According to Preacher Sibisi, there are all in all about seven million Nazaretha members. This growth, plus the church's "success in creating a religious presence which is distinctively African," causes Gunner to see it as a "force to be reckoned with in social, religious and political terms" (2002: 1).[1]

Isaiah Shembe forged his church by blending Christian and African forms, many of the latter being downgraded and prohibited in mainstream churches (Fig. 145.2). As Kunene argues, "Whilst keeping the Christian principles the Shembites have retained some of the traditional customs and practices" (1961: 197–198). Isaiah Shembe's theology clashed with that of the nonconformist missionaries, whose intention it was, as the Comaroffs have noted, "to "civilise" the native by remaking his person and his context; by reconstructing his habit and habitus; by taking back the savage mind from Satan, who had emptied it of all traces of spirituality and reason" (1991: 238). Some of the main issues of conflict were African song and dance, polygamy, and recognition of the ancestors. One of the most significant expressive forms in the Nazaretha Church is *umgidi* (the sacred dance) in which the performers' dress includes loinskins, headties and other attire made from animal hides. The sacred dance itself involves the beating of cowhide drums and the singing of hymns or songs composed by Isaiah Shembe, and is arguably an improvization on the dances that took place in pre-colonial society, which were labelled "uncivilised" and "anti-Christian."

The hymns that were in existence in isiZulu prior to Isaiah Shembe and during his time were mainly translations of the English hymns from Orthodox churches; something which Bengt Sunkler has lamented:

> One of the most striking – and disconcerting – examples of the White man's dominance even in his spiritual matters over his Zulu co-religionists is this fact that Zulu Christians have not felt led to express their new faith in the composing of songs and hymns of their own, whereas this is quite common in certain Mission Churches in East Africa. (1948: 193)

B. W. Vilakazi has lauded P. T. Gumede and N. Luthuli for their contributions in the field of hymns in isiZulu for their *Amagama okuhlabelela* (American Board Hymnbook), even though their influence was limited. But Vilakazi's statement that, "After these two there is a great break, up to now. The field of hymns seems to be dead" (Quoted in Gerard 1971: 187), is very disturbing considering

[1] While the Church has been marginalised in the past, this has changed in the new South African context. The Nazaretha Church has been visited by all the Presidents of the new South Africa and is highly regarded for its upholding of African values. For instance, in 2004, some African Americans came to South Africa supposedly in search of their African roots and the KwaZulu-Natal government sent them to EBuhleni where a cow was slaughtered for them and they took part in the sacred dance. But, it also needs to be noted that the popularity with the politicians has to do with the fact that as more and more people join the Church, the politicians see potential voters in them.

Fig. 145.2 Prophet Isaiah Shembe (1870–1935), as he appears in a church hymnal

the fact that he was aware of Shembe's hymns but seems to dismiss them as he dismisses the Nazaretha Church itself in the following statement:

> The followers of the zionist movement have incorporated into their services most of the first-fruit ceremonial observencies, in the purification of the priest or king, colourful dresses and community singing, mixed with dancing, consisting largely of rhythmical raising of the feet, thundering stamp upon the ground, and a series of grotesque shuffles, interspersed with vigorous leaps by leaders of the groups. This has an attraction for the average Native, and he eagerly supports such a movement, for he has an active part to play, besides the priest. (Quoted in Gerard 1971: 192)

Even though Vilakazi holds this hostile view of the Nazaretha Church, which was common among the educated *kholwa* (believers) of his time, and does not see Shembe's hymns as bridging the gap he has noted in the field of hymns, but he did acknowledge Shembe's achievement. "Shembe composed for his sect a number of songs. If one reads through a typical one, the VIIIth, one notices the African atmosphere pervading it, in its poetic figures of speech" (1946: 175). So it seems that for Vilakazi these were mere songs, not hymns, because the people who created them and performed them did not conform to the orthodox Christian way of life.

Isaiah Shembe's contribution was unique in the sense that he created his own original hymns, not ones based on the hymns of other churches. In doing so he drew on pre-colonial African song and the Bible, and used the genre of the hymns to articulate his own feelings and ideas. Gerard has remarked that, "[i]t is in his hymnal poetry that Shembe makes his unique contribution to Zulu literary history" (1971: 188). Reading these hymns as "literary texts of extraordinary power and vision" (Brown 1999: 197), I am interested in the way in which Shembe utilises the Bible in making his contribution and in creating his literary texts.

Isaiah Shembe was a self-taught musician. Bongani Mthethwa, in an interview with Carol Muller, has described Isaiah Shembe as "a great maskanda musician of his time" (Muller 2010, Track 5). He maintains that Shembe would "lock himself inside the room and sing a song and rehearse how to crystallise it, as it were, into dance steps, and then take that and teach it to the people" (Track 5). Shembe was also, still in the words of Bongani Mthethwa, "an excellent maskanda concertina player"; was able to play any instrument he laid his hands on. In the 1920s he started an instrumental ensemble in which the African flutes and the makweyana were played (Track 5).

The hymns that Shembe composed are now published in the Church's hymnal, which was first published by Johannes Galilie Shembe in 1940. Isaiah Shembe created 219 out of the 242 hymns in the hymnal that J. G. Shembe published in 1940, and that is still used, with some minor alterations, in the EBuhleni Sect of the iBandla lamaNazaretha.[2] The translation of the hymnal has recently been published by Carol Muller (2010). Isaiah Shembe was not educated, so he could not write his own hymns. He had scribes, normally young women members of his church, who would write down the hymns as they "came" to him. I am putting "came" in scare quotes because the belief that is generally accepted that Isaiah Shembe's hymns came with the messengers of heaven is partly true and I have argued elsewhere (Sithole 2011) that this view deprives Shembe of his creativity and agency. Sundkler has reported that:

> [Isaiah Shembe] would hear a woman's voice, often a girl's voice, singing new and unexpected words. He could not see her, but as he woke up from a dream or walked along the path in Zululand, meditating, he heard that small voice, that clear voice, which gave him a new hymn. He had to write down the new words, while humming and singing the tune which was born with the words. (1976: 186)

But there are a number of hymns that seem to me to challenge this view as they are related to certain events that actually happened and Shembe created the hymns

[2] The iBandla lamaNazaretha split in 1977 when J. G. Shembe (Successor to Isaiah Shembe) passed away. J.G.'s son Londa Shembe remained in eKuphakameni with a smaller group of followers while a larger grout left with J.G.'s brother, Amos, to create a new home, EBuhleni. Londa added more hymns to the hymnal while Amos did not.

as reflections to those events. A close and contextual reading of the hymns themselves negates the above understanding of the creation of the hymns. Hymn No. 3 is a case in point: it is based on Shembe's prayer before he and his followers took on a journey to Nhlangakazi in 1923. He had been informed by the authorities that he could no longer undertake his pilgrimage to Nhlangakazi without prior permit by the magistrate. When the time had come to go and there was no permit, Shembe decided to defiantly embark on his journey, and before he went, he said a prayer that was to form this hymn:

Nkosi Nkosi bubusise	Lord, Lord bless
Lobu buNazaretha	This Nazaretha Church
Uchoboze izitha zabo	Crush its enemies
Zingabu vukeli.	That they don't rise against it.
Vuka Vuka wena Nkosi	Wake up, wake up, Oh Lord
Mabulwelwe nguwe	Be the one that fights for it
Uzuhambe phambi kwabo	Travel ahead of it
Zingabuvukeli.	So they don't rise against it.
Noma siya entabeni	Even as we travel to the mountain
Owasikhethela yona	You chose for us
Ethiwa yiNhlangakazi	Called Nhlangakazi
Bungakhubeki.	Let it not falter.

Clearly, this hymn speaks to Shembe's situation at the time of composing it. The "enemies" that are spoken about are the state, the police and the missionaries (with the black believers), all those responsible for his predicament. He was here deliberating about his problem of being prohibited to undertake a journey he had undertaken for the last 5 years or so. This then can only have come from his mind, not brought by some spirits. Sundkler's quote above should be understood in terms of the composition of the interview from which it came, viz., a white scholar interviewing a black religious leader (of a church that had for many years been subjected to scrutiny by the state) in a country ruled unfairly by a white government. The statement could have been invoked by Isaiah Shembe and J. G. Shembe to channel people's attention away from the political nature of some of the hymns.

145.4 Isaiah Shembe and the Notion of the Black Messiah

There is no doubt that Shembe spent some time in his life reflecting on his own identity not just as an African, Zulu or human being, but also as a religious figure of great prominence. It is also apparent that in his endeavours to understand and articulate his own self he found the Bible to be very useful. That he was aware of his (healing) abilities and wisdom is clear from his self delineation in his assertion that, "[i]f you had educated him in your schools you would have taken pride in him. But that God may demonstrate his wisdom, he sent Shembe, a child, so that he may speak like the wise and the educated" (Gunner 1986: 182). Nellie Wells, "Special Representative" of *The Natal Mercury* and Isaiah Shembe's ardent admirer who wrote a transcript for a proposed film on Shembe, reports that

> Everywhere chapels, churches and schools were emptied as Shembe approached and the people crowded to listen with great joy. Immorality so prevalent among Christians who were living under false economic conditions was driven to shame, mental snobbery was pricked like a bubble, and a simple folk wearing skins or next to nothing accepted the gospel with great joy and were baptised, then the missionaries waged warfare, because, as they said, Shembe was undoing much if not most of what they had done. (Quoted in Mpanza 1999: 56)

The popularity that Shembe commanded and the conflict with the missionaries, the educated elites, and the State which resulted from such popularity made it important for Shembe to deal with the question of his identity. On a number of occasions he was interrogated with regards to his identity and praxis. In 1921 he was invited to give his life story to then assistant Magistrate C.N.C Barrett and in 1923 he was interviewed by Magistrate Charles McKenzie in Ndwedwe Court (Papini 1999: 254). He was also interviewed in 1929 by Carl Faye (see Papini 1999) and in April 1931 by the Native Economic Commission. Clearly, in these interviews he told his interrogators what they would like to hear, as in the case of the 1923 interview he said something far removed from his many texts addressed to his followers:

> God has given me certain work to do amongst my people. I therefore realise that God has also placed the authorities over us, and those who disregard or defy the Government, disregard the will of God. (Quoted in Papini 1999: 255)

While this is clearly a reference to Romans 13, it also echoes Jesus' words in Luke 20 verse 25, "Give unto Caesar what is Caesar's and to God what is God's." However, in reality, this is not what Shembe encouraged and taught his followers to do. I stated elsewhere (see Sithole 2010), following on James Scott's formulation, that in the Nazaretha Church there is a marked difference between the "public transcript" and the "hidden transcript." According to these concepts, the above statement in the interview would belong to the "public transcript" while there are a number of other texts (hymns and sermons) that would constitute a "hidden transcript" in which Shembe addresses his followers away from the ears of the authorities (Scott 1990). In these texts he teaches defiance, rather than submission.

The kind of wisdom and powers that Shembe was believed and believed himself to have commanded are close to those which Jesus was believed to have commanded; Shembe was aware of this fact. Thus, in hymn 34 he appropriates the story of Jesus's birth and uses it to claim that what Jesus was for the Israelites which Shembe is for the Zulus or even Africans:

Kwafika izazi	The wise men came
Ziphuma empumalanga	They came from the East
Zathi uphi lowo	They said where is that one
Oyinkosi yabaJuda.	Who is the Lord of the Jews

Chorus:

Kunjalo-ke namhlanje	It is like that today
Emagqumeni as'Ohlange	In the hills of Ohlange
Nawe-ke Betlehema	And you Bethlehem
Muzi wakwaJuda	Village of Judah
Awusiye omncinyane	You are not the smaller one
Kunababusi bakwaJuda.	Than the rulers of Judah.
Chorus	
Lanyakaza iJerusalema	Jerusalem was shaken
Bathi niyayizwa lendaba	They said do you hear this news
Evele phakathi kwethu	That happened amongst us
Efike nezazi	It came with the wise men
Sibutheleni abafundisi	Call for us the priests
Bahlole imibhalo.	To examine the scriptures.
Chorus	
Bafike bathi yebo	They came and said yes
Kulotshiwe kanjalo	It is written so
Nawe Kuphakama	And you Kuphakama
Magquma as'Ohlange	Hills of Ohlange
Chorus	
Awusiye omncinyane	You are not the smaller one
Kunababusi bakwaJuda	Than the rulers of Judah
Kuyakuvela kuwe	From you shall come forth
AbaProfithi	Prophets
Abayakusindisa	Who will save
Umuzi was'Ohlange.	The village of Ohlange.
Chorus	

One of the reasons why Shembe read the Bible so extensively as stated above was so that he could use it to defend himself against the people (black church leaders, the missionaries and the state) who seemed troubled by his work. He called on the educated and knowledgeable people to examine the scripture and hoped that they would realize – by identifying the similarities between himself and other prophets – that Shembe himself was a prophet and was sent by God. Thus he begins the above hymn at the time when the wise men were going to see and salute the young Jesus. Shembe emphasises that the wise men came with the question, "Where is that one who is the king of the Jews?" This challenges his enemies to enquire about who and what he is, instead of simply dismissing him as "the madman, son of Mayekisa" (Papini 1999: 251). Rather than following the story of Jesus as it is told in the Bible, at this point he inserts the chorus which makes a statement about himself and places this narrative in the context of South Africa, using a location occupied by only black people (*Ohlange*) to emphasize the fact that he was sent to "save" black people in South Africa.

What Isaiah Shembe does with the Bible here is what West calls "re-membering" the Bible. West argues that ordinary people (as opposed to trained biblical scholars) have their own tools for "reading" the Bible. Their 'reading' of the Bible is "more akin to "rewriting" than reading in any scholarly sense" (2003: 78). They reinterpret the Bible (sometimes in the way antagonistic to that of the missionaries) and give it new meaning relevant to them and their contexts. As West goes on to argue, while the ordinary African interpreters of the Bible do not rewrite the Bible as such,

> they are [also] not as transfixed and fixated by the text as their textually trained pastors and theologians… The Bible they work with is always an already "re-membered" "text" – a text, both written and oral, that has been dismembered, taken apart, and then re-membered (2003: 78).

The text that is commonly used or appropriated in the Nazaretha Church to confirm Isaiah Shembe's messianic position is that of Deutoronomy 18:18. While this text was used by Isaiah Shembe himself (as can be seen in his writings) and Church members even today use it, an example of its use by Amos Shembe (Isaiah Shembe's son and third leader of the Church according to the Ebuhleni sect) will suffice here. In an undated cassette, Amos (known to members as iNyanga YeZulu / Moon of Heaven) preached a sermon in which he dismissed Jesus as white and calls for a black prophet for black people. He told the story of a woman who wanted her children to go to church. The children asked who was to be worshipped and the mother told them it was Jesus. When they asked who Jesus was, and were told that he was white, the children were very angered and disappointed.

> "Kanti mama sihlupheka kangaka nje sikhonz'umlungu?" Wadumal umam 'esepheth 'incwad'ey 'esontweni… Zathi "Siyabuza?…Kant'uJesu wumlungu?....Abelungu basihlupha kangaka nje sikhonza bona?…Kwakwenzenjani?Kwakonakelen'e-Afrika mama? Kwakwenzenjan'-uNkulunkul 'engavez 'umuntu…" Ngiyoniveze…ves 18 Deutoronomy 18… wa…zibuz-'ingan'ukuth'akamvezanga ngan'uNkulunkulu uma umunt 'omnyama kuyisizwe ngempela kushukuthi siqalekisiwe thina mama?…Bafundele phela ngoba naz'ingane zath 'akufundwe lapho…(**Ngiyakubavezela umprofethi kubafowabo. Abe njengabo. Ngiyakubeka amazwi ami emlonyeni wakhe… Uyakukhuluma konke akutshelwa yimina… Uyakuthi ongawalaleliyo amazwi…**)Eya! Ngiyakubavezela…Abantu!… Umprofethi…Abantu!… Onjengabo…Abantu!…Hhayi ilokhu u Sheti…iNdiya min 'alifani nam. UHhay'u Ferguson! Hhay'uFerguson umlungu…no…no…no! Ngiyovez'umuntu kwabakubo …Akushiw' ukuthi kwabakubo? …Ehhe! Kwabakub 'onjengabo abantu…
>
> TRANSLATION: "Is it true, mother, that we worship a white man as we suffer like this?" The woman was disappointed as she held her books on her way to church…They [the children] said, "We are asking! Is Jesus a white man? The whites abuse us like this but we worship them? What happened? What went wrong in Africa mother? Why didn't God raise a person?",,, I will raise for you… verse 18 Deutoronomy 18… he… the children [were] asking, "Why didn't God raise a person? If black people are a nation it means we are cursed mother"…Do read for them because the children said we should read there[3]…(**I will raise the prophet among your brothers. He will be like them. I will put my words into his mouth. He will speak to them what I tell him and if anyone does not listen to the words**) I will raise for them…the people…the prophet…the people…from their race…the

[3] In the Nazaretha Church there is a tradition of reading the Bible where a person preaching a sermon will not read the Bible for himself, but will ask another person in the congregation. While one person reads from the bible, the preacher would reiterate what is read, sometimes changing some words and emphasizing whatever he feels like emphasizing. Here, I have used the parentheses and bold to show the voice of the person reading for Amos Shembe.

> people…he will be like them…the people. Not the…not Shetty… the Indian does not look like me… Not Ferguson…Not Ferguson he is white.. .No! No! No! I will raise a person from his race…Is it not said that from his race? Yes! From his race who is like them…

The above text does not suggest who is the Messiah for the black people, but the audience need not be told. They see the "chosen" one sitting in front of them. While Amos Shembe states clearly and openly his relationship to Jesus, that he was white and, therefore, represented the whites, he states in a subtle way Shembe's position and his relationship to God. He does this through singing verse two of hymn No. 239 (Fig. 145.3):

Ufikile abakhuluma ngaye	He has arrived, the one they spoke about
Aba profithi	The Prophets
Babazani wemadoda	Praise, you men
Babazani zizwe nonke	Praise, you nations

The singing of this hymn serves two purposes here, a common tendency in Nazaretha sermon performance. It is firstly a way of engaging with the audience, allowing them to be active participants in the performance. Secondly, it is a subtle way of making a point: that Shembe (and here he has his father, Isaiah Shembe in mind, even though he was himself believed to be a messiah) is the prophet the children confronted their mother about. It means God has actually done what the children were crying that He had not done.

Fig. 145.3 Members of the Nazaretha Church in EBuhlemi, near Durban, are at the end of the Sabbath service; the sermon has been preached and they are singing the closing hymn, July 2010 (Photo by Nkosinathi Sithole)

Now back to the hymns. How exactly does Isaiah Shembe "re-member" the story of Jesus's birth in his hymns? Shembe does this by appropriating a written story and making it a predominantly oral one by renewing it "in a sung context" (Watson 2009: 330) and thus allowing it to be received communally. This rendering of the text in oral form also "Africanises" it because, as Ruth Finnegan has noted, "Africa is celebrated above all for the treasure of her voiced and auditory arts, and as the home of oral literature, orature and orality, and the genesis and inspiration of the voiced traditions of the great diaspora" (2007: 1). The text itself is not 'borrowed' as it is from the Bible. Isaiah Shembe omits certain parts and emphasizes others as he sees appropriate for his own purpose. That the story of Jesus's birth was important to Isaiah Shembe is exemplified by the fact that there are two hymns based on this story. Another hymn telling the story of Jesus' birth is No. 152 "*Jerusalema Betlehema*" (Jerusalem Bethlehem). Unlike hymn No. 34 where the speaker simply narrates the story, here in the first stanza Isaiah Shembe uses two crucial locales in the story of Jesus's birth, Jerusalem and Bethlehem, as the addressees. But here too the wise men take the position as the second most important characters in the story. Shembe did not see Jesus as important in and of himself, but what was important was what he did and what he stood for. That is why the name Jesus is not mentioned in this stanza. What is emphasised is the fact that he is the Saviour (*uMsindisi*):

Jerusalema Betlehema	Jerusalem Bethlehem
Izazi zisitshelile	The wise men told us
Umsindisi uselezelwe	The Saviour is born
Kwelasebetlehema.	In the land of Bethlehem.

What Jesus does is important because of the people for whom it is done. The significance of the Saviour is dependent or reliant on the people who are saved, and I think that is why Shembe downplays or omits the name of Jesus in his re-membering of the biblical story of Jesus's birth. The text on which both hymn No. 34 and hymn No. 152 are based is Matthew 2. Interestingly, the name of Jesus, which Shembe avoids completely in hymn No. 34 and only mentions in the very last line of hymn No. 152, is the very first one in Matthew 2:

> UJesu se e zelwe eBetlehema la seJudia, emihleni kaHerodi inkosi, kwafika eJerusalema izazi zivela empumalanga, zithi, "U pi lowo o zelwe e inkosi yabaJuda na? Sabona inkanyezi yakhe sisempumalanga, size kukhuleka kuye."
>
> TRANSLATION: When Jesus had been born in Bethlehem of Judea, in the time of King Herod, the wise men came to Jerusalem from the East, they said, "Where is the one born king of the Jews? We saw his star while we were in the East, we have come to salute him."

The way in which Isaiah Shembe re-members this text (Matthew 2) in each of the two hymns mentioned above is different. Hymn No. 152 seems to be simply a re-telling of the narrative, albeit with certain omissions and alterations of some parts of the text. Hymn No. 34 in contrast is unapologetically claiming the story of Jesus for the amaNazaretha, for the Africans. In hymn No. 34, Isaiah Shembe inserts the chorus that claims that what happened in Jerusalem has happened in Ohlange too. For instance, stanza 3 talks about the stirring of the people in Jerusalem because of the news that the king of the Jews had been born, and a call is made that the ministers should examine the scripture to ascertain if what was happening was in accordance with what had happened or had been prophesied before. This call is Isaiah Shembe's own and is not part of Matthew 2. Through it Shembe is arguing that he and his work should be judged and examined in terms of what the scripture says. This is probably in response to what Elizabeth Gunner refers to as the 'police surveillance' and to the "Makholwas" (Believers), for whom Shembe's appeal was "very alarming and threatening" (1988: 215).

But what is even more telling is the fact that although the call to examine the scripture is with regards to Jesus' birth and Jerusalem, in Shembe's text, the ministers' examination shows that eKuphakameni (not Bethlehem), in the hills of Ohlange (not of Judah), is not smaller than the rulers of Judah, as the biblical text says about Jerusalem. Here Shembe is claiming that the black people of Africa are God's chosen people as the Israelites were. He says in another of his hymns (No. 101):

Akusiyo iJerusalema kuphela	It is not Jerusalem alone
Owayithandayo.	That you loved.

As stated earlier, hymn No. 152 does not deviate a great deal from the biblical text on which it is based. Here Shembe does not make a statement about ubuNazaretha, at least not openly. It seems to me that even though hymn No. 34 comes before hymn No. 152 in the hymn book, it is the latter that was composed first. In "Jerusalem Bethlehem" Shembe repeats the Matthew 2 text, although with a great deal of selection and very little alteration of the way in which the text is presented. I think when he composed this hymn, Shembe was still coming to terms with his own identity and spirituality, and this text appealed to him so much that he wanted it to be part of his church's repertoire. He then worked on the story itself, selecting some parts and leaving out others without bringing in as many of his own ideas as he does in hymn No. 34. This is not to say that this hymn is without creativity on Shembe's part. While the act of selecting itself is a creative process, Shembe presents the story in his own way. As mentioned above, the addressees of this hymn are "Jerusalem and Bethlehem." It is to these two that the persona narrates the story. And while the beginning of hymn No. 34 is based on the biblical text itself, "the wise men came," in hymn No. 152 the same idea is presented, but here "the wise men told us" (l. 2) that the saviour is born.

"Jerusalem and Bethlehem" are the addressees only up to stanza 3. In stanza 4 Shembe returns to the text as it is in Matthew 2. This is the only stanza which is repeated in full in hymn No. 34, and it is the one that is based on Mathew 2: 6:

Nawe-ke Betlehema	And you Bethlehem
Awusiye omuncinyane	You are not the smaller one
Kunababusi bakwaJuda	Than the rulers of Judah
Wena Bethlehema.	You, Bethlehem.

What seems to attract Shembe to this verse is the idea that the presence of a strong religious leader in a country (as Jesus was to be) raises even the political status of that country. I mentioned elsewhere (see Sithole 2011) that what interested Shembe in the Zulu kingdom was Shaka's idea of a strong, unified black nation, and that Shembe saw himself as filling the position of a leader who could bring about that unity, though differently from Shaka by not using force in order to bring those people together. In short, Shembe saw himself as a Black messiah and he used the Bible to confirm such a claim.

However, Isaiah Shembe's imagined relationship with Jesus was more complex than simply viewing him as an equal who was for the Jews what he himself was for the Zulus/Blacks. In hymn No. 82, Shembe praises Jesus' perseverance in the face of adversity, and the fact that Jesus did not resort to violence of any kind when his enemies abused him.

Mangibenjengawe Nkosi	Let me be like you, Lord
Ekukhonzeni kwakho	In your worshipping
Awesabanga lutho	You feared nothing
Noma bekuhlupha.	Even when they ill-treated you.

Mangibenjengawe Nkosi	Let me be like you, Lord
Noma bekuhlupha	Even when they ill-treated you
Nasekufeni kwakho	Even in your death
Awushongo lutho.	You said nothing.

That Jesus died for all the people's sins Shembe accepts and celebrates in Hymn No. 132:

Bamenywa ngowayelenga emthini	They are invited by the one who hung on a tree
Esiphambanweni	On the cross
Emthini wokudelwa	On the tree of the forsaken
Umhlaba wazamazama	The earth quaked

Wayevuma ezami izono	He was confessing my own sins
Emthini wokudelwa	On the tree of the forsaken
…..	
Mina ngaphunyuzwa	As for me, I was relieved
Ngowayelenga emthini	By the one who hung on a tree
Senizwile wozani.	Now that you have heard, come.
Ezami zagqitshwa kanye naye	[My sins] were buried with him
Ethuneni, senizwile, wozani.	In the grave, you have heard, come.

145.5 Conclusion

Isaiah Shembe's appropriation or engagement with the Bible depends on what he intends to accomplish with a particular hymn he is composing. With the hymns on the birth of Jesus, he tries to negotiate his identity as a powerful healer and religious leader who is constantly violated by the state, the missionaries, and the Black educated elites. While this may seem to be pointing to a claim to be a Black Christ, it is in fact meant to iterate that his predicaments are the same as those faced by Jesus. This becomes clearer in the later hymns (at least according to the order of presentation in the hymnal) where he praises Jesus and wishes to be like him, and when he rejoices in the knowledge that Jesus died for his sins. He thus calls on the people to join him in worshipping Jesus.

References

Achebe, C. (1989). *Hopes and impediments: Selected essays*. New York: Doubleday.

Boer, R. (2001). *Introduction: Vanishing mediators? Semia Special Issue: A Vanishing Mediator? The presence/absence of the Bible in postcolonialism*, 88 (pp. 1–12).

Brown, D. (1999). Orality and Christianity: The hymns of Isaiah Shembe and the Church of the Nazarites. In D. Brown (Ed.), *Oral literature and performance in southern Africa* (pp. 195–219). Oxford/Cape Town: James Currey/David Philip.

Brown, D. (2006). *To speak of this land: Identity and belonging in South Africa and beyond.* Pietermaritzburg: University of KwaZulu-Natal Press.

Brown, D. (2008a). Modern prophets produce a new Bible: Christianity, Africanness and the poetry of Nontsizi Mgqwetho. *Current Writing, 20*(2), 77–91.

Brown, D. (2008b). Introduction. *Current Writing, 20*(2), 1–6.

Brown, D. (Ed.). (2009). *Religion and spirituality in South Africa: New perspectives.* Pietermaritzburg: University of KwaZulu-Natal Press.

Comaroff, J., & Comaroff, J. (1991). *Of revelation and revolution: Christianity, colonialism and consciousness in South Africa*. Chicago: University of Chicago Press.

Dube, M. (2000). *Postcolonial feminist interpretation of the Bible*. St. Louis: Chalice Press.

Finnegan, R. (2007). *The oral and beyond: Doing things with words in Africa*. Oxford/Chicago/Pietermaritzburg: Currey/University of Chicago Press/University of KwaZulu-Natal Press.

Gerard, A. S. (1971). *Four African literatures: Zulu, Sotho, Xhosa and Amharic*. Berkeley: University of California Press.

Gunner, E. (1986). The word, the book and the Zulu Church of Nazareth. In R. A. Whitaker & E. R. Sienaet (Eds.), *Oral tradition and literacy: Changing visions of the word* (pp. 179–188). Durban: University of Natal Documentation & Research Centre.

Gunner, E. (1988). Power house, prison house: An oral genre and its use in Isaiah Shembe's Nazareth Baptist Church. *Journal of Southern African Studies, 14*(2), 204–227.

Gunner, E. (2002/2004). *The man of heave and the beautiful ones of God: Writings from Ibandla lamaNazaretha, a Nazaretha Church.* Leiden/London/Köln/Peitermaritzburg: Brill/University of KwaZulu-Natal Press.

Kunene, R. M. (1961). *An analytical survey of the Zulu poetry both traditional and modern*. Master of Arts thesis, Zulu Department, University of Natal.

Mpanza, M. (1999). *Izwi lezulu*. Empangeni: Excellentia Publishers.

Muller, C. A. (Ed.). (2010). *Shembe hymns* (B. Mthethwa, Trans.). Pietermaritzburg: University of KwaZulu-Natal Press.

Papini, R. (1999). Carl Faye's transcript of Isaiah Shembe's testimony of his early life and calling. His early life. *Journal of Religion in Africa, 29*(3), 243–284.
Peel, J. D. Y. (2003) [2000]. *Religious encounter and the making of the Yoruba*. Bloomington: Indiana University Press.
Punt, J. (2010, October 29–31). *What is meant by Postcolonial Biblical Studies?* Paper presented at a workshop on Religion, Spirituality and the Postcolonial, University of Stellenbosch.
Roberts, E. (1936). *Shembe: The man and his work*. Johannesburg: University of the Witwatersrand. Unpublished master's thesis, African Studies Department.
Scott, J. (1990). *Domination and the arts of resistance: Hidden transcripts*. New Haven: Yale University Press.
Sithole, N. (2010). Aquiescence or resistance? The role of the AIC in the struggle against apartheid. *Journal of Theology for Southern Africa, 139*(2), 104–119.
Sithole, N. (2011). *Performance, power and agency: Isaiah Shembe's hymns and the sacred dance in the Church of the Nazarites*. Ph.D. thesis, English Studies Department, University of KwaZulu-Natal, Pietermaritzburg.
Sundkler, B. G. M. (1961) [1948]. *Bantu prophets in South Africa*. London/New York/Toronto: Oxford University Press.
Sundkler, B. G. M. (1976). *Zulu Zion and some Swazi Zionists*. Oxford: Oxford University Press.
Vilakazi, B. W. (1946). *The oral and written literature in Nguni*. D. Litt. thesis, Department of Bantu Studies, University of the Witwatersrand, Johannesburg.
Watson, J. R. (2009). Eighteenth century hymn writers. In R. Lemon, E. Mason, J. Roberts, & C. Rowland (Eds.), *The Blackwell companion to the Bible in English literature* (pp. 329–344). Oxford: Blackwell Publishing.
West, G. (2000). Mapping African biblical interpretation: A tentative sketch. In G. West & M. Dube (Eds.), *The Bible in Africa: Transactions, trajectories and trends* (pp. 29–53). Boston/Leiden: Brill.
West, G. (2003) (1999). *The academy of the poor: Towards a dialogical reading of the Bible*. Sheffield/Pietermaritzburg: Sheffield Academic Press/Cluster Publications.
West, G. (2006). Reading Shembe 'Remembering' the Bible: Isaiah Shembe's instructions on adultery. *Neotestamenica, 40*(1), 157–184.
West, G. (2007). The Bible and the female body in Ibandla lamaNazaretha: Isaiah Shembe and Jephthah's daughter. *Old Testament Essays, 20*(2), 489–509.
Young, R. J. C. (2001). *Postcolonialism: An historical introduction*. Oxford: Blackwell.

Chapter 146
Music as Catechesis and Cultural Transformation in the East African Revival

Anna Swynford

146.1 Introduction

Perhaps nothing is more closely associated with the East African Revival than the song "Tukutendereza Yesu" (translated, "We praise you, Jesus"). Revival accounts suggest that it was sung at every gathering. The East African Revival is usually traced back to a meeting in 1929 between a young British medical doctor, Joe Church, and an African government worker, Simeoni Nsibambi. Beginning with their prayers, preaching and evangelism, revival spread throughout Kenya, Rwanda, Tanzania and Uganda, lasting three generations. It is significant that theirs was a partnership of black and white since the revival itself relied on both cultures; the themes of the revival display an African enculturation of nineteenth century British and American revival theology. Although the revival was very much a movement propagated and led by Africans, the theological underpinnings reflect nineteenth century Keswick theology, which focused on personal repentance, reliance on Jesus' blood and holy living through the power of the Holy Spirit.

Unlike other twentieth century revival movements, this movement mostly stayed within established structures. Although it began among Anglicans, it spread to other "Western" churches and indigenous Christian groups, acting as a catalyst for change within denominations, rather than creating the necessity for new structures. In this way, African Christians gained a new agency in formerly white ecclesial structures and in the process significantly changed the shape of Christianity in East Africa.

First hand testimonies of the revival from "Balokole" (a term those in the movement used to describe those who had been "saved") recount hours of singing, in many instances lasting late into the night. Hymn-singing informally became the main vehicle for catechesis, spreading the doctrinal foci of the movement to a largely

A. Swynford (✉)
Duke Divinity School, Duke University, Durham, NC 27705, USA
e-mail: anna.swynford@gmail.com

S.D. Brunn (ed.), *The Changing World Religion Map*,
DOI 10.1007/978-94-017-9376-6_146

illiterate population. The songs, mostly translations of hymns from nineteenth century American and British revivals, quickly became indigenized, reflecting their African context even more than their Western origins. African adaptations of Western hymnody, a mainstay of the East African revival, not only reflected but also helped propagate the revival and its theological particularities. In addition, it significantly transformed the practice of worship in East African churches.

146.1.1 Worship as Catechesis

This paper implicitly relies on recent work by theologians and sociologists who have examined how and why practices are formational. Dorothy Bass defines Christian practices as the "things Christian people do together over time in response to and in the light of God's active presence for the life of the world" (Volf and Bass 2001: 5). Christian practices are inherently social – they emerge from and occur in community – and are always a response to God's initiative. Worship, of which music plays a vital role, is the pinnacle of Christian practices. Although the common working assumption is that Christian practices flow out of Christian beliefs, theologian Miroslav Volf challenges this, arguing, "In most cases, Christian practices come first and Christian beliefs follow – or rather, beliefs are already entailed in practices, so that their explicit espousing becomes a matter of bringing to consciousness what is implicit in the engagement of practices themselves" (Volf and Bass 2001: 256). In other words, when people engage in Christian practices, they are already living as if the world exists in a certain way and they simply learn to articulate these preexisting beliefs.

Philosopher James K. A. Smith (2009), in his book *Desiring the Kingdom*, shifts the locus of catechesis from beliefs to practices. He counters the Enlightenment view of human beings as primarily thinking beings, and instead defines human beings by their desires and loves. Smith believes humans are at the core motivated by what they love and desire, which stem from their view of the "good life." The desire for a certain vision of life is largely instilled in humans by practices, especially practices that cater to the imagination. He writes:

> Discipleship and formation are less about erecting an edifice of Christian knowledge than they are a matter of developing a Christian know-how that intuitively "understands" the world in the light of the fullness of the gospel. And insofar as an understanding is implicit in practice, the practices of Christian worship are crucial – the sine qua non – for developing a distinctly Christian understanding of the world. (Smith 2009: 68)

Smith's focus on developing Christians who "understand the world in the light of the fullness of the gospel" is centered on the Christian practice of worship.

These authors help us to see that music sung in worship is not simply an embodiment of theology, but influences and reinforces belief itself. Therefore, music in the East African Revival was not a mere accompaniment or response to teaching, but a profound vehicle for discipleship that helped propagate the revival across multiple generations. I will explicitly examine music as theological text in the second half of this essay.

146.1.2 Background of Hymnody in East Africa

In order to understand revival music in twentieth century East Africa, we must first examine a brief history of hymnody in the region. In the eighteenth and nineteenth centuries when Anglican missionaries from England brought the gospel to far away countries, understandably they brought with them English customs and culture. It was not uncommon for Christian converts to adapt English dress and participate in English worship. The various factors for this importation of culture ranged from pragmatism (missionaries did not know native music, language or culture) to cultural elitism. Nowhere is this cultural influence seen more than in the use of Western hymnody in newly founded African churches.

As soon as possible, missionaries began translating Western hymns into native languages. In 1897 Anglican missionaries translated hymns into Kiswahili and called the songbook the "Nyimbo Standard." Unfortunately, the translations in this and other hymnbooks proved problematic for a number of reasons. First of all, most Africans did not easily assimilate the complicated harmony and melody lines found in many Western hymns. One Kenyan scholar explains: "The chorale, with its varied chord selection and rapidly changing harmonic structure, presents almost insurmountable difficulties to those who are not initiated to Western music" (Okeyo 2008: 33). Secondly, the translations did not take into account the tonal nature of most East African languages. As a result the translations often sounded forced, not reflecting the natural speech contours of the target language (Okeyo 2008: 27).

A.M. Jones, in his book, *African Hymnody in Christian Worship*, closely examines the reasons, both pragmatic and theoretical, that missionaries introduced this music to African converts:

> The more musical ones naturally desired to hand on to the Africans the riches of our Western Christian musical culture; the less musical would want to teach the hymns and hymn-tunes of which they themselves were fond. No one at that time knew anything much about either African music or about the subtler aspects of African languages. (Jones 1976: 7)

Adding to the subtle cultural elitism of most missionaries was their sheer ignorance of African music and languages. Missionaries and African converts alike feared the old tribal associations of African music and they wondered if African music was "good enough" for glorifying God. Perhaps most importantly, most African Christians themselves did not desire to associate with their old pagan musical forms (Jones 1976. 7).

Evidence suggests that translated hymns were usually sung in a style that would have been familiar to most nineteenth century British Anglicans. Unfortunately, this did not translate into robust and fervent hymn-singing. H. H. Osborn, in his book, *The Living Legacy of the East African Revival*, counts lackluster worship as one of the main problems with African Christianity before the revival:

> One of the reasons why Christians prayed for Revival in the early days was the apparent 'deadness' of many services as expressed in the seemingly lifeless following of 'orders of service' and dull singing of hymns. All that changed when there was revival. (Osborn 2006: 55)

146.2 Music in the East African Revival

146.2.1 The Presence and Function of Singing

First-hand accounts of the East African Revival point to hymn-singing as one of its characteristic features. Music and revival spirituality seem inextricably linked. In the centuries preceding the East African Revival a great outpouring of music often accompanied revivals in other parts of the world, from the Wesleyan hymns of eighteenth century Britain to the Sankey hymns that grounded Moody's revivals in late nineteenth-century America. Hymn singing, and in particular the revival chorus "Tukutendereza Yesu," which I will discuss in detail later in this paper, became synonymous with the movement (Jenkins 2011: 45).

Osborn outlines two distinct purposes of music in the Revival: "The communication and embedding in the mind of biblical truth to those who often were illiterate or for whom the printed form was difficult to obtain, and the expression of the joy of salvation and praise and worship to God for His greatness and the great things He was doing" (Osborn 2006: 57). These two main grounds for singing, encompassing both doxology and theology, impacted the choice of hymns and how and when they were sung.

Although hymnody has been used to teach theological truths worldwide, ethno-musicologist Roberta King sees the singing of hymns as particularly pedagogically important in Africa. She explains, "Oral dynamics inherent in African song foster an environment for theologizing and embedding the Christian faith in the soil of Africa. Among their multiple functions, songs provide opportunities for processing life's vagaries, difficulties, and daily interactions" (King et al. 2008: 120). Especially crucial at a time when the majority of Christians in Africa were not literate, music enabled theological truths to be memorized and internalized.

And yet far from being a mere teaching tool, music in the revival grew out of a fervent love of God and a desire to express thanks and praise. Music was always a response to the Gospel. When non-Christians were present at revival meetings, the priority was given to preaching, explanation of "the word" and personal testimonies. Once the gospel was presented, however, those who were "saved," including the newly saved, would often sing for hours, thanking God for cleansing them from their sin.

Memoirs of the East African Revival often mention all-night hymn-singing. Joe Church, one of the founders of the East African Revival, wrote in his diary in 1936: "There is little time to write. The night singing has been more intense; all night long there are groups singing and praying. It was very noisy on the 15th so I went down at 12:30 a.m. and sat for two hours with them, to try to help them" (Church 1981: 131). In response to this all-night zeal, missionaries at the outpost tried to stop the "disruption," but they were unsuccessful. In another instance non-Christians complained to the High Commissioner, who introduced a curfew banning singing after 10 p.m., but due to continued fervor and lack of timepieces, the singing continued until far after the curfew (Church 1981: 163).

Revival services and singing did not take the place of normal Sunday liturgies. Instead, spontaneous revival meetings often followed directly after a Sunday service.

During the years of the revival, Joe Church recounts, "Anglican services all over the country were followed by revival meetings outside church, spontaneous and yet keeping to a certain orderly pattern, where the saved ones greeted one another warmly with singing, handshakes and embraces" (Church of the Province of Kenya 1994). This spontaneous meeting would be integrally connected with the liturgical service that preceded it. Church writes later: "Sometimes a crowd would collect outside the church to go over the points of the sermon again, testifying to what had been a blessing to them, and then an appropriate hymn would be started up and sung with tremendous fervor, especially if some important person had just been saved" (Church 1981: 131).

146.2.2 The Style and Origin of Music

The music of the East African Revival could best be summed up as indigenized versions of American and British revival hymnody. Most of the selections came from *Golden Bells*, a 1925 hymnal that was published in both the United Kingdom and the United States. One edition of the hymnal has the subtitle: "a hymnal for revival services, Sunday schools, home singing, church services, etc." The music in the hymnal consists mostly of nineteenth century revival hymns ("Golden Bells | Hymnary.org," n.d.).

Revival founder Joe Church reminisces early in his memoir, "We talked about the old evangelical hymns that God has so greatly used in the days of Moody and how they seemed to have the same appeal to Africans" (Church 1981: 22). Various aspects of revival hymnody made it a natural fit for Kenyan and Ugandan Christians. Sam Mac Okeyo, in his article on Kenyan liturgical music, draws on Don Hustad's analysis of American gospel hymnody to explain its resonance with Africans. Harmony was simple, with chords usually limited to the tonic, dominant and subdominant. In addition, "The melodies were designed to be 'catchy'. The refrain rarely found in earlier hymns became their essential components. They were bound in repetition both in the lyrics and hymn tune" (Okeyo 2008: 33). Okeyo sees all of these aspects as coinciding closely with the African folk musical idiom. Because of this, they were quickly accepted and embraced by African Christians.

Hymns that naturally resonated with African sensibilities became more African as Balokole adopted them for their own use. Africans transformed hymns that were originally introduced by white missionaries. "They naturally and unconsciously permeated Revival Christianity with patterns of worship learned from their own tradition, and which were remodeled on their contact with British Revivalism" (Wild 1999: 420). Both the music itself and the way in which they were sung underwent changes. Significantly, Africans changed the rhythm of most of the songs they sang. Joe Church gives an example of how syncopation began to appear in hymn tunes:

> For example the hymn, "My hope is built on nothing less than Jesus' blood and righteousness," by Edward Mote, with the three/four time of William Bradbury's tune, was sometimes sung

> nearly all night, more and more syncopated until the Africanised six/eight time completely took the place of the original and incidentally seemed to many of us to fit the vernacular much better. (Church 1981: 131)

This quotation highlights two changes: rhythmic modifications in the actual music and a different way of approaching hymnody that was uniquely African. While Christians in England might sing "My hope is built on nothing less" once and be finished in 5 min, in this story the African Christians sang the same song almost all night. In addition, drumming, hand-clapping and dance began to accompany hymns. Church writes of celebratory singing after a person had been saved: "Gathering around the rejoicing person, arms would be waved, fervent handshakes and embraces given and the singing would be accompanied by swaying of the body and beating time. Of course there was fear that it would be associated with pagan dancing" (Church 1981 131).

The negative associations connected to pagan forms of music and dancing that I mentioned in the first section of this paper still remained, yet that no longer stopped Christians from assimilating them into Christian worship. Church tells a story that explicitly shows this redemption in practice, "This was harvest time in Ruanda, the beginning of the long dry season, and it was quite normal for the hundreds of homesteads around Lake Mohasi to meet and rejoice in a heathen way to the beat of the drums and clapping of hands. But this time it was to sing praises to the Giver of all life whom they had learned to love" (Church 1981: 131). Drums, swaying and clapping of hands, all associated previously with pagan religion, now accompanied praises to God.

This indigenization of Western music progressed to the extent that many songs transcended their now forgotten western origins. The iconic anthem of the revival, "Tukutendereza Yesu," is a Lugandan translation and modification of the chorus to the hymn, "Cleansing Blood," originally used in the revivals of Moody and Sankey. In its current form, however, it is unrecognizable from the original. This is clear in a 2010 video of the song recorded at a service to commemorate the beginning of the revival (Powerful East African Revival Testimony 2010) (Fig. 146.1). It is impossible to make a clear distinction between imported Western music and indigenous hymnody because of this natural process of indigenization.

Not all revival hymns, however, have direct connections to Western hymnody. The creative outpouring of the revival surpassed mere adaption to include composition. While many of these hymns have since been lost or forgotten, a few, like "Grace of God" by Emmanual Sibomana and others mentioned later in this chapter, have survived in hymnbooks and in Revival accounts (Osborn 2006: 104; Ward 1991: 132).

New outpourings of the Spirit often lead to new expressions of worship, as with the East African Revival. And it is fitting that in a revival characterized by the partnership between African Christians and western missionaries, where black and white worked together and openly confessed their sins to one another, that music would also be a blend of British/American sources and African style and sensibility.

Fig. 146.1 A 2010 gathering near Lake Mohasi, Ruanda celebrating the East African Revival (Photo © 2010 Chris Cairns)

146.3 The Theological Content and Focus

The songs that became central to the East African Revival reflect the main foci of the revival itself: confession/repentance, the centrality of Jesus and his blood, the Holy Spirit, and mission. In his analysis of the revival, MacMaster Jacobs summarizes the revival thus:

> Come to Jesus with your sins; repent and be cleansed by the blood of Jesus Christ; live in the immediacy of the presence of Jesus, and walk in open fellowship with the brothers and sisters; absorb yourself in the Word of God by life-changing Bible study; allow Jesus Christ to do good deeds through you by the enabling of the Holy Spirit; and witness with word, life and action that Jesus Christ is the head of the individual and of the body of believers. (MacMaster and Jacobs 2006: 21)

The experience of confession and repentance played a primary role in revival gatherings. To the Balokole, being a Christian meant forgiveness of sin through the power of Christ's blood and "walking in the light" through the power of the Holy Spirit. Music inhabited a pivotal place in the liturgy of conversion and in one's ongoing reliance on Christ's blood and the Holy Spirit.

146.3.1 *Confession and Repentance*

Repentance and public confession grounded and fueled the revival. Mark Shaw, a theology professor in Kenya, describes two main themes in the preaching of revival co-founder Nsibambi:

> The first theme was echoed in the oft-repeated phrase: Ekibi kibi nyo, "Sin is Sin." The ugliness and destruction of sin was so clear to him. But of equal emphasis was the softly spoken ekissa, Luganda for mercy. The bridge between dealing with sin and receiving mercy was public confession of such sins as "debt, dishonesty, immorality, and hatred of Europeans." (Shaw 2003: 30)

Public confession became the place where sin and mercy came together.

At revival gatherings after the word was preached and testimonies spoken, convicted attendees often publicly confessed their sins. This public display of repentance helped bring non-Christians or formerly lackluster Christians into the community of the Balokole. If the gathered deemed the confession sincere, new converts would be embraced with the chanting of the aforementioned "Tukutendereza Yesu." "Precious Saviour I will Praise thee/ Thine, and only Thine, I am;/For the Cleansing Blood has reached me; Glory, Glory to the Lamb" (Karanja 2011: 219). The new convert's confession showed that they, too, needed the blood of Christ to cleanse them from sin.

Not all public confessions were met with this song, however. In John Karanja's account of confession in the East African Revival, he writes, "if a brother or sister was perceived to have withheld or stumbled over vital information that was known to some brethren, his/her testimony was greeted with a chorus in which God lovingly warned the sinner: You cover your sin/ But a friend sees you/ When death comes/ Then you'll realise/ That sin is not a friend" (Karanja 2011: 220).

In the music itself we see a radical dichotomy between the saved and damned. The aforementioned warning is made even more explicit in an old Rukiya hymn, which begins:

> You who delight in worldly pleasures You will never feel satisfied
> For even your ancestors Were not satisfied
> You eat and drink You dress yourselves well
> You play and laugh But you forget that there is death. (Ward 1991: 132)

This song may at times have been directed to professed Christians, who in the eyes of the Balokole, were not "walking in the light," repenting of their sins and actively following Jesus. The distinction between the saved and the unsaved often stemmed from outward actions; both of these hymns emphasize public sin and worldly pleasures. Upright living, including sexual purity and abstention from alcohol, was essential to following Christ.

146.3.2 *Centrality of Jesus and His Blood*

Confession and repentance occupied a central place in meetings because awareness of personal sin led to reliance on Jesus as the only savior from sin and death. The key theme of the revival, the centrality and sufficiency of Jesus, was often reflected in the music used in revival services. The hymn, "I'm not ashamed to own my Lord," an Isaac Watts text translated into Kikuyu, often followed communal repentance (Church of the Province of Kenya 1994: 85). The second verse clearly claims Jesus as one's only hope:

> Jesus, my God! I know His Name,
> His Name is all my trust;
> Nor will He put my soul to shame,
> Nor let my hope be lost. (Watts n.d.)

Sung in the context of repentance, the words remind the singers that it is not repentance or holy living in itself that saves, but only Jesus' name. Other songs, like the Fanny Crosby hymn, "Pass me not, O gentle Savior," petition God for continual need and assistance:

> Pass me not, O gentle Savior,
> Hear my humble cry;
> While on others Thou art calling,
> Do not pass me by. (Crosby n.d.)

These songs, and others like them, point to the necessity of a personal, saving relationship with Jesus, and look to Jesus as the only savior.

Bishop Bill Butler, describing salvation and assurance of conversion, wrote in 1945 from Kenya, "The Blood of Jesus has made Victory gloriously available to me and entering experimentally into that victorious life is dependent on just one thing: my willingness to keep saying "Yes" to Jesus!" (Hooper 2007: 87). Even more prevalent than general songs about Christ's sufficiency were songs about the efficacy of Christ's blood. Two of the songs most referenced in revival accounts are the hymns "Nothing but the blood of Jesus" and "My hope is built on nothing less." Both emphasize that it is only because of Jesus' blood that one can be cleansed from sin and saved from death. To the questions "What can wash away my sin?" and "What can make me whole again?" the prior hymn responds, "Nothing but the blood of Jesus." The latter hymn likewise places all hope on "Jesus' blood and righteousness."

Another hymn mentioned in revival accounts, "There is life for a look at the crucified one," explicitly makes the connection between repentance and Christ's blood. Verse three clarifies that it is not one's repentance that saves, but the blood of Jesus. It begins, "It is not thy tears of repentance or prayers, But the blood, that redeemeth the soul" (Hull n.d.).

Philip Jenkins explains this fascination with the blood of Jesus, seen both in these hymns and in "Tukutendereza Yesu."

> We have to remember the setting – a society in which animal sacrifice remains commonplace and in which people have a deep appreciation for the value of the blood that is the life. In such a setting, concepts of atonement and sacrifice have an intuitive power that they lack in the modern West. (Jenkins 2011: 45)

Jenkins posits that the atonement, a doctrine that has fallen out of favor in much of the West, carries a deeper and more intuitive meaning in much of Africa. This highlights how Western hymns, when translated to an African context, take on new layers of meaning and significance that they did not have in their original contexts.

146.3.3 The Holy Spirit

The intense focus on Jesus and his blood paralleled a focus on the Holy Spirit. Scriptural teachings on the Holy Spirit not only helped begin the East African revival, but also were foundational to its theology and experience. Joe Church's study of the Scofield Bible before the revival began focused on the need for indwelling by the Holy Spirit. Even before Simeoni Nisbambi met Joe Church, he learned of the power of the Holy Spirit, both intellectually and experientially. Nsibambi recounts:

> There were a number of tracts, booklets and books from Britain and elsewhere which helped me to understand the ministry of the Holy Spirit. For a whole year I gave myself to the study of Scriptural materials with prayer. The reading of my Bible led me to complete commitment, and God filled me with the Holy Spirit. (Hooper 2007: 78)

In one of Church's early meetings at the beginning of the revival, he said to those around him, "God has shown us that the Church has many, many Christians but they are all dead in sin. They are very many, but they are very dry" (Farrimond 2011: 143). He connected this dryness to lack of the Holy Spirit by immediately leading the gathered in the song, "Spirit of the living God come afresh on me." This simple revival song, written in the 1920s, petitions the Holy Spirit to "melt me, mold me, use me, fill me." Unlike later Pentecostal movements, the reception of the Holy Spirit in the revival was not connected with ecstatic signs and wonders, but power to live faithful lives, "walking in the light."

146.3.4 Mission

Those who had been saved by the blood of Jesus and empowered with the Holy Spirit were not content to keep this to themselves. Instead, believers sought to bring others into the fold. Despite the presence of mass evangelistic movements, the revival mainly relied on personal testimony and evangelism, that "each one – teach one " (King 1968: 162). This began with Nsibambi himself. One irate missionary once approached Church and asked what he had done with Nsibambi. When Church inquired at the problem, she responded, "Oh, he's gone mad, and is going around everywhere asking people if they are saved. He's just left my gardener" (Shaw 2003: 29). This model of personal evangelism pervaded the entire revival.

In addition to actively evangelizing, the Balokole also sang about their mission. After revivalist Blasio died early in the movement, gathered Christians sang his favorite hymns, "Onward Christian Soldiers" and "Work for the night is coming." The first hymn utilizes military language to emphasize that it is Christ that leads his people into battle; mission is first and foremost God's mission. The second hymn emphasizes the urgency of the task:

> Work, for the night is coming,
> Work through the morning hours…
> Work, for the night is coming,
> When man's work is done. (Church 1981: 123)

Missions had exigency because the Balokole saw that the chasm between the saved and the unsaved was the difference between life and death. The Rukiya hymn quoted in the first section explicitly warns of the consequences of sin. The lavish lifestyle people revel in won't prevent them, "From being taken to Gehena." The song ends:

> We the saved ones, We have seen it
> And never shall we stop Talking about it
> And when you see it After your death,
> Do not regret, For it was you who rebelled. (Ward 1991: 32)

The saved spoke out of the overflowing joy of their own experience ("we have seen it and never shall we stop talking about it"), but also out of fear for the fate of their friends. Positive exhortations to accept Jesus were always connected with warnings for those who refused, because the decision had life and death consequences.

146.4 Conclusion

Through the hymns mentioned above, and others, Christians in the East African Revival "sang their faith." The palpable joy they experienced in Christ's forgiveness led to an outpouring of singing, which in turn reinforced their faith and led them to share with others the good news found in Jesus' blood. Philip Jenkins posits African hymnody as providing the reason behind the success of the revival and the strength of African Christianity in the twenty-first century:

> Watching the triumph of a hymn like "Tukutendereza Yesu" and its many modern counterparts makes us question the dismissive view often heard in the West about the weak theological underpinnings of Christian belief in the newer churches – the idea that African Christianity is a mile wide but an inch deep. We might draw a close analogy to the 18th-century revivals in Britain or America, when immersion in the great hymns of Whitefield and the Wesleys more than made up for the lack of formal theological education among ordinary believers. The popular religious music of that time created a faith a mile deep, much as the modern hymns are doing in churches across Africa and Asia. (Jenkins 2011: 45)

Although it is impossible to prove, the evidence suggests that music played a large role in the ongoing breadth and depth of the East African Revival. It is uncontested that the movement massively transformed Christianity in East Africa. Unlike many revivals, this revival stayed mostly within established structures. Therefore, the Anglican Church, which before the revival was largely small and lifeless, became a vibrant locus of revival Christianity.

Not only did the East African Revival renew the church, but it also transformed the cultural expression of Christianity in East Africa. While Christians continued to sing translations of Western hymnody during the revival as they did before, the type of songs and the manner in which they were sung changed drastically. In this collision of cultures, Africans adapted hymns to better fit their sensibilities and musical cultures. Drumming, clapping and dancing became an integral part of hymn singing. In addition, East Africans began composing their own music, something that had not yet been accomplished within the Anglican Church before the revival.

As I have shown, music was central, not peripheral to the revival. Unfortunately, little scholarly work has examined the place of music in the movement. Although I have chosen to foreground the theological and religious implications of this music, there are many other avenues yet to be explored. It would be especially fruitful to study the differences in music styles between the East African Revival and other revival movements that were mostly outside of church structures. In addition, although scholars have looked at the relationship between the revival and the contemporary movements for independence, it would be interesting to see how music was used differently in each of these movements. Because music leaves a small archival footprint, much of this research needs to be done while those who remember the movement are still alive to tell their stories.

References

Church, J. E. (1981). *Quest for the highest: An autobiographical account of the East African revival*. Exeter: Paternoster Press.

Church of the Province of Kenya. (1994). *Rabai to Mumias: A short history of the Church of the Province of Kenya, 1844–1994*. Nairobi: Uzima.

Crosby, F. (n.d.). Pass me not, O gentle Savior. *Hymnary.org*. Retrieved February 26, 2013, from www.hymnary.org/text/pass_me_not_o_gentle_savior

Farrimond, K. (2011). Personal loyalties: Revival, Church and mission in the lives of early revival leaders in Busoga. In K. Ward & E. Wild-Wood (Eds.), *The East African revival: Histories and legacies* (pp. 89–104). Burlington: Ashgate Pub. Ltd.

Golden Bells | Hymnary.org. (n.d.). Retrieved April 27, 2012, from www.hymnary.org/hymnal/GB1923

Hooper, E. (2007). The theology of Trans-Atlantic evangelicalism and its impact on the East African revival. *Evangelical Review of Theology, 31*(1), 71–89.

Hull, A. (n.d.). Life for a look. *Hymnary.org*. Retrieved February 26, 2013, from www.hymnary.org/text/there_is_life_for_in_a_look_at_the_cruci

Jenkins, P. (2011). Tukutendereza Yesu. *Christian Century, 128*(2), 45.

Jones, A. M. (1976). *African Hymnody in Christian worship: A contribution to the history of its development*. Gwelo, Rhodesia: Mambo Press.

Karanja, J. (2011). Confession and cultural dynamism in the East African revival. In K. Ward & E. Wild-Wood (Eds.), *The East African revival: Histories and legacies* (pp. 143–152). Burlington: Ashgate Pub. Ltd.

King, N. Q. (1968). The East African revival movement and evangelism. *Ecumenical Review, 20*(2), 159–162.

King, R. R., Kidula, J. N., Krabill, J. R., & Oduro, T. (2008). *Music in the life of the African church*. Waco: Baylor University Press.

MacMaster, R. K., & Jacobs, D. R. (2006). *A gentle wind of God: The influence of the East Africa revival*. Waterloo: Herald Press.

Okeyo, S. M. (2008). The categories and substances of liturgical music of the Christian church of Kenya between 1900–2000. In *An anthology of Christian music worship styles in Kenya* (pp. 27–40). Nairobi: Permanent Presidential Music Commission.

Osborn, H. H. (2006). *The living legacy of the East African revival*. Eastbourne: Apologia Publications.

Powerful East African Revival Testimony. (2010). Filmed by Chris Cairns. Retrieved from www.youtube.com/watch?v=oTzqnqfRck8&feature=youtube_gdata_player
Shaw, M. (2003). A hunger for holiness. *Christian History, 22*(3), 28.
Smith, J. K. A. (2009). *Desiring the kingdom: Worship, worldview, and cultural formation*. Grand Rapids: Baker Academic.
Volf, M., & Bass, D. C. (Eds.). (2001). *Practicing theology: Beliefs and practices in Christian life* (Edition Unstated.). San Francisco: Wm. B. Eerdmans Publishing Company.
Ward, K. (1991). Tukutendereza Yesu: The Balokole revival in Uganda. In Z. Nthamburi (Ed.), *From mission to church: A handbook of Christianity in East Africa* (pp. 113–144). Nairobi: Uzima Press.
Watts, I. (n.d.). I'm not ashamed to own my Lord. *Hymnary.org*. Retrieved February 26, 2013, from www.hymnary.org/text/im_not_ashamed_to_own_my_lord
Wild, E. (1999). "Walking in the Light" the liturgy of fellowship in the early years of the East African revival. In R. N. Swanson (Ed.), *Continuity and change in Christian worship* (pp. 419–431). Rochester: Boydell Press.

Chapter 147
Zen Buddhism and Music: Spiritual Shakuhachi Tours to Japan

Kiku Day

147.1 Introduction

This paper deals with organized tours to Japan for players of the *shakuhachi*—a Japanese vertical, notched, oblique bamboo flute, which from the seventeenth to the late nineteenth century was played by *komusō* monks of Fuke sect, a subsect of Rinzai Zen Buddhism. The instrument was used by the monks as a tool for spiritual training and for begging (*takuhatsu*).

The tours discussed here are arranged by North American *shakuhachi* "authorities," both of whom studied in Japan during the course of their *shakuhachi* training. The aspect that is particularly interesting is the link to religion and spirituality, which are important points of attraction for both of the tours. The tour we examine in greater detail is the "Shakuhachi Roots Pilgrimage" led by Alcvin Ramos (USA/Canada), which has been a yearly event since 2003 (Shakuhachi Roots 2013a). Ramos meditates and considers himself a Buddhist, although he is not an ordained Buddhist monk. The other tour to be discussed, albeit in less detail, is the "Annual Japan Tenri Tours," a 17-day trip to Japan, which has been organized yearly by Ronnie Nyôgetsu Reishin Seldin (USA) since 1980 (Nyogetsu 2013). Seldin is a member of the Tenri kyō or sect, a religion closely related to Shintoism (although monotheistic), which was established in the nineteenth century (Jansen 2000: 712).

The reason for focusing on Ramos' tour is that religion and the experience of spirituality is explicitly at the heart of the "Shakuhachi Roots Pilgrimage," whereas the "Tenri Tours" are more of a leisure trip, within which spirituality is only one of several focal points of visiting Japan as a *shakuhachi* player. Both tours take participants to

K. Day (✉)
Ethnomusicology, Aarhus University, Elme Alle 24, 8766 Nr Snede, Aarhus, Denmark
e-mail: kikuday@gmail.com

S.D. Brunn (ed.), *The Changing World Religion Map*,
DOI 10.1007/978-94-017-9376-6_147

Japan and provide them with experiences, including public performances for Japanese audiences, that connect them with the homeland and the history of their chosen instrument.

The music played and studied while on tour is discussed only briefly below, as generally speaking it consists of traditional *honkyoku* (original or fundamental pieces) music and has been widely documented. The term refers to the solo pieces played by *komusō* monks, who have roots in the Edo period (1603–1868), used for spiritual training or religious mendicancy.

The performances and the journeys themselves possess strong ritualistic elements. Here I explore how the participants utilize the performances as strategies of resistance in order to manipulate and modify *a priori* attitudes towards ethnical ownership and authenticity of *shakuhachi* music. Through their performances they authenticate their connection to an esoteric Buddhist music tradition.

147.2 Brief History of the Shakuhachi

Scholars now widely agree that the *shakuhachi* was introduced into Japan from China via the Korean peninsula during the Nara period (710–794) as one of the instruments in the *gagaku* (court) ensemble (Tsukitani et al. 1994: 105). The earliest extant examples of the *shakuhachi* are found at the Shōsōin, a repository built in 756, which contains eight *shakuhachi* used in the ceremony performed for the consecration of the Great Buddha of Tōdaiji temple in 752 (Tsukitani 2008). These *shakuhachi* have five finger holes in the front plus a thumbhole and produce a heptatonic (seven note) scale as probably used contemporaneously in China. When the *gagaku* ensemble was reorganized in the mid-tenth century, the *shakuhachi* fell into desuetude and no references to the instrument appear in surviving historical documents until the thirteenth century, by which time it had become a five-holed flute. During the early seventeenth century, *shakuhachi*-playing monks organized themselves within an institutional setting under the *Fukeshū* or Fuke sect, a subsect of Rinzai Zen. The monks of the Fuke sect were termed *komusō* or "priests of nothingness." The sect was granted special privileges by the Tokugawa government (the *de facto* rulers of Japan during the Edo period) in the seventeenth century, including monopoly rights over the use of the *shakuhachi* and passes that allowed them to travel to any part of Japan (Berger and Hughes 2001). According to the rules of the sect, the *shakuhachi* was to be used exclusively as *hōki*, or sacred tool, for the purpose of spiritual training and for *takuhatsu* (religious mendicancy). This served as the legal basis for establishment of the Fuke sect, which only admitted men of the *samurai* (military nobility) class and *rōnin* (*samurai* with no master to serve) as members of the order (Takahashi 1990).

In all, Nakatsuka Chikuzen lists 77 Fuke temples scattered around Japan during the Edo period. Three of the most important were Myōanji in Kyoto and Ichigetsuji and Reihōji in the Kantō region, the area around Edo (present-day Tokyo). Each temple developed its own corpus of music, which together comprise the repertoire of approximately 150 *honkyoku* from the Edo period known today. Interaction among the temples, including musical exchange, took place by means of *komusō* monks who wandered from temple to temple (Shimura 2002). Music other than *honkyoku* was referred to as *gaikyoku* (outer pieces) or *rankyoku* (disorderly pieces).

The Edo government was overthrown in 1868 and replaced by the new Meiji government (1868–1912), which abolished the Fuke sect in October 1871 and prohibited begging between 1872 and 1881 (Takahashi 1990: 127). These two political measures completely transformed the *shakuhachi* world in Japan.

147.3 Shakuhachi and Spirituality Today

According to Tsukitani Tsuneko and Shimura Satoshi, after the abolition of the Fuke sect, the *shakuhachi* was to follow two distinct paths, one secular and one religious (Tsukitani 2008: 152; Shimura 2002: 705). The *shakuhachi* entered the stage as a musica--ather than a religiou--nstrument played not to achieve enlightenment, but to entertain the public. The instrument began to be played as a hobby and in ensembles. It was also used to accompany *min'yō* (folk songs).

Today, the Tozan and Kinko schools, the schools that are descended directly from the Fuke-sect tradition, dominate the world of *shakuhachi* in Japan. Both schools are secular and begin teaching new students with *gaikyoku* or ensemble pieces, and—in the case of the Kinko School—only accomplished players are allowed to enter the realm of *honkyoku*. The Tozan School, the largest school of *shakuhachi* in Japan, does not teach *honkyoku* at all, although the term is used; the so-called "*honkyoku*" taught at the Tozan School are pieces composed mostly by the founder of the school.

Thus the transmission strategies and teaching methods that came to dominate Japan not only differed from those of the monks of the Edo period, but also from those of the Myōan players, who continued to play *shakuhachi* as a spiritual praxis and therefore only played *honkyoku*. This loose grouping of players, who followed the heritage of the *komusō* monks, became utterly marginalized in the *hōgaku* (Japanese traditional music) world as religious eccentrics and musical amateurs.

The construction of the instrument itself changed and skillful makers began to make *shakuhachi* with a rebuilt and lined bore, which enabled better pitch accuracy and larger volume (Maru 1922). Strong counter-reactions to these changes arose, as in the cases of two of the most charismatic figures in the *shakuhachi* world, Watazumi Dōsō (1910–1992) and Nishimura Kokū (1915–2002), both players of the unlined *shakuhachi* who regarded their playing as a means of spiritual training in the traditional manner (Keister 2004: 107). They called their instruments *hotchiku* and *kyotaku*, respectively, in order to differentiate them from the modern mainstream *shakuhachi*. During the twentieth century, these two players became increasingly well known outside Japan. Watazumi Dōsō became an icon of Zen Buddhist followers in *avant-garde* milieu—not least in the U.S. Watazumi travelled to the United States several times. He was invited to teach at the Zen Mountain Monastery and at the Creative Music Studio in New York, where he met other musicians, including John Cage, Steve Lacy, and Pauline Oliveros.

In Volume One of *The Annals of The International Shakuhachi Society,* published in 1990, a chapter with several subchapters is devoted to each of the two main schools of *shakuhachi* Kinko and Tozan. Another chapter is devoted to Chikuho *ryū* or school, and all four subchapters are written by Riley K. Lee, an American who happens to be a player of this school. In the last chapter, titled "Miscellaneous Shakuhachi Schools" one subchapter is devoted to Watazumi, an edited transcript of a lecture / demonstration at the Creative Music Studio, Woodstock, N.Y. in 1981 (Mayers 1990: 189–191). The short bibliographical note explains:

> Watazumido-so Roshi is a leading Zen Priest who specializes in his own style of Honkyoku, played on Shakuhachi of enormous length and diameter. A unique character who defies classification and whose spiritual theories are expressed in forms difficult to comprehend. (Mayers 1990: 191)

However, in Volume Two, published in 2004, the whole first section, which comprises some 8.9 % of the entire volume, is devoted to Watazumi, which reflects the enormous increase in knowledge and interest in this particular player and his Zen philosophy which had occurred in the meantime. Watazumi, his lifestyle, and his playing (which is not considered music but spiritual praxis) are described in terms such as "Zen only happens when the tool and the person cannot be divided," "To attempt to do Zen immediately does away with Zen." "The attitude of NO ZEN is the beginning of facing in the direction of Zen," and:

> The intellectual playthings that people attempt as Zen have nothing to do with Zen. They are only specious (that which is false though appearing deceptively true–pseudo). It may sound like Zen, feel like "blowing Zen," or be the cute splashes of ink on paper, these are all ZERO ZEN. (Karhu 2004: 14)

The editor of *The Annals*, Dan Mayers writes: "Watazumido epitomised Zen, and faced the same problem as did Confucius in the need to form a personal philosophy. In this he succeeded, as anyone who associated with him will attest" (Mayers 2004a, b: 25). In the eyes of these non-Japanese *shakuhachi* aficionados, Watazumi was elevated to the status of a Confucius.

Nishimura Kokū's playing disseminated via very different channels, probably partly due to his less extravagant and sensational philosophy and personality. Many came into contact with Nishimura's concept of *kyotaku* at Osho International Meditation Resort in Pune, India, where several of Nishimura's senior disciples were regular visitors (personal conversation with Noiri Kyosui, Darcy O'Byrne and Tilo Budach) as they were followers of Osho (Bhagwan Shree Rajneesh, 1931–1990). Nishimura's CD, "Kyotaku" has become iconic partly due to his photo on the CD cover (Fig. 147.1), where he resembled an old hermit or sage with a long white beard, playing on a long *kyotaku*.

In the CD sleeve, Nishimura's life as a *komusō*, his founding of the Tani sect of flute players, woodcarver, painter, and a black belt six *dan* in karate is described. The story of *shakuhachi* playing going back to the Zen Master Fuke, which today is known to be a myth, is described as historical fact. The difference in the instruments is explained as following:

> When the shogun-epoch came to an end around 1868, the shakuhachi began to change. It became more like the flutes in the Western world. The shakuhahi was made smoother and

Fig. 147.1 Nishimura Kokū (Photo from the front cover of the CD Kyotaku, 1998, reproduced by the courtesy of Nishimura Koryū and Agar Noiri, used with permission)

more symmetrical on the inside and could now be disassembled. These developments in the structure of the instrument affected the way in which the flute was played. It has become more and more difficult to find original kyotaku flutes. The original kyotaku flutes are played by breathing, rather than blowing, to get the desired sound. Today only the bamboo flutes made by Kokū Nishimura may be called "kyotaku." Each kyotaku made by the master receives his special stamp of authenticity. (Nishimura 1998: 3)

The myth building in English-language sources surrounding these two figures, who remained marginalized in Japan, created a great interest in them among non-Japanese *shakuhachi* players, which also helped promoting the *shakuhachi* as a spiritual instrument among people in the West interested in Zen Buddhism. And it is intriguing to note that while Nishimura and Watazumi's playing styles and philosophy differed greatly and have been propagated via very different channels, they were both eventually embraced by Westerners seeking spirituality.

147.4 Shakuhachi Tours to Japan

Emphasis on the spiritual aspect of *shakuhachi* on the part of non-Japanese players has without doubt influenced the itinerary of the tours. Both tours visit sacred sites and provide experiences of a sacred Japan "off the beaten track."

The analysis presented here is based on theories of rituals and performance as applied to material available and gathered online and in the course of telephone, Skype conversations, and email interviews with both the leaders of and participants in the tours. I also address the psychological importance that the deep connection most of the participants come to feel with the bamboo, the soil of Japan, and *shakuhachi* history and lineages during the course of the tours plays for their music, and how it enrich the lives of those who take part in them.

147.4.1 Tour 1: The Shakuhachi Roots Pilgrimage

The Shakuhachi Roots Pilgrimage is extensively documented online. It is packaged, promoted and sold on the Internet as a spiritual tour to the homeland of the *shakuhachi*. Ramos, the organizer, writes on his website, blog, and on *shakuhachi* fora on the Internet. He posts promotion before the tours, descriptive posts while on tour, and provides links to photo reportage. Among the target groups for the performances on tour are thus those who follow the tours online on blogs, websites, and web fora.

The Shakuhachi Roots Pilgrimage visits temples—with or without a *shakuhachi* connection—worships at temples, shrines and holy sites, performs Buddhist and Shintō rites, visits people who work within or cultivate traditional arts, and performs retreats at Buddhist temples such as Eiheiji temple, Tekishinjuku or Rengejōin. A large part of the trip is dedicated to bamboo harvesting and the making of one's own *shakuhachi*—just as the *komusō* monks used to do during the Edo period or as Nishimura and Watazumi did in the twentieth century.

On 9th August 2006, Ramos posts the following on the shakuhachi forum regarding the schedule of the tour (Andersson 2006):

> …visit Musashi Miyamoto's cave, experience the onsens (hotsprings), visit the newest Komus temple in Kumamoto, and meditate at the Komus Temple, Saik-ji (Hakata Ichoken). Then we will also participate in our obligatory visit to Tsukuba Shinto Shrine in Mie prefecture to offer shakuhachi to the gods of music and entertainment and do waterfall purification as well as visit shakuhachi maker and kyd master, Taro Miura. Then we'll visit Koya-san, center of esoteric buddhism…

The attraction and admiration felt by non-Japanese *shakuhachi* players towards *rōnin* such as the famous Miyamoto Musashi (1584–1645), known for his excellent swordsmanship, is intriguing, given the fact that they have a negative image among most Japanese as ruthless troublemakers. After the visit to the cave (Fig. 147.2), the offering of *shakuhachi* [music] at the Shinto shrine Tsukuba, is named, followed by a Shintō rite—waterfall purification or *misogi*. The *misogi* is a firmly established event on Shakuhachi Roots Pilgrimage, having been performed on each tour. And finally, *kyūdō* or archery and Koyasan, which Ramos connects with esoteric Buddhism. The eclectic character of the Shakuhachi Roots Pilgrimage where Buddhism, in the shape of esotericism and the Fuke sect are mixed with Shintō rites and Japanese fine arts associated with Zen Buddhism, is worthy of attention; it is clear

Fig. 147.2 Alcvin Ramos conducting the Shinto rite misogi (Photo courtesy of Darren Stone, used with permission)

that the tour does not accurately present the *shakuhachi*'s history and current situation, but rather offers a broad experience of a particular and to some degree imaginary aspect of Japan.

On Ramos' new website, which was relaunched in January 2013, the tour is described as follows:

> This trip is different than most other tours of Japan in that you will be entering the experience already with the spirit of Japan within you, being a shakuhachi player. It is geared mainly for shakuhachi players and other musicians of Japanese music living and studying outside of Japan with the intention of deepening one's practice by connecting with the root culture and experiencing the spiritual and artistic traditions of Japan… It is a very special trip for a very specific group of people; and it is this connection to shakuhachi that will be the key to opening doors in Japan that no ordinary person can access.

From the above, it can be seen that which religion of Japan the participants of the tour experience is of secondary importance, so long as they have a spiritual experience of traditional Japan—or rather of *their* imagined traditional Japan. It is a tour for selected participants who already physically embody Japan through breathing when playing *shakuhachi*. As Ramons puts it: It is a "3-week journey into the heart of Japan… It is a true pilgrimage, where one has the opportunity to visit sacred sites, study with master teachers" (Shakuhachi Roots 2013b). As the Japanese ethnomusicologist Seyama Tôru observes, "Playing the *shakuhachi* is…playing the history of the instrument" (Seyama 1998: 76). And indeed the Shakuhachi Roots Pilgrimage is to no small degree about experiencing the spiritual history of the instrument. As playing and experiencing the history of the *shakuhachi* are well documented online, I find it reasonable here to view the music played by the *shakuhachi* players on tour in Japan, the tour itself, and the context within which the travel takes place as performances—the ritualistic aspects of which enable the players to occupy a space within the history of *shakuhachi*. Performance is a concept that defies definition. Here we are talking about a postmodern performance practice, a broader concept than that of theatrical or music performance (Schechner 2005: 4–6): a "communal activity set apart from the everyday through varying degree of formalism and performativity" (Harris and Norton 2002: 1) and which can include both secular and religious or sacred activities. I view the activities such as *misogi* and playing in Myōanji temple in front of *shakuhachi* ancestors as performances which seek to negotiate a space within the spiritual world of *shakuhachi* for the performers (Fig. 147.3). The audience is not merely defined by the Japanese people present during these activities, as the largest audience is, in fact, to be found on the Internet worldwide.

Fig. 147.3 Darren Stone playing shakuhachi facing the altar where the statue of Kakushin (1207–1298), the founder of Fuke sect at Myōanji temple (Photo courtesy of Darren Stone, used with permission)

Fig. 147.4 Alcvin Ramos playing in a temple service at Shōfukuji temple (Photo courtesy of Darren Stone, used with permission)

In an interview conducted with Ramos in 2006, he describes the aim of the Shakuhachi Roots Pilgrimage as giving the opportunity to others to connect more deeply with Japan through the *shakuhachi*. When asked what makes Shakuhachi Roots Pilgrimage a pilgrimage, Ramos answered that his own *shakuhachi* experience has been a religious experience, which made him want to explore its roots. And as Japan is the root of *shakuhachi*, it, too is considered sacred. As holy places are visited in order to play the *shakuhachi* as a sacred religious practice, the tour thus becomes a pilgrimage. However, he emphasizes the sharing of experiences with the participants as the main goal and this sharing makes the tour experience as a pilgrimage even stronger for Ramos himself. Doing *taketori* or bamboo harvesting is also in Ramos' terms "a deep *shakuhachi* experience," and he explained that the tour is demanding for body, mind, and spirit, and is to be regarded as a kind of *shugyō* or religious practice (Fig. 147.4).

147.4.2 Tour 2: Annual Japan Tenri Tours

The Annual Japan Tenri Tours, led by one of the North American grandfathers of *shakuhachi* Ronnie Nyōgetsu Reishin Seldin from New York City is more a leisure trip with a duration of 17 days. The participants here take lessons with, in addition to Seldin himself, such players as Kurahashi Yōdō, the son of Seldin's teacher, and Aoki Reibo (Living National Treasure). They visit temples, shrines, museums, soak in hot springs, watch a *kabuki* (Japanese traditional theatre) play their instruments and dine well (the tour costs US$3,400 all inclusive except air fares and 1-week Japan Railway pass). They also visit and play at Myōanji temple, so important in the history of the *shakuhachi*, and also perform elsewhere depending on the year. One interesting aspect of Annual Japan Tenri Tours is, as Seldin himself is a member of the Tenri Kyō sect, the participants listen to several *besseki* lectures lasting 90 min, which cover the basic teachings and narratives of the Tenri sect. The *besseki* is viewed as a preparation, after which a person may receive the sacrament of the *sazuke*, a divine intervention to relieve pain caused by illness. Thus, although spirituality and religion is not the main focus of the Annual Japan Tenri Tours, it is nonetheless an important aspect through the performance at Myōanji, the visits to temples and shrines, and the *besseki*.

147.5 The Role of Religion and Spirituality

As we have seen, in the context of the Shakuhachi Roots Pilgrimage, the *shakuhachi* is spiritual. Simply playing it is a spiritual act. However, it is not only the *shakuhachi* that is spiritual—as the whole experience of Japan becomes spiritual due to its being the root or origin of Ramos' sacred instrument of choice. During the tour, this spirituality is experienced at first hand by the participants. Below are two descriptions of tour stops from the Shakuhachi Roots Pilgrimage 2006, the first from the Shinto shrine where *misogi* is conducted:

> The Tsubaki Grand Shrine in Suzuka… [was] established in 3 B.C. (in the 27th year of 11th Emperor Suinin)—is one of the oldest and most prestigious shrines in Japan. Tsubaki Daihogu is famous for its Konryu-Myojin waterfall where misogi (a waterfall purification) is practiced…

The feeling was quite magical as we walked under the giant torii and into the temple grounds down a long gravel road lined with candle-lit lamps leading to the main shrine. The wind was still blowing wildly through the ancient trees which added to the mysterious atmosphere. Iwasaki-san the Shinto priest led us through the front of shrine… into changing rooms where we changed into our minimal misogi outfits (fundoshi [loincloth] and hachimaki [headband—a symbol of perseverance]). Then we all met in front to the waterfall where we all did a kind of warm up composed of various calisthenics and chanting to prepare our bodies and minds for the coming experience. After the warm-up we followed Iwasaki-san down the torch-lined staircase leading to the waterfall. After ceremonially spraying salt and sake into the air Iwasaki entered the water and under the waterfall. Then it was my turn. After clapping two times and slicing the air with my fingers, I entered the roaring waterfall, while chanting Shinto mantra, positioning my body so that the water hit my upper back and shoulders. After about 30 s I heard Iwasaki-san's signal to exit the fall. Everyone took turns to experience the waterfall then we all changed back into our clothes and met in the main hall where we sat in meditation for several minutes.

The following passage describes a stay in the Zen Buddhist dōjō (training center) in Musashi Koganei, Tokyo:

> … Adam instructed us to clean the dojo before meditation. We swept the floors and tatami [floor] mats, toweled down the floors and wiped all the windows clean. Then the head teacher instructed us in their particular form of meditation. Before we actually meditated, the regular dojo members did 30 minutes of a special misogi training for advanced students. Then we all went up to the second floor where the meditation hall was, took cushions, folded them under us, and sat on the cushions in half lotus or full lotus position. We sat for 1 hour in the austere cold of early morning. Then we followed the senior students to the kitchen where they prepared us a breakfast of green tea, miso soup and rice… After the break, was our next meditation session. Before sitting we all were handed bokuto (wooden swords) and took our places in the meditation hall and everyone did 1000 strokes of the sword as each person took turns counting to 100. Then we sat again for an hour. Afterwards we… went out into the yard where we swept and raked leaves for a few hours. Then it was lunch time. We had soba, umeboshi, and tea. Meals were a solemn affair. We were supposed to follow every movement of the senior students as close as possible. Very minimal talking. Only seiza [position sitting on one's heels] sitting is permitted.

These examples throw the participants and reader into exotic and unfamiliar situations which require courage, and offer an authenticity which cannot be experienced by simply visiting Japan. Here we receive an image of an exclusive and deeply spiritual Japan available only to the few. Although the activities described above have nothing directly to do with the *shakuhachi*, experiencing a spiritual Japan seems to be one of the key points of the tours and provides the participants with a feeling of a deep connection with the instrument.

147.6 Analysis of the *Shakuhachi* Tours

147.6.1 Tours as Experienced by People

The participants always described Shakuhachi Roots Pilgrimage as having been extraordinary and according to some, even life-changing experiences. "The trip was great because it was both off the beaten track and behind closed doors," as one of the participants wrote (email correspondence, June 2006).

The two most important experiences for the participants on the tour seem to be bamboo harvesting and learning from Japanese teachers in Japan. As Darren Stone, who has participated in the Shakuhachi Roots Pilgrimage several times, wrote:

> The single biggest thing to connect me to my hocchiku (and subsequent ones I've made) was to see and touch and participate in the bamboo harvest. Seeing my bamboo go from in-the-ground, to cutting it, cleaning and burning, sunning, watching the colors change, and finishing them into several poorly-tuned (but personal and playable) flutes just 8 months later… This was really important to me. (email correspondence, Feb. 2006).

Learning from a Japanese teacher in Japan—even if Kurahashi Yōdō, for example, has taught more students in the U.S. than in Japan—is much appreciated by the participants. It makes them feel that they have learned from "the real thing," which they regard as an important aspect of their *shakuhachi* studies. In the course of the interviews, participants indicated the feelings of authenticity they experience when learning from a Japanese teacher rather than the teacher they had in their respective countries.

All of the participants explained that they have had a long-time interest in Japanese culture, and in particularly Zen Buddhism and martial arts such as *aikido*. It was through these disciplines they encountered the *shakuhachi*.

When asked about what they expected from the tour, most answered they had no expectations and tried to remain open to new experiences. However, a few were disappointed by what they saw as "westernization" in Japan.

147.6.2 Spirituality and Shakuhachi in the West and Japan

Meditating at a Zen Buddhist *dōjō*, *misogi*, attending a secret ceremony commemorating the famous Zen Buddhist monk and poet, Ikkyū Sōjun (1394–1481), performing *kyūdō*, and experiencing Koyasan (a UNESCO World Heritage Site) all add to the drawing power of the tour and render it an authentic Japanese experience. Zen Buddhism, as it is known in the West, has been branded as an authentic, pure, rational, and individualistic spirituality that enjoys a commonality with modern scientific discourse. Mainly due to the works of Suzuki Shunryū (1904–1971), it has been transformed from a religion in Japan to a spiritual movement in the West, not least in the U.S. The spirituality that the tour seeks to find in Japan is one that has been transformed in the West. As Jørn Borup (2013) notes "Zen" has become an icon, a floating signifier through the use of the concept as "be Zen" or "find your Zen." Here, it is no longer just a religion, but also a state of being or something one possesses. Borup furthermore points out that in Japan, there are basically no associations between Zen Buddhism and spirituality. In Japan Zen Buddhism is rather an institutionalized religion or, for the majority of the Japanese, a funerary institution. Zen Buddhism as an individual spirituality is a Western interpretation.

For *shakuhachi* players the individuality of (their version of) Zen connects well with the stories of highly individualistic persons such as Watazumi and Nishimura. Most *shakuhachi* players in the Western countries have no larger group of which to be a part. Although they may identify themselves with a larger school in Japan, the fact is that most would never become an equal member of such a group. Thus players are often isolated figures with a special interest, and rebellious individuals such as Watazumi become iconic figures with whom they can identify themselves.

The Western interest in Japanese religion, in particular Zen Buddhism, and arts have led to a coupling of these two aspects to the degree that they have become almost inseparable (Yamada 2009: 214). In Yamada's view, the Japanese feel flattered when non-Japanese people express an interest in their culture, despite it being a Western-appropriated view of Japanese culture. As a result of this feedback from the West, knowledge of the raw and less refined *shakuhachi* of Watazumi and Nishimura has increased in Japan during the past decade. The Myōan group of *shakuhachi* players is no longer as isolated as before and is invited in to festivals and other larger events, although it remains a curiosity.

Returning to Seyama Tōru's thesis that "playing the *shakuhachi* is...playing the history of the instrument" (1998: 76), it can be seen that, due to the fact that participating in the tradition and history of the *shakuhachi* and in Japanese culture in its entirety is of such importance, the ritualistic aspects of the performances permit participants in the tour to claim a place in the history of *shakuhachi*; the participants perform for the ancestors of the *shakuhachi* lineages—the *komusô* monks. By praying to the *shakuhachi* ancestors as if they were their own, the participants signal that they are, indeed, heirs to this tradition. Thus the negotiation of a multi-cultural *shakuhachi* lineage has begun. The music itself serves as the vehicle which enables the rituals—and with the online documentation of the rituals, the negotiation takes place. Indeed, I view the *shakuhachi* offerings to the ancestors as one of the key factors that serve to make the tour as a whole into a performance, due to the heavy online coverage through the materialization of photos, daily blogs, and descriptions of the tours.

On the Roots Pilgrimage 2005 blog Ramos discusses playing at Myōanji temple, the most important of the *shakuhachi* temples, which he regarded as playing a crucial role in the continuation of the tradition inherited from the *komusō* monks:

> After placing a monetary offering in an envelope upon the altar, and bowing deeply, we all played the piece Tamuke as our honkyoku offering to the spirits of the shakuhachi ancestors... Our vibrations will merge with the ancestors who are watching us and playing with us, guiding us on our Path...

Tamuke is not dissimilar to the Western requiem, as the piece is often played for newly departed souls. Playing and thereby offering *shakuhachi* music to the *shakuhachi* ancestors within the Japanese tradition of visiting the family grave to make offerings to the ancestors, can be viewed as participants in the tour linking themselves and the spirits of the *shakuhachi* past—Japanese players, the *komusô* monks of the Fuke sect—into one great *shakuhachi* family. Visits at Shinto shrines are described in the following terms: "We offered shakuhachi honkyoku within the inner sanctuary of the Honden after offering tamagushi (sacred branch offering of the Yu tree) to connect more with the Kamisama (God)" (2005 blog, accessed Sept. 26, 2006).

Connecting Shakuhachi Roots Pilgrimage participants to *shakuhachi* ancestors and Shinto Gods seems analogous to the authenticity process that the Meiji emperor (1852–1912) had to perform when he became the head of state after the Tokugawa regime was overthrown. He performed pilgrimages throughout Japan and prayed at old graves, assumed to be the tombs of ancient emperors, in support of his claim to a direct imperial rule due him as the heir of the Gods (McClain 2002: 198). By performing a pilgrimage around Japan to pay tribute to the ancient emperors, he signaled to the inhabitants of the country that he was indeed the direct descendant of these mystical figures. In the case of the Shakuhachi Roots Pilgrimage tour, the private act of playing the *shakuhachi* for one's own enlightenment or spiritual transcendence (which was partly the aims of the *komusō* monks) has changed into a public one, carrying new meanings not only for the players, but also for the international *shakuhachi* community which is watching online.

Membership in the *shakuhachi* tradition depends on how successfully one can define oneself as an insider or outsider to the tradition. Riley Lee writes that the definition of an 'insider' most likely would begin with membership in the instrument's original ethnic group, meaning being Japanese:

> One is either inside or outside. Since one is a Japanese only if born a Japanese, people who are gaijin [foreigner] are by definition complete and permanent outsiders. Gaijin shakuhachi players are likewise never 'insiders' to the shakuhachi tradition in the minds of many Japanese. (Lee 1992: 16)

Be that as it may, playing for the *shakuhachi* ancestors can be viewed as an initial negotiation of that *emic/etic* status. The use of Japanese religious rites can be viewed as a means to negotiate an inclusion into an *emic* stance in the *shakuhachi* world. However, as the Internet world is highly separated into English-speaking *shakuhachi* players on the one hand and Japanese-speaking ones on the other, the question becomes among whom is the negotiation taking place and for which audience?

Another aspect of the performance is the experience of playing in Japan in front of a Japanese audience or in temples, which participants reported in interviews to be a sacred gesture. I see this as something of a rite of passage. Many of the interviewed answered that performing in Japan was an important step to them. Here I sense a feeling of having crossed a borderline. It is almost as stepping into adulthood when one has dared to play with or in front a Japanese audience—not Japanese people living in North America, but with "real" Japanese in Japan. As one participant explained:

> I have played with Japanese musicians in the States; but in doing so, they have often tended to adopt Western rehearsal habits and mannerisms. So, to play with musicians in their native land, and in the land to which the music is native, is quite different—more authentic.

The tours thus enable the participants to take a big psychological step. Were a *shakuhachi* player to go to Japan on his or her own, it would require serious involvement with a *shakuhachi* school and religious institutions, and a much longer stay to attain this degree of involvement. As seen by North Americans, the *shakuhachi* tour is a convenient way of opening the door into the spiritual world of Japan—a spiritual world, however, which has been appropriated by the West.

147.7 Conclusion

As rituals have the capacity to restructure time, the participants experience—through the rituals they participate in on the tour (both religious and secular)—a collapse of the spatial and temporal boundaries obtaining between the *shakuhachi* and themselves. By bonding with the *shakuhachi* ancestors and gods of Japan, through their rituals of playing, donating offerings, and praying, they transcend the limitations of the present and secure for themselves a "past" as a member of the *shakuhachi* lineage, thereby transforming both past and present and their futures as authenticated *shakuhachi* players. Rituals also encompass control of a space and have the ability to restructure space and mark out a new space of belonging; thus the participants make the space of the "homeland of the *shakuhachi*" their own. In this manner the rituals serve to connect the participants both physically and spiritually to the land of the *shakuhachi* and the bamboo and to its sacred places. The rituals are conducted with seriousness, care, and respect. The Shakuhachi Roots Pilgrimage is a pilgrimage and the participants pilgrims; on Ramos' blog each new participant is entered as a "new pilgrim" after signing up for the tour (http://alcvin.ca/japantrip/hello-world, accessed 10.03.13).

By incorporating vigorous religious practice, not performed by most Japanese players, Shakuhachi Roots Pilgrimage, in my view, can be seen as "investing old ritual forms with new meanings" (Jones 1999: 33). The rituals transform *a priori* attitudes of ethnic ownership of the music. By participating in ritual acts in Japan, the players challenge the boundaries of authenticity and resist being excluded due to ethnic categorization. By performing far more rituals than most Japanese players, they manifest themselves as insiders in a tradition which was once highly exclusive. They therewith transform the realm of the *shakuhachi* into an internationally open tradition and create a new sense of community as one big *shakuhachi* family.

The *shakuhachi* tours are closely linked to images of the Other. The participants possess a strong sense of the Japan that they desire to see and not to see, and thus they invent their own image of "Japan"—although during interviews most participants said that they tried not to have specific expectations with regard to the tour. However, most of them had, in fact, studied Buddhism and ancient Japan before taking the tour, thus inevitably constructing their own images of the Japan they were to experience. It is striking, for example, to see how important the images of traditional Japan are for the participants:

> I live in Manhattan. I try to bring back old Japan. I play the shakuhachi. My apartment reflects Japanese lifestyle: There is not too much technology [sic!]. I keep it simple! I keep objects that I brought back from Japan or that remind me of Japan.

However, the image of Japan as traditional, secret, and mysterious is a Western-constructed image, promoted, not least, by the transformation of Zen Buddhism in the West. Participating in a *shakuhachi* tour not only acquaints players with the history of the music, but also with a lost exotic Japan, exemplified by the *komusō*, their wanderings from temple to temple, far from lives of contemporary Japanese (Fig. 147.5).

The difference in focus between these the Annual Japan Tenri Tours which has continued with the same leader since 1980 and Shakuhachi Roots Pilgrimage which commenced in 2003 shows, in my view, the changes in the approach to the *shakuhachi* among non-Japanese players. While both *shakuhachi* tours emphasize visits to such sacred places as temples or shrines, in Shakuhachi Roots Pilgrimage, which started 23 years after the Annual Japan Tenri Tours, the role of the *shakuhachi* as a religious instrument and experiencing spirituality is far more emphasized. And the spiritual aspect of the tour is clearly based upon a Western-appropriated Zen Buddhism and spirituality.

Fig. 147.5 The participants in front of the famous stone with the inscription Suizen (lit., blowing Zen) at Myōanji temple (Photo courtesy of Darren Stone, used with permission)

The Shakuhachi Roots Pilgrimage is packaged, promoted, and sold on the Internet as a spiritual tour to the homeland of the *shakuhachi*, in which rites are performed for those who follow the tours online on blogs, websites, and web fora. No matter how we interpret the tours, it is certain that they enrich the participants' lives and that they return inspired

References

Andersson. (2006). www.shakuhachiforum.com. Accessed 10 Aug 2006.

Berger, D. P., & Hughes, D. W. (2001). Shakuhachi. In S. Stanley & J. Tyrrell (Eds.), *The new Grove dictionary of music and musicians* (Vol. 12, pp. 831–836). London: Macmillan Publishers.

Borup, J. (2013, March 13). *Zen and spirituality: The Easternization of the East*. Seminar, Aarhus University.

Harris, R., & Norton, B. (2002). Introduction: Ritual music and communism. *British Journal of Ethnomusicology, 11*(1), 1–8.

Jansen, M. B. (2000). *The making of modern Japan*. Cambridge, MA: Harvard University Press.

Jones, S. (1999). Chinese ritual music under Mao and Deng. *British Journal of Ethnomusicology, 8*, 27–66.

Karhu, C. W. (2004). Zen, Zero Zen, No Zen and Watazumidozen. In D. Mayers (Ed.), *Annals of the ISS vol. 2* (pp. 14–15). Hong Kong: ISS.

Keister, J. (2004). The *Shakuhachi* as a spiritual tool: A Japanese Buddhist instrument in the west. *Asian Music, 35*(2), 99–131.

Lee, R. (1992). *Yearning for the bell: A study of transmission in the shakuhachi honkyoku tradition*. Thesis, University of Sydney.

Maru, S. (1922). Shakuhachi seisakuhō (尺八製作法 [Methods of shakuhachi making]). *Sankyoku, 14*, 45–47.
Mayers, D. E. (Ed.). (1990). *Annals of the ISS vol. 1*. Hong Kong: ISS.
Mayers, D. E. (Ed.). (2004a). The philosophy of Watazumido Doso Roshi. In D. Mayers (Ed.), *Annals of the ISS vol. 2* (p. 25). Hong Kong: ISS.
Mayers, D. E. (Ed.). (2004b). *Annals of the ISS vol. 2*. Hong Kong: ISS.
McClain, J. L. (2002). *Japan: A modern history*. New York: W.W. Norton & Company.
Nishimura Kokū. (1998). (Recording). *Kyotaku*. CD. Kyotaku: NG 2845.
Nyogetsu. (2013). www.nyogetsu.com/events. Accessed 16 Mar 2013.
Schechner, R. (2005). *Performance theory*. London: Routledge.
Seyama, T. (1998). The re-contextualisation of the *Shakuhachi* (*Syakuhati*) and its music from traditional/classical into modern/popular. *The World of Music, 40*(2), 69–83.
Shakuhachi Roots. (2013a). http://alcvin.ca/japantrip. Accessed 7 Mar 2013.
Shakuhachi Roots. (2013b). http://alcvin.ca/japantrip/alcvinryuzen-ramos. Accessed 10 Mar 2013.
Shimura, S. (2002). Chamber music for *Syakuhati*. In R. Provine, Y. Tokumaru, & J. Witzleben (Eds.), *The Garland encyclopedia of world music, volume 7: East Asia: China, Japan, and Korea* (pp. 701–703). New York: Routledge.
Takahashi, T. (1990). *Tozan-ryu: And innovation of the shakuhachi tradition from the Fuke-shu to secularism*. PhD dissertation, The Florida State University.
Tsukitani, T. (2008). The shakuhachi and its music. In A. Tokita & D. Hughes (Eds.), *The Ashgate companion to Japanese music* (pp. 145–168). London: Ashgate.
Tsukitani, T., Seyama, T., & Shimura, S. (1994). The Shakuhachi: The instrument and its music, change and diversification. *Contemporary Music Review, 8*(2), 105.
Yamada, S. (2009). *Shots in the dark: Japan, Zen, and the West* (E. Hartman, Trans.). Kyoto: International Research Center for Japanese Studies.

Chapter 148
The Festival of World Sacred Music: Creating a Destination for Tourism, Spirituality, and the Other

Deborah Justice

148.1 Introduction

Clad in the scanty skins and loincloths of traditional Zulu warriors, the nine performers of Colenso Abafana rhythmically stomped on stage holding spears and shields. Zulu-language chanting, percussive stamping, athletic leaping, clapping, spear clanging, and the sheer tribal rawness of it all captivated the audience. The dancers' bare chests glistened under the searing Moroccan sun as their leader, a shaman, stepped forward to address the wide-eyed crowd:

> Good afternoon you all. How are you? My name is Victor Mkhisé, and I am for Colenso Abafana from South Africa. I am very happy to be with you and I think you will be happy because we bring our song from South Africa, traditional song in our culture, I think you will be happy and because our song is sung in Zulu <pause, to build tension> and in English! I think you will be happy. (Field recording, Colenso Abafana 2001)

The audience broke into wild cheering and applause. In the happy compromise of ethnotourism, the concert-goers could experience exotic natives *and* understand them. The audience demanded multiple encores and *Colenso Abafana* was hailed as one of the most successful acts of the 2001 Fes Festival of Sacred Music[1] (Fig. 148.1).

Founded in 1994, the Fes Festival of World Sacred Music has grown into one of the largest touristic events centered on sacred music. By 2001, roughly 30,000 ticket holders were visiting the Fes Festival annually (Lynch 2000: 79) and the United Nations designated the festival as one of the major events contributing "in remarkable fashion" to dialogue between civilizations (Fes Festival 2013). The festival has

[1] For more on audience reaction to Colenso Abafana's 2001 Fes Festival Performance, see Passelègue 2001; Fez Festival 2001.

D. Justice (✉)
Department of Art and Music Histories, Syracuse University, Syracuse, NY 13244-1200, USA
e-mail: deborah.ruth.justice@gmail.com

S.D. Brunn (ed.), *The Changing World Religion Map*,
DOI 10.1007/978-94-017-9376-6_148

Fig. 148.1 Colenso Abafana at the Fes Festival of World Sacred Music 2001 (Photo by Deborah Justice)

continued to grow, in both ticket sales and influence.[2] Beautiful programs, trilingually produced in French, English, and Arabic, showcase the festival's development (Fig. 148.2). The production values have continually increased as the festival has expanded.

Promising the lure of adventure, travel, and cross-cultural contact, discretely and innocuously packaged to facilitate Western consumption, events like the Fes Festival of World Sacred Music, the World Festival of Sacred Music-Los Angeles (founded by the Dalai Lama), the World Sacred Music Festival in Olympia, Washington, and numerous smaller festivals acknowledge that the recontextualization of staging sacred performance raises issues of authenticity. Their carefully crafted display, as performance scholar and folklorist Barbara Kirshenblatt-Gimblett explains, "enables playful participation in a zone of repudiation once it has been insulated from the possibility of anyone going native" (1998: 8). As world sacred music enjoys increased global relevance, festivals have joined the industry of "ethnotourism" by enabling people to experience the excitement of "authentic" cross-cultural contact while retaining a sense of control and safety, to maintain a degree of distance from the sensitive cultural and religious subject matter.

Official Fes Festival travel agency, Sarah Tours, claims that "the festival is born out of two elements: the encounter of people from diverse areas and a rich variety of specialties, and its site, Fez,[3] a spiritual metropolis" (Sarah Tours brochure

[2] See listing of festivals in the Resources section at the end of this chapter.

[3] In this paper, I use the spelling "Fes" because it transliterates more directly from Arabic and because it is the spelling used by the Festival of World Sacred Music. However, the alternate spelling "Fez," guide is used frequently as well.

Fig. 148.2 Fes Festival posters, clockwise from upper left, 1999, 2001, 2008, 2013 (Source: © The Spirit of Fes Foundation, www.fesfestival.com/2013/en/fes.php?ld_rub=37)

2001). The company's website advertises the potential for sacred experience via a post-modern pilgrimage to Fes:

> ... travelers from all over the world will meet in the holy city of Fez.... A sacred place and a noble event where leading musicians of world caliber will share sacred music from the spiritual traditions of both East and West. Join in the spirit of this unique multi-cultural event and experience the beauty and majesty of the world's most moving sacred music. (Sarah Tours 2002)

The language in this advertisement suggests the Fes Festival as an appropriate, even glorious, extension of Fes' innate spirituality.

The Fes Festival gains authority by building upon the city's spiritual geography. The festival program rejoices in its home in "a city that holds the memory of what once was the culture and tradition of the Sacred. Tied to this past, the *medina* (old city) of Fez still continues to live by the rhythm of the call to prayer and religious celebrations" (Abdelfettah Bouzoubâa 2001: 22). Situated at the crossroads of Europe, sub-Saharan Africa, and the Middle East, the UNESCO world heritage site enjoys a heritage as a cultural and spiritual hub. As invading Arab armies settled on the Mediterranean coast, native Berber tribes concentrated in the mountains. When the Christian reconquest of Spain forced the Muslims and Jews from Andalucia, these displaced populations took refuge in Morocco's northern cities. The oldest of the four imperial cities of Morocco, Fes has been routinely graced with the presence of the royal family. Founded by national hero and Muslim saint Moulay Idriss II in 809, this "the spiritual heart of Morocco" boasts a medina (old city quarter) radiating from Zaouia Moulay Idriss II, a popular shrine and pilgrimage destination. A few meters from the intricately ornamented fourteenth century Madrassa al-Attarine, stands the magnificent Qaraouine Mosque. Founded in 859. Later centuries of French colonialism have added their influence to the mix, notably in the adoption of French as the cosmopolitan *lingua franca*. In sum, Fes' spiritual lineage equals the city's colorful cultural pedigree (Fig. 148.3).

Fig. 148.3 Bab Boujeloud, "The Blue Gate" into the medina (old city) of Fes (Photo by Bjørn Christian Tørrissen, http://en.wikipedia.org/wiki/File:Fes_Bab_Bou_Jeloud_2011.jpg)

While the spiritual venue of Fes remains the same, the festival roster changes annually to feature performers from a variety of world regions and religions. A number of Moroccan groups representing traditional regional spiritual practices tend to anchor the festival vis-à-vis local context, providing a balance to the rich palate of international acts. Similarly, Mediterranean artists have a strong presence: in 2013, 14 of the 21 artists represented Mediterranean countries/traditions; From the founding of the festival to 2013, the gathering aims "to harness the arts and spirituality in the service of human and social development, and the relationship between peoples and cultures" (Fes Festival 2013). In this eclectic setting, festival-goers' varying levels of familiarity with the cultural and spiritual traditions being presented embody what Kirshenblatt-Gimblett describes as "convergence between practical limitations on what audiences can be expected to know or learn and avant-garde principles of reception that, at least theoretically, require minimal preparation or expert foreknowledge and may even benefit from their absence" (1998: 237). In practice, this means that festival guests' expectations may lead to experience of the musics as texts unto themselves, but unfamiliarity may also prove to be a cultural and aesthetic barrier.

This essay explores the Festival of World Sacred Music as a location between the exotic and the accessible. With varying degrees of mediation and explanation, these festivals present the culturally and religiously specific for mass consumption, revisiting ethnomusicologist Lois Ib al-Faruqi's fundamental question, "what makes 'religious music' religious?" (al-Faruqi 1983). Recontextualization of religious ritual on the secular stage raises tensions between performing authenticity and presenting accessibility to challenge festival patrons, producers, and performers.

We often speak of featured acts on a festival stage as "presentations of" a particular culture or religious tradition. This language reveals a poignant awareness of the extraordinary elements of festival performance and acknowledges that festivals mark a liminal space and a "time out of time." An intensification of cultural and spiritual awareness often occurs within this temporally and physically demarcated sphere.

The geography of world sacred music festivals poses related questions of authenticity. How does the festival present itself? Do producers construe the colorful costumes, attention to "authentic" performance details, and strong sense of communitas created by festival performance as encapsulating community ideals? Conversely, does the heightened attention to providing "authentic" performance become a contrived display intended for cultural outsiders and tourists? Categorizing an event as "overly religious and proselytizing," "too pluralistic," or an "incoherent conglomeration of spirituality" says as much about the commentator's preconceived frame than the content of the festival itself. The various goals of these festivals in promoting arts agendas, highlighting ethnic minorities, proselytizing, worship, and increasing tourism shape their content and presentation.

Adding the element of sacred performance further complicates issues of performance authenticity. What are the implications when a festival advertises itself as a forum for spiritual discovery, as the majority of large international world sacred music festivals such as the Fes Festival do? How do featured artists present their music as related to the experience of the sacred? What is the impact of the festival frame in encouraging religious specificity while also encouraging an overarching atmosphere of spiritual pluralism that validates myriad individual approaches to the sacred?

Drawing upon interdisciplinary discourse running through performance studies, folklore, history, sociology, and ethnomusicology, I use the Fes Festival as a case study to show how sacred music festivals interrupt mundane geography to create poignant points of entrée into ecumenical spiritual experience. This understanding of festivals of world sacred music problematizes ethnographic musical objects as created, not found, through processes of "detachment and contextualization" (Kirshenblatt-Gimblett 1998: 3). The created festival product in turn offers "in a concentrated form, at a designated place and time, what the tourist would otherwise search out in the diffuseness of everyday life, with no guarantee of ever finding it" (Falassi 1987: 59). With its intersection of tourism, festival, and sacred experience as special and removed from mundane, everyday realities, the festival of world sacred music has created a new spiritual destination within a globalizing world.

148.2 Festivals as Time Out of Time, Space Out of Space

An integral part of organized society, festivals function similarly to music, the sacred, and the exotic by interrupting habitual human experience of time and space. In his work on the festival construct, folklorist Alessandro Falassi claims that festival activity relates to everyday life, yet:

> …at festival times, people do something they normally do not; they abstain from something they normally do; they carry to the extreme behaviors that are usually regulated by measure; they invert patterns of daily social life. Reversal, intensification, trespassing, and abstinence are the four cardinal points of festive behavior. (1987: 3)

By interrupting ordinary temporal experience, the festival creates an atmosphere of possibility and exploration. This "time out of time" (Falassi 1987: 4) not only allows, but advocates, experimentation with the unknown, as well as with that which is known, but unacceptable in everyday life.

While many festivals may draw from local populations, the Fes Festival of World Sacred Music represents a different kind of event, that of decontextualized post-tourism spirituality. These festivals promise to combine physical and spiritual geographies to convey layers of meaning that visitors do not find in everyday life. The congruence of place and content is important in a world described by historian Maxine Feifer as "post-touristic" (1985). Her term refers to an increasingly globalized world—at least in the affluent West or global north—where very little left on earth that is not accessible in some form. Sociologists Lash and Urry assert that "people are tourists most of the time, whether they are literally mobile or only experience simulated mobility through the incredible fluidity of multiple signs and electronic images" (1994: 259). Sounds, both sacred and secular, play a key role in making the global locally accessible, as ethnomusicologist Stephen Feld describes in his work on disconnects between sources of sound, their distribution, and consumption. He finds that schizophonia (decontextualization) and schismogenesis (transcontextualization and imitation) form the backbone of discourses and commodification practices of "World Music" and "World Beat" (1994). Fellow ethnomusicologist Timothy Taylor's study of marketing music demonstrates how decontextualized cultural symbols have become accepted, and even expected: "A touch of the exotic as local color … is giving way to a more pervasive series of representations and icons of Otherness as a cultural dominant" (1999: 180). Cultural recontextualization thus becomes a normative practice as the lines between cultures are blurred via increasingly frequent crossings.

Within their post-touristic worlds, many Western Fes Festival audience members spoke of a keen awareness that the festival's spiritual aspects were especially designed to highlight the event as a time and space set apart from increasingly mundane globalized existence. Cross-cultural festivals of world sacred music, claims ethnomusicologist Timothy Taylor describe their distinctive "kind of informational capital… associated only with certain highly placed social groups, even though it is theoretically available to anyone" (2000: 179). For non-Moroccans attending the festival, the cost of transportation, tickets, and lodging run high. Although locals may be interested in attending the Fes Festival (Passelègue 2001), financial realities (and a small army of security guards and ushers) initially barred them from the festival, despite its slogans of universal brotherhood and inclusion. In 2001, admission to a single concert cost the average working Moroccan 4 days' wages (*CIA Factbook* 2002). By 2013, the price in dollars remained roughly the same for official festival concerts and some free concerts have been added to appeal to a more local audience. Nevertheless, despite rhetoric of egalitarian accessibility and tolerant pluralism, with most of the visitors to the Fes Festival flying into Morocco from overseas, the audience demographics at the Fes Festival indeed represent the elite economic groups able to take part in discourses of post-tourism. The relatively high levels of exoticism and spiritual content battle new realities of the exotic as normal. In other words, building upon cultural studies scholar Marianna Torgovnick, festivalization of the global sacred creates a new geography to compensate for devaluing proximity when previously exotic sounds and rituals "can no longer serve as a locus for our powerful longings precisely because they have entered our own normative conditions of urban life" (1990: 192).

148.3 Transforming Function via Forum

> The love brings people together—love for your brother, for what you do. It overrides disagreements. Some people do it [perform] for the love of God, for the love of clapping, for the love of glory, for the love of money. But it is the love that brings them together. And if I represent what I stand for, then God is love, and God's love will come through. (T. Paschall, 11-3-2001, personal communication)

If "carnival represented is carnival tamed" (Kirshenblatt-Gimblett 1998: 77), then festivals, by their very nature, are unable to present a truly "authentic" experience. Folklorist Daniel Sheehy calls attention to modifications in content and delivery as performers "transcontextualize" (1992: 220) music that stems from religious tradition into a public-sector festival frame. On the whole, however, festival organizers, patrons, and promoters find that the experience of community created by music uniting people in a type of touristic pilgrimage nevertheless retains value. They tend to voice sentiments similar to New York Public Radio producer and Fes Festival sound man John Schafer's description of the festival as an invitation:

> I take the performance for what it is: a re-creation of a sacred rite, often of great subtlety and antiquity, in a "performance" setting. It is usually not the rite itself; rather, it's usually an introduction, and perhaps an invitation to explore more deeply. (J. Schaefer, 1-4-2002, personal communication)

Sacred world music festival participants acknowledge the demarcated nature of the festival environment such that the self-conscious perception of the festival frame preserves the integrity of the represented tradition, while packaging the performance in an accessible format. In detaching musical and religious practices from their original contexts, interfaith festivals of sacred music have created a new international geography of sacred consumption.

The selection of certain musics as legitimate for presentation plays a substantial role not only audience perceptions of the music, but in reinterpreting performance for the musicians as well. Presented by pluralistic festivals as culturally-situated couriers of the spiritually generic, sacred musicians must balance "authentic" performance practices with at least a degree of stage appeal. The process of festivalization—with its marketing, profits, fame, and the demands of tolerant pluralism—impacts performers' motivation, redefines performance practice, and problematizes proselytizing.

148.3.1 A Concert, Not a Crusade

The Fes Festival of World Sacred Music, and many similar ecumenical musical showcases, emphasize the cultural aspects of performance and a generic spiritual appeal. Musicians are framed as emissaries of the culturally specific and the spiritually ambiguous, such that pluralism is promoted but overt proselytization is discouraged: "Instead of trying to unify world visions, it would certainly be more profitable to see how the specificities and competences (sic) unique to each culture (and in the end to each experience) join to enrich this world debate" (Fes Festival Program 2001: 10). Fes' location in Morocco additionally impacts musicians' incorporation of missions into their performance. According to the United States Bureau of Consular Affairs:

> Islam is the state religion of Morocco. The Moroccan Government does not interfere with public worship by the country's Christian or Jewish minorities. However, … some activities, such as proselytizing or encouraging conversion to the Christian faith … are prohibited. It is illegal for a Muslim to convert to Christianity. In the past, American citizens have been detained or arrested and expelled for discussing or trying to engage Moroccans in debate about Christianity. (Morocco- Consular Information Sheet, March 10, 2002)

Even within the Fes Festival's relatively tolerant climate, musicians must temper their performances in conformity to local law. Although musical expressions stemming from Christian traditions should be authentic and thereby effective to a certain extent, they should not be too convincing in terms of advancing a conversion to Christianity. While they may represent a tradition that uses music as a way to reach people—for example, sonically pulling worshippers forward to (re)dedicate their lives to Christ in an alter call in evangelical Christianity—both the general pluralism and the specific rebuke against Muslim-Christian conversion tempers what these practitioners are allowed to present at Fes.

Modifications made by the Eddie Hawkins Gospel Singers to their performance at the Fes Festival illustrate these concerns of staging the religiously specific. The ensemble's website espouses their music as ministry:

> We, as Christians, are fishers of men. It is our job to "CATCH" anyone and everyone in God's "SAFETY NET" by sharing the gospel with "WHOSOEVER" will listen. It is not our job to determine who can receive the gospel of Jesus Christ. It is our job to share the gospel of Jesus Christ with everyone who wants to hear it. Thus, everyone is welcome to hear God's Word regardless of income, title, position or circumstance. GOD does the rest. (What We Believe 2002)

Despite the specificity of their statement of faith, their appearance at the 2001 Fes festival lacked testimonial and prayerful elements often found in their stage shows (Eddie Hawkins Gospel Singers 2001).

While all festivals of sacred world music are not constrained by national religious laws, they do prescribe to the egalitarian pluralism of post-modernism. Since festivals such as the Fes Festival and the Dalai Lama's World Festival of Sacred Music function on the pretext of promoting human experience of the sacred, the performers at these multi-religious venues are faced with a unique challenge. While providing an "authentic," audience-friendly show of the exotic other, artists must package their approach to the sacred in a universally appealing manner that remains true to its religious convictions.

If conversion is not the primary emphasis for performance at an ecumenical world sacred music festival, how do artists present their motivations? Fes Festival artists fall into three categories: those who base their performance of the sacred in (1) specific cultural appeal, (2) cosmopolitan appeal, and (3) historical appeal. When I attended the festival in 2001, of the 16 featured groups (Table 148.1), 10 cast themselves as primarily cultural, 3 catered to the cosmopolitan atmosphere, and 3 grounded their performances in historical recreation.

148.3.2 Culturally Based Performance

Culturally specific performance ensembles contributed their distinctive localized flavors to the overall spiritual stew of the festival. Three Moroccan spiritual brotherhoods—The Hmadcha of Fes, Les Mouloudiyats du Maroq, and the Dakka of Taroudant—presented abbreviated versions of their groups' religious ceremonies. From the East, Afroz Bano sang the Bahjan and Thumris of Rajastan and Abida

Table 148.1 Performers at the 2001 Fes Festival of World Sacred Music

Performer/Group	Country/Region of origin	Type of music	Primary type of performance appeal	Percent
The Hmadcha of Fes	Morocco	Moroccan ceremonial	Cultural	62.5 %
Les Mouloudiyats du Maroq	Morocco	Moroccan ceremonial	Cultural	
Dakka of Taroudant	Morocco	Moroccan ceremonial	Cultural	
Afroz Bano	Rajasthan	Bahjan and Thumris, devotional	Cultural	
Abida Paveen	Pakistan	Ghazal, devotional	Cultural	
Sheikh al-Tuni	Upper Egypt	Islamic devotional	Cultural	
Al-Kindi Ensemble	Syria	Arabic classical	Cultural	
Luzmila Carpio	Bolivia	Sacred songs and chants of the Andes	Cultural	
Eddie Hawkins Singers	United States	Gospel	Cultural	
Colenso Abafana	South Africa	Zulu song and chant	Cultural	
Sister Marie Keyrouz and the Ensemble of Peace	Lebanon	Mediterranean spirituality "traditional oriental orchestra"	Cosmopolitan	18.75 %
Abby Lincoln	United States	Jazz	Cosmopolitan	
Enrique Morente	Spain	Fusion of new music and traditional Flamenco style	Cosmopolitan	
Micrologus	Italy	Lay religious songs of sixteenth century Italy	Historical	18.75 %
Jordi Savall and the Hesperion XXI Ensemble	Spain	Pre-1800 European and Hispanic repertoire	Historical	
Ensemble Naguila	France	Jewish and Muslim Andalucian	Historical	

Source: Deborah Justice

Paveen contributed Pakistani *ghazals*.[4] Sheikh al-Tuni of Upper Egypt and the Al-Kindi Ensemble of Syria represented traditional Levantine Islam. Luzmila Carpio's voice soared in the sacred chants of the Andes. The Eddie Hawkins Singers brought American gospel music, while the Zulus of Colenso Abafana come fully dressed in traditional tribal garb. Although many of these musicians give dynamic and innovative performances, they choose to identify themselves as representative examples of their particular religious musical traditions.

[4] Ghazals are a genre of Islamic devotional song.

These performers base their identities upon specific cultural and religious authenticity. Since decontextualized spiritual and cultural fragments are becoming increasingly common in everyday experience, these festival performers chose to recreate a stable reference point for situating their cultural evocations of the divine. Such firm assertions of a distinctive identity become particularly important within the context of a multinational, multi-religious festival of sacred music. In large-scale events such as the Fes Festival, the discourse of pluralism, of unity in diversity, runs the risk of what might be termed the "'banality of difference,' whereby the proliferation of variations has the neutralizing effect of rendering difference (and conflict) inconsequential" (Kirshenblatt-Gimblett 1998: 76). While elements of the unfamiliar delineate different performance traditions, the sheer number and exoticism of the religious traditions presented frequently appeal to society's tendency toward generalizing the "other." If our desire for something outside of everyday life is "by nature and effect inexact or composite," we run the risk of creating a mash-up that "habitually and sometimes willfully infuses the attributes of different societies" (Torgovnick 1990: 22). To combat this sense of all-inclusive "global pop" or "world beat," some performers base their musical identities and presentation of the sacred within culturally specific media and continuously assert themselves, both over time and within specific festival appearances, in order to counter their relegation among the generic, irrelevant "other."

148.3.3 Pluralism Based Performance

In contrast, another category of Fes Festival musicians embraces exactly this all-encompassing world music and spirituality by catering their performances to the pluralistic philosophy of the gathering. The performance content and styles of Sister Marie Keyrouz, Abby Lincoln, and Enrique Morente, and to some extent the Al-Kindi Ensemble, lend themselves to the "ethnographic humanism ... cosmopolitan, progressive, and democratic" values of large festivals of world sacred music (Kirshenblatt-Gimblett 1998: 217). While these performers ground their repertoire in traditional musics, they draw inspiration and performance practices liberally from a variety of sources. Their final artistic product is a syncretic, innovative composition based that appeals to the broad-based spirituality of the festival construct.

Fes Festival organizers exclaim that Sister Marie Keyrouz "perfectly personifies synthesis of Mediterranean spirituality" (Fes Festival Program 2001: 32). The Lebanese nun's Paris-based musical "Ensemble of Peace" features both Christian and Muslim musicians. The "traditional oriental orchestra's" appearance at the Fes Festival was billed as furthering Keyrouz' mission of "fostering Eastern and Western dialogue." A doctor of musicology and skilled performer in many genres,[5] Keyrouz

[5] "Cultivating Milanese and Gregorian chant with equal finesse as the Byzantine or Syro-Maronite style of the Christian Oriental Church, or even Gounod, Bruckner or Schubert (as in a recent recording)..." (Ibid).

espouses her message as: "a prayer for love and peace which helps to break down barriers and helps to expand boundaries-fostering an understanding that brings solitary individuals and regions together." This message of pluralistic tolerance, expressed through a fusion of sacred styles, dovetails with the Fes Festival's aims.

Such musical innovation and fusion, however, raises the ever-present issue of authenticity. By venturing beyond the safety of tradition, artists like Keyrouz run the risk of adopting and epitomizing Torgovnick's inexact, composite exoticism in their approach to experience of the sacred. By including a grand piano in her "oriental orchestra," Sister Keyrouz chooses to limits her repertoire to *maqamat*, Arabic modes, without semitones, or forces unorthodox modification of the *maqamat* to accommodate the piano's intervals. In addition to these modal variations, my interviews with participating musicians revealed musical cosmopolitanism at work. When I talked with the pianist after the show, he explained that he hailed from a jazz background and frequently tried to "spice things up a bit" by infusing chromatic runs and jazzy riffs into his playing. Such musical invocation can be seen subjectively either as positive development of within the tradition, or as a negative departure from it.

In refusing to adhere to a single musical traditions or religious approach to the divine, these musicians create a genre of post-modern sacred music that reflects the cross-cultural and cross-confessional festival context. Sharing the post-tourist's penchant for introspection and evaluation, these performers are keenly aware of the cultural and musical influences of their music. Writing "Misa Flamenca" Spanish artist Enrique Morente purposefully incorporated the liturgy of the Catholic mass, his knowledge of classical music theory, and "new musical trends without losing touch with the traditional Flamenco style" (Fes Festival Program 2001: 48). The result embodies a syncretic, self-determining musical pursuit of the experience of the sacred aligned with the personalized, reception-based spirituality of the Fes Festival of World Sacred Music.

148.3.4 Historically Based Performance

The third group of performers uses the spiritual connotations of history as an alternative basis of legitimacy and identity. As British novelist L.P. Hartley (1953: 1) famously penned, "The past is a foreign country; they do things differently there." Positioning themselves in opposition to the modern mainstream, this group of performers appropriates musics of the past, infusing them with new potential for experiencing timeless aspects of the sacred within the paradigm of post-modern spirituality. Ethnomusicologist Tamara Livingston suggests understanding revivalists as sharing an "overt cultural and political agenda" that prompts them to "align themselves with a particular historical lineage, and offer a cultural alternative in which legitimacy is grounded in reference to authenticity and historical fidelity" (Livingston 1999: 66). These performers strive to communicate their experiences of the sacred what they perceive as the purity, authenticity, and "otherness" of the past (Figs. 148.4 and 148.5).

Fig. 148.4 Les Mouloudiyats du Maroq at the Fes Festival of World Sacred Music 2001 (Photo by Deborah Justice)

Fig. 148.5 Jordi Savall and the Hesperion XXI Ensemble at the Fes Festival of World Sacred Music 2001 (Photo by Deborah Justice)

Historically-based ensembles constituted 18.75 % of the performing groups at the 2001 Fes Festival. Micrologus,[6] Jordi Savall and the Hesperion XXI

[6]An Italian ensemble that recreates the orally transmitted laudes and other lay religious songs of sixteenth century Italy. According to the 2001 Fes Festival program, "The group bases its interpretations on solid historical, organolgical and iconographical research, as well as comparative folklore studies" (54).

Ensemble,[7] and Ensemble Naguila[8] billed themselves as performing "sacred music," but emphasized their repertoire's historical roots over its spiritual content. Hesperion XXI described itself as "brought together by the study and interpretation of ancient music through a modern premise, as well as a deep respect for the immense richness of the pre-1800 European and Hispanic repertoire" (Fes Festival Program 2001: 40).[9] The musicians seemed to share a "flow" experience of emotion and musical intensity with their listeners, as both groups swayed, closed their eyes, and clapped. Although early music may initially seem far removed from the post-modern pursuit of the sacred, the festival of world sacred music demonstrates how casting the past as a place of authenticity, purity, and return supports festival-goers' post-touristic rejection of everyday banality.

148.4 Locating the Festival World Sacred Music in the Geography of Post-tourism

Larger international festivals of world sacred music, such as the Fes Festival, advertise their potential for encouraging experience of the sacred, "The World Sacred Music Festival is an offering of shared blissful moments, through the medium of devotional music from around the globe, known to all human hearts and understood as a universal sacred language of the soul" (Sarah Tours 2001: 21). Told that they can indeed experience the sacred through all of these musics, festival-goers are given the opportunity to reconcile their own religious beliefs and practices with the diverse array of featured sacred musical traditions.

While much of the music featured at a festival of world sacred music generally functions within ritual as religious music, often performed for an actively participating congregation of a single faith, the festival setting transforms this insider's active experience of the sacred and the religious into a staged spectacle that permits, but does not demand, active audience participation and experience of the sacred. Most festivals of world sacred music emphasize their *potential* for promoting experience of the sacred, and many festival-goers attend the event with a goal of spiritual experience in mind. The festival removes much of the cultural and religious framework for the represented sacred musical traditions, leaving something of a contextual vacuum. Unencumbered by highly enculturated ritual settings, festival-goers impose their own meanings and experiences of the sacred from a mélange of featured tradi-

[7] Led by Spaniard Jordi Savall, Hesperion XXI is "brought together by the study and interpretation of ancient music through a modern premise, as well as a deep respect for the immense richness of the pre-1800 European and Hispanic Repertoire" (Fes Festival Program: 40).

[8] Formed in France, Ensemble Naguila "reunites Jewish and Muslim musicians in the spirit of nurturing peaceful coexistence and a shared musical quest to preserve these traditions" of Andalucian Spain and post-Reconquista 'Ala al-Andalus in Morocco (Fes Festival Program: 50).

[9] Regrettably, the Festival program neglects to elaborate upon the details of Hesperion XXI's "modern premise" of interpretation.

tions. The nonconfrontational nature of the world sacred music festival setting encourages individuality to the point that one cannot speak of a "shared" festival experience (Kirshenblatt-Gimblett 1998: 247). Ultimately, the impact of festivalization of sacred world music rests upon the festival-goer's personal approach to the experience such that, "meaning derives not from the original context of the fragments but from their juxtaposition in a new context" (Kirshenblatt-Gimblett 1998: 3). Interaction with the sacred remains a highly individualized phenomenon, resting entirely upon the individual's level of participation, mental openness, and suspension of disbelief and skepticism.

Armed with increasing technology, upward mobility, and disposable income, consumers pursue a newer kind of cultural and educational capital, defined by Timothy Taylor as "*global informational capital*: distinction, knowledge, the ability to travel, learn, and discriminate" (2000: 174). Post-tourists embrace the confidence that this global information capital elevates their musical and spiritual experiences. These consumers often express sentiments that, as a result of enlightened introspection, their "orientation to travel is far more sophisticated than that of mere 'tourists'" (Lash and Urry 1994: 308). Building on Feifer's "post-tourism" model, Lash and Urry contend that reflexive self-awareness is the defining mark of recent discourse, "Indeed, most important of all, the post-tourist knows that he/she is a tourist, and that tourism is merely a series of games and multiple texts and no single authentic experience. The post-tourist is ironic and cool, self-conscious and role-distanced" (1994: 275–276). Following a concert at the 2001 Fes Festival that she described as particularly moving, one Australian woman, voiced the sentiments of many when she looked disappointedly for a merchandise table and exclaimed, "You'd think they'd at least have t-shirts!"

By extension, Fes Festival attendees largely understand themselves as participating in a constructed event that is at once authentic and artificial. By staging sacred music outside of its original ritual context, festivals of world sacred music project the exotic and the spiritual in a new physical and conceptual destination. Marianna Torgovnick explains that such ambiguity encourages exploration that simultaneously allures and frightens:

> Our interest in the primitive meshes thoroughly, in ways that we have only begun to understand, with our passion for clearly marked and definable beginning and ending that will make what comes between them coherent narrations. A significant motivation for primitivism in modernism, and perhaps especially in postmodernism, is a new version of the idyllic, utopian primitive…the wish for physical, psychological, and social integrity as a birthright, within … cultural traditions that both connect to the past and allow for a changing future. (1990: 245)

Embracing the ambiguity and flexible roles of post-modernism, the post-tourist festival-goer faces the difficult task of casting off cynicism, preconceptions, and "knowledge" in order to experience meaning. To varying degrees, he or she recognizes that the sacred music presented for mass enjoyment comes from enculturated religious practices and that its nature is transformed by presentation on a public stage. Embracing the reflexivity and ambiguity of post-tourism, the festival-goer may relish the power and autonomy of defining his/her own role in the recontextualized performance.

The inviting forum of the world sacred music festival encourages all of its participants, producers, performers, and audience members alike to join in this pilgrimage of sorts on a post-modern search for identity. By positioning themselves vis-à-vis the festival's definition of the exotic and authentic, audience members, performers, and patrons negotiate boundaries of cultural and spiritual identity:

> ...the West was once much more convinced of the illusion of Otherness it created. Now everything is mixed up, and the Other controls some of the elements in the mix ... the "them" is much more like us now, and us, often garbed in clothing and living amid objects that evoke "their" traditional forms of life. (1990: 38)

The festival of world sacred music carves out a unique space in the sacred landscape, one that affords participants the chance to try on certain self-identities and spiritualties within the tolerantly ambiguous pluralism of a post-modern, post-touristic sacred geography.

148.5 Resources

148.5.1 Audio from the Fes Festival

- B'ismillah: Highlights from the Fes Festival of World Sacred Music. (1997) Sounds True. AF MM 00339D. Compact disc.
- Hamdulillah; Fes Festival of World Sacred Music, Volume II (1998) Sounds True. AF MI08D-2.101, W3987. Compact disc.
- Traces de Lumiere: Fes Festival of World Sacred Music. (2005) Buda Musique. BUD601205 Compact Disc.
- Under the Moroccan Sky. (2001) Sounds True. AF B00005B53M Compact Disc.

148.5.2 Video from the Fes Festival

- *Sawt-e-Sarmad.* (2008) CreateSpace. B001L4L9G8. DVD.
- *Sound of the Soul: The Fez Festival of World Sacred Music.* (2005) Alive Mind. DVD.

148.5.3 Selected Listing of Festivals of World Sacred Music

- Fes Festival of World Sacred Music. Fes, Morocco. www.fesfestival.com
- Le Festival Gnaoua et des Musiques du Monde d'Essaouira. Esaouira, Morocco. www.festival-gnaoua.net/en/
- World Festival of Sacred Music. Los Angeles, California, USA. www.festivalor-sacredmusic.org
- World Sacred Music Festival. Olympia, Washington, USA. 222.oly-wa.us/olysacredmusic

References

al-Faruqi, Lois Ibsen. (1983). What makes 'religious music' religious? In J. Irwin (Ed.), *Sacred sound* (pp. 21–34). Chicago: Scholars Press.
Bouzoubâa, A. (2001). The Fez festival: The meeting with a city. In *Fes festival of world sacred music* [Concert program] (p. 22). Fes.
CIA Factbook: Morocco: Economy. (2002). Retrieved April 2, 2002, from www.odci.gov/cia/publications/factbook/
Colenso, A. (2001, June 3). Field recording. *Fes Festival of World Sacred Music*.
Eddie Hawkins Gospel Singers. (2001, June 10). Field recording: *Fez Festival of World Sacred Music*.
Falassi, A. (Ed.). (1987). Festival: Definition and morphology. In A. Fallassi (Ed.), *Time out of time: Essays on the festival* (pp. 1–12). Albuquerque: University of New Mexico Press.
Feifer, M. (1985). *Tourism in history: From imperial Rome to the present*. New York: Stein and Day.
Fes Festival of World Sacred Music. (2001). *Festival program*. Morocco: Fes.
Fes Festival of World Sacred Music. (2013). Retrieved March 8, 2013, from www.fesfestival.com/2013/indexen.php
Fez Festival of World Sacred Music. (2001). Retrieved December 2, 2001, from www.fezfestival.org/
Hamdulillah: Fes Festival of World Sacred Music, Volume II. (1998). Sounds true. AF MI08D-2.101, W3987. Compact disc.
Hartley, L. P. (1953). *The go-between*. London: Hamish Hamilton.
Kirshenblatt-Gimblett, B. (1998). *Destination culture: Tourisms, museums, and heritage*. Los Angeles: University of California Press.
Lash, S., & Urry, J. (1994). *Economies of sings and space*. London: Sage Publications.
Livingston, T. E. (1999). Music revivals: Towards a general theory. *Ethnomusicology, 43*(1), 66–81.
Lynch, D. (2000). *Staging the sacred in Morocco: The Fes festival of world sacred macred Music*. MA thesis, Department of Anthropology, University of Texas, Austin.
Morocco-Consular Information Sheet. (2002). Retrieved March 10, 2002, from http://travel.state.gov/morocco.html
Passelègue, V. (2001). Culture shock: Sufi Poetry, Sephardic Chants and Zulu Rhythms in Fes. *Vibrations*. Retrieved June 9, 2001, from http://rfimusique.com/gb/Datas_Articles/1206101162706.html
Sarah Tours. (2001). Retrieved December 2, 2001, from www.morocco-fezfestival.com/5thm_s99.html
Sarah Tours. (2002). Retrieved January 20, 2002, from www.sarahtours.com
Sheehy, D. (1992). Crossover dreams: The folklorist and the folk arrival. In R. Baron & N. R. Spitzer (Eds.), *Public folklore* (pp. 217–229). Washington, DC: Smithsonian Institution.
Taylor, T. (2000). World music in television ads. *American Music, 18*, 162–192.
Torgovnick, M. (1990). *Gone primitive*. Chicago: University of Chicago Press.
What We Believe. (2002). Retrieved March 3, 2002, from www.lovecenter.org/WhatWeOuterFrame.htm

Chapter 149
More Than Meets the Ear: The Agency of Hindustani Music in the Lives and Careers of John Coltrane and George Harrison

Kevin D. Kehrberg

> *But there are cultures of this earth which do not depend on either the clock or causality, cultures where thought, value, and even life itself are not primarily linear, cultures in which right-brain thought processes are predominant. (Jonathan D. Kramer, The Time of Music (1988))*

149.1 Introduction

During the 1960s, western interest in the music of India rose to unprecedented heights. For Ravi Shankar—who unequivocally remains the best-known Indian musician to westerners—1967 marked the apex of this rise:

> As far as superstardom and popularity went, I was absolutely at the height of it! I appeared on numerous TV talk shows, including Johnny Carson (twice), David Frost, Joey Bishop, Dick Cavett and Mike Wallace. I couldn't walk in the street without being mobbed, with fans following me shouting, "Hi Ravi!" or even, "Hi Rav!" It was unbelievable—though I was enjoying it, I won't deny that. I was so busy at the time that I couldn't think properly. Everything was happening at once: moving so fast, giving concerts and making appearances, I don't know where the next two years went. It was as if I was completely in the hands of my managers, with all the dates fixed by them. Everything was a blur. (Shankar and Craske 1997: 204)

As he had been touring in the West for over a decade by 1967, Shankar was able to watch this explosion unfold, and he was certainly a prime factor in presaging its emergence.[1] What ultimately shifted the West's interest in Indian music (and thus in the sitarist himself) from marginal to monumental at this time was Shankar's recent and blossoming association with George Harrison. The Beatles lead guitarist was part of a musical group that by the mid-1960s had reached popularity levels that no one dreamed existed, and the eye of the public was on its every move.

[1] Shankar had made the acquaintance of violinist Yehudi Menuhin as early as 1952, and he recorded an album with a group of Western jazz musicians as early as 1961.

K.D. Kehrberg (✉)
Department of Music, Warren Wilson College, Asheville, NC 28815, USA
e-mail: kkehrberg@warren-wilson.edu

S.D. Brunn (ed.), *The Changing World Religion Map*,
DOI 10.1007/978-94-017-9376-6_149

Indeed, through the medium of popular music, the western world was exposed to this music from the East. However, rock was not the only popular music vehicle through which it arrived. Some authors claim that by the last half of the 1950s, jazz musicians had already taken notice of Indian music and were entertaining and realizing ways of grafting its principles of improvisation with their own (Berendt 1984; Harrison 2001).[2] As early as 1961, Shankar recorded an album entitled *Improvisations* that featured himself collaborating with western jazz musicians. This was years before rock began experimenting with Indian music, and by the time of Harrison and Shankar's first meeting in 1966, many jazz musicians were already looking at Indian musical practice as a way of informing and enhancing their own art.

Although both North (Hindustani) and South (Karnatic) Indian classical music contain many similarities (the latter is the root of the former), the exchanges with western popular music in the 1960s primarily involved Hindustani musical style.[3] With an instrumental tradition dominated by the lyrical sitar and the resonant patterns of the tabla drums along with a much greater emphasis on improvisation, North Indian classical music contained elements that were perhaps more striking for the uninitiated jazz and/or rock musician. Through Shankar's performances, the Hindustani style certainly received more exposure in the West.

Despite the superstar status obtained by Shankar in the late 1960s, writers have shown that—at least in the rock world—the draw of Indian music faded fast, becoming a worn-out fad after only a few years. Although it was a consuming interest for a time, it was, nevertheless, one that lacked serious understanding of the art form itself or the culture it represented (Lal 1995; Bellman 1998).[4] As far as creating individuals who were delving into the substance *behind* the music—the meanings, philosophies, and religious significances—the same was true for the jazz world at large. However, musical appropriation in jazz often occurs within a much different conceptual framework. For many jazz musicians, once a new musical implement has been added to their toolbox, there is no reason to discard it. As a part of their growing vocabulary, it becomes one among many methods of creating their art. Rather than seen as a fad, it is joined to a larger body of concepts and approaches upon which they draw.

Regardless of opinions involving Indian music's purportedly faddish stint on the West's pop music radar screen and whether or not a certain artist used its elements in an "authentic" way, two musical figures garner mention in almost any discussion of this exchange. They are John Coltrane and the aforementioned

---

[2] Harrison posits that it was Shankar's concerts in the West during the late 1950s that sparked the interest of American jazz musicians in Indian music, revealing possibly another aspect of the sitarist's involvement in the music's spread.

[3] For this reason, discussions of any facet of "Indian classical music" in this study, unless otherwise noted, refer to the North Indian style.

[4] Lal gives the cover art for Jimi Hendrix's 1967 album *Axis: Bold as Love*, which depicts his band as "a three-headed Hindu deity," as an example of the "superficial interest" that marred much of Indian music's popularity during this time (p. 94).

George Harrison. The inconspicuous importance of these two musicians in discourse covering the relationships of Indian music to western popular music are undeniable, and as such they appear almost without fail in literature on the subject. In addition, they both are members of that small group of artists who went much deeper than just the music of India, forming enduring relationships with the culture, religion, and philosophy that lie at its root.

Both Coltrane (1926–1967) and Harrison (1943–2001) are considered pioneers in the incorporation of Indian musical elements into their respective musical genres. Both also became profoundly and permanently transformed in personal and spiritual ways as a result of their study of and experimentation with Indian music. Not surprisingly, these transformations reshaped their artistic goals to new heights of social consciousness, producing in both a strong desire for their music to be a force for good and an agent for social change on one hand and a form of praise to God on the other. These ideals became the top priority for Coltrane and Harrison, far outshining hopes for personal recognition and/or commercial gain through their music (although, paradoxically, these were still achieved on astonishing levels for both).

A comparison of John Coltrane and George Harrison reveals striking parallels in the ways that each man's career and life were affected by their contact with Indian music. This study examines these parallels, paying close attention to how each musician/composer was personally transformed as well as how each developed a subsequent desire to transform others with their music. It investigates the different manners in which Coltrane and Harrison implemented Indian musical elements into their own music as a means to accomplish their artistic goals. In doing so, it provides case studies of Indian music's ability to sonically embody—through such properties as modality, time, and the use of drones—the philosophies and thinking that are inseparable from it in its native cultural context. Both Coltrane and Harrison had distinctly different approaches in their efforts to raise audience awareness and reach listeners on a deeper level with their music. A probe into the success and reception history of these approaches evaluates the effectiveness of each in fulfilling his objectives. Finally, this study considers Coltrane's and Harrison's appropriation of Indian music and philosophy within the context of their careers. How do the demands and expectations of the respective pop-culture worlds of Coltrane, a black American jazz musician, and Harrison, a white British rocker, reflect themselves in each man's relationship with this eastern music and culture?

As a result of these queries, portraits emerge of two artists who, during the third quarter of the twentieth century, found spiritual renewal by looking for musical inspiration from a particular source outside of their cultures. From roughly that time forward, their desire to share their experience consumed their careers in different ways: Coltrane's method was more understated; Harrison was, initially, very forthright. While their careers reached their respective peaks within a decade or so of each other, both operated in considerably different popular- and musical-cultural worlds. Their use of Indian musical elements reflects this. Harrison was coping with questions of identity and independence amidst the reality of being permanently associated with the greatest pop group of all time. Coltrane was dealing with the rising volatility of the Civil Rights movement and the conflicting ideals of unification and divisiveness entwined within his elevation as a symbol of black nationalism. Through it all, the spirituality acquired through their studies of Indian music helped both men sustain a social-centered consciousness in their art and persevere in their creative pursuits.

149.2 John Coltrane

The history of the confluence of western jazz and Indian music begins much earlier than Coltrane's forays. It essentially begins during the final decades of the British Raj, when big bands of the Swing Era both traveled to and arose within the borders of India. Although swing was certainly popular there, it operated primarily within the confines of the colonial elite with little, if any, musical amalgamation occurring between the music of India and that of swing.[5] In the American scene, however, Coltrane is generally regarded as the first to produce a recording—"My Favorite Things" (1960)—that genuinely attempts to incorporate Indian musical elements and sensibilities with jazz (Berendt 1984: 13; Pickney 1989–1990: 41).

Certain biographical details concerning the great saxophonist imply that he was entertaining the potential of such an exchange long before he attempted it. As early as 1956, Coltrane reportedly told composer David Amram that "the improvisational principle of Indian ragas could be very important for jazz" (Berendt 1984: 13). In 1957 he was part of Thelonius Monk's group for a short time which included bassist Ahmed Abdul-Malik who, according to biographer Bill Cole, "was one of the early African American musicians to show serious interest in the music of the East" and whose "association with Trane continued Trane's own interest in Eastern music." (Cole 1976: 64). Cole also claims that the modal music that Miles Davis's group began experimenting with in 1958, particularly the composition "Milestones," resonated with Coltrane's new awareness of Indian music and its system of *ragas*.[6] Perhaps most significant of these precursory events was Coltrane's involvement in the recording of the legendary Miles Davis album *Kind of Blue* (1959), which represents a zenith in the use of modal concepts in jazz.[7]

In the same year that *Kind of Blue* was recorded, Coltrane also completed his second significant album as a leader, *Giant Steps* (released in 1960). On this album, the composition "Naima" (purportedly Coltrane's favorite of his original works)

---

[5] According to author Warren R. Pickney, Jr., Rudy Cotton was an exception to this convention (Pickney 1989–1990). Cotton was a saxophonist from the territory of Goa in western India who recorded with Indian swing bands during the 1930s and 1940s and continued playing into the 1970s.

[6] Modal jazz, which undoubtedly took part of its inspiration from the modal sound that dominates the native music of many cultures (including India), focuses the improvisation on select scales as opposed to a series of several chords progressing at a fixed rate.

[7] Some authors even cite the music of *Kind of Blue* as a precedent to Coltrane's "My Favorite Things" in the emergence of Indian elements in jazz (Harrison 2001). However, whether enough elements are there for the Davis album to warrant this distinction is questionable.

reveals even more hints of eastern musical influence. In the words of Coltrane himself, "The tune is built on suspended chords over an Eb pedal tone on the outside. On the inside—the channel—the chords are suspended over a Bb pedal tone" (Hentoff 1998a) According to Cole, this tonic-dominant relationship represents "the drone from which improvisations are developed, just as in the music of India" (Cole 1976: 110).

During this period of the late 1950s, explorations into various religions and philosophies—including those of India—paralleled and complemented Coltrane's multicultural musical interests of this time. In 1957, he decided to break the cycle of drug and alcohol abuse that had plagued his life up to that point and experienced a "spiritual awakening" in the process (Coltrane 1964). This awakening had been spurred on by an emerging interest in Christianity, Islam, and other spiritual worldviews, including the practice of meditation. Coltrane's exposure to eastern writings had also begun at least as early as that year, and by 1958 he was experimenting with yoga and vegetarianism (Cole 1976: 97). This also marked the year that Bill Evans, the pianist in Miles Davis's group at the time, introduced Coltrane to the writings of J. Krishnamurti, the influential twentieth-century Indian spiritual philosopher.

In 1960, Coltrane had his first major "hit" with a recording of the popular tune "My Favorite Things." The song also represents Coltrane's first notable amalgamation of jazz and Indian musical elements. His use of the soprano saxophone is one example; he was primarily a tenor player up to this point. Authors have noted the striking resemblance of his piercing, penetrating, and nasal soprano tone to the North Indian shenai (Farrell 1988: 189–205). At almost 14 min in length, the performance far exceeds in duration the modal experiments of *Kind of Blue*, providing a temporal affinity to the lengthy improvisations that characterize Hindustani music. Furthermore, the drone sound that Coltrane was perhaps after in "Naima" is much more prominent in "My Favorite Things." The double bass practically strums an open fifth (E and B) the entire time, changing *only* for the dominant chord of the melody's final cadence. This drone also dominates the solo sections where, other than infrequent statements of the melody, Coltrane discards the form in favor of a static vamp.

Coltrane continued to perform and record with his soprano into the early 1960s, and in 1961 he recorded another soprano composition titled "India," signaling his growing affinity for the foreign land's culture and music.[8] In 1964, Coltrane recorded *A Love Supreme*, a landmark album which declared his increasing dedication to producing music that was grounded in his spirituality and devotion to God. Complete with a poem of praise of the same title, Coltrane's album notes open with the lines "ALL PRAISE BE TO GOD TO WHOM ALL PRAISE IS DUE" (Coltrane 1964). The music itself is completely modal, featuring virtually no harmonic changes, and heavily improvised. The final minutes of the opening piece, "Acknowledgement," contain a mantra over the words "a love supreme."

---

[8] It should be noted that this composition also served to honor the new daughter, named India, of one of Coltrane's good friends, Calvin Massey (Cole 1976: 146).

From this point forward, Coltrane's studio albums and compositions are almost completely sacred in nature. His compositions include such titles as "Vigil," "Ascension," "Dear Lord," "Prayer and Meditation: Day," "Prayer and Meditation: Evening," "Prayer and Meditation: 4 AM," "Dearly Beloved," "Amen," "Joy," and "Selflessness"—all recorded in 1965. In this same year Coltrane recorded a piece simply called "Om," which one biographer has labeled as "among the finest—if not *the* finest—playing that Trane did in his career" (Cole 1976: 179). "Om," nearly 29 min long, is also unique because it marks Coltrane's most extensive use of various outside instruments—gongs, bells, wooden flutes, a thumb piano, cymbals—to lend the music a decidedly non-western sound. In addition, the musicians chant selected verses from the Bhagavad-Gita to open and close the piece. In November 1965, Coltrane went into the studio again and recorded *Meditations*. After an opening track entitled "The Father and the Son and the Holy Ghost," he offers listeners musical representations of "Compassion," "Love," "Consequences," and "Serenity." 1965 marked an extremely productive year for Coltrane, and although some important live recordings were made after this year, his studio output seriously declined. Only a handful of studio works were recorded from 1966 until his death in 1967.

Naturally, this music reflected Coltrane's personal transformation into an individual with a profound interest in spirituality in general and its relationship to music in particular. One admirer who met the great saxophonist in 1960 observed that he traveled with three or four Bibles and would talk to younger musicians about the benefits of meditation (Cole 1976: 121). Such an interest in matters both spiritual and musical led Coltrane to dig deeper into Hindustani music. In November 1965, he finally had the opportunity to meet Ravi Shankar, an individual whose music Coltrane highly respected (he named his second son, born in August of that year, Ravi). Evidently the two had been corresponding via mail for some time, and they set up a meeting in New York. (Thomas 1975: 199–200). They spent much of their time together discussing Indian music and its spiritual basis. Ravi recalls:

> He was intrigued by Indian music and asked me all about it. As a jazz artist, he was amazed by our different system of improvisation within the framework and discipline of fixed melody forms, by the complexity of our talas, and more than anything by how we can create such peace, tranquility and spirituality in our music. (Shankar and Craske 1997: 178)

According to Shankar, he and Coltrane had three or four instructional sessions such as this one. The two even discussed a plan for the saxophonist to come study with Shankar for an extended period of time, but Coltrane's untimely death from liver cancer prevented its fruition.

By 1965, Coltrane had been studying such Hindu religious texts as the Bhagavad-Gita (used in "Om") and the writings of Ramakrishna. His meditation practices had become an important part of his daily routine, as evidenced by some of his works from 1965 as well as the album *Meditations*. In discussing the music of the latter, Coltrane remarked:

> There is never any end…There are always new sounds to imagine, new feelings to get at. And always, there is the need to keep purifying these feelings and sounds so that we can

really see what we've discovered in its pure state. So that we can see more and more clearly what we are. In that way, we can give to those who listen the essence, the best of what we are. But to do that at each stage, we have to keep on cleaning the mirror. (Hentoff 1996)

One might presume that Coltrane was joining the trend in spiritual awareness via eastern thinking that had begun to sweep western popular culture during this time. But coming from a jazz musician in the 1960s, Coltrane's worldview was uncommon. In the words of one writer, jazz musicians up to that time "only talked about music—and if they talked about anything else it was about chicks and grass and booze" (Berendt 1984: 14). Ravi Shankar makes a similar observation:

Coltrane's approach was different from that of most jazz musicians I had seen or met over the years... [He was] so clean, well-mannered and humble...He had apparently given up drugs and drink, become a vegetarian and taken to reading Ramakrishna's books. For a jazz musician to go to the other extreme, especially in those days, was a pleasant surprise. (Shankar and Craske 1997: 176–78)

Respect for and contemplation of nature had also become an important facet of Coltrane's worldview in the 1960s. In the last years of his life, Coltrane maintained a sizeable garden at his residence in which he spent considerable time. In India, nature is sometimes worshipped as God's "Universal Form" (Greene 2006: 226). One particular comment of Coltrane's shows how he wove reflections on nature into his musical philosophy. "All a musician can do is to get closer to the sources of nature, and so feel that he is in communion with the natural laws. Then he can feel that he is interpreting them to the best of his ability, and can try and convey that to others" (Cole 1976: 11).

Coltrane's remark here about conveying these interpretations "to others" is noteworthy. It reflects the social obligation that had begun to manifest itself in him and his music ever since his spiritual awakening in 1957, at which time he "humbly asked [God] to be given the means and privilege to make others happy through music" (Coltrane 1964). This compulsion became prominent in the 1960s, paralleling his deep investigations into the philosophies of Christianity and other religions, including Islam, Hinduism, and Sufism. The core of Coltrane's faith was Christian, but he drew heavily on the meditative practices of eastern thought and maintained a very pluralistic view of religion, at one point claiming that he "believed in all religions" (Hentoff 1996). In this way, he was able to extend his mission of social change through his music to anybody and everybody, regardless of whether or not they professed the same tenets of faith.

Perhaps the most vivid indications of Coltrane's growing social consciousness are his own words on the subject. In a 1962 interview, he said, "I know that I want to produce beautiful music, music that does things to people that they need. Music that will uplift and make them happy: those are the qualities I'd like to produce" (Cole 1976: 11). From his comments concerning *Meditations*, Coltrane reiterates this purpose, "My goal in meditating on this [force for unity in life] through music...remains the same. And that is to uplift people, as much as I can. To inspire them to realize more and more their capacities for living meaningful lives" (Hentoff

1996). In one of his final interviews in the summer of 1966, Coltrane outlines his opinions on the power of music to transform (and his own ambition to do so) with even greater clarity:

> I think I can truthfully say that in music I make or I have tried to make a conscious attempt to change what I've found...In other words, I've tried to say, "Well, this I feel, could be better, in my opinion, so I will try to do this to make it better." [W]hen there's something we think could be better, we must make an effort to try and make it better. So it's the same socially, musically, politically, and in any department of our lives. I think music is an instrument. It can create the initial thought patterns that can change the thinking of people. (Kofsky 1970: 227)

What sort of change is Coltrane after? Later on in the interview he states very simply, "I want to be a force for real good... I know that there are bad forces, forces put here that bring suffering to others and misery to the world, but I want to be the force which is truly for good" (Kofsky 1970: 241).

Coltrane wanted to harness this force in a concrete fashion. He envisioned establishing an educational institution where young jazz musicians could come for not only practical music instruction, but also instruction in the "theoretical and philosophical aspects of jazz" (Cole 1976: 195). Unfortunately, his premature death left this and surely other humanitarian goals unrealized, precipitating a continuous dialogue ever since that ponders the legacy of the visionary musician had he lived longer.

149.3 George Harrison

Like John Coltrane, George Harrison garners the attention of most writers as the major figure in pioneering a fusion of rock music with Indian musical elements. This distinction, however, mainly grows out of the fact that the Beatles were the first rock group to record and commercially release a song that featured the sound of a sitar. Multiple authors have made strong arguments showing that while the Beatles were the first group to achieve this particular milestone, other groups were the first to appropriate certain structural elements of Indian music (Bellman 1998: 294–297; Harrison 2001).[9] Regardless, due to the overwhelming popularity of the Beatles, they undoubtedly deserve credit as the musical group most responsible for elevating public interest in Indian music during the 1960s.

Despite the debate surrounding who was first, none of the early precedents possesses a release date before 1965, by which time the Indian/jazz exchange was arguably well under way. The extent to which jazz inspired rock and folk musicians to experiment with Indian musical elements is unclear, although it certainly played a role in the case of artists like David Crosby and the Byrds. Crosby introduced the music of John Coltrane and Ravi Shankar to the Byrds, an

[9] Both Bellman and F. Harrison propose the Yardbirds and the Kinks as possible precursors to the Beatles in the Indian/rock exchange.

American rock band, and their 1966 release "Eight Miles High" actually incorporates a quotation from Coltrane's playing on "India" (Harrison 2001). According to the Byrds's Roger McGuinn:

> The first break is a direct quote from a Coltrane phrase, and throughout the rest of the song, we try to emulate the scales and modes that Trane was using. Especially his spiritual feeling, which got me into transcendental meditation not long afterwards. (Thomas 1975: 199)

Evidently, Coltrane was accomplishing his musical objectives with at least some listeners. As for Harrison, however, I have uncovered no direct connection between him and the saxophonist, although it is doubtful that he was completely unaware of Coltrane's work.

Harrison's introduction to Indian music and ultimately Indian culture and religion was through the sitar. His first encounter with the instrument in 1965 was quite happenstance. There was one being used on the set of the movie *Help!* and the curious Harrison picked it up, strummed it, and remembered thinking, "This is a funny sound" (Bellman 1998: 293). This first impression sparked the young guitarist's interest enough that he decided to use one on the recording of "Norwegian Wood" in October of that year. Incidental as the playing was, it marks a milestone in rock history as the first recording to contain the instrument.

In June of the next year, Harrison met Ravi Shankar for the first time in London and soon began sitar lessons with him. Their first lesson affected the young guitarist deeply. "I felt like I wanted to walk out of my home that day and take a one-way ticket to Calcutta," he later remarked (Greene 2006: 63). He did leave for India in September 1966, spending his time studying the sitar with Shankar, staying with him on a private houseboat on a lake in Kashmir, and being introduced to the teachings of important Hindu thinkers. The experience was life changing for Harrison:

> One book ... that was a great influence on me was *Raja Yoga* by Swami Vivekananda. Vivekananda says right at the beginning: "Each soul is potentially divine. The goal is to manifest that divinity. Do this through work and yoga and prayer, one or all of these means, and be free. Churches, temples, rituals and dogmas are but secondary details." As soon as I read that, I thought, "That's what I want to know!"...[It] was very important for me. That's the essence of yoga and Hindu philosophy. (Shankar and Craske 1997: 195)

Harrison's newfound enthusiasm for Indian music (and philosophy) immediately began impacting his contribution to the Beatles' repertoire. Unlike "Norwegian Wood," the Indian-influenced Beatles works from this time forward—for example, "Love You Too" (1966) and "The Inner Light" (1968)—were more in line stylistically with the way such musical elements were used in their native context. Of these, most writers acknowledge "Within You, Without You" (1967) as the Beatles' quintessential amalgamation of pop and Indian music (Farrell 1988; Bellman 1998).

According to Gerry Farrell, "Within You, Without You" takes a variety of features of Indian music—drones, the sitar, certain vocal and instrumental nuances (glides, slurs, etc.), additive rhythms, the use of talas, modal melodies, question-answer passages—and successfully weaves them together to produce a sophisticated whole. In addition, "the spiritual sentiments of the lyrics are reminiscent of *bhajans* [ancient Hindu devotional songs] and other religious songs"

(Farrell 1988: 195). Though nowhere near a true representation of Indian classical music (although uninformed fans held it up as such), the song's intelligent manipulations of Indian musical forms, features, and philosophy render it the Beatles' most mature attempt at hybridization.

In the late 1960s, Harrison's adoption of Indian religion and philosophy was becoming more and more crystallized, and the new practices and beliefs had helped him curtail his drug use. He became convinced that it was particularly the study, practice, and hearing of Indian music that provided the path to reach his inner self. He even remarked to an American reporter that "through the musical you reach the spiritual" (Greene 2006: 84). Harrison desperately desired his fellow band mates to experience the same spiritual transformation and renewal, and he convinced them to accompany him on a 3-month Transcendental Meditation retreat in Rishikesh, India, in 1968. The trip was, in short, unsuccessful, and Harrison recognized that if he was going to realize aspirations of combining his music with his new world view, he would have to do it without the other Beatles.

In 1969, Harrison began his association with a group of Hare Krishna devotees hoping to start a temple in London. He now felt a responsibility to use his status as way to assist others and set a good example. One of the first musical fruits of this growing sense of obligation was the release of the single "Hare Krishna Mantra." It combined a Sanskrit prayer mantra sung and chanted by his new Krishna friends with a Beatles-esque drumbeat and Harrison playing a harmonium part. Biographer Joshua Greene elaborates on the social consciousness that underlined Harrison's decision to record a Krishna mantra:

> Chanting God's names steadied his mind, and George felt purified of anger and greed. Beyond wanting that peace for himself, he felt a calling to reach people who had no knowledge of spiritual life. Recording mantras would be his way of giving back something for all he had received. The world would acknowledge that gift or not…. But at least he would see to it that as many people as possible heard the chanting. (Greene 2006: 146)

"Hare Krishna Mantra" was an unexpected success, selling 70,000 copies on the first day and going to number one on many European charts. Harrison was successfully accomplishing his goal of spreading the spiritual teachings of Krishna through his music.

The success of "Hare Krishna Mantra" coupled with the disbanding of the Beatles in 1970 gave Harrison the opportunity to focus on his new musical purpose with even greater exuberance.[10] Early in the year he released "Govinda," a pop music setting of what is considered the earliest Hindu poem; the single was another top-ten hit in Europe. 1970 also marked the publication of *KRSNA Book*, a Sanskrit devotional scripture that had been translated by Harrison's guru. Harrison was the sole financial source for its publishing costs, and the book became a best seller. In

[10] Greene is the major source for the biographical information in this paragraph and the three that follow.

November, George's three-record masterpiece *All Things Must Pass* was released to critical reviews that hailed it as the best of the Beatles members' solo albums to date. Containing the super hit "My Sweet Lord" (also based on a Sanskrit mantra), the album musically embodied Harrison's spiritual quest thus far. His use of Indian musical elements now mainly consisted of finding ways to incorporate mantras—either new original ones or those involving ancient Sanskrit—into his brand of pop music settings.

Harrison's goal of implementing his music as an agent for raised awareness and social change continued to reach new heights in the early 1970s. His Concert for Bangladesh, a benefit for the war-torn country that marked the first rock event of its kind, was a complete success. His second acclaimed solo album, *Living in the Material World* (1972), equaled his last as a profound statement of his spirituality and continued his practice of incorporating sacred Hindu texts into the music.

In 1974, however, Harrison's streak of successes came to sudden halt. His desire to "give back something" that had originally sparked his decision to record "Hare Krishna Mantra" remained a priority, and he was still emotionally riding the success of the Concert for Bangladesh. Eager to get back on stage and continue his mission, he hastily completed his third solo album, *Dark Horse* (1974), so that it would be released in time to publicize his upcoming American tour. Reviews of the album were the poorest that Harrison would receive, and the subsequent *Dark Horse* tour was a failure. His voice gave out, and his proselytizing was beginning to border on the extreme for his audiences. He exhorted them to chant "Krishna" and purportedly waved pictures of the deity over his head during performances. For one biographer, Harrison's worst offense during this tour was how he changed his best-loved songs to reflect his spiritual convictions. "Now his guitar 'gently *smiled,*' and in his life he 'loved *God* more'" (Greene 2006: 214).

Harrison was devastated and disillusioned by the *Dark Horse* debacle, and he did not tour again for 20 years. Although he continued his religious practices to varying degrees during this time, his subsequent albums showed much more restraint in their spiritual overtones. In his personal life he became a zealous gardener, drawing spiritual strength from maintaining the natural beauty of his large estate. He revived his career in the late 1980s and 1990s through the success of his own pop releases as well as collaborations with other individuals. In his later years, Harrison admitted the folly of engaging in such energetic proselytizing during his youth, but he remained committed to his goal of sharing the music of his faith with the world. Now, however, he accomplished this mainly through his work in helping Ravi Shankar produce albums of Indian classical music. Despite the loss of momentum after the *Dark Horse* tour, Harrison's early solo works continue to draw new listeners year after year, and in that sense his initial musical objective to uplift and change has never ceased, even after his death in 2001.

149.4 A Comparative Look at Coltrane and Harrison

After examining the ways in which Indian music and the philosophies behind it impacted the lives of John Coltrane and George Harrison, some interesting parallels begin to emerge. Both men were transformed personally by first searching and experimenting musically. In the areas of substance abuse, both were aided in their abstinence by their adherence to eastern philosophy and meditation. Both formed close bonds with Ravi Shankar (although Harrison's bond was certainly closer). Both were deeply religious but also were strongly pluralistic in their worldviews. There are some more obscure connections that are equally compelling. There was a Krishna devotee who was part of the group that helped record "Hare Krishna Mantra," "Govinda," and *All Things Must Pass* who was a former pianist for saxophonist Pharoah Sanders, a member of Coltrane's last working band (Greene 2006: 84). Harrison and Coltrane both became avid gardeners in the last years of their lives.

Perhaps the most significant parallel is that both men became consumed with a desire to positively transform others and endeavored to do so by implementing certain characteristics of Hindustani music into their own art. The method behind each musician's presentation of their art, however, was different. Coltrane chose to let the music primarily speak for itself. He was not one to preach or try and persuade with words, except for perhaps in his liner notes and other remarks about his music. He did not concertedly attempt to draw attention to himself. In response to a question about whether or not he would like to communicate some extra-musical understanding to his audiences, Coltrane replied:

> Sure, I feel this, and this is one of the things I am concerned about now. I just don't know how to go about this. I want to find out just how I should do it. I think it's going to have to be very subtle; you can't ram philosophies down anybody's throat, and the music is enough! That's philosophy. (Kofsky 1970: 241)

Harrison—at least during the late 1960s and early 1970s—was much more up front. He was not afraid to talk to reporters about the power of music to spiritually transform. He held press conferences for the release of his devotional music ("Hare Krishna Mantra"). This unabashed forthrightness culminated in the *Dark Horse* tour, where it seemed to reach a level of intolerance for his audiences.

Perhaps not surprisingly, both approaches yielded somewhat mixed outcomes. Coltrane's strictly musical performances certainly produced results. According to former band mate Art Davis, "People would be shouting, like you go to church, a holy roller church…You could hear people screaming" (Hentoff 1998b). Today, there is a Saint John Coltrane Church in San Francisco that has formed a liturgy out of his poem "A Love Supreme" and other quotations (see www.coltranechurch.org). Harrison's Indian devotional singles and initial solo albums sold incredible numbers of units and led many to contemplate their spirituality. His widely publicized Concert for Bangladesh also raised awareness and depicted a sense of worldly brotherhood. Along with these successes, however, both artists had critics and detractors.

The different ways in which Coltrane and Harrison utilized Indian musical elements in their works are also telling. In his study of Indian musical properties in western popular music and jazz, Gerry Farrell concludes that jazz generally gravitates toward incorporating the deeper structural and formal elements of Indian music (for example, scales, concepts of improvisation, forms, time spans, static harmony, additive rhythm) while rock tends to emphasize more surface-oriented traits, including specific instruments, ornamentations, and other characteristic sounds (Farrell 1988: 203). Coltrane and Harrison each utilize both categories to some degree, but for the most part they reflect this general assessment. Harrison's use of mantras in his post-Beatles solo work (which lies outside Farrell's scope) also maintains this distinction. As a text that is repeated throughout, the mantra definitely assumes a place on the surface of those particular recordings. It is not embedded into the structural aspects of the song.

This can help explain the more "faddish" history of the India/rock exchange as opposed to the enduring use of Indian musical elements in the jazz world. The surface elements, if only treated as such, represent simple colorations—they can become familiar and lose their vitality after only a short time. Structural elements, however, can inform the composition or performance of a piece that bears little resemblance in "sound" to Indian music. These deeper, structural elements may also be more effective agents in drawing western rock and jazz musicians to the philosophies behind the music of India.

Joachim Berendt has examined how the use of modal scales, if approached properly, automatically engages the natural chords of the overtone series. The music produced contains sounds that are emanating from the "endless scale" made up of the overtones, thus producing a sonic "metaphor for eternity" (Berendt 1984: 14–15). For music that relies heavily on modal scales (like that of India) this metaphor manifests itself in spiritual connotations. However, the strength of western chords and harmony is so strong that, once introduced, it overpowers the subtle properties of modal music.

In his book *The Time of Music*, Jonathan D. Kramer proposes a perception of musical time that he calls "nonlinearity" (as opposed to western "linearity"). Kramer defines nonlinearity as "the determination of some characteristic(s) of music in accordance with implications that arise from principles or tendencies governing an entire piece or section" (Kramer 1988: 20). Kramer elaborates this concept further by saying "a work's or section's nonlinearity is present from its beginning. The dynamic of comprehending a work's nonlinearity is learning its immutable relationships" (Kramer 1988: 21). Furthermore, he explains how the temporal qualities that characterize a certain culture's music will be reflected in other, non-musical aspects (for example, language, worldview, calendar cycles) of that culture. In particular, several non-European cultures bear traditions of predominantly nonlinear music, which correspond with other nonlinear cultural patterns and ways of life. As examples, Kramer points to the music/culture of Bali and that of the Trobriand Islands.

Although Indian classical music does contain linear aspects (for example, the *Alap-Jor-Gat* succession, the general melodic contour of the *Alap* improvisation), several structural elements of Indian classical music display a stronger emphasis on what Kramer calls "nonlinear" temporality. The importance of the drone, the long length of the works, the lack of harmonic change, the use of a single mode and *raga*, and the use of repetition serve as major examples that betray a significant nonlinear element. Culturally embedded in this nonlinearity is the meditative and self-realizing philosophy that is integral to the music's true understanding. Thus, by simply listening to this music, a westerner begins to be exposed to the philosophies that lie behind it.

In this way, certain concrete elements of Indian music played an active role in attracting both John Coltrane and George Harrison to dig much deeper into its culture. Other, more circumstantial reasons were a significant factor as well. In the case of George Harrison, he was struggling to establish his own identity outside of the shadow of the Beatles. While the group was together, John Lennon and Paul McCartney monopolized its creative output and suppressed Harrison's own contributions at times. Harrison was desperate for a way to assert his individuality, and the sitar and the world of Indian culture provided him with a vehicle to do so. From this perspective, Harrison's predilection for adopting more surface-oriented elements of the music acquires new meaning. He embraced the culture whole-heartedly, and today is arguably identified just as much with India and its music as he is with the Beatles.

For Coltrane, he was struggling with the dichotomy of being upheld by black nationalists as a symbol of "the new black music." Although the movement strove to unify blacks, in many ways it was divisive as well. The following exchange from a 1966 Coltrane interview helps illustrate his concerns (Kofsky 1970: 226):

> Kofsky: Have you ever noticed—since you've played all over the United States and in all kinds of circumstances—have you ever noticed that the reaction of an audience varies or changes if it's a black audience or a white audience or a mixed audience? Have you ever noticed that the racial composition of the audience seems to determine how the people respond?
>
> Coltrane: Well, sometimes, yes, and sometimes, no.
>
> Kofsky: Any examples?
>
> Coltrane: Sometimes it might appear to be one; you might say…it's hard to say, man. Sometimes people like it or don't like it, no matter what color they are.
>
> Kofsky: You don't have any preferences yourself about what kind of an audience you play for?
>
> Coltrane: Well, to me, it doesn't matter. I only hope that whoever is out there listening, they enjoy it; and if they're not enjoying it, I'd rather not hear.

The uneasiness that exudes form Coltrane's responses to these questions discloses his desire for unity among all peoples, regardless of skin color. Establishing a strong link with Indian (and other) musical and cultural constructs meant transcending political and racial barriers to create music that represented a bond of brotherhood. By taking certain structural elements and fashioning them to the core of his musical vision (extended improvisation, modal playing, technical mastery, etc.), he was establishing a sense of unity that went below the surface level—below the level of race. It counteracted the troubling aspects of black nationalism.

149.5 Conclusion

As a result of their encounters with Hindustani music, both John Coltrane and George Harrison were transformed personally, professionally, and spiritually. These transformations refocused their creative vision to situate the audience's spiritual and emotional edification as a top priority along with God's glorification. Consequently, their aspirations for personal fame and recognition shifted as well. The approaches of both men were different, as were their methods of interlacing Indian musical elements with their own brands of popular music, but their objectives bore striking similarities. In this sense, the careers of a black American jazz musician and a white British rocker combine to form a compelling case study of the agency of Hindustani music in diffusing the spiritual attitudes and principles that lie at its core. Finally, it comes full circle back to Ravi Shankar, the great Indian sitarist who profoundly touched both of these western musical icons. Perhaps his words best condense the agency of Hindustani music in the hands of a master:

> There is more to [Hindustani music] than exciting the senses of the listeners with virtuosity and loud crash-bang effects. My goal has always been to take the audience along with me deep inside, as in meditation, to feel the sweet pain of trying to reach out for the Supreme, to bring tears to the eyes, and to feel totally peaceful and cleansed. (Greene 2006: 63)

For musicologists, ethnomusicologists, and other specialists who research the intersections between music and spirituality within specific culture groups, perhaps this study can serve as evidence of the rich potential available to those willing to explore links across geographical, cultural, and religious boundaries. As a result of globalization and modernization, sacred music today must negotiate a multitude of shifting transnational and transcultural contexts, producing a collective field that is ripe for investigation. For example, music that is popular among the Islamic diaspora is settling in new places as the faith continues to grow and prosper. Contemporary Christian music now reaches a rapidly growing body of evangelical Protestants in such regions as East and Southeast Asia, and it is particularly exceptional at absorbing and integrating local popular music styles. The revival of nineteenth-century shape-note singing in the United States is now spreading to the United Kingdom and Europe. Each of these examples offers various avenues of inquiry, and together they only scratch the surface. The time has come to advance and expand our understandings of sacred music in a globalized world.

References

Bellman, J. (1998). Indian romances in the British invasion, 1965–1968. In J. Bellman (Ed.), *The exotic in western music* (pp. 292–306). Boston: Northeastern University Press.
Berendt, J. E. (1984). Jazz and world music. *Jazz Educators Journal, 16*(4), 12–16, 82–84.
Cole, B. (1976). *John Coltrane*. New York: Schirmer Books.
Coltrane, J. (1964). Liner notes to Coltrane, J. In *A love supreme*. Impulse AS-77.

Farrell, G. (1988). Reflecting surfaces: The use of elements from Indian music in popular music and jazz. *Popular Music, 7*(2), 189–205.
Greene, J. M. (2006). *Here comes the sun: The spiritual and musical journey of George Harrison*. Hoboken: Wiley.
Harrison, F. W. (2001, April). West meets east: Or how the sitar came to be heard in western pop music. *Soundscapes: Journal on Media Culture, 4*. Retrieved: July 29, 2012, from http://www.icce.rug.nl/~soundscapes/VOLUME04/West_meets_east.shtml
Hentoff, N. (1996). Liner notes to Coltrane, J. In *Meditations*. Impulse A-9110, 1966. Reissue, 1996.
Hentoff, N. (1998a). Liner notes to Coltrane, J. In *Giant steps*. Atlantic R2-75203, 1960. Reissue, 1998.
Hentoff, N. (1998b). Liner notes to Coltrane, J. In *My favorite things*. Atlantic R2-75204, 1961. Reissue, 1998.
Kofsky, F. (1970). *Black nationalism and the revolution in music*. New York: Pathfinder Press.
Kramer, J. D. (1988). *The time of music: New meanings, new temporalities, new listening strategies*. New York: Schirmer Books.
Lal, A. (1995). Rock & raga: The Indo-West musical interface. *Rasa: The Indian Performing Arts in the Last Twenty-Five Years, 1*, 89–98.
Pickney, W. R. (1989–1990). Jazz in India: Perspectives on historical development and musical acculturation. *Asian Music, 21*(1), 35–77.
Shankar, R., & Craske, O. (1997). *Rage Mala: The autobiography of Ravi Shankar*. New York: Welcome Rain Publishers.
Thomas, J. C. (1975). *Chasin' the Trane*. New York: Da Capo.

Chapter 150
Eating, Drinking and Maintenance of Community: Jewish Dietary Laws and Their Effects on Separateness

Stanley Waterman

150.1 Preface

A religion is a set of beliefs concerning the cause, nature, and purpose of the universe. As a rule, religion involves faith, which is, in other words, unquestioningly accepting certain givens as true. In particular, it concerns obedience to some supernatural power, which believers regard as controlling human destiny. Each religion normally contains a moral code that governs the conduct of human affairs and additionally, almost all religions involve devotional and ritual observances and practices to some degree.

One of the consequences of any successful religion is that sooner or later it tends towards institutionalization, so much so that it may be almost axiomatic to state that without institutionalized rules and regulations it becomes difficult for any religion to hold its adherents together and in check.

Religious beliefs are most frequently expressed through a set of behaviors and practices. Some of these are active on a personal level whereas others, though governing personal behaviors and practices, are designed to create a feeling and spirit of community, setting off members of one religion—or sect or sub-religion—from another. The customs and rituals of a religion's adherents are usually determined and administered by human intermediaries who regard themselves, and are generally accepted, as the approved interpreters of the sets of rules and regulations (religious laws) that govern the group's and the individuals' lifestyles. Over and above their role as interpreters of religious laws, this activity by religious functionaries is part of ongoing attempts to safeguard their existing strength, and to maintain control of their authority over current believers and to attract newcomers (that is, to engage in practices of conversion).

S. Waterman (✉)
Department of Geography and Environmental Studies, University of Haifa, Haifa, Israel
e-mail: stanleywaterman@me.com

S.D. Brunn (ed.), *The Changing World Religion Map*,
DOI 10.1007/978-94-017-9376-6_150

Over the centuries and the world over food has had significant roles to play in religious practice and observance. This has led some to ask what makes food so influentially potent, unleashing such keen memories and powerful forces of cohesion or revulsion (Holtzman 2006). As Holtzman noted, the most compelling answer is that the sensuality of eating transmits powerful mnemonic cues, principally through smells and tastes.

More often than not, the effects of food on religious practice and on perceptions related to religion have been as negative rather than positive forces. As Adam Gopnik (2011: 94) has so aptly put it, "Before modern times, taste in food particularly was enforced by the two greatest of enforcers: faith and famine, and their not-so-nice foster parent, fear." Strictures and prohibitions against the consumption of specific food substances are commoner than positive exhortations to eat or drink other foods. This is because it has been found that food and food consumption are efficient media for enhancing the perception of difference between groups and of the members of these groups—of dividing the pure from the impure, the sacred from the profane—and this is, of course, an important factor in a world in which there is vigorous competition over the justification of religious beliefs and practices.

Translated into religious injunctions rather than just cultural differences, practices relating to which foods are proper for consumption and which are inappropriate, customs regarding the care and attention devoted to the preparation of food, and the existence of food taboos become powerful elements in distinguishing members of one religion from another.[1]

In this chapter I examine the food consumption practices of Jews with clear emphasis on Jewish dietary laws and their effects on Jewish community and separation.

150.2 Introduction

> I will buy with you, sell with you, talk with you, walk with you, and so following; but I will not eat with you, drink with you, nor pray with you. (The Merchant of Venice (I, iii, 35–39))

In Shakespeare's *The Merchant of Venice*, the Jewish moneylender Shylock declines an invitation from Bassanio to have dinner with Bassanio's friend, the merchant Antonio. By refusing this solicitation, Shylock simultaneously asserts a inclination to do business with Christians and an unwillingness to interact with them socially. Thus, in this famous infamous instance, the social segregation of Jews in sixteenth century Venice is expressed through the medium of food and an eating

[1] As Gopnik, again, puts it (2011: 94): "Dietary restrictions are a big part of most religious practice; kosher and halal rules are the most famous of these, and though various attempts have been made to justify them on rational grounds .... Even the rabbis and mullahs admit now that the purpose of food laws is to create a form of symbolic solidarity that keeps a tribe or faith together. There is nothing so powerful to keep you eating with your family on this side of the river as disgust at what the people on the other side think tastes good. Faith shapes food..."

taboo. Although Shakespeare portrays much personal antipathy between Antonio and Shylock (Antonio lends money without interest, thereby affecting Shylock's business and that of other—mostly Jewish—Venetian usurers) it is Shylock, masquerading as the Jewish people *in toto*, who bears an historic grudge against all Christians. Whatever else, Shakespeare voices the differences between Christians and Jews and seems to indicate that this separation is less the ghettoization of the Jews by the Christians and more the desire of Jews *to remain apart*, separated and segregated from the Christian majority.

In this dramatization of Jewish difference and segregation, what are the guidelines and rules that dictated Shylock's refusal to dine with Antonio and Bassanio? Underlying Shylock's behavior is a litany of do's and don'ts concerning which foods are permissible and which are forbidden to Jews to eat and drink and how such foods should be prepared. In common parlance, these are referred to as the laws of *kashruth*—from the Hebrew word *kasher*, meaning "fit" or "proper" (for consumption). The concept [of *kasher* (kosher)] is hard to pin down. The word itself is a late Hebrew word that does not occur in the Pentateuch. In English, kosher is 'fit' in the sense of proper or suitable but is mostly ceremonial (Masoudi 1993). Although the words *kashruth* and *kasher* are usually understood in the context of food, they can equally well be used to describe other religiously fit items such as a *sukkah* (a temporary living and eating booth used by religious Jews to celebrate the autumn Festival of Tabernacles) or a *mezzuzah*, an icon affixed to doorposts.

150.3 Pork and Pigs

As Bahloul (1995) has noted "Meat is sacred" and much of the long litany of obligations and prohibitions concern meat, its preparation and consumption. Although there are many foods forbidden to observant Jews and others that they are encouraged to consume at specific times, the iconic prohibition, the one that most people associate with Jews, is the taboo concerning pork, that is, meat from pigs. In fact, so closely are Jews historically associated with their reluctance to eat pork that Jews and the pig have become almost inseparable (Rosenblum 2010).[2]

In his classic work on meat-eating taboos around the world, Frederick Simoons (1994) discussed the origins of the pork prohibition. Though Simoons dealt with meat taboos worldwide, I summarize his findings for just pork in the Middle East and Mediterranean. He noted (1994: 13) that although the main center of pork

[2] Rosenblum (p. 60) notes that by refusing to eat pig, Jews were never able to ingest Romanness and thus could never truly become Roman. ... and that over time, the practice of refusing to ingest pork came to be viewed as a distinctly Jewish one, leading to the marshaling of the pig in anti-Semitic tropes throughout the medieval and modern periods. This explains, for example, the otherwise incomprehensible pejorative term marrano, deriving from the Arabic muḥarram, meaning "forbidden, anathematized," and meaning "pig" in 15th century Spanish, for a Portuguese or Spanish Jew who converted to Christianity during the time of the Inquisition, but secretly practiced Judaism (Rosenblum 2010: 61).

avoidance is the Middle East where the Muslims have a strong anathema to pork, this avoidance cannot simply be explained in terms of Islam and even today non-Muslim groups there entertain strong feelings against it. There was longstanding anti-pork feeling in the region long before Islam though the pig had been domesticated and used a supplementary source of meat protein in prehistoric times.[3]

The Hebrew attitude to the pig and the prohibition in Leviticus are quite clear: as the pig is an unclean animal its flesh should be avoided. In Talmudic times, the breeding or keeping of swine was banned. During the Seleucid period in the second century BCE pork avoidance became a symbol of Jewish religion (1994: 21) and when Islam adapted and adopted Jewish dietary practices, the ban on pork consumption spread with it. Simoons (1994: 64) also noted that because the ancient Hebrews are so closely associated with the ban on pork in Western thinking, it is assumed that it originated with them, although there is no scientific consensus about this. In general, in the Mediterranean and Near East, pigs and pork are restricted to Christian areas along the northern Mediterranean whereas in Muslim regions and Israel, they are banned. Even among Christian minorities in Muslim and Jewish societies, there is reticence to use the pig or society places obstacles on its use.[4]

The underlying question seldom asked is why there is a pork taboo. Simoons recorded five principal hypotheses for Jewish dietary laws. First, they are arbitrary and can only be understood by the deity or strict believers. Second, they arose because of hygienic concerns about danger from illness or disease. Third, Jewish dietary laws are symbolic, setting off proper from errant behavior. Fourth, they originated in rejection by the Hebrews of the cultic practices of alien peoples. Finally, he also raises the issue that there may be an economic, environmental or ecological explanation (1994: 65). If the first hypothesis is accepted, there is little point in searching further. However, the others are worth pursuing.

The most commonly proffered hypothesis is that the ancient Hebrew rejected pork on hygienic grounds: it decays rapidly in the high temperatures of the Middle East; the scavenging pig, eating everything and physically dirty, might be dangerous to eat; eating pork can cause trichinosis, a parasitic disease. Simoons rejects the

---

[3] Some antipathy to the pig appears to have taken root in Egypt and Mesopotamia around 1200 BCE. Herodotus noted that though the pig was not banned completely, most Egyptians did not eat pork. Nevertheless he noted annual sacrifices where pork was eaten and that the poor who could not afford pig sacrificed baked pig-shaped dough instead. In Anatolia, the Hittites kept pigs, which appear to have survived solely on scavenging. Thus, although tolerated, pigs were considered unclean, as they were in Sumer. There, as in Babylonia and Assyria their flesh was eaten on certain days only and young undefiled animals were used in rites of healing and exorcism; a similar attitude existed in Anatolia and lasted until a later period. In contrast, pork was a preferred food among the Greeks and the Romans and, as Christianity developed and spread, pork-eating migrated with it. (Simoons 1994: 16–24).

[4] Anecdotally, there is a description of a "ceremony" the late Eli Landau, a well-known amateur cook and medical doctor prepared a porchetta feast in a gourmet restaurant in Tel Aviv. Landau had come to celebrate the launch of his new book, the first collection of pork recipes ever published in Hebrew. However, as Israel's major publishers had banded together and jointly refused to print it, Landau published the book himself—and even then, some of the chain bookstores declined to sell it (Vered 2010).

decay idea as nonsense; pigs are eaten in other hot regions with no apparent problems. The trichinosis hypothesis is also hard to accept because its cause is not pig-specific, was relatively uncommon in the ancient Middle East and the relationship between undercooked meat and *Trichinella* roundworms was only discovered in the mid-nineteenth century. Simoons does not rule out the possibility that the Hebrews or some other group using commonsense methods associated the first-stage symptoms (vomiting, diarrhea, fever, and general malaise) with eating pork, but that such an association is not easy to make. However, he concludes (1994: 71) that the hygienic hypothesis is false because the core concept of the Hebrew ban on pigs is the defiling nature of the pig itself; this objection also applies to the economic-ecological premise.[5]

In the long run, Simoons favors the symbolic and cultic hypothesis. The Levitical bans give only one reason: holiness—the need to emulate God's nature and separation from the impurities of pagans. And it is this desire to remain apart from other ethnic groups that was the prime instigator of the Hebrew ban on pork, a situation magnified when Antiochus made the pig central to his policy of forcing the Jews to abandon their faith. In Simoons' view, Western observers need to consider areas outside the Near East where pork taboos also occur and the widespread revulsion of the pig and its eating habits among many peoples.

150.4 Origins

Pigs apart, the basic list of food permissions and proscriptions are set out in Leviticus. Mammals must both ruminate and have cloven hooves. Possessing just one of these characteristics is insufficient to render the animal fit to eat; animals such as camels or horses are rejected as they do not have cloven hooves and animals such as rabbits do not have hooves at all. Pigs have cloven hooves but do not ruminate, allowing it to become the ultimate culturally non-kosher (*tareff, treiffah*) animal. Several of these animals are specifically excluded by name (Leviticus, 11: 3–8 and in Deuteronomy, Chapter 14). Birds of prey, too, are forbidden food and they are specifically listed in Deuteronomy 14: 12–18—alongside bats, which are mammals. Fish, in order to be acceptable, must have both fins and scales thereby ruling out shellfish (Leviticus, 11: 9–12). And just in case of any misunderstanding,

[5] The anthropologist Carleton Coon suggested that where the environment was relatively untouched by humans pigs were able to forage for acorns, beechnuts and truffles (cited in Simoons 1994: 71–72). Once human activity increased, however, the pig was seen as a competitor to humans and was displaced as a favored animal. Under such conditions a person keeping pigs would be displaying his wealth and disturbing the ecological balance of the environment, leading to the pig's displacement. However, there is no such evidence for this in the Bible or in the archeology, as there is little verification relating climatic or environmental deterioration to a ban on eating pork. That the pig was abandoned because of an unwillingness to display wealth is also questionable, as it appears that pork was not an expensive delicacy.

it is stated three times in the chapter.[6] Notwithstanding, debate rages as to when scales are really scales and in some cases, the size of scales changes as the fish matures, thereby affecting the potential consumption of certain fish such as turbot or swordfish. In addition, all insects and any crawling creature, with the exception of one type of locust—irrelevant today for almost everyone—are forbidden.

This is the basic list of forbidden food, the bottom layer, the basic set of prohibitions (and permissions). However, other strata above this fundamental layer further restrict the freedoms of observant Jews in all that concerns food. And whereas the fundamental layer relates specifically to animals and flesh not all these additional directives relate to meat and meat products. Whereas most of the bottom stratum is based on direct biblical injunction, most of the upper echelons are based on interpretations of biblical injunctions, which have, over the centuries, become laws unto themselves.

In addition to the non-kosher animals listed in Leviticus and Deuteronomy, there are prohibitions on eating meat from any permitted animal that has not been slaughtered according to the laws of *shechitah* or from an animal with a significant anatomical or physiological defect. There is a clear prohibition against consuming blood, initiating designated procedures for its removal in general and even more specific procedures for blood-rich liver; certain fats from the abdomens of cattle, goats and sheep must be removed in order to render the meat fit to eat. In addition, the sciatic nerve area in the hindquarters of the animal is also generally not consumed; its labor-intensive removal requires specially trained personnel, making it an expensive operation. As a consequence, it was common for the entire hindquarters to be sold on to the non-kosher market.

In addition to these prohibitions on animal products, fruit from a tree cannot be consumed in the first 3 years after planting (Leviticus 19: 23–24) and there are also similar prohibitions on newly grown grain.

As well as these regulations derived directly from the Old Testament, there are others whose origins are more oblique and which have been variously interpreted over the years. The most widespread of these is the prohibition against mixing meat (including fowl) and dairy products, derived from the biblical injunction not to "cook a kid in its mother's milk" (Exodus, 23:19, 34:26 and Deuteronomy, 14:21). Over the centuries, this has led to a slew of regulations in which not only are dairy and meat foods separated from one another, but also the vessels in which they are cooked and from which they are consumed. Vessels for meat and dairy products must not come into any contact with one another. This means keeping separate sets of cookware, flatware and dishes and, in more extreme cases, discrete cupboards, ovens and even kitchens—all of which is compounded during the festival of

[6] 9 These you may eat, of all that are in the waters. Everything in the waters that has fins and scales, whether in the seas or in the streams—such you may eat. … 10 … that does not have fins and scales … they are detestable to you … 12 Everything in the waters that does not have fins and scales is detestable to you.

Passover (see below). In an effort to keep these products strictly separated from one another, customs have developed whereby, for instance, no dairy products are consumed for a stipulated time after having eaten meat or fowl. The strictest view requires at least 6 h between the two. Most Orthodox Jews in the West wait 3 h but the custom among observant Dutch Jews is to wait 1 h, thus highlighting the symbolic nature of this stricture. There are also other differences of opinion, such as about hard cheeses. Because some cheeses are made with a curdling agent from the walls of a calf's stomach, some Jews view the cheese as a prohibited mixture of milk and meat. Others believe that all cheeses, hard and soft, are kosher (Masoudi 1993). In a modern world where people eat outside the home, this gives rise to separate meat or dairy (or vegetarian) restaurants—or at least strictly segregated non-contiguous areas within restaurants or dining areas.

Passover imposes further dietary obligations on observant Jews. Exodus 12:15 stipulates: "Seven days you shall eat unleavened bread; on the first day you shall remove leaven from your houses, for whoever eats leavened bread from the first day until the seventh day shall be cut off from Israel." All leavened products or leavening agents (chametz) may be simply depleted, removed, destroyed or transferred ("sold") to gentiles. The objective is that no Jew possess any *chametz* for the whole week of Passover and that any food product that might contain even a minutia of *chametz* not be ingested. All dishes and appliances in the household, which might have been in contact with *chametz* during the year must be ritually cleansed or "sold" or stored away and sealed, not used during the festival.

However, as with so many things Jewish, there are differences of opinion over what exactly constitutes *chametz*, with the general rule to allow what has become general practice in individual communities. As a consequence, it is traditional among all communities not only to refrain from eating leavened bread but also to consume unleavened flatbread (*matzah*) instead. In addition, Ashkenazi communities generally ban the use of pulses (*qitniyot*) and rice whereas Sephardi and other non-European Jewish communities permit them.[7]

How these demands and proscriptions arose and why they continue to apply has been the source of much discussion among Jews and others (see Severson 2010). For believers, the reasons are plain and require no elaboration. They are simply there, to be followed explicitly. For doubting traditionalists, if the rules do not stand up fully to logical analysis, (i.e., make sense) they constitute part of what is done to maintain group cohesion and thus for the sake of group continuity to be followed. However, true skeptics may ask "Why on earth ...?" and wish to know if some innate logic was involved in their origin and whether this is sufficient reason for encouraging or forbidding the consumption of one food or another.

[7]This may cause confusion and even give rise to absurdity and farce, especially in Israel, when a supermarket product might have a label from a Sephardi rabbinical authority permitting its consumption for Passover while the opposite side of the same package or jar might bear a label from an Ashkenazi authority prohibiting its use—all of which is a reminder of differences past and present, real or imagined.

150.5 Institutionalization and Modernity

What is obvious from this cursory survey is that observant Jews have felt a need to protect themselves from transgression by consuming forbidden foods. Thus, over the years, this has brought about widespread organization and regulation of things to do with preparing, storing and retailing food.

A fundamental aspect concerning the kashrut status of meat and fowl, besides ascertaining whether or not the animal is an approved species, is that it should be slaughtered according to accepted practice (*shechitah*). Observant Jews regard shechitah as an act of respect and compassion for the animals whereby the trachea, esophagus, carotid arteries and jugular veins are severed using a very sharp blade and the blood allowed to drain. A *shochet*, the person practicing *shechitah*, normally a religious Jew, expertly trained and licensed, is the only person permitted to perform this operation, which must be carried out without pressing, pausing, piercing, tearing or allowing the knife to be covered.

In traditional pre-modern Jewish communities, the *shochet* would have been a local person, either from the village or township or at least from the region, covering several settlements. The *shochet* often doubled as a *mohel* or ceremonial circumciser whose status in the local community was almost equal to that of the rabbi. As a local person, or at least a person known to all, the community he served trusted the *shochet* and there was seldom doubt that the animals selected and the means by which they had been slaughtered conformed with Jewish religious law.

However, the meeting of Jews with modernity—first with the Enlightenment in Central and Western Europe and later with the unprecedented freedoms accorded Jews in North America—gave rise to issues that had hardly existed before. Nevertheless, some semblance of the Old World continued into the New, as Norman Ravvin (1997: 4) recounted:

> my … grandfather was a shochet … a necessary figure in many Jewish communities, but in Canadian terms a rather atavistic sort of professional man, a disappearing breed. Ironically, [his] knowledge of … shchita … […] got him work in Canada and entitled him to a visa. … [He was] a travelling rabbi and shochet on the Canadian prairie ….

But even here, one gets the impression that the itinerant, though known, was not a local and because distances between settlements were longer and the areas covered vaster, the potential for "suspicious" activities became greater.[8]

The potential for losing trust in *shochetim* on the Canadian prairies perhaps did not often come to fruition, but the travelling slaughterer was just the tip of the iceberg. In towns and cities, the picture is different and the need for kosher supervision is long-standing. In New Amsterdam as early as 1660, a Portuguese Jew applied for a license

[8] In the modern age, as the philosopher Onora O'Neill (2002a) has written, "Every day we read of untrustworthy action by politicians and officials, by hospitals and exam boards, by companies and schools. … we cannot have guarantees that everyone will keep trust. Elaborate measures to ensure that people keep agreements and do not betray trust must, in the end, be backed by—trust. At some point we just have to trust. There is, I think, no complete answer to the old question: 'who will guard the guardians?"

to sell kosher meat. The first recorded complaint against a *shochet* was lodged in 1771 and the first court license revocation against a kosher butcher took place in 1796.[9]

From the middle of the nineteenth century several attempts were made by American Jewish laymen and *shochetim* to improve *kashrut* standards, but it was not until the Union of Orthodox Rabbis became involved in 1924 that modern kosher supervision was born. As more products were produced in manufacturing plants rather than households, kosher product certification became more necessary to protect those for whom *kashrut* is important. Although initially necessary to ensure that the animals and plants sold met with the strictures of *kashrut* (type of animal or plant, slaughter, etc.), it increasingly came to mean a guarantee that manufactured foodstuffs did not contain even the smallest traces of forbidden substances. Although in recent times the heightened regulation of food manufacturers in general has placed ever-increasing demands on producers to label accurately and truthfully the contents of their products, this was not always the case. Consequently, Ⓤ, the symbol of the Union of Orthodox Rabbis, became an icon for observant Jews in America.

As the manufacture of kosher food spread throughout the United States, the resources of the New York-based certifying organization began to be stretched, so that another certifying agency was needed. This led to the establishment of the O/K Laboratories, which was also based in New York. Over the years more such agencies were established throughout the country and worldwide. One source now lists 143 such agencies worldwide, with over half in the United States alone (Table 150.1). Each of these agencies uses a symbol, supposedly to distinguish it readily from the others (Fig. 150.1). In addition to the agencies, individual rabbis have entered the field of kashrut certification, using their own kosher symbol or even just a plain “K” to designate a product’s kosher status. However, rather than clarify matters, it has only supplemented the confusion and, as an unforeseen consequence of the kosher business, it has helped separate observant Jews into “camps” that will trust only one agency but not others, or one that people have heard from others “can be trusted.”

As well as indicating the diffusion of the manufacturing and preparation of food products for observant Jewish populations and the spread of these populations, this explosion of agencies and individuals with the power and say-so to authorize the fitness of food products also signifies something more disturbing: a breakdown of trust in which some people refuse to recognize the authority of a certifying agency or individual and/or only have confidence in another. In the words of Onora O’Neill (2002b, Section 1), “… this high enthusiasm for ever more complete openness and transparency has done little to build or restore public trust. On the contrary, trust seemingly has receded as transparency has advanced.”

This perspective indicates that some segments of the observant Jewish population only consume food products if they have been certified as fit by a favored kashrut agency or individual. In essence, all other food products are taboo. Although most observant Jews are willing to accept the authority of the longest-standing and moderate agencies—that is, to trust them—or of agencies related to a “higher” (that is, stricter) authority, groups who are more particular will only eat at the tables of those

[9] History of Kosher Certification. www.kosherquest.org (Viewed 6 March 2012).

Table 150.1 Kashruth authority locations

Location	Number
New York	18
Israel	12
California	9
Canada	8
France	7
Texas	6
United Kingdom	6
Massachusetts, Pennsylvania, Australia	5 each
Florida, Maryland, Ohio, Argentina	4 each
Illinois, New Jersey, Austria, The Netherlands	3 each
Arizona, Colorado, Virginia, Washington, Brazil, Germany, Italy, Mexico	2 each
Connecticut, District of Columbia, Delaware, Georgia, Indiana, Kansas, Minnesota, Missouri, Rhode Island, Wisconsin, Philippines, Colombia, Belgium, Czech Republic, Hong Kong, India, Russia, South Africa	1 each
Total	143

Data source: www.hanefesh.com

who accept the same authority. This, lack of trust *in extremis*, can produce absurd scenarios in which two strictly religious Jews paint themselves into different kosher corners from which they can only eat separately.

The obsessive hair-splitting of some strictly religious groups and individuals leads to irrational—and distressing—situations where members of the same family choose not to eat with one another because they cannot agree over the fitness of the rabbi officiating over the fitness of the food! This is not much different to a situation in which a Jew concerned over kashrut will not eat at the table of one who cares little or not at all. As Mars (1997: 189) observed, "Commensality, eating food together, has been considered both a manifestation and a symbol of social solidarity and of community" and any diminution in commensality is an indication of a commensurate shrinking of social cohesion and social capital. Mars also pointed to a case in his research in the Welsh city of Swansea where a Jew from an observant background became strictly Orthodox and subsequently refused to eat in the house of his observant brother because he was not satisfied with the kashrut supervision proffered by the (Orthodox) United Synagogue. Unfortunately, such situations are not uncommon.

It is not just cynicism on my part that suggests a justification for suspecting false play in the world of kashrut certification, for kosher food is a multi-billion dollar industry.[10]

[10] Boland and Geisler (2006/2012) reported that U.S. supermarkets have 25,000 kosher product lines and that even large food manufacturers such as Coca-Cola, Kraft and General Mills have some food products certified kosher. Moreover, the kosher food market does not cater just for Jews, one estimate suggesting that just 44 % of the kosher market is Jewish. More than a quarter of all kosher consumers are non-Jewish health-conscious consumers, kosher having become associated with being clean and safe.

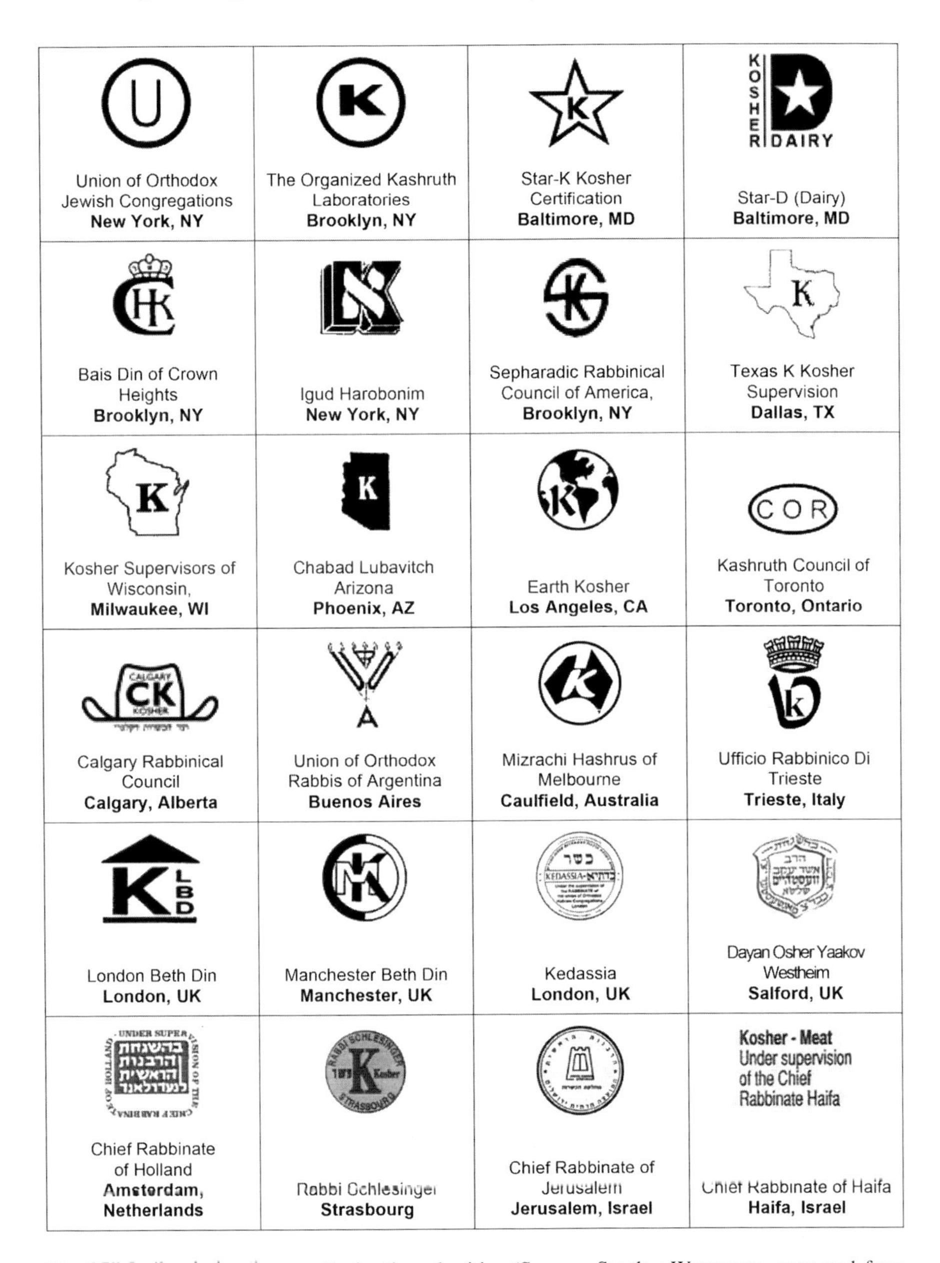

Fig. 150.1 Symbols of some Kashruth authorities (Source: Stanley Waterman, extracted from www.hanefesh.com)

From the beginning, there has been a need in the United States to enact kosher food laws. New York State enacted the first in 1915 in response to the

> chaotic state of the kosher food industry—its charlatans, profiteers and outright crooks—which, coupled with the huge influx of immigrants ... unfamiliar with local circumstances, made any assurance of kashruth all but impossible.... (Masoudi 1993)

Masoudi noted that between 1985 and 1988 over 240 kosher food violations were referred to the New York Attorney General.[11] However, legal disputes involving kashrut are complex and in order to regulate kosher food constitutionally, Masoudi records, a state must not define the term "kosher" at all and because the term is essentially religious, courts are not fit to resolve disputes over its meaning. Consequently, in his opinion, the kosher food laws in most states violate the Constitution. Laws that select a particular definition of "kosher" favored by a particular sect as the state's definition fail on the basis of denominational preferences. Those that select a definition without reference to a particular sect's views fail on grounds of the First Amendment, that is, impeding the free exercise of religion. In order to protect its consumers from kosher food fraud, no state may go further than to require the sellers to disclose their certifying organization or the procedures followed in preparing their food.

150.6 Secularity and Jewish Dietary Laws

From the preceding discussion, one should not assume that all Jews actually consume or demand kosher food. Where many people in the "West" have adopted secular lifestyles, religious proscriptions have been transformed into cultural or ethnic preferences in good circumstances whereas in worse situations they have been jettisoned entirely.

In a nominally Orthodox Jewish community such as that in Leeds in the United Kingdom, just under two-thirds (63 %) of a survey sample of almost 1,500 Jewish households purchased meat from a kosher butcher. Nevertheless, 78 % of the same sample reported that they ate non-kosher food outside the home occasionally or frequently, a typically Anglo-Jewish situation in which many people retain certain Jewish religious customs while being simultaneously lax toward many others (Waterman 2003: 8). In the United States it is common to find "kosher-style" delis and restaurants in which food has been transformed from being ritually fit to eat into being an ethnic food catering not only to the tastes and preferences of nostalgic but non-Orthodox or non-observant Jews, but also to inquisitive gentiles. As the

[11] For example, a New York State Division of Kosher Law Enforcement inspector, on a surprise visit to Commack Kosher Deli & Market on Long Island, inspected the factory-sealed packages of kosher turkey thighs in the refrigerator. The department subsequently declared that these were in violation of section 201-a(2) of the New York Agriculture and Market Law because the thighs were not individually marked "soaked and salted," one of the key steps in making meat officially kosher. The Department of Agriculture and Markets fined Commack $11,100. (Marsh 2002).

prize-winning secular Jewish-British novelist Howard Jacobson (1993: 34) colorfully put it during a visit to New York:

> Even without cheesecake and sour cream, I was in heaven. This was not like eating kosher in England. Not surprising, since the food here was not so much kosher as kosher-ish—ersatz kosher, kosher freed from the rules and restrictions which usually make you wish you were in another restaurant, eating something that didn't have matzo balls in it, or that wasn't mashed and strained and puréed, as though for Jewish babies. And there was no wizened little watchdog from the Beth Din sitting in a corner either—no ancient guardian of the Orthodox palate, such as you find in English kosher delis, hovering like a chaperon over the promiscuity of your digestive system, spoiling your appetite, and making you feel that eating is not a pleasure but a penance, a mortification of the duodenum which you misperform at your peril.

But secularism does not just mean freedom to ignore the constraints of the Jewish religion with regard to what they consume for modernity has also brought about a heightened awareness of what has come to be known as "animal welfare." Apparently, some methods of animal slaughter are considered to be more "humane" than others and, it is little surprise that *shechitah* and *halal* forms of killing animals are regarded in some countries as less humane methods than those which require pre-stunning of animals. Predictably, this has led to charges of anti-Semitism on the part of Jews towards legislators favoring the banning of *shechitah* and other forms of ritual slaughter and equally surely, to the actual proscription of these forms of slaughter in some countries—not just countries in which anti-Semitism was rife as in Nazi Germany or Fascist Italy but in modern democracies where animal rights are taken seriously such as Sweden and New Zealand (which had a Jewish Prime Minister at the time the ban came into force in 2010). Thus Jews in those countries in which there are ongoing debates about proscribing *shechitah* are fighting to protect what they regard as one of their fundamental rights. Illogically, even many Jews for whom *kashrut* is an anachronism are prepared to give their co-religionists (or co-ethnies) support on this issue. Ironically, their natural allies in this struggle are the Muslim communities (even though some Muslim authorities are more lenient over the stunning issue than their stricter counterparts), which have the numbers to back up their demands. Even more paradoxical, of course, is that there are many people around who regard all forms of animal killing to be inhumane and who wish that this be terminated, much as in earlier history human sacrifice and cannibalism were relinquished.

150.7 Conclusion

Food and food consumption has traditionally been a sure way of separating Jews from the peoples among whom they have lived. There have been many mechanisms for effecting such separation, the most common being prohibitions on the consumption of specific animals and plants, directives and guidelines on the preparation of permitted foods so that they are fit for eating, and in the modern world, institutionalizing the certification procedures for such foods. Conventionally, in the period before the Enlightenment, these proved reasonably efficient in achieving their goals, which, in essence, was the preservation of Jewish peoplehood.

However, the emergence of different customs in diverse Jewish communities and the institutionalization designed to safeguard observant Jews from consuming prohibited food have served to highlight differences among Jews themselves. This has only been emphasized further by the secularization of many Jews in modern societies, undermining the role played by *kashrut* in setting off Jews from the population in general and further widening gaps between observant Jews and others. On the other hand, attempts to ban *shechitah* are regarded as an infringement of the rights of an ethnic group and, almost perversely, brought Jews closer as members of the same ethnic group.

References

Bahloul, J. (1995). Food practices among Sephardic immigrants in contemporary France: Dietary laws in urban society. *Journal of the American Academy of Religion, 53*(3), 485–496.

Boland, M., & Geisler, M. (2006/2012). *Kosher industry profile.* Ames: Agricultural Marketing Resource Center, Iowa State University. www.agmrc.org/markets__industries/food/kosher_industry_profile.cfm. Viewed May 4 2012.

Gopnik, A. (2011). *The table comes first.* New York: A.A. Knopf.

Holtzman, J. D. (2006). Food and memory. *Annual Review of Anthropology, 35*, 361–378.

Jacobson, H. (1993). *Roots Schmoots: Journeys among Jews.* London: Penguin Books.

Mars, L. (1997). Food and disharmony: Commensality among Jews. *Food and Foodways, 7*(3), 189–202.

Marsh, K. (2002, September/October). Busting Chops. *Legal Affairs.* www.legalaffairs.org/issues/September-October2002/story_marsh_sepoct2002.msp. Viewed July 18 2012.

Masoudi, G. F. (1993). Kosher food regulation and religion clauses in the First Amendment. *University of Chicago Law Review, 50*, 667–696.

O'Neill, O. (2002a). Spreading suspicion. Lecture 1, BBC Reith Lectures 2002, A question of trust. www.bbc.co.uk/radio4/reith2002/lecture1.shtml

O'Neill, O. (2002b). Trust and transparency. Lecture 4, BBC Reith Lectures 2002, A question of trust. www.bbc.co.uk/radio4/reith2002/lecture4.shtml

Ravvin, N. (1997). *A house of words: Jewish writing, identity, and memory.* Montreal: McGill-Queen's University Press.

Rosenblum, J. D. (2010). Why do you refuse to eat pork? Jews, food, and identity in Roman Palestine. *Jewish Quarterly Review, 100*(1), 95–100.

Severson, K. (2010, January 13). For some, 'Kosher' equals pure. *New York Times.* http://www.nytimes.com/2010/01/13/dining/13kosh.html?pagewanted=all&_r=0. Viewed 20 September 2014.

Simoons, F. J. (1994). *Eat not this flesh: Food avoidances from prehistory to the present.* Madison: University of Wisconsin Press.

Vered, R. (2010, Summer). Prescribing pork in Israel. *Gastronomica, 10*, 19–22.

Waterman, S. (2003). *The Jews of Leeds in 2001: Portrait of a community.* London: The Institute for Jewish Policy Research.

Chapter 151
The Shrines of Sport: Sacred Space and the World's Athletic Venues

Arthur Remillard

151.1 Introduction

While visiting Anchorage, Alaska, George Sheehan—a medical doctor and popular distance running author—met a skeptical reporter who asked, "Is running your religion?" Pausing thoughtfully, Sheehan replied, "Running is not a religion, it is a place" (Sheehan n.d.). To explain this, he channeled insights from Henri Nouwen, a Belgian priest and noted author. Wearied by his unceasing routine of writing, traveling, and speaking, Nouwen retreated to a Trappist monastery in Genesee, New York, for 7 months. Solitude, reflection, prayer, and manual labor allowed the priest to reconnect with himself and his faith. Prior to his departure, though, Nouwen worried that he would slip back in to old habits, losing everything that he had gained. He sought council from a sagacious abbot, who instructed Nouwen bring Genesee home with him. In other words, he needed to infuse the habits of monastery life, such as regular prayer, with the habits of his normal life (1981). With Nouwen in mind, Sheehan likened running to "a monastery—a retreat, a place to commune with God and yourself, a place for psychological and spiritual renewal" (Sheehan). What prayer was to the priest, running was to Sheehan.

Although we tend to classify sports as a distinctly secular activity, for many who engage in the realm of play, it carries a certain religious weight. As religion scholar David Chidester posits, popular culture is rife with the "traces of transcendence, the sacred, and the ultimate." Comparison is a means by which we can uncover these "traces." In other words, just as Sheehan saw a parallel between a rigorous prayer life and the rhythms of distance running, we might discover other forms of religious life lingering in the world of sports. Comparison, Chidester notes, is a rather old tool in the religion scholar's toolbox. When Christopher Columbus landed in the

A. Remillard (✉)
Department of Religious Studies, St. Francis University, Loretto, PA 15940, USA
e-mail: aremillard@francis.edu

S.D. Brunn (ed.), *The Changing World Religion Map*,
DOI 10.1007/978-94-017-9376-6_151

Caribbean in 1492, he branded the indigenous people atheists because they had no formal church or creed. Columbus's definition of religion, though, effectively began and ended with his Catholic worldview. Missionaries soon revised Columbus's conclusions after comparing the "familiar metaphors" of Catholicism to "the strange beliefs" of their Indian counterparts. For Chidester, popular culture—which includes sports—represents a largely uncharted religious domain, awaiting a new map to better define and categorize it (2005: 10, 50).

This task lies at the heart of the following essay, an examination of "sacred ground" at sporting venues around the world. These places are not empty vessels, meant merely to house games. Rather, these places contain symbolic potency, they are special, set apart from the ordinary. Here, throngs of people cheer their heroes, honor their homeland, and even mourn their dead. Examining these places, then, offers insights on how people identify themselves at their core, where what they are meets what they aspire to be. Put another way, a study of the sacred places of sports is also a study of the people who inhabit them.

151.2 Understanding Sacred Space

In ancient Roman temples, the *sacrum* was the area inside the temple, while the *profanum* referred to everything outside of the temple. The *sacrum* was reserved, set apart, designed to focus the attention of adherents on the pagan gods. The philosopher of religion Mircea Eliade used this image for describing the sacred more generally. "The sacred," he asserted, "always manifests itself as a reality of a wholly different order from 'natural' realities." But Eliade stopped short of defining the sacred, certain that this numinous phenomenon resisted containment by mere prose. Instead, Eliade used negative terminology, referring to the sacred as "wholly other" or "wholly different from the profane" (1959: 10, 11).

Scholars have since built upon Eliade's notion of the sacred as "wholly other," while setting aside his conclusion that these places have a *sui generis*, nameless power. Instead, they have followed the lead of figures like Jonathan Z. Smith, emphasizing the role that individuals and groups play in "producing" the sacred (Nelson 2006). Smith argues that words like "sacred" and "profane" do not exist in on their own, but rather "in relation" (1982: 55). A sacred place becomes sacred when people interpret it that way, organizing it in contrast to other places deemed profane. Creations of the "wholly other," then, bear the stamp of those who make it so.

David Chidester and Edward Linenthal fall within Smith's intellectual lineage, bringing attention to the role of human agency in generating sacred space. Moreover, the scholars introduce three "defining features" of sacred space. First, sacred space is "ritual space," or "a location for formalized, repeatable symbolic performances." Rituals thicken the sacred air of a location through enacting the central narratives and symbols of a community. Additionally, the rules of ritual often only make sense within a sacred place, since these geographic points of origin relate directly to the group's collective identity. Next, sacred space is "significant space," in that "it focuses crucial questions about what it means to be a human being in a meaningful world." In a sacred place, people affirm, define, and defend their worldviews, the lens through which they interpret existence. In doing so, they engage the ultimate questions of life, death, and meaning. Finally, sacred space "is inevitably contested space, a site of negotiated contests over the legitimate ownership of sacred symbols." As individuals and communities generate sacred meanings, they do so in contrast and competition with others. The lines drawn around the sacred, in other words, create a world of "Us," but also a world of "Them." Thus, a sacred place might have multiple interpretations, making it into a river of competing discourses (1995: 9, 12, 15).

This essay employs this tripartite schema of sacred space as a structure for examining the interplay of religion, sports, and place. Overlap between these categories is inevitable, but the case studies sufficiently illustrate each dimension. This essay *does not* address "religion" and sports. Rather, it is about the adjectival "religious" experience of sports. The former evokes images of church structures and formalized creeds. But the latter ventures into the ambiguous realm of wonder, awe, inspiration, and transcendence—categories associated with traditional religions, but not entirely owned by them. When people walk into a soccer stadium or ball park, they continue to be Christians, Jews, Muslims, Buddhists, atheists, or seekers. At the same time, contact with these places awakens religious sensibilities. As Gary Laderman states, "God or no God, play can animate religious energies that bind communities of fans, athletes, and teams together around idols that are worshipped in ways that, for some, create shared experiences and memories as impressive and meaningful as any other sacred encounters in this life" (2009: 62). What follows, then, is an investigation of the creative ways that people do religious things, speak religious words, and have religious feelings—all in seemingly secular places.

151.3 Ritual Space

"One of the most easily identifiable ritual actions associated with baseball," writes religion scholar Joseph Price, "is the journey of pilgrimage" (2006: 144). As anthropologist Victor Turner has explained, a pilgrimage involves purposeful travel to a "center out there," a sacred place with a certain spiritual allure. Through ritualized travel, the pilgrim sustains a hope for enlightenment, certain that contact with the objects and structures along the route and at its terminus will enable some degree of transformation (1973). For Price, the Hall of Fame in Cooperstown, New York, is baseball's "center out there." "Enshrined there," he elaborates, "are tributes to the mythic heroes and their incredible achievements in the game" (2006: 146–147).

Any true baseball pilgrim would certainly want to see the "Sacred Ground" exhibit, which showcases over 200 artifacts from baseball's storied past (Fig. 151.1). As visitors enter, they walk under an impressive stone arch to a ticket booth that had serviced Yankee Stadium for 50 years. On the booth, a sign announces,

Fig. 151.1 Sacred Grounds exhibit at the National Baseball Hall of Fame and Museum (Photo by Milo Stewart, Jr., 2005, used with permission)

"Ballparks are baseball's sacred ground. The total ballpark experience goes beyond hot dogs, luxury boxes and video scoreboards. Ballparks provide the stage for the game, a frame for memories of games past, and the promise of future games enjoyed with family and friends" (Mock n.d.). From here, visitors amble past six themed sections: "Fans," "Ballpark Business," "Evolution of the Ballpark," "The Stadium World," "Reverence," and "Ballpark Entertainment." Throughout the 1,800 ft^2 (167 m^2) exhibit are famous baseball bats, bumper stickers, score cards, rings, trophies, and bleacher seats. Interactive exhibits allow visitors to see, hear, and even smell these ballparks. Dale Petroskey, a former president of the Hall of Fame, referenced his own baseball autobiography to describe his love of the exhibit. "Growing up in Detroit, Tiger Stadium was like a second home to me," he recalled. "It was where my friends and family came together. It was where we celebrated, where we built memories, and saw bits of history play out on a field of green. This exhibit captures that magical connection that forms between fans and their ballpark" (Holmes 2005).

For those who ritually process to and through the Hall of Fame, this "magical connection" orients their attention to the sacred realm of "America's pastime." No doubt, in America, the game has strong ties to what it means to be America (Evans 2002). Baseball, though, is a global sport with multiple "centers out there." Each year in Japan, players and fans flock to Koshien Stadium near Osaka for the National High School Baseball Championship Tournament. To gauge the importance of this stadium, consider that after teams play their final games here the players return to the field and scoop up containers of dirt. "The dirt of Koshien is sacred for Japanese baseball players," observed one tournament staffer. "High school baseball players think the dirt of Koshien is a precious thing, a treasure" (Baker 2012).

As one journalist explained, the ritualized collection of dirt connects players with "something bigger than the outcome of any one particular game" (Baker 2012). The "something bigger" here—its "wholly otherness"—relates to the transcendent fusion of baseball and Japanese identity. The tournament began in 1915. Organizers theorized that baseball could be a valuable tool for teaching the Japanese youth about mental, spiritual, and physical discipline. Additionally, they aspired to reinforce a broad social value known as *wa*, or social harmony. Put simply, the concept of *wa* valorizes the forsaking individual interests for the good of the community. Accordingly, the "sacrifice bunt" is a common sight at Koshien, since it is a microcosmic performance of this defining social value (Whiting 1990).

Soon after its introduction, the tournament became *the* "universal Japanese experience." For 2 weeks each spring, fans stream to Osaka wearing regionally distinct colors, brass bands play celebratory tunes, and professional talent scouts watch for the next great player. The great player at Koshein often embodies what the Japanese call the "fighting spirit." As one fan summarized, "Koshien is a big festival. ... Only it's dedicated to spirit and guts." In 1969, pitcher Koji Ohta threw four consecutive complete games, with the third going for 18 innings. While his final game was a 4-2 loss, he was still elevated to hero status for his show of resilience (Whiting 1990: 247, 262, 246). In 1998, Daisuke Matsuzaka threw 250 pitches and 17 innings to win a quarterfinal game. The next day, he played outfield with his pitching arm wrapped in bandages. And the day after that, he returned to the mound only to throw a no-hitter and win the championship. "Nobody will ever forget what Daisuke did," remarked one journalist. "It was an unbelievable three days" (Larimer 2000).

Despite the spirited atmosphere at Koshien, there is little in the way of unrestrained rowdiness. Crowds and players generally live out the inscription on centerfield, "Without principles, nothing can be done" (Belson 2011). People do not challenge umpires, even when they make bad calls. If an errant pitch strikes a batter, the pitcher immediately gestures a bow to apologize. And when players enter and exit the field, they bow to centerfield, because, as Robert Whiting says, "this stadium is a very special repository of the spirit—like a cathedral. One should pay it utmost respect" (1990: 258).

The rituals performed within this sacred place reinforce key elements of Japanese identity. In the face of tragedy, these rituals functioned as a healing rite. Four months after the March 2011 earthquake and tsunami hit northeastern Japan, a team from Tohoku High School arrived at Koshien Stadium. The school itself was relatively unharmed. But many players had lost relatives and experienced property damage. After the disaster, the team debated whether or not to play, but reasoned that it was an opportunity to honor their region and to display their "fighting spirit." Tohoku lost in the first round, but the crowd gave them a resounding cheer before and after the game. As the team manager remarked, "there are no words to describe how it feels to have people cheer for you" (Belson 2011).

151.4 Significant Space

In the wake of unthinkable destruction, the rituals of Koshien made an unstable world a little bit more stable. In this context, the stadium assumed a new meaning, one related to the realities of human loss. Death, writes philosopher Thomas Attig, "disrupts the continuity of our life stories" (1996: 149). To manage this disruption, individuals and groups develop ritual acts as a means of making this transition meaningful. Funerals, prayer services, burial rites—each seek to reorder worldviews in light of loss. These acts generally occur in designated spaces, sacred environments like a church, mosque, or burial ground (Grimes 2000: 217–284). For the Japanese, a baseball stadium became a shelter for mourning. In one African country, a soccer stadium served a similar function.

Soccer came to Zambia by way of the British, who had colonized what was then known as Northern Rhodesia in 1911. When Zambia gained independence in 1964, the British left, but soccer remained. In fact, the sport became a consuming passion. Yet, on the world stage, their national teams had little success. Then, in 1988, Zambia finished tied for fifth at the Seoul Olympics. Many of those players stayed with the team as the 1994 World Cup neared, where Zambia looked to be a strong contender (Darby 2005: 130–131).

On April 27, 1993, however, 18 members of the Zambian national soccer team and 12 of their support crew died in an airplane crash off the coast of Gabon in Africa. They were headed to Senegal for a World Cup qualifier game. The tragedy left an indelible mark on the Zambian psyche. "For that entire week I wasn't worth anything," remarked one soccer fan. "I didn't want to talk with anyone. I just wanted to be alone." When the bodies returned, a national day of mourning was declared and a funeral service was held at Independence Stadium, the largest sporting venue in Zambia. While approximately 30,000 mourners sat inside the stadium, roughly 100,000 more gathered outside. All watched as the deceased were buried next to the stadium. In the coming weeks and months, their graves became objects of veneration, drawing streams of mourners. Officials soon posted a security detail to guard the graves. "This is sacred ground," one guard remarked. "These are our national heroes" (Montville 1993).

Pangs of sorrow soon turned to hopes for redemption. "From the ashes of disaster," proclaimed a Zambian soccer announcer, "our soccer program is headed for glory, glory hallelujah." Zambia hastily assembled a team and forged ahead with their World Cup ambitions. A mere 10 weeks after the crash, the fresh team faced Morocco at Independence Stadium. Down 1-0 early, the Zambian crowd almost all collectively gazed upon the graves outside the stadium, beckoning the departed for supernatural assistance. Zambia scored twice and won. The word "miracle" was in no short supply. With momentum on their side, the team had a remarkable run, beating South Africa and shooting a tie with Zimbabwe. "After the tragedy and all, this team has been sort of the focal point for the entire country" the team's coach remarked. Young people across the country were excited to watch and play. "If you ever could put together the financing," he speculated, "some African country is going to come very close to winning the World Cup very soon" (Montville 1993).

Zambia's last test was in Casablanca against Morocco. A win or tie would get to them to the World Cup in America. Players visited the graves once more before leaving. One remarked, "In Africa we believe that the spirits must be satisfied. If someone dies, everything must be done properly for that person. Everything has been done properly here" (Montville 1993). Despite a valiant effort, Zambia lost. But the nation used soccer to make meaning of a tragedy, and affirm what it meant to be Zambian.

This bundling of nationhood and soccer also defines the symbolic contours of Mexico City's Aztec Stadium. Built in 1966 in anticipation of the 1968 Olympics, Aztec Stadium is a massive structure, able to hold 126,000 fans. As one sportswriter noted, it intended to show "that Mexico is no longer a stepchild in the family of nations" (Smith 1968). In addition to the Olympics, the stadium hosted the World Cup twice, in 1970 and 1986. Describing the atmosphere of the 1970 event, one journalist invoked the Spanish word *la locura*, "a mixture of madness and folly." Music, crowds, and heightened emotions ruled the landscape. And while police tried to maintain some sense of order in the city, near Aztec Stadium, "chaos" reigned. Pedestrians jammed the streets, cars struggled to inch ahead, and crowds waved Mexican flags yelling, "Meh-he-co, Meh-he-co" (Maule 1970).

With this pervasive sense of national pride brimming, hurt feelings resulted when the English team arrived with their own food, fearing that the local fare would be unsanitary. "We consider them our *huespedes*," remarked one merchant. "You know, when you have guest in your house, is for him everything of the best. The best to eat, the best to drink, the best of your courtesy." To have this hospitality turned away was a dishonor. So the locals reacted by serenading their guests with clanging pots and pans in the early hours of the morning. And when team England played Brazil, Mexicans cheered for their fellow Latin Americans, and jeered the opposition. Mexico eventual lost in the quarterfinal, but locals delighted in watching Brazil continue their dominance, enabled by the superhuman play of their star, Pelé (Maule 1970).

In 1986, the World Cup returned to Mexico City and so too did *la locura*. Argentinian Diego Maradona left a longstanding mark on the event. In the championship game against England, he scored two goals, both of which would be given titles. The "Hand of God" goal came when he slipped the ball into the goal using both his head and hand. Referees did not notice the hand ball, so the goal counted. When questioned afterward, Maradona quipped that the ball was struck both by "the head of Maradona" and by "the hand of God" (Wells 2008). The second goal came near the end of the game. This one was legitimate and it gave Argentina a 2-1 victory. A plaque at Aztec Stadium commemorates this "Goal of the Century" (Adams 2013).

While venerated in Mexico and Argentina, however, Maradona has no such reputation in England, where fans only remember the "Hand of the Devil" goal (Teale 2009). Crafty gamesmanship for one group was perceived as cheating to another. Thus, Aztec Stadium conjures competing memories, derived from different historical interpretations of this single goal.

151.5 Contested Space

The phrase "moral geography" refers to the intersection of moral and spatial order. As cultural geographer Tim Cresswell has noted, people inscribe on to public places moral prescriptions about who can use this space, and how it ought to be used. Often, the exact parameters of these rules emerge only when an outside group misuses the space, and is deemed "out-of-place" by the majority (1996). So space is not morally neutral. Rather, decisions made about its design and usage carries moral weight. As the "Hand of God/Devil" goal indicates, the world of sports has its own, unique moral geography where events, symbols, and structures are subject to figurative tug-of-wars between competing interests.

Writing in 1907, golf enthusiast A. J. Robertson voiced his concern over the "congestion" at the Saint Andrews Old Course, which in his view was the best course in the world. But the Old Course was under threat by a new "evil," namely, the golfer with little regard for the game. Some men play in large groups and women "who ought to know better" wander the course "staring off gaily." Meanwhile, "good players" wait and are often unable to complete a round. Robertson offered a solution: charge a fee for playing the Old Course, which at the time was free. He doubted this would happen. Robertson grumbled that townspeople would not do anything to dissuade tourism, their financial lifeblood. He found this short-sighted. For Robertson, the "sacred ground" of the Old Course had to be preserved for the "genuine golfer" and not for the "cheap tourist and tripper" (1907: 572).

For Robertson, the moral geography of the Old Course demanded that only a qualified player shadow its sacred links. Plenty of other golfers have expressed reverence for the Old Course. "The hair on the back of your neck stands up when you are here," marveled Padraig Harrington, as he prepared for the 150th anniversary of the British Open in 2010. "The setting, the history, all the things that have happened here in the town and on the golf course." He continued, "It's spine chilling. There is no other place in the world like it" (DiMeglio 2010). A former superintendent once reported, "I've seen American visitors come to Saint Andrews, walk down the steps to the No. 1 tee and bow down and kiss the turf." He also claimed to have witnessed a man drive a ball off of the first tee, then leave. "I hit a golf ball at Saint Andrews," the player stated. "Now, I can die when I go back to the states tomorrow with my wife and children" ("Sacred ground," 1970: A10).

History is one significant part of the Old Course's allure. It claims to be the oldest existing golf course, dating back to the fourteenth century. And while most contemporary courses are designed with golfers in mind, the Old Course evolved over time and adapted to the local surroundings. One legend says that the bunkers formed when animals sought shelter on the course during inclement weather. This at least accounts for the course's randomly placed bunkers, which number over 100. Other anomalies include shared fairways, blind shots, and double greens that measure in yards rather than feet. It is a unique golf course that scarcely looks like a golf course (Fig. 151.2). In 1946, professional golfing great Sam Snead arrived at the Old Course for the first time, surveyed the scene, and wondered where the course

Fig. 151.2 Valley of Sin (Photo by Kevin Murray, kevinmurraygolfphotography.com, used with permission)

was. To Snead's astonishment, it was right in front of him. "Until you play it," Snead marveled, "St. Andrews looks like the sort of real estate you couldn't give away" (Shackelford 2003: 15).

For the golfing faithful, the Old Course draws its sacredness from its historic roots, and seeming unchanging landscape. But the course has changed, ever-so slightly, especially in the mid-nineteenth century as professional golf began to develop. Custodians smoothed greens, strengthened bunkers, and made fairways slightly more distinguishable. The game itself also changed. The first standardized golf balls were made of small leather sacks packed with goose or chicken feathers. In 1848, Allan Robertson became the first official custodian of the Old Course and he fancied himself a craftsman of these "featheries." Innovation, though, was on the horizon. The gutta-percha ball was made from the dried sap of the Sapodilla tree, found in Malaysia. These balls were more durable, cheaper, and hit further than the old feathery ball. Robertson was not impressed. He called them "the filth," and tossed any gutta that he found into the fireplace at Saint Andrews. Then, in 1850, he witnessed his longtime protégé, Tom Morris, approach the final green experimenting with a gutta-percha ball. Enraged, Robertson fired Morris on the spot, even though the two had stuffed featheries for 10 years together (Shackelford 2003: 17–18).

"Old Tom" Morris eventually found his way back to the Old Course in 1864, after Robertson had left. Morris made some of the most significant changes to the course, to include adding two new greens, a move that made some traditionalists uneasy. Still, Morris left an indelible mark at the final hole, which is named after him. He relocated the green just beyond a massive depression that he called the "Valley of Sin." On this hole, hitting a short approach puts a golfer in definite peril. In 1933, Leo Deigel had a chance to win the British Open until his ball landed in the valley. He finished with a bogey, missing a playoff by one stroke. A similar fate befell Doug Sanders in 1970. On the final day of regulation, his ball bounced into the depression, resulting in a tie with Jack Nicklaus who beat Sanders in a playoff the next day (MacKie 1995: 159).

This final hole, then, has an infamous history of failure, one that golf enthusiasts recite in order to reflect upon the humbling nature of their game. But not everyone stands in awe. When the British Open returned to the Old Course in 2010, one journalist dared to call the 18th a "terrible finishing hole." Despite the notorious "Valley of Sin," he did not think that it was challenging enough. He believed that an additional fairway bunker might produce slightly more drama. Of course, the journalist knew that this was a rhetorical exercise, since modifying the Old Course like this would be "sacrilege" (Achenbach 2010).

While the symbolic competition at the Old Course described here relates to the game itself, the 1960 Olympic Marathon brings a geopolitical edge to this conversation. In 1935, Italy invaded Ethiopia and, along the way, looted the Obelisk of Axum. Constructed in the fourth century, the 179-ft (55 m) funerary stele had been a distinguishing feature of Axum, the birthplace of the Ethiopian Orthodox Church. While religiously significant for Ethiopians, it was a mere war treasure for Mussolini and his troops. British and Ethiopian forces managed to defeat the Italians in 1941, but the Obelisk remained in Italy until 2005 (Pankhurst 1999). Ethiopian Abebe Bikila would help to re-interpret this relic en route to winning a gold medal.

Bikila grew up the son of a shepherd in the small village of Jatto, just north of Addis Abada, the capital of Ethiopia. There, he tramped about barefoot, minding his herds, and playing a game called *ganna*—a soccer-like game where goals are separated by several miles rather than yards. By 1951, his family moved to Addis Abada and he began serving with the Imperial Army. Bikila soon met Onni Niskanen, the director of physical education for the Ministry of Education. Niskanen believed that Ethiopia could gain national stature by excelling on the international athletic stage. He saw promise in Bikila and recruited him for Ethiopia's marathon team. By 1960, with two suits and $150, Bikila and his teammate, Abebe Wakjira, traveled to Rome (Judah 2008).

While physically fit and ready to run, one question remained—their shoes. Both runners had been accustomed to running barefoot. But they worried that their unshod feet might reinforce negative stereotypes about Africans. They tried shoes just before the race, only to develop blisters. So when the starting gun sounded, the Ethiopian runners took off in their orange shorts, green shirts, and bare feet. The press knew very little about Bikila, but took note of the runner as he nudged ahead to the front pack. By the final miles, Bikila was in the lead with Moroccan Rhadi Ben Abdessselem. Bikila's stride looked effortless, even as he negotiated the jagged cobblestone streets of Rome. One journalist described him as "running so lightly that his feet scarcely see to touch the ground." Then, at 39.9 km, the runners passed the Obelisk of Axum. Here, Bikila surged ahead, separating from the Moroccan for the remainder of the race. The crowd was stunned as the barefoot Bikila strode into the stadium in a new world record of 2:15:16. The first African to win a medal at the Olympics, he returned to Ethiopia a conquering hero (Judah 2008).

Bikila again won marathon gold in the 1964 Olympics, this time wearing shoes. He continued running competitively until 1968, when a car accident left him paralyzed. Bikila died on October 25, 1973, at the age of 41. Over 75,000 people attended his funeral, including the Emperor who declared it a national day of mourning. Above his grave stands a massive bronze statue, depicting him running barefoot through the streets of Rome (Robinson 2008).

151.6 Conclusion

As Tim Cresswell explains, "Place is the raw material for the creative production of identity rather than an *a priori* label of identity" (2004: 39). Physical locations, in other words, are not static products used solely for pragmatic ends. They are instead pregnant with meaning. As this essay has shown, the world's sporting venues are indeed "raw material" for the formation of collective identity. Stadium seats, grassy fields, and cobblestone roads serve as platforms for ritual performances, for making life meaningful, and for defining and redefining the vital symbols of a community. In acquainting ourselves with these sacred places, we also acquaint ourselves with the people inside them.

References

Achenbach, J. (2010, July 11). St. Andrews needs better finishing hole. *Golf Week*. Retrieved May 30, 2012, from http://golfweek.com/news/2010/jul/11/st-andrews-needs-better-finishing-hole/

Attig, T. (1996). *How we grieve: Relearning the world*. New York: Oxford University Press.

Adams, M. (2013, July 11). *Great soccer games played at the Estadio Azteca Stadium. Sporting Life 360*. Retrieved September 22, 2014, from http://www.sportinglife360.com/index.php/great-soccer-games-played-at-the-estadio-azteca-stadium-2-180/

Baker, G. (2012, March 25). Japan's love of baseball runs deep. *Seattle Times*. Retrieved April 30, 2012, from http://seattletimes.nwsource.com/html/mariners/2017834159_marijapan25.html

Belson, K. (2011, July 15). A call to play ball in Japan followed much fretting. *New York Times*. Retrieved May 5, 2012, from http://nyti.ms/neSz4N

Chidester, D. (2005). *Authentic fakes: Religion and American popular culture*. Berkeley: University of California Press.

Chidester, D., & Linenthal, E. T. (1995). Introduction. In D. Chidester & E. T. Linenthal (Eds.), *American sacred space* (pp. 1–42). Bloomington: Indiana University Press.

Cresswell, T. (1996). *In place/out of place: Geography, ideology, and transgression*. Minneapolis: University of Minnesota Press.

Cresswell, T. (2004). *Place: A short introduction*. Malden: Blackwell.

Darby, P. (2005). A context of vulnerability: The Zambian air disaster, 1993. In P. Darby, M. Johnes, & G. Mellor (Eds.), *Soccer and disaster: International perspectives* (pp. 124–140). New York: Routledge.

DiMeglio, S. (2010, July 15). History, mythology combine at St. Andrews. *USA Today*. Retrieved May 15, 2012, from http://usat.ly/bU8O40

Eliade, M. (1959). *The sacred and the profane: The nature of religion*. New York: Harcourt, Brace, and Co.

Evans, C. H. (2002). Baseball as civil religion: The genesis of an American creation story. In C. H. Evans & W. R. Herzog (Eds.), *The faith of fifty million: Baseball, religion, and American culture* (pp. 13–33). Louisville: Westminster John Knox.

Grimes, R. L. (2000). *Deeply into the bone: Re-inventing rites of passage*. Berkeley: University of California Press.

Holmes, D. (2005). *Treading on 'sacred ground.'* Retrieved May 1, 2012, from http://mlb.mlb.com/news/article.jsp?ymd=20050619&content_id=1096259&vkey=news_mlb&c_id=mlb

Judah, T. (2008, July 25). The glory trail. *The Guardian*. Retrieved May 20, 2012, from www.guardian.co.uk/books/2008/jul/26/books.sport

Laderman, G. (2009). *Sacred matters: Celebrity worship, sexual ecstasies, the living dead, and other signs of religious life in the United States*. New York: The New Press.

Larimer, T. (2000, September 9). Daisuke Matsuzaka. *Time*. Retrieved March 30, 2012, from http://ti.me/8lf0Yh

MacKie, K. (1995). *Golf at St. Andrews*. Gretna: Pelican.

Maule, T. (1970, June 22). Soccer is a frenzy. *Sports Illustrated*. Retrieved April 20, 2012, from http://sportsillustrated.cnn.com/vault/article/magazine/MAG1083735/index.htm

McPhee, J. (2010, September 6). Linksland and bottle. *The New Yorker*. Retrieved May 1, 2012, from http://nyr.kr/bZR1qd

Mock, J. (n.d.). *Visiting sacred ground*. Retrieved April 30, 2012, from www.baseballparks.com/hofexh.asp

Montville, L. (1993, October 18). Triumph on sacred ground. *Sports Illustrated*. Retrieved May 15, 2012, from http://sportsillustrated.cnn.com/vault/article/magazine/MAG1138563/index.htm

Nelson, L. P. (2006). Introduction. In L. P. Nelson (Ed.), *American sanctuary: Understanding sacred spaces* (pp. 1–14). Bloomington: Indiana University Press.

Nouwen, H. (1981). *The Genesee diary*. New York: Doubleday.

Pankhurst, R. (1999). Ethiopia, the Aksum Obelisk, and the return of Africa's cultural heritage. *African Affairs, 98*(391), 229–239.

Price, J. L. (2006). *Rounding the bases: Baseball and religion in America*. Macon: Mercer University Press.

Robertson, A. J. (1907). Congested St. Andrews. *Country Life 22*, 563.

Robinson, S. (2008, August 6). Barefoot in Rome. *Time*. Retrieved June 1, 2012, from http://ti.me/8829tl

"Sacred ground" of Golf World. (1970, February 10). *Cape Girardeau southeast Missourian*. p. A10.

Shackelford, G. (2003). *Grounds for golf: The history and fundamentals of golf course design*. New York: Macmillan.

Sheehan, G. (n.d.). *Is running a religion?* Retrieved March 20, 2012, from www.georgesheehan.com/essays/essay46.html

Smith, F. R. (1968, July 15). Ancient contests, shining arenas. *Sports Illustrated*. Retrieved April 15, 2012, from http://sportsillustrated.cnn.com/vault/article/magazine/MAG1081378/index.htm

Smith, J. Z. (1982). *Imagining religion: From Babylon to Jonestown*. Chicago: University of Chicago Press.

Teal, N. (2009, November 20). *Nick Maradona goal 'Hand of the Devil'*. Arsenal.com. Retrieved September 22, 2014, from http://www.arsenal.com/news/news-archive/wenger-maradona-goal-hand-of-the-devil

Turner, V. (1973). The center out there: Pilgrim's goal. *History of Religions, 12*(3), 191–230.

Wells, T. (2008, January 31). Maradona: I hold my hands up. *The Sun* (London). Retrieved May 15, 2012, from www.thesun.co.uk/sol/homepage/news/745800/Diego-Maradona-Football-legend-finally-says-sorry-for-that-goal.html

Whiting, R. (1990). *You gotta have wa*. New York: Vintage.

Chapter 152
Religion's Future and the Future's Religions Through the Lens of Science Fiction

James F. McGrath

152.1 Introduction

There have been those who have doubted that religion had a future, or that the future would have religion, while others have expressed confidently – as French author André Malraux is reported to have said about the twenty-first century – that the future will either be religious or not at all. That people have very different visions of the future is not surprising. If we turn our attention to science fiction, the genre most closely associated with imagining the future, we do not find as many different futures as there are authors; we find a greater number even than that, as authors explore several different possible futures across a number of stories and series. And as different as those futures are, so likewise the place of religions in them differs.

Science fiction has often been a means for exploring cultural diversity projected onto the vista of a distant future – or one that is said to have existed "A long time ago, in a galaxy far, far away." In our time, increased globalization and multiculturalism have brought the actual science fiction stories of human beings from a wider variety of cultures into one another's orbits, although to a lesser degree than perhaps one might have hoped. And as some cultures have remained the dominant exporters of science fiction (as of other sorts of literature, television, and film), without the flow being reciprocal, we are reminded of the legacy of colonialism, one that has been central to countless sci-fi stories featuring new and powerful arrivals who wish to take over our homes and lands, and recounting the struggle of human beings to resist.

Even as authors have explored imagined worlds and hypothetical futures, the palette available to them through their own experience always provides some of the dominant colors they use. The future is always viewed through the lens of the

J.F. McGrath (✉)
Department of Philosophy and Religion, Butler University, Indianapolis, IN 46028, USA
e-mail: jfmcgrat@butler.edu

S.D. Brunn (ed.), *The Changing World Religion Map*,
DOI 10.1007/978-94-017-9376-6_152

present. And thus the futures authors envisage may feature one or more particular human religions, or none at all, or some distorted or combined version of them, whether because of the author's limited experience, or because the author creatively imagined how the passing of centuries or millennia might change them. What religious trends are imagined, if any, tell us at least as much about how religion seems to us in the present, as what the future of religion may be.

In our time, religion and science fiction may relate in any number of ways. For some, science fiction depicts the hope for a future of religious tolerance, or for a future free from religion. For some, science fiction itself takes on religious overtones, as people congregate periodically, donning sacred costumes to recite or reenact their sacred stories and find inspiration within them for their daily lives. For still others, the converse has occurred, and a traditional religion has taken on science-fictional overtones, taking the author of Revelation's obscure but recognizable references to the Roman power of his time, and substituting in their place an apocalyptic vision of the future involving global conspiracy, implanted microchips, and nuclear war. And of course, science fiction can give birth to, or be incorporated into, new religions, whether one thinks of the Raelians (UFO-oriented religion), or Scientology, or the Church of Jesus Christ of Latter Day Saints. Scientology's founder L. Ron Hubbard was an author of science fiction, while several Mormons have expressed their faith through science fiction, including Orson Scott Card and Glen L. Larson, the producer of the original Battlestar Galactica series.

If there is something that all treatments of religion in science fiction have in common, it is that they approach these themes and subjects by means of story. In what follows, I will survey a selection of the range of specific motifs through which sci-fi has explored the terrain of religion, contributing in its own way to the changing map of world religion both in the present and with respect to the future as human beings today envisage it might be.

152.2 Time Travel and the Nature of Time

Philosophers have long debated whether God is within or outside of time, and indeed, whether notions such as being outside of time make any sort of sense. The progress of modern physics has led to comparable explorations of time, with models such as that of "block time" being developed to account for the idea, required by the theory of relativity, that time may be different for different persons or objects moving at different speeds. This view of time serves as the basis for the scenarios explored in science fiction, in which advanced technology enables people to travel backwards and forwards in time, since the past and future must in some sense persist and still be "there" for it to be possible to travel to them.

The relevance of this to religion does not relate only to the model of time that is necessary for time travel to be feasible, which happens to be the same one that makes possible the envisioning of a deity residing outside of time. The biggest point of intersection between time travel and religion is in connection with those religious traditions which place great emphasis on certain historical moments. What would happen if time travelers were to visit those moments? Would they find the claims of sacred texts to be confirmed or denied? And would the very presence of time travelers at those moments perhaps contribute to, or interfere with and even change, those events – and if so, what might the consequences be?

In the story "Let's Go to Golgotha!" Garry Kilworth imagines a world in which time travelers can go to witness historical events, provided they dress appropriately and act according to the predetermined historical script to avoid changing history. But one family which goes to first century Judaea discovers – as they blend into a crowd shouting "Crucify him!" – that the crowd consists entirely of time-traveling tourists, and no one from the local population. In a similar (but far more satirical) vein is Michael Moorcock's famous story, "Behold the Man." In it, a time traveler goes back to the time of Jesus to find him, only to discover that there is a man named Jesus in Nazareth, but who is mentally disabled. Yet finding himself there, the time traveler steps into the shoes of the Jesus of the Gospels, and in essence becomes him. Even on Doctor Who, although there has never been an episode telling a story about the Doctor traveling to the time of Jesus, there have been references to his having done so. In "Voyage of the Damned," a Christmas special, the Doctor mentions that he was there for the original Christmas, and that he "got the last room," suggesting (however tongue-in-cheek) that the Doctor contributed to the literal factuality of the traditional Christmas story. (The version with an "inn," although traditional, has been challenged by scholars not only on historical, but also on cultural grounds). And in the Easter special "Planet of the Dead" there is an Easter reference, with the Doctor beginning to say "What really happened is…" before getting cut off. And so that show tantalizingly hints at the potential of time travel scenarios to confirm and even contribute to the unfolding of a story as traditionally known, or to witness or contribute to a revisionist version. Yet this creates paradoxes, since the nature of time would seem to have to be more flexible for history to be changeable than is required for time travel to be possible in the first place.

Be that as it may, closed causal loops of the sort Kilworth envisages – people traveling through time to see something ending up causing what they went to witness – also raise another question for religion. In the distant future, might our science advance to such a point that it could bring a universe into existence? If so, could that universe, in conjunction with technology that allows for time travel, be our own universe? On a smaller scale, could the activity of time traveling fundamentalists with advanced technology, determined to prove the Bible correct by ensuring that things happen exactly as written on its pages, actually *be* the inspiration for the stories, and in essence take on the role of and become the deity whose presumed existence initially motivated their interference in history in the first place? Time travel scenarios thus raise for the first time in this chapter the question of what a "god" might be within a science fictional universe. Time travelers are but one possibility, as we shall see.

152.3 Technologies

Time travel in most instances involves the use of technology, but it tends to be the means to an end, the way of enabling the telling of stories set in the distant future or past, rather than an end in itself. But there are technologies which have at times become the focus of stories, and science fiction has proven particularly adept at exploring ethical issues related to those possible future technologies. Mary Shelley's *Frankenstein* is usually considered the first ever work of science fiction, and it had as one of its central themes that of the scientist playing God. That the relationship between technology and divinity would be explored is articulated fairly explicitly in the novel's alternative title: *or, The Modern Prometheus*.

Every future technology human beings have succeeded in imagining has been incorporated into a science fiction story, and sometimes this has been done intentionally with the aim of exploring philosophical and ethical issues. The short story "Where Am I?" by Daniel Dennett is a philosophical thought experiment, but it is also a science fiction tale. Sometimes, technology has been posited that may not be realistic, for practical reasons. The transporter on Star Trek is a famous example, being introduced because it would have been too expensive to film a shuttle landing sequence for every episode. But realistic or not, such fictional technology still allows for the exploration of philosophical thought experiments. If someone is essentially disintegrated in one location, and a precise copy is assembled in another, can we say that the same person has made the trip, or is that copy simply that, a duplicate? Star Trek has raised this issue explicitly on occasion, such as when a transporter mishap produced a duplicate of Cmd. Riker, as featured in the *Next Generation* episode "Second Chances." This question relates to some religions quite directly, if they posit an afterlife in which human beings who have died and ceased to exist will be recreated and resurrected. Will those so raised be the same person, or merely copies?

There are other technologies which likewise produce copies of people, or of personhood. These include cloning, the creation of androids, and artificial intelligence. It has sometimes been asked whether clones would have souls, and stories have been told of human clones made specifically for the purpose of having spare parts for the aging or injured original (as in *The Island*), or as a body to transfer one's memories into, and thus cheat death (as in the movie *The Sixth Day*). But since nature itself produces clones – otherwise known as identical twins – the matter of clones might seem relatively straightforward – or at least, no more complex than the case of identical twins. The identical twin even more than the artificially-created clone raises issues for the religious viewpoint which claims that the soul is present in a human being from conception. If that is so, then do identical twins only share one soul between them?

In the Star Wars prequel *Attack of the Clones*, and in the spin off cartoon series *The Clone Wars*, we encounter a scenario in which clones are mass produced to provide a powerful and obedient army. Even in shows aimed primarily at a younger audience, one finds significant issues being raised, pertaining to the rights of such human fighting machines, the genetic manipulation involved in creating them, and the conditioning that prepares them for battle. If one can modify human beings to make ideal soldiers, perhaps mass producing them, does that give one proprietary rights over the product? Even if one proceeds entirely from scratch, rather than beginning with human DNA and modifying it, would the result be any less a person, and any less worthy of rights? A central question is whether personhood itself correlates with certain rights, or whether our notion of *human* rights may prove too restrictive to encompass certain futuristic scenarios involving modifications to and variations on the persons we know of today – including the clone, the cyborg, and perhaps also the extraterrestrial.

Such lines of inquiry continue beyond the realm of biological entities, at least in science fiction. The possibility of computers being programmed to have sentience, or having sentience emerge spontaneously in complex artificial intelligences, raises the possibility of there being non-biological persons. Whether they can ever have a soul is a question that we may never be able to answer (not least because the precise meaning of the terminology of the "soul" remains hotly debated even when it comes to human beings, as well as because of the different views of personhood espoused by various religious traditions). But whether machines could ever deserve legal rights is a question that we may find ourselves having to answer in the not too distant future in real life. Both the legal and the spiritual status of such entities has been explored in science fiction – on television shows like *Star Trek: The Next Generation* (whose Cmd. Data is the subject of a trial to determine his status in the episode "The Measure of a Man"), the rebooted *Battlestar Galactica,* and *Terminator: The Sarah Connor Chronicles,* and literary explorations such as Isaac Asimov's famous *I, Robot* collection and Jack McDevitt's "Gus." These stories have explored not only the ethical questions raised for humans by AIs and androids, but also the possibility of ethics, religiosity and spirituality being features of machine intelligences themselves. As entities at least in the first instance modeled on our own selves, created "in our image and in our likeness," there is no reason to expect that they would sidestep entirely all aspects of the multifaceted ethical and spiritual lives that characterize their human creators. Whether as a result of direct programming by us or through emulation of us, such machines are liable to mirror, to at least some extent, everything we hold dear and everything we find most disappointing about ourselves as human beings.

Drugs have not had the prominence in recent science fiction as they did half a century ago, in an era typified by pharmacological experimentation inducing altered states of consciousness. Huxley's *Brave New World* nonetheless remains strikingly timely in its depiction of a society sedated into a pharmacologically-induced contentment, and Frank Herbert's *Dune* series seems similarly up-to-date in its concern for the environment, and its depiction of a future upon which Islam is one of the religions that has had a profound long-term influence. Both those stories explore the role of controlled substances on human existence, and indeed the power that comes with controlling those substances which human beings desire. The *Dune* novels in particular explore the potential for religion to be used by powerful organizations to manipulate whole worlds, and to inspire movements seeking to escape the control of such powers, which take human history in unexpected directions, led on by mahdis or messiahs.

152.4 Manipulating and Controlling Nature

The *Dune* books are rare in being set in a future where certain elements of technology have been largely eliminated – the Butlerian Jihad, with its obvious religious overtones, is mentioned on multiple occasions as having driven computers and any "thinking machines" from legality and thus from widespread use. The relationship between what nature has produced – whether human brains or planetary ecosystems – and what human beings have developed to supplement, supplant, enhance, or alter what nature itself has produced – is one of the major focuses of an entire genre of science fiction, and the case could perhaps be made that it is the most universal feature of the genre. Whether one thinks of artificial as opposed to biological organisms and minds, machines that allow us to travel through space and/or time in ways that we could not on our own, weapons that allow people to be vaporized or whole worlds to be blasted into debris, or anything else that typifies science fiction worlds, the relationship of humanity and its creations to "nature" is a perennial theme.

It could be argued that the dichotomy is a false one. After all, we ourselves are products of nature, and the technologies we develop simply use the minds that nature has endowed us with to utilize the raw materials that nature has provided us with in a creative and innovative fashion. Nevertheless, the potential of science to unbalance nature and send it spinning out of control, or to control nature and regulate it, remains a constant topic of interest in sci-fi, and rightly so, however problematic the dichotomy may be on one level. Our potential to blow ourselves up is both the backdrop to the original *Planet of the Apes* movie, and the ending to the story of life on Earth in a sequel to that film. And while science fiction has constantly invented new technologies, it has all the while continued to point our attention to the harm that technology can do. The movie *Avatar* envisaged a world which represents a much more literal "Mother Earth" than the Gaia hypothesis envisaged, a world on which all life it quite literally interconnected, or at least can be.

If in *Avatar* there were organisms that could plug into their natural world to commune with their planet as a unified living entity, movies like the *Matrix* films and *eXistenZ*, and novels such as William Gibson's *Neuromancer*, have explored the possibility of creating *virtual* worlds. The two are once again not inherently diametrically opposed. Humans might shift certain activities into a virtual world as a step towards preservation of the natural world, or could create (or be enslaved in) a virtual world that resurrects a likeness of a natural world long since devastated. Technology in science fiction may transport living things and in some cases may resurrect them. Technology may bring life, may destroy it, and in some instances may do both simultaneously, as in the famous instance of the Genesis device that is the focus of the second and third Star Trek movies. The possibility of terraforming a world – taking a barren world and making it habitable – represents the promise of technology to support nature at its most spectacular. The possibility of turning a habitable world into a dead wasteland represents the horror of technology's potential to undermine nature at its most terrifying.

152.5 Extraterrestrials

Even without terraforming, science fiction regularly expects the cosmos to offer other places where life can and has flourished, and thus where intelligent beings not entirely unlike ourselves have evolved. Stories about the alien, the extraterrestrial, allow us to explore not only hopeful scenarios related to interstellar cooperation, which parallel our hopes for international and interpersonal friendship and community, but also bring attention to the darker side of human existence, as the other is feared frequently because of the extent to which that other resembles ourselves, or in ways that parallel our fears and hatred towards other human beings.

Few science fiction stories are more familiar than those in which one or more alien spacecraft, more powerful than anything on Earth, appear in the skies above our planet. In some instances, they immediately begin firing weapons to which we have no equivalent. In others, they speak words of peace and friendship, but hide ulterior motives. Both scenarios reflect the ways that human nations and empires with more advanced technology have invaded the lands of other people, and have either immediately opened fire, or have offered gentle words and nevertheless taken possession of whatever they wished. And in stories in which a kind alien has ended up in our midst, as in the classic film *E. T.*, the situations may have been somewhat reversed, but instances of the use of power to dominate and control by force appear nonetheless. Perhaps this is because we find it difficult to tell an interesting and compelling story that does not involve conflicts of interests. But perhaps it is because the future and the alien that we imagine are based on our own history and experience, and finds its inspiration within our own selves, with all that is beautiful and all that is disturbing therein being projected onto the pages or screens as we tell our stories.

152.6 Communication and Communion

In some stories, the hurdles to interplanetary communication are overcome by means of technologies that are implausible, even ludicrously so. The "Babel fish" of Douglas Adams' *Hitchhiker's Guide to the Galaxy* pokes fun at Star Trek's notion of a universal translator. Such devices must be posited in order to tell the stories that we wish to, since it would frustrate us if story after story involved encounters in which communication remained impossible. But the reality is that we have terrestrial languages that we have still not managed to decode, whether the script of the Indus Valley civilization or the precise meaning of much of Etruscan. That we could fail to make sense of the communication of other human beings, and yet create a device that can instantly make sense of one from another world, seems at best far-fetched. Of course, science fiction is characterized by the use of technology and science that does not currently exist as the backdrop for the stories it tells. If we were to exclude any story that involves technology which is unrealistic to us at present, we would in fact eliminate a great deal of sci-fi, and perhaps prematurely. At present, not only the universal translator but also the warp drive and the time machine and the light saber appear unlikely to ever become realistic. But the same was said of other things that people once imagined, such as sending human beings to the moon. The whole point of science fiction is to imagine, and it is right to imagine that, if we ever manage to make contact with other sentient beings, or they make contact with us, we will also find ways to communicate.

Contact, based on a story by Carl Sagan, envisages the use of mathematics as a universal language which could be used to make a first step towards conveying information. But when Ellie Arroway travels in a device the instructions for which human beings managed to decode and build, there seems to be no actual issue of language, and indeed the one being from another world who addresses her takes on a familiar human form. The possibility there may be not merely technologies, but ways of existing and being alive, that we can scarcely imagine, renders the limits of current technology but one part of a broader puzzle that sci-fi explores.

152.7 Life and Living

The opening monologue of the television series *Star Trek* spoke of "strange new worlds" and seeking out "new life and new civilizations." Often, depending on constraints not only of the imagination but of budget and special effects capabilities, the non-human entities that have been encountered in science fiction have still had a vaguely human form, with relatively minor modifications to indicate otherness, whether the addition of green skin color or of pointiness to the ears. When such constraints have not been present or were less stringent, far more different forms of life have been imagined. Beings of silicon which seem like rocks or monsters to us but have intelligence and feelings, as in the Star Trek episode "Devil in the Dark." Beings that have left bodily existence behind, as in Ray Bradbury's story "The Fire Balloons." Science fiction has even explored the possibility that the gods of old might have been powerful aliens mistaken by humans for something supernatural, as well as the possibility that the future for humanity and other species might resemble something akin to what has traditionally been meant by divinity. While at this point, religion and science fiction have sometimes been felt to collide and conflict, when explorations of these topics have been thoughtful and nuanced, the situation seems to more appropriately be viewed as one of overlap and intersection.

Science fiction has explored as fiction stories which some have treated as fact. One thinks, in this category, or Erich von Däniken and Zecharia Sitchin, who claimed quite seriously that the gods of human antiquity were in fact alien space travelers. It is interesting that this viewpoint, relegated to the ludicrous fringe when one takes it seriously, remains an incredibly popular scenario to explore in stories, as long as it is clearly being regarded as fiction. This may have something to do with the fact that science fiction often represents an updating of classic streams of storytelling with their roots deep in human antiquity. While some love fantasy, some science fiction, and some both, the distinction between the two genres is not the plausibility of the scenarios, but whether something *allegedly* scientific is posited as providing an explanatory framework. And thus entities like Apollo and Q can inhabit the largely secular Star Trek universe because, as Clarke's Law states, "Any sufficiently advanced technology is indistinguishable from magic." One could consider it to be a corollary of that principle that any sufficiently advanced being, or being in possession of sufficiently advanced technology, is indistinguishable from a god.

Whether telling stories of gods, near gods, godlike beings, or something very different from any human conception of a god, science fiction certainly has sought to explore the limits of the definition of "life." From entities of pure energy akin to spirits, to zombies which seem to have everything but spirit, and from artificially-created life forms to hybrid cyborgs to parallel versions of ourselves inhabiting a parallel universe, if there is a domain in which science fiction has been constrained only by the limits of the human imagination, it is in the life forms that have been made to populate our fictional worlds. As for which of them if any could legitimately be termed "deities," that can often be difficult to say, and is obviously a matter of perspective. Indeed, when humans have been depicted as visiting worlds where the advanced technology available to humans is not yet known, those human beings have on occasion themselves been mistaken for gods.

Science fiction has explored not only what it means to be a life form, but also what living might be like in our own future. Life without end has long been the offer held out to human beings by certain religions. The possibility that science could provide greatly extended lifespans, if not in fact immortality, may seem like the ultimate encroachment of science onto religion's territory. Science fiction, exploring such possibilities, might be said to not merely offer scientific gods, but also scientific salvation. This is sometimes accomplished through medical treatments performed on humans or by drugs, which raises the specter of whether life-extending technology will be available to all or only the rich. Indeed, some stories have imagined clones – exact copies of human beings, complete with personalities and personhood – being created solely to provide spare parts for the originals of whom they are identical copies. In other cases, human beings must transcend biological bodily existence altogether in order to achieve such immortality, whether by transferring our thoughts into artificial and immortal bodily replicas of themselves, or by transferring their thoughts into a machine. While the *Matrix* films envisaged a scenario in which artificial intelligences enslaved us within an artificial world, it is also possible to imagine human beings choosing of their own free will to enter and exist forever within an artificial world, rather than die at the end of finite biological lifespans in the real world. And indeed, when a complete artificial world can be created, the precise definition of a "real" world, and what distinguishes it from an artificial one, becomes blurry – as does, once again, the role of gods and mortals in bringing entire universes into existence.

Science fiction stories – like all human storytelling – give voice to our deepest hopes and fears. Sometimes we envisage a future in which there is unlimited energy, and there is no shortage of food or of opportunity. Sometimes we envisage post-apocalyptic situations in which, whether through natural disaster or our own misuse of technology, life on Earth is brought near to extinction, and some resort to cannibalism in order to survive (as in *The Book of Eli* and *The Road*). We hope that science will provide possibilities and resolve problems. We fear that in the hands of those who do not use it wisely or kindly, it will result in horrors reminiscent of scenes from the Book of Revelation. And whether telling or reading either kind of story, we usually recall that ultimately whether the future will be closer to one or the other largely depends on us.

152.8 Travel

Astronomy has revealed to us increasingly intriguing images of other planets and moons besides our own, of beautiful stellar nurseries and enormous colored clouds of gas. But must we only view them from afar? The possibility of travel to the celestial realm has been a subject of human stories long before there was science fiction, in particular within religious literature. As we have attained an actual glimpse of what is up there, even though the expected host of angels, and then of Martians, were not to be seen, that has only increased our desire to find ways to journey further beyond our world and explore the cosmos. In the long run, this may in fact be something that our survival depends on, as our sun's lifespan is finite, even if for most of us still unimaginably long. As a species, all our eggs are in one basket, on one planet. The history of our species will have an inevitable end to its story unless we can set forth and make new homes for ourselves elsewhere. But even were that not the case, curiosity would be more than sufficient to make us desire to venture out into space.

But is it feasible? We have sent probes throughout our solar system, but human beings have been no further than our moon, and only a small handful of us have done even that. To travel to planets around other stars would require enormous amounts of time – or the ability to travel faster than light, which physicists continue to discouragingly say is impossible. We have not allowed this to hinder our imagination, but this is one point at which it disheartens us to think that we may never overcome the obstacles that stand in the way of our turning what we imagine into reality.

As long as feasibility is not a factor, however, fictional science has provided countless methods of achieving interstellar travel – from clearly technological warp drives and hyperdrives, to the tongue-in-cheek improbability drive of Douglas Adams' novels, to the use of spice in Frank Herbert's *Dune* series, spice being a drug that somehow allows not merely mind-altering but space-altering. Likewise what seems like magical teleportation is achieved through stargates and transporters, allowing a person to be transported instantaneously to another place. Whether the person who appears on the other end is oneself or merely a copy is a philosophical question worth pondering before stepping into such a device, if continuity of existence is important to you, and not merely having an exact copy of you see things that you could never have without the help of such technology – a point we mentioned earlier. Yet while some technologies of transportation present hurdles of fundamental physics, others have probably not been realized for more practical reasons. Until we find ways to utilize the vehicles we currently have with a much higher degree of safety, it would be unwise of anyone to finally provide us with flying cars, hoverbikes, or jet packs.

It is not only through space that we have imagined technology transporting us. Time travel is also a popular device in science fiction. In some instances, the warping of the fabric of spacetime inadvertently catapults travelers through space backwards or forwards in time as well. But often travel to the past or future has been the aim in itself. H. G. Wells told what is arguably the most famous story involving a time machine, and since then Wells himself has appeared occasionally as a character in other stories involving comparable technology. Since science fiction already allows the telling of stories about the future, what does time travel add? The possibility of returning to one's own time, having seen what the future will be like, armed with knowledge that might perhaps allow the course of history to be changed. As we have already discussed time travel in an earlier section, we need not dwell on it here. But it is one more dimension of the cosmos that we have envisaged ourselves traveling through.

152.9 Multicultural and Multiplanetary Diversity

Science fiction has long been a global phenomenon. Recent years have seen an increased flow of information from one part of the world to another, enhanced by technology, although where the political sway of former imperial powers has waned, the possibility of a different sort of cultural imperialism, accomplished by the spread of culture through the aptly-named world wide web, still remains. Nevertheless, the flow of science fiction has by no means been unidirectional, and even within this or that cultural setting, authors of science fiction stories have regularly used the encounter of fictional species to explore the more Earthbound matter of intercultural interaction. Every scenario of colonization, of making peace, of waging war, of misunderstanding, of spreading disease with deadly consequences – in short, every situation that has been witnessed as a reality in human history when societies have encountered one another for the first time has been projected onto the interplanetary canvas of sci-fi.

In some instances, the stories have reflected our cultural prejudices. Aliens have often stepped in to replace the traditional role of "Indians" in equivalent shoot-outs involving space cowboys with ray guns. And where this extremely ethnocentric outlook has been avoided, it has still proven hard for authors to avoid painting the alien as a monolithic and monochrome "other" which, however much prejudice may be combatted in the telling of stories, still reflects a single species which seems to have a single culture, language, and religion, and thus mirrors our impression of the foreign human societies of which they are a symbol, much more than the likely realities of an entire *planet* akin to our own. Our species here on Earth speaks a wide array of unrelated languages, practices a variety of religions, and espouses a diverse range of values, and it is realistic to expect that the inhabitants of other planets would be similar in this regard.

152.10 Conclusion: Mapping Religion and Science Fiction

Although science fiction envisages a wide array of futures, it views them from the perspective of our own experience. As a result, those futures are often remarkably similar across a range of human cultures. Some stories have been told by the imperial conquerors, and some by the colonized, but both have managed to tell stories that explore humanity setting forth to conquer others, or of our being conquered ourselves or resisting such conquest. Some stories have been told first within a society that developed new technology and exported it to other places, while others have originated in a society that experienced the cultural quakes resulting from that technology's arrival, but in both cases, authors have imagined futures in which technology solves problems and in which technology causes all sorts of new ones, either giving expression to or undermining our most cherished values, or both.

And so as scholars seek to explore the way that the phenomenon of religion is changing and developing on a global scale, and the way the scholarly study of religion is seeking to plot, map, analyze, study and account for what is unfolding, the study of religion in science fiction can provide a useful tool. Science fiction brings into the foreground our current hopes, fears, expectations, and dreams about the future. It envisages what we fear might be, what we hope will be, and what we hope will never be, all based on trajectories present in the actual world the authors and fans of science fiction currently inhabit. As such, it provides an important barometer to the state of popular culture and broader cultural trends. Maps quickly become outdated, and so too do scholarly "maps" of the aspects of human life and history that scholars study. Even science fiction quickly becomes dated, as assumptions about life on Mars and it having a breathable atmosphere, or references to the "wireless," seem quaint and antiquated rather than something realistic in the present or future. But by turning our attention to the intersection of religion and science fiction, we can hope to gain an understanding of important aspects of the phenomenon of religion and the directions in which it is heading in our own time, however ephemeral or otherwise our scholarly reflections on these may or may not prove to be. Trends and currents change, but the very fact of the recycling of ancient gods and literary motifs in generation after generation of science fiction, and the popularity of remakes and the "reboot" of older stories and series, shows that, if the longevity of the specific visions and the popularity of specific science fictional worlds can hardly be guessed, the phenomenon of science fiction itself is one that we can expect to persist, and to continue to provide a glimpse of important components of our world, and of its thinking, cultures, and values.

Bibliography

Adams, D. (1979). *A Hitchhiker's guide to the galaxy*. London: Pan Books.
Asimov, I. (1950). *I, Robot*. New York: Gnome.
Bertonneau, T., & Paffenroth, K. (2006). *The truth is out there: Christian faith and the classics of TV science fiction*. Grand Rapids: Brazos.
Bradbury, R. (1951). The fire balloons. In *The illustrated man*. Garden City: Doubleday.
Caputo, J. D. (2001). *On religion*. New York: Routledge.
Clarke, A. C. (1972). *Profiles of the future: An inquiry into the limits of the possible*. New York: Harper and Row.
Cowan, D. E. (2010). *Sacred space: The quest for transcendence in science fiction film and television*. Waco: Baylor University Press.
Crome, A., & McGrath, J. F. (Eds.). (2013). *Time and relative dimensions in faith: Religion and doctor who*. London: Darton, Longman, and Todd.
Dennett, D. (1978). Where am I? In *Brainstorms*. Montgomery: Bradford Books.
Gibson, W. (1984). *Neuromancer*. New York: Ace Books.
Hatch, R. (Ed.). (2006). *So say we all: An unauthorized collection of thoughts and opinions on Battlestar Galactica*. Dallas: BenBella.
Herbert, F. (1965). *Dune*. Philadelphia/New York: Chilton Books.
Hopkinson, N., & MeHan, U. (Eds.). (2004). *So long been dreaming: Postcolonial science fiction and fantasy*. Vancouver: Arsenal Pulp.
Huxley, A. (1932). *Brave new world*. London: Chatto & Windus.
Kilworth, G. (1975). Let's go to Golgotha. In *The Gollancz/Sunday Times best SF stories*. London: Granada.
McDevitt, J. (1991). Gus. In A. M. Greenley & M. Cassutt (Eds.), *Sacred visions* (pp. 1–25). New York: TOR.
McGrath, J. F. (2002). Religion but not as we know it: Spirituality and sci-fi. In C. K. Robertson (Ed.), *Religion as entertainment* (pp. 153–172). New York: Peter Lang.
McGrath, J. F. (Ed.). (2011). *Religion and science fiction*. Eugene: Pickwick Publications.
Moorcock, M. (1966). Behold the man. *New Worlds* 166. London: Compact.
Porter, J., & McLaren, D. (Eds.). (1999). *Star Trek and sacred ground: Explorations of Star Trek, religion, and American culture*. Albany: SUNY.
Sagan, C. (1985). *Contact*. New York: Pocket Books.
Shelley, M. (1823). *Frankenstein: Or, a modern prometheus*. London: Whittaker.
von Däniken, E. (1970). *Chariots of the Gods?* New York: Putnam.
Wagner, J., & Lundeen, J. (1998). *Deep space and sacred time: Star Trek in the American mythos*. Westport: Praeger.
Wells, H. G. (1893). *The time machine*. New York: Henry Holt & Co.

Part XII
Organizations

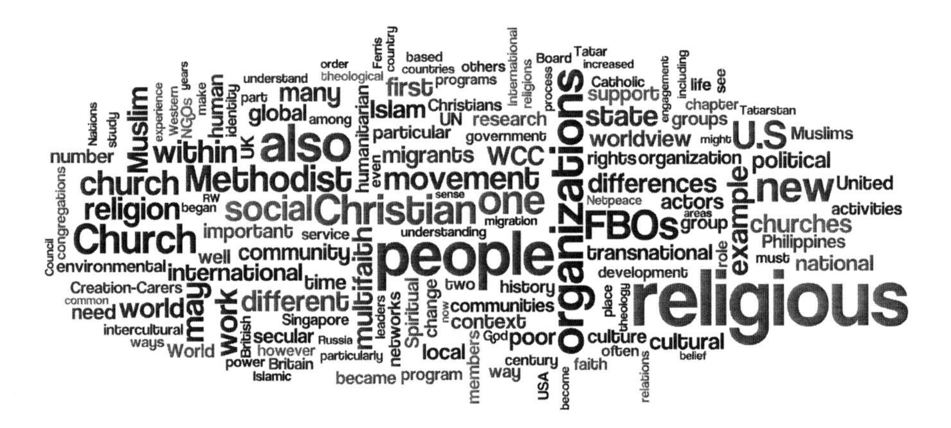

Chapter 153
Global Reach and Global Agenda: The World Council of Churches

Katharina Kunter

153.1 Introduction

Today historians interpret the twentieth century as an age of extreme violence, wars, destruction, ideologies and dictatorships. At the same time, it opened new forms of transnational communication, set up global organizations and networks, and connected at the same time the local to the universal. Just as secular history, the history of Christianity was embedded between these two extremes of the short twentieth century (Lehmann 2012; Kunter and Schjørring 2008). As a result of these developments Christian transnational cooperation as well as interdenominational cooperation increased, amounting to the heyday of the Ecumenical movement.

Notwithstanding the schism and splits that have been a part of Christian history since its onset, the Great Schism of 1045 between the Eastern and the Western Church became a crucial symbol for a divided church, just as the Reformation in the sixteenth century that finally resulted in a new theological doctrine and the separation of Protestantism from the Roman Catholic Church. Although in reality these different Christian denominations and churches face each other irreconcilably, they are united by the shared belief that one day all these separations will be overcome and all Christians will be one in Christ, as it is written in John 17:21. Contemporaries of the Ancient World called this vision of a worldwide Christian community *oíkumene*, after the Greek word *oíkos* which referred to the whole inhabited world on earth.

Initially promoted by young Protestant men in Europe and North America, the idea of a worldwide Christian community grew since the middle of the nineteenth century and gave birth to several new ecumenical networks and organizations. The World Council of Churches (WCC) consisting of Protestants, Anglicans and Orthodox, emerged from these organizations. It was founded in 1948 in Amsterdam

---

K. Kunter (✉)
Faculty of Theology, University of Bochum, Bochum, Germany
e-mail: katharina.kunter@gmx.de

S.D. Brunn (ed.), *The Changing World Religion Map*,
DOI 10.1007/978-94-017-9376-6_153

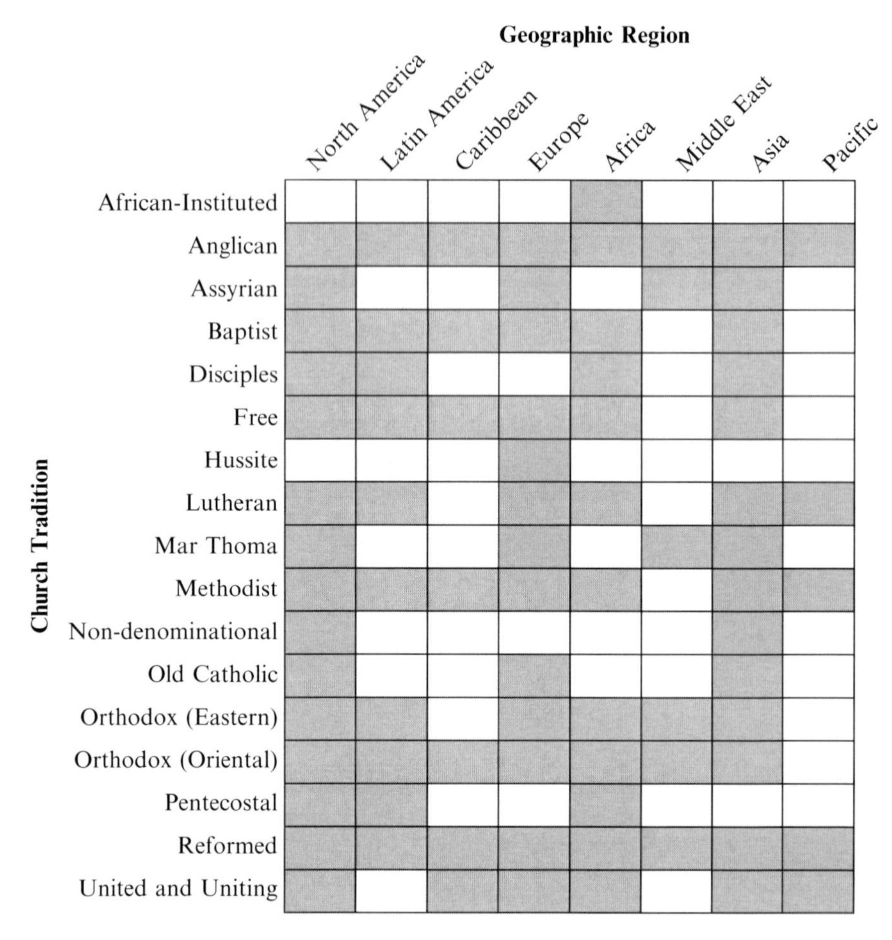

Fig. 153.1 Church tradition and geographic regions of members of the World Council of Churches (Source: World Council of Churches, used with permission)

and has been located in Geneva, Switzerland ever since. During the postwar period, this Christian organization developed into the most influential international church board besides the Roman Catholic Church. But while the Vatican is an independent state under international law, the WCC has not so much possibilities to influence international politics because it is just a transnational umbrella association (Rittberger and Zangl 2003). Despite the ongoing, and often justified, criticisms about the theology and policy of the WCC, it is still the only non-Catholic Christian organization with a worldwide, interdenominational membership (Kunter 2012; Richter 2011; Bremer 2003; Greschat 2000; Joppien 2000; Besier et al. 1999). With 349 member churches in more than 140 countries on all six continents, the WCC today represents more 572 million Christians. More than a quarter of the member churches (132 million) are from Africa, nearly the half (287 million) from Europe, and a fifth (62, 6 million) from Asia[1] (Figs. 153.1 and 153.2). Based on the assumption

[1] Data from 2006, see WCC (Ed.) (2006). Zeugnis – Einheit – Dienst. Geneva: WCC Publications.

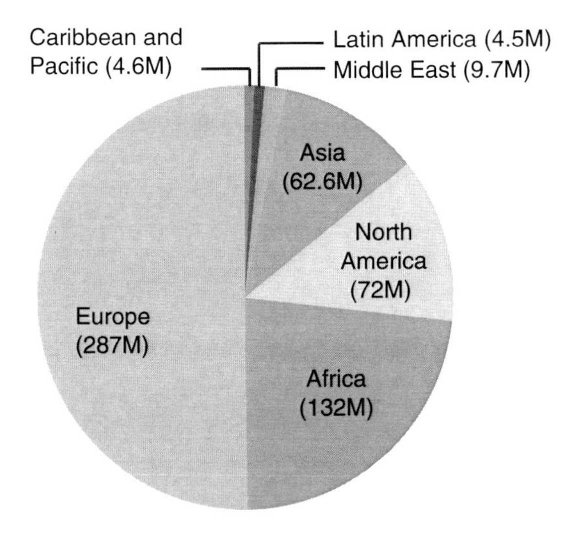

Fig. 153.2 Regional percentage of the 572 million members of the World Council of Churches (Source: World Council of Churches, used with permission)

that this worldwide scope of the WCC goes back to the 1960s and early 1970s and went hand in hand with decolonization and the rise of the organizations such as the United Nations, this article explores three sides of this ecumenical transformation. First it looks back on the forerunners of the WCC and the period of the first globalization; second it discusses when and in which way the WCC adopted and represented the topics and policies of international organizations such as the United Nations, and third it shows how the WCC has been transformed from a formerly Western organization into a global community and also the conflicts and tensions that this transformation has caused.

While the majority of studies on the history of WCC are written from a very European or Western perspective this chapter explores the WCC from a global historical perspective. Using new approaches of the study of Global History, the term "Globalization" is applied to this historical process. Hence, the WCC in the "long sixties" achieved not only global structures and global members, but also a global consciousness. This framework was formed by the thought that all people are living in one world – and therefore responsible for each other and everybodies' conditions of life. This view implied, according to Roland Robertson (Robertson 1992), the synchrony between the global and the local. It will be shown that especially in the 1960s and 1970s the WCC and its conferences conferences and programs formed the global platform which interacted directly with its local bases, viz, the parishes and local communities of their member churches.

153.2 Beginnings of the Ecumenical Movement and the First Phase of Globalization

The origins of the Ecumenical movement can be traced back into the middle of the nineteenth century when English and the American revivalism called for a renewed Christendom. Inspired young men sought to cross national boundaries and to win people all over the world for Christianity. Aiming for conversion and mission in the inner city especially, they believed in the decision of the individual for Christ as their personal savior. Their vision was a movement for Christ, a worldwide community that encouraged the incorporation of Christ in the daily practical and spiritual lives of its members. The *Young Men's Christian Association* (YMCA) was one of the first of these initiatives. It was founded in 1844 and focused on young men in English towns and cities. The YMCA turned out to be the first transnational ecumenical movement (Mjagkij 2003; Davidann 1998; Xing 1996). Within a short time, it spread from Great Britain to Australia, India, Japan, China and to the United States as well as into other countries of Europe including Germany, Switzerland, The Netherlands, Belgium and France. The first world conference of the YMCA took place in Paris in 1855. With its mission statement and its conference style of delegates from national YMCA groups, it set a standard for all later ecumenical movements and bodies. The *Young Women's Christian Association* (YWCA), the corresponding organization aiming at women, soon spread to Europe, to the United States, Japan, China and India. The great missionary dedication of the YMCA and the YMCA inspired the foundation of other Christian youth movements. Among those was the World Student Christian Federation (WSCF, founded in 1895) which became another forerunner of the Protestant ecumenical movement (Selles 2011; Evans 2003; Lehtonen 1998; Potter and Wieser 1996). Organized in the same way as the YMCA, it shared at first the idea of evangelization and the personal decision for Christ, but then stepped outside of the conservative piety of the revival tradition and became much more liberal and modern. For example, many American members joined the Social Gospel, a socially progressive leftist movement. Women were also allowed to become members of the WCSF. From 1895 onwards John R. Mott (1865–1955), a famous Methodist preacher from the United States, and Clara Ruth Rouse (1872–1956), a former English missionary in India, led the WSCF for more than 20 years. During this period the WCSF grew into a real international ecumenical movement, with more than 3,000 groups with 300,000 members worldwide in the 1920s. The WSCF formed the base from which future leading figures of the WCC like the general secretaries Willem Visser't Hooft from the Netherlands or Philipp Potter from the Caribbean received their theological and cultural imprint.

The YMCA and WSCF developed into the main transnational actors during this first period of global ecumenical movements. Many of their members were also delegates at the first World Mission Conference, held from June 14–23, 1910 in Edinburgh in Scotland. It was the biggest ecumenical church gathering at that time, with 1,400 representatives coming from 159 different Protestant Missionary Societies worldwide (Kunter 2011; Stanley 2009; Hogg 1952; Gairdner 1910). Mott summarized the conference as "The most notable gathering in the worldwide expansion of Christianity ever held, not only in missionary annals, but in all Christian annals" (Hopkins 1979: 342). Its vision was a renewed and dynamic, steadily growing Christendom that finds its way from the "West" into the rest of the world.

Fuelled by the great optimism of the new century, the quote "The evangelization of the world in this generation" by conference chairman and later Peace Nobel prize winner John R. Mott soon became an aphorism of the new aspiration towards a globally interlinked Christendom.[2] But despite this global motto, the Edinburgh conference was primarily an event of the Protestant West. Only 17 of the 1,400 participants came from countries of the so-called Third World, even if the many international working-places of the participants suggested a highly global representation of World Protestantism. Paternalistic ideas of mission and the expansion of the Western culture into the Non-Western world dominated the discussions.

Nevertheless, the Edinburgh conference became a milestone in ecumenical history that led to the development of new international friendships and networks.

153.3 The Moral, Theological and Political Foundation of the WCC

The First World War severed many of these new connections. A new start was made in 1938 in Utrecht in the Netherlands. Members of the ecumenical groups "Faith and Order" and "Life and Work" decided to establish the World Council of Churches (WCC) with its headquarter in Geneva as an international institution for all Protestant churches. However, any concrete plans to realize this institution were stopped by the outbreak of World War II. Yet, there was an inner circle around future General secretary of the WCC Willem Visser't Hooft (1900–1985) who were still in operation, helping Jews and other refugees, and passing secret information from the German resistance to the Allies.

After the war in 1946, a number of prominent Anglo-American members of the WCC and the International Missionary Council founded the Commission of Churches on International Affairs (CCIA) (Nurser 2005; Greschat 1998; Nolde 1974; Fey 1968). A leading figure of this commission was John Forster Dulles, the later United States Secretary of State. He had already chaired the "Commission to Study the Bases of a Just and Durable Peace during Wartimes" in the 1940s which had tried to promote ideas of an international world order, decolonization, disarmament and democracy during the war. The CCIA continued in this spirit this policy, but focused particularly on human rights and religious liberty. Its members maintained close relations with high ranking politicians and followed carefully the work of the United Nations from its beginning. They also participated in the drafting stage of the Universal Declaration of Human Rights. With their statement on religious liberty of 1947, the CCIA contributed the final version of Article 18 on

[2] It was originally the title of Mott's bestselling book: Mott, John R. (1900). The evangelization of the world in this generation. New York: Student Volunteer Movement for Foreign Missions.

freedom of thought, conscience and religion that was included in the final version of the Human Rights Declaration from 10th December 1948.

The first meeting of the WCC in autumn 1948 in Amsterdam was based in many ways on the moral and political groundwork of the UN (WCC 1949; t'Hooft 1982). Many delegates were further inspired by the thoughts presented by Scottish missionary Joseph H. Oldham and Swiss theologian Karl Barth on the role of the church in a world in disorder. These two directions resulted in the theological and socio-ethical concept of a responsible society. This concept stressed the Christian responsibility towards God and Men in a free society, appreciated the role of the UN, and demanded the implementation of the universal human rights, thus setting the course for a global political engagement of the churches – although, like in Edinburgh 30 years before, the founding members in Amsterdam came largely from North America and Western Europe. Only 30 out of the 147 member churches were from Africa, Asia and Latin America.

But while the WCC was still dominated by Western ideas and staff, it endeavored to reduce slowly its political activities in postwar Europe. The close relationship between the WCC, the CCIA and the UN was strengthened when the UN accredited the CCIA as one of the first NGOs (Kunter 2002) (Fig. 153.3). This gave the WCC

Fig. 153.3 CCIA staff on United Nations Balcony D (Photo from World Council of Churches, used with permission)

access to more information, influence and cooperation within other organizations of the UN like the Economic and Social Council (ECOSOC), the Food and Agriculture Organization (FAO), the International Labour Organization (ILO), the United Nations Educational, Scientific and Cultural Organization (UNESCO) and the United Nations International Children's Emergency (UNICEF). In order to coordinate its work and to communicate more effectively with the UN, the CCIA maintained offices in New York, Geneva, London and Paris. Today, only two of those offices still exist: the headquarters in Geneva and the office in New York, which was renamed CCIA/WCC United Nations Headquarter Liaison Office in 1960. All these connections provided incentives for the practical work of the WCC and the CCIA. Just as during World War II, international refugee work soon emerged as one of the central areas of the humanitarian work of the churches. In 1949, the CCIA, troubled by the situation of the Palestinian refugees after the 1948 Arab-Israeli War, inquired about their future in the UN General Assembly and called for a broader and more comprehensive approach to the refugee problem which was not only limited to European refugees and displaced persons (DPs). Its intervention was successful. After the foundation of the UNHCR in 1951, the WCC and the Lutheran World Federation succeeded in procuring more funding for UN refugee work and in improving the international standards for those working with refugees.

153.4 The Second Phase of Globalization: Decolonization and De-Westernization

Decolonization, the Cold War, and the growing wealth of the West ended the postwar period. From the 1950s to the middle of the 1970s a new period of stability started, which marked at the same time a period of tension and dynamic changes that had an impact on politics and society worldwide, the "Golden Age" as Eric Hobsbawm labeled these years with a particular focus on the Western hemisphere. As for Christian churches, this period posed a double challenge. On the one hand they had to face pluralism, the growing secularization and the loss of Christian tradition and piety in Europe and partly in North America. On the other hand, the growing number of countries in the Third World[3] led to a change in the self-conception of the WCC as a truly ecumenical organization (Kunter and Schilling 2013). New and independent Christian churches had been formed in countries of Africa and Asia. They demanded for themselves the same privileges within the WCC as the old colonial churches, which were still frequently associated with the Western missionary churches. As a result of these tensions the World Missionary Conference in Ghana 1958 decided to abolish all distinctions between "old" and "young" churches (Orchard 1959). Yet, equality of the "young" churches of the South within the ecumenical movement became one of the crucial issues in international ecumenical meetings.

[3] The term Third World is here used as a historical term of the contemporaries and the sources. It referred mostly to the developing countries in Africa and Asia. In the ecumenical movement it also included Latin America.

It was at this peak of decolonization when the third General Assembly of the WCC convened in 1961, just one year after 17 former African colonies had gained their independence in what has been called the "African Year" 1960 (Kalter 2010). For the first time in the history of the WCC, the assembly did not take place in Europe or North America, but in one of the starting points of decolonization, viz., in the Indian capital New Delhi (t'Hooft 1962). The atmosphere was one of change and new beginnings, evidenced in the large number of more than 1,000 participants and great media interest. Only a few of the "great old men" (t'Hooft 1973) who had given shape to the ecumenical movement before World War II were present in New Delhi. None of them was in the leadership, except the general secretary Visser't Hooft, representing now the older generation. Another indication of the deep transformation within the WCC was the growing number of member churches from Africa, Asia and Latin America. Over time, their proportion increased from 37 % at the second general assembly in Evanston in 1954 to 54 % at the fourth general assembly in Nairobi in 1975 (Kunter and Schilling 2014). In part, these changes were due to the "Rapid Social Change Program" that was established after the second general assembly of the WCC that was held at the same time as the Third World Conference in Bandung. The program was intended to devote more attention to the young churches and their socio-political challenges (Albrecht 1961). Together with two other important developments, the program paved the way for a wider geographical and interdenominational expansion of the WCC. The first of these two developments was to integrate the International Missionary Council, which was an independent organization until then, into the WCC. Behind this step stood a new concept of mission, as the term could not be understood anymore to describe the traditional European missionary movements, but as mission coming from and going to all six continents (Müller-Krüger 1964). On the other hand, the integration of 23 churches as new members was an important agenda item. Among these, there were eleven new churches from Africa, five from Asia, and three from Latin America and the Caribbean.[4] Yet, the most controversial matter in the media was the entry into the WCC of the Russian Orthodox Church and of three other Orthodox Churches of Eastern Europe. The topic of debate was in how far the WCC might be instrumentalized by Moscow. But despite the criticism that the membership of the Russian Orthodox Church in the WCC provoked, its inclusion marked not only a geographical, but also an interdenominational expansion of the WCC (Fig. 153.4).

One of the new voices representing the Third World in New Delhi was the Indian theologian Madathilparampil Mammen (M. M.) Thomas. He came from a Protestant-Orthodox Indian church, had been in contact with the ecumenical movement via the Indian World Student Christian Federation and was the only representative of the Third World who had already participated in the preparations for the Amsterdam assembly (Thomas 1990). His biography personified the new revolutionary hopes for global transformation of Christianity that the ecumenical movement meant for

[4]The new member churches are listed in Visser 't Hooft (Ed.) (1962). Neu-Delhi 1961. Dokumentarbericht über die Dritte Vollversammlung des Ökumenischen Rates der Kirchen. Stuttgart: Evangelischer Missionsverlag, 16–17.

Fig. 153.4 Presidents of the World Council of Churches, 1961 (Photo from World Council of Churches, used with permission)

many Christians: He came from the Indian independence movement, was connected with Mahatma Gandhi and was for, a while, was a member of the communist party. Nevertheless he was a devoted intellectual Christian, convinced that Christians had to fight for good conditions for a humane life. Thomas and other speakers saw the assembly in New Delhi as a turning point in the ecumenical history, or, as he later wrote: "New Delhi 1961 was the Assembly which began the conversion of the WCC from being a movement largely of West European Protestant churches to being a truly *world* movement" (Thomas 1990: 252).

153.5 De-Westernization of the WCC

After the New Delhi assembly, the World Conference for Church and Society held in Geneva in 1966 set another milestone. Already half of the 450 participants came from countries of the Third World. They addressed topics that started to change the ecumenical discourse. For example, controversial discussions arose around the "theology of revolution" and the "Program to Combat Racism" (PCR) that was implemented in 1969. Its aim was to support different liberation movements, especially in Southern Africa by providing humanitarian help. Yet, there was a dissent between the different member churches of the WCC about the extent of this support and who was going to steer it.

However, the election of the new third General Secretary of the WCC in 1972 received much more publicity than the program to combat racism. 51-year-old Methodist Philip A. Potter came from the Caribbean island Dominica, which was at that time still under British colonial rule and became the first non-white General Secretary after the Dutch Willem A. Visser't Hooft and the U.S.-American Eugene Carson Blake (Jagessar 1997). As Potter was considered a symbol that equality of the Third World churches had now been achieved, his election was associated with tremendous hopes for a wide renewal and of inner reforms as well as the ecumenical movement. But Potter's ecumenical career was not only a sign that the Western hemisphere had lost its hold on the South, but also showed that he was neither an ecumenical outsider nor a newcomer, but rather an old stager (Kunter 2011). With support from both former General Secretaries, he had gone through all the important stages that qualified him for an international leadership position. Starting with the Christian World Federation, he had been youth speaker at the WCC General Assemblies in Amsterdam (1948) and Evanston (1954). At the age of 33, he moved to Geneva in 1954, where he worked in the youth department of the WCC and between 1958 and 1961 as its director. After a couple of years at the Methodist Missionary Society in London he returned to Geneva in 1967 as director of the new commission for World Mission and Evangelism, the former International Missionary Council. In this position Philip Potter was also responsible for the organization of the 9th World Mission Conference, which took place from 20th December 1972 to 13th January 1973 in Bangkok, Thailand (Potter 1973).

More than any other conference that had preceded it, this ecumenical meeting saw the final abolishment of the Western understanding of mission that was a direct result of the meeting in Edinburgh in 1910. The main themes of the Bangkok conference were today's understanding of mission and salvation; and more than 300 participants from 69 countries came together to explore this theme (Kunter and Schilling 2014; Kunter 2011). The department for World Mission and Evangelism under Potter endeavored to set up a constructive, communicative atmosphere that ensured that the young churches felt included. They, therefore, decided to deviate from the structure that was typically considered a Western style with papers, lectures and discussions of experts that normally excluded other styles of cultural expressions. Instead, the WCC prepared a conference program that gave more room to discussions and personal talks in small groups. Long plenary speeches of Western participants were avoided and only voices from the South should be heard in the plenary. Several highly emotional appeals were presented to the participants, for example by the former Indonesian independence fighter and church leader T.B. Simatupang who proclaimed the end of the four hundred era years of dominance of Western Culture und Western Church. This leitmotif dominated the whole conference. Also, several theological concepts underwent a reinterpretation: "Salvation" was now seen as liberation from unjust political and social situations, "Mission" as an inner renewal of the churches and their ecumenical relations. The old idea of mission that had been born in 1910 – that mission started from the West and would then change into a worldwide Mission – was dead.

Many European and North American Protestants saw the World Mission Conference of Bangkok as a fundamental watershed in the history of ecumenism (Kunter and Schilling 2014; Kunter 2011). Nearly all national newspapers reported details about the missionary conference. The Schweizer Tageblatt wrote for example

Fig. 153.5 World Conference on Salvation Today, Bangkok, 1973 (Right: P. Potter) (Photo from World Council of Churches, used with permission)

"Das kopfnickende Negerlein ist tot" (*"the headnodding little negroe is dead"*) and the German newspaper "Die Zeit"captioned: "Das Heil der Heiden – am Ende abendländisch-christlicher Vorherrschaft" (*"Salvation of the Heath – at the End of occidental-christian Hegemony")* (WCC 272.018). Conservative and evangelical Protestants from Germany and the United States criticized the new ecumenical consensus and the "Humanization" of the Gospel. The end of the Western missionary concept would be the total surrender of the Christian belief in the future, they claimed (Beyerhaus 1973). Evangelical groups who former joined the International Missionary Council consequently left the WCC in the aftermath (Fig. 153.5).

But the De-Westernization of the WCC was not only a matter of theology and geography. When a member of the East German Protestant Churches (GDR) summarized Bangkok with the words: "Salvation is not coming from the West" Adler 1973) it became clear that the North-south conflict was embedded into the Cold War and the antagonism between East and West too.

153.6 Politicized Issues: Racism and Human Rights

Against this background during the 1960s and 1970s new alliances developed between the First, Second and Third World within the WCC. They went together with new theological ideas and political activities. Anti-colonialism, anti-racism, socialism and the theology of liberation replaced in many aspects the concept of the responsible society and the traditional anticommunism of the WCC, which was focused mainly on Eastern Europe. Here the contribution of leftwing Protestantism of Latin America was remarkable (Schilling 2013) as the WCC wanted to be politically and economically independent from the United States and pleaded for democratic socialism which broke off the issue of socialism out of the East-west-Conflict in Europe. Similarly the liberal idea of human rights underwent a change. For the general assembly of the WCC in 1968 in Uppsala in Sweden racial discrimination became now the top priority of any political activity in the ecumenical movement (Godall 1968). Hence, the WCC implemented after Uppsala the *Programme to Combat Racism* (PCR) (Adler 1974). It later was implemented as a steady part of the WCC's structure. Its aim was to identify and establish ecumenical policies and programs that would substantially contribute to the liberation of victims of racism. Special attention was paid to white racism and on the struggle against institutional racism, because they were based on social, economic and political power structures. A new approach in the ecumenical human rights discourse came up (Albers 2014). The collective human rights of all men (***sic***.) were stressed in opposite to the classical Western individual human rights. That also meant the right to enable people to liberate themselves out of unjust structures – in a pinch with violence.

From now on social and human rights stood in the center of the ecumenical policy (Kunter 2000). Most prominent it included the support for the Anti-Apartheid movement in South Africa from up the late 1970s to the end of the 1980s. But the change towards a more global and integrative approach of human rights had its problems too. It brought on the one hand more alliances with representatives of the churches in the socialist countries of Middle and Eastern Europe, especially with those who were loyal to their state and the communist ideology. Stressing social and human rights therefore soon became a highly politicized matter of the Cold War. In many cases there was opposition to Western liberal individual rights and polarized discussions and representatives. Or to state it more simply: Being against racism and apartheid was left, accusing human rights violations in Eastern Europe was right.

On the other hand this ecumenical globalization of the human right discourse followed similar developments on the international level, especially in the UN. Less controversial were the changes in the ecumenical efforts on behalf of education, although it followed in the same way the demands of the Third World representatives. A new office for education was set up, prominently represented with the Brazil educator Paulo Freire and his "Paedagogy of the Oppressed" (Strümpfel 2010). And another important arena of WCC's work was deeply influenced by the challenges through the Third World, viz., humanitarian help. Instead of the old idea of paternalistic "diaconical help" the slogan was now: "Justice, not charity" (Hookway et al. 2002).

153.7 Conclusions

Throughout the "long sixties" the WCC would run through a phase of changes and challenges. During this decade the WCC transformed from a mainly Anglo-American network to a modern, non-governmental international organization.

Fig. 153.6 Eucharistic celebration with United Liturgy for East Africa, 1975 (Photo from World Council of Churches, used with permission)

Voices of the younger churches in Asia and Africa had more participation. While the North-south conflict came into the center, the East-west conflict moved to the periphery of the ecumenical debates. This globalization of the WCC had four elements: first the denominational extension – to Protestants and Anglicans came also Orthodox and Pentecostals (Fig. 153.6); second, the geographical expansion from West and North to East and South; third, the integration of more secular analysis and, fourth, and last but not least, the participation of laypeople, especially women and young people.

The WCC functioned as a catalyst, because it transported global issues and trends into its worldwide membership. At the same time the challenges of the globalization led to a renewal of its structures, policies and mentalities under the paradigm of globalization.

References

Adler, E. (1973). Das Heil kommt nicht vom Westen. Neue Zeit, 8.2.

Adler, E. (1974). *Small beginning: An assessment of the first five years of the programme to combat racism*. Geneva: WCC Publications.

Albers, C. (2014). Der ÖRK und die Menschenrechte im Kontext von Kaltem Krieg und Dekolonialisierung. In K. Kunter & A. Schilling (Eds.), *Globalisierung und Kirchen. Der Őkumenische Rat der Kirchen und die "Entdeckung" der Dritten Welt*. Göttingen: Vandenhoeck & Ruprecht.

Albrecht, P. (1961). *The churches and rapid social change*. London: USA SCM Press.

Besier, G., Boyens, A., & Lindemann, G. (1999). *Nationaler Protestantismus und ökumenische Bewegung. Kirchliches Handeln in Kalten Krieg 1945–1990*. Berlin: Duncker & Humblot.

Beyerhaus, P. (1973). *Bangkok'73. Anfang oder Ende der Weltmission? Ein gruppendynamisches Experiment*. Bad Liebenzell: Verlag der Liebenzeller Mission.
Bremer, T. (2003). Die ökumenische Bewegung während des Kalten Krieges - Eine Rückschau. *Theologische Revue, 119*, 1177–1190.
Davidann, J. T. (1998). *A world of crises: The American YMCA in Japan 1890–1930*. Bethlehem: Leigh University Press.
Evans, S. M. (2003). *Journeys that opened up the world: Women, student Christian movements and social justice*. Piscataway: Rutgers University Press.
Fey, H. (Ed.). (1968). *Geschichte der ökumenischen Bewegung 1948–1968*. Göttingen: Vandenhoeck & Ruprecht.
Gairdner, W. H. T. (1910). *Edinburgh 1910: An account and interpretation of the World Missionary Conference*. New York: Layman's Missionary Movement.
Godall, N. (1968). *The Uppsala report 1968. Official report of the fourth assembly of the WCC, Uppsala. July 4–10*. Geneva: WCC Publications.
Greschat, M. (1998). Verantwortung für den Menschen. Protestantische Aktivitäten für Menschenrechte und Religionsfreiheit in und nach dem Zweiten Weltkrieg. In B. Jendorff & G. Schmalenberg (Eds.), *Politik - Religion - Menschenwürde*. Gießen: Gießen Schriften.
Greschat, M. (2000). Ökumenisches Handeln der Kirchen in den Zeiten des Kalten Krieges. *Őkumenische Rundschau, 49*, 7–11.
Hogg, W. R. (1952). *Ecumenical foundations: A history of the International Missionary Council and its nineteenth-century background*. New York: Wipf & Stock Publishers.
Hookway, E., Francis, C., Blyth, M., & Belopopsky, A. (2002). *From inter-church aid to jubilee: A brief history of ecumenical diakonia in the World Council of Churches*. Geneva: WCC Publications.
Hopkins, C. H. (1979). *John R. Mott 1865–1955. A biography*. Grand Rapids: Eerdmans.
Jagessar, J. (1997). *Full life for all. The work and theology of Philip A. Potter: A historical survey and systematic analysis of major themes*. Zoetermeer: Boekencentrum.
Joppien, J. J. (Ed.). (2000). *Der Őkumenische Rat der Kirchen in den Konflikten des Kalten Krieges. Kontexte-Kompromisse-Konkretionen*. Lembeck: Frankfurt am Main.
Kalter, C. (2010). Aufbruch und Umbruch: Das "Afrika-Jahr" vor einem halben Jahrhundert. In Zeitgeschichte-online June 2010. www.zeitgeschichte-online.de/Themen-Kalter-06-2010 (12.9.2012).
Kunter, K. (2000). Die Schlussakte von Helsinki und die Diskussion in ŐRK um die Verletzung der Religionsfreiheit in Ost-und Mitteleuropa 1975–1977. *Őkumenische Rundschau, 49*, 43–51.
Kunter, K. (2002). Vereinte Nationen/Völkerbund. *Theologische Realenzyklopädie, 34*, 657–662.
Kunter, K. (2011). The End of Protestant Western Mission: The World Mission Conference in Bangkok in 1972–1973. In A. Laine & A. Laitenen (Eds.), *Yliopisto, Kirkko ja Yhteiskunta. Aila Lauha juhlairja* (pp. 122–127). Helsinki: Suomen kirkkohistoriallinen seura.
Kunter, K. (2012). Der Őkumenische Rat der Kirchen in Kalten Krieg. *Religion und Gesellschaft, 40*, 12–13.
Kunter, K., & Schilling, A. (Eds.). (2014). *Globalisierung und Kirchen. Der Őkumenische Rat der Kirchen und die "Entdeckung" der Dritten Welt*. Göttingen: Vandenhoeck & Ruprecht.
Kunter, K., & Schjørring, J. H. (Eds.). (2008). *Changing relations between churches in Europe and Africa. The internationalization of Christianity and politics in the 20th century*. Wiesbaden: Harrasowitz Verlag.
Lehmann, H. (2012). *Das Christentum in 20. Jahrhundert: Fragen, Probleme, Perspektiven*. Leipzig: Evangelische Verlangsanstalt.
Lehtonen, R. (1998). *Story of the storm. The ecumenical student movement in the turmoil of the revolution, 1968–1973*. Grand Rapids: Eerdmans.
Mjagkij, N. (2003). *Light in the darkness: African Americans and YMCA 1952–1946*. Lexington: University Press of Kentucky.
Müller-Krüger, T. (Ed.). (1964). *In sechs Kontinenten, Dokumente der Weltmissionskonferenz. Mexiko 1963*. Stuttgart: Evangelischer Missionsverlag.

Nolde, F. (1974). Ökumenisches Handeln in internationalen Angelegenheiten. In H. Fey (Ed.), *Geschichte der ökumenischen Bewegung 1948–1968* (pp. 344–375). Göttingen: Vandenhoeck & Ruprecht.

Nurser, J. S. (2005). *For all peoples and all nations: The ecumenical church and human rights*. Washington: Georgetown University Press.

Orchard, R. (1959). *The Ghana Assembly of the International Missionary Council*. New York: Friendship Press.

Potter, P. (1973). *Das Heil der Welt heute. Ende oder Beginn der Weltmission? Dokumente der Weltmissionskonferenz Bangkok 1973*. Stuttgart: Kreuzverlag.

Potter, P., & Wieser, T. (1996). *Seeking and serving the truth. The first hundred years of the World Student Christian Federation*. Geneva: World Council of Churches Publishing.

Richter, H. (2011). Der Protestantismus und das linksrevolutionäre Pathos. Der Ökumenische Rat der Kirchen in Genf im Ost-West-Konflikt in den 1960er und 1970er Jahren. *Geschichte und Gesellschaft, 36*, 408–436.

Rittberger, V., & Zangl, B. (2003). *Internationale Organisationen – Politik und Geschichte. Europäische und weltweite Zusammenarbeit*. Opladen: VS Verlag für Sozialwissenschaften.

Robertson, R. (1992). *Globalization: Social theory and global culture*. London: Sage.

Schilling, A., & Schilling, A. (2014). Demokratischer Sozialismus, Humanisierung und Befreiung: Der Beitrag Lateinamerikas zur Globalisierung der Ökumene. In K. Kunter (Ed.), *Globalisierung und Kirchen. Der Őkumenische Rat der Kirchen und die "Entdeckung"der Dritten Welt*. Göttingen: Vandenhoeck & Ruprecht.

Selles, J. J. (2011). *The World Student Christian Federation, 1985–1925. Motives, methods and influential women*. Eugene: Pickwick Publications.

Stanley, B. (2009). *The World Missionary Conference*. Grand Rapids: Eerdmans/Edinburgh.

Strümpfel, A. (2010). Őkumenisches Lernen im Exil. Zu den Auswirkungen der Mitabeit Paulo Freires im Őkumenischen Rat der Kirchen. In A. S. Sören (Ed.), *Lernen für das Leben. Perspektiven ökumenischen Lernens und ökumenischer Bildung* (pp. 33–48). Frankfurt am Main: Lembeck.

t'Hooft, W. V. (Ed.). (1962). *Neu Delhi 1961. Dokumentarbericht über die Dritte Vollversammlung des Őkumenischen Rates der Kirchen*. Stuttgart: Evangelischer Mission Verlag.

t'Hooft, W. F. (1973). *Memoirs*. Geneva: WCC Publications.

t'Hooft, W. V. (1982). *The genesis and formation of the World Council of Churches*. Geneva: WCC Publications.

Thomas, M. M. (1990). *My ecumenical journey 1947–1975*. Trivandrum: Ecumenical Publishing Center Private Ltd.

WCC (Ed.). (1949). *Man's disorder and God's design* (The Amsterdam Assemblies series). New York: Harper.

WCC (Ed.). (2006). *Zeugnis-Einheit-Dienst*. Geneva: WCC Publications.

WCC Archive. 272.018.

Xing, J. (1996). *Baptized in the fire or revolution: The American social gospel and the YMCA in China 1919–1937*. Bethlehem: Leigh University Press.

Chapter 154
Religious Presence in the Context of the United Nations Organization: A Survey

Karsten Lehmann

154.1 Introduction

The multifold contributions that form this edited volume, present a dynamic sketch of what Stanley Brunn describes in his introduction as the "world religion map." In doing so, they open up an interesting perspective on processes, the Academic Study of Religions tends to analyze along the lines of globalization-theory.[1] From this point of view, scholars such as José Casanova and Hans G. Kippenberg have been emphatically underlining the significance of globalization-processes for the history of religions as well as the strengthening of religions in global public spaces:

> For the 'world religions', globalization offers the opportunity to become for the first time truly world religions, that is, global. But it also threatens them with deterritorialization. [... O]ne could say that the world religions, through the linking of electronic mass media and mass migration, are being reconstituted as deterritorialized religions 'at large' or as global ummahs. (Casanova 2008: 116; also, Kippenberg 2009: 193)

This paper adds a fascinating – though frequently neglected – aspect to these processes by providing a description of what has been characterized as a 'religious presence' in the context of the United Nations Organization (UN).[2] Since its formation, this organization has been confronted again and again with "religion" – whether through continuing conflicts in the Middle East, the international engagement of the Holy See, or the debates about religious freedom. At the same time, various

[1] See for example: Peter Beyer. (2006). Religions in Global Society. London/New York: Routledge and. Oliver Roy. (2002). L'Islam mondialisé. Paris: Editions du Seuil.

[2] In order to get an idea of the theoretical implications of these developments, see Karsten Lehmann (2013). Shifting boundaries between the religious and the secular religions organizations in global public space. Journal of Religion in Europe, 6 (1), in print.

K. Lehmann(✉)
Senior Lecturer, Science des Religions, Bayreuth University, Bd de Pérolles 90, Büro D421, CH-1700 Fribourg, Switzerland
e-mail: karsten.lehmann@kaiciid.org

S.D. Brunn (ed.), *The Changing World Religion Map*,
DOI 10.1007/978-94-017-9376-6_154

religiously affiliated organizations and movements – early on – perceived the work of the United Nations (and particularly its Economic and Social Council, ECOSOC) as an important context for their own activities – thus delegating representatives, contributing surveys or formal statements, and participating in the major campaigns of the UN (for example, the Development Decades, the Decades to Combat Racism and Racial Discrimination, or the Decades for the Eradication of Colonialism).[3]

On the basis of these general observations, the paper at hand elaborates upon three major points:

1. The ways in which the UN serves as a stage for a segment of religiously affiliated organizations.
2. The changing role of these organizations in an emerging global public.
3. The processes of reconstruction that are triggered by these developments.

Following this line of thought, the article begins with general remarks on the place of religion in the context of the United Nations. There follows a survey of the processes that a group of organizations has gone through, that label themselves as Religious Non-Governmental Organizations (RNGOs). A third section focuses upon the activities of the World Council of Churches (WCC), with its particular links to the context of the United Nations. The article concludes with more general remarks that reflect upon the establishment of religions in the United Nations framework.

154.2 The Place of Religions in the Context of the United Nations

In order to get a first idea of the place of religions in the context of the UN, it must be pointed out that the role and significance of the United Nations itself has been controversial since its establishment on October 24, 1945 (as the successor organization to the League of Nations). The spectrum of attitudes ranges from the desire for a "civilization" of world politics by a "world parliament" to objections of wide-ranging and ongoing corruption, powerlessness, and inefficiency (see Rittberger et al. 1997 and Kennedy 2007). The only certainty appears to be that the United Nations is a comprehensive federation of states with – at present – 193 member states, whose various commissions and sub-organizations collect information, formulate opinions, and promulgate declarations or (lawfully binding) conventions.

[3] In past years, two political science introductions have been published: Eric O. Hanson. (2006). Religion and politics in the international system today. Cambridge, New York, Melbourne: Cambridge University Press. Jeffrey Haynes. (2007). An introduction to international relations and religion. London, New York: Pearson Education. See also the discussion in Karsten Lehmann (2010a). Interdependenzen zwischen Religionsgemeinschaften und internationaler Politik: Religionswissenschaftliche Anmerkungen zu politikwissenschaftlichen Religionskonzeptionen. Zeitschrift für international Beziehungen, 17, (3), 75–99.

Against this background the general debates about religion in the context of the United Nations follow a particular pattern.

154.2.1 The General Debates About Religion in the Context of the United Nations

In order to understand the pattern of these debates, one has to start with the observation that the United Nations has developed no particular policy on religion. As soon as one browses through the main documents of the UN to find explicit references to "religion," it becomes clear that the organizations and committees of the UN tend to express opinions about religious issues only in exceptional cases and then only on a largely formal plane (for example in relation to the usage rights of religious buildings in Palestine).[4] And this is true with respect to disputes about the general *policy* of the United Nations (in the General Assembly, the Economic and Social Council (ECOSOC), or the Security Council) as well as to official analyses of international conflicts. Thus the theme of religion has, over long stretches, been largely marginal(ized), while topics such as peace, development, or economic policy were much more prevalent.

For most of the UN's history, the systematic place of religion-related debates was linked to the establishment of religious freedom as it was expressed in 1948 in the Universal Declaration of Human Rights (UDHR). Documents such as the *"Declaration on the Elimination of All Forms of Intolerance and Discrimination Based on Religion or Belief"* (1981) and the *"Resolution on the Elimination of All Forms of Religious Intolerance"* (1993) form milestones of the respective debates. In conjunction with other UN-documents, they constitute the framework of a discussion that finally led – among other things – to the establishment of the office of a *"UN Special Rapporteur on Freedom of Religion or Belief"* – an honorary office that up to now was staffed by personalities that shaped the international debates about religious freedom through country-reports and -missions (for juridical implications see Wiener 2007).

This particular approach to religions has been changing towards the end of the 1990s and the beginning of the 2000s. (Lincoln 2003; Favret-Saada 2010)[5] Kofi Annan was the first Secretary-General of the United Nations who expressed himself prominently on the theme of religion (Kille 2007). In 2001 (and this means of course in the aftermath of the 9/11-attacks) he dedicated around one-third of his Nobel Prize reception speech to the relationship between religions. His central point here was the conviction that:

[4] See, for example, the well-known Resolution 194 of the General Assembly of December 11, 1948.

[5] For a UN-critical essay that summarizes this process pointedly, see Jeanne Favret-Saada. (2010), Jeux d'ombres sur la scène de l'ONU: Droit humains et laicité. Paris: Editions de l'Olivier.

> Each of us has the right to take pride in our particular faith or heritage. But the notion that what is ours is necessarily in conflict with what is theirs is both false and dangerous. It has resulted in endless enmity and conflict [...]. It need not be so. People of different religions and cultures live side by side in almost every part of the world, and most of us have overlapping identities which unite us with very different groups. (UN Press Release SG/SM/8071)

In the same year the General Assembly, in Resolution 59/142, started to charge the Secretary-Generals to report upon the theme of *Promotion of religious and cultural understanding, harmony, and cooperation*. In a way, this became the starting-point for an increasing interest in inter-religious and inter-cultural dialogue inside the UN-system and other organizations leading up to the "UN Year of Dialogue among Civilizations" (2001), the "High-level Dialogue on Interreligious and Intercultural Understanding and Cooperation for Peace" (2007), or the ongoing attempts to initiate a "UN Decade of Interreligious, and Intercultural Dialogue, Understanding and Cooperation for Peace."

All these developments help to understand the organizational place of religions in the framework of the UN.

154.2.2 The Organizational Place of Religions in the Framework of the United Nations

If we direct our attention onto institutional aspects, several states or state confederations come into view that present themselves as having particular links to religion. As a paradigmatic case, one might name the Holy See, which in the UN has had the status of a *permanent observer* since 1964. But there is, of course, also the *Organization of the Islamic Cooperation* (OIC), that is based upon the ideal of the Muslim *ummah* (the religious community of all Muslims) and is, – as a regional confederation of states – officially connected with the United Nations.[6] Furthermore, some states (such as Egypt, the USA, or Russia) frequently describe themselves as "representatives" or "strongholds" of specific religious traditions.

As far as statistics are concerned, these examples are, however, more the exception than the rule. Numerically speaking, a much larger space is occupied by those religiously affiliated organizations that this article labels as Religious Non-Governmental Organizations (RNGOs) if they comply with two requirements:

- an explicit affiliation with particular religious traditions (be it in terms of explicit references to religious symbols or by formal links to organizations that refer to such symbols), and
- a successful application for the formal status of a Non-Governmental Organization/NGO with the Economic and Social Council/ECOSOC of the UN (based upon Art. 71 of the UN-Charter).

[6] One of the few depictions: Abdullah al Ahsan. (1988). The Organization of the Islamic Conference: An introduction to an Islamic political institution. Herndon [Islamization of Knowledge].

By focusing on those RNGOs, the following pages will deal with a group of organizations that went through highly complex processes – both on the part of the UN and the various religious communities – in order to cooperate with the UN:

To begin with the side of the UN: a special NGO committee (first established in 1946) is in charge of the accreditation-process. Since 1981, this committee consists of 19 representatives of UN member states (Resolution 1981/50) that are chosen for 4 years according to geographic distribution: 5 members from Africa, 4 members from Asia, 2 members from eastern Europe, 4 members from Latin America and the Caribbean, and 4 members from western Europe. Under the conditions of the Cold War, this composition frequently led to debates that were able to delay the accreditation of religiously-affiliated NGOs considerably. Only with the end of these ideological differences and the increasing accreditation of regional and national NGOs did the variety of accreditations become possible that characterize the present situation. A process that can be observed in a similar manner in other international organizations, insofar as they seek the official accreditation of NGOs.

At least equally complex are the processes that lead organizations or movements to striving for accreditation with the UN. In these cases, the decisions for or against an application are decisively shaped by the formal structures of the respective organizations. Frequently, a traditional identification with specific policy areas of the UN (for example the "peace policy" with *Society of Friends/Quaker* or the "social policy" with *Caritas Internationalis*) serves as the basis for cooperation. Accreditation can also be determined by the initiative of single members (such as with the *Sisters of Mercy* or the *Franciscans International*) or by decisions that are taken in the wake of centrally-determined strategy developments (as with *Kolping International*). In the case of a positive result of the procedure, often committees are created (such as the *Commission of the Churches for International Affairs*, the *Baha'i International Community,* or the *Quaker UN Offices*) that specialize on the cooperation with the United Nations.

The development of successful accreditations thus depends on contingent influences that cannot all be listed in detail here. As far as this article is concerned, it should be sufficient to highlight the very significance of the fact that single religious organizations perceive the formal cooperation with the UN as a viable way to follow their own interests. For a scholar of religions it is furthermore revealing that the concept of the "religious non-governmental organization" (RNGO) has developed into a self-description that is becoming more and more prevalent (for a case analysis, see Lehmann 2012). Quite a number of religious organizations that wish to be present in the context of international politics do no longer call themselves "churches," "orders," or "religious communities." Rather they employ the term "RNGO" in order to emphasize their established place within the UN.[7]

---

[7]This status provides these organizations not only with access to informational materials and contact with specific committees. It also opens a wide spectrum of informal forms of influence: Peter M. Schulze. (2000). Nichtstaatliche Organisationen (NGOs). In Helmut Volger (Ed.). Lexikon der Vereinten Nationen. (pp. 397–405) München Wien: R. Oldenbourg Verlag. Also see

Keeping this very complex setting in mind, the following section will sketch the overall developments of the accredited RNGOs in the context of the United Nations.

154.3 Accredited RNGOs in the Context of the United Nations

So far, we know only little about the overall situation of the RNGOs as well as the changes inside the particular organizations (Knox 2002).[8] To date, the most detailed study is Geoffrey Knox's volume from 2002 entitled "Religion and Public Policy at the UN." Since its publication, authors such as Julia Berger or Marie Juul Petersen added a number of interesting articles to the literature (Berger 2003; Boehle 2007; Karam 2010; Petersen 2010; Lehmann 2010b), and in the near future there are more results to be expected. At the moment three recent research-projects are just about to publish their findings.[9]

All these researchers approach their analyses from quite different starting-points. On the one hand, they use different notions of NGO. Some authors are trying to deal with all non-state actors present within the UN context (despite their formal status) and others focus on the embeddedness of the organizations into civil society (with their respective structures and ideals). On the other hand, the complexity of the attribute "religious" triggers different research-strategies. So far, none of these researchers goes so far as to apply a functional definition of religion, but even the more descriptive, substantial approaches are confronted with categorical difficulties. Some of the UN-related NGOs most active in the field of "religion" describe themselves, for example, as secular, whereas the activities of some of those NGOs with direct links to religious communities have no practical religious emphasis what so ever. So let us continue with some references to the formal concept of RNGO that is applied in the article at hand.

Kersten Martens. (2005). NGOs and the United Nations: Institutionalization, professionalization and adaption. London: Palgrave Macmillan.

[8] Of further interest is the French-language debate in: Bruno Duriez, François Mabille & Kathy Rousselet (Eds.). (2007). Les ONG confessionnelles: Religions et action internationale. Paris: Editions L'Harmattan.

[9] The article at hand presents some of the results of one of these projects. Under the title RNGOs at the UNO it is a cooperation between the "Institut zur Erforschung der religiösen Gegenwartskultur (IrG), the Observatoire des Religions en Suisse (ORS), and the Berkley Center for Religion, Peace, and World Affairs" by Karsten Lehmann. The other two are: Religious Non-Governmental Organizations and the United Nations in New York and Geneva by Jeremy Carrette, Hugh Miall & Evelyn Bush, which focuses on present-day developments and Josef Boehle, Religions, Civil Society and International Institutions which primarily analyzes attempts to foster inter-religious dialogue.

154.3.1 A Formal Concept of RNGO

As mentioned above, the present article favors a concept that is based upon (a) the status of formal accreditation as well as (b) the explicit links to systems of symbols described as religious. Consequently, the following considerations are based upon an analysis of the official ECOSOC-list of NGOs (with general and special status).[10] In the vast majority of the NGOs listed in this document, the identification with particular systems of religious symbols becomes evident as soon as one has a closer look at the official names. Most of them highlight either the nexus to a particular religious tradition (e.g. the manifold organizations that describe themselves as International Catholic or Worldwide Muslim organizations), or to a personality or a concept that is of religious significance (for example, the *"Imam Al-Sadr Foundation," "Inner Trip Reiykai International*" or the *"International Kolping Society"*).

There are, however, NGOs that make it more difficult to identify them as RNGOs and the discussion of these exceptional cases reveals already a great deal about the specific character of the RNGOs in general. In some cases the main character of the organizations in question has shifted over time. There are NGOs that in the beginning were dominated by founding figures with a high religious commitment that seems to have decreased throughout the history of the respective organization. Moreover, it is not always easy to draw a line between the dominant religious culture of a particular region and the character of the organizations embedded into this culture. The Christian affiliation of the members of American-based conservative NGOs does, for example, not necessarily imply that these NGOs fall into the above category. This becomes particularly difficult on the fringes of religious traditions or in those cases that refer to a rather general spiritual commitment.

In order to cope with these problems, the author applies a rather strict concept of RNGOs. Most generally speaking, he follows the rule that (a) explicit references to systems of religious symbols or religious institutions in official statements override the mere self-descriptions of the NGOs and that (b) individual convictions have to be undergirded by official statements or organizational links. This implies for example, that the following section excludes humanist NGOs with an explicit atheist agenda (for example, the *"International Humanist and Ethical Union"*) or human rights NGOs dealing with religious freedom among their main fields of activity (for example, "*Cairo Institute for Human Rights Studies*" or the "*Philippine Human Rights Information Center*"). They also exclude humanitarian organizations such as *"CARE – Cooperative for Assistance and Relief Everywhere"* that once had a religious background but nowadays presents itself as secular organization. On this basis, it becomes possible to identify a number of general trends among the RNGOs.

[10] Unfortunately, this list provides us only with (a) those NGOs still active and (b) the date of their accreditation with the most exclusive status. Even after detailed research, this might still add a small amount of inaccuracy.

154.3.2 General Trends Among RNGOs

In order to describe these trends, it is helpful to begin with the observation that in 2011 over 1,000 NGOs (with special or general status) are formally accredited to the ECOSOC (document E/2011/INF/4). Around one-fifth of these NGOs (224, to be exact) either refer explicitly to systems of religious symbols or maintain – on a structural level – relations with religious communities. In no sense can we understand these RNGOs as a homogeneous phenomenon. Rather they appear to constitute an extremely pluralistic setting that is characterized by multifold (and sometimes contradicting) trends:

First, it is noticeable that the number of RNGOs has steadily increased. Some of the oldest organizations can even trace their first activities back to the League of Nations. In the first years after the establishing of the United Nations (1945–1949) around 10 RNGOs were accredited. Between the 1950s and the 1970s the number of accreditations increased from 6 in the 1950s to 14 in the 1980s. After the end of the Cold War their number massively increased once again. In the 1990s around 70 religiously affiliated NGOs received their special or general status, and in the 2000s more than 100 RNGOs went through the respective processes. And over all, these accreditation rates are slightly higher than the general increase in accredited NGOs.

Second, there are significant differences with regard to the religious traditions that anchor the RNGOs. Over half stem from the Christian realm (around 130). The second largest group is connected to various traditions of Islam (31), followed by inter-religious organizations (15) and Jewish groupings (12), as well as religious organizations with a Buddhist (9) or Hindu background (7). During the 1940s, 1950s and 1960s, the Christian and Jewish RNGOs dominated the activities inside the UN-context. The number of accreditations of Muslim, Buddhist and Hindu NGOs rose only in the following decades without having reached the Christian numbers, yet.

Third, we must recall that RNGOs are often supported by specific religious milieus. Among the nearly 50 Catholic NGOs, for example, there are welfare groups founded in the middle of the nineteenth century ("*Brothers of Charity*," "*School Sisters of Notre Dame*," "*Sisters of Mercy*"), the Industrial Revolution ("*Kolping International*," "*Young Christian Workers*"), or the post-war period ("*Caritas International*" and "*Emmaus International Association*"). Moreover, Catholic umbrella organizations of single regional, professional, or interest groups ("*Pax Romana*," "*International Union of the Press*," "*International Commission of Catholic Prison Pastoral Care*") are present in the United Nations, which mostly arose in the first half of the twentieth century. Finally, there is a number of Catholic NGOs that developed out of specific spiritual traditions within the Roman Catholic Church (for example, "*Franciscans International*," "*Catholics for Choice*," and "*Community of St. Egidio*").

Finally, in the frame of this volume, the geographical distribution is of particular interest. Unfortunately, it is not always easy to identify for example the headquarters of a particular organization. This problem notwithstanding, a first approximation shows that the vast majority of the RNGOs have their headquarters either in Europe or America (in this means primarily in the U.S.). There is however, also a significant number of RNGOs (predominantly with a Muslim, Buddhist, and Hindu background) that have their organizational center in Asia. Furthermore, it is particularly interesting to see that seven out of the 31 Muslim NGOs officially accredited to ECOSOC designate headquarters in Europe (for example, "Islamic Relief" or "Islamic World Studies").

On the basis of these trends, a differentiated network of representatives of religious NGOs has developed in the UN-context that cooperate closely with each other. Among the oldest cooperations is the *"International Catholic Organizations Information Center"* (and its brother- or sister-organization in Geneva), established in New York in 1946. In the 1960s the women's section of *"American Baptists"* founded the *"Church Center for the United Nations (CCUN),"* an 11-story building situated directly across from the United Nations-headquarters. In this building are the offices of representatives of around 40 Christian and inter-confessional NGOs. The *"Committee of Religious Non-Governmental Organizations at the United Nations"* is an example for an inter-religious group of NGO representatives that has been meeting regularly since 1972. Among the most recent networks of this kind is the *"Tripartite Forum on Interfaith Cooperation for Peace"* which met for the first time in June 2005.

All the descriptions must, however, remain only superficial as long as they neglect the processes inside the respective organizations. In order to analyze these internal processes, one has to turn to detailed case-studies. Accordingly, the next section will focus upon the engagement of the World Council of Churches (WCC) within the UN.

154.4 Exemplary Case of an RNGO: The World Council of Churches (WCC)

The decision to highlight this exemplary case stems above all from two characteristics: First, the WCC can look back upon a very long tradition of engagement in the UN-context and belongs to the first NGOs accredited by ECOSOC. Secondly, the WCC was and is (after its official foundation at the first General Assembly in Amsterdam/1948) frequently perceived as "the representative" of Protestant and Orthodox churches in general (Fitzgerald 2004; Fey 1970; Briggs et al. 2004). Following Peter Beyer it is even possible to identify particular links between the WCC and the UN that make it particularly worthwhile to have a closer look at this case:

> Its [the WCC's] similarity and difference with respect to the latter [the UN] is instructive. On the one hand, the WCC, like the UN, had global pretensions from the very beginning, even though it was in its earlier period largely the expression of only the large, mainline or national Western churches. […] On the other hand, the WCC is also unlike the United Nations in critical respects, the most important of which is that, without the artifice of

> territorial states with precise and contiguous boundaries that cover the entire globe, Christianity, like religion more generally, has no equivalent of territorial-based sovereignty, and therefore no handy mechanism for symbolizing the inclusion of all its subdivisions. (Beyer 2006: 153)

As soon as one has a closer look at the archives of the WCC, it becomes possible to identify a close institutional and contentual cooperation between the WCC and the UN. At the institutional level the *"Commission of the Churches for International Affairs (CCIA)"* is foremost. It was established in 1946 (as a Commission of the *"International Missionary Council (IMR)*" and of the World Council of Churches) and officially accredited in 1948 by the UN as a non-governmental organization. Furthermore, on the contentual level, the decision-making councils and representatives of the WCC (such as the General Assemblies, the Executive Committees, or the General Secretaries) have published various commentaries, statements, and demands on a wide range of policies linked to the UN, whereby the spectrum reaches from semi-official letters to official resolutions.[11]

Against this general background, the following section focuses upon the developments that led to the WCC's activities within the context of the human rights debates.[12]

154.4.1 WCC Engagement in Human Rights in the 1940s and 1950s

A central basis for WCC's UN involvement was first of all the concept of "Christendom," with which Christian theologians (primarily in England and the U.S.) formulated a world-wide Christian claim around the turn of the twentieth century. With John Nurser (2005), we can deduce two developments from this self-understanding that were decisive for the WCC's early UN engagement in general and its human rights aspects in particular:

---

[11] An overview of these contributions of the WCC is accessible via four volumes edited by the CCIA between 1992 and 2007 under the title The Churches in International Affairs. Further insight is provided by: Ans J von der Bent. (1986). Christian response in a world of crises: A brief History of the WCC's Commission of the Churches on International Affairs. Geneva: WCC Publishing.

[12] At this point we enter a field that hitherto has attracted little attention either in studies of the history of the WCC in analyses of the development of human rights within the UN. Roger Normand, Sarah Zaidi. (2008). Human rights at the UN: The political history of universal justice. Bloomington: Indiana University Press [United Nations Intellectual History Project Series]. Mary Ann Glendon. (2001). A world made new: Eleanor Roosevelt and the Universal Declaration of Human Rights. New York: Random House. Peter Willetts (Ed). (1996). The conscience of the world: The influence of non-governmental organizations on the UN system. Washington: Brookings Institution Press. William Korey. (1998). NGOs and the Universal Declaration of Human Rights: A curious grapevine. New York, Basenstoke: Palgrave Macmillian. Furthermore, in Karlsruhe members of the DFG-Project of Katharine Kunter, "Globalized Christianity" are at work in this field: www.rz.uni-karlsruhe.de/~geschichte/index.php?page=forschung (last access 2012).

- First, Nurser emphasizes the lengthy debates that were carried out since the beginning of the 1920s within Anglo-American Protestantism and that contributed to an increasing involvement of the early ecumenical movement in the area of human rights. He sees the WCC as part of an avant-garde that early on in the twentieth century committed itself to human rights.[13]
- Furthermore, Nurser recapitulates how the members of the CCIA (and particularly its first long-time director O. Frederick Nolde) actively intervened in the UN's early human rights debates and thus shaped the formulation of the Preamble to the UN Charter and the content of the Universal Declaration of Human Rights. (Nolde 1970)

It is, however, crucial that these early activities have begun to change fundamentally since the 1960s. The respective developments refer to highly complex processes that allow, on the one hand, conclusions regarding the inner constitution of the WCC, and that can, on the other hand, serve as a basis for an understanding of its activities within a UN context.

154.4.2 Changes in WCC Engagement Beginning in the 1960s

In order to assess these changes, one has to keep in mind that in the background stand not only structural and contentual developments within the WCC (Hudson 1977, 1969), but also much wider changes that affect, for example, the UN as well as the wider NGO-community. Two points should be emphasized here:

First, the main officers of the CCIA strove to anchor human rights within the WCC and its member churches. These efforts did, however, not always garner full support even within the WCC headquarters (in Geneva).[14] They were even more difficult to mediate to a number of member churches, especially in Eastern Europe and Africa, insofar as at the time human rights were perceived as chiefly Western and secular values. In this situation, the protagonists inside the CCIA developed an argumentative strategy that placed the human right of religious freedom in the center, in order to deduce from there the significance of other human rights (World Council of Churches 1945). This argumentative model changed in the course of the 1970s and 1980s. In this period the churches of Africa and Latin America not only gained influence, but also the so-called new social movements in general (such as the civil rights movement, the peace movement etc.). They all discovered the "language of human rights" as a possibility for furthering their interests. Since then, references to human rights shape broad areas of the WCC and of its member churches, gaining even more importance after the end of the Cold War.

[13]The classical contemporary text of the debate: M. Searle Bates. (1945). Religious liberty: An inquiry. London, New York: Kessinger Reprint.

[14]This led to the establishment of more and more working groups that dealt with this theme. See the boxes of WCC archives in Geneva: 428:3.01-428.3.06 (religious liberty) and 428.3.23-428.3.25 (human rights).

Secondly, it is interesting to see how the forms of primary and personal contacts, at first so important for the day-to-day activities of the CCIA members, increasingly lost importance around the end of the 1960s. National and international political partners that CCIA officers could depend on in the 1950s lost influence, so that the WCC position founded on these partners was significantly weakened within the UN framework (Koshy 1994). So, in the 1960s, there followed a search for new paths of contentual and structural cooperation, which finally led to the much more pronounced integration of the CCIA in the so-called "NGO Community" and thus to more intensive cooperation with other religious and secular NGOs (such as *"Pax Romana"* or *"Amnesty International"*) (Martens 2005). In other words, the 1960s saw a development during which the contours of the WCC and its UN-engagement changed significantly.

From this background, the internal dimension of WCC engagement at the UN in general and the CCIA-activities in particular can be summarized along the lines of three major points:

- The above considerations should have made it quite clear that the WCC is an umbrella organization of Christian churches that bundles and represents even opposing interests in a global setting.[15] Its long-time support of human rights represents on the one hand a civil-society engagement of its member churches. On the other hand it is only comprehensible if the early (personal and financial) dominance of the Anglo-American churches, the increasing influence of African and Latin American churches, and the interests of various WCC committees are taken into account.
- Despite these factors, the WCC has entered into international relations as an independent actor and has, among other things, emphatically influenced UN human rights policy. The experts of the CCIA shaped this engagement from the 1940s into the 1960s, but then stepped more and more into the background. In the meantime theme-related NGO networks dominate activities in the field of human rights.
- Engagement within the UN has also shaped the WCC itself. Its support for collective human rights in the 1940s and 1950s (as e.g. formulated in the Human Rights Declaration of the United Nations) established early on a semantic link to human rights inside the WCC, to which reference could be made by member churches during the 1970s. That engagement placed the WCC, however, also in the framework of other NGOs and thus formed the basis for cooperation that includes partners with an explicit secular background as well as partners from other religious traditions.

Of course, the work of the CCIA and the WCC is a very special case that must not hastily be generalized. Unfortunately, this article does not allow to expand upon the specific position of the CCIA among the RNGOs in any sufficient way. On the basis of the above descriptions it should, however, be possible to give some further insights on the establishment of religions in public space.

---

[15] In the background here is a theological discussion that is thoroughly controversial even within the WCC. See Reinhard Frieling. (1992), Der Weg des ökumenischen Gedankens. Göttingen: Vandenhoeck & Ruprecht [Zugänge zur Kirchengeschichte].

154.5 Further Thoughts on the Establishment of Religions in Public Space

With reference to the introductory question of the evolution of the structural place and the content direction of UN-accredited religious non-governmental organizations, we may present from the previous discussion three results:

First, the foregoing discussion has shown that religious organizations (above all from the Christian and Jewish areas) were present early on at the UN and endeavored to influence its activities. In encountering the sociologist of religion José Casanova, we can speak here of a civil-societal deprivatization of religions (Casanova 1994, 2008), whereby it must be stressed that inside the UNO context this process began already in the 1940s. Several RNGOs began very quickly to use the option of UN accreditation and to establish themselves within this new context, whereby they usually could reach back to already established structures.

Secondly, it has become clear that these processes are shaped to a high degree by the specific UN context. This is true firstly for the structures built within the religious organizations, in order to become active in the UN framework. These are structures that aim both at contacts with UN members as well as networking with these representatives. It is also true for contentual debates that emerged in exchange with UN organs. Both observations imply that processes of civil-social deprivatization can evolve very differently and that it is dangerous to draw hasty conclusions.[16]

Thirdly, we have shown to what degree these institutionalization processes are shaped by the specifics of various religious traditions. Thus Catholic and Protestant NGOs, for example, have found very different ways of networking among themselves within the UN context. While the Catholic NGOs built up a network with the *International Catholic Organisations Information Center* that was quasi established around the representative of the Holy See, the Protestant NGOs act much more independently of each other, whereby the *World Council of Churches* plays a coordinating role in the UN. These differences become even clearer with reference to the comparatively late establishment of Muslim and Buddhist NGOs in the UN.

Acknowledgements This essay is based on research and archival studies that the author carried out in the context of a project supported by the Deutsche Forschungsgemeinschaft (DFG). A first version was presented in 2009 in the context of a project workshop at the University of Leipzig. The author is particularly thankful for the valuable suggestions of my colleague Dr. Ansgar Jödicke (University of Fribourg) as well as the highly professional translation of an earlier version of this paper by Professor Michael Jones at the University of Kentucky.

[16]This is an objection that Casanova, in his most recent publications, has begun to explore with reference to globalization theory: José Casanova. (2009), Europas Angst vor der Religion. Berlin: Berlin University Press.

References

Ahsan, Abdullah al. (1988). *The Organization of the Islamic Conference: An introduction to an Islamic political institution* (Islamization of knowledge). Herndon.

Bates, M. S. (1945). *Religious liberty: An inquiry*. London/New York: Kessinger Reprint.

Berger, J. (2003). Religious nongovernmental organizations, – An exploratory analysis. *International Journal of Voluntary and Nonprofit Organizations, 14*(2), 15–39.

Beyer, P. (2006). *Religions in global society*. London/New York: Routledge.

Boehle, J. (2007). Religions, civil society and the UN system. *Studies in Interreligious, Dialogue, 17*(1), 20–30.

Briggs, J., Mercy, A. O., & Georges, T. (Eds.). (2004). *A history of the ecumenical movement. Vol. 3, 1968–2000*. Geneva: WCC Publishing.

Casanova, J. (1994). *Public religions in the modern world*. Chicago/London: University of Chicago Press.

Casanova, J. (2008). Public religions revisited. In H. de Vries (Ed.), *Religion beyond a concept* (pp. 101–119). New York: Fordam University Press.

Duriez, B., Mabille, F., & Rousselet, K. (Eds.). (2007). *Les ONG confessionnelles: Religions et action internationale*. Paris: Editions L'Harmattan.

Favret-Saada, J. (2010). *Jeux d'ombres sur la scène de l'ONU: Droit humains et laicité*. Paris: Editions de l'Olivier.

Fey, H. C. (Ed.). (1970). *A history of the ecumenical movement. Vol. 2, 1948–1968*. Geneva: WCC Publishing.

Fitzgerald, T. E. (2004). *The ecumenical movement: An introductory history* (Contributions to the study of religion). Westport/London: Greenwood Publishing Group.

Frieling, R. (1992). *Der Weg des ökumenischen Gedankens* (Zugänge zur Kirchengeschichte). Göttingen: Vandenhoeck & Ruprecht.

Glendon, M. A. (2001). *A world made new: Eleanor Roosevelt and the Universal Declaration of Human Rights*. New York: Random House.

Hanson, E. O. (2006). *Religion and politics in the international system today*. Cambridge/New York/Melbourne: Cambridge University Press.

Haynes, J. (2007). *An introduction to international relations and religion*. Harlow/London/New York: Pearson Education.

Hudson, D. (1969). *The ecumenical movement in world affairs*. London: Weidenfeld & Nicolson.

Hudson, D. (1977). *The World Council of Churches in international affairs*. Leighton: Faith Press.

Karam, A. (2010). Concluding thoughts on religion and the United Nations: Redesigning the culture of development. *CrossCurrents, 60*(3), 462–474.

Kennedy, P. (2007). *The parliament of man: The past, present and future of the United Nations*. New York: Knopf Doubleday Publishing Group.

Kille, K. J. (Ed.). (2007). *The UN Secretary General and moral authority: Ethics and religion in international leadership*. Washington: Georgetown University Press.

Kippenberg, H. G. (2009). Religiöse Gemeinschaftlichkeit im Zeitalter der Globalisierung. In G. M. Hoff (Ed.), *Weltordnungen* (pp. 193–221). Insbruck/Wien: Tyrolia-Verlag.

Knox, G. (Ed.). (2002). *Religion and public policy at the UN* (Religion counts report). Washington: GPO.

Korey, W. (1998). *NGOs and the Universal Declaration of Human Rights: "A Curious Grapevine"*. New York/Basingstoke: Palgrave MacMillan.

Koshy, N. (1994). *Churches in the World of Nations: International politics and the mission and the ministry of the churches*. Geneva: WCC Publishing.

Lehmann, K. (2010a). Interdependenzen zwischen Religionsgemeinschaften und internationaler Politik: Religionswissenschaftliche Anmerkungen zu politikwissenschaftlichen Religionskonzeptionen. *Zeitschrift für international Beziehungen, 17*(3), 75–99.

Lehmann, K. (2010b). Zur Etablierung von Religionen im Kontext der Vereinten Nationen: Ein Überblick. *Religion – Staat – Gesellschaft, 11*(2), 33–52.

Lehmann, K. (2012). "Etablierung als Nicht-Regierungsorganisation": Zum frühen Engagement der Quäker im Kontext der UNO. *Religion – Staat – Gesellschaft, 13*(2), 35–52.
Lehmann, K. (2013). Shifting Boundaries between the Religious and the Secular: Religious organizations in global public space. *Journal of Religion in Europe. 6*(1), 201–228.
Lincoln, B. (2003). *Holy terrors: Thinking about religion after September 11*. Chicago/London: University of Chicago Press.
Martens, K. (2005). *NGOs and the United Nations: Institutionalization, professionalization and adaption*. London: Palgrave MacMillan.
Nolde, O. F. (1970). *The Churches and the nations*. Philadelphia: Fortress Press.
Normand, R., & Zaidi, S. (2008). *Human rights at the UN: The political history of universal justice* (United Nations intellectual history project series). Bloomington: University Press.
Nurser, J. (2005). *For all the peoples and all nations: Christian churches and human rights* (Advancing human rights). Geneva: WCC Publishing.
Petersen, M. J. (2010). International religious NGOs at the United Nations: A study of a group of religious organizations. *Journal of Humanitarian Assistance* 2010. http://sites.tufts.edu/jha/archives/847. Last access: 2012.
Rittberger, V., Mogler, M., & Zangl, B. (1997). *Vereinte Nationen und Weltordnung: Zivilisierung der internationalen Politik?* Opladen: Leske + Budrich.
Roy, O. (2002). *L'Islam mondialisé*. Paris: Editions du Seuil.
Schulze, P. M. (2000). Nichtstaatliche Organisationen (NGOs). In H. Volger (Ed.), *Lexikon der Vereinten Nationen* (pp. 397–405). München Wien: R. Oldenbourg Verlag.
von der Bent, A. J. (1986). *Christian response in a world of crises: A brief history of the WCC's Commission of the Churches on International Affairs*. Geneva: WCC Publishing.
Wiener, M. (2007). *Das Mandat des UN-Sonderberichterstatters über Religions- und Weltanschauungsfreiheit: Instituionelle, prozedurale und materielle Rechtsfragen* (Schriften zum Staats- und Völkerrecht). Frankfurt a.M.: Lang.
Willetts, P. (Ed.). (1996). *"The conscience of the world:" The influence of non-governmental organizations on the UN System*. Washington: Brookings Institution Press.
World Council of Churches (Ed.). (1945). *International affairs – Christians in the struggle for world community: An ecumenical survey prepared under the auspices of the World Council of Churches*. New York.

Chapter 155
Preparing Professional Interculturalists for Interfaith Collaboration

Naomi Ludeman Smith

155.1 World Headlines Report Increased Conflict Due to Religion

Conflict over religious differences makes news headlines almost every day. The Pew Research Forum on Religion & Public Life reported in September 2012 that of the 197 countries studied between mid-2009 and mid-2010, "restrictions on religion rose in each of the five major regions of the world—including in the Americas and sub-Saharan Africa, the two regions where overall restrictions previously had been declining" (Full Report 2012: 9). The Government Restrictions Index reported that "63 % of countries had increased … while 25 % decreased" (2012: 12). For the Social Hostilities Index, "49 % of countries had increases in hostilities … while 32 % had decreased" (2012: 13). What is the nature of these hostile incidences? The study found "increases in crimes, malicious acts and violence motivated by religious hatred or bias, as well as increased government interference with worship or other religious practices" (2012: 10). In other words, though the news headlines might indicate that these incidents are isolated squabbles due to local issues, the Pew research tells a fuller story. These incidents are happening nation-wide in an increasing number of countries and connected to decisions made by a majority of the nation's people.

These findings also indicate a correlation between the rise of government restrictions and the rise of people committing hostilities in support and protest to these restrictions (2012: 20). For example, on 29 November 2009, the BBC NEWS headline read "Swiss voters back ban on minaret." The Swiss government opposed this ban and advised against it, though in the end they conceded to the majority

N. Ludeman Smith (✉)
Learning and Women's Initiatives, 570 North Albany Street,
St. Paul, MN 55104, USA
e-mail: naomi.ludeman.smith@peace-catalyst.net

S.D. Brunn (ed.), *The Changing World Religion Map*,
DOI 10.1007/978-94-017-9376-6_155

voice and it became law. For 57 % of the people voting for the ban, the BBC article reported that the most common reasons were that "minarets are a sign of Islamization" and that voters were "worried about rising immigration—and with it the rise of Islam." According to the article, at the time of the vote there were 400,000 Muslims living in Switzerland and a total of four minarets. A leader of the Muslim community in Switzerland was quoted as saying that "this will cause major problems because during this campaign mosques were attacked, which we never experienced in 40 years in Switzerland."

On 6 August 2012, an AsianNews.it headline offered an example of religious hostilities on the rise due to political elections (Hariyadi 2012). The news source reported that 6,000 Catholics were kept from the parish of St. John the Baptist in Waru, Indonesia. Though a legally sufficient number of local residents gave written permission for this minority religious group to worship together in the building, officials sealed off the area due to the lack of the village chief's signature on the necessary petition. *The Jakarta Post* headline 4 months later read "Politicians fuel raging [religious] hatred and intolerance" (Aritonang 2012). The National Commission on Human Rights of Indonesia (Koomas HAM) asserts that Indonesian politicians across the country are using religious popular sentiment against religious minorities to win votes for local and regional elections. This local human rights watchdog organization "warned that religious tension will escalate as the country headed toward election year" (Aritonang 2012).

Indonesia's Koomas HAM leaders believe that local politicians, governments and religious leaders are conspiring against religious tolerance. Indonesia's Deputy National Police chief explains that the police

> could do little in the face of politics. … We are conscious of human rights. … However, we are very much influenced by politics. The country's leadership and public policies determine our responses when dealing with religious conflicts.

In light of the world religion map shifts, of particular interest in this story is that the religious hostilities are not isolated to the Christian minority up against the Muslim majority, nor is it only coming from within the harbors of Indonesia. The persecution is targeted at any minority religious group that is seen as being politically beneficial to deny freedom. According to the *Jakarta Post*, Muslim minority sects such as the Ahmadiyah are also Indonesian politicians' pawns. The Ahmadiyah are known for leading peaceful movements to propagate their particular beliefs. In addition, the Komnas HAM chairman, Otto Nur Abdullah, asserts that a rising ultraconservative branch of Sunni Islam coming from the Middle East, the Wahhabists, has also fueled this increased intolerance. "This strand of Islamic thinking," claims Otto, "had been adopted by local politicians to boost their credibility. … Indonesia is a target for such transnational Islamic thought."

And so the stories continue, as they have since ages past, of government and social unrest rising and falling due to religious intolerance. People with religious differences just cannot seem to get along. Is there any hope for our world as we experience increased international and inter-religious interaction, and interdependence and globalization? I am relieved to report that there are rays of hope with a growing number of people and organizations that are seeking interfaith collaboration instead of hostilities. The headlines are posted, but they are not making it to the popular news sources. Collections of people with diverse belief systems are taking advantage, in the best sense of the word, of their commonalities and their differences for innovative solutions to address some of our world's greatest struggles. Informing people and organizations in these interfaith endeavors is a relatively new category of professionals called interculturalists.

Though interculturalists have been about their work for a very long time, their number and recognition as a formal discipline is growing, along with their body of specialized research, grounded theories, tools and professional societies. They identify themselves by several different titles because they work in a range of disciplines in the academy, as well as the marketplace, government, non-government organizations, the military, and in far reaching corners of the world. Can they overturn the increasing inter-religious threats and violence and put an end to them? Most would sadly concede that this is unrealistic. Still, professional interculturalists are a hopeful and needed resource to prepare and inform those who are working in the aisles of interfaith collaboration to rewrite the world's headlines.

155.2 The Interculturalist and the Interfaith Collaborator

What does crossing cultures have to do with crossing religious differences? The key word here is differences. When we engage with people with a different belief system from our own, we often confront behaviors, feelings, beliefs and thinking that can create psychological, epistemological and moral dissonance in us. There is nothing wrong with this dissonance; it is how we respond to it that can end in the horrific stories and injustices that news headlines are reporting. If we are not practiced in productive ways to respond to these intense experiences of differentness, we can act and think in hostile ways toward individuals and groups. When that religious group is a minority in a particular setting, such as Indonesia, then issues of power can be at play.

We have these same responses of tension when we cross cultural borders. In fact, many would agree that religious groups develop the same characteristics as a culture group, often referred to as a sub-culture of the larger culture (Himells 1997, Hiebert et al. 1999; Muck 2005). This is why interfaith collaborators and peacemakers can tap into the knowledge and training techniques that interculturalists use to prepare and support people who are crossing cultural borders. Both are sojourners of a type, trying to traverse the complexities of their inner and outer life in relationship to those who are different from themselves. Most interculturalists assert that ethnic and religious sojourners can develop what they refer to as intercultural competencies to not just exist in these cross-religious, cross-cultural settings, but thrive in them to make a better life for all people.

155.2.1 Foundational Theories

What are the foundational theories that inform interculturalists to work with people and organizations from across the world in collaborative and innovative ways? Interculturalists identify Milton Bennett's Developmental Model of Intercultural Sensitivity as one of the leading conceptual frameworks and theories (Berardo 2008: 28). As a result of a plethora of research to explore Bennett's theory, Mitchell Hammer (2011, 2012) revised Bennett's theory and named it the Intercultural Development Continuum. These two cultural adaptation theories offer insight to how people can sequentially develop their affective, behavioral and cognitive skills to work effectively across cultural and belief system differences and similarities (Bennett 1993, 2004, 2012; Abu-Nimer 2001; Bennett and Bennett 2004; Bennett and Castiglioni 2004; Bennett 2009a; Stuart 2012).

The Developmental Model of Intercultural Sensitivity and the Intercultural Development Continuum work from a constructivist paradigm. It is based on personal and professional development models that describe the cross-cultural sojourner's worldview orientation as it relates to the phenomenology of differences. Bennett describes the major worldview orientation shifts as a developmental journey from ethnocentrism toward ethnorelativism, becoming increasingly more interculturally competent with each shift. The models are based on the psychological organizing concept of differentiation, as this is our overwhelming subjective experience from which we confront different ways of feeling, behaving, believing and thinking as a way to make sense of our life experience. This is what makes up our worldview. When we are exposed and engaged in relationships with people from different religious beliefs from our own, our worldview is often challenged and we must differentiate what is regarded as our norm from what is new and often strange, and then decide how to respond to make meaning of the diverse and sometimes opposing perceptions and patterns of reality.

155.3 The Changing World Religion Map and the Interculturalist

With this changing world religion map, the task and challenge for the interculturalists, then, is to effectively prepare themselves and those involved in interfaith collaboration to understand and develop the intercultural skills and mindsets needed to do their work. It is a risky and personal business to challenge people to deeply reflect on the core of their existence and assumptions about life as they are confronting these complex social, economic, political and historical situations where reconciliation is needed. Given the current rise in religiously motivated hostilities, the people and their stories on our shifting map need to tap into professional interculturalists' resources for the sake of peaceful coexistence and innovations to critically integrate the diverse assets that the people of the world have to offer.

This chapter will explore who interculturalists are, the phenomenon they confront and the theories and tools that they use to work through these challenges. Then we will journey to a place on the map where a peacemaker/interculturalist is applying his knowledge and skills for interfaith collaboration: The Middle East.

155.4 Worldview Development, Belief Systems and the Professional Interculturalist

Who is an interculturalist and what do they do? As a way to answer this question, a precursor explanation is the phenomenon that the interculturalist and interfaith collaborator are confronting. This is the phenomenon of worldview dissonance related to differences, specifically belief system differences. Simplistically stated, a central reason that we cannot seem to get along when we must live next to people with religious differences is that many of us are not practiced in the skills needed. This dynamic skill is to respectfully shift our own frame of reference to that of another and then back again while staying in tact with our own psychological and epistemological well-being and with those in our own and other communities of faith. Understanding this phenomenon and supporting people who want to be able to make these shifts is the work of the interculturalist. They can encourage global-minded worldview development. The manifestation of these kinds of shifts is intercultural skill development.

Sojourners often experience challenges to their worldviews during their cross-religious exchanges. This experience is referred to as a critical incident because it is a particular or a collection of experiences that forces a person to reconsider how to make sense of a new and often conflicting possible organization of reality. If the person shifts his or her answer from what was previously held, a worldview shift is the result. Interculturalists can prepare sojourners to anticipate these critical incidents and intervene in the midst of them. They can offer ways to process the experiences that supports increased intercultural competencies and peaceful resolutions. They can do so because at the core of their expertise is an understanding of the intricacies of worldview development.

155.4.1 Worldview Development

Anthropologist Paul Hiebert (2008) offers an explanation to understand how our worldviews shift or develop, the functions worldviews perform in our daily lives, consciously and unconsciously, and the potential for worldview transformation. He defines worldview as "'the foundational cognitive, affective, and evaluative assumptions and frameworks groups of people make about the nature of reality which they use to order their lives.' It encompasses people's images or maps of the reality of all things they use for living their lives" (2008: 25–26).

A worldview is not merely a vision of life, but a vision for how life is to be lived. People experience life's daily events, filtering and testing them through their worldview (beliefs, feelings, values) as informed, reinforced and codified by their unique cultural context and the members of that culture (Kraft 1979; Marsella 2005; Hiebert 2008). This is the epistemological exercise of defining what is real. People then make decisions based on this dynamic processing and then behave in a way that a cultural context supports.

Contributing to Hiebert's explanation, Marsella (2005) describes worldviews as:

> Cultural templates for negotiating reality [that] emerge from our in-born human effort after meaning, an effort that reflexively provokes us to describe, understand, predict, and control the world about us through ordering of stimuli into complex belief and meaning systems that can guide behavior. Our brain not only responds to stimuli, it also organizes, connects and symbolizes them, and in this process, it generates patterns of explicit and implicit meanings and purposes that promote survival, growth, and development. This process occurs through socialization and often leads us to accept the idea that our constructed realities are in fact realities. The "relativity" of the process and product is ignored in favor of the "certainty" provided by the assumption that our way of life is correct, righteous, and indisputable (e.g., ethnocentricity). (Marsella 2005: 658–659)

Thus, the dynamic organization of reality is shared with a community and experienced and confirmed within a cultural context.

155.4.2 Beliefs and Belief Systems

One of the functions of a worldview is to make sense of other worldviews, which include beliefs and belief systems. People filter the new experiences through their worldviews, explains Hiebert (2008), "to select those that fit our culture and reject those that do not. It also helps us to reinterpret those we adopt so that they fit our overall cultural pattern" (2008: 30). Kraft (1979) affirms that this worldview monitoring function "lies at the heart of a culture, providing the basic model(s) for bridging the gap between the 'objective' reality outside people's heads and the culturally agreed upon perception of that reality inside their heads" (2008: 56). Our worldviews provide a "psychological assurance that the world is truly as we see it and a sense of peace and belonging in the world in which we live. People experience a worldview crisis when there is a gap between their worldview and their experience of reality" (Hiebert 2008: 30).

These explanations shed light on why engagement across religious differences, especially when cultural borders are also crossed, presents humans with psychological dissonance. During these cross-religious experiences, a sojourner's worldview—the norms and places of comfort—is challenged and he or she must decide how to affectively, behaviorally and cognitively make sense of the differences, testing his or her truth, the authorities on truth and its reliability according to the known context that is reinforced by members in the culture. In the inter-religious experience, dissonance creates a disorienting dilemma (Allan 2003; Milstein 2005; Hunter 2008). This is the critical incident.

155.4.3 The Critical Incident and Transformative Worldview Development

This dilemma forms the critical incident that creates epistemological stress. It requires a response and presents the potential for a worldview shift if the sojourner is willing to take the risk and has the preparation and tools to process the critical incident. Because of this incident, the sojourner must negotiate reality, which involves trying to make meaning and shifting meaning perspectives. This involves learning from differences and questioning how to integrate them with his or her own norms and places of comfort. This presents the sojourner with the unique opportunity for deep transformation to take place that can lead to increased intercultural sensitivity and competency (Kraft 1979; Mezirow 2000; Friedman and Berthoin Antal 2005; Marsella 2005; Friedman and Rholes. 2007; Hiebert 2008).

Considerable research has explored the psychological, sociological, spiritual tension and shifts when we are in the midst of experiences that challenge our organizational patterns of reality. Interculturalists have drawn from this body of research to develop and test, for example, coping strategies, models and pedagogy to prepare sojourners for the experience of engaging alterity (Grove and Torbiörn 1993; Rogers and Ward 1993; Weaver 1993; Ward and Kennedy 1999; Ward and Rana-Deuba 2000; Bennett and Bennett 2004; Cushner and Karim 2004; Savicki et al. 2004; Lewis Hall et al. 2006; Sandage and Harden 2011; Sandage et al. 2014; Sandage and Jankowski 2012).

The Immersion Myth An example of research that interculturalists are engaged in is to explore beliefs and preparation practices that educators, mission agencies and corporate, government and military human resource staff—anyone sending people abroad to live and work—have employed for decades. Specifically, researchers have identified a commonly held assumption that people have relied on to teach intercultural, inter-religious skills: the Immersion Myth. This immersion myth perpetuates the belief that if we just spend time immersed in cross-cultural settings with some language preparation and some information about the cultural behaviors and regional history, we will successfully adapt to our new environment and our intercultural skills will naturally develop. Intercultural education research, however, has all but dismissed this myth (Pedersen 2009, 2010; Sample 2009; Vande Berg et al. 2009, 2012). The finding has also served as a catalyst to develop and test sophisticated theories and tools to more precisely explain what is happening in the cross-cultural, cross-religious experience (Paige and Vande Berg 2012: 29–60).

This is an example of the interculturalists' academic and practitioner endeavors, further establishing it as a field unto itself. It is also this kind of cross-discipline research that informs interculturalists about the power and potential of the critical incident for transformational worldview shifts and an emphasis on developmental models and experiential/constructivist paradigms for reflective learning (Vande Berg, Paige and Hemming Lou 2012: 15–25).

155.4.4 *Transformative Worldview Shifts Through Interfaith Exploration*

As intercultural educators and researchers learn more about the complexity of the critical incident in cross-cultural experiences, the conversation is shifting. It is shifting from how to prepare sojourners for effective adjustment to how to prepare and support sojourns for epistemological transformation. The goal has shifted from preparing people to think of cultural differences not as a problem to solve, but as a resource for solving problems (Mezirow 2000; Parks Daloz 2000; Abu-Nimer 2001; Marsella 2005; Asay et al. 2006; Hage et al. 2006; Hoff 2008; Hunter 2008; Savicki 2008; Bennett 2009b). This pattern of thinking, argues Friedman and Berthoin Antal (2005), "enables people to discover differing views of reality, making it more likely that they will create common understandings and generate collaborative action" (2005: 70). Scholars and practitioners investigating this transformative goal would describe a person with this competency as one living out of an ethnorelative worldview orientation. He or she has a multicultural identity and demonstrates it through interculturally sensitive behaviors that become consistent competencies.

How does a person shift to this transformed orientation? Milton Bennett (1993) observes that: "intercultural sensitivity is not natural. It is not part of our primate past" (1993: 21). He goes on to say that "the concept of fundamental difference in cultural worldview is the most problematic and threatening idea that many of us ever encounter. Learners (and teachers) employ a wide range of strategies to avoid confronting the implications of such differences" (1993: 22). Confrontation with religious and cultural differences and the emotions it produces often forces people to choose between isolating and withdrawing from the affective, behavioral, and cognitive messiness or engaging in the complexity of negotiating the inevitable worldview tensions that result. This is why it has not worked well to depend solely on immersion into the interfaith context as a training method to develop the necessary skills.

While knowing history and human proclivity, good-willed people are known to celebrate the similarities of different people groups. Often, though, this is where they stop because they do not know what to do with the challenge of differences. Others tolerate differences with little depth or sophisticated knowledge or skills of how to process the differences. They do not know how to build positive and productive relationships with people different from themselves. A person with intercultural competence, then, is someone who knows how to process the critical experiences and ideas rather than ignore them. The sojourner has developed the skills and capacity to negotiate the religious and cultural similarities and differences. He or she has learned "to be emotionally resilient in responding to the challenges and frustrations" when humans engage with different worldviews and the belief systems within those worldviews (Paige 1993: 1).

Despite the reality and challenge of this sojourn across differences, people do still have the growing conviction that people can and must develop the knowledge and skills to be collaborative and productive intercultural world citizens. The question is how humans develop intercultural knowledge, skills and attitudes so as to engage peacefully and productively with people who hold distinctly different worldviews. To achieve this goal, interculturalists are honing sophisticated methods for deliberate and highly reflective training that moves far beyond the immersion myth. The work of the interculturalist, then, is to prepare and intervene in relationships with people in these critical experiences to develop the intercultural competencies needed to peacefully engage across religious and cultural differences.

155.5 Developing Intercultural Competency for Interfaith Collaboration

To develop intercultural competencies for interfaith collaboration, sojourners must be willing to engage deeply with others who have different worldviews from their own and then have the courage to confront the core assumptions of their own worldviews. This kind of intimate and epistemological engagement is referred to as alterity. When sojourners engage in alterity, they are inevitably engaging with others' beliefs and belief systems.

Beliefs and belief systems are intricate to and foundational in one's worldview and its development (Hiebert et al. 1999: 40). They are "bodies of knowledge that emerge in response to key questions [about life's experiences] and agreed-upon methods to find answers" (Hiebert et al. 1999: 39). Most often humans develop and share these belief systems with a group of people and together they hold a collective belief regarding what is truth. This often is how a religion is started and maintained. Religion can be understood as "beliefs about the ultimate nature of things, as deep feelings and motivations, and as fundamental values and allegiances" (35). Religion or belief systems, then, are at the core of one's worldview. When they are challenged by other beliefs and belief systems, the confrontation results in deep epistemological tension.

In the midst of the cross-religious experience, sojourners must explore how their culture influences their worldview beliefs and the behavior and values that express their beliefs. The reason is that "the worldview lies at the very heart of culture, touching, interacting with, and strongly influencing every other aspect of the culture" (Kraft 1979: 53). Therefore, sojourners must reflect on their tension and their culture's influence on their worldview and religious beliefs in order to face the assumptions that often support a worldview. Here is where the tension and turbulence begins. Upon reflection, sojourners usually challenge the unfounded beliefs that their culture supports and must face the question of what to do with the contrasting beliefs. They often question what is truth. They must decide if they are going to allow new beliefs and patterns of thinking to influence and shift their worldview perspective.

155.5.1 Psychology and Theology Integrated Exploration

What happens psychologically and spiritually when people confront contrasting worldviews and the challenge to shift their paradigm? Psychologist Steven Sandage and theologian LeRon Shults integrated their areas of expertise to explore this question. In their book *Transforming Spirituality* (2006) they conclude that "spiritual transformations involve profound changes in self-identity and meaning in life, often following periods of significant stress and emotional turbulence" (Shults and Sandage 2006: 19). Interfaith sojourners experience much internal spiritual and psychological tension. They confront the crucible of how people from a different religion make sense of life. Sojourners' norms of self-identity and reality are challenged. They challenge the spiritual and psychological wellbeing and emotions that are more intense than the normal, everyday experience. How one processes these critical incidents directly determines their outcome. This is where the professional interculturalist can intervene to support a positive outcome and growth toward increased intercultural competence for people in inter-religious experiences.

155.5.2 Intercultural Competence

Intercultural competence, then, is to know how to effectively process the intense affective, behavioral and cognitive critical incident. It is the skill to detect differences, recognize and value our own belief system within our worldview and accept as real the worldview of people of another religion, even if we do not fully understand, share or agree with the other worldview. The cross-religious sojourner does not have to accept the other religion as the "true" religion, but accept and respect the reality that others also hold their convictions as truth. People with intercultural competence also recognize that exploring others' contrasting beliefs can offer insight into life's big questions, serving as a resource rather than a problem. To develop and practice these competencies, however, is no easy task.

Much of people's experience and understanding of how to get along with those who are different from themselves is to tolerate these differences. Intercultural competence requires more than just tolerance. Developing or practicing tolerance toward people with differences rarely leads people to explore their own worldview or open themselves to others' ways of experiencing life. Tolerance is a way of protecting oneself from alterity, from the risk of deeply experiencing and relating to differentness. A person who seeks to tolerate another person's religious differences rarely seeks to understand why there are the differences, especially if there is tension. Practicing only tolerance fails to teach people how to adapt one's behavior and thinking in order to successfully relate to the perceived differentness of others.

155.5.3 Conflict Resolution, Religious Peacemaking and Intercultural Development

The Mennonite community is historically and currently known for moving beyond tolerance to places of understanding and relationship. The result is their reputation for active engagement in peace and justice activities around the world. They themselves come from a history of being oppressed by the religious majority. Fortunately, they have surfaced as peacemakers rather than continuing the cycle of moving from the oppressed to the oppressor.

A crucial conviction of their tested ways of turning conflict into peace is the practice of alterity, the same finding that Shults and Sandage (2006) asserted and Bennett's (1993, 2004, 2012) and Hammer's (2009, 2012) intercultural development theories support. In Marc Gopin's *Between Eden and Armageddon: The Future of world religions, violence and peacemaking* (2000), he explores how the study of religion and the study of conflict resolution confirms the peacemakers/interculturalists' mindsets and approaches to effectively confront religiously motivated hostilities.

In the discussion of various paradigms of religious peacemaking, Gopin highlights two critical and integrated differences between how, for example, "instrumentalist and consequentialist methods of process and outcome evaluations" and "the Mennonite religious response is different" (2000: 155). The first is the crucial emphasis on alterity. "Mennonite spirituality guides the peacemaker to build relationships in the conflict situation," Gopin observes (2000: 154). "Evaluation is not the problem here," Gopin clarifies. "It is the reduction of the human moment of relation to its instrumentality that is problematic for Mennonite peacemakers and, undoubtedly, many other peacemakers" (2000: 154). Too often, Gopin asserts, conflict resolution methods overlook the "very real clash of cultures" (2000: 155). With a focus on relationships with the people who are in conflict, the mediator and people in conflict must also consider culture as the second crucial component. Gopin explains that "the cultivation of relationship between human beings and the careful attention to the style of interaction and the character that one brings to that relationship are the essential elements of [Mennonite] peacemaking activity, because they are the primary moments of religious experience and discovery of the Other in the world" (2000: 156). Culture norms are at the core of the style of interaction, or communication, and what interculturalists consider as factors to notice and then interpret meaning from within the cultural context.

There is a growing body of intercultural research, training and tools to build self-awareness of intercultural communication styles and managing conflict (Ting-Toomey and Oetzel 2001; Lebaron and Pillay 2006; Hammer 2009). To do the kind of careful and nuanced work that Mennonites do in their inter-religious peacemaking efforts, they must gain knowledge and skills through self-awareness, reflection and experiential practice. They develop intercultural skills that come from ethnorelative worldview mindsets about differences between people and the groups to which they are members.

Therefore, to further develop intercultural skills a person must reach deeply into one's self-identity to consider how culture influences these things *while in relationship* with another person who is distinctly different. Of course, the ideal situation would be that both people are working toward this end of finding meaningful and peace-filled relationships amidst their commonalities and differences. We all know that this is not always the case, however. This is where the peacemaker/interculturalist can bring or intervene in the relationship to offer the knowledge and skills to evaluate the intercultural paradigm from which people are working and meet them where they are at in their perspectives of the Other. "'Faith alone'," one facilitator noted, 'is not enough'" (Abu-Nimer 2011: 568).

An interculturally competent person, then, must develop the virtues of worldview respect and cultural humility. They must develop "self and other" cultural awareness. This means that the effective interfaith collaborator must be able to respectfully and patiently advocate for one's own convictions, knowing how to appropriately express those convictions in the particular cultural context. The professional interculturalist's challenge is to soundly support the peacemaking sojourner to develop these competencies through the critical experiences. To do this effectively through these experiences is the challenge. Milton Bennett (1993, 2012) and Mitchell Hammer's (2009, 2012) cultural adaptation theories have proven to be effective international tools in the interculturalist's belt to constructively build the competencies needed to address these crucible experiences.

155.6 The Developmental Model of Intercultural Sensitivity and the Intercultural Development Continuum

Professional interculturalists often cite Milton Bennett's Developmental Model of Intercultural Sensitivity (DMIS) (1993, 2004, 2012) as a theory and framework from which to understand and support cross-cultural sojourners' transformational development toward a multicultural identity and productive global citizenship (Table 155.1). An accompanying measurement inventory that grew out of the DMIS is one of the leading tools interculturalists use to identify placement of people and their worldview orientation regarding differences. Bennett and colleague Mitchell Hammer (Hammer et al. 2003) originally authored this tool: the Intercultural Development Inventory (IDI). It is now in its third version with Hammer as its sole author and owner.

The IDI is used and available in thirty different countries and languages and tested for cultural bias. It is used to support personal and professional development in cross-cultural and racially diverse settings. At the writing of this chapter, over 1,400 professional interculturalists have completed the required qualifying seminar to prepare and authorize them to distribute and use the tool. In addition, these interculturalists have published over 60 articles and book chapters and more than 42

Table 155.1 Bennett's developmental model of intercultural sensitivity, DMIS

Ethnocentric			Ethnorelative		
The experience of one's own beliefs and behaviors (culture) as "just the way things are"			The experience of one's beliefs and behaviors (culture) as one's organization of reality among many viable possibilities (other cultures)		
Denial →	Defense →	Minimization →	Acceptance →	Adaptation →	Integration
Blissful ignorance	Negative reaction	Emphasizing similarity	Curiosity, respect	Adjusting	Becoming bicultural
Lack of awareness of cultural difference and/or lack of interest in cultural difference	Feeling threatened	One's own culture is experienced as universal	One's own culture experienced as one among many	Perception and behavior appropriate to another culture	One's experience of self is expanded to include movement in and out of different cultural contexts
Other cultures are not noticed, or construed vaguely ("foreigner" or "immigrant," "Asian" or "African")	Us/them polarization	Subsuming cultural differences into familiar categories (projecting one's culture onto others)	Discriminating differences among cultures	Ability to "look through the others' eyes"	Cultural marginality—sense of identity at the margins of two or more cultures (central to none)
	Positive stereotyping of one's own culture, negative stereotyping of others	Sympathy—doing unto others as you would have them do unto you	Experiencing others as different but equally human	Movement toward Empathy—doing unto others as they would have you do unto them	
	Note Reversal			Not assimilation, but expansion of your repertoire of beliefs and behavior	

Source: Magnuson (2008), used with permission

recorded dissertations have used the instrument (Hammer 2012, 2013). Only those who are trained and authorized by IDI, LLC can produce individual and group reports from the 50-item inventory, showing the developmental placement along a continuum (Hammer 2012, 2013).

As a result of extensive psychometric testing of the IDI, Hammer developed a revised version of the DMIS and IDI to reflect the most recent findings about the phenomenon of the development of our worldview orientations related to differences. Hammer named the revised theory the Intercultural Development Continuum (IDC) (Hammer et al. 2003; Hammer 2011). Essentially, the two constructivist theories are very similar in the foundational nature and description of the stages. The primary difference is that Hammer dropped Bennett's final stage due to a lack of sufficient findings to support the reliability and validity of the measurement of it (Hammer 2012).

As stated previously, the DMIS and the IDC are personal-growth models to describe a sojourner's worldview orientation as it relates to the phenomenology of differences. It describes stages of growth and encourages movement toward increased intercultural sensitivity and an ethnorelativist or intercultural worldview and identity. Bennett based the model on the psychological organizing concept of differentiation, as this is sojourners' overwhelming subjective experiences from which they confront differences. The sojourner's worldview is challenged and he or she must differentiate what is regarded as the norm from what is new and strange, and then decide how to respond to make meaning of reality. This is the epistemological task of negotiating reality (Friedman and Berthoin Antal 2005; Marsella 2005). The challenge of cultural differences stem from humans' proclivity to avoid making accurate meaning from different perspectives and the deep worldview shift. To do so means inviting psychological dissonance.

155.6.1 Monocultural to Intercultural Mindsets for Responding to Differences

In the Intercultural Development Continuum, Hammer (2011, 2012) posits a spectrum of five stages in which sojourners and interculturalists can describe how we respond to worldview differences (Fig. 155.1). Bennett and Hammer theorize that sojourners move through these stages sequentially as we expand our capacity to cope and adjust to an increasingly more complex and sophisticated understanding of the influence that culture has on worldview development. If carefully attended to, this expanded capacity to cope can develop intercultural sensitivity and lead to increased intercultural competency. Hammer (2013) defines intercultural competence as "the capability to accurately understand and adapt behavior to cultural difference and commonality."

Hammer's first set of two stages begins with a monocultural or ethnocentric mindset. Ethnocentric means that people make sense of life through one ethnic lens of their own culturally influenced worldview. The theory assumes that all people from all cultures initially experience life from an ethnocentric orientation, as this is what most people primarily know and what their cultural community reinforces (Marsella 2005: 659).

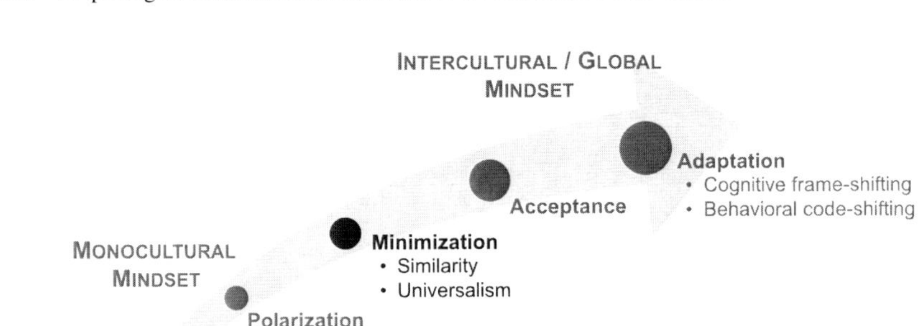

Fig. 155.1 Intercultural development continuum: the DMIS (Source: Modified from Hammer 2009 and Bennett 1986)

A transitional stage called Minimization is between the monocultural and intercultural mindsets. It is one of Hammer's revised descriptions on the Intercultural Development Continuum from Bennett's DMIS. This Minimization stage reflects a place where people "experience a certain degree of success in navigating unfamiliar cultural practices" (Hammer 2012: 122). If the experience demands that they navigate differences, however, they do not yet have the skills to do so with consistent success. Understanding the distinction and paradigm shift between these two perspectives—monocultural and intercultural—is essential to intercultural sensitivity development.

Hammer's next set of two stages reflects an intercultural or ethnorelativist perspective. A person with an intercultural mindset and identity will have developed the capacity to differentiate with increasing skill the characteristics of differences and make meaning of life through multiple ethnic lenses resulting from culturally influenced worldviews. The intercultural identity recognizes the bias of a single worldview due to culture and expands the possibilities of how a person might interpret and make sense of life's experience, both one's own and others. The transformative intercultural worldview would also embrace these differences as resources to make meaning (Parks Daloz 2000; Friedman and Berthoin Antal 2005; Marsella 2005; Sandage et al. 2008).

Denial The first phase in both Bennett and Hammer's theories is called "Denial." Simply stated, a person denies that culture matters as a way to make sense of differences. The conviction is that everyone holds, or should hold, the same worldview. As such, there are no discernable differences due to culture. Therefore, a person does not recognize that culture influences beliefs, belief systems, behaviors, feelings, and so forth, and, idealistically, everyone should feel, believe, think and behave the same in order to peacefully live together and to live a good life.

Polarization The second phase in the developmental model Hammer calls "Polarization." This ethnocentric worldview finds that a way to make sense of differences is to defend a particular worldview as the best or true way of thinking, acting and feeling. One can think of it as an "us and them" description of differences. The results are that sojourners can find themselves feeling threatened and defending their own culture as better, as a way to make sense of the differences that they confront. They can also do the reverse, a mindset conveniently referred to as Reversal in this Polarization orientation. People find themselves defending the other culture as better. In both cases, the polarized position is that one worldview is better than another. They do not yet have the skill or enough capacity to equalize their critique and to evaluate from multiple frames of reference the virtues and vices of the similarities and the differences.

Minimization Hammer's third phase is "Minimization" and a transitional mindset between monocultural and intercultural orientations. People in minimization agree that there are cultural differences. They minimize, however, the differences and emphasize the universal similarities and under emphasize the impact of culture on deep and complex worldview components. This perspective believes that humanity does have surface differences due to culture. What lies below these surface differences, however, are universal similarities or essential human values, standards and conditions that we all (should) have in common. The orientation offers consideration and sympathy to more than one worldview, but assumes that, for example, what one considers as good is what all others should also consider as good and as a universal truth.

A person with a Minimization mindset would say, for example, that there is a single answer and description about what is true love and how love is authentically felt and shown, and that it is universally received and valued in the same manner. Another example of this perspective is the Golden Rule: Do to others as you would have them do to you. The tension with this ethnocentric perspective occurs when people are crossing ethnic and religious borders. What one person's worldview in a particular culture understands or values as good, a person with a worldview in another culture may not understand or experience as "good." Thus, to encourage worldview development toward an intercultural mindset, the interculturalist might suggest that the sojourner consider the Platinum Rule: Do to others as they would want done to them. To apply the Platinum Rule the sojourner must learn from people from the other culture what and how they ascribe meaning to behavior, symbols and other particulars of the culture. A primary way for this to happen is through alterity.

The tendency of this Minimization orientation, then, is to organize cultural differences into categories familiar to what one knows in one's own culture while at the same time transitioning to multiple explanations for other topics or situations in which differences present themselves. The developmental goal would be to develop greater self-awareness of one's own culture and to probe deeper to understand the values, perceptions and beliefs about other cultural and religious differences.

Acceptance The fourth mindset is "Acceptance." This position believes that humans are different and similar and accepts that how people judge the value or truth of these differences is connected to the cultural context in which humans reside. It accepts these differences and values them in their context, seeking to understand the patterns of thinking and the cultural belief systems and values that influence the differences.

This phase and theory, however, does not mean that with an acceptance of the differences the sojourner diminishes his or her own belief system and convictions in order to be an effective cross-cultural sojourner. Neither Bennett nor Hammer promote that people become chameleons, relativists or uncritical interculturalists. "A state of ethnorelativism does not imply an ethical agreement with all differences nor a disavowal of stating (and acting upon) a preference for one worldview over another," qualifies Bennett (1993: 46). "Acceptance does not mean agreement" (2004: 69). Rather, Bennett (1993) understands that people who become savvy cross-cultural sojourners develop the habit of asking questions and reflecting on what they are observing and learning, processing the differences that challenge their own worldview:

> It is naïve to think that intercultural sensitivity and competence is always associated with liking other cultures or agreeing with their values or ways of life. In fact, the uncritical agreement with other cultures is more characteristic of the ethnocentric condition of Reversal, particularly if it is accompanied by a critical view of your own culture. Some cultural differences may be judged negatively–but the judgment is not ethnocentric unless it is associated with simplification, or withholding equal humanity. (1993: 46)

In other words, people can seek to understand beliefs, feelings and behaviors from multiple perspectives. People do not, however, have to exchange their entire worldview with each new exposure to a new way of making meaning. In fact, Milstein (2005) found in his study that subjects who rated their intercultural experience as a challenge significantly increased their self-efficacy (2005: 228). As the subjects embraced the challenge of cultural differences and dissonance, they believed themselves more capable of communicating their belief and in their abilities "to organize and execute the actions required to manage prospective situations and produce given attainments" (2005: 222). This increased self-efficacy belief enabled them to become more engaged in processing the epistemological sojourn challenges (2005: 235). Bennett warns, however, that "to figure out how to maintain ethical commitment in the face of such relativity" is a major challenge to resolve in the Acceptance orientation (2004: 69).

People with an Acceptance mindset, then, accept the contextual influence of both the other worldview and their own. The testing and filtering function of their worldview accepts that culture deeply influences worldview. With this orientation, sojourners gain the skill to interact with these perceived differences productively and to communicate and collaborate effectively with others from a distinctly different worldview and cultural experience. Curiosity about and respect for cultural differences expand, as well as the skill to differentiate between multiple cultures, especially one's own. The relational challenges that can come from alterity, however, become epistemologically and spiritually transformative (Sandage et al. 2008). This is why developing the skill of reflective cultural self-awareness to recognize and understand the foundation of one's own worldview is critical to understanding the religion of another in a particular cultural context.

Adaptation The final mindset in Hammer's intercultural sensitivity theory is "Adaptation." This view supports the skill of actually adapting one's ways of thinking, acting and feeling to another's worldview and culture. It is more than just being able to eat food from another culture and not wake up in the night with indigestion. It is the skill of being able, for example, to take on the patterns of thinking similar to people in another culture. A person in the Adaptation orientation develops the skill to intentionally shift affectively, behaviorally and cognitively from different cultural orientations.

Both Bennett's Developmental Model of Intercultural Sensitivity and Hammer's Intercultural Development Continuum offer sojourners theoretical frameworks by which to recognize and describe their worldview perspectives. It also offers interculturalists who are preparing and intervening with sojourners a theoretical framework by which to explain and observe the phenomenon. The accompanying measurement tool, the Intercultural Development Inventory, assists the interculturalist to articulate developmental goals for how to develop inter-religious and intercultural competencies, such as listening for accuracy, empathy, curiosity, coping skills for epistemological ambiguity, and so forth.

155.7 The Interculturalist's Work and Challenge

The intercultural trainer's challenge is to more deeply understand the phenomenon and definition of the religious tension experienced during the critical incidents in cross-religious interaction. Cultural adaptation theories such as the DMIS and IDC are an excellent foundation for this exploration. They also have their limitations. Professional interculturalists who work in settings in which religious differences and hostilities are present must further investigate the possibilities and limitations. In the meantime, they should continue to move forward with the intention and hope that with increased intercultural competency could come increased interfaith collaboration. Of course, all involved must be willing to participate in this challenge of worldview shifts to try to understand humanity's diverse ways to make sense of life through diverse religions and their systems.

To answer the questions Who is an interculturalist? and What do they do?, is to understand some of the leading frameworks and theories that are the foundation of their work. Bennett's Developmental Model of Intercultural Sensitivity and Hammer's Intercultural Development Continuum are two of those theories. They offer insight to the phenomenon of making sense of differences, differences of culture and differences of religion. This phenomenon is personal and holistic in that it addresses our affective, behavioral, spiritual and cognitive experiences. The troubling headlines of religious conflict point to why we need interculturalists' growing expertise to support the delicate attempts at interfaith collaboration. We do not need to think long about where on the map the people and places are who need interventions informed by the interculturalist's expertise and study. One of the places at the top of the current headlines is the Middle East.

155.7.1 Jews, Jesus and Mohammed: Applying Intercultural Tools in Interfaith Conflict

The field of Peace Studies is also investigating the spiritual and religious dimensions related to the development of intercultural competence. Employing Bennett's Developmental Model of Intercultural Sensitivity (DMIS) as his training approach and assessment framework for interreligious groups and conflict resolution in the Middle East, Mohammed Abu-Nimer investigated the "importance and uniqueness of religious attitudes and settings in intercultural exchanges" (2001: 688). This was the first of many subsequent uses that Abu-Nimer has and continues to apply the DMIS, and now the IDC, as a way to support interfaith collaboration.

With this particular interreligious group of Israeli Jews and Palestinian Christians and Muslims, Abu-Nimer observed that the respondents' belief systems significantly influenced their ability to engage with and develop toward the next worldview orientation in Bennett's model. What surfaced when Abu-Nimer integrated spirituality with the affective, behavioral and cognitive skill sets was the deep dynamic that spirituality and religious beliefs play in an effort to increase interreligious tolerance, understanding and dialogue. His findings suggest that if the intercultural trainer's goal is to increase intercultural competence, a sojourner's preparation and support requires the integration of spirituality and religion in order to process this deep epistemological dynamic of worldview orientation and development.

Abu-Nimer deliberately included in his objectives the exploration of "how religion had helped to construct [the participants'] world-view and how it shapes their value system" (689). Specifically, and at the suggestion of the participants, they integrated as a central and interconnected component in the DMIS the role of spirituality in the development of the cognitive, affective and behavioral skills skillsets to support intercultural competency development. Using a triangle as the image, they placed the respective skillset at each corner of the triangle and placed "spiritual" in the center, suggesting that spirituality influences and interconnects each of the skillsets.

Abu-Nimer used the DMIS with the following objective:

> To increase the participants' awareness of how limited their tolerance and interreligious interaction is (both cognitively and behaviorally) and to explore the group's vision of interreligious relations and dialogue. Thus, participants are asked to identify their attitudes toward other religious groups using the proposed developmental model. (2001: 697)

Overall, the DMIS did prove useful to support the training objective to increase interreligious competency.

Abu-Nimer also noted several limitations in his use of the DMIS. Some participants concluded that Reversal would mean they would need to convert to another religion. Reversal is a posture of giving superiority to the other culture, or religion for Abu-Nimer's participants. This is a significant challenge to anyone's worldview. He explained it in this way:

> Conversion might involve denigrating one's previous culture or insisting on the superiority of the new one; however, conversion in an interreligious interaction is often perceived as an abandonment of the current faith and adoption of a new faith, and denigration or superiority are not necessarily components of this response, as suggested by many participants. (2001: 698)

From a spiritual and religious experience, this development is not just a worldview shift but a worldview abandonment. Abu-Nimer notes that his participants' response might suggest a low level of religious differentiation, not being able to feel loyal to one's own convictions while at the same time seeking to understand the beliefs of another religion. Still, the challenge is significant. Abu-Nimer asserts that Bennett's DMIS discussion "does not address this distinction sufficiently" (697). Abu-Nimer's conclusions suggests that specific attention must be given to how to negotiate this affective response as sojourners confront the cognitive challenge of belief system differences. Interculturalists need to study more deeply how to support sojourners' processing of spiritual and religious beliefs if the goal is to support movement toward transformative worldview development.

Abu-Nimer also found that most participants disagreed with the idea of the Acceptance worldview orientation, understanding it to promote a relativist framework whereby there are no absolute standards of right and wrong. While he conceded that it could be that the participants were not yet able to shift to understand or experience another's worldview as a possible viable worldview explanation, he also offered an alternative explanation. Simply stated, in the interreligious settings, "the moral and spiritual dimensions of the identity add more difficulty to the persons' ability to move from an ethnocentric to an ethnorelative stage (than in a cultural or non-spiritual setting)" (2001: 699).

Some might conclude that Abu-Nimer's observations mean that religious people are unable to develop the same levels of intercultural competence. This is a faulty conclusion. Research is beginning to demonstrate that this sort of challenging worldview shift requires interculturalists to intervene. Sojourners need challenge and support that addresses sojourners' developmental and transformative reflective processing of spiritual and religious sensibilities (Ludeman Smith 2010). The IDC accompanied with The IDI can help determine the kinds of challenge and support that can encourage worldview shifts toward an ethnorelative worldview. These tools are not enough, however. Training methods, reflective processes, experiential design and support before, during and after the experiences are critical to transformation toward an intercultural mindset (Ludeman Smith 2013)

155.8 Conclusion

Though rarely found are the news headlines telling stories about interfaith interculturalists like Abu-Nimer, the work is happening with growing awareness. For example, in 2013 the Parents Circle Families Forum in Israel and Palestine, a grassroots organization of bereaved parents who seek reconciliation instead of revenge, received an honorable mention in competition for the United Nations Alliance of Civilization and the BMW Group's prized Intercultural Innovation Award. This was the Parents Circle Families Forum eleventh award in 5 years for its effective and active efforts to bring healing across interfaith and enemy lines. Professional interculturalists trained in increasingly more sophisticated and research supported theories, tools, and training methods are applying their expertise in these settings of increased religious intolerance. Still, as Abu-Nimer (2011) reported, more professionally trained people in these intercultural methods are needed to meet the increasing needs for interfaith and multifaith peace efforts (2011: 568).

There will always be limitations in any theory and method applied to actual contexts, as Abu-Nimer found. Professional interculturalists are responding in increasingly more sophisticated ways to this critique through their research and international networks. An example is the rigorous psychometric analysis that Hammer applied to version two of the Intercultural Development Inventory (IDI) and its supporting constructivist theory. The result was a revised theory description in 2009 that he named the Intercultural Development Continuum and a third version of the IDI (Hammer 2011, 2012). This kind of reflective and responsive activity is characteristic of the field.

Professional interculturalist societies and their networks, such as the Society for Intercultural Education, Training and Research (SIETAR), are growing in membership and in the number of new societies being added around the world. SIETAR—USA developed a "Living Code of Ethical Behavior" by which its signing members commit to adhere to in their practice (Thacker 2012; Ludeman Smith 2013). People around the world are finding the theories, tools and methods effective. Especially telling is that the authors of the peer-reviewed articles are coming from all over the world. Thus, an increasing number of Professional interculturalists in the contexts of the conflict can offer developmental approaches to answer this call for worldview shifts that manifest mindsets and skills equipped to meet the conflicts and collaboration of the people on this changing world religion map. Professional interculturalists can help create the needed world headlines that tell of how interfaith collaborators are creating, together, new solutions to our world's greatest needs *because* of our religious and cultural diversity and commonalities.

References

Abu-Nimer, M. (2001). Conflict resolution, culture, and religion: Toward a training model of interreligious peacebuilding. *Journal of Peace Research, 38*(6), 685–704.

Abu-Nimer, M. (2011, October). Religious leaders in the Israeli-Palestinian conflict: From violent incitement to nonviolence resistance. *Peace & Change, 36*(4), 556–580.

Allan, M. (2003). Frontier crossings: Cultural dissonance, intercultural learning and the multicultural personality. *Journal of Research in International Education, 2*(1), 83–110.

Aritonang, M. S. (2012, December 21). Politicians fuel raging hatred and tolerance. *The Jakarta Post*. Retrieved February 27, 2013, from http://www.thejakartapost.com/news/2012/12/21/politicians-fuel-raging-hatred-and-intolerance.html

Asay, S. M., Younes, M. N., & James Moore, T. (2006). The cultural transformation model: Promoting cultural competence through international study experiences. In R. R. Harmon (Ed.), *International family studies: Developing curricula and teaching tools* (pp. 84–99). Binghamton: Haworth Press, Inc.

Bennett, M. J. (1986). A developmental approach to training for intercultural sensitivity. *International Journal of Intercultural Relations, 10*(2), 179–196.

Bennett, M. J. (1993). Toward ethnorelativism: A developmental model of intercultural sensitivity. In R. M. Paige (Ed.), *Education for the intercultural experience* (2nd ed., pp. 21–71). Yarmouth: Intercultural Press, Inc.

Bennett, M. J. (2004). Becoming interculturally competent. In J. Wurzel (Ed.), *Toward multiculturalism: A reader in multicultural education* (2nd ed., pp. 62–77). Newton: Intercultural Resource Corporation.

Bennett, J. M. (2009a). Transformative training: Designing programs for culture learning. In M. A. Moodian (Ed.), *Contemporary leadership and intercultural competence: Exploring the cross-cultural dynamics within organizations* (pp. 95–110). Thousand Oaks: Sage.

Bennett, M. F. (2009b). Religious and spiritual diversity in the workplace. In M. A. Moodian (Ed.), *Contemporary leadership and intercultural competence: Exploring the cross-cultural dynamics within organizations* (pp. 45–57). Thousand Oaks: Sage.

Bennett, M. J. (2012). Paradigmatic assumptions and a developmental approach to intercultural learning. In M. Vande Berg, R. M. Paige, & K. Hemming Lou (Eds.), *Student learning abroad: What our students are learning, what they're not, and what we can do about it* (pp. 90–114). Sterling: Stylus Publishing, LLC.

Bennett, J. M., & Bennett, M. J. (2004). Developing intercultural sensitivity: An integrative approach to global and domestic diversity. In D. Landis, J. M. Bennett, & M. J. Bennett (Eds.), *Handbook of intercultural training* (3rd ed., pp. 147–165). Thousand Oaks: Sage.

Bennett, M. J., & Castiglioni, I. (2004). Developing intercultural sensitivity: An integrative approach to global and domestic diversity. In D. Landis, J. M. Bennett, & M. J. Bennett (Eds.), *Handbook of intercultural training* (3rd ed., pp. 249–265). Thousand Oaks: Sage.

Berardo, K. (2008). *The intercultural profession in 2007: Profiles, practices & challenges*. Unpublished paper. Retrieved March 3, 2013, from http://www.culturosity.com/pdfs/Intercultural%20Profession%20Report.%20Berardo.%202008.pdf

Cushner, K., & Karim, A. U. (2004). Study abroad at the university level. In D. Landis, J. M. Bennett, & M. J. Bennett (Eds.), *Handbook of intercultural training* (3rd ed., pp. 289–308). Thousand Oaks: Sage.

Friedman, V. J., & Berthoin Antal, A. (2005). Negotiating reality: A theory of action approach to intercultural competence. *Management Learning, 36*(1), 69–86.

Friedman, M., & Rholes, W. S. (2007). Successfully challenging fundamentalist beliefs results in increased death awareness. *Journal of Experimental Social Psychology, 34*, 794–801.

Gopin, M. (2000). *Between Eden and Armageddon: The Future of world religions, violence, and peacemaking*. Oxford: Oxford University Press.

Grove, C., & Torbiörn, I. (1993). A new conceptualization of intercultural adjustment and the goals of training. In R. M. Paige (Ed.), *Education for the intercultural experience* (2nd ed., pp. 73–108). Yarmouth: Intercultural Press.

Hage, S. M., Hopson, A., Siege, M., Payton, G., & DeFanti, E. (2006). Multicultural training in spirituality: An interdisciplinary review. *Counseling and Values, 50*, 217–234.

Hammer, M. R. (2009). Solving problems and resolving conflict using the intercultural conflict style model and inventory. In M. A. Moodian (Ed.), *Contemporary leadership and intercultural competence: Exploring the cross-cultural dynamics within organizations* (pp. 219–232). Thousand Oaks: Sage.

Hammer, M. R. (2011). Additional cross-cultural validity testing of the Intercultural Development Inventory. *International Journal of Intercultural Relations, 35*, 474–487.

Hammer, M. R. (2012). The Intercultural Development Inventory: A New frontier in assessment and development of intercultural competence. In M. Vande Berg, R. M. Paige, & K. Hemming Lou (Eds.), *Student learning abroad: What our students are learning, what they're not, and what we can do about it* (pp. 115–136). Sterling: Stylus Publishing, LLC.

Hammer, M. R. (2013). *What is the Intercultural Development Inventory (IDI)?* Retrieved March 2, 2013 from http://www.idiinventory.com/about.php

Hammer, M. R., Bennett, M. J. & Wiseman, R. (2003). The Intercultural Development Inventory: A measure of intercultural sensitivity. In R. M. Paige (Guest Editor), Special issue on the Intercultural Development Inventory. *International Journal of Intercultural Relations*, *27*, 421–443.

Hariyadi, M. (2012, August 14). *Bogor, after 5 years radical Muslims force closure of Saint John the Baptist Parish Church*. AsianNews. IT. Retrieved February 27, 2013, from http://www.asianews.it/news-en/Bogor,-after-6-years-radical-Muslims-force-closure-of-Saint-John-the-Baptist-Parish-Church-25549.html#
Hiebert, P. G. (2008). *Transforming worldviews*. Grand Rapids: Baker Academic.
Hiebert, P. G., Shaw, D., & Tiénou, T. (1999). *Understanding Folk Religion*. Grand Rapids: Baker Books.
Hinnells, J. R. (1997). *A new handbook of living religions*. London: Penguin.
Hoff, J. G. (2008). Growth and transformation outcomes in international education. In V. Savicki (Ed.), *Developing intercultural competence and transformation: Theory, research and application in international education* (pp. 53–73). Sterling: Stylus Publishing, LLC.
Hunter, A. (2008). Transformative learning in international education. In V. Savicki (Ed.), *Developing intercultural competence and transformation: Theory, research and application in international education* (pp. 92–107). Sterling: Stylus Publishing, LLC.
Kraft, C. H. (1979). *Christianity in culture*. Maryknoll: Orbis Books.
Lebaron, M., & Pillay, V. (2006). *Conflict across cultures*. Boston: Intercultural Press.
Lewis Hall, M. E., Edwards, K. J., & Hall, T. W. (2006). The role of spiritual and psychological development in the cross-cultural adjustment of missionaries. *Mental Health, Religion and Culture, 9*(2), 193–208.
Ludeman Smith, N. (2010). *Sojourn through spiritual and religious tension: A Quantitative study of intercultural competence and worldview development*. Unpublished doctoral dissertation, Bethel University & Seminary, Department of Transformational Leadership, St. Paul, MN.
Ludeman Smith, N. (2013, March). Ethics Work Group Minutes, March 5, 2013. SIETAR—USA. Unpublished document.
Ludeman Smith, N. (2013, Summer). (Re) Considering a critical ethnorelative worldview goal and pedagogy for global and biblical demands in Christian higher education. In *Christian Scholar's Review, Special Issue*.
Magnuson, D. (2008). *Developmental Model of Intercultural Sensitivity (DMIS): How we can experience cultural differences*. Unpublished handout.
Marsella, A. J. (2005). Culture and conflict: Understanding, negotiating, and reconciling conflicting constructions of reality. *International Journal of Intercultural Relations, 29*, 651–673.
Mezirow, J. (2000). Learning to think like an adult: Core concepts of transformation theory. In J. Mezirow & Associates (Eds.), *Learning as transformation: Critical perspectives on a theory in progress* (pp. 3–33). San Francisco: Wiley.
Milstein, T. (2005). Transformation abroad: Sojourning and the perceived enhancement of self-efficacy. *International Journal of Intercultural Relations, 29*, 217–238.
Muck, T. C. (2005). *How to study religion*. Wilmore: Wood Hill Books.
Paige, R. M. (1993). On the nature of intercultural experiences and intercultural education. In R. M. Paige (Ed.), *Education for the intercultural experience* (pp. 1–19). Yarmouth: Intercultural Press, Inc.
Paige, R. M., & Vande Berg, M. (2012). Why students are and are not learning. In M. Vande Berg, R. M. Paige, & K. Hemming Lou (Eds.), *Student learning abroad: What our students are learning, what they're not, and what we can do about it* (pp. 29–58). Sterling: Stylus Publishing, LLC.
Parks Daloz, L. A. (2000). Transformative learning for the common good. In J. Mezirow & Associates (Eds.), *Learning as transformation: Critical perspectives on a theory in progress* (pp. 103–123). San Francisco: Wiley.
Pedersen, P. J. (2009). Teaching towards an ethnorelative worldview through psychology study abroad. *Intercultural Education, 20*(supplement 1–2), 73–86.
Pedersen, P. J. (2010). Assessing intercultural effectiveness outcomes in a year-long study abroad program. *International Journal of Intercultural Relations, 34*, 70–80.
Rogers, J., & Ward, C. (1993). Expectation-experience discrepancies and psychological adjustment during cross-cultural reentry. *International Journal of Intercultural Relations, 17*, 185–196.

Sample, S. G. (2009, October). *Intercultural development and the international curriculum*. Paper presented at the IDI Conference, Minneapolis, MN.
Sandage, S. J., & Harden, M. G. (2011). Relational spirituality, differentiation of self, and virtue as predictors of intercultural development. *Mental Health, Religion & Culture, 14*(8), 819–838.
Sandage, S. J. & Jankowski, P. J. (2012). Spirituality, social justice, and intercultural competence: Mediator effects for differentiation of self. *International Journal of Intercultural Relations*.
Sandage, S. J., Jensen, M. L., & Jass, D. (2008). Relational spirituality and transformation: Risking intimacy and alterity. *Journal of Spiritual Formation and Soul Care, 1*(2), 182–206.
Sandage, S. J., Crabtree, S. & Schweer, M. (2014). Differentiation of self and social justice commitment mediated by hope. *Journal of Counseling & Development*.
Savicki, V. (2008). Experiential and affective education for international educators. In V. Savicki (Ed.), *Developing intercultural competence and transformation: Theory, research and application in international education* (pp. 74–107). Sterling: Stylus Publishing, LLC.
Savicki, V., Downing-Burnette, R., Heller, L., Binder, F., & Suntinger, W. (2004). Contrasts, changes, and correlates in actual and potential intercultural adjustment. *International Journal of Intercultural Relations, 29*, 311–329.
Shults, F. L., & Sandage, S. J. (2006). *Transforming spirituality: Integrating theology and psychology*. Grand Rapids: Baker Academic.
Stuart, D. K. (2012). Taking stage development theory seriously. In M. Vande Berg, R. M. Paige, & K. Hemming Lou (Eds.), *Student learning abroad: What our students are learning, what they're not, and what we can do about it* (pp. 61–89). Sterling: Stylus Publishing, LLC.
Swiss voters back ban on minarets. (2009, November 29). BBC NEWS. Retrieved February 23, 2013, from http://news.bbc.co.uk/go/pr/fr/-/2/hi/europe/8385069.stm
Thacker, M. (Ed.). (2012). *Eye on ethics: A conversation for professional interculturalists*. Portland: SIETAR—USA.
The Parents Circle Families Forum. (2013). Retrieved March 4, 2013, from http://www.theparentscircle.org/
The Pew Forum on Religion & Public Life. (2012). Retrieved February 22, 2013, from The Pew Research Center Government/Rising-Tide-of-Restrictions-on-Religion-findings.aspx
Ting-Toomey, S., & Oetzel, J. G. (2001). *Managing intercultural conflict effectively*. Thousand Oaks: Sage.
Vande Berg, M., Connor-Linton, J., & Paige, R. M. (2009). The Georgetown Consortium Project: Interventions for student learning abroad. *Frontiers: The Interdisciplinary Journal of Study Abroad, 17*, 1–76.
Vande Berg, M., Paige, R. M., & Hemming Lou, K. (2012). Student learning abroad: Paradigms and assumptions. In M. Vande Berg, R. M. Paige, & K. Hemming Lou (Eds.), *Student learning abroad: What our students are learning, what they're not, and what we can do about it* (pp. 3–26). Sterling: Stylus Publishing, LLC.
Ward, C., & Kennedy, A. (1999). The measurement of sociocultural adaptation. *International Journal of Intercultural Relations, 23*(4), 659–677.
Ward, C., & Rana-Deuba, A. (2000). Home and host culture influences on sojourner adjustment. *International Journal of Intercultural Relations, 24*, 291–306.
Weaver, G. R. (1993). Understanding and coping with cross-cultural adjustment stress. In R. M. Paige (Ed.), *Education for the intercultural experience* (2nd ed., pp. 137–167). Yarmouth: Intercultural Press.

Chapter 156
Multifaith Responses to Global Risks

Anna Halafoff

156.1 Introduction

While a politics of fear has dominated the turn of the twenty-first century, it has paradoxically led to a growing interest in collaborative solutions to counter global risks, such as terrorism and more recently climate change, which address the root causes of problems (Bauman 2006; Beck 2006). The rise of the multifaith movement during this period is exemplary of these peacebuilding strategies—aimed at addressing risks and advancing common security—at local and global levels.

A rise in multifaith engagement occurred in the U.S. following the tragic events of September 11, 2001 (Eck 2001: xiii–xix; Brodeur 2005: 42; McCarthy 2007: 85; Niebuhr 2008: xxii, 5–7, 10–11) and the number of multifaith and multi-actor peacebuilding networks, including both religious and state actors, also increased in the UK and Australia as a result of September 11 and the 7 and 21 July 2005 London bombings (Braybrooke 2007: 1, 13; Pearce quoted in Bharat and Bharat 2007: 245–246; Bouma et al. 2007: 6, 57–60). For example, there were 24 multifaith councils in the U.S. in 1980 and 500 by 2006 (National Council of Churches 1980, Pluralism Project 2006 cited in McCarthy 2007: 85). Similarly, the UK had 27 local interfaith organisations in 1987, while in 2007 the number had grown to 200 (Pearce quoted in Bharat and Bharat 2007: 245–246).

These twenty-first century developments form part of a long history of multifaith initiatives that well predate the events of September 11, 2001. This chapter draws on existing literature and on the author's *Netpeace* study, which consisted of 56 semi-structured interviews with leading multifaith practitioners in Australia, the United Kingdom and the United States, conducted in 2007 and 2008. Respondents' remarks are tagged including the participant's surname, the year of the interview

A. Halafoff (✉)
Centre for Citizenship and Globalisation, Deakin University,
Melbourne Burwood Campus, 221 Burwood Highway, Burwood, VIC 31215, Australia
e-mail: anna.halafoff@deakin.edu.au

S.D. Brunn (ed.), *The Changing World Religion Map*,
DOI 10.1007/978-94-017-9376-6_156

and the country in which they reside (Netpeace: Patel 2007, USA). This study identified four principle aims of the multifaith movement and proposed a new theoretical framework termed *netpeace,* which recognizes the interconnectedness of global problems and solutions and the capacity of multi-actor peacebuilding networks, including religious and state actors, to overcome the most pressing risks of our times (Halafoff 2010).

This chapter presents a brief history of the multifaith movement and then applies the *netpeace* framework to two more recent multifaith developments, viz., the 2009 Melbourne Parliament of World's Religions (PWR) and the recent establishment of the new multifaith organization Groundswell. In so doing, it argues that as the multifaith movement's focus has shifted away from terrorism to climate change and the global financial crisis, the movement has become more inclusive. It also suggests that this growing inclusivity, which encompasses non-religious as well as religious actors, combined with what could perhaps be described as a swing against conservative religiosity, is leading the movement in a more socially progressive direction, returning in a sense to its progressive roots.

156.2 A Brief History of the Multifaith Movement

156.2.1 The 1893 Parliament of the World's Religions

The 1893 PWR is widely acknowledged as the beginning of the global multifaith movement. The 1893 PWR was held as part of a World Columbian Exposition celebrating Christopher Columbus's "discovery" of America and has been widely and rightly criticized as a flawed model of multifaith relations due to its Christian bias and "civilizing mission." The international gathering of the 1893 PWR was made possible by increased opportunities for travel and communication, which continued to escalate throughout the twentieth century, thus further enabling the global expansion of the multifaith movement. The high level of representation and enthusiastic participation of Jewish leaders at the 1893 PWR established a tradition of ecumenical relations among Catholics, Protestants and Jews in America (Braybrooke 1992: 7–9, 18, 22, 27–30, 39–42, 309). In addition, due to the spread of the British Empire throughout Asia, a fascination with Eastern philosophy was prevalent in certain segments of Western societies in the late nineteenth century (Croucher 1989: 6–1; Braybrooke 1996: 10; McCarthy 2007: 15). While Ralph Waldo Emerson, Henry David Thoreau and the Theosophical Society had all been influential in introducing Americans to Hindu and Buddhist thought, the 1893 PWR provided the first opportunity for Americans to have direct contact with Hindu and Buddhist teachers (Eck 2001: 96–97, 180–184). These teachers, in addition to providing first-hand explanations of their religious and philosophical traditions, utilized the PWR as a platform to challenge Christian universalism and exclusivity propagated by British and American missionaries (Braybrooke 1992: 25; Eck 2001: 182–185).

Although invited, the caliph of Turkey refused to send Muslim representatives to the first PWR. One of the only Muslim participants at the first PWR was American Mohammed Russell Alexander Webb who had converted to Islam while posted as America's consul general in the Philippines. Webb publicly acknowledged the negative stereotypes associated with Islam in America, yet also articulated his confidence that once Americans had a true understanding of Islam they would learn to appreciate it (Eck 2001: 234–235). The relatively small Muslim presence at the 1893 PWR illustrates that the first bridges to be built in global multifaith engagement were largely among Hindus, Buddhists, Jews and Christians. Indigenous people were excluded from the main assembly of the 1893 PWR (Brodeur 2005: 44) and it was not until a century later that Indigenous and Muslim participants began to play a prominent role in multifaith initiatives in Western societies.

The 1893 PWR's Declaration emphasized the need for more understanding among religions, and also of the need for common action in response to social "problems … and questions connected with Temperance, Labour, Education, Wealth and Poverty" and to "bring nations of the earth into a more friendly fellowship in the hope of securing permanent international peace" (Barrows 1893: 18 quoted in Braybrooke 1992: 324). Therefore, despite its many flaws the 1893 PWR is said to have established "a normative model" for multifaith encounters conducted in a spirit of openness and respect for diversity, with a new emphasis on non-proselytizing, promoting understanding between faith traditions and, significantly, on peacebuilding (McCarthy 2007: 18; Brodeur 2005: 43).

156.2.2 Twentieth Century Multifaith Initiatives

Following the 1893 PWR, multifaith congresses and conferences were held in the U.S. the UK and Europe and several multifaith organizations were also established in the U.S. and the UK. While the First World War (WWI) and WWII restricted global multifaith engagement, several multifaith congresses were convened in America and Europe before WWI and between WWI and WWII. At the end of WWII multifaith congresses and conferences resumed in the U.S., the UK and Europe (Braybrooke 1992: 49–56, 67–72, 95–108, 114). In response to the atrocities of WWII the imperative to address global risks and injustices intensified within the multifaith movement alongside other social movements of this period. After WWII multifaith activities were focused largely on bilateral initiatives of Jewish–Christian relations in response to the tragedy of the Holocaust (Braybrooke 1992: 175–215; Niebuhr 2008: 126–127; Netpeace: Dupuche 2008, AUS; Shashoua 2008, UK; Summers 2008, AUS; Voll 2007, USA). Jewish–Christian dialogue enabled communities to confront and address gross injustices committed against Jews and to unite in common action to prevent the recurrence of such atrocities.

A new concern over peace and nuclear disarmament also arose within the multifaith movement in response to the atrocities committed in Hiroshima and Nagasaki (Braybrooke 1992: 52, 56, 59). During the Cold War, the nuclear threat became a major focus of multifaith engagement and numerous multifaith initiatives were coordinated to protest against the Vietnam War in the 1960s and 1970s (Brodeur 2005: 45–6).

A growing awareness of the global environmental crisis and the role that religious and spiritual traditions can play in either inflaming or ameliorating this problem was also steadily building alongside other social movements in the 1960s and 1970s (Braybrooke 1992: 148–149; Rockefeller and Elder 1992: 1; Netpeace: Harper 2008, USA; Kearns 2008, USA). As religious and multifaith organizations began to focus much of their attention on social problems including civil rights, the Vietnam War, poverty and gender inequality, extending this concern to the rights of nonhuman life emerged as a continuation of this pattern (Nash 1996: 220; Keller and Kearns 2007: 2; Tucker 2007: 496). Buddhist philosophy also played a central role in the environmental movement from the 1960s onwards, particularly due to the doctrines of interdependent arising, impermanence, non-violence and respect for all beings (Rockefeller 1992: 156; Nash 1996: 215). Greater awareness of the environmental crisis also led Western societies to turn to Indigenous, especially Native American, traditions for inspiration and guidance on how best to live in harmony with the natural environment (Braybrooke 1992: 242; Rockefeller and Elder 1992: 6; Nash 1996: 214).

The period from the 1960s to the 1980s heralded a new era of social movements, including anti-war, women's and environmental movements, which arose to defend the "lifeworld" from the negative effects of capitalist modernity and to advance equal rights for all (Habermas 1981: 35). William Sims Bainbridge (1997: 3) defines social movements as "collective human attempts to create or to block change" and religious movements as "a relatively organized attempt by a number of people to cause or prevent change in a religious organization or in religious aspects of life." Fred Kniss and Paul D. Numrich (2007: 227) describe the multifaith movement in particular as "a decentralized social movement of individuals, groups and organizations seeking to foster mutual respect and understanding across religions in order to achieve positive individual, social, cultural, and civic change." The multifaith movement is made up of numerous local, national and global multifaith organizations, networks and actors, who meet at local, regional and global multifaith conferences, festivals and events. As will be explained in more detail throughout this chapter, the multifaith movement has always been committed to social change on issues such as protecting the environment, nuclear disarmament, poverty alleviation, peacebuilding and post-conflict reconstruction. This commitment aligns the multifaith movement with other social movements, which emerged at the turn of the twenty-first century.

Another contributing factor behind the growth of the multifaith movement in the late twentieth century was the rise in immigration to Western societies during the 1960s and 1970s (Eck 2001: 1–4; Bouma 2006: 52–53, 64; Weller 2008: 32–42; Netpeace: Aly 2008, AUS; Dellal 2008, AUS; Ficca 2007, USA; Mogra 2007, UK; Murdoch 2008, UK; Patel 2007, USA; Voll 2007, USA). These developments restructured the focus of multifaith engagement in Western societies after WWII from the predominance of Catholics, Protestants and Jews to the inclusion of Hindus, Buddhists, Sikhs and Muslims (McCarthy 2007: 7).

156.2.3 Multifaith Initiatives in the 1990s

Following on from the increase in multifaith activity from the 1960s through to the 1980s, there was a dramatic rise of multifaith engagement in the early 1990s (Baldock 1997: 197; Eck 2001: 370; Kirkwood 2007: xiv). In 1993 over 7,000 people from diverse faith traditions including Buddhists, Jains, Zoroastrians, Hindus, Muslims, Sikhs, Protestants, Catholics, Jews, Taoists, Wiccans, Baha'is and Indigenous peoples from all over the world assembled in Chicago to participate in the PWR (Eck 2001: 366–8). The 1993 PWR is acknowledged as the real beginning of a global multifaith movement (Netpeace: Knitter 2007, USA; Patel 2007, USA), in so far as this event raised the profile (Netpeace: Braybrooke 2007, UK) and visibility (Netpeace: Gibbs 2007, USA) of international multifaith engagement. Therefore, while the 1893 PWR was seen as the beginning of the global multifaith movement, the 1993 PWR is said to have signified its "coming of age" (Braybrooke 1992: 7–8).

Multifaith activity increased dramatically in the 1990s, as evidenced in the renewal and/or foundation of many of the world's largest multifaith organizations (Netpeace: Knitter 2007, USA; Patel 2007, USA) including the Tanenbaum Centre for Interreligious Understanding (TCIU) (Netpeace: Dubensky 2007, USA), the United Religions Initiative (URI) (Netpeace: Gibbs 2007, USA), the International Interfaith Centre at Oxford (IIC) (Netpeace: Braybrooke 2007, UK) and the Interfaith Centre of New York (ICNY) (Netpeace: Breyer 2007, USA). The World Conference of Religions for Peace also grew substantially under the new leadership of William F. Vendley during this period (Netpeace: Patel 2007, USA).

However, the euphoria and optimism of the 1990s was soon to be overshadowed by other international events. After the end of the Cold War, a series of "new wars" erupted in which the new actors were not states, but rather cultural and/or religious movements claiming power based on identity politics (Kaldor 1999: 1–2, 6). While an increased awareness of interconnectedness created a perception of "oneness" among an emerging global citizenry, it also heightened differences, thereby threatening ontological security (Kaldor 1999: 2–4, 6; Beckford 2003: 109). Growing fears around losing identity and power as a result of processes of globalization resulted in a global reassertion of "*introverted* forms of nationalism," of ethnic and religious identities, fostering aggressive intolerances (Beck 2006: 4). As a result, a global resurgence of religion and a rise of religious fundamentalisms categorized the turn of the twenty-first century (Marty and Appleby 1992). The fifth WCRP Assembly declaration of 1989 (1989: 4 quoted in Braybrooke 1992: 156) foresaw these events by ending with the words: "'Lead us from fear to trust.' Lead us from common terror to common security." Lead us from common terror to common security.'" This statement reflected both growing concerns regarding the global rise of religious extremism and the recognition that religions could play a positive role in countering religiously motivated violence.

As Muslim communities were frequently at the center of crisis events in the late 1980s and 1990s, such as the *The Satanic Verses* controversy in the UK, the 1995

Oklahoma City bombings and the First Gulf War, a rise of Islamophobia, fermented by divisive discourses emanating from state actors and the media, spread throughout Western societies (Weller 2008: 155, 163–167, 194–195; Eck 2001: 2, 8, 296–300, 303, 306). As a result of these events, diverse faith communities joined together with Muslim communities to show solidarity and joint outrage in response to hate crimes that targeted Muslims. Muslim communities became proactive in countering negative stereotypes, often through multifaith activities, and new multifaith alliances were formed especially among Christians, Muslims and Jews in the USA, the UK and Australia (Eck 2001: 341–347, 374; Bharat and Bharat 2007: 236) (Netpeace: Abu-Nimer 2008, USA; Braybrooke 2007, UK; Dupuche 2008, AUS; Ficca 2007, USA; Hassan 2008, AUS; Jones 2008, AUS; Ozalp 2008, AUS; Postma 2008, AUS; Shashoua 2008, UK).

It is also important to note that since the 1990s a new emphasis on youth engagement also developed within the multifaith movement, evident in a rise of youth programs within the major multifaith organizations and the creation of the Interfaith Youth Core (IFYC) founded in 1999 in the U.S. (Brodeur and Patel 2006: 4; Bharat and Bharat 2007: 190) (Netpeace: Patel 2007, USA).

156.2.4 Twenty-First Century Multifaith Initiatives

As briefly described above, there was a dramatic increase in multifaith engagement in the U.S., the UK and Australia in response to the events of September 11 as multifaith initiatives were implemented as peacebuilding and counter-terrorism strategies in response to this crisis event (Eck 2001: xiii–xix; Brodeur 2005: 42; Halafoff 2006: 10–12; Bouma et al. 2007: 6, 22–26, 55, 57–60; Braybrooke quoted in Bharat and Bharat 2007: 225; Kirkwood 2007: v–vi; McCarthy 2007: 85; Pearce quoted in Bharat and Bharat 2007: 245–246; Niebuhr 2008: xxii, 5–7, 10–11). The events of September 11 also lent more urgency to multifaith engagement (Netpeace: Smock 2007, USA; Dellal 2008, AUS) and intensified the need for understanding among diverse communities (Netpeace: Lacey 2008, AUS). Consequently, the multifaith movement gathered strength (Netpeace: Braybrooke 2007, UK) and a greater awareness of the importance of cooperation and understanding across faith communities emerged at both the local and global level.

In addition, subsequent to September 11, multifaith engagement was identified as a potential solution to counter religious extremism, which suddenly propelled it to 'the centre stage of world attention' (Brodeur 2005: 42; Braybrooke, quoted in Bharat and Bharat 2007: 225). Indeed, Sr. Joan Kirby (Netpeace: 2007, USA) exclaimed how after September 11, "it's as if the wave is cresting. There's such enormous interest in interfaith dialogue and cooperation" from the grassroots all the way to the United Nations!' Although September 11 and the need to counter religiously motivated terrorism created "a new sense of urgency" for multifaith engagement, it is critical to recognize that this momentum was building in the multifaith movement well before 2001, and particularly during the 1990s, as described above (Niebuhr 2008: xxii; Smock 2002: 3).

The 2002 Bali bombings and the 2005 London bombings also led to an increase of multifaith initiatives in the UK and Australia (Braybrooke 2007: 1, 13; Capie quoted in Bharat and Bharat 2007: 233–234; Cahill et al. 2004: 86–88; Bouma et al. 2007: 6, 55, 57–59; Bouma 2008: 13) (Netpeace: Blundell 2008, AUS; Cass 2008, UK; Hassan 2008, AUS; Mogra 2007, UK; Murdoch 2008, UK; Pearce 2007, UK). In particular, the "home-grown" nature of the terrorists involved in the London bombings at once increased fears and prejudices and also provided greater impetus for multifaith engagement in both the UK and Australia (Netpeace: Cass 2008, UK; Hassan 2008, AUS; Murdoch 2008, UK). Multifaith networks that were well established after September 11 also enabled quicker and more effective responses to these crisis events (Netpeace: Gibbs 2007, USA; Pearce 2007, UK).

Muslim communities also suddenly became more "visible" after September 11, thereby heightening awareness of Islam and of Muslims in Western societies (Netpeace: Harper 2008, USA; Shashoua 2008, UK; Ramey 2007, USA) and a new imperative to include Muslim communities in multifaith activities arose within the multifaith movement (Netpeace: Kearns 2008, USA; Pearce 2007, UK; Postma 2008, AUS). Multifaith initiatives provided a platform for Muslim communities to differentiate themselves from terrorists, to dispel negative stereotypes of Muslims and to affirm their commitment to non-violent principles (Netpeace: Mogra 2007, UK; Toh 2008, AUS). September 11 was thereby seen as a notable turning point for Muslim multifaith engagement in that Muslim communities became more proactive in initiating dialogue and educational activities to dispel misconceptions and to promote the peacebuilding aspects of Islam (Netpeace: Braybrooke 2007, UK; Dellal 2008, AUS; Gibbs 2007, USA; Hassan 2008, AUS; Mogra 2007, UK; Ozalp 2008, AUS; Ramey 2007, USA; Smock 2007, US; Woodlock 2008, AUS).

In addition, following the commercial release of *An Inconvenient Truth* in 2006, a renewed focus on environmental issues emerged within the multifaith movement. While environmental issues began to occupy a prominent place in the public sphere during the 1990s (Rockefeller 1992: 167), the so-called War on Terror, combined with a widespread apathy and denial of climate change across various sectors including conservative religious groups, marginalized the issue of climate change from the public mind (Kearns and Keller 2007: xii; Tucker 2007: 496) (Netpeace: Kearns 2008, USA). However, by the mid-2000s the global risk of climate change began to eclipse the global risk of terrorism, as the most prominent perceived threat to public security. Consequently, the multifaith movement shifted its focus again toward environmental issues.

156.2.5 Multi-actor Peacebuilding Networks

Toward the end of the twentieth century, the processes of globalization and the increase of interest in religion led to a new awareness among non-religious organizations of the need to partner with religious actors in response to common concerns. This new emphasis on collaboratively countering global risks resulted in the formation of multi-actor peacebuilding networks in which faith-based actors played an increasingly significant role alongside non-religious actors such as state, UN, NGO and Inter-Governmental Organization (IGO) actors, on issues of common security. In an increasingly interdependent world, issues such as climate change, HIV-AIDS and economic inequalities required global solutions in which faith communities and their theologies and philosophies had an important role to play (Eck 2001: 380; Bharat and Bharat 2007: 243). Significant multi-actor peacebuilding initiatives of this period included: a World Day of Prayer for Peace, hosted by the Vatican in Assisi in 1986 (Braybrooke 1992: 141) and sponsored by the World Wildlife Fund (WWF) (Rockefeller and Elder 1992: 10–11); The Global Forum, founded by the ToU, which held meetings in Tarrytown, New York State (1985), Oxford (1988) and Moscow (1990) (Braybrooke 1992: 109–110); and the World Faiths Development Dialogue (WFDD), established in 1998 by James Wolfensohn, the then President of the World Bank, and the then Archbishop of Canterbury George Carey, to create partnerships between faith and development actors to address the problems of poverty (Bharat and Bharat 2007: 247; Marshall and Keough 2004: 3; Wolfensohn 2004: xii) (Netpeace: Braybrooke 2007, UK; Marshall 2007, USA).

This momentum for cooperation between religious, state and NGO actors continued to build in the early twenty-first century and intensified after the events of September 11. Notable developments included: The Millennium World Peace Summit of Religious and Spiritual Leaders convened at the United Nations in 2000 (Eck 2001: 380; Smock 2002: 3); the creation of a Faith Zone and *A Shared Act of Reflection and Commitment by Faith Communities* at the Palace of Westminster, coordinated by the UK Government, the Archbishop of Canterbury's Office and the Interfaith Network of the UK as part of the Millennium celebrations in London (Weller, 2002: 138; Braybrooke 2007: 8); the Interreligious and International Peace Council (IIPC), first held in New York in 2003; the Tripartite Forum on Interfaith Cooperation for Peace (Tripartite Forum), which includes UN and NGO faith-based actors (Petrovsky 2003: 49; Williams in Bharat and Bharat 2007: 281) (Netpeace: Kirby 2007, USA; Knitter 2007, USA; Ramey 2007, USA); the World Economic Forum's (WEF) Council of 100 Leaders (C-100) founded in 2004 (Netpeace: Braybrooke 2007, UK; Marshall 2007, USA); and the meeting of religious leaders that took place in Russia prior to the G8 meeting in 2006/7 (Netpeace: Marshall 2007, USA).

Academia also displayed a strong interest in religion at the turn of the twenty-first century and particularly after the events of September 11 (Netpeace: Abu-Nimer 2008, USA; Amatullah 2007, USA; Dellal 2008, AUS; Dupuche 2008, AUS; Glasman 2007, UK; Laing 2007, UK; Smock 2007, USA; von Hippel 2007, USA). The Pluralism Project was established at Harvard University in Boston (Eck 2001: 17, 12, 20), while multifaith centres were established at the University of Derby in the UK (Bharat and Bharat 2007: 68–69) and Monash University and Griffith University in Australia (Baldock 1997: 198). The Religion and Peacemaking Program, established at the United States Institute of Peace in Washington in 2000, also brought religious and state actors together on issues of common security (Netpeace: Smock 2007, USA).

In addition, since September 11, state actors in Australia and the UK have increasingly initiated and financially supported multifaith activities with a focus on

Fig. 156.1 InterAction: Multifaith youth network, founded in 2009 in Melbourne, Australia (Photo by Anna Halafoff)

social inclusion and countering radicalization (Brodeur 2005: 42; Halafoff 2006: 11–12, 2007; Bouma et al. 2007: 69–74, 111–112; Braybrooke 2007: 1, 13; Bouma 2008: 13; Weller 2008: 198–199) (Netpeace: Aly 2008, AUS; Blundell 2008, AUS; Braybrooke 2007, UK; Camilleri 2008, AUS; Dellal 2008, AUS; Dupuche 2008, AUS; Hirst 2008, AUS; Jones 2008, AUS; Keyes 2007, UK; Lacey 2008, AUS; Pascoe 2008, AUS; Pearce 2007, UK; Postma 2008, AUS; Ozalp 2008, AUS; Toh 2008, AUS). A heightened emphasis on multifaith youth initiatives also emerged in the USA, UK and Australia (Fig. 156.1) with a focus on countering extremism and home-grown terrorism following September 11 and the London bombings (Netpeace: Cass 2008, AUS; Dellal 2008, AUS; Epstein 2007, USA; Mogra 2007, UK; Shashoua 2008, UK; Young 2008, AUS). There was also a significant rise of women's multifaith initiatives subsequent to the events of September 11 as women's interfaith networks were formed in the USA, Australia, the UK after September 11 and the Bali and London bombings Lohre's (2007: 11) (Netpeace: Lacey 2008, AUS; Murdoch 2008, UK).

156.3 Netpeace and the Principle Aims of the Multifaith Movement

By examining historical and contemporary multifaith developments and building on Patrice Brodeur's (2005) characteristics of the multifaith movement, Halafoff (2010) identified four principal aims of the multifaith movement as follows:

1. developing understanding of diverse faiths and of the nature of reality;
2. challenging exclusivity and normalising pluralism;

3. addressing global risks and injustices; and
4. creating multi-actor peacebuilding networks for common security.

While John Arquilla's and David Ronfeldt's (2001: 15) observation that "it takes networks to fight networks" achieved almost axiomatic status in much contemporary scholarship on counter-terrorism that emerged at the turn of the twenty-first century, Halafoff (2010) argued that *netpeace* is a preferable option to netwar for countering global risks such as terrorism and climate change. The concept of *netpeace* acknowledges the interconnectedness of global problems and solutions, and particularly the capacity of critical and collaborative networks, including state, non-state and religious actors co-committed towards common good, to solve the world's most pressing issues.

However, two key issues emerged from the *Netpeace* study that raise some concerns about the multifaith movement's future directions. Firstly, one of the greatest differences between multifaith engagement in Australia, the UK and U.S. at the time of the *Netpeace* study occurred at the level of government participation in multifaith initiatives, which was, and continues to be, far closer in Australia, than in the U.S.[1] and even in the UK. As described above, both Victorian and UK governments prioritized working collaboratively with religiously diverse communities to prevent extremism and to promote social cohesion in response to September 11 and the London bombings. However, in the UK, state bodies did not establish multifaith networks, which instead remained autonomously run by community organisations that received increased state support. Conversely, in Victoria, Australia, local councils and the Victorian State Government created networks, in collaboration with faith communities, within local and state government structures (Halafoff 2010: 197).

Due to the prominent role that communities have played within these networks and the success of these initiatives, especially as they have enabled deliberative processes between religious communities and state actors, these multifaith and multi-actor networks have been widely praised in previous studies (Halafoff 2006; Bouma et al. 2007). Actor perspectives gathered in the course of the *Netpeace* study were also largely positive regarding this increased collaboration between religious and state actors, thereby contributing to evidence documenting the success of these initiatives. However, both locally and globally some concerns were raised regarding the growing proximity between governments and religious actors, as they partner on issues of common security, particularly whether increased state financial support of multifaith initiatives might impede the critical voice of religious actors in the public sphere (Halafoff 2010: 197).

[1] Interviews for the Netpeace study were conducted prior to the Obama administration's establishment of the Centre for Faith-Based and Community Initiatives, which brings diverse religious groups in much closer proximity with state actors than previously in the USA.

Secondly, in recent times, more conservative faith communities have joined the multifaith movement as a new emphasis on practical action in place of dialogue has increased the scope of the movement's activities. On the positive side, this has led conservative actors in some cases to become more respectful of, and open-minded about, religious diversity. Conversely, the *Netpeace* study argued that the multifaith movement may be in danger of compromising its peacebuilding potential, especially in relation to promoting minority rights, such as gender equality, by giving conservative voices a platform within a movement that has been traditionally progressive. The *Netpeace* study concluded by stating that both of these issues required further investigation (Halafoff 2010: 197, 205).

156.4 Recent Multifaith Developments

Two recent multifaith developments, namely the 2009 Melbourne Parliament of World's Religions and the recent establishment of the new multifaith organization Groundswell, indicate that as the multifaith movement's focus has shifted away from terrorism to climate change and economic crisis, the movement has actually become more inclusive. They also suggest that this growing inclusivity, which encompasses non-religious as well as religious actors, combined with what could perhaps be described as a swing against conservative religiosity, is leading the movement in a more socially progressive direction, returning in a sense to its progressive roots.

156.4.1 The 2009 Melbourne Parliament of Religions

The 2009 Parliament was held in Melbourne on the 3–9th December, a city praised for its commitment to promoting positive multicultural and multifaith relations. 6,500 participants, representing over 200 religions from over 80 countries participated in over 600 programs and a number of cultural and sacred performances and events (CPWR 2010: 2). These "religions" included Indigenous ways of knowing, earth-based spiritualities, major religious traditions, new religious movements and diverse spiritualities as the organizers repeatedly stressed that everyone was welcome to attend and participate in the Melbourne PWR.

The 2009 Parliament's organizers described how since September 11, 2001, diverse spiritual and religious traditions began to be recognized as "influential and constructive forces" at the grassroots and global levels, addressing "global concerns with global solutions" creating "new approaches, resources and partnerships" to

Fig. 156.2 "Poverty must no longer be with us," a panel at Melbourne Parliament of World Religions (Photo by Anna Halafoff)

counter these common crises (CPWR 2010: 1). The 2009 Parliament's main theme was 'Making a World of Difference: Hearing each Other, Healing the Earth'. Its subthemes were: Healing the Earth with Care and Concern; Indigenous People; Overcoming Poverty in an Unequal World (Fig. 156.2); Securing Food and Water for All People; Building Peace in the Pursuit of Justice; Creating Social Cohesion in Village and City; and Sharing Wisdom in the Search for Inner Peace. It also included a Youth Program coordinated by a Youth Committee (PoR, 2009a, b) (Fig. 156.3).

These sub-themes align the 2009 PWR with the main aims of the multifaith movement described above, yet the order of these sub-themes demonstrates that environmental and Indigenous issues rose to the forefront of the movement during this period, whereas social cohesion (and the emphasis on countering radicalization) while remaining important was less prominent by 2009. And the imperative for inner reflection, while still noted, assumed a lower priority.

As global risks and concerns shifted away from terrorism towards global warming, perhaps the issue of climate change, more than any other issue, has assisted in demonstrating the interconnected nature of crises and their solutions, given that everyone is affected by climate variability. This foregrounding of environmental concerns led to expanding the conversation, not only beyond the Abrahamic faiths but also to Indigenous communities, a range of new religious movements, new and ancient spiritualities and also non-religious actors. The

Fig. 156.3 Council for a Parliament of the World's Religions (Image from http://www.parliamentofreligions.org/index.cfm?n=7&sn=115, used with permission)

2009 Parliament also generated substantially more media attention than any previous Parliament, as over 160 media outlets and nearly 200 journalists covered its proceedings in the local and international press (CPWR 2010: 12). Moreover it received substantial financial support from local, state and federal Australian governments, community organizations and local and global foundations (CPWR 2010: 20–21). Therefore the multifaith movement continued to expand during this period, not only in terms of the numbers and diversity of participants, but also the level of state and public support it received and also its media profile.

156.4.2 Groundswell

Another new feature of twenty-first century multifaith engagement has been the impetus to include non-religious actors in multifaith peacebuilding initiatives. These relatively recent developments were spearheaded by the Interfaith Youth Core (IFYC) in Chicago and by Harvard University's Humanist Chaplain Greg

Epstein (Netpeace: Epstein 2007, USA). More recently "faithiest" Chris Stedman, who previously worked at the IFYC and then became the Interfaith and Community Service Fellow for Harvard's Humanist Chaplaincy has described the benefits of partnering between religious and non-religious actors on service projects and social justice initiatives (Stedman forthcoming).

Groundswell, a new socially progressive multifaith movement headed by Sikh filmmaker and activist Valarie Kaur (ValarieKaur 2012) also stresses the importance of partnerships between religious and non-religious people and communities. Groundswell includes "the secular, the seeking, and people of faith," mobilizing people 'to take strategic social action around justice issues of the hour' (Auburn Seminary 2012). These issues include campaigning for LGBTI dignity and equality, and ending youth sex trafficking (Groundswell 2012) (Fig. 156.4).

Ten years after September 11, Groundswell's emergence indicates that significant changes are afoot within the multifaith movements' next generation. As noted above, in the early 2000s the progressive nature of the late twentieth century multifaith movement was in danger of being compromised as the multifaith movement expanded to include more conservative actors. Groundswell appears to be reversing this trend by taking a publicly progressive stance on LGBTI and young people's rights. While this might risk alienating more conservative religious people and groups, Groundswell seems determined to campaign for equal rights of all minorities, aligning it with the multifaith movements original aims.

156.5 Conclusions

The multifaith movement has always been responsive to crisis events and the focus of the movement has shifted accordingly. The focus of the movement also influences which actors have a central place at the table, so to speak, and, therefore, determines not only who is included in the conversation, but also the nature of the dialogue.

Briefly examining the history of the multifaith movement, and the recent 2009 Parliament of World's Religions and the formation of Groundswell in 2011, suggests that while the multifaith movements aims have remained relatively consistent over the past 120 years, it appears to be entering a new more inclusive era, expanding to involve Indigenous ways of knowing, earth based spiritualities, major religions, new religious movements and non-religious actors. This presents new opportunities and challenges for the multifaith movement as it may well have to change its labeling/branding to reflect these new developments and also increasingly address the tensions between religious freedoms and respecting the rights of others.

These observations require further investigation in order to be substantiated, yet offer some evidence that the multifaith movement continues to be responsive to external events and crises, and also perhaps of the movements' underlying socially progressive agenda.

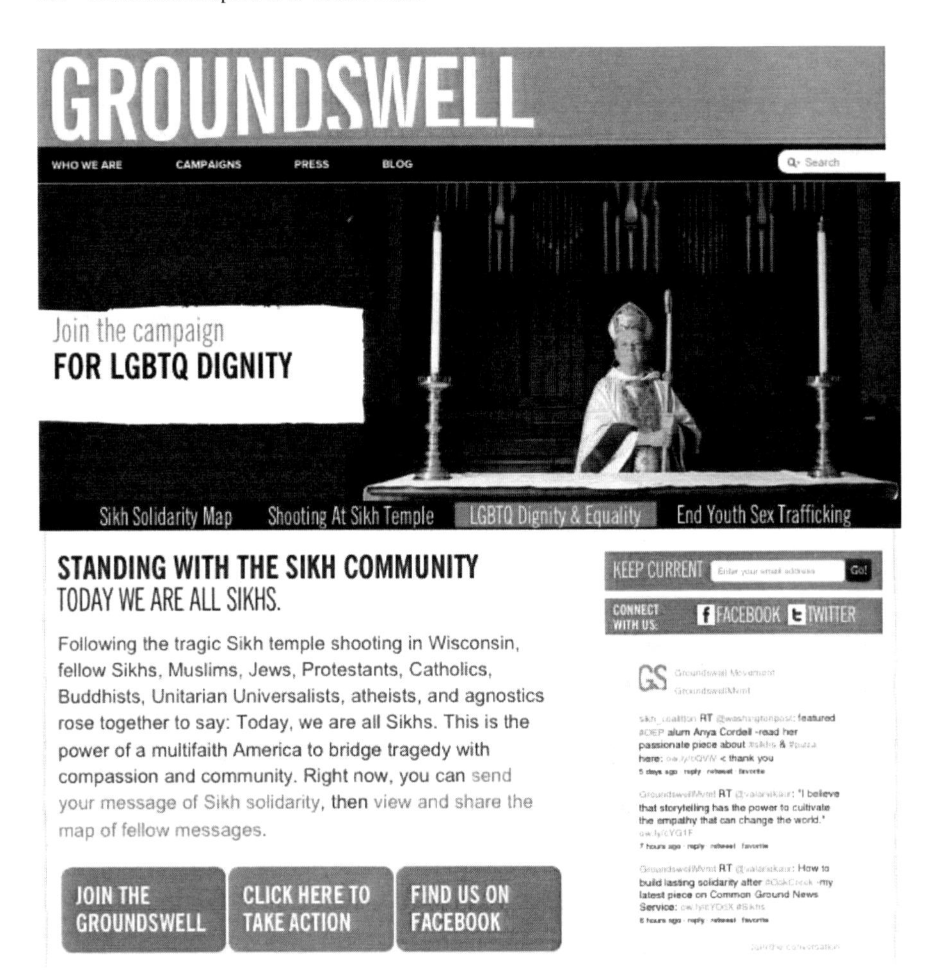

Fig. 156.4 Groundswell website (Image from Groundswell, used with permission)

References

Arquilla, J., & Ronfeldt, D. (2001). The advent of netwar (Revisited). In J. Arquilla & D. Ronfeldt (Eds.), *Networks and netwars: The future of terror, crime, and militancy* (pp. 1–25). Santa Monica: RAND.

Auburn Seminary. (2012). *Religion and justice: Groundswell*. Retrieved July 20, 2012, from www.auburnseminary.org/religion-and-justice

Bainbridge, W. S. (1997). *The sociology of religious movements*. New York: Routledge.

Baldock, J. (1997). Responses to religious plurality in Australia. In G. D. Bouma (Ed.), *Many religions, all Australian: Religious settlement, identity and cultural diversity* (pp. 193–204). Melbourne: The Christian Research Association.

Barrows, J. H. (Ed.). (1893). *The world's parliament of religions*. Chicago: The Parliament Publishing Co.

Bauman, Z. (2006). *Liquid fear*. Cambridge: Polity Press.
Beck, U. (2006). *The cosmopolitan vision*. Cambridge: Polity Press.
Beckford, J. A. (2003). *Social theory & religion*. Cambridge: Cambridge University Press.
Bharat, S., & Bharat, J. (2007). *A global guide to interfaith: Reflections from around the world*. Winchester: O Books.
Bouma, G. D. (2006). *Australian soul: Religion and spirituality in the twenty-first century*. Cambridge: Cambridge University Press.
Bouma, G. D. (2008). The challenge of religious revitalization and religious diversity to social cohesion in secular societies. In B. S. Turner (Ed.), *Religious diversity and civil society: A comparative analysis* (pp. 13–25). Oxford: The Bardwell Press.
Bouma, G. D., Pickering, S., Halafoff, A., & Dellal, H. (2007). *Managing the impact of global crisis events on community relations in multicultural Australia*. Brisbane: Multicultural Affairs Queensland.
Braybrooke, M. (1992). *Pilgrimage of hope: One hundred years of global interfaith dialogue*. London: SCM Press Ltd.
Braybrooke, M. (1996). *A wider vision: A history of the world congress of faiths*. Oxford: Oneworld.
Braybrooke, M. (2007). *Interfaith witness in a changing world: The world congress of faiths, 1996–2006*. Abingdon: Braybrooke Press.
Brodeur, P. (2005). From the margins to the centers of power: The increasing relevance of the global interfaith movement. *Cross Currents, 55*(1), 42–53.
Brodeur, P., & Patel, E. (2006). Introduction: Building the interfaith youth movement. In E. Patel & P. Brodeur (Eds.), *Building the interfaith youth movement: Beyond dialogue to action* (pp. 1–14). Oxford: Rowman & Littlefield Publishers, Inc.
Cahill, D., Bouma, G. D., Dellal, H., & Leahy, M. (2004). *Religion, cultural diversity and safeguarding Australia*. Canberra: Department of Immigration, Multicultural and Indigenous Affairs.
Council for a Parliament of the World's Religions (CPWR). (2010). *Parliament of the world's religions report 2009*. Melbourne: Council for a Parliament of the World's Religions. Retrieved July 20, 2012, from www.parliamentofreligions.org/_includes/files/reports/PWR2009-Report.pdf
Croucher, P. (1989). *A history of Buddhism in Australia 1848–1988*. Kensington: New South Wales University Press.
Eck, D. L. (2001). *A new religious America: How a "Christian country" has become the world's most religiously diverse nation*. New York: HarperOne.
Groundswell. (2012). Groundswell Home Page. Retrieved July 20, 2012, from www.groundswell-movement.org/
Habermas, J. (1981). New social movements. *Telos, 49*, 33–37.
Halafoff, A. (2006) UnAustralian values. In *UNAustralia*. The Cultural Studies Association of Australasia Annual Conference Proceedings. Canberra: University of Canberra.
Halafoff, A. (2010). *Netpeace: The multifaith movement and common security.* PhD Thesis, Melbourne: Monash University.
Kaldor, M. (1999). *New & old wars: Organized violence in a global era*. Cambridge: Polity Press.
Keller, C., & Kearns, L. (2007). Introduction: Grounding theory – Earth in religion and philosophy. In L. Kearns & C. Keller (Eds.), *Ecospirit: Religions and philosophies for the earth* (pp. 1–20). New York: Fordham University Press.
Kirkwood, P. (2007). *The quiet revolution: The emergence of interfaith consciousness*. Sydney: ABC Books.
Kniss, F., & Numrich, P. D. (2007). *Sacred assemblies and civic engagement: How religion matters for America's newest immigrants*. New Brunswick: Rutgers University Press.
Lohre, K. (2007). Women's interfaith initiatives in the United States post 9/11. *Interreligious Insight, 5*(2), 11–23.

Marshall, K., & Keough, L. (2004). *Mind, heart, and soul in the fight against poverty*. Washington, DC: The World Bank.

Marty, M. E., & Appleby, R. S. (1992). *The glory and the power: The fundamentalist challenge to the modern world*. Boston: Beacon.

McCarthy, K. (2007). *Interfaith encounters in America*. New Brunswick: Rutgers University Press.

Nash, R. (1996). The greening of religion. In R. S. Gottlieb (Ed.), *This sacred earth: Religion, nature and environment* (pp. 194–229). New York: Routledge.

National Council of Churches, Committee on Regional & Local Ecumenism. (1980). *Survey of Interfaith/Inter-religious Councils in the United States*. www.nain.org/library/councils.htm

Netpeace Study participants: Prof. Mohammed Abu-Nimer, Director, Peacebuilding and Development Institute, International Peace and Conflict Resolution Program, School of International Service, American University, Washington, USA; Waleed Aly, Lecturer, Global Terrorism Research Centre, School of Social and Political Inquiry, Monash University, Melbourne, Australia; Nurah Amatullah, Executive Director of the Muslim Women's Institute for Research and Development, New York, USA; Dr. Patricia Blundell, Coordinator Chaplaincy, Griffith University, Brisbane, Australia; Rev. Professor Marcus Braybrooke, President, World Congress of Faiths, Oxford, UK; Rev. Dr. Chloe Breyer, Executive Director, The Interfaith Centre of New York, New York, USA; Prof. Joseph Camilleri, Director, Centre for Dialogue, La Trobe University, Melbourne, Australia; Josh Cass, Youth Worker, Encounter, London, UK; Dr. Bulent (Hass) Dellal, Executive Director, Australian Multicultural Foundation, Melbourne, Australia; Joyce S. Dubensky, Executive Vice-President and CEO, Tanenbaum Centre for Interreligious Understanding, New York, USA; Fr. Dr. John Dupuche, Chair, Catholic Interfaith Committee of the Catholic Archdiocese of Melbourne, Melbourne, Australia; Greg M. Epstein, Humanist Chaplain at Harvard University, Cambridge (interviewed in Chicago, USA at the Interfaith Youth Core Conference), USA.; Rev. Dirk Ficca, Executive Director, Council for a Parliament of the World's Religions, Chicago (interviewed in Melbourne, Australia), USA; Rev. Charles Gibbs, Executive Director, United Religions Initiative, San Francisco, USA; Dr. Maurice Glasman, Director, Faith and Citizenship Program, London Metropolitan University, London, UK; Rev. Fletcher Harper, Executive Director, GreenFaith, New Brunswick, USA; Sherene Hassan, Interfaith Officer, Islamic Council of Victoria, Melbourne, Australia; Di Hirsh, Interfaith and Intercultural Chair, National Council of Jewish Women of Australia, Melbourne, Australia; Jeremy Jones, Co-Chair of the Australian National Dialogue of Christians, Muslims and Jews, Sydney, Australia; Assoc. Prof. Laurel Kearns, Associate Professor, Sociology of Religion and Environmental Studies, Drew Theological School and Graduate Division of Religion, Drew University, Madison, USA; Simon Keyes, Director, St. Ethelburga's Centre for Reconciliation and Peace, London, UK; Sr. Joan Kirby, United Nations Representative, The Temple of Understanding, New York, USA; Prof. Paul F. Knitter, Paul Tillich Professor of Theology, World Religions and Culture, Union Theological Seminary, New York, USA; Josie Lacey, Convener, Women's Interfaith Network, Sydney, Australia; Catriona Laing, Project Manager of The Cambridge Inter-Faith Programme, Faculty of Divinity, University of Cambridge (interviewed in London, UK), UK; Katherine Marshall, Senior Fellow, Berkley Center for Religion, Peace and World Affairs, Georgetown University, Washington, USA; Shaykh Ibrahim Mogra, Chair, Interfaith Relations Committee, Muslim Council of Britain, London, UK; Alison Murdoch, Director, Foundation for Developing Compassion and Wisdom, London, UK; Mehmet Ozalp, Chief Executive Officer, Affinity Intercultural Foundation, Sydney, Australia; Dr. Susan Pascoe, Chair, Australian National Commission for UNESCO, Canberra (interview conducted in Melbourne), Australia; Dr. Eboo Patel, Executive Director, Interfaith Youth Core, Chicago, USA; Brian Pearce, Director, The Inter Faith Network for the UK, London, UK; Maureen Postma, General Secretary, Victorian Council of Churches, Melbourne, Australia; Ibrahim Abdil-Mu'id Ramey, Director, Human and Civil Rights Division, Muslim American Society Freedom (MAS Freedom), Washington, USA; Stephen Shashoua, Director, The Three Faiths Forum, London, UK; Dr. David R. Smock, Vice-President, Centre for Mediation and Conflict Resolution, Religion and Peacebuilding

Center of Innovation, United States Institute of Peace, Washington, USA; Rev. Helen Summers, Director, The Interfaith Centre of Melbourne, Melbourne, Australia; Prof. Toh Swee-Hin, Director, Multi-Faith Centre, Griffith University, Brisbane; Prof. John O. Voll, Associate Director, Prince Alwaleed Bin Talal Center for Muslim-Christian Understanding, Georgetown University, Washington, USA; Dr. Karin von Hippel, Director, Post-Conflict Reconstruction Project, Centre for Strategic and International Studies, Washington, USA; Rachel Woodlock, Researcher, Centre for Islam and the Modern World, School of Social and Political Inquiry, Monash University, Melbourne, Australia; Elizabeth Young, Student and Multifaith Youth Worker, Flinders University, Adelaide (interviewed in Melbourne), Australia.

Niebuhr, G. (2008). *Beyond tolerance: Searching for interfaith understanding in America*. New York: Viking.

Parliament of Religions (PoR). (2009a). *Themes*. Retrieved July 30, 2012, from www.parliamentofreligions.org/index.cfm?n=7&sn=18

Parliament of Religions (PoR). (2009b). *Youth*. Retrieved July 30, 2012, from www.parliamentofreligions.org/index.cfm?n=7&sn=5

Pluralism Project. (2006). *Resources by tradition: Interfaith*. The President and Fellows of Harvard College & Diana L. Eck, The Pluralism Project. www.pluralism.org/resources/tradition/index.php

Rockefeller, S. C. (1992). Faith and community in an ecological age. In S. C. Rockefeller & J. C. Elder (Eds.), *An interfaith dialogue, spirit and nature: Why the environment is a religious issue* (pp. 139–172). Boston: Beacon.

Rockefeller, S. C., & Elder, J. C. (1992). Introduction. In S. C. Rockefeller & J. C. Elder (Eds.), *An interfaith dialogue, spirit and nature: Why the environment is a religious issue* (pp. 1–14). Boston: Beacon.

Smock, D. R. (2002). Introduction. In D. R. Smock (Ed.), *Interfaith dialogue and peacebuilding* (pp. 3–12). Washington, DC: United States Institute of Peace Press.

Stedman, C. (forthcoming). *Faitheist: How an atheist found common ground with the religious*. Boston: Beacon Press.

Tucker, M. E. (2007). Ethics and ecology: A primary challenge of the dialogue of civilizations. In L. Kearns & C. Keller (Eds.), *Ecospirit: Religions and philosophies for the earth* (pp. 495–503). New York: Fordham University Press.

ValarieKaur. (2012). ValarieKaur Home Page. Retrieved July 20, 2012, from www.valariekaur.com/

Weller, P. (2008). *Religious diversity in the UK: Contours and issues*. London: Continuum.

Wolfensohn, J. D. (2004). Foreword. In K. Marshall & L. Keough (Eds.), *Mind, heart, and soul in the fight against poverty* (pp. xi–xiii). Washington, DC: The World Bank.

World Conference of Religions for Peace (WCRP). (1989). *Melbourne declaration*. Geneva: WCRP.

Chapter 157
Effecting Environmental Change: The Challenges of Transnational Environmental Faith-Based Organizations

Deborah Lee and Lily Kong

157.1 Introduction: Religion and Environment

Religion has a potential and sometimes significant role in shaping religious adherents' environmental beliefs and even actions (see, for example, Glacken 1967; Kay 1989; Cooper and Palmer 1998; Jamieson 2001; McFague 2008). Kong (2010) highlights two broad ways in which this link has been studied. One body of research takes a discursive approach by focusing on the religious belief systems that underpin human-nature relations. Examples include the religious traditions of St Francis of Assisi and St Benedict in Christianity which preach the importance of living in harmony with the natural world and the wise use of natural resources (Livingstone 2002: 351), the notion of *ahimsa* (prohibition of violence to all living things) that influences Hinduism, Buddhism and Jainism, as well as other injunctions in religions such as Islam (Akhtaruddin 1997). Much of this work has tended to focus on analysis of theological texts and what they say about the environment, rather than on actual human behavior and how that may have religious influences (Proctor 2006). In fact, while selected religious teachings may advocate environmental respect, such prescriptions may not translate into behavior (Mawdsley 2006: 383). A second body of research thus focuses on praxis – examining how "religion" (both as institution and belief) impacts the environment. As a social institution, religious institutions (such as churches or religiously-inspired NGOs) may give attention to environmental causes, with common religious beliefs forming the basis for gathering adherents, and the religious institutional framework forming the basis on which they may "negotiate their ways through religious ideology, organizational structures/strictures, and different international cultures and contexts" (Kong 2010: 767). At the same time, the religious symbolism of places may provide the rationale for

D. Lee (✉) • L. Kong
Department of Geography, National University of Singapore,
AS2, 1 Arts Link, Singapore 11 7570, Singapore
e-mail: lee.debs@gmail.com; lilykong@nus.edu.sg

S.D. Brunn (ed.), *The Changing World Religion Map*,
DOI 10.1007/978-94-017-9376-6_157

environmental protection and conservation (for example, sacred forests in India) (Tomalin 2009) or sacred cities like Medina for Muslims (Akhtaruddin 1997).

On the other hand, scholars have also disputed the view that religion has positive impacts on the environment. The classic study that propagates this view is White's (1967: 1205) study of Western Christianity, which, he argues, promulgates "an implicit faith in perpetual progress" which leads human beings to destroy the environment.

In more recent times, both practical and academic interest has grown around the question of how religions can be engaged to help address environmental problems, particularly those that are transnational in nature, such as climate change. Attention has thus turned to transnational religious networks and their potential to effect change, given how they are both extensive in scope and grounded in the everyday lives of religious adherents. Transnational environmental faith-based organisations (FBO), in particular, offer potential to effect change, and thus deserve closer scrutiny and understanding. The growing interest in transnational environmental FBOs is an extension of the interest in non-faith based environmental NGOs and their effects (see, for example, Princen and Fingers 1994; Jänicke 2006). But the boundary-crossing and religious nature of transnational environmental FBOs potentially offers something more. As Gardner (2002: 5) argues, religions "have the ear of multitudes of adherents, often possess strong financial and institutional assets, and are strong generators of social capital," so religion can be a strong motivating factor for environmental action. At the same time, their transnational structure also enables them to mobilize individuals in different contexts with the common aim of environmental protection (Kong 2010).

157.2 Study Objectives

While the number of environmental FBOs has been increasing since the 1980s (Kearns 1997; Gardner 2002), these organizations are starting to take on a transnational form with a *modus operandi* distinct from organizations whose operations remain within national boundaries. Given the potential that transnational environmental FBOs have in effecting positive environmental change, we are motivated by a desire to understand how they work to enlarge their network.

Our research is motivated by three specific objectives. First, we seek to understand how transnational environmental FBOs endeavour to create partnerships in a new destination. Second, we wish to examine whether and how the (religious) identities of transnational actors are accepted, negotiated and resisted by local actors and the implications of such transnational-local interactions for the embedding of the transnational environmental FBO. Finally, we seek to understand how historically and geographically contingent factors in a locality may influence the relative success in the embedding and development of a transnational organization's activities. Factors such as a locality's history of religious participation in "secular" activities, the "policing" of the boundaries of such activities by the state or secular groups, and the history of FBO success in effecting change, may all have impacts on the entry of a new transnational environmental FBO by presenting opportunities for learning and cooperation, or greater competition for resources.

We examine the case of a transnational environmental FBO (Creation-Carers)[1] as it seeks to establish a national movement in a new country – Singapore. Through an examination of its nascent efforts, we aim to understand the opportunities and challenges that this organization faces. While this remains an analysis of a specific FBO with its particular transnational networks seeking to embed and develop in a distinct destination, we hope that our analysis will be helpful in understanding more broadly the conditions which confront transnational environmental FBOs as they seek to widen their networks and operational localities, and the strategies which they might use in enlarging their transnational presence.

The chapter will be divided into four sections hereafter. The first will introduce the study context, presenting readers with the background about Creation-Carers and the situation in secular Singapore regarding the relationship between religion and the state, and the place of FBOs in the larger polity. The second will introduce the concept of embeddedness – borrowed from economic geography – as a useful framework for understanding the efforts of Creation-Carers in Singapore. The third will constitute an analysis of the empirical material collected from interviews with key actors, and the final section will conclude with broader observations about the relationship between religious institutions and environmental change.

157.3 Study Context

Founded in 1983 and headquartered in England, Creation-Carers is a Christian organization that focuses on environmental action. The main aim for Creation-Carers is environmental protection, using Biblical scriptures as guiding principles. Its activities include community garden farming where visitors learn about sustainable agriculture and study Biblical scriptures pertaining to environmental action, and restoring environmentally-damaged wetlands as a symbolic way of enacting the renewal of the earth that Creation-Carers actors believe God will do in the future.

Creation-Carers International, a network of Creation-Carers national movements, was established in 2001. As of 2012, there were 19 such national movements, led by founder/director Joseph Scott.[2] The Creation-Carers International team headquartered in England manages finances and administration, as well as national movements, which are bound by the 'Five Commitments' – (to be) Christian, (to support) Conservation, (to develop) Community, (to be) Cross-cultural and (to seek) Cooperation. The Creation-Carers' International Trustees govern Creation-Carers

[1] All names of organizations and interviewees have been replaced with pseudonyms.

[2] These 19 countries are Brazil, Bulgaria, Canada, Czech Republic, Finland, France, Ghana, India, Kenya, Lebanon, Netherlands, New Zealand, Peru, Portugal, South Africa, Switzerland, Uganda, UK and USA.

International, and the International Council of Reference promotes Creation-Carers through their professional work. Trustees possess local experience and are well-known in their respective professions (such as in marketing or fund-raising), while the International Council of Reference comprises world-renowned theologians and scientists, some of whom are faculty at theological seminaries or hold key positions in the International Union for Conservation of Nature.

Besides the 19 other countries in which Creation-Carers is currently located, the organization sought to establish a national movement in Singapore in 2010. Prior to that, since 2008, projects had begun to be introduced in Singapore in the hope of leading up to the establishment of a national movement in the city-state. It is important to understand the specific context of Singapore, in order to fully appreciate the promises and challenges of this intention.

Singapore is characterised by a high degree of religious heterogeneity, with the population comprising in 2010 Chinese religionists (44.2 %); Christians (18.3 %); Muslims (14.7 %); and Hindus (5.1 %). In addition, 0.7 % of the population adhere to other religions, while 17 % have no religion (Singapore Department of Statistics 2010). Because of such variation, the state has adopted a secular position characterised by four specific tenets. Singapore is a secular state in the sense that there is no official state religion. The state allows for freedom of worship. There is official commitment to multi-culturalism[3] where all cultural groups and in this instance, all religious groups, are treated fairly without prejudice to any group in particular, whether they are majority or minority groups. In turn, religion and politics must be kept strictly separate. Religious groups should not venture into politics and political parties should not use religious sentiments to gather popular support. The Maintenance of Religious Harmony Act is designed to ensure that this is so. It allows the relevant Minister to issue prohibition orders should any individual engage in any of four categories of harmful conduct. These are where a person causes feelings of enmity or hatred between different religious groups; if, under the guise of religion or propagating religious activity, one carries out political activities for promoting a political cause or the cause of any political party; carrying out subversive activities under the guise of propagation of religion; and instigating and provoking feelings of disloyalty or hatred against the President or the government.

The clear separation between religion and politics, however, does not mean that religion is necessarily viewed as opposed to national interests in this island-state. Instead, as Hill and Lian (1995) highlight, religion is considered an important part of Singapore's nation-building strategy, though in specified ways. This was clearly articulated in the late 1980s by the then Prime Minister Lee Kuan Yew, who

[3] Various versions of this concept have been discussed by sociologists. For example, Benjamin (1976: 115) discussed the concept of "multiracialism" in the context of Singapore as the "ideology that accords equal status to the cultures and ethnic identities of the various 'races' that are regarded as comprising the population of a plural society." Siddique (1989) discussed the "4Ms": multiracialism, multilingualism, multiculturalism and multireligiosity. To her, multireligiosity "acknowledges a societal situation in which a number of religions are practised, but none is officially recognized as paramount." It is distinguished by a "religious populism supportive of moral order" (Siddique 1989: 565).

emphasised that religious groups should look after the spiritual, moral, and social well-being of their followers but should leave the economic and political needs of people to nonreligious groups such as political parties (Lee 1988). As Goh (2005: 43) highlights:

> In Singapore's tightly-governed socio-political climate, there are fewer opportunities for the churches to play a socially active role, in contrast to places like Hong Kong and the Philippines where the churches often play a significant role in social activism, civil rights demonstrations, and support for political candidates.

Whether it is such official circumscription or a history of social service provision perhaps best exemplified by Christian groups from an inherited colonial past (Mathews 2008), the involvement of religious groups in non-religious areas in Singapore has largely been focused on social provision, such as education, healthcare, eldercare, and aftercare. Indeed, they play an important role in Singapore society, and all the major religious groups engage in such social service provision.

Another dimension of the influence of Singapore's Christian community can be seen from Singapore's status as a hub of Christian activities – attributable to both religious and extra-religious reasons. Singapore has been dubbed by renowned Christian leaders Reverend Billy Graham and Reverend Yong Gi Cho as the "Antioch of Asia," 'alluding to a multi-ethnic city…that was the cradle of Christianity in the first century A.D.' (DeBernardi 2008: 120–121). Renowned Christian organizations such as the Overseas Missionary Fellowship have established their headquarters or regional offices in Singapore. Goh (2005: 44) argues that one reason for the high concentration of transnational Christian organisations is that they "follow their secular counterparts in choosing Singapore for its stability, strategic location and religious tolerance." These organisations depend on transnational networks, developed largely as a result of the Singapore government's attempts at making Singapore a "hub" for telecommunications, businesses and tourism to remain connected with their operations in other countries (Goh 2009). Such linkages attract other transnational Christian organisations seeking to locate in Singapore, including Creation-Carers.

The efforts to establish in Singapore have taken a variety of forms. First, a Creation-Carers International staff, Hannah Koo, and a volunteer, Thomas Lu, are Singaporeans who organized the first Creation-Carers activity in Singapore together with Singapore Theological Seminary – a conference concerning Christian environmentalism. From 2005 to 2008, Koo and Lu worked with Creation-Carers Canada and produced films for Creation-Carers International. Koo and Lu have successfully completed several activities in Singapore since late 2008. They delivered talks on Christian environmentalism to several Christian tertiary student groups and churches. They also organized a major two-day Creation-Carers Conference in July 2009, in partnership with Christ College Singapore, one of 14 local theological schools. Joseph Scott, the founder/director of Creation-Carers, delivered the keynote lecture and there was a plenary session comprising panellists representing three theological schools in Singapore. There was also a strategy session at the end of the second day. The objectives were twofold: to brainstorm the future of Creation-Carers in Singapore and a networking session for (potential) local partners.

Not all Creation-Carers activities have been successfully completed. Some projects – such as a nature walk for a local church – were not realized because of changes in the church's schedule. Other planned projects, such as a community garden at a home for disadvantaged children, suffered from a lack of funds. Yet other projects, such as plans for an environmental education business, were unsuccessful due to a lack of partners.

To understand the opportunities and challenges facing such a new transnational environmental FBO as it seeks to establish in Singapore, we will examine the relationships its key leaders were able to tap on and develop and the networks they were able to access (or not). To provide a frame through which the empirical observations can be discussed, we introduce some conceptual tools in the next section.

157.4 Embeddedness, Social Capital and Trust

'Embeddedness', social capital and trust are useful concepts to frame our empirical discussion. They are useful lenses for understanding the opportunities and obstacles that Creation-Creators faced in the nascent stages of establishing a presence in Singapore. 'Embeddedness' is a concept developed out of economic geography and economic sociology where the concern is with where entrepreneurs are located within network relations and how these relations are influenced by historical factors that continue to shape entrepreneurial activity (Sunley 2008). Another interest centres on how government policies can help to attract certain segments of the network to particular cities in order to create new network hubs (Henderson et al. 2002).

Hess (2004) surfaces three important dimensions of embeddedness: *network, societal and territorial embeddedness.* Network embeddedness comprises both relational and structural dimensions. The relational dimension focuses on an individual actor's relations with other actors, while the structural dimension is concerned with broader structures that networks are situated in, such as government policies. Studies of network embeddedness are concerned with how relations are constructed, whether the relations are resilient, the role of trust-formation in these relations, and the ways in which broader structures affect the development of such relations. The second dimension of embeddedness is societal embeddedness. It is concerned with how an actor's social and personal background enables or constrains the embedding process. The third dimension of embeddedness – territorial embeddedness – shifts the focus from the actors to the locations, particularly, the conditions and processes operating in the destination location, and the ways in which they impact on the degree of anchoring of the transnational organization in a new destination.

Especially in relation to understanding the relational aspects of network embeddedness, the concepts of social capital and trust are further helpful frames for analyzing our empirical data. Various scholars have sought to develop the concept of "social capital," each critiquing and building on earlier efforts. Without going into the debates about shortcomings of various conceptualisations, we highlight here those salient ideas that are helpful to our understanding of the case at hand. Putnam's (1993) conceptualisations of social capital highlight two useful ideas. One is the role of social networks, or, as Jenkins (2009) has described it – the "friends of friends" connections. Another is Putnam's focus on trust, which 'denotes a wider facility for co-operative behaviour' (Hems and Tonkiss 2000: 6), particularly for civil societal groups. Mark Granovetter, who wrote the 1985 seminal paper on embeddedness, argues for the need to "recognise the importance of concrete personal relations and networks of relations in generating trust, in establishing expectations, and in creating and enforcing norms" (Coleman 2000: 15). Social capital is a matter of '*enforceable* trust' (Portes 1998: 8, emphasis his). In other words, trust is not a feel-good effect of social capital, but is instead a modality of power with the ability to enact changes. This acknowledgement that power relations are involved points to the need to recognize another theorist of social capital – Pierre Bourdieu – who recognises that patterns of resource access 'tend to reproduce existing distributions of power and capitals' (Bebbington 2009: 165). Bourdieu thus acknowledges that structures and institutional behavior (including state structures and religious institutions, for instance) influence the development of social capital.

Given these conceptual scaffolds, we turn now to empirical analyses of our particular case. We organize our discussion in three sections, according to each type of embeddedness discussed above.

157.5 Network Embeddedness: Leveraging Institutional and Personal Relationships

In this section, we focus on an early effort on the part of Creation-Carers to organize an activity and introduce their ideas more broadly in Singapore. This was done through connecting with major seminary leaders in the organization of an environmental conference. We use the notion of network embeddedness with its related ideas of social capital and trust to understand how the partnership was forged.

The partnership with major seminary leaders in Singapore in the organization of an environmental conference was successful in seeding the faith-based environmental work in Singapore principally because of the web of personal relationships centered on trust in a key Canadian institution associated with Creation-Carers and a key leader of Creation-Carers International (Joseph Scott).

When approached to be part of the organizing efforts, Christ College Singapore (CCS) was ready to make the commitment because several seminary leaders in Singapore, including CCS leaders were part of a network of direct and indirect relationships with individuals and institutions associated with Creation-Carers and the conference in question. In particular, Scott, as leader of Creation-Carers International, and a plenary speaker identified for the conference, was a key actor within an evangelical Christian seminary in Canada (Sovereign College) that local seminary leaders had a network of links with. These local leaders had either graduated from, taught at, been taught by, had heard talks by or heard of Scott (and others linked to Scott via the same means as they themselves). For those who had been at Sovereign College, trust in the institution was based on their own experience of the institution (see Uslaner 2008). For those who had not been at Sovereign College but knew others who had been or who knew Scott, the "friends of friends" (Jenkins 2009) network became important in developing a sense of trust. As Jenkins asserts: "[y]ou don't actually have to know someone personally to be in the same network, although you do know *about* them and, through mediators, *potentially* know them" (2009: 149, original emphasis). For those who had heard (of) Scott, his reputation and social capital afforded him credibility and agency. In this sense, Scott could tap into the convergence of myriad transnational ties to help facilitate his organization's embedding process in Singapore. All these networks, built on social capital and trust, and in turn contributing to them, are well summarized by one of the local seminary leaders, who says:

> It is really more personal. It is relational. I know Josiah Eliot [my doctoral dissertation supervisor at Sovereign College]; we are good friends. If he recommends someone, it should be good. Josiah mentioned that Joseph Scott teaches a course at Sovereign College. I know Sovereign [College] would not just invite anyone who has no experience in the area he is talking about. So for me, it is a matter of trust. This person I know knows him, so I know he knows. (personal interview)

In seeking to embed in a new destination, transnational organizations, including transnational environmental FBOs, will stand a higher chance of success if they can draw on a network of relationships, both institutional and personal, direct and indirect. In this regard, Creation-Carers found some appropriate transnational networks that were helpful in their cause, and Singapore's long history of having its citizens train overseas – whether in secular or religious education – helped to create some of these networks.

157.6 Societal Embeddedness: Foreign Actors, Postcolonial Consciousness

Societal embeddedness refers to the ways in which the social and personal background of key actors can either enable or constrain the embedding process. In this section, we examine the ways in which certain aspects of the background of key actors involved in Creation-Carers International enable its embedding while other dimensions became constraining.

Creation-Carers' identity as an organization that is part of a global Evangelical Protestant community and supported by numerous key Evangelical leaders in the worldwide community has enabled the organization to gain "insider" status in Singapore's Christian community and hence facilitated its embedding process. Further, the personal background of some of the key actors helped to open the doors. First, Scott himself was a former Anglican clergyman. This helped conference participants to identify with him as fellow members of a transnational denominational network. As one Creation-Carers supporter in Singapore shared:

> It is very good to have the founder come ... he is an Anglican minister. I was thinking, how nice if he meets some Anglican vicars because people tend to have the mindset that if this is an Anglican minister, [as an Anglican] I am more likely to hear [him out].

Second, the personal backgrounds of members of Creation-Carers' International Council of Reference improved the likelihood of their reception. Some of these individuals already had an established reputation in Singapore. For instance, one of the key actors in the Council was a highly influential Evangelical leader who had been named one of *Time Magazine's* 100 most influential people. Another is the Secretary for Social Engagement of the International Evangelical Youth Fellowship, travelling frequently to different parts of the world to deliver talks on Christian social involvement, and is a highly respected individual. Yet others are renowned in their field of expertise – mainly theological education or scientific research. Their endorsements, like that of Scott's, helped the Creation-Carers attain "insider" status. Many would-be supporters of the FBO in Singapore believed it to be a trustworthy organization because respected thinkers and leaders supported it. Such social and personal background is crucial to the process of trust-formation, especially in transnational settings where meetings between actors may be brief, and trust built on understanding derived from interaction is difficult to achieve. Borrowing from Routledge and Cumbers (2009), these individuals enabled a certain degree of societal embedding because they act as "imagineers" – transnational NGO actors who explain the concept of an NGO's network (the "imaginary") to other (potential) members. By virtue of who they are, they enable the fostering of trust among new members (Bandy and Smith 2005). Scott acknowledges that, more ideally, those who decide to subscribe to and support Creation-Carers should do so because they believe in the work of Creation-Carers, but he recognizes the power of influential endorsers:

> For many people, it is sufficient for them to know that those thought leaders are on board. And they find that easier than thinking for themselves. It shouldn't be like that but it is. (personal communication)

On the other hand, the "insider" status is not always guaranteed because many of the key actors are viewed as Westerners leading a "Western" organisation with "foreign" theology. This has constrained Creation-Carers' embedding process. The process of fostering trust in a new local context, even for key actors with social capital has to contend with the history of Christianity and Christian identities in Singapore, which arrived with the British colonials. While Scott's influence portrayed a coherent organisational identity, the effect was ambivalent because of the exclusion of other identities, particularly local voices. Further, having "Western" imagineers as key representatives perpetuates the idea that Creation-Carers is a "Western" organisation bringing 'Western' values in, rather than a transnational FBO seeking to establish local movements with local leadership. As one potential Creation-Carers partner observed:

> I am not too keen on connecting with Creation-Carers so fast because they are a Western organisation and they have their resources…We have been, for the past seven years, raising the flag for Asian environmentalism. Why would we want all the good work to go under a Western organisation? (personal interview)
>
> I don't fully agree with [him]. He is not from Singapore but comes to us speaking about integral missions, not knowing the issues we face. And then he says 'oh the issues in Singapore are for you to think through,' but that is unsatisfactory. I know he is a strong supporter of Creation-Carers, but he does not know the environmental challenges we face in urban Singapore…so how does integral mission feature in Singapore? I don't know. (personal interview, Creation-Carers' potential partner)

The possibility that the development of Creation-Carers in Singapore may be met with less enthusiasm than the leaders desire may be attributed at least in part to a sense of postcolonial independence, in which local actors recognize the embedding process is not always a "positive, benign form of development based on collaboration and equality" (Hess 2009: 428). Societal embeddedness is thus a factor of the social and personal backgrounds of key actors, which can enhance the opportunities for a transnational organization to embed, or conversely, constrain its ability to do so.

157.7 Territorial Embeddedness: The Uniqueness of Local Conditions

Territorial embeddedness focuses on the conditions operating in the destination location, and the ways in which they impact on the transnational organization's anchoring in a new destination. Three conditions unique to Singapore affect Creation-Carers' ability to embed. The first has to do with the clustering of transnational Christian organizations in Singapore. The second is related to the circumscription of religions to social service provision and as the guardian of (human) morality. The third is the overt and conscientious commitment to multi-religiosity, often interpreted as equal official treatment of all religious groups.

In Christian circles, Singapore is known as the "Antioch of Asia" due to the presence of many transnational and local Christian organisations (DeBernardi 2008: 120–121). The potential to learn and partner with these organizations is a positive factor leading Creation-Carers to embed in Singapore. As a transnational organization operating mainly in non-Asian contexts, Creation-Carers actors need to learn from other organisations about operating in Asia. This can be seen in Creation-Carers' partnership with Transformational Development, a Singapore-based Christian FBO that engages in development projects in Asia. Transformational Development was the first group that Koo contacted when seeking partnerships in its first activities. Working together was instrumental in ensuring its success, as the local group already had a constituency and audience.

Conversely, the potential benefits of locating in a cluster are reduced due to competition for limited resources between Creation-Carers and already-existing parachurch organisations and churches. Creation-Carers is a parachurch organization whose members are generally drawn from different churches. Like other such organisations, "[it relies] upon a large pool of lay people who are willing to volunteer time and energy" (Sng 2003: 262). It thus draws resources from churches and other parachurch organisations, including Christian volunteers, participants and audiences. This suggests that locating in a cluster is not necessarily beneficial as actors within it may already be engaged in cluster-related activities.

The problems that Creation-Carers faces may be described as the "dilemma of embeddedness" (Grabher 1993). On the one hand, too little embeddedness may lead to erosion of support for an organisation. On the other hand, too much embeddedness "may pervert networks into cohesive coalitions against more radical innovations" (Grabher 1993: 26). At the time of writing, Creation-Carers is weakly embedded in Singapore in this regard, with almost no resource base which its actors can rely on.

Beyond the specific condition of clustering, another unique circumstance in Singapore is the way in which the state defines the role of religion to be the provision of social service and as the guardian of (human) morality. Because of this clear delineation of activities, alternatives are rarely articulated. Consequently, when other forms of Christian social involvement are proposed, such as in the case of Creation-Carers' call for Christian environmentalism, Christians themselves sometimes consider it a deviation from the norm, deeming the hybrid religious-secular nature of Creation-Carers' organizational aims as transgressing the state-inscribed ambit of religious activities in Singapore and hence inappropriate.

Finally, a third condition in Singapore is the conscientious official stance of multi-religiosity which propagates equal treatment of all religions. Often, such equality is interpreted as equal absence rather than equal presence (Kong 2006). As a result, Mathews (2008: 544) observes that in Singapore, a primary reason for Christian-related "agencies downplaying their religious...identification is the need to adapt to various sensitivities of operating in a secular state and multi-religious social setting." In the context of collaboration with para-religious organisations such as Creation-Carers, some state-related institutions have been reluctant to be seen as supporting the activities of a particular religious group more than others. For instance, Creation-Carers actors found it difficult to rent a venue from a government agency for its activities and de-emphasized its religious character as a way of finding access. One conference event entailed a change of title, from "Climate change, health and faith – calling Christians in medicine" to "Climate change, health and faith – the challenge facing the healthcare community."

The preceding discussion demonstrates the continued importance of the "national" in influencing "transnational" organizations, especially with regards to the role religious actors are allowed to play and the commitment to parity of treatment which has led to non-engagement of government agencies. In this sense, "transnational actions are bounded...(by) policies and practices of territorially-based sending and receiving local and national states and communities" (Guarnizo and Smith 1998: 10). The territory then, with its existing relationships between actors such as the state and religious organizations, plays a crucial role in influencing the embedding process of transnational organizations.

157.8 Conclusions

As calls for environmental protection become increasingly urgent in light of warnings about climate change and environmental degradation, and as religion continues to play a significant role in the everyday lives of many, organizations such as Creation-Carers have the potential to play an increasingly important role in improving the environment. This chapter has explored the opportunities and obstacles that a transnational environmental FBO encounters as it tries to embed itself in a new location, to spread its approach to environmental protection. In particular, we have shown how such organizations will have to confront a complex of factors, some enabling and others constraining. Strong relationship networks – both first hand relationships and indirect ones (through "friends of friends") – are enabling, in the same way that the social and personal backgrounds of key actors in the organisation can engender social capital and trust. At the same time, the existence of other similar organisations in the new destination can be welcoming and afford learning and collaboration opportunities.

On the other hand, in a postcolonial context such as Singapore, the social and personal background of external actors has the potential to arouse caution rather than enthusiasm, particularly if they are perceived to be, at best, "Western" and thus ignorant of local contexts, or worse, imperialistic in attitude. Further, receiving communities, with their specific contexts and conditions, including historical relations between state and religion and among religions, as well as currently applicable government laws and policies, will set boundaries on how successfully transnational organizations can embed in the new location.

In studies of transnational religion, religious teachings and networks are generally thought to play a unifying role, bringing together religious actors in different locations in shared beliefs, practices and actions. Yet, as our analysis of transnational religious organizations shows, the processes of establishment, embedding and evolution across national boundaries are shaped by the particularities of the local context. Transnational religion alone, by virtue of shared religious beliefs, is insufficient to ensure the successful embedding of transnational organizations in different destinations. The strategies for establishment, embedding and evolution will necessarily be different from location to location.

References

Akhtaruddin, A. (1997). *Islam and the environmental crisis*. London: Ta Ha Publishers.
Bandy, J., & Smith, J. (Eds.). (2005). *Coalitions across borders*. Oxford: Rowman and Littlefield.
Bebbington, A. (2009). Social capital. In R. Kitchin & N. Thrift (Eds.), *International encyclopaedia of human geography* (pp. 165–170). Amsterdam: Elsevier.
Benjamin, G. (1976). The cultural logic of Singapore's 'Multiculturalism'. In R. Hassan (Ed.), *Singapore: Society in transition* (pp. 115–133). Kuala Lumpur: Oxford University Press.

Coleman, J. (2000). Social capital in the creation of human capital. In P. Dasgupta & I. Serageldin (Eds.), *Social capital: A multifaceted perspective* (pp. 13–30). Washington, DC: The World Bank.

Cooper, D., & Palmer, J. (Eds.). (1998). *Spirit of the environment: Religion, value and environmental concern*. London: Routledge.

DeBernardi, J. (2008). Global Christian culture and the Antioch of Asia. In A. Lai (Ed.), *Religious diversity in Singapore* (pp. 116–141). Singapore: Institute of Southeast Asian Studies.

Gardner, G. (2002). Invoking the spirit: Religion and spirituality in the quest for a sustainable world. *Worldwatch Paper, 164*, 5–62.

Glacken, C. (1967). *Traces on the Rhodian shore: Nature and culture in western thought from ancient times to the end of the eighteenth century*. Berkeley: University of California Press.

Goh, R. (2005). *Christianity in southeast Asia*. Singapore: Institute of Southeast Asian Studies.

Goh, R. (2009). Christian identities in Singapore: Religion, race and culture between state controls and transnational flows. *Journal of Cultural Geography, 26*(1), 1–23.

Grabher, G. (1993). Rediscovering the social in the economics of interfirm relations. In G. Grabher (Ed.), *The embedded firm: On the socioeconomics of industrial networks* (pp. 1–31). London: Routledge.

Guarnizo, L., & Smith, M. (1998). The locations of transnationalism. In L. Guarnizo & M. Smith (Eds.), *Transnationalism from below* (pp. 3–34). New Brunswick: Transaction Publishers.

Hems, L., & Tonkiss, F. (2000). Introduction. In F. Tonkiss & A. Passey (Eds.), *Trust and civil society* (pp. 1–11). London: Macmillan Press Ltd.

Henderson, J., Dicken, P., Hess, M., Coe, N., & Yeung, H. (2002). Global production networks and the analysis of economic development. *Review of International Political Economy, 9*(3), 436–464.

Hess, M. (2004). 'Spatial' relationships? Towards a reconceptualisation of embeddedness. *Progress in Human Geography, 28*(2), 165–186.

Hess, M. (2009). Embeddedness. In R. Kitchin & N. Thrift (Eds.), *International encyclopaedia of human geography* (pp. 423–428). Amsterdam: Elsevier.

Hill, M., & Lian, K. F. (1995). *The politics of nation building and citizenship in Singapore*. London: Routledge.

Jamieson, D. (Ed.). (2001). *A companion to environmental philosophy*. Malden: Blackwell Publishers.

Jänicke, M. (2006). The environmental state and environmental flows: The need to reinvent the nation-state. In G. Spaargaren, A. Mol, & F. Buttel (Eds.), *Governing environmental flows: Global challenges to social theory* (pp. 83–106). Cambridge, MA: MIT Press.

Jenkins, R. (2009). The ways and means of power: Efficacy and resources. In S. Clegg & M. Haugaard (Eds.), *The SAGE handbook of power* (pp. 140–156). London: Sage.

Kay, J. (1989). Human dominion over nature in the Hebrew Bible. *Annals of the Association of American Geographers, 79*(2), 214–232.

Kearns, L. (1997). Noah's ark goes to Washington: A profile of evangelical environmentalism. *Social Compass, 44*(3), 349–366.

Kong, L. (2006). Religion and spaces of technology: Constructing and contesting nation, transnation and place. *Environment and Planning A, 38*, 903–918 (Special issue on Geographies and Politics of Transnationalism.).

Kong, L. (2010). Global shifts, theoretical shifts: Changing geographies of religion. *Progress in Human Geography, 34*(6), 755 776.

Lee, K. Y. (1988). Politics and religion best be kept separate. *Speeches, 12*(6), 13.

Livingstone, D. (2002). Ecology and environment. In G. B. Ferngren (Ed.), *Science and religion: A historical introduction* (pp. 345–355). Baltimore: The John Hopkins University Press.

Mathews, M. (2008). Saving the city through good works: Christian involvement in social services. In A. E. Lai (Ed.), *Religious diversity in Singapore* (pp. 524–533). Singapore: Institute of Southeast Asian Studies.

Mawdsley, E. (2006). Hindu nationalism, neo-traditionalism and environmental discourses in India. *Geoforum, 37*, 380–390.
McFague, S. (2008). *A new climate for theology: God, the world, and global warming*. Minneapolis: Fortress Press.
Portes, A. (1998). Social capital: Its origins and applications in modern sociology. *Annual Review of Sociology, 24*, 1–24.
Princen, T., & Fingers, M. (Eds.). (1994). *Environmental NGOs in world politics: Linking the local and the global*. London: Routledge.
Proctor, J. (2006). Religion as trust in authority: Theocracy and ecology in the United States. *Annals of the Association of American Geographers, 96*(1), 188–196.
Putnam, R. D. (1993). The prosperous community. *The American Prospect, 4*(13), 1–11.
Routledge, P., & Cumbers, A. (2009). *Global justice networks: Geographies of transnational solidarity*. Manchester: Manchester University Press.
Siddique, S. (1989). Singaporean identity. In K. S. Sandhu & P. Wheatley (Eds.), *Management of success: The moulding of modern Singapore* (pp. 563–577). Singapore: Institute of Southeast Asian Studies.
Singapore Department of Statistics. (2010). *Census of population 2010: Demographic characteristics, education, language and religion*. Department of Statistics, Ministry of Trade & Industry, Republic of Singapore.
Sng, B. (2003.) [1980]. *In his good time: The story of the church in Singapore* (3rd ed.). Singapore: Bible Society of Singapore.
Sunley, P. (2008). Relational economic geography: A partial understanding or a new paradigm? *Economic Geography, 84*(1), 1–26.
Tomalin, E. (2009). *Biodivinity and biodiversity: The limits of religious environmentalism*. Aldershot: Ashgate.
Uslaner, E. (2008). Trust as a moral value. In D. Castiglione, J. Van Deth, & G. Wolleb (Eds.), *The handbook of social capital* (pp. 101–122). Oxford: Oxford University Press.
White, L., Jr. (1967). The historical roots of our ecological crisis. *Science, 155*(3767), 1203–1207.

Chapter 158
Mapping Methodism: Migration, Diversity and Participatory Research in the Methodist Church in Britain

Lia Dong Shimada and Christopher Stephens

> *Our servant came up and said, "Sir, there is no travelling today. Such a quantity of snow has fallen in the night that the roads are quite filled up." I told him, "At least we can walk twenty miles a day, with our horses in our hands." So in the name of God we set out. The northeast wind was piercing as a sword and had driven the snow into such uneven heaps that the main road was impassable. However, we kept on, afoot or on horseback, till we came to the White Lion at Grantham. (John Wesley's journal entry from 18 February 1747)*

158.1 Introduction

In the city of Bristol, in southwest England, the Reverend John Wesley rests immortalized before the oldest Methodist chapel in the world. Lifesize, he sits astride a horse, an open book in his right hand. There could be no more appropriate image of Wesley in this historic place than as a preacher on horseback.

The founder of Methodism is known for his tireless journeying across the English countryside. Wesley's example was to shape the denomination from its inception. Movement and travel are integral parts of the itinerant ministry that defines Methodist activity and shapes the ethos of the Methodist Church to this day. These early historical images – of preachers crossing the country on horseback and seeing even a tree stump as suitable space for preaching – pre-figured a widespread international missionary movement. Facilitated by the tides of British imperialism, Methodist missionaries carried the denomination and its theologies to the far corners of the globe. In its heyday, the "Mother Church" in Britain gave rise to Methodism in places that stretched from Fiji to Zimbabwe, from Brazil to Korea.

L.D. Shimada (✉)
Conflict Mediator, Methodist Church in Britain, London NW1 5JR, UK
e-mail: lshimada@alum.wellesley.edu

C. Stephens
Southlands College, University of Roehampton, London SW15 5SL, UK
e-mail: christopher.stephens@roehampton.ac.uk

S.D. Brunn (ed.), *The Changing World Religion Map*,
DOI 10.1007/978-94-017-9376-6_158

The early twenty-first century, however, marks a vastly different era in British society – one profoundly reshaped by globalization and migration, which has been increasing since the post-war years and has now reached unprecedented and unanticipated levels. This changing context has brought radical change and significant challenges for all Christian denominations in these isles. The Methodist Church is no exception. As Britain has moved to a post-imperial age, so has the British Methodist Church responded to a changing world. Once upon a time, "overseas districts" answered to the Mother Church, witnessing in their subsidiary status to the zeal and reach of generations of Methodist missionaries. Over several decades in the twentieth century, however, these districts became autonomous Methodist Churches in their own right; the last of these districts evolved into full independence as "The Methodist Church, The Gambia" in 2008. In place of its imperially-marked missionary approach, today the Methodist Church in Britain is one partner of many in the global network of Methodist Churches. Moreover, the migration of diverse people into Britain – particularly from its former colonies – is redefining the missionary nature of the Methodist Church in Britain. The concept of "reverse mission" has entered into common parlance, as people from Methodist Churches abroad observe the increasing secularism of British society and see here a fertile ground for their own missionary activity.

As the UK becomes increasingly diverse and global, the Methodist Church finds itself confronting once-unimagined opportunities and challenges. As a denomination, Methodism grapples with potent questions of history, faith, culture, identity and belonging; with questions of geography and theology in an ever-shifting world. As churchgoers from former British colonies migrate to the United Kingdom, how are post-imperial dynamics finding expression in the Methodist Church in Britain? How are the shifting cultural geographies of the Methodist Church shaping its theologies? How are distinctly Methodist theologies being invoked, reshaped and deployed in response to an increasingly diverse Church? What are the implications for the identity (indeed, identi*ties*) of British Methodism? The Methodist Church in Britain is only beginning to understand the impact of these issues on its future life.

In this chapter, we explore these questions in the context of participation, framing our discussion around the rich theoretical seam of participatory research. In the social sciences, research conventions have long upheld the values of neutrality, objectivity and approval of experts in the field. In contrast, participatory research has evolved as an alternative system that values local, indigenous and multiple types of knowledge (Reason and Bradbury 2002). In their approach, participatory methodologies emphasize participation and collaboration, placing high value on people's experiences and harnessing their knowledge for the purpose of political action. Gatenby and Humphries (2000: 89) describe participatory research as a "form of praxis aimed at social change." Unlike other approaches, participatory research is unique in the high value it places on reflexive practice (for example, Kesby 2007). In this way, the participatory approach is particularly useful as a lens for exploring forms of identity and subjectivity (Parr 2007; Cameron and Gibson 2005). These are themes that resonate strongly for questions of world religions.

Although participatory research is far more established in the social sciences than in theological studies, some theologians have begun to explore the terrain of "participatory theology" (see Conde-Frazier 2006; Hermans 2003; Lienemann-Perrin 2004; Cameron et al 2010).

In this chapter, we seek to facilitate a conversation between *participatory research* and *participatory theology*, grounded in practice. The chapter begins with a brief history of the Methodist Church, emphasizing the role of migration in shaping denominational identity. In the second section, we discuss contemporary efforts to engage the Methodist Church in Britain with its shifting dimensions of cultural diversity. The third section is devoted to recent research activities of the Methodist Church, including an initiative to map denominational diversity. Finally, we explore the scope for participatory mapping to illuminate the challenges and opportunities that diversity brings to the life of Methodism, and point toward diverse geographies and theologies of identity and belonging.

158.2 A Brief History of Methodism

John Wesley had no intention of founding a new denomination. As the son of the vicar of Epworth, Lincolnshire, the Anglican ministry was a logical and expected vocational path. In the eighteenth century, the Church of England was a bastion for established order, and John Wesley's early ministry was no different. As he later reflected, "I had been all my life…so tenacious of every point relating to decency and order that I should have thought the saving of souls almost a sin if it had not been done in a church." (*Journal*, 17th March 1739). However, while studying at Oxford University, Wesley became introduced to new theological thinking and a new approach to Christian ministry that would ultimately, and dramatically, reshape his ministry. In his journal entry for 24 May 1738, Wesley famously described the moment he became assured of God's love: "I felt my heart strangely warmed." From this assurance flowed Wesley's deep commitment to share this message with those who needed to hear it most. Since many working-class people of the day often felt excluded from churches, "field preaching" became Wesley's way of working. By emphasizing God's forgiveness and love for all, Wesley was able to reach many who felt alienated from the established Church of England. His sermons were experienced as deeply personal messages to those who heard them, including the increasing number of lay preachers whom Wesley trained.

For Wesley, preaching to the masses became a theological imperative – "the logical outcome of his new convictions" (Davies 1985: 59). Readings of Wesley's account of this time have even attributed to him a providential view of the weather in sanctioning his method of preaching (Rack 2002). As many working-class people felt excluded from churches, field preaching became a key feature of Wesley's revival movement, with increasing numbers of preachers trained to join Wesley in this activity of missional outreach. Wesley himself went on to spend his life traveling the length and breadth of Britain, preaching to crowds on village greens and at the pitheads of mines and quarries. The various entries in his diaries lead to a common scholarly consensus that, during his lifetime, Wesley travelled an estimated 250,000 miles and preached 40,000 times.[1]

It is useful to examine briefly the theological rationale behind this shift, which made preaching to the uneducated a missionary imperative resulting from Wesley's developing theological convictions. The desire to preach to the masses in any available setting emanated from Wesley's rejection of a firm Calvinistic view of predestination, common even in the revivalists of his own age. There can be no doubt that, for Wesley, the practicalities of his preaching activity were a result of theological conviction: the rejection of predestination, the role of the individual in his or her salvation through their response to the saving grace of God, and the pressing need for the preacher to bring the message of salvation to all in order to illicit that response. In particular, the Wesleyan theological position emphasized the need to pay attention to those who live without opportunities for theological education and regular, appropriate ministry. By the time the Church of England and the Methodist movement had become so organizationally and ideologically distinct that a formal division took place (see Baker 1970), Methodist denominational identity had begun to demonstrate a clear and identifiable shape that emphasized the assurance of grace for all, social justice and a broadly Arminian theology.[2] John Wesley's published sermons became the doctrinal standard of the Methodist Church, and remain so today.

All this has left two defining marks on the Methodist Church in Britain. The first is a fundamentally and crucially itinerant ministry that emerges from Methodist theology. As Wesley's networks of meetings developed across the country into a formalized and, eventually, independent body of Methodist chapels, they came to see themselves as a network – a "Connexion" – of societies that gather as a worshipping community. Today, the Methodist Church in Britain continues to conceptualize itself as a Connexion – a structure that deliberately and explicitly interconnects as one every small worshipping society across Britain. In this way, the tiny cluster of chapels in the Shetland Islands holds the same status on any map of ministerial activity as the vast array of churches in London. The concept of connexionalism is deeply rooted in Methodist theological identity. As Carter (1998: 59) describes, "connexionalism implies harmony, balance, mutual respect and deference between the members and leaders of the Church, a balance" (Fig. 158.1).

In Methodism, the language of "Church" or "denomination" is held alongside that of "movement." The self-consciously connexional nature of the Methodist Church explicitly requires that its ministers remain itinerant. Within this context, Methodist ministers (at least in theory) have no personal choice over the location of their ministry. It is a ministry of movement, with up to 20 % of all Methodist ministers changing "station" each September, often long distances from their previous

[1] We note with interest the belief that "Wesley probably paid more for turn-pikes than any other man in England, for no other person travelled so much" (Southey 1858: 293).

[2] For further reading on this theme, see Rupp (1965); Oden (2008); Rack (2002); Baker (1970); Turner (2005).

Fig. 158.1 The British Methodist Connexion (Map by Thyra Hawkes for the Methodist Church in Britain, © Groundwork London, 2012, used with permission)

location, and leading to almost all ministers moving to a new location within the space of 5 years. Ministers do not apply for these roles, but are sent by the Church to the areas where their skills and gifts are most needed.

A second defining quality of British Methodism has been a consistent desire to send preachers, ministers and mission workers around the world – an international approach that began in Wesley's era. As Birtwhistle (1983: 1) observes, "(t)he very nature of the Methodist Revival made it impossible that its energies could be confined to one small country." Driving this movement was, of course, the theology of connexionalism: "Connexional means being ecumenical in the widest sense, pushing the boundaries at every point" (Drake 2004: 138). The reach of Methodism was extraordinary; by the twentieth century, it had crossed oceans and continents. In an unfortunate yet informative simile, Hempton (2005: 19) compares Methodists to parasites that rode the wider British and Anglican imperial effort. Following imperial expansion, Methodists found headway in the majority of lands where the Church of England laid claim, and likewise struggled where it struggled.

In the twentieth century, as empires collapsed and imperialism became associated less with progress and development and more with oppression and racism, so the Christian missionary movement that had developed – at least pragmatically – off the back of imperialism required major rethinking. Indeed, as missionaries themselves became involved in nationalist movements in the countries in which they served, so their sending Churches – Methodist included – were required to listen to a new generation of missionary thinking (Hempton 2005; Koss 1975). Today, Methodist districts abroad no longer exist; each is now its own autonomous Church, and none is managed administratively from Britain. The last vestiges of the British Church's international reach lie only in the small British territories of Gibraltar and Malta; due to the national identities of their residents, they exist only as extensions of British Methodist home districts. Just as the British Empire has evolved into only a small collection of overseas protectorates under direct control from Britain, existing essentially as "Britain abroad" and dwarfed by the vast and largely independent Commonwealth, so the Methodist Church has taken on a similar nature. It has become a federation – a connexion of connexions, so to speak – that reflects the breadth and reach of those early missionaries. The tides, however, have reversed, carrying migrants from those places back to Britain, and in the process creating a new landscape of diversity for the Methodist Church in Britain.

158.3 The Challenge of Diversity

British Methodism cherishes its participation in the "World Methodist Church." In recent times, however, sharper questions are emerging that test this in practice. To draw from real examples, in a quiet rural village – in a tiny chapel in which Wesley himself may have preached – an ordained minister from Sierra Leone (or Singapore, or Brazil) may struggle to serve a community long accustomed to a certain type of white, British minister. What does "world ministry" mean in this context? For ears used to hearing sermons from ministers trained at Bristol, Cambridge or Birmingham, what challenges and opportunities does an unfamiliar accent and a different set of cultural references present from the pulpit? In addition to new types of ministers, global migration has also brought thousands of lay people through the doors of Methodist buildings. Across Britain, non-English-speaking congregations (Cantonese, Korean, Farsi, Urdu, to name a few) make their homes in Methodist churches. In cities like London and Birmingham, Ghanaians and Zimbabweans may gather for special services that draw on cultural expressions of Methodism familiar to their home context. Such gatherings offer space for these communities to worship in ways that are decidedly absent from the mainstream British Church on a typical Sunday morning. How might these gatherings – and the larger narratives behind them – engage constructively with the life of the local church? More broadly, how can British Methodism learn not just to accommodate, but to embrace its increasing ethnic, linguistic and cultural diversity?

These are important questions that test the space between the aspiration of global Methodism, and the reality of tensions that may be felt most acutely in the everyday rhythms of local congregations. Heightening these questions is the sheer breadth of theological perspectives accompanying British Methodism's increasing ethnic, linguistic and cultural diversity. Over two centuries, the Methodist Church in Britain has navigated its identity as a mainstream Christian denomination that can encompass a reasonably wide theological spectrum. Today, the denomination is witnessing a theological sea change, as the tides of global migration bring diverse strands of Methodism "home." For those early missionaries, who went forth with a unified Methodist message, the contemporary, post-colonial world would be unrecognizable, not least in the variety of Methodist theologies now present in Britain (see Walls 1996).

Although the demographics of the Methodist Church in Britain are shifting rapidly, the Church struggles to reflect its increasing diversity in its leadership structures. Throughout the Methodist Church in Britain, those who inhabit key decision-making bodies tend to be male, white, and middle-aged or older. In 2010, the Methodist Conference, the supreme governance body of the Church, reported that over 70 % of its membership was over 50 years old and only 9 % were under 40; only 6 % registered a physical or mental disability or impairment; and only 11 % identified their ethnicity as anything other than white, despite the compulsory inclusion in the membership of that body of representatives from partner "World Churches."[3] Not surprisingly, institutional practices have evolved within the Church that support and affirm the status quo. The discrepancy between the Church's demographic diversity and the relative visual homogeneity of its decision-makers is by no means a new realization. In the 1980s, two landmark publications reshaped the landscape of British Methodism. A report titled *A Tree God Planted* highlighted the difficulties faced by black members who sought to offer their distinctive contributions. The report concluded that British Methodism made

[3] The report on the membership of the Conference can be found at www.methodist.org.uk/statisticsformission

insufficient effort to encourage them to participate in the leadership and ministry of the Church. Two years later, at its annual meeting, the Methodist Conference received a report titled *Faithful and Equal*. This report affirmed the theological stance that racism is in direct contradiction to Christian values, and urged the Methodist Church "to more faithfully respond to Britain as a multi-cultural society in which all receive equal opportunity and treatment." In response to this charge, the Committee for Racial Justice was formed, devoting itself to leading the Church in this work.

In June 2009, the Methodist Church in Britain committed the resources – organizational, financial, and above all human – to a 3-year initiative called *Belonging Together*. This project, which was administered through the administrative headquarters in collaboration with a connexionally-representative steering group and three regional partner districts, began officially in September 2010. *Belonging Together* sought to engage the Methodist Church in Britain with its ethnic, linguistic and cultural diversity. Under the rubric of "Unity in Diversity," *Belonging Together* built on the work of the Committee for Racial Justice, positioning it alongside wider discussions about equalities and diversity in the Methodist Church in Britain.

The 12 aims of *Belonging Together* addressed major questions related to wide-ranging areas of policy and practice in the Methodist Church in Britain. For example, how might Methodist theological colleges embed cultural diversity in their training programs? What are the experiences of people from under-represented backgrounds who offer themselves for leadership roles? How is "leadership" defined culturally in the diverse communities that, today, comprise the Methodist Church in Britain? What kind of conversations, opportunities and support should the Church be facilitating for children and young people from diverse backgrounds? How can an ethos of diversity embed itself sustainably in the life, learning and ministries of the Methodist Church?

Above all, what does the increasing diversity reveal about the identity – organizational, cultural and theological – of the Methodist Church in Britain? What has long been considered "the norm" is now in flux, with the dawning realization that new voices and ways of worship are re-shaping the landscape of British Methodism. In the next section, we look at how the Methodist Church in Britain is attempting to listen to these voices, in order to understand the shifting identities of the Church (Fig. 158.2).

158.4 Participation in Research

All of the major Christian denominations in Britain currently have, in one form or another, research or development staff serving their central administrative headquarters. However, this does not necessarily lead to recognition of the importance of such resources. Within the Methodist Church in Britain, the concept of a staff group dedicated to producing research that could inform strategy or policy in more than one area of Church life, or for more than one major project, emerged only in 2008.

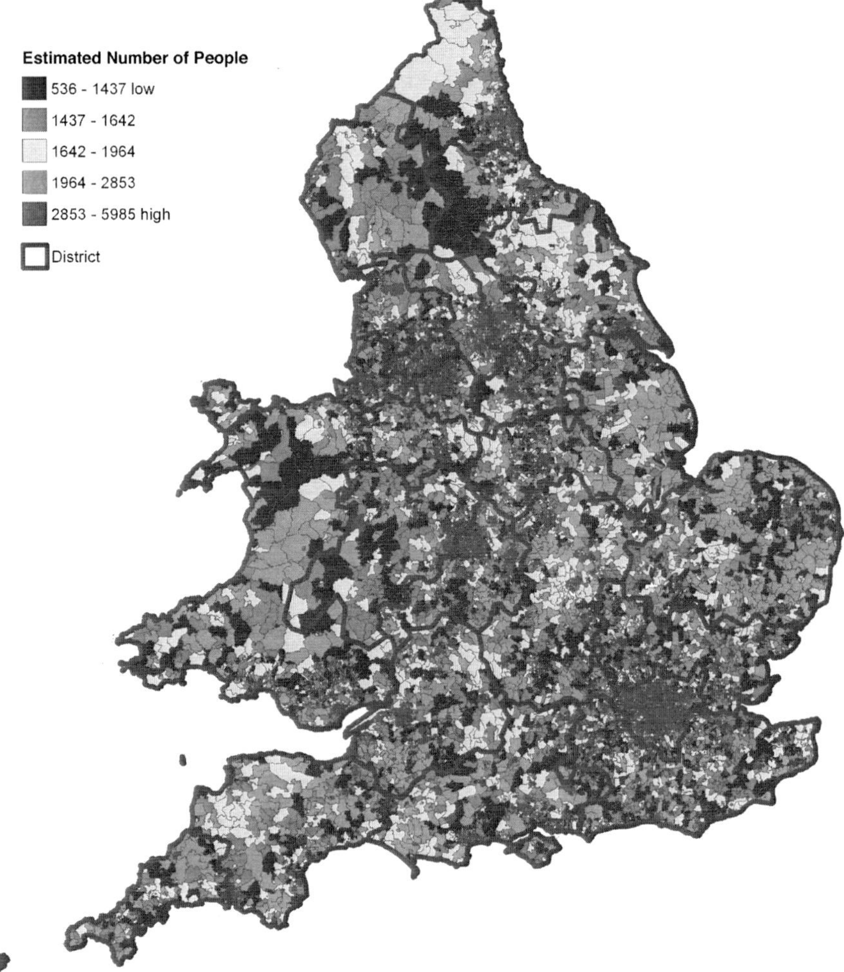

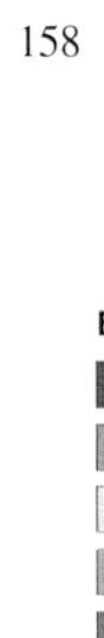

Fig. 158.2 Methodist district population density (Map by Thyra Hawkes for the Methodist Church in Britain, © Groundwork London, 2012, used with permission)

Even within the context of such staff teams, approaches to research have been restricted in Methodist churches to a reasonably narrow set of questions, inevitably limited by resources and by time. Counting has dominated as the primary research activity, of which counting for specific and pre-decided purposes has been the norm. As with most organizational research, key individuals or committees largely steer the work of Church-based research from the desire to answer the specific questions that seem most pressing or important to their work.

Emerging from this context, and seeking to challenge the established norm, a deliberately distinct approach shaped the research aims of *Belonging Together*. As an initiative emerging out of a belief that the Church and its churches need new ways to listen and to respond to the diversity that now comprises Methodist membership, the research which *Belonging Together* commissioned does not start with any specific question to answer or problem to solve. Instead, it called for the development of professional research, which will ultimately inform the progression of policy and strategy within the Church. This research involves listening actively for the question, then encouraging members of local churches to participate in framing the question and responding to it. The final set of outputs will be largely unknown until the close of the research period because the fundamental significance of the work lies in its process, rather than the discovery of a solution to an organizationally-defined problem. Participation characterizes the research, particularly in the fluid development of the research framework. In this way, the research remains open to exploring diversity of opinion, experience and knowledge across the reach of British Methodism.

Although it is fundamentally a Connexion, the Methodist Church in Britain is held together within a tight and clear organizational structure. The advantage of this relatively small and centralized administrative structure is that the Church can decide to look intentionally at itself. As a result, of all British Christian denominations, the Methodist Church has the largest and richest data set relating to its diversity.

Historically, the Methodist Church has valued research as an organizational activity, in its many phases and guises. Records of the annual Methodist Conference go back well into the nineteenth century and show a marked propensity towards counting in the Church. By the end of the twentieth century, the varied numerical reporting types seen in reports to the Conference had developed into the relatively consistent triennial reports. These reports, known as "Statistics for Mission," are administered centrally, under the close direction of the Methodist Conference. Previously, the concerns of these reports were to gather a small but seemingly critical data set of those things which had traditionally marked the presence of Methodism within society. Membership numbers, weekly attendance figures, youth group activities and churches' involvements in baptisms, weddings and funerals were presented as a full demonstration of the "health" of the Church.

In 2008, a significant reorganization of the Church's central administrative team took place, which had a critical impact on how the Methodist Church recognizes its markers of health. This reorganization was driven largely by an intentional response to the interests, concerns, fears and energies of the Methodist people, who had put pressure on the Church's leaders to develop a more responsive and flexible administrative function. As a result, the Statistics for Mission process became properly regarded as a research tool for mission planning. For the first time, it embodied a Church-wide acknowledgement that membership and attendance numbers were, increasingly, a shallow representation of denominational life. Indeed, the numbers which had been collected formally by the Methodist Church until this time were insufficient on their own as markers of health in a contemporary context. The British Methodist Church needed to find out more if it was truly to know itself, and its local churches were vocal in ensuring that change to data collection reflected the experience of life on the ground.

From October 2009, the Statistics for Mission process expanded more than it had done at any point in the history of Methodism. Individual churches and congregations were asked to provide more information about the people they served, the activities they led or encouraged, and the approaches they took. At the same time, the formal process became, for the first time, a self-consciously participatory, grassroots-led exercise in data collection. Every church member, minister and district officer was asked to participate in shaping the questions that were asked: What would they like to ask about our churches? Why is this important? How would they respond? How might the reporting of this information best serve the Church's mission needs, in their local context? The result was a massive increase in the data available to members of the Church, in the breadth of data available, and in the methods of reporting. Because these developments were driven by energy from church ministers and members themselves, it was matched by a rapid and marked increase in interest and investment in the Statistics for Mission process from those asked to submit the required information. By 2011, over 97 % of all Methodist churches were regularly providing detailed responses to the annual October census – another exceptional statistic among British Christian denominations. A striking but unsurprising result of this shift was the critical position given to ethnic and linguistic diversity as locally-recognized markers of church life. Every data collection exercise since 2009 has been shaped by a range of questions proposed by Methodist members across the Connexion, which seek diverse expressions of cultural and national identity within the Methodist Church in Britain.

Although the statistical record of a denomination can never provide a full account of its life and activities, the early years of development in gathering data on the linguistic and ethnic diversity of Methodist churchgoers has been dramatic (Stephens 2011).[4] For example, in October 2009, the first year of the augmented data collection, 89 Methodist churches reported running worship or fellowship activities in a language other than English or Welsh. The following year, this figure increased to 125 churches. To date, the Statistics for Mission returns indicate that over 40 languages are spoken in Methodist worship or fellowship across the Connexion (Figs. 158.3, 158.4, and 158.5).

Yet the task of learning about the identity of the Methodist Church is more than just an exercise in numbers. Running alongside the Statistics for Mission process is a pilot initiative that will reshape *how* the Methodist people choose to identify. Previously, the Methodist Church (like many of its counterparts, both secular and religious) has drawn upon the UK census format to collect information about diversity. The census form – a standard tick-box – is problematic for a number of reasons, not least its inflexible and increasingly outdated options for denoting ethnic background (see McKenney and Bennett 1994; Kaneshiro et al. 2011; Weller 2004; Christopher 2011). In 2011, the Methodist Church piloted a new diversity monitoring form for people seeking to train for the ministry. In lieu of the standardized tick-boxes, the respondents were invited, anonymously, to describe in their own words a range of diversity dimensions: How would you describe your ethnic identity?

[4] For the full reporting of Methodist Statistics for Mission, visit www.methodist.org.uk/statisticsformission

Fig. 158.3 Churches that offer worship or fellowship in a language other than English or Welsh to serve a distinct language group, 2009–2010 (Map by Thyra Hawkes for the Methodist Church in Britain, © Groundwork London, 2012, used with permission)

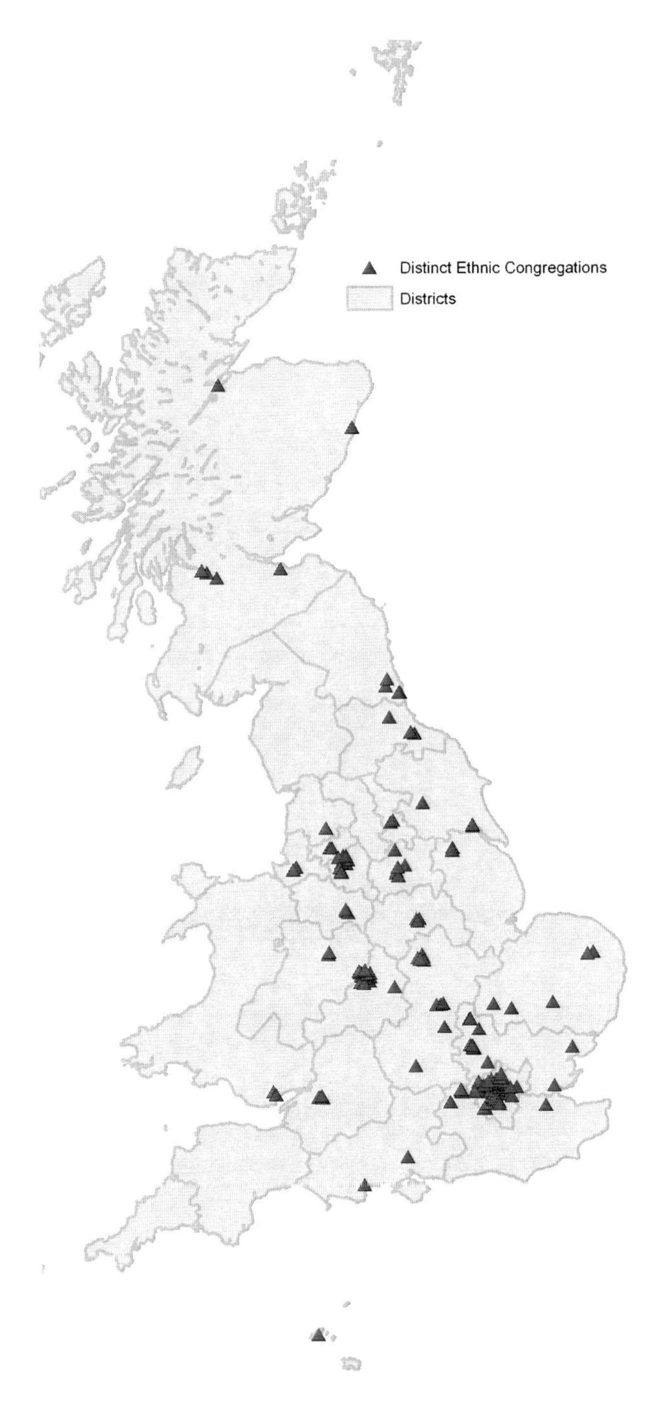

Fig. 158.4 Churches that offer worship or fellowship in English to serve a distinct language or ethnic group, 2009–2010 (Map by Thyra Hawkes for the Methodist Church in Britain, © Groundwork London, 2012, used with permission)

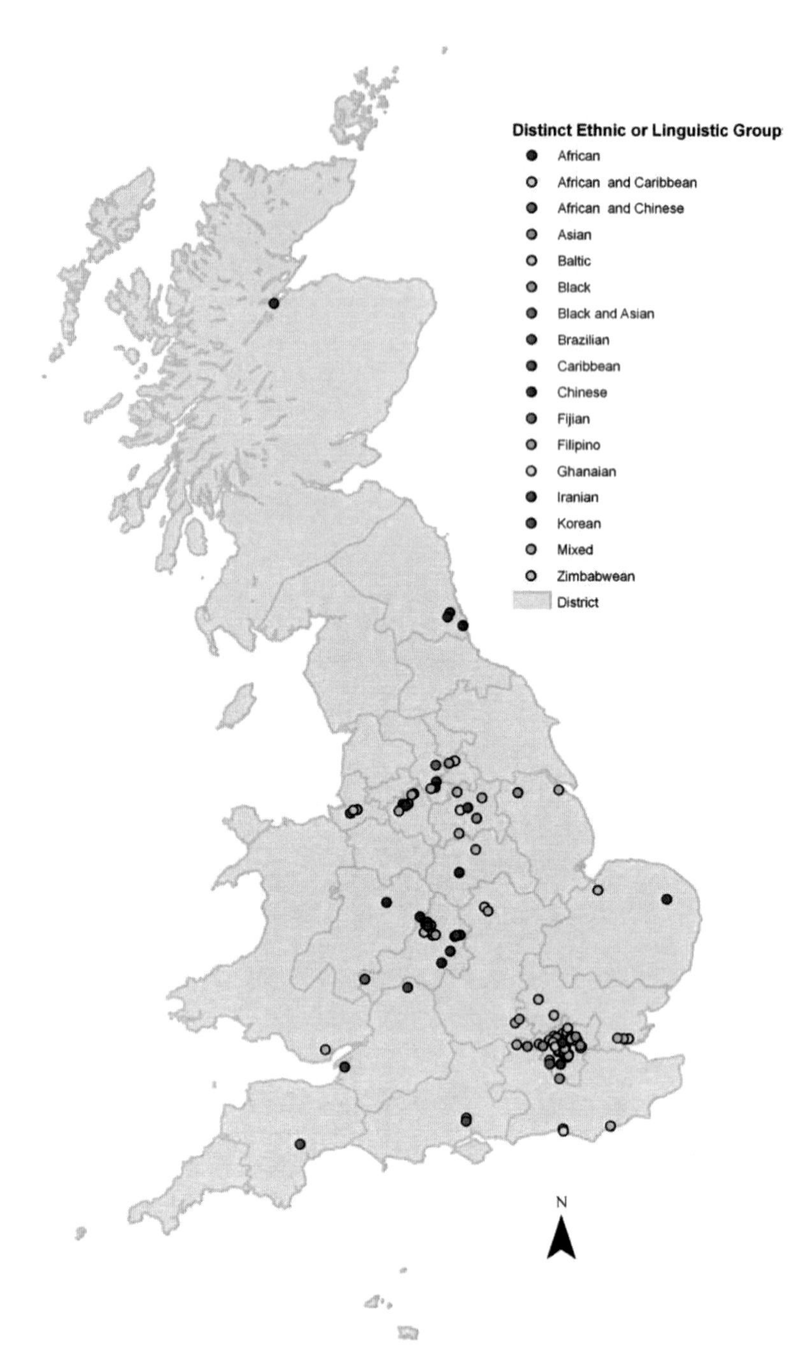

Fig. 158.5 Churches that offer group activities for a distinct ethnic or linguistic group, 2010–2011 (Map by Thyra Hawkes for the Methodist Church in Britain, © Groundwork London, 2012, used with permission)

How would you describe your gender identity? Your sexual orientation? Your disability status? The pilot exercise, which continued into 2012–2013, yielded fascinating initial findings, which are currently being analyzed. Ultimately, this research exercise will expand the ways in which Methodism identifies itself and, crucially, it will offer new language to describe the identities of the Methodist people. Theologically, it has at its center the desire to see within every human individual a unique manifestation of the image of God.

However, where participatory research truly takes flight is in the Methodist Church in Britain's new mapping initiative, which draws on the rich – and growing – body of academic work on participatory mapping (for example, Aberley 1993; Al-Kodmany 2001; Dower 1996; Wood 2005). Both inside and outside academia, the power of participatory mapping has been expressed and demonstrated in broader organizational contexts. As Chambers (2006: 1) observes: "The means and medium of mapping, whether ground, paper or GIS, and the style and mode of facilitation, influence who takes part, the nature of outcomes and power relationships." Offen's study (2003) explored the relationship between the mapping process and an identity politics of place, addressing the transformational potential for the mapping process. For the Methodist Church in Britain, it is this transformational potential that drives the invocation of mapping as both research product and participatory process.

Having seen the potential of maps as a reporting tool for statistical reports, in 2010 the Church commissioned a large piece of development work that sought to empower individuals within local churches to develop their own Methodist maps. The desire to involve the breadth of Methodist membership in the development of the Church's mapping emerged from the same desire to encourage local churches to define new questions in the Statistics for Mission data collection. The starting point for the new Methodist mapping initiative is the assumption that relevant knowledge lies not with the central administrative offices, but with individual members and churchgoers across the British Connexion. The mapping initiative provides an online portal and mapping tool which anyone can use, and onto which all the available data the Church possesses can be added in a clear and easily understood form. It is a dynamic online map, which allows Methodist information to be presented geographically alongside other data available from national archives and partner organizations. In this context, mapping becomes a participatory exercise, whereby any Methodist can be the decision maker and any map can be made from available data. Thus, the power to tell the Methodist story is vested in the local users, who can discover, report and demonstrate what they believe to be the critical stories that need to be told.

Central to these stories are the diverse identities that comprise the British Methodist Connexion. It is no coincidence that the recent emphasis on mapping is emerging through and alongside wider conversations about cultural diversity in the Methodist Church. As well as providing a spatial depiction of the diversity of the Methodist Church, the maps which Methodist members create are pointing to areas of need in further research on migration, culture, identity and belonging in the contemporary Church. For example, a map of Methodist Fijian fellowship groups in Britain (Fig. 158.6) reveals a strong geographic association with British military

Fig. 158.6 Fijian fellowship groups (Map by Thyra Hawkes for the Methodist Church in Britain, © Groundwork London, 2012, used with permission)

sites. This map captures a sense of the migratory dynamics of British post-imperial history, with citizens from the British Commonwealth migrating to the UK to serve the military that, historically, expanded Britain's imperial reach and trailed Methodist missionaries in its wake. The map invites speculation: How are these Fijian Methodist individuals locating themselves in the wider connexional geography of Methodist denominational identity?

The answer to this question, and others like it, may lie in part with the maps themselves. In July 2012, the Methodist mapping initiative was launched with a deliberate invitation for the people of the British Connexion to contribute their own voices to the online map. They were asked: *What speaks to your cultural identities as Methodist communities?* Each day now sees new entries added by local users, who upload photographs, sound files, video clips, and links to discussion boards. Methodists can now express themselves not just by name in location, but in picture, song and theological reflection. In doing so, they challenge the British Methodist Church's preconceived ideas of itself as a singular, stable and relatively homogenous worshipping community.

In giving voice to diverse voices, the Methodist mapping initiative will help the Church to come to new understanding of itself and its mission in the twenty-first century. The creation of a multi-sensory map – where recordings of Fijian women at British military bases, video clips of the choirs of Zimbabwean fellowships, and debate and discussion about the identity of local church-based youth projects all appear as demonstrations of identity – frames the Methodist Church's participatory research program as an explicitly theological discourse. Already, the mapping initiative is re-shaping new theologies of inclusion, pushing the Methodist Church in Britain to value the divine in the sheer breadth of its human diversity.

158.5 Conclusion

> I look upon all the world as my parish thus far I mean, that, in whatever part of it I am, I judge it meet, right, and my bounden duty, to declare unto all that are willing to hear, the glad tidings of salvation. – John Wesley's journal entry from 11 June 1739

When John Wesley wrote these words, he had little idea how relevant they would be in the years and centuries to come. Although he may have glimpsed the possibilities for the Methodist Church, the extent of its eventual global reach would have astonished him. Methodism took the vigor of British non-conformist faith to the far reaches of the world. Today, Wesleyan theology   Methodism   remains one of the dominant expressions of Christianity throughout the world.

For the Methodist Church in Britain, however, this history of global triumph is no longer the dominant story. The independence of Methodist Churches across the world has taken place alongside a background of decline in traditional measures of Church 'health' in Britain itself. Such is the severity of this decline that the current General Secretary of the Methodist Church in Britain describes the Church as "in dire straits" (Atkins 2007: 2). Methodism in Britain has, for several decades, followed suit with the Church of England and most other Protestant denominations in the country, witnessing a broad numerical decline. Membership numbers, average attendance and even youth activities, for which the Methodist Church was once widely regarded, have suffered at worrying rates. As the expectation that one simply went to church as part of a normal weekly routine has disappeared from any definition of popular British society, so have the firm identities of the non-conformist groups in Britain, as distinct from the established Church of England, become blurred in a post-denominational and post-ecclesial age. Duly, the possibility of extinction has entered the conceptual framework of discourse in the Church; its specter thus has a significant impact on the Church's strategic direction.[5]

In this context, the influx of new communities to Britain, and particularly to its urban centres, can be seen as a critical breath of life for those churches which have embraced those communities and found new ways of serving them in a Methodist context. For the denomination more broadly, just as new immigrant communities have brought to Britain a new injection of life to the nation's Roman Catholic and Orthodox Churches, so also have Methodist churches and congregations benefited from the arrival of culturally distinct churchgoers. As the number of Methodist congregations in Britain continues to decline, the number of languages spoken in formal worship in British Methodism is far more than it has ever been, as is the number of ethnic and cultural groups to which Methodist churches intentionally minister.

This leaves a Methodist Church which is fundamentally different from that of Wesley's day, and from that of even 20 years ago. In this context, the Church has forced itself to re-think what it means to be the Methodist Church *in Britain*. This involves far more than a readjustment of what "mission abroad" might entail, provoking Methodists throughout the British Connexion to question how well the Church's structures, processes and decisions reflect the denomination's shifting identities. To respond to these queries, the Church has adopted a radically different approach to research and to data collection needs to be taken. Already, the participatory mapping infrastructure is inviting questions that will drive further research studies: How is the legacy of colonialism expressing itself in the identities of contemporary British Methodism? What conversations are taking shape around conflict and reconciliation related to the denominations increasing diversity? How might the new emphases around participation drive the expression of diasporic theologies particular to Methodism in Britain?

In its research approach and emphases, the Methodist Church in Britain strives to embody doctrinal imperatives of service in humility, drawing on developments in approaches to participatory research and mapping within an explicitly theological framework. Participatory mapping within the life of the Methodist Church expresses the recognition that, in order to serve and to minister,, the fullest expressions of the identities of Methodists in Britain needs to be known, understood and celebrated.

[5] For a useful study of church decline and its causes, see Francis and Richter (2007).

Acknowledgement All maps in this chapter were produced by Thyra Covell, © Methodist Church 2012© Groundwork London 2012. Contains Ordnance Survey data © Crown copyright and database right 2012. The maps were produced on ESRI AarcView using data from ONS. We wish to thank Thyra for producing them.

References

Aberley, D. (Ed.). (1993). *Boundaries of home: Mapping for local empowerment*. Philadelphia: New Society Publishers.

Al-Kodmany, K. (2001). Online tools for public participation. *Government Information Quarterly, 18*, 329–341.

Atkins, M. (2007). *Resourcing renewal: Shaping churches for the emerging future*. Peterborough: Inspire.

Baker, F. (1970). *John Wesley and the Church of England*. London: Epworth Press.

Birtwhistle, N. A. (1983). Methodist missions. In R. Davies, A. R. George, & G. Rupp (Eds.), *A history of the Methodist Church in Britain* (Vol. 3, pp. 1–116). London: Epworth Press.

Cameron, J., & Gibson, K. (2005). Participatory action research in a poststructuralist vein. *Geoforum, 36*(3), 315–331.

Cameron, H., Bhatti, D., Duce, C., Sweeney, J., & Watkins, C. (2010). *Talking about God in practice. Theological action research and practical theology*. London: SCM Press.

Carter, D. (1998). Some Methodist principles of ecumenism: A discussion article. *Epworth Review, 25*(4), 53–63.

Chambers, R. (2006). Participatory mapping and Geographic Information Systems: Whose map? Who is empowered and who disempowered? Who gains and who loses? *The Electronic Journal of Information Systems in Developing Countries, 25*(2), 1–11.

Christopher, A. J. (2011). Questions of language in the commonwealth census. *Population, Space and Place, 17*(5), 534–549.

Conde-Frazier, E. (2006). Participatory action research: Practical theology for social justice. *Religious Education, 101*(3), 321–329.

Davies, R. E. (1985). *Methodism* (2nd ed.). Peterborough: Epworth Press.

Dower, M. (1996). Forward. In S. Clifford & A. King (Eds.), *From place to PLACE: Maps and parish maps*. London: Common Ground.

Drake, P. (2004). Joining the dots: Methodist membership and connectedness. In C. Marsh, B. Beck, A. Shier-Jones, & H. Wareing (Eds.), *Unmasking Methodist theology* (pp. 131–141). New York/London: Continuum.

Francis, L. J., & Richter, P. (2007). *Gone for good? Church leaving and returning in the 21st century*. Peterborough: Epworth.

Gatenby, B., & Humphries, M. (2000). Feminist participatory action research: Methodological and ethical issues. *Women's Studies International Forum, 23*(1), 89–105.

Hempton, D. (2005). *Methodism: Empire of the spirit*. New Haven: Yale University Press.

Hermans, C. A. M. (2003). *Participatory learning: Religious education in a globalizing society*. Boston: Brill.

Kaneshiro, B., Geling, O., Geller, K., & Millar, L. (2011). The challenges of collecting data on race and ethnicity in a diverse, multiethnic state. *Hawaii Medical Journal, 70*(8), 168–171.

Kesby, M. (2007). Spatialising participatory approaches: The contribution of geography to a mature debate. *Environment and Planning A, 39*, 2813–2831.

Koss, S. (1975). Wesleyan and empire. *The Historical Journal, 18*(1), 105–118.

Lienemann-Perrin, C. (2004). The biblical foundations for a feminist and participatory theology of mission. *International Review of Mission, 93*(368), 17–34.

McKenney, N. R., & Bennett, C. E. (1994). Issues regarding data on race and ethnicity: The Census Bureau experience. *Public Health Rep, 109*(1), 16–25.

Oden, T. C. (2008). *Doctrinal standards in the Wesleyan tradition*. Nashville: Abingdon.
Offen, K. H. (2003). Narrating place and identity, or mapping Miskitu land claims in northeastern Nicaragua. *Human Organization, 62*(4), 382–392.
Parr, H. (2007). Mental health, nature work and social inclusion. *Environment and Planning D: Society and Space, 25*(3), 537–561.
Rack, H. D. (2002). *Reasonable enthusiast: John Wesley and the rise of Methodism*. London: Epworth Press.
Reason, P., & Bradbury, H. (2002). Introduction: Inquiry and participation in search of a world worthy of human aspiration. In P. Reason & H. Bradbury (Eds.), *Handbook of action research* (pp. 1–14). London: Sage.
Rupp, G. (1965). Introductory essay. In R. Davies & G. Rupp (Eds.), *A history of the Methodist Church in Britain, Volume 1* (pp. xxvi–xxxix). London: Epworth Press.
Southey, R. (1858). *The life of Wesley and the rise and progress of Methodism in two volumes* (Vol. 1). Longman, Brown, Green, Longmans and Roberts: London.
Stephens, C. W. B. (2011). *Methodist statistics: Are we yet alive?* London: Methodist Publishing.
Walls, A. (1996). *The missionary movement in Christian history: Studies in the transmission of faith*. Edinburgh: Orbis.
Weller, P. (2004). Identity, politics, and the future(s) of religion in the UK: The case of the religion questions in the 2001 decennial census. *Journal of Contemporary Religion, 19*(1), 3–21.
Wood, J. (2005). "How green is my valley?" Desktop geographic information systems as a community-based participatory mapping tool. *Area, 37*(2), 159–170.

Chapter 159
Territoriality and the Muslim Spiritual Boards of Post-Soviet Russia

Matthew A. Derrick

159.1 Introduction

The estimated 20 million Muslims of Russia (Walker 2005: 247) comprise an internally diverse, highly fragmented *umma* (Islamic community). Their fragmented situation foremost relates to the manner in which religion and ethnicity are intertwined and territorialized within the county's complex federal structure. First, it is important to recognize that, in the context of Russia, who is considered a Muslim – or an Orthodox Christian, for that matter – is not defined by rigorous adherence to religious rules, but instead is a question of self-identification that highly correlates to ethno-nationality. Just as most ethnic Russians consider themselves to be Orthodox Christians, regardless of active religious observance or conviction, Tatars, Bashkirs, Chechens, and other cultural groups in Russia consider themselves to be Muslims by fate of belonging to an ethno-national group traditionally associated with Islam.[1] Second, it is important to understand how ethno-national identity is territorialized within Russia's complex federal structure. The Russian Federation is a hierarchy of 83 administrative units that can be separated into two basic categories: (1) ethnically defined republics and autonomous regions and (2) non-ethnic regions (Fig. 159.1). The former, designated as the historic homelands of titular (that is, non-Russian) populations, have enjoyed certain cultural privileges, such as native language rights, and some attributes of quasi-statehood, including in the post-Soviet era their own presidents, parliaments, constitutions, and flags. The latter, presumably inhabited almost exclusively by ethnic Russians, enjoy no special status.[2]

[1] Indeed, as indicated in a recent poll, 90 % of Russia's citizenry who self-identify as Muslims do not attend mosques (Malashenko 2009: 321).

[2] The ethno-federal structure of contemporary Russia is a Soviet legacy, designed to give the country's minorities a semblance of statehood. The Bolsheviks created internal autonomous

M.A. Derrick (✉)
Department of Geography, Humboldt State University, Arcata, CA 95521, USA
e-mail: mad632@humboldt.edu

S.D. Brunn (ed.), *The Changing World Religion Map*,
DOI 10.1007/978-94-017-9376-6_159

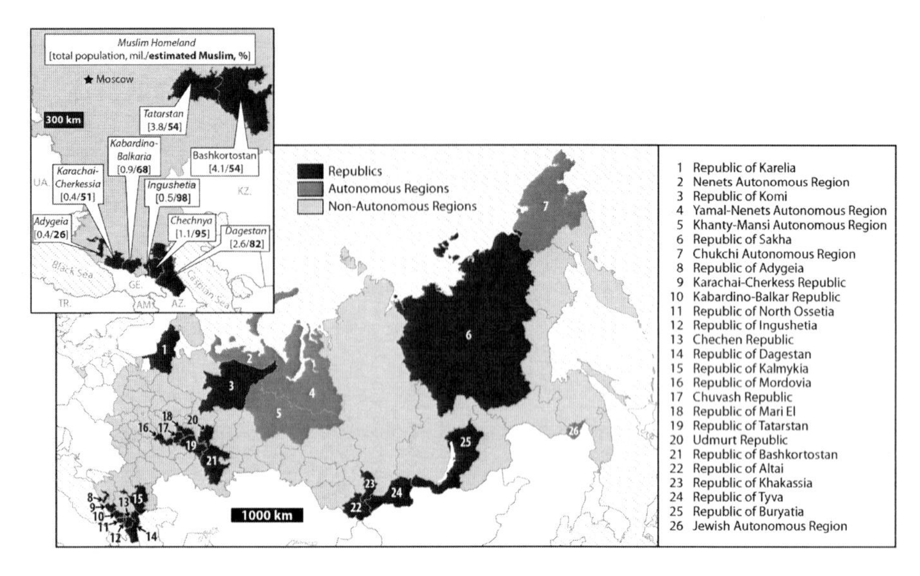

Fig. 159.1 The Russian Federation, showing republics designated as historic homelands of Muslim groups (Map by Matthew Derrick with Stephen Mangum)

The most important of the autonomous republics are those located in two areas, the Middle Volga Basin and the North Caucasus, which are designated as the historic homelands of Muslim groups. Tatarstan and Bashkortostan, combining to make the Middle Volga region Russia's largest Muslim area, are recognized as the homelands of the Tatars and Bashkirs, closely related ethnic groups of Turkic heritage. The first Muslims to be brought into the Russian Empire in the middle of the sixteenth century were Sunni Muslims who traditionally follow the Hanafi *madhhab* (juridical school of Islamic thought) (Mukhametshin 2006). In the North Caucasus, which Russia began colonizing in the latter part of the eighteenth century, six autonomous republics are recognized as the historic homelands of Muslim peoples: Dagestan, Chechnya, Ingushetia, Kabardino-Balkaria, Karachai-Cherkessia, and Adygeia. While this bloc claims fewer total Muslims than the Middle Volga, the North Caucasus hosts far greater ethno-religious complexity. Dagestan alone is home to dozens of Muslim nationalities, none forming an outright majority, speaking an array of languages. Sunni Islam is most widespread in the North Caucasus, yet its *ulema* (Islamic juridical scholars) are divided between the Shafi and Hanafi *madhhabs*. There are also pockets of Shiites in the region, and Sufism is widely prac-

republics, which, while distinguishing them from the 15 union republics, were intended to give Russia's minorities a semblance of statehood. The hierarchy of constituent units that was created after the 1917 revolution loosely mirrored the Marxist-Leninist conception of nationhood, which made a distinction between a "nation" (*natsiia*) and a "nationality" (*natsional'nost'* or *narodnost'*). A nation was viewed by Bolshevik theoreticians as a more developed cultural entity than a nationality. Therefore, nations warranted more self-government (Stoliarov 2003: 62).

ticed throughout the North Caucasus, historically associated with the mountain peoples' resistance to the Russian Empire (Makarov and Mukhametshin 2003: 134). But, again, Sufism in the North Caucasus is split into numerous *tariqats* (brotherhoods) that often correspond to clan or tribal loyalties.

The country's *umma* is further fragmented institutionally into dozens of different Muslim Spiritual Boards, hierarchical ecclesiastical organizations that have played an integral role in shaping Islamic revivalism in post-Soviet Russia. With roots in the Tsarist Empire, Muslim Spiritual Boards today are charged with monitoring the activities of the approximately 5,000 officially registered mosques – up from only 179 in the 1980s (Yemelianova 2003a: 54) – guiding learning in the country's burgeoning network of Islamic education centers, producing and disseminating religious literature, as well as a fulfilling a host of other tasks. Although Russia is constitutionally a secular country, state authorities maintain a heavy hand in the Muslim Spiritual Boards, relying on the Muslim organizations to communicate and reinforce "official" Islam. Official, or "traditional to Russia," understandings of Islam may vary among the different institutions and the diverse cultural and geographical contexts (Makarov and Mukhametshin 2003). In practice, however, official Islam as conceptualized and propagated via Russia's Spiritual Board system is above all defined by its "loyalty to the state" (Malashenko 2007: 92), the state in its modern form being an explicitly territorial, discretely bounded entity. Indeed, leading clerics of the Muslim Spiritual Boards are of one voice with governmental authorities in their support for official, *indigenous to Russia* Islam as a bulwark against religious fundamentalism (not to mention Islamic extremism and radicalism), which they frame as a foreign import that is wholly unsuitable to the conditions of the poly-confessional, multiethnic Russian Federation. In backing native/official Islam against nonnative/unofficial Islam, Russia's Muslim Spiritual Boards produce an inside/outside bordering effect in which preferred notions of faith and group identity are territorialized through institutional discourses and practices.

The question of territory has been at the heart of Islamic revivalism in post-Soviet Russia. Muslim-majority regions, most notably Chechnya and Tatarstan, drew on incipient religious rebirths in the early 1990s to fuel a "parade of sovereignties" (Kahn 2000) that threatened Russia's territorial integrity. These sovereignty campaigns, as examined in this chapter, to a significant degree account for the current institutional fragmentation of the Muslim Spiritual Board system. Whereas there were only two discretely delimited Muslim Spiritual Boards in Russia – one for each of the country's major Muslim regions – until the fall of the USSR, Muslim leaders of sovereignty movements seceded from these macro-scale institutions and reterritorialized Spiritual Boards to the political borders of regions designated as their historic homelands. Preferred understandings of Islam mixed with preferred senses of territory and were transmitted through the rescaled ecclesiastical institutions. A similar dynamic is at play today, only a different sense of territory – corresponding to that of the country as a unified polity – has been communicated and reinforced via the Muslim Spiritual Board system amid the political-geographical recentralization that has taken place since 2000.

The fundamental purpose of this chapter is to examine the relationship between territoriality and Russia's Muslim Spiritual Board system, showing how shifting territorial circumstances have influenced the character of religion propagated by this Islamic institution. To that end, the remainder of this chapter is comprised of four parts, beginning with overview of the concept of territoriality and the role played by institutions such as the Muslim Spiritual Board in socializing individuals and groups as members of distinct territories. The second section consists of an outline of the historical development of Russia's Muslim Spiritual Board system, showing how the establishment and subsequent development of the institution has been influenced by changes to the territorial structuration of the Russian state in the Tsarist, Soviet, and post-Soviet eras. Through the case study of the Muslim Spiritual Board of Tatarstan, one of Russia's most powerful Muslim-majority regions, part three illustrates how the institution, as well as the character of Islam it has supported, has been conditioned by post-Soviet Russia's political-territorial transformation. The chapter concludes with a discussion of the implications for Russia and how a consideration of the role institutions play in shaping territorial identities can serve as a counterweight to essentialist accounts of Islam.

159.2 Territorializing Identity

Over the past decade Russia's Muslim Spiritual Board system has attracted the attentions of a number of scholars (for example, Campbell 2005; Crews 2003; Matsuzato 2007; Matsuzato and Sawae 2010; Norihiro 2006; Yemelianova 2003b). Common to their studies, both in the historical and contemporary context, has been an acknowledgement of the role played by the ecclesiological institution in mediating relations between the state and Muslims, and in turn conditioning Muslim group identity and the character of Islam practiced in Russia. Missing from these analyses, however, has been an explicit consideration of territory, which in its predominant modern expression – the nation-state – presumes that cultural boundaries are more or less coterminous with political boundaries (Murphy 2005), and territoriality. Territoriality is "a primary geographical expression of power," according to geographer Robert Sack (1986: 3), who defines the term more fully as "the attempt by an individual or group to affect, influence, or control people, phenomena, and relationships, by delimiting and asserting control over a geographic area."

Sack emphasizes that territoriality is not an essential part of human behavior, unlike with animals, but rather a learned "strategy" in which power relations are reified through a threefold process of (1) *classification* of space, for example, "homeland" vs. "foreign land;" (2) *communication* of a sense of place, for example. the erection of boundaries; and (3) *enforcement* of control, for example, policing, surveillance (Sack 1986: 32–33). Following from this dynamic, the usefulness of territoriality as a control-oriented strategy lies in its efficiency. Power relations are depersonalized by moving attention away from individuals and to the entire extent of bounded space. Hence, territoriality shapes identity in its role of defining group membership, literally who is considered "in" and who is "out" of place. In the context of the territorial state, territoriality and its accompanying bordering processes are employed in the construction of the nation to "identify who is included and excluded" by accentuating "external difference or internal unity" (Herb 2004: 144).

Geographer Anssi Paasi builds on Sack's conceptualization of territoriality with his idea of "spatial socialization," which he defines as a constant "process" in which individuals and groups

> are socialized as members of specific territorially bounded spatial entities, participate in their reproduction and "learn" collective territorial identities, narratives of shared traditions and inherent spatial images (e.g. visions regarding boundaries, regional divisions, regional identities, etc.), which may be, and often are, contested. (Paasi 2009: 226)

Territories and identities, according to Paasi, are co-constructed through boundaries separating "us" from "them" via the process of Othering. These borders are not just the physical borders located at the edges of states. Rather, the borders of Paasi's spatial socialization, both discursive and material, are "spread" – however unevenly – throughout territories and permeate everyday life (Paasi 2008: 113). These lines of inclusion and exclusion, in their ubiquity, represent hidden power relations that are communicated foremost through institutions, including national school systems, politics, popular culture, government, media outlets, and multiple others via practices and discourses that serve to "nationalize everyday life" (Paasi 1999).

The question of territory clearly needs to be brought into the study of state-aligned Muslim organizations, along with the role they play in shaping Islamic identity. This chapter works from the basic understanding that the Muslim Spiritual Board system represents an expression of territoriality, influencing the character of Islam in Russia via institutionalized discourses and practices that communicate and reinforce a preferred sense of territory. Russia's *umma* is spatially socialized through the Spiritual Boards, learning collective territorial identities. This process is uneven, contested, and resisted. Preferred senses of territory that are reproduced through the institution are not static. Indeed, sense of territory within the context of Russia, which has undergone two major political-territorial upheavals in the past century, is particularly fluid and complex.

159.3 The Changing Territoriality of Russia's Muslim Spiritual Board System

Following her earlier decrees on religious tolerance, Catherine II ("the Great") established Russia's first Muslim Spiritual Board in the city of Ufa (currently the capital of Bashkortostan) in 1788[3] as a means to integrate the Muslims of the Middle Volga into the administrative structures of the Russian Empire (Campbell 2005).

[3] The Muslim Spiritual Board was relocated in the city of Orenburg (currently located in the Orenburg oblast of the Middle Volga region) in 1796, but returned to Ufa in 1802 (Azamatov and Usmanov 1999).

The creation of the Spiritual Board came as a response to Muslim revolts that resulted from previous waves of forced Christianization; it was also formed in anticipation of the empire's continued eastward expansion, a process in which Moscow would employ Volga-area Muslims as a "civilizing force" to promote Russia's interests among the "culturally less-developed" Central Asians (Yemelianova 2003a: 25). Roughly replicating the hierarchical administrative structure of the Russian Orthodox Church, the Muslim Spiritual Board (also known as the Muftiate) was headed by a state-appointed mufti to register and manage cadres imams and mullahs. Rank-and-file clergy in turn formed a bridge between Muslims and the state, monitoring mosque activity, overseeing the operation of Islamic schools, and fulfilling a range of other social functions (Zagidullin 2007).

The formation of the Muftiate, the first example of what geographer Jouni Häkli (2001) terms the "state-centered construction of society" for the empire's Muslims, precipitated a profound transformation in the character of Islam and group identity. The institutional standardization of religious doctrine produced an "Islamic orthodoxy" (Crews 2006) that, while formally based on the Hanafi *madhhab*, was Russian in the territorial sense and above all loyal to the state. *Abyzes* (elders) and *inshans* (respected Sufi leaders), who previously served as informal clergy in villages that were for all intents and purposes autonomous of state structures, were sidelined as the guardians of Muslim tradition; their "folk" Islam, mixed as it was with local and ethnic traditions, was decried by the Spiritual Board as being riddled with *bid'ah* (illicit innovation) (Mukhametshin 2006: 136). In the late nineteenth century and into the twentieth century the Muftiate directed similar accusations at modernist religious reformers (known as Jadidists, from the Arabic word for "renewal") in their midst, who were often accused of pan-Islamism as the state became the arbiter over the meaning of Islamic tradition in the empire.

The institutional authority of the Muslim Spiritual Board, a scalar shift upward away from the autonomy of localities, cultivated a regional Muslim identity associated with and loyal to the state (worship included blessing the tsar in recognition of his sovereignty), an imperial sense of territory, but it also kept Muslims of Russia segregated from Christians and other non-Muslims. The institutional segregation of Muslims contributed to what historian Robert Crews has termed the "confessionalization" of population and empire (2006: 8) – citizenship based not on an inclusive all-Russian (*rossiiskii*) territorial identity, but instead based on exclusive religious affiliation. Nonetheless, this religious-based segregation gave Russia's Muslims a significant degree of autonomy and room for negotiation in their relations with the state. While the relations between the empire and its Muslims, as institutionalized by the Spiritual Board, have been described as confederative (Norihiro 2006), they comprised a non-territorial confederation.

Against the background of broader and more intense institutional atheism, the Bolsheviks retained the Muftiate as a means of controlling Soviet Muslims by molding an Islam that was "fundamentally loyal to the Soviet state and generally pliant" (Ro'i 2000: 155). However, with the sweeping political-territorial transformation that resulted in the creation of Muslim homelands – the Union Republics of Central Asia, as well as the Autonomous Soviet Socialist Republics inside Russia (which that today comprise the country's ethnic republics) – the state now provided a secular education for Russia's Muslims, developing their ethno-national group identities while integrating them institutionally and socially with non-Muslims; via these ethno-nationally defined territories the state assumed jurisdiction over most social functions formerly coordinated by the Spiritual Board. By the late 1920s the Spiritual Board was permitted to perform only narrowly defined religious rituals as the state began persecuting clergy and destroying mosques. By the mid-1930s, after a decade of militant atheism, the institution was deprived of a mufti; its clergy and network of temples decimated, the Spiritual Board existed in name only (Usmanova et al. 2010: 40–43).

The Spiritual Board received a new lease on life during World War II when Stalin revived the institution in a bid to foster state patriotism among the Volga Muslims and mobilize them in the military campaign. He also ordered the creation of three new Muftiates: the Muslim Spiritual Board of the North Caucasus, based in Buinaksk (later Makhachkala); the Spiritual Board of Transcaucasia in Baku; and the Spiritual Board of Central Asia in Tashkent (Fig. 159.2).

The new Muftiates, according to historian Galina Yemelianova, "were designed to tighten state control over Soviet Muslims who were regarded by German commanders as a potential fifth column" (2003b: 139–140). At the war's conclusion the state returned to its antireligious policies, but all four Muslim Spiritual Boards – now discretely territorialized to cover the entire extent of the USSR – were nonetheless retained.

In the latter decades of the USSR's existence, the Muftiates' activities were severely limited and tightly controlled by Soviet authorities. Friday sermons at the few remaining official mosques, for example, had to be approved beforehand by the Council for Religious Affairs (CRA), a state organ that policed the Spiritual Boards and often forced Muslim religious leaders to teach their parishes in contradiction of their beliefs (Matsuzato and Sawae 2010: 346). Increasingly associated

Fig.159.2 Four Soviet-era Muftiates (Map by Matthew Derrick with Stephen Mangum)

with atheist Soviet regime, the dwindling number of official mosques attracted only a small stream of the faithful, and Islam in Russia once again became something akin to the "folk" religion that predated Catherine's establishment of the Muslim Spiritual Board, practiced in secret or led by untrained, semiliterate clergy who could carry out only basic rituals (Usmanova et al. 2010: 46–47).

In spite of its associations with the atheist regime, the institution of the Muslim Spiritual Board did not disappear alongside the USSR. On the contrary, since the collapse of the Soviet polity the four Soviet-era Muftiates have been succeeded by dozens of new Spiritual Boards. The Muslim Spiritual Board of Central Asia was promptly divided into five independent Mufitates coterminous with the political borders of the newly sovereign states of Central Asia (Kazakhstan, Kyrgyzstan, Uzbekistan, Tajikistan, and Turkmenistan); each of the new Muslim organizations in Central Asia has been employed by state leaders in their respective post-Soviet nation- and state-building projects.[4] In the Russian Federation the two Soviet-era Muftiates have given way to more than 40 new Spiritual Boards (Silant'ev 2008: 18–63). Unlike its previous discrete territoriality, the post-Soviet administrative-territorial arrangement of the Muftiate system is multi-scalar and overlapping, with

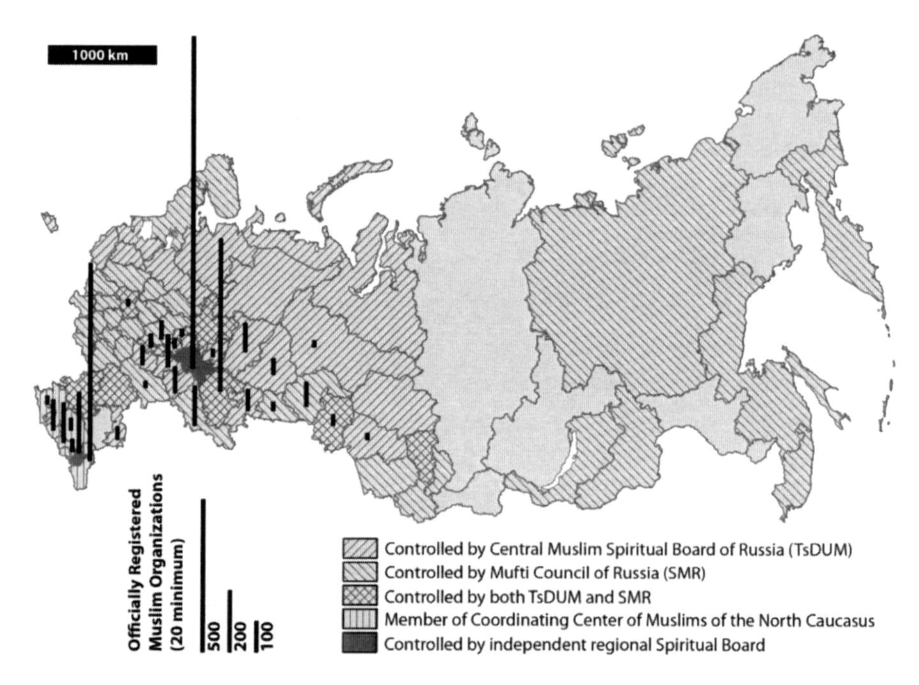

Fig. 159.3 The current administrative-territorial arrangement of Russia's Muslim Spiritual Board system, showing regional distribution of mosques (Map by Matthew Derrick with Stephen Mangum)

[4] Khalid (2007) provides an excellent examination of post-Soviet Islamic revivals in Central Asia.

two main national-level Spiritual Boards competing for the allegiances of regional-level Spiritual Boards and mosques (Fig. 159.3), as well as the Kremlin's favor.

The first national-level Islamic organization is the Central Muslim Spiritual Board of Russia (*Tsentral'noe dukhovnoe upravlenie musul'man Rossii* – TsDUM). As the successor organization of the Muslim Spiritual Board of the European Part of the USSR and Siberia (*Dukhovnoe upravlenie musul'man evropeiskoi chasti SSSR i Sibiri* – DUMES), TsDUM maintains its headquarters in the city of Ufa. It is a highly centralized, rigidly hierarchical organization that claims the loyalties of about 20 regional-level Muftiates, controlling a total of about 800 officially registered mosques (Silant'ev 2008: 28). TsDUM is led by Mufti Talgat Tadzhudin, who has held his post since 1980 and in 1990 was elected supreme mufti for life. Like his predecessors in the Tsarist Empire and the Soviet Union, Tadzhudin is a protector of a traditionalist, conservative "Islamic orthodoxy." A *gosudarstvennik* (in favor of a strong, unified Russia), he maintained his position as "state mufti" after the fall of the USSR, was integrated into Yeltsin's political establishment, and unambiguously opposed the territorial aspirations of Muslim-majority regions such as Tatarstan and Chechnya (Yemelianova 2003b: 146).

The second national-level organization, the Mufti Council of Russia (*Sovet muftiev Rossii* – SMR), was formed in the mid-1990s in opposition to TsDUM. Headed by Mufti Ravil' Gainutdin and headquartered in Moscow, SMR is distinguished by the decentralized character of its administrative structure (Hunter 2004: 57). More than 30 regional Muftiates, totaling about 1,000 mosques, are aligned with SMR (Silant'ev 2008: 41). Although free of associations with the "Islamic orthodoxy" that was propagated in previous eras, Gainutdin promotes a "Eurasian Islam," claiming that the religion as it has developed over the centuries in the geographical context of Russia mixes eastern and western traditions and, therefore, is particularly suited for life in a poly-confessional, multiethnic state (1999: 78).

A number of important regions – including most of the ethnic republics designated as the historic homelands of Muslim ethno-national groups – are not directly aligned with either of the national-level Muftiates. First is a collective of associate regions with membership in the Coordinating Center of Muslims of the North Caucasus (*Koordinatsionnyi tsentr musul'man Severnogo Kavkaza* – KTsMSK). This bloc includes the Spiritual Boards of Dagestan, Ingushetia, Kabardino-Balkaria, Karachai-Cherkessia and Stavropol, North Ossetia, and Adygea and Krasnodar. About 1,000 officially registered mosques, nearly 70 % of them in Dagestan alone, are contained within this territorial grouping (Silant'ev 2008: 51–53).

Two additional Muslim-majority republics host independent Muslim Spiritual Boards. The first is Chechnya. Home to nearly 200 officially registered mosques, Chechnya was an associate member of KTsMSK until announcing its independence in April 2012 (Malashenko 2012). The region's leading cleric, Sultan Mirzaev, formally justified his Spiritual Board's exit from KTsMSK with the claim that the organization was ineffective in combating the growth of religious extremism in the North Caucasus. Yet, because the decision was made with the full backing of Chechnya's strongman president, Ramzan Kadyrov (who also claims religious authority in the republic), questions of politics clearly were also at play (Malashenko

2012). The second independent Muftiate is the Muslim Spiritual Board of the Republic of Tatarstan (*Dukhovnoe upravlenie Respubliki Tatarstan* – DUMRT). Founded in 1992, DUMRT today oversees all but only a few of approximately 1,200 mosques within Tatarstan's borders (CRA 2009); this figure distinguishes DUMRT as Russia's largest centralized religious organization.

A number of explanations have been put forth to elucidate the institutional fragmentation of Russia's *umma* that has been effected since the demise of the Soviet Union, including the personal ambitions of muftis and the outside interference of powerful political and financial backers (for example, Bukharaev 1996; Tul'skii 2004; Yemelianova 2003b) – and similar arguments have been issued to explain why reunification of the institution continues to prove elusive (for example, Tul'skii 2004; Mel'nikov 2009). While these issues may indeed play a role, a more salient factor has been the political-territorial restructuration of the Russian Federation over the past two decades. Powerful Muslim-majority regions such as Chechnya and Tatarstan most aggressively pursued sovereignty in the 1990s. It was in these regions that the fragmentation of Russia's Spiritual Board system started, and it is these regions that most jealously guard the independence of their Muftiates today. To gain some insight into the relationship between territoriality and the institutions of Islam, as well as its implications for politics and Muslim group identity in post-Soviet Russia, the following section examines the conditions that gave rise to the formation of an independent Muftiate in Tatarstan in the early 1990s and important developments to the institution thereafter.

159.4 Case Study: The Muslim Spiritual Board of the Republic of Tatarstan

The formation of the Muslim Spiritual Board of the Republic of Tatarstan (DUMRT) was conditioned by two interrelated factors. First was the political-territorial decentralization of the Russian Federation that followed the USSR's collapse. Second was the response to those forces by the Tatar national movement, which, after Tatarstan's 1990 sovereignty declaration, viewed an independent Muftiate as an important institution for the redevelopment of Tatar national consciousness and a key attribute of sovereign statehood. The national movement, led by the centrist Tatar Social Center (*Tatarskii obschestvennyi tsentr* – TOTs) as well as the more radical Ittifak ("Alliance"), discursively framed DUMES, the all-Russian Muftiate based in Ufa, as a colonial instrument of a Moscow-centered state striving to recreate a "unitary-totalitarian type of government" (TOTs 1991: 134). The national organizations also rejected the traditionalist "Islamic orthodoxy" propagated by DUMES as anti-intellectual and backward-looking, limiting the Tatars' national development; instead they put their support behind a modernizing "Tatar Islam," understood then as a revived version of Jadidism, the indigenous liberal religious reform movement developed by Tatar intellectuals in the latter part of the nineteenth century (Iskhakov 1992; Mukhametshin 1994).

Fig. 159.4 Kazan Kremlin (Photo by Matthew A. Derrick)

Allied with the Tatar political leadership, the national organizations took the initiative in seceding from DUMES in August 1992 and establishing a new Muftiate headquartered in Kazan (Fig. 159.4), the capital of Tatarstan. Gabdullah Galiullin, one of the new generation of young imams and allied with TOTs and Ittifak (Malashenko 1992: 3), was elected supreme mufti of DUMRT. In justifying Tatarstan's exit from DUMES, Gailiullin said, "We want to address the problems not included in those dealt with by the Spiritual Board and its headquarters in Ufa. What we find acceptable is not often acceptable for the Spiritual Board operating in Russia" (quoted in PostFaktum 1992: 1). In making a political-geographical distinction between Russia and Tatarstan, Galiullin also drew on the historical memory of the Muslim Kazan Khanate and the loss of Tatar statehood, echoing the primary political discourse underpinning Tatarstan's sovereignty claim, declaring, "They [that is, Russians] have kept our ancestors and us down under their crosses since 1552" (quoted in Galimov 1992: 1).

The formation of DUMRT was followed by the establishment of an *Ulema* Council, which consisted of academics, intelligentsia, and national leaders associated with TOTs and Ittifak. The *Ulema* Council's support for a modernizing Tatar Islam that embraced *ijtihad* (independent interpretation of the Koran and Sunnah) and secular learning was fundamentally shaped by political-territorial aspirations, as evidenced in Rashad Amirkhanov's take on the Jadidists' legacy of religious reform. He wrote that their "Islamic teachings, which stimulated a progressive reformation of the Muslim's life" a century earlier, were needed by Tatars today

> to fill the ideological vacuum, formed as a result of the crisis of a totalitarian ideology, and return to the people its faith in spiritual values. Islam is to educate a generation of thinkers and believers on the basis of humanistic principles. The fate of our people is largely deter-

> mined by the ability of the new generation to use this historical chance given to us by the crisis of an imperial state and follow the struggle for national sovereignty through to the end. (1996: 28)

Amirkhanov maintained that the "function" of Islam foremost was to serve the Tatar nation and the nascent Tatar state against what he viewed as a revanchist, innately imperialist power opposed to a people's right to self-determination (1996: 28). The neo-Jadidism promulgated by DUMRT, although grounded in universal "humanistic principles," was defined foremost against the colonial Russian Other.

The national movement spoke in a single voice for the formation of an independent Muftiate of Tatarstan, but the divide between the moderate and radical wings widened and hardened over the terms of the 1994 bilateral treaty with Moscow, which ambiguously acknowledged Tatarstan as a "State … united with the Russian Federation" (Treaty 1994: Preamble). The division in the national movement, based on the treaty's glaring elision of the question of Tatarstan's sovereignty, in turn affected the character of DUMRT. The centrist leadership of TOTs put its weight behind the political leadership's policy of "gradualism" (Malik 1994), understanding that the case for Tatarstan's sovereignty would be built methodically over time through institutions, notably among them DUMRT. Viewing the Muftiate as part of a long-term nation- and state-building project that was based on the 1994 bilateral treaty, TOTs declared that "the Islamic community and the Spiritual Board of Tatarstan, not yielding to the influence of any political party, organization, or group, must serve the entire nation and the state" (1996: 1).

Mufti Galiullin, however, became more associated with Ittifak, whose leader, Fauiziia Bairamova, decried the 1994 treaty because, as she contended, it "took everything from us, most of all our national aspirations, our aspirations for independence" (2004). Allied with Ittifak, the head of DUMRT nurtured a confrontational relationship with Tatarstan's political establishment (Khabibullin 2008: 80–81), echoing the radicals' accusation of betraying the cause of the Tatars' independent statehood in favor of melding a state-defined national polity – a multiethnic "Tatarstani people" that included both Tatars and Russians of the republic.[5] Thus, with conflicts arising over diverging understandings of homeland, Galiullin opposing the political elite and a number of imams and parishioners following his lead, DUMRT enjoyed no regular support, material or otherwise, from the government (Yakupov 2005: 76–78).

The future shape of DUMRT was decisively thrown into question in late 1996, when Galiullin's allies in Ittifak took part in the formation of an oppositional political party called Muslims of Tatarstan (Mukhametshin 2005: 176). The party's creation coincided with a distinct shift in the character of Islam propagated by the radicals. Bairamova abandoned her previous support for Jadidism, claiming that the Tatars' religious reform had "destroyed Islam from within" (1997: 22). She wrote,

---

[5] Tatars comprise just over half of Tatarstan's population, while ethnic Russians make up nearly 40 %; other Orthodox peoples comprise most of the remainder.

> If the essence of Jadidism (reformism) is reduced to the renewal of religion under the guise of progress, pushing the nation toward infidels and battling against fundamental Islam, then the absolute unbelief of the Tatar people can be viewed as a telling result of Jadidism.

Bairamova's repudiation of Jadidism, along with the explicit politicization of Islam that came with her participation in the political party Muslims of Tatarstan, represented a decisive split in the Tatar national movement. Whereas the national movement, theretofore, collectively supported the nationalization and modernization of Islam under the banner of Jadidism, seeking its propagation through DUMRT, Ittifak now called for the Islamization of the Tatar nation, a return to explicit faith based on the Koran, not universal humanistic values. The radicals' shifting register that incorporated an ambiguous "fundamental Islam," however, was not based on any theological argumentation. Rather, it was driven by territorial grievances coming out of the 1994 treaty. While Mufti Galiullin publicly maintained his adherence to the principles of Jadidism and also worked with the moderates from TOTs who largely comprised the *Ulema* Council (Interfax 1996), his alliance with Ittifak not only brought him into opposition with the political authorities, but also contributed to a renewed division in Tatarstan's *umma*, reflective of the hardened split in the national movement.

For most of the 1990s, Tatar President Mintimer Shaimiev remained behind the scenes in his involvement with DUMRT, allowing the national leadership to guide the Muftiate's development. However, by the latter part of the decade the Tatar national movement and its involvement in Spiritual Board had become a liability to the republic's sovereignty project, which, since the 1994 treaty, was dependent on cultivating state (not ethnic) nationalism. With its mufti allied with nationalists who maintained a vision of Tatarstan as the Tatars' independent homeland, DUMRT had become an oppositional organization rather than an institution that fulfilled its envisaged purpose of contributing to Tatarstan's regime of territorial legitimation (Murphy 2005). Facing these challenges to the post-treaty territorial status quo, Shaimiev mobilized resources to assure the election of his hand-picked mufti, Gusman Iskhakov, a more pliant imam and a conservative traditionalist who had developed good relations with Tatarstan's political elite, at the Muslim Congress of Tatarstan in February 1998 (Todres 1998; Sanin 1998; Vetlugin 1998). Shaimiev justified his role in engineering the election:

> Although separate from state, religion is not separate from society. Therefore the state recognizes the necessity of constructive cooperation with religion in solving many socially important problems. Foremost that means the consolidation of society. (Shaimiev 1998: 3)

With the state and the new mufti together claiming to serve the unity of the people the "multinational Tatarstani people," not necessarily the Tatar nation DUMRT abandoned its previous dedication to religious reform and officially rededicated itself to the promotion of a so-called "traditional" Tatar Islam based on fealty "to the religious school (*madhhab*) of Imam Abu Hanifa" (Ustav 1998: 38–39).

The 1998 reformation of the Muftiate signaled an end to the direct influence of the Tatar national movement on the institution's development. That the Kazan-based government was able to engineer these changes to DUMRT without a significant backlash is indicative of both the national movement's success and its declining importance. The creation of DUMRT was only one part of a much larger program spearheaded by TOTs, Ittifak, and other organizations to redevelop Tatar national culture and to bolster Tatarstan's sovereignty claim. Although the latter of these twin goals was only nebulously formalized in the 1994 bilateral treaty, Tatarstan nonetheless in many ways operated as an independent state in the latter part of the 1990s (Stepanov 2000; Derrick 2008). As for the former, the Tatar language was taught almost universally in the republic's schools as a required course, and Tatarstan appeared to be on the way to accomplishing the stated goal of functional bilingualism (Cashaback 2008). In short, although mosques had multiplied throughout the region, the broader Tatar national revival proceeded apace through a variety of secular institutions. With its primary goals seemingly having been met, the influence of the Tatar national movement was already in sharp decline by the latter part of the 1990s (Iskhakov 1997: 105–118).

Unlike the neo-Jadidism inchoately formulated by the national intelligentsia for much of the 1990s, traditional Tatar Islam as advocated by DUMRT today requires active observance and practice of prescribed religious principles and rituals. Yet the religious requirements of this official Islam do not detract from its national orientation. If reformist Tatar Islam was cast foremost as an attribute of national culture by the national leadership of the first part of the 1990s, the post-1998 Spiritual Board represents Islam as the very core of the Tatar nation. Mufti Iskhakov made this point clear following his election: "Islam, tightly interwoven with the customs and moral codes of our people, defines our national identity" (1998: 27). In this understanding, religion and national identity cannot be separated; Islam is the very foundation of Tatar national identity, not simply a cultural attribute of the nation.

The national movement of the 1990s hailed Jadidism for its ostensible modernizing qualities, which they associated with Europe, their geo-civilizational orientation and model for Tatarstan. In contrast, the post-1998 DUMRT emphasizes the Tatars' specific local and regional geo-history, which its leaders claim has formed a distinctive Tatar Islam over the centuries. Mufti Iskhakov stressed that Tatars have developed as a unique Muslim people, their identity and religion fundamentally having been shaped by their relative isolation from the broader Islamic world and their adaptation over the centuries to living within a non-Muslim state (1998: 26). Because the Hanafi *madhhab* is distinguished by its "particular tolerance toward other religions and the capacity to coexist with culturally dissimilar peoples," the mufti contended, traditional Tatar Islam is "suited to our local conditions" (1998: 27). The Tatars' ancestors could recognize the authority of a Christian ruler and dwell among Orthodox Russians, he said, but only their strict observance of long-established religious traditions saved them from total assimilation.

Mufti Iskhakov's emphasis on the tolerant nature of Tatar traditionalism was codified in DUMRT's 1998 charter. A primary goal of the organization was "preserving inter-confessional and inter-national [that is, interethnic] peace and accord in society" (Ustav 1998: 38). This aim was calibrated to complement Shaimiev's concept of a "multinational Tatarstani people" (Derrick 2008: 81–83), thereby enhancing the legitimacy of Tatarstan's sovereignty claim and contributing to the maintenance of the territorial status quo. Whereas the Tatar president formerly discussed mutual tolerance as expressed between the Tatar and Russian "peoples" [*narody*] in the republic, following DUMRT's reformation he shifted focus to the tolerance he insisted exists between Islam and Orthodoxy in the region, claiming that "the kind of inter-confessional relations that have formed in Tatarstan are a model for the rest of Russia" (quoted in Zargishiev 1999: 4). Thus, DUMRT and the Kazan-based state synchronized discursive borders of inclusion/exclusion, framing social relations within the republic as qualitatively different (better) than in the rest of Russia, reinforcing notions of a certain tolerance arising from a centuries-long shared history and geography between Muslim Tatars and Orthodox Russians within the contemporary borders of Tatarstan.

DUMRT's promulgation of tolerant traditionalism was originally intended to counter previously institutionalized representations of Tatar Islam defined against an innately imperial Russian Other and thereby assist the political leadership in cultivating a state nationalism that would contribute to Tatarstan's pretensions to sovereign statehood. With the political-territorial recentralization of the Russian Federation that has been effected under Vladimir Putin's watch, however, DUMRT's message of simple tolerance has shifted to one of mutual respect and even a significant degree of similarity between the two main religions practiced in the Middle Volga region. Former first deputy Mufti Valiulla Yakupov (2009), for instance, hailed the "experience of peaceful coexistence of the two traditional religions" – Russian Orthodoxy and what he called "orthodox Islam" (*ortodoksal'nyi Islam*) – as being characterized by

> not mere tolerance, but in fact deep respect for each other. The two religions cooperate in Tatarstan, complement each other. Tatars and Russians have lived side by side in the Volga area for centuries, our religions have evolved in a close dialogue with each other. And therefore our faiths actually have a lot in common.[6]

A short distance separated the discursive *konsolidatsiia* ("consolidation") of a "multinational Tatarstani people" shaped by Shaimiev after the 1994 treaty and Putin's discourse of the *konsolidatsiia* of a "multinational Russian people" who are united by the fate of sharing "religions traditional to Russia" (Putin 2003). In short, official Islam promoted by DUMRT today, by the former deputy mufti's own accord, is a rescaled version of the Islamic orthodoxy promoted and reinforced by the Spiritual Board in Tsarist and Soviet Russia.

Considering that DUMRT is recognized as the single centralized religious organization in Tatarstan and the institution supports only Tatar traditionalism based on the Hanafi *madhhab*, all other Islamic movements in the republic are *de facto* understood as unofficial. However, the most important target of the Spiritual Board's censure is Wahhabism. DUMRT representatives decry Wahhabism as an aggressive "sect" that, in its pretensions to universalism, poses a distinct threat to the inter-confessional harmony in Tatarstan and, moreover, to Tatar culture itself (Yakupov

---

[6] This message was repeated in multiple interviews by the author with imams at mosques in Kazan and other parts of Tatarstan. For his part, Yakupov has gone so far to identify "theological reasons for the compatibility of Islam and Orthodoxy in Tatarstan" (2006: 433–435).

2006). Yet Wahhabism became a salient feature of social and political discourse in Tatarstan only in August 1999, after then-Prime Minister Putin reignited warfare in Chechnya. The presence of some foreign militants, including Arab *mujahideen*, fighting against Russian troops allowed Moscow to frame the conflict as a campaign to liberate Chechens from the clutch of international Islamic extremism and thereby restore the country's territorial integrity (Ware 2005). The proliferation of Wahhabism in the North Caucasus, along with the justifications proffered by Moscow in its bombardment of Grozny, served as a warning to DUMRT and the political establishment of Tatarstan to cleanse the republic's Muslim communities of, as Mufti Iskhakov put it weeks after the renewed warfare in Chechnya, "alien, nonnative doctrines" that have the potential of "destroying society's traditional soil" (1999: 18). Since the start of the second Chechen war, accusations of Wahhabism have become a political weapon of sorts, deployed first by opponents of DUMRT outside of Tatarstan and eventually by different factions within Tatarstan's Muslim community itself. Anxieties of Chechenization and the political power in allegations of being friendly to religious fundamentalism endure as drivers behind DUMRT's policing of the republic's mosques.

DUMRT's campaign to keep "nonnative" versions of Islam out of mosques under its purview, however, is not simply a prophylactic measure. With the Putin-era recentralization having denuded Tatarstan of much of its formal autonomy, the temple is seen today as one of the last bastions of the post-Soviet revival of Tatar culture. The Muftiate's leadership is insistent that Tatar remain the language of official Islam in the mosques of Tatarstan, as has historically been the case in the region. All *medresses* (Islamic secondary schools) in Tatarstan, in addition to providing a religious education, also require *shakirds* (students at *medresses*) to take courses on the history of the Tatar people, Tatar language, and Tatar literature (Khabutdinov 2005). Additionally, many mosques now offer similar courses for non-*shakirds*. Because the number of Tatar-language schools has precipitously dropped, as has the number of hours devoted to the Tatar language in ethnically mixed schools in the republic (Suleymanova 2010),[7] Islam is once again said to "fulfill the task of preserving the Tatar nation" from all-out assimilation (Yakupov 2006: 25).

DUMRT's campaign against unofficial Islam also aids in fulfilling an important political function by ensuring that Tatarstan, although disabused of its pretensions to sovereign statehood, retains a significant degree of informal autonomy in its relations with the Moscow. Controlling some 1,200 mosques – about one-quarter of all Muslim communities in Russia – DUMRT is the largest Spiritual Board in Russia (CRA 2009; Silant'ev 2008: 217). As such, DUMRT contributes to a relationship between Russia and Tatarstan that in many ways can be described as a confessional confederation, which, unlike in the Tsarist era, today is territorial. While no longer the capital of sovereign Tatarstan, Kazan can rightly claim the title of Russia's "Muslim capital" (Garaev 2009), ensuring itself a significant role in the country's push to expand diplomatic and economic relations with Muslim countries.

---

[7] Additionally, the Tatar language is being squeezed out of state-sponsored broadcast media.

To ensure that mosques remain spaces where official Islam is followed, DUMRT relies on a number of reinforcement mechanisms. First, since its 1998 reformation the Spiritual Board maintains a close relationship with the Council for Religious Affairs (CRA) (Nabiev 2002), a state agency that, much as was the case in the Soviet era, is associated with internal security and intelligence forces and retains its duties in monitoring mosques in the republic. Second, the territorial-administrative arrangement of the Spiritual Board aids in reinforcing the dictates of the institution's central apparatus. Each of the region's mosques is arranged within one of 45 regional subdivisions (*mukhtasibaty*) that are coterminous with the administrative-geographical districts (*raiony*) of the republic. Each of the 45 regional subdivisions is in turn arranged within one of nine meso-scale regions (*khaziiaty*), each headed by an imam-*khaziiat* who reports to the supreme mufti of the central apparatus. Third, DUMRT reinforces official Islam through accreditation and licensing procedures. Since 2000 all religious schools in the region have been subject to the centralized institution (Mukhametshin 2005), which has implemented a standardized educational program that trains imams and mullahs steeped in the traditions of the Hanafi *madhhab*. Although each Islamic community has the right to select its own clergy, imams and mullahs must first be approved by the central apparatus of DUMRT (Yakupov 2009). In addition to policing, administrative, and accrediting oversight, the Spiritual Board also produces an array of Islamic literature that helps communicate and reinforce preferred religious expressions and practices, while at the same time making sure that religious literature legally deemed as "extremist" is kept out of all mosques under its jurisdiction.

For the nationalist organizations that led the formation of DUMRT in 1992, the political-territorial transformation that has been effected since 2000 has served to intensify the Islamist register to which they had begun shifting by the latter part of the 1990s. Ittifak's transformation from a "radical nationalist" to a "radical Islamist" organization was already underway before the 1998 reformation of DUMRT. TOTs shifted to a more Islamist register only after the federal center had unambiguously launched its campaign against Tatarstan's sovereignty claim, as witnessed in a resolution accepted at the organization's 2002 congress. Stating that "religion and the national movement are elements of a single social system, tightly intertwined, and complement each other," TOTs echoed its earlier assertions. But the declaration went on to urge Muslim Tatars

> to actively fight for freedom and independence of peoples and nations in this world, not in the life hereafter. Those believers who fall for the false exhortations of the allies of colonial policies commit a great sin, attempting to maintain neutrality, not differentiating good from evil, just preparing themselves for the next life … [A]ccording to the canons of Islam, only the Muslim who in this life actively fights for his rights and the rights of his people for freedom and independence will be permitted entrance through the heavenly gates … The fight for national and state sovereignty of the Tatar people and the Republic of Tatarstan is the sacred duty of the Muslim in this life, prescribed to him by the Koran. (TOTs 2002: 24)

TOTs offered a thinly veiled denunciation of DUMRT's leaders for being "allies of colonialism" who maintain "neutrality" in the face of perceived injustice, an accusation repeated in a more recent declaration in which the nationalist organization paints the Muftiate as an imperial instrument "for the control of Muslim clergy and the brainwashing of Muslims" (TOTs 2011). Like the leadership of Ittifak, TOTs has moved to a position in which negotiations with "colonial rulers" are now viewed as futile. Only an active fight "in this life" will secure Tatarstan as the Tatars' namesake nation-state; to fight for independence is a "sacred duty" laid out in the Koran. While this argument was framed in a larger struggle going on in the Muslim world, TOTs understands the *umma* as fundamentally divided into distinct peoples and nations, each with a right to its own sovereign homeland. Thus, TOTs' shift to an Islamist register, like Ittifak's, is the result of unfulfilled territorial demands.

159.5 Conclusion

State patronage of traditional Tatar Islam via Tatarstan's Muslim Spiritual Board, although initiated before the aggressive political-territorial recentralization of the Russian Federation, has been part of a countrywide effort in state- and nation-(re) building that began with the renewed conflagration in Chechnya. Under Putin's watch the state has supported versions of Islam deemed "traditional to Russia" in almost every Muslim-majority region of the federation, with "traditional" being defined foremost by perceived loyalty to the state, while simultaneously "categorically rejecting the legal existence of Islamic opposition and mercilessly suppressing any appearance of political protest in religious form" (Malashenko 2008: 2). This approach has strong parallels to the way the empire, beginning with Catherine II and continuing through the Soviet era, dealt with Islam. This chapter's examination of the role played by DUMRT in the Islamic revival in post-Soviet Tatarstan points out the need for additional in-depth scholarly investigations into how "traditional" Islam is represented, produced, reproduced, and contested via institutions and symbols in other Muslim-majority regions of Russia. Each of the country's Muslim regions is unique in its ethnic makeup and historical experience with the empire. Therefore, empirically rich comparative studies could enhance our understanding of the changing place of Islam in contemporary Russia, a federation that in many respects appears to be well on its way to becoming a unitary state (Oversloot 2009).

Yet this chapter carries a more basic implication for scholarly research on Islam's social expression. While Islamic revivals today may be conditioned by sectarian differences, language, history, socio-economic development, and other factors, all cases of Islamic revival take place within a modern political-territorial order that assumes states represent a defined "people," understood as a nation. Two prevalent paradigms in the social scientific interrogation of Islam, however, preclude serious consideration of the role territory plays in shaping Islamic identity. The first paradigm, common in the academic International Relations (IR) community, may be termed "Islamic exceptionalism," which views Islam as fundamentally incompatible with the modern political-territorial order. IR neorealism *à la* Samuel Huntington (1996) is most notorious for asserting the *umma*'s innate aversion to its territorial division, but advocates of IR's cultural turn, although more sympathetic to

Islam, similarly argue that Muslims are averse to Western notions of national community (for example, Pasha 2003; Mandaville 2011). A second paradigm, termed the "comparative fundamentalisms" model, views the current upswing in Islamic fundamentalism as part of a global shift toward stricter forms of faith that is witnessed among all major world religions as a result of the "crisis of the nation-state" brought about by globalization (Castells 2010: 21).

As suggested in this chapter's case study, neither of these paradigms, in positing Islam's innate aversion to the nation-state or a globalization-fuelled crisis of the nation-state giving rise to Islamic fundamentalism, would likely provide a plausible account of the Tatars' post-Soviet Islamic revival. From the beginning, the Tatars' religious renaissance was tightly connected with the revival of national culture and the pursuit of sovereign statehood. It is virtually impossible to separate Islam from Tatar national identity – the question is the relative weight religion assumes in the balance of Tatar national identity, depending on the political-territorial circumstances. The "pure Islam" espoused by the more radical Tatar nationalists was not a result of globalization, but came in tandem with intensified grievances and territorial demands. Neither paradigm would be able to differentiate expressions of the revival of Tatar traditionalism from expressions of Wahhabism in Tatarstan, as both phenomena would be grouped together under a single heading: Islamic Fundamentalism. For those seeking to understand the nature and significance of Islam as a social and political force in the contemporary world, the task at hand is to go beyond the perception that Muslim identities and practices are somehow incompatible with the nation-state and interrogate the ideas and ideologies underpinning the modern political-territorial order that condition Islam's social and political expression.

References

Amirkhanov, R. (1996). Tatarskaia natsional'naia ideologiia: Istoriia i sovremennost'. *Panorama-Forum, 1*(4), 27–38.

Azamatov, D., & Usmanov, K. (1999). *Orenburgskoe magametanskoe dukhovnoe sobranie v kontse XVIII-XIX vv*. Ufa: Izdatel'stvo Gilem.

Bairamova, F. (1997). O tatarizme i musul'manstve. *Tatarstan, 8*, 22–24.

Bairamova, F. (2004). Author's interview with Bairamova, Naberezhnye Chelny, 24 April.

Bukharaev, R. (1996). Islam in Russia: Crisis of leadership. *Religion, State & Society, 24*(2/3), 167–182.

Campbell, A. (2005). The autocracy and the Muslim clergy in the Russian Empire (1850s-1917). *Russian Studies in History, 44*(2), 8–29.

Cashaback, D. (2008). Assessing asymmetrical federal design in the Russian Federation: A case study of language policy in Tatarstan. *Europe-Asia Studies, 60*(2), 249–275.

Castells, M. (2010). *The power of identity*. West Sussex: Wiley-Blackwell.

CRA. (2009). Data provided to the author from the Council for Religious Affairs, Kazan.

Crews, R. (2003). Empire and the confessional state: Islam and religious politics in nineteenth-century Russia. *The American Historical Review, 108*(1), 50–83.

Crews, R. (2006). *For prophet and tsar: Islam and empire in Russia and Central Asia*. Cambridge: Harvard University Press.

Derrick, M. (2008). Revisiting "sovereign" Tatarstan. *Journal of Central Asian and Caucasian Studies, 3*(6), 75–103.

Gainutdin, R. (1999). Islamskoe vozrozhdenie v Rossii. *Mir Islama, 1*(1), 71–78.

Galimov, V. (1992). Chernoe i beloe? *Izestiia Tatarstana*, 3 September, 1.

Garaev, D. (2009). Esli my khotim byt' musul'manskim tsentrom – nado aktivnee rabotat'. *IslamInfo, 4*(36), 3.

Häkli, J. (2001). In the territory of knowledge: State-centered discourses and the construction of society. *Progress in Human Geography, 25*(3), 403–422.

Herb, G. (2004). Double vision: Territorial strategies in the construction of national identities in Germany, 1949-1979. *Annals of the Association of American Geographers, 94*(1), 140–164.

Hunter, S. (2004). *Islam in Russia: The politics of identity and security*. Armonk: M.E. Sharpe.

Huntington, S. (1996). *The clash of civilizations and the remaking of the world order*. New York: Touchstone.

Interfax. (1996). *Islam v Rossii na poroge novogo raskola – muftii Tatarii*. Retrieved October 8, 2012, from www.interfax-religion.ru/?act=archive&div=14384

Iskhakov, D. (1992). Neformal'nye ob'edineniia v sovremennom tatarskom obschestve (istoriia formirovaniia, struktura, programmnye polozheniia). In D. Iskhakov & R. Musina (Eds.), *Sovremennye natsional'nye protsessy v Respublike Tatarstan* (Vol. 1, pp. 5–51). Kazan: Academy of Sciences of Tatarstan.

Iskhakov, D. (1997). *Problemy stanovleniia i transformatsiia tatarskoi natsii*. Kazan: Academy of Sciences of Tatarstan.

Iskhakov, G. (1998 [1999]). Obraschenie k musul'manam Tatarstana vo imia Allakha milostivogo, miloserdnogo. In V. Yakupov (Ed.), *Materialy pervogo (ob'edinitel'nogo) s'ezda musul'man Republiki Tatarstan* (pp. 26–28). Kazan: Iman.

Iskhakov, G. (1999). *Islam za mir (doklad na nauchno-prakticheskoi konferentsii 1.10.1999)*. Kazan: Iman.

Kahn, J. (2000). The parade of sovereignties: Establishing the vocabulary of the new Russian federalism. *Post-Soviet Affairs, 16*(1), 58–89.

Khabibullin, R. (2008). *Istoriia Islama u Tatar*. Kazan: Iman.

Khabutdinov, A. (2005). *Islam v Tatarstane v pervye gody novogo tysiacheletiia*. Retrieved October 8, 2012, from www.idmedina.ru/books/islamic/?455

Khalid, A. (2007). *Islam after communism: Religion and politics in Central Asia*. Berkeley: University of California Press.

Makarov, D., & Mukhametshin, R. (2003). Official and unofficial Islam. In H. Pilkington & G. Yemelianova (Eds.), *Islam in post-Soviet Russia: Public and private faces* (pp. 117–163). London: RoutledgeCurzon.

Malashenko, A. (1992). Muftii, mully, imamy i politika. *Nezavisimaia Gazeta*, 24 October, 3

Malashenko, A. (2007). Gosudarstvo i islam v possovetskoi Rossii. *Svobodnaia Mysl', 3*(1574), 92–104.

Malashenko, A. (2008). Islam and state in Russia. *Russian Analytical Digest, 44*, 1–5.

Malashenko, A. (2009). Islam in Russia. *Social Research, 76*(1), 321–358.

Malashenko, A. (2012). *All is not quiet in Russian Islam*. Retrieved October 8, 2012, from http://carnegieendowment.org/2012/05/16/all-is-not-quiet-in-russian-islam/az90

Malik, H. (1994). Tatarstan's treaty with Russia: Autonomy or independence? *Journal of South Asian and Middle Eastern Studies, 18*(2), 1–36.

Mandaville, P. (2011). Transnational Muslim solidarities and everyday life. *Nations and Nationalism, 17*(1), 7–24.

Matsuzato, K. (2007). Muslim leaders in Russia's Volga-Urals: Self-perceptions and relations with regional authorities. *Europe-Asia Studies, 59*(5), 779–805.

Matsuzato, K., & Sawae, F. (2010). Rebuilding a confessional state: Islamic ecclesiology in Turkey, Russia and China. *Religion, State & Society, 38*(4), 331–360.

Mel'nikov, A. (2009). *Mirazh islamskogo edinstva*. Retrieved October 8, 2012, from http://religion.ng.ru/events/2009-12-23/1_islam.html

Mukhametshin, R. (1994). Dinamika islamskogo faktora v obschestvennom soznanii tatar XVI-XX vv (istoriko-sotsiologicheskii ocherk). In D. Iskhakov & R. Musina (Eds.), *Sovremennye Natsional'nye Protsessy v Respublike Tatarstan* (Vol. 2, pp. 99–115). Kazan: Academy of Sciences of Tatarstan.

Mukhametshin, R. (2005). *Islam v obschestvennoi i politicheskoi zhizni tatar i Tatarstana v XX veke*. Kazan: Tatar Book Publishing.

Mukhametshin, R. (2006). Islam v tatarskom obschestve: Poisk putei vyzhivaniia. In R. Mukhametshin (Ed.), *Islam i musul'manskskaia kul'tura v Srednem Povolzh'e: Istoriia i sovremennost'* (pp. 133–141). Kazan: Academy of Sciences of the Republic of Tatarstan.

Murphy, A. (2005). Territorial ideology and interstate conflict. In C. Flint (Ed.), *The geography of war and peace: From death camps to diplomats* (pp. 280–296). Oxford: Oxford University Press.

Nabiev, R. (2002). *Islam i gosudarstvo: Kul'turno-istoricheskaia evoliutsiia musul'manskoi religii na evropeiskom vostoke*. Kazan: Kazan State University.

Norihiro, N. (2006). Molding the Muslim community through the tsarist administration: Mahalla under the jurisdiction of the Orenburg Mohammedan Spiritual Assembly after 1905. *Acta Slavica Iaponica, 23*(1), 101–123.

Oversloot, H. (2009). The merger of federal subjects of the Russian Federation during Putin's presidency and after. *Review of Central and East European Law, 34*(2), 119–135.

Paasi, A. (1999). Nationalizing everyday life: Individual and collective identities as practice and discourse. *Geography Research Forum, 19*, 4–21.

Paasi, A. (2008). Territory. In J. Agnew, K. Mitchell, & G. Toal (Eds.), *A companion to geography* (pp. 109–122). Malden: Blackwell Publishing.

Paasi, A. (2009). Bounded spaces in a "borderless world:" Border studies, power and the anatomy of territory. *Journal of Power, 2*(2), 213–234.

Pasha, M. (2003). Fractured worlds: Islam, identity, and international relations. *Global Society, 17*(2), 111–120.

PostFaktum. (1992). Sozdanny 29 maia v Kazani islamskyi tsentr Tatarstana ne poluchil podderzhki muftia Talgata Tadjudtina. *PostFaktum*, 1 June, 1.

Putin, V. (2003). *Poslanie federal'nomu sobraniiu Rossiiskoi Federatsii*. Retrieved October 8, 2012, from www.kremlin.ru/text/appears/2003/05/44623.shtml

Ro'i, Y. (2000). *Islam in the Soviet Union: From the Second World War to Gorbachev*. New York: Columbia University Press.

Sack, R. (1986). *Human territoriality: Its theory and history*. Cambridge: Cambridge University Press.

Sanin, S. (1998). Musul'mane tak i ne pomiriatsia? *Kazanskoe Vremia*, 5 February, 1.

Shaimiev, M. (1998). O polozhenii v respublike i osnovnykh napravleniiakh sotsial'no-ekonomicheskoi politiki. *Respublika Tatarstan*, 14 February, 3–4.

Silant'ev, R. (2008). *Islam v sovremennoi Rossii: Entsiklopediia*. Moscow: Algoritm.

Stepanov, V. (2000). Ethnic tensions and separatism in Russia. *Journal of Ethnic and Migration Studies, 26*(2), 305–332.

Stoliarov, M. (2003). *Federalism and the dictatorship of power*. London: Routledge.

Suleymanova, D. (2010). International language rights norms in dispute over latinization reform in the Republic of Tatarstan. *Caucasian Review of International Affairs, 4*(1), 43–56.

Todres, V. (1998). Troemuftie v Tatarii. *Vecherniaia Kazan'*, 4 March, 3

TOTs. (1991 [1998]). Programma (platform) Vsesoiuznogo tatarskogo obschestvennogo tsentra In D. Iskhakov (Ed.), *Suverennyi Tatarstan. Tom II* (pp. 133–173). Moscow: Russian Academy of Sciences.

TOTs. (1996). Zaiavlenie. *Izvestiia TOTs*, 2, 1.

TOTs. (2002). *Materialy sed'mogo koryltaia (s'ezda) VTOTs*. Kazan: TOTs.

TOTs. (2011). *Otkrytoe pis'mo Shaimievu M.Sh.* Retrieved October 8, 2012, from http://open-letter.ru/letter/27704

Treaty. (1994). *Treaty between the Russian Federation and the Republic of Tatarstan on delimitation of jurisdictional subjects and mutual delegation of authority between the state bodies of the Russian Federation and the state bodies of the Republic of Tatarstan*. Retrieved October 8, 2012, from www.kcn.ru/tat_en/tatarstan/treaty.htm

Tul'skii, M. (2004). *Prichiny raskola musul'manskikh organizatsii Rossii*. Retrieved October 8, 2012, from www.ca-c.org/journal/2004-04-rus/02.tulprimru.shtml

Usmanova, D., Minnullin, I., & Mukhametshin, R. (2010). Islamic education in Soviet and post-Soviet Tatarstan. In M. Kemper, R. Motika, & S. Reichmuth (Eds.), *Islamic education in the Soviet Union and successor states* (pp. 21–66). London: Routledge.

Ustav DUMRT. (1998 [2006]). Ustav Dukhovnogo upravleniia musul'man Respubliki Tatarstan. In V. Yakupov (Ed.), *Ustavnaia Literatura Dukhovnyx Upravlenii Tatarstana* (pp. 64–84). Kazan, Russia: Iman.

Vetlugin, A. (1998). Stanet li M.Sh. aiatolloi Shaimini? *Vecherniaia Kazan'*, 13 February 13, 1.

Walker, E. (2005). Islam, territory, and contested space in post-Soviet Russia. *Eurasian Geography and Economics, 46*(4), 247–271.

Ware, R. (2005). Mythology and political failure in Chechnya. In R. Sakwa (Ed.), *Chechnya: From past to future* (pp. 79–115). London: Anthem Press.

Yakupov, V. (2005). *Islam v Tatarstane v 1990-e gody*. Kazan: Iman.

Yakupov, V. (2006). *K prorocheskomu islamu*. Kazan: Iman.

Yakupov, V. (2009). Author's interview with Yakupov, Kazan, 22 April.

Yemelianova, G. (2003a). Islam in Russia: An historical perspective. In H. Pilkington & G. Yemelianova (Eds.), *Islam in post-Soviet Russia: Public and private faces* (pp. 15–60). London: RoutledgeCurzon.

Yemelianova, G. (2003b). Russia's *umma* and its muftis. *Religion, State & Society, 31*(2), 139–150.

Zagidullin, I. (2007). *Islamskie instituty v Rossisskoi Imperii*. Kazan: Tatar Book Publishing.

Zargishiev, M. (1999). Prezident Tatarstana Mintimer Shaimiev: "Vera prisutstvuet v kazhdom iz nas." *Respublika Tatarstana*, 4 February, 4.

Chapter 160
A Needs-Based GIS Approach to Accessibility and Location Efficiency of Faith-Based Social Programs

Jason E. VanHorn and Nathan A. Mosurinjohn

160.1 Introduction

Faith-based or religious oriented congregations along with the spectrum of non-profits and public services offer a variety of programs to local communities (Cnaan et al. 2004). Program offerings are numerous and meet a variety of age, ethnic, and racial groups (Chaves 2004; Hungerman 2008). Some such as Polson (2008) contend that nearly 90 % of congregations in the United States provide some level of social program to the local community. If this is true, then the old adage that *'there is a church on every corner'* means that there is a tremendous amount of community activity centered on faith-based congregational work in various cities and towns throughout the US.

National survey-based data such as the National Congregations Study 2006–2007 can help to sift out totals of social service programs by congregations, but what is missing is a measure of congregational capacity to meet needs within a community. What is capacity? In regards to health, Goodman et al. (1998) argue that community capacity is a condition for effective local health program promotion and success. They view capacity as a form of empowerment and the potential of a community to affect change. The authors provide helpful constructs to capture capacity dimensions that include: citizen participation; leadership; various skills within the community; resources and social networks; community perceptions and history; real power by citizens that function within common values; and finally the ability to critically reflect on ideas and decisions. Applying this framework to congregations, we might look at congregational capacity along the same lines. Who is

J.E. VanHorn (✉)
Department of Geography, Calvin College, Grand Rapids, MI 49546, USA
e-mail: jev35@calvin.edu

N.A. Mosurinjohn
Center for Social Research, Calvin College, Grand Rapids, MI 49546, USA
e-mail: nam3@calvin.edu

S.D. Brunn (ed.), *The Changing World Religion Map*,
DOI 10.1007/978-94-017-9376-6_160

in the congregation and who are the leaders? What skills are present and what resources (people, social networks, and geographic proximities) exist to maximize the allocation of those resources of time, talents, and finances? Who outside the congregation is involved, have they been effective, and what kinds of synergy can be incorporated to meet common goals?

Evaluating congregational communities on the basis of the criteria above can be helpful to establish levels of potential service. Goodman et al. (1998) focus on characteristics of people within a community to say potential capacity either is high or low. This can be difficult to quantify, however, and it does not give an accurate picture of what community needs are actually being met. Examining the landscape of congregational activity is useful to allow for comparison between congregational programs and their prospective populations – people who might benefit from the programs. With the frequency and size of unknowns present in an analysis of potential capacity (e.g. who are the leaders in the congregational social network, etc.), we suggest rethinking congregational service potential within a framework of analysis that examines comparative levels of present service to needs. Knowing high or low capacity in general might allow congregations to decide if they should implement a program (i.e. "we have the capacity to do that program"), but we suggest an alternate view of capacity – that congregations use analytical tools that examine the ability of congregations to meet need in an area.

How can comparisons be examined to know how well congregational programs are meeting the need within an area? Raw totals or ratio statistical measures that look at the number of people served over time or some manner of program evaluation is a start. These types of numbers are helpful to compare how well congregations are serving compared to each other. However, what is missing is the ability to know how well needs are being met within a community. Matching the need with the level of service provides a statistical measure that provides basis to answer questions of gaps in service coverage in a community. Where could investment be made to improve efficiency? What would be the impact of a congregation taking on a new program in another part of the city? What places are over-served in the community along a certain program?

What is currently missing in the literature is a clear method for comparison across congregational communities or non-profit service organization which addresses the questions of program potential and how that interacts with need. To fill the gap, we have developed a methodology to give objective comparisons between congregations so that measurement may be made that answers to what degree congregational capacity exists and how well are congregations doing to meet the need in an area. We argue that the best way to understand how service and need interact is to develop a spatial statistical model. Inclusion of distance as an integral part of measuring service and need allows analysis of relationships across space and is consequently a better venue for comparison among congregations then just simple a-spatial statistical ratios. To demonstrate the effectiveness and utility of spatial modeling, we have used the Kent County Congregations Study in Grand Rapids, Michigan in conjunction with US Census data to build a congregational capacity model using Geographic Information Systems (GIS).

160.2 Data

The Kent County Congregations Study (KCCS) is a local survey in Grand Rapids, Michigan conducted in 2007 by the Center for Social Research at Calvin College and supported by the Doug and Maria DeVos Foundation. Hernández et al. (2008) describe some of the key findings that the Grand Rapids community is much higher than national averages in congregational worship attendance and that congregations in the area provide a large degree of social programs in the community. Congregations in Kent County were found to provide social programs with an annual replacement value between \$95 million and \$118 million. Cnaan and Newman (2010) point out that this study is consistent with other studies that religious congregations are spatially distributed throughout the urban and rural area and are often more visible than other non-profit organizations. The KCCS researchers constructed a full census as opposed to a sampling scheme of all congregations. Hernández et al. (2008) indicate that the 2007 census included 717 congregations, but later in 2009 in preparation for another survey round, the census numbers were increased to 760 due to better counting (e.g. more than one congregation in a building) and to the emergence of new congregations (Carlson 2011, Personal communication). The congregational data and the locations of all congregations were made available for this study and are available for download through the Association of Religion Data Archives.

160.3 Conceptual Framework and Methodology

In order to geographically analyze the data to determine congregational service capacities the congregations themselves were situated in digital geographic space using ArcGIS 10 (ESRI 2010). To tease out the concept of congregational capacity power and potential to affect change we have come up with a term we call congregational service power (CSP). The definition of CSP is the amount of service power a particular congregation has with regard to a program that serves the community. In a sense, it is the service market for a congregation and the strength of that market area. To construct the CSP, we draw on the work of Huff (1964). In this seminal work on accessibility, he describes a geographic trade region as a place where probabilities can be established for potential customers living in that region for a set of products based on interaction. He describes this mathematically as:

$$T_j = \sum_{i=1}^{n} P_{ij} \times C_i \tag{160.1}$$

where:

T_j = *the trade area for a firm j, which is the total number of customers likely to patronize*
P_{ij} = *the probability of an individual within region i to shop at j*
C_i = *the total number of customers with region i*

The probability portion of the model is (Huff and Black 1997):

$$P_{ij} = \frac{A_i \times d_{ij}^{-\gamma}}{\sum_{i=1}^{n} A_i \times d_{ij}^{-\gamma}} \tag{160.2}$$

Where:

P_{ij} = the probability of a person choosing to shop at location i traveling from j
A_i = the attractiveness of store i
$d_{ij}^{-\gamma}$ = distance decay between i and j

The Huff model is a market area measure that provides theoretical modeling for consumer's spatial behavior from the vantage point of the facility attraction. Notice distance is included in this model; the decay of distance helps up determine how people are less attracted to a place because it is too far away. Other models have been developed primarily from the work of Kwan (Kwan 1998, 1999, 2002; Kwan and Hong 1998; Kwan et al. 2003) that look at market accessibility from the vantage point of the time-space constraints of individuals.

Based on the type of survey data we have gathered from congregations in the KCCS study we build from Huff's work to first create the concept of congregational service markets (CSM) as:

$$CSM_{ij} = \frac{A_i \times d_{ij}^{-\gamma}}{\sum_{i=1}^{n} A_i \times d_{ij}^{-\gamma}} \tag{160.3}$$

Where:

CSM_{ij} = the probability of a person in area j, going to a congregational program at location i
A_i = the attractiveness of the congregational program at i (Measured using a factor or factors of the congregation – size, volunteers, budget, etc…)
$d_{ij}^{-\gamma}$ = distance decay between i and j

For this study, we used the congregational budget as the attractiveness value for the congregation. Better measures could be used, such as specifics regarding the program budget, number of program workers, or/and the number of people served, however this data was not available. In some cases, the budget for some congregations was imputed by survey staff workers (Carlson 2011, personal communication). However, we still believe that program budget offers a reasonable estimate for how attractive a program is and how far reaching it might be. The magnitude of distance decay could not be verified, which is an exceedingly important parameter for model specification (Skov-Petersen 2001), but it requires survey information on travel distances and willingness of people to travel for certain services, which was not available. To approximate the distance decay exponent in Eq. 160.3, we used a value of 2.0, which is a commonly used value

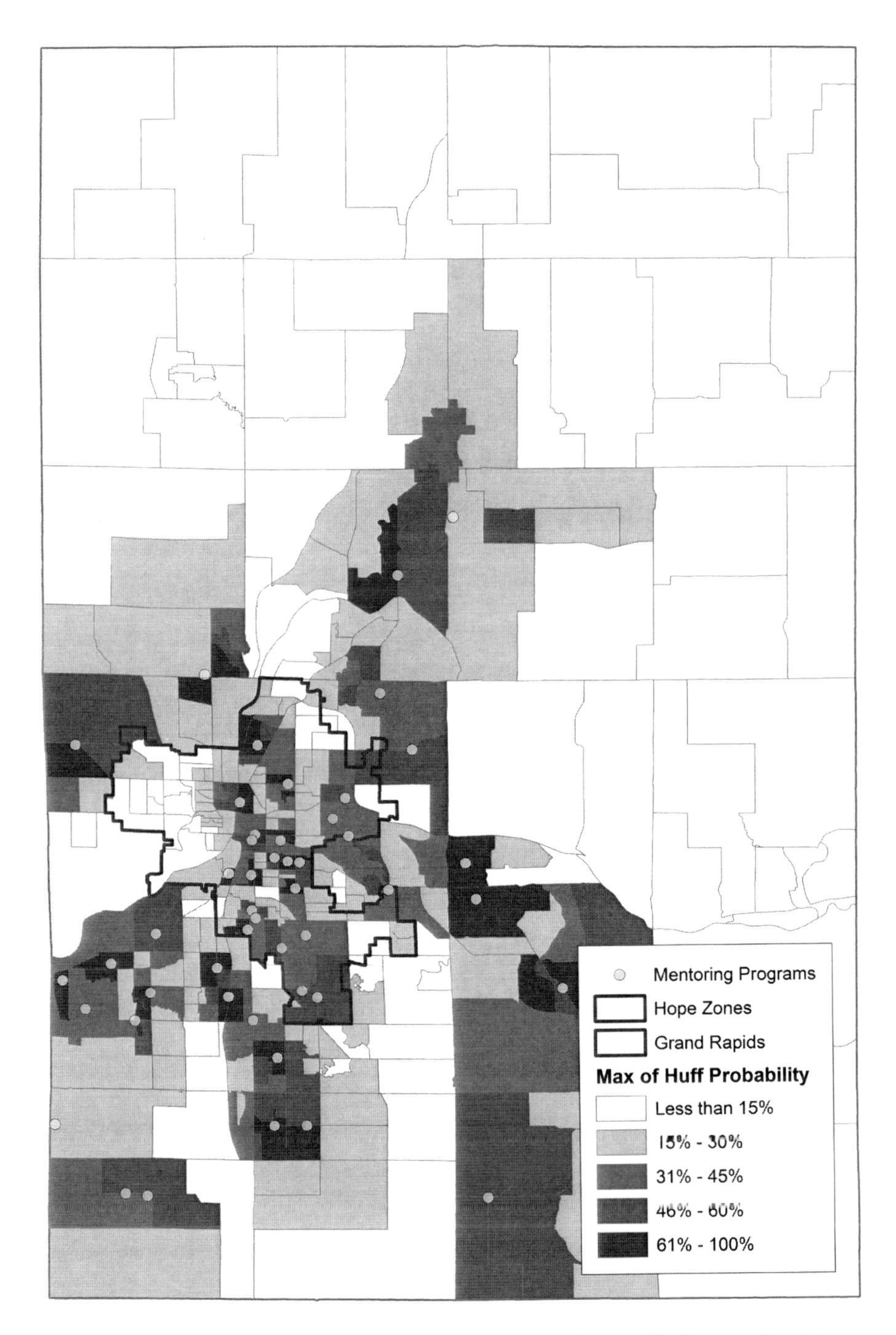

Fig. 160.1 Market areas of mentoring programs using the Maximum of Huff Probability (Map by Jason E. VanHorn and Nathan A. Mosurinjohn)

(Haggett 1965) but by no means compensates for the vast differences in social geographies across the country. The results of the CSM are presented in Fig. 160.1. The map shows the market areas of all mentoring programs at congregations within the KCCS study. In this example, all congregations with a mentoring program were compared with the other congregations with that specific program (mentoring) to delineate the geographic markets. Researchers (Amato 1994; Bank et al. 1993) have demonstrated increased poverty, stresses, and parent-child authority issues are associated with single parent households for mothers who are the only parent. Mentoring organizations like Big Brothers Big Sisters use adult volunteers to model healthy relationships for children from single-parent families (Abbott et al. 1997). We used single parent households as the population of these markets as they are a good indicator of the need of mentoring programs from available census data. These were aggregated at a census block group level. Markets are shown using the variable Maximum of Huff Probability (MHP). Huff Probability (calculated as CSM in Eq. 160.3) shows the likelihood that the population of a certain geographic area will choose a specific mentoring service over others, based on distance and attractiveness. As seen in the equation above, attractiveness is derived from congregational budget size. It would be possible to construct a more realistic value of attractiveness if variables such as program budget, denomination, and race are taken into consideration.

After finding all Huff Probabilities between each program and each block group, we attributed the maximum value derived by the model within each block group. This shows the decisiveness of a population when choosing services. A population with a high MHP has a high likelihood of choosing one specific program, while a low MHP indicates relative indifference. Using these values, we are able to visually construct market areas for each program. Areas with relatively high MHP are in the market of the nearest, most attractive program. As the indifference increases away from the program, its market influence dwindles. The Hope Zones shown are areas identified by the Doug and Maria DeVos Foundation as focus zones to improve the lives of residents within the zones, especially children (Carlson 2011, personal communication). These areas are considered to be where children are in most need of improved services and thus are representative of areas in the geographic region of particular note. The Hope Zones show areas of both high and low levels of indifference.

To create a map showing the areas that have high and low geographies of service for the need that exists, first we needed to define and represent both access to services and need in each block group. The block group level of geography was chosen because of its relative granular size, especially in high density areas, while still having detailed demographic data. In this case, we examined where adequate offerings of mentoring programs were located. To approximate access to service, we used a portion of Eq. 160.4 above to create congregational service power (CSP):

$$CSP_j = \sum_{i=1}^{n} A_i \times d_{ij}^{-\gamma} \tag{160.4}$$

Where:

CSP_j = *the congregation service power for congregation j*
A_i = *the attraction of a program within region i to go to participate in a program at j*
$d_{ij}^{-\gamma}$ = *distance decay between i and j*

Equation 160.4 provides the output of service power without delineating the market boundaries for congregations so that all congregations are measured based on their attractiveness by program as a measure of their budget.

To determine need for mentoring opportunities, we chose the variable "number of single-parent households," which was obtained from US Census data. The census data is from Census 2000, as the 2010 block group data was not yet released at the time the study was completed. By totaling the number of single-parent households by block group, we can better determine the population from which the mentoring programs are likely to draw from. Figure 160.2 shows the need by block group as a percentage of single parent households.

As a slight alternative to the $d_{ij}^{-\gamma}$, we first built a spatial weights matrix (SWM), which would give a weight, ranging from zero to one, to the spatial relationship between every block group centroid and every religious congregation in the study area. ArcGIS Geostatistical Analyst provides the ability to automate this procedure. We again used the value 2.0 as an exponent of distance decay in specifying the SWM. The result of the distance decay then indexed to a value between zero and one. Multiplying the congregation budget by distance decay allowed larger programs to have a larger spatial influence than smaller ones. Figure 160.3 shows CSP, or the attraction of block groups to congregations with a mentoring program. A correct interpretation of the low values would be that the block group is not very attracted to the congregation for the mentoring program because of distance. In this case, larger programs have a larger spatial influence than smaller ones.

To answer the questions posed earlier like where are gaps in service coverage in a community, where investment could be made to improve efficiency of programs, what would be the impact of a congregation taking on a new program in another part of the city, or what places are over-served in the community along a certain program, we have created a combination of Figs. 160.2 and 160.3 which gives us a service to need ratio shown in Fig. 160.4. In this map, dark green areas have a relatively large amount of service for the amount of need, or single parent households, in that area. The opposite is true for light green areas. The yellow circles are the mentoring programs reported in the KCCS study, with their size being represented by the budget of the congregation, which was used as a proxy for program influence.

160.4 Analysis and Discussion

Identifying the service-to-need ratio by block group provides basis to answer two key questions. The first question is to what degree are congregations doing well with a particular program (in this case mentoring programs) to meet the apparent

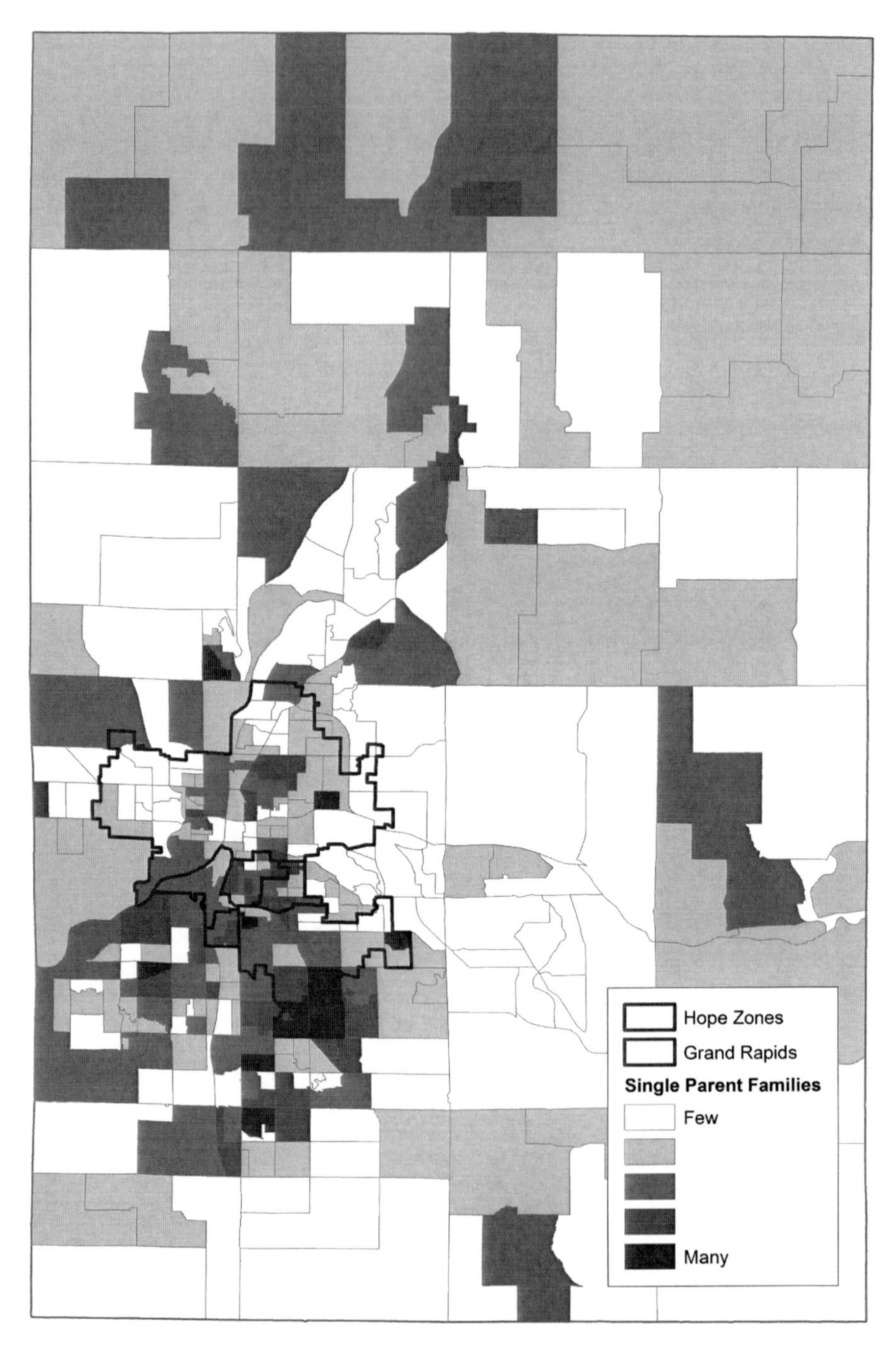

Fig. 160.2 Need by block group as a percentage of single parent households (Map by Jason E. VanHorn and Nathan A. Mosurinjohn)

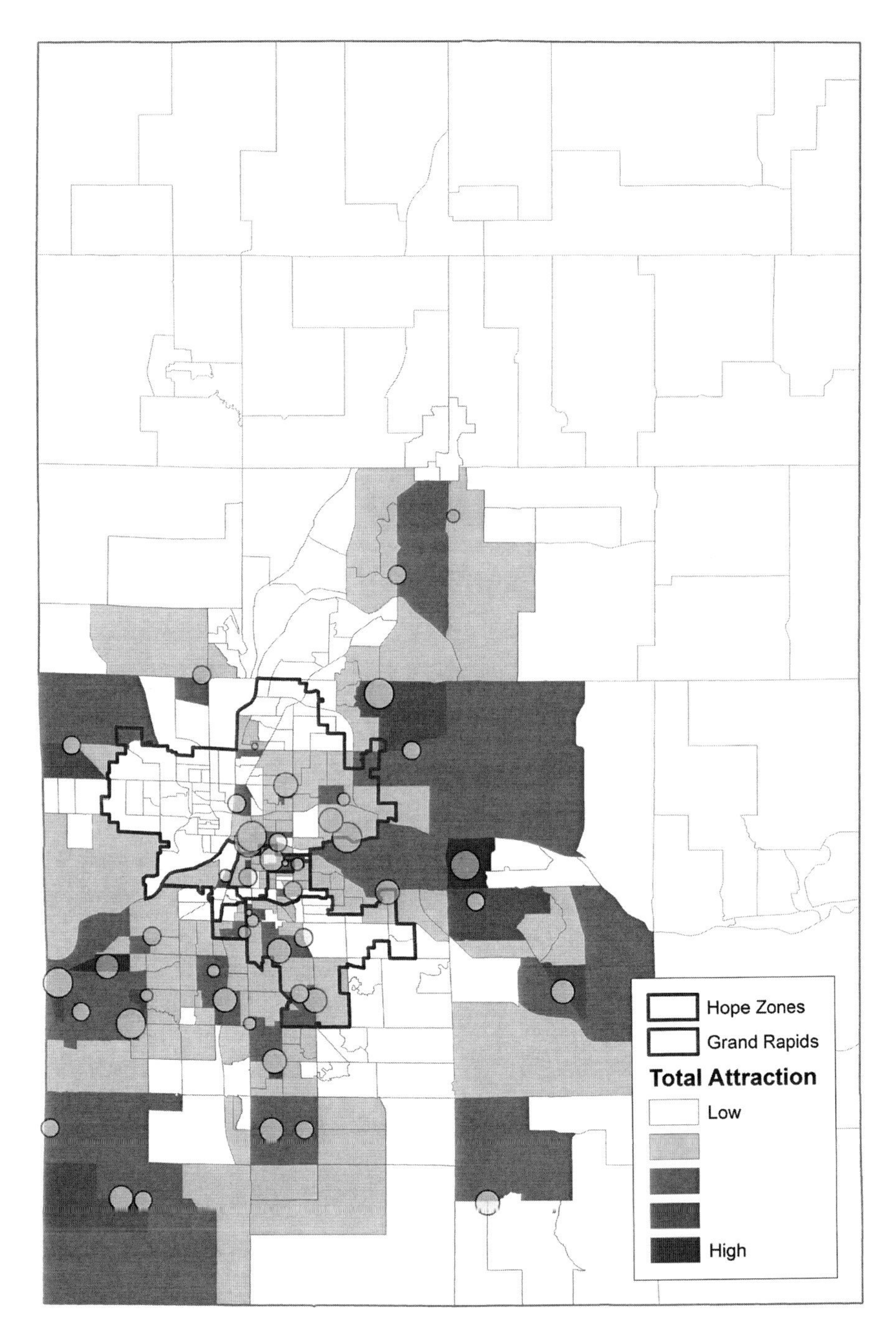

Fig. 160.3 Attraction of block groups to congregations with a mentoring program (Map by Jason E. VanHorn and Nathan A. Mosurinjohn)

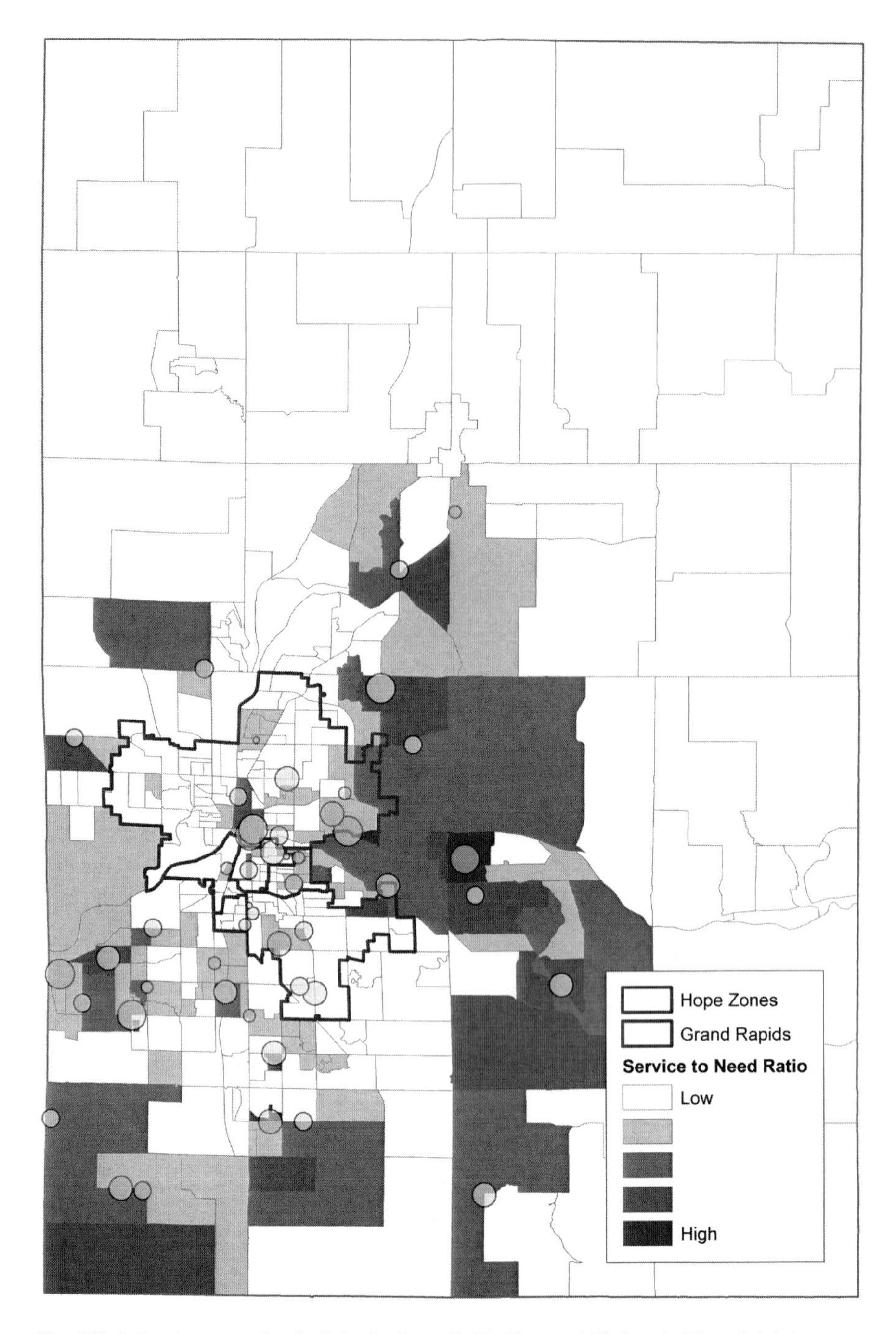

Fig. 160.4 Service to need ratio (Map by Jason E. VanHorn and Nathan A. Mosurinjohn)

need near to them? As is shown in Fig. 160.4, the darkest areas indicate the places where service is high and likely meets or exceeds the current need. The lighter two categories in the map are where most concern can be focused. The lighter categories in the map are places where, even if there is a program nearby, there is a relatively low amount of service in relation to the need. That is not to say that an organization other than a congregation is not at work in that area (e.g. schools might offer a free mentoring program). However, from the standpoint of the congregational view alone, these areas are the areas of weakest service compared with need.

It is worth noting here that the service-to-need ratios within the study are only relative to each other. That is, a low ratio would only indicate that the level of service as compared to the need is only low relative to other locations in the county. If Kent County, for some reason, was overzealous in its administration of mentoring programs, it could be that the lower levels of service have more appropriate priorities. Community organizers should look for neighborhoods they know are preforming well in offering the service in question, and compare the rest of the county to that area.

The second key question answered by the methodology is where should increases in support be made to congregations already invested in a particular program *or* where should new program locations by considered? Focusing on the Hope Zones, clearly there are several areas of concern and bolstering certain congregations, such as the one furthest west in the West Hope Zone, which has a comparatively lower congregational budget, could be a potential area of greater investment where activities along this particular program are already occurring. As well, we could also investigate the South Hope Zone to see what other existing congregations are present and what their capacity is based on congregational budget or some other parameter. With this modeling approach it is possible to model the effects of an additional congregation providing a mentoring service to that Hope Zone. The methodology approach outlined above provides a measure to foundations, non-profits, and others as a decision-support mechanism to help answer questions and provide answers based on specific focused agendas to help people in need along a variety of themes such as mentoring, health-care, education, and more.

Although modification of the Huff model provides new avenues for research of congregational capacity it is not without limitations. There can be situations, where regardless of the attractiveness or nearness of a service, individuals daily geographies are so limited that people cannot take advantage of the service, or are in a situation where they are not knowledgeable about programs because of various circumstances (such as work). In an effort then to provide a more accurate view of capacity, certain assumptions must be made.

One assumption this approach is dependent on is that people within a block group are more similar than those further away. For instance, since we used single parent households as a key market factor of who might be most likely to send their children to mentoring programs at congregational institutions, we have made the assumption that single parent households, in general, are similar within this geographic region.

In addition, due to limited data, there are a few parts of this model which could be optimized in the future, as mentioned above. Primarily, the parameters of attraction and distance decay can be improved. Knowing these variables will help give a more accurate picture of how service and need interact geographically in Kent County. First, we assumed that program capacity was comparable to congregation budget. One could imagine a large congregation with a small mentoring program. The model also assumes that the program takes place at the congregation. If a congregation sends volunteers to mentor at a nearby (or far away) school, this will not be adequately reflected.

Lastly, congregations are not the only organizations which provide social service programs, and mentoring programs specifically. Schools, community groups, and other non-profit organizations all provide these services as well, and including them in a model such as this would serve to give a more accurate representation of social service coverage. Different types of organizations would serve slightly different populations and have varying degrees of spatial influence, so consideration of these differences would help community organizers even more. Given these limitations, there is potential strength with this model to provide a visual and statistical output that can help organizations interested in improving congregational capacity. By identifying markets, congregations can work to focus attention on collaboration and cooperation within markets, knowing who is working to improve the lives of people along certain program lines (e.g. mentoring). Service power, determined with congregational budget or some other parameter, can provide a base-line measure for foundations or others interested in establishing or increasing support of existing congregations dedicated to work within a particular geographic area. More importantly, however, is market delineation and service power potential is a tool to identify the gaps in service along various programs.

160.5 Conclusion

We have shown using the KCCS study how congregational communities can use an analytical approach to answering questions around program service efficiency using a GIS-based statistical approach. The Huff model provides an elegant choice for delineation of the markets because it uses both an attractiveness measure and a distance decay parameter to provide a realistic component of choice. People make decisions about distance, perhaps more than they know, where certain services or amenities will be accessed if and only if it is within a reasonable distance. The modification of the Huff model accommodates that consumer choice and provides a visual and statistical measure for congregational capacity. Our model shows how social service programs interact with each other and the people they serve, and gives a spatial statistical representation of how needs are being met. Future efforts should be made to provide better weighting parameters that curb the limitations mentioned above.

Acknowledgements The authors would like to thank the Douglas & Maria DeVos Foundation and the Board of Trustees at Calvin College in support of this work, as well as, reviewers who offered helpful comments on this work.

References

Abbott, D. A., Meredith, W. H., SelfKelly, R., & Davis, M. E. (1997). The influence of a Big Brothers program on the adjustment of boys in single-parent families. *Journal of Psychology, 131*(2), 143–156.
Amato, P. R. (1994). Father-child relations, mother-child relations, and offspring psychological well-being in early adulthood. *Journal of Marriage and the Family, 56*(4), 1031–1042.
Bank, L., Forgatch, M. S., Patterson, G. R., & Fetrow, R. A. (1993). Parenting practices of single mothers – Mediators of negative contextual factors. *Journal of Marriage and Family, 55*(2), 371–384.
Chaves, M. (2004). *Congregations in America*. Cambridge, MA: Harvard University Press.
Cnaan, R. A., & Newman, E. (2010). The safety net and faith-based services. *Journal of Religion & Spirituality in Social Work: Social Thought, 29*(4), 321–336.
Cnaan, R. A., Sinha, J. W., & McGrew, C. C. (2004). Congregations as social service providers: Services, capacity, culture, and organizational behavior. *Administration in Social Work, 28*(3–4), 47–68.
ESRI. (2010). *ArcGIS*. Redlands: Environmental Systems Research Institute, Inc.
Goodman, R. M., Speers, M. A., McLeroy, K., Fawcett, S., Kegler, M., Parker, E., et al. (1998). Dimensions of community capacity to provide a basis for measurement. *Health Education & Behavior, 25*(3), 258–278.
Haggett, P. (1965). *Locational analysis in human geography*. New York: Wiley.
Hernández, E. I., Carlson, N., Medeiros-Ward, N., Stek, A., & Verspoor, L. (2008). *Gatherings of hope: How religious congregations contribute to the quality of life in Kent County*. Grand Rapids: Calvin College Center for Social Research.
Huff, D. L. (1964). Defining and estimating a trading area. *The Journal of Marketing, 28*(3), 34–38.
Huff, D. L., & Black, W. (1997). The Huff model in retrospect. *Applied Geographic Studies, 1*, 83–93.
Hungerman, D. M. (2008). Race and charitable church activity. *Economic Inquiry, 46*(3), 380–400.
Kwan, M.-P. (1998). Space-time and integral measures of individual accessibility: A comparative analysis using a point-based framework. *Geographical Analysis, 30*, 191–216.
Kwan, M.-P. (1999). Gender and individual access to urban opportunities: A study using space-time measures. *The Professional Geographer, 51*, 210–227.
Kwan, M.-P. (2002). Time, information technologies and the geographies of everyday life. *Urban Geography, 23*(5), 471–482.
Kwan, M.-P., & Hong, X. (1998). Network-based constraints-oriented choice set formation using GIS. *Geographical Systems, 5*, 139–162.
Kwan, M.-P., Murray, A. T., O'Kelly, M. E., & Tiefelsdorf, M. (2003). Recent advances in accessibility research: Representation, methodology and applications. *Journal of Geographical Systems, 5*, 129–138.
Polson, E. C. (2008). The inter-organizational ties that bind: Exploring the contributions of agency-congregation relationships. *Sociology of Religion, 69*(1), 45–65.
Skov-Petersen, H. (2001, June 25–27). *Estimation of distance-decay parameters – GIS-based indicators of recreational accessibility*. In ScanGIS'2001: The 8th Scandinavian Research Conference on Geographical Information Science, Norway.

Chapter 161
Welcome the Stranger or Seal the Borders? Conflicting Religious Responses to Migrants

Thia Cooper

161.1 Introduction

> If one attitude can be said to characterize America's regard for immigration over the past two hundred years it is the belief that while immigration was unquestionably a wise and prescient thing in the case of one's parents or grandparents, it really ought to stop now. Succeeding generations of Americans have persuaded themselves that the country faced imminent social dislocation, and eventual ruin, at the hands of the grasping foreign hordes pouring into its ports or across its borders. (Bryson 1996: 145–146)

Responses to immigration have shifted throughout U.S. history. The question is: Who should be included and who should be excluded? As a nation, we have been inconsistent with regard to immigration- welcoming and rejecting migrants alternately and sometimes simultaneously. Host[1] responses to migrants[2] have been to exclude them or to expect them to assimilate. Exclusion results from the fear that the migrants' diversity threatens "United Statesians."[3] The feared groups have ranged from Irish and Italian to Asian and Latin American and from Catholic to Jewish, Hindu and Muslim. Even when not excluded physically, migrants are often excluded culturally if they do not assimilate. "The most pot image here is the melting pot … where differences dissolve into the common pot" (Eck 2001: 54). This model assumes three things: (1) Migrants should give up their distinctive characteristics: cultural, religious, economic etc.; (2) Migrants are willing to do so;

[1] I use host to refer to U.S. citizens.

[2] I will use the term migrant throughout this essay. Eithne Luibheid suggests this term because it "makes no distinction among legal immigrants, refugees, asylum seekers, or undocumented immigrants" (Zavella 2011: xiii). Migrants remain the same people although they may shift categories.

[3] I use United Statesian rather than American since America is comprised of two continents and many countries. American then applies to all in the western hemisphere.

T. Cooper (✉)
Department of Religion, Gustavus Adolphus College, St. Peter, MN 56082, USA
e-mail: tcooper@gustavus.edu

S.D. Brunn (ed.), *The Changing World Religion Map*,
DOI 10.1007/978-94-017-9376-6_161

(3) Migrants can do so. The hosts on the other hand need not change. Even if one argued for the first and second components (countered by multiculturalists), the third piece is not always possible. For example, a dark skinned black African cannot become "white." The host group can change its notion of "white" to include another group, as with Italians.[4] Yet, many racial, religious, ethnic, and other groups remain marginalized from the "dominant" culture.

Further, the dominant narrative of U.S. immigration history, handed down by white Anglo-Saxon Protestants does not describe accurately what occurred or why. In our history of immigration the group that holds power economically and politically considers only itself and forces or rejects migration based on its own desires. There has been a consistent conflict between needing economic labor and not wanting racial or religious change. Migrants were primarily considered as labor, used when necessary, whitened and converted if possible, and when not needed or when they failed to assimilate were rejected and excluded.

As the nation coalesced, citizenship was restricted to "whites" but any person could reside in the country. Migration into the U.S. was only fully restricted in 1965.[5] The restrictions maintained privilege for the dominant group, initially "White Anglo-Saxon Protestants" and today the more amorphous "Americans," prioritizing the 'national' society and economy over humanitarian needs. Yet, conflicts between economy, politics, race and religion continue because a migrant is more than a quantity of labor.

This chapter will explore responses to migration in U.S. history and then responses of faith-based hosts and migrants today. The current question is: should we keep our privilege enshrined in law or prioritize humans? While hosts debate the merits of both, for migrants, the answer is simple, humans; and for Christian migrants, Christ is the migrant and Christ is with the migrant.

161.2 Immigration, Religion and Race: A Kaleidoscopic History

Every United Statesian has migrant ancestors. Some were the original migrants: Native Americans, who later migrants excluded from citizenship until 1924. Others were forced to migrate through slavery, war or colonialism. And finally, some "chose" to migrate to this country, whether for family, study, economics, or adventure.

The English were not the first migrants, nor the majority to enter what would become the U.S.. First, "Native" Americans migrated to this continent. Five hundred years ago, these indigenous Americans numbered about 2–2.5 million.[6] The Pueblo cultures were at their height around 1100 CE and the Iroquois Confederacy

[4] This could occur with the Latino/a community, an ethnicity made up of a variety of races.

[5] See Appendix 2 for the details of the varieties of restrictions beginning in 1875.

[6] These original inhabitants were considered "aliens" and were only granted citizenship in 1924.

was negotiated in 1451. (Hoerder 2011: 3) The Norse briefly migrated, around 1000 CE, although probably only into Canada. After 1492, Spanish Catholics arrived and then the French. According to Zolberg, net immigration during this time was 501,000 Europeans (Zolberg 2006: 26). The new colonists excluded the Native Americans and migrants including Quakers, Jews and Catholics (Eck 2001: 48).

Protestants were initially a minority in number and territory of what would become the U.S. Yet Protestantism (and anti-Catholicism) became the dominant narrative in public discourse. The English migrants identified themselves as "Christian" ... after about 1680, "white" began to differentiate colonists from African slaves (Tan 2008: 123). White Anglo-Saxon Protestant was at the top of the hierarchy.

The largest group of migrants in these early years was forced into this country, slaves from West Africa. The 1808 law ending importation of slaves was the first migration law. Some enslaved Africans practiced African traditional religions, later becoming Christian. Others practiced Islam leading their slaveowners to label them Arab,[7] although they were from the same parts of Africa as other slaves. Nearly half of the 1,000,000 residents during the two centuries of African slavery were not initially Christian. (Daniels 2002: 61)

Other migrants were also "non-white." In the 1720s and 1730s as Germans migrated into Pennsylvania, Ben Franklin among others thought that Germans would never integrate. Franklin argued, "Why should *Pennsylvania*, founded by the English, become a colony of *Aliens*, who will shortly be so numerous as to Germanize us instead of our Anglifying them, and will never adopt our language or customs, any more than they can acquire our complexion" (Isbister 1996: 7). By 1790, Congress passed a law that limited citizenship to "whites" with "good moral character" (Schrag 2010: 23). The definition of white slowly expanded over time and Germans would be included.

In the 1800s, most migrants were Catholic, causing strong anti-migrant sentiment. First, Mexico gave up the rights to California, New Mexico and Texas at the end of the U.S.-Mexico War and nearly 100,000 Mexican Catholic citizens living in these areas could become U.S. citizens. The Mexicans did not cross the border; it crossed them (Soerens and Hwang 2009: 52–53).

The U.S. government now had land, but not enough people. It wanted migrants to improve the economy (Wilson 2009: 23). Before, during and after the Civil War states actively recruited migrants to settle land because of the Homestead Act and as troops were fighting. Despite the fact that 25 % of the Union Army was foreign-born and the government thought them important to the economy, they were not wanted culturally.[8]

More than two million people, many Catholics from Ireland and Germany, migrated in the 1850s. By the 1850 census, Catholicism was already the largest denomination in the U.S. (Matovina 2012: 15). For many Protestants, Catholicism

[7] In fact, the majority of our Arab migrants have been Christian throughout our country's history.

[8] The situation of migrants in the army has been true throughout our history.

was "an invading enemy, audaciously conspiring, under the mask of holy religion, against the liberties of our country." (William Brownlee qtd in Soerens and Hwang 2009: 51) The one million member American Party (the Know Nothing party) was both anti-immigrant and anti-Catholic. (Soerens and Hwang 2009: 50–51) "'Whiteness' became an aspirational goal as these European migrants sought to assimilate into the "whiteness" norm that defined U.S. society. In the process, they had to shed their distinctive native languages, cultural norms, and traditional customs" (Tan 2008: 124). And maybe their Catholicism too? They were eventually successful in becoming white.

From the 1850s onward, Chinese men from Canton migrated to the West Coast, as male labor was requested by the mine owners and others (Soerens and Hwang 2009: 53). They had to leave their wives and families behind, making the ratio of Chinese 100 males to one female in 1890 (Lee 1995: 15). Once the need for their labor dissipated, the men too were excluded. The Chinese Exclusion Act in 1882 barred Chinese migration; the first time it was possible to migrate illegally. The Chinese, non-white and non-Christian, were deemed unable to assimilate.

While race seemed immutable, religion did not. Protestant clergy worked amongst Chinese men on the West Coast, attempting to convert them, and defending their rights. Protestants also welcomed the second wave of European migrants, trying to convert the masses of Italian Catholics and Russian Jews who arrived in the late nineteenth century. "The change came about as Christians heeded the call of evangelical leaders such as Howard Grose, who extolled the arrival of the "incoming millions" as "an opportunity" to "carry the gospel to [foreigners] in our own land"" (Soerens and Hwang 2009: 60). Perhaps evangelization would bring assimilation.

Anti-immigration sentiment, however, abounded. In 1891, the White League lynched 11 Italian prisoners. In 1900, the Chinese Exclusion Act was expanded to the Japanese as well. The migrants were good for the U.S. economy as the country shifted toward manufacturing. However, the perception was that migrants harmed the economy and culture (Soerens and Hwang 2009: 61).

The concepts of whiteness and Christianness shifted over time. African Americans would not be considered white, although they were no longer enslaved by 1870. However, it was less clear for other groups. The U. S. courts shifted in the definition of whiteness. For example, Armenians were first considered Asian and later white (Takaki 1989: 15). However, by 1923, race was thought to be obvious. *United States v. Bhagat Singh Thind* argued:

> What we now hold is that the words "free white persons" are words of common speech, to be interpreted in accordance with the understanding of the common man. … It is a matter of familiar observation and knowledge that the physical group characteristics of the Hindus render them readily distinguishable from the various groups of persons in this country commonly recognized as white.... What we suggest is merely racial difference, and it is of such character and extent that the great body of our people instinctively recognize it and reject the thought of assimilation. (261 U.S. 204, 214–215 in Tan 2008: 131)

Hindu was non-white and non-Christian. Note here that the argument is over who IS white, not whether "whiteness" should be privileged.

This racism formed the basis of the 1924 immigration law. The law decreased immigration from nations such as Italy, and almost eliminated immigration from eastern Europe, Asia, and Africa, any country without people already in the U.S. (Daniel 2010: 64). One now had to get permission from the U.S. government (a visa) before entering and permission was limited.[9] The Border Patrol was set up and the prospect of "Illegal" immigration, by other than Chinese, began. Congressman Albert Johnson, a Republican from Washington State and an author of the act, said: Our capacity to maintain our cherished institutions stands diluted by a stream of alien blood, with all its inherited misconceptions respecting the relationships of the governing power to the governed. ... The day of indiscriminate acceptance of all races, has definitely ended (Daniels 2002: 283–284). The assumption was that the USA should remain white and the whitest people were from Northern Europe. Race clipped nation here: a racially Chinese Person from England was excluded for being Chinese. No Asians or Africans were allowed.[10] Asians continued to be "aliens ineligible to citizenship" (Ngai 2004: 9). Italians and Irish were suspect enough to reduce but not eliminate their migration. However, Native Americans were finally given citizenship.

This new law did not apply to white migrants from the Western Hemisphere, so Mexicans and Canadians continued to migrate. Canada became the leading donor country along with Germany and there was no anti-Canadian backlash[11] (Ramirez 2011: 86). Throughout the 1900s, however, Mexicans were alternately welcomed as labor was needed (WWI, WWII, postwar) and rejected when labor was not needed (Great Depression).[12] Some argued for Mexico to be included in the 1924 law but the southwestern states successfully argued they needed Mexican agricultural labor. Mexicans became associated with "illegal" immigration although they were considered "white" and could become citizens. Over this century, Mexicans shifted from "white" to "non-white" in the public imagination. The southwest also shifted in the public imagination from a frontier to a region with specific borders and began to be included in the dominant narrative of the U.S..

Religion continued to be a source of hatred. Catholics and Jews were targeted by the Ku Klux Klan (made up of white Protestants), in the early twentieth century (Alba et al. 2009: 1). Further, the resentment of migrants, particularly Jews, was repeated throughout WWII and after. A ship with 900 Jews fled Nazi Germany and requested that President Roosevelt allow them to enter the U.S.. Roosevelt said no, apparently worried about the next election. Eventually, the ship had to return to Nazi-occupied Europe (Eck 2001: 61). Only slowly did assimilation expand with

---

[9] Unless one is Chinese or Japanese, arguing one's ancestors migrated legally before 1924 is a misnomer, as all migration was legal.

[10] Only from the late 1980s onward, has there been a substantial increase in African migrants to the U.S., due to conflicts in the Democratic Republic of Congo, Sierra Leone, Somalia, Liberia, Rwanda, Burundi, and the Sudan. In 2003, more than 1,000,000 African-born migrants were in the U.S., compared to 230,000 in 1990 (Olupona and Gemignani 2007: 1–2).

[11] The 1900 Census showed 22 % of Canada's population was living in the U.S. (Ramirez 2011: 76).

[12] A similar back and forth occurred with Filipino migrants.

regard to religion and Catholics and Jews became part of what is now called the "Judeo-Christian" tradition.

The rejection was racial too. For example, when Japan bombed Pearl Habor in 1941, Japanese Buddhist ministers and community leaders were immediately rounded up by the FBI (Eck 2001: 177). German-Americans, however, did not face this discrimination. They were "white Christian Americans."

From 1950, the debate about whether and how to change immigration restrictions began. The current law was criticized from a racial, economic, and humanitarian perspective. In 1953 the National Council of Churches of Christ argued for a new law to increase "producers and consumers" (Ngai 2004: 250). The U.S. economy would improve with more migrants, maintaining and increasing our privilege.

Others argued against the racism. Herbert Lehman, former governor of New York and a public voice in favor of immigration reform in the 1950s, argued it "is based on the same discredited racial theories from which Adolph Hitler developed the infamous Nuremberg laws. This System is based on the hypothesis that persons of Anglo-Saxon birth are superior to other nationalities and therefore better qualified to be admitted into the United States, and to become Americans" (Ngai 2004: 242). Coming on the heels of U.S. reticence to take refugees during and after WWII, this allegation hit home.

Still others focused on humanity globally. The first and second Hart bills introduced in 1962 and 1963 tried to shift the focus to the world's most needy people. Neither time was it passed. As Ngai notes, "the view that the nation's interests could be considered in conjunction with the interests of other nations failed to rally mass support" (2004: 254).

A new law was passed in 1965, which promoted family reunification rather than national origin.[13] "Sixty-three religious, labor, ethnic, and charitable organizations … 'hail [ed] the passage of the bill…,' which they said 'finally established an immigration policy consistent with our national philosophy that all men are entitled to equal opportunity regardless of race or place of birth" (Ngai 2004: 259–260.) It retained numerical restriction and added an economic component to get the best workers the U.S. needed to migrate. Economic privilege was to be maintained although overt racism was not.

Not all were pleased with the new law. The National Association of Evangelicals, opposed the opening of immigration policy. However, since 1965 millions of Asian and Hispanic migrants have joined evangelical churches (Soerens and Hwang 2009: 61). Conversion is occurring but still not "assimilation." Race and ethnicity still keep migrants marginalized.

We have had a racial (Asian) and ethnic (Latino/a) shift in the migrant population toward those considered "non-white." Latin Americans have grown from 4 % of migrants (1940s) to 33 % in the 1980s and make up 53 % of the foreign-born in the U.S. as of 2010. Asians rose from 3 % of migrants (1940s) to almost 50 % in the

---

[13] 1951–1960, 60 % of legally admitted migrants were European or Canadian, with the top five countries: Germany, Canada, Mexico, the UK and Italy. 1971–1980, the top five countries were Mexico, the Philippines, Korea, Cuba, and India.

1980s and make up 28 % of the foreign-born in the USA as of 2010. It remains to be seen whether one of these groups will become "white" or if another marker will appear.

There was a category of refugees to partly address humanitarian concerns. This quickly became politicized. The U.S. received large numbers of refugees from Asian Communist countries (Carnes and Yang 2004: 17). Many Vietnamese (30 % Catholic) came after the fall of Saigon. Political turmoil and an economic downturn in the 1980s brought millions of Koreans to the U.S. as well.[14] (Min 2010: 48) We prioritized those escaping "atheist" communist countries, those looking for economic and religious freedom.

We did not prioritize those repressed by non-communist governments. Civil wars in Central America and economic hardship throughout Latin America in the 1980s brought millions of migrants. Most were undocumented because the U.S. supported governments and armed forces of El Salvador and Guatemala and would not acknowledge the humanitarian crisis (Daniel 2010: 125–126).

Unlike earlier in U.S. history, faith communities took the government to task on this inequality, forming a Sanctuary Movement. Jews, Catholics, Lutherans, Methodists, Presbyterians and Quakers were involved, among others. By 1986, 567 churches and synagogues gave sanctuary to refugee families. Some people were prosecuted and convicted, but by 1989 they achieved the basic aims of the movement. It was a major effort to create a humanitarian policy and briefly it worked. However, the immigration law itself remains restrictive and focused on our privilege.

Where previously we discussed our privilege in terms of race, religion or politics, it now seems more bound up in economic discussions, with the other three underlying. We do not want those who threaten our economy or our national safety. The two appear to be bound up with each other, although there is actually no connection.

In the past 10 years there have been two foci with regard to immigration, both wrapped around anti-terrorism: the border and Arab Muslims. After 9/11, the U.S. detained hundreds of Middle Eastern and South Asian noncitizens. Nearly 800 had closed deportation proceedings. The FBI took an average of 80 days to clear innocent detainees, even after it "became clear that many [hundreds] of the September 11 detainees had no immediately apparent nexus to terrorism" (Kerwin 2008: 202–203). The Arab Muslim migrants were not white and not part of the Judeo-Christian tradition and easily became "aliens."

Oddly, this became bound up with protecting our borders from "illegals," although those involved in the 9/11 terror attacks entered the country legally. Somehow Mexican migrants have become non-white, associated with terror, and can be excluded. Those detained after 9/11 unjustly and those who die crossing the border do not elicit humanitarian responses. They are considered casualties of protecting U.S. privilege.

Even if the border were successfully sealed, undocumented migrants would still exist. Almost half the undocumented enter legally and later overstay their visas. Further, the border is permeable and cannot be sealed. Known as the "Tortilla

[14] Most Korean migrants are affiliated with Korean Protestant churches. (Min 2010: 3)

Curtain," the U.S.-Mexico border has more than one million crossings a day. Twenty years ago the border was mostly open to those who wanted to walk across it.[15] The building of walls began in the mid-1990s and sped up after 9/11. Migrants now cross in remote and dangerous desert regions. The underlying strategy was that deaths of migrants attempting the hazardous crossing would deter others from crossing. It has increased the number of deaths. About 2,000 people died along the border between 1995 and 2005. In contrast, more than 1,300 died from 2004 to 2009 on the Arizona border alone (Daniel 2010: 110). However, increased border enforcement has turned migration into a more permanent journey, as migrants stay rather than returning home and facing the risk of having to cross again at some other time.

During 2008, there was a crackdown on undocumented migrants, with many raids rounding up and deporting suspected undocumented workers. A Sanctuary Movement, smaller in size, but similar to that in the 1980s emerged calling for an end to the raids.[16] Christians and Jews were involved in the protests.

At the same time, ignorance and misinformation continue. Lou Dobbs, a CNN broadcaster, stated that undocumented migrants made up a third of the inmates in our prison system. In fact, they make up around 2 % (Schrag 2010: 196). And earlier this year in Missouri, a white U.S. couple was able to adopt a child whose Guatemalan mother had been arrested as undocumented, and put in detention while awaiting deportation. Separated from her mother, the child was taken care of by relatives (Bauer 2012). A similar situation occurred in 2007 (De la Torre 2009: 87).

A 2009 Task Force on U.S. Immigration Policy, whose chairs were Jeb Bush and Thomas F. McLarty III, argued that Congress should pass comprehensive immigration reform: (1) to attract and retain talented and ambitious immigrants; (2) to secure America's borders and strongly discourage employers from hiring illegal workers, and (3) to legalize many undocumented (Haass 2009: x). It would have a partly humanitarian focus, while at the same time protecting U.S. economic and political privilege. As of writing (August 2012), the measures have not been introduced.

161.3 The State of Migration Today

In 2005 the country became less than 50 % Protestant for the first time since the colonial period. What many white Protestants feared in the 1800s is now coming true. We do have less than 50 % Protestant and a large number of Catholics. Yet, white Protestant fear has now turned to Islam, although Muslims remain <1 % of our population, and the vast majority of those who practice Islam migrate from Asian countries, not the Middle East. We also associate this fear with the need to patrol our borders, although those crossing our borders without documents are

[15] In 1993, in San Ysidro, California, the U.S. government built a fence from Vietnam-era landing mats. The steel grooves run horizontally, creating a ladder (Daniel 2010: 114).

[16] See for example, www.sanctuaryphiladelphia.org; www.oregonsanctuary.org; and www.newsanctuarynyc.org

migrants looking for work, usually Catholic or Evangelical Christians. We conflate race, religion and politics in order to keep our economic privilege.

Globally, many people are being displaced from their homes. Nationally, our immigration policies appear to be failing to deal with this pressure. Over 3,000,000 people wait between 4 and 20 years to join U.S. relatives and more people than ever die crossing the border because they cannot migrate legally (Suarez-Orozco et al. 2011: xii).

In addition, those most in need often cannot migrate. It is not the poorest who reach our shores. In order to be able to migrate, first, a migrant has to know where they can migrate to and how to do so: an access to information component. Second, they need to find affordable transport: a monetary and access component. Third, they must be able to leave their place of origin: a mobility and permissions component. And, finally, they must be able to get through the borders of the new country, which requires all of the above components.

The top five countries sending migrants to the U.S. in the early 2000s were Mexico, the Philippines, China, India, and the Dominican Republic. Of an estimated 37 million migrants, 35 % are U.S. citizens, 33 % are Lawful Permanent Residents and 2 % have temporary resident status. About 31 % have no legal status (Soerens and Hwang 2009: 29). About 56 % of unauthorized migrants come from Mexico while the remaining 44 % come from Asia, Europe, Latin America, Canada and Africa (Soerens and Hwang 2009: 37).

The majority of our migrants are Christian. About half of migrants come from Latin America and self-identify as Christian.[17] Latino/as constitute more than half of U.S. Catholics under the age of 25 and about 25 % of U.S. Catholics are Latino/a. (Matovina 2012: vii) Asian Americans constitute the second-fastest-growing racial group in the U.S.[18] Over 60 % of Asian Americans who have a religious identification are Christian and are the second-fastest-growing group after Latino/a Christians.

161.4 Responses Today

161.4.1 Christian Hosts

Migration is not a crime. The crime is that which causes migration. (Kerwin 2009: 97)

Underneath the question of the right to exclude is the question of maintaining our privilege. Within the Christian tradition[19] the argument is sometimes expressed as

[17] Sixty-four percent of the U.S. Hispanic population is of Mexican descent with Puerto Ricans at 9 % (Rivera 2009: 137).

[18] As of 2000, Asian Americans make up 4.2 % (11.9 million) of the total U.S. population. Chinese Americans comprise 2.7 million (Tan 2008: 3).

[19] A variety of Christian and other traditions speak about immigration. I use mainly Lutheran and Catholic contributions here for the sake of brevity. See, for example, ELCA 1998 and Pontifical Council 2003.

God's law versus civil law. Yet within the argument of upholding civil law is also an argument for tightening restrictions. While not overtly racially or religiously biased, those two biases still sometimes underlie our responses. Are United Statesians more important than non-United Statesians? Who are we allowed to prioritize and who can we discriminate against and why?

From one perspective, protecting those within U.S. borders, at least those who "belong" there, comes first. Some want migration slowed or stopped to keep the U.S. as it is. From a faith perspective, God is seen as the God of the U.S. rather than the God of all peoples. This particular nation has been inspired by God and needs to protect its borders. This type of argument tends to emerge from Evangelical Protestant churches. However, some Evangelicals are also found at the other end of the spectrum. So too while many Catholic are at the other end of the spectrum, some U.S. Catholic bishops contend that border enforcement is a necessary first step before the consideration of any other measures (Matovina 2012: 202–203).

From a more extreme perspective of enshrining privilege in law, there are some citizens who want to exclude some born in the U.S. from citizenship. Currently, if born here, one is a U.S. citizen. The Fourteenth Amendment states that: "All persons born or naturalized in the United States, and subject to the jurisdiction thereof, are citizens of the United States and of the State wherein they reside." Some would like the U.S. born children of undocumented migrant parents to not automatically have citizenship, leaving those children stateless. (Soerens and Hwang 2009: 104)

From a more moderate perspective, many Christians are concerned that more migration may negatively affect impoverished U.S. citizens (Soerens and Hwang 2009: 94–95). Migrants should be welcomed only if they increase U.S. prosperity. And so economic arguments abound to assess whether or not migrants add to the economy.[20] Here too, some worry migration will harm the natural environment (Soerens and Hwang 2009: 96–97).

Some hope to blend God's law and civil law for a more humanitarian policy. "Christians are generally bound to submit to the rule of law" (Soerens and Hwang 2009: 13). The Catholic Church has been at the forefront of the calls for immigration reform. Some non-Catholics critique the prioritization of migration, arguing the Catholic Church 'uses' it to promote themselves (Schrag 2010: 179). Yet, Catholic tradition does recognize the importance of national sovereignty. "Catholic teaching recognizes the authority and duty of a state to ensure the orderly entry of migrants and to exclude and remove persons whose presence would not advance the common good. But sovereignty does not immunize states from responsibility for receiving persons who are forced to cross borders to realize their rights" (Kerwin 2009: 102). There is a call for more humanitarian laws where sovereignty is respected but so too are humans.

---

[20] For example, roughly 75 % of undocumented migrants have false Social Security numbers, which generate $6–7 billion of Social Security withholdings each year (Daniel 2010: 53). Even Lawful Permanent Residents are not eligible for food stamps or for any cash benefit until they have been Lawful Permanent Residents for at least 5 years (Soerens and Hwang 2009: 42).

Toward the other end of the spectrum, some Christians argue that all humans are equal and deserve equal privilege and protection, according to God's law. Civil law should shift to a fully humane perspective. For example, countering the argument that the New Testament tells Christians to obey civil law, some argue that neither Paul nor Peter would have advocated that Christians should obey the laws that made Christianity illegal in the Roman Empire (Daniel 2010: 42).

Catholic tradition emphasizes the notion of the common good "the sum total of social conditions which allow people, either as groups or as individuals, to reach their fulfillment" (Kerwin 2009: 101). Communities suffer when any member is excluded. "The present laws make it impossible for most people to immigrate legally, while the economic or political situations in their home countries make it exceedingly difficult to stay put there" (Soerens and Hwang 2009: 108). This is not upholding the common good.

In thinking through economics, two questions have been asked: what does the economy do FOR people? And what does the economy do TO people? This perspective argues that it is unfair to prioritize United Statesians over others. Reverend Paul Empie of the National Lutheran Council argued even before the change to the 1965 law that the U.S. should, "resist the insistent pressure of groups of our citizens for a quick increase in their standard of living without regard to our relations to a world society" (Ngai 2004: 253).

In thinking globally, immigration and helping those in other countries both are important. Those who prioritize humanity argue that "the right to migrate is among the most basic human rights ... because God has created the good things of this world to be shared by all men and women – not just a privileged few" (Matovina 2012: 200–201). For example, the narrative of the Hebrew Bible is one of migration. Exodus 23:9 fits today as it did back then: "You shall not oppress a resident alien; you know the heart of an alien, for you were aliens in Egypt." From the Christian New Testament, others argue that Joseph and Mary were a refugee family and Jesus was born as undocumented. "They are 'the archetype of every refugee family' and are 'the models and protectors of every migrant, alien and refugee of whatever kind'" (Kerwin 2008: 193). Further, Christianity itself spread through migration.

Those allowed to migrate should not be those who help our economy, but the neediest, the poorest, refugees. Before the 1965 law, many faith-based organizations argued this. Ngai notes that a coalition of Jewish organizations argued "those areas of the world whose peoples are most urgently in need of resettlement and most deserving of assistance have among the lowest of the quotas" (2004: 253). Human beings are made in the image of God. "Do you see migrants as a risk, or do you see migrants at risk?" (Kerwin 2009: 93). People are human beings, not commodities. The Jewish and Christian sacred texts do not suggest migrants are to be accepted only when they are economically beneficial (Soerens and Hwang 2009: 136–137).

Empie noted the attempt to keep privilege is not new. He critiqued the nation's "use of force... in obtaining various parts of this nation from the Indians, the British, the Mexicans. Having thus inherited the continent their descendents continued to use force to keep most other people out" (Ngai 2004: 253). This behavior should end. The U.S. should base its policy on "the interlocking and mutual interests of all nations with regard to the immigration of peoples, the interaction of culture, and the respect of universal human rights" (Ngai 2004: 253). Others echo his words today.

This perspective also wants to help the undocumented and poor within our own borders. For example, "we have more than a million agricultural workers, 70–80 % of whom are undocumented, and we only have five thousand visas (for workers)" (Daniel 2010: 69–70). Right now, there is no way for an undocumented migrant to "repent" and "fix" undocumented status except to leave the country. In the 1980s, when pressed by the Sanctuary Movement, the U.S. Department of Justice agreed to: "(1) end all deportations to El Salvador and Guatemala; (2) give all refugees from those countries work permits and temporary protective status, (3) reform the political asylum process at the INS [Immigration and Naturalization Service]" (De la Torre 2009: 173). It was a humanitarian shift on a temporary and small-scale. Legalization is being called for again but it is still piecemeal and does not solve the problems that push people out of their home countries in the first place, or attend to our immigration policy as a whole.

Some Evangelicals have moved in this direction too (see Carroll 2008). According to one pastor: "Which is better, to come from São Paulo on a tourist visa and overstay your visa, or to come from Britain in 1690 on a boat when nobody had to have a visa, kill a few Indians and steal their land?" (Freston 2008: 265). Others are somewhere in the middle: "One pastor told me that the American co-pastor in his church… was horrified when he found out that 70 % of the Brazilians were illegal. But this same Brazilian pastor concluded that the Berlin Wall, the old wall of shame, had fallen, but the new wall of shame was the one along the border with Mexico" (Freston 2008: 266).

Note that this conversation goes beyond immigration to how we help or harm other countries. In addition to a new immigration policy, Empie argued years ago, the U.S. should use "our resources in such a way as to help such persons in the countries where they are" (Ngai 2004: 253).[21] Economic and political policy decisions in the U.S. have effects on other countries. For example, the U.S. imports goods from other countries, made with labor cheaper than in the U.S., leaving those laborers with a lower standard of living than they would have in the U.S. While we import the goods, we do not import the labor and let people make a higher living wage here. Further, with regard to pollution, not only do we create more pollution and waste than poorer countries, but many richer countries export some of their waste, especially hazardous waste to poorer countries.

From this perspective, the problem is not the border. We live in a globalized world where borders are crossed constantly by people and things. The issue is 'how' to cross the border both in terms of receiving and giving. Many see Christianity as full of border crossings. For example, in Christ there is the crossing of the border between human and divine. There is the crossing of borders between human beings

[21] The 2009 U.S. Immigration Policy Task Force has a section on Development and states that "The Task Force believes that economic development inside sending countries is the best way to discourage mass emigration" (Bush and McCarty 2009: 34).

of different cultures. Further, there is the crossing of borders between our citizenship on earth and citizenship in heaven (Groody 2009a).

Some faith communities have come together to support Comprehensive Immigration Reform: "(1) Border protection policies consistent with humanitarian values...; (2) reforms in family-based immigration to reduce backlogs...; (3) creation of legal avenues for workers and their families...; and (4) earned legalisation of undocumented immigrants" (Soerens and Hwang 2009: 141–142). It works toward the middle ground of the spectrum I described in this section.

161.4.2 Christian Migrants

What do migrants want? Their voice is always silenced in the debate. Those not resident in the U.S. have no say in what immigration policy is, although it directly affects many of them. I suggest one faith-based aspect here, but many others need to be heard. It is not easy to migrate, to leave home, community and country. Migrants are categorized as labor; professionals; entrepreneurs; refugees, asylees; those who come for family, study, fun or adventure.

The first stage of migration is leaving one's home. Most migrants leave their country to improve their economic situation. They may suffer from poverty, violence, war, prejudice, and so forth. In particular, economic or political conflict in the home country drive people out. It is also important to note here that forced migration still occurs. Today it is labeled human trafficking rather than slavery. It occurs most frequently within the sex trade but there is also forced migration for cheap labor in sweatshops, etc.[22]

They are drawn to the U.S. because there is a wide gap between the rich and poor of the world and the U.S. is on the richer end of the spectrum. The U.S. offers jobs and economic improvement, freedom and sometimes family. Migrants can make a living off their labor unlike in their home countries. And there are more jobs available in the host country. A demand for labor exists. If migrants can find this out, they can try to get here.

The next stage is the travel, including acquiring permission. There are many challenges here. If permission is not available, as is true for the majority of people on the planet, then they need to decide whether the need is great enough to risk illegal entry. Further, travel must be available and affordable. Again, the poorest do not migrate. For example, Brazilian migrants need to earn $10,000 to pay a smuggler when they cannot get a visa to enter legally. The travel itself can be arduous too.

Upon arrival, the migrants need to settle into life in the U.S. It is difficult economically, politically, and culturally, even when one has the correct documents. It is even more difficult if one tries to live in the shadows and remain unnoticed.

---

[22] See www.humantrafficking.org/countries/united_states_of_america. See also Batstone 2007, in particular the introduction, conclusion and the chapter on the USA.

And yet, despite these difficulties or maybe in part because of them, the experience of migration seems to hold a spiritual component for many migrants. I highlight just one aspect here: "The Migrant as Jesus Christ." There is a statue outside a church in El Paso, Texas that captures this sense of Christ. "It depicts not one of the usual images of Christ, the triumphant savior or the bleeding crucified one, but the peasant Christ, with a rough-woven serape over his shoulders, a walking stick in one hand, a small bag in other" (Journalist Paul Wilkes quoted in Bevans 2008: 89).

Many migrants see Christ as a migrant, making the journey with them. God is on their side as they try to meet their basic needs. They are called by God to migrate; some describe themselves as pilgrims. They are kept alive on the journey by God. "(Josette's) comment that "Jesus came with us on the boat" indicates a firm belief that God enters the world of poor migrants just as he entered the world in the person of Jesus Christ 2,000 years ago" (Mooney 2009: 50). And once they arrive, they are helped to survive in the U.S. by God. Many migrants articulate that they become more faithful, more religious after the journey. For example, it has been documented in the Mexican migrant community, the Brazilian migrant community and in the Hmong community (see Freston 2008; Hillmer 2010; Daniel 2010).

Similar to the hosts quoted in the previous section, migrants see the birth of Jesus in light of his family's migration. Further, the crucified, broken and bleeding Christ mirrors their troubles as they journey. Some theologians have called the poorest and marginalized, "crucified people," echoing this interpretation (Groody 2009b).

This is not unlike the view of other migrants throughout our history beginning with the Puritan narrative that became the dominant migration narrative. They articulated that God was with them, on their side. This notion can be one that brings us together or can be one we use to exclude another when we are no longer the migrant but the new host. The key is whether this will be used exclusively to privilege one group, or inclusively to welcome all.

Conclusion

There are many areas that need further research and thinking. One gap in the literature is on the narrative of African migration to the U.S. throughout history. Another gap is thinking through U.S. migration abroad, what U.S. citizens expect when they want to move to another country and how this compares or contrasts to migration into the U.S. From a faith perspective, thinking through the need for hosts to change as migrants enter is critical, as is widening the notion of assimilation and perhaps moving beyond that term altogether. Examining the rationale for enshrining economic privilege is also important. Right now, it tends to be an assumption rather than a perspective with a solid theological basis.

In essence, among Christian hosts there is a deepening divide between those who want to welcome the stranger and view immigration as part of our globalizing and impoverished world and those who want to submit God's law

(continued)

to civil law and see national borders as crucial to our protection. Yet from the migrant perspective, God is on the side of the migrant.

How these faith responses will frame actual legislation remains to be seen. But it is critical that the notion of borders be examined from the perspective of privilege. While the U.S. would no longer legally exclude another race or religion to privilege our culture, racism and fear of other religions still seem to lurk beneath despite our best intentions. The notion of what it means to assimilate still dominates the debate. What would a broader notion of U.S. culture look like? And in terms of protecting our privilege, in 100 years time will we have changed our policy of economic exclusion? Will we look back on this time period with shame? Or will we hide it under a different narrative as so often happens with our migration history now?

Appendices

Appendix 1: Glossary

Asylum Seeker: A person applying for refugee status who has already arrived in the U.S. See refugee.

Diversity Visa: 1990- present. They can be applied for by anyone from an under-represented country who has completed high school or has at least 2 years of work experience in a skilled profession. There is a lottery and those chosen must demonstrate that they meet requirements for admission.

Employment-based visa: Migrants may enter if an employer supports a visa application to meet the labor demands of the U.S. economy. Labor shortages in jobs with unique skills or further education are prioritized.

Eventual Voluntary Departure (EVD): This status can be granted when conditions in the sending country make it dangerous for migrants to return. Since 1960, EVD had been granted to Afghans, Cambodians, Chileans, Cubans, Czechs, Dominicans, Ethiopians, Hungarians, Iranians, Lebanese, Poles, Romanians, Ugandans, and Vietnamese. (Garcia 2005: 162)

Expatriate: A migrant living in a country other than that of their birth, who aims to be only temporarily in the new country.

Family-based visa: There are four categories, listed here in order of preference: (1) unmarried adult children of U.S. citizens (average 6 year wait); (2) spouses and children of Lawful Permanent Residents; (3) married sons and daughters of U.S. citizens (average 8 year wait); (4) brothers and sisters of U.S. citizens (11 year average wait).

Lawful Permanent Residents: Lawful Permanent Residents have the right to work, and their status never expires, although it must be renewed every 10 years.

Lawful Permanent Residents have a Diversity, Employment-based, or Family-based Visa.

Non-immigrant visa: Migrants come to the U.S. on a temporary basis, usually either as tourists, business travelers, temporary workers or students. Nonimmigrant visas generally have a specific expiration date. Approximately 45 % of the estimated 11–12 million people who are undocumented came on a valid visitor visa but overstayed. (Soerens and Hwang 2009: 67) They crossed the border legally and later became unauthorized.

Refugee: "A refugee or asylum seeker is a person unable or unwilling to return to his or her home country because of persecution or a well-founded fear of persecution on account of religion, race, nationality, membership in a particular social group, or political opinion." A refugee applies from abroad. (Carnes and Yang 2004: 17–18)

U.S. citizens: U.S. citizens include those born in the U.S. and naturalized citizens, who must first have been Lawful Permanent Residents and swear allegiance to the U.S. among other requirements.

Appendix 2: Rough Timeline

"Native" Americans came across the Aleutian ice bridge from 30,000 BCE.

The Norsemen crossed the Atlantic in the early centuries CE.

Early 1500s: People from the regions of Spain arrived in Florida, South Carolina and New Mexico.

1565: The Spanish founded the first permanent European settlement within the current borders of the 50 states at St. Augustine, Florida four decades before Jamestown. (Matovina 2012: 7)

1587: Asians land on U.S. soil. Luzon Indians came ashore in California with Captain Pedro de Unamuno. (Lee 1995: 20)

1600s–1830s: More Africans reached the Americas than Europeans. About 645,000 Africans were brought into the U.S. as slaves (Soerens and Hwang 2009: 49–50). Daniels (2002: 61) states 427,300.

1790: The Naturalization Act of 1790 gave citizenship to "free white persons" only.

1798: The Alien and Sedition Acts of 1798 gave the government the right to deport migrants whom it believed to be a danger to the USA or who came from countries at war with the U.S.

1820–1860: The First European Wave: In the 1840s, a massive famine in Ireland caused many Irish to come to the U.S. After 1848, German immigration spiked as people fled government instability. Over five million migrants arrived between 1820 and 1860, about 20 times more than the previous 44 years. By 1860, 13.2 % of the population of the U.S. was foreign born.

1848: The Treaty of Guadalupe Hidalgo: From 1810 to 1853, the U.S. acquired Florida, Texas, New Mexico, Arizona, California, Nevada, Utah, and parts of Colorado, Kansas, Oklahoma, and Wyoming. This included more than half of

Mexico and more than doubled the size of the U.S. (Gonzalez 1990: 31–32) Congress declared war against Mexico May 13, 1846. Mexico signed a peace treaty February 2, 1848.

1848s–1860s: Chinese began to migrate. By 1870, the census counted 60,000 Chinese in the USA.

1875: The Page Act, the first federal law restricting immigration, ended the trafficking of prostitutes and forced laborers from "China, Japan and other Oriental countries." Chinese women were considered prostitutes.

1882–1902: In 1882, President Arthur signed the Chinese Exclusion Act. In 1888, the Scott Act prohibited Chinese workers from returning unless they had relatives in the U.S. or owned land worth $1,000. A second and third Exclusion Acts were passed in 1892 and 1902.

1882: Immigration law began to exclude the: "lunatic, idiot, or any person unable to take care of himself or herself without becoming a public charge" (Soerens and Hwang 2009: 49).

1880–1920: About 23.4 million migrants entered mainly Italians, Poles and Russian people of Jewish descent. About 2.5 million were Jewish. Beginning in 1892, they were processed through Ellis Island in New York. Thirteen to 15 % of the population was foreign born.

1880s: Japanese laborers began to arrive.

1894: The Immigrant Restriction League was founded by recent Harvard graduates.

1898: Spanish-American War: The U.S. annexed Puerto Rico and Cuba.

1900: Filipino laborers began to arrive.

1903: Immigration Act bars polygamists and anarchists. Koreans began to arrive.

1907: Indians began to arrive.

1907: The Gentlemen's Agreement between the U.S. and Japan bars Japanese migration to the U.S.

1913: Alien ownership of California land forbidden. Japanese were ineligible for citizenship. (Lee 1995).

1917: Congress ended migration from Asia and required basic literacy for migrants over 16.

1917: Jones-Shafroth Act granted Puerto Ricans U.S. Citizenship.

1922: President Harding signed a law to set the number of visas extended to each country at 3 % of the total number of people born in that country as of the 1910 U.S. census.

1924: The quota was revised to 2 % of each migrant community- as measured in the 1890 census, which had the effect of preferencing Anglo-Saxons.

1924–1929: About 285,000 Mexicans enter the U.S.

1930s: After the Great Depression, the U.S. establishes a repatriation program for Mexican migrants.

1934: Filipinos may no longer enter the U.S. freely due to Philippine Independence.

1941: Japan bombs Pearl Harbor.

1942: President Roosevelt established military defense zones to exclude persons considered a threat to the U.S. military effort. 117,000 Japanese-Americans were interned. 70,000 were U.S. citizens. (Lee 1995: 20)

1943: The Chinese Exclusion Act was repealed.

1942–1964: The Bracero Program begins. The U.S. government grants visas to as many as 400,000 Mexican workers each year.

1945–1965: The continental Puerto Rican population grew from about 13,000 to 1,000,000+.

1945–1946: Legislation permitted foreign spouses and children of U.S. servicemen to enter the U.S. The Luce-Celler Act allows exceptions to the restriction of Asian migration.

1948 (and 1950) Displaced Persons Act: War refugees may enter the U.S.

1952: The McCarran-Walter Act changed the quotas to one-sixth of 1 % of each ethnic group based on the 1920 census. Changes included: Highly skilled migrants whose services are needed; A minimum of 100,000 visas annually for each Asian country and migrants of any nationality can become citizens.

1952: The Texas Proviso: Congress made it a felony to harbor (but not employ) the undocumented.

1953: Refugee Relief Act grants permanent residence to 200,000+ refugees.

1954: Operation Wetback: The Eisenhower administration deported about 1.1 million Mexican and Mexican Americans, including U.S.-born citizens of Latina/o descent.

1959: The Cuban Revolution: nearly 500,000 Cubans migrated.

1965: Hart-Celler Immigration Act. President Johnson abolished national immigration quotas and allowed foreign-born residents to sponsor the immigration of family members. (1) Unmarried adult children of US citizens; (2) spouses and unmarried adult children of permanent residents; (3) members of the professions and scientists and artists of exceptional ability; (4) married children of US citizens; (5) brothers and sisters of U.S. citizens over age 21; (6) skilled and unskilled workers in occupations for which labor is in short supply; (7) refugees from communist countries, or the Middle East.

1975: Indochina Refugee Act accepts refugees from Indochina.

1975–1980s: After the fall of Saigon in 1975, 130,000 Vietnamese arrived in the US. Further waves entered over the next decade. For example, 1978–1982, 300,000 ‘boat people’ who had been sheltered in various refugee camps- mainly in Thailand, the Philippines, and Hong Kong arrived in the US.

1979–1985: More than 150,000 Cambodians came to the U.S.

1980: Mariel Boat Lift: 10,000 Cubans migrated.

1980s: With unrest in El Salvador, Nicaragua and Guatemala, thousands of refugees arrived.

1982: Start of the Sanctuary Movement.

1986: Immigration Act (IRCA) provides legal residency to 3,000,000 undocumented migrants who resided in the USA prior to 1982, expands the Border Patrol, imposes sanctions on employers who knowingly hired the undocumented; and makes it a crime to work without documents.

1990: The immigration ceiling was raised to 700,000 for the next 3 years, followed by 675,000 for each subsequent year. The act removed homosexuality as a reason for exclusion and increased the quota for employment-based skilled laborers, and created a lottery program that randomly assigned a number of visas, known as the Diversity Visa.

1994: Operation Hold the Line, Gatekeeper, and Safeguard: The Clinton administration enacted initiatives to close the popular urban crossing cities along the border.

1996: President Clinton doubled the number of Border Patrol agents; eliminated due process for migrants; tightened claims for asylum; increased penalties for the undocumented; and toughened legislation against smugglers of migrants.

2000: 11 % of the population was foreign-born, with 43 % having arrived since 1990. (Soerens and Hwang 2009: 42)

2002: The Homeland Security Act. After 9/11, this act created the U.S. Immigration and Custom Enforcement (ICE), increased the number of INS agents, tightened systems to track migrants and instructed government agencies to share information about migrants.

2005–2006: Acts enabled further construction of a fence along the Mexican-U.S. border, prohibited the undocumented from obtaining a driver's license; and toughened asylum regulations. By 2009, the border along Mexico/California, Arizona, and New Mexico was almost entirely fenced off.

References

Alba, R., Raboteau, A. J., & De Wind, J. (2009). Introduction: Comparisons of migrants and their religions, past and present. In R. Alba, A. J. Baboteau, & J. DeWind (Eds.), *Immigration and religion in America: Comparative and historical perspectives* (pp. 1–24). New York: New York University Press.

Batstone, D. (2007). *Not for sale: The return of the global slave trade – And how we can fight it*. New York: HarperOne.

Bauer, L. (2012). Missouri judge denies Guatemalan mother custody of her son. *Kansas City Star*. www.kansascity.com/2012/07/18/3711223/judge-says-boy-should-stay-with.html. Accessed 15 Aug 2012.

Bevans, S. (2008). Mission *among* migrants, mission *of* migrants: Mission of the church. In D. G. Groody & G. Campese (Eds.), *A promised land, a perilous journey: Theological perspectives on migration* (pp. 89–106). Notre Dame: University of Notre Dame Press.

Bryson, B. (1996). *Made in America: An informal history of the English language in the United States*. New York: William Morrow Paperbacks.

Bush, J., & McLarty, T. F. (Chairs.). (2009). *U.S. immigration policy* (Independent task force report, no. 63). New York: Council on Foreign Relations.

Carnes, T., & Yang, F. (2004). Introduction. In T. Caarnes & F. Yang (Eds.), *Asian American religions: The making and remaking of borders and boundaries* (pp. 1–37). New York: New York University Press.

Carroll, R. M. D. (2008). *Christians at the border: Immigration, the church, and the Bible*. Grand Rapids: Baker.

Daniel, B. (2010). *Neighbor: Christian encounters with "illegal" immigration*. Louisville: Westminster John Knox Press.

Daniels, R. (2002). *Coming to America: A history of immigration and ethnicity in American life* (2nd ed.). New York: Harper Perennial.

De la Torre, M. A. (2009). *Trails of hope and terror: Testimonies on immigration*. Maryknoll: Orbis Books.

Eck, D. (2001). *A new religious America: How a "Christian country" has become the world's most religiously diverse nation*. New York: HarperSanFrancisco.
Evangelical Lutheran Church in America (ELCA). (1998). *Immigration*. www.elca.org/What-We-Believe/Social-Issues/Messages/Immigration.aspx. Accessed 15 Aug 2012.
Freston, P. (2008). The religious field among Brazilians in the United States. In C. Jonet-Pastre & L. J. Braga (Eds.), *Becoming Brazuca: Brazilian immigration to the United States* (pp. 255–270). Cambridge, MA: Harvard University Press.
Garcia, M. C. (2005). 'Dangerous times call for risky responses': Latino immigration and sanctuary, 1981–2001. In G. Espinosa & J. Miranda (Eds.), *Latino religions and civic activism in the United States* (pp. 159–176). New York: Oxford University Press.
Gonzalez, J. L. (1990). *Mañana: Christian theology from a Hispanic perspective*. Nashville: Abingdon.
Groody, D. G. (2009a). Crossing the divide: Foundations of a theology of migration and refugees. *Theological Studies, 70*, 638–667.
Groody, D. G. (2009b). Jesus and the undocumented immigrant: A spiritual geography of a crucified people. *Theological Studies, 70*, 298–316.
Haass, R. N. (2009). Foreword. In J. Bush & T. F. McLarty III (Eds.), *U.S. immigration policy* (Independent task force report, no. 63, pp. ix–xi). New York: Council on Foreign Relations.
Hillmer, P. (2010). *A people's history of the Hmong*. St Paul: Minnesota Historical Society Press.
Hoerder, D. (2011). Introduction: Migration, people's lives, shifting and permeable borders. The North American and Caribbean societies in the Atlantic world. In D. Hoerder & N. Faires (Eds.), *Migrants and migration in modern North America: Cross-border lives, labor markets, and politics* (pp. 1–48). Durham: Duke University Press.
Isbister, J. (1996). *The immigration debate: Remaking America*. West Hartford: Kumarian Press.
Kerwin, D. (2008). The natural rights of migrants and newcomers: A challenge to U.S. law and policy. In D. G. Goody & G. Campese (Eds.), *A promised land, a perilous journey: Theological perspectives on migration* (pp. 192–209). Notre Dame: University of Notre Dame Press.
Kerwin, D. (2009). Rights: The common good, and sovereignty in service of the human person. In D. Kerwin & J. M. Gerschutz (Eds.), *And you welcomed me: Migration and Catholic social teaching* (pp. 93–122). Lanham: Lexington Books.
Lee, J. Y. (1995). *Marginality: The key to multicultural theology*. Minneapolis: Fortress Press.
Matovina, T. (2012). *Latino Catholicism: Transformation in America's largest church*. Princeton: Princeton University Press.
Min, P. G. (2010). *Preserving ethnicity through religion in America: Korean Protestants and Indian Hindus across generations*. New York: New York University Press.
Mooney, M. A. (2009). *Faith makes us live: Surviving and thriving in the Haitian diaspora*. Berkeley: University of California Press.
Ngai, M. M. (2004). *Impossible subjects: Illegal aliens and the making of modern America*. Princeton: Princeton University Press.
Olupona, J. K., & Gemignani, R. (2007). Introduction. In J. K. Olupona & R. Gemignani (Eds.), *African immigrant religions in America* (pp. 1–26). New York: New York University Press.
Pontifical Council for the Pastoral Care of the Migrants and Itinerant People. (2003). *Starting afresh from Christ. Towards a renewed pastoral care for migrants and refugees*. Final document of the fifth world congress on the pastoral care of migrants and refugees, Rome, 17–22 November 2003. www.vatican.va/roman_curia/pontifical_councils/migrants/documents/rc_pc_migrants_doc_2004001_Migrants_Vcongress_%20findoc_en.html
Ramirez, B. (2011). Through the northern borderlands: Canada-U.S. migrations in the nineteenth and twentieth centuries. In D. Hoerder & N. Faires (Eds.), *Migrants and migration in modern North America: Cross-border lives, labor markets and politics* (pp. 76–98). Durham: Duke University Press.
Rivera, L. R. (2009). *El Cristo migrante*/The migrant Christ. In H. J. Recinos & H. Magallanes (Eds.), *Jesus in the Hispanic community: Images of Christ from theology to popular religion* (pp. 135–154). Louisville: Westminster John Knox Press.

Schrag, P. (2010). *Not fit for our society: Immigration and nativism in America*. Berkeley: University of California Press.
Soerens, M., & Hwang, J. (2009). *Welcoming the stranger: Justice, compassion & truth in the immigration debate*. Downers Grove: IVP Books.
Suarez-Orozco, M. M., Louie, V., & Suro, R. (2011). Preface. In M. M. Suarez-Orozco, V. Louie, & R. Suro (Eds.), *Writing immigration: Scholars and journalists in dialogue* (pp. ix–xxiv). Berkeley: University of California Press.
Takaki, R. (1989). *Strangers from a different shore: A history of Asian Americans*. Boston: Little, Brown.
Tan, J. Y. (2008). *Introducing Asian American theologies*. Maryknoll: Orbis Books.
Wilson, T. D. (2009). *Women's migration networks in Mexico and beyond*. Albuquerque: University of New Mexico Press.
Zavella, P. (2011). *I'm neither here nor there: Mexicans' quotidian struggles with migration and poverty*. Durham: Duke University Press.
Zolberg, A. R. (2006). *A nation by design: Immigration policy in the fashioning of America*. New York: Russell Sage.

Chapter 162
Evangelical Geopolitics: Practices of Worship, Justice and Peacemaking

Nick Megoran

162.1 Introduction

In his disturbing yet ultimately hopeful study of life under the Chilean military dictatorship of the 1970s, William Cavanaugh describes how the Catholic Church gradually developed a theology and practice of resistance to torture (Cavanaugh 1998). In the early days of the Pinochet regime, the church's emphasis on social action over poverty as the primary mode of public engagement left it ill-equipped to challenge the regime's practices of torture. However, argues Cavanaugh, in time the church cultivated practices of opposition that centered around the Eucharist. This 'Eucharistic counter-politics' contrasted the torture that Christ endured to the torture of the regime, and created alternative communities that resisted and exposed state violence.

Cavanaugh's research is a reminder that Christian political imaginations, geopolitical visions, and the practices of engagement that result from them, are created through and structured by practices of worship. As Bernd Wannenwetsch argues in his post-liberal critique of traditional theological ethical reflection, worship is central to the formation of Christian political ethics (Wannenwetsch 2004: 8). This chapter extends the growing interest in the geopolitics of evangelical Christianity by considering practices of worship that produce progressive political programs.

This emphasis on uncovering progressive evangelical geopolitics is an important counterbalance to popular impressions of evangelicals as being politically conservative or espousing an otherworldliness that is detached from reality. This conception is epitomised by Harold Camping, an 89-year old US Christian radio broadcaster who predicted that the world would end on May 21, 2011 (Tenety 2011). His claim was based on an idiosyncratic reading of Biblical prophecy, and led some of his

N. Megoran (✉)
Department of Geography, University of Newcastle-upon-Tyne, Newcastle NE1 7RU, UK
e-mail: nick.megoran@newcastle.ac.uk

S.D. Brunn (ed.), *The Changing World Religion Map*,
DOI 10.1007/978-94-017-9376-6_162

relatively small band of followers to divest themselves of worldly goods in anticipation. Journalists followed their preparations with a mixture of fascination and scorn, and took obvious pleasure in their bewilderment when the predicted events failed to materialise (Harris 2011).

That media outlets in the U.S. and UK should so widely report a non-story like the world's continued existence, or the predictions of an obscure geriatric businessman that it was about to end, is remarkable. The story revisits a certain stereotype of U.S. evangelical Christians as wealthy, other-worldly, conservative, and irrational. Reporters barely mentioned that most Christian communities have other ways of thinking about end-times theology that inform very different political engagements.

In this chapter I argue that something analogous is in danger of occurring within the "geopolitics of religion" literature that has emerged over the past half-decade. Although I welcome study of the geopolitical significance of religion, I am concerned at its relatively narrow focus on right-wing militaristic readings of the end-times theology. Following Susan Harding's argument about the field of the anthropology of Christianity, I suggest that the emerging geopolitics of Christianity is constructing fundamentalist/evangelical Christians as our "repugnant cultural other." Although the critical geopolitical scrutiny of how evangelical theologies inform bellicose political practices is crucial, it is equally important to recover how people use the same traditions to inform more progressive social activism. This chapter considers practices by evangelical Christians that frame geopolitical interventions and imaginations in South Africa, the Middle East, and the UK that are marked by the pursuit of justice and peace through nonviolent means. Following a discussion of the recent geopolitics' literature on evangelicals, it does this by considering examples of three important evangelical practices of worship: preaching, prayer walking, and prayer meetings.

162.2 Evangelicals – Critical Geopolitics' "Repugnant Other?"

In 1961 Freeman lamented the "impatience" of many geographers with religion, even though it was clearly such an important geopolitical factor (Freeman 1961: 206). This same point about the neglect of the geopolitics of religion was echoed by a number of geographers almost half a century later (Dijkink 2006; Megoran 2006a). Scholars of geopolitics are at last taking religion seriously. Dittmer and Sturm's focused, coherent, theoretically-informed and empirically-diverse collection on the geopolitics of American Evangelical end-times prophecy belief (Dittmer and Sturm 2010) is ample demonstration of progress. A whole chapter is devoted to "Evangelicals" in the *Ashgate Research Companion to Critical Geopolitics* (Dittmer 2013).

However, much of this work depicts religion in general, Christians in particular, and evangelicals specifically, as indelibly mendacious. This is the case from Dalby's 1990 asides about evangelicals reading prophecy to oppose arms reduction talks

(Dalby 1990), to Sturm and Dittmer's summary in their book that "Evangelicals often support highly violent foreign policies" (Sturm and Dittmer 2010: 4). Sturm (2006) and Dittmer (2008) consider interpretations of the Book of Revelation in U.S. popular culture and their significance within right-wing, pro-war constituencies and formal politics. In his own chapter of the edited book, Dittmer uses internet ethnographies to show how some evangelicals suspected that the 2008 U.S. Presidential winner Barack Obama was "the Antichrist" (Dittmer 2010). Across the Atlantic, Sidaway mentions how some anti-EU Britons read the union as the beast of Revelation (Sidaway 2006: 2–4). Catholicism has received less treatment but fares little better: Ó Tuathail explores the "Jesuit anti-Communism" of Fr. Edmund Walsh's Cold War "spiritual geopolitics" (Ó Tuathail 2000), and Agnew considers the geopolitics of the church under Pope Benedict XVI as based on a "Hobbesian-Stalinist model" of its place in global affairs (Agnew 2010: 56). Exceptions do exist (Gerhardt 2008; Gallaher 2010), but although each of the above studies represents by itself a valuable and important contribution to our understanding of geographies of religion, together they present Evangelical Christians as warlike, bigoted, racist, credulous, irrational, conspiratorially paranoid, and right-wing. No doubt some are, but Gallaher is at least partially correct in stating that "most commentators on the evangelical phenomenon (whether in the media or academia) are extrapolating the views of a few well known" figures to the entire movement (Gallaher 2010: 229). Critical geography should be wary about creating new Others.

The narrow focus of the geopolitics of religion is of interest not simply because of what it says about a certain group of Christians, but because of what it says about "us." Here we can learn from the vibrant field of the anthropology of Christianity. A foundational text was Susan Harding's 1991 article in *Social Research*, "Representing fundamentalism." Harding argues that fundamentalist/evangelical Christians have become anthropology's "repugnant cultural other." Seen as antithetical to modernity, "which emerges as the positive term in an escalating string of oppositions between supernatural belief and unbelief, literal and critical, backward and progressive, bigoted and tolerant," they constitute us as the modern subject (Harding 1991: 374). Our study of the geopolitics of religion may be doing likewise. Here the anthropology of religion can point a way forwards. A recent review article in *Anthropological Forum* shows how this literature may point us forwards, enjoining scholars to attend to diverse local forms of Christianity (McDougall 2009: 188). I suggest that one way to advance the study of the geographical study of evangelicals is to do just that.

In this chapter, I will outline three ways in which Evangelical Christians have engaged in practices with more pacific geopolitical implications. The first is the preaching of South African anti-Apartheid activist and church leader Allan Boesak; the second is a "prayer walk" by U.S. missionaries in the Middle East to deliver an apology for the First Crusade; and the third is UK evangelical prayer meetings that acted as a way to reflect on and repent over nationalism and racism. These divergent evangelical activities inform geopolitical visions and sustain political practices quite different from those considered in the literature thus far.

162.3 Preaching: Allan Boesak and the Re-reading of Revelation

More so than for Christian traditions that emphasise sacramental or liturgical worship, the performance of the sermon is a crucial element in evangelical practices of worship. To cite a well-known example, many of the addresses of Rev Martin Luther King Jr in the civil rights campaign were sermons delivered in his or another church at the height of a particular element of the civil rights (King 1981).. This section considers the sermons of a black leader of the South African anti-apartheid movement who drew on King's legacy, Rev Allan Boesak. In particular, it explores his exposition of apocalyptic scripture to critique apartheid and to succour its opponents.

Much of the work on evangelical geopolitics outlined above has looked at geopolitical interpretations of the biblical book of Revelation. Arranged as the final book of the New Testament canon, it is ascribed to "John," exiled to the island of Patmos "because of the word of God and the testimony of Jesus."[1] Revelation is an example of the genre of contemporary apocalyptic Jewish writing. This genre uses extraordinarily vivid imagery and language to depict cataclysmic events, moments that may present substantial ruptures in human history. All texts are open to multiple interpretations; but history has shown that Revelation is more so than others.[2]

The geopolitical analyses cited above have largely considered how Revelation is mapped onto present geopolitics by evangelical Christians using the theological paradigm of "Dispensationalism premillennialism" (or simply "premillennalism"). Although its antecedents are in the earliest Christian times, dispensational premillennialism was developed in the nineteenth century by John Nelson Darby. It is a highly literalistic interpretation of texts like Revelation and holds that at the end of the present "dispensation" the church would be "secretly raptured," or safely removed to heaven in an instant. Following this, earth's remaining inhabitants would be subject to the great "tribulation" (including disease, war, and famine), before the salvation of the Jewish people and Christ's return to rule the earth from Jerusalem for 1,000 years of peace. The end of this "millennium" would witness a final decisive battle with evil in the form of "Babylon" and its minions (Boyer 1992: chapter 3; Weber 1987: 1–27). Premillennialism understands many passages of the book of Revelation as literal predictions rather than symbols. It thus lends itself to the identification of specific events in our time with what are understood to be biblical predictions. It is premillennialist circles – and often extreme and marginal elements of them, such as Harold Camping – that have been the focus of most of the scholarship on geopolitics and religion cited above.

In contrast to premillennialism, Kovacs and Rowland's illuminating compendium of readings of Revelation over Christian history shows that this approach of

[1] Revelation 1: 9, Bible. All Bible references are to the New International Version, London Hodder and Stoughton 1973.

[2] This argument is developed more fully in Megoran 2013.

seeing Revelation as coded references to specific future events and people is only one of a broad range of approaches (Kovacs and Rowland 2004). Richard Bauckham, an influential scholar of Revelation, dismisses this approach of seeing Revelation as coded future predictions, insisting instead on the need to understand the social, political, cultural and religious resonances of its symbols (Bauckham 1993: 19). These, he posits, are primarily Roman and particularly the Roman empire of Nero and Domitian that was brutally persecuting Christians. In Bauckham's interpretation, the strategy of Revelation, he suggests, is to create a symbolic world for its readers to enter in order to "redirect their imaginative response to the world," and thereby break the bounds which Roman power and ideology set on the world (Bauckham 1993: 129).

Ben Witherington's respected commentary on Revelation fleshes this out. For example, Revelation chapter 17 depicts "Babylon," mother of harlots, as a woman on a city built on seven hills, deceiving the world, draining it of wealth, and drunk on the blood of the saints. Rome was, famously, a city built on seven hills. Witherington observes that a coin minted under Vespasian, and still in circulation under Domitian (when Revelation may have been written) depicted the female divinity Roma sitting on seven hills. Witherington writes that "John's depiction may owe something to this coin, but one must bear in mind that he is doing a deliberate parody of such images that involves comic exaggeration of features, such as we see today in political cartoons" (Witherington 2003: 218–219). Numerous such references can be drawn from Revelation, by particular attention to the textual and visual discourses of Roman imperial propaganda and emperor worship. The cult of the worship of Ceasar as emperor, as "the son of god" who had brought "peace" and "salvation" to the earth, was increasingly seen as integral to Roman life and welfare. Witherington thus asserts that Revelation's proclamation of Jesus (rather than Caesar Domitian) as Lord is a profoundly political act (Witherington 2003: 162). This is an articulation of a politics of resistance that is non-violent because "conquering takes place through dying not killing" (Witherington 2003: 174).

This idea that Revelation is not merely an anti-imperial text, but one that posits *non-violent* resistance to empire is gaining increasing currency. Patricia McDonald reads Revelation's the battle scenes between angelic and demonic forces as examples of "nonviolent conquering" (McDonald 2004: 265). Likewise Mark Bredin argues, "The Jesus of Revelation is a Revolutionary of peace," who defeats his enemies by dying at their hands, who fights violence with nonviolence, and who stands for all humankind rather than projecting violence against the evildoer who must be eradicated (Bredin 2003: 223). In his attempt to "aid the recovery of a spiritual reading of geopolitics in our time," Michael Northcott argues that contemporary premillennialist right-wing U.S. readings of the apocalypse are a product of the Constantinian shift in the fourth century when "Christianity was turned from its non-violent and anti-imperial origins into an imperial cult" (Northcott 2004). The task, then, is "saving Christianity from empire" (Nelson-Pallmeyer 2005).

A practical outworking of this theology can be seen in the life and work of Allan Boesak, was a Black theologian, church minister and influential anti-apartheid activist in South Africa (Ackerman and Duval 2000: 348). He argued for the importance of social context for the doing of theology by Black Christians in Africa and North America (Boesak 1977). His African liberation theology incorporated a rigorous theorization of violence, with a clear vision of a just and peaceful future obtainable by peaceful means, and practical application in the messy context of a localized struggle. For example, in 1983 he discovered that he was the target of a foiled assassination plot by the *Afrikaner Weerstandsbeweging* (Akrikaner Resistance Movement). In response, he declared that their actions were explicable when "racist laws, racist structures, racist attitudes emphasize in a thousand ways the sub-human status of black people in South Africa" (Boesak 1983a: 46). He rejected the option of "cheap reconciliation" which "denies justice, and which compromises the God-given dignity of black people" (46–47), and instead averred his commitment to "true peace" that would only occur with the dismantling of apartheid, and was to be pursued through nonviolent civil disobedience.

His sermons repeatedly turn to Revelation. For Boesak, prophecy is "much less predicting the future that contradicting the present" (Boesak 1983b: 29) and Revelation contradicts Apartheid as much as it contradicted Rome. For example, in 1978 the so-called 'Information Scandal' plunged the South African government into turmoil when it was revealed that it had been using illegal means to sell the government's Apartheid policy to the outside world. Boesak reads this through Revelation as God unveiling and judging the immoral workings of empire (Boesak 1987).

Boesak's 1987 book C*omfort and Protest,* is a commentary on Revelation, or, perhaps more accurately, a commentary on the Apartheid regime performed through a reading of the book of Revelation. He identifies specific Apartheid policies and official proclamations, comparing them to the Rome/Babylon of Revelation. In closing his book with a discussion of the final chapter of Revelation, Boesak quotes a poem written by Martial praying for the safe homecoming of the Emperor Domitian:

Thou, morning star,
Bring on the day!
Come and expel our fears,
Rome begs that Caesar
May soon appear

"The church smiles at this last desperate attempt at power and glory," writes Boesak, observing that the final chapter of Revelation hails not Domitian, but the Lord Jesus Christ as "the bright morning star" (Revelation 22:16). As John surveys the heavenly Jerusalem, the city of justice and peace, descending to earth to reunite symbolically God and humanity, he implores, "Come, Lord Jesus." Boesak (1987: 137–138) creates a liturgical prayer from this:

For the pain and tears and anguish must end… Come, Lord Jesus
For there must be an end to the struggle when the unnecessary dying is over… Come, Lord Jesus
For the patterns of this world must change… Come, Lord Jesus
For hate must turn to love, fear must turn to joy…Come, Lord Jesus
For war must cease and peace must reign… Come, Lord Jesus.

Apocalyptic scripture presents the empire is a force whose wickedness must be taken seriously, yet insists that, for all its might and resources, its claims to represent civilisation or God's agency in history are nothing but lies. It does not have the final word in the human story, indeed, it will be overthrown spectacularly. Kovel argues that the apocalyptic mindset's ability to conceive of the wholesale overthrow of an unjust order has made Revelation a key text in revolutionary history, from anti-slavery struggles to Communism (Kovel 2007). The emerging geopolitics of religion needs to uncover radical and revolutionary readings of apocalyptic scripture, not just reactionary ones. It is crucial to place these readings not only as part of academic theological discourse, but as sermons embedded in specific struggles for peace and justice.

162.4 Prayer Walking: Crusade Apologies

The second evangelical practice of worship that we will consider is 'prayer walking,' an activity that came to assume importance in certain strands of evangelical church life in the 1980s. Although evangelicalism commonly eschews the idea of "sacred places," prayer walking emerged from a theology that emphasised 'spiritual warfare' against 'territorial spirits.' This theology posited that the work of the church could be enhanced by specifically praying in certain places. One product of this was the "March for Jesus" movement, that saw sometimes thousands of Christians at a time processing through British streets with songs and prayer (Kendrick 1992). An unusual example of this movement, that grew out of it and took it in new directions, was the Reconciliation Walk, a prayer walk in apology for the first Crusade that took its participants overland from France and Germany to Jerusalem.[3] This phenomenon will be considered in this section as a particular form of engagement with the geopolitics of the Israel-Palestine question.

A recurrent criticism of the geopolitics of U.S. and UK evangelicalism is its often uncritical support of Israel in recent Arab-Israeli disputes. For geopolitical analysis, the key place of premillennalism is important in explaining this, as it predicted the 'ingathering' of the Jews to Palestine and a 'great tribulation' before Christ's return (Boyer 1992). By the 1970s premillennialism was firmly entrenched in the U.S. as the prevalent evangelical end-times doctrine. What began as an apparently obscure theological debate gained important geopolitical significance with the creation of the state of Israel in 1948.

The belief that the emergence of the modern state of Israel is a fulfilment of Biblical prophecy (Sizer 2004; Clark 2007) has led to the condition whereby 'the vast majority of [American] evangelicals instinctively believe that vigorous support

[3] It might be more correct to say that the Reconciliation Walk grew along with the March for Jesus. As Graham Kendrick records in his history of March for Jesus, Lynn Green – the originator of the Reconciliation Walk – was involved with prayer walking and March for Jesus in Britain from their earliest stages (Kendrick 1992, chapter 4).

for Israel is the only appropriate response to the conflicts in the Middle East' (Burge 2003: 236). Ruether and Ruether argue that "Christian fundamentalist support for Israel is not simply a matter of apocalyptic theories; it is a matter of garnering major economic and political support behind an expansionist vision of the State of Israel" (Ruether and Reuther 2002: 182). Likewise, evangelicals in this camp view "the native Arab population in generally negative terms," rarely calling for their equal treatment (Weber 1987: 206–207). The attitude of this "Christian Zionism" towards Muslims is part of a wider geopolitical vision that is antagonistic towards Muslims. Thus, for popular U.S. theologian Don Carson, the "war on terror" is "a civilisational struggle between the world of Islam and West" (Carson 2002: chapter 4).

The geopolitical impact of U.S. premillennialism is an example of the central assumption of geopolitical study: the views we hold about the world have real impacts upon the way we act in it. Having identified the most common geopolitical implications of U.S. evangelicalism in regard to the Arab-Israeli conflicts, this chapter will now consider the ways in which the Reconciliation Walk, contested this geopolitical vision and moved towards an alternative one.

On November 27, 1095, Pope Urban II preached a sermon to crowds of clergy and laity attending a church council in the French town of Clermont-Ferrand. Calling on Europe to unite and defend itself against Muslim attacks on Christian territory and pilgrims, his speech initiated what became known to history as "the First Crusade." The First Crusade culminated on July 15, 1099, when Jerusalem fell and Jewish and Muslim defenders and residents were massacred. On November 27, 1995, evangelical Christians gathered in Clermont-Ferrand to launch "The Reconciliation Walk" (henceforth 'RW'). This involved thousands of largely American and European Christians retracing the routes of the First Crusade and apologising to Jews, Eastern Christians, and Muslims for the Crusades.[4] Although based in England, the RW was largely a project of the influential U.S.-based global evangelical Christian mission agency, Youth With A Mission (henceforth "YWAM"). The RW culminated in Jerusalem on July 15, 1999, when a formal apology was issued to, and received by, Muslim, Orthodox and Jewish leaders in the city.

This juxtaposition of parallel journeys between Clermont-Ferrand and Jerusalem undoubtedly appears bizarre. But it reveals how a geopolitical vision can be transformed through engagement. I conducted research on the RW by interviewing its key organisers and originators, and through accessing its archives at a UK regional YWAM centre in Harpenden, England.[5]

According to its original promotional literature, the RW was designed to "… make a major contribution to peace between the peoples of Christianity, Islam and Judaism" (Reconciliation Walk no date). By apologising to Jews, Muslims and Eastern Christians, it would assist in "defusing the legacy of the Crusades" (Reconciliation Walk 1996). An important cognate idea was that "defusing" this

[4] The text of the apology is available at the RW's website: www.crusades-apology.org/Crusades%20 Project/turkpres.htm (accessed December 2009).

[5] For a fuller investigation of the Reconciliation Walk, see Megoran, N. (2010) 'Towards a geography of peace: pacific geopolitics and evangelical Christian Crusade apologies'. Transactions of the Institute of British Geographers, 35(3), 382–398.

legacy would remove an obstacle to the conversion of Muslims to Christianity, as one RW leader put it, removing barriers "between the Islamic world and evangelism" (Cathy Nobles, interview, Harpenden, 4/08/2006).

The RW was officially launched with a day of prayer in Clermont-Ferrand on November 27 1995. Although it might appear to be a form of traditional pilgrimage, in fact its more immediate context was the practice of "prayer walks" developed by charismatic Christians in the 1970s and 1980s. Lynn Green (see below, one of the RW leaders) was involved in the "March for Jesus," a series of prayer walks in UK cities in the 1980s, and in an interview with me indicated that this practice was in his mind as he planned the RW.

In the spring of 1996, a few small groups of walkers retraced Crusader routes along the Rhine and Danube and via Italy and the Balkans, praying for peace and reconciliation and focussing on presenting an apology to Jewish communities, the first targets of the Crusaders. In the summer of 1996, the first teams began to arrive in Istanbul. From then on, greater numbers of larger teams joined the RW for often short periods of time (2 weeks). Following an induction, teams then fanned out across different parts of Turkey, meeting people en route in public spaces such as cafes, shops, and parks, and being invited to homes. At the same time, leaders and teams held official meetings with religious and civic leaders, when framed copies of the apology were presented and discussions held in well-publicised meetings that attracted sometimes significant media coverage. The walk then continued down the Levant through Syria, Lebanon, and to Israel/Palestine. A similar format was followed in these countries. Around 3000 people from over 30 countries and a variety of Protestant denominations took part.

The RW was a project with great sensitivity to geography. Following the exact routes that Crusaders took, and reaching places on the anniversary of their arrival, was considered a vital aspect of the historical authenticity and contemporary spiritual effectiveness of the project.

The genesis of the RW is traceable to a small number of US citizens working for evangelical Christian mission agencies in the late 1980s and early 1990s. These were principally Lynn Green, currently the International Chairman of YWAM, Cathy Nobles, a YWAM leader based in Harpenden, England, and Matthew Hand of the U.S. Lutheran Orient Mission Society. In interviews, all three identified themselves as having been influenced to some degree by the Christian Zionism of the US Christian Right at some point in their lives. I identified above two geopolitical implications of the theological position of the U.S. Christian Right: support of the state of Israel, and antipathy to Muslims. The remaining sections will examine how the RW came to question these positions and articulate alternatives.

162.4.1 The Reconciliation Walk and Israel

Those interviewed all spoke about having more-or-less developed Christian Zionist backgrounds, which they had come to re-evaluate and reject through participation in the RW. Key to this was the impact of actually meeting people in the Middle East.

Cathy Nobles cited the example or a 2006 "RW follow-up trip composed of 'mostly Zionist-type Christians from the US, UK and elsewhere' she had taken to Israel/ Palestine. They began the trip on the city wall in East Jerusalem, and suddenly came across a group of angry Palestinian teenagers chanting and shouting. It transpired that an Israeli air-raid had hit the wrong target, massacring families in Gaza. The Israeli authorities arrived and, without even attempting to calm down or disperse the crowd peacefully, began beating and arresting them. Cathy Nobles relayed that the visitors were shocked to see the actions of the state they supported" (Cathy Nobles, conversation, Harpenden, 11/11/2006).

The importance of face-to-face meetings with local people in transforming views was also stressed by Lynn Green. I asked him whether, bearing in mind how critically he spoke of Christian Zionism, there was a political goal to the RW (that is, of changing these views). He replied that he didn't think they "set out with any kind of political goal," but that they became aware from their own experiences that meeting people face-to-face and "humbling ourselves and taking a message of apology over something that is such an open wound amongst these peoples" changes people in a way that "will always have political implications." These implications were that those who went on the walk "would be much more reticent about supporting militaristic action, for the expansion of the borders of Israel, for example" whereas previous to their participation he thought that many would have had "no qualms" about that:

> Once you know the people, and they are people instead of images on the television screen, it does change your politics. That's the tragedy, isn't it, you know that oftentimes the people making the decisions, the people who wield the power, don't actually know the people that they are deciding about as people, they know them as images on the television screen, strategic objectives in some sort of geopolitical scenario. (interview, 5/08/2006)

By bringing British and American Christians into contact with those suffering as a result of Christian Zionism, the RW sought to transform the theological and thus geopolitical visions of those taking part. As the next section shows, this was the case not only with regard to Israel, but also Islam and Muslims.

162.4.2 The Reconciliation Walk and Muslims

Just as participation in the RW transformed its leaders' theological understandings and thus geopolitical visions towards Israel, so it shifted their attitudes towards Muslims. Crucial to this shift was the impact of meeting Muslims in the context of apologizing to them. In many cases, this was the simple overcoming of prejudices and stereotypes by folk whom participants would instinctively have been afraid of. For example, Cathy Nobles spoke about her fears of meeting and giving the apology to a "rough guy, head of a fundamentalist, reactionary group" in Beruit. She was surprised that he not only did "not look fundamentalist" but was "sweet, welcoming, loving." On a different occasion she was hosted to a sumptuous breakfast by

Istanbul's deputy mayor. She was struck by his suggestion that Christians and Muslims were on the same side facing overlapping moral concerns, such as pornography and secularisation (interview, 4/08/2006).

More striking than simply the realization that Muslims who appear frightening may in fact be pleasant and personable, or the identification of common perspectives on social morality, was an appreciation of the spirituality of Muslims. This was more disconcerting for those involved. Cathy Nobles (interview, 4/08/2006) spoke about "meeting very godly Muslims," people who:

> as far as works of godliness are far exceeding us, in their understanding of God, in the way that they treated their neighbour, you just had to envy and marvel at what they knew.

This raised some uncomfortable questions for her, challenging her preconception that "they know nothing and I bring everything to the table." It led her to a theological position which she identifies as "the openness of God," that if the Holy Spirit is working in all the world then "you should be finding truth in other cultures." Lynn Green recounted a similar transformation. He described his pre-RW perspective as one of "Western superiority," the:

> unconscious thinking was that God was at work amongst us, and outside of us it was kind of like a vacuum, a spiritual vacuum, and people were out there sitting in total darkness, and they had no understanding of God, or what understanding they had was heretical. (Interview, 5/08/2006)

He said that meeting Muslim groups in Turkey such as the Alevi and Mevlana, and learning of many more where "the fundamental message of Jesus has been proclaimed, that God is love, and we're called to love our neighbour." he saw so much evidence that the Holy Spirit was already at work. He professed to have read many books on this theology, "but the penny never dropped," and what he found was "totally unexpected" – "a sort of spiritual communion with some Muslims." He underlined how unexpected and disconcerting this was for him, by adding to his previous quote that "even saying those words I realise that I'm going to be branded a heretic by a number of Western Christians, but there it is."

Cathy Nobles did not understand this merely as human interaction, but as God speaking to her through it. Through these experiences, she concluded that:

> God confronted me that in falling in love with the Jews, that I had hatred for the Muslims… I heard God speak to me that "I am in love with these people, I am passionate about these people."

The implication of this for Cathy Nobles was that if God loved Muslims, she ought to too – but had not done so.

There were geopolitical as well as theological implications of this shift. RW leaders spoke about their desire to challenge visions of global space that pitted the West/Christianity against an Islam that they perceived to be dangerous. Lynn Green said that he had been increasingly

> dismayed at the tendency in the American Christian press for some of the best known spokespeople to go right along with the secular media perspective of casting Islam as the

> enemy, and not coming to grips with love of enemies, forgiveness, and getting the idol of nationalism in its proper place. (interview, 5/08/2006)

Based on her experiences in Turkey, and listening to Turkish recipients of the apology, Cathy Nobles said she became aware that "we have carried prejudice towards Turks," and "this whole image of a Christian Europe up against Turkey, two empires colliding, it really doesn't have a whole lot to do with Jesus" (interview, 3/08/2006).

Thus, the act of meeting Muslims and giving them an apology challenged assumptions and prejudices on numerous levels. The RW leaders came to people their imaginative landscapes of the Middle East with real Muslims rather than frightening stereotypes, spiritual illiterates, or dangerous enemies of the state of Israel. These personal engagements challenged the antagonistic views on Islam common within American evangelicalism. They questioned those geopolitical visions that posited clashing civilizations or re-enacted crusades. They created spaces for more peaceful and transformative interactions with Muslims. This is remarkable enough, but as we shall see in the next section, the RW did not only contribute to changes in views on Islam. It also led to changed understandings of evangelicalism's nature and its role in the Middle East.

162.4.3 The Reconciliation Walk and U. S. Evangelicalism

The RW challenged deeply-held US evangelical views about Israel and Muslims. At the same time, it transformed understandings of Christianity and Christian mission. It began by seeing the Crusades as a discreet historical episode whose spiritual legacy could be "defused" in order to facilitate conversion. However, according to interviewees, recipients often expressed thanks for the apology, but at the same time pointed to ongoing perceived Western and Christian injustices. Thus the RW leaders came to see the Crusades as emblematic of a "Crusader spirit" of arrogant superiority that infects subsequent Christianity (and Westernism) down to and including contemporary evangelicalism. Cathy Nobles told me that she frequently observed participants going through RW training and then presenting the message, who came to realise (as Cathy Nobles herself had) that "I had a lot of Crusader in me and the way I live my faith." This "Crusader spirit"

> comes into any place that we've got Manifest Destiny going, with the Northern Irish and the Scots feeling its their promised land, South Africans felt like that with their promised land, America is the promised land, so there's that same ethos. I still think that's the over-riding spirit we need to get out of the church. (interview, 3/08/2006)

The reference to "Manifest Destiny" in the U.S. context is clearly geopolitical, indexing the idea that the US has a unique, divinely-given destiny that justifies – or rather demands – power projection outside its own borders. Lynn Green observes that:

> there is still a great big chunk of, especially US American evangelicalism, that is just so firmly and closely identified with conservative politics there, including a deep belief in the

efficacy of redemptive violence, and the idea that a lot of problems in the world can be solved militarily. (interview, 5/08/2006)

He describes this as being generally seen as part of the "whole package" of being an evangelical Christian, that if someone "holds steadfastly to the basics of the evangelical faith" they ought naturally to hold to this right-wing conservative political position, and cannot understand why "you don't cheer when the U.S. military goes to war." He reported that "I'm just no longer convinced" by the idea that this conservative theological position should be wedded to the right-wing position on foreign policy, concluding that "it's not a whole package." He explained that through involvement in the RW he had come to realise that about U.S. evangelicalism and reject it (interview 5/08/2006). As he put it to my question about how involvement in the RW changed him and his understanding of God, "it changed me in that I began to see the gospel completely differently" (interview, 5/08/2006).

This would appear to be the chief impact of the RW on the understanding of Christianity itself by Lynn Green and Cathy Nobles – that reconciliation "is the core of the gospel" (interview, 4/08/2006). This understanding leads her as an American Christian to see an important Christian task as being "to challenge power and weaponry in the age we're living in." In relation to the "war on terror," it means "to get people, instead of being reactionary against Islam" to ask

'why is this happening?', are we asking the right question of why these people [the 9/11 attackers] feel so passionately about why they kill themselves, and attack us in this way, and is there something that we can change, especially as Christians.

This is a significant movement away from the geopolitical perspective that sees the U.S. as a righteous innocent violated by a pathologically evil world of Islam. Concomitantly, it involves a withdrawal from the position that Christians should back the U.S. government as it responds with military force to its enemies, and as it offers uncritical support of the state of Israel.

This research concludes that participation in the Reconciliation Walk occasioned significant shifts in the geopolitical visions of its leaders. What began as a tactic to facilitate conversion led to a wholesale rejection of a theology of Christian Zionism and its associated foreign geopolitical vision and foreign policy agenda. It also precipitated movement towards a very different theology and practice of engagement with Muslims. Due to the influence of these people within U.S. evangelicalism, this potentially has significant broader implications – particularly as they have identified spreading the lessons of the RW within the movement as a key ongoing goal.[6] The Reconciliation Walk demonstrates how evangelical Christian practices such as prayer walks can produce politically-progressive geopolitical practices and visions and prove transformative for those engaged in them. The final section will consider the long-term effects of walk on participants.

[6] For more on these implications, see Megoran, Towards a geography of peace.

162.5 Prayer and Repentance: Reconciliation Walk Follow-Up Meetings

Prayer and repentance is an important element of everyday evangelical Christian worship. Following the completion of the Reconciliation Walk (RW) in 1999, Lynn Green and Cathy Nobles renamed it, "Reconciliation Walk… the Journey Continues" and the project has continued to undertake a number of activities. Building on relationships formed up to 1999, it takes groups of largely British and American Christians to Lebanon and Israel/Palestine, where they meet Christians, Jews and Muslims, and receive teaching about reconciliation. It also conducts work on Christian-Muslim dialogue in the UK. These activities are supported through bi-monthly prayer meetings at the YWAM "base" in Harpenden, and it is these prayer meetings that are considered in this final section of the chapter.

These meetings serve as a means for Lynn Green and Cathy Nobles to report back on and discuss ongoing activities, to have them prayed for and to bring together veterans of the walk. Most of the people who regularly attend the meetings had been inspired by the original RW to continue peacemaking/reconciliation activities in their own communities in southern England: for example, in local Christian-Muslim dialogue groups, or promoting alternative perspectives in their own congregations. Meetings tend to have 20–30 people present, most of them late middle aged white Britons. From 2006 until the time of writing, I have attended one or two of these a year to conduct an ongoing ethnographic study through participation (Megoran 2006b) in the continuing work of the RW. In this section I describe two of these meetings and show how prayer for the work of the RW abroad served to inform and rework geopolitical understandings of participants (Megoran 2010).

The RW prayer meeting in June 2006 coincided with the football world cup. The chapel on the YWAM base had been taken over to screen England's first game, against Paraguay, and the RW meeting took place in the sunshine outside the chapel.[7] Perhaps appropriately, prayer and discussion that day revolved around the topic of nationalism. Lynn Green began by reading from the Old Testament book of Jeremiah, chapter 28. This passage depicts the confrontation between the prophet Jeremiah and the royal court prophet, Hananiah, over the subjugation of the Kingdom of Judah to the Babylonian empire of King Nebuchadnezzar (reigned c. 605–562 BC). Hananiah prophesied that God would "break the yoke of the king of Babylon," whereas Jeremiah contended that Babylon was God's instrument of punishing Judah for its idolatry and oppression of the vulnerable. In Lynn Green's reading of this encounter, Hananiah

> identified the interests of the elite with God's interests, and said that God was on their side…. Many evangelicals make that mistake today as well. But God's plan is to bring a nation out of nations, a people out of peoples, a race out of races – and we make that harder the more that we identify with a certain nation, as we drag our principalities and powers into the mixture.

[7] Reconciliation Walk prayer day, Highfield Oval, Harpenden, 10/06/2006.

Lynn Green identified nation states, in Jeremiah's day or ours, as 'principalities and powers.' Although he did not expand, this reprises an obscure Biblical idea that states and empires are entities that have spiritual identities that influence those who inhabit them and that are ultimately answerable to God for their conduct. It has been developed by radical theologians such as Walter Wink (1998) and William Stringfellow in their critiques of violent state power as demonic. A particularly influential reading of "the powers" is offered by Mennonite theologian John Howard Yoder, whom Green explicitly identified in this meeting as an influence on his thought. After mentioning Yoder, Green continued (Yoder 1994 (1972)) "These principalities and powers hate this new nation, because they have to bend their knees to it, because their power is challenged when old enemies of different nations reconcile."

This led into a discussion about how Jeremiah contested the nationalist assumptions of his own day, and what it would similarly mean for Christian peacemakers to challenge the spirit of nationalism today. This discussion was forced to a halt, with much mirth, when the British national anthem blared out from the nearby chapel before the match began! "Should we stand up?" joked Cathy Nobles. Lynn Green made the obvious point about nationalism and the church, and led the group into a time of prayer. He began by praying, "Jesus Christ is Lord, and one day every tongue will confess this and every knee will bow, you reign over every principality and power… not by might, nor by power, but by my Spirit."

Further discussion and prayer that afternoon majored on nationalism: the choice between identifying with a warring nation state, or with the people of God as a new nation. Prayers were framed in a way that sought to stand outside identification with the foreign policy interests and objectives of the participants' nation states. Thus there was a discussion of the killing that week of Al-Qaeda in Iraq leader Abu Musab al-Zarqawi by a U.S. air strike. People voiced feeling uncomfortable at the celebration of his death by U.S. leaders, with Cathy Nobles beginning a prayer on the topic with the words, "We know that you sorrow at a life that brought such destruction." Other matters discussed and prayed for included the RW/YWAM's work in helping Italian Catholics to rethink the legacy of the Italian church's support of Mussolini's African wars, promoting U.S.-Libyan relations, developing existing relationships with Palestinian Islamists, and supporting programmes to engage young British Christians with the topic of reconciliation.

A recurrent motif was the idea of nationalism as a demonic power that can warp the ways that unsuspecting Christians think and act in the world, and that it must therefore be combated through prayer. Thus, for example, the leaders updated the group on the efforts of Messianic Jewish congregations (that is, Jews who accept Jesus Christ as the messiah and maintain a strong Jewish identity) and Palestinian churches to work together. In praying for them, Lynn Green prayed for good "Arab-Messianic relations" and prayed against "the divisions that the powers and authorities want to introduce" – on the last count specifically praying 'against the gods of nationalism.' Humorously, and with heavy ironic reference to the football tournament, Cathy Nobles concluded the meeting with a final prayer and the words, "God bless England!" to laughter all round.

If nationalism emerged as the theme of the June 2006 meeting, then a RW prayer day held in August that year was dominated by that of racial prejudice.[8] This was prefigured in discussion even before the leaders of the meeting began speaking. The meeting began with me being introduced. In order to avoid covert research and that I might approximate some level of informed consent, I explained my work, and how studying the RW had been transformative for me, and that I would be taking notes unless anyone objected (as the leader, Lynn Green, joked, "be careful what you pray – it may end up in a textbook!") My introduction prompted one of the participations, Georgina,[9] a late middle-aged white British woman, offered her experiences of how participating in the RW had led to a "seismic shift" in her thinking, going from being "very pro-Israel" to a radically different view of the Middle East conflicts and of theology.

Lynn Green began by saying:

> Today, we are going to pray for unity in the body of Christ. Our prayers are blunted when we don't pray with a sense of empathy towards people on both sides of this. It is important for us to reclaim the "two-eyed" perspective on the Middle East.

He explained that YWAM had had a 'word of knowledge' (understood as a direct revelation of God to the organisation) in the early 1990s that the organisation was looking at the Middle East with only "one eye," that is, through the lens of the state of Israel as the fulfilment of Biblical prophecy and thus deserving of partisan political support. He then proceeded to provide an account of the Arab-Israeli conflicts seen from an Israeli perspective, and asked if anyone identifies with that perspective.

Again, Georgina spoke up, saying that she did: 'we were taught to pray and rejoice in the return of Jesus to the land, even though we thought that this would involve a big battle. It was uncaring.' Lynn Green commented that although he now see[s] this end-times theology as "misguided," he exhorted humility and cautioned against judgementalism towards those who hold it. He went on to present an Arab/Palestinian perspective that foregrounded Palestinian grievances and historical narratives.[10] He concluded that, 'Our task is therefore to pray for both, empathise with them, remember that in Christ there is neither Jew nor Palestinian.' Following more open discussion, this led into an extended time of prayer for Palestinians and Israelis.

This was followed by a tea break, but the resumption showed that praying for others to understand the narratives of their enemies was productive of an introspective search for the same prejudice in oneself. Thus Lynn Green began by saying:

> I was speaking to Rudolph at the tea break, and as he reminded me, when we pray it is important to address the causes of sin and see if we have any remnants of that same sin in us. We have less provocation, but the same source may be there.

[8] Reconciliation Walk prayer day, Highfield Oval, Harpenden, 6/08/2006.

[9] Apart from Reconciliation Walk leaders, the names of other people cited have been replaced by pseudonyms.

[10] This approach of presenting parallel narratives of the same events was one that epitomized what the Peace Research Institute of the Middle East would later call the 'dual narrative approach' (Adwan and Bar-On 2012: x).

Rudolph, a late middle-aged white British man, then confessed:

> I was walking along a street with my wife the other day, and started grumbling about all the non-whites and the takeover of the street by people of different nationalities. I had also been in a hospital waiting room, and there were lots of translators waiting for people who had not turned up. I suddenly noticed that my wife was not by my side, and looked back, and she had stopped a few yards behind me. She said that she was suddenly convicted by the Holy Spirit who said to her, 'Whose street is this anyway?' Answer: it is God's street. We were very convicted of our prejudice.

Lynn Green invited the group to "wait on the Lord" to impress upon us awareness of similar sin: "purify us, and give us the gift of repentance," he prayed. There followed a time of silent prayer and meditation, which was productive of a number of similar prayers of confession. Angela, again a former participant of the RW from a similar demographic as the other participants mentioned, spoke up that she "really hate[s] the smell of curry," it makes her feel "physically sick." She continued that she was walking down the street past a curry house recently when she smelt the cooking, and felt ill, and said to herself that she hoped she "never had a neighbour move in who cooked that stuff." This story was presented as a prayer of repentance to God, complete with asking for forgiveness. Another participant told of a recent golfing outing with his friends, when he saw a couple of golfers on a parallel course who appeared to be of Pakistani background – an unusual sight on his golf course. He recounted that here was some confusion over whose ball was whose, and one of the other golfers kicked the speaker's ball into the longer grass to examine it. The speaker recounted that he was angry because that simply wasn't golfing etiquette, and he found himself saying to himself that it was because these people were foreign, they didn't understand. Again, this confession was accompanied by prayers of repentance.

The Reconciliation Walk prayer meetings would be understood in their own terms to exist to seek divine assistance and guidance for the organisation's activities. But they also have an obvious geopolitical significance. They act as a forum for a group of committed individuals to critically reflect on international relations in the light of their experiences on past prayer walks, and to support each other in pursuing activities that many in evangelical circles would be wary of. The combination of discussion and prayer challenges geopolitical designations of Christian-Muslim and US/UK-Middle Eastern relations as indelibly hostile. These alternative framings, and the supporting community in which they are generated and articulated, sustain both leaders and participants in what they would understand as their ongoing activities for Christian peacemaking.

162.6 Conclusion: Worship, Geopolitics and Peace

"The recent emergence of critical geopolitical study of evangelicalism is important and welcome. However by focusing largely on right-wing and militaristic evangelical geopolitics, it risks stereotyping this community as indelibly reactionary. It is thus also important to mine evangelicalism for its pacific geopolitics" (Megoran 2010).

One way to approach this is to explore not simply texts, but practices of worship. As Wannenwetsch contends, “For Christians, the experience of worship is seminal. It is here that they experience the presence of the acting and judging God in a formative way; and here, at the same time, a reflective ethics will emerge among them” (Wannenwetsch 2004:5). By considering preaching on apocalyptic scripture in Apartheid South Africa, prayer walking in the Middle East, and prayer meetings in the UK, this chapter has shown how more pacific practices of thinking about and doing geopolitics emerge from evangelical Christian worship.

Acknowledgements I am very grateful to Lynn Green and Cathy Nobles of the Reconciliation Walk for assistance with research for this chapter. Material for Sect. 162.3 has been drawn from “Radical politics and the Apocalypse: Activist readings of Revelation,” forthcoming in *Area*. Material for Sect. 162.4 has been reworked from my article originally published as 2010 “Towards a geography of peace: pacific geopolitics and evangelical Christian Crusade apologies,” *Transactions of the Institute of British Geographers, 35*(3), 382–398. I would like to thank Wiley for permission to reuse material in both cases: especially as not all journal publishers grant this without charging a fee nowadays.

References

Ackerman, P., & Duval, J. (2000). *A force more powerful: A century of nonviolent conflict*. New York: St Martin’s Press.
Adwan, S., & Bar-On, D. (2012). The dual-narrative approach: Jewish-Israeli and Palestinian pupils learn the history of the other party in the conflict. In S. Adwan, D. Bar-On, & E. Naveh (Eds.), *Side by side: Parallel histories of Israel-Palestine* (pp. ix–xviii). New York: The New Press.
Agnew, J. (2010). Deus Vult: The geopolitics of the Catholic Church. *Geopolitics, 15*(1), 39–61.
Bauckham, R. (1993). *The theology of the book of Revelation*. Cambridge: Cambridge University Press.
Boesak, A. (1977). *Farewell to innocence: A socio-ethical study of black theology and power*. New York: Orbis.
Boesak, A. (1983a). …And even his own life…. In *Walking on thorns: The call to Christian obedience* (pp. 42–49). Geneva: World Council of Churches.
Boesak, A. (1983b). Into the fiery furnace. In *Walking on thorns: The call to Christian obedience* (pp. 26–34). Geneva: World Council of Churches.
Boesak, A. (1987). *Comfort and protest: Reflections on the apocalype of John of Patmos*. Philadelphia: Westminister.
Boyer, P. (1992). *When time shall be no more: Prophecy belief in modern American culture*. London: Harvard University Press.
Bredin, M. (2003). *Jesus, revolutionary of peace: A nonviolent Christology in the book of Revelation*. Carlisle: Paternoster Press.
Burge, G. (2003). *Whose Land? Whose promise? What Christians are not being told about Israel and the Palestinians*. Cleveland: Pilgrim Press.
Carson, D. (2002). *Love in hard places*. Wheaton: Crossway Books.
Cavanaugh, W. (1998). *Torture and eucharist: Theology, politics, and the body of Christ*. Oxford: Blackwell.
Clark, V. (2007). *Allies for Armageddon: The rise of Christian Zionism*. London: Yale University Press.
Dalby, S. (1990). *Creating the second cold war: The discourse of politics*. London: Pinter.

Dijkink, G. (2006). When geopolitics and religion fuse: A historical perspective. *Geopolitics, 11*(2), 192–208.
Dittmer, J. (2008). The geographical pivot of the end of history: Evangelical geopolitical imaginations and audience interpretation of *Left Behind. Political Geography, 27*, 280–300.
Dittmer, J. (2010). Obama, son of perdition? Narrative rationality and the role of the 44th President of the United States in the end-of-days. In J. Dittmer & T. Sturm (Eds.), *Mapping the end times: American evangelical geopolitics and apocalyptic visions* (pp. 73–95). Farnham: Ashgate.
Dittmer, J. (2013). Evangelicals. In K. Dodds, M. Kuus, & J. Sharp (Eds.), *The Ashgate research companion to critical geopolitics*. Farnham: Ashgate.
Dittmer, J., & Sturm, T. (2010). *Mapping the end times: American evangelical geopolitics and apocalyptic visions*. Farnham: Ashgate.
Freeman, T. W. (1961). *A hundred years of geography, the 'hundred years' series*. London: Gerald Duckworth & Co.
Gallaher, C. (2010). Between Armageddon and hope: Dispensational premillennialism and evangelical missions in the Middle East. In J. Dittmer & T. Sturm (Eds.), *Mapping the end times: American evangelical geopolitics and apocalyptic visions* (pp. 209–232). Farnham: Ashgate.
Gerhardt, H. (2008). Geopolitics, ethics, and the evangelicals' commitment to Sudan. *Environment and Planning D: Society and Space, 26*(5), 911–928.
Harding, S. F. (1991). Representing fundamentalism: The problem of the repugnant cultural other. *Social Research, 58*(2), 373–393.
Harris, P. (2011, May 22). World doesn't end: California prophet had no Plan B. *The Observer*. www.guardian.co.uk. Accessed 20 June 2011.
Kendrick, G. (1992). *Public praise: Celebrating Jesus on the streets of the world*. Altamonte Springs: Creation House.
King, M. L., Jr. (1981). *Strength to love*. Philadelphia: Fortress Press.
Kovacs, J., & Rowland, C. (2004). *Revelation: The apocalypse of Jesus Christ*. Oxford: Blackwell.
Kovel, J. (2007). Facing end-time. *Capitalism Nature Socialism, 18*(2), 1–6.
McDonald, P. (2004). *God and violence: Biblical resources for living in a small world*. Scottdale: Herald Press.
McDougall, D. (2009). Rethinking Christianity and anthropology: A review article. *Anthropological Forum, 19*(2), 185–194.
Megoran, N. (2006a). God on our side? The Church of England and the geopolitics of mourning 9/11. *Geopolitics, 11*(4), 561–579.
Megoran, N. (2006b). For ethnography in political geography: Experiencing and re-imagining Ferghana Valley boundary closures. *Political Geography, 25*(6), 622–640.
Megoran, N. (2010). Towards a geography of peace: Pacific geopolitics and evangelical Christian Crusade apologies. *Transactions of the Institute of British Geographers, 35*(3), 382–398.
Megoran, N. (2013). Radical politics and the Apocalypse: Activist readings of Revelation. *Area, 45*, 141–147.
Nelson-Pallmeyer, J. (2005). *Saving Christianity from empire*. London: Continuum.
Northcott, M. (2004). *An angel directs the storm: Apocalyptic religion and American empire*. London: I.B. Tauris.
Ó Tuathail, G. (2000). Spiritual geopolitics: Fr. Edmund Walsh and Jesuit anti-communism. In K. Dodds & D. Atkinson (Eds.), *Geopolitical traditions: A century of geopolitical thought* (pp. 187–210). London: Routledge.
Reconciliation Walk. (1996). *The Reconciliation Walk: Defusing the bitter legacy of the Crusades*. Harpenden: Reconciliation Walk.
Reconciliation Walk. The Reconciliation Walk – who are we? *Undated pamphlet*.
Ruether, R., & Reuther, H. (2002). *The wrath of Jonah: The crisis of religious nationalism in the Israeli-Palestinian conflict*. Minneapolis: Fortress Press.
Sidaway, J. (2006). On the nature of the beast: Re-charting political geographies of the European Union. *Geografiska Annaler Series B, 88*(2), 1–14.
Sizer, S. (2004). *Christian Zionism: Road-map to Armageddon?* Leicester: Inter-Varsity Press.

Sturm, T. (2006). Prophetic eyes: The theatricality of Mark Hitchcock's premillennial geopolitics. *Geopolitics, 11*(2), 231–255.
Sturm, T., & Dittmer, J. (2010). Mapping the end times. In J. Dittmer & T. Sturm (Eds.), *Mapping the end times: American evangelical geopolitics and apocalyptic visions* (pp. 1–23). Farnham: Ashgate.
Tenety, E. (2011, January 3). May 21, 2011: Harold Camping says the end is near. *The Washington Post*. www.washingtonpost.com. Accessed 20 June 2011.
Wannenwetsch, B. (2004). *Political worship*. Oxford: Oxford University Press.
Weber, T. (1987). *Living in the shadow of the second coming: American Premillennialism, 1875–1982*. London: University of Chicago Press.
Wink, W. (1998). *The powers that be: Theology for a new millennium*. London: Galilee Doubleday.
Witherington, B. (2003). *Revelation*. Cambridge: Cambridge University Press.
Yoder, J. H. (1994) (1972). *The politics of Jesus: Vicit Agnus Noster* (3rd ed.). Carlisle: Paternoster Press.

Chapter 163
From the Church of the Powerful to the Church of the Poor: Liberation Theology and Catholic Praxis in the Philippines

William Holden

The King will say to them in reply, 'Amen, I say to you, whatever you did for one of these least brothers of mine, you did for me.' (Matthew 25: 40)

163.1 Introduction

The Republic of the Philippines is an archipelago of approximately 7,100 islands located in Southeast Asia (Fig. 163.1). Unlike its Buddhist (Taiwan and Vietnam) and Muslim (Indonesia and Malaysia) neighbors the Philippines is an overwhelmingly Christian country wherein 95 % of the population are Christian and the most well established religious domination is the Roman Catholic Church with 81 % of all Filipinos (Central Intelligence Agency 2011). The church is a profoundly important institution in the Philippines. Environmental Science for Social Change (1999: 95) described the church as "a vitally important part of the life and history of the Filipino nation; it is, in a sense, the soul of the nation; more than any other it has shaped the ethos of the nation." Among its members, "confidence in the church is overwhelmingly higher than in other groups and organizations" (Nicolas et al. 2011: 94). Indeed, some have gone so far as to declare the church to have "omnipresence in Philippine culture" (Nicolas et al. 2011: 97). This chapter examines one of the most important aspects of the church in the Philippines: its commitment to the poor and marginalized members of that archipelago's society.

W. Holden (✉)
Department of Geography, University of Calgary, Calgary, AB T2N 1N4, Canada
e-mail: wnholden@ucalgary.ca

S.D. Brunn (ed.), *The Changing World Religion Map*,
DOI 10.1007/978-94-017-9376-6_163

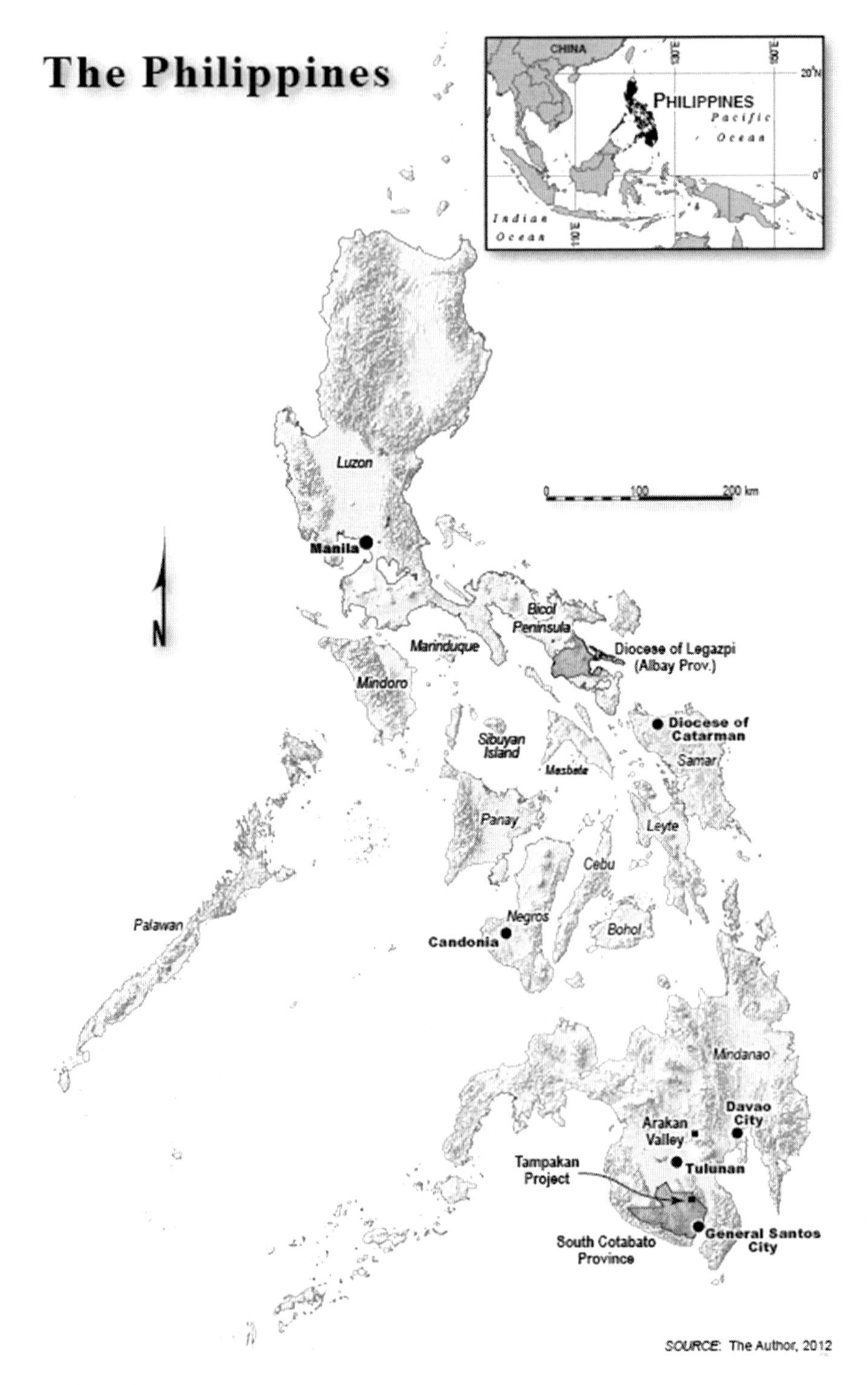

Fig. 163.1 The Philippines, an archipelago of 7,100 islands in Southeast Asia (Map by William Holden 2012)

163.2 Historical Background

163.2.1 Vatican II and the Church of the Poor

Throughout much of history the Roman Catholic Church has been aligned with the rich and powerful and, with few exceptions, showed little concern for the poor and marginalized. The church traditionally justified this orientation by using an approach known as *the distinction of planes*, which argued there were two planes of existence: *the sacred plane*, the concern of the church, and the *secular plane*, the concern of secular society (Holden and Jacobson 2011). Any potential destabilizing influences emerging from a discussion of Jesus' love for the poor in the scriptures were blunted by making it abundantly clear that any poverty being referred to was spiritual poverty and not material poverty (Holden and Jacobson 2011).

By the 1960s, however, it became increasingly apparent to many in the church that the demands of the modern world required it to make changes or risk becoming irrelevant. The 1960s was a time of freedom, experimentation, and revolution wherein all old certainties, including those of religion, became questioned (Nicolas et al. 2011). During the second Vatican Council (1962 to 1965), the church became determined to bring itself up to date and it took on the crucial role of addressing issues of poverty, economics, and social justice (Ramos-Llana 2011). This led to a shift in Catholic teaching away from a purely spiritual understanding of salvation and towards a greater commitment to challenge unjust social structures (Holden and Jacobson 2011).

It was in Latin American where the church first responded to Vatican II with the most notable degree of rapidity and the concept of liberation theology, an interpretation of Christianity from the experience of the poor, began to emerge (Holden and Jacobson 2011). When Vatican II described the role of the church as service to the world, Latin American theologians placed this within the context of their world of poverty and quickly came to see this as a vehicle for social involvement (Holden and Jacobson 2011). It was in this context that the *Confederacion Episcopal Latina America* (Latin American Episcopal Confederation or CELAM) met in Medellin, Colombia in August of 1968 and placed the moral weight of the church on the side of the poor. One of the most important organizers of this conference was the Peruvian priest Gustavo Gutierrez. In July of 1968, (1 month before Medellin) Gutierrez presented a paper in Chimbote, Peru entitled *Hacia una Teologia de la Liberacion* (Towards a Theology of Liberation) and this paper presented liberation theology as a theological rationale for doing pastoral work among the poor. With the presentation of this paper, with the publication of Gutierrez's 1971 book *A Theology of Liberation*, and with publication of works by other writers (such as Leonardo Boff and Jon Sobrino) the concept of liberation theology began to emerge. At its core is the concept of the preferential option for the poor. The poor are to be treated with a preference; they are to be taken into account first. Liberation theology also entails their participation; the poor must be afforded a choice about what happens to them.

From a theological perspective, one of the most important aspects of liberation theology is the eschatology of Matthew 25 (Berryman 1987). In that Chapter of the gospel Jesus renders his final judgment and the standard for salvation is whether one has lived a just life by rendering practical material aid for others, particularly the less fortunate. As Berryman (1987: 55) wrote, "The criterion is not whether one considers oneself Christian or not- one might even be an atheist- but whether one has served the needs of others."

As a result of Vatican II, and the emergence of liberation theology, the Roman Catholic Church transformed itself from an institution supportive of the established social order to an institution containing a sector engaged in activism on behalf of the poor that may be referred to as "the church of the poor." By no means does the entirety (or even majority) of the church constitute this church of the poor but those who do play a role disproportionate to their numbers, particularly since they are the ones in direct contact with the poor (Holden and Jacobson 2011).

163.2.2 Emergence of the Church of the Poor in the Philippines

The Philippines has a long history of Roman Catholicism dating back to the Spanish colonization of the sixteenth century. However, during the Spanish colonial period (1568 to 1898) the church, as Tan (2002: 17) wrote, "arrogated unto itself the political prerogatives and authority of the state" and the long years of oppressive domination by Catholic friars came to be known as *La Frailocracia* (the rule by the friars). Indeed, the church was so dominant during Spanish colonial times that Tan (2002: 17) referred to this period as "four centuries of brainwashing by catechetical methods" regimenting "the compliant population into devotion to various icons of Spanish religiosity."

During the Spanish colonial period the church "taught a mystified and otherworldly version of Christianity to indoctrinate and subdue the masses for their conquerors" (Nadeau 2008: 30). Filipinos were made to believe that all bad things happening to them in this world were the wrath of God (Roque and Garcia 1993). As a result of the extensive use of the distinction of planes during the Spanish colonial period a low degree of secularization developed in Filipino society; as Picardal (1995: 44) wrote, "the culture of modernity and secularization that has dominated the West has not made any deep impact on the Filipino culture so far."

With a heavy reliance upon a distinction of planes approach to spirituality it was not surprising that throughout much of history the church in the archipelago became aligned with the rich and powerful and acted in near total disregard of the poor and marginalized (Nadeau 2008). One of the best examples of the was the fact that during the 1930s and 1940s many members of the Catholic hierarchy in the islands were supporters of Spain's General Franco, "particularly the Spanish priests of the Dominican hierarchy running the University of Santo Tomas" (Constantino 1975: 386). Indeed, the church engaged in displays of reactionary behavior as recently as the 1950s when Father John Delaney, the parish priest for the Catholic community at the University of the Philippines Diliman, launched a crusade "to 'cleanse' the campus of 'atheists'" (Ordonez 2008: 38).

Nevertheless, by the 1960s, with profound changes occurring worldwide within the church as a result of Vatican II, the church in the Philippines began to make a commitment to act as a church of the poor (Holden 2009; Holden and Nadeau 2010). What accelerated this commitment more than anything else were the worsening conditions encountered by the poor under the dictatorship of President Ferdinand Marcos (1972–1986). During the 1970s, many Filipino lay persons, sisters, and priests began to develop a critical awareness of the depredations of the Marcos regime and this caused them to act on behalf of the poor (Danenberg et al. 2007). "Marcos," wrote Gaspar (2004: 98), "exhibited the characteristics of biblical figures [that] oppressed God's people from the Pharaoh to Herod." At this time English translations of Gutierrez's writing were beginning to circulate in the archipelago and these came to "define the church's mission as facilitating humankind's 'total liberation'" (Jones 1989: 208). Activism on behalf of the poor received an additional impetus when Bishops such as Bishop Antonio Fortich, Bishop Julio Xavier Labayen, and Bishop Francisco Claver sided with the poor articulating their concern for them despite the oppressive conditions prevailing under martial law (Youngblood 1990). Indeed, in 1973, Bishop Labayen, Bishop of the Prelature of Infanta, specifically called for the church to "listen to the voice of the many poor" (Quinones et al. 2005: 3). Eventually, as a result of the 1991 Second Plenary Council of the Philippines (PCP II), the Church, as a whole, decided that it must become a "Church of the Poor" (Catholic Bishops' Conference of the Philippines 1992: 48). As Ramos-Llana (2011: 61) wrote:

> Essentially the vision of the church in the Philippines in PCP II is to become a Church of the Poor- a church whose members are 'in solidarity with the poor,' and who 'collaborate with the poor themselves and with others to uplift the poor from their poverty.'

What is interesting about the diffusion of the principles of the church of the poor throughout the archipelago is that these principles did not spread from a core area (namely Manila on the island of Luzon) to periphery areas (outer islands). Instead, these principles spread from the island of Mindanao to the rest of the archipelago (Holden 2009; Holden and Nadeau 2010).

Mindanao has always been regarded in the Philippines as a "frontier area" or as "the land of promise." During the 1950s and 1960s the government of the Philippines, an institution dominated by landowners, encouraged landless people to migrate from Luzon and the Visayan islands to Mindanao to forestall agrarian unrest on Luzon and in the Visayas. In total, the migration of Christians to Mindanao was so extensive that the proportion of Mindanao's population that was Christian increased from 22 %, in 1918, to 82 % in 1990 (Holden and Nadeau 2010). Consequently, the church on Mindanao, in the late 1960s and early 1970s, became a frontier church with newly established dioceses filled with people who had moved from elsewhere. These dioceses were staffed by young bishops who were receptive to new ideas and who needed priests; in many cases they turned to foreign missionaries to staff their parishes and many of these had been exposed to liberation theology in Latin America and this acted as a conduit carrying concepts from there to the Philippines. Liberation theology found fertile ground in Mindanao as it had a migrant population that was more open to change; Mindanao's Christians had moved from elsewhere and, as they were willing to move, they became more receptive to new things. Eventually, throughout the 1970s and into the 1980s, the precepts of the church of the poor began to diffuse throughout the archipelago from Mindanao to Negros, Samar, and ultimately Luzon. Manila, despite its role as the capital city of the Philippines, serves no role as a center of progressive ecclesial thought. Instead Manila, the capital city of an archipelagic nation with different regional cultures and languages, acts largely as a refuge for the poor and marginalized seeking employment and as a departure point for overseas Filipino workers leaving the country to support their families by remitting money from abroad (Tyner 2009).

163.2.3 Endurance of the Church of the Poor in the Philippines

While some writers (De Oliveira Ribeiro 1999; Kater 2001) have argued that liberation theology is waning in parts of its original Latin American hearth, the church in the Philippines remains committed to the poor. There are two major explanations for the endurance of the church of the poor in the Philippines.

The first explanation is the legacy of the Marcos dictatorship during which the church developed a strong sense of social justice as it acted against Marcos (Youngblood 1990). However, when the (supposed) "People Power" revolution of 1986, which replaced Marcos with Corazon Aquino, failed to solve the underlying problems of Filipino society the church continued with its activism on behalf of the poor eventually institutionalizing this (Gaspar 2004). This has led to many of the principles of the progressive church being taught in seminaries across the archipelago. In Davao City, Brother Karl Gaspar, a member of the Redemptorists (and a political prisoner during martial law), is a professor at Saint Alphonsus Theologate (Gaspar 2006). Brother Karl, a member of the Ecumenical Association of Third World Theologians (an association of people committed to the liberation of Third World peoples), was heavily influenced by liberation theology and this was the source of his passion for justice. In St. Joseph's Parish, in Dumingag in the Diocese of Pagadian on the island of Mindanao, Father Riolito Ramos is the Parochial Vicar (Ramos 2005). Father Riolito attended Saint Mary's Theologate in Ozamis City, which was created to train priests to suit the signs of the times on Mindanao, and was heavily influenced by liberation theology. While a seminarian Father Riolito visited the poor and from these visits, taken into conjunction with his training, developed a heavy emphasis on social justice.

The second explanation of the endurance of the church of the poor is one offered by Gaspar (2010) and is based upon the influences of the indigenous belief systems prevailing in the archipelago before the introduction of Islam (in the fourteenth century) and the introduction of Christianity (in the sixteenth century). The indigenous belief systems lacked any form of distinction between the spiritual plane and the secular plane and were based upon reciprocal relations between people and the spirit world. The pre-Islamic and pre-Hispanic residents of the archipelago would engage in religious rituals and ceremonies to show their respect to the spirits and, in exchange for this, they would receive good health and prosperity. These belief systems had what Gaspar (2010: 103) called a "this-worldly orientation" and lacked any discussion of enduring suffering on this world in exchange for a better life in an after world. As a result of the deeply engrained influences of these belief systems within the Filipino people they were already pre-disposed to the progressive church and its lack of a distinction of planes approach to spirituality. In the word of Gaspar (2010: 352):

> Our indigenous knowledge systems, practices and spirituality which have a 'this-worldly' perspective are wellsprings that can quench our thirst for justice, peace, mutuality and solidarity. The seeds of a transformation orientation of our spirituality were there already long before conquest and conversion and despite the impact of colonization these never left the collective psyche of our people.

163.2.4 The Role of Women in the Church of the Poor in the Philippines

It bears stressing that the church of the poor in the Philippines is by no means a patriarchal institution and many of its most important members are women. In addition to the countless female lay workers who work in the church many nuns are also engaged in issues of social justice. Consider Sister Susan Bolanio, who became a nun in 1971 and took her exposure courses during the Marcos dictatorship and, through this, learnt about liberation theology and the preferential option for the poor (Bolanio 2005). Sister Susan was influenced by Gutierrez and Boff and regards liberation theology as a "progressive understanding of the Bible" (Bolanio 2005: interview). To Sister Susan, God is the god of the poor, God hears the groaning of the poor, the poor should be the subject of engagement in their liberation from injustice, and humans are to be sustained both spiritually and physically. Consider as well Sister Crescencia Lucero (Lucero 2007). Sister Crescencia is the Executive Director of the Task Force Detainees of the Philippines (TFDP), a nongovernmental organization (NGO) engaged in human rights advocacy created by the Association of Major Religious Superiors of the Philippines in 1974 that had, by 1993, become "the largest human rights NGO both in the Philippines and in the developing world" (Clarke 1998: 159). Sister Crescencia became a nun in 1969 and was influenced by liberation theology, which she regards as a rereading of the gospel from the perspective of the poor.

163.3 The Church of the Poor in the Philippines Today

163.3.1 The Church of the Poor: A Multidimensional Undertaking

The Philippines is a country with widespread poverty (Fig. 163.2). In 2006, the official poverty rate in the archipelago stood at 33 % (World Bank 2010). Where poverty becomes particularly acute is in rural areas where people are involved in agriculture. While the national poverty rate may have stood at 33 % in 2006 the rural poverty rate for the same year stood at 46 % and 71 % of all poor people lived in rural areas (World Bank 2010). In the Philippines today the church is committed to act on behalf of the poor and marginalized through a number of different methods. Some of these methods consist of programs that the church conducts to assist the poor while others consist of policy stances it has adopted. What is notable about the church's commitment to the poor is the extent to which it frequently comes into conflict with the neoliberal agenda adopted by the government of the Philippines.

The term "neoliberalism" refers to a set of economic polices emphasizing free trade, privatization, deregulation, and the retreat of the state from matters of wealth redistribution and social service provision (Ward and England 2007). A term frequently used in conjunction with neoliberalism is "globalization," the tendency for

Fig. 163.2 Urban poor in Barangay Calumpang, General Santos City (Photo by William Holden 2007)

economic interdependencies to occur on a global scale. Although activities have occurred on a global scale for years, neoliberalism, with its heavy emphasis on free trade, has led to such an amplification of globalization that Ward and England (2007: 12) have taken to calling globalization the "international face of neoliberalism."

The Philippines has long been reputed to be among the most accommodating in Asia to the prescriptions of neoliberalism (Holden 2011). During the presidency of Gloria Macapagal-Arroyo (2001 to 2010) the government consolidated its commitment to neoliberalism by aggressively encouraging foreign investment in various economic sectors and by entering into free trade agreements with other countries (Holden 2011). Neoliberalism is widely regarded as a set of polices prioritizing efficiency over equity and growth over poverty alleviation and the church, with its commitment to the poor, often finds itself in contradistinction to the government's neoliberal agenda. Attention now turns to the methods implemented by the church to act on behalf of the poor.

163.3.2 Diocesan Social Action Centers

In 1966 the Catholic Bishops Conference of the Philippines (CBCP), the collective body of all bishops in the archipelago, created the National Secretariat for Social Action (NASSA), which later became the National Secretariat for Social Action, Justice and Peace (NASSA-JP) (Ramos-Llana 2011). After the creation of NASSA, Social Action Centers (SACs) were created in each diocese in the Philippines "to respond to the socioeconomic needs of the poor" (Ramos-Llana 2011: 62). Sister Susan Bolanio was the Social Action Director of the Diocese of Marbel from 1992 to 1995 and again from 2003 to 2004. During this time Sister Susan was substantially involved in activism on behalf of the poor (Bolanio 2005).

One of the most vibrant SACs in the Philippines is that in the Diocese of Legazpi, in the Bicol Peninsula (see Fig. 163.1). This SAC was created by the Diocese of Legazpi in 1972 in response to the call of Vatican II for greater church involvement in social issues; it acts to implement programs improving the conditions of the poor in the province of Albay (Ramos-Llana 2011). The mission of this SAC is social transformation through empowered leaders and sustainable communities living out the gospel values and it engages in a number of programs (Fig. 163.3) to assist the poor such as: its child abuse prevention and intervention unit (CAPIU) delivering services to child abuse survivors; its health program, which has created a parish pharmacy providing access to low cost medicines; its good governance program, aimed at combating corruption in the Province of Albay; a variety of livelihood programs, such as livestock rearing and bio-intensive gardening; and its disaster management program, which seeks to train communities (in an area highly vulnerable to typhoons, volcanic eruptions, landslides, and floods) how to prevent, cope with and rebuild after calamities (Ramos-Llana 2011). The SAC of the Diocese of Legazpi has been described as "a model social action center," living up to the call of PCP II for a truly functioning and effective social action apostolate responding to the needs of the poor (Ramos-Llana 2011: 71).

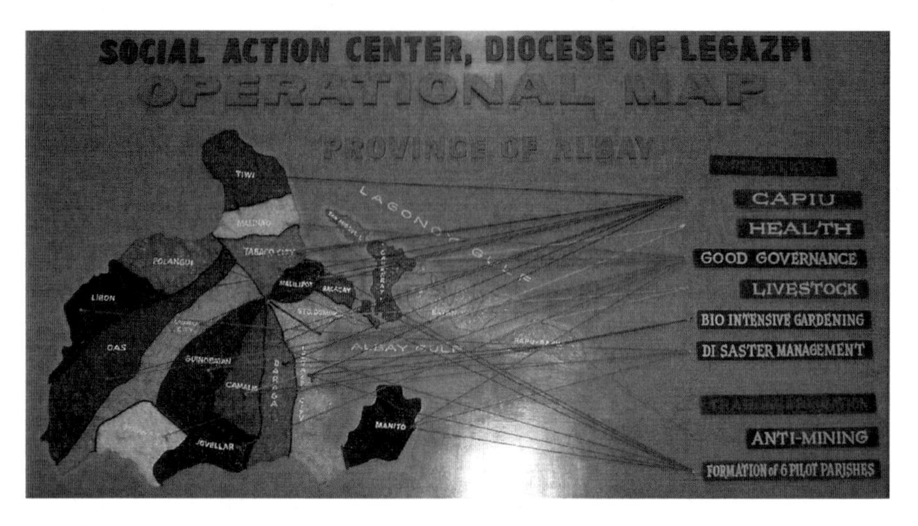

Fig. 163.3 Programs carried out by the Social Action Center of the Diocese of Legazpi (Photo by William Holden 2009)

These SACs can perform a vital function linking the grassroots of the church at the parish level in the *barangays* (villages) of the Philippines with national wide church organizations, such as NASSA-JP, international church organizations, such as Misereor (the German Catholic Bishops' Organization for Development Cooperation), and with national and international nongovernmental organizations (NGOs) active in a variety of activism.

163.3.3 The Basic Ecclesial Community Movement

The Basic Ecclesial Community (BEC) movement is one of the most tangible programs by which the church acts to assist the poor (Fig. 163.4) and Carroll (2011: 17) describes the BEC as:

> A community of believers, at the grassroots level, which meets regularly, under the leadership of a lay minister, to express their faith in common worship, to discern on their common living of the faith, to plan and act on common decisions regarding their life of faith, in community as community.

The term "basic" refers to both the size and the social location of the BECs (Holden 2009; Holden and Nadeau 2010). They are small communities, consisting of from 40 to 200 families organized on a parish-by-parish basis, and most BECs consist of the "small people," the poor and the marginalized. Members of all social classes may join but, overwhelmingly, they are a movement of the poor since middle class and upper class people have eschewed participation in them. They are called "ecclesial" communities because this emphasizes their place within the church; they are a way of being a church that is realized, located and experienced at the grassroots.

Fig. 163.4 Basic Ecclesial community members from Our Mother of Perpetual Help Parish, Davao City (Photo by William Holden 2007)

The word "community" emphasizes their communitarian nature; these are not societies or associations but are communities whose members live in close spatial proximity to each other and who regularly interact with each other. The role of the BECs as communitarian organizations stands in sharp contrast with the individualistic, selfish, privatized and competitive style marking modern western culture, particularly under neoliberalism (Holden 2009).

The BECs provide livelihood programs for their members such as handicraft production, food processing, garment making, soap making, cooperative stores, communal farming and livestock rearing programs (Holden 2009; Holden and Nadeau 2010). Given the high cost of pharmaceutical drugs, the low quality of public health care and the biodiversity prevalent in the Philippines, an important component of the BEC movement is the provision of herbal medicine (Holden 2009; Holden and Nadeau 2010). Many BEC members use herbal medicine and members will often gather together to make herbal medicines. Some BECs have arranged for doctors to provide seminars on how to develop herbal medicine; this serves the twofold purpose of providing people with something they can sell to others in their *barangay* as well as an affordable form of medicine. The BECs also engage in numerous programs designed to help protect the environment such as organic farming, solid waste management and tree planting, (Holden 2009).

One of the more interesting programs carried out by the BECs to improve the condition of the poor are its peace zones (Avruch and Jose 2007; Gaspar 2010; Holden 2009; Picardal 2005). The Philippines has experienced armed insurgent groups defying the authority of the state and the most widespread of these is the New People's Army (NPA) the armed wing of the Communist Party of the Philippines (CPP). Since 1969, the conflict between the NPA and the Armed Forces of the Philippines (AFP) has taken over 40,000 lives and it is one of the longest running Maoist insurgencies in the world (Holden 2011). In a peace zone, BEC members make it clear that they are autonomous from both the AFP and the NPA and then refuse to give any assistance to either side. Two of the earliest peace zones were the ones in Candonia, on the island of Negros, and in Tulunan, on the island of Mindanao (see Fig. 163.1) (Avruch and Jose 2007; Picardal 2005). What is interesting about peace zones is that both the AFP and the NPA decry them claiming that they obey them while the other side does not and, consequently, the other side gains an advantage from them. In May 2006, for example, Lieutenant Colonel Cesar Idio, commander of the 25th Infantry Battalion, held a press conference and declared that the NPA was using Tulunan, North Cotabato as a refuge area and as a base for recruitment (Holden 2009). The fact that both the AFP and the NPA are opposed to peace zones could well be a sign that these are making it more difficult for both sides to engage in hostilities thus creating peace.

There are many who have criticized the BEC movement for not doing enough to help the poor and for being largely stalled at a purely liturgical level (Holden 2009; Holden and Nadeau 2010). Perhaps the greatest success of the BEC movement is the way it actively involves Catholics in their church and, in doing so, prevents their conversion to fundamentalist Christian churches, which have experienced "remarkable membership growth throughout the Philippines in recent decades" (Eder 1999: 160). While some of these protestant churches (most notably the United Church of Christ in the Philippines) have engaged in social justice issues, few of them care little, if at all, about the material well being of people and few of them have articulated an alternative vision of society and the social order such as that advanced by the BEC movement (Holden 2009; Holden and Nadeau 2010). These churches operate with an exclusively spiritual, and highly individualistic, orientation and in some ways resemble the pre-Vatican II distinction of planes approach used by the Roman Catholic Church. Ultimately it seems that the criticism of the BEC movement is largely focused on the rapidity of the program's development not on its direction. The program is still largely a work in progress but it shows that there are programs through which the poor can help themselves.

163.3.4 Ecclesial Activism for Land Reform

One of the most tangible examples of the commitment of the church to the poor and marginalized is its activism for land reform (Marin 2011). Land reform, often referred to as "agrarian reform," is defined by Putzel (1992) as "programs, usually introduced by the state, that have the intention of redistributing agricultural land to its tillers and providing them with secure property rights and the means to earn an adequate living." In the Philippines the agricultural sector is responsible for 40 % of all employment and in this sector there is widespread landlessness with many Filipino farmers living a serf-like existence on lands owned by members of the archipelago's oligarchic elite (Borras 2007).

In the Philippines the current land reform statute is the Comprehensive Agrarian Reform Law (CARL), or Republic Act 6657, which was signed into effect by President Corazon Aquino in 1988. This statute created the Comprehensive Agrarian Reform Program (CARP) and this "was not a law universally favored by the church" and "CARL was considered by more progressive forces in the church to be insufficiently redistributive" (Marin 2011: 45). Land reform outcomes under CARP are well below those in other countries and the extent of "redistributive land reform outcome is far below the official claims in government statistics" (Borras 2007: 287). Obstacles to the implementation of CARP include landlord opposition, a lack of will by the Department of Agrarian Reform (DAR), the government agency responsible for its implementation, and even the commission of acts of violence against land reform beneficiaries with thirty-four peasant leaders being killed between 1998 and 2006 (Marin 2011).

"The support of the Philippine Catholic Church," wrote Marin (2011: 43), "for the plea of the rural poor for agrarian reform emanates from the experiences of her personnel in the rural areas, as well as from the inspiration they have gathered from Catholic social teaching." In 2007, the CBCP wrote a pastoral letter expressing its "concern over the deteriorating dignity of the rural poor who have become the greatest victims of the country's unjust economic system" (Marin 2011: 43). Later that year, Archbishop Antonio Ledesma, of the Archdiocese of Cagyan de Oro, wrote "that the church was perturbed about the incomplete implementation of the program and the consequent persistence of an inequitable distribution of the nation's wealth and of other social injustices" (Marin 2011: 43). Then, in 2008, as CARP came up for an extension in Congress, the bishops said special masses for the farmers, created venues for dialogue between farmers and legislators, wrote letters advocating for CARP extension with reform (CARPER), and even partook in protests (Marin 2011). When CARP was extended in 2009 seventy-four bishops signed a statement opposing what they saw as the toothless resolution professing to extend it and urged Congress to restore the power of the state to expropriate land for redistribution (Marin 2011). The church is clearly an active participant in matters of land reform and it is not content to remain on the sidelines refraining from becoming involved in a concern of secular society. As Marin (2011: 50) wrote:

> The church is considered to be among the forces influential in agrarian issues in recent years, particularly in getting CARPER passed. After listening to and empathizing with the people, and carrying out both spiritual duties and secular activities in support of the farmers' cause, she does not intend to stop now and rest on her laurels.

163.3.5 Ecclesial Opposition to the Extrajudicial Killings

In the Philippines, Supreme Court Administrative Order No. 25-2007 defines extrajudicial killings as "killings due to the political affiliation of the victims; the method of attack; and involvement or acquiescence of state agents in the commission of the killings" (Parreno 2010: 39). Across the archipelago there has been a wave of extrajudicial killings wherein environmental activists, labor organizers, human rights activists, and peasants campaigning for land reform, have been assassinated (Holden 2011). Most of the killings have followed a pattern wherein the victims were shot during broad daylight by men riding motorcycles; after being shot nothing was taken from them and they were left to die where they had been shot (Holden 2011). The nature of these killings indicates that the assailants had little fear of any police or government reaction and there is widespread consensus that these killings can be attributed to the government (in general) and to the AFP (in particular) as opposed to merely being acts of violent crime (Holden 2011).

The church has not stood by silently while the extrajudicial killings have occurred. The TFDP has been one of the foremost organizations in the Philippines campaigning against them and in 2006 the CBCP issued a pastoral letter wherein it condemned them stating: "It is not right that people be killed simply because they have different 'political beliefs' or are suspected of being 'subversive'" (Catholic Bishops Conference of the Philippines 2006: 1). The church has also campaigned against the extrajudicial killings at the grassroots. In the Diocese of Legazpi, for example, the SAC has collaborated with various NGOs to campaign against extrajudicial killings and enforced disappearances (Ramos-Llana 2011). An important component of this collaboration has been the involvement of the SAC in the *Andurog Kan Derechos* (Support for Rights), a "quick reaction team organized to provide a venue for reporting, responding to, documenting, and monitoring cases of extrajudicial killings and enforced disappearances in Albay" (Ramos-Llana 2011: 68).

The issue of extrajudicial killings is, unfortunately, one area where the church has also been the victim as well as the advocate and there have been recent incidents where members of the church have been killed for their activism. In September 2009, Father Cecilio Lucero, the chair of the committee on human rights of the Diocese of Catarman (see Fig. 163.1), was killed in an ambush while travelling in the Province of Northern Samar (Gaspar 2010). Two days before his death, soldiers from the Philippine Army's 63rd Infantry Battalion had visited Father Cecilio at his convent regarding his human rights activism (Karapatan 2009). In October 2011, Father Fausto Tentorio, an Italian missionary priest, was killed inside the parish compound in Arakan Valley, North Cotabato (see Fig. 163.1). After he was killed two men on a motorcycle were seen driving away and, across the street, members of the Philippine Army's Fifth Special Forces were stationed engaging in intelligence gathering work (Karapatan 2011). Father Fausto had been deemed by the AFP to be a supporter of the NPA due to his opposition to mining and militarization in North Cotabato (Karapatan 2011). These attacks on members of the church demonstrate that social activism is not a casual activity and it can have fatal consequences.

163.3.6 Ecclesial Opposition to Mining

The church has been heavily involved in advocacy pertaining to environmental issues such as climate change, illegal logging, and opposition to nuclear power (Balane 2011; Catholic Bishops Conference of the Philippines 2009; Fruto et al 2006). However, the most salient example of the church's environmentalism (and arguably the most salient example of its commitment to act on behalf of the poor) is its opposition to mining (Fig. 163.5). As Karaos (2011: 53) wrote:

> One social issue on which the Catholic Church has not remained silent in the face of what she perceives as injustices committed against the poor is that of mining. On this particular issue, the Philippine bishops have given credible witness to PCP II's vision of a Church of the Poor. They have spoken out strongly against large-scale mining in a series of pastoral statements, marshaled the organizational resources and networks of the Church behind their advocacy, and shown remarkable solidarity with their poor and marginalized constituents.

Starting during the 1990s the government of the Philippines has engaged in efforts to encourage large-scale mining carried out by multinational corporations (Holden and Jacobson 2011). Mining is an activity with a substantial potential for environmental harm and the church's primary objection to mining is that it will

Fig. 163.5 Father Edwin Garriguez, Executive Secretary of the National Secretariat for Social Action, Justice, and Peace at Anti-Mining Mobilization, Metro Manila (Photo by William Holden 2005)

degrade the environment depended upon by subsistence farmers and subsistence fisherfolk (Holden and Jacobson 2011). In 1998, 2006, and 2008 the CBCP declared its opposition to mining in pastoral letters (Karaos 2011). The church has also become involved in an organization known as the Anti-Mining Campaign, which has enjoined the current administration of President Benigno Aquino III (son of former President Corazon Aquino) to declare a moratorium on large-scale mining and to repudiate the Macapagal-Arroyo administration's policy of aggressively promoting large-scale mining as a development strategy (Karaos 2011).

Consider Bishop Pedro Arigo, the Bishop of the Apostolic Vicariate of Palawan, who issued a pastoral letter calling for a ban on new mining operations on the island of Palawan (see Fig. 163.1) in October 2006 (Argio 2009). Bishop Argio was ordained as a priest in 1963 during Vatican II, read the writings of Gutierrez, and was influenced by Matthew 25. From such an interpretation of the Bible, Bishop Pedro came to see Christ as identified with the poor and marginalized and this constitutes the basis of his objection to mining on Palawan, which will disrupt the environment depended upon by Palawan's poor and, in doing so, deprive them of their livelihoods.

On of the most effective examples of the church's opposition to mining comes from the Diocese of Marbel, in the Province of South Cotabato (see Fig. 163.1), where a joint Australian, Swiss, and Filipino corporation has been attempting to develop a large-scale copper mine, known as the Tampakan Project, using open pit mining. However, after the Diocese of Marbel brought substantial pressure on the government of South Cotabato the latter banned open pit mining in June of 2010 (MindaNews 2010). As a result of this ban the Department of Environment and Natural Resources, the national government agency responsible for regulating mining, refused to grant the Tampakan Project environmental clearance (Sarmiento 2012). Baring a repudiation of the ban implemented by the Province of South Cotabato, this mine, the largest undeveloped copper deposit in Southeast Asia, will not be able to proceed. This, perhaps more than anything else, demonstrates the seriousness of the church's opposition to mining; as an unnamed exploration company president stated: "[In the Philippines], NGOs, peasants and church groups override [the] government constantly. You can spend millions developing a property in the Philippines, only to have it swept away by peasants, lobby groups [and] churches" (Fraser Institute 2011: 49).

163.4 Concluding Discussion

The church in the Philippines is an institution firmly committed to act on behalf of the poor and marginalized members of society. This commitment emanates from the institutionalization within the church of one of the central tenets of liberation theology: the preferential option for the poor. There are, of course, some contradictions inherent in the church's advocacy on behalf of the poor: the church is economically affluent, yet remains much involved in the politics of power and prestige; it claims to value participation and democracy, while acting hierarchically and secretly

(Gaspar 2010). The church's commitment to the poor also does not mean that the church has become a radicalized institution (or even a liberalized institution) as the church in many ways remains deeply conservative and there is perhaps no better example of this than its adamant opposition to government proposals to reduce population growth (Carroll 2011; Dionisio 2011b).

In 2011 the Philippines had a population of almost 102 million people and, with a 2011 population growth rate of 1.9 % the population can be expected to double in 37 years (Central Intelligence Agency 2011). This means that in the year 2048 the Philippines will have a population of almost 204 million people living on only 300,000 square kilometers (115,839 square miles) of land, only 19 % of which is arable (Central Intelligence Agency 2011).With such high population growth the Reproductive Health (RH) Bill has been introduced in Congress. This bill, if passed into law, will ensure government funding of family planning methods, such as contraception, while retaining a strict prohibition of abortion. The church has been an adamant opponent of the RH Bill (Fig. 163.6). When a group of Ateneo de

Fig. 163.6 Poster offering free intrauterine device removal by the Archdiocesan Center for Family and Life Apostolate Foundation of the Archdiocese of Davao (Photo by William Holden 2007)

Manila professors came out with a statement in 2008 saying that a Catholic could, in conscience, support the RH Bill they received harsh criticism from the hierarchy (Carroll 2011). In 2011, a group of theologians at the Loyola School of Theology issued a paper advocating "critical and constructive engagement in the discussions over the RH Bill" (Dionisio 2011b: 27). This paper caused Father Jaime Achacoso, a member of Opus Dei, to suggest "the necessity of some form of ecclesiastical sanction against the school" (Dionisio 2011b: 28). Father Jaime also suggested that "the energies devoted by the bishops to opposing the environmental pollution caused by large-scale mining ought more properly to be directed against the doctrinal pollution being brought about by irresponsible Catholic theologizing in Catholic universities" (Dionisio 2011b: 28).

While this opposition to a reproductive health bill (retaining a strict prohibition of abortion) shows that the church, in many ways, remains a deeply conservative institution in Filipino society, one must not lose sight of how far this institution has come. From being a staunch ally of the rich and powerful the church today is committed to acting on behalf of the poor. It does this through: its social action centers, its basic ecclesial communities, its activism for land reform, its advocacy for human rights, and through its opposition to large-scale mining. Today, "the Roman Catholic Church in the Philippines has grown in her engagement with pressing and persistent problems of the nation" (Dionisio 2011a: 8).

References

Argio, Bishop P. (2009, December 15). *Bishop, Apostolic Vicariate of Palawan*, Personal Interview, Puerto Princesa City, Philippines.

Avruch, K., & Jose, R. S. (2007). Peace zones in the Philippines. In L. E. Hancock & C. Mitchell (Eds.), *Zones of peace* (pp. 51–69). Bloomfield: Kumarian Press.

Balane, W. I. (2011). Bukidnon Bishop Assails "Climate Sins," Calls For Hope Amid Tragedy. *MindaNews*. Retrieved 20 December, 2011, from www.mindanews.com/top-stories/2011/12/20/bukidnon-bishop-assails-%E2%80%98climate-sins%E2%80%99-calls-for-hope-amid-tragedy/

Berryman, P. (1987). *Liberation theology: Essential facts about the revolutionary movement in Latin America and beyond*. Philadelphia: Temple University Press.

Bolanio, S. O. Sister. (2005, May 24). *Sister, Oblates of Notre Dame*, Personal Interview, Davao City, Philippines.

Borras, S. M. (2007). *Pro-poor land reform: A critique*. Ottawa: University of Ottawa Press.

Carroll, J. J. (2011). Bishop Claver's vision of the church. In E. R. Dionisio (Ed.), *Becoming a church of the poor: Philippine Catholicism after the Second Plenary Council* (pp. 11–23). Quezon City: John J. Carroll Institute on Church and Social Issues.

Catholic Bishops Conference of the Philippines. (1992). *Acts and Decrees of the Second Plenary Council of the Philippines*. Pasay City: Pauline Publishing House.

Catholic Bishops Conference of the Philippines. (2006). *Let us keep human life sacred*. Manila: Catholic Bishops Conference of the Philippines.

Catholic Bishops Conference of the Philippines. (2009). *No to Bataan nuclear power plant: A pastoral statement*. Manila: Catholic Bishops Conference of the Philippines.

Central Intelligence Agency. (2011). *The World Factbook*. Retrieved January 18, 2012, from https://www.cia.gov/library/publications/the-world-factbook/geos/rp.html

Clarke, G. (1998). Human rights non-governmental organizations in the Philippines: A case study of task force detainees of the Philippines. In G. S. Silliman & L. G. Noble (Eds.), *Organizing for democracy: NGOs, civil society, and the Philippine State* (pp. 157–192). Honolulu: University of Hawaii Press.

Constantino, R. (1975). *The Philippines: A past revisited.* Quezon City: Tala Publishing Service.

Danenberg, T., Ronquillo, C., de Mesa, J., Villegas, E., & Piers, M. (2007). *Fired from within: Spirituality in the social movement.* Manila: Institute of Spirituality in Asia.

De Oliveira Ribeiro, C. (1999). Has liberation theology died? Reflections on the relationship between community life and the globalization of the economic system. *The Ecumenical Review, 51*(3), 304–314.

Dionisio, E. R. (2011a). Introduction. In E. R. Dionisio (Ed.), *Becoming a church of the poor: Philippine Catholicism after the Second Plenary Council* (pp. 1–10). Quezon City: John J. Carroll Institute on Church and Social Issues.

Dionisio, E. R. (2011b). Who speaks for the church? Catholic cacophony in the reproductive health debate. In E. R. Dionisio (Ed.), *Becoming a church of the poor: Philippine Catholicism after the Second Plenary Council* (pp. 24–41). Quezon City: John J. Carroll Institute on Church and Social Issues.

Eder, J. F. (1999). *A generation later: Household strategies and economic change in the rural Philippines.* Honolulu: University of Hawaii Press.

Environmental Science for Social Change. (1999). *Mining revisited.* Quezon City: Environmental Science for Social Change.

Fraser Institute. (2011). *Fraser institute annual survey of mining companies 2010–2011.* Vancouver: Fraser Institute.

Fruto, R. O., Picardal, A. L., & Gaspar, K. M. (2006). *Being sent: Redemptorist missions in Mindanao (1975–2005).* Quezon City: Claretian Publications.

Gaspar, K. M. (2004). *To be poor and obscure: The spiritual sojourn of a Mindanawon.* Manila: Center for Spirituality.

Gaspar, K. M. (2006, January 5). *Redemptorist Brother*, Personal Interview, Davao City, Philippines.

Gaspar, K. M. (2010). *The masses are messiah: Contemplating the Filipino soul.* Manila: Institute of Spirituality in Asia.

Holden, W. N. (2009). Post modern public administration in the land of promise: The basic ecclesial community movement of Mindanao. *Worldviews: Environment, Culture, Religion, 13*(2), 180–218.

Holden, W. N. (2011). Neoliberalism and state terrorism in the Philippines: The fingerprints of Phoenix. *Critical Studies on Terrorism, 4*(3), 331–350.

Holden, W. N., & Jacobson, R. D. (2011). Ecclesial opposition to nonferrous metals mining in Guatemala and the Philippines: Neoliberalism encounters the church of the poor. In S. Brunn (Ed.), *Engineering the earth: The impacts of Megaengineering projects* (pp. 383–411). Dordrecht: Kluwer.

Holden, W. N., & Nadeau, K. M. (2010). Philippine liberation theology and social development in anthropological perspective. *Philippine Quarterly of Culture and Society, 38*(2), 89–129.

Jones, G. R. (1989). *Red revolution: Inside the Philippine Guerrilla movement.* Boulder: Westview Press.

Karaos, A. M. A. (2011). The Church and the environment: Prophets against the mines. In E. R. Dionisio (Ed.), *Becoming a church of the poor: Philippine Catholicism after the Second Plenary Council* (pp. 53–60). Quezon City: John J. Carroll Institute on Church and Social Issues.

Karapatan. (2009). *Karapatan 2009 report on the human rights situation in the Philippines.* Quezon City: Karapatan.

Karapatan. (2011). *Karapatan 2011 report on the human rights situation in the Philippines.* Quezon City: Karapatan.

Kater, J. L. (2001). Whatever happened to liberation theology? New directions for theological reflection in Latin America. *Anglican Theological Review, 30*(4), 735–773.

Lucero, C. (2007, May 29). *Sister, executive director, task force detainees of the Philippines*, Personal Interview, Quezon City, Philippines.
Marin, G. R. R. (2011). A church for agrarian reform. In E. R. Dionisio (Ed.), *Becoming a church of the poor: Philippine Catholicism after the Second Plenary Council* (pp. 42–52). Quezon City: John J. Carroll Institute on Church and Social Issues.
MindaNews. (2010). *South Cotabato Provincial Board Bans Open-Pit Mining*. Retrieved January 30, 2012, from http://mindanews.com/main/2010/06/04/southcot-provincial-board-bans-open-pit-mining/
Nadeau, K. M. (2008). *The history of the Philippines*. Westport: Greenwood Press.
Nicolas, G. M., Batomalaque, L. A., & Rabacal, G. A. G. (2011). Qualified confidence: Church, state, and public opinion. In E. R. Dionisio (Ed.), *Becoming a church of the poor: Philippine Catholicism after the Second Plenary Council* (pp. 90–101). Quezon City: John J. Carroll Institute on Church and Social Issues.
Ordonez, E. A. (2008). Recrudescence. In B. Lumbera, J. Taguiwalo, R. Tolentino, R. Guillermo, & A. Alamon (Eds.), *Serve the people: Ang Kasaysayan ng Radikal na Kilusan sa Unibersidad ng Pilipinas* (pp. 37–40). Quezon City: IBON Books.
Parreno, A. A. (2010). *Report on the Philippine extrajudicial killings (2001-August 2010)*. San Francisco: The Asia Foundation.
Picardal, A. L. (1995). *Basic ecclesial communities in the Philippines: An ecclesiological perspective*. Doctoral dissertation, Faculty of Theology, Pontifical Gregorian University, Rome.
Picardal, A. L. (2005). BECs in the Philippines: Renewing and transforming. In J. G. Healey & J. Hinton (Eds.), *Small Christian communities today: Capturing the new moment* (pp. 117–122). Maryknoll: Orbis Books.
Putzel, J. (1992). *A captive land: The policies of agrarian reform in the Philippines*. London: Catholic Institute for International Relations.
Quinones, M., Marino, N. V., Zijl, A., & Geertman, W. (2005). *Historical consciousness of the reaching out in solidarity between the basic sectors and the Church of Infanta: Journeying towards the Church of the poor*. Angeles City: Pima Press.
Ramos, R. Father. (2005, June 15). *Priest, St. Joseph's Parish, Diocese of Pagadian*, Personal Interview, Dipolog, Philippines.
Ramos-Llana, M. (2011). The Church of the poor in the province of Albay: The Diocesan Social Action Center of Legazpi. In E. R. Dionisio (Ed.), *Becoming a church of the poor: Philippine Catholicism after the Second Plenary Council* (pp. 61–72). Quezon City: John J. Carroll Institute on Church and Social Issues.
Roque, C. R., & Garcia, M. I. (1993). The ecology of rebellion: Economic inequality, environmental degradation and civil strife in the Philippines. *Solidarity, 139*(140), 88–120.
Sarmiento, B. S. (2012). *DENR denies SMI's ECC application: SMI to appeal for reconsideration*. Retrieved January 30, 2012, from www.mindanews.com/environment/2012/01/13/denr-denies-smi%E2%80%99s-ecc-application-smi-to-appeal-for-reconsideration
Tan, S. K. (2002). *The Filipino-American War, 1899–1913*. Quezon City: University of the Philippines Press.
Tyner, J. A. (2009). *The Philippines: Mobilities, identities, globalization*. London: Routledge.
Ward, K., & England, K. (2007). Introduction: Reading neoliberalization. In K. Ward & K. England (Eds.), *Neoliberalization: States, networks, peoples* (pp. 1–22). Oxford: Blackwell.
World Bank. (2010). *Philippines: Fostering more inclusive growth*. Washington, DC: World Bank.
Youngblood, R. L. (1990). *Marcos against the church: Economic development and political repression in the Philippines*. Ithaca: Cornell University Press.

Chapter 164
Faith Based Organizations and International Responses to Forced Migration

Sarah Ann Deardorff Miller

164.1 Introduction

While significant scholarship has demonstrated the ways in which faith and religion can affect refugees and other forced migrants, either as a cause of forced migration, a coping mechanism while living in exile, or as a way to negotiate one's surroundings (Fiddian-Qasmiyeh 2011b: 430), studies on faith-based organizations working with refugees are only recently gaining traction as a worthy area of scholarship. Indeed, since 2008, there have been a range of publications, most notably the recent *Journal of Refugee Studies* 2011 "Special Issue: Faith-Based Humanitarianism in Contexts of Forced Displacement." Other recent scholarship has looked more broadly at faith-based humanitarian organizations in the context of development or humanitarian relief work more generally, as scholars like Michael Barnett and Janice Stein demonstrate in their book, *Sacred Aid: Faith and Humanitarianism* (2012). These trends represent interest from a range of disciplines—from politics to international relations to sociology, human geography and anthropology. This seemingly up-and-coming issue area, however, still holds many unanswered questions. This chapter will focus largely on Christian international faith-based organizations (FBOs)[1] and their role in humanitarian assistance, looking both broadly at humanitarian aid and specifically in relation to forced migration. It will begin with the starting assumption that faith-based organizations are unique and different from secular ones, first outlining how and why this assumption can be taken as true. It will then examine how these differences can work as assets or challenges to humanitarian

[1] Recognizing that insufficient scholarship exists on other FBOs and calling for more research of non-western, non-Christian FBOs.

S.D. Miller (✉)
Researcher, Refugee Studies Centre, 3 Mansfield Road, Oxford OX1 3 TB, UK
e-mail: sarah.deardorff@gmail.com

S.D. Brunn (ed.), *The Changing World Religion Map*,
DOI 10.1007/978-94-017-9376-6_164

work, looking particularly at the refugee context, and demonstrating the ways in which strengths largely outweigh the challenges. Finally, it will identify important questions and gaps in the research, calling for further scholarship in response to these issues.

164.2 Unpacking Faith Based Organizations (FBOs)

164.2.1 Definitions, Concepts and a Brief Introduction to the Literature

Recent scholarship has provided useful baseline definitions of FBOs, which this chapter will draw upon in order to maintain consistency within the literature. It is worth noting, however, that there is significant variation in how some understand FBOs, and that more of the literature deals with Western, Christian FBOs than other regions and religions. Several possible explanations for this may exist, including the fact that the meager data and reporting that does exist tends to come from larger, western, Christian FBOs. Thus, a disproportionate amount of attention is paid to these organizations simply because information on them is more readily available. This alone presents an enormous gap that needs to be addressed, as will be discussed later. Regarding definitions, however, the recent *JRS* Special Issue, understands an FBO as "…any organization that derives inspiration from and guidance for its activities from the teachings and principles of faith or from a particular interpretation or school of thought within a faith" (Clarke and Jennings 2008: 6; Fiddian-Qasmiyeh 2011b: 430). In this respect, FBOs are said to derive their "organizational identity and mission from a particular religion or spiritual tradition" (Palmer 2011: 97), but are distinct from the faith community whose ethos guides their work, insofar as their programs and projects are guided to fulfill a particular function, such as responding to humanitarian needs arising from forced migration (Fiddian-Qasmiyeh 2011b: 430). Elizabeth Ferris (2005: 312) notes that FBOs are characterized by one or more of the following:

> …affiliation with a religious body; a mission statement with explicit reference to religious values; financial support from religious sources; and/or a governance structure where selection of board members or staff is based on religious belief or affiliation and/or decisions-making processes based on religious values.

She admits, however, that the term is problematic, and that it would be better to understand differences of secular and FBOs as a continuum than a dichotomy (Ferris 2011: 622). Ager and Ager (2011), among others like Barnett and Stein (2012), also emphasize that it can be problematic to use secular organizations as the neutral object of study, arguing that functional secularism frames the discourse of contemporary humanitarianism, and unintentionally marginalizes religious language, practice and experience, making it difficult to engage with the dynamics of faith, particularly in relation to displaced populations. While this chapter is focused more directly on the organizations working internationally (not the individuals experiencing displacement or humanitarian crisis and their psychological state or coping mechanisms), their point is relevant, demonstrating that scholar bias can affect the trajectory of how FBOs are studied and understood.

Despite these attempts to describe FBOs, trying to define FBOs is very difficult. Like secular organizations, FBOs are highly heterogeneous, particularly with respect to forced migration: "…they can be small-scale local-level religious congregations to national inter-denominational coalitions and networks to international humanitarian agencies associated with particular religions; and they have diverse histories, motivations, fund-raising mechanisms and modes of operation" (Fiddian-Qasmiyeh 2011b; Ferris 2011: 621). Indeed, generalizing FBOs in one way or another would be a mistake, as they are very diverse.[2] Ferris writes:

> The growing number of humanitarian organizations or NGOs and their incredible variety makes generalizations impossible. Refugee-serving NGOs include small organizations staffed by volunteers and housed in church basements as well as organizations with annual budgets close to US $1 billion per year—about the same as the United Nations High Commissioner for Refugees. Some NGOs, particularly faith-based organizations, have large constituencies numbering in the hundreds of millions. Others are membership organizations whose members contribute funds and volunteer their time. Like many of their secular counterparts, most faith-based organizations are involved in a wide range of activities, including long-term development and advocacy for justice as well as humanitarian assistance. (2005: 312)

Ferris also highlights that there are differences within the Christian community of FBOs, including those linked directly to churches; those with Christian values, but without formal links to churches; and those that work internationally, versus locally or nationally. FBOs of other religions are likely to be just as diverse, and there is an urgent need for greater attention and scholarship on non-Christian and non-western FBOs. In sum, viewing NGOs or IOs with faith connections may best be understood via a spectrum, rather than binary categories.

Historically, FBOs have always been known as important actors in humanitarian, development and emergency assistance (Parsitau 2011: 493; Ferris 2011: 609). Ferris provides an excellent overview of some historical insight, noting, "Long before international humanitarian law was formalized in treaty law, individuals and faith communities provided assistance to those afflicted by natural disaster, persecution, uprooting and war" (Ferris 2005: 313). She continues to explain that themes

[2] Broad generalizations are something the author is constantly mindful of, seeking to guard against; one must also be cautious of seeing secular and FBOs as a binary, which would be too simplistic. Clarke (2006: 835) categorizes FBOs using five different categories of functions: faith-based representative organizations; faith-based charitable or development organizations; faith-based socio-political organizations; faith-based missionary organizations; faith-based radical, illegal or terrorist organizations (see Orji 2011: 474). Goldsmith et al. (2006: 3) provide another analysis, dividing U.S. faith-based operations into four levels (local/regional/national ecumenical/interfaith coalitions; incorporated non-profits independent or affiliated with congregations; organizations or projects sponsored by religious organizations; and relief operations by religious congregations) (in Orji 2011: 480). Barnett and Stein also explore the strategic reasons for categorizing and organization as secular or religious (Barnett and Stein 2012: 9).

of justice for the poor, marginalized and the alien are central to Hebrew scriptures, and that there is a long history of persecuted people seeking sanctuary in temples and cities of refuge and, "…in the later medieval period, monasteries were often places of refuge and hospitality for strangers" (Ferris 2005: 313; see also Marfleet 2011). Mission societies focused on evangelism in the eighteenth and nineteenth centuries and were key in raising awareness in individual congregations; to that end, they were also involved in lobbying and advocacy, not limiting themselves to charity or relief (Ferris 2005: 314). These acts brought public attention and international awareness for governments to respond, and reflect different historical moments when faith communities have been more or less political (2005: 314).[3] Ferris also notes an important trend in the 1980s, whereby secular and faith-based organizations were encouraged to decrease direct involvement abroad and support the development of indigenous NGOs or local institutions, in the context of different conceptualizations of development and capacity-building (2005: 316). In spite of this, however, some FBOs continue to be major players globally, with larger budgets than some of the government ministries with whom they work (Ferris 2005: 311).

Barnett and Stein (2012) also provide a useful historical perspective on the role of religion in developing concepts of humanitarianism of today. They note that "… religious discourses and organizations helped to establish humanitarianism in the early nineteenth century, and it is only a slight exaggeration to say 'no religion, no humanitarianism'" (2012: 4). Indeed, early roots of humanitarianism shared significant overlap with mission ideas. Barnett and Stein note, however, that there have been shifts back and forth with how "religious" humanitarianism has been during different historical periods:

> Over the course of the nineteenth century, however, many religious organizations began to downplay their interest in conversion in favor of improving the lives of the local peoples; they became less reliant on the "good book" and more reliant on the public health manual. By the end of the twentieth century, many religious organizations were beginning to work with secular agencies and use secularized international legal principles and international institutions to further their goals. Missionaries, for instance, were quite involved in the campaign to establish international human rights conventions during the interwar years. Then, after World War Two, Western governments became the chief funders of humanitarian action and increasingly favored secular agencies such as CARE. Once-avowedly religious organizations such as World Vision International and Catholic Relief Services downplayed their religious identity. Much like the rest of the world, it seemed as if humanitarianism was succumbing to the pull and power of secularism. (2012: 4)

They go on to write, however, that despite this appearance, religion remained "front and center influencing humanitarianism," and FBOs, especially Christians in the West, continued to expand (2012: 5). They also provide deeper analysis of the concepts of "sacred" and "profane" to understand how religion and humanitarianism have been interlinked.

---

[3] It is worth noting that faith communities encompass FBOs and individual congregations/churches/worship bodies, even though this chapter is focused largely on FBOs alone.

164.2.2 Context

The broader context in which humanitarian-focused FBOs operate overlaps with the work of many secular organizations, and thus provides some basis for comparison and analysis. Ferris, for example, explains how coordination remains a weak area among secular and FBOs: "The fact is that coordination of NGO work implies a loss of the 'sovereignty' which most NGOs are reluctant to give up. In view of the competitive environment for raising funds, it is important for international NGOs to demonstrate their presence in a given emergency—even when it might be more cost-effective to channel funds through an already operational partner" (2005: 322). All NGOs, faith-based or not, feel the pressure of an environment in which they must compete for funds. Similarly she emphasizes the constant struggle of using humanitarian assistance as an instrument of foreign policy, and even the emergence of "for profit" humanitarian players, such as military or private contractors (2005: 323). These developments can affect funding, mandate and project decisions of secular and faith-based organizations. In addition, FBOs have the added burden of distinguishing themselves from other FBOs that may hold very different beliefs and priorities. Obviously FBOs of different faiths are less likely to be conflated, but FBOs of the same religion may easily be lumped into one general category, even if they are quite different. For example, an FBO with Christian roots but focused on humanitarian relief may not want to be identified with a highly evangelical group focused on spreading the Gospel over feeding hungry people or providing clean water (2005: 323). It may be seen as bad public relations, or worse, jeopardize their work with a government, local village, other organizations, or the population they are serving. On some level this effort may occur with all organizations at work, secular or faith-based, but the need to differentiate FBOs within religions is an added challenge for FBOs. Indeed, more broadly speaking, FBOs may operate in a context where they need to justify their motives to a greater extent, given that authorities may be suspicious of their motivations over, say, Doctors without Borders or the International Red Cross/Red Crescent.

164.3 A Few Relevant Theoretical Approaches

Because FBOs have been studied from a range of disciplines, one would imagine that various theoretical lenses have been applied. However, few have drawn on specific theories to describe the nature of FBOs and their role and influence in the places in which they work. While any number of frameworks could be applied, a natural theoretical fit from a political science/international relations perspective may be to employ transnational non-state actor perspectives within neoliberal institutionalist or constructivist approaches. Moving away from focusing on the state as the main actor, transnational literature sheds light on how non-state actors (particularly networks) can have influence across state lines. Risse et al. (1999), for example, look specifically at the abilities of transnational non-state actors (like international FBOs) to link up with domestic actors and partners, in turn affecting policy or state behavior. They argue that the diffusion of international norms in the human rights area depends on the establishment and the sustainability of networks among domestic and transnational actors who manage to link up with international regimes and alert Western public opinion and Western governments (1999: 4). Case studies look closely at how norms and ideas influence state actions, and explore the conditions under which networks of domestic and transnational actors are able to change domestic structures themselves. Their constructivist approach thus allows a fuller understanding of how non-state actors and transnational actors can shape politics (Betts 2009: 33). This might be a natural theory to apply to the study of FBOs for a better understanding of their influence vis-à-vis the states in which they work.

Keck and Sikkink (1998) have also contributed directly to this line of reasoning, discussing the importance of "transnational advocacy networks" which rally around a "principled issue" for the diffusion of international human rights norms (see also Keohane and Nye 1971). They argue that these networks build links among civil society, states, and IOs, and multiply the channels of access to the international system (Keck and Sikkink 1998: 1). In some areas they bring the international to the domestic by making resources more available:

> By thus blurring the boundaries between a state's relations with its own nationals and the recourse both citizens and states have to the international system, advocacy networks are helping to transform the practice of national sovereignty. (1998: 1–2)

They claim that because the networks are motivated by values more than resources, they can go beyond policy change to even change the nature and terms of the debate by "framing" issues and targeting accordingly (1998: 2). Thus, transnational networks are complex agents who not only participate in politics, but also shape them, and thus "…bridge the increasingly artificial divide between international and national realm" (1998: 4). These networks, then, "…participate in domestic and international politics simultaneously" (1998: 4). Grace Skogstad's (Ed.) *Policy Paradigms, Transnationalism, and Domestic Politics* (2011) also demonstrates how transnational actors can be sources of norms (2011: 17).

These theories are generally executed within the discipline of political science and international relations, however they shed light on interesting claims about the role and capacities of international organizations. It is even more interesting to see where and how such theories might be applied to international FBOs in particular, as will be considered below via an analysis of the facets of FBOs. Additional studies may also consider employing constructivist approaches that examine the organizational nature of international FBOs, exploring whether they can be prone to the same bureaucratic pathologies as some other international organizations (see Barnett and Finnemore 1999), and how they develop and change over time.

164.4 Why FBOs Are Unique to Other Organizations

Cautious that not all FBOs are the same and that they are better viewed on a spectrum than via any generalizations or binary analysis with secular organizations, there are some broader facets that make FBOs unique. Indeed, it is a question that FBOs themselves also wonder (for example, Ferris 2011: 614 quotes a leader of Christian Aid asking, "…is Christian Aid just an Oxfam with hymn books?"). Building on the definitions, concepts, history and theory above, this section will unpack some of the differences that make FBOs, of varying places on the spectrum, unique to secular organizations. Ferris identifies two facets that set FBOs apart from most other organizations:

> …they are motivated by their faith and they have a constituency which is broader than humanitarian concerns. For believers, to be a Jew or a Muslim or a Christian implies a duty to respond to the needs of the poor and the marginalized. (Ferris 2005: 316)

These differences also infer a few other important characteristics that make FBOs unique, including their relationship to those they serve and how they are perceived by others (media, for example, or others outside their religion). It may also be the case that they more often consist of transnational networks of other religious affiliates, lending them different political sway than other actors, although this has yet to be fully studied.

164.4.1 Motives

Stemming from the working definitions above, it is clear that FBOs are more likely to be working from different motives than other organizations. Themes of hospitality, welcoming the stranger, exile, assisting the poor and marginalized and loving one's neighbor are common to most major religions, and especially Christianity, which this chapter looks at more closely. As mentioned above, themes of "sanctuary" and "refuge" resonate throughout Christian history, and one might even consider how moral authority is attached to humanitarian work of faith or secular bases, particularly in light of "humanitarian space" and "responsibility to protect" themes in recent years. Barnett and Stein analyze notions of humanitarianism and religion in light of Durkheim's "sacred" and "profane" (with "sacred" being "superior in dignity and power to profane things," and "profane" being "everyday" things (2012: 15). They write, "Religious orders are often known for their lifelong commitment to the marginalized and vulnerable, believing that by serving the poor they are serving God" (Barnett and Stein 2012: 20). Ferris also writes how "Both Christians and Muslims believe that 'there's a witness of faith through charity that is a way of life and expression of obedience to God'"

(Ferris 2005: 324)[4] and that for many, it is hard to conceive of a humanitarian gesture that is outside the scope of religion, noting Islamic societies, for example, which seek to integrate all aspects of physical life with spiritual life, a theme common to some Christian organizations as well. These motives can also make FBOs incredibly useful:

> [People of faith are] usually the people on the front lines of need and human assistance. They go there motivated purely out of love for their human brothers and sisters (…). The faith-based mechanism is a lot of times the easiest mechanism for the government to use to reach those people who are not usually reached, and, therefore, more in need. (Ferris 2005: 324)[5]

Wilson (2011) also depicts these unique motives, writing on the concept of hospitality among FBOs in the politics of asylum in Australia. She considers how they might hold extra leverage, noting that religion helped develop many concepts of hospitality invoked today. She writes, "Hospitality has a long association with asylum and sanctuary practices in various religious and secular traditions and is used with particular reference to strangers and foreigners in the philosophical and political writings of Kant, Levinas (1981) and Derrida (2000); (see also Baker 2009; Bretherton 2010; Gauthier 2007; Marfleet 2011; Pohl 2006, 1999)" (2011: 550).

164.4.2 Network and Donor Base

Among the most distinctive qualities of FBOs is their donor base, often the foundation of the transnational network. First and foremost, this relates to finances; FBOs can tap into a different base, one that is often a "built-in" tradition throughout history. Indeed, Abby Stoddard writes that in the U.S.,

> [t]he widespread practice among evangelicals of tithing (giving 10 % of income to church-sponsored charity) makes them a potentially much more lucrative source of private relief and development funding than the average US private donor, who directs only roughly 1 % of donations to foreign causes' (Ferris 2005: 323)[6]

Many Christians in the U.S. context, for example, expect their churches to be involved in international mission, and see that as an extension of their own place in the church. Ferris writes that because of this, FBOs tend to have at least some unrestricted funding, giving them operational freedom unlike organizations that rely completely on government funds for their resources (Ferris 2005: 617).[7] Thus, to

[4] Citing "International faith-based initiatives: Can they work?" Available at http://woodstock.georgetown.edu/resources/articles/International-Faith-Based-Initiatives.html (last visited 6 June 2012).

[5] Citing Linda Shovlain of the Center for Faith-Based and Community Initiatives of the USAID.

[6] Citing Abby Stoddard, "With Us or Against Us? NGO Neutrality on the Line, Humanitarian Practice Network," December 2003.

[7] However, many are still in competition with other secular NGOs (and even military or private contractors) for funds as well, and can thus be subject to turf wars like so many other organizations.

some extent—and likely a greater extent than secular organizations—FBOs can count on local church communities to be an automatic place to seek funds. Certainly variation may occur, but it is an automatic donor base to tap. FBOs aligned with a particular denomination (Lutheran World Relief or Presbyterian Disaster Assistance, for example), can also expect funds (though certainly unpredictable and sometimes meager as any other organization might face as well) from their specific denomination and even individual churches. In many cases, mission committees may designate special offerings for such organizations.[8]

Beyond funding, FBOs can also appeal to this "automatic constituency" with information. Indeed, they can spread the word with an already-established database of people throughout the world who can be reached with information about an FBOs cause, work, or a recent crisis worthy of attention. These networks are broad and geographically-spread, and because of them FBOs are more likely to be "…'well-equipped' to offer trained and experienced leadership, money and grass-roots participants and they have pre-existing communication channels (from weekly bulletins and address lists to synods) and enterprise tools (from telephones and facilities for printing publicity material to legal advice) which can be drawn upon" (Snyder 2011b: 578). There is also a sense of connection between locals, spanning the globe—for example, a church in California may feel an automatic connection to a church in Nigeria. Indeed, Ferris writes that this built-in vast global network links people to one another besides funding and programming (Ferris 2011: 617). Of course technology makes this even easier—churches now have Facebook pages, leaders of FBOs can "tweet" to church members, and people can spread the message through blogs and email forwards. Thus, "…religion can be tapped to raise emotion for a religious community and national identity" (Keyes 1979 in Horstmann 2011: 516). Though certainly not always successful, FBOs tend to have a useful database of "constituents" who are ready to receive information; this social organization alone is an asset coveted by any communications officer.[9]

164.4.3 Reach and Scope

Finally, closely linked to the preceding section, FBOs are unique in the scope of reach that they have throughout their network. Though certainly larger secular organizations can span the globe, reaching highly remote areas with numerous field offices, FBOs can be connected in a different way. Namely, being able to connect with local churches on the ground can, in some cases, give them an angle in that other international organizations might not have. In other words, if both an FBO and a secular organization are new to a country, the FBO might have an easier time

[8] This is not to say that funds are easy to come by, particularly when congregations may be underfunded or unwilling to provide funds beyond small local projects.

[9] It is also worth noting that Barnett and Stein (2012) unpack some of the different ways religious labels can be used in strategy.

connecting because they can go to already established churches or congregations to connect locally. In Barnett and Stein's volume, Taithe (2012: 20) writes,

> …missionaries have always played the 'long game' and missions, whether by choice or circumstance, have been closer to the people than their modern, secular contemporaries…

Barnett and Stein even argue that religious organizations are more accountable to local populations because they are more firmly embedded in the local community (2012: 21) and are "less likely to have a traveling class of expatriate professionals who move from one 'emergency' to another" (2012: 22). Ferris (2005) also notes that FBOs can, in some cases, be the first to respond on the ground if they already have local offices or support. In the case of refugees entering Tanzania, for example, Tanganyika Christian Refugee Service (connected to the international Lutheran World Federation) reminds its donors that it was able to mobilize and respond to refugee needs instantly because they were already working in those regions, whereas larger, international organizations took days to get expatriate staff in place. Ferris (2005) reminds readers that these groups seldom receive media attention, but that they may be among the best-situated to respond to some crises. Damaris Seleina Parsitau also demonstrates how faith can be used to help integrate displaced persons into new circumstances—again, something that FBOs might be better situated for (2011: 494).

164.5 The Uniqueness of FBOs: Challenge or Asset?

The preceding section considered ways in which FBOs—along the diverse spectrum in which they may fall and the many ways they can operate as a transnational network within different state environments—are unique to secular organizations, which are also very diverse. It examined three broad categories: motivations, donor base/network, and reach/scope. This section will consider the ways in which these differences represent challenges or assets to their work. It will avoid normative judgments, but will try to provide better context for the debate, outlining where these highlighted differences benefit or challenge their work. Ultimately, it will argue that the differences represent far more strengths than challenges, and that FBOs are well-positioned to respond in ways that secular organizations may fall short. This does not mean that FBOs are "better" than secular organizations—indeed, secular organizations may outperform FBOs in other ways—but that they may be able to fill in gaps of needs that may otherwise go unaddressed.

164.5.1 In Light of Different Motivations

Working from different motivations is certainly unique, but can be both a challenge and an asset to the work of FBOs, and again, it greatly matters what "type" of FBO one is examining. It is also important to note that a number of FBOs hire workers of different faiths (or no faith), and even those that do require employees to believe the same principles cannot guarantee the personal motives of every employee. In other words, one cannot be certain or measure whether every World Vision employee, for example, is motivated solely by his or her faith in Jesus Christ. Regardless, as Snyder notes, a voice of a religious viewpoint may carry moral authority that secular organizations may not. In looking at Christian FBOs, churches may use their voice of moral authority to highlight certain issues or even push for social change. In one case, Snyder writes how churches can help to "settle" refugees who have arrived, introducing them to their new community, and "unsettle" an established population's attitudes and government policies (Snyder 2011a, b: 556). Similarly this moral authority has been used to mobilize and raise awareness among broader populations, as has been seen in the United States with the Sanctuary Movement (Marfleet 2011), or in Australia with churches speaking out against detention (Wilson 2011), to name a few examples.

These different motivations can also be an asset to FBO's work in helping to reframe and conceptualize humanitarian work. Ferris (2005), Barnett and Stein (2012), and others have indicated that humanitarian and human rights of today have their roots in religion, demonstrating that motivations to respond to an "aching world" and the assumption of a voice of moral authority have brought about positive elements. Even today FBOs help to frame and reframe the terms of humanitarian issues broadly, and relating to the inclusion of forced migrants; from how organizations relate to individuals to the vocabulary used toward them. Wilson, for example, examines how religious conceptions of hospitality undergird discourses of protection relating to asylum seekers and refugees (2011: 551), and notes that religion underlies human rights as well (also see Ager and Ager 2011: 551). Verbs like "accompany" and "serve" evoke a different relationship to individuals with whom they work than some secular discourse. Similarly Ferris (2011) writes that FBOs moved away from the "donor-recipient" vocabulary in favor or "partnership" approaches as they sought to move away from Northern churches overpowering Southern churches (Ferris 2011: 618), noting an odd benefit from a missionary past in the 1960s, where power dynamics were questioned. Likewise such motives may predispose employees of FBOs to "care" and "love" those with whom they work in a different way, that is, if they are drawing upon religious themes in their faith tradition.[10] FBOs such as the Free Burma Rangers working with Karen refugees in Thailand, for example, draw upon a number of themes of exile and displacement in the Bible in comparing the suffering of refugees and internally displaced persons to biblical stories.[11]

Working from different motivations can also pose a challenge to FBOs' work. Ferris notes that classic humanitarian principles of neutrality and objectivity may clash with FBOs' motivations of justice and solidarity with the poor (2011: 618).

[10] That is not to imply, however, that those working for secular organizations cannot also "care" and "love" those they are serving.

[11] There are, of course, additional examples where the invocation of religion can be a method of political or social manipulation (see for example, Fiddian-Qasmiyeh 2011a, b, c; Horstmann 2011).

Likewise, FBOs are subject to rifts in their own faith communities about theological issues—ranging anywhere debates about Israel and Palestine to the use of condoms in Africa. Some FBOs may also receive a lot of money from governments, and may not be as independent as they proclaim (Ferris 2011). More common criticisms may also apply: some FBOs, particularly those on the more extreme end of the spectrum, may view "evangelism" and "salvation" as more important than the provision of medical supplies. Certainly many FBOs would adamantly disagree with such practices, but the diversity among FBOs exemplifies these clashing priorities. Likewise while religious communities have been vehicles of social change in some historical moments, the reality is that they have also stood in the way at other points in history. These challenges are, of course, in addition to the challenges that all NGOs or IOs face, including but not limited to competition for donor funds, turf wars, and difficulties with coordination.

164.5.2 In Light of a Different Donor Base/Network

Receiving funds and resources from a different donor base, such as church congregations and denominations in the Christian Western tradition, many FBOs have an advantage in that they have a separate place to seek funds—an automatic group that is already well-established. They may also have a significant leg up on secular organizations trying to gain funds from the same faith communities, as faith communities may prefer to give to an organization that shares their beliefs over one that is secular. Church mission committees, for example, will recommend FBOs worthy of support to broader church members; they may even take special offerings or participate in special programs to raise funds. These networks are a unique element to the international community, and remain largely understudied. Indeed, transnational advocacy networks have received significant attention in recent years, but little research has been carried out on the multi-layer, diversified religious-based networks that FBOs can tap into. Horstmann, for example, draws on Castells social network approach (understanding the world as "reconstituting itself around a series of networks strung around the globe based on advanced communication technologies") (Horstmann 2011: 516). This different donor base can also be an asset for mobilizing political or social campaigns to pressure government decisions, or to carrying out massive grassroots campaigns to assist a population, be it via an "underground railroad" of some sort, as with the Sanctuary Movement, or with simply sharing church building space to teach job training classes, language classes, or community meeting space (Snyder 2011b: 576) Indeed, "…from 'potluck dinners to job referrals', being part of a religious community provides avenues for social advancement, leadership, community service and respect, as well as social, cultural and socio-economic roles" (Hirschman 2007: 414 in Snyder 2011b: 576).

However, this alternative donor base/network does have some strings attached. As with any donor base for faith or secular organizations, givers may grow tired of supporting certain projects, or may simply find their budgets tighter depending on economic conditions or willingness to give. In addition, FBOs may feel beholden to carry out projects in line with donor priorities, again, a similar problem experienced by all organizations. FBOs may also feel like every cause has to be justified according to that faith community's priorities, limiting their scope in some cases (that is, a focus on child poverty or anti-abortion may push an FBO to ignore other needs, such as income-generation activities). They may also have fewer resources to tap into as some church membership decreases, or as some churches trend away from international "mission" and focus on helping people locally. These shifts are not just haphazard; they represent deep theological movements within religious communities, and are a reflection of religious landscape today. Finally, in some cases, FBOs are holding their work to the emotionally charged level of spirituality, righteousness and God, invoking a heftiness to their actions that, in some cases, may leave little room for debate or discussion. Invoking moral authority may, in more extreme cases, omit the possibility of "agreeing to disagree" and cause tension and infighting more often than in other organizations.[12]

164.5.3 Reach and Scope

As noted above, FBOs can have a different reach than other organizations. In the best of cases, being tied to religion may grant them more access, enabling them the cover to work more freely and to be seen as spiritual rather than political entities. A Christian organization may gain greater trust from a government than a non-Christian one in a predominantly Christian country, or a Muslim organization more trust than a secular one in a Muslim-majority country. Greater access and freedom of operation can be enormous assets in a humanitarian response. FBOs may also reach more people in need by tapping into the larger diaspora of the community of faith. Indeed, Ferris notes that churches can be excellent providers of information (Ferris 2005: 320), and that in some cases many can continue working long after NGOs have withdrawn expatriate staff because they have local connections on the ground (Ferris 2005: 321). Religious or secular global networks, like the International Council of Voluntary Agencies (ICVA), InterAction and the Steering Committee for Humanitarian Response (SCHR)[13] can also foster their unique reach and scope. FBOs may also dabble in a range of issue areas, or may even overlap between human rights work and humanitarian work (Ferris 2005: 321). Ferris

[12] Of course this is not to say that many FBOs do not have healthy debates and approach difficult humanitarian issues with humility—indeed, many are exemplary in how they operate in this manner.

[13] Created in 1972, the Steering Committee for Humanitarian Response (SCHR) is an alliance for voluntary action of ACT Alliance, Care International, Caritas Internationalis, the International Committee of the Red Cross, the International Federation of Red Cross and Red Crescent Societies, Lutheran World Federation, Oxfam International, Save the Children and World Vision International. See www.humanitarianinfo.org/iasc/pageloader.aspx?page=content-about-schr for more. Accessed 29 July 2012.

writes: "In practice, it is not possible to entirely separate the work of humanitarian and human rights NGOs, as they often overlap. Faith-based organizations, with their commitment to justice, tend to further blur the distinctions" (2005: 321). This can be both an asset and a challenge, and may cause them to be viewed differently. In blurring boundaries of humanitarian work and human rights, some with a greater focus on evangelism may also blur proselytizing and assistance, in a worst case scenario being seen as trying to "sneak in" conversion under the auspices of aid. Though somewhat rare, all it takes are a few horror stories of forced altar calls in return for food to raise concern about the intentions of all FBOs. There are past stories of a small number of FBO selecting beneficiaries based on faith, a practice that violates the humanitarian principles of neutrality and impartiality, in some cases detracting from the work of FBOs (see, for example, Ferris 2011: 617).

164.6 Shifts

There are significant shifts occurring among FBOs, some of which will directly affect the challenges and assets mentioned above, and all of which pertain to the unique characteristics of FBOs. Ferris, for example, emphasizes trends of professionalization as one area of change. In particular FBOs are increasingly trying to distance themselves from missionary activity, employing more secular professionals and adhering to high professional standards to be taken seriously by others in the international humanitarian community, including other organizations, donors and governments (Ferris 2011: 614). She notes that the word "professionalism" is highly charged, and has a clear Northern/Western bias (2011: 619), implying in some cases that "faith-based" is the opposite of "professionalism" (2011: 619). Whether this is fair or not, numerous partnerships continue to develop between secular organizations, and it is certainly the case that some large FBOs are more comfortable associating with secular organizations than evangelicals in their own religious traditions. Ferris writes, "A fifty-year tradition of 'inter-church aid' is being replaced in many quarters by professional programs to eradicate poverty and respond to emergencies" (2005: 319). FBOs are also increasingly encouraged to depoliticize their views (2011: 615). Similar to secular organizations, Ferris notes that FBOs operate in a context that is increasingly about emergency response and one where governments are more and more willing to contract funds to NGOs than before (2005: 317). She writes, "The impact of increasing governmental resources, the shift toward emergency response, the expanding role of the media in shaping humanitarian response, and the proliferation of NGOs has accentuated competition between NGOs…At the same time, greater media attention to emergencies coupled with growing donor requirements for accountability has stepped up the pressure on international NGOs to implement more professional programs" (2005: 318). Obtaining more government funds and under greater scrutiny, FBOs are having to response to a greater degree of measurable results and reporting requirements; more stringent standards than they were likely to have faced when only dealing with funds from congregations.

Many church-based organizations have also signed the NGO/Red Cross and Red Crescent Code of Conduct for Humanitarian Work (2005: 319), and there is also a greater push to foster more local NGO responses as a way to balance North/South power imbalances (2005: 318).

Similarly, as many scholars point out, the pendulum shifts regarding how religion is perceived and used strategically (Barnett and Stein 2012; Fiddian-Qasmiyeh 2011a, b, c). Though just one example and certainly not generalizable for all FBOs, Horstmann (2011) considers how faith traditions can be invoked via networks to push for a specific cause. He looks specifically at FBOs' alliance with Karen Baptist networks to explore the nexus of being stateless and mobilized by Christian missionary movements, and examines FBOs like the Thai-Burma Border Consortium (TBBC) (see www.tbbc.org/) and Free Burma Rangers (see www.freeburmarangers.org). He writes of global alliances that fuel their cause, writing, "The Karen and their Christian partner organizations have woven the images of the atrocities perpetrated by the Burmese military into a powerful narrative about social suffering and Christian liberation. In the propaganda material that is shown in videos, magazines, and on websites, the suffering of the Karen people provides the platform for the heroic efforts of Karen Christian relief teams that provide humanitarian aid to the wounded" (2011a, b, c: 515). He notes that they are able to reach places that other humanitarian workers cannot, and that they are able to fund and share their message far more widely because of these alliances. He also writes, however, that this relationship places humanitarian NGOs in an ethical dilemma, as they are supposed to stay neutral, but in this case have been drawn into privileging the Christian Baptist Karen. Thus he argues that FBOs in the camps on the Thai-Burma border understand their support as spiritual engagement in which humanitarian aid and proselytizing are intertwined; this in turn makes these FBOs politically biased, something he fears may fuel more conflict (2011a, b, c: 515). Even training materials from the Free Burma Rangers evoke powerful imagery. One leadership training program available on the Free Burma Rangers website is entitled "Fighting the Serpent," a title meant to evoke images of fighting evil. At the same time, the training focuses on aiding internally displaced persons of any ethnicity or religious persuasion in a humanitarian way, focusing on medicine, food, water and protection.[14]

In a different vein, Barnett and Stein also point out that globalization is changing the way FBOs operate, and that even as some churches lose membership, FBOs continue to operate with strength. For example, they write of a surge in some giving

[14] A story incorporated in the training material discusses a young woman who protected Jews fleeing Nazis. It reads, "Stefania especially was in a situation very similar to FBR relief teams accompanying IDPs. If you find yourself in a desperate situation when you are on an FBR relief team, do what Stefania did. Do not give in to fear or despair even if there is no human way out. Even if it looks for sure like you and the IDPs with you will die in a matter of hours or in minutes, do not give in to fear and do not give up. Pray together. Ask God for help, and be ready to act on an answer if and when it comes. And if you do die, know that you die for love and that you die in love and that the serpent has not won and will not." Linking these narratives provides a powerful religious undertone of righteousness. Full training literature available from www.freeburmarangers.org/wp-content/uploads/2011/02/FightingTheSerpent.pdf.

to FBOs since the 1990s: "American churches have increased their giving to overseas ministries by almost 50 % over the past decade and, according to recent figures, gave nearly $3.7 billion..." (2012: 5). They attribute some of this to increased affluence among American Christians, but also emphasize the effects of globalization bringing greater global engagement on issues that were previously foreign (2012: 6). They note that "...just as Christianity is globalizing, so too is Christian-based humanitarianism" (2012: 6).[15] Some Christians may even feel closer and more solidarity with Christians on the other side of the world than they did before. Megachurches may even see "new lands of opportunity for activism and missionary work," considering how they might work directly in establishing global ministries rather than working through intermediaries (2012: 6). This may require former "middlemen" like the World Council of Churches or Action by Churches Together to reinvent themselves, as churches find themselves doing their own projects directly, rather than going through these organizations.

164.7 Gaps and Further Questions

In addition to calls for further research on non-western and non-Christian FBOs, a number of theoretical questions emerge, some of which are listed here:

- If FBOs operate from different motivations and donor bases, do they also, in a sense, play by different rules? Does that make them more or less predictable; more or less useful to work with?
- What differences and similarities emerge between Christian, Muslim, Buddhist and other FBOs—there are so many differences within each one, is it even possible to compare across religions?
- In refugee situations in particular, and humanitarian situations in general, there is often such a conglomeration of organizations (secular NGOs, FBOs, the UN and other "implementing partners"). Is it reasonable to expect quality coordination and cooperation with so many different actors, each with different motives?
- How is "professionalization" changing the humanitarian work of FBOs? Can FBOs get away with positions or behaviors that others cannot (that is, are they held to a different standard, such as that they might not be required to maintain political neutrality in humanitarian assistance)?
- How can greater cooperation between FBOs and secular organizations be attained?
- What other gaps exist, and what might different disciplines have to offer this area of study?
- On a deeper level, how much do motives really matter? Do some human rights or humanitarian secular organizations take on "religious" undertones with human rights rhetoric being the "religion"? For example, Barnett and Stein write, "...all

[15] They go on to discuss other religions as well.

humanitarian organizations are faith-based—but they are faith-based in different ways" (Barnett and Stein 2012: 23). Is this true?
- Barnett and Stein hold that religious organizations may be more likely to behave consistently with their principles. Is this the case? (Barnett and Stein 2012: 20).

164.8 Conclusions

This chapter has examined FBOs in light of their unique differences to secular organizations, noting all the while the problematic notions tied to terms, definitions, generalizations and binaries. It has considered motives, donor bases and networks, and the reach and scope of FBOs, particularly in light of challenges and strengths attributed with each characteristic, ultimately arguing that FBOs may have the potential to fill gaps unaddressed by other organizations, but that further coordination and research is needed. It has looked specifically at refugee examples, and predominantly at Christian FBOs, and has considered international relations theoretical lenses, such as constructivist and neoliberal institutionalist transnational network perspectives. Certainly there is much more work to be done. Indeed, Ferris emphasizes that most of what is known about FBOs comes from within their own assessments, and more scholarly research is needed, particularly to avoid generalizing and "homogenizing" them (Ferris 2011: 621). Barnett and Stein also echo calls for further research more broadly and with respect to religious organizations, writing that "The humanitarian sector is data-poor, frustrating any attempt to talk about trends, patterns and dynamics" (Barnett and Stein 2012: 10). Thus, the field is ripe for further research from various disciplines, including politics and international relations (which may apply transnational network or regime theoretical approaches), sociological or anthropological studies (perhaps examining the organizational or bureaucratic constructs of FBOs and the power and authority they project) or research on intersections with political and civil society, and each other. Indeed, scholarship in these areas is only beginning to unlock doors of potential research. But it can go much further than intellectual interest; understanding the role and capacities of FBOs may help reveal better ways of responding to the needs of those facing humanitarian crises, something that FBOs and secular organizations alike could always improve on.

References

Ager, A., & Ager, J. (2011). Faith and the discourse of secular humanitarianism. *Journal of Refugee Studies, 24*(3), 456–472.
Baker, G. (2009). Cosmopolitan as hospitality: Revisiting identity and difference in cosmopolitanism. *Alternatives, 34*, 107–128.
Barnett, M. (2011). *Empire of humanity: A history of humanitarianism.* Ithaca: Cornell University Press.

Barnett, M., & Finnemore, M. (1999). The politics, power, and pathologies of international organization. *International Organization, 53*(4), 699–732.
Barnett, M., & Stein, J. (2012). *Sacred aid: Faith and humanitarianism*. Oxford: Oxford University Press.
Betts, A. (2009). *Forced migration and global politics*. West Sussex: Wiley-Blackwell.
Bretherton, L. (2010). *Christianity and contemporary politics*. Oxford: Wiley-Blackwell.
Clarke, G. (2006). Faith matters: Faith-based organisations, civil society and international development. *Journal of International Development, 18*(6), 835–848.
Clarke, G., & Jennings, M. (2008). *Development, civil society and faith-based organizations: Bridging the sacred and the secular*. London: Palgrave Macmillan.
Derrida, J. (2000). Foreigner question in *of Hospitality: Anne Dufourmantelle invites Jacques Derrida to respond* (R. Bowlby, Trans.). Stanford: Stanford University Press.
Durkheim, É. (2008). *The elementary forms of religious life* (C. Cosman, Trans.). New York: Oxford University Press.
Eby, J., Iverson, E., Smyers, J., & Kekic, E. (2011). The faith community's role in refugee resettlement in the United States. *Journal of Refugee Studies, 24*(3), 586–605.
Ferris, E. (2005). Faith-based and secular humanitarian organizations. *International Review of the Red Cross, 87*(858), 311–325.
Ferris, E. (2011). Faith and humanitarianism: It's complicated. *Journal of Refugee Studies, 24*(3), 606–625.
Fiddian-Qasmiyeh, E. (Ed.). (2011a). Special issue: Faith-based humanitarianism in contexts of forced displacement. *Journal of Refugee Studies, 24*(3).
Fiddian-Qasmiyeh, E. (2011b). Introduction: Faith-based humanitarianism in the contexts of forced displacement. *Journal of Refugee Studies, 24*(3), 429–439.
Fiddian-Qasmiyeh, E. (2011c). The pragmatics of performance: Putting 'faith' in aid in the Sahrawi refugee camps. *Journal of Refugee Studies, 24*(3), 533–547.
Gauthier, D. J. (2007). Levinas and the politics of hospitality. *History of Political Thought, 28*(1), 158–180.
Georgetown University. International faith-based initiatives: Can they work? Available at http://woodstock.georgetown.edu/resources/articles/International-Faith-Based-Initiatives.html. Last visited 6 June 2012.
Goldsmith, S., Eimicke, W., & Pineda, C. (2006). *Faith-based organizations versus their secular counterparts: A primer for local officials*. Available at www.innovations.harvard.edu/cache/documents/11120.pdf. Accessed 29 July 2012.
Groody, D., & Campese, G. (Eds.). (2008). *Promised land, a perilous journey: Theological perspectives on migration*. Notre Dame: Notre Dame University Press.
Hirschman, C. (2007). The role of religion in the origins and adaptation of immigrant groups in the United States. In A. Portes & J. DeWind (Eds.), *Rethinking migration: New theoretical and empirical perspectives* (pp. 391–418). New York/Oxford: Berghahn.
Horstmann, A. (2011). Ethical dilemmas and identifications of faith-based humanitarian organizations in the Karen refugee crisis. *Journal of Refugee Studies, 24*(3), 513–532.
Keck, M., & Sikkink, K. (1998). *Activists beyond borders: Advocacy networks in international politics*. Ithaca: Cornell University Press.
Keohane, R., & Nye, J. S., Jr. (Eds). (1971). Transnational relations and world politics: An introduction. *International Organization, 25*(3), 329–349.
Levinas, E. (1981). *Totality and infinity: An essay on exteriority* (M. Lingis, Trans.). London: Martinus Nijhoff.
Marfleet, P. (2011). Understanding 'sanctuary:' Faith and traditions in asylum. *Journal of Refugee Studies, 24*(3), 440–455.
Orji, N. (2011). Faith-based aid to people affected by conflict in Jos, Nigeria: An analysis of the role of Christian and Muslim organizations. *Journal of Refugee Studies, 24*(3), 473–492.
Palmer, V. (2011). Analysing cultural proximity: Islamic relief worldwide and Rohingya refugees in Bangladesh. *Development in Practice, 21*(1), 96–108.

Parsitau, D. (2011). The role of faith and faith-based organizations among internally displaced persons in Kenya. *Journal of Refugee Studies, 24*(3), 493–512.

Pohl, C. D. (1999). *Making room: Recovering hospitality as a Christian tradition*. Grand Rapids/Cambridge: William B. Eerdmans Publishing.

Pohl, C. D. (2006). Responding to strangers: Insights from the Christian tradition. *Studies in Christian Ethics, 19*(1), 81–101.

Risse, T., Ropp, S., & Sikkink, K. (1999). *The power of human rights: International norms and domestic change*. Cambridge: Cambridge University Press.

Risse-Kappen, T. (Ed.). (1995). *Bringing transnational relations back in: Non-state actors, domestic structures and international institutions*. Cambridge: Cambridge University Press.

Skogstad, G. (Ed.). (2011). *Policy paradigms, transnationalism, and domestic politics*. Toronto: University of Toronto Press.

Snyder, J. (2011a). *Religion and international relations theory*. New York: Columbia University Press.

Snyder, S. (2011b). Un/settling angels: Faith-based organizations and asylum-seeking in the UK. *Journal of Refugee Studies, 24*(3), 565–585.

Stoddard, A. (2003, December). With us or against us? NGO neutrality on the line. *Humanitarian Practice Network*.

Taithe, B. (2012). Pyrrhic victories? French Catholic missionaries, modern expertise, and secularizing technologies. In M. Barnett & J. Stein (Eds.), *Sacred aid: Faith and humanitarianism* (pp. 166–187). Oxford: Oxford University Press.

Wilson, E. (2011). Much to be proud of, much to be done: Faith-based organizations and the politics of asylum in Australia. *Journal of Refugee Studies, 24*(3), 548–564.

Printed by Books on Demand, Germany